Encyclopedia of Algorithms

Ming-Yang Kao

Editor

Encyclopedia of Algorithms

Second Edition

Volume 1

A–F

With 379 Figures and 51 Tables

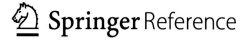

Editor
Ming-Yang Kao
Department of Electrical Engineering
and Computer Science
Northwestern University
Evanston, IL, USA

ISBN 978-1-4939-2863-7   ISBN 978-1-4939-2864-4   (eBook)
ISBN 978-1-4939-2865-1   (print and electronic bundle)
DOI 10.1007/ 978-1-4939-2864-4

Library of Congress Control Number: 2015958521

This Springer imprint is published by SpringerNature

The registered company is Springer Science+Business Media LLC New York

Preface

The Encyclopedia of Algorithms provides researchers, students, and practitioners of algorithmic research with a mechanism to efficiently and accurately find the names, definitions, and key results of important algorithmic problems. It also provides further readings on those problems.

This *encyclopedia* covers a broad range of algorithmic areas; each area is summarized by a collection of entries. The entries are written in a clear and concise structure so that they can be readily absorbed by the readers and easily updated by the authors. A typical encyclopedia entry is an in-depth mini-survey of an algorithmic problem written by an expert in the field. The entries for an algorithmic area are compiled by area editors to survey the representative results in that area and can form the core materials of a course in the area.

This 2nd edition of the encyclopedia contains a wide array of important new research results. Highlights include works in tile self-assembly (nanotechnology), bioinformatics, game theory, Internet algorithms, and social networks. Overall, more than 70 % of the entries in this edition and new entries are updated.

This reference work will continue to be updated on a regular basis via a live site to allow timely updates and fast search. Knowledge accumulation is an ongoing community project. Please take ownership of this body of work. If you have feedback regarding a particular entry, please feel free to communicate directly with the author or the area editor of that entry. If you are interested in authoring a future entry, please contact a suitable area editor. If you have suggestions on how to improve the Encyclopedia as a whole, please contact me at kao@northwestern.edu. The credit of this Encyclopedia goes to the area editors, the entry authors, the entry reviewers, and the project editors at Springer, including Melissa Fearon, Michael Hermann, and Sylvia Blago.

About the Editor

Ming-Yang Kao is a Professor of Computer Science in the Department of Electrical Engineering and Computer Science at Northwestern University. He has published extensively in the design, analysis, and applications of algorithms. His current interests include discrete optimization, bioinformatics, computational economics, computational finance, and nanotechnology. He serves as the Editor-in-Chief of Algorithmica.

He obtained a B.S. in Mathematics from National Taiwan University in 1978 and a Ph.D. in Computer Science from Yale University in 1986. He previously taught at Indiana University at Bloomington, Duke University, Yale University, and Tufts University. At Northwestern University, he has served as the Department Chair of Computer Science. He has also cofounded the Program in Computational Biology and Bioinformatics and served as its Director. He currently serves as the Head of the EECS Division of Computing, Algorithms, and Applications and is a Member of the Theoretical Computer Science Group.

For more information, please see www.cs.northwestern.edu/~kao

Area Editors

Algorithm Engineering

Giuseppe F. Italiano* Department of Computer and Systems Science, University of Rome, Rome, Italy

Department of Information and Computer Systems, University of Rome, Rome, Italy

Rajeev Raman* Department of Computer Science, University of Leicester, Leicester, UK

Algorithms for Modern Computers

Alejandro López-Ortiz David R. Cheriton School of Computer Science, University of Waterloo, Waterloo, ON, Canada

Algorithmic Aspects of Distributed Sensor Networks

Sotiris Nikoletseas Computer Engineering and Informatics Department, University of Patras, Patras, Greece

Computer Technology Institute and Press "Diophantus", Patras, Greece

Approximation Algorithms

Susanne Albers* Technical University of Munich, Munich, Germany

Chandra Chekuri* Department of Computer Science, University of Illinois, Urbana-Champaign, Urbana, IL, USA

Department of Mathematics and Computer Science, The Open University of Israel, Raanana, Israel

Ming-Yang Kao Department of Electrical Engineering and Computer Science, Northwestern University, Evanston, IL, USA

Sanjeev Khanna* University of Pennsylvania, Philadelphia, PA, USA

Samir Khuller* Computer Science Department, University of Maryland, College Park, MD, USA

---

* Acknowledgment for first edition contribution

Average Case Analysis

Paul (Pavlos) Spirakis* Computer Engineering and Informatics, Research and Academic Computer Technology Institute, Patras University, Patras, Greece

Computer Science, University of Liverpool, Liverpool, UK

Computer Technology Institute (CTI), Patras, Greece

Bin Packing

Leah Epstein Department of Mathematics, University of Haifa, Haifa, Israel

Bioinformatics

Miklós Csűrös Department of Computer Science, University of Montréal, Montréal, QC, Canada

Certified Reconstruction and Mesh Generation

Siu-Wing Cheng Department of Computer Science and Engineering, Hong Kong University of Science and Technology, Hong Kong, China

Tamal Krishna Dey Department of Computer Science and Engineering, The Ohio State University, Columbus, OH, USA

Coding Algorithms

Venkatesan Guruswami* Department of Computer Science and Engineering, University of Washington, Seattle, WA, USA

Combinatorial Group Testing

Ding-Zhu Du Computer Science, University of Minnesota, Minneapolis, MN, USA

Department of Computer Science, The University of Texas at Dallas, Richardson, TX, USA

Combinatorial Optimization

Samir Khuller* Computer Science Department, University of Maryland, College Park, MD, USA

Compressed Text Indexing

Tak-Wah Lam Department of Computer Science, University of Hong Kong, Hong Kong, China

Compression of Text and Data Structures

Gonzalo Navarro Department of Computer Science, University of Chile, Santiago, Chile

Computational Biology

Bhaskar DasGupta Department of Computer Science, University of Illinois, Chicago, IL, USA

Tak-Wah Lam Department of Computer Science, University of Hong Kong, Hong Kong, China

Computational Counting

Xi Chen Computer Science Department, Columbia University, New York, NY, USA

Computer Science and Technology, Tsinghua University, Beijing, China

Computational Economics

Xiaotie Deng AIMS Laboratory (Algorithms-Agents-Data on Internet, Market, and Social Networks), Department of Computer Science and Engineering, Shanghai Jiao Tong University, Shanghai, China

Department of Computer Science, City University of Hong Kong, Hong Kong, China

Computational Geometry

Sándor Fekete Department of Computer Science, Technical University Braunschweig, Braunschweig, Germany

Computational Learning Theory

Rocco A. Servedio Computer Science, Columbia University, New York, NY, USA

Data Compression

Paolo Ferragina* Department of Computer Science, University of Pisa, Pisa, Italy

Differential Privacy

Aaron Roth Department of Computer and Information Sciences, University of Pennsylvania, Levine Hall, PA, USA

Distributed Algorithms

Sergio Rajsbaum Instituto de Matemáticas, Universidad Nacional Autónoma de México (UNAM) México City, México

Dynamic Graph Algorithms

Giuseppe F. Italiano* Department of Computer and Systems Science, University of Rome, Rome, Italy

Department of Information and Computer Systems, University of Rome, Rome, Italy

Enumeration Algorithms

Takeaki Uno National Institute of Informatics, Chiyoda, Tokyo, Japan

Exact Exponential Algorithms

Fedor V. Fomin Department of Informatics, University of Bergen, Bergen, Norway

External Memory Algorithms

Herman Haverkort Department of Computer Science, Eindhoven University of Technology, Eindhoven, The Netherlands

Game Theory

Mohammad Taghi Hajiaghayi Department of Computer Science, University of Maryland, College Park, MD, USA

Geometric Networks

Andrzej Lingas Department of Computer Science, Lund University, Lund, Sweden

Graph Algorithms

Samir Khuller* Computer Science Department, University of Maryland, College Park, MD, USA

Seth Pettie Electrical Engineering and Computer Science (EECS) Department, University of Michigan, Ann Arbor, MI, USA

Vijaya Ramachandran* Computer Science, University of Texas, Austin, TX, USA

Liam Roditty Department of Computer Science, Bar-Ilan University, Ramat-Gan, Israel

Dimitrios Thilikos AlGCo Project-Team, CNRS, LIRMM, France

Department of Mathematics, National and Kapodistrian University of Athens, Athens, Greece

Graph Drawing

Seokhee Hong School of Information Technologies, University of Sydney, Sydney, NSW, Australia

Internet Algorithms

Edith Cohen Tel Aviv University, Tel Aviv, Israel

Stanford University, Stanford, CA, USA

I/O-Efficient Algorithms

Herman Haverkort Department of Computer Science, Eindhoven University of Technology, Eindhoven, The Netherlands

Kernels and Compressions

Gregory Gutin Department of Computer Science, Royal Holloway, University of London, Egham, UK

Massive Data Algorithms

Herman Haverkort Department of Computer Science, Eindhoven University of Technology, Eindhoven, The Netherlands

Mathematical Optimization

Ding-Zhu Du Computer Science, University of Minnesota, Minneapolis, MN, USA

Department of Computer Science, The University of Texas at Dallas, Richardson, TX, USA

Mechanism Design

Yossi Azar* Tel-Aviv University, Tel Aviv, Israel

Mobile Computing

Xiang-Yang Li* Department of Computer Science, Illinois Institute of Technology, Chicago, IL, USA

Modern Learning Theory

Maria-Florina Balcan Department of Machine Learning, Carnegie Mellon University, Pittsburgh, PA, USA

Online Algorithms

Susanne Albers* Technical University of Munich, Munich, Germany

Yossi Azar* Tel-Aviv University, Tel Aviv, Israel

Marek Chrobak Computer Science, University of California, Riverside, CA, USA

Alejandro López-Ortiz David R. Cheriton School of Computer Science, University of Waterloo, Waterloo, ON, Canada

Parameterized Algorithms

Dimitrios Thilikos AlGCo Project-Team, CNRS, LIRMM, France

Department of Mathematics, National and Kapodistrian University of Athens, Athens, Greece

Parameterized Algorithms and Complexity

Saket Saurabh Institute of Mathematical Sciences, Chennai, India

University of Bergen, Bergen, Norway

Parameterized and Exact Algorithms

Rolf Niedermeier* Department of Mathematics and Computer Science, University of Jena, Jena, Germany

Institut für Softwaretechnik und Theoretische Informatik, Technische Universität Berlin, Berlin, Germany

Price of Anarchy

Yossi Azar* Tel-Aviv University, Tel Aviv, Israel

Probabilistic Algorithms

Sotiris Nikoletseas Computer Engineering and Informatics Department, University of Patras, Patras, Greece

Computer Technology Institute and Press "Diophantus", Patras, Greece

Paul (Pavlos) Spirakis* Computer Engineering and Informatics, Research and Academic Computer Technology Institute, Patras University, Patras, Greece

Computer Science, University of Liverpool, Liverpool, UK

Computer Technology Institute (CTI), Patras, Greece

Quantum Computing

Andris Ambainis Faculty of Computing, University of Latvia, Riga, Latvia

Radio Networks

Marek Chrobak Computer Science, University of California, Riverside, CA, USA

Scheduling

Leah Epstein Department of Mathematics, University of Haifa, Haifa, Israel

Scheduling Algorithms

Viswanath Nagarajan University of Michigan, Ann Arbor, MI, USA

Kirk Pruhs* Department of Computer Science, University of Pittsburgh, Pittsburgh, PA, USA

Social Networks

Mohammad Taghi Hajiaghayi Department of Computer Science, University of Maryland, College Park, MD, USA

Grant Schoenebeck Computer Science and Engineering, University of Michigan, Ann Arbor, MI, USA

Stable Marriage Problems, k-SAT Algorithms

Kazuo Iwama Computer Engineering, Kyoto University, Sakyo, Kyoto, Japan

School of Informatics, Kyoto University, Sakyo, Kyoto, Japan

String Algorithms and Data Structures

Paolo Ferragina* Department of Computer Science, University of Pisa, Pisa, Italy

Gonzalo Navarro Department of Computer Science, University of Chile, Santiago, Chile

Steiner Tree Algorithms

Ding-Zhu Du Computer Science, University of Minnesota, Minneapolis, MN, USA

Department of Computer Science, The University of Texas at Dallas, Richardson, TX, USA

Sublinear Algorithms

Andrew McGregor School of Computer Science, University of Massachusetts, Amherst, MA, USA

Sofya Raskhodnikova Computer Science and Engineering Department, Pennsylvania State University, University Park, State College, PA, USA

Tile Self-Assembly

Robert Schweller Department of Computer Science, University of Texas Rio Grande Valley, Edinburg, TX, USA

VLSI CAD Algorithms

Hai Zhou Electrical Engineering and Computer Science (EECS) Department, Northwestern University, Evanston, IL, USA

Contributors

Karen Aardal Centrum Wiskunde & Informatica (CWI), Amsterdam, The Netherlands

Department of Mathematics and Computer Science, Eindhoven University of Technology, Eindhoven, The Netherlands

Ittai Abraham Microsoft Research, Silicon Valley, Palo Alto, CA, USA

Adi Akavia Department of Electrical Engineering and Computer Science, MIT, Cambridge, MA, USA

Réka Albert Department of Biology and Department of Physics, Pennsylvania State University, University Park, PA, USA

Mansoor Alicherry Bell Laboratories, Alcatel-Lucent, Murray Hill, NJ, USA

Noga Alon Department of Mathematics and Computer Science, Tel-Aviv University, Tel-Aviv, Israel

Srinivas Aluru Department of Electrical and Computer Engineering, Iowa State University, Ames, IA, USA

Andris Ambainis Faculty of Computing, University of Latvia, Riga, Latvia

Christoph Ambühl Department of Computer Science, University of Liverpool, Liverpool, UK

Nina Amenta Department of Computer Science, University of California, Davis, CA, USA

Amihood Amir Department of Computer Science, Bar-Ilan University, Ramat-Gan, Israel

Department of Computer Science, Johns Hopkins University, Baltimore, MD, USA

Spyros Angelopoulos Sorbonne Universités, L'Université Pierre et Marie Curie (UPMC), Université Paris 06, Paris, France

Anurag Anshu Center for Quantum Technologies, National University of Singapore, Singapore, Singapore

Alberto Apostolico College of Computing, Georgia Institute of Technology, Atlanta, GA, USA

Vera Asodi Center for the Mathematics of Information, California Institute of Technology, Pasadena, CA, USA

Peter Auer Chair for Information Technology, Montanuniversitaet Leoben, Leoben, Austria

Pranjal Awasthi Department of Computer Science, Princeton University, Princeton, NJ, USA

Department of Electrical Engineering, Indian Institute of Technology Madras, Chennai, Tamilnadu, India

Adnan Aziz Department of Electrical and Computer Engineering, University of Texas, Austin, TX, USA

Moshe Babaioff Microsoft Research, Herzliya, Israel

David A. Bader College of Computing, Georgia Institute of Technology, Atlanta, GA, USA

Michael Bader Department of Informatics, Technical University of Munich, Garching, Germany

Maria-Florina Balcan Department of Machine Learning, Carnegie Mellon University, Pittsburgh, PA, USA

Hideo Bannai Department of Informatics, Kyushu University, Fukuoka, Japan

Nikhil Bansal Eindhoven University of Technology, Eindhoven, The Netherlands

Jérémy Barbay Department of Computer Science (DCC), University of Chile, Santiago, Chile

Sanjoy K. Baruah Department of Computer Science, The University of North Carolina, Chapel Hill, NC, USA

Surender Baswana Department of Computer Science and Engineering, Indian Institute of Technology (IIT), Kanpur, Kanpur, India

MohammadHossein Bateni Google Inc., New York, NY, USA

Luca Becchetti Department of Information and Computer Systems, University of Rome, Rome, Italy

Xiaohui Bei Division of Mathematical Sciences, School of Physical and Mathematical Sciences, Nanyang Technological University, Singapore, Singapore

József Békési Department of Computer Science, Juhász Gyula Teachers Training College, Szeged, Hungary

Djamal Belazzougui Department of Computer Science, Helsinki Institute for Information Technology (HIIT), University of Helsinki, Helsinki, Finland

Aleksandrs Belovs Computer Science and Artificial Intelligence Laboratory, MIT, Cambridge, MA, USA

Aaron Bernstein Department of Computer Science, Columbia University, New York, NY, USA

Vincent Berry Institut de Biologie Computationnelle, Montpellier, France

Randeep Bhatia Bell Laboratories, Alcatel-Lucent, Murray Hill, NJ, USA

Andreas Björklund Department of Computer Science, Lund University, Lund, Sweden

Eric Blais University of Waterloo, Waterloo, ON, Canada

Mathieu Blanchette Department of Computer Science, McGill University, Montreal, QC, Canada

Markus Bläser Department of Computer Science, Saarland University, Saarbrücken, Germany

Avrim Blum School of Computer Science, Carnegie Mellon University, Pittsburgh, PA, USA

Hans L. Bodlaender Department of Computer Science, Utrecht University, Utrecht, The Netherlands

Sergio Boixo Quantum A.I. Laboratory, Google, Venice, CA, USA

Paolo Boldi Dipartimento di Informatica, Università degli Studi di Milano, Milano, Italy

Glencora Borradaile Department of Computer Science, Brown University, Providence, RI, USA

School of Electrical Engineering and Computer Science, Oregon State University, Corvallis, OR, USA

Ulrik Brandes Department of Computer and Information Science, University of Konstanz, Konstanz, Germany

Andreas Brandstädt Computer Science Department, University of Rostock, Rostock, Germany

Department of Informatics, University of Rostock, Rostock, Germany

Gilles Brassard Université de Montréal, Montréal, QC, Canada

Vladimir Braverman Department of Computer Science, Johns Hopkins University, Baltimore, MD, USA

Tian-Ming Bu Software Engineering Institute, East China Normal University, Shanghai, China

Adam L. Buchsbaum Madison, NJ, USA

Costas Busch Department of Computer Science, Lousiana State University, Baton Rouge, LA, USA

Jaroslaw Byrka Centrum Wiskunde & Informatica (CWI), Amsterdam, The Netherlands

Department of Mathematics and Computer Science, Eindhoven University of Technology, Eindhoven, The Netherlands

Jin-Yi Cai Beijing University, Beijing, China

Computer Sciences Department, University of Wisconsin–Madison, Madison, WI, USA

Mao-cheng Cai Chinese Academy of Sciences, Institute of Systems Science, Beijing, China

Yang Cai Computer Science, McGill University, Montreal, QC, Canada

Gruia Calinescu Department of Computer Science, Illinois Institute of Technology, Chicago, IL, USA

Colin Campbell Department of Physics, Pennsylvania State University, University Park, PA, USA

Luca Castelli Aleardi Laboratoire d'Informatique (LIX), École Polytechnique, Bâtiment Alan Turing, Palaiseau, France

Katarína Cechlárová Faculty of Science, Institute of Mathematics, P. J. Šafárik University, Košice, Slovakia

Nicolò Cesa-Bianchi Dipartimento di Informatica, Università degli Studi di Milano, Milano, Italy

Amit Chakrabarti Department of Computer Science, Dartmouth College, Hanover, NH, USA

Deeparnab Chakrabarty Microsoft Research, Bangalore, Karnataka, India

Erin W. Chambers Department of Computer Science and Mathematics, Saint Louis University, St. Louis, MO, USA

Chee Yong Chan National University of Singapore, Singapore, Singapore

Mee Yee Chan Department of Computer Science, University of Hong Kong, Hong Kong, China

Wun-Tat Chan College of International Education, Hong Kong Baptist University, Hong Kong, China

Tushar Deepak Chandra IBM Watson Research Center, Yorktown Heights, NY, USA

Kun-Mao Chao Department of Computer Science and Information Engineering, National Taiwan University, Taipei, Taiwan

Bernadette Charron-Bost Laboratory for Informatics, The Polytechnic School, Palaiseau, France

Ioannis Chatzigiannakis Department of Computer Engineering and Informatics, University of Patras and Computer Technology Institute, Patras, Greece

Shuchi Chawla Department of Computer Science, University of Wisconsin–Madison, Madison, WI, USA

Shiri Chechik Department of Computer Science, Tel Aviv University, Tel Aviv, Israel

Chandra Chekuri Department of Computer Science, University of Illinois, Urbana-Champaign, Urbana, IL, USA

Department of Mathematics and Computer Science, The Open University of Israel, Raanana, Israel

Danny Z. Chen Department of Computer Science and Engineering, University of Notre Dame, Notre Dame, IN, USA

Ho-Lin Chen Department of Electrical Engineering, National Taiwan University, Taipei, Taiwan

Jianer Chen Department of Computer Science, Texas A&M University, College Station, TX, USA

Ning Chen Division of Mathematical Sciences, School of Physical and Mathematical Sciences, Nanyang Technological University, Singapore, Singapore

Xi Chen Computer Science Department, Columbia University, New York, NY, USA

Computer Science and Technology, Tsinghua University, Beijing, China

Siu-Wing Cheng Department of Computer Science and Engineering, Hong Kong University of Science and Technology, Hong Kong, China

Xiuzhen Cheng Department of Computer Science, George Washington University, Washington, DC, USA

Huang Chien-Chung Chalmers University of Technology and University of Gothenburg, Gothenburg, Sweden

Markus Chimani Faculty of Mathematics/Computer, Theoretical Computer Science, Osnabrück University, Osnabrück, Germany

Francis Y.L. Chin Department of Computer Science, University of Hong Kong, Hong Kong, China

Rajesh Chitnis Department of Computer Science, University of Maryland, College Park, MD, USA

Minsik Cho IBM T. J. Watson Research Center, Yorktown Heights, NY, USA

Rezaul A. Chowdhury Department of Computer Sciences, University of Texas, Austin, TX, USA

Stony Brook University (SUNY), Stony Brook, NY, USA

George Christodoulou University of Liverpool, Liverpool, UK

Marek Chrobak Computer Science, University of California, Riverside, CA, USA

Chris Chu Department of Electrical and Computer Engineering, Iowa State University, Ames, IA, USA

Xiaowen Chu Department of Computer Science, Hong Kong Baptist University, Hong Kong, China

Julia Chuzhoy Toyota Technological Institute, Chicago, IL, USA

Edith Cohen Tel Aviv University, Tel Aviv, Israel

Stanford University, Stanford, CA, USA

Jason Cong Department of Computer Science, UCLA, Los Angeles, CA, USA

Graham Cormode Department of Computer Science, University of Warwick, Coventry, UK

Derek G. Corneil Department of Computer Science, University of Toronto, Toronto, ON, Canada

Bruno Courcelle Laboratoire Bordelais de Recherche en Informatique (LaBRI), CNRS, Bordeaux University, Talence, France

Lenore J. Cowen Department of Computer Science, Tufts University, Medford, MA, USA

Nello Cristianini Department of Engineering Mathematics, and Computer Science, University of Bristol, Bristol, UK

Maxime Crochemore Department of Computer Science, King's College London, London, UK

Laboratory of Computer Science, University of Paris-East, Paris, France

Université de Marne-la-Vallée, Champs-sur-Marne, France

Miklós Csürös Department of Computer Science, University of Montréal, Montréal, QC, Canada

Fabio Cunial Department of Computer Science, Helsinki Institute for Information Technology (HIIT), University of Helsinki, Helsinki, Finland

Marek Cygan Institute of Informatics, University of Warsaw, Warsaw, Poland

Artur Czumaj Department of Computer Science, Centre for Discrete Mathematics and Its Applications, University of Warwick, Coventry, UK

Bhaskar DasGupta Department of Computer Science, University of Illinois, Chicago, IL, USA

Constantinos Daskalakis EECS, Massachusetts Institute of Technology, Cambridge, MA, USA

Mark de Berg Department of Mathematics and Computer Science, TU Eindhoven, Eindhoven, The Netherlands

Xavier Défago School of Information Science, Japan Advanced Institute of Science and Technology (JAIST), Ishikawa, Japan

Daniel Delling Microsoft, Silicon Valley, CA, USA

Erik D. Demaine MIT Computer Science and Artificial Intelligence Laboratory, Cambridge, MA, USA

Camil Demetrescu Department of Computer and Systems Science, University of Rome, Rome, Italy

Department of Information and Computer Systems, University of Rome, Rome, Italy

Ping Deng Department of Computer Science, The University of Texas at Dallas, Richardson, TX, USA

Xiaotie Deng AIMS Laboratory (Algorithms-Agents-Data on Internet, Market, and Social Networks), Department of Computer Science and Engineering, Shanghai Jiao Tong University, Shanghai, China

Department of Computer Science, City University of Hong Kong, Hong Kong, China

Vamsi Krishna Devabathini Center for Quantum Technologies, National University of Singapore, Singapore, Singapore

Olivier Devillers Inria Nancy – Grand-Est, Villers-lès-Nancy, France

Tamal Krishna Dey Department of Computer Science and Engineering, The Ohio State University, Columbus, OH, USA

Robert P. Dick Department of Electrical Engineering and Computer Science, University of Michigan, Ann Arbor, MI, USA

Walter Didimo Department of Engineering, University of Perugia, Perugia, Italy

Ling Ding Institute of Technology, University of Washington Tacoma, Tacoma, WA, USA

Yuzheng Ding Xilinx Inc., Longmont, CO, USA

Michael Dom Department of Mathematics and Computer Science, University of Jena, Jena, Germany

Riccardo Dondi Università degli Studi di Bergamo, Bergamo, Italy

Gyorgy Dosa University of Pannonia, Veszprém, Hungary

David Doty Computing and Mathematical Sciences, California Institute of Technology, Pasadena, CA, USA

Ding-Zhu Du Computer Science, University of Minnesota, Minneapolis, MN, USA

Department of Computer Science, The University of Texas at Dallas, Richardson, TX, USA

Hongwei Du Department of Computer Science and Technology, Shenzhen Graduate School, Harbin Institute of Technology, Shenzhen, China

Ran Duan Institute for Interdisciplinary Information Sciences, Tsinghua University, Beijing, China

Devdatt Dubhashi Department of Computer Science, Chalmers University of Technology, Gothenburg, Sweden

Gothenburg University, Gothenburg, Sweden

Adrian Dumitrescu Computer Science, University of Wisconsin–Milwaukee, Milwaukee, WI, USA

Iréne Durand Laboratoire Bordelais de Recherche en Informatique (LaBRI), CNRS, Bordeaux University, Talence, France

Stephane Durocher University of Manitoba, Winnipeg, MB, Canada

Pavlos Efraimidis Department of Electrical and Computer Engineering, Democritus University of Thrace, Xanthi, Greece

Charilaos Efthymiou Department of Computer Engineering and Informatics, University of Patras, Patras, Greece

Michael Elkin Department of Computer Science, Ben-Gurion University, Beer-Sheva, Israel

Matthias Englert Department of Computer Science, University of Warwick, Coventry, UK

David Eppstein Donald Bren School of Information and Computer Sciences, Computer Science Department, University of California, Irvine, CA, USA

Leah Epstein Department of Mathematics, University of Haifa, Haifa, Israel

Jeff Erickson Department of Computer Science, University of Illinois, Urbana, IL, USA

Constantine G. Evans Division of Biology and Bioengineering, California Institute of Technology, Pasadena, CA, USA

Eyal Even-Dar Google, New York, NY, USA

Rolf Fagerberg Department of Mathematics and Computer Science, University of Southern Denmark, Odense, Denmark

Jittat Fakcharoenphol Department of Computer Engineering, Kasetsart University, Bangkok, Thailand

Piotr Faliszewski AGH University of Science and Technology, Krakow, Poland

Lidan Fan Department of Computer Science, The University of Texas, Tyler, TX, USA

Qizhi Fang School of Mathematical Sciences, Ocean University of China, Qingdao, Shandong Province, China

Martín Farach-Colton Department of Computer Science, Rutgers University, Piscataway, NJ, USA

Panagiota Fatourou Department of Computer Science, University of Ioannina, Ioannina, Greece

Jonathan Feldman Google, Inc., New York, NY, USA

Vitaly Feldman IBM Research – Almaden, San Jose, CA, USA

Henning Fernau Fachbereich 4, Abteilung Informatikwissenschaften, Universität Trier, Trier, Germany

Institute for Computer Science, University of Trier, Trier, Germany

Paolo Ferragina Department of Computer Science, University of Pisa, Pisa, Italy

Johannes Fischer Technical University Dortmund, Dortmund, Germany

Nathan Fisher Department of Computer Science, Wayne State University, Detroit, MI, USA

Abraham Flaxman Theory Group, Microsoft Research, Redmond, WA, USA

Paola Flocchini School of Electrical Engineering and Computer Science, University of Ottawa, Ottawa, ON, Canada

Fedor V. Fomin Department of Informatics, University of Bergen, Bergen, Norway

Dimitris Fotakis Department of Information and Communication Systems Engineering, University of the Aegean, Samos, Greece

Kyle Fox Institute for Computational and Experimental Research in Mathematics, Brown University, Providence, RI, USA

Pierre Fraigniaud Laboratoire d'Informatique Algorithmique: Fondements et Applications, CNRS and University Paris Diderot, Paris, France

Fabrizio Frati School of Information Technologies, The University of Sydney, Sydney, NSW, Australia

Engineering Department, Roma Tre University, Rome, Italy

Ophir Frieder Department of Computer Science, Illinois Institute of Technology, Chicago, IL, USA

Hiroshi Fujiwara Shinshu University, Nagano, Japan

Stanley P.Y. Fung Department of Computer Science, University of Leicester, Leicester, UK

Stefan Funke Department of Computer Science, Universität Stuttgart, Stuttgart, Germany

Martin Fürer Department of Computer Science and Engineering, The Pennsylvania State University, University Park, PA, USA

Travis Gagie Department of Computer Science, University of Eastern Piedmont, Alessandria, Italy

Department of Computer Science, University of Helsinki, Helsinki, Finland

Gábor Galambos Department of Computer Science, Juhász Gyula Teachers Training College, Szeged, Hungary

Jianjiong Gao Computational Biology Center, Memorial Sloan-Kettering Cancer Center, New York, NY, USA

Jie Gao Department of Computer Science, Stony Brook University, Stony Brook, NY, USA

Xiaofeng Gao Department of Computer Science, Shanghai Jiao Tong University, Shanghai, China

Juan Garay Bell Laboratories, Murray Hill, NJ, USA

Minos Garofalakis Technical University of Crete, Chania, Greece

Olivier Gascuel Institut de Biologie Computationnelle, Laboratoire d'Informatique, de Robotique et de Microélectronique de Montpellier (LIRMM), CNRS and Université de Montpellier, Montpellier cedex 5, France

Leszek Gąsieniec University of Liverpool, Liverpool, UK

Serge Gaspers Optimisation Research Group, National ICT Australia (NICTA), Sydney, NSW, Australia

School of Computer Science and Engineering, University of New South Wales (UNSW), Sydney, NSW, Australia

Maciej Gazda Department of Mathematics and Computer Science, Eindhoven University of Technology, Eindhoven, The Netherlands

Raffaele Giancarlo Department of Mathematics and Applications, University of Palermo, Palermo, Italy

Gagan Goel Google Inc., New York, NY, USA

Andrew V. Goldberg Microsoft Research – Silicon Valley, Mountain View, CA, USA

Oded Goldreich Department of Computer Science, Weizmann Institute of Science, Rehovot, Israel

Jens Gramm WSI Institute of Theoretical Computer Science, Tübingen University, Tübingen, Germany

Fabrizio Grandoni IDSIA, USI-SUPSI, University of Lugano, Lugano, Switzerland

Roberto Grossi Dipartimento di Informatica, Università di Pisa, Pisa, Italy

Lov K. Grover Bell Laboratories, Alcatel-Lucent, Murray Hill, NJ, USA

Xianfeng David Gu Department of Computer Science, Stony Brook University, Stony Brook, NY, USA

Joachim Gudmundsson DMiST, National ICT Australia Ltd, Alexandria, Australia

School of Information Technologies, University of Sydney, Sydney, NSW, Australia

Rachid Guerraoui School of Computer and Communication Sciences, EPFL, Lausanne, Switzerland

Heng Guo Computer Sciences Department, University of Wisconsin–Madison, Madison, WI, USA

Jiong Guo Department of Mathematics and Computer Science, University of Jena, Jena, Germany

Manoj Gupta Indian Institute of Technology (IIT) Delhi, Hauz Khas, New Delhi, India

Venkatesan Guruswami Department of Computer Science and Engineering, University of Washington, Seattle, WA, USA

Gregory Gutin Department of Computer Science, Royal Holloway, University of London, Egham, UK

Michel Habib LIAFA, Université Paris Diderot, Paris Cedex 13, France

Mohammad Taghi Hajiaghayi Department of Computer Science, University of Maryland, College Park, MD, USA

Sean Hallgren Department of Computer Science and Engineering, The Pennsylvania State University, University Park, State College, PA, USA

Dan Halperin School of Computer Science, Tel-Aviv University, Tel Aviv, Israel

Moritz Hardt IBM Research – Almaden, San Jose, CA, USA

Ramesh Hariharan Strand Life Sciences, Bangalore, India

Aram W. Harrow Department of Physics, Massachusetts Institute of Technology, Cambridge, MA, USA

Prahladh Harsha Tata Institute of Fundamental Research, Mumbai, Maharashtra, India

Herman Haverkort Department of Computer Science, Eindhoven University of Technology, Eindhoven, The Netherlands

Meng He School of Computer Science, University of Waterloo, Waterloo, ON, Canada

Xin He Department of Computer Science and Engineering, The State University of New York, Buffalo, NY, USA

Lisa Hellerstein Department of Computer Science and Engineering, NYU Polytechnic School of Engineering, Brooklyn, NY, USA

Michael Hemmer Department of Computer Science, TU Braunschweig, Braunschweig, Germany

Danny Hendler Department of Computer Science, Ben-Gurion University of the Negev, Beer-Sheva, Israel

Monika Henzinger University of Vienna, Vienna, Austria

Maurice Herlihy Department of Computer Science, Brown University, Providence, RI, USA

Ted Herman Department of Computer Science, University of Iowa, Iowa City, IA, USA

John Hershberger Mentor Graphics Corporation, Wilsonville, OR, USA

Timon Hertli Department of Computer Science, ETH Zürich, Zürich, Switzerland

Edward A. Hirsch Laboratory of Mathematical Logic, Steklov Institute of Mathematics, St. Petersburg, Russia

Wing-Kai Hon Department of Computer Science, National Tsing Hua University, Hsin Chu, Taiwan

Seokhee Hong School of Information Technologies, University of Sydney, Sydney, NSW, Australia

Paul G. Howard Akamai Technologies, Cambridge, MA, USA

Peter Høyer University of Calgary, Calgary, AB, Canada

Li-Sha Huang Department of Computer Science and Technology, Tsinghua University, Beijing, China

Yaocun Huang Department of Computer Science, The University of Texas at Dallas, Richardson, TX, USA

Zhiyi Huang Department of Computer Science, The University of Hong Kong, Hong Kong, Hong Kong

Falk Hüffner Department of Math and Computer Science, University of Jena, Jena, Germany

Thore Husfeldt Department of Computer Science, Lund University, Lund, Sweden

Lucian Ilie Department of Computer Science, University of Western Ontario, London, ON, Canada

Sungjin Im Electrical Engineering and Computer Sciences (EECS), University of California, Merced, CA, USA

Csanad Imreh Institute of Informatics, University of Szeged, Szeged, Hungary

Robert W. Irving School of Computing Science, University of Glasgow, Glasgow, UK

Alon Itai Technion, Haifa, Israel

Giuseppe F. Italiano Department of Computer and Systems Science, University of Rome, Rome, Italy

Department of Information and Computer Systems, University of Rome, Rome, Italy

Kazuo Iwama Computer Engineering, Kyoto University, Sakyo, Kyoto, Japan

School of Informatics, Kyoto University, Sakyo, Kyoto, Japan

Jeffrey C. Jackson Department of Mathematics and Computer Science, Duquesne University, Pittsburgh, PA, USA

Ronald Jackups Department of Pediatrics, Washington University, St. Louis, MO, USA

Riko Jacob Institute of Computer Science, Technical University of Munich, Munich, Germany

IT University of Copenhagen, Copenhagen, Denmark

Rahul Jain Department of Computer Science, Center for Quantum Technologies, National University of Singapore, Singapore, Singapore

Klaus Jansen Department of Computer Science, University of Kiel, Kiel, Germany

Jesper Jansson Laboratory of Mathematical Bioinformatics, Institute for Chemical Research, Kyoto University, Gokasho, Uji, Kyoto, Japan

Stacey Jeffery David R. Cheriton School of Computer Science, University of Waterloo, Waterloo, ON, Canada

Madhav Jha Sandia National Laboratories, Livermore, CA, USA

Zenefits, San Francisco, CA, USA

David S. Johnson Department of Computer Science, Columbia University, New York, NY, USA

AT&T Laboratories, Algorithms and Optimization Research Department, Florham Park, NJ, USA

Mark Jones Department of Computer Science, Royal Holloway, University of London, Egham, UK

Tomasz Jurdziński Institute of Computer Science, University of Wrocław, Wrocław, Poland

Yoji Kajitani Department of Information and Media Sciences, The University of Kitakyushu, Kitakyushu, Japan

Shahin Kamali David R. Cheriton School of Computer Science, University of Waterloo, Waterloo, ON, Canada

Andrew Kane David R. Cheriton School of Computer Science, University of Waterloo, Waterloo, ON, Canada

Mamadou Moustapha Kanté Clermont-Université, Université Blaise Pascal, LIMOS, CNRS, Aubière, France

Ming-Yang Kao Department of Electrical Engineering and Computer Science, Northwestern University, Evanston, IL, USA

Alexis Kaporis Department of Information and Communication Systems Engineering, University of the Aegean, Karlovasi, Samos, Greece

George Karakostas Department of Computing and Software, McMaster University, Hamilton, ON, Canada

Juha Kärkkäinen Department of Computer Science, University of Helsinki, Helsinki, Finland

Petteri Kaski Department of Computer Science, School of Science, Aalto University, Helsinki, Finland

Helsinki Institute for Information Technology (HIIT), Helsinki, Finland

Hans Kellerer Department of Statistics and Operations Research, University of Graz, Graz, Austria

Andrew A. Kennings Department of Electrical and Computer Engineering, University of Waterloo, Waterloo, ON, Canada

Kurt Keutzer Department of Electrical Engineering and Computer Science, University of California, Berkeley, CA, USA

Mohammad Reza Khani University of Maryland, College Park, MD, USA

Samir Khuller Computer Science Department, University of Maryland, College Park, MD, USA

Donghyun Kim Department of Mathematics and Physics, North Carolina Central University, Durham, NC, USA

Jin Wook Kim HM Research, Seoul, Korea

Yoo-Ah Kim Computer Science and Engineering Department, University of Connecticut, Storrs, CT, USA

Valerie King Department of Computer Science, University of Victoria, Victoria, BC, Canada

Zoltán Király Department of Computer Science, Eötvös Loránd University, Budapest, Hungary

Egerváry Research Group (MTA-ELTE), Eötvös Loránd University, Budapest, Hungary

Lefteris Kirousis Department of Computer Engineering and Informatics, University of Patras, Patras, Greece

Jyrki Kivinen Department of Computer Science, University of Helsinki, Helsinki, Finland

Masashi Kiyomi International College of Arts and Sciences, Yokohama City University, Yokohama, Kanagawa, Japan

Kim-Manuel Klein University Kiel, Kiel, Germany

Rolf Klein Institute for Computer Science, University of Bonn, Bonn, Germany

Adam Klivans Department of Computer Science, University of Texas, Austin, TX, USA

Koji M. Kobayashi National Institute of Informatics, Chiyoda-ku, Tokyo, Japan

Stephen Kobourov Department of Computer Science, University of Arizona, Tucson, AZ, USA

Kirill Kogan IMDEA Networks, Madrid, Spain

Christian Komusiewicz Institute of Software Engineering and Theoretical Computer Science, Technical University of Berlin, Berlin, Germany

Goran Konjevod Department of Computer Science and Engineering, Arizona State University, Tempe, AZ, USA

Spyros Kontogiannis Department of Computer Science, University of Ioannina, Ioannina, Greece

Matias Korman Graduate School of Information Sciences, Tohoku University, Miyagi, Japan

Guy Kortsarz Department of Computer Science, Rutgers University, Camden, NJ, USA

Nitish Korula Google Research, New York, NY, USA

Robin Kothari Center for Theoretical Physics, Massachusetts Institute of Technology, Cambridge, MA, USA

David R. Cheriton School of Computer Science, Institute for Quantum Computing, University of Waterloo, Waterloo, ON, Canada

Ioannis Koutis Computer Science Department, University of Puerto Rico-Rio Piedras, San Juan, PR, USA

Dariusz R. Kowalski Department of Computer Science, University of Liverpool, Liverpool, UK

Evangelos Kranakis Department of Computer Science, Carleton, Ottawa, ON, Canada

Dieter Kratsch UFM MIM – LITA, Université de Lorraine, Metz, France

Stefan Kratsch Department of Software Engineering and Theoretical Computer Science, Technical University Berlin, Berlin, Germany

Robert Krauthgamer Weizmann Institute of Science, Rehovot, Israel

IBM Almaden Research Center, San Jose, CA, USA

Stephan Kreutzer Chair for Logic and Semantics, Technical University, Berlin, Germany

Sebastian Krinninger Faculty of Computer Science, University of Vienna, Vienna, Austria

Ravishankar Krishnaswamy Computer Science Department, Princeton University, Princeton, NJ, USA

Danny Krizanc Department of Computer Science, Wesleyan University, Middletown, CT, USA

Piotr Krysta Department of Computer Science, University of Liverpool, Liverpool, UK

Gregory Kucherov CNRS/LIGM, Université Paris-Est, Marne-la-Vallée, France

Fabian Kuhn Department of Computer Science, ETH Zurich, Zurich, Switzerland

V.S. Anil Kumar Virginia Bioinformatics Institute, Virginia Tech, Blacksburg, VA, USA

Tak-Wah Lam Department of Computer Science, University of Hong Kong, Hong Kong, China

Giuseppe Lancia Department of Mathematics and Computer Science, University of Udine, Udine, Italy

Gad M. Landau Department of Computer Science, University of Haifa, Haifa, Israel

Zeph Landau Department of Computer Science, University of California, Berkelely, CA, USA

Michael Langberg Department of Electrical Engineering, The State University of New York, Buffalo, NY, USA

Department of Mathematics and Computer Science, The Open University of Israel, Raanana, Israel

Elmar Langetepe Department of Computer Science, University of Bonn, Bonn, Germany

Ron Lavi Faculty of Industrial Engineering and Management, Technion, Haifa, Israel

Thierry Lecroq Computer Science Department and LITIS Faculty of Science, Université de Rouen, Rouen, France

James R. Lee Department of Computer Science and Engineering, University of Washington, Seattle, WA, USA

Stefano Leonardi Department of Information and Computer Systems, University of Rome, Rome, Italy

Pierre Leone Informatics Department, University of Geneva, Geneva, Switzerland

Henry Leung Department of Computer Science, The University of Hong Kong, Hong Kong, China

Christos Levcopoulos Department of Computer Science, Lund University, Lund, Sweden

Asaf Levin Faculty of Industrial Engineering and Management, The Technion, Haifa, Israel

Moshe Lewenstein Department of Computer Science, Bar-Ilan University, Ramat-Gan, Israel

Li (Erran) Li Bell Laboratories, Alcatel-Lucent, Murray Hill, NJ, USA

Mengling Li Division of Mathematical Sciences, Nanyang Technological University, Singapore, Singapore

Ming Li David R. Cheriton School of Computer Science, University of Waterloo, Waterloo, ON, Canada

Ming Min Li Computer Science and Technology, Tsinghua University, Beijing, China

Xiang-Yang Li Department of Computer Science, Illinois Institute of Technology, Chicago, IL, USA

Vahid Liaghat Department of Computer Science, University of Maryland, College Park, MD, USA

Jie Liang Department of Bioengineering, University of Illinois, Chicago, IL, USA

Andrzej Lingas Department of Computer Science, Lund University, Lund, Sweden

Maarten Löffler Department of Information and Computing Sciences, Utrecht University, Utrecht, The Netherlands

Daniel Lokshtanov Department of Informatics, University of Bergen, Bergen, Norway

Alejandro López-Ortiz David R. Cheriton School of Computer Science, University of Waterloo, Waterloo, ON, Canada

Chin Lung Lu Institute of Bioinformatics and Department of Biological Science and Technology, National Chiao Tung University, Hsinchu, Taiwan

Pinyan Lu Microsoft Research Asia, Shanghai, China

Zaixin Lu Department of Mathematics and Computer Science, Marywood University, Scranton, PA, USA

Feng Luo Department of Mathematics, Rutgers University, Piscataway, NJ, USA

Haiming Luo Department of Computer Science and Technology, Shenzhen Graduate School, Harbin Institute of Technology, Shenzhen, China

Rune B. Lyngsø Department of Statistics, Oxford University, Oxford, UK

Winton Capital Management, Oxford, UK

Bin Ma David R. Cheriton School of Computer Science, University of Waterloo, Waterloo, ON, Canada

Department of Computer Science, University of Western Ontario, London, ON, Canada

Mohammad Mahdian Yahoo! Research, Santa Clara, CA, USA

Hamid Mahini Department of Computer Science, University of Maryland, College Park, MD, USA

Veli Mäkinen Department of Computer Science, Helsinki Institute for Information Technology (HIIT), University of Helsinki, Helsinki, Finland

Dahlia Malkhi Microsoft, Silicon Valley Campus, Mountain View, CA, USA

Mark S. Manasse Microsoft Research, Mountain View, CA, USA

David F. Manlove School of Computing Science, University of Glasgow, Glasgow, UK

Giovanni Manzini Department of Computer Science, University of Eastern Piedmont, Alessandria, Italy

Department of Science and Technological Innovation, University of Piemonte Orientale, Alessandria, Italy

Madha V. Marathe IBM T.J. Watson Research Center, Hawthorne, NY, USA

Alberto Marchetti-Spaccamela Department of Information and Computer Systems, University of Rome, Rome, Italy

Igor L. Markov Department of Electrical Engineering and Computer Science, University of Michigan, Ann Arbor, MI, USA

Alexander Matveev Computer Science and Artificial Intelligence Laboratory, MIT, Cambridge, MA, USA

Eric McDermid Cedar Park, TX, USA

Catherine C. McGeoch Department of Mathematics and Computer Science, Amherst College, Amherst, MA, USA

Lyle A. McGeoch Department of Mathematics and Computer Science, Amherst College, Amherst, MA, USA

Andrew McGregor School of Computer Science, University of Massachusetts, Amherst, MA, USA

Brendan D. McKay Department of Computer Science, Australian National University, Canberra, ACT, Australia

Nicole Megow Institut für Mathematik, Technische Universität Berlin, Berlin, Germany

Manor Mendel Department of Mathematics and Computer Science, The Open University of Israel, Raanana, Israel

George B. Mertzios School of Engineering and Computing Sciences, Durham University, Durham, UK

Julián Mestre Department of Computer Science, University of Maryland, College Park, MD, USA

School of Information Technologies, The University of Sydney, Sydney, NSW, Australia

Pierre-Étienne Meunier Le Laboratoire d'Informatique Fondamentale de Marseille (LIF), Aix-Marseille Université, Marseille, France

Ulrich Meyer Department of Computer Science, Goethe University Fankfurt am Main, Frankfurt, Germany

Daniele Micciancio Department of Computer Science, University of California, San Diego, La Jolla, CA, USA

István Miklós Department of Plant Taxonomy and Ecology, Eötvös Loránd University, Budapest, Hungary

Shin-ichi Minato Graduate School of Information Science and Technology, Hokkaido University, Sapporo, Japan

Vahab S. Mirrokni Theory Group, Microsoft Research, Redmond, WA, USA

Neeldhara Misra Department of Computer Science and Automation, Indian Institute of Science, Bangalore, India

Joseph S.B. Mitchell Department of Applied Mathematics and Statistics, Stony Brook University, Stony Brook, NY, USA

Shuichi Miyazaki Academic Center for Computing and Media Studies, Kyoto University, Kyoto, Japan

Alistair Moffat Department of Computing and Information Systems, The University of Melbourne, Melbourne, VIC, Australia

Mark Moir Sun Microsystems Laboratories, Burlington, MA, USA

Ashley Montanaro Department of Computer Science, University of Bristol, Bristol, UK

Tal Mor Department of Computer Science, Technion – Israel Institute of Technology, Haifa, Israel

Michele Mosca Canadian Institute for Advanced Research, Toronto, ON, Canada

Combinatorics and Optimization/Institute for Quantum Computing, University of Waterloo, Waterloo, ON, Canada

Perimeter Institute for Theoretical Physics, Waterloo, ON, Canada

Thomas Moscibroda Systems and Networking Research Group, Microsoft Research, Redmond, WA, USA

Yoram Moses Department of Electrical Engineering, Technion – Israel Institute of Technology, Haifa, Israel

Shay Mozes Efi Arazi School of Computer Science, The Interdisciplinary Center (IDC), Herzliya, Israel

Marcin Mucha Faculty of Mathematics, Informatics and Mechanics, Institute of Informatics, Warsaw, Poland

Priyanka Mukhopadhyay Center for Quantum Technologies, National University of Singapore, Singapore, Singapore

Kamesh Munagala Levine Science Research Center, Duke University, Durham, NC, USA

J. Ian Munro David R. Cheriton School of Computer Science, University of Waterloo, Waterloo, ON, Canada

Joong Chae Na Department of Computer Science and Engineering, Sejong University, Seoul, Korea

Viswanath Nagarajan University of Michigan, Ann Arbor, MI, USA

Shin-ichi Nakano Department of Computer Science, Gunma University, Kiryu, Japan

Danupon Nanongkai School of Computer Science and Communication, KTH Royal Institute of Technology, Stockholm, Sweden

Giri Narasimhan Department of Computer Science, Florida International University, Miami, FL, USA

School of Computing and Information Sciences, Florida International University, Miami, FL, USA

Gonzalo Navarro Department of Computer Science, University of Chile, Santiago, Chile

Ashwin Nayak Department of Combinatorics and Optimization, and Institute for Quantum Computing, University of Waterloo, Waterloo, ON, Canada

Amir Nayyeri Department of Electrical Engineering and Computer Science, Oregon State University, Corvallis, OR, USA

Jesper Nederlof Technical University of Eindhoven, Eindhoven, The Netherlands

Ofer Neiman Department of Computer Science, Ben-Gurion University of the Negev, Beer Sheva, Israel

Yakov Nekrich David R. Cheriton School of Computer Science, University of Waterloo, Waterloo, ON, Canada

Jelani Nelson Harvard John A. Paulson School of Engineering and Applied Sciences, Cambridge, MA, USA

Ragnar Nevries Computer Science Department, University of Rostock, Rostock, Germany

Alantha Newman CNRS-Université Grenoble Alpes and G-SCOP, Grenoble, France

Hung Q. Ngo Computer Science and Engineering, The State University of New York, Buffalo, NY, USA

Patrick K. Nicholson Department D1: Algorithms and Complexity, Max Planck Institut für Informatik, Saarbrücken, Germany

Rolf Niedermeier Department of Mathematics and Computer Science, University of Jena, Jena, Germany

Institut für Softwaretechnik und Theoretische Informatik, Technische Universität Berlin, Berlin, Germany

Sergey I. Nikolenko Laboratory of Mathematical Logic, Steklov Institute of Mathematics, St. Petersburg, Russia

Sotiris Nikoletseas Computer Engineering and Informatics Department, University of Patras, Patras, Greece

Computer Technology Institute and Press "Diophantus", Patras, Greece

Aleksandar Nikolov Department of Computer Science, Rutgers University, Piscataway, NJ, USA

Nikola S. Nikolov Department of Computer Science and Information Systems, University of Limerick, Limerick, Republic of Ireland

Kobbi Nisim Department of Computer Science, Ben-Gurion University, Beer Sheva, Israel

Lhouari Nourine Clermont-Université, Université Blaise Pascal, LIMOS, CNRS, Aubière, France

Yoshio Okamoto Department of Information and Computer Sciences, Toyohashi University of Technology, Toyohashi, Japan

Michael Okun Weizmann Institute of Science, Rehovot, Israel

Rasmus Pagh Theoretical Computer Science, IT University of Copenhagen, Copenhagen, Denmark

David Z. Pan Department of Electrical and Computer Engineering, University of Texas, Austin, TX, USA

Peichen Pan Xilinx, Inc., San Jose, CA, USA

Debmalya Panigrahi Department of Computer Science, Duke University, Durham, NC, USA

Fahad Panolan Institute of Mathematical Sciences, Chennai, India

Vicky Papadopoulou Department of Computer Science, University of Cyprus, Nicosia, Cyprus

Fabio Pardi Institut de Biologie Computationnelle, Laboratoire d'Informatique, de Robotique et de Microélectronique de Montpellier (LIRMM), CNRS and Université de Montpellier, Montpellier cedex 5, France

Kunsoo Park School of Computer Science and Engineering, Seoul National University, Seoul, Korea

Srinivasan Parthasarathy IBM T.J. Watson Research Center, Hawthorne, NY, USA

Apoorva D. Patel Centre for High Energy Physics, Indian Institute of Science, Bangalore, India

Matthew J. Patitz Department of Computer Science and Computer Engineering, University of Arkansas, Fayetteville, AR, USA

Mihai Pătraşcu Computer Science and Artificial Intelligence Laboratory (CSAIL), Massachusetts Institute of Technology (MIT), Cambridge, MA, USA

Maurizio Patrignani Engineering Department, Roma Tre University, Rome, Italy

Boaz Patt-Shamir Department of Electrical Engineering, Tel-Aviv University, Tel-Aviv, Israel

Ramamohan Paturi Department of Computer Science and Engineering, University of California at San Diego, San Diego, CA, USA

Christophe Paul CNRS, Laboratoire d'Informatique Robotique et Microélectronique de Montpellier, Université Montpellier 2, Montpellier, France

Andrzej Pelc Department of Computer Science, University of Québec-Ottawa, Gatineau, QC, Canada

Jean-Marc Petit Université de Lyon, CNRS, INSA Lyon, LIRIS, Lyon, France

Seth Pettie Electrical Engineering and Computer Science (EECS) Department, University of Michigan, Ann Arbor, MI, USA

Marcin Pilipczuk Institute of Informatics, University of Bergen, Bergen, Norway

Institute of Informatics, University of Warsaw, Warsaw, Poland

Michał Pilipczuk Institute of Informatics, University of Warsaw, Warsaw, Poland

Institute of Informatics, University of Bergen, Bergen, Norway

Yuri Pirola Università degli Studi di Milano-Bicocca, Milan, Italy

Olivier Powell Informatics Department, University of Geneva, Geneva, Switzerland

Amit Prakash Microsoft, MSN, Redmond, WA, USA

Eric Price Department of Computer Science, The University of Texas, Austin, TX, USA

Kirk Pruhs Department of Computer Science, University of Pittsburgh, Pittsburgh, PA, USA

Teresa M. Przytycka Computational Biology Branch, NCBI, NIH, Bethesda, MD, USA

Pavel Pudlák Academy of Science of the Czech Republic, Mathematical Institute, Prague, Czech Republic

Simon J. Puglisi Department of Computer Science, University of Helsinki, Helsinki, Finland

Balaji Raghavachari Computer Science Department, The University of Texas at Dallas, Richardson, TX, USA

Md. Saidur Rahman Department of Computer Science and Engineering, Bangladesh University of Engineering and Technology, Dhaka, Bangladesh

Naila Rahman University of Hertfordshire, Hertfordshire, UK

Rajmohan Rajaraman Department of Computer Science, Northeastern University, Boston, MA, USA

Sergio Rajsbaum Instituto de Matemáticas, Universidad Nacional Autónoma de México (UNAM), México City, México

Vijaya Ramachandran Computer Science, University of Texas, Austin, TX, USA

Rajeev Raman Department of Computer Science, University of Leicester, Leicester, UK

M.S. Ramanujan Department of Informatics, University of Bergen, Bergen, Norway

Edgar Ramos School of Mathematics, National University of Colombia, Medellín, Colombia

Satish Rao Department of Computer Science, University of California, Berkeley, CA, USA

Christoforos L. Raptopoulos Computer Science Department, University of Geneva, Geneva, Switzerland

Computer Technology Institute and Press "Diophantus", Patras, Greece

Research Academic Computer Technology Institute, Greece and Computer Engineering and Informatics Department, University of Patras, Patras, Greece

Sofya Raskhodnikova Computer Science and Engineering Department, Pennsylvania State University, University Park, PA, USA

Rajeev Rastogi Amazon, Seattle, WA, USA

Joel Ratsaby Department of Electrical and Electronics Engineering, Ariel University of Samaria, Ariel, Israel

Kaushik Ravindran National Instruments, Berkeley, CA, USA

Michel Raynal Institut Universitaire de France and IRISA, Université de Rennes, Rennes, France

Ben W. Reichardt Electrical Engineering Department, University of Southern California (USC), Los Angeles, CA, USA

Renato Renner Institute for Theoretical Physics, Zurich, Switzerland

Elisa Ricci Department of Electronic and Information Engineering, University of Perugia, Perugia, Italy

Andréa W. Richa School of Computing, Informatics, and Decision Systems Engineering, Ira A. Fulton Schools of Engineering, Arizona State University, Tempe, AZ, USA

Peter C. Richter Department of Combinatorics and Optimization, and Institute for Quantum Computing, University of Waterloo, Waterloo, ON, Canada

Department of Computer Science, Rutgers, The State University of New Jersey, New Brunswick, NJ, USA

Liam Roditty Department of Computer Science, Bar-Ilan University, Ramat-Gan, Israel

Marcel Roeloffzen Graduate School of Information Sciences, Tohoku University, Sendai, Japan

Martin Roetteler Microsoft Research, Redmond, WA, USA

Heiko Röglin Department of Computer Science, University of Bonn, Bonn, Germany

José Rolim Informatics Department, University of Geneva, Geneva, Switzerland

Dana Ron School of Electrical Engineering, Tel-Aviv University, Ramat-Aviv, Israel

Frances Rosamond Parameterized Complexity Research Unit, University of Newcastle, Callaghan, NSW, Australia

Jarek Rossignac Georgia Institute of Technology, Atlanta, GA, USA

Matthieu Roy Laboratory of Analysis and Architecture of Systems (LAAS), Centre National de la Recherche Scientifique (CNRS), Université Toulouse, Toulouse, France

Ronitt Rubinfeld Massachusetts Institute of Technology (MIT), Cambridge, MA, USA

Tel Aviv University, Tel Aviv-Yafo, Israel

Atri Rudra Department of Computer Science and Engineering, State University of New York, Buffalo, NY, USA

Eric Ruppert Department of Computer Science and Engineering, York University, Toronto, ON, Canada

Frank Ruskey Department of Computer Science, University of Victoria, Victoria, BC, Canada

Luís M.S. Russo Departamento de Informática, Instituto Superior Técnico, Universidade de Lisboa, Lisboa, Portugal

INESC-ID, Lisboa, Portugal

Wojciech Rytter Institute of Informatics, Warsaw University, Warsaw, Poland

Kunihiko Sadakane Graduate School of Information Science and Technology, The University of Tokyo, Tokyo, Japan

S. Cenk Sahinalp Laboratory for Computational Biology, Simon Fraser University, Burnaby, BC, USA

Michael Saks Department of Mathematics, Rutgers, State University of New Jersey, Piscataway, NJ, USA

Alejandro Salinger Department of Computer Science, Saarland University, Saarbücken, Germany

Sachin S. Sapatnekar Department of Electrical and Computer Engineering, University of Minnesota, Minneapolis, MN, USA

Shubhangi Saraf Department of Mathematics and Department of Computer Science, Rutgers University, Piscataway, NJ, USA

Srinivasa Rao Satti Department of Computer Science and Engineering, Seoul National University, Seoul, South Korea

Saket Saurabh Institute of Mathematical Sciences, Chennai, India

University of Bergen, Bergen, Norway

Guido Schäfer Institute for Mathematics and Computer Science, Technical University of Berlin, Berlin, Germany

Dominik Scheder Institute for Interdisciplinary Information Sciences, Tsinghua University, Beijing, China

Institute for Computer Science, Shanghai Jiaotong University, Shanghai, China

Christian Scheideler Department of Computer Science, University of Paderborn, Paderborn, Germany

André Schiper EPFL, Lausanne, Switzerland

Christiane Schmidt The Selim and Rachel Benin School of Computer Science and Engineering, The Hebrew University of Jerusalem, Jerusalem, Israel

Markus Schmidt Institute for Computer Science, University of Freiburg, Freiburg, Germany

Dominik Schultes Institute for Computer Science, University of Karlsruhe, Karlsruhe, Germany

Robert Schweller Department of Computer Science, University of Texas Rio Grande Valley, Edinburg, TX, USA

Shinnosuke Seki Department of Computer Science, Helsinki Institute for Information Technology (HIIT), Aalto University, Aalto, Finland

Pranab Sen School of Technology and Computer Science, Tata Institute of Fundamental Research, Mumbai, India

Sandeep Sen Indian Institute of Technology (IIT) Delhi, Hauz Khas, New Delhi, India

Maria Serna Department of Language and System Information, Technical University of Catalonia, Barcelona, Spain

Rocco A. Servedio Computer Science, Columbia University, New York, NY, USA

Comandur Seshadhri Sandia National Laboratories, Livermore, CA, USA

Department of Computer Science, University of California, Santa Cruz, CA, USA

Jay Sethuraman Industrial Engineering and Operations Research, Columbia University, New York, NY, USA

Jiří Sgall Computer Science Institute, Charles University, Prague, Czech Republic

Rahul Shah Department of Computer Science, Louisiana State University, Baton Rouge, LA, USA

Shai Shalev-Shwartz School of Computer Science and Engineering, The Hebrew University, Jerusalem, Israel

Vikram Sharma Department of Computer Science, New York University, New York, NY, USA

Nir Shavit Computer Science and Artificial Intelligence Laboratory, MIT, Cambridge, MA, USA

School of Computer Science, Tel-Aviv University, Tel-Aviv, Israel

Yaoyun Shi Department of Electrical Engineering and Computer Science, University of Michigan, Ann Arbor, MI, USA

Ayumi Shinohara Graduate School of Information Sciences, Tohoku University, Sendai, Japan

Eugene Shragowitz Department of Computer Science and Engineering, University of Minnesota, Minneapolis, MN, USA

René A. Sitters Department of Econometrics and Operations Research, VU University, Amsterdam, The Netherlands

Balasubramanian Sivan Microsoft Research, Redmond, WA, USA

Daniel Sleator Department of Computer Science, Carnegie Mellon University, Pittsburgh, PA, USA

Michiel Smid School of Computer Science, Carleton University, Ottawa, ON, Canada

Adam Smith Computer Science and Engineering Department, Pennsylvania State University, University Park, State College, PA, USA

Dina Sokol Department of Computer and Information Science, Brooklyn College of CUNY, Brooklyn, NY, USA

Rolando D. Somma Theoretical Division, Los Alamos National Laboratory, Los Alamos, NM, USA

Wen-Zhan Song School of Engineering and Computer Science, Washington State University, Vancouver, WA, USA

Bettina Speckmann Department of Mathematics and Computer Science, Technical University of Eindhoven, Eindhoven, The Netherlands

Paul (Pavlos) Spirakis Computer Engineering and Informatics, Research and Academic Computer Technology Institute, Patras University, Patras, Greece

Computer Science, University of Liverpool, Liverpool, UK

Computer Technology Institute (CTI), Patras, Greece

Aravind Srinivasan Department of Computer Science, University of Maryland, College Park, MD, USA

Venkatesh Srinivasan Department of Computer Science, University of Victoria, Victoria, BC, Canada

Gerth Stølting Department of Computer Science, University of Aarhus, Århus, Denmark

Jens Stoye Faculty of Technology, Genome Informatics, Bielefeld University, Bielefeld, Germany

Scott M. Summers Department of Computer Science, University of Wisconsin – Oshkosh, Oshkosh, WI, USA

Aries Wei Sun Department of Computer Science, City University of Hong Kong, Hong Kong, China

Vijay Sundararajan Broadcom Corp, Fremont, CA, USA

Wing-Kin Sung Department of Computer Science, National University of Singapore, Singapore, Singapore

Mario Szegedy Department of Combinatorics and Optimization, and Institute for Quantum Computing, University of Waterloo, Waterloo, ON, Canada

Stefan Szeider Department of Computer Science, Durham University, Durham, UK

Tadao Takaoka Department of Computer Science and Software Engineering, University of Canterbury, Christchurch, New Zealand

Masayuki Takeda Department of Informatics, Kyushu University, Fukuoka, Japan

Kunal Talwar Microsoft Research, Silicon Valley Campus, Mountain View, CA, USA

Christino Tamon Department of Computer Science, Clarkson University, Potsdam, NY, USA

Akihisa Tamura Department of Mathematics, Keio University, Yokohama, Japan

Tiow-Seng Tan School of Computing, National University of Singapore, Singapore, Singapore

Shin-ichi Tanigawa Research Institute for Mathematical Sciences (RIMS), Kyoto University, Kyoto, Japan

Eric Tannier LBBE Biometry and Evolutionary Biology, INRIA Grenoble Rhône-Alpes, University of Lyon, Lyon, France

Alain Tapp Université de Montréal, Montréal, QC, Canada

Stephen R. Tate Department of Computer Science, University of North Carolina, Greensboro, NC, USA

Gadi Taubenfeld Department of Computer Science, Interdiciplinary Center Herzlia, Herzliya, Israel

Kavitha Telikepalli CSA Department, Indian Institute of Science, Bangalore, India

Barbara M. Terhal JARA Institute for Quantum Information, RWTH Aachen University, Aachen, Germany

Alexandre Termier IRISA, University of Rennes, 1, Rennes, France

My T. Thai Department of Computer and Information Science and Engineering, University of Florida, Gainesville, FL, USA

Abhradeep Thakurta Department of Computer Science, Stanford University, Stanford, CA, USA

Microsoft Research, CA, USA

Justin Thaler Yahoo! Labs, New York, NY, USA

Sharma V. Thankachan School of CSE, Georgia Institute of Technology, Atlanta, USA

Dimitrios Thilikos AlGCo Project-Team, CNRS, LIRMM, France

Department of Mathematics, National and Kapodistrian University of Athens, Athens, Greece

Haitong Tian Department of Electrical and Computer Engineering, University of Illinois at Urbana-Champaign, Urbana, IL, USA

Ioan Todinca INSA Centre Val de Loire, Universite d'Orleans, Orléans, France

Alade O. Tokuta Department of Mathematics and Physics, North Carolina Central University, Durham, NC, USA

Laura Toma Department of Computer Science, Bowdoin College, Brunswick, ME, USA

Etsuji Tomita The Advanced Algorithms Research Laboratory, The University of Electro-Communications, Chofu, Tokyo, Japan

Csaba D. Tóth Department of Computer Science, Tufts University, Medford, MA, USA

Department of Mathematics, California State University Northridge, Los Angeles, CA, USA

Luca Trevisan Department of Computer Science, University of California, Berkeley, CA, USA

John Tromp CWI, Amsterdam, The Netherlands

Nicolas Trotignon Laboratoire de l'Informatique du Parallélisme (LIP), CNRS, ENS de Lyon, Lyon, France

Jakub Truszkowski Cancer Research UK Cambridge Institute, University of Cambridge, Cambridge, UK

European Molecular Biology Laboratory, European Bioinformatics Institute (EMBL-EBI), Wellcome Trust Genome Campus, Hinxton, Cambridge, UK

Esko Ukkonen Department of Computer Science, Helsinki Institute for Information Technology (HIIT), University of Helsinki, Helsinki, Finland

Jonathan Ullman Department of Computer Science, Columbia University, New York, NY, USA

Takeaki Uno National Institute of Informatics, Chiyoda, Tokyo, Japan

Ruth Urner Department of Machine Learning, Carnegie Mellon University, Pittsburgh, USA

Jan Vahrenhold Department of Computer Science, Westfälische Wilhelms-Universität Münster, Münster, Germany

Daniel Valenzuela Department of Computer Science, Helsinki Institute for Information Technology (HIIT), University of Helsinki, Helsinki, Finland

Marc van Kreveld Department of Information and Computing Sciences, Utrecht University, Utrecht, The Netherlands

Rob van Stee University of Leicester, Leicester, UK

Stefano Varricchio Department of Computer Science, University of Roma, Rome, Italy

José Verschae Departamento de Matemáticas and Departamento de Ingeniería Industrial y de Sistemas, Pontificia Universidad Católica de Chile, Santiago, Chile

Stéphane Vialette IGM-LabInfo, University of Paris-East, Descartes, France

Sebastiano Vigna Dipartimento di Informatica, Università degli Studi di Milano, Milano, Italy

Yngve Villanger Department of Informatics, University of Bergen, Bergen, Norway

Paul Vitányi Centrum Wiskunde & Informatica (CWI), Amsterdam, The Netherlands

Jeffrey Scott Vitter University of Kansas, Lawrence, KS, USA

Berthold Vöcking Department of Computer Science, RWTH Aachen University, Aachen, Germany

Tjark Vredeveld Department of Quantitative Economics, Maastricht University, Maastricht, The Netherlands

Magnus Wahlström Department of Computer Science, Royal Holloway, University of London, Egham, UK

Peng-Jun Wan Department of Computer Science, Illinois Institute of Technology, Chicago, IL, USA

Chengwen Chris Wang Department of Computer Science, Carnegie Mellon University, Pittsburgh, PA, USA

Feng Wang Mathematical Science and Applied Computing, Arizona State University at the West Campus, Phoenix, AZ, USA

Huijuan Wang Shandong University, Jinan, China

Joshua R. Wang Department of Computer Science, Stanford University, Stanford, CA, USA

Lusheng Wang Department of Computer Science, City University of Hong Kong, Hong Kong, Hong Kong

Wei Wang School of Mathematics and Statistics, Xi'an Jiaotong University, Xi'an, Shaanxi, China

Weizhao Wang Google Inc., Irvine, CA, USA

Yu Wang Department of Computer Science, University of North Carolina, Charlotte, NC, USA

Takashi Washio The Institute of Scientific and Industrial Research, Osaka University, Ibaraki, Osaka, Japan

Matthew Weinberg Computer Science, Princeton University, Princeton, NJ, USA

Tobias Weinzierl School of Engineering and Computing Sciences, Durham University, Durham, UK

Renato F. Werneck Microsoft Research Silicon Valley, La Avenida, CA, USA

Matthias Westermann Department of Computer Science, TU Dortmund University, Dortmund, Germany

Tim A.C. Willemse Department of Mathematics and Computer Science, Eindhoven University of Technology, Eindhoven, The Netherlands

Ryan Williams Department of Computer Science, Stanford University, Stanford, CA, USA

Tyson Williams Computer Sciences Department, University of Wisconsin–Madison, Madison, WI, USA

Andrew Winslow Department of Computer Science, Tufts University, Medford, MA, USA

Paul Wollan Department of Computer Science, University of Rome La Sapienza, Rome, Italy

Martin D.F. Wong Department of Electrical and Computer Engineering, University of Illinois at Urbana-Champaign, Urbana, IL, USA

Prudence W.H. Wong University of Liverpool, Liverpool, UK

David R. Wood School of Mathematical Sciences, Monash University, Melbourne, VIC, Australia

Damien Woods Computer Science, California Institute of Technology, Pasadena, CA, USA

Lidong Wu Department of Computer Science, The University of Texas, Tyler, TX, USA

Weili Wu College of Computer Science and Technology, Taiyuan University of Technology, Taiyuan, Shanxi Province, China

Department of Computer Science, California State University, Los Angeles, CA, USA

Department of Computer Science, The University of Texas at Dallas, Richardson, TX, USA

Christian Wulff-Nilsen Department of Computer Science, University of Copenhagen, Copenhagen, Denmark

Mingji Xia The State Key Laboratory of Computer Science, Chinese Academy of Sciences, Beijing, China

David Xiao CNRS, Université Paris 7, Paris, France

Dong Xu Bond Life Sciences Center, University of Missouri, Columbia, MO, USA

Wen Xu Department of Computer Science, The University of Texas at Dallas, Richardson, TX, USA

Katsuhisa Yamanaka Department of Electrical Engineering and Computer Science, Iwate University, Iwate, Japan

Hiroki Yanagisawa IBM Research – Tokyo, Tokyo, Japan

Honghua Hannah Yang Strategic CAD Laboratories, Intel Corporation, Hillsboro, OR, USA

Qiuming Yao University of Missouri, Columbia, MO, USA

Chee K. Yap Department of Computer Science, New York University, New York, NY, USA

Yinyu Ye Department of Management Science and Engineering, Stanford University, Stanford, CA, USA

Anders Yeo Engineering Systems and Design, Singapore University of Technology and Design, Singapore, Singapore

Department of Mathematics, University of Johannesburg, Auckland Park, South Africa

Chih-Wei Yi Department of Computer Science, National Chiao Tung University, Hsinchu City, Taiwan

Ke Yi Hong Kong University of Science and Technology, Hong Kong, China

Yitong Yin Nanjing University, Jiangsu, Nanjing, Gulou, China

S.M. Yiu Department of Computer Science, University of Hong Kong, Hong Kong, China

Makoto Yokoo Department of Information Science and Electrical Engineering, Kyushu University, Nishi-ku, Fukuoka, Japan

Evangeline F.Y. Young Department of Computer Science and Engineering, The Chinese University of Hong Kong, Hong Kong, China

Neal E. Young Department of Computer Science and Engineering, University of California, Riverside, CA, USA

Bei Yu Department of Electrical and Computer Engineering, University of Texas, Austin, TX, USA

Yaoliang Yu Machine Learning Department, Carnegie Mellon University, Pittsburgh, PA, USA

Raphael Yuster Department of Mathematics, University of Haifa, Haifa, Israel

Morteza Zadimoghaddam Google Research, New York, NY, USA

Francis Zane Lucent Technologies, Bell Laboratories, Murray Hill, NJ, USA

Christos Zaroliagis Department of Computer Engineering and Informatics, University of Patras, Patras, Greece

Norbert Zeh Faculty of Computer Science, Dalhousie University, Halifax, NS, Canada

Li Zhang Microsoft Research, Mountain View, CA, USA

Louxin Zhang Department of Mathematics, National University of Singapore, Singapore, Singapore

Shengyu Zhang The Chinese University of Hong Kong, Hong Kong, China

Zhang Zhao College of Mathematics Physics and Information Engineering, Zhejiang Normal University, Zhejiang, Jinhua, China

Hai Zhou Electrical Engineering and Computer Science (EECS) Department, Northwestern University, Evanston, IL, USA

Yuqing Zhu Department of Computer Science, California State University, Los Angeles, CA, USA

Department of Computer Science, The University of Texas at Dallas, Richardson, TX, USA

Sandra Zilles Department of Computer Science, University of Regina, Regina, SK, Canada

Aaron Zollinger Department of Electrical Engineering and Computer Science, University of California, Berkeley, CA, USA

Uri Zwick Department of Mathematics and Computer Science, Tel-Aviv University, Tel-Aviv, Israel

A

Abelian Hidden Subgroup Problem

Michele Mosca
Canadian Institute for Advanced Research,
Toronto, ON, Canada
Combinatorics and Optimization/Institute for
Quantum Computing, University of Waterloo,
Waterloo, ON, Canada
Perimeter Institute for Theoretical Physics,
Waterloo, ON, Canada

Keywords

Abelian hidden subgroup problem; Abelian stabilizer problem; Quantum algorithms; Quantum complexity; Quantum computing

Years and Authors of Summarized Original Work

1995; Kitaev
2008; Mosca

Problem Definition

The Abelian hidden subgroup problem is the problem of finding generators for a subgroup K of an Abelian group G, where this subgroup is defined implicitly by a function $f : G \rightarrow X$, for some finite set X. In particular, f has the property that $f(v) = f(w)$ if and only if the cosets (we are assuming additive notation for the group operation here.) $v + K$ and $w + K$ are equal. In other words, f is constant on the cosets of the subgroup K and distinct on each coset.

It is assumed that the group G is finitely generated and that the elements of G and X have unique binary encodings. The binary assumption is only for convenience, but it is important to have unique encodings (e.g., in [22] Watrous uses a quantum state as the unique encoding of group elements). When using variables g and h (possibly with subscripts), multiplicative notation is used for the group operations. Variables x and y (possibly with subscripts) will denote integers with addition. The boldface versions $\mathbf{x}$ and $\mathbf{y}$ will denote *tuples* of integers or binary strings.

By assumption, there is computational means of computing the function f, typically a circuit or "black box" that maps the encoding of a value g to the encoding of $f(g)$. The theory of reversible computation implies that one can turn a circuit for computing $f(g)$ into a reversible circuit for computing $f(g)$ with a modest increase in the size of the circuit. Thus, it will be assumed that there is a reversible circuit or black box that maps $(g, \mathbf{z}) \mapsto (g, \mathbf{z} \oplus f(g))$, where $\oplus$ denotes the bit-wise XOR (sum modulo 2), and $\mathbf{z}$ is any binary string of the same length as the encoding of $f(g)$.

Quantum mechanics implies that any reversible gate can be extended linearly to a unitary operation that can be implemented in the model of quantum computation. Thus, it is assumed that there is a quantum circuit or

© Springer Science+Business Media New York 2016
M.-Y. Kao (ed.), *Encyclopedia of Algorithms*,
DOI 10.1007/978-1-4939-2864-4

black box that implements the unitary map $U_f : |g\rangle |\mathbf{z}\rangle \mapsto |g\rangle |\mathbf{z} \oplus f(g)\rangle$.

Although special cases of this problem have been considered in classical computer science, the general formulation as the hidden subgroup problem seems to have appeared in the context of quantum computing, since it neatly encapsulates a family of black-box problems for which quantum algorithms offer an exponential speedup (in terms of query complexity) over classical algorithms. For some explicit problems (i.e., where the black box is replaced with a specific function, such as exponentiation modulo N), there is a conjectured exponential speedup.

Abelian Hidden Subgroup Problem:

Input: Elements $g_1, g_2, \ldots, g_n \in G$ that generate the Abelian group G. A black box that implements $U_f : |m_1, m_2, \ldots, m_n\rangle |\mathbf{y}\rangle \mapsto |m_1, m_2, \ldots, m_n\rangle |f(g) \oplus \mathbf{y}\rangle$ where $g = g_1^{m_1} g_2^{m_2} \ldots g_n^{m_n}$ and K is the hidden subgroup corresponding to f.
Output: Elements $h_1, h_2, \ldots, h_l \in G$ that generate K.

Here we use multiplicative notation for the group G in order to be consistent with Kitaev's formulation of the Abelian stabilizer problem. Most of the applications of interest typically use additive notation for the group G.

It is hard to trace the precise origin of this general formulation of the problem, which simultaneously generalizes "Simon's problem" [20], the order-finding problem (which is the quantum part of the quantum factoring algorithm [18]), and the discrete logarithm problem.

One of the earliest generalizations of Simon's problem, order-finding problem, and discrete logarithm problem, which captures the essence of the Abelian hidden subgroup problem, is the *Abelian stabilizer problem* which was solved by Kitaev using a quantum algorithm in his 1995 paper [14] (and also appears in [15, 16]).

Let G be a group acting on a finite set X. That is, each element of G acts as a map from X to X in such a way that for any two elements $g, h \in G$, $g(h(z)) = (gh)(z)$ for all $z \in X$. For a particular element $z \in X$, the set of elements that fix z (i.e., the elements $g \in G$ such that $g(z) = z$) form a subgroup. This subgroup is called the stabilizer of z in G, denoted $St_G(z)$.

Abelian Stabilizer Problem:

Input: Elements $g_1, g_2, \ldots, g_n \in G$ that generate the group G. An element $z \in X$. A black box that implements $U_{(G,X)} : |m_1, m_2, \ldots, m_n\rangle |z\rangle \mapsto |m_1, m_2, \ldots, m_n\rangle |g(z)\rangle$ where $g = g_1^{m_1} g_2^{m_2} \ldots g_n^{m_n}$.
Output: Elements $h_1, h_2, \ldots, h_l \in G$ that generate $St_G(z)$.

Let f_z denote the function from G to X that maps $g \in G$ to $g(z)$. One can implement U_{f_z} using $U_{(G,X)}$. The hidden subgroup corresponding to f_z is $St_G(z)$. Thus, the Abelian stabilizer problem is a special case of the Abelian hidden subgroup problem.

One of the subtle differences (discussed in Appendix 6 of [12]) between the above formulation of the Abelian stabilizer problem and the Abelian hidden subgroup problem is that Kitaev's formulation gives a black box that for any $g, h \in G$ maps $|m_1, \ldots, m_n\rangle |f_z(h)\rangle \mapsto |m_1, \ldots, m_n\rangle |f_z(hg)\rangle$, where $g = g_1^{m_1} g_2^{m_2} \ldots g_n^{m_n}$. The algorithm given by Kitaev is essentially one that estimates eigenvalues of shift operations of the form $|f_z(h)\rangle \mapsto |f_z(hg)\rangle$. In general, these shift operators are not explicitly needed, and it suffices to be able to compute a map of the form $|\mathbf{y}\rangle \mapsto |f_z(h) \oplus \mathbf{y}\rangle$ for any binary string $\mathbf{y}$.

Generalizations of this form have been known since shortly after Shor presented his factoring and discrete logarithm algorithms (e.g., [23] presents the hidden subgroup problem for a large class of finite Abelian groups or more generally in [11] for finite Abelian groups presented as a product of finite cyclic groups. In [17] the natural Abelian hidden subgroup algorithm is related to eigenvalue estimation.)

Other problems which can be formulated in this way include:

Deutsch's Problem:

Input: A black box that implements $U_f : |x\rangle |b\rangle \mapsto |x\rangle |b \oplus f(x)\rangle$, for some function f that maps $\mathbb{Z}_2 = \{0, 1\}$ to $\{0, 1\}$.

Output: "constant" if $f(0) = f(1)$ and "balanced" if $f(0) \neq f(1)$.

Note that $f(x) = f(y)$ if and only if $x - y \in K$, where K is either $\{0\}$ or $\mathbb{Z}_2 = \{0, 1\}$. If $K = \{0\}$, then f is $1-1$ or "balanced," and if $K = \mathbb{Z}_2$, then f is constant [5,6].

Simon's Problem:

Input: A black box that implements U_f : $|\mathbf{x}\rangle |\mathbf{b}\rangle \mapsto |\mathbf{x}\rangle |\mathbf{b} \oplus f(\mathbf{x})\rangle$ for some function f from $\mathbb{Z}_2^n$ to some set X (which is assumed to consist of binary strings of some fixed length) with the property that $f(\mathbf{x}) = f(\mathbf{y})$ if and only if $\mathbf{x} - \mathbf{y} \in K = \{\mathbf{0}, \mathbf{s}\}$ for some $\mathbf{s} \in \mathbb{Z}_2^n$.
Output: The "hidden" string $\mathbf{s}$.

The decision version allows $K = \{\mathbf{0}\}$ and asks whether K is trivial. Simon [20] presents an efficient algorithm for solving this problem and an exponential lower bound on the query complexity. The solution to the Abelian hidden subgroup problem is a generalization of Simon's algorithm.

Key Results

Theorem (ASP) There exists a quantum algorithm that, given an instance of the Abelian stabilizer problem, makes $n + O(1)$ queries to $U_{(G,X)}$ and uses poly(n) other elementary quantum and classical operations, with probability at least $\frac{2}{3}$ output values $h_1, h_2, \ldots, h_l$ such that $St_G(z) = \langle h_1 \rangle \oplus \langle h_2 \rangle \oplus \cdots \langle h_l \rangle$.

Kitaev first solved this problem (with a slightly higher query complexity, because his eigenvalue estimation procedure was not optimal). An eigenvalue estimation procedure based on the Quantum Fourier Transform achieves the $n + O(1)$ query complexity [5].

Theorem (AHSP) There exists a quantum algorithm that, given an instance of the Abelian hidden subgroup problem, makes $n + O(1)$ queries to U_f and uses poly(n) other elementary quantum and classical operations, with probability at least $\frac{2}{3}$ output values $h_1, h_2, \ldots, h_l$ such that $K = \langle h_1 \rangle \oplus \langle h_2 \rangle \oplus \cdots \langle h_l \rangle$.

In some cases, the success probability can be made 1 with the same complexity, and in general the success probability can be made $1 - \epsilon$ using $n + O(\log(1/\epsilon))$ queries and poly$(n, \log(1/\epsilon))$ other elementary quantum and classical operations.

Applications

Most of these applications in fact were known before the Abelian stabilizer problem or hidden subgroup problem were formulated.

Finding the order of an element in a group: Let a be an element of a group H (which does *not* need to be Abelian), and let r be the smallest positive integer so that $a^r = 1$.

Consider the function f from $G = \mathbb{Z}$ to the group H where $f(x) = a^x$ for some element a of H. Then $f(x) = f(y)$ if and only if $x - y \in r\mathbb{Z}$. The hidden subgroup is $K = r\mathbb{Z}$ and a generator for K gives the order r of a.

Finding the period of a periodic function: Consider a function f from $G = \mathbb{Z}$ to a set X with the property that for some positive integer r, we have $f(x) = f(y)$ if and only if $x - y \in r\mathbb{Z}$. The hidden subgroup of f is $K = r\mathbb{Z}$ and a generator for K gives the period r.

Order finding is a special case of period finding and was also solved by Shor's algorithm [18].

Discrete Logarithms: Let a be an element of a group H (which does *not* need to be Abelian), with $a^r = 1$, and suppose $b = a^k$ from some unknown k. The integer k is called the *discrete logarithm of b to the base a*. Consider the function f from $G = \mathbb{Z}_r \times \mathbb{Z}_r$ to H satisfying $f(x_1, x_2) = a^{x_1} b^{x_2}$. Then $f(x_1, x_2) = f(y_1, y_2)$ if and only if $(x_1, x_2) - (y_1, y_2) \in \{(t, -tk), t = 0, 1, \ldots, r - 1\}$ which is the subgroup $\langle (1, -k) \rangle$ of $\mathbb{Z}_r \times \mathbb{Z}_r$. Thus, finding a generator for the hidden subgroup K will give the discrete logarithm k. Note that this algorithm works for H equal to the multiplicative group of a finite field, or the additive group of points on an elliptic curve, which are groups that are used in public-key cryptography.

Recently, Childs and Ivanyos [3] presented an efficient quantum algorithm for finding **discrete logarithms in semigroups**. Their algorithm makes use of the quantum algorithms for period finding and discrete logarithms as subroutines.

Hidden Linear Functions: Let σ be some permutation of $\mathbb{Z}_N$ for some integer N. Let h be a function from $G = \mathbb{Z} \times \mathbb{Z}$ to $\mathbb{Z}_N$, $h(x, y) = x + ay \mod N$. Let $f = \sigma \circ h$. The hidden subgroup of f is $\langle (-a, 1) \rangle$. Boneh and Lipton [1] showed that even if the linear structure of h is hidden (by σ), one can efficiently recover the parameter a with a quantum algorithm.

Self-Shift-Equivalent Polynomials: Given a polynomial P in l variables $X_1, X_2, \ldots, X_l$ over $\mathbb{F}_q$, the function f that maps $(a_1, a_2, \ldots, a_l) \in \mathbb{F}_q^l$ to $P(X_1 - a_1, X_2 - a_2, \ldots, X_l - a_l)$ is constant on cosets of a subgroup K of $\mathbb{F}_q^l$. This subgroup K is the set of shift-self-equivalences of the polynomial P. Grigoriev [10] showed how to compute this subgroup.

Decomposition of a Finitely Generated Abelian Group: Let G be a group with a unique binary representation for each element of G, and assume that the group operation, and recognizing if a binary string represents an element of G or not, can be done efficiently.

Given a set of generators $g_1, g_2, \ldots, g_n$ for a group G, output a set of elements $h_1, h_2, \ldots, h_l$, $l \leq n$, from the group G such that $G = \langle g_1 \rangle \oplus \langle g_2 \rangle \oplus \ldots \oplus \langle g_l \rangle$. Such a generating set can be found efficiently [2] from generators of the hidden subgroup of the function that maps $(m_1, m_2, \ldots, m_n) \mapsto g_1^{m_1} g_2^{m_2} \ldots g_n^{m_n}$.

This simple algorithm directly leads to an algorithm for computing the class group and class number of a quadratic number field, as pointed out by Watrous [22] in his paper that shows how to compute the order of solvable groups. Computing the class group of a more

general number field is a much more difficult task: this and related problems have been successfully tackled in a series of elegant work summarized in ▶ Quantum Algorithms for Class Group of a Number Field.

Such a decomposition of Abelian groups was also applied by Friedl, Ivanyos, and Santha [9] to test if a finite set with a binary operation is an Abelian group, by Kedlaya [13] to compute the zeta function of a genus g curve over a finite field $\mathbb{F}_q$ in time polynomial in g and q, and by Childs, Jao, and Soukharev [4] in order to construct elliptic curve isogenies in subexponential time.

Discussion: What About Non-Abelian Groups?

The great success of quantum algorithms for solving the Abelian hidden subgroup problem leads to the natural question of whether it can solve the hidden subgroup problem for non-Abelian groups. It has been shown that a polynomial number of queries suffice [8]; however, in general there is no bound on the overall computational complexity (which includes other elementary quantum or classical operations).

This question has been studied by many researchers, and efficient quantum algorithms can be found for some non-Abelian groups. However, at present, there is no efficient algorithm for most non-Abelian groups. For example, solving the HSP for the symmetric group would directly solve the graph automorphism problem.

Cross-References

▶ Quantum Algorithm for Factoring
▶ Quantum Algorithm for Solving Pell's Equation
▶ Quantum Algorithms for Class Group of a Number Field

Recommended Reading

1. Boneh D, Lipton R (1995) Quantum cryptanalysis of hidden linear functions (extended abstract). In: Proceedings of 15th Annual International Cryptology Conference (CRYPTO'95), Santa Barbara, pp 424–437

2. Cheung K, Mosca M (2001) Decomposing finite Abelian groups. Quantum Inf Comput 1(2):26–32
3. Childs AM, Ivanyos G (2014) Quantum computation of discrete logarithms in semigroups. J Math Cryptol 8(4):405–416
4. Childs AM, Jao D, Soukharev V (2010) Constructing elliptic curve isogenies in quantum subexponential time. preprint. arXiv:1012.4019
5. Cleve R, Ekert A, Macchiavello C, Mosca M (1998) Quantum algorithms revisited. Proc R Soc Lond A 454:339–354
6. Deutsch D (1985) Quantum theory, the Church-Turing principle and the universal quantum computer. Proc R Soc Lond A 400:97–117
7. Deutsch D, Jozsa R (1992) Rapid solutions of problems by quantum computation. Proc R Soc Lond Ser A 439:553–558
8. Ettinger M, Høyer P, Knill E (2004) The quantum query complexity of the hidden subgroup problem is polynomial. Inf Process Lett 91:43–48
9. Friedl K, Ivanyos G, Santha M (2005) Efficient testing of groups. In: Proceedings of the 37th Annual ACM Symposium on Theory of Computing (STOC'05), Baltimore, pp 157–166
10. Grigoriev D (1997) Testing shift-equivalence of polynomials by deterministic, probabilistic and quantum machines. Theor Comput Sci 180: 217–228
11. Høyer P (1999) Conjugated operators in quantum algorithms. Phys Rev A 59(5):3280–3289
12. Kaye P, Laflamme R, Mosca M (2007) An introduction to quantum computation. Oxford University Press, Oxford, UK
13. Kedlaya KS (2006) Quantum computation of zeta functions of curves. Comput Complex 15:1–19
14. Kitaev A (1995) Quantum measurements and the Abelian stabilizer problem. quant-ph/9511026
15. Kitaev A (1996) Quantum measurements and the Abelian stabilizer problem. In: Electronic Colloquium on Computational Complexity (ECCC), vol 3. http://eccc.hpi-web.de/report/1996/003/
16. Kitaev A. Yu (1997) Quantum computations: algorithms and error correction. Russ Math Surv 52(6):1191–1249
17. Mosca M, Ekert A (1998) The hidden subgroup problem and eigenvalue estimation on a quantum computer. In: Proceedings 1st NASA International Conference on Quantum Computing & Quantum Communications, Palm Springs. Lecture notes in computer science, vol 1509. Springer, Berlin, pp 174–188
18. Shor P (1994) Algorithms for quantum computation: discrete logarithms and factoring. In: Proceedings of the 35th Annual Symposium on Foundations of Computer Science (FOCS'94), Santa Fe, pp 124–134
19. Shor P (1997) Polynomial-time algorithms for prime factorization and discrete logarithms on a quantum computer. SIAM J Comput 26:1484–1509
20. Simon D (1994) On the power of quantum computation. In: Proceedings of the 35th IEEE Symposium on the Foundations of Computer Science (FOCS'94), Santa Fe, pp 116–123
21. Simon D (1997) On the power of quantum computation. SIAM J Comput 26:1474–1483
22. Watrous J (2000) Quantum algorithms for solvable groups. In: Proceedings of the 33rd ACM Symposium on Theory of Computing (STOC'00), Portland, pp 60–67
23. Vazirani U (1997) Berkeley lecture notes. Fall. Lecture 8. http://www.cs.berkeley.edu/~vazirani/qc.html

Abstract Voronoi Diagrams

Rolf Klein
Institute for Computer Science, University of Bonn, Bonn, Germany

Keywords

Abstract Voronoi diagram; Bisector; Computational geometry; Metric; Voronoi diagram

Years and Authors of Summarized Original Work

2009; Klein, Langetepe, Nilforoushan

Problem Definition

Concrete Voronoi diagrams are usually defined for a set S of sites p that exert influence over the points z of a surrounding space M. Often, influence is measured by distance functions $d_p(z)$ that are associated with the sites. For each p, its *Voronoi region* is given by

$$\mathrm{VR}(p, S)$$
$$= \{ z \in M \, ; \, d_p(z) < d_q(z) \text{ for all } q \in S \setminus \{p\}\},$$

and the *Voronoi diagram* $V(S)$ of S is the decomposition of M into Voronoi regions; compare the entry ▸ Voronoi Diagrams and Delaunay Triangulations of this Encyclopedia.

Quite different Voronoi diagrams result depending on the particular choices of space, sites, and distance measures; see Fig. 1. A great number of other types of Voronoi diagrams can be found in the monographs [1] and [14]. In each

Abstract Voronoi Diagrams, Fig. 1 Voronoi diagrams of points in the Euclidean and Manhattan metric and of disks (or additively weighted points) in the Euclidean plane

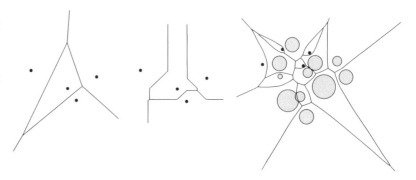

case, one wants to quickly compute the Voronoi diagram, because it contains a lot of distance information about the sites. However, the classical algorithms for the standard case of point sites in the Euclidean plane do not apply to more general situations.

To free us from designing individual algorithms for each and every special case, we would like to find a unifying concept that provides structural results and efficient algorithms for generalized Voronoi diagrams. One possible approach studied in [5,6] is to construct the lower envelope of the 3-dimensional graphs of the distance functions $d_p(z)$, whose projection to the XY-plane equals the Voronoi diagram.

Key Results

A different approach is given by *abstract Voronoi diagrams* that are not based on the notions of sites and distance measures (as their definitions vary anyway). Instead, AVDs are built from *bisecting curves* as primary objects [7].

Let $S = \{p, q, r, \ldots\}$ be a set of n indices, and for $p \neq q \in S$, let $J(p, q) = J(q, p)$ denote an unbounded curve that bisects the plane into two unbounded open domains $D(p, q)$ and $D(q, p)$. We require that each $J(p, q)$ is mapped to a closed Jordan curve through the north pole, under stereographic projection to the sphere. Now we define Voronoi regions by

$$\text{VR}(p, S) := \bigcap_{q \in S \setminus \{p\}} D(p, q)$$

and the abstract Voronoi diagram by

$$V(S) := \mathbf{R}^2 \setminus \bigcup_{p \in S} \text{VR}(p, S).$$

The system **J** of the curves $J(p, q)$ is called *admissible* if the following axioms are fulfilled for every subset T of S of size three.

A1. Each Voronoi region $\text{VR}(p, T)$ *is pathwise connected.*

A2. Each point of $\mathbf{R}^2$ *lies in the closure of a Voronoi region* $\text{VR}(p, T)$.

These combinatorial properties should not be too hard to check in a concrete situation because only triplets of sites need to be inspected. Yet, they form a strong foundation, as was shown in [8]. The following fact is crucial for the proof of Theorem 1. It also shows that AVDs can be seen as lower envelopes of surfaces in dimension 3.

Lemma 1 *For all* p, q, r *in* S, *we have* $D(p, q) \cap D(q, r) \subset D(p, r)$. *Consequently, for each point* $z \in \mathbf{R}^2$ *not contained in any curve of* **J**, *the relation*

$$p <_z q \; :\Leftrightarrow \; z \in D(p, q)$$

is an ordering of the sites in S *at* z.

Theorem 1 *If* **J** *is admissible, then axioms A1 and A2 hold for all subsets* T *of* S. *Moreover, the abstract Voronoi diagram* $V(S)$ *is a planar graph of size* $O(n)$.

Of the classical algorithm for constructing Voronoi diagrams, the randomized incremental construction method works best for abstract Voronoi diagrams [8, 10].

Theorem 2 *If* **J** *is admissible, then* $V(S)$ *can be constructed in an expected number of* $O(n \log n)$ *many steps and in expected linear space.*

Here basic operations like computing an intersection of two bisecting curves are counted as one step.

Applications

To show that a concrete type of Voronoi diagram is under the roof of abstract Voronoi diagrams, one needs to prove that its bisector system is admissible.

Let d be a metric in the plane that enjoys the following properties. Each d-disk contains a Euclidean disk and vice versa; for any two points a, b, there exists a point c different from a and b such that $d(a, b) = d(a, c) + d(c, b)$ holds; for any two points a, b, their metric bisector

$$B_d(a, b) = \{z \in \mathbf{R}^2; d(a, z) = d(b, z)\}$$

is itself a curve that maps to a closed Jordan curve through the north pole by stereographic projection to the sphere, or, in case $B_d(p, q)$ contains 2-dimensional pieces, its boundary consists of two such curves.

The first two properties ensure that any two points can be connected by a d-straight path along which d-distances add up. The third condition ensures that we can choose from $B_d(p, q)$ suitable bisecting curves. Let us call metric d *very nice* if also a fourth condition is fulfilled. Given three points a, p_0, p_1, there exist d-straight paths from a to p_0 and from a to p_1 that have only point a in common, or each d-straight path from a to p_i contains p_{1-i} for $i = 0$ or $i = 1$. All *convex distance functions (gauges)* are very nice.

Theorem 3 *Very nice metrics have admissible point bisector curves.*

Other applications of AVDs include points with additive weights, both the regular and the Hausdorff Voronoi diagram of disjoint convex sites with respect to a convex distance function, and some types of city Voronoi diagrams; see [1] for further details.

Generalizations

How to *dynamize* abstract Voronoi diagrams has been studied in [12]. Special cases of *3-dimensional* abstract Voronoi diagrams have been discussed in [11]; they include all convex distance functions whose unit spheres are ellipsoids. It is well known that for the vertices of a *convex polygon*, the Voronoi diagram can be constructed in linear time. This result has been generalized to AVDs in [9] and [4]. In [3] the *path-connectedness* of abstract Voronoi regions (axiom A1) has been relaxed. If a region of three sites can have up to s connected components, the abstract Voronoi diagram can still be constructed in expected time $O(s^2 n \sum_{j=3}^{n} m_j/j)$, where m_j denotes the average number of faces per region in any subdiagram of j sites from S.

In an *order-k Voronoi diagram*, all points of space M are placed in one region that shares the same k nearest sites in S. For $k = n - 1$, this concept has been generalized to furthest site abstract Voronoi diagrams in [13]. Here the furthest (or inverse) region of $p \in S$ is the intersection of all domains $D(q, p)$, where $q \in S \setminus \{p\}$. If all regular Voronoi regions are nonempty, then the furthest site AVD is a tree of size $O(n)$, even though some regions may be disconnected.

General order-k abstract Voronoi diagrams have been studied in [2]. If all regular Voronoi regions are nonempty and if bisecting curves are in general position, a tight upper complexity bound of $2k(n-k)$ can be shown. Fortunately, the nonemptiness of the regular regions need only be tested for all subsets of S of size 4.

Cross-References

▶ Voronoi Diagrams and Delaunay Triangulations

Recommended Reading

1. Aurenhammer F, Klein R, Lee DT (2013) Voronoi diagrams and Delaunay triangulations. World Scientific, Singapore
2. Bohler C, Cheilaris P, Klein R, Liu CH, Papadopoulou E, Zavershynskyi M (2013) On the complexity of higher order abstract Voronoi diagrams. In: Proceedings of the 40th international colloquium on automata languages and programming, Riga. Lecture notes in computer science, vol 7965, pp 208–219
3. Bohler C, Klein R (2013) Abstract Voronoi diagrams with disconnected regions. In: Proceedings of the 24th international symposium on algorithms and computation, Hong Kong. Lecture notes in computer science, vol 8283, pp 306–316
4. Bohler C, Klein R, Liu CH (2014) Forest-like abstract Voronoi diagrams in linear time. In: 26th Canadian conference on computational geometry, Halifax
5. Boissonnat JD, Wormser C, Yvinec M (2006) Curved Voronoi diagrams. In: Boissonnat JD, Teillaud M (eds) Effective computational geometry for curves and surfaces. Mathematics and visualization. Springer, Berlin/New York, pp 67–116
6. Edelsbrunner H, Seidel R (1986) Voronoi diagrams and arrangements. Discret Comput Geom 1:387–421
7. Klein R (1989) Concrete and abstract Voronoi diagrams. Lecture notes in computer science, vol 400. Springer, Berlin/New York
8. Klein R, Langetepe E, Nilforoushan Z (2009) Abstract Voronoi diagrams revisited. Comput Geom Theory Appl 42(9):885–902
9. Klein R, Lingas A (1994) Hamiltonian abstract Voronoi diagrams in linear time. In: Proceedings of the 5th international symposium on algorithms and computation, Beijing. Lecture notes in computer science, vol 834, pp 11–19
10. Klein R, Mehlhorn K, Meiser S (1993) Randomized incremental construction of abstract Voronoi diagrams. Comput Geom Theory Appl 3:157–184
11. Lê NM (1995) Randomized incremental construction of simple abstract Voronoi diagrams in 3-space. In: Proceedings of the 10th international conference on fundamentals of computation theory, Dresden. Lecture notes in computer science, vol 965, pp 333–342
12. Malinauskas KK (2008) Dynamic construction of abstract Voronoi diagrams. J Math Sci 154(2): 214–222
13. Mehlhorn K, Meiser S, Rasch R (2009) Furthest site abstract Voronoi diagrams. Int J Comput Geom Appl 11:583–616
14. Okabe A, Boots B, Sugihara K, Chiu SN (2000) Spatial tessellations: concepts and applications of Voronoi diagrams. Wiley, Chichester

Active Learning – Modern Learning Theory

Maria-Florina Balcan and Ruth Urner
Department of Machine Learning, Carnegie Mellon University, Pittsburgh, PA, USA

Keywords

Active learning; Computational complexity; Learning theory; Sample complexity

Years and Authors of Summarized Original Work

2006; Balcan, Beygelzimer, Langford
2007; Balcan, Broder, Zhang
2007; Hanneke
2013; Urner, Wulff, Ben-David
2014; Awashti, Balcan, Long

Problem Definition

Most classic machine learning methods depend on the assumption that humans can annotate all the data available for training. However, many modern machine learning applications (including image and video classification, protein sequence classification, and speech processing) have massive amounts of unannotated or unlabeled data. As a consequence, there has been tremendous interest both in machine learning and its application areas in designing algorithms that most efficiently utilize the available data while minimizing the need for human intervention. An extensively used and studied technique is *active learning*, where the algorithm is presented with a large pool of unlabeled examples (such as all images available on the web) and can interactively ask for the labels of examples of its own choosing from the pool, with the goal to drastically reduce labeling effort.

Formal Setup

We consider *classification problems* (such as classifying images by who is in them or classifying emails as spam or not), where the goal is to predict a label y based on its corresponding input vector x. In the standard machine learning formulation, we assume that the data points (x, y) are drawn from an unknown underlying distribution D_{XY} over $X \times Y$; X is called the feature (instance) space and $Y = \{0, 1\}$ is the label space. The goal is to output a hypothesis function h of small error (or small $0/1$ loss), where $\text{err}(h) = \mathbb{P}_{(x,y) \sim D_{XY}}[h(x) \neq y]$. In the passive learning setting, the learning algorithm is given a set of labeled examples $(x_1, y_1), \ldots, (x_m, y_m)$ drawn i.i.d. from D_{XY} and the goal is to output a hypothesis of small error by using only a polynomial number of labeled examples. In the *realizable* case [10] (PAC learning), we assume that the true label of any example is determined by a deterministic function of the features (the so-called target function) that belongs to a known concept class C (e.g., the class of linear separators, decision trees, etc.). In the *agnostic* case [10, 13], we do not make the assumption that there is a perfect classifier in C, but instead we aim to compete with the best function in C (i.e., we aim to identify a classifier whose error is not much worse than $opt(C)$, the error of the best classifier in C). Both in the realizable and agnostic settings, there is a well-developed theory of Sample Complexity [13], quantifying in terms of the so-called *VC-dimension* (a measure of complexity of a concept class) how many training examples we need in order to be confident that a rule that does well on training data is a good rule for future data as well.

In the active learning setting, a set of labeled examples $(x_1, y_1), \ldots, (x_m, y_m)$ is also drawn i.i.d. from D_{XY}; the learning algorithm is permitted direct access to the sequence of x_i values (unlabeled data points), but has to make a label request to obtain the label y_i of example x_i.

The hope is that we can output a classifier of small error by using many fewer label requests than in passive learning by actively directing the queries to informative examples (while keeping the number of unlabeled examples polynomial).

It has been long known that, in the realizable case, active learning can sometimes provide an exponential improvement in label complexity over passive learning. The canonical example [6] is learning threshold classifiers ($X = [0, 1]$ and $C = \{\mathbf{1}_{[0,a]} \mid a \in [0, 1]\}$). Here we can actively learn with only $\tilde{O}(\log(1/\epsilon))$ label requests by using a simple binary search-like algorithm as follows: we first draw $N = \tilde{O}((1/\epsilon) \log(1/\delta))$ unlabeled examples, then do binary search to find the transition from label 1 to label 0, and with only $O(\log(N))$ queries we can correctly infer the labels of all our examples; we finally output a classifier from C consistent with all the inferred labels. By standard VC-dimension based bounds for supervised learning [13], we are guaranteed to output an ϵ-accurate classifier. On the other hand, for passive learning, we provably need $\Omega(1/\epsilon)$ labels to output a classifier of error at most ϵ with constant probability, yielding the exponential reduction in label complexity.

Key Results

While in the simple threshold concept class described above active learning always provides huge improvements over passive learning, things are more delicate in more general scenarios. In particular, both in the realizable and in the agnostic case, it has been shown that for more general concept spaces, in the worst case over all data-generating distributions, the label complexity of active learning equals that of passive learning. Thus, much of the literature was focused on identifying non-worst case, natural conditions about the relationship between the data distribution and the target, under which active learning provides improvements over passive. Below, we discuss three approaches, under which active learning has been shown to reduce the label complexity: disagreement-based techniques, margin-based techniques and cluster-based techniques.

Disagreement-Based Active Learning

Disagreement-based active learning was the first method to demonstrate the feasibility of *agnostic active learning* for general concept classes. The general algorithmic framework of disagreement-based active learning in the presence of noise was introduced with the A^2 algorithm by Balcan et al. [2]. Subsequently, several researchers have proposed related disagreement-based algorithms with improved sample complexity, e.g., [5, 8, 11].

At a high level, A^2 operates in rounds. It maintains a set of candidate classifiers from the concept class C and in each round queries labels aiming to efficiently reduce this set to only few high-quality candidates. More precisely, in round i, A^2 considers the set of surviving classifiers $C_i \subseteq C$, and asks for the labels of a few random points that fall in the *region of disagreement* of C_i. Formally, the region of disagreement of a set of classifiers C_i is $\text{DIS}(C_i) = \{x \in X \mid \exists f, g \in C_i : f(x) \neq g(x)\}$. Based on these queried labels from $\text{DIS}(C_i)$, to obtain C_{i+1}, the algorithm then throws out hypotheses that are suboptimal. The key ingredient is that A^2 *only* throws out hypotheses, for which it is *statistically confident* that they are suboptimal.

Balcan et al. [2] show that A^2 provides exponential improvements in the label sample complexity in terms of the $1/\epsilon$-parameter when the noise rate η is sufficiently small, both for learning thresholds and for learning homogeneous linear separators in R^d, one of the most widely used and studied classes in machine learning. Following up on this, Hanneke [9] provided a generic analysis of the A^2 algorithm that applies to *any concept class*. This analysis quantifies the label complexity of A^2 in terms of the so-called *disagreement coefficient* of the class C. The disagreement coefficient is a distribution-dependent sample complexity measure that quantifies how fast the region of disagreement of the set of classifiers at distance r of the optimal classifier collapses as a function r. In particular, [9] showed that

the label complexity of the A^2 algorithm is $O\left(\theta^2\left(\frac{v^2}{\epsilon^2}+1\right)(d \log(1/\epsilon)+\log(1/\delta)) \log(1/\epsilon)\right)$, where v is the best error rate of a classifier in C, d is the VC-dimension of C, and θ is the disagreement-coefficient. As an example, for homogeneous linear separators, we have $\theta = \theta(\sqrt{d})$ under uniform marginal over the unit ball. Here, the disagreement-based analysis yields a label complexity of $\tilde{O}\left(d^2 \frac{v^2}{\epsilon^2} \log(1/\epsilon)\right)$ in the agnostic case and $\tilde{O}\left(d^{3/2} \log(1/\epsilon)\right)$ in the realizable case.

Margin-Based Active Learning

While the disagreement-based active learning line of work provided the first general understanding of the sample complexity benefits with active learning for arbitrary concept classes, it suffers from two main drawbacks: (1) methods and analyses developed in this context are often suboptimal in terms of label complexity, since they take a conservative approach and query even points on which there is only a small amount of uncertainty, (2) the methods are computationally inefficient. *Margin-based* active learning is a technique that overcomes both the above drawbacks for learning homogeneous linear separators under log-concave distributions. The technique was first introduced by Balcan et al. [3] and further developed by Balcan et al. [4], and Awasthi et al. [1].

At a high level, like disagreement-based methods, the margin-based active learning algorithm operates in rounds, in which a number of labels are queried in some subspace of the domain and a set of candidate classifiers for the next round is identified. The crucial idea to reduce the label complexity is to design a *more aggressive querying strategy* by carefully choosing where to query instead of querying in all of the current disagreement region. Concretely, in round k the algorithm has a *current hypothesis* w_k, and the set of candidate classifiers for the next round consists of all homogeneous halfspaces that lie in a *ball of radius r_k around w_k* (in terms of their angle with w_k). The algorithm then queries points for labels near the decision boundary of w_k; that is, it only queries points

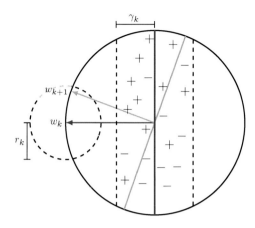

Active Learning – Modern Learning Theory, Fig. 1
The margin-based active learning algorithm after iteration k. The algorithm samples points within margin γ_k of the current weight vector w_k and then minimizes the hinge loss over this sample subject to the constraint that the new weight vector w_{k+1} is within distance r_k from w_k

that are within a *margin* γ_k of w_k; see Fig. 1. To obtain w_{k+1}, the algorithm finds a loss minimizer among the current set of candidates with respect to the queried examples of round k. In the realizable case, this is done by $0/1$-loss minimization. In the presence of noise, to obtain a computationally efficient procedure, the margin-based technique minimizes a convex surrogate loss.

Balcan et al. [3] and Balcan and Long [4] showed that by localizing aggressively, namely by setting the margin parameter to $\gamma_k = \Theta(\frac{1}{2^k})$, one can actively learn with only $\tilde{O}(d \log(1/\epsilon))$ label requests in the realizable case, when the underlying distribution is isotropic log-concave. A key idea of their analysis is to decompose, in round k, the error of a candidate classifier w as its error outside margin γ_k of the current separator plus its error inside margin γ_k, and to prove that for the above parameters, a small constant error inside the margin suffices to reduce the overall error by a constant factor. For the constant error inside the margin only $\theta(d)$ labels need to be queried, and since in each round the overall error gets reduced by a constant factor, $O(\log(1/\epsilon))$ rounds suffice to reduce the error to ϵ, yielding the label complexity of $\tilde{O}(d \log(1/\epsilon))$. Passive learning here provably requires $\Omega(d/\epsilon)$ labeled examples. Thus, the

dependence on $1/\epsilon$ is exponentially improved, but without increasing the dependence on d (as in the disagreement-based method for this case, see above).

Building on this work, [1] gave the first *polynomial-time* active learning algorithm for learning linear separators to error ϵ in the presence of agnostic noise (of rate $O(\epsilon)$) when the underlying distribution is an isotropic log-concave distribution in R^d. They proposed to use a normalized hinge loss minimization (with normalization factor τ_k) for selecting the next classifier w_{k+1} in round k. Awasthi et al. [1] show that by setting the parameters appropriately (namely, $\tau_k = \Theta(1/2^k)$ and $r_k = \Theta(1/2^k)$), the algorithm again achieves error ϵ using only $O(\log(1/\epsilon))$ rounds, with $O(d^2)$ label requests per round. This yields a query complexity of poly$(d, \log 1/\epsilon)$. The key ingredient for the analysis of this computationally efficient version in the noisy setting is proving that by constraining the search for w_{k+1} to vectors within a ball of radius r_k around w_k, the hinge-loss acts as a sufficiently faithful proxy for the $0/1$-loss.

A recent work [14] proposes an elegant generalization of [3, 4] to more general concept spaces and shows an analysis that is always tighter than disagreement-based active learning (though their results are not computationally efficient).

Cluster-Based Active Learning
The methods described above (disagreement-based and margin-based active learning) use active label queries to efficiently identify a classifier from the concept class C with low error. An alternative approach to agnostic active learning is to design active querying methods that efficiently find a (approximately) correct labeling of the unlabeled input sample. Here, "correct labeling" refers to the hidden labels y_i in the sample $(x_1, y_1), \ldots, (x_m, y_m)$ from the distribution D_{XY} (as defined in the formal setup section). The so labeled sample can then be used as input to a passive learning algorithm to learn an arbitrary concept class.

Cluster-based active learning is a method for the latter approach and was introduced by Dasgupta and Hsu [7]. The idea is to use a hierarchical clustering (cluster tree) of the unlabeled data, and check the clusters for label homogeneity by starting at the root of the tree (the whole data set) and working towards the leaves (single data points). The label homogeneity of a cluster is estimated by choosing data points for label query uniformly at random from the cluster. If a cluster is considered label homogeneous (with sufficiently high confidence), all remaining unlabeled points in that cluster are labeled with the majority label. If a cluster is detected to be label heterogeneous, it is split into its children in the cluster tree and processed later. The key insight in [7] is that since the cluster tree is fixed before any labels were seen, the induced labeled subsample of a child cluster can be considered a sample that was chosen uniformly at random from the points in that child-cluster. Thus, the algorithm can reuse labels from the parent cluster without introducing any sampling bias. The label efficiency of this paradigm crucially depends on the quality of input hierarchical clustering. Intuitively, if the cluster tree has a small pruning with label homogeneous clusters, the procedure will make only few label queries.

Urner et al. [12] proved label complexity reductions with this paradigm under a distributional assumption. They analyze a version (PLAL) of the above paradigm that uses hierarchical clusterings induced by *spatial trees* on the domain $[0, 1]^d$ and provide label query bounds in terms of the *Probabilistic Lipschitzness* of the underlying data-generating distribution. Probabilistic Lipschitzness quantifies a marginal-label relatedness in the sense of close points being likely to have the same label. For a distribution with deterministic labels ($\Pr[Y = 1 \mid X = x] \in \{0, 1\}$ for all x), the Probabilistic Lipschitzness is a function ϕ that bounds, as a function of λ, the mass of points x for which both labels 0 and 1 occur in the ball $B_\lambda(x)$.

Urner et al. [12] show that, independently of the any data assumptions, (with probability $1 - \delta$) PLAL labels a $(1 - \epsilon)$-fraction of the input points correctly. They further show that using PLAL as a preprocedure, if the data-generating distribution has deterministic labels and its Probabilistic Lipschitzness is bounded by $\phi(\lambda) = \lambda^n$ for some

$n \in \mathbb{N}$, then classes C of bounded VC-dimension on domain $X = [0, 1]^d$ can be learned with $\tilde{O}\left(\left(\frac{1}{\epsilon}\right)^{\frac{n+2d}{n+d}} \right)$ many labels, while any passive proper learner (i.e., a passive learner that outputs a function from C) requires to see $\Omega(1/\epsilon^2)$ many labels. Further, [12] show that PLAL can be used to reduce the number of labels needed for nearest neighbor classification (i.e., labeling a test point by the label of its nearest point in the sample) from $\Omega\left(\left(\frac{1}{\epsilon}\right)^{1 + \frac{d-1}{n}} \right)$ to $\tilde{O}\left(\left(\frac{1}{\epsilon}\right)^{1 + \frac{d^2}{n(n+d)}} \right)$.

Cross-References

▶ PAC Learning

Recommended Reading

1. Awasthi P, Balcan M-F, Long PM (2014) The power of localization for efficiently learning linear separators with noise. In: Proceedings of the 46th annual symposium on the theory of computing (STOC), New York
2. Balcan MF, Beygelzimer A, Langford J (2006) Agnostic active learning. In: Proceedings of the 23rd international conference on machine learning (ICML), Pittsburgh
3. Balcan M-F, Broder A, Zhang T (2007) Margin based active learning. In: Proceedings of the 20th annual conference on computational learning theory (COLT), San Diego
4. Balcan M-F, Long PM (2013) Active and passive learning of linear separators under log-concave distributions. In: Proceedings of the 26th conference on learning theory (COLT), Princeton
5. Beygelzimer A, Hsu D, Langford J, Zhang T (2010) Agnostic active learning without constraints. In: Advances in neural information processing systems (NIPS), Vancouver
6. Cohn D, Atlas L, Ladner R (1994) Improving generalization with active learning. In: Proceedings of the 11th international conference on machine learning (ICML), New Brunswick
7. Dasgupta S, Hsu D (2008) Hierarchical sampling for active learning. In: Proceedings of the 25th international conference on machine learning (ICML), Helsinki
8. Dasgupta S, Hsu DJ, Monteleoni C (2007) A general agnostic active learning algorithm. In: Advances in neural information processing systems (NIPS), Vancouver
9. Hanneke S (2007) A bound on the label complexity of agnostic active learning. In: Proceedings of the

24th international conference on machine learning (ICML), Corvallis
10. Kearns MJ, Vazirani UV (1994) An introduction to computational learning theory. MIT, Cambridge
11. Koltchinskii V (2010) Rademacher complexities and bounding the excess risk in active learning. J Mach Learn 11:2457–2485
12. Urner R, Wulff S, Ben-David S (2013) Plal: cluster-based active learning. In: Proceedings of the 26th conference on learning theory (COLT), Princeton
13. Vapnik VN (1998) Statistical learning theory. Wiley, New York
14. Zhang C, Chaudhuri K (2014) Beyond disagreement-based agnostic active learning. In: Advances in neural information processing systems (NIPS), Montreal

Active Self-Assembly and Molecular Robotics with Nubots

Damien Woods
Computer Science, California Institute of Technology, Pasadena, CA, USA

Keywords

Molecular robotics; Rigid-body motion; Self-assembly

Years and Authors of Summarized Original Work

2013; Woods, Chen, Goodfriend, Dabby, Winfree, Yin
2013; Chen, Xin, Woods
2014; Chen, Doty, Holden, Thachuk, Woods, Yang

Problem Definition

In the theory of molecular-scale self-assembly, large numbers of simple interacting components are designed to come together to build complicated shapes and patterns. Many models of self-assembly, such as the abstract Tile Assembly Model [6], are cellular automata-like crystal growth models. Indeed such models have given rise to a rich theory of self-assembly as described elsewhere in this encyclopedia. In biological organisms we frequently see much more sophisticated growth processes, where self-assembly is combined with active molecular components that change internal state and even molecular motors that have the ability to push and pull large structures around. Molecular engineers are now beginning to design and build molecular-scale DNA motors and active self-assembly systems [2]. We wish to understand, at a high level of abstraction, the ultimate computational capabilities and limitations of such molecular-scale rearrangement and growth. The nubot model, put forward in [8], is akin to an asynchronous nondeterministic cellular automaton augmented with nonlocal rigid-body movement. Unit-sized monomers are placed on a 2D triangular grid. Monomers undergo state changes, appear, and disappear using local rules, as shown in Fig. 1. However, there is also a nonlocal aspect to the model: rigid-body movement that comes in two forms, movement rules and random agitations.

A *movement rule r*, consisting of a pair of monomer states A, B and two unit vectors, is a programmatic way to specify unit-distance translation of a set of monomers in one step. See Fig. 2 for an example. If A and B are in a prescribed orientation, one is nondeterministically chosen to move unit distance in a prescribed direction. The rule r is applied in a rigid-body fashion: roughly speaking, if A is to move right, it pushes anything immediately to its right and pulls any monomers that are bound to its left which in turn push and pull other monomers, all in one step. The rule may not be applicable if it is blocked (i.e., if movement of A would force B to also move), which is analogous to the fact that an arm cannot push its own shoulder. The other, somewhat related, form of movement is called *agitation*: at every point in time, every monomer on the grid may move unit distance in any of the six directions, at unit rate for each (monomer, direction) pair. An agitating monomer will push or pull any monomers that it is adjacent to, in a way that preserves rigid-body structure and all in

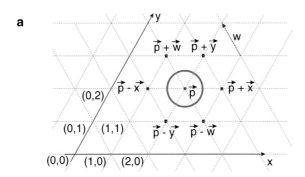

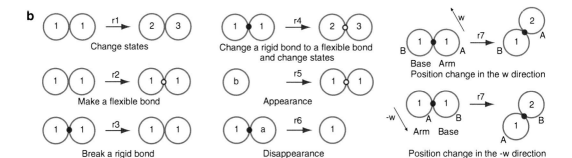

Active Self-Assembly and Molecular Robotics with Nubots, Fig. 1 Overview of the nubot model. (**a**) A nubot configuration showing a single nubot monomer on the triangular grid. (**b**) Examples of nubot monomer rules. Rules r1–r6 are local cellular automaton-like rules, whereas r7 effects a nonlocal movement that may translate other monomers as shown in Fig. 2. Monomers continuously undergo agitation, as shown in Fig. 3. A flexible bond is depicted as an empty *red circle* and a rigid bond is depicted as a *solid red disk* (from [8])

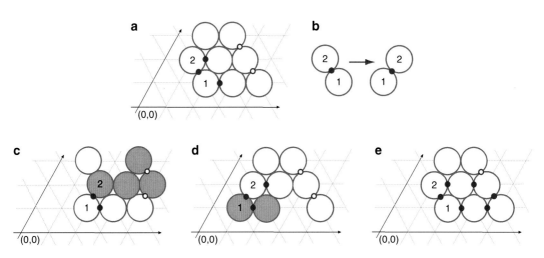

Active Self-Assembly and Molecular Robotics with Nubots, Fig. 2 Movement rule. (**a**) Initial configuration. (**b**) Movement rule with one of two results depending on the choice of arm or base. (**c**) Result if the monomer with state 2 is the arm or (**d**) monomer with state 1 is the arm. The shaded monomers are the movable set. The affect on rigid (*filled red disks*), flexible (*hollow red circles*), and null bonds is shown. (**e**) A configuration for which the movement rule is blocked: movement of 1 or 2 would force the other to move; hence the rule is not applicable (from [3])

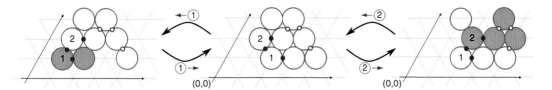

Active Self-Assembly and Molecular Robotics with Nubots, Fig. 3 Example agitations. Starting from the centre configuration, there are **48** possible agitations (8 monomers, 6 directions each) each with equal probability. The right configuration is the result of agitation of the monomer in state 2 in direction $\rightarrow$, the left is the result of the agitation of the monomer in state 1 in direction $\leftarrow$. The shaded monomers are the agitation set – monomers that are moved by the agitation (from [4])

one step as shown in Fig. 3. Unlike movement, agitations are never blocked. Rules are applied asynchronously and in parallel. Taking its time model from stochastic chemical kinetics, a nubot system evolves as a continuous time Markov process.

For intuition, we describe motion in terms of pushing and pulling. However movement and agitation are actually intended to model a nanoscale environment with diffusion, Brownian motion, convection, turbulent flow, cytoplasmic streaming, and other uncontrolled inputs of energy that interact monomers in all directions, moving large molecular assemblies in a random fashion (i.e., agitation) and allowing motors to simply latch and unlatch large assemblies into position (i.e., the movement rule).

Key Results

Assembling simple structures, namely, lines and squares, has proven to be a fruitful way to explore the power of the nubot model for a few reasons. Firstly, it helps us develop a number of techniques and intuitions for the model. Secondly, lines and squares get used again and again in more general results that show the full power of the model. Thirdly, the efficiency of assembling simple shapes has been a de facto benchmark problem for a number of self assembly models (although this benchmark often does not give the full story). In a variety of models, such as the abstract Tile Assembly Model, cellular automata, and some robotics models, it takes time $\Omega(n)$ to assemble a length n line. In the nubot model this

is achieved in merely $O(\log n)$ expected time and $O(\log n)$ states.

Theorem 1 ([8]) *For each $n \in \mathbb{N}$, there is a set of nubot rules $\mathcal{N}_n^{\text{line}}$ such that starting from a single monomer $\mathcal{N}_n^{\text{line}}$ assembles a length n line in $O(\log n)$ expected time, $n \times 2$ space, and $O(\log n)$ states.*

One can trade time for states by giving a slightly slower method with fewer states:

Theorem 2 ([3]) *There is a set of nubot rules $\mathcal{N}^{\text{line}}$ such that for each $n \in \mathbb{N}$, from a line of $O(\log n)$ "binary" monomers (each in state 0 or 1), $\mathcal{N}^{\text{line}}$ assembles a length n line in $O(\log^2 n)$ expected time, $n \times O(1)$ space, and $O(1)$ states.*

An $n \times n$ square can be built by growing a horizontal line and then n vertical lines, showing that assembly of squares with nubots is exponentially faster than the $\Theta(n)$ expected time seen in the abstract Tile Assembly Model [1]:

Theorem 3 ([8]) *For each $n \in \mathbb{N}$, there is a set of nubot rules $\mathcal{N}_n^{\text{square}}$ such that starting from a single monomer, $\mathcal{N}_n^{\text{square}}$ assembles a $n \times n$ square in $O(\log n)$ expected time, $n \times n$ space, and $O(\log n)$ states.*

The results above, and all of those in [3, 8], crucially make use of the rigid-body movement rule: the ability for a single monomer to control the movement of large objects quickly and at a time and place of the programmer's choosing. However, in a molecular-scale environment, molecular motion is happening in a largely uncontrolled and fundamentally random manner, all of the time. The *agitation nubot* model does not have the movement rule, but instead permits such uncontrolled random agitation (movement). Although this form of movement is challenging to control in a precise manner, the following result shows we can use it to achieve sublinear expected time growth of a length n line in only $O(n)$ space:

Theorem 4 ([4]) *There is a set of nubot rules* $\mathcal{N}_{\text{line}}$*, such that* $\forall n \in \mathbb{N}$*, starting from a line of* $\lfloor \log_2 n \rfloor + 1$ *monomers, each in state* 0 *or* 1*,* $\mathcal{N}_{\text{line}}$ *in the agitation nubot model assembles an* $n \times 1$ *line in* $O(n^{1/3} \log n)$ *expected time,* $n \times 5$ *space, and* $O(1)$ *monomer states.*

For a square we can do much better, achieving polylogarithmic expected time:

Theorem 5 ([4]) *There is a set of nubot rules* $\mathcal{N}_{\text{square}}$*, such that* $\forall n \in \mathbb{N}$*, starting from a line of* $\lfloor \log_2 n \rfloor + 1$ *monomers, each in state* 0 *or* 1*,* $\mathcal{N}_{\text{square}}$ *in the agitation nubot model assembles an* $n \times n$ *square in* $O(\log^2 n)$ *expected time,* $n \times n$ *space, and* $O(1)$ *monomer states.*

This section concludes with three results on general-purpose computation and shape construction with the nubot model. First we have a computability-theoretic result: any finite computable connected shape can be quickly self-assembled.

Theorem 6 ([8]) *An arbitrary connected computable 2D shape of size* $\leq \sqrt{n} \times \sqrt{n}$ *can be assembled in expected time* $O(\log^2 n + t(|n|))$ *using* $O(s + \log n)$ *states. Here,* $t(|n|)$ *is the time required for a program-size* s *Turing machine to compute, given a pixel index as a binary string of length* $|n| = \lfloor \log_2 n \rfloor + 1$*, whether or not the pixel is present in the shape.*

For complicated computable shapes the construction for Theorem 6 necessarily requires computation workspace outside of the shape's bounding box. The next result is of a more resource-bounded style and, roughly speaking, states that 2D patterns with efficiently computable pixel colors can be assembled using nubots in merely polylogarithmic expected time while staying inside the pattern's bounding box.

Theorem 7 ([8]) *An arbitrary finite computable 2D pattern of size* $\leq n \times n$*, where* $n = 2^p$*,* $p \in \mathbb{N}$*, with pixels whose color is computable on a polynomial time* $O(|n|^{\ell})$ *(inputs are binary strings of length* $|n| = O(\log n)$*), linear space* $O(|n|)$*, program-size* s *Turing machine, can be assembled in expected time* $O(\log^{\ell+1} n)$*, with* $O(s + \log n)$ *monomer states and without growing outside the pattern borders.*

The results cited so far can be used to compare the nubot model to other models of self-assembly and tell us that nubots build shapes and patterns in a fast parallel manner. The next result quantifies this parallelism in terms of a well-known parallel model from computational complexity theory: NC is the class of problems solved by uniform polylogarithmic depth and polynomial-size Boolean circuits.

Theorem 8 ([3]) *For each language* $L \in \text{NC}$*, there is a set of nubot rules* $\mathcal{N}_L$ *that decides* L *in polylogarithmic expected time, constant number of monomer states, and polynomial space in the length of the input string of binary monomers (in state* 0 *or* 1*). The output is a single binary monomer.*

This result stands in contrast to sequential machines like Turing machines, that cannot read all of an n-bit input string in polylogarithmic time, and "somewhat parallel" models like cellular automata and the abstract Tile Assembly Model, that cannot have all of n bits influence a single output bit decision in polylogarithmic time [5]. Thus, adding the nubot rigid-body movement primitive to an asynchronous nondeterministic cellular automaton drastically increases its parallel processing abilities.

Open Problems

Some future research directions are discussed here and in [3, 4, 8]. It remains as future work to look at other topics such as fault tolerance, self-healing, dynamical tasks, or systems that continuously respond to the environment.

The Complexity of Assembling Lines

Theorem 1 states that a line can be grown in expected time $O(\log n)$, space $O(n) \times O(1)$, and $O(\log n)$ states, and Theorem 2 trades time for states to get expected time $O(\log^2 n)$, space $O(n) \times O(1)$, and $O(1)$ states. What is the complexity (expected time $\times$ states) of assembling a line in the nubot model? Is it possible to meet the lower bound of expected time $\times$ states $= \Omega(\log n)$? In this problem, the input should be a set of monomers with space $\times$ states $= O(\log n)$.

Computational Power

Theorem 8 gives a lower bound on the computational complexity of the nubot model. What is the exact power of polylogarithmic expected time nubots? The answer may differ on whether we begin from a small collection of monomers (as in Theorem 8) or a large prebuilt structure. One challenge, for the upper bound, involves finding better Turing machine space, or circuit depth, bounds on computing multiple applications of the movable set on a large nubots grid.

Synchronization and Composition of Nubot Algorithms

Synchronization is a method to quickly send signals using nonlocal rigid-body motion [3, 8]. The nubot model is asynchronous, but synchronization can be used to set discrete stages, or checkpoints, during a complicated construction. This in turn facilitates composition of nubot algorithms (run algorithm 1, synchronize, run algorithm 2, synchronize, etc.) and many of the results cited here use it for exactly that reason. However, synchronization-less constructions often exhibit a kind of independence where growth proceeds everywhere in parallel, without waiting on signals from distant components. Such systems are highly distributed, easy to analyze, and perhaps more amenable to laboratory implementation. Intuitively, this seems like the right way to program molecules. The proof of Theorem 7 does not use synchronization which shows that without it a very general class of (efficiently) computable patterns can be grown and indeed the proof gives methods to compose nubot algorithms without resorting to synchronization.

It remains as future work to formalize both this notion of synchronization-less "independence" and what we mean by "composition" of nubot algorithms. What conditions are necessary and sufficient for composition of nubot algorithms? What classes of shapes and patterns can be assembled using without synchronization or other forms of rapid long-range communication?

Agitation Versus the Movement Rule

Is it possible to simulate the movement rule using agitation? More formally, is it the case that for each nubot program $\mathcal{N}$, there is an agitation nubot program $\mathcal{A}_{\mathcal{N}}$, that acts just like $\mathcal{N}$ but with some $m \times m$ scale-up in space, and a k factor slowdown in time, where m and k are (constants) independent of $\mathcal{N}$ and its input? As motivation, note that every self-assembled molecular-scale structure was made under conditions where random jiggling of monomers is a dominant source of movement! Our question asks if we can *programmably exploit* this random molecular motion to build structures quicker than without it.

Intrinsic Universality and Simulation

Is the nubot model intrinsically universal? Specifically, does there exist a set of monomer rules U, such that any nubot system $\mathcal{N}$ can be simulated by "seeding" U with a suitable initial configuration? Here the simulation should have a spatial scale factor m that is a function of the number of states in the simulated system $\mathcal{N}$. Is the agitation nubot model intrinsically universal? Our hope would be that simulation could be used to tease apart the power of different notions of movement (e.g., to understand if nubot-style movement is weaker or stronger than other notions of robotic movement), in the way it has been used to characterize and separate the power of other self-assembly models [7].

Brownian Nubots

With nubots, under agitation, or multiple parallel movement rules, larger objects move faster. This is intended to model an environment with uncontrolled and rapid fluid flows. But in Brownian motion, larger objects move slower: what is the power of nubots with such a rate model, for example, with rate equal to object size? Although assembly in such a model may be slower than with the usual model, many of the same programming principles should apply, and indeed it will still be possible to assemble objects in a parallel distributed fashion.

Cross-References

▶ Combinatorial Optimization and Verification in Self-Assembly
▶ Intrinsic Universality in Self-Assembly
▶ Patterned Self-Assembly Tile Set Synthesis
▶ Randomized Self-Assembly
▶ Robustness in Self-Assembly
▶ Self-Assembly at Temperature 1
▶ Self-Assembly of Fractals
▶ Self-Assembly of Squares and Scaled Shapes
▶ Self-Assembly with General Shaped Tiles
▶ Temperature Programming in Self-Assembly

Acknowledgments A warm thanks to all of my coauthors on this topic and especially to Erik Winfree and Chris Thachuk for their helpful comments. The author is supported by NSF grants 0832824, 1317694, CCF-1219274, and CCF-1162589.

Recommended Reading

1. Adleman LM, Cheng Q, Goel A, Huang MD (2001) Running time and program size for self-assembled squares. In: STOC 2001: proceedings of the 33rd annual ACM symposium on theory of computing, Hersonissos. ACM, pp 740–748
2. Bath J, Turberfield A (2007) DNA nanomachines. Nat Nanotechnol 2:275–284
3. Chen M, Xin D, Woods D (2013) Parallel computation using active self-assembly. In: DNA19: the 19th international conference on DNA computing and molecular programming. LNCS, vol 8141. Springer, pp 16–30. arxiv preprint arXiv:1405.0527
4. Chen HL, Doty D, Holden D, Thachuk C, Woods D, Yang CT (2014) Fast algorithmic self-assembly of simple shapes using random agitation. In: DNA20: the 20th international conference on DNA computing and molecular programming. LNCS, vol 8727. Springer, pp 20–36. arxiv preprint arXiv:1409.4828
5. Keenan A, Schweller R, Sherman M, Zhong X (2014) Fast arithmetic in algorithmic self-assembly. In: UCNC: the 13th international conference on unconventional computation and natural computation. LNCS, vol 8553. Springer, pp 242–253. arxiv preprint arXiv:1303.2416 [cs.DS]
6. Winfree E (1998) Algorithmic self-assembly of DNA. PhD thesis, California Institute of Technology
7. Woods D (2015) Intrinsic universality and the computational power of self-assembly. Philos Trans R Soc A: Math Phys Eng Sci 373(2046). doi:10.1098/rsta.2014.0214, ISBN:1471-2962, ISSN:1364-503X
8. Woods D, Chen HL, Goodfriend S, Dabby N, Winfree E, Yin P (2013) Active self-assembly of algorithmic shapes and patterns in polylogarithmic time. In: ITCS'13: proceedings of the 4th conference on innovations in theoretical computer science. ACM, pp 353–354. Full version: arXiv:1301.2626 [cs.DS]

Adaptive Partitions

Ping Deng[1], Weili Wu[1,4,5], Eugene Shragowitz[2], and Ding-Zhu Du[1,3]
[1]Department of Computer Science, The University of Texas at Dallas, Richardson, TX, USA
[2]Department of Computer Science and Engineering, University of Minnesota, Minneapolis, MN, USA
[3]Computer Science, University of Minnesota, Minneapolis, MN, USA
[4]College of Computer Science and Technology, Taiyuan University of Technology, Taiyuan, Shanxi Province, China
[5]Department of Computer Science, California State University, Los Angeles, CA, USA

Keywords

Technique for constructing approximation

Years and Authors of Summarized Original Work

1986; Du, Pan, Shing

Problem Definition

Adaptive partition is one of major techniques to design polynomial-time approximation algorithms, especially polynomial-time approximation schemes for geometric optimization problems. The framework of this technique is to put the input data into a rectangle and partition this rectangle into smaller rectangles by a sequence of cuts so that the problem is also partitioned into smaller ones. Associated with each adaptive partition, a feasible solution can be constructed recursively from solutions in smallest rectangles to bigger rectangles. With dynamic programming, an optimal adaptive partition is computed in polynomial time.

Historical Note

The adaptive partition was first introduced to the design of an approximation algorithm by Du et al. [4] with a guillotine cut while they studied the minimum edge-length rectangular partition (MELRP) problem. They found that if the partition is performed by a sequence of guillotine cuts, then an optimal solution can be computed in polynomial time with dynamic programming. Moreover, this optimal solution can be used as a pretty good approximation solution for the original rectangular partition problem. Both Arora [1] and Mitchell et al. [12,15] found that the cut does not need to be completely guillotine. In other words, the dynamic programming can still run in polynomial time if subproblems have some relations but the number of relations is small. As the number of relations goes up, the approximation solution obtained approaches the optimal one, while the run time, of course, goes up. They also found that this technique can be applied to many geometric optimization problems to obtain polynomial-time approximation schemes.

Key Results

The MELRP was proposed by Lingas et al. [10] as follows: Given a rectilinear polygon possibly with some rectangular holes, partition it into rectangles with minimum total edge length. Each hole may be degenerated into a line segment or a point.

There are several applications mentioned in [10] for the background of the problem: process control (stock cutting), automatic layout systems for integrated circuit (channel definition), and architecture (internal partitioning into offices). The *minimum edge-length* partition is a natural goal for these problems since there is a certain amount of waste (e.g., sawdust) or expense incurred (e.g., for dividing walls in the office) which is proportional to the sum of edge lengths drawn. For very large-scale integration (VLSI) design, this criterion is used in the MIT placement and interconnect (PI) system to divide the routing region up into channels – one finds that this produces large "natural-looking" channels with a minimum of channel-to-channel interaction to consider.

They showed that while the MELRP in general is nondeterministic polynomial-time (NP)-hard, it can be solved in time $O(n^4)$ in the hole-free case, where n is the number of vertices in the input rectilinear polygon. The polynomial algorithm is essentially a dynamic programming based on the fact that there always exists an optimal solution satisfying the property that every cut line passes through a vertex of the input polygon or holes (namely, every maximal cut segment is incident to a vertex of input or holes).

A naive idea to design an approximation algorithm for the general case is to use a forest connecting all holes to the boundary and then to solve the resulting hole-free case in $O(n^4)$ time. With this idea, Lingas [9] gave the first constant-bounded approximation; its performance ratio is 41.

Motivated by a work of Du et al. [6] on application of dynamic programming to optimal routing trees, Du et al. [4] initiated an idea of adaptive partition. They used a sequence of guillotine cuts to do rectangular partition; each guillotine cut breaks a connected area into at least two parts. With dynamic programming, they were able to show that a minimum-length guillotine rectangular partition (i.e., one with minimum total length among all guillotine partitions) can be computed in $O(n^5)$ time. Therefore, they suggested using the minimum-length guillotine rectangular partition to approximate the MELRP and tried to analyze the performance ratio. Unfortunately, they failed to get a constant ratio in general and only obtained an upper bound of 2 for the performance ratio in an NP-hard special case [7]. In this special case, the input is a rectangle with some points inside. Those points are holes. The following is a simple version of the proof obtained by Du et al. [5].

Theorem 1 *The minimum-length guillotine rectangular partition is an approximation with performance ratio 2 for the MELRP.*

Proof Consider a rectangular partition P. Let $proj_x(P)$ denote the total length of segments on a horizontal line covered by vertical projection of the partition P.

A rectangular partition is said to be covered by a guillotine partition if each segment in the rectangular partition is covered by a guillotine cut of the latter. Let $guil(P)$ denote the minimum length of the guillotine partition covering P and $length(P)$ denote the total length of rectangular partition P. It will be proved by induction on the number k of segments in P that

$$guil(P) \leq 2 \cdot length(P) - proj_x(P).$$

For $k = 1$, one has $guil(P) = length(P)$. If the segment is horizontal, then one has $proj_x(P) = length(P)$ and hence

$$guil(P) = 2 \cdot length(P) - proj_x(P).$$

If the segment is vertical, then $proj_x(P) = 0$ and hence

$$guil(P) < 2 \cdot length(P) - proj_x(P).$$

Now, consider $k \geq 2$. Suppose that the initial rectangle has each vertical edge of length a and each horizontal edge of length b. Consider two cases.

Case 1. There exists a vertical segment s having length greater than or equal to $0.5a$. Apply a guillotine cut along this segment s. Then the

remainder of P is divided into two parts, P_1 and P_2, which form rectangular partition of two resulting small rectangles, respectively. By induction hypothesis,

$$guil(P_i) \leq 2 \cdot length(P_i) - proj_x(P_i)$$

for $i = 1, 2$. Note that

$$guil(P) \leq guil(P_1) + guil(P_2) + a,$$
$$length(P) = length(P_1) + length(P_2)$$
$$+ length(s),$$
$$proj_x(P) = proj_x(P_1) + proj_x(P_2).$$

Therefore,

$$guil(P) \leq 2 \cdot length(P) - proj_x(P).$$

Case 2. No vertical segment in P has length greater than or equal to $0.5a$. Choose a horizontal guillotine cut which partitions the rectangle into two equal parts. Let P_1 and P_2 denote rectangle partitions of the two parts, obtained from P. By induction hypothesis,

$$guil(P_i) \leq 2 \cdot length(P_i) - proj_x(P_i)$$

for $i = 1, 2$. Note that

$$guil(P) = guil(P_1) + guil(P_2) + b,$$
$$length(P) \geq length(P_1) + length(P_2),$$
$$proj_x(P) = proj_x(P_1) = proj_x(P_2) = b.$$

Therefore,

$$guil(P) \leq 2 \cdot length(P) - proj_x(P).$$

Gonzalez and Zheng [8] improved this upper bound to 1.75 and conjectured that the performance ratio in this case is 1.5.   □

Applications

In 1996, Arora [1] and Mitchell et al. [12, 14, 15] found that the cut does not necessarily have to be completely guillotine in order to have a polynomial-time computable optimal solution for such a sequence of cuts. Of course, the number of connections left by an incomplete guillotine cut should be limited. While Mitchell et al. developed the m-guillotine subdivision technique, Arora employed a "portal" technique. They also found that their techniques can be used for not only the MELRP, but also for many geometric optimization problems [1–3, 12–15].

Open Problems

One current important submicron step of technology evolution in electronics interconnects has become the dominating factor in determining VLSI performance and reliability. Historically a problem of interconnects design in VLSI has been very tightly intertwined with the classical problem in computational geometry: Steiner minimum tree generation. Some essential characteristics of VLSI are roughly proportional to the length of the interconnects. Such characteristics include chip area, yield, power consumption, reliability, and timing. For example, the area occupied by interconnects is proportional to their combined length and directly impacts the chip size. Larger chip size results in reduction of yield and increase in manufacturing cost. The costs of other components required for manufacturing also increase with the increase of the wire length. From the performance angle, longer interconnects cause an increase in power dissipation, degradation of timing, and other undesirable consequences. That is why finding the minimum length of interconnects consistent with other goals and constraints is such an important problem at this stage of VLSI technology.

The combined length of the interconnects on a chip is the sum of the lengths of individual signal nets. Each signal net is a set of electrically connected terminals, where one terminal acts as a driver and other terminals are receivers of electrical signals. Historically, for the purpose of finding an optimal configuration of interconnects, terminals were considered as points on the plane, and a routing problem for individual nets was formulated as a classical Steiner minimum tree problem. For a variety of reasons, VLSI technology implements only rectilinear wiring on the set of parallel planes, and, consequently, with few exceptions, only a rectilinear version of the Steiner tree is being considered in the VLSI domain. This problem is known as the RSMT.

Further progress in VLSI technology resulted in more factors than just length of interconnects gaining importance in selection of routing topologies. For example, the presence of obstacles led to reexamination of techniques used in studies of the rectilinear Steiner tree, since many classical techniques do not work in this new environment. To clarify the statement made above, we will consider the construction of a rectilinear Steiner minimum tree in the presence of obstacles.

Let us start with a rectilinear plane with obstacles defined as rectilinear polygons. Given n points on the plane, the objective is to find the shortest rectilinear Steiner tree that interconnects them. One already knows that a polynomial-time approximation scheme for RSMT without obstacles exists and can be constructed by adaptive partition with application of either the portal or the m-guillotine subdivision technique. However, both the m-guillotine cut and the portal techniques do not work in the case that obstacles exist. The portal technique is not applicable because obstacles may block the movement of the line that crosses the cut at a portal. The m-guillotine cut could not be constructed either, because obstacles may break down the cut segment that makes the Steiner tree connected.

In spite of the facts stated above, the RSMT with obstacles may still have polynomial-time approximation schemes. Strong evidence was given by Min et al. [11]. They constructed a polynomial-time approximation scheme for the problem with obstacles under the condition that the ratio of the longest edge and the shortest edge of the minimum spanning tree is bounded by a constant. This design is based on the classical nonadaptive partition approach. All of the above make us believe that a new adaptive technique can be found for the case with obstacles.

Cross-References

▶ Metric TSP
▶ Rectilinear Steiner Tree
▶ Steiner Trees

Recommended Reading

1. Arora S (1996) Polynomial-time approximation schemes for Euclidean TSP and other geometric problems. In: Proceedings of the 37th IEEE symposium on foundations of computer science, pp 2–12
2. Arora S (1997) Nearly linear time approximation schemes for Euclidean TSP and other geometric problems. In: Proceedings of the 38th IEEE symposium on foundations of computer science, pp 554–563
3. Arora S (1998) Polynomial-time approximation schemes for Euclidean TSP and other geometric problems. J ACM 45:753–782
4. Du D-Z, Pan, L-Q, Shing, M-T (1986) Minimum edge length guillotine rectangular partition. Technical report 0241886, Math. Sci. Res. Inst., Univ. California, Berkeley
5. Du D-Z, Hsu DF, Xu K-J (1987) Bounds on guillotine ratio. Congressus Numerantium 58:313–318
6. Du DZ, Hwang FK, Shing MT, Witbold T (1988) Optimal routing trees. IEEE Trans Circuits 35:1335–1337
7. Gonzalez T, Zheng SQ (1985) Bounds for partitioning rectilinear polygons. In: Proceedings of the 1st symposium on computational geometry
8. Gonzalez T, Zheng SQ (1989) Improved bounds for rectangular and guillotine partitions. J Symb Comput 7:591–610
9. Lingas A (1983) Heuristics for minimum edge length rectangular partitions of rectilinear figures. In: Proceedings of the 6th GI-conference, Dortmund. Springer
10. Lingas A, Pinter RY, Rivest RL, Shamir A (1982) Minimum edge length partitioning of rectilinear polygons. In: Proceedings of the 20th Allerton conference on communication, control, and computing, Illinois
11. Min M, Huang SC-H, Liu J, Shragowitz E, Wu W, Zhao Y, Zhao Y (2003) An approximation scheme for the rectilinear Steiner minimum tree in presence of obstructions. Fields Inst Commun 37:155–164
12. Mitchell JSB (1996) Guillotine subdivisions approximate polygonal subdivisions: a simple new method for the geometric k-MST problem. In: Proceedings of the 7th ACM-SIAM symposium on discrete algorithms, pp 402–408
13. Mitchell JSB (1997) Guillotine subdivisions approximate polygonal subdivisions: part III – faster polynomial-time approximation scheme for geometric network optimization, manuscript, State University of New York, Stony Brook
14. Mitchell JSB (1999) Guillotine subdivisions approximate polygonal subdivisions: part II – a simple polynomial-time approximation scheme for geometric k-MST, TSP, and related problem. SIAM J Comput 29(2):515–544
15. Mitchell JSB, Blum A, Chalasani P, Vempala S (1999) A constant-factor approximation algorithm for the geometric k-MST problem in the plane. SIAM J Comput 28(3):771–781

Additive Spanners

Shiri Chechik
Department of Computer Science, Tel Aviv University, Tel Aviv, Israel

Keywords

Approximate shortest paths; Spanners; Stretch factor

Years and Authors of Summarized Original Work

2013; Chechik

Problem Definition

A spanner is a subgraph of a given graph that faithfully preserves the pairwise distances of that graph. Formally, an (α, β) spanner of a graph $G = (V, E)$ is a subgraph H of G such that for any pair of nodes x, y, $\mathbf{dist}(x, y, H) \leq \alpha \cdot \mathbf{dist}(x, y, G) + \beta$, where $\mathbf{dist}(x, y, H')$ for a subgraph H' is the distance of the shortest path from s to t in H'. We say that the spanner is *additive* if $\alpha = 1$, and if in addition $\beta = O(1)$, we say that the spanner is *purely additive*. If $\beta = 0$, we say that the spanner is *multiplicative*; otherwise, we say that the spanner is *mixed*.

Key Results

This section presents a survey on spanners with a special focus on additive spanners.

Graph spanners were first introduced in [12,13] in the late 1980s and have been extensively studied since then.

Spanners are used as a key ingredient in many distributed applications, e.g., synchronizers [13], compact routing schemes [6, 14, 17], broadcasting [11], etc.

Much of the work on spanners considers multiplicative spanners. A well-known theorem on multiplicative spanners is that one can efficiently construct a $(2k - 1, 0)$ spanner with $O\left(n^{1+1/k}\right)$ edges [2]. Based in the girth conjecture of Erdős [10], this size-stretch ratio is conjectured to be optimal.

The problem of additive spanners was also extensively studied, but yet several key questions remain open. The girth conjecture does not contradict the existence of $(1, 2k-2)$ spanners of size $O\left(n^{1+1/k}\right)$, or in fact any (α, β) spanners of size $O\left(n^{1+1/k}\right)$ such that $\alpha + \beta = 2k - 1$ with $\alpha \geq 1$ and $\beta > 0$.

The first construction for purely additive spanners was introduced by Aingworth et al. [1]. It was shown in [1] how to efficiently construct a $(1, 2)$ spanner, or a 2-*additive* spanner for short, with $O\left(n^{3/2}\right)$ edges (see also [8, 9, 16, 19] for further follow-up). Later, Baswana et al. [3, 4] presented an efficient construction for 6-additive spanners with $O\left(n^{4/3}\right)$ edges. Woodruff [21] further presented another construction for 6-additive spanners with $\tilde{O}\left(n^{4/3}\right)$ edges with improved construction time. Chechik [7] recently presented a new algorithm for $(1, 4)$-additive spanners with $\tilde{O}\left(n^{7/5}\right)$ edges. These are the only three constructions known for purely additive spanners. Interestingly, Woodruff [20] presented lower bounds for additive spanners that match the girth conjecture bounds. These lower bounds do not rely on the correctness of the conjecture. More precisely, Woodruff [20] showed the existence of graphs for which any spanner of size $O\left(k^{-1}n^{1+1/k}\right)$ must have an additive stretch of at least $2k - 1$.

In the absence of additional purely additive spanners, attempts were made to seek sparser spanners with nonconstant additive stretch. Bollobás et al. [5] showed how to efficiently construct a $\left(1, n^{1-2\delta}\right)$ spanner with $O\left(2^{1/\delta}n^{1+\delta}\right)$ edges for any $\delta > 0$. Later, Baswana et al. in [3,4] improved this additive stretch to $\left(1, n^{1-3\delta}\right)$, and in addition, Pettie [15] improved the stretch to $\left(1, n^{9/16-7\delta/8}\right)$ (the latter is better than the former for every $\delta < 7/34$).

Chechik [7] recently further improved the stretch for a specific range of δ. More specifically, Chechik presented a construction for additive spanners with $\tilde{O}\left(n^{1+\delta}\right)$ edges and $\tilde{O}\left(n^{1/2-3\delta/2}\right)$ additive stretch for any $3/17 \leq \delta < 1/3$. Namely, [7] decreased the stretch for this range to the root of the best previously known additive stretch.

Sublinear additive spanners, that is, spanners with additive stretch that is sublinear in the distances, were also studied. Thorup and Zwick [19] showed a construction of a $O\left(kn^{1+1/k}\right)$ size spanner such that for every pair of nodes s and t, the additive stretch is $O\left(d^{1-1/k} + 2^k\right)$, where $d = \mathbf{dist}(s, t)$ is the distance between s and t. This was later improved by Pettie [15] who presented an efficient spanner construction with $O\left(kn^{1+\frac{(3/4)^{k-2}}{7-2(3/4)^{k-2}}}\right)$ size and $O\left(kd^{1-1/k} + k^k\right)$ additive stretch, where $d = \mathbf{dist}(s, t)$. Specifically, for $k = 2$, the size of the spanner is $O\left(n^{6/5}\right)$ and the additive stretch is $O\left(\sqrt{d}\right)$.

Chechik [7] slightly improved the size of Pettie's [15] sublinear additive spanner with additive stretch $O(\sqrt{d})$ from $O\left(n^{1+1/5}\right)$ to $\tilde{O}\left(n^{1+3/17}\right)$.

Open Problems

A major open problem in the area of additive spanners is on the existence of purely additive spanners with $O\left(n^{1+\delta}\right)$ for any $\delta > 0$. In particular, proving or disproving the existence of a spanner of size $O\left(n^{4/3-\epsilon}\right)$ for some constant ϵ with constant or even polylog additive stretch would be a major breakthrough.

Recommended Reading

1. Aingworth D, Chekuri C, Indyk P, Motwani R (1999) Fast estimation of diameter and shortest paths (without matrix multiplication). SIAM J Comput 28(4):1167–1181
2. Althöfer I, Das G, Dobkin D, Joseph D, Soares J (1993) On sparse spanners of weighted graphs. Discret Comput Geom 9:81–100
3. Baswana S, Kavitha T, Mehlhorn K, Pettie S (2005) New constructions of (α, β)-spanners and purely additive spanners. In: Proceedings of the 16th symposium on discrete algorithms (SODA), Vancouver, pp 672–681
4. Baswana S, Kavitha T, Mehlhorn K, Pettie S (2010) Additive spanners and (α, β)-spanners. ACM Trans Algorithm 7:1–26. A.5
5. Bollobás B, Coppersmith D, Elkin M (2003) Sparse distance preservers and additive spanners. In: Proceedings of the 14th ACM-SIAM symposium on discrete algorithms (SODA), Baltimore, pp 414–423
6. Chechik S (2013) Compact routing schemes with improved stretch. In: Proceedings of the 32nd ACM symposium on principles of distributed computing (PODC), Montreal, pp 33–41
7. Chechik S (2013) New additive spanners. In: Proceedings of the 24th symposium on discrete algorithms (SODA), New Orleans, pp 498–512
8. Dor D, Halperin S, Zwick U (2000) All-pairs almost shortest paths. SIAM J Comput 29(5):1740–1759
9. Elkin M, Peleg D (2004) $(1 + \epsilon, \beta)$-spanner constructions for general graphs. SIAM J Comput 33(3):608–631
10. Erdős P (1964) Extremal problems in graph theory. In: Theory of graphs and its applications. Methuen, London, pp 29–36
11. Farley AM, Proskurowski A, Zappala D, Windisch K (2004) Spanners and message distribution in networks. Discret Appl Math 137(2):159–171
12. Peleg D, Schäffer AA (1989) Graph spanners. J Graph Theory 13:99–116
13. Peleg D, Ullman JD (1989) An optimal synchronizer for the hypercube. SIAM J Comput 18(4):740–747
14. Peleg D, Upfal E (1989) A trade-off between space and efficiency for routing tables. J ACM 36(3):510–530
15. Pettie S (2009) Low distortion spanners. ACM Trans Algorithms 6(1):1–22
16. Roditty L, Thorup M, Zwick U (2005) Deterministic constructions of approximate distance oracles and spanners. In: Proceedings of the 32nd international colloquium on automata, languages and programming (ICALP), Lisbon, pp 261–272
17. Thorup M, Zwick U (2001) Compact routing schemes. In: Proceedings of the 13th ACM symposium on parallel algorithms and architectures (SPAA), Heraklion, pp 1–10
18. Thorup M, Zwick U (2005) Approximate distance oracles. J ACM 52(1):1–24
19. Thorup M, Zwick U (2006) Spanners and emulators with sublinear distance errors. In: Proceedings of the 17th ACM-SIAM symposium on discrete algorithms (SODA), Miami, pp 802–809
20. Woodruff DP (2006) Lower bounds for additive spanners, emulators, and more. In: Proceedings of the 47th IEEE symposium on foundations of computer science (FOCS), Berkeley, pp 389–398
21. Woodruff DP (2010) Additive spanners in nearly quadratic time. In: Proceedings of the 37th international colloquium on automata, languages and programming (ICALP), Bordeaux, pp 463–474

Adwords Pricing

Tian-Ming Bu
Software Engineering Institute, East China
Normal University, Shanghai, China

Keywords

Adword auction; Convergence; Dynamics; Equilibrium; Mechanism design

Years and Authors of Summarized Original Work

2007; Bu, Deng, Qi

Problem Definition

The model studied here is the same as that which is first presented in [10] by Varian. For some keyword, $\mathcal{N} = \{1, 2, \ldots, N\}$ advertisers bid $\mathcal{K} = \{1, 2, \ldots, K\}$ advertisement slots ($K < N$) which will be displayed on the search result page from top to bottom. The higher the advertisement is positioned, the more conspicuous it is and the more clicks it receives. Thus for any two slots $k_1, k_2 \in \mathcal{K}$, if $k_1 < k_2$, then slot k_1's click-through rate (CTR) c_{k_1} is larger than c_{k_2}. That is, $c_1 > c_2 > \cdots > c_K$, from top to bottom, respectively. Moreover, each bidder $i \in \mathcal{N}$ has privately known information, v^i, which represents the expected return of per click to bidder i.

According to each bidder i's submitted bid b^i, the auctioneer then decides how to distribute the advertisement slots among the bidders and how much they should pay for per click. In particular, the auctioneer first sorts the bidders in decreasing order according to their submitted bids. Then the highest slot is allocated to the first bidder, the second highest slot is allocated to the second bidder, and so on. The last $N - K$ bidders would lose and get nothing. Finally, each winner would be charged on a per-click basis for the next bid in the descending bid queue. The losers would pay nothing.

Let b_k denote the kth highest bid in the descending bid queue and v_k the true value of the kth bidder in the descending queue. Thus if bidder i got slot k, i's payment would be $b_{k+1} \cdot c_k$. Otherwise, his payment would be zero. Hence, for any bidder $i \in \mathcal{N}$, if i were on slot $k \in \mathcal{K}$, his/her utility (payoff) could be represented as

$$u_k^i = (v^i - b_{k+1}) \cdot c_k$$

Unlike one-round sealed-bid auctions where each bidder has only one chance to bid, the adword auction allows bidders to change their bids any time. Once bids are changed, the system refreshes the ranking automatically and instantaneously. Accordingly, all bidders' payment and utility are also recalculated. As a result, other bidders could then have an incentive to change their bids to increase their utility and so on.

Definition 1 (Adword Pricing)

INPUT: The CTR for each slot, each bidder's expected return per click on his/her advertising

OUTPUT: The stable states of this auction and whether any of these stable states can be reached from any initial states

Key Results

Let $\mathbf{b}$ represent the bid vector $(b^1, b^2, \ldots, b^N)$. $\forall i \in \mathcal{N}$, $\mathcal{O}^i(\mathbf{b})$ denotes bidder i's place in the descending bid queue. Let $\mathbf{b}^{-i} = (b^1, \ldots, b^{i-1}, b^{i+1}, \ldots, b^N)$ denote the bids

of all other bidders except i. $\mathcal{M}^i(\mathbf{b}^{-i})$ returns a set defined as

$$\mathcal{M}^i(\mathbf{b}^{-i}) = \arg \max_{b^i \in [0, v^i]} \left\{ u_{\mathcal{O}^i(b^i, \mathbf{b}^{-i})}^i \right\} \quad (1)$$

Definition 2 (Forward-Looking Best-Response Function) Given $\mathbf{b}^{-i}$, suppose $\mathcal{O}^i(\mathcal{M}^i(\mathbf{b}^{-i}), \mathbf{b}^{-i}) = k$, then bidder i's forward-looking response function $\mathcal{F}^i(\mathbf{b}^{-i})$ is defined as

$$\mathcal{F}^i(\mathbf{b}^{-i})$$
$$= \begin{cases} v^i - \frac{c_k}{c_{k-1}}(v^i - b_{k+1}) & 2 \le k \le K \\ v^i & k = 1 \text{ or } k > K \end{cases}$$
$$(2)$$

Definition 3 (Forwarding-Looking Equilibrium) A forward-looking best-response function-based equilibrium is a strategy profile $\hat{\mathbf{b}}$ such that

$$\forall i \in \mathcal{N}, \hat{\mathbf{b}}^i \in \mathcal{F}^i(\hat{\mathbf{b}}^{-i})$$

Definition 4 (Output Truthful (Kao et al., 2006, Output truthful versus input truthful: a new concept for algorithmic mechanism design, unpublished) [7]) For any instance of adword auction and the corresponding equilibrium set $\mathcal{E}$, if $\forall \mathbf{e} \in \mathcal{E}$ and $\forall i \in \mathcal{N}$, $\mathcal{O}^i(\mathbf{e}) = \mathcal{O}^i(v^1, \ldots, v^N)$, then the adword auction is *output truthful* on $\mathcal{E}$.

Theorem 1 *An adword auction is output truthful on $\mathcal{E}_{forward-looking}$.*

Corollary 1 *An Adword auction has a unique forward-looking Nash equilibrium.*

Corollary 2 *Any bidder's payment under the forward-looking Nash equilibrium is equal to his/her payment under the VCG mechanism for the auction.*

Corollary 3 *For adword auctions, the auctioneer's revenue in a forward-looking Nash equilibrium is equal to his/her revenue under the VCG mechanism for the auction.*

Definition 5 (Simultaneous Readjustment Scheme) In a simultaneous readjustment scheme, all bidders participating in the auction will use forward-looking best-response function $\mathcal{F}$ to update their current bids simultaneously, which turns the current stage into a new stage. Then based on the new stage, all bidders may update their bids again.

Theorem 2 *An adword auction may not always converge to a forward-looking Nash equilibrium under the simultaneous readjustment scheme even when the number of slots is 3. But the protocol converges when the number of slots is 2.*

Definition 6 (Round-Robin Readjustment Scheme) In the round-robin readjustment scheme, bidders update their biddings one after the other, according to the order of the bidder's number or the order of the slots.

Theorem 3 *An adword auction may not always converge to a forward-looking Nash equilibrium under the round-robin readjustment scheme even when the number of slots is 4. But the protocol converges when the number of slots is 2 or 3.*

---

1 Readjustment Scheme: Lowest-First$(K, j, b_1, b_2, \cdots, b_N)$

1: **if** $(j = 0)$ **then**
2:     exit
3: **end if**
4: Let i be the ID of the bidder whose current bid is b_j (and equivalently, b^i).
5: Let $h = \mathcal{O}^i(\mathcal{M}^i(\mathbf{b}^{-i}), \mathbf{b}^{-i})$.
6: Let $\mathcal{F}^i(\mathbf{b}^{-i})$ be the best response function value for Bidder i.
7: Re-sort the bid sequence. (So h is the slot of the new bid $\mathcal{F}^i(\mathbf{b}^{-i})$ of Bidder i.)
8: **if** $(h < j)$ **then**
9:     call Lowest-First $(K, j, b_1, b_2, \cdots, b_N)$,
10: **else**
11:     call Lowest-First$(K, h-1, b_1, b_2, \cdots, b_N)$
12: **end if**

Theorem 4 *Adword auctions converge to a forward-looking Nash equilibrium in finite steps with a lowest-first adjustment scheme.*

Theorem 5 *Adword auctions converge to a forward-looking Nash equilibrium with probability 1 under a randomized readjustment scheme.*

Applications

Online adword auctions are the fastest growing form of advertising. Many search engine companies such as Google and Yahoo! make huge profits on this kind of auction. Because advertisers can change their bids anytime, such auctions can reduce the advertisers' risk. Further, because the advertisement is only displayed to those people who are really interested in it, such auctions can reduce the advertisers' investment and increase their return on investment.

For the same model, Varian [10] focuses on a subset of Nash equilibria, called *Symmetric Nash Equilibria*, which can be formulated nicely and dealt with easily. Edelman et al. [8] study *locally envy-free* equilibria, where no player can improve his/her payoff by exchanging bids with the player ranked one position above him/her. Coincidently, locally envy-free equilibrium is equal to symmetric Nash equilibrium proposed in [10]. Further, the revenue under the forward-looking Nash equilibrium is the same as the lower bound under Varian's symmetric Nash equilibria and the lower bound under Edelman et al.'s locally envy-free equilibria. In [6], Cary et al. also study the dynamic model's equilibria and convergence based on the *balanced bidding strategy* which is actually the same as the *forward-looking best-response function* in [4]. Cary et al. explore the convergence properties under two models, a *synchronous* model which is the same as *simultaneous readjustment scheme* in [4] and an *asynchronous* model which is the same as *randomized readjustment scheme* in [4].

In addition, there are other models for adword auctions. Abrams [1] and Bu et al. [5] study the model under which each bidder could submit his/her daily budget, even the maximum number of clicks per day, in addition to the price per click. Both [9] and [3] study bidders' behavior of bidding on several keywords. Aggarwal et al. [2] studies the model where the advertiser not only submits a bid but additionally submits which positions he/she is going to bid for.

Open Problems

The speed of convergence still remains open. Does the dynamic model converge in polynomial time under randomized readjustment scheme? Even more, are there other readjustment scheme that converge in polynomial time?

Cross-References

▶ Multiple Unit Auctions with Budget Constraint
▶ Position Auction

Recommended Reading

1. Abrams Z (2006) Revenue maximization when bidders have budgets. In: Proceedings of the seventeenth annual ACM-SIAM symposium on discrete algorithms (SODA-06), Miami, pp 1074–1082
2. Aggarwal G, Feldman J, Muthukrishnan S (2006) Bidding to the top: VCG and equilibria of position-based auctions. In: Proceedings of the 4th international workshop on approximation and online algorithms (WAOA-2006), Zurich, pp 15–28
3. Borgs C, Chayes J, Etesami O, Immorlica N, Jain K, Mahdian M (2006) Bid optimization in online advertisement auctions. In: 2nd workshop on sponsored search auctions (SSA2006), in conjunction with the ACM conference on electronic commerce (EC-06), Ann Arbor
4. Bu TM, Deng X, Qi Q (2008) Forward looking nash equilibrium for keyword auction. Inf Process Lett 105(2):41–46
5. Bu TM, Qi Q, Sun AW (2008) Unconditional competitive auctions with copy and budget constraints. Theor Comput Sci 393(1–3):1–13
6. Cary M, Das A, Edelman B, Giotis I, Heimerl K, Karlin AR, Mathieu C, Schwarz M (2007) Greedy bidding strategies for keyword auctions. In: Proceedings of the 8th ACM conference on electronic commerce (EC-2007), San Diego, pp 262–271
7. Chen X, Deng X, Liu BJ (2006) On incentive compatible competitive selection protocol. In: Proceedings of the 12th annual international computing and combinatorics conference (COCOON06), Taipei, pp 13–22
8. Edelman B, Ostrovsky M, Schwarz M (2007) Internet advertising and the generalized second price auction: selling billions of dollars worth of keywords. Am Econ Rev 97(1):242–259
9. Kitts B, Leblanc B (2004) Optimal bidding on keyword auctions. Electron Mark Spec Issue Innov Auction Mark 14(3):186–201
10. Varian HR (2007) Position auctions. Int J Ind Organ 25(6):1163–1178

Algorithm DC-Tree for *k*-Servers on Trees

Marek Chrobak
Computer Science, University of California, Riverside, CA, USA

Keywords

Competitive analysis; K-server problem; On-line algorithms; Trees

Years and Authors of Summarized Original Work

1991; Chrobak, Larmore

Problem Definition

In the *k-Server Problem*, one wishes to schedule the movement of *k*-servers in a metric space $\mathbb{M}$, in response to a sequence $\varrho = r_1, r_2, \ldots, r_n$ of *requests*, where $r_i \in \mathbb{M}$ for each i. Initially, all the servers are located at some initial configuration $X_0 \subseteq \mathbb{M}$ of k points. After each request r_i is issued, one of the k-servers must move to r_i. A *schedule* specifies which server moves to each request. The *cost* of a schedule is the total distance traveled by the servers, and our objective is to find a schedule with minimum cost.

In the *online* version of the k-Server Problem, the decision as to which server to move to each request r_i must be made before the next request r_{i+1} is issued. In other words, the choice of this server is a function of requests $r_1, r_2, \ldots, r_i$. It is quite easy to see that in this online scenario, it is not possible to compute an optimal schedule for each request sequence, raising the question of how to measure the accuracy of such online algorithms. A standard approach to doing this is based on competitive analysis. If $\mathcal{A}$ is an online k-server algorithm denote by $cost_{\mathcal{A}}(\varrho)$ the cost of the schedule produced by $\mathcal{A}$ on a request sequence ϱ, and by $opt(\varrho)$ the cost of the optimal schedule. $\mathcal{A}$ is called *R-competitive* if $cost_{\mathcal{A}}(\varrho) \leq R \cdot opt(\varrho) + B$, where B is a constant that may depend on $\mathbb{M}$ and X_0. The smallest such R is called the *competitive ratio* of $\mathcal{A}$.

The k-Server Problem was introduced by Manasse, McGeoch, and Sleator [7, 8], who proved that there is no online R-competitive algorithm for $R < k$, for any metric space with at least $k + 1$ points. They also gave a 2-competitive algorithm for $k = 2$ and formulated what is now known as the *k-Server Conjecture*, which postulates that there exists a k-competitive online algorithm for all k. Koutsoupias and Papadimitriou [5, 6] proved that the so-called *Work-Function Algorithm* has competitive ratio at most $2k - 1$, which to date remains the best upper bound known.

Efforts to prove the k-Server Conjecture led to discoveries of k-competitive algorithms for some restricted classes of metric spaces, including Algorithm DC-Tree for trees [3] presented in this entry. (See [1, 2, 4] for other examples.) A *tree* is a metric space defined by a connected acyclic graph whose edges are treated as line segments of arbitrary positive lengths. This metric space includes both the tree's vertices and the points on the edges, and the distances are measured along the (unique) shortest paths.

Key Results

Let $\mathbb{T}$ be a tree, as defined above. Given the current server configuration $S = \{s_1, \ldots, s_k\}$, where s_j denotes the location of server j, and a request point r, the algorithm will move several servers, with one of them ending up on r. For two points $x, y \in \mathbb{T}$, let $[x, y]$ be the unique path from x to y in $\mathbb{T}$. A server j is called *active* if there is no other server in $[s_j, r] - \{s_j\}$, and j is the minimum-index server located on s_j (the last condition is needed only to break ties).

Algorithm DC-Tree. On a request r, move all active servers, continuously and with the same speed, towards r, until one of them reaches the request. Note that during this process some active servers may become inactive, in which case they halt. Clearly, the server that will arrive at r is the one that was closest to r at the time when r was issued.

More formally, denoting by s_j the variable representing the current position of server j, the algorithm serves r as follows:

while $s_j \neq r$ for all j **do**
   let $\delta = \frac{1}{2} \min_{i<j} \{d(s_i, s_j) + d(s_i, r)$
      $-d(s_j, r)\}$
   move each active server s_j by distance
      δ towards r

The example below shows how DC-Tree serves a request r (Fig. 1).

The competitive analysis of Algorithm DC-Tree is based on a potential argument. The cost of Algorithm DC-Tree is compared to that of an adversary who serves the requests with her own servers. Denoting by A the configuration of the adversary servers at a given step, define the potential by $\Phi = k \cdot D(S, A) + \sum_{i<j} d(s_i, s_j)$, where $D(S, A)$ is the cost of the minimum matching between S and A. At each step, the adversary first moves one of her servers to r. In this substep the potential increases by at most k times the increase of the adversary's cost. Then, Algorithm DC-Tree serves the request. One can show that then the sum of Φ and DC-Tree's cost does not increase. These two facts, by amortization over

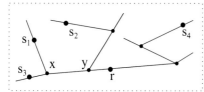 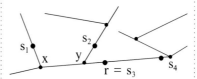

Algorithm DC-TREE for *k*-Servers on Trees, Fig. 1
Algorithm DC-TREE serving a request on *r*. The config-uration before *r* is issued is on the left; the configuration after the service is completed is on the right. At first, all

servers are active. When server 3 reaches point *x*, server 1 becomes inactive. When server 3 reaches point *y*, server 2 becomes inactive

the whole request sequence, imply the following result [3]:

Theorem 1 ([3]) *Algorithm* DC-TREE *is k-competitive on trees.*

Applications

The *k*-Server Problem is an abstraction of various scheduling problems, including emergency crew scheduling, caching in multilevel memory systems, or scheduling head movement in 2-headed disks. Nevertheless, due to its abstract nature, the *k*-server problem is mainly of theoretical interest.

Algorithm DC-TREE can be applied to other spaces by "embedding" them into trees. For example, a uniform metric space (with all distances equal 1) can be represented by a star with arms of length $1/2$, and thus Algorithm DC-TREE can be applied to those spaces. This also immediately gives a *k*-competitive algorithm for the *caching problem*, where the objective is to manage a two-level memory system consisting of a large main memory and a cache that can store up to *k* memory items. If an item is in the cache, it can be accessed at cost 0, otherwise it costs 1 to read it from the main memory. This caching problem can be thought of as the *k*-server problem in a uniform metric space where the server positions represent the items residing in the cache. This idea can be extended further to the *weighted caching* [4], which is a generalization of the caching problem where different items may have different costs. In fact, if one can embed a metric space $\mathbb{M}$ into a tree with distortion bounded by δ, then Algorithm DC-TREE yields a δk-competitive algorithm for $\mathbb{M}$.

Open Problems

The *k*-Server Conjecture – whether there is a *k*-competitive algorithm for *k*-servers in any metric space – remains open. It would be of interest to prove it for some natural special cases, for example the plane, either with the Euclidean or Manhattan metric. (A *k*-competitive algorithm for the Manhattan plane for $k = 2, 3$ servers is known [1], but not for $k \geq 4$).

Very little is known about online *randomized* algorithms for *k*-servers. In fact, even for $k = 2$ it is not known if there is a randomized algorithm with competitive ratio smaller than 2.

Cross-References

▶ Deterministic Searching on the Line
▶ Generalized Two-Server Problem
▶ Metrical Task Systems
▶ Online Paging and Caching
▶ Work-Function Algorithm for *k*-Servers

Recommended Reading

1. Bein W, Chrobak M, Larmore LL (2002) The 3-server problem in the plane. Theor Comput Sci 287: 387–391
2. Borodin A, El-Yaniv R (1998) Online computation and competitive analysis. Cambridge University Press, Cambridge
3. Chrobak M, Larmore LL (1991) An optimal online algorithm for *k* servers on trees. SIAM J Comput 20:144–148
4. Chrobak M, Karloff H, Payne TH, Vishwanathan S (1991) New results on server problems. SIAM J Discret Math 4:172–181

5. Koutsoupias E, Papadimitriou C (1994) On the k-server conjecture. In: Proceedings of the 26th symposium on theory of computing (STOC). ACM, Montreal, pp 507–511
6. Koutsoupias E, Papadimitriou C (1995) On the k-server conjecture. J ACM 42:971–983
7. Manasse M, McGeoch LA, Sleator D (1988) Competitive algorithms for online problems. In: Proceedings of the 20th symposium on theory of computing (STOC). ACM, Chicago, pp 322–333
8. Manasse M, McGeoch LA, Sleator D (1990) Competitive algorithms for server problems. J Algorithms 11:208–230

Algorithmic Cooling

Tal Mor
Department of Computer Science, Technion – Israel Institute of Technology, Haifa, Israel

Keywords

Cooling; Data compression; Nuclear magnetic resonance; Quantum computing; Spin cooling; State initialization

Years and Authors of Summarized Original Work

1999; Schulman, Vazirani
2002; Boykin, Mor, Roychowdhury, Vatan, Vrijen

Problem Definition

The fusion of concepts taken from the fields of quantum computation, data compression, and thermodynamics has recently yielded novel algorithms that resolve problems in nuclear magnetic resonance and potentially in other areas as well, algorithms that "cool down" physical systems.

- A leading candidate technology for the construction of quantum computers is nuclear magnetic resonance (NMR). This technology has the advantage of being well established for other purposes, such as chemistry and medicine. Hence, it does not require new and exotic equipment, in contrast to ion traps and optical lattices, to name a few. However, when using standard NMR techniques, (not only for quantum computing purposes) one has to live with the fact that the state can only be initialized in a very noisy manner: The particles' spins point in mostly random directions, with only a tiny bias towards the desired state.

The key idea of Schulman and Vazirani [27] is to combine the tools of both data compression and quantum computation, to suggest a *scalable* state initialization process, a "molecular-scale heat engine." Based on Schulman and Vazirani's method, Boykin, Mor, Roychowdhury, Vatan, and Vrijen [4] then developed a new process, "heat-bath algorithmic cooling," to significantly improve the state initialization process, by opening the system to the environment. Strikingly, this offered a way to put to good use the phenomenon of decoherence, which is usually considered to be the villain in quantum computation. These two methods are now sometimes called "closed-system" (or "reversible"), algorithmic cooling, and "open-system" algorithmic cooling, respectively.

- The far-reaching consequence of this research lies in the possibility of reaching beyond the potential implementation of remote-future quantum computing devices. An efficient technique to generate ensembles of spins that are highly polarized by external magnetic fields is considered to be a Holy Grail in NMR spectroscopy. Spin-half nuclei have steady-state polarization biases that increase inversely with temperature; therefore, spins exhibiting polarization biases above their thermal-equilibrium biases are considered *cool*. Such cooled spins present an improved signal-to-noise ratio if used in NMR spectroscopy or imaging.

Existing spin-cooling techniques are limited in their efficiency and usefulness. Algorithmic cooling is a promising new spin-cooling approach that employs data compression methods in *open systems*.

It reduces the entropy of spins to a point far beyond Shannon's entropy bound on reversible entropy manipulations, thus increasing their polarization biases. As a result, it is conceivable that the open-system algorithmic cooling technique could be harnessed to improve on *current uses of NMR* in areas such as chemistry, material science, and even medicine, since NMR is at the basis of MRI – magnetic resonance imaging.

Basic Concepts

Loss-Less In-Place Data Compression

Given a bit string of length n, such that the probability distribution is known and far enough from the uniform distribution, one can use data compression to generate a shorter string, say of m bits, such that the entropy of each bit is much closer to one. As a simple example, consider a four-bit string which is distributed as follows: $p_{0001} = p_{0010} = p_{0100} = p_{1000} = 1/4$, with p_i the probability of the string i. The probability of any other string value is exactly zero, so the probabilities sum up to one. Then, the bit string can be compressed, via a lossy compression algorithm, into a 2-bit string that holds the binary description of the location of "1" in the above four strings. One can also envision a similar process that generates an output which is of the same length n as the input, but such that the entropy is compressed via a loss-less, in-place, data compression into the last two bits. For instance, logical gates that operate on the bits can perform the permutation $0001 \rightarrow 0000$, $0010 \rightarrow 0001$, $0100 \rightarrow 0010$, and $1000 \rightarrow 0011$, while the other input strings transform to output strings in which the two most significant bits are not zero; for instance, $1100 \rightarrow 1010$. One can easily see that the entropy is now fully concentrated on the two least significant bits, which are useful in data compression, while the two most significant bits have zero entropy.

In order to gain some intuition about the design of logical gates that perform entropy manipulations, one can look at a closely related scenario which was first considered by von Neumann.

He showed a method to extract fair coin flips, given a biased coin; he suggested taking a pair of biased coin flips, with results a and b, and using the value of a conditioned on $a \neq b$. A simple calculation shows that $a = 0$ and $a = 1$ are now obtained with equal probabilities, and therefore, the entropy of coin a is increased in this case to 1. The opposite case, the probability distribution of a given that $a = b$, results in a highly determined coin flip, namely, a (conditioned) coin flip with a higher bias or lower entropy. A gate that flips the value of b if (and only if) $a = 1$ is called a controlled NOT gate. If after applying such a gate $b = 1$ is obtained, this means that $a \neq b$ prior to the gate operation; thus, now the entropy of a is 1. If, on the other hand, after applying such a gate $b = 0$ is obtained, this means that $a = b$ prior to the gate operation; thus, the entropy of a is now lower than its initial value.

Spin Temperature, Polarization Bias, and Effective Cooling

In physics, two-level systems, namely, systems that possess only binary values, are useful in many ways. Often it is important to initialize such systems to a pure state "0" or to a probability distribution which is as close as possible to a pure state "0." In these physical two-level systems, a data compression process that brings some of them closer to a pure state can be considered as "cooling." For quantum two-level systems, there is a simple connection between temperature, entropy, and population probability. The population difference between these two levels is known as the polarization bias, ϵ. Consider a single spin-half particle – for instance, a hydrogen nucleus – in a constant magnetic field. At equilibrium with a thermal heat-bath, the probability of this spin to be up or down (i.e., parallel or antiparallel to the field direction) is given by $p_\uparrow = \frac{1+\epsilon}{2}$ and $p_\downarrow = \frac{1-\epsilon}{2}$. The entropy H of the spin is $H(\text{single} - \text{bit}) = H(1/2 + \epsilon/2)$ with $H(P) \equiv -P \log_2 P - (1 - P) \log_2(1 - P)$ measured in bits. The two pure states of a spin-half nucleus are commonly written as $|\uparrow\rangle \equiv$ "0" and $|\downarrow\rangle \equiv$ "1"; the $|\rangle$ notation will be clarified elsewhere. (Quantum Computing entries in this encyclopedia.) The polarization bias of the spin at thermal equilibrium is given by $\epsilon = p_\uparrow - p_\downarrow$. For such a physical System, the bias is obtained via a quantum statistical mechanics argument, $\epsilon = \tanh\left(\frac{\hbar \gamma B}{2K_B T}\right)$, where $\hbar$ is Planck's constant, B is the magnetic field, γ is the particle-dependent gyromagnetic constant,(This constant, γ, is thus responsible for the difference in equilibrium polarization bias [e.g., a hydrogen nucleus is 4 times more polarized than a carbon isotope ^{13}C nucleus, but about 10^3 less polarized than an electron spin].) K_B is Boltzman's coefficient, and T is the thermal heat-bath temperature. For high temperatures or small biases $\epsilon \approx \frac{\hbar \gamma B}{2K_B T}$; thus, the bias is inversely proportional to the temperature. Typical values of ϵ for spin-half nuclei at room temperature (and magnetic field of $\sim 10\,T$) are 10^{-5}–10^{-6}, and therefore, most of the analysis here is done under the assumption that $\epsilon \ll 1$. The spin temperature at equilibrium is thus $T = \frac{Const}{\epsilon}$, and its (Shannon) entropy is $H = 1 - (\epsilon^2/\ln 4)$.

A spin temperature out of thermal equilibrium is still defined via the same formulas. Therefore, when a system is moved away from thermal equilibrium, achieving a greater polarization bias is equivalent to cooling the spins, *without cooling the system*, and to decreasing their entropy. The process of increasing the bias (reducing the entropy) without increasing the temperature of the thermal bath is known as "effective cooling." After a typical period of time, termed the thermalization time or relaxation time, the bias will gradually revert to its thermal equilibrium value; yet during this process, typically in the order of seconds, the effectively cooled spin may be used for various purposes as described in section "Applications."

Consider a molecule that contains n adjacent spin-half nuclei arranged in a line; these form the bits of the string. These spins are initially at thermal equilibrium due to their interaction with the environment. At room temperature, the bits at thermal equilibrium are not correlated to their neighbors on the same string: more precisely, the correlation is very small and can be ignored. Furthermore, in a liquid state one can also neglect the interaction between strings (between molecules). It is convenient to write the

probability distribution of a single spin at thermal equilibrium using the "density-matrix" notation

$$\rho_\epsilon = \begin{pmatrix} p_\uparrow & 0 \\ 0 & p_\downarrow \end{pmatrix} = \begin{pmatrix} (1+\epsilon)/2 & 0 \\ 0 & (1-\epsilon)/2 \end{pmatrix}, \tag{1}$$

since these two-level systems are of a quantum nature (namely, these are quantum bits – qubits) and, in general, can also have states other than just a classical probability distribution over "0" and "1." The classical case will now be considered, where ρ contains only diagonal elements, and these describe a conventional probability distribution. At thermal equilibrium, the state of $n = 2$ uncorrelated qubits that have the same polarization bias is described by the density matrix $\rho_{init}^{\{n=2\}} = \rho_\epsilon \otimes \rho_\epsilon$, where $\otimes$ means tensor product. The probability of the state "00," for instance, is then $(1+\epsilon)/2 \times (1+\epsilon)/2 = (1+\epsilon)^2/4$, etc. Similarly, the initial state of an n-qubit system of this type, at thermal equilibrium, is

$$\rho_{init}^{\{n\}} = \rho_\epsilon \otimes \rho_\epsilon \otimes \cdots \otimes \rho_\epsilon. \tag{2}$$

This state represents a thermal probability distribution, such that the probability of the classical state "000...0" is $P_{000...0} = (1+\epsilon_0)^n/2^n$, etc. In reality, the initial bias is not the same on each qubit,(Furthermore, individual addressing of each spin during the algorithm requires a slightly different bias for each.) but as long as the differences between these biases are small (e.g., all qubits are of the same nucleus), these differences can be ignored in a discussion of an idealized scenario.

Key Results

Molecular-Scale Heat Engines

Schulman and Vazirani (SV) [27] identified the importance of in-place loss-less data compression and of the low-entropy bits created in that process: physical two-level systems (e.g., spin-half nuclei) may be similarly cooled by data compression algorithms. SV analyzed the cooling of such a system using various tools of data compression. A loss-less compression of an n-bit binary string distributed according to the thermal equilibrium distribution, Eq. 2, is readily analyzed using information-theoretical tools: In an ideal compression scheme (not necessarily realizable), with sufficiently large n, all randomness – and hence all the entropy – of the bit string is transferred to $n - m$ bits; the remaining m bits are thus left, with extremely high probability, at a known deterministic state, say the string "000...0." The entropy H of the entire system is $H(\text{system}) = nH(\text{single} - \text{bit}) = nH(1/2 + \epsilon/2)$. Any compression scheme cannot decrease this entropy; hence, Shannon's source coding entropy bound yields $m \leq n[1 - H(1/2 + \epsilon/2)]$. A simple leading-order calculation shows that m is bounded by (approximately) $\frac{\epsilon^2}{2\ln 2}n$ for small values of the initial bias ϵ. Therefore, with typical $\epsilon \sim 10^{-5}$, molecules containing an order of magnitude of 10^{10} spins are required to cool a single spin close to zero temperature.

Conventional methods for NMR quantum computing are based on unscalable state initialization schemes [7, 14] (e.g., the "pseudo-pure-state" approach) in which the signal-to-noise ratio falls exponentially with n, the number of spins. Consequently, these methods are deemed inappropriate for future NMR quantum computers. SV [27] were first to employ tools of information theory to address the scaling problem; they presented a compression scheme in which the number of cooled spins scales well (namely, a constant times n). SV also demonstrated a scheme approaching Shannon's entropy bound, for very large n. They provided detailed analyses of three cooling algorithms, each useful for a different regime of ϵ values.

Some ideas of SV were already explored a few years earlier by Sørensen [29], a physical chemist who analyzed effective cooling of spins. He considered the entropy of several spin systems and the limits imposed on cooling these systems by polarization transfer and more general polarization manipulations. Furthermore, he considered spin-cooling processes in which only unitary operations were used, wherein unitary matrices are applied to the density matrices; such operations are realizable, at least from a conceptual point of view. Sørensen derived a stricter bound on unitary cooling, which today bears his name.

Yet, unlike SV, he did not infer the connection to data compression or advocate compression algorithms.

SV named their concept "molecular-scale heat engine." When combined with conventional polarization transfer (which is partially similar to a SWAP gate between two qubits), the term "reversible polarization compression (RPC)" is more descriptive.

Heat-Bath Algorithmic Cooling

The next significant development came when Boykin, Mor, Roychowdhury, Vatan, and Vrijen (hereinafter referred to as BMRVV), invented a new spin-cooling technique, which they named *Algorithmic cooling* [4] or more specifically heat-bath algorithmic cooling in which the use of controlled interactions with a heat bath enhances the cooling techniques much further. Algorithmic cooling (AC) expands the effective cooling techniques by exploiting entropy manipulations in *open systems*. It combines RPC steps (When the entire process is RPC, namely, any of the processes that follow SV ideas, one can refer to it as reversible AC or closed-system AC, rather than as RPC.) with fast relaxation (namely, thermalization) of the *hotter spins*, as a way of pumping entropy outside the system and cooling the system *much beyond Shannon's entropy bound*. In order to pump entropy out of the system, AC employs regular spins (here called computation spins) together with rapidly relaxing spins. The latter are auxiliary spins that return to their thermal equilibrium state very rapidly. These spins have been termed "reset spins," or, equivalently, reset bits. The controlled interactions with the heat bath are generated by polarization transfer or by standard algorithmic techniques (of data compression) that transfer the entropy onto the reset spins which then lose this excess entropy into the environment.

The ratio $R_{\text{relax}-\text{times}}$, between the relaxation time of the computation spins and the relaxation time of the reset spins, must satisfy $R_{\text{relax}-\text{times}} \gg 1$. This condition is vital if one wishes to perform many cooling steps on the system to obtain significant cooling.

In a pure information-theoretical point of view, it is legitimate to assume that the only restriction on ideal RPC steps is Shannon's entropy bound; then the equivalent of Shannon's entropy bound, when an ideal open-system AC is used, is that all computation spins can be cooled down to zero temperature, that is to $\epsilon = 1$. Proof: repeat the following till the entropy of all computation spins is exactly zero: (i) push entropy from computation spins into reset spins and (ii) let the reset spins cool back to room temperature. Clearly, each application of step (i), except the last one, pushes the same amount of entropy onto the reset spins, and then this entropy is removed from the system in step (ii). Of course, a realistic scenario must take other parameters into account such as finite relaxation-time ratios, realistic environment, and physical operations on the spins. Once this is done, cooling to zero temperature is no longer attainable. While finite relaxation times and a realistic environment are system dependent, the constraint of using physical operations is conceptual.

BMRVV therefore pursued an algorithm that follows some physical rules; it is performed by unitary operations and reset steps and still bypass Shannon's entropy bound, by far. The BMRVV cooling algorithm obtains significant cooling beyond that entropy bound by making use of very long molecules bearing hundreds or even thousands of spins, because its analysis relies on the law of large numbers.

Practicable Algorithmic Cooling

The concept of algorithmic cooling then led to practicable algorithms [13] for cooling *small molecules*. In order to see the impact of practicable algorithmic cooling, it is best to use a different variant of the entropy bound. Consider a system containing n spin-half particles with total entropy higher than $n - 1$, so that there is no way to cool even one spin to zero temperature. In this case, the entropy bound is a result of the compression of the entropy into $n - 1$ fully random spins, so that the remaining entropy on the last spin is minimal. The entropy of the remaining single spin satisfies

$H(\text{single}) \geq 1 - n\epsilon^2/\ln 4$; thus, at most, its polarization can be improved to

$$\epsilon_{\text{final}} \leq \epsilon \sqrt{n} \,. \tag{3}$$

The practicable algorithmic cooling (PAC), suggested by Fernandez, Lloyd, Mor, and Roychowdhury in [13], indicated potential for a near-future application to NMR spectroscopy. In particular, it presented an algorithm named PAC2 which uses any (odd) number of spins n, such that one of them is a reset spin, and $(n - 1)$ are computation spins. PAC2 cools the spins such that the coldest one can (approximately) reach a bias amplification by a factor of $(3/2)^{(n-1)/2}$. The approximation is valid as long as the final bias $(3/2)^{(n-1)/2} \epsilon$ is much smaller than 1. Otherwise, a more precise treatment must be done. This proves an exponential advantage of AC over the best possible reversible AC, as these reversible cooling techniques, e.g., of [27, 29], are limited to improve the bias by no more than a factor of $\sqrt{n}$. PAC can be applied for small n (e.g., in the range of 10–20), and therefore, it is potentially suitable for near-future applications [9, 13, 19] in chemical and biomedical usages of NMR spectroscopy.

It is important to note that in typical scenarios, the initial polarization bias of a reset spin is higher than that of a computation spin. In this case, the bias amplification factor of $(3/2)^{(n-1)/2}$ is relative to the larger bias, that of the reset spin.

Exhaustive Algorithmic Cooling

Next, AC was analyzed, wherein the cooling steps (reset and RPC) are repeated an arbitrary number of times. This is actually an idealization where an unbounded number of reset and logic steps can be applied without error or decoherence, while the computation qubits do not lose their polarization biases. Fernandez [12] considered two computation spins and a single reset spin (the least significant bit, namely, the qubit at the right in the tensor-product density-matrix notation) and analyzed optimal cooling of this system. By repeating the reset and compression exhaustively, he realized that the bound on the final biases of the three spins is approximately $\{2, 1, 1\}$ in units of ϵ, the polarization bias of the reset spin.

Mor and Weinstein generalized this analysis further and found that $n - 1$ computation spins and a single reset spin can be cooled (approximately) to biases according to the Fibonacci series: $\{\ldots 34, 21, 13, 8, 5, 3, 2, 1, 1\}$. The computation spin that is further away from the reset spin can be cooled up to the relevant Fibonacci number F_n. That approximation is valid as long as the largest term times ϵ is still much smaller than 1. Schulman then suggested the "partner pairing algorithm" (PPA) and proved the optimality of the PPA among all *classical and quantum* algorithms. These two algorithms, the Fibonacci AC and the PPA, led to two joint papers [25, 26], where upper and lower bounds on AC were also obtained. The PPA is defined as follows: repeat these two steps until cooling sufficiently close to the limit: (a) RESET, applied to a reset spin in a system containing $n - 1$ computation spins and a single (the LSB) reset spin, and (b) SORT, a permutation that sorts the 2^n diagonal elements of the density matrix by decreasing order, so that the MSB spin becomes the coldest. Two important theorems proven in [26] are:

(1) Lower bound: When $\epsilon 2^n \gg 1$ (namely, for long enough molecules), Theorem 3 in [26] promises that $n - \log(1/\epsilon)$ cold qubits can be extracted. This case is relevant for scalable NMR quantum computing.
(2) Upper bound: Section 4.2 in [26] proves the following theorem: No algorithmic cooling method can increase the probability of any basis state to above $\min\{2^{-n} e^{2^n \epsilon}, 1\}$, wherein the initial configuration is the completely mixed state (the same is true if the initial state is a thermal state).

More recently, Elias, Fernandez, Mor, and Weinstein [9] analyzed more closely the case of $n < 15$ (at room temperature), where the coldest spin (at all stages) still has a polarization bias much smaller than 1. This case is most relevant for near-future applications in NMR spectroscopy. They generalized the Fibonacci-AC to algorithms yielding higher-term Fibonacci series, such as the tribonacci (also known as 3-term Fibonacci series), $\{\ldots 81, 44, 24, 13, 7, 4, 2, 1, 1\}$, etc. The ultimate limit of these multi-term Fibonacci series is obtained when each term in the series is the sum of all previous terms. The resulting series is precisely the exponential series $\{\ldots 128, 64, 32, 16, 8, 4, 2, 1, 1\}$, so the coldest spin is cooled by a factor of 2^{n-2}. Furthermore, a leading-order analysis of the upper bound mentioned above (Section 4.2 in Ref. [26]) shows that no spin can be cooled beyond a factor of 2^{n-1}; see Corollary 1 in [9].

Other Results

For several other theoretical results dealing with relevant algorithms and with the connection to thermodynamics, see [11, 15, 17, 21]. For several popular "News and Views" discussions of AC in Nature, see [18, 22, 24].

Applications

The two major far-future and near-future applications are already described in section "Problem Definition." It is important to add here that although the specific algorithms analyzed so far for AC are usually classical, their practical implementation via an NMR spectrometer must be done through analysis of universal quantum computation, using the specific gates allowed in such systems. Therefore, AC could yield the first near-future application of quantum computing devices.

AC may also be useful for cooling various other physical systems; for several examples (theoretical and experimental), see [2, 16, 28, 30, 31], since state initialization is a common problem in physics in general and in quantum computation in particular.

Open Problems

A main open problem in practical AC is technological; can the ratio of relaxation times be increased so that many cooling steps may be applied onto relevant NMR systems? Other methods, for instance, a spin-diffusion mechanism [3, 23], may also be useful for various applications.

Another interesting open problem is whether the ideas developed during the design of AC can also lead to applications in classical information theory.

Last but not least, in the context of building scalable quantum computers, it is interesting to study if AC can become a practical tool for advancing the non-conventional model of quantum computing called the one pure qubit (or one clean qubit) model as suggested in [1, 8] and to study if AC can be useful for designing fault-tolerant quantum computers as suggested in [20].

Experimental Results

Various ideas of AC had already led to several experiments using 3–4 qubit quantum computing devices in NMR (AC used in other systems was mentioned earlier in section "Applications"):

(1) An experiment [6] that implemented a single RPC step.
(2) Two experiments [5, 10] in which entropy-conservation bounds (which apply in any closed system) were bypassed. The second one [10] was done on bio-molecules – amino acids.
(3) A full AC experiment [3] that includes the initialization of three carbon nuclei to the bias of a hydrogen spin, followed by a single compression step on these three carbons. This work was later on extended also to multi-cycle AC [23].

Cross-References

Quantum computing entries such as ▶ Quantum Algorithm for Factoring, ▶ Quantum Algorithm for the Parity Problem and ▶ Quantum Key Distribution. Data compression entries such as ▶ Dictionary-Based Data Compression.

Recommended Reading

1. Ambainis A, Schulman LJ, Vazirani U (2006) Computing with highly mixed states. J ACM 53: 507–531
2. Bakr WS, Preiss PM, Tai ME, Ma R, Simon J, Greiner M (2011) Orbital excitation blockade and algorithmic cooling in quantum gases. Nature 480:500–503
3. Baugh J, Moussa O, Ryan CA, Nayak A, Laflamme R (2005) Experimental implementation of heat-bath algorithmic cooling using solid-state nuclear magnetic resonance. Nature 438:470–473
4. Boykin PO, Mor T, Roychowdhury V, Vatan F, Vrijen R (2002) Algorithmic cooling and scalable NMR quantum computers. Proc Natl Acad Sci USA 99:3388–3393
5. Brassard G, Elias Y, Fernandez JM, Gilboa H, Jones JA, Mor T, Weinstein Y, Xiao L (2005) Experimental heat-bath cooling of spins. Submitted to Euro Phys Lett PLUS. See also arXiv:0511156 [quant-ph]. See a much improved version in arXiv:1404.6885 [quant-ph], 2014
6. Chang DE, Vandersypen LMK, Steffen M (2001) NMR implementation of a building block for scalable quantum computation. Chem Phys Lett 338: 337–344
7. Cory DG, Fahmy AF, Havel TF (1997) Ensemble quantum computing by NMR spectroscopy. Proc Natl Acad Sci USA 94:1634–1639
8. Datta A, Flammia ST, Caves CM (2005) Entanglement and the power of one qubit. Phys Rev A 72:042316
9. Elias Y, Fernandez JM, Mor T, Weinstein Y (2007) Optimal algorithmic cooling of spins. Isr J Chem 46:371–391
10. Elias Y, Gilboa H, Mor T, Weinstein Y (2011) Heat-bath cooling of spins in two amino acids. Chem Phys Lett 517:126–131
11. Elias Y, Mor T, Weinstein Y (2011) Semioptimal practicable algorithmic cooling. Phys Rev A 83:042340
12. Fernandez JM (2004) De computatione quantica. Ph.D. Dissertation, University of Montreal, Montreal
13. Fernandez JM, Lloyd S, Mor T, Roychowdhury V (2004) Practicable algorithmic cooling of spins. Int J Quantum Inf 2:461–477
14. Gershenfeld NA, Chuang IL (1997) Bulk spin-resonance quantum computation. Science 275:350–356
15. Kaye P (2007) Cooling algorithms based on the 3-bit majority. Quantum Inf Proc 6:295–322
16. Ladd TD, Goldman JR, Yamaguchi F, Yamamoto Y, Abe E, Itoh KM (2002) All-Silicon quantum computer. Phys Rev Lett 89:017901
17. Linden N, Popescu P, Skrzypczyk P (2010) How small can thermal machines be? The smallest possible refrigerator. Phys Rev Lett 105:130401

18. Lloyd S (2014) Quantum optics: cool computation, hot bits. Nat Photon 8:90–91
19. Mor T, Roychowdhury V, Lloyd S, Fernandez JM, Weinstein Y (2005) Algorithmic cooling. US patent No. 6,873,154
20. Paz-Silva GA, Brennen GK, Twamley J (2010) Fault tolerance with noisy and slow measurements and preparation. Phys Rev Lett 105:100501
21. Rempp F, Michel M, Mahler G (2007) Cyclic cooling algorithm. Phys Rev A 76:032325
22. Renner R (2012) Thermodynamics: the fridge gate. Nature 482:164–165
23. Ryan CA, Moussa O, Baugh J, Laflamme R (2008) Spin based heat engine: demonstration of multiple rounds of algorithmic cooling. Phys Rev Lett 100:140501
24. Schulman LJ (2005) Quantum computing: a bit chilly. Nature 438:431–432
25. Schulman LJ, Mor T, Weinstein Y (2005) Physical limits of heat-bath algorithmic cooling. Phys Rev Lett 94:120501
26. Schulman LJ, Mor T, Weinstein Y (2007) Physical limits of heat-bath algorithmic cooling. SIAM J Comput (SICOMP) 36:1729–1747
27. Schulman LJ, Vazirani U (1999) Molecular scale heat engines and scalable quantum computation. In: Proceedings of the 31st ACM STOC (Symposium on Theory of Computing), Atlanta, pp 322–329
28. Simmons S, Brown RM, Riemann H, Abrosimov NV, Becker P, Pohl HJ, Thewalt MLW, Itoh KM, Morton JJL (2011) Entanglement in a solid-state spin ensemble. Nature 470:69–72
29. Sørensen OW (1989) Polarization transfer experiments in high-resolution NMR spectroscopy. Prog Nucl Magn Reson Spectrosc 21:503–569
30. Twamley J (2003) Quantum-cellular-automata quantum computing with endohedral fullerenes. Phys Rev A 67:052318
31. Xu JS, Yung MH, Xu XY, Boixo S, Zhou ZW, Li CF, Aspuru-Guzik A, Guo GC (2014) Demon-like algorithmic quantum cooling and its realization with quantum optics. Nat Photonics 8:113–118

Algorithmic Mechanism Design

Ron Lavi
Faculty of Industrial Engineering and Management, Technion, Haifa, Israel

Years and Authors of Summarized Original Work

1999; Nisan, Ronen

Problem Definition

Mechanism design is a sub-field of economics and game theory that studies the construction of social mechanisms in the presence of selfish agents. The nature of the agents dictates a basic contrast between the social planner, that aims to reach a socially desirable outcome, and the agents, that care only about their own private utility. The underlying question is how to incentivize the agents to cooperate, in order to reach the desirable social outcomes.

In the Internet era, where computers act and interact on behalf of selfish entities, the connection of the above to algorithmic design suggests itself: suppose that the input to an algorithm is kept by selfish agents, who aim to maximize their own utility. How can one design the algorithm so that the agents will find it in their best interest to cooperate, and a close-to-optimal outcome will be outputted? This is different than classic distributed computing models, where agents are either "good" (meaning obedient) or "bad" (meaning faulty, or malicious, depending on the context). Here, no such partition is possible. It is simply assumed that all agents are utility maximizers. To illustrate this, let us describe a motivating example:

A Motivating Example: Shortest Paths

Given a weighted graph, the goal is to find a shortest path (with respect to the edge weights) between a given source and target nodes. Each edge is controlled by a selfish entity, and the weight of the edge, w_e is private information of that edge. If an edge is chosen by the algorithm to be included in the shortest path, it will incur a cost which is minus its weight (the cost of communication). Payments to the edges are allowed, and the total utility of an edge that participates in the shortest path and gets a payment p_e is assumed to be $u_e = p_e - w_e$. Notice that the shortest path is *with respect to the true weights of the agents, although these are not known to the designer.*

Assuming that each edge will act in order to maximize its utility, how can one choose the path and the payments? One option is to ignore the strategic issue all together, ask the edges to simply report their weights, and compute the shortest path. In this case, however, an edge dislikes being selected, and will therefore prefer to report a very high weight (much higher than its true weight) in order to decrease the chances of being selected. Another option is to pay each selected edge its reported weight, or its reported weight plus a small fixed "bonus". However in such a case all edges will report lower weights, as being selected will imply a positive gain.

Although this example is written in an algorithmic language, it is actually a mechanism design problem, and the solution, which is now a classic, was suggested in the 1970's. The chapter continues as follows: First, the abstract formulation for such problems is given, the classic solution from economics is described, and its advantages and disadvantages for algorithmic purposes are discussed. The next section then describes the new results that algorithmic mechanism design offers.

Abstract Formulation

The framework consists of a set A of alternatives, or outcomes, and n players, or agents. Each player i has a valuation function $v_i : A \to \Re$ that assigns a value to each possible alternative. This valuation function belongs to a domain V_i of all possible valuation functions. Let $V = V_1 \times \cdots \times V_n$, and $V_{-i} = \prod_{j \neq i} V_j$. Observe that this generalizes the shortest path example of above: A is all the possible $s - t$ paths in the given graph, $v_e(a)$ for some path $a \in A$ is either $-w_e$ (if $e \in a$) or zero.

A *social choice function* $f : V \to A$ assigns a socially desirable alternative to any given profile of players' valuations. This parallels the notion of an algorithm. A *mechanism* is a tuple $M = (f, p_1, \ldots, p_n)$, where f is a social choice function, and $p_i : V \to \Re$ (for $i = 1, \ldots, n$) is the price charged from player i. The interpretation is that the social planner asks the players to reveal their true valuations, chooses the alternative according to f as if the players have indeed acted truthfully, and in addition rewards/punishes the players with the prices. These prices should

induce "truthfulness" in the following strong sense: no matter what the other players declare, it is always in the best interest of player i to reveal her true valuation, as this will maximize her utility. Formally, this translates to:

Definition 1 (Truthfulness) M is "truthful" (in dominant strategies) if, for any player i, any profile of valuations of the other players $v_{-i} \in V_{-i}$, and any two valuations of player i $v_i, v'_i \in V_i$,

$$v_i(a) - p_i(v_i, v_{-i}) \geq v_i(b) - p_i(v'_i, v_{-i})$$

where $f(v_i, v_{-i}) = a$ and $f(v'_i, v_{-i}) = b$.

Truthfulness is quite strong: a player need not know anything about the other players, even not that they are rational, and still determine the best strategy for her. Quite remarkably, there exists a truthful mechanism, even under the current level of abstraction. This mechanism suits all problem domains, where the social goal is to maximize the "social welfare":

Definition 2 (Social welfare maximization) A social choice function $f : V \to A$ maximizes the social welfare if $f(v) \in \text{argmax}_{a \in A} \sum_i v_i(a)$, for any $v \in V$.

Notice that the social goal in the shortest path domain is indeed welfare maximization, and, in general, this is a natural and important economic goal. Quite remarkably, there exists a general technique to construct truthful mechanisms that implement this goal:

Theorem 1 (Vickrey–Clarke–Groves (VCG)) *Fix any alternatives set A and any domain V, and suppose that $f : V \to A$ maximizes the social welfare. Then there exist prices p such that the mechanism (f, p) is truthful.*

This gives "for free" a solution to the shortest path problem, and to many other algorithmic problems. The great advantage of the VCG scheme is its generality: it suits *all* problem domains. The disadvantage, however, is that the method is tailored to social welfare maximization. This turns out to be restrictive, especially for algorithmic and computational settings, due to several reasons: (i) different algorithmic goals: the algorithmic literature considers a variety of goals, including many that cannot be translated to welfare maximization. VCG does not help us in such cases. (ii) computational complexity: even if the goal is welfare maximization, in many settings achieving exactly the optimum is computationally hard. The CS discipline usually overcomes this by using approximation algorithms, but VCG will not work with such algorithm – reaching exact optimality is a necessary requirement of VCG. (iii) different algorithmic models: common CS models change "the basic setup", hence cause unexpected difficulties when one tries to use VCG (for example, an online model, where the input is revealed over time; this is common in CS, but changes the implicit setting that VCG requires). This is true even if welfare maximization is still the goal.

Answering any one of these difficulties requires the design of a non-VCG mechanism. What analysis tools should be used for this purpose? In economics and classic mechanism design, average-case analysis, that relies on the knowledge of the underlying distribution, is the standard. Computer science, on the other hand, usually prefers to avoid strong distributional assumptions, and to use worst-case analysis. This difference is another cause to the uniqueness of the answers provided by algorithmic mechanism design. Some of the new results that have emerged as a consequence of this integration between Computer Science and Economics is next described. Many other research topics that use the tools of algorithmic mechanism design are described in the entries on Adword Pricing, Competitive Auctions, False Name Proof Auctions, Generalized Vickrey Auction, Incentive Compatible Ranking, Mechanism for One Parameter Agents Single Buyer/Seller, Multiple Item Auctions, Position Auctions, and Truthful Multicast.

There are two different but closely related research topics that should be mentioned in the context of this entry. The first is the line of works that studies the "price of anarchy" of a given

system. These works analyze *existing* systems, trying to quantify the loss of social efficiency due to the selfish nature of the participants, while the approach of algorithmic mechanism design is to understand how new systems should be designed. For more details on this topic the reader is referred to the entry on Price of Anarchy. The second topic regards the algorithmic study of various equilibria computation. These works bring computational aspects into economics and game theory, as they ask what equilibria notions are reasonable to assume, if one requires computational efficiency, while the works described here bring game theory and economics into computer science and algorithmic theory, as they ask what algorithms are reasonable to design, if one requires the resilience to selfish behavior. For more details on this topic the reader is referred (for example) to the entry on Algorithms for Nash Equilibrium and to the entry on General Equilibrium.

Key Results

Problem Domain 1: Job Scheduling

Job scheduling is a classic algorithmic setting: n jobs are to be assigned to m machines, where job j requires processing time p_{ij} on machine i. In the game-theoretic setting, it is assumed that each machine i is a selfish entity, that incurs a cost p_{ij} from processing job j. Note that the payments in this setting (and in general) may be negative, offsetting such costs. A popular algorithmic goal is to assign jobs to machines in order to minimize the "makespan": $\max_i \sum_{j \text{ is assigned to } i} p_{ij}$. This is different than welfare maximization, which translates in this setting to the minimization of $\sum_i \sum_{j \text{ is assigned to } i} p_{ij}$, further illustrating the problem of different algorithmic goals. Thus the VCG scheme cannot be used, and new methods must be developed.

Results for this problem domain depend on the specific assumptions about the structure of the processing time vectors. In the *related machines* case, $p_{ij} = p_j / s_i$ for any $i\, j$, where the p_j's are public knowledge, and the only secret parameter of player i is its *speed*, s_i.

Theorem 2 ([3, 22]) *For job scheduling on related machines, there exists a truthful exponential-time mechanism that obtains the optimal makespan, and a truthful polynomial-time mechanism that obtains a 3-approximation to the optimal makespan.*

More details on this result are given in the entry on Mechanism for One Parameter Agents Single Buyer. The bottom line conclusion is that, although the social goal is different than welfare maximization, there still exists a truthful mechanism for this goal. A non-trivial approximation guarantee is achieved, even under the additional requirement of computational efficiency. However, this guarantee does not match the best possible without the truthfulness requirement, since in this case a PTAS is known.

Open Question 1 *Is there a truthful PTAS for makespan minimization in related machines?*

If the number of machines is fixed then [2] give such a truthful PTAS.

The above picture completely changes in the move to the more general case of *unrelated machines*, where the p_{ij}'s are allowed to be arbitrary:

Theorem 3 ([13, 30]) *Any truthful scheduling mechanism for unrelated machines cannot approximate the optimal makespan by a factor better than $1 + \sqrt{2}$ (for deterministic mechanisms) and $2 - 1/m$ (for randomized mechanisms).*

Note that this holds regardless of computational considerations. In this case, switching from welfare maximization to makespan minimization results in a strong impossibility. On the possibilities side, virtually nothing (!) is known. The VCG mechanism (which minimizes the total social cost) is an m-approximation of the optimal makespan [32], and, in fact, nothing better is currently known:

Open Question 2 *What is the best possible approximation for truthful makespan minimization in unrelated machines?*

What caused the switch from "mostly possibilities" to "mostly impossibilities"? Related machines is a single-dimensional domain (players hold only one secret number), for which truthfulness is characterized by a simple monotonicity condition, that leaves ample flexibility for algorithmic design. Unrelated machines, on the other hand, are a multi-dimensional domain, and the algorithmic conditions implied by truthfulness in such a case are harder to work with. It is still unclear whether these conditions imply real mathematical impossibilities, or perhaps just pose harder obstacles that can be in principle solved. One multi-dimensional scheduling domain for which possibility results are known is the case where $p_{ij} \in \{L_j, H_j\}$, where the "low" 's and "high" 's are fixed and known. This case generalizes the classic multi-dimensional model of restricted machines ($p_{ij} \in \{p_j, \infty\}$), and admits a truthful 3-approximation [27].

Problem Domain 2: Digital Goods and Revenue Maximization

In the E-commerce era, a new kind of "digital goods" have evolved: goods with no marginal production cost, or, in other words, goods with unlimited supply. One example is songs being sold on the Internet. There is a sunk cost of producing the song, but after that, additional electronic copies incur no additional cost. How should such items be sold? One possibility is to conduct an *auction*. An auction is a one-sided market, where a monopolistic entity (the auctioneer) wishes to sell one or more items to a set of buyers.

In this setting, each buyer has a privately known value for obtaining one copy of the good. Welfare maximization simply implies the allocation of one good to every buyer, but a more interesting question is the question of revenue maximization. How should the auctioneer design the auction in order to maximize his profit? Standard tools from the study of revenue-maximizing auctions (This model was not explicitly studied in classic auction theory, but standard results from there can be easily adjusted to this setting.) suggest to simply declare a price-per-buyer, determined by the probability distribution of the buyer's value, and make a take-it-or-leave-it offer.

However, such a mechanism needs to know the underlying distribution. Algorithmic mechanism design suggests an alternative, worst-case result, in the spirit of CS-type models and analysis.

Suppose that the auctioneer is required to sell all items in the same price, as is the case for many "real-life" monopolists, and denote by $F(\vec{v})$ the maximal revenue from a fixed-price sale to bidders with values $\vec{v} = v_1, \ldots v_n$, *assuming that all values are known*. Reordering indexes so that $v_1 \geq v_2 \geq \cdots \geq v_n$, let $F(\vec{v}) = \max_i i \cdot v_i$. The problem is, of-course, that in fact *nothing* about the values is known. Therefore, a truthful auction that extracts the players' values is in place. Can such an auction obtain a profit that is a constant fraction of $F(\vec{v})$, for any $\vec{v}$ (i.e., in the worst case)? Unfortunately, the answer is provably no [17]. The proof makes use of situations where the entire profit comes from the highest bidder. Since there is no potential for competition among bidders, a truthful auction cannot force this single bidder to reveal her value.

Luckily, a small relaxation in the optimality criteria significantly helps. Specifically, denote by $F^{(2)}(\vec{v}) = \max_{i \geq 2} i \cdot v_i$ (i.e., the benchmark is the auction that sells to at least two buyers).

Theorem 4 ([17, 20]) *There exists a truthful randomized auction that obtains an expected revenue of at least $F^{(2)}/3.25$, even in the worst-case. On the other hand, no truthful auction can approximate $F^{(2)}$ within a factor better than 2.42.*

Several interesting formats of distribution-free revenue-maximizing auctions have been considered in the literature. The common building block in all of them is the random partitioning of the set of buyers to random subsets, analyzing each set separately, and using the results on the other sets. Each auction utilizes a different analysis on the two subsets, which yields slightly different approximation guarantees. Aggarwal et al. [1] describe an elegant method to derandomize these type of auctions, while losing another factor of 4 in the approximation. More details on this problem domain can be found in the entry on Competitive Auctions.

Problem Domain 3: Combinatorial Auctions

Combinatorial auctions (CAs) are a central model with theoretical importance and practical relevance. It generalizes many theoretical algorithmic settings, like job scheduling and network routing, and is evident in many real-life situations. This new model has various pure computational aspects, and, additionally, exhibits interesting game theoretic challenges. While each aspect is important on its own, obviously only the integration of the two provides an acceptable solution.

A combinatorial auction is a multi-item auction in which players are interested in *bundles* of items. Such a valuation structure can represent substitutabilities among items, complementarities among items, or a combination of both. More formally, m items (Ω) are to be allocated to n players. Players value subsets of items, and $v_i(S)$ denotes i's value of a bundle $S \subseteq \Omega$. Valuations additionally satisfy: (i) monotonicity, i.e., $v_i(S) \leq v_i(T)$ for $S \subseteq T$, and (ii) normalization, i.e., $v_i(\emptyset) = 0$. The literature has mostly considered the goal of maximizing the social welfare: find an allocation $(S_1, \ldots, S_n)$ that maximizes $\sum_i v_i(S_i)$.

Since a general valuation has size exponential in n and m, the representation issue must be taken into account. Two models are usually considered (see [11] for more details). In the *bidding languages* model, the bid of a player represents his valuation is a concise way. For this model it is NP-hard to approximate the social welfare within a ratio of $\Omega(m^{1/2-\epsilon})$, for any $\epsilon > 0$ (if "single-minded" bids are allowed; the exact definition is given below). In the *query access* model, the mechanism iteratively queries the players in the course of computation. For this model, any algorithm with polynomial communication cannot obtain an approximation ratio of $\Omega(m^{1/2-\epsilon})$ for any $\epsilon > 0$. These bounds are tight, as there exist a deterministic $\sqrt{m}$-approximation with polynomial computation and communication. Thus, for the general valuation structure, the computational status by itself is well-understood.

The basic incentives issue is again well-understood: VCG obtains truthfulness. Since

VCG requires the exact optimum, which is NP-hard to compute, the two considerations therefore clash, when attempting to use classic techniques. Algorithmic mechanism design aims to develop new techniques, to integrate these two desirable aspects.

The first positive result for this integration challenge was given by [29], for the special case of "single-minded bidders": each bidder, i, is interested in a specific bundle S_i, for a value v_i (any bundle that contains S_i is worth v_i, and other bundles have zero value). Both v_i, S_i are private to the player i.

Theorem 5 ([29]) *There exists a truthful and polynomial-time deterministic combinatorial auction for single-minded bidders, which obtains a $\sqrt{m}$-approximation to the optimal social welfare.*

A possible generalization of the basic model is to assume that each item has B copies, and each player still desires at most one copy from each item. This is termed "multi-unit CA". As B grows, the integrality constraint of the problem reduces, and so one could hope for better solutions. Indeed, the next result exploits this idea:

Theorem 6 ([7]) *There exists a truthful and polynomial-time deterministic multi-unit CA, for $B \geq 3$ copies of each item, that obtains $O(B \cdot m^{1/(B-2)})$-approximation to the optimal social welfare.*

This auction copes with the representation issue (since general valuations are assumed) by accessing the valuations through a "demand oracle": given per-item prices $\{p_x\}_{x \in \Omega}$, specify a bundle S that maximizes $v_i(S) - \sum_{x \in S} p_x$.

Two main drawbacks of this auction motivate further research on the issue. First, as B gets larger it is reasonable to expect the approximation to approach 1 (indeed polynomial-time algorithms with such an approximation guarantee do exist). However here the approximation ratio does not decrease below $O(\log m)$ (this ratio is achieved for $B = O(\log m)$). Second, this auction does not provide a solution to the original setting, where $B = 1$, and, in general for small

B's the approximation factor is rather high. One way to cope with these problems is to introduce randomness:

Theorem 7 ([26]) *There exists a truthful-in-expectation and polynomial-time randomized multi-unit CA, for any $B \geq 1$ copies of each item, that obtains $O(m^{1/(B+1)})$-approximation to the optimal social welfare.*

Thus, by allowing randomness, the gap from the standard computational status is being completely closed. The definition of truthfulness in expectation is the natural extension of truthfulness to a randomized environment: the *expected* utility of a player is maximized by being truthful.

However, this notion is strictly weaker than the deterministic notion, as this implicitly implies that players care only about the expectation of their utility (and not, for example, about the variance). This is termed "the risk-neutrality" assumption in the economics literature. An intermediate notion for randomized mechanisms is that of "universal truthfulness": the mechanism is truthful given any fixed result of the coin toss. Here, risk-neutrality is no longer needed. Dobzinski et al. [15] give a universally truthful CA for $B = 1$ that obtains an $O(\sqrt{m})$-approximation. Universally truthful mechanisms are still weaker than deterministic truthful mechanisms, due to two reasons: (i) It is not clear how to actually create the correct and exact probability distribution with a deterministic computer. The situation here is different than in "regular" algorithmic settings, where various derandomization techniques can be employed, since these in general does not carry through the truthfulness property. (ii) Even if a natural randomness source exists, one cannot improve the quality of the actual output by repeating the computation several times (using the the law of large numbers). Such a repetition will again destroy truthfulness. Thus, exactly because the game-theoretic issues are being considered in parallel to the computational ones, the importance of determinism increases.

Open Question 3 *What is the best-possible approximation ratio that deterministic and*

truthful combinatorial auctions can obtain, in polynomial-time?

There are many valuation classes, that restrict the possible valuations to some reasonable format (see [28] for more details). For example, sub-additive valuations are such that, for any two bundles $S, T, \subseteq \Omega$, $v(S \cup T) \le v(S) + v(T)$. Such classes exhibit much better approximation guarantees, e.g., for sub-additive valuation a polynomial-time 2-approximation is known [16]. However, no polynomial-time truthful mechanism (be it randomized, or deterministic) with a constant approximation ratio, is known for any of these classes.

Open Question 4 *Does there exist polynomial-time truthful constant-factor approximations for special cases of CAs that are NP-hard?*

Revenue maximization in CAs is of-course another important goal. This topic is still mostly unexplored, with few exceptions. The mechanism [7] obtains the same guarantees with respect to the optimal revenue. Improved approximations exist for multi-unit auctions (where all items are identical) with budget constrained players [12], and for unlimited-supply CAs with single-minded bidders [6].

The topic of Combinatorial Auctions is discussed also in the entry on Multiple Item Auctions.

Problem Domain 4: Online Auctions

In the classic CS setting of "online computation", the input to an algorithm is not revealed all at once, before the computation begins, but gradually, over time (for a detailed discussion see the many entries on online problems in this book). This structure suits the auction world, especially in the new electronic environments. What happens when players arrive over time, and the auctioneer must make decisions facing only a subset of the players at any given time?

The integration of online settings, worst-case analysis, and auction theory, was suggested by [24]. They considered the case where players arrive one at a time, and the auctioneer must provide an answer to each player *as it arrives*,

without knowing the future bids. There are k identical items, and each bidder may have a distinct value for every possible quantity of the item. These values are assumed to be marginally decreasing, where each marginal value lies in the interval $[\underline{v}, \bar{v}]$. The private information of a bidder includes both her valuation function, and her arrival time, and so a truthful auction need to incentivize the players to arrive on time (and not later on), and to reveal their true values. The most interesting result in this setting is for a large k, so that in fact there is a continuum of items:

Theorem 8 ([24]) *There exists a truthful online auction that simultaneously approximates, within a factor of $O(\log(\bar{v}/\underline{v}))$, the optimal offline welfare, and the offline revenue of VCG. Furthermore, no truthful online auction can obtain a better approximation ratio to either one of these criteria (separately).*

This auction has the interesting property of being a "posted price" auction. Each bidder is not required to reveal his valuation function, but, rather, he is given a price for each possible quantity, and then simply reports the desired quantity under these prices.

Ideas from this construction were later used by [10] to construct *two*-sided online auction markets, where multiple sellers and buyers arrive online.

This approximation ratio can be dramatically improved, to be a constant, 4, if one assumes that (i) there is only one item, and (ii) player values are i.i.d from some fixed distribution. No a–priori knowledge of this distribution is needed, as neither the mechanism nor the players are required to make any use of it. This work, [19], analyzes this by making an interesting connection to the class of "secretary problems".

A general method to convert online algorithms to online mechanisms is given by [4]. This is done for one item auctions, and, more generally, for one parameter domains. This method is competitive both with respect to the welfare and the revenue.

The revenue that the online auction of Theorem 8 manages to raise is competitive only with respect to VCG's revenue, which may be far from optimal. A parallel line of works is concerned with revenue maximizing auctions. To achieve good results, two assumptions need to be made: (i) there exists an unlimited supply of items (and recall from section "Problem Domain 2: Digital Goods and Revenue Maximization" that $F(v)$ is the offline optimal monopolistic fixed-price revenue), and (ii) players cannot lie about their arrival time, only about their value. This last assumption is very strong, but apparently needed. Such auctions are termed here "value-truthful", indicating that "time-truthfulness" is missing.

Theorem 9 ([9]) *For any $\epsilon > 0$, there exists a value-truthful online auction, for the unlimited supply case, with expected revenue of at least $(F(v))/(1 + \epsilon) - O(h/\epsilon^2)$.*

The construction exploits principles from learning theory in an elegant way. Posted price auctions for this case are also possible, in which case the additive loss increases to $O(h \log \log h)$. Hajiaghayi et al. [19] consider fully-truthful online auctions for revenue maximization, but manage to obtain only very high (although fixed) competitive ratios. Constructing fully-truthful online auctions with a close-to-optimal revenue remains an open question. Another interesting open question involves multi-dimensional valuations. The work [24] remains the only work for players that may demand multiple items. However their competitive guarantees are quite high, and achieving better approximation guarantees (especially with respect to the revenue) is a challenging task.

Advanced Issues

Monotonicity

What is the general way for designing a truthful mechanism? The straight-forward way is to check, for a given social choice function f, whether truthful prices exist. If not, try to "fix" f. It turns out, however, that there exists a more structured way, an *algorithmic* condition that will imply the *existence* of truthful prices. Such a condition shifts the designer back to the familiar territory of algorithmic design. Luckily, such a condition do exist, and is best described in the abstract social choice setting of section "Problem Definition":

Definition 3 ([8, 23]) A social choice function $f: V \to A$ is "weakly monotone" (W-MON) if for any i, $v_{-i} \in V_{-i}$, and any $v_i, v_i' \in V_i$, the following holds. Suppose that $f(v_i, v_{-i}) = a$, and $f(v_i', v_{-i}) = b$. Then $v_i'(b) - v_i(b) \geq v_i'(a) - v_i(a)$.

In words, this condition states the following. Suppose that player i changes her declaration from v_i to v_i', and this causes the social choice to change from a to b. Then it must be the case that i's value for b has increased in the transition from v_i to v_i' no-less than i's value for a.

Theorem 10 ([35]) *Fix a social choice function $f: V \to A$, where V is convex, and A is finite. Then there exist prices p such that $M = (f, p)$ is truthful if and only if f is weakly monotone.*

Furthermore, given a weakly monotone f, there exists an explicit way to determine the appropriate prices p (see [18] for details).

Thus, the designer should aim for weakly monotone algorithms, and need not worry about actual prices. But how difficult is this? For single-dimensional domains, it turns out that W-MON leaves ample flexibility for the algorithm designer. Consider for example the case where every alternative has a value of either 0 (the player "loses") or some $v_i \in \Re$ (the player "wins" and obtains a value v_i). In such a case, it is not hard to show that W-MON reduces to the following monotonicity condition: if a player wins with v_i, and increases her value to $v_i' > v_i$ (while v_{-i} remains fixed), then she must win with v_i' as well. Furthermore, in such a case, the price of a winning player must be set to the infimum over all winning values.

Impossibilities of truthful design

It is fairly simple to construct algorithms that satisfy W-MON for single-dimensional domains, and a variety of positive results were obtained for such domains, in classic mechanism design, as well as in algorithmic mechanism design. But how hard is it to satisfy W-MON for multi-dimensional domains? This question is yet unclear, and seems to be one of the challenges of algorithmic mechanism design. The contrast between single-dimensionality and multi-dimensionality appears in all problem domains that were surveyed here, and seems to reflect some inherent difficulty that is not exactly understood yet. Given a social choice function f, call f *implementable* (in dominant strategies) if there exist prices p such that $M = (f, p)$ is truthful. The basic question is then *what forms of social choice functions are implementable*.

As detailed in the beginning, the welfare maximizing social choice function is implementable. This specific function can be slightly generalized to allow weights, in the following way: fix some non-negative real constants $\{w_i\}_{i=1}^n$ (not all are zero) and $\{\gamma_a\}_{a \in A}$, and choose an alternative that maximizes the *weighted* social welfare, i.e., $f(v) \in \text{argmax}_{a \in A} \sum_i w_i v_i(a) + \gamma_a$. This class of functions is sometimes termed "affine maximizers". It turns out that these functions are also implementable, with prices similar in spirit to VCG. In the context of the above characterization question, one sharp result stands out:

Theorem 11 ([34]) *Fix a social choice function $f : V \rightarrow A$, such that (i) A is finite, $|A| \geq 3$, and f is onto A, and (ii) $V_i = \Re^A$ for all i. Then f is implementable (in dominant strategies) if and only if it is an affine maximizer.*

The domain V that satisfies $V_i = \Re^A$ for all i is term an "unrestricted domain". The theorem states that, if the domain is unrestricted, at least three alternatives are chosen, and the set A of alternatives is finite, then nothing besides affine maximizers can be implemented!

However, the assumption that the domain is unrestricted is very restrictive. All the above example domains exhibit some basic combina-torial structure, and are therefore restricted in some way. And as discussed above, for many restricted domains the theorem is simply not true. So what *is* the possibilities – impossibilities border? As mentioned above, this is an unsolved challenge. Lavi, Mu'alem, and Nisan [23] explore this question for Combinatorial Auctions and similar restricted domains, and reach partial answers. For example:

Theorem 12 ([23]) *Any truthful combinatorial auction or multi-unit auction among two players, that must always allocate all items, and that approximates the welfare by a factor better than 2, must be an affine maximizer.*

Of-course, this is far from being a complete answer. What happens if there are more than two players? And what happens if it is possible to "throw away" part of the items? These questions, and the more general and abstract characterization question, are all still open.

Alternative solution concepts

In light of the conclusions of the previous section, a natural thought would be to re-examine the *solution concept* that is being used. Truthfulness relies on the strong concept of dominant strategies: for each player there is a unique strategy that maximizes her utility, no matter what the other players are doing. This is very strong, but it fits very well the worst-case way of thinking in CS. What other solution concepts can be used? As described above, randomization, and truthfulness-in-expectation, can help. A related concept, again for randomized mechanisms, is truthfulness with high probability. Another direction is to consider mechanisms where players cannot improve their utility *too much* by deviating from the truth-telling strategy [21].

Algorithm designers do not care so much about actually reaching an equilibrium point, or finding out what will the players play – the major concern is to guarantee the optimality of the solution, taking into account the strategic behavior of the players. Indeed, one way of doing this is to guarantee a good equilibrium point. But there is no reason to rule out mechanisms where several acceptable strategic choices for the players exist, provided that the approximation will be achieved *in each of these choices.*

As a first attempt, one is tempted to simply let the players try and improve the basic result by allowing them to lie. However, this can cause unexpected dynamics, as each player chooses her lies under some assumptions about the lies of the others, etc. etc. To avoid such an unpredictable situation, it is important to insist on using rigorous game theoretic reasoning to explain exactly why the outcome will be satisfactory.

The work [31] suggests the notion of "feasibly dominant" strategies, where players reveal the possible lies they consider, and the mechanism takes this into account. By assuming that the players are computationally bounded, one can show that, instead of actually "lying", the players will prefer to reveal their true types plus all the lies they might consider. In such a case, since the mechanism has obtained the true types of the players, a close-to-optimal outcome will be guaranteed.

Another definition tries to capture the initial intuition by using the classic game-theoretic notion of undominated strategies:

Definition 4 ([5]) A mechanism M is an "algorithmic implementation of a c-approximation (in undominated strategies)" if there exists a set of strategies, D, such that (i) M obtains a c-approximation for any combination of strategies from D, in polynomial time, and (ii) For any strategy not in D, there exists a strategy in D that weakly dominates it, and this transition is polynomial-time computable.

By the second condition, it is reasonable to assume that a player will indeed play *some* strategy in D, and, by the first condition, it does not matter what tuple of strategies in D will actually be chosen, as any of these will provide the approximation. This transfers some of the burden from the game-theoretic design to the algorithmic design, since now a guarantee on the approximation should bu provided for a larger range of strategies. Babaioff et al. [5] exploit this notion to design a deterministic CA for multi-dimensional players that achieves a close-to-optimal approximation guarantee. A similar-in-spirit notion, although a weaker one, is the notion of "Set-Nash" [25].

Applications

One of the popular examples to a "real-life" combinatorial auction is the spectrum auction that the US government conducts, in order to sell spectrum licenses. Typical bids reflect values for different spectrum ranges, to accommodate different geographical and physical needs, where different spectrum ranges may complement or substitute one another. The US government invests research efforts in order to determine the best format for such an auction, and auction theory is heavily exploited. Interestingly, the US law guides the authorities to allocate these spectrum ranges in a way that will maximize *the social welfare*, thus providing a good example for the usefulness of this goal.

Adword auctions are another new and fast-growing application of auction theory in general, and of the new algorithmic auctions in particular. These are auctions that determine the advertisements that web-search engines place close to the search results they show, after the user submits her search keywords. The interested companies compete, for every given keyword, on the right to place their ad on the results' page, and this turns out to be the main source of income for companies like Google. Several entries in this book touch on this topic in more details, including the entries on Adwords Pricing and on Position Auctions.

A third example to a possible application, in the meanwhile implemented only in the academic research labs, is the application of algorithmic mechanism design to pricing and congestion control in communication networks. The existing fixed pricing scheme has many disadvantages, both with respect to the needs of efficiently allocating the available resources, and with respect to the new opportunities of the Internet companies to raise more revenue due to specific types of traffic. Theory suggests solutions to both of these problems.

Cross-References

▶ Adwords Pricing
▶ Competitive Auction
▶ False-Name-Proof Auction
▶ Generalized Vickrey Auction
▶ Incentive Compatible Selection
▶ Position Auction
▶ Truthful Multicast

Recommended Reading

The topics presented here are detailed in the textbook [33]. Section "Problem Definition" is based on the paper [32], that also coined the term "algorithmic mechanism design". The book [14] covers the various aspects of combinatorial auctions.

1. Aggarwal G, Fiat A, Goldberg A, Immorlica N, Sudan M (2005) Derandomization of auctions. In: Proceedings of the 37th ACM symposium on theory of computing (STOC'05)
2. Andelman N, Azar Y, Sorani M (2005) Truthful approximation mechanisms for scheduling selfish related machines. In: Proceedings of the 22nd international symposium on theoretical aspects of computer science (STACS), pp 69–82
3. Archer A, Tardos É (2001) Truthful mechanisms for one-parameter agents. In: Proceedings of the 42nd annual symposium on foundations of computer science (FOCS), pp 482–491
4. Awerbuch B, Azar Y, Meyerson A (2003) Reducing truth-telling online mechanisms to online optimization. In: Proceedings of the 35th ACM symposium on theory of computing (STOC'03)
5. Babaioff M, Lavi R, Pavlov E (2006) Single-value combinatorial auctions and implementation in undominated strategies. In: Proceedings of the 17th symposium on discrete algorithms (SODA)
6. Balcan M, Blum A, Hartline J, Mansour Y (2005) Mechanism design via machine learning. In: Proceedings of the 46th annual symposium on foundations of computer science (FOCS'05)
7. Bartal Y, Gonen R, Nisan N (2003) Incentive compatible multiunit combinatorial auctions. In: Proceedings of the 9th conference on theoretical aspects of rationality and knowledge (TARK'03)
8. Bikhchandani S, Chatterjee S, Lavi R, Mu'alem A, Nisan N, Sen A (2006) Weak monotonicity characterizes deterministic dominant-strategy implementation. Econometrica 74:1109–1132
9. Blum A, Hartline J (2005) Near-optimal online auctions. In: Proceedings of the 16th symposium on discrete algorithms (SODA)
10. Blum A, Sandholm T, Zinkevich M (2006) Online algorithms for market clearing. J ACM 53(5):845–879
11. Blumrosen L, Nisan N (2005) On the computational power of iterative auctions. In: Proceedings of the 7th ACM conference on electronic commerce (EC'05)
12. Borgs C, Chayes J, Immorlica N, Mahdian M, Saberi A (2005) Multi-unit auctions with budget-constrained bidders. In: Proceedings of the 7th ACM conference on electronic commerce (EC'05)
13. Christodoulou G, Koutsoupias E, Vidali A (2007) A lower bound for scheduling mechanisms. In: Proceedings of the 18th symposium on discrete algorithms (SODA)
14. Cramton P, Shoham Y, Steinberg R (2005) Combinatorial auctions. MIT
15. Dobzinski S, Nisan N, Schapira M (2006) Truthful randomized mechanisms for combinatorial auctions. In: Proceedings of the 38th ACM symposium on theory of computing (STOC'06)
16. Feige U (2006) On maximizing welfare when utility functions are subadditive. In: Proceedings of the 38th ACM symposium on theory of computing (STOC'06)
17. Goldberg A, Hartline J, Karlin A, Saks M, Wright A (2006) Competitive auctions. Games Econ Behav 55(2):242–269
18. Gui H, Muller R, Vohra RV (2004) Characterizing dominant strategy mechanisms with multidimensional types. Working paper
19. Hajiaghayi M, Kleinberg R, Parkes D (2004) Adaptive limited-supply online auctions. In: Proceedings of the 6th ACM conference on electronic commerce (EC'04)
20. Hartline J, McGrew R (2005) From optimal limited to unlimited supply auctions. In: Proceedings of the 7th ACM conference on electronic commerce (EC'05)
21. Kothari A, Parkes D, Suri S (2005) Approximately-strategy proof and tractable multi-unit auctions. Decis Support Syst 39:105–121
22. Kovács A (2005) Fast monotone 3-approximation algorithm for scheduling related machines. In: Proceedings of the 13th annual European symposium on algorithms (ESA), pp 616–627
23. Lavi R, Mu'alem A, Nisan N (2003) Towards a characterization of truthful combinatorial auctions. In: Proceedings of the 44rd annual symposium on foundations of computer science (FOCS'03)
24. Lavi R, Nisan N (2004) Competitive analysis of incentive compatible on-line auctions. Theor Comput Sci 310:159–180
25. Lavi R, Nisan N (2005) Online ascending auctions for gradually expiring items. In: Proceedings of the 16th symposium on discrete algorithms (SODA)

26. Lavi R, Swamy C (2005) Truthful and near-optimal mechanism design via linear programming. In: Proceedings of the 46th annual symposium on foundations of computer science (FOCS), pp 595–604
27. Lavi R, Swamy C (2007) Truthful mechanism design for multi-dimensional scheduling via cycle monotonicity. Working paper
28. Lehmann B, Lehmann D, Nisan N (2006) Combinatorial auctions with decreasing marginal utilities. Games Econ Behav 55(2):270–296
29. Lehmann D, O'Callaghan L, Shoham Y (2002) Truth revelation in approximately efficient combinatorial auctions. J ACM 49(5):577–602
30. Mu'alem A, Schapira M (2007) Setting lower bounds on truthfulness. In: Proceedings of the 18th symposium on discrete algorithms (SODA)
31. Nisan N, Ronen A (2000) Computationally feasible VCG mechanisms. In: Proceedings of the 2nd ACM conference on electronic commerce (EC'00)
32. Nisan N, Ronen A (2001) Algorithmic mechanism design. Games Econ Behav 35:166–196
33. Nisan N, Roughgarden T, Tardos E, Vazirani V (2007) Algorithmic game theory. Cambridge University Press (expected to appear)
34. Roberts K (1979) The characterization of implementable choice rules. In: Laffont JJ (ed) Aggregation and revelation of preferences. North-Holland, pp 321–349
35. Saks M, Yu L (2005) Weak monotonicity suffices for truthfulness on convex domains. In: Proceedings of the 6th ACM conference on electronic commerce (ACM-EC), pp 286–293

Algorithms for Combining Rooted Triplets into a Galled Phylogenetic Network

Jesper Jansson[1] and Wing-Kin Sung[2]
[1]Laboratory of Mathematical Bioinformatics, Institute for Chemical Research, Kyoto University, Gokasho, Uji, Kyoto, Japan
[2]Department of Computer Science, National University of Singapore, Singapore, Singapore

Keywords

Dense set; Galled phylogenetic network; Phylogenetic tree; Polynomial-time approximation algorithm; Rooted triplet

Years and Authors of Summarized Original Work

2006; Jansson, Sung
2006; Jansson, Nguyen, Sung
2006; He, Huynh, Jansson, Sung
2010; Byrka, Gawrychowski, Huber, Kelk
2011; van Iersel, Kelk

Problem Definition

A *phylogenetic tree* is a binary, rooted, unordered tree whose leaves are distinctly labeled. A *phylogenetic network* is a generalization of a phylogenetic tree formally defined as a rooted, connected, directed acyclic graph in which (1) each node has outdegree at most 2; (2) each node has indegree 1 or 2, except the root node which has indegree 0; (3) no node has both indegree 1 and outdegree 1; and (4) all nodes with outdegree 0 are labeled by elements from a finite set L in such a way that no two nodes are assigned the same label. Nodes of outdegree 0 are referred to as *leaves* and are identified with their corresponding elements in L. Nodes with indegree 2 are called *reticulation nodes*. For any phylogenetic network N, let $\mathcal{U}(N)$ be the undirected graph obtained from N by replacing each directed edge by an undirected edge. N is said to be a *galled phylogenetic network* (*galled network*, for short) if all cycles in $\mathcal{U}(N)$ are node-disjoint. Galled networks are also known in the literature as *topologies with independent recombination events* [15], *galled-trees* [6], and *level-1 phylogenetic networks* [2, 5, 7, 9, 10, 14].

A phylogenetic tree with exactly three leaves is called a *rooted triplet*. The unique rooted triplet on a leaf set $\{x, y, z\}$ in which the lowest common ancestor of x and y is a proper descendant of the lowest common ancestor of x and z (or equivalently, where the lowest common ancestor of x and y is a proper descendant of the lowest common ancestor of y and z) is denoted by $xy|z$. For any phylogenetic network N, the rooted triplet $xy|z$ is said to be *consistent* with N if N contains three leaves labeled by x, y, and z

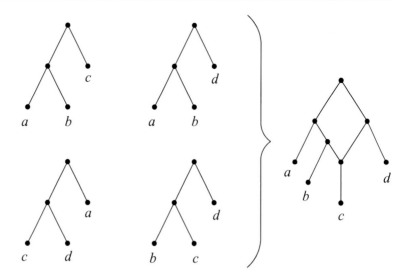

Algorithms for Combining Rooted Triplets into a Galled Phylogenetic Network, Fig. 1 A dense set $T = \{ab|c, ab|d, cd|a, bc|d\}$ of rooted triplets with leaf set $\{a, b, c, d\}$ and a galled phylogenetic network that is consistent with T. Note that this solution is not unique

as well as two internal vertices w and z such that there are four directed paths of nonzero length from w to a, from w to b, from z to w, and from z to c that are vertex-disjoint except for in the vertices w and z. A set T of rooted triplets is *consistent* with N if every rooted triplet in T is consistent with N. See Fig. 1 for an example.

Denote the set of leaves in any phylogenetic network N by $\Lambda(N)$, and for any set T of rooted triplets, define $\Lambda(T) = \bigcup_{t_i \in T} \Lambda(t_i)$. A set T of rooted triplets is *dense* if for each $\{x, y, z\} \subseteq \Lambda(T)$, at least one of the three possible rooted triplets $xy|z$, $xz|y$, and $yz|x$ belongs to T. Observe that if T is dense, then $|T| = \Theta(|\Lambda(T)|^3)$. Jansson and Sung introduced the following problem in [10].

Problem 1 Given a set T of rooted triplets, output a galled network N with $\Lambda(N) = \Lambda(T)$ such that N and T are consistent, if such a network exists; otherwise, output *null*.

A natural optimization version of Problem 1 is:

Problem 2 Given a set T of rooted triplets, output a galled network N with $\Lambda(N) = \Lambda(T)$ that is consistent with the maximum possible number of rooted triplets belonging to T.

A generalization of Problem 1 studied by He et al. in [8] involves *forbidden* rooted triplets and is defined as follows.

Problem 3 Given two sets T and $\mathcal{F}$ of rooted triplets, output a galled network N with $\Lambda(N) = \Lambda(T) \cup \Lambda(\mathcal{F})$ such that (1) N and T are consistent and (2) N is not consistent with any rooted triplet belonging to $\mathcal{F}$; if no such network exists, output *null*.

Below, we write $L = \Lambda(T)$ and $n = |L|$.

Key Results

As shown in [11], Problem 1 can be solved in (optimal) $O(|T|) = O(n^3)$ time for dense inputs:

Theorem 1 ([11]) *Given any dense set T of rooted triplets with leaf set L, a galled network consistent with T (if one exists) can be constructed in $O(n^3)$ time, where $n = |L|$.*

The algorithm referred to in Theorem 1 was extended by van Iersel and Kelk [14] as follows.

Theorem 2 ([14]) *Given any dense set T of rooted triplets with leaf set L, a galled network consistent with T (if one exists) that contains as few reticulation nodes as possible can be constructed in $O(n^5)$ time, where $n = |L|$.*

For the more general case of nondense inputs, Problem 1 becomes harder:

Theorem 3 ([11]) *The problem of determining if there exists a galled network that is consistent* *with an input nondense set $\mathcal{T}$ of rooted triplets is NP-hard.*

Since not all sets of rooted triplets are consistent with a galled network, it is of interest to consider Problem 2. It follows from Theorem 3 that Problem 2 is also NP-hard for nondense inputs, and this motivates polynomial-time approximation algorithms. Say that an algorithm for Problem 2 is an *f-approximation algorithm* if it always returns a galled network N such that $\frac{N(\mathcal{T})}{|\mathcal{T}|} \geq f$, where $N(\mathcal{T})$ is the number of rooted triplets in $\mathcal{T}$ that are consistent with N. Define the nonlinear recurrence relation $S(n) = \max_{1 \leq a \leq n}\left\{\binom{a}{3}+2\cdot\binom{a}{2}\cdot(n-a)+a\cdot\binom{n-a}{2}+S(n-a)\right\}$ for $n > 0$ and $S(0) = 0$. It was shown in [4] that $\lim_{n\to\infty}\frac{S(n)}{3\binom{n}{3}} = \frac{2(\sqrt{3}-1)}{3} \approx 0.488033\ldots$ and that $\frac{S(n)}{3\binom{n}{3}} > \frac{2(\sqrt{3}-1)}{3} \approx 0.488033\ldots$ for all $n > 2$. The following theorem was proved by Byrka et al. in [2].

Theorem 4 ([2]) *There exists an $\frac{S(n)}{3\binom{n}{3}}$-approximation algorithm for Problem 2 that runs in $O(n^3 + n|\mathcal{T}|)$ time.*

A matching negative bound is:

Theorem 5 ([11]) *For any $f > \lim_{n\to\infty}\frac{S(n)}{3\binom{n}{3}}$, there exists a set $\mathcal{T}$ of rooted triplets such that no galled network can be consistent with at least a factor of f of the rooted triplets in $\mathcal{T}$. (Thus, no f-approximation algorithm for Problem 2 is possible.)*

For Problem 3, Theorem 3 immediately implies NP-hardness by taking $\mathcal{F} = \emptyset$. The following positive result is known for the optimization version of Problem 3.

Theorem 6 ([8]) *There exists an $O(|L|^2|\mathcal{T}|(|\mathcal{T}| +|\mathcal{F}|))$-time algorithm for inferring a galled network N that guarantees $|N(\mathcal{T})| - |N(\mathcal{F})| \geq \frac{5}{12}\cdot(|\mathcal{T}| - |\mathcal{F}|)$, where $L = \Lambda(\mathcal{T}) \cup \Lambda(\mathcal{F})$.*

Finally, we remark that the analogous version of Problem 1 of inferring a phylogenetic *tree* consistent with all the rooted triplets in an input set (when such a tree exists) can be solved in polynomial time with a classical algorithm by Aho et al. [1] from 1981. Similarly, for Problem 2, to infer a phylogenetic tree consistent with as many rooted triplets from an input set of rooted triplets as possible is NP-hard and admits a polynomial-time 1/3-approximation algorithm, which is optimal in the sense that there exist certain inputs for which no tree can achieve a factor larger than 1/3. See, e.g., [3] for a survey of known results about maximizing rooted triplet consistency for trees. On the other hand, more complex network structures such as the *level-k phylogenetic networks* [5] permit a higher percentage of the input rooted triplets to be embedded; in the extreme case, if there are no restrictions on the reticulation nodes at all, then a sorting network-based construction yields a phylogenetic network that is trivially consistent with every rooted triplet over L [10]. A number of efficient algorithms for combining rooted triplets into higher level networks have been developed; see, e.g., [2,7,14] for further details and references.

Applications

Phylogenetic networks are used by scientists to describe evolutionary relationships that do not fit the traditional models in which evolution is assumed to be treelike. Evolutionary events such as horizontal gene transfer or hybrid speciation (often referred to as *recombination events*) which suggest convergence between objects cannot be represented in a single tree but can be modeled in a phylogenetic network as internal nodes having more than one parent (i.e., reticulation nodes). The phylogenetic network is a relatively new tool, and various fast and reliable methods for constructing and comparing phylogenetic networks are currently being developed.

Galled networks form an important class of phylogenetic networks. They have attracted special attention in the literature [5, 6, 15] due to their biological significance (see [6]) and their simple, almost treelike, structure. When the number of recombination events is limited and most of the recombination events have occurred recently, a galled network may suffice to accurately describe the evolutionary process under study [6]. The motivation behind the rooted triplet approach taken here is that a highly accurate tree for each cardinality-three subset of the leaf set can be obtained through maximum likelihood-based methods or Sibley-Ahlquist-style DNA-DNA hybridization experiments (see [13]). The algorithms mentioned above can be used as the merging step in a divide-and-conquer approach for constructing phylogenetic networks analogous to the quartet method paradigm for inferring unrooted phylogenetic trees [12] and other supertree methods. We consider dense input sets in particular because this case can be solved in polynomial time.

Open Problems

The approximation factor given in Theorem 4 is expressed in terms of the number of rooted triplets in the input $\mathcal{T}$, and Theorem 5 shows that it cannot be improved. However, if one measures the quality of the approximation in terms of a galled network N_{OPT} that is consistent with the maximum possible number of rooted triplets from $\mathcal{T}$, Theorem 4 can be far from optimal. An open problem is to determine the polynomial-time approximability and inapproximability of Problem 2 when the approximation ratio is defined as $\frac{N(\mathcal{T})}{N_{OPT}(\mathcal{T})}$ instead of $\frac{N(\mathcal{T})}{|\mathcal{T}|}$.

Another research direction is to develop fixed-parameter polynomial-time algorithms for Problem 1. The level of the constructed network, the number of allowed reticulation nodes, or some measure of the density of the input set of rooted triplet might be suitable parameters.

URLs to Code and Data Sets

A Java implementation of the algorithm for Problem 1 referred to in Theorem 2 (coded by its authors [14]) is available at http://skelk.sdf-eu.org/marlon.html. See also http://skelk.sdf-eu.org/lev1athan/ for a Java implementation of a polynomial-time heuristic described in [9] for Problem 2.

Cross-References

► Directed Perfect Phylogeny (Binary Characters)
► Distance-Based Phylogeny Reconstruction (Fast-Converging)
► Distance-Based Phylogeny Reconstruction: Safety and Edge Radius
► Perfect Phylogeny (Bounded Number of States)
► Phylogenetic Tree Construction from a Distance Matrix

Acknowledgments JJ was funded by the Hakubi Project at Kyoto University and KAKENHI grant number 26330014.

Recommended Reading

1. Aho AV, Sagiv Y, Szymanski TG, Ullman JD (1981) Inferring a tree from lowest common ancestors with an application to the optimization of relational expressions. SIAM J Comput 10(3):405–421
2. Byrka J, Gawrychowski P, Huber KT, Kelk S (2010) Worst-case optimal approximation algorithms for maximizing triplet consistency within phylogenetic networks. J Discret Algorithms 8(1):65–75
3. Byrka J, Guillemot S, Jansson J (2010) New results on optimizing rooted triplets consistency. Discret Appl Math 158(11):1136–1147
4. Chao K-M, Chu A-C, Jansson J, Lemence RS, Mancheron A (2012) Asymptotic limits of a new type of maximization recurrence with an application to bioinformatics. In: Proceedings of the 9th annual conference on theory and applications of models of computation (TAMC 2012), Beijing, pp 177–188
5. Choy C, Jansson J, Sadakane K, Sung W-K (2005) Computing the maximum agreement of phylogenetic networks. Theor Comput Sci 335(1):93–107
6. Gusfield D, Eddhu S, Langley C (2003) Efficient reconstruction of phylogenetic networks with constrained recombination. In: Proceedings of computational systems bioinformatics (CSB2003), Stanford, pp 363–374
7. Habib M, To T-H (2012) Constructing a minimum phylogenetic network from a dense triplet set. J Bioinform Comput Biol 10(5):1250013
8. He Y-J, Huynh TND, Jansson J, Sung W-K (2006) Inferring phylogenetic relationships avoiding forbidden rooted triplets. J Bioinform Comput Biol 4(1):59–74
9. Huber KT, van Iersel L, Kelk S, Suchecki R (2011) A practical algorithm for reconstructing level-1 phylogenetic networks. IEEE/ACM Trans Comput Biol Bioinform 8(3):635–649

10. Jansson J, Sung W-K (2006) Inferring a level-1 phylogenetic network from a dense set of rooted triplets. Theor Comput Sci 363(1):60–68
11. Jansson J, Nguyen NB, Sung W-K (2006) Algorithms for combining rooted triplets into a galled phylogenetic network. SIAM J Comput 35(5):1098–1121
12. Jiang T, Kearney P, Li M (2001) A polynomial time approximation scheme for inferring evolutionary trees from quartet topologies and its application. SIAM J Comput 30(6):1942–1961
13. Kannan S, Lawler E, Warnow T (1996) Determining the evolutionary tree using experiments. J Algorithms 21(1):26–50
14. van Iersel L, Kelk S (2011) Constructing the simplest possible phylogenetic network from triplets. Algorithmica 60(2):207–235
15. Wang L, Zhang K, Zhang L (2001) Perfect phylogenetic networks with recombination. J Comput Biol 8(1):69–78

All Pairs Shortest Paths in Sparse Graphs

Seth Pettie
Electrical Engineering and Computer Science (EECS) Department, University of Michigan, Ann Arbor, MI, USA

Keywords

Quickest route; Shortest route

Years and Authors of Summarized Original Work

2004; Pettie

Problem Definition

Given a communications network or road network, one of the most natural algorithmic questions is how to determine the shortest path from one point to another. The *all pairs* shortest path problem (APSP) is, given a directed graph $G = (V, E, l)$, to determine the distance and shortest path between every pair of vertices, where $|V| = n, |E| = m$, and $l : E \rightarrow \mathbb{R}$ is the edge length (or weight) function. The output is in the form of two $n \times n$ matrices: $D(u, v)$ is the distance from u to v and $S(u, v) = w$ if (u, w) is the first edge on a shortest path from u to v. The APSP problem is often contrasted with the *point-to-point* and *single source* (SSSP) shortest path problems. They ask for, respectively, the shortest path from a given source vertex to a given target vertex and all shortest paths from a given source vertex.

Definition of Distance

If l assigns only non-negative edge lengths then the definition of distance is clear: $D(u, v)$ is the length of the minimum length path from u to v, where the length of a path is the total length of its constituent edges. However, if l can assign negative lengths then there are several sensible notations of distance that depend on how negative length cycles are handled. Suppose that a cycle C has negative length and that $u, v \in V$ are such that C is reachable from u and v reachable from C. Because C can be traversed an arbitrary number of times when traveling from u to v, there is no shortest path from u to v using a finite number of edges. It is sometimes assumed a priori that G has no negative length cycles; however it is cleaner to define $D(u, v) = -\infty$ if there is no finite shortest path. If $D(u, v)$ is defined to be the length of the shortest *simple* path (no repetition of vertices) then the problem becomes NP-hard. (If all edges have length -1 then $D(u, v) = -(n-1)$ if and only if G contains a Hamiltonian path [7] from u to v.) One could also define distance to be the length of the shortest path without repetition of edges.

Classic Algorithms

The Bellman-Ford algorithm solves SSSP in $O(mn)$ time and under the assumption that edge lengths are non-negative, Dijkstra's algorithm solves it in $O(m + n \log n)$ time. There is a well known $O(mn)$-time shortest path preserving transformation that replaces any length function with a non-negative length function. Using this transformation and n runs of Dijkstra's algorithm gives an APSP algorithm running in $O(mn + n^2 \log n) = O(n^3)$ time. The Floyd-Warshall algorithm computes APSP in a more direct manner, in $O(n^3)$ time. Refer to [4] for a description of these algorithms. It is known that APSP on complete graphs is asymptotically equivalent to (min, +) matrix multiplication [1], which can be computed by a non-uniform algorithm that performs $O(n^{2.5})$ numerical operations [6].

Integer-Weighted Graphs

Much recent work on shortest paths assume that edge lengths are integers in the range $\{-C, \ldots, C\}$ or $\{0, \ldots, C\}$. One line of research reduces APSP to a series of standard matrix multiplications. These algorithms are limited in their applicability because their running times scale linearly with C. There are faster SSSP algorithms for both non-negative edge lengths and arbitrary edge lengths. The former exploit the power of RAMs to sort in $o(n \log n)$ time and the latter are based on the *scaling* technique. See Zwick [20] for a survey of shortest path algorithms up to 2001.

Key Results

Pettie's APSP algorithm [12] adapts the *hierarchy* approach of Thorup [16] (designed for undirected, integer-weighted graphs) to general real-weighted directed graphs. Theorem 1 is the first improvement over the $O(mn + n^2 \log n)$ time bound of Dijkstra's algorithm on arbitrary real-weighted graphs.

Theorem 1 *Given a real-weighted directed graph, all pairs shortest paths can be solved in* $O(mn + n^2 \log \log n)$ *time.*

This algorithm achieves a logarithmic speedup through a trio of new techniques. The first is to exploit the necessary similarity between the SSSP trees emanating from nearby vertices. The second is a method for computing discrete approximate distances in real-weighted graphs. The third is a new hierarchy-type SSSP algorithm that runs in $O(m + n \log \log n)$ time when given suitably accurate approximate distances.

Theorem 1 should be contrasted with the time bounds of other hierarchy-type APSP algorithms [11, 14, 16].

Theorem 2 ([14], 2005) *Given a real-weighted undirected graph, APSP can be solved in* $O(mn \log \alpha(m, n))$ *time.*

Theorem 3 ([16], 1999) *Given an undirected graph $G(V, E, l)$, where ℓ assigns integer edge lengths in the range $\{-2^{w-1}, \ldots, 2^{w-1} - 1\}$, APSP can be solved in $O(mn)$ time on a RAM with w-bit word length.*

Theorem 4 ([13], 2002) *Given a real-weighted directed graph, APSP can be solved in polynomial time by an algorithm that performs $O(mn \log \alpha(m, n))$ numerical operations, where α is the inverse-Ackermann function.*

A secondary result of [12, 14] is that no *hierarchy-type* shortest path algorithm can improve on the $O(m + n \log n)$ running time of Dijkstra's algorithm.

Theorem 5 *Let G be an input graph such that the ratio of the maximum to minimum edge length is r. Any hierarchy-type SSSP algorithm performs $\Omega(m + \min\{n \log n, n \log r\})$ numerical operations if G is directed and $\Omega(m + \min\{n \log n, n \log \log r\})$ if G is undirected.*

Applications

Shortest paths appear as a subproblem in other graph optimization problems; the minimum weight perfect matching, minimum cost flow, and minimum mean-cycle problems are some examples. A well known commercial application of shortest path algorithms is finding efficient routes on road networks; see, for example, Google Maps, MapQuest, or Yahoo Maps.

Open Problems

The longest standing open shortest path problems are to improve the SSSP algorithms of Dijkstra's and Bellman-Ford on *real*-weighted graphs.

Problem 1 Is there an $o(mn)$ time SSSP or point-to-point shortest path algorithm for arbitrarily weighted graphs?

Problem 2 Is there an $O(m) + o(n \log n)$ time SSSP algorithm for directed, non-negatively weighted graphs? For undirected graphs?

A partial answer to Problem 2 appears in [14], which considers undirected graphs. Perhaps the most surprising open problem is whether there is any (asymptotic) difference between the complexities of the all pairs, single source, and point-to-point shortest path problems on arbitrarily weighted graphs.

Problem 3 Is point-to-point shortest paths easier than all pairs shortest paths on arbitrarily weighted graphs?

Problem 4 Is there a truly subcubic APSP algorithm, i.e., one running in time $O(n^{3-\epsilon})$? In a recent breakthrough on this problem, Williams [19] gave a new APSP algorithm running in $n^3/2^{\Theta(\sqrt{\log n/\log \log n})}$ time. Vassilevska Williams and Williams [17] proved that a truly subcubic algorithm for APSP would imply truly subcubic algorithms for other graph problems.

Experimental Results

See [5, 8, 15] for recent experiments on SSSP algorithms. On sparse graphs the best APSP algorithms use repeated application of an SSSP algorithm, possibly with some precomputation [15]. On dense graphs cache-efficiency becomes a major issue. See [18] for a cache conscious implementation of the Floyd-Warshall algorithm.

The trend in recent years is to construct a linear space data structure that can quickly answer exact or approximate point-to-point shortest path queries; see [2, 6, 9, 10].

Data Sets

See [5] for a number of U.S. and European road networks.

URL to Code

See [5].

Cross-References

▶ All Pairs Shortest Paths via Matrix Multiplication
▶ Single-Source Shortest Paths

Recommended Reading

1. Aho AV, Hopcroft JE, Ullman JD (1975) The design and analysis of computer algorithms. Addison-Wesley, Reading
2. Bast H, Funke S, Matijevic D, Sanders P, Schultes D (2007) In transit to constant shortest-path queries in road networks. In: Proceedings of the 9th workshop on algorithm engineering and experiments (ALENEX), San Francisco
3. Chan T (2007) More algorithms for all-pairs shortest paths in weighted graphs. In: Proceedings of the 39th ACM symposium on theory of computing (STOC), San Diego, pp 590–598
4. Cormen TH, Leiserson CE, Rivest RL, Stein C (2001) Introduction to algorithms. MIT, Cambridge
5. Demetrescu C, Goldberg AV, Johnson D (2006) 9th DIMACS implementation challenge – shortest paths. http://www.dis.uniroma1.it/~challenge9/
6. Fredman ML (1976) New bounds on the complexity of the shortest path problem. SIAM J Comput 5(1):83–89
7. Garey MR, Johnson DS (1979) Computers and intractability: a guide to NP-completeness. Freeman, San Francisco
8. Goldberg AV (2001) Shortest path algorithms: engineering aspects. In: Proceedings of the 12th international symposium on algorithms and computation (ISAAC), Christchurch. LNCS, vol 2223. Springer, Berlin, pp 502–513
9. Goldberg AV, Kaplan H, Werneck R (2006) Reach for A*: efficient point-to-point shortest path algorithms. In: Proceedings of the 8th workshop on algorithm engineering and experiments (ALENEX), Miami
10. Knopp S, Sanders P, Schultes D, Schulz F, Wagner D (2007) Computing many-to-many shortest paths using highway hierarchies. In: Proceedings of the 9th workshop on algorithm engineering and experiments (ALENEX), New Orleans
11. Pettie S (2002) On the comparison-addition complexity of all-pairs shortest paths. In: Proceedings of the 13th international symposium on algorithms and computation (ISAAC), Vancouver, pp 32–43

12. Pettie S (2004) A new approach to all-pairs shortest paths on real-weighted graphs. Theor Comput Sci 312(1):47–74
13. Pettie S, Ramachandran V (2002) Minimizing randomness in minimum spanning tree, parallel connectivity and set maxima algorithms. In: Proceedings of the 13th ACM-SIAM symposium on discrete algorithms (SODA), San Francisco, pp 713–722
14. Pettie S, Ramachandran V (2005) A shortest path algorithm for real-weighted undirected graphs. SIAM J Comput 34(6):1398–1431
15. Pettie S, Ramachandran V, Sridhar S (2002) Experimental evaluation of a new shortest path algorithm. In: Proceedings of the 4th workshop on algorithm engineering and experiments (ALENEX), San Francisco, pp 126–142
16. Thorup M (1999) Undirected single-source shortest paths with positive integer weights in linear time. J ACM 46(3):362–394
17. Vassilevska Williams V, Williams R (2010) Subcubic equivalences between path, matrix, and triangle problems. In Proceedings of the 51st IEEE symposium on foundations of computer science (FOCS), Los Alamitos, pp 645–654
18. Venkataraman G, Sahni S, Mukhopadhyaya S (2003) A blocked all-pairs shortest paths algorithm. J Exp Algorithms 8
19. Williams R (2014) Faster all-pairs shortest paths via circuit complexity. In Proceedings of the 46th ACM symposium on theory of computing (STOC), New York, pp 664–673
20. Zwick U (2001) Exact and approximate distances in graphs – a survey. In: Proceedings of the 9th European symposium on algorithms (ESA), Aarhus, pp 33–48. See updated version at http://www.cs.tau. ac.il/~zwick/

All Pairs Shortest Paths via Matrix Multiplication

Tadao Takaoka
Department of Computer Science and Software Engineering, University of Canterbury, Christchurch, New Zealand

Keywords

Algorithm analysis; All pairs shortest path problem; Bridging set; Matrix multiplication; Shortest path problem; Two-phase algorithm; Witness

Years and Authors of Summarized Original Work

2002; Zwick

Problem Definition

The all pairs shortest path (APSP) problem is to compute shortest paths between all pairs of vertices of a directed graph with nonnegative real numbers as edge costs. Focus is given on shortest distances between vertices, as shortest paths can be obtained with a slight increase of cost. Classically, the APSP problem can be solved in cubic time of $O(n^3)$. The problem here is to achieve a sub-cubic time for a graph with small integer costs.

A directed graph is given by $G = (V, E)$, where $V = \{1, \ldots, n\}$, the set of vertices, and E is the set of edges. The cost of edge $(i, j) \in E$ is denoted by d_{ij}. The (n, n)-matrix D is one whose (i, j) element is d_{ij}. It is assumed for simplicity that $d_{ij} > 0$ and $d_{ii} = 0$ for all $i \neq j$. If there is no edge from i to j, let $d_{ij} = \infty$. The cost, or distance, of a path is the sum of costs of the edges in the path. The length of a path is the number of edges in the path. The shortest distance from vertex i to vertex j is the minimum cost over all paths from i to j, denoted by d_{ij}^*. Let $D^* = \{d_{ij}^*\}$. The value of n is called the size of the matrices.

Let A and B are (n, n)-matrices. The three products are defined using the elements of A and B as follows: (1) Ordinary matrix product over a ring $C = AB$ (2) Boolean matrix product $C = A \cdot B$ (3) Distance matrix product $C = A \times B$, where

$$(1)\ c_{ij} = \sum_{k=1}^{n} a_{ik} b_{kj}, \quad (2)\ c_{ij} = \bigvee_{k=1}^{n} a_{ik} \wedge b_{kj},$$

$$(3)\ c_{ij} = \min_{1 \le k \le n} \{a_{ik} + b_{kj}\}.$$

The matrix C is called a product in each case; the computational process is called multiplication, such as distance matrix multiplication. In those three cases, k changes through the entire set $\{1, \ldots, n\}$. A partial matrix product of A and B is defined by taking k in a subset I of V. In other words, a partial product is obtained by multiplying a vertically rectangular matrix, $A(*, I)$, whose columns are extracted from A corresponding to the set I, and similarly a horizontally rectangular matrix, $B(I, *)$, extracted from B with rows corresponding to I. Intuitively, I is the set of check points k, when going from i to j.

The best algorithm [11] computes (1) in $O(n^\omega)$ time, where $\omega = 2.373$. This was recently achieved as improvement from $\omega = 2.376$ in [4] after more than two decades of interval. We use $\omega = 2.376$ to describe Zwick's result in this article. Three decimal points are carried throughout this article. To compute (2), Boolean values 0 and 1 in A and B can be regarded as integers and use the algorithm for (1), and convert nonzero elements in the resulting matrix to 1. Therefore, this complexity is $O(n^\omega)$. The witnesses of (2) are given in the witness matrix $W = \{w_{ij}\}$ where $w_{ij} = k$ for some k such that $a_{ik} \wedge b_{kj} = 1$. If there is no such k, $w_{ij} = 0$. The witness matrix $W = \{w_{ij}\}$ for (3) is defined by $w_{ij} = k$ that gives the minimum to c_{ij}. If there is an algorithm for (3) with $T(n)$ time, ignoring a polylog factor of n, the APSP problem can be solved in $\tilde{O}(T(n))$ time by the repeated squaring method, described as the repeated use of $D \leftarrow D \times D$ $O(\log n)$ times.

The definition here of computing shortest paths is to give a witness matrix of size n by which a shortest path from i to j can be given in $O(\ell)$ time where ℓ is the length of the path. More specifically, if $w_{ij} = k$ in the witness matrix $W = \{w_{ij}\}$, it means that the path from i to j goes through k. Therefore, a recursive function $\text{path}(i, j)$ is defined by $(\text{path}(i, k), k, \text{path}(k, j))$ if $\text{path}(i, j) = k > 0$ and nil if $\text{path}(i, j) = 0$, where a path is defined by a list of vertices excluding endpoints. In the following sections, k is recorded in w_{ij} whenever k is found such that a path from i to j is modified or newly set up by paths from i to k and from k to j. Preceding results are introduced as a framework for the key results.

Alon-Galil-Margalit Algorithm

The algorithm by Alon, Galil, and Margalit [1] is reviewed. Let the costs of edges of the given graph be ones. Let $D^{(\ell)}$ be the ℓ-th approximate matrix for D^* defined by $d_{ij}^{(\ell)} = d_{ij}^*$ if $d_{ij}^* \leq \ell$, and $d_{ij}^{(\ell)} = \infty$ otherwise. Let A be the adjacency matrix of G, that is, $a_{ij} = 1$ if there is an edge (i, j), and $a_{ij} = 0$, otherwise. Let $a_{ii} = 1$ for all i. The algorithm consists of two phases. In the first phase, $D^{(\ell)}$ is computed for $\ell = 1, \ldots, r$, by checking the (i, j) element of $A^\ell = \{a_{ij}^\ell\}$. Note that if $a_{ij}^\ell = 1$, there is a path from i to j of length ℓ or less. Since Boolean matrix multiplication can be computed in $O(n^\omega)$ time, the computing time of this part is $O(rn^\omega)$.

In the second phase, the algorithm computes $D^{(\ell)}$ for $\ell = r, \lceil \frac{3}{2}r \rceil, \lceil \frac{3}{2}\lceil \frac{3}{2}r \rceil \rceil, \ldots, n'$ by repeated squaring, where n' is the smallest integer in this sequence of ℓ such that $\ell \geq n$. Let $T_{i\alpha} = \{j \mid d_{ij}^{(\ell)} = \alpha\}$ and $I_i = T_{i\alpha}$ such that $|T_{i\alpha}|$ is minimum for $\lceil \ell/2 \rceil \leq \alpha \leq \ell$. The key observation in the second phase is that it is only needed to check k in I_i whose size is not larger than $2n/\ell$, since the correct distances between $\ell + 1$ and $\lceil 3\ell/2 \rceil$ can be obtained as the sum $d_{ik}^{(\ell)} + d_{kj}^{(\ell)}$ for some k satisfying $\lceil \ell/2 \rceil \leq d_{ik}^{(\ell)} \leq \ell$. The meaning of I_i is similar to I for partial products except that I varies for each i. Hence, the computing time of one squaring is $O(n^3/\ell)$. Thus, the time of the second phase is given with $N = \lceil \log_{3/2} n/r \rceil$ by $O\left(\sum_{s=1}^{N} n^3/((3/2)^s r)\right) = O(n^3/r)$. Balancing the two phases with $rn^\omega = n^3/r$ yields $O(n^{(\omega+3)/2}) = O(n^{2.688})$ time for the algorithm with $r = O(n^{(3-\omega)/2})$.

Witnesses can be kept in the first phase in time polylog of n by the method in [2]. The maintenance of witnesses in the second phase is straightforward.

When a directed graph G whose edge costs are integers between 1 and M is given, where M is a positive integer, the graph G can be expanded to G' by creating up to $M - 1$ new vertices for each vertex and replacing each edge by up to M edges with unit cost. Obviously, the problem for G can be solved by applying the above algorithm to G', which takes $O\left((Mn)^{(\omega+3)/2}\right)$ time. This time is sub-cubic when $M < n^{0.116}$. The maintenance of witnesses has an extra polylog factor in each case.

For undirected graphs with unit edge costs, $\tilde{O}(n^\omega)$ time is known in Seidel [9].

Takaoka Algorithm

When the edge costs are bounded by a positive integer M, a better algorithm can be designed than in the above as shown in Takaoka [10]. Romani's algorithm [7] for distance matrix multiplication is reviewed briefly.

Let A and B be (n, m) and (m, n) distance matrices whose elements are bounded by M or infinite. Let the diagonal elements be 0. A and B are converted into A' and B' where $a'_{ij} = (m + 1)^{M-a_{ij}}$, if $a_{ij} \neq \infty$, 0, otherwise, and $b'_{ij} = (m + 1)^{M-b_{ij}}$, if $b_{ij} \neq \infty$, 0, otherwise.

Let $C' = A'B'$ be the product by ordinary matrix multiplication and $C = A \times B$ be that by distance matrix multiplication. Then, it holds that

$$c'_{ij} = \sum_{k=1}^{m} (m + 1)^{2M-(a_{ik}+b_{kj})},$$

$$c_{ij} = 2M - \lfloor \log_{m+1} c'_{ij} \rfloor.$$

This distance matrix multiplication is called (n, m)-Romani. In this section, the above multiplication is used with square matrices, that is, (n, n)-Romani is used. In the next section, the case where $m < n$ is dealt with.

C can be computed with $O(n^\omega)$ arithmetic operations on integers up to $(n + 1)^M$. Since these values can be expressed by $O(M \log n)$ bits and Schönhage and Strassen's algorithm [8] for multiplying k-bit numbers takes $O(k \log k \log \log k)$ bit operations, C can be computed in $O(n^\omega M \log n \log(M \log n) \log \log (M \log n))$ time, or $\tilde{O}(Mn^\omega)$ time.

The first phase is replaced by the one based on (n, n)-Romani and the second phase is modified based on path lengths, not distances.

Note that the bound M is replaced by ℓM in the distance matrix multiplication in the first phase. Ignoring polylog factors, the time for the first phase is given by $\tilde{O}(n^\omega r^2 M)$. It is assumed that M is $O(n^k)$ for some constant k. Balancing this complexity with that of second phase, $O(n^3/r)$, yields the total computing time of $\tilde{O}(n^{(6+\omega)/3} M^{1/3})$ with the choice of $r = O(n^{(3-\omega)/3} M^{-1/3})$. The value of M can be almost $O(n^{0.624})$ to keep the complexity within sub-cubic.

Key Results

Zwick improved the Alon-Galil-Margalit algorithm in several ways. The most notable is an improvement of the time for the APSP problem with unit edge costs from $O(n^{2.688})$ to $O(n^{2.575})$. The main accelerating engine in Alon-Galil-Margalit [1] was the fast Boolean matrix multiplication and that in Takaoka [10] was the fast distance matrix multiplication by Romani, both powered by the fast matrix multiplication of square matrices.

In this section, the engine is the fast distance matrix multiplication by Romani powered by the fast matrix multiplication of rectangular matrices given by Coppersmith [3] and Huang and Pan [5]. Suppose the product of (n, m) matrix and (m, n) matrix can be computed with $O(n^{\omega(1,\mu,1)})$ arithmetic operations, where $m = n^\mu$ with $0 \leq \mu \leq 1$. Several facts such as $O(n^{\omega(1,1,1)}) = O(n^{2.376})$ and $O(n^{\omega(1,0.294,1)}) = \tilde{O}(n^2)$ are known. To compute the product of (n, n) square matrices, $n^{1-\mu}$ matrix multiplications are needed, resulting in $O(n^{\omega(1,\mu,1)+1-\mu})$ time, which is reformulated as $O(n^{2+\mu})$, where μ satisfies the equation $\omega(1, \mu, 1) = 2\mu + 1$. Also, the upper bound of $\omega(1, \mu, 1)$ is given by

$$\omega(1, \mu, 1) = 2, \text{ if } 0 \leq \mu \leq \alpha$$

$$\omega(1, \mu, 1) = 2 + (\omega - 2)$$
$$(\mu - \alpha)/(1 - \alpha), \text{ if } \alpha \leq \mu \leq 1$$

The best known value for μ, when [12] was published, was $\mu = 0.575$, derived from the above formulae, $\alpha > 0.294$ and $\omega < 2.376$. So, the time becomes $O(n^{2.575})$, which is not as good as $O(n^{2.376})$. Thus, we use the algorithm for rectangular matrices in the following.

The above algorithm for rectangular matrix multiplication is incorporated into (n, m)-Romani with $m = n^\mu$ and $M = n^t$, and the computing time of $\tilde{O}(Mn^{\omega(1,\mu,1)})$. The next step is how to incorporate (n, m)-Romani into the APSP algorithm. The first algorithm is a monophase algorithm based on repeated squaring, similar to the second phase of the algorithm in [1]. To take advantage of rectangular matrices in (n, m)-Romani, the following definition of the bridging set is needed, which plays the role of the set I in the partial distance matrix product in section "Problem Definition."

Let $\delta(i, j)$ be the shortest distance from i to j, and $\eta(i, j)$ be the minimum length of all shortest paths from i to j. A subset I of V is an ℓ-bridging set if it satisfies the condition that if $\eta(i, j) \geq \ell$, there exists $k \in I$ such that $\delta(i, j) = \delta(i, k) + \delta(k, j)$. I is a strong ℓ-bridging set if it satisfies the condition that if $\eta(i, j) \geq \ell$, there exists $k \in I$ such that $\delta(i, j) = \delta(i, k) + \delta(k, j)$ and $\eta(i, j) = \eta(i, k) + \eta(k, j)$. Note that those two sets are the same for a graph with unit edge costs.

Note that if $(2/3)\ell \leq \mu(i, j) \leq \ell$ and I is a strong $\ell/3$-bridging set, there is a $k \in I$ such that $\delta(i, j) = \delta(i, k) + \delta(k, j)$ and $\mu(i, j) = \mu(i, k) + \mu(k, j)$. With this property of strong bridging sets, (n, m)-Romani can be used for the APSP problem in the following way. By repeated squaring in a similar way to Alon-Galil-Margalit, the algorithm computes $D^{(\ell)}$ for $\ell = 1, \lceil \frac{3}{2} \rceil$, $\lceil \frac{3}{2} \lceil \frac{3}{2} \rceil \rceil, \ldots, n'$, where n' is the first value of ℓ that exceeds n, using various types of set I described below. To compute the bridging set, the algorithm maintains the witness matrix with extra polylog factor in the complexity. In [12], there are three ways for selecting the set I. Let $|I| = n^r$ for some r such that $0 \leq r \leq 1$.

1. Select $9n \ln n/\ell$ vertices for In from V at random. In this case, it can be shown that the algorithm solves the APSP problem with high probability, i.e., with $1 - 1/n^c$ for some constant $c > 0$, which can be shown to

be 3. In other words, I is a strong $\ell/3$-bridging set with high probability. The time T is dominated by (n, m)-Romani. It holds that $T = \tilde{O}(\ell Mn^{\omega(1,r,1)})$, since the magnitude of matrix elements can be up to ℓM. Since $m = O(n \ln n/\ell) = n^r$, it holds that $\ell = \tilde{O}(n^{1-r})$, and thus $T = O(Mn^{1-r}n^{\omega(1,r,1)})$. When $M = 1$, this bound on r is $\mu = 0.575$, and thus $T = O(n^{2.575})$. When $M = n^t \geq 1$, the time becomes $O(n^{2+\mu(t)})$, where $t \leq 3 - \omega = 0.624$ and $\mu = \mu(t)$ satisfies $\omega(1, \mu, 1) = 1 + 2\mu - t$. It is determined from the best known $\omega(1, \mu, 1)$ and the value of t. As the result is correct with high probability, this is a randomized algorithm.

2. Consider the case of unit edge costs here. In (1), the computation of witnesses is an extra thing, i.e., not necessary if only shortest distances are needed. To achieve the same complexity in the sense of an exact algorithm, not a randomized one, the computation of witnesses is essential. As mentioned earlier, maintenance of witnesses, that is, matrix W, can be done with an extra polylog factor, meaning the analysis can be focused on Romani within the $\tilde{O}$-notation. Specifically, I is selected as an $\ell/3$-bridging set, which is strong with unit edge costs. To compute I as an $O(\ell)$-bridging set, obtain the vertices on the shortest path from i to j for each i and j using the witness matrix W in $O(\ell)$ time. After obtaining those n^2 sets spending $O(\ell n^2)$ time, it is shown in [12] how to obtain a $O(\ell)$-bridging set of $O(n \ln n/\ell)$ size within the same time complexity. The process of obtaining the bridging set must stop at $\ell = n^{1/2}$ as the process is too expensive beyond this point, and thus, the same bridging set is used beyond this point. The time before this point is the same as that in (1) and that after this point is $\tilde{O}(n^{2.5})$. Thus, this is a two-phase algorithm.

3. When edge costs are positive and bounded by $M = n^t > 0$, a similar procedure can be used to compute an $O(\ell)$-bridging set of $O(n \ln n/\ell)$ size in $\tilde{O}(\ell n^2)$ time. Using the bridging set, the APSP problem can be solved in $\tilde{O}(n^{2+\mu(t)})$ time in a similar way to (1).

The result can be generalized into the case where edge costs are between $-M$ and M within the same time complexity by modifying the procedure for computing an ℓ-bridging set, provided there is no negative cycle. The details are shown in [12].

Applications

The eccentricity of a vertex v of a graph is the greatest distance from v to any other vertices. The diameter of a graph is the greatest eccentricity of any vertices. In other words, the diameter is the greatest distance between any pair of vertices. If the corresponding APSP problem is solved, the maximum element of the resulting matrix is the diameter.

Open Problems

Recently, LeGall [6] discovered an algorithm for multiplying rectangular matrices with $\omega(1, 0.530, 1) < 2.060$, which gives the upper bound $\mu < 0.530$. This improves the complexity of APSP with unit edge costs from $O(n^{2.575})$ by Zwick to $O(n^{2.530})$ in the same framework as that of Zwick in this article. Two major challenges are stated here among others. The first is to improve the complexity of $\tilde{O}(n^{2.530})$ for the APSP with unit edge costs.

The other is to improve the bound of $M < O(n^{0.624})$ for the complexity of the APSP with integer costs up to M to be sub-cubic.

Cross-References

▶ All Pairs Shortest Paths in Sparse Graphs

Recommended Reading

1. Alon N, Galil Z, Margalit O (1991) On the exponent of the all pairs shortest path problem. In: Proceedings of the 32th IEEE FOCS, San Juan, pp 569–575. Also JCSS 54 (1997), pp 255–262
2. Alon N, Galil Z, Margalit O, Naor M (1992) Witnesses for Boolean matrix multiplication and for shortest paths. In: Proceedings of the 33th IEEE FOCS, Pittsburgh, pp 417–426
3. Coppersmith D (1997) Rectangular matrix multiplication revisited. J Complex 13:42–49
4. Coppersmith D, Winograd S (1990) Matrix multiplication via arithmetic progressions. J Symb Comput 9:251–280
5. Huang X, Pan VY (1998) Fast rectangular matrix multiplications and applications. J Complex 14:257–299
6. Le Gall F (2012) Faster algorithms for rectangular matrix multiplication. In: Proceedings of the 53rd FOCS, New Brunswick, pp 514–523
7. Romani F (1980) Shortest-path problem is not harder than matrix multiplications. Inf Proc Lett 11:134–136
8. Schönhage A, Strassen V (1971) Schnelle Multiplikation Großer Zahlen. Computing 7:281–292
9. Seidel R (1992) On the all-pairs-shortest-path problem. In: Proceedings of the 24th ACM STOC, Victoria, pp 745–749. Also JCSS 51 (1995), pp 400–403
10. Takaoka T (1998) Sub-cubic time algorithms for the all pairs shortest path problem. Algorithmica 20:309–318
11. Williams VV (2012) Multiplying matrices faster than Coppersmith-Winograd. In: Proceedings of the 44th symposium on theory of computing, ACM STOC, New York, pp 887–898
12. Zwick U (2002) All pairs shortest paths using bridging sets and rectangular matrix multiplication. J ACM 49(3):289–317

All-Distances Sketches

Edith Cohen
Tel Aviv University, Tel Aviv, Israel
Stanford University, Stanford, CA, USA

Keywords

Distance distribution; Distinct counting; Graph analysis; Influence; Nodes similarity; Summary structures

Years and Authors of Summarized Original Work

1997, 2014; Cohen
2002; Palmer, Gibbons, Faloutsos
2004, 2007; Cohen, Kaplan

Problem Definition

All-distances sketches (The term *least element lists* was used in [3]; the terms MV/D lists and Neighborhood summaries were used in [6].) are randomized summary structures of the distance relations of nodes in a graph. The graph can be directed or undirected, and edges can have uniform or general nonnegative weights.

Preprocessing Cost
A set of sketches, ADS(v) for each node v, can be computed efficiently, using a near-linear number of edge traversals. The sketch sizes are well concentrated, with logarithmic dependence on the total number of nodes.

Supported Queries
The sketches support approximate distance-based queries, which include:

- *Distance distribution*: The query specifies a node v and value $d \geq 0$ and returns the cardinality $|N_d(v)|$ of the d-neighborhood of v $N_d(v) = \{u \mid d_{vu} \leq d\}$, where d_{uv} is the shortest path distance from u to v. We are interested in estimating $|N_d(v)|$ from ADS(v).

  A related property is the *effective diameter* of the graph, which is a quantile of the distance distribution of all node pairs; we are interested in computing an estimate efficiently.
- *Closeness centrality* (distance-decaying) is defined for a node v, a monotone nonincreasing function $\alpha(x) \geq 0$ (where $\alpha(+\infty) \equiv 0$), and a nonnegative function $\beta(u) \geq 0$:

$$C_{\alpha,\beta}(v) = \sum_u \alpha(d_{vu})\beta(u) . \quad (1)$$

The function α specifies the decay of relevance with distance and the function β weighs nodes based on metadata to focus on a topic or property of relevance. Neighborhood cardinality is a special case obtained using $\beta \equiv 1$ and $\alpha(x) = 1$ if $x \leq d$ and $\alpha(x) = 0$ otherwise. The number of reachable nodes from v is

obtained using $\alpha(x) \equiv 1$. Also studied were exponential decay $\alpha(x) = 2^{-x}$ with distance [10], the (inverse) harmonic mean of distances $\alpha(x) = 1/x$ [1, 14], and general decay functions [4,6]. We would like to estimate $C_{\alpha,\beta}(v)$ from ADS(v).

- *Closeness similarity* [8] relates a pair of nodes based on the similarity of their distance relations to all other nodes.

$$\mathrm{SIM}_{\alpha,\beta}(v, u) = \frac{\sum_j \alpha(\max\{d_{u,j}, d_{v,j}\})\beta(j)}{\sum_j \alpha(\min\{d_{u,j}, d_{v,j}\})\beta(j)} . \quad (2)$$

We would like to estimate $\mathrm{SIM}_{\alpha,\beta}(v, u) \in [0, 1]$ from ADS(v) and ADS(u).
- *Timed influence* of a seed set S of nodes depends on the set of distances from S to other nodes. Intuitively, when edge lengths model transition times, the distance is the "elapsed time" needed to reach the node from S. Influence is higher when distances are shorter:

$$\mathrm{INF}_{\alpha,\beta}(S) = \sum_j \alpha(\min_{i \in S} d_{ij})\beta(j) . \quad (3)$$

We would like to estimate $\mathrm{INF}_{\alpha,\beta}(S)$ from the sketches $\{\mathrm{ADS}(v) \mid v \in S\}$.
- *Approximate distance oracles:* For two nodes v, u, use ADS(u) and ADS(v) to estimate $d_{u,v}$.

Key Results

We provide a precise definition of ADSs, overview algorithms for scalable computation, and finally discuss estimators.

Definition
The ADS of a node v, ADS(v), is a set of node ID and distance pairs (u, d_{vu}). The included nodes are a sample of the nodes reachable from v. The sampling is such that the inclusion probability of a node is inversely proportional to its Dijkstra rank (nearest neighbor rank). That is, the probability that the ith closest node is sampled is proportional to $1/i$.

The ADSs are defined with respect to random mappings/permutations r of the set of all nodes and come in three flavors: bottom-k, k-mins, and k-partition. The integer parameter k determines a tradeoff between the sketch size and computation time on one hand and the information and estimate quality on the other. For simplicity, we assume here that distances d_{vu} are unique for different u (using tie breaking). We use the notation $\Phi_{<u}(v)$ for the set of nodes that are closer to node v than node u and $\pi_{vu} = 1 + |\Phi_{<u}(v)|$ for the *Dijkstra rank* of u with respect to v (u is the π_{vu} closest node from v). For a set N and a numeric function $r : N$, the function $k_r\text{th}(N)$ returns the kth smallest value in the range of r on N. If $|N| < k$, then we define $k_r\text{th}(N)$ to be the supremum of the range of r.

A *bottom-k* ADS [3,7] is defined with respect to a single random permutation r. ADS(v) includes a node u if and only if the rank $r(u)$ is one of the k smallest ranks among nodes that are at least as close to v:

$$u \in \text{ADS}(v) \iff r(u) < k_r\text{th}(\Phi_{<u}(v)). \quad (4)$$

A *k-partition* ADS (implicit in) [2] is defined with respect to a random partition BUCKET : $V \to [k]$ of the nodes to k subsets and a random permutation r. ADS(v) includes u if and only if u has the smallest rank among nodes in its bucket that are at least as close to v.

$$u \in \text{ADS}(v) \iff r(u)$$
$$< \min \left\{ r(h) \mid \begin{matrix} \text{BUCKET}(h) = \text{BUCKET}(u) \\ \wedge\, h \in \Phi_{<u}(v) \end{matrix} \right\}.$$

A *k-mins* ADS [3, 15] is k bottom-1 ADSs, defined with respect to k independent permutations.

It is often convenient to specify the ranks $r(j)$ and (for k-partition ADSs) the bucket BUCKET(j) using random hash functions, so they are readily available from the node ID. The same randomization is used for all nodes, which results in the sketches being coordinated. This means that a node sampled in one sketch is more likely to be included in other sketches. Coordination is an artifact of scalable computation of the sketches

but also facilitates more accurate similarity and influence queries.

Relation to MINHASH Sketches
All-distances sketches are related to MINHASH sketches: ADS(v) is the union of the MINHASH sketches of the neighborhoods $N_d(v)$, for all possible values of d. We explain how a MINHASH sketch of a neighborhood $N_d(v)$ can be obtained from ADS(v). From a k-mins ADS, we obtain a k-mins MINHASH sketch of $N_d(v)$, which includes for each of the k permutations r the value $x \leftarrow \min_{u \in N_d(v)} r(u)$. Note that x is the minimum rank of a node of distance at most d in the respective bottom-1 ADS defined for r. The k minimum rank values $x^{(t)}$ $t \in [k]$ we obtain from the different permutations are the k-mins MINHASH sketch of $N_d(v)$. We now consider obtaining a bottom-k MINHASH sketch of $N_d(v)$ from a bottom-k ADS(v). The MINHASH sketch of $N_d(v)$ includes the k nodes of minimum rank in $N_d(v)$, which are also the k nodes of minimum rank in ADS(v) within distance at most d. Finally, a k-partition MINHASH sketch of $N_d(v)$ is similarly obtained from a k-partition ADS by taking, for each bucket $i \in [k]$, the smallest rank value of a node in bucket i that is in $N_d(v)$. This is also the smallest value in ADS(v) over nodes in bucket i that have distance at most d from v.

Direction
For directed graphs, influence, centrality, and closeness similarity queries can be defined with respect to either forward or reversed distances. Accordingly, we can separately consider for each node v the *forward* ADS and the *backward* ADS of each node, which are defined respectively using forward or reverse paths from v.

Node Weights
All the ADS flavors can be extended to be with respect to specified node weights $\beta(v) \geq 0$ [3,4]. This makes queries formulated with respect to the weights more efficient.

Algorithms
There are two meta-approaches for scalable computation of an ADS set. The first approach,

PRUNEDDIJKSTRA (Algorithm 1), performs pruned applications of Dijkstra's single-source shortest paths algorithm (BFS when unweighted) [3, 7]. The second approach, DP (implicit in [2, 15]), applies to unweighted graphs and is based on dynamic programming or Bellman-Ford shortest paths computation. LOCALUPDATES (Algorithm 2) [4] extends DP to weighted graphs. LOCALUPDATES is node-centric and is appropriate for MapReduce or similar platforms, but can incur more overhead than PRUNEDDIJKSTRA. The algorithms are presented for bottom-k sketches, but both approaches can be easily adopted to work with all three ADS flavors.

Algorithm 1: ADS set for G via PRUNEDDI-JKSTRA

1 **for** u *by increasing* $r(u)$ **do**
2 | Perform Dijkstra from u on G^T (the transpose graph)
3 | **foreach** *scanned node* v **do**
4 | | **if** $|\{(x, y) \in \text{ADS}(u) \mid y < d_{vu}\}| = k$ **then**
5 | | | prune Dijkstra at v
6 | | **else**
7 | | | $\text{ADS}(v) \leftarrow \text{ADS}(v) \cup \{(r(u), d_{vu})\}$

Both PRUNEDDIJKSTRA and DP can be performed in $O(km \log n)$ time (on unweighted graphs) on a single processor in main memory, where n and m are the number of nodes and edges in the graph. These algorithms maintain a partial ADS for each node, as entries of node ID and distance pairs. $\text{ADS}(v)$ is initialized with the pair $(v, 0)$. The basic operation we use is *edge relaxation*: when relaxing (v, u), $\text{ADS}(v)$ is updated using $\text{ADS}(u)$. For bottom-k, the relaxation modifies $\text{ADS}(v)$ when $\text{ADS}(u)$ contains a node i such that $r(i)$ is smaller than the kth smallest rank among nodes in $\text{ADS}(v)$ with distance at most $d_{ui} + w_{vu}$ from v. Both PRUNEDDIJKSTRA and DP perform relaxations in an order which guarantees that inserted entries are part of the final ADS, that is, there are no other nodes that are both closer and have lower rank: PRUNEDDIJKSTRA iterates over all nodes

in increasing rank, runs Dijkstra's algorithm from the node on the transpose graph, and prunes at nodes when the ADS is not updated. DP performs iterations, where in each iteration, all edges (v, u), such that $\text{ADS}(v)$ was updated in the previous step, are relaxed. Therefore, entries are inserted by increasing distance.

Algorithm 2: ADS set for G via LOCALUP-DATES

    // Initialization
1 **for** u **do**
2 | $\text{ADS}(u) \leftarrow \{(r(u), 0)\}$
    // Propagate updates (r, d) at node u
3 **if** (r, d) *is added to* $\text{ADS}(u)$ **then**
4 | **foreach** $y \mid (u, y) \in G$ **do**
5 | | send $(r, d + w(u, y))$ to y
    // Process update (r, d) **received** at u
6 **if** *node* u *receives* (r, d) **then**
7 | **if** $r < \text{k}_x^{\text{th}}\{(x, y) \in \text{ADS}(u) \mid y < d\}$ **then**
8 | | $\text{ADS}(u) \leftarrow \text{ADS}(u) \cup \{(r(v), d)\}$
    // Clean-up $\text{ADS}(u)$
9 | **for** *entries* $(x, y) \in \text{ADS}(u) \mid y > d$ *by increasing* y **do**
10 | | **if** $x > \text{k}_h^{\text{th}}\{(h, z) \in \text{ADS}(u) \mid z < y\}$ **then**
11 | | | $\text{ADS}(u) \leftarrow \text{ADS}(u) \setminus (x, y)$

Estimation

Distance Distribution

Neighborhood cardinality queries for a node v, and $d \geq 0$ can be estimated with a small relative error from $\text{ADS}(v)$. The generic estimator extracts a MINHASH sketch of the neighborhood $N_d(v)$ from $\text{ADS}(v)$ and applies a MINHASH cardinality estimator to this sketch. This approach was used in [3, 7, 15]. A nearly optimal estimator, the Historic Inverse Probability (HIP) estimator [4], has a factor 2 improvement in variance by using all information in the ADS instead of just the MINHASH sketch. HIP works by considering for each entry (u, d) in the sketch, the *HIP threshold probability*, which is the probability, under randomly drawn rank for the node u, but fixing ranks of all other nodes, that the entry is included in the sketch. The entry then obtains an adjusted weight that is the inverse of the HIP threshold probability. Neighborhood cardinality can be estimated by the sum of the adjusted weights of ADS entries that fall in the neighborhood.

Closeness Centrality (Distance-Decaying)

$C_{\alpha,\beta}(v)$ can be estimated from $\text{ADS}(v)$ with a small relative error when the set of ADSs is computed with respect to β. Estimators using MINHASH sketches were given in [6]. The tighter HIP estimator in [4] simply sums, over entries $(u, d) \in \text{ADS}(v)$, the product of the adjusted weight of the entry and $\alpha(d)\beta(u)$.

Closeness Similarity and Influence

When the sketches are computed with respect to β, the closeness similarity of two nodes u and v can be estimated from $\text{ADS}(u)$ and $\text{ADS}(v)$ within a small additive ϵ [8]. The influence of a set of nodes S can be estimated from $\{\text{ADS}(v) \mid v \in S\}$ to within a small relative error [9]. These estimators are instances of the L^* estimator [5] applied with the HIP inclusion probabilities [4].

Approximate Distance Oracles

An upper bound on the distance of two nodes u, v can be computed from $\text{ADS}(u)$ and $\text{ADS}(v)$ [8]. This is done by looking at the minimum, over nodes h that are in the intersection $\text{ADS}(u) \cap \text{ADS}(v)$, of $d_{uh} + d_{hv}$. When the graph is undirected, the oracle has worst-case quality guarantees that match the distance oracle of [16] (oracle time can be improved by looking only at nodes in the few ADS entries that correspond to $k = 1$). We note that observed quality in practice (using the full oracle) tends to have a small relative error [8].

Applications

Massive graphs, with billions of edges, are prevalent and model web graphs and social networks. Centralities, similarities, influence, and distances are basic data analysis tasks on these graphs. ADSs are a powerful tool for scalable analysis of very large graphs.

Extensions

An ADS can be viewed as a MINHASH sketch constructed from a stream, where all updates are recorded. This means that the HIP estimator [4] can be applied for distinct counting on streams, obtaining improved performance over estimators applied to the MINHASH sketch alone [11, 12].

In a graph context, $\text{ADS}(v)$ is a recording of all updates to a MINHASH sketch obtained by sweeping through nodes in increasing distance from v. More generally, we can construct ADS for other settings and apply the same estimation machinery. One example is Euclidean distances [6, 13]. Another example is constructing a *combined ADS* of multiple graphs for the application of timed influence oracles [9].

Cross-References

▶ All-Distances Sketches can be viewed as an extension of ▶ Min-Hash Sketches. They are also ▶ Coordinated Sampling.

Recommended Reading

1. Boldi P, Vigna S (2014) Axioms for centrality. Internet Math 10:222–262
2. Boldi P, Rosa M, Vigna S (2011) HyperANF: approximating the neighbourhood function of very large graphs on a budget. In: WWW, Hyderabad
3. Cohen E (1997) Size-estimation framework with applications to transitive closure and reachability. J Comput Syst Sci 55:441–453
4. Cohen E (2014) All-distances sketches, revisited: HIP estimators for massive graphs analysis. In: PODS. ACM. http://arxiv.org/abs/1306.3284
5. Cohen E (2014) Estimation for monotone sampling: competitiveness and customization. In: PODC. ACM. http://arxiv.org/abs/1212.0243, full version http://arxiv.org/abs/1212.0243
6. Cohen E, Kaplan H (2007) Spatially-decaying aggregation over a network: model and algorithms. J Comput Syst Sci 73:265–288. Full version of a SIGMOD 2004 paper
7. Cohen E, Kaplan H (2007) Summarizing data using bottom-k sketches. In: PODC, Portland. ACM

8. Cohen E, Delling D, Fuchs F, Goldberg A, Goldszmidt M, Werneck R (2013) Scalable similarity estimation in social networks: closeness, node labels, and random edge lengths. In: COSN, Boston. ACM
9. Cohen E, Delling D, Pajor T, Werneck RF (2014) Timed influence: computation and maximization. Manuscript. ArXiv 1410.6976
10. Dangalchev C (2006) Residual closeness in networks. Phisica A 365:556–564
11. Flajolet P, Martin GN (1985) Probabilistic counting algorithms for data base applications. J Comput Syst Sci 31:182–209
12. Flajolet P, Fusy E, Gandouet O, Meunier F (2007) Hyperloglog: the analysis of a near-optimal cardinality estimation algorithm. In: Analysis of algorithms (AOFA), Juan des Pins
13. Guibas LJ, Knuth DE, Sharir M (1992) Randomized incremental construction of Delaunay and Voronoi diagrams. Algorithmica 7:381–413
14. Opsahl T, Agneessens F, Skvoretz J (2010) Node centrality in weighted networks: generalizing degree and shortest paths. Soc Netw 32. http://toreopsahl.com/2010/03/20/
15. Palmer CR, Gibbons PB, Faloutsos C (2002) ANF: a fast and scalable tool for data mining in massive graphs. In: KDD, Edmonton
16. Thorup M, Zwick U (2001) Approximate distance oracles. In: Proceedings of the 33th annual ACM symposium on theory of computing, Crete, pp 183–192

Alternate Parameterizations

Neeldhara Misra
Department of Computer Science and
Automation, Indian Institute of Science,
Bangalore, India

Keywords

Above guarantee; Complexity ecology of parameters; Dual parameters; Structural parameterization

Years and Authors of Summarized Original Work

2013; Fellows, Jansen, Rosamond
2014; Lokshtanov, Narayanaswamy, Raman, Ramanujan, Saurabh
2014; Marx, Pilipczuk

Problem Definition

A parameterized problem is a language $L \subseteq \Sigma^* \times \mathbb{N}$. Such a problem is said to be *fixed-parameter tractable* if there is an algorithm that decides if $(x, k) \in L$ in time $f(k)|X|^{O(1)}$. For attacking an intractable problem within the multivariate algorithmic framework, a necessary first step is to identify some reasonable parameters. The relevance of an FPT algorithm will depend on the quality of the choice of parameters. The first objective is of a practical concern: the choice of parameter should not "cheat," that is, it should be a choice that leads to tractability in the context of instances that are relevant to real-world applications. On the other hand, the parameter should also lend a perspective that is useful to the algorithm designer, usually by providing additional structural insights, thereby making an otherwise unwieldy problem manageable. Finally, the parameter itself should be accessible, in the sense that it should either typically accompany the input or be easy to compute from the input.

For a combinatorial optimization problem, the size of the desired solution is a natural parameter. For a minimization problem, it is usually reasonable to assume that this parameter is also small in practice. For a maximization problem, the *dual* parameter, which is the difference from the best possible upper bound on the optimum, is also a natural choice. For example, consider the problem of satisfying at least k clauses of a CNF formula. Here, the standard parameter would be k, while the dual parameter would be $(m - k)$: in other words, can we satisfy *all but k* clauses in the formula? In the rest of this section, we broadly describe the other possibilities for parameters.

Structural Parameterizations

Structural parameters are a considered attempt at acknowledging that various aspects of an instance influence its complexity. A classic example is ML-type checking [11], which is an NP-complete problem but can be resolved in time $O(2^k)n^{O(1)}$, where k is the maximum nesting depth of the input program. Fortunately, nesting depths of most programs are no more than four or five; the algorithm proposed is entirely adequate for real-world instances.

Since every problem context is inherently suggestive of several possible parameters, we are only able to describe a few illustrative examples. In the context of graph problems, *width parameters* such as treewidth, cliquewidth, and rankwidth have enjoyed immense success. The notion of *treewidth* is particularly popular because of a number of real-world instances that are known to exhibit small treewidth, and on the other hand, the theoretical foundations of algorithms on graphs of bounded treewidth are extremely well established and actively developed (see, e.g., [3, Chapter 7]). An analogous notion for treewidth for directed graphs remains elusive, although several proposals with varying merits exist in the literature [7].

Special graph classes, such as interval graphs, chordal graphs, planar graphs, and so on, have been extensively studied, and most hard problems turn out to be tractable on these classes. For an arbitrary graph, one might hope that the tractability carries over if the graph is "close enough" to being, say, a chordal graph. An increasingly popular program involves considering distance-to-$\mathcal{C}$ parameterizations, where $\mathcal{C}$ is a class of graphs on which the problem of interest is easily solvable. For instance, we might let k be the size of a smallest subset of vertices whose removal makes the input graph a member of $\mathcal{C}$. Other measures of closeness, using operations like addition, removal, or contraction of edges, are also frequently considered.

In the context of satisfiability and constraint satisfaction also, the notion of distance from tractable subclasses has garnered much attention in recent times. This is formalized by the notion of *backdoors*, which are subsets of variables whose "removal" makes the formula tractable (for instance, one of the Schaefer classes). Much work has been done with backdoors as parameters for determining satisfiability, and we refer the reader to [9].

Above or Below Guarantee Parameterizations

Consider the standard parameterization of VERTEX COVER: given a graph $G = (V, E)$ on n vertices, decide whether G has a vertex cover of size at most k. The best-known algorithm for vertex cover is due to Chen et al. [1] and runs in time $O(1.2852^k + kn)$. Observe that if G has a matching of size μ, then any vertex cover also has size at least μ. In particular, if G has a perfect matching, then all vertex covers of G have $\Omega(n)$ edges, and even the FPT algorithms for the standard parameter will be obliged to spend time that is exponential in n.

Mahajan and Raman [13] consider the following alternative parameterization: does G have a vertex cover of size at most $\mu + k$? Note that the parameter k here is the size of the vertex cover *above* the matching size. Since the matching size is a guaranteed lower bound on the vertex cover size, this problem is referred to as the ABOVE GUARANTEE VERTEX COVER problem. Just as one can parameterize above guaranteed values, one can consider parameterizations below guaranteed values. A classic example is the following variant of VERTEX COVER: given a planar graph $G = (V, E)$ on n vertices and an integer parameter k, does G admit a vertex cover of size $\lfloor 3n/4 \rfloor - k$?

Key Results

One of the earliest attempts at parameterizing by the size of the vertex cover was made in [4]. Various graph layout problems were considered, where it turned out that a small vertex cover led to a very convenient structure for formulating a linear program. For many of these problems, it is not known if they are FPT parameterized by treewidth, and some are hard even on graphs of bounded treewidth (indeed, BANDWIDTH is NP-hard even on trees). This justifies the need for a stronger structural parameter, and in these examples, vertex cover turned out to be a very fruitful parameterization.

These examples led to a broader theme, namely, the "complexity ecology of parameters" program, which was proposed in [5]. The theoretical foundations of this program were further established and surveyed in [6]. An immediate concern is that of how one formalizes the structural parameterization in question. Along the lines of [6], we distinguish the following possible objectives from the formalization:

1. The complexity of verifying and then exploiting a bounded parameter value
2. The complexity of exploiting structure that is guaranteed, but not given explicitly (a "promise" problem)
3. The complexity of exploiting structure that is explicitly provided along with the input, as a "witness"

The first setting is the most general but also the most computationally restrictive, as it puts the burden of discovering the structure also on the algorithm. This definition makes the study of parameters like bandwidth and cliquewidth prohibitive, as these are hard to determine even in the parameterized framework. On the other hand, while the other two notions are increasingly relaxed, the premise of a promise or the availability of witnesses in real-world witnesses remains a concern. We point the reader to [6] for a detailed discussion of the precise formalisms and their respective merits and trade-offs.

One of the major theoretical themes with alternate parameterizations is the exercise of identifying *meta-theorems* that explain the influence of the parameter over a large class of problems, usually specified in an appropriate logic. A cornerstone result of this kind is Courcelle's theorem, establishing that a problem expressible in Monadic Second Order Logic is FPT when parameterized by treewidth and the size of the formula [2]. Several generalizations of Courcelle's theorem have since been proposed, and many of them are surveyed in [10]. More recent work also establishes a similar result in the context of kernelization and parameters like vertex cover [8].

There is a rich literature that evidences the growing consideration of alternate parameterizations for optimization problems in varied

contexts. As a concluding example, we turn our attention to [14], which serves to illustrate the scale at which it is possible to execute an exercise in understanding a question from several perspectives. Given two graphs H and G, the SUBGRAPH ISOMORPHISM problem asks if H is isomorphic to a subgraph of G. In [14], a framework is developed involving ten relevant parameters for each of H and G (such as treewidth, maximum degree, number of components, and so on). The generic question addressed in this work is if the problem admits an algorithm with running time:

$$f_1(p_1, p_2, \ldots, p_\ell) \cdot n^{f_2(p_{\ell+1}, \ldots, p_k)},$$

where each of $p_1, \ldots, p_k$ is one of the ten parameters depending only on H or G. We refer the reader to Figure 1 in [14] for a concise tabulation of the results. Notably, *all* combinations of questions (the number of which runs into the billions) are answered by a set of 28 of positive and negative results.

There are many examples of problems that are parameterized away from guaranteed bounds. We note that parameterizing vertex cover above the LP optimum has attracted considerable interest because a number of fundamental problems including Above Guarantee Vertex Cover, Odd Cycle Transversal, Split Vertex Deletion, and Almost 2-SAT reduce to this problem. Indeed, for many of these problems, the fastest algorithms at the time of this writing are obtained by reducing these problems to vertex cover parameterized above the LP optimum [12].

Open Problems

We direct the reader to the excellent survey [6] for several open problems concerning specific combinations of parameters for particular problems. In an applied context, an interesting possibility is to investigate if parameters can be learned from large samples of data.

Cross-References

▶ Kernelization, Constraint Satisfaction Problems Parameterized above Average
▶ Kernelization, Max-Cut Above Tight Bounds
▶ Kernelization, MaxLin Above Average
▶ Kernelization, Permutation CSPs Parameterized above Average
▶ LP Based Parameterized Algorithms
▶ Parameterization in Computational Social Choice
▶ Parameterized SAT
▶ Treewidth of Graphs

Recommended Reading

1. Chen J, Kanj IA, Jia W (2001) Vertex cover: further observations and further improvements. J Algorithms 41(2):280–301
2. Courcelle B (1990) The monadic second-order logic of graphs I: recognizable sets of finite graphs. Inf Comput 85:12–75
3. Cygan M, Fomin FV, Kowalik L, Lokshtanov D, Marx D, Pilipczuk M, Pilipczuk M, Saurabh S (2015) Parameterized algorithms. Springer, Cham. http://www.springer.com/us/book/9783319212746
4. Fellows MR, Lokshtanov D, Misra N, Rosamond FA, Saurabh S (2008) Graph layout problems parameterized by vertex cover. In: 19th international symposium on algorithms and computation (ISAAC). Lecture notes in computer science, vol 5369. Springer, Berlin, pp 294–305
5. Fellows MR, Lokshtanov D, Misra N, Mnich M, Rosamond F, Saurabh S (2009) The complexity ecology of parameters: an illustration using bounded max leaf number. ACM Trans Comput Syst 45:822–848
6. Fellows MR, Jansen BMP, Rosamond FA (2013) Towards fully multivariate algorithmics: parameter ecology and the deconstruction of computational complexity. Eur J Comb 34(3):541–566
7. Ganian R, Hlinený P, Kneis J, Meister D, Obdrzálek J, Rossmanith P, Sikdar S (2010) Are there any good digraph width measures? In: IPEC, vol 6478. Springer, Berlin, pp 135–146
8. Ganian R, Slivovsky F, Szeider S (2013) Meta-kernelization with structural parameters. In: 38th international symposium on mathematical foundations of computer science, MFCS 2013. Lecture notes in computer science, vol 8087. Springer, Heidelberg, pp 457–468
9. Gaspers S, Szeider S (2012) Backdoors to satisfaction. In: Bodlaender HL, Downey R, Fomin FV, Marx D (eds) The multivariate algorithmic revolution and beyond. Lecture notes in computer science, vol 7370. Springer, Berlin/Heidelberg, pp 287–317
10. Grohe M, Kreutzer S (2011) Methods for algorithmic meta theorems. In: Grohe M, Makowsky J (eds) Model theoretic methods in finite combinatorics. Contemporary mathematics, vol 558. American Mathematical Society, Providence, pp 181–206
11. Henglein F, Mairson HG (1991) The complexity of type inference for higher-order typed lambda calculi. J Funct Program 4:119–130
12. Lokshtanov D, Narayanaswamy NS, Raman V, Ramanujan MS, Saurabh S (2014) Faster parameterized algorithms using linear programming. ACM Trans Algorithms 11(2):15:1–15:31
13. Mahajan M, Raman V (1999) Parameterizing above guaranteed values: maxsat and maxcut. J Algorithms 31(2):335–354
14. Marx D, Pilipczuk M (2014) Everything you always wanted to know about the parameterized complexity of subgraph isomorphism (but were afraid to ask). In: 31st international symposium on theoretical aspects of computer science (STACS), Lyon, pp 542–553

Alternative Performance Measures in Online Algorithms

Alejandro López-Ortiz
David R. Cheriton School of Computer Science, University of Waterloo, Waterloo, ON, Canada

Keywords

Bijective analysis; Diffuse adversary; Loose competitiveness; Relative interval analysis; Relative worst-order ratio; Smoothed analysis

Years and Authors of Summarized Original Work

2000; Koutsoupias, Papadimitriou
2005; Dorrigiv, López-Ortiz

Problem Definition

While the competitive ratio [19] is the most common metric in online algorithm analysis and it has led to a vast amount of knowledge in the field, there are numerous known applications in which the competitive ratio produces unsatisfactory results. Far too often, it leads to unrealistically pessimistic measures including the failure to distinguish between algorithms that have vastly differing performance under any practical characterization in practice. Because of this there, has been extensive research in alternatives to the competitive ratio, with a renewed effort in the period from 2005 to the present date.

The competitive ratio metric can be derived from the observation that an online algorithm, in essence, computes a partial solution to a problem using incomplete information. Then, it is only natural to quantify the performance drop due to this absence of information. That is, we compare the quality of the solution obtained by the online algorithm with the one computed in the presence of full information, namely, that of the offline optimal OPT, in the *worst case*. More formally,

Definition 1 An online algorithm $\mathcal{A}$ is said to have (asymptotic) competitive ratio c if $\mathcal{A}(\sigma) \leq c \cdot \text{OPT}(\sigma) + b$ for all input sequences σ and fixed constants b and c.

The early literature considered only algorithms with constant competitive ratio, and all others are termed as algorithms with *unbounded* competitive ratio. However, it is easy to extend this definition to a $C(n)$-competitive algorithm as follows:

Definition 2 An online algorithm $\mathcal{A}$ is said to have (asymptotic) competitive ratio $C(n)$ if $\mathcal{A}(\sigma) \leq C(n) \cdot \text{OPT}(\sigma) + b$ for all σ and a fixed constant b. When $b = 0$, $C(n)$ is termed the *absolute* competitive ratio.

A natural expectation would be that the performance of OPT reflects both knowledge of the future and the inherent structure of the specific instance being solved, and hence, an online algorithm with optimal competitive ratio must handle most if not all instances in an efficient manner. Unfortunately, for most problems, the worst-case nature of the competitive ratio leads to algorithms of varying degrees of sophistication having the same equally bad competitive ratio. As a consequence the competitive ratio leads to "equiva-

lence" for online algorithms with vastly differing performance in practice.

In the next sections we discuss the main alternatives to and refinements of the competitive ratio and highlight their relative benefits and drawbacks.

Key Results

Relative Worst-Order Ratio

The relative worst-order ratio [8, 10, 11] combines some desirable properties of two earlier measures, namely, the max/max ratio [6] and the random order ratio [15]. Using this measure we can directly compare two online algorithms. Informally, for a given sequence, it considers the worst-case ordering of that sequence for each algorithm and compares their behavior as a ratio on these orderings. Then it finds among all sequences (not just reorderings) the one that maximizes the ratio above in the worst-case performance.

Let $\mathcal{A}$ and $\mathcal{B}$ be online algorithms for an online minimization problem and let $\mathcal{A}(I)$ be the cost of $\mathcal{A}$ on an input sequence $I = (i_1, i_2, \ldots, i_n)$. Denote by I_σ the sequence obtained by applying a permutation σ to I, i.e., $I_\sigma = (i_{\sigma_1}, \ldots, i_{\sigma_n})$. Define $\mathcal{A}_W(I) = \min_\sigma \mathcal{A}(I_\sigma)$.

Definition 3 ([11]) Let $S_1(c)$ and $S_2(c)$ be statements about algorithms $\mathcal{A}$ and $\mathcal{B}$ defined in the following way:

$S_1(c)$: There exists a constant b such that $\mathcal{A}_W(I) \leq c \cdot \mathcal{B}_W(I) + b$ for all I.

$S_2(c)$: There exists a constant b such that $\mathcal{A}_W(I) \geq c \cdot \mathcal{B}_W(I) - b$ for all I.

The relative worst-order ratio $WR_{\mathcal{A},\mathcal{B}}$ of an online algorithm $\mathcal{A}$ to algorithm $\mathcal{B}$ is defined if $S_1(1)$ or $S_2(1)$ holds. In this case $\mathcal{A}$ and $\mathcal{B}$ are said to be comparable. If $S_1(1)$ holds, then $WR_{\mathcal{A},\mathcal{B}} = \sup\{r \mid S_2(r)\}$, and if $S_2(r)$ holds, then $WR_{\mathcal{A},\mathcal{B}} = \inf\{r \mid S_1(r)\}$.

$WR_{\mathcal{A},\mathcal{B}}$ can be used to compare the qualities of $\mathcal{A}$ and $\mathcal{B}$. If $WR_{\mathcal{A},\mathcal{B}} = 1$, then these two algorithms have the same quality with respect to this measure. The magnitude of difference between $WR_{A,B}$ and 1 reflects the difference between the behavior of the two algorithms. For a minimization problem, A is better than B with respect to this measure if $WR_{A,B} < 1$ and vice versa. Boyar and Favrholdt showed that the relative worst-order ratio is transitive [8].

Note that we can also compare the online algorithm A to an optimal offline algorithm OPT. The *worst-order ratio* of A is defined as $WR_A = WR_{A,OPT}$. For some problems, OPT is the same for all order of requests on a given input sequence, and hence, the worst-order ratio is the same as the competitive ratio. However, for other problems such as paging the order does matter for OPT.

In [10], three online algorithms (FIRST-FIT, BEST-FIT, and WORST-FIT) for two variants of the seat reservation problem [9] are compared using the relative worst-order ratio. The relative worst-order ratio when applied to paging algorithms can be used to differentiate LRU which is strictly better than FWF with respect to the worst-order ratio, while they have the same competitive ratio [11]. Similarly, [11] proposes a new paging algorithm, retrospective LRU (RLRU), and shows that it is better than LRU under this measure while not under the competitive ratio.

Loose Competitiveness

Loose competitiveness was first proposed in [22] and later modified in [25]. It attempts to obtain a more realistic measure by observing that first, in many real online problems, we can ignore those input sequences on which the online algorithm incurs a cost less than a certain threshold and, second, many online problems have a second resource parameter (e.g., size of cache, number of servers) and the input sequences are independent of these parameters. In contrast, in competitive analysis, the adversary can select sequences tailored against those parameters. For example, for caching the worst-case input with competitive ratio k can only be constructed by the adversary if it is aware of the size k of the cache. However, in practice the competitive ratios of many online paging algorithms have

been observed to be constant [25], i.e., independent of k.

In loose competitiveness we consider an adversary that is oblivious to the parameter by requiring it to give a sequence that is bad for most values of the parameter rather than just a specific bad value of the parameter. Let $A_k(I)$ denote the cost of an algorithm A on an input sequence I, when the parameter of the problem is k.

Definition 4 ([25]) An algorithm A is (ϵ, δ)-loosely c-competitive if, for any input sequence I and for any n, at least $(1 - \delta)n$ of the values $k \in \{1, 2, \ldots, n\}$ satisfy $A_k(I) \leq \max\{c \cdot OPT_k(I), \epsilon |I|\}$.

Therefore, we ignore the input sequences I which cost less than $\epsilon |I|$. Also we require the algorithm to be good for at least $(1 - \delta)$ fraction of the possible parameters. For each online problem, we can select the appropriate constants ϵ and δ. The following result shows that by this modification of the competitive analysis, we can obtain paging algorithms with constant performance ratios.

Theorem 1 ([25]) *Every k-competitive paging algorithm is (ϵ, δ)-loosely c-competitive for any $0 < \epsilon, \delta < 1$, and $c = (e/\delta)\ln(e/\epsilon)$, where e is the base of the natural logarithm.*

Diffuse Adversary Model

The diffuse adversary model [16] tries to refine the competitive ratio by restricting the set of legal input sequences. In the diffuse adversary model, the input is generated according to a distribution belonging to a member of a class Δ of distributions.

Definition 5 Let A be an online algorithm for a minimization problem and let Δ be a class of distributions for the input sequences. Then A is c-competitive against Δ, if there exists a constant b, such that $\mathcal{E}_{I \in D} A(I) \leq c \cdot \mathcal{E}_{I \in D} OPT(I) + b$, for every distribution $D \in \Delta$, where $A(I)$ denotes the cost of A on the input sequence I and the expectations are taken over sequences that are picked according to D.

In other words, for a given algorithm A, the adversary selects the distribution D in Δ that leads to its worst-case performance in that family. If Δ is highly restrictive, then A knows more about the distribution of input sequences and the power of adversary is more constrained. When Δ contains all possible distributions, then the competitive analysis against Δ is the same as the standard competitive ratio.

Computing the actual competitive ratio of both deterministic and randomized paging algorithms against Δ_ϵ is studied in [23, 24]. An estimation of the optimal competitive ratio for several algorithms (such as LRU and FIFO) within a factor of 2 is given. Also it is observed that around the threshold $\epsilon \approx 1/k$, the best competitive ratios against Δ_ϵ are $\Theta(\ln k)$. The competitive ratios rapidly become constant for values of ϵ less than the threshold. For $\epsilon = \omega(1/k)$, i.e., values greater than the threshold, the competitive ratio rapidly tends to $\Theta(k)$ for deterministic algorithms while it remains unchanged for randomized algorithms.

Note that we can also model locality of reference using the diffuse adversary model by considering only those distributions that are consistent with distributions obeying a locality of reference principle. In particular Dorrigiv et al. showed that for the list update problem MTF is optimal in expected cost under any probability distribution that has locality of reference monotonicity, i.e., a recently accessed item has equal or larger probability of being accessed than a less recently accessed item [14].

Bijective Analysis

Bijective analysis and average analysis [3] build upon the framework of locality of reference by [1]. These models directly compare two online algorithms without appealing to the concept of the offline "optimal" cost. In addition, these measures do not evaluate the performance of the algorithm on a single "worst-case" request, but instead use the cost that the algorithm incurs on each and all request sequences. Informally, bijective analysis aims to pair input sequences for two algorithms A and B using a bijection in such a way that the cost of A on input σ is no more than the cost of B on the image of σ, for all request sequences σ of the same

length. In this case, intuitively, A is no worse than B. On the other hand, average analysis compares the average cost of the two algorithms over all request sequences of the same length. For an online algorithm A and an input sequence σ, let $A(\sigma)$ be the cost incurred by A on σ. Denote by I_n the set of all input sequences of length n.

We say that an online algorithm A is *no worse* than an online algorithm B according to bijective analysis if there exists an integer $n_0 \geq 1$ so that for each $n \geq n_0$, there is a bijection $b : I_n \leftrightarrow I_n$ satisfying $A(\sigma) \leq B(b(\sigma))$ for each $\sigma \in I_n$. We denote this by $A \preceq_b B$.

We say that an online algorithm A is *no worse* than an online algorithm B according to average analysis if there exists an integer $n_0 \geq 1$ so that for each $n \geq n_0$, $\sum_{I \in I_n} A(I) \leq \sum_{I \in I_n} B(I)$. We denote this by $A \preceq_a B$.

Under both bijective analysis and average analysis alone, all *lazy* algorithms (including LRU and FIFO, but not FWF) are in fact strongly equivalent. This is evidence of an inherent difficulty to separate these algorithms in any general unrestricted setting. Their superiority is seemingly derived from the well-known observation that input sequences for paging and several other problems show locality of reference [12,13]. This means that when a page is requested, it is more likely to be requested in the near future. Therefore, several models for paging with locality of reference have been proposed.

Hence, the need to combine bijective analysis with an assumption of locality of reference model such as concave analysis. In this model a request sequence has high locality of reference if the number of distinct pages in a window of size n is small.

Using this measure Angelopoulos et al. [3] show that LRU is never outperformed in any possible subpartition on the request sequence space induced by concave analysis, while it always outperforms *any other paging algorithm* in at least one subpartition of the sequence space. This result proves separation between LRU and all other algorithms and provides theoretical backing to the observation that LRU is preferable in practice. This is the first deterministic theoretical model to provide full separation between LRU and all other algorithms. Recently this result was strengthened by Angelopolous and Schweitzer [2] where they showed that the separation also holds under the stricter bijective analysis (as opposed to average analysis) using the concave analysis framework.

Smoothed Competitiveness

Some algorithms that have very bad *worst-case* performance behave very well in practice. One of the most famous examples is the simplex method. This algorithm has a very good performance in practice but it has exponential worst-case running time. *Average case* analysis of algorithms can somehow explain this behavior, but sometimes there is no basis to the assumption that the inputs to an algorithm are random.

Smoothed analysis of algorithms [21] tries to explain this intriguing behavior without assuming anything about the distribution of the input instances. In this model, we randomly perturb (smoothen) the input instances according to a probability distribution f and then analyze the behavior of the algorithm on these perturbed (smoothed) instances. For each input instance $\check{I}$, we compute the neighborhood $N(\check{I})$ of $\check{I}$ which contains the set of all perturbed instances that can be obtained from $\check{I}$. Then we compute the expected running time of the algorithm over all perturbed instances in this neighborhood. The smoothed complexity of the algorithm is the maximum of this expected running time over all the input instances. Intuitively, an algorithm with a bad worst-case performance can have a good smoothed performance if its worst-case instances are isolated. Spielman and Teng show [21] that the simplex algorithm has polynomial smoothed complexity. Several other results are known about the smoothed complexity of the algorithms [4, 7, 18, 20].

Becchetti et al. [5] introduced *smoothed competitive analysis* which mirrors competitive analysis except that we consider the cost of the algorithm on randomly perturbed adversarial sequences. As in the analysis of the randomized online algorithms, we can have either an *oblivious adversary* or an *adaptive adversary*. The smoothed competitive ratio of an online algorithm $\mathcal{A}$ for a minimization problem can be formally defined as follows.

Definition 6 ([5]) The *smoothed competitive ratio* of an algorithm $\mathcal{A}$ is defined as

$$c = \sup_{\check{I}} \mathcal{E}_{I \leftarrow N(\check{I})} \left[\frac{\mathcal{A}(I)}{OPT(I)} \right],$$

where the supremum is taken over all input instances $\check{I}$ and the expectation is taken over all instances I that are obtainable by smoothening the input instance $\check{I}$ according to f in the neighborhood $N(\check{I})$.

In [5], they use the smoothed competitive ratio to analyze the MULTI-LEVEL FEEDBACK(MLF) algorithm for processor scheduling in a time-sharing multitasking operating system. This algorithm has very good practical performance, but its competitive ratio is very bad and obtains strictly better ratios using the smooth competitive analysis than with the competitive ratio.

Search Ratio

The search ratio belongs to the family of measures in which the offline OPT is weakened. It is defined only for the specific case of geometric searches in an unknown terrain for a target of unknown position. Recall that the competitive ratio compares against an all-knowing OPT; indeed, for geometric searches in the competitive ratio framework, the OPT is simply a shortest path algorithm, while the online search algorithm has intricate methods for searching. The search ratio instead considers the case where OPT knows the terrain but not the position of the target. That is, the search ratio compares two search algorithms, albeit one more powerful than the other. By comparing two instances of like objects, the search ratio can be argued to be a more meaningful measure of the quality of an online search algorithm. Koutsopias et al. show that searching in trees results the same large competitive ratio regardless of the search strategy, yet under the search ratio framework, certain algorithms are far superior to others [17].

Cross-References

▶ Online Paging and Caching

Recommended Reading

1. Albers S, Favrholdt LM, Giel O (2005) On paging with locality of reference. JCSS 70(2): 145–175
2. Angelopoulos S, Schweitzer P (2009) Paging and list update under bijective analysis. In: Proceedings of the twentieth annual ACM-SIAM symposium on discrete algorithms (SODA 2009), New York, 4–6 Jan 2009, pp 1136–1145
3. Angelopoulos S, Dorrigiv R, López-Ortiz A (2007) On the separation and equivalence of paging strategies. In: Proceedings of the eighteenth annual ACM-SIAM symposium on discrete algorithms (SODA 2007), New Orleans, 7–9 Jan 2007, pp 229–237
4. Banderier C, Mehlhorn K, Beier R (2003) Smoothed analysis of three combinatorial problems. In: Proceedings of the 28th international symposium on mathematical foundations of computer science 2003 (MFCS 2003), Bratislava, 25–29 Aug 2003, vol 2747, pp 198–207
5. Becchetti L, Leonardi S, Marchetti-Spaccamela A, Schafer G, Vredeveld T (2003) Average case and smoothed competitive analysis of the multi-level feedback algorithm. In: IEEE (ed) Proceedings of the 44th symposium on foundations of computer science (FOCS 2003), Cambridge, 11–14 Oct 2003, pp 462–471
6. Ben-David S, Borodin A (1994) A new measure for the study of on-line algorithms. Algorithmica 11:73–91
7. Blum A, Dunagan J (2002) Smoothed analysis of the perceptron algorithm for linear programming. In: Proceedings of the thirteenth annual ACM-SIAM symposium on discrete algorithms, San Francisco, 6–8 Jan 2002, pp 905–914
8. Boyar J, Favrholdt LM (2003) The relative worst order ratio for on-line algorithms. In: Proceedings of the 5th Italian conference on algorithms and complexity (CIAC 2003), Rome, 28–30 May 2003,
9. Boyar J, Larsen KS (1999) The seat reservation problem. Algorithmica 25(4):403–417
10. Boyar J, Medvedev P (2004) The relative worst order ratio applied to seat reservation. In: Proceedings of the 9th Scandinavian workshop on algorithm theory (SWAT 2004), Humlebaek, 8–10 July 2004
11. Boyar J, Favrholdt LM, Larsen KS (2005) The relative worst order ratio applied to paging. In: Proceedings of the sixteenth annual ACM-SIAM symposium on discrete algorithms (SODA 2005), Vancouver, 23–25 Jan 2005. ACM, pp 718–727
12. Denning PJ (1968) The working set model for program behaviour. CACM 11(5):323–333
13. Denning PJ (1980) Working sets past and present. IEEE Trans Softw Eng SE-6(1):64–84
14. Dorrigiv R, López-Ortiz A (2012) List update with probabilistic locality of reference. Inf Process Lett 112(13):540–543
15. Kenyon C (1996) Best-fit bin-packing with random order. In: Proceedings of the seventh annual ACM-SIAM symposium on discrete algorithms, Atlanta, 28–30 Jan 1996, pp 359–364
16. Koutsoupias E, Papadimitriou C (2000) Beyond competitive analysis. SIAM J Comput 30: 300–317
17. Koutsoupias E, Papadimitriou C, Yannakakis M (1996) Searching a fixed graph. In: Proceedings of the 23rd international colloquium on automata, languages and programming (ICALP96), Paderborn, 8–12 July 1996, vol 1099, pp 280–289
18. Manthey B, Reischuk R (2005) Smoothed analysis of the height of binary search trees. Schriftenreihe der Institut für Informatik und Mathematik A-05-17, Universität zu Lübeck, Lübeck. http://www.tcs.uni-luebeck.de/pages/manthey/publica tions/SearchTrees-SIIM-A-05-17.pdf
19. Sleator DD, Tarjan RE (1985) Amortized efficiency of list update and paging rules. Commun ACM 28:202–208
20. Spielman DA, Teng SH (2003) Smoothed analysis of termination of linear programming algorithms. Math Program 97(1–2):375–404
21. Spielman DA, Teng SH (2004) Smoothed analysis of algorithms: why the simplex algorithm usually takes polynomial time. J ACM 51(3):385–463
22. Young NE (1994) The k-server dual and loose competitiveness for paging. Algorithmica 11(6):525–541
23. Young NE (1998) Bounding the diffuse adversary. In: Proceedings of the ninth annual ACM-SIAM symposium on discrete algorithms, San Francisco, 25–27 Jan 1998, pp 420–425
24. Young NE (2000) On-line paging against adversarially biased random inputs. J Algorithms 37(1):218–235
25. Young NE (2002) On-line file caching. Algorithmica 33(3):371–383

Amortized Analysis on Enumeration Algorithms

Takeaki Uno
National Institute of Informatics, Chiyoda, Tokyo, Japan

Keywords

Amortized analysis; Delay; Enumeration tree; Recursion

Years and Authors of Summarized Original Work

1998; Uno

Problem Definition

Let $\mathcal{A}$ be an enumeration algorithm. Suppose that $\mathcal{A}$ is a recursive type algorithm, i.e., composed of a subroutine that recursively calls itself several times (or none). Thus, the recursion structure of the algorithm forms a tree. We call the subroutine or the execution of the subroutine an *iteration*. We here assume that an iteration does not include the computation done in the recursive calls generated by itself. We regard a series of subroutines of different types as an iteration if they form a nested recursion. We simply write the set of all iterations of an execution of $\mathcal{A}$ by $\mathcal{X}$.

When an iteration X recursively calls an iteration Y, X is called the *parent* of Y, and Y is called a *child* of X. The *root iteration* is that with no parent. For non-root iteration X, its parent is unique and is denoted by $P(X)$. The set of the children of X is denoted by $C(X)$. The parent-child relation between iterations forms a tree structure called a *recursion tree*, or an *enumeration tree*. An iteration is called a *leaf iteration* if it has no child and an *inner iteration* otherwise.

For iteration X, an upper bound of the execution time (the number of operations) of X is denoted by $T(X)$. Here we exclude the computation for the output process from the computation time. We remind that $T(X)$ is the time for local execution time and thus does not include the computation time in the recursive calls generated by X. For example, when $T(X) = O(n^2)$, $T(X)$ is written as cn^2 for some constant c. T^* is the maximum $T(X)$ among all leaf iterations X. Here, T^* can be either constant or a polynomial of the input size. If X is an inner iteration, let $\overline{T}(X) = \sum_{Y \in C(X)} T(Y)$.

Key Results

We explain methods to amortize the computation time of iterations that only requires a local condition and give simple algorithms which achieves nontrivial time complexity. On enumeration algorithms, it is very hard to grasp the global structures of the computation and the recursion tree that is coming from the hardness of estimating the number of iterations in a branch. Instead of that, we approach from local amortization from parent and children. When we go deep in a recursion tree, the number of iterations tends to increase exponentially, and the size of the input of each iteration often decreases on the other hand. Motivated by this observation, we amortize the computation time by moving the computation time of each iteration to its children from the top to bottom, so that the long computation time on upper levels is diffused.

Amortization by Children

Suppose that each iteration X takes $O((|C(X)| + 1)T)$ time. Note that this implies that a leaf iteration takes $O(T)$ time. Then, the total computation time of the algorithm is $O(T \sum_{X \in \mathcal{X}} |C(X)| + 1) = O(T(|\mathcal{X}| + \sum_{X \in \mathcal{X}} |C(X)|)) = O(T|\mathcal{X}|)$, since any iteration is a child of at most one iteration. Hence, an iteration takes $O(T)$ time on average. Let us see an example on the following algorithm for enumerating all subsets of $\{1, \ldots, n\}$.

We can confirm that the algorithm correctly enumerates all subsets without duplications, and an iteration X takes $O(|C(X)|)$ time, except for the output process. Without amortization, the time complexity is $O(n)$ for each iteration, but the above amortization reduces it to $O(1)$. Note that the output process is shortened by outputting each subset by the difference from the previously output subset, and by this the accumulated computation time for output process is also bounded by $O(1)$ for each subset. This amortization technique is common in many algorithms. Further, in the enumeration of spanning trees, the time

---

Algorithm EnumSubset (S, x):

1 **output** S
2 **for** $i := x + 1$ **to** n; **call** EnumSubset $(S \cup \{i\}, i + 1)$

---

complexity is amortized by not only the children but also the grandchildren [3]. More sophisticated amortization is used in [1, 2] for path connecting given two vertices and subtrees of size k.

Push-Out Amortization

When the computation time of an iteration X is not proportional $|C(X)|$, the above amortization does not work. In such cases, *push-out* amortization [4–6] can work. We amortize the computation time by charging the computation time of iterations near by the root of the recursion to those in bottom levels, by recursively moving the computation time from an iteration to its children from top to down. The move is done in the following push-out rule.

Push-out rule (PO rule): Suppose that iteration X receives a computation time of $S(X)$ from its parent; thus X has computation time of $S(X) + T(X)$ in total. Then, we fix $\frac{\beta}{\alpha-1}(|C(X)|+1)T^*$ of the computation time to X and charge (push-out) the remaining computation time of quantity $S(X) + T(X) - \frac{\beta}{\alpha-1}(|C(X)| + 1)T^*$ to its children. Each child Z of X receives computation time proportional to $T(Z)$, that is, $S(Z) = (S(X) + T(X) - \frac{\beta}{\alpha-1}(|C(X)| + 1)T^*)\frac{T(Z)}{T(X)}$.

After the moves in this rule from the top to bottom of the recursion tree, each inner iteration has $O((|C(X)| + 1)T^*)$ computation time,

thus $O(T^*)$ time per iteration. Moreover, when the following push-out condition holds for any non-leaf iteration X, each leaf iteration receives computation time of $O(T^*)$ from its parent; thus the computation time per iteration is bounded by $O(T^*)$. Suppose that $\alpha > 1$ and $\beta \geq 0$ are two constants.

Push-Out Condition (PO Condition)

$\overline{T}(X) \geq \alpha T(X) - \beta(|C(X)| + 1)T^*$

Intuitively, this means that $\overline{T}(X) \geq \alpha T(X)$ holds after the assignment of the computation time of $\alpha\beta(|C(X)| + 1)T^*$ to children and the remaining to itself, the inequation. Thus, the computation time of one level of recursion intuitively increases as the depth, unless there are not so many leaf iterations. These suggest that the total computation time spent by middle-level iterations is relatively short compared to that by leaf iterations.

Theorem 1 *If any inner iteration of an enumeration algorithm satisfies PO condition, the amortized computation time of an iteration is* $O(T^*)$.

Proof We state by induction that when we charge computation time with PO rule, from the root iteration to the leaf iterations, each iteration X satisfies $S(X) \leq T(X)/(\alpha - 1)$. The root iteration satisfies this condition. Suppose that an iteration X satisfies it. Then, for any child Z of X, we have

$$S(Z) = (S(X) + T(X) - \frac{\beta}{\alpha - 1}(|C(X)| + 1)T^*)\frac{T(Z)}{\overline{T}(X)}$$

$$\leq (T(X)/(\alpha - 1) + T(X) - \frac{\beta}{\alpha - 1}(|C(X)| + 1)T^*)\frac{T(Z)}{\overline{T}(X)}$$

$$= \frac{\alpha T(X) - \beta(|C(X)| + 1)T^*}{\overline{T}(X)} \times \frac{T(Z)}{\alpha - 1}.$$

Therefore, any leaf iteration receives $O(T^*)$ time from its parent, and the statement holds.

Since PO condition is satisfied, $\overline{T}(X) \geq \alpha T(X) - \beta(|C(X)| + 1)T^*$. Thus,

$$\frac{\alpha T(X) - \beta(|C(X)| + 1)T^*}{\overline{T}(X)} \frac{T(Z)}{\alpha - 1} \leq \frac{T(Z)}{\alpha - 1}. \square$$

Matching Enumeration

Let us see an example of designing algorithms so that push-out amortization does work. The problem is the enumeration of matchings in an undirected graph $G = (V, E)$. A *matching* is an edge set $M \subseteq E$ such that any vertex is incident to at most one edge in M. A straightforward way to enumerate all matchings is to choose an edge e and enumerate matchings including e and enumerate matchings not including e, recursively. This algorithm yields the time complexity of $O(|V|)$ for each matching.

We here consider another way for the enumeration. We choose a vertex v of the maximum degree and partition the problem into enumeration of matchings including e_1, matchings including $e_2, \ldots$, matchings including e_k, and matchings including none of $e_1, \ldots, e_k$. Here $e_1, \ldots, e_k$ are the edges incident to v. Since any matching has at most one edge incident to v, this algorithm is complete and makes no duplication. The algorithm is described as follows. Note that $G \setminus \{v\}$ denotes the graph obtained by removing vertex v and edges incident to v from G.

Algorithm EnumMatching $(G = (V, E), M)$:

1 **if** $E = \emptyset$ **then output** M; **return**
2 choose a vertex v having the maximum degree in G
3 **call** EnumMatching $(G \setminus \{v\}, M)$
4 **for** each edge $e = (v, u)$, **call** EnumMatching $(G \setminus \{u, v\}, M \cup \{e\})$

$G \setminus \{u, v\}$ is obtained from $G \setminus \{u', v\}$ in $O(d(u) + d(u'))$ time, where $d(u)$ and $d(u')$ are the degrees of u and u', respectively. From this, the computation time in step 4 is bounded by the sum of degrees of all vertices adjacent to v. Here $T(X) = c|E|$ for some c, except for the output process. Note that $|E|$ is the number of edges in the graph given to X.

The input graph of the child generated in step 3 has $|E| - d(v)$ edges and that in step 4 has $|E| - d(v) - d(u) + 1$ edges. Thus, when $d(v) < |E|/4$, we have $\overline{T}(X) = c((|E| - d(v)) + (|E| - d(v) - d(u) + 1)) \geq 1.25c|E|$. When $d(v) \geq |E|/4$, $|C(X)| \geq |E|/4$. Thus, PO condition holds by setting $\alpha = 1.25$ and choosing a certain β. The output process can be shorten as the subset enumeration.

Theorem 2 *Matchings of a graph can be enumerated in $O(1)$ time for each matching.*

Elimination Ordering

An *elimination ordering* is a sequence of elements obtained by iteratively removing an element from an object G with keeping a property satisfied, until the object will be empty. Examples are perfect elimination ordering and perfect sequence. The former is the removal sequence of simplicial vertices from a chordal graph, and the latter is the removal sequence of cliques from a connected chordal graph. Elimination orderings can be enumerated by a simple algorithm as follows.

Algorithm EnumElim (G, S):

1 **if** $G = \emptyset$ **then output** S; **return**
2 **for** each element e of G that can be removed, **call** EnumElim $(G \setminus \{e\}, S \cup \{e\})$

Here we assume that $T(X) = \text{poly}(|G|)$ except for output process. The decision problem of removing an element from G is naturally considered to be solved in $O(\text{poly}(|G|))$ time; thus this assumption is natural.

Theorem 3 *If any G of size larger than some constant c has at least two removable elements, elimination orderings are enumerated in $O(1)$ time for each.*

Proof The statement means that each iteration has at least two children, if its computation time is not constant. For sufficiently large constant δ, we always have $\text{poly}(|G|) \leq 2\text{poly}(|G| - 1)$ for any $|G| > \delta$. This implies that PO condition always holds for these iterations. $\square$

Recommended Reading

1. Birmele E, Ferreira RA, Grossi R, Marino A, Pisanti N, Rizzi R, Sacomoto G (2013) Optimal listing of cycles and st-paths in undirected graphs. SODA 2013:1884–1896
2. Ferreira RA, Grossi R, Rizzi R (2011) Output-sensitive listing of bounded-size trees in undirected graphs. ESA 2011:275–286
3. Shioura A, Tamura A, Uno T (1997) An optimal algorithm for scanning all spanning trees of undirected graphs. SIAM J Comput 26:678–692

4. Uno T (1998) New approach for speeding up enumeration algorithms. LNCS 1533:287–296
5. Uno T (1999) A new approach for speeding up enumeration algorithms and its application for matroid bases. LNCS 1627:349–359
6. Uno T (2014) A new approach to efficient enumeration by push-out amortization. arXiv:1407.3857

AMS Sketch

Graham Cormode
Department of Computer Science, University of Warwick, Coventry, UK

Keywords

Euclidean norm; Second-moment estimation; Sketch; Streaming algorithms

Years and Authors of Summarized Original Work

1996; Alon, Matias, Szegedy

Problem Definition

Streaming algorithms aim to summarize a large volume of data into a compact summary, by maintaining a data structure that can be incrementally modified as updates are observed. They allow the approximation of particular quantities. The AMS sketch is focused on approximating the sum of squared entries of a vector defined by a stream of updates. This quantity is naturally related to the Euclidean norm of the vector and so has many applications in high-dimensional geometry and in data mining and machine learning settings that use vector representations of data.

The data structure maintains a linear projection of the stream (modeled as a vector) with a number of randomly chosen vectors. These random vectors are defined implicitly by simple hash functions, and so do not have to be stored explicitly. Varying the size of the sketch

changes the accuracy guarantees on the resulting estimation. The fact that the summary is a linear projection means that it can be updated flexibly, and sketches can be combined by addition or subtraction, yielding sketches corresponding to the addition and subtraction of the underlying vectors.

Key Results

The AMS sketch was first proposed by Alon, Matias, and Szegedy in 1996 [1]. Several refinements or variants have subsequently appeared in the literature, for example, in the work of Thorup and Zhang [4]. The version presented here works by using hashing to map each update to one of t counters rather than taking the average of t repetitions of an "atomic" sketch, as was originally proposed. This hash-based variation is often referred to as the "fast AMS" summary.

Data Structure Description

The AMS summary maintains an array of counts which are updated with each arriving item. It gives an estimate of the ℓ_2-norm of the vector v that is induced by the sequence of updates. The estimate is formed by computing the norm of each row and taking the median of all rows. Given parameters ε and δ, the summary uses space $O(1/\varepsilon^2 \log 1/\delta)$ and guarantees with probability of at least $1 - \delta$ that its estimate is within relative ε-error of the true ℓ_2-norm, $\|v\|_2$.

Initially, v is taken to be the zero vector. A stream of updates modifies v by specifying an index i to which an update w is applied, setting $v_i \leftarrow v_i + w$. The update weights w can be positive or negative.

The AMS summary is represented as a compact array C of $d \times t$ counters, arranged as d rows of length t. In each row j, a hash function h_j maps the input domain U uniformly to $\{1, 2, \ldots t\}$. A second hash function g_j maps elements from U uniformly onto $\{-1, +1\}$. For the analysis to hold, we require that g_j is *four-wise* independent. That is, over the random choice of g_j from the set of all possible hash functions, the probability that any four distinct items from the domain that get mapped to $\{-1, +1\}^4$ is uniform: each of the 16 possible outcomes is equally likely. This can be achieved by using polynomial hash functions of the form $g_j(x) = 2((ax^3 + bx^2 + cx + d \mod p) \mod 2) - 1$, with parameters a, b, c, d chosen uniformly from the prime field p.

The sketch is initialized by picking the hash functions to be used and initializing the array of counters to all zeros. For each update operation to index i with weight w (which can be either positive or negative), the item is mapped to an entry in each row based on the hash functions h and the update applied to the corresponding counter, multiplied by the corresponding value of g. That is, for each $1 \leq j \leq d$, $h_j(i)$ is computed, and the quantity $wg_j(i)$ is added to entry $C[j, h_j(i)]$ in the sketch array. Processing each update therefore takes time $O(d)$, since each hash function evaluation takes constant time.

The sketch allows an estimate of $\|v\|_2^2$, the squared Euclidean norm of v, to be obtained. This is found by taking the sums of the squares of the rows of the sketch and in turn finding the median of these sums. That is, for row j, it computes $\sum_{k=1}^{t} C[j, k]^2$ as an estimate and takes the median of these d estimates. The query time is linear in the size of the sketch, $O(td)$, as is the time to initialize a new sketch. Meanwhile, update operations take time $O(d)$.

The analysis of the algorithm follows by considering the produced estimate as a random variable. The random variable can be shown to be correct in expectation: its expectation is the desired quantity, $\|v\|_2^2$. This can be seen by expanding the expression of the estimator. The resulting expression has terms $\sum_i v_i^2$ but also terms of the form $v_i v_j$ for $i \neq j$. However, these "unwanted terms" are multiplied by either $+1$ or -1 with equal probability, depending on the choice of the hash function g. Therefore, their expectation is zero, leaving only $\|v\|_2$. To show that it is likely to fall close to its expectation, we also analyze the variance of the estimator and use Chebyshev's inequality to argue that with constant probability, each estimate is close to the desired value. Then, taking the median of sufficient repetitions

amplifies this constant probability to be close to certainty.

This analysis shows that the estimate is between $(1 - \varepsilon)\|v\|_2^2$ and $(1 + \varepsilon)\|v\|_2^2$. Taking the square root of the estimate gives a result that is between $(1 - \varepsilon)^{1/2}\|v\|_2$ and $(1 + \varepsilon)^{1/2}\|v\|_2$, which means it is between $(1 - \varepsilon/2)\|v\|_2$ and $(1 + \varepsilon/2)\|v\|_2$.

Note that since the updates to the AMS sketch can be positive or negative, it can be used to measure the Euclidean distance between two vectors v and u: we can build an AMS sketch of v and one of $-u$ and merge them together by adding the sketches. Note also that a sketch of $-u$ can be obtained from a sketch of u by negating all the counter values.

Applications

The sketch can also be applied to estimate the inner product between a pair of vectors. A similar analysis shows that the inner product of corresponding rows of two sketches (formed with the same parameters and using the same hash functions) is an unbiased estimator for the inner product of the vectors. This use of the summary to estimate the inner product of vectors was described in a follow-up work by Alon, Matias, Gibbons, and Szegedy [2], and the analysis was similarly generalized to the fast version by Cormode and Garofalakis [3]. The ability to capture norms and inner products in Euclidean space means that these sketches have found many applications in settings where there are high-dimensional vectors, such as machine learning and data mining.

URLs to Code and Data Sets

Sample implementations are widely available in a variety of languages.

C code is given by the MassDAL code bank: http://www.cs.rutgers.edu/~muthu/massdalcode-index.html.

C++ code given by Marios Hadjieleftheriou is available at http://hadjieleftheriou.com/sketches/index.html.

Cross-References

▶ Count-Min Sketch

Recommended Reading

1. Alon N, Matias Y, Szegedy M (1996) The space complexity of approximating the frequency moments. In: ACM symposium on theory of computing, Philadelphia, pp 20–29
2. Alon N, Gibbons P, Matias Y, Szegedy M (1999) Tracking join and self-join sizes in limited storage. In: ACM principles of database systems, New York, pp 10–20
3. Cormode G, Garofalakis M (2005) Sketching streams through the net: distributed approximate query tracking. In: International conference on very large data bases, Trondheim
4. Thorup M, Zhang Y (2004) Tabulation based 4-universal hashing with applications to second moment estimation. In: ACM-SIAM symposium on discrete algorithms, New Orleans

Analyzing Cache Behaviour in Multicore Architectures

Alejandro López-Ortiz[1] and Alejandro Salinger[2]
[1]David R. Cheriton School of Computer Science, University of Waterloo, Waterloo, ON, Canada
[2]Department of Computer Science, Saarland University, Saarbücken, Germany

Keywords

Cache; Chip multiprocessor; Multicore; Online algorithms; Paging

Years and Authors of Summarized Original Work

2010; Hassidim
2012; López-Ortiz, Salinger

Problem Definition

Multicore processors are commonly equipped with one or more levels of cache memory, some of which are shared among two or more cores. Multiple cores compete for the use of shared caches for fast access to their program's data, with the cache usage patterns of a program running on one core, possibly affecting the cache performance of programs running on other cores.

Paging

The management of data across the various levels of the memory hierarchy of modern computers is abstracted by the paging problem. Paging models a two-level memory system with a small and fast memory – known as cache – and a large and slow memory. Data is transferred between the two levels of memory in units known as pages. The input to the problem is a sequence of page requests that must be made available in cache as they are requested. If the currently requested page is already present in the cache, then this is known as a *hit*. Otherwise a *fault* occurs, and the requested page must be brought from slow memory to cache, possibly requiring the eviction of a page currently residing in the cache. An algorithm for this problem must decide, upon each request that results in a fault with a full cache, which page to evict in order to minimize the number of faults. Since the decision of which page to evict must be taken without information of future requests, paging is an online problem.

The most popular framework to analyze the performance of online algorithms is competitive analysis [10]: an algorithm A for a minimization problem is said to be c-competitive if its cost is at most c times that of an optimal algorithm that knows the input in advance. Formally, let $A(r)$ and $\mathrm{OPT}(r)$ denote the costs of A and the optimal algorithm OPT on an input r. Then A is c-competitive if for all inputs r, $A(r) \leq c \cdot \mathrm{OPT}(r) + \beta$, where β is a constant that does not depend on r. The infimum of all such values c is known as A's *competitive ratio*.

Traditional paging algorithms, like least recently used (LRU), evict the page currently in cache that was least recently accessed, or first-in-first-out (FIFO), evict the page currently in the cache that was brought in the earliest, have an optimal competitive ratio equal to the cache size. Other optimal eviction policies are flush-when-full (FWF) and Clock (see [2] for definitions).

Paging in Multicore Caches

The paging problem described above can be extended to model several programs running simultaneously with a shared cache. For a multicore system with p cores sharing one cache, the multicore paging problem consists of a set r of p request sequences $r_1, \ldots r_p$ to be served with one shared cache of size k pages. At any timestep, at most p requests from different processors can arrive and must be served in parallel. A paging algorithm must decide which pages to evict when a fault occurs on a full cache.

The general model we consider for this problem was proposed by Hassidim [6]. This model defines the fetching time τ of a page as the ratio between a cache miss and a cache hit. A sequence of requests that suffers a page fault must wait τ timesteps for the page to be fetched into the cache, while other sequences that incur hits can continue to be served. In addition, paging algorithms can decide on the schedule of request sequences, choosing to serve a subset of the sequences and delay others. In this problem, the goal of a paging algorithm is to minimize the makespan. López-Ortiz and Salinger [8] proposed a slightly different model in which paging algorithms are not allowed to make scheduling decisions and must serve requests as they arrive. Furthermore, instead of minimizing the makespan, they propose two different goals: minimize the number of faults and decide if each of the sequences can be served with a number of faults below a given threshold. We consider both these settings here and the following problems:

Definition 1 (Min-Makespan) Given a set r of request sequences $r_1, \ldots, r_p$ to be served with a cache of size k, minimize the timestep at which the last request among all sequences is served.

Definition 2 (Min-Faults) Given a set r of requests $r_1, \ldots, r_p$ to be served with a cache of size k, minimize the total number of faults when serving r.

Definition 3 (Partial-Individual-Faults) Given a set r of requests $r_1, \ldots, r_p$ to be served with a cache of size k, a timestep t, and a bound b_i for each sequence, decide whether r can be served such that at time t the number of faults on r_i is at most b_i for all $1 \leq i \leq p$.

Key Results

Online Paging

For both the models of Hassidim and López-Ortiz and Salinger, no online algorithm has been shown to be competitive, while traditional algorithms that are competitive in the classic paging setting are not competitive in the multicore setting. Hassidim shows that LRU and FIFO have a competitive ratio in the Min-Makespan problem of $\Theta(\tau)$, which is the worst possible for any online algorithm in this problem.

In the following, k is the size of the shared cache of an online algorithm, and h is the size of the shared cache of the optimal offline.

Theorem 1 ([6]) *For any $\alpha > 1$, the competitive ratio of LRU (or FIFO) is $\Theta(\tau/\alpha)$, when $h = k/\alpha$. In particular, if we give LRU a constant factor resource augmentation, the ratio is $\Theta(\tau)$. There is a setting with this ratio with just $\lceil \alpha \rceil + 1$ cores.*

The bad competitive ratio stems from the ability of the offline algorithm to schedule sequences one after the other one so that each sequence can use the entire cache. Meanwhile, LRU or FIFO will try to serve all sequences simultaneously, not having enough cache to satisfy the demands of any sequence. A similar result is shown in [8] for the Min-Faults problem, even in the case in which the optimal offline cannot explicitly schedule the input sequences. In this case, given a set of request sequences that alternate periods of high and low cache demand, the optimal offline algorithm can delay some sequences through faults in order to align periods of high demands of some sequences with periods of low demands of others and with a total cache demand below capacity. As in the previous lower bound, traditional online algorithms will strive to serve all sequences simultaneously, incurring only faults in periods of high demand.

Theorem 2 ([8]) *Let A be any of LRU, FIFO, Clock, or FWF, let $p \geq 4$, let n be the total length of request sequences, and assume $\tau > 1$. The competitive ratio of A is at least $\Omega(\sqrt{n\tau/k})$ when the optimal offline's cache is $h \geq k/2 + 3p/2$. If A has no resource augmentation, the competitive ratio is at least $\Omega(\sqrt{n\tau p/k})$.*

These results give light about the characteristics required by online policies to achieve better competitive ratios. López-Ortiz and Salinger analyzed paging algorithms for Min-Faults , separating the cache partition and the eviction policy aspects. They defined partitioned strategies as those that give a portion of the cache to each core and serve the request sequences with a given eviction policy exclusively with the given part of the cache. The partition can be static or dynamic. They also define shared strategies as those in which all requests are served with one eviction policy using a global cache. The policies considered in Theorems 1 and 2 above are examples of shared strategies.

If a cache partition is determined externally by a scheduler or operating system, then traditional eviction policies can achieve a good performance when compared to the optimal eviction policy with the same partition. More formally,

Theorem 3 *Let A be any marking or conservative paging algorithm and B be any dynamically conservative algorithm [9] (these classes include LRU, FIFO, and Clock). Let S and D be any static and dynamic partition functions and let OPT_s and OPT_d denote the optimal eviction policies given S and D, respectively. Then, for all inputs r, $A(r) \leq k \cdot OPT_s(r)$ and $B(r) \leq pk \cdot OPT_d(r)$.*

The result above relies on a result by Peserico [9] which states that dynamically conservative policies are k-competitive when the size of the cache varies throughout the execution of the cache instance.

When considering a strategy as a partition plus eviction policy, it should not be a surprise that a strategy involving a static partition cannot be competitive. In fact, even a dynamic partition that does not change the sizes of the parts assigned to its cores often enough cannot be competitive. There are sequences for which the optimal static partition with the optimal paging policy in each part can incur a number of faults that is arbitrarily large compared to an online shared strategy using LRU. A similar result applies to dynamic partitions that change a sublinear number of times. These results suggest that in order to be competitive, an online strategy needs to be either shared or partitioned with a partition that changes often.

Offline Paging

We now consider the offline multicore paging problem. Hassidim shows that Min-Makespan is NP-hard for $k = p/3$ and [7] extends it to arbitrary k and p. In the model without scheduling of [8], Partial-Individual-Faults, a variant of the fault minimization problem, is also shown to be NP-hard. It is not known, however, whether Min-Faults is NP-hard as well. Interestingly, Partial-Individual-Faults remains NP-hard when $\tau = 1$ (and hence a fault does not delay the affected sequence with respect to other sequences). In contrast, in this case, Min-Faults can be solved simply by evicting the page that will be requested furthest in the future, as in classic paging. On the positive side, the following property holds for both Min-Makespan and Min-Faults (on disjoint sequences).

Theorem 4 *There exist optimal algorithms for Min-Makespan and Min-Faults that, upon a fault, evict the page that is the furthest in the future for some sequence.*

This result implies that multicore paging reduces to determining the optimal dynamic partition of the cache: upon a fault, the part of the cache of one sequence is reduced (unless this sequence is the same as the one which incurred the fault), and the page whose next request is furthest is the future in this sequence should be evicted.

Finally, in the special case of a constant number of processors p and constant delay τ, Min-Makespan admits a polynomial time approximation scheme (PTAS), while Min-Faults and Partial-Individual-Faults admit exact polynomial time algorithms.

Theorem 5 ([6]) *There exists an algorithm that, given an instance of Min-Makespan with optimal makespan m, returns a solution with makespan $(1 + \epsilon)m$. The running time is exponential on p, τ, and $1/\epsilon$.*

Theorem 6 ([8]) *An instance of Min-Faults with p requests of total length n, with $p = O(1)$ and $\tau = O(1)$ can be solved in time $O(n^{k+p}\tau^p)$.*

Theorem 7 ([8]) *An instance of Partial-Individual-Faults with p requests of total length n, with $p = O(1)$ and $\tau = O(1)$, can be solved in time $O(n^{k+2p+1}\tau^{p+1})$.*

Other Models

Paging with multiple sequences with a shared cache has also been studied in other models [1,3–5], even prior to multicores. In these models, request sequences may be interleaved; however, only one request is served at a time and all sequences must wait upon a fault affecting one sequence.

In the application-controlled model of Cao et al. [3], each process has full knowledge of its request sequence, while the offline algorithm also knows the interleaving of requests. As opposed to the models in [6, 8], the interleaving is fixed and does not depend on the decisions of algorithms. It has been shown that for p sequences and a cache of size k, no online deterministic algorithm can have a competitive ratio better than $p + 1$ in the case where sequences are disjoint [1] and $\frac{p}{2} \log\left(\frac{4(k+1)}{3p}\right)$ otherwise [7]. On the other hand, there exist algorithms with competitive ratios $2(p + 1)$ [1, 3] and $\max\{10, p + 1\}$ [7] for the disjoint case, and $2p(\ln(ek/p) + 1)$ [7] for the shared case.

Open Problems

Open problems in multicore paging are finding competitive online algorithms, determining the exact complexity of Min-Faults, obtaining approximation algorithms for Min-Makespan for a wider range of parameters, and obtaining faster exact offline algorithms for Min-Faults and Partial-Individual-Faults. Another challenge in multicore paging is concerned with modeling the right features of the multicore architecture while enabling the development of meaningful algorithms. Factors to consider are cache coherence, limited parallelism in other shared resources (such as bus bandwidth), different cache associativities, and others.

Cross-References

▶ Online Paging and Caching

Recommended Reading

1. Barve RD, Grove EF, Vitter JS (2000) Application-controlled paging for a shared cache. SIAM J Comput 29:1290–1303
2. Borodin A, El-Yaniv R (1998) Online computation and competitive analysis. Cambridge University Press, New York
3. Cao P, Felten EW, Li K (1994) Application-controlled file caching policies. In: Proceedings of the USENIX summer 1994 technical conference – volume 1 (USTC'94), Boston, pp 171–182
4. Feuerstein E, Strejilevich de Loma A (2002) On-line multi-threaded paging. Algorithmica 32(1):36–60
5. Fiat A, Karlin AR (1995) Randomized and multi-pointer paging with locality of reference. In: Proceedings of the 27th annual ACM symposium on theory of computing (STOC'95), Las Vegas. ACM, pp 626–634
6. Hassidim A (2010) Cache replacement policies for multicore processors. In: Yao ACC (ed) Innovations in computer science (ICS 2010), Tsinghua University, Beijing, 2010, pp 501–509
7. Katti AK, Ramachandran V (2012) Competitive cache replacement strategies for shared cache environments. In: International parallel and distributed processing symposium (IPDPS'12), Shanghai, pp 215–226
8. López-Ortiz A, Salinger A (2012) Paging for multicore shared caches. In: Proceedings of the 3rd innovations in theoretical computer science conference (ITCS'12), Cambridge. ACM, pp 113–127

9. Peserico E (2013) Elastic paging. In: SIGMETRICS, Pittsburgh. ACM, pp 349–350
10. Sleator DD, Tarjan RE (1985) Amortized efficiency of list update and paging rules. Commun ACM 28(2):202–208

Analyzing Cache Misses

Naila Rahman
University of Hertfordshire, Hertfordshire, UK

Keywords

Cache analysis

Years and Authors of Summarized Original Work

2003; Mehlhorn, Sanders

Problem Definition

The problem considered here is *multiple sequence access via cache memory*. Consider the following pattern of memory accesses. k sequences of data, which are stored in disjoint arrays and have a total length of N, are accessed as follows:

for $t := 1$ to N do

    select a sequence $s_i \in \{1, \ldots k\}$

    work on the current element of sequence s_i

    advance sequence s_i to the next element.

The aim is to obtain exact (not just asymptotic) closed form upper and lower bounds for this problem. Concurrent accesses to multiple sequences of data are ubiquitous in algorithms. Some examples of algorithms which use this paradigm are distribution sorting, k-way merging, priority queues, permuting, and FFT. This entry summarizes the analyses of this problem in [5, 8].

Caches, Models, and Cache Analysis

Modern computers have hierarchical memory which consists of registers, one or more levels of caches, main memory, and external memory devices such as disks and tapes. Memory size increases, but the speed decreases with distance from the CPU. Hierarchical memory is designed to improve the performance of algorithms by exploiting temporal and spatial locality in data accesses.

Caches are modeled as follows. A cache has m *blocks* each of which holds B data elements. The capacity of the cache is $M = mB$. Data is transferred between one level of cache and the next larger and slower memory in blocks of B elements. A cache is organized as $s = m/a$ *sets* where each set consists of a blocks. Memory at address xB, referred to as memory block x, can only be placed in a block in set $x \bmod s$. If $a = 1$, the cache is said to be *direct mapped*, and if $a = s$, it is said to be *fully associative*.

If memory block x is accessed and it is not in cache, then a *cache miss* occurs, and the data in memory block x is brought into cache, incurring a performance penalty. In order to accommodate block x, it is assumed that the least recently used (LRU) or the first used (FIFO) block from the cache set $x \bmod s$ is evicted, and this is referred to as the *replacement strategy*. Note that a block may be evicted from a set, even though there may be unoccupied blocks in other sets.

Cache analysis is performed for the number of cache misses for a problem with N data elements. To read or write N data elements, an algorithm must incur $\Omega(N/B)$ cache misses. These are the *compulsory* or *first reference misses*. In the multiple sequence access via cache memory problem, for given values of M and B, one aim is to find the largest k such that there are $O(N/B)$ cache misses for the N data accesses. It is interesting to analyze cache misses for the important case of direct mapped cache and for the general case of set-associative caches.

A large number of algorithms have been designed on the external memory model [11], and these algorithms optimize the number of data transfers between main memory and disk. It seems natural to exploit these algorithms to minimize cache misses, but due to the limited associativity of caches, this is not straightforward. In the external memory model, data transfers are under programmer control, and the multiple sequence access problem has a trivial solution. The algorithm simply chooses $k \leq M_e/B_e$, where B_e is the block size and M_e is the capacity of the main memory in the external memory model. For $k \leq M_e/B_e$, there are $O(N/B_e)$ accesses to external memory. Since caches are hardware controlled, the problem becomes nontrivial. For example, consider the case where the starting addresses of $k > a$ equal length sequences map to the ith element of the same set, and the sequences are accessed in a round-robin fashion. On a cache with an LRU or FIFO replacement strategy, all sequence accesses will result in a cache miss. Such pathological cases can be overcome by randomizing the starting addresses of the sequences.

Related Problems

A very closely related problem is where accesses to the sequences are interleaved with accesses to a small working array. This occurs in applications such as distribution sorting or matrix multiplication.

Caches can emulate external memory with an optimal replacement policy [3, 10]; however, this requires some constant factor more memory. Since the emulation techniques are software controlled and require modification to the algorithm, rather than selection of parameters, they work well for fairly simple algorithms [6].

Key Results

Theorem 1 ([5]) *Given an a-way set-associative cache with m cache blocks, $s = m/a$ cache sets, cache blocks size B, and LRU or FIFO replacement strategy. Let U_a denote the expected number of cache misses in any schedule of N sequential accesses to k sequences with starting addresses that are at least $(a + 1)$-wise independent:*

$$U_1 \leq k + \frac{N}{B}\left(1 + (B - 1)\frac{k}{m}\right), \tag{1}$$

$$U_1 \geq \frac{N}{B}\left(1 + (B - 1)\frac{k - 1}{m + k - 1}\right), \tag{2}$$

$$U_a \leq k + \frac{N}{B}\left(1 + (B - 1)\left(\frac{k\alpha}{m}\right)^a + \frac{1}{m/(k\alpha) - 1} + \frac{k - 1}{s - 1}\right) \tag{3}$$

$$\text{for } k \leq \frac{m}{\alpha},$$

$$U_a \leq k + \frac{N}{B}\left(1 + (B - 1)\left(\frac{k\beta}{m}\right)^a + \frac{1}{m/(k\beta) - 1}\right) \tag{4}$$

$$\text{for } k \leq \frac{m}{2\alpha},$$

$$U_a \geq \frac{N}{B}\left(1 + (B - 1)P_{\text{tail}}\left(k - 1, \frac{1}{s}, a\right)\right) - kM, \tag{5}$$

$$U_a \geq \frac{N}{B}\left(1 + (B - 1)\left(\frac{(k - a)\alpha}{m}\right)^a\left(1 - \frac{1}{s}\right)^k\right) - kM, \tag{6}$$

where $\alpha = \alpha(a) = a/(a!)^{1/a}$, $P_{\text{tail}}(n, p, a) = \sum_{i \geq a} \binom{n}{i} p^i (1-p)^{n-i}$ is the cumulative binomial probability, and $\beta := 1 + \alpha(\lceil ax \rceil)$ where $x = x(a) = \inf\{0 < z < 1 : z + z/\alpha(\lceil az \rceil) = 1\}$.

Here $1 \leq \alpha < e$ and $\beta(1) = 2$, $\beta(\infty) = 1 + e \approx 3.71$. This analysis assumes that an adversary schedules the accesses to the sequences. For the lower bound, the adversary initially advances sequence s_i for $i = 1 \ldots k$ by X_i elements, where the X_i is chosen uniformly and independently from $\{0, M - 1\}$. The adversary then accesses the sequences in a round-robin manner.

The k in the upper bound accounts for a possible extra block that may be accessed due to randomization of the starting addresses. The $-kM$ term in the lower bound accounts for the fact that cache misses cannot be counted when the adversary initially winds forwards the sequences.

The bounds are of the form $pN + c$, where c does not depend on N and p is called the *cache miss probability*. Letting $r = k/m$, the ratio between the number of sequences and the number of cache blocks, the bounds for the cache miss probabilities in Theorem 1 become [5]

$$p_1 \leq (1/B)(1 + (B - 1)r), \tag{7}$$

$$p_1 \geq (1/B)\left(1 + (B - 1)\frac{r}{1 + r}\right), \tag{8}$$

$$p_a \leq (1/B)(1 + (B - 1)(r\alpha)^a + r\alpha + ar)$$
$$\text{for } r \leq \frac{1}{\alpha}, \tag{9}$$

$$p_a \leq (1/B)(1 + B - 1)(r\beta)^a + r\beta \text{ for } r \leq \frac{1}{2\beta} \tag{10}$$

$$p_a(1/B)\left(1 + (B - 1)(r\alpha)^a \left(1 - \frac{1}{s}\right)^k\right). \tag{11}$$

The $1/B$ term accounts for the compulsory or first reference miss, which must be incurred in

order to read a block of data from a sequence. The remaining terms account for *conflict misses*, which occur when a block of data is evicted from cache before all its elements have been scanned. Conflict misses can be reduced by restricting the number of sequences. As r approaches zero, the cache miss probabilities approach $1/B$. In general, inequality (4) states that the number of cache misses is $O(N/B)$ if $r \leq 1/(2\beta)$ and $(B - 1)(r\beta)^a = O(1)$. Both of these conditions are satisfied if $k \leq m/\max(B^{1/a}, 2\beta)$. So, there are $O(N/B)$ cache misses provided $k = O(m/B^{1/a})$.

The analysis shows that for a direct-mapped cache, where $a = 1$, the upper bound is a factor of $r + 1$ above the lower bound. For $a \geq 2$, the upper bounds and lower bounds are close if $(1 - 1/s)^k \approx$ and $(\alpha + \alpha)r \ll 1$, and both these conditions are satisfied if $k \ll s$.

Rahman and Raman [8] obtain closer upper and lower bounds for average case cache misses assuming the sequences are accessed uniformly randomly on a direct-mapped cache. Sen and Chatterjee [10] also obtain upper and lower bounds assuming the sequences are randomly accessed. Ladner, Fix, and LaMarca have analyzed the problem on direct-mapped caches on the independent reference model [4].

Multiple Sequence Access with Additional Working Set

As stated earlier in many applications, accesses to sequences are interleaved with accesses to an additional data structure, a *working set*, which determines how a sequence element is to be treated. Assuming that the working set has size at most sB and is stored in contiguous memory locations, the following is an upper bound on the number of cache misses:

Theorem 2 ([5]) *Let U_a denote the bound on the number of cache misses in Theorem 1 and define $U_0 = N$. With the working set occupying w conflict-free memory blocks, the expected number of cache misses arising in the N accesses to the sequence data, and any number of accesses to the working set, is bounded by $w + (1 - w/s)U_a + 2(w/s)U_{a-1}$.*

On a direct-mapped cache, for $i = 1, \ldots, k$, if sequence i is accessed with probability p_i independently of all previous accesses and is followed by an access to element i of the working set, then the following are upper and lower bounds for the number of cache misses:

Theorem 3 ([8]) *In a direct-mapped cache with m cache blocks, each of B elements, if sequence i, for $i = , \ldots, k$, is accessed with probability p_i and block j of the working set, for $j = , \ldots, k/B$, is accessed with probability P_j, then the expected number of cache misses in N sequence accesses is at most $N(p_s + p_w) + k(1 + 1/B)$, where*

$$p_s \le \frac{1}{B} + \frac{k}{mB} + \frac{B-1}{mB}$$

$$\sum_{i=1}^{k} \left(\sum_{j=1}^{k/B} \frac{p_i P_j}{p_i + P_j} + \frac{B-1}{B} \sum_{j=1}^{k} \frac{p_i p_j}{p_i + p_j} \right),$$

$$p_w \le \frac{k}{B^2 m} + \frac{B-1}{mB} \sum_{i=1}^{k/B} \sum_{j=1}^{k} \frac{P_i p_j}{P_i p_j}.$$

Theorem 4 ([8]) *In a direct-mapped cache with m cache blocks each of B elements, if sequence i, for $i = 1, \ldots, k$, is accessed with probability $p_i \ge 1/m$, then the expected number of cache misses in N sequence accesses is at least $Np_s + k$, where*

$$p_s \ge \frac{1}{B} + \frac{k(2m-k)}{2m^2} + \frac{k(k-3m)}{2Bm^2} - \frac{1}{2Bm} - \frac{k}{2B^2 m} + \frac{B(k-m) + 2m - 3k}{Bm^2} \sum_{i=1}^{k} \sum_{j=1}^{k} \frac{(p_i)^2}{p_i + p_j}$$

$$+ \frac{(B-1)^2}{B^3 m^2} \sum_{i=1}^{k} p_i \left[\sum_{j=1}^{k} \frac{p_i(1 - p_i - p_j)}{(p_i + p_j)^2} - \frac{B-1}{2} \sum_{j=1}^{k} \sum_{l=1}^{k} \frac{p_i}{p_i + p_j + p_l - p_j p_l} \right] - O(e^{-B}).$$

The lower bound ignores the interaction with the working set, since this can only increase the number of cache misses.

In Theorems 3 and 4, p_s is the probability of a cache miss for a sequence access, and in Theorem 3, p_w is the probability of a cache miss for an accesses to the working set.

If the sequences are accessed uniformly randomly, then using Theorems 3 and 4, the ratio between the upper and lower bound is $3/(3 - r)$, where $r = k/m$. So for uniformly random data, the lower bound is within a factor of about $3/2$ of the upper bound when $k \le m$ and is much closer when $k \ll m$.

Applications

Numerous algorithms have been developed on the external memory model which access multiple sequences of data, such as merge sort, distribution sort, priority queues, and radix sorting. These analyses are important as they allow initial parameter choices to be made for cache memory algorithms.

Open Problems

The analyses assume that the starting addresses of the sequences are randomized, and current approaches to allocating random starting addresses waste a lot of virtual address space [5]. An open problem is to find a good online scheme to randomize the starting addresses of arbitrary length sequences.

Experimental Results

The cache model is a powerful abstraction of real caches; however, modern computer architectures have complex internal memory hierarchies, with registers, multiple levels of caches, and *translation lookaside buffers* (TLB). Cache miss penalties are not of the same magnitude as the cost of disk accesses, so an algorithm may perform better by allowing conflict misses to increase in order to reduce computation costs and compulsory misses, by reducing the number of passes over the data. This means that in practice, cache analysis is used to choose an initial value of k which is then fine-tuned for the platform and algorithm $[1, 2, 6, 7, 9, 12, 13]$.

For distribution sorting, in [6], a heuristic was considered for selecting k, and equations for approximate cache misses were obtained. These equations were shown to be very accurate in practice.

Cross-References

► Cache-Oblivious Model
► Cache-Oblivious Sorting
► External Sorting and Permuting
► I/O-Model

Recommended Reading

1. Bertasi P, Bressan M, Peserico E (2011) Psort, yet another fast stable sorting software. ACM J Exp Algorithmics 16:Article 2.4
2. Bingmann T, Sanders P (2013) Parallel string sample sort. In: Proceedings of the 21st European symposium on algorithms (ESA'13), Sophia Antipolis. Springer, pp 169–180
3. Frigo M, Leiserson CE, Prokop H, Ramachandran S (1999) Cache-oblivious algorithms. In: Proceedings of the 40th annual symposium on foundations of computer science (FOCS'99), New York. IEEE Computer Society, Washington, DC, pp 285–297
4. Ladner RE, Fix JD, LaMarca A (1999) Cache performance analysis of traversals and random accesses. In: Proceedings of the 10th annual ACM-SIAM symposium on discrete algorithms (SODA'99), Baltimore. Society for Industrial and Applied Mathematics, Philadelphia, pp 613–622
5. Mehlhorn K, Sanders P (2003) Scanning multiple sequences via cache memory. Algorithmica 35:75–93
6. Rahman N, Raman R (2000) Analysing cache effects in distribution sorting. ACM J Exp Algorithmics 5:Article 14
7. Rahman N, Raman R (2001) Adapting radix sort to the memory hierarchy. ACM J Exp Algorithmics 6:Article 7
8. Rahman N, Raman R (2007) Cache analysis of non-uniform distribution sorting algorithms. http://www.citebase.org/abstract?id=oai:arXiv.org:0706.2839. Accessed 13 Aug 2007 Preliminary version in: Proceedings of the 8th annual European symposium on algorithms (ESA'00), Saarbrücken. Lecture notes in computer science, vol 1879. Springer, Berlin/Heidelberg, pp 380–391 (2000)
9. Sanders P (2000) Fast priority queues for cached memory. ACM J Exp Algorithmics 5:Article 7
10. Sen S, Chatterjee S (2000) Towards a theory of cache-efficient algorithms. In: Proceedings of the 11th annual ACM-SIAM symposium on discrete algorithms (SODA'00), San Francisco. Society for Industrial and Applied Mathematics, pp 829–838
11. Vitter JS (2001) External memory algorithms and data structures: dealing with massive data. ACM Comput Surv **33**, 209–271
12. Wassenberg J, Sanders P (2011) Engineering a Multicore Radix Sort, In: Proceedings of the 17th international conference, Euro-Par (2) 2011, Bordeaux. Springer, pp 160–169
13. Wickremesinghe R, Arge L, Chase JS, Vitter JS (2002) Efficient sorting using registers and caches. ACM J Exp Algorithmics 7:9

Applications of Geometric Spanner Networks

Joachim Gudmundsson[1,2], Giri Narasimhan[3,4], and Michiel Smid[5]
[1]DMiST, National ICT Australia Ltd, Alexandria, Australia
[2]School of Information Technologies, University of Sydney, Sydney, NSW, Australia
[3]Department of Computer Science, Florida International University, Miami, FL, USA
[4]School of Computing and Information Sciences, Florida International University, Miami, FL, USA
[5]School of Computer Science, Carleton University, Ottawa, ON, Canada

Keywords

Approximation algorithms; Cluster graphs; Dilation; Distance oracles; Shortest paths; Spanners

Years and Authors of Summarized Original Work

2002; Gudmundsson, Levcopoulos, Narasimhan, Smid
2005; Gudmundsson, Narasimhan, Smid
2008; Gudmundsson, Levcopoulos, Narasimhan, Smid

Problem Definition

Given a geometric graph in d-dimensional space, it is useful to preprocess it so that distance queries, exact or approximate, can be answered efficiently. Algorithms that can report distance queries in constant time are also referred to as "distance oracles." With unlimited preprocessing time and space, it is clear that exact distance oracles can be easily designed. This entry sheds light on the design of approximate distance oracles with limited preprocessing time and space for the family of geometric graphs with constant dilation.

Notation and Definitions

If p and q are points in $\mathcal{R}^d$, then the notation $|pq|$ is used to denote the Euclidean distance between p and q; the notation $\delta_G(p, q)$ is used to denote the Euclidean length of a shortest path between p and q in a geometric network G. Given a constant $t > 1$, a graph G with vertex set S is a t-spanner for S if $\delta_G(p, q) \leq t|pq|$ for any two points p and q of S. A t-spanner network is said to have *dilation* (or *stretch factor*) t. A $(1 + \varepsilon)$-approximate shortest path between p and q is defined to be any path in G between p and q having length Δ, where $\delta_G(p, q) \leq \Delta \leq (1+\varepsilon)\delta_G(p, q)$. For a comprehensive overview of geometric spanners, see the book by Narasimhan and Smid [14].

All networks considered in this entry are simple and undirected. The model of computation used is the traditional algebraic computation tree model with the added power of indirect addressing. In particular, the algorithms presented here do not use the non-algebraic floor function as a unit-time operation. The problem is formalized below.

Problem 1 (Distance Oracle) Given an arbitrary real constant $\epsilon > 0$, and a geometric graph G in d-dimensional Euclidean space with constant dilation t, design a data structure that answers $(1 + \epsilon)$-approximate shortest path length queries in constant time.

The data structure can also be applied to solve several other problems. These include (a) the problem of reporting approximate distance queries between vertices in a planar polygonal domain with "rounded" obstacles, (b) query versions of *closest pair* problems, and (c) the efficient computation of the approximate dilations of geometric graphs.

Survey of Related Research

The design of efficient data structures for answering distance queries for general (non-geometric) networks was considered by Thorup and Zwick [17] (unweighted general graphs), Baswanna and Sen [3] (weighted general graphs, i.e., arbitrary metrics), and Arikati et al. [2] and Thorup [16] (weighted planar graphs).

For the geometric case, variants of the problem have been considered in a number of papers (for a recent paper, see, e.g., Chen et al. [5]). Work on the approximate version of these variants can also be found in many articles (for a recent paper, see, e.g., Agarwal et al. [1]). The focus of this entry is the results reported in the work of Gudmundsson et al. [10–13]. Similar results on distance oracles were proved subsequently for *unit disk graphs* [7]. Practical implementations of distance oracles in geometric networks have also been investigated [15].

Key Results

The main result of this entry is the existence of approximate distance oracle data structures for geometric networks with constant dilation (see Theorem 4 below). As preprocessing, the network is "pruned" so that it only has a linear number of edges. The data structure consists of a series of "cluster graphs" of increasing coarseness, each of which helps answer approximate queries for pairs of points with interpoint distances of different scales. In order to pinpoint the appropriate cluster graph to search in for a given query, the data structure uses the bucketing tool described below. The idea of using cluster graphs to speed up geometric algorithms was first introduced by Das and Narasimhan [6] and later used by Gudmundsson et al. [9] to design an efficient algorithm to compute $(1 + \varepsilon)$-spanners. Similar ideas were explored by Gao et al. [8] for applications to the design of mobile networks.

Pruning

If the input geometric network has a superlinear number of edges, then the preprocessing step for the distance oracle data structure involves efficiently "pruning" the network so that it has only a linear number of edges. The pruning may result in a small increase of the dilation of the spanner. The following theorem was proved by Gudmundsson et al. [12].

Theorem 1 *Let* $t > 1$ *and* $\varepsilon' > 0$ *be real constants. Let* S *be a set of* n *points in* $\mathcal{R}^d$, *and let* $G = (S, E)$ *be a* t-*spanner for* S *with* m *edges. There exists an algorithm to compute in* $O(m + n \log n)$ *time, a* $(1 + \varepsilon')$-*spanner of* G *having* $O(n)$ *edges and whose weight is* $O(wt(MST(S)))$.

The pruning step requires the following technical theorem proved by Gudmundsson et al. [12].

Theorem 2 *Let* S *be a set of* n *points in* $\mathcal{R}^d$, *and let* $c \geq 7$ *be an integer constant. In* $O(n \log n)$ *time, it is possible to compute a data structure* $D(S)$ *consisting of:*

1. *A sequence* $L_1, L_2, \ldots, L_\ell$ *of real numbers, where* $\ell = O(n)$, *and*
2. *A sequence* $S_1, S_2, \ldots, S_\ell$ *of subsets of* S *such that* $\sum_{i=1}^{\ell} |S_i| = O(n)$,

such that the following holds. For any two distinct points p *and* q *of* S, *it is possible to compute in* $O(1)$ *time an index* i *with* $1 \leq i \leq \ell$ *and two*

points x *and* y *in* S_i *such that (a)* $L_i / n^{c+1} \leq |xy| < L_i$ *and (b) both* $|px|$ *and* $|qy|$ *are less than* $|xy| / n^{c-2}$.

Despite its technical nature, the above theorem is of fundamental importance to this work. In particular, it helps to deal with networks where the interpoint distances are not confined to a polynomial range, i.e., there are pairs of points that are very close to each other and very far from each other.

Bucketing

Since the model of computation assumed here does not allow the use of floor functions, an important component of the algorithm is a "bucketing tool" that allows (after appropriate preprocessing) constant-time computation of a quantity referred to as BINDEX, which is defined to be the floor of the logarithm of the interpoint distance between any pair of input points.

Theorem 3 *Let* S *be a set of* n *points in* $\mathcal{R}^d$ *that are contained in the hypercube* $(0, n^k)^d$, *for some positive integer constant* k, *and let* ε *be a positive real constant. The set* S *can be preprocessed in* $O(n \log n)$ *time into a data structure of size* $O(n)$, *such that for any two points* p *and* q *of* S, *with* $|pq| \geq 1$, *it is possible to compute in constant time the quantity* $\text{BINDEX}_\varepsilon(p, q) = \lfloor \log_{1+\varepsilon} |pq| \rfloor$.

The constant-time computation mentioned in Theorem 3 is achieved by reducing the problem to one of answering least common ancestor queries for pairs of nodes in a tree, a problem for which constant-time solutions were devised most recently by Bender and Farach-Colton [4].

Main Results

Using the bucketing and the pruning tools, and using the algorithms described by Gudmundsson et al. [13], the following theorem can be proved.

Theorem 4 *Let* $t > 1$ *and* $\varepsilon > 0$ *be real constants. Let* S *be a set of* n *points in* $\mathcal{R}^d$, *and let* $G = (S, E)$ *be a* t-*spanner for* S *with* m *edges. The graph* G *can be preprocessed into a data structure of size* $O(n \log n)$ *in time* $O(mn \log n)$, such that for any pair of query points $p, q \in S$, it is possible to compute a $(1 + \varepsilon)$-approximation of the shortest path distance in G between p and q in $O(1)$ time. Note that all the big-Oh notations hide constants that depend on d, t, and ε.

Additionally, if the traditional algebraic model of computation (without indirect addressing) is assumed, the following weaker result can be proved.

Theorem 5 *Let S be a set of n points in $\mathcal{R}^d$, and let $G = (S, E)$ be a t-spanner for S, for some real constant $t > 1$, having m edges. Assuming the algebraic model of computation, in $O(m \log \log n + n \log^2 n)$ time, it is possible to preprocess G into a data structure of size $O(n \log n)$, such that for any two points p and q in S, a $(1 + \varepsilon)$-approximation of the shortest-path distance in G between p and q can be computed in $O(\log \log n)$ time.*

Applications

As mentioned earlier, the data structure described above can be applied to several other problems. The first application deals with reporting distance queries for a planar domain with polygonal obstacles. The domain is further constrained to be t-rounded, which means that the length of the shortest obstacle-avoiding path between any two points in the input point set is at most t times the Euclidean distance between them. In other words, the visibility graph is required to be a t-spanner for the input point set.

Theorem 6 *Let $\mathcal{F}$ be a t-rounded collection of polygonal obstacles in the plane of total complexity n, where t is a positive constant. One can preprocess $\mathcal{F}$ in $O(n \log n)$ time into a data structure of size $O(n \log n)$ that can answer obstacle-avoiding $(1+\varepsilon)$-approximate shortest path length queries in time $O(\log n)$. If the query points are vertices of $\mathcal{F}$, then the queries can be answered in $O(1)$ time.*

The next application of the distance oracle data structure includes query versions of *closest pair*

problems, where the queries are confined to specified subset(s) of the input set.

Theorem 7 *Let $G = (S, E)$ be a geometric graph on n points and m edges, such that G is a t-spanner for S, for some constant $t > 1$. One can preprocess G in time $O(m + n \log n)$ into a data structure of size $O(n \log n)$ such that given a query subset S' of S, a $(1 + \varepsilon)$-approximate closest pair in S' (where distances are measured in G) can be computed in time $O(|S'| \log |S'|)$.*

Theorem 8 *Let $G = (S, E)$ be a geometric graph on n points and m edges, such that G is a t-spanner for S, for some constant $t > 1$. One can preprocess G in time $O(m + n \log n)$ into a data structure of size $O(n \log n)$ such that given two disjoint query subsets X and Y of S, a $(1 + \varepsilon)$-approximate bichromatic closest pair (where distances are measured in G) can be computed in time $O((|X| + |Y|) \log(|X| + |Y|))$.*

The last application of the distance oracle data structure includes the efficient computation of the approximate dilations of geometric graphs.

Theorem 9 *Given a geometric graph on n vertices with m edges, and given a constant C that is an upper bound on the dilation t of G, it is possible to compute a $(1 + \varepsilon)$-approximation to t in time $O(m + n \log n)$.*

Open Problems

Two open problems remain unanswered:

1. Improve the space utilization of the distance oracle data structure from $O(n \log n)$ to $O(n)$.
2. Extend the approximate distance oracle data structure to report not only the approximate distance but also the approximate shortest path between the given query points.

Cross-References

▶ Geometric Spanners
▶ Sparse Graph Spanners

Recommended Reading

1. Agarwal PK, Har-Peled S, Karia M (2000) Computing approximate shortest paths on convex polytopes. In: Proceedings of the 16th ACM symposium on computational geometry, Hong Kong, pp 270–279
2. Arikati S, Chen DZ, Chew LP, Das G, Smid M, Zaroliagis CD (1996) Planar spanners and approximate shortest path queries among obstacles in the plane. In: Proceedings of the 4th annual european symposium on algorithms, Barcelona. Lecture notes in computer science, vol 1136. Springer, Berlin, pp 514–528
3. Baswana S, Sen S (2004) Approximate distance oracles for unweighted graphs in $\tilde{O}(n^2)$ time. In: Proceedings of the 15th ACM-SIAM symposium on discrete algorithms, Philadelphia, pp 271–280
4. Bender MA, Farach-Colton M (2000) The LCA problem revisited. In: Proceedings of the 4th Latin American symposium on theoretical informatics, Punta del Este. Lecture notes in computer science, vol 1776. Springer, Berlin, pp 88–94
5. Chen DZ, Daescu O, Klenk KS (2001) On geometric path query problems. Int J Comput Geom Appl 11:617–645
6. Das G, Narasimhan G (1997) A fast algorithm for constructing sparse Euclidean spanners. Int J Comput Geom Appl 7:297–315
7. Gao J, Zhang L (2005) Well-separated pair decomposition for the unit-disk graph metric and its applications. SIAM J Comput 35(1):151–169
8. Gao J, Guibas LJ, Hershberger J, Zhang L, Zhu A (2003) Discrete mobile centers. Discret Comput Geom 30:45–63
9. Gudmundsson J, Levcopoulos C, Narasimhan G (2002) Fast greedy algorithms for constructing sparse geometric spanners. SIAM J Comput 31:1479–1500
10. Gudmundsson J, Levcopoulos C, Narasimhan G, Smid M (2002) Approximate distance oracles for geometric graphs. In: Proceedings of the 13th ACM-SIAM symposium on discrete algorithms, San Francisco, pp 828–837
11. Gudmundsson J, Levcopoulos C, Narasimhan G, Smid M (2002) Approximate distance oracles revisited. In: Proceedings of the 13th international symposium on algorithms and computation, Osaka. Lecture notes in computer science, vol 2518. Springer, Berlin, pp 357–368
12. Gudmundsson J, Narasimhan G, Smid M (2005) Fast pruning of geometric spanners. In: Proceedings of the 22nd symposium on theoretical aspects of computer science, Stuttgart. Lecture notes in computer science, vol 3404. Springer, Berlin, pp 508–520
13. Gudmundsson J, Levcopoulos C, Narasimhan G, Smid M (2008) Approximate distance oracles for geometric spanners. ACM Trans Algorithms 4(1):article 10
14. Narasimhan G, Smid M (2007) Geometric spanner networks. Cambridge University, Cambridge
15. Sankaranarayanan J, Samet H (2010) Query processing using distance oracles for spatial networks. IEEE Trans Knowl Data Eng 22(8):1158–1175
16. Thorup M (2004) Compact oracles for reachability and approximate distances in planar digraphs. J ACM 51:993–1024
17. Thorup M, Zwick U (2001) Approximate distance oracles. In: Proceedings of the 33rd annual ACM symposium on the theory of computing, Crete, pp 183–192

Approximate Dictionaries

Venkatesh Srinivasan
Department of Computer Science, University of Victoria, Victoria, BC, Canada

Keywords

Cell probe model; Data structures; Static membership

Years and Authors of Summarized Original Work

2002; Buhrman, Miltersen, Radhakrishnan, Venkatesh

Problem Definition

The Problem and the Model

A static data structure problem consists of a set of data D, a set of queries Q, a set of answers A, and a function $f : D \times Q \to A$. The goal is to store the data succinctly, so that any query can be answered with only a few probes to the data structure. *Static membership* is a well-studied problem in data structure design [2, 6, 9, 10, 16, 17, 23].

Definition 1 (Static Membership) In the static membership problem, one is given a subset S of at most n keys from a universe $U = \{1, 2, \ldots, m\}$. The task is to store S so that queries of the form "Is u in S?" can be answered by making few accesses to the memory.

A natural and general model for studying any data structure problem is the *cell probe model* proposed by Yao [23].

Definition 2 (Cell Probe Model) An (s, w, t) cell probe scheme for a static data structure problem $f : D \times Q \to A$ has two components: a storage scheme and a query scheme. The storage scheme stores the data $d \in D$ as a Table $T[d]$ of s cells, each cell of word size w bits. The storage scheme is deterministic. Given a query $q \in Q$, the query scheme computes $f(d, q)$ by making at most t probes to $T[d]$, where each probe reads one cell at a time, and the probes can be adaptive. In a deterministic cell probe scheme, the query scheme is deterministic. In a randomized cell probe scheme, the query scheme is randomized and is allowed to err with a small probability.

Buhrman et al. [3] study the complexity of the static membership problem in the *bitprobe* model. The bitprobe model is a variant of the cell probe model in which each cell holds just a single bit. In other words, the word size w is 1. Thus, in this model, the query algorithm is given bitwise access to the data structure. The study of the membership problem in the bitprobe model was initiated by Minsky and Papert in their book *Perceptrons* [16]. However, they were interested in *average*-case upper bounds for this problem, while this work studies *worst*-case bounds for the membership problem.

Observe that if a scheme is required to store sets of size at most n, then it must use at least $\lceil \log \sum_{i \leq n} \binom{m}{i} \rceil$ number of bits. If $n \leq m^{1 - \Omega(1)}$, this implies that the scheme must store $\Omega(n \log m)$ bits (and therefore use $\Omega(n)$ cells). The goal in [3] is to obtain a scheme that answers queries, uses only constant number of bitprobes, and at the same time uses a table of $O(n \log m)$ bits.

Related Work

The static membership problem has been well studied in the cell probe model, where each cell is capable of holding one element of the universe. That is, $w = O(\log m)$. In a seminal paper, Fredman, Komlós, and Szemerédi [10] proposed a scheme for the static membership problem in

the cell probe model with word size $O(\log m)$ that used a constant number of probes and a table of size $O(n)$. This scheme will be referred to as the FKS scheme. Thus, up to constant factors, the FKS scheme uses optimal space and number of cell probes. In fact, Fiat et al. [9], Brodnik and Munro [2], and Pagh [17] obtain schemes that use space (in bits) that is within a small additive term of $\lceil \log \sum_{i \leq n} \binom{m}{i} \rceil$ and yet answer queries by reading at most a constant number of cells. Despite all these fundamental results for the membership problem in the cell probe model, very little was known about the bitprobe complexity of static membership prior to the work in [3].

Key Results

Buhrman et al. investigate the complexity of the static membership problem in the bitprobe model. They study

- Two-sided error randomized schemes that are allowed to err on positive instances as well as negative instances (i.e., these schemes can say "No" with a small probability when the query element u is in the set S and "Yes" when it is not).
- One-sided error randomized schemes where the errors are restricted to negative instances alone (i.e., these schemes never say "No" when the query element u is in the set S);
- Deterministic schemes in which no errors are allowed.

The main techniques used in [3] are based on 2-colorings of special set systems that are related to r-cover-free family of sets considered in [5, 7, 11]. The reader is referred to [3] for further details.

Randomized Schemes with Two-Sided Error

The main result in [3] shows that there are randomized schemes that use *just one bitprobe* and yet use space close to the information theoretic lower bound of $\Omega(n \log m)$ bits.

Theorem 1 *For any $0 < \epsilon \leq \frac{1}{4}$, there is a scheme for storing subsets S of size at most n of a universe of size m using $O\left(\frac{n}{\epsilon^2} \log m\right)$ bits so that any membership query "Is $u \in S$?" can be answered with error probability at most ϵ by a randomized algorithm which probes the memory at just one location determined by its coin tosses and the query element u.*

Note that randomization is allowed only in the query algorithm. It is still the case that for each set S, there is exactly one associated data structure $T(S)$. It can be shown that deterministic schemes that answer queries using a single bitprobe need m bits of storage (see the remarks following Theorem 4). Theorem 1 shows that, by allowing randomization, this bound (for constant ϵ) can be reduced to $O(n \log m)$ bits. This space is within a constant factor of the information theoretic bound for n sufficiently small. Yet, the randomized scheme answers queries using a single bitprobe.

Unfortunately, the construction above does not permit us to have sub-constant error probability and still use optimal space. Is it possible to improve the result of Theorem 1 further and design such a scheme? Buhrman et al. [3] shows that this is not possible: if ϵ is made sub-constant, then the scheme must use more than $n \log m$ space.

Theorem 2 *Suppose $\frac{n}{m^{1/3}} \leq \epsilon \leq \frac{1}{4}$. Then, any two-sided ϵ-error randomized scheme which answers queries using one bitprobe must use space $\Omega\left(\frac{n}{\epsilon \log(1/\epsilon)} \log m\right)$.*

Randomized Schemes with One-Sided Error

Is it possible to have any savings in space if the query scheme is expected to make only one-sided errors? The following result shows it is possible if the error is allowed only on negative instances.

Theorem 3 *For any $0 < \epsilon \leq \frac{1}{4}$, there is a scheme for storing subsets S of size at most n of a universe of size m using $O\left(\left(\frac{n}{\epsilon}\right)^2 \log m\right)$ bits so that any membership query "Is $u \in S$?" can*

be answered with error probability at most ϵ by a randomized algorithm which makes a single bitprobe to the data structure. Furthermore, if $u \in S$, the probability of error is 0.

Though this scheme does not operate with optimal space, it still uses significantly less space than a bitvector. However, the dependence on n is quadratic, unlike in the two-sided scheme where it was linear. Buhrman et al. [3] shows that this scheme is essentially optimal: there is necessarily a quadratic dependence on $\frac{n}{\epsilon}$ for any scheme with one-sided error.

Theorem 4 *Suppose $\frac{n}{m^{1/3}} \leq \epsilon \leq \frac{1}{4}$. Consider the static membership problem for sets S of size at most n from a universe of size m. Then, any scheme with one-sided error ϵ that answers queries using at most one bitprobe must use $\Omega\left(\frac{n^2}{\epsilon^2 \log(n/\epsilon)} \log m\right)$ bits of storage.*

Remark 1 One could also consider one-probe one-sided error schemes that only make errors on positive instances. That is, no error is made for query elements *not* in the set S. In this case, [3] shows that randomness does not help at all: such a scheme must use m bits of storage.

The following result shows that the space requirement can be reduced further in one-sided error schemes if more probes are allowed.

Theorem 5 *Suppose $0 < \delta < 1$. There is a randomized scheme with one-sided error $n^{-\delta}$ that solves the static membership problem using $O\left(n^{1+\delta} \log m\right)$ bits of storage and $O\left(\frac{1}{\delta}\right)$ bitprobes.*

Deterministic Schemes

In contrast to randomized schemes, Buhrman et al. show that deterministic schemes exhibit a time-space tradeoff behavior.

Theorem 6 *Suppose a deterministic scheme stores subsets of size n from a universe of size m using s bits of storage and answers membership queries with t bitprobes to memory. Then, $\binom{m}{n} \leq \max_{i \leq nt} \binom{2s}{i}$.*

This tradeoff result has an interesting consequence. Recall that the FKS hashing scheme is a data structure for storing sets of size at most n from a universe of size m using $O(n \log m)$ bits, so that membership queries can be answered using $O(\log m)$ bitprobes. As a corollary of the tradeoff result, [3] show that the FKS scheme makes an optimal number of bitprobes, within a constant factor, for this amount of space.

Corollary 1 *Let $\epsilon > 0, c \geq 1$ be any constants. There is a constant $\delta > 0$ so that the following holds. Let $n \leq m^{1-\epsilon}$ and let a scheme for storing sets of size at most n of a universe of size m as data structures of at most $cn \log m$ bits be given. Then, any deterministic algorithm answering membership queries using this structure must make at least $\delta \log m$ bitprobes in the worst case.*

From Theorem 6, it also follows that any deterministic scheme that answers queries using t bitprobes must use space at least $\Omega\left(tm^{1/t}n^{1-1/t}\right)$ in the worst case. The final result shows the existence of schemes which almost match the lower bound.

Theorem 7 *1. There is a nonadaptive scheme that stores sets of size at most n from a universe of size m using $O\left(ntm^{\frac{2}{t+1}}\right)$ bits and answers queries using $2t + 1$ bitprobes. This scheme is non-explicit.*
2. There is an explicit adaptive scheme that stores sets of size at most n from a universe of size m using $O\left(m^{1/t}n \log m\right)$ bits and answers queries using $O(\log n + \log \log m) + t$ bitprobes.

Power of Few Bitprobes

In this section, we highlight some of recent results for this problem subsequent to [3] and encourage the reader to read the corresponding references for more details. Most of these results focus on the power of deterministic schemes with a small number of bitprobes.

Let $S(m, n, t)$ denote the minimum number of bits of storage needed by a deterministic scheme that answers queries using t (adaptive) bitprobes. In [3], it was shown that $S(m, n, 1) = m$ and $S(m, n, 5) = o(m)$ for $n = o\left(m^{1/3}\right)$ (Theorem 7, Part 1). This leads us to a natural question:

Is $S(m, n, t) = o(m)$ for $t = 2, 3$, and 4 and under what conditions on n?

Initial progress for the case $t = 2$ was made by [18] who considered the simplest case: $n = 2$. They showed that $S(m, 2, 2) = O\left(m^{2/3}\right)$. It was later shown in [19] that $S(m, 2, 2) = \Omega\left(m^{4/7}\right)$. The upper bound result of [18] was improved upon by the authors of [1] who showed that $S(m, n, 2) = o(m)$ if $n = o(\log m)$. Interestingly, a matching lower bound was shown recently in [12]: $S(m, n, 2) = o(m)$ only if $n = o(\log m)$.

Strong upper bounds were obtained by [1] for the case $t = 3$ and $t = 4$. They showed that $S(m, n, 3) = o(m)$ whenever $n = o(m)$. They also showed that $S(m, n, 4) = o(m)$ for $n = o(m)$ even if the four bitprobes are nonadaptive. Recently, it was shown in [14] that $S(m, 2, 3) \leq 7m^{2/5}$. This work focuses on explicit schemes for $n = 2$ and $t \geq 3$.

Finally, we end with two remarks. Our problem for the case $n = \Theta(m)$ has been studied by Viola [22]. A recent result of Chen, Grigorescu, and de Wolf [4] studies our problem in the presence of adversarial noise.

Applications

The results in [3] have interesting connections to questions in coding theory and communication complexity. In the framework of coding theory, the results in [3] can be viewed as constructing locally decodable source codes, analogous to the locally decodable channel codes of [13]. Theorems 1–4 can also be viewed as giving tight bounds for the following communication complexity problem (as pointed out in [15]): Alice gets $u \in \{1, \ldots, m\}$, Bob gets $S \subseteq \{1, \ldots, m\}$ of size at most n, and Alice sends a single message to Bob after which Bob announces whether $u \in S$. See [3] for further details.

Recommended Reading

1. Alon N and Feige U (2009) On the power of two, three and four probes. In: Proceedings of SODA'09, New York, pp 346–354

2. Brodnik A, Munro JI (1994) Membership in constant time and minimum space. In: Algorithms ESA'94: second annual European symposium, Utrecht. Lecture notes in computer science, vol 855, pp 72–81. Final version: Membership in constant time and almost-minimum space. SIAM J Comput 28(5):1627–1640 (1999)
3. Buhrman H, Miltersen PB, Radhakrishnan J, Venkatesh S (2002) Are bitvectors optimal? SIAM J Comput 31(6):1723–1744
4. Chen V, Grigorescu E, de Wolf R (2013) Error-correcting data structures. SIAM J Comput 42(1):84–111
5. Dyachkov AG, Rykov VV (1982) Bounds on the length of disjunctive codes. Problemy Peredachi Informatsii 18(3):7–13 [Russian]
6. Elias P, Flower RA (1975) The complexity of some simple retrieval problems. J Assoc Comput Mach 22:367–379
7. Erdős P, Frankl P, Füredi Z (1985) Families of finite sets in which no set is covered by the union of r others. Isr J Math 51:79–89
8. Fiat A, Naor M (1993) Implicit $O(1)$ probe search. SIAM J Comput 22:1–10
9. Fiat A, Naor M, Schmidt JP, Siegel A (1992) Non-oblivious hashing. J Assoc Comput Mach 31:764–782
10. Fredman ML, Komlós J, Szemerédi E (1984) Storing a sparse table with $O(1)$ worst case access time. J Assoc Comput Mach 31(3):538–544
11. Füredi Z (1996) On r-cover-free families. J Comb Theory Ser A 73:172–173
12. Garg M, Radhakrishnan J (2015) Set membership with a few bit probes. In: Proceedings of SODA'15, San Diego, pp 776–784
13. Katz J, Trevisan L (2000) On the efficiency of local decoding procedures for error-correcting codes. In: Proceedings of STOC'00, Portland, pp 80–86
14. Lewenstein M, Munro JI, Nicholson PK, Raman V (2014) Improved explicit data structures in the bit-probe model. In: Proceedings of ESA'14, Wroclaw, pp 630–641
15. Miltersen PB, Nisan N, Safra S, Wigderson A (1998) On data structures and asymmetric communication complexity. J Comput Syst Sci 57:37–49
16. Minsky M, Papert S (1969) Perceptrons. MIT, Cambridge
17. Pagh R (1999) Low redundancy in static dictionaries with $O(1)$ lookup time. In: Proceedings of ICALP '99, Prague. Lecture notes in computer science, vol 1644, pp 595–604
18. Radhakrishnan J, Raman V, Rao SS (2001) Explicit deterministic constructions for membership in the bitprobe model. In: Proceedings of ESA'01, Aarhus, pp 290–299
19. Radhakrishnan J, Shah S, Shannigrahi S (2010) Data structures for storing small sets in the bitprobe model. In: Proceedings of ESA'10, Liverpool, pp 159–170
20. Ruszinkó M (1984) On the upper bound of the size of r-cover-free families. J Comb Theory Ser A 66:302–310
21. Ta-Shma A (2002) Explicit one-probe storing schemes using universal extractors. Inf Process Lett 83(5):267–274
22. Viola E (2012) Bit-probe lower bounds for succinct data structures. SIAM J Comput 41(6):1593–1604
23. Yao ACC (1981) Should tables be sorted? J Assoc Comput Mach 28(3):615–628

Approximate Distance Oracles with Improved Query Time

Christian Wulff-Nilsen
Department of Computer Science, University of Copenhagen, Copenhagen, Denmark

Keywords

Approximate distance oracle; Graphs; Query time; Shortest paths

Years and Authors of Summarized Original Work

2013; Wulff-Nilsen

Problem Definition

This problem is concerned with obtaining a compact data structure capable of efficiently reporting approximate shortest path distance queries in a given undirected edge-weighted graph $G = (V, E)$. If the query time is independent (or nearly independent) of the size of G, we refer to the data structure as an *approximate distance oracle* for G. For vertices u and v in G, we denote by $d_G(u, v)$ the shortest path distance between u and v in G. For a given *stretch* parameter $\delta \geq 1$, we call the oracle δ-*approximate* if for all vertices u and v in G, $d_G(u, v) \leq \tilde{d}_G(u, v) \leq \delta d_G(u, v)$, where $\tilde{d}_G(u, v)$ is the output of the query for u and v. Hence, we allow estimates to be stretched by a factor up to δ but not shrunk.

Key Results

A major result in the area of distance oracles is due to Thorup and Zwick [4]. They gave, for every integer $k \geq 1$, a $(2k - 1)$-approximate distance oracle of size $O\left(kn^{1+1/k}\right)$ and query time $O(k)$, where n is the number of vertices of the graph. This is constant query time when k is constant. Corresponding approximate shortest paths can be reported in time proportional to their length. Mendel and Naor [3] asked the question of whether query time can be improved to a universal constant (independent also of k) while keeping both size and stretch small. They obtained $O\left(n^{1+1/k}\right)$ size and $O(1)$ query time at the cost of a constant-factor increase in stretch to $128k$. Unlike the oracle of Thorup and Zwick, Mendel and Naor's oracle is not path reporting.

In [5], it is shown how to improve the query time of Thorup-Zwick to $O(\log k)$ without increasing space or stretch. This is done while keeping essentially the same data structure but applying binary instead of linear search in so-called bunch structures that were introduced by Thorup and Zwick [4]; the formal definition will be given below. Furthermore, it is shown in [5] how to improve the stretch of the oracle of Mendel and Naor to $(2 + \epsilon)k$ for an arbitrarily small constant $\epsilon > 0$ while keeping query time constant (bounded by $1/\epsilon$). This improvement is obtained without an increase in space except for large values of k close to $\log n$ (only values of k less than $\log n$ are interesting since the Mendel-Naor oracle has optimal $O(n)$ space and $O(1)$ query time for larger values). Below, we sketch the main ideas in the improvement of Thorup-Zwick and of Mendel-Naor, respectively.

Oracle with $O(\log k)$ Query Time

The oracle of Thorup and Zwick keeps a hierarchy of sets of sampled vertices $V = A_0 \supseteq A_1 \supseteq A_2 \ldots \supseteq A_k = \emptyset$, where for $i = 1, \ldots, k - 1$, A_i is obtained by picking each element of A_{i-1} independently with probability $n^{-1/k}$. Define $p_i(u)$ as the vertex in A_i closest to u. The oracle precomputes and stores for each vertex $u \in V$ the bunch B_u, defined as

$$B_u = \bigcup_{i=0}^{k-1} \{v \in A_i \setminus A_{i+1} | d_G(u, v) $$

$$< d_G(u, p_{i+1}(u))\}.$$

See Fig. 1 for an illustration of a bunch. The distance $d_G(u, v)$ for each $v \in B_u$ is precomputed as well.

Now, to answer a query for a vertex pair (u, v), the oracle performs a linear search through bunches B_u and B_v. Pseudocode is given in Fig. 2. It is clear that query time is $O(k)$, and it can be shown that the estimate output in line 6 has stretch $2k - 1$.

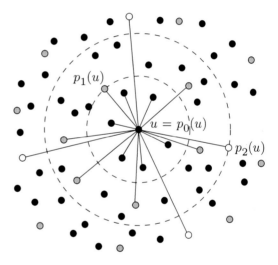

Approximate Distance Oracles with Improved Query Time, Fig. 1 A bunch B_u in a complete Euclidean graph with $k = 3$. *Black vertices* belong to A_0, *grey vertices* to A_1, and *white vertices* to A_2. Line segments connect u to vertices of B_u

---

Algorithm $\text{dist}_k(u, v)$

1. $w \leftarrow p_0(u); \ j \leftarrow 0$
2. while $w \notin B_v$
3. $\quad j \leftarrow j + 1$
4. $\quad (u, v) \leftarrow (v, u)$
5. $\quad w \leftarrow p_j(u)$
6. return $d_G(w, u) + d_G(w, v)$

Approximate Distance Oracles with Improved Query Time, Fig. 2 Answering a distance query, starting at sample level i

We can improve query time to $O(\log k)$ by instead doing binary search in the bunch structures. A crucial property of the Thorup-Zwick oracle is that every time the test in line 2 succeeds, $d_G(u, p_j(u))$ increases by at most $d_G(u, v)$, and this is sufficient to prove $2k - 1$ stretch. In particular, if the test succeeds two times in a row, $d_G(u, p_{j+2}(u)) - d_G(u, p_j(u)) \leq 2d_G(u, v)$, where j is even. If we can check that $d_G(u, p_{j'+2}(u)) - d_G(u, p_{j'}(u)) \leq 2d_G(u, v)$ for all smaller even indices j', we may start the query algorithm at index j instead of index 0. Since we would like to apply binary search, pick j to be (roughly) $k/2$. It suffices to check only one inequality, namely, the one with the largest value $d_G(u, p_{j'+2}(u)) - d_G(u, p_{j'}(u))$. Note that this value depends only on u and k, so we can precompute the index j' with this largest value. In the query phase, we can check in $O(1)$ time whether $d_G(u, p_{j'+2}(u)) - d_G(u, p_{j'}(u)) \leq 2d_G(u, v)$. If the test succeeds, we can start the query at j, and hence, we can recurse on indices between j and $k - 1$. Conversely, if the test fails, it means that the test in line 2 fails for either j' or $j' + 1$. Hence, the query algorithm of Thorup and Zwick terminates no later than at index $j' + 1$, and we can recurse on indices between 0 and $j' + 1$. In both cases, the number of remaining indices is reduced by a factor of at least 2. Since each recursive call takes $O(1)$ time, we thus achieve $O(\log k)$ query time.

Since the improved oracle is very similar to the Thorup-Zwick oracle, it is path reporting, i.e., it can report approximate paths in time proportional to their length.

Oracle with Constant Query Time

The second oracle in [5] can be viewed as a hybrid between the oracles of Thorup-Zwick and of Mendel-Naor. An initial estimate is obtained by querying the Mendel-Naor oracle. This estimate has stretch at most $128k$, and it is refined in subsequent iterations until the desired stretch $(2 + \epsilon)k$ is obtained. In each iteration, the current estimate is reduced by a small constant factor greater than 1 (depending on ϵ). Note that after a constant number of iterations, the estimate will be below the desired stretch, but it needs to be

ensured that it is not below the shortest path distance.

In each iteration, the hybrid algorithm attempts to start the Thorup-Zwick query algorithm at a step corresponding to this estimate. If this can be achieved, only a constant number of steps of this query algorithm need to be executed before the desired stretch is obtained. Conversely, if the hybrid algorithm fails to access the Thorup-Zwick oracle in any iteration, then by a property of the bunch structures, it is shown that the current estimate is not below the shortest path distance. Hence, the desired stretch is again obtained.

An important property of the Mendel-Naor oracle needed above is that the set d_{MN} of all different values the oracle can output has size bounded by $O\left(n^{1+1/k}\right)$. This is used in the hybrid algorithm as follows. In a preprocessing step, values from d_{MN} are ordered in a list $\mathcal{L}$ together with additional values corresponding to the intermediate estimates that the hybrid algorithm can consider in an iteration. Updating the estimate in each iteration then corresponds to a linear traversal of part of $\mathcal{L}$. Next, each vertex p_i of each bunch structure B_u of the Thorup-Zwick oracle is associated with the value in the list closest to $d_G(u, p_i)$. For each element of $\mathcal{L}$, a hash table is kept for the bunch vertices associated with that element. It can be shown that this way of linking the oracle of Thorup-Zwick and Mendel-Naor achieves the desired.

Applications

The practical need for efficient algorithms to answer the shortest path (distance) queries in graphs has increased significantly over the years, in large part due to emerging GPS navigation technology and other route planning software. Classical algorithms like Dijkstra do not scale well as they may need to explore the entire graph just to answer a single query. As road maps are typically of considerable size, obtaining compact distance oracles has received a great deal of attention from the research community.

Open Problems

A widely believed girth conjecture of Erdős [2] implies that an oracle with stretch $2k - 1$, size $O\left(n^{1+1/k}\right)$, and query time $O(1)$ would be optimal. Obtaining such an oracle (preferably one that is path reporting) is a main open problem in the area. Some progress has recently been made: Chechik [1] gives an oracle (not path reporting) with stretch $2k - 1$, size $O\left(kn^{1+1/k}\right)$, and $O(1)$ query time.

Recommended Reading

1. Chechik S (2014) Approximate distance oracles with constant query time. In: STOC, New York, pp 654–663
2. Erdős P (1964) Extremal problems in graph theory. In: Theory of graphs and its applications (Proceedings of the symposium on smolenice, 1963). Czechoslovak Academy of Sciences, Prague, pp 29–36
3. Mendel M, Naor A (2007) Ramsey partitions and proximity data structures. J Eur Math Soc 9(2):253–275. See also FOCS'06
4. Thorup M, Zwick U (2005) Approximate distance oracles. J Assoc Comput Mach 52:1–24
5. Wulff-Nilsen C (2013) Approximate distance oracles with improved query time. In: Proceedings of the 24th ACM-SIAM symposium on discrete algorithms (SODA), New Orleans, pp 202–208

Approximate Matching

Ran Duan
Institute for Interdisciplinary Information Sciences, Tsinghua University, Beijing, China

Keywords

Approximation algorithms; Maximum matching

Years and Authors of Summarized Original Work

1973; Hopcroft, Karp
1980; Micali, Vazirani
2014; Duan, Pettie

Problem Definition

In graph theory, a *matching* in a graph is a set of edges without common vertices, while a *perfect matching* is one in which all vertices are associated with matching edges. In a graph $G = (V, E, w)$, where V is the set of vertices, E is the set of edges, and $w : E \to \mathbb{R}$ is the edge weight function, the *maximum matching* problem determines the matching M in G which maximizes $w(M) = \sum_{e \in M} w(e)$. Note that a maximum matching is not necessarily perfect. The *maximum cardinality matching (MCM)* problem means the maximum matching problem for $w(e) = 1$ for all edges. Otherwise, it is called the *maximum weight matching (MWM)*.

Algorithms for Exact MWM

Although the maximum matching problem has been studied for decades, the computational complexity of finding an optimal matching remains quite open. Most algorithms for graph matchings use the concept of augmenting paths. An alternating path (or cycle) is one whose edges alternate between M and $E \setminus M$. An alternating path P is *augmenting* if P begins and ends at free vertices, that is, $M \oplus P \stackrel{\text{def}}{=} (M \setminus P) \cup (P \setminus M)$ is a matching with cardinality $|M \oplus P| = |M| + 1$. Therefore, the basic algorithm finds the maximum cardinality matching by finding an augmenting path in the graph and adding it the matching each time, until no more augmenting paths exist. The running time for the basic algorithm will be $O(mn)$ where $m = |E|$ and $n = |V|$. The major improvement over this for bipartite graphs is the Hopcroft-Karp algorithm [10]. It finds a maximal set of vertex disjoint shortest augmenting paths in each step and shows that the length of shortest augmenting paths will increase each time. The running time of the Hopcroft-Karp algorithm is $O(m\sqrt{n})$. Its corresponding algorithm for general graphs is given by Micali and Vazirani [14].

For the maximum weight matching (MWM) and maximum weight perfect matching (MWPM), the most classical algorithm is the Hungarian algorithm [12] for bipartite graphs and the Edmonds algorithm for general graphs [6, 7].

For fast implementations, Gabow and Tarjan [8] gave bit-scaling algorithms for MWM in bipartite graphs running in $O(m\sqrt{n}\log(nN))$ time, where the edge weights are integers in $[-N,\ldots,N]$. Then, they also gave its corresponding algorithm for general graphs [9]. Extending [15], Sankowski [18] gave an $O(Nn^{\omega})$ MWM algorithm for bipartite graphs (here, $\omega \leq 2.373$ denotes the exponential of the complexity of fast matrix multiplication (FMM) [2, 20]), while Huang and Kavitha [11] obtained a similar time bound for general graphs. We can see these time complexities are still far from linear, which shows the importance of fast approximation algorithms.

Approximate Matching

Let a δ-MWM be a matching whose weight is at least a δ fraction of the maximum weight matching, where $0 < \delta \leq 1$, and let δ-MCM be defined analogously.

It is well known that the *greedy* algorithm – iteratively chooses the maximum weight edge not incident to previously chosen edges – produces a $\frac{1}{2}$-MWM. A straightforward implementation of this algorithm takes $O(m\log n)$ time. Preis [3,17] gave a $\frac{1}{2}$-MWM algorithm running in linear time. Vinkemeier and Hougardy [19] and Pettie and Sanders [16] proposed several $\left(\frac{2}{3} - \epsilon\right)$-MWM algorithms (see also [13]) running in $O(m\log\epsilon^{-1})$ time; each is based on iteratively improving a matching by identifying sets of short weight-augmenting paths and cycles.

Key Results

Approximate Maximum Cardinality Matching

In fact, the Hopcroft-Karp algorithm [10] for bipartite graphs and Micali-Vazirani [14] algorithm for general graphs both imply a $(1 - \epsilon)$-MCM algorithm in $O(\epsilon^{-1}m)$ time. We can search for a maximal set of vertex disjoint shortest augmenting paths for k steps, and the matching obtained is a $\left(1 - \frac{1}{k+1}\right)$-MCM.

Theorem 1 ([10, 14]) *In a general graph G, the $(1 - \epsilon)$-MCM algorithm can be found in time $O(\epsilon^{-1}m)$.*

Approximate Maximum Weighted Matching

In 2014, Duan and Pettie [5] give the first $(1 - \epsilon)$-MWM algorithm for arbitrary weighted graphs whose running time is linear. In particular, we show that such a matching can be found in $O(m\epsilon^{-1}\log\epsilon^{-1})$ time, improving a preliminary result of $O(m\epsilon^{-2}\log^3 n)$ running time by the authors in 2010 [4]. This result leaves little room for improvement. The main results are given in the following two theorems:

Theorem 2 ([5]) *In a general graph G with integer edge weights between $[0, N]$, a $(1 - \epsilon)$-MWM can be computed in time $O(m\epsilon^{-1}\log N)$.*

Theorem 3 ([5]) *In a general graph G with real edge weights, a $(1 - \epsilon)$-MWM can be computed in time $O(m\epsilon^{-1}\log\epsilon^{-1})$.*

Unlike previous algorithms of approximation ratios of $1/2$ [3, 17] or $2/3$ [16, 19], the new algorithm does not find weight-augmenting paths and cycles directly, but follows a primal-dual relaxation on the linear programming formulation of MWM. This relaxed complementary slackness approach relaxes the constraint of the dual variables by a small amount, so that the iterative process of the dual problem will converge to an approximate solution much more quickly. While it takes $O(\sqrt{n})$ iterations of augmenting to achieve a perfect matching, we proved that we only need $O(\log N/\epsilon)$ iterations to achieve a $(1 - \epsilon)$-approximation. Also, we make the relaxation "dynamic" by shrinking the relaxation when the dual variables decrease by one half, so that finally the relaxation is at most ϵ times the edge weight on each matching edge and very small on each nonmatching edge, which gives an approximate solution.

Applications

Graph matching is a fundamental combinatorial problem that has a wide range of applications in many fields, and it can also be building blocks of other algorithms, such as the Christofides algorithm [1] for approximate traveling salesman problem. The approximate algorithm for maximum weight matching described above has linear running time, much faster than the Hungarian algorithm [12] and Edmonds [6,7] algorithm. It is also much simpler than the Gabow-Tarjan scaling algorithms [8,9] of $\tilde{O}(m\sqrt{n})$ running time. Thus, it has a great impact both in theory and in real-world applications.

Cross-References

▸ Assignment Problem
▸ Maximum Matching

Recommended Reading

1. Christofides N, GROUP CMUPPMSR (1976) Worst-case analysis of a new heuristic for the travelling salesman problem. Defense Technical Information Center, http://books.google.de/books?id=2A7eygAACAAJ
2. Coppersmith D, Winograd T (1987) Matrix multiplication via arithmetic progressions. In: Proceedings of 19th ACM symposium on the theory of computing (STOC), New York, pp 1–6
3. Drake D, Hougardy S (2003) A simple approximation algorithm for the weighted matching problem. Inf Process Lett 85:211–213
4. Duan R, Pettie S (2010) Approximating maximum weight matching in near-linear time. In: Proceedings 51st IEEE symposium on foundations of computer science (FOCS), Las Vegas, pp 673–682
5. Duan R, Pettie S (2014) Linear-time approximation for maximum weight matching. J ACM 61(1):1:1–1:23. doi:10.1145/2529989, http://doi.acm.org/10.1145/2529989
6. Edmonds J (1965) Maximum matching and a polyhedron with 0, 1-vertices. J Res Nat Bur Stand Sect B 69B:125–130
7. Edmonds J (1965) Paths, trees, and flowers. Can J Math 17:449–467
8. Gabow HN, Tarjan RE (1989) Faster scaling algorithms for network problems. SIAM J Comput 18(5):1013–1036
9. Gabow HN, Tarjan RE (1991) Faster scaling algorithms for general graph-matching problems. J ACM 38(4):815–853
10. Hopcroft JE, Karp RM (1973) An $n^{5/2}$ algorithm for maximum matchings in bipartite graphs. SIAM J Comput 2:225–231
11. Huang CC, Kavitha T (2012) Efficient algorithms for maximum weight matchings in general graphs with small edge weights. In: Proceedings of the twenty-third annual ACM-SIAM symposium on discrete algorithms, SODA'12, Kyoto. SIAM, pp 1400–1412. http://dl.acm.org/citation.cfm?id=2095116.2095226
12. Kuhn HW (1955) The Hungarian method for the assignment problem. Nav Res Logist Q 2:83–97
13. Mestre J (2006) Greedy in approximation algorithms. In: Proceedings of the 14th conference on annual European symposium, Zurich, vol 14. Springer, London, pp 528–539. doi:10.1007/11841036_48, http://portal.acm.org/citation.cfm?id=1276191.1276239
14. Micali S, Vazirani V (1980) An $O(\sqrt{|V|} \cdot |E|)$ algorithm for finding maximum matching in general graphs. In: Proceedings of 21st IEEE symposium on foundations of computer science (FOCS), Syracuse, pp 17–27
15. Mucha M, Sankowski P (2004) Maximum matchings via Gaussian elimination. In: Proceedings of 45th symposium on foundations of computer science (FOCS), Rome, pp 248–255
16. Pettie S, Sanders P (2004) A simpler linear time $2/3 - \epsilon$ approximation to maximum weight matching. Inf Process Lett 91(6):271–276
17. Preis R (1999) Linear time $1/2$-approximation algorithm for maximum weighted matching in general graphs. In: Proceedings of 16th symposium on theoretical aspects of computer science (STACS), Trier. LNCS, vol 1563, pp 259–269
18. Sankowski P (2006) Weighted bipartite matching in matrix multiplication time. In: Proceedings of 33rd international symposium on automata, languages, and programming (ICALP), Venice, pp 274–285
19. Vinkemeier DED, Hougardy S (2005) A linear-time approximation algorithm for weighted matchings in graphs. ACM Trans Algorithms 1(1):107–122
20. Williams VV (2012) Multiplying matrices faster than Coppersmith-Winograd. In: Proceedings of the 44th symposium on theory of computing, STOC'12, New York. ACM, New York, pp 887–898. doi:10.1145/2213977.2214056, http://doi.acm.org/10.1145/2213977.2214056

Approximate Regular Expression Matching

Gonzalo Navarro
Department of Computer Science, University of Chile, Santiago, Chile

Keywords

Regular expression matching allowing errors or differences

Years and Authors of Summarized Original Work

1995; Wu, Manber, Myers

Problem Definition

Given a *text string* $T = t_1 t_2 \ldots t_n$ and a *regular expression* R of length m denoting language, $\mathcal{L}(R)$ over an alphabet Σ of size σ, and given a *distance function* among strings d and a *threshold* k, the *approximate regular expression matching (AREM)* problem is to find all the text positions that finish a so-called approximate occurrence of R in T, that is, compute the set $\{ j, \exists i, 1, \leq i \leq j, \exists P \in \mathcal{L}(R), d(P, t_i, \ldots, t_j) \leq k \}$ T, R, and k are given together, whereas the algorithm can be tailored for a specific d.

This entry focuses on the so-called weighted edit distance, which is the minimum sum of weights of a sequence of operations converting one string into the other. The operations are insertions, deletions, and substitutions of characters. The weights are positive real values associated to each operation and characters involved. The weight of deleting a character c is written $w(c \rightarrow \varepsilon)$, that of inserting c is written $w(\varepsilon \rightarrow c)$, and that of substituting c by $c \neq c\prime$ is written $w(c \rightarrow c\prime)$. It is assumed $w(c \rightarrow c) = 0$ for all $c \in \Sigma \cup \varepsilon$ and the triangle inequality, that is, $w(x \rightarrow y) + w(y \rightarrow z) \geq w(x \rightarrow z)$ for any $x, y, z, \in \Sigma \cup \{\varepsilon\}$. As the distance may be asymmetric, it is also fixed that $d(A, B)$ is the cost of converting A into B. For simplicity and practicality, $m = o(n)$ is assumed in this entry.

Key Results

The most versatile solution to the problem [3] is based on a graph model of the distance computation process. Assume the regular expression R is converted into a nondeterministic finite automaton (NFA) with $O(m)$ states and transitions using Thompson's method [8]. Take this automaton as a directed graph $G(V, E)$ where edges are labeled by elements in $\Sigma \cup \{\varepsilon\}$. A directed and

weighted graph $\mathcal{G}$ is built to solve the AREM problem. $\mathcal{G}$ is formed by putting $n + 1$ copies of $G, G_0, G_1, \ldots, G_n$ and connecting them with weights so that the distance computation reduces to finding shortest paths in $\mathcal{G}$.

More formally, the nodes of $\mathcal{G}$ are $\{v_i, v \in V, 0 \leq i \leq n\}$, so that v_i is the copy of node $v \in V$ in graph G_i. For each edge $u \xrightarrow{c} v$ in $E, c \in \Sigma \cup \varepsilon$, the following edges are added to graph $\mathcal{G}$:

$$u_i \rightarrow v_i, \quad \text{with weight } w(c \rightarrow \varepsilon),$$

$$0 \leq i \leq n.$$

$$u_i \rightarrow u_{i+1}, \quad \text{with weight } w(\varepsilon \rightarrow t_{i+1}),$$

$$0 \leq i \leq n.$$

$$u_i \rightarrow v_{i+1}, \quad \text{with weight } w(c \rightarrow t_{i+1}),$$

$$0 \leq i \leq n.$$

Assume for simplicity that G has initial state s and a unique final state f (this can always be arranged). As defined, the shortest path in $\mathcal{G}$ from s_0 to f_n gives the smallest distance between T and a string in $\mathcal{L}(R)$. In order to adapt the graph to the AREM problem, the weights of the edges between s_i and s_{i+1} are modified to be zero.

Then, the AREM problem is reduced to computing shortest paths. It is not hard to see that $\mathcal{G}$ can be topologically sorted so that all the paths to nodes in G_i are computed before all those to G_{i+1}. This way, it is not hard to solve this shortest path problem in $O(mn\log m)$ time and $O(m)$ space. Actually, if one restricts the problem to the particular case of *network expressions*, which are regular expressions without Kleene closure, then G has no loops and the shortest path computation can be done in $O(mn)$ time, and even better on average [2].

The most delicate part in achieving $O(mn)$ time for general regular expressions [3] is to prove that, given the types of loops that arise in the NFAs of regular expressions, it is possible to compute the distances correctly within each G_i by (a) computing them in a topological order of G_i without considering the *back edges* introduced by Kleene closures, (b) updating path costs by using the back edges once, and (c) updating path costs once more in topological order ignoring back edges again.

Theorem 1 (Myers and Miller [3]) *There exists an $O(mn)$ worst-case time solution to the AREM problem under weighted edit distance.*

It is possible to do better when the weights are integer-valued, by exploiting the unit-cost RAM model through a four-Russian technique [10]. The idea is as follows. Take a small subexpression of R, which produces an NFA that will translate into a small subgraph of each G_i. At the time of propagating path costs within this automaton, there will be a counter associated to each node (telling the current shortest path from s_0). This counter can be reduced to a number in $[0, k + 1]$, where $k + 1$ means "more than k." If the small NFA has r states, $r\lceil \log_2(k + 2)\rceil$ bits are needed to fully describe the counters of the corresponding subgraph of G_i. Moreover, given an initial set of values for the counters, it is possible to precompute all the propagation that will occur within the same subgraph of G_i, in a table having $2^{r\lceil\log_2(k + 2)\rceil}$ entries, one per possible configuration of counters. It is sufficient that $r < \alpha\log_{k + 2}n$ for some $\alpha < 1$ to make the construction and storage cost of those tables $o(n)$. With the help of those tables, all the propagation within the subgraph can be carried out in constant time. Similarly, the propagation of costs to the same subgraph at $G_{i + 1}$ can also be precomputed in tables, as it depends only on the current counters in G_i and on text character $t_{i + 1}$, for which there are only σ alternatives.

Now, take all the subtrees of R of maximum size not exceeding r and preprocess them with the technique above. Convert each such subtree into a leaf in R labeled by a special character a_A, associated to the corresponding small NFA A. Unless there are consecutive Kleene closures in R, which can be simplified as $R^{* *} = R^{*}$, the size of R after this transformation is $O(m/r)$. Call $R\prime$ the transformed regular expression. One essentially applies the technique of Theorem 1 to $R\prime$, taking care of how to deal with the special leaves that correspond to small NFAs. Those leaves are converted by Thompson's construction into two nodes linked by an edge labeled a_A.

When the path cost propagation process reaches the source node of an edge labeled a_A with cost c, one must update the counter of the initial state of NFA A to c (or $k + 1$ if $c > k$). One then uses the four-Russians table to do all the cost propagation within A in constant time and finally obtain, at the counter of the final state of A, the new value for the target node of the edge labeled a_A in the top-level NFA. Therefore, all the edges (normal and special) of the top-level NFA can be traversed in constant time, so the costs at G_i can be obtained in $O(mn/r)$ time using Theorem 1. Now one propagates the costs to $G_{i + 1}$, using the four-Russians tables to obtain the current counter values of each subgraph A in $G_{i + 1}$.

Theorem 2 (Wu et al. [10]) *There exists an $O(n + mn/\log_{k + 2}n)$ worst-case time solution to the AREM problem under weighted edit distance if the weights are integer numbers.*

Applications

The problem has applications in computational biology, to find certain types of motifs in DNA and protein sequences. See [1] for a more detailed discussion. In particular, PROSITE patterns are limited regular expressions rather popular to search protein sequences. PROSITE patterns can be searched for with faster algorithms in practice [7]. The same occurs with other classes of complex patterns [6] and network expressions [2].

Open Problems

The worst-case complexity of the AREM problem is not fully understood. It is of course $\Omega(n)$, which has been achieved for $m\log(k + 2) = O(\log n)$, but it is not known how much can this be improved.

Experimental Results

Some experiments are reported in [5]. For small m and k, and assuming all the weights are 1 (except $w(c \rightarrow c) = 0$), bit-parallel algorithms of worst-case complexity $O(k\,n(m/\log n)^2)$ [4,9] are the fastest (the second is able to skip some text characters, depending on R). For arbitrary integer weights, the best choice is a more complex bit-parallel algorithm [5] or the four-Russians based one [10] for larger m and k. The original algorithm [3] is slower, but it is the only one supporting arbitrary weights.

URL to Code

A recent and powerful software package implementing AREM is *TRE* (http://laurikari. net/tre), which supports edit distance with different costs for each type of operation. Older packages offering efficient AREM are *agrep* [9] (https://github.com/Wikinaut/agrep) for simplified weight choices and *nrgrep* [4] (http:// www.dcc.uchile.cl/~gnavarro/software).

Cross-References

▶ Approximate String Matching is a simplification of this problem, and the relation between graph $\mathcal{G}$ here and matrix C there should be apparent.
▶ Regular Expression Matching is the simplified case where exact matching with strings in $\mathcal{L}(R)$ is sought.

Recommended Reading

1. Gusfield D (1997) Algorithms on strings, trees and sequences. Cambridge University Press, Cambridge
2. Myers EW (1996) Approximate matching of network expressions with spacers. J Comput Biol 3(1):33–51
3. Myers EW, Miller W (1989) Approximate matching of regular expressions. Bull Math Biol 51:7–37
4. Navarro G (2001) Nr-grep: a fast and flexible pattern matching tool. Softw Pract Exp 31:1265–1312
5. Navarro G (2004) Approximate regular expression searching with arbitrary integer weights. Nord J Comput 11(4):356–373
6. Navarro G, Raffinot M (2002) Flexible pattern matching in strings – practical on-line search algorithms for texts and biological sequences. Cambridge University Press, Cambridge
7. Navarro G, Raffinot M (2003) Fast and simple character classes and bounded gaps pattern matching, with applications to protein searching. J Comput Biol 10(6):903–923
8. Thompson K (1968) Regular expression search algorithm. Commun ACM 11(6):419–422
9. Wu S, Manber U (1992) Fast text searching allowing errors. Commun ACM 35(10):83–91
10. Wu S, Manber U, Myers EW (1995) A subquadratic algorithm for approximate regular expression matching. J Algorithms 19(3):346–360

Approximate String Matching

Gonzalo Navarro
Department of Computer Science, University of Chile, Santiago, Chile

Keywords

Inexact string matching; Semiglobal or semilocal sequence similarity; String matching allowing errors or differences

Years and Authors of Summarized Original Work

1980; Sellers
1989; Landau, Vishkin
1999; Myers
2003; Crochemore, Landau, Ziv-Ukelson
2004; Fredriksson, Navarro

Problem Definition

Given a *text string* $T = t_1 t_2 \ldots t_n$ and a *pattern string* $P = p_1 p_2 \ldots p_m$, both being sequences over an alphabet Σ of size σ, and given a *distance function* among strings d and a *threshold* k, the *approximate string matching (ASM)* problem is to find all the text positions that finish the so-called approximate occurrence of P in T, that is, compute the set $\{j, \exists i, 1 \leq i \leq j, d(P, t_i \ldots t_j) \leq k\}$. In the sequential version of the problem, T, P, and k are given together, whereas the algorithm can be tailored for a specific d.

The solutions to the problem vary widely depending on the distance d used. This entry focuses on a very popular one, called *Levenshtein distance* or *edit distance*, defined as the minimum number of character insertions, deletions, and substitutions necessary to convert one string into the other. It will also pay some attention to other common variants such as *indel distance*, where only insertions and deletions are permitted and is the dual of the *longest common subsequence* lcs ($d(A, B) = |A| + |B| - 2 \cdot lcs(A, B)$), and *Hamming distance*, where only substitutions are permitted.

A popular generalization of all the above is the *weighted edit distance*, where the operations are given positive real-valued weights and the distance is the minimum sum of weights of a sequence of operations converting one string into the other. The weight of deleting a character c is written $w(c \rightarrow \varepsilon)$, that of inserting c is written $w(\varepsilon \rightarrow c)$, and that of substituting c by $c' \neq c$ is written $w(c \rightarrow c')$. It is assumed $w(c \rightarrow c) = 0$ and the triangle inequality, that is, $w(x \rightarrow y) + w(y \rightarrow z) \geq w(x \rightarrow z)$ for any $x, y, z, \in \sum \cup \{\varepsilon\}$. As the distance may now be asymmetric, it is fixed that $d(A, B)$ is the cost of converting A into B. Of course, any result for weighted edit distance applies to edit, Hamming, and indel distances (collectively termed *unit-cost edit distances*) as well, but other reductions are not immediate.

Both worst- and average-case complexity are considered. For the latter, one assumes that pattern and text are randomly generated by choosing each character uniformly and independently from Σ. For simplicity and practicality, $m = o(n)$ is assumed in this entry.

Key Results

The most ancient and versatile solution to the problem [13] builds over the process of computing weighted edit distance. Let $A = a_1 a_2 \ldots a_m$ and $B = b_1 b_2 \ldots b_n$ be two strings. Let $C[0 \ldots m, 0 \ldots n]$ be a matrix such that $C[i, j] = d(a_1 \ldots a_i, b_1 \ldots b_j)$. Then, it holds $C[0, 0] = 0$ and

$$C[i, j] = \min(C[i - 1, j]$$
$$+ w(a_i \rightarrow \varepsilon), C[i, j - 1] + w(\varepsilon \rightarrow b_j),$$
$$C[i - 1, j - 1] + w(a_i \rightarrow b_j)),$$

where $C[i, -1] = C[-1, j] = \infty$ is assumed. This matrix is computed in $O(mn)$ time and $d(A, B) = C[m, n]$. In order to solve the approximate string matching problem, one takes $A = P$ and $B = T$ and sets $C[0, j] = 0$ for all j, so that the above formula is used only for $i > 0$.

Theorem 1 (Sellers 1980 [13]) *There exists an $O(mn)$ worst-case time solution to the ASM problem under weighted edit distance.*

The space is $O(m)$ if one realizes that C can be computed column-wise and only column $j - 1$ is necessary to compute column j. As explained, this immediately implies that searching under unit-cost edit distances can be done in $O(mn)$ time as well. In those cases, it is quite easy to compute only part of matrix C so as to achieve $O(kn)$ average-time algorithms [14].

Yet, there exist algorithms with lower worst-case complexity for weighted edit distance. By applying a Ziv-Lempel parsing to P and T, it is possible to identify regions of matrix C corresponding to substrings of P and T that can be computed from other previous regions corresponding to similar substrings of P and T [5].

Theorem 2 (Crochemore et al. 2003 [5]) *There exists an $O(n + mn/\log_\sigma n)$ worst-case time solution to the ASM problem under weighted edit distance. Moreover, the time is $O(n + mnh/\log n)$, where $0 \leq h \leq \log \sigma$ is the entropy of T.*

This result is very general, also holding for computing weighted edit distance and local similarity (see section on "Applications"). For the case of edit distance and exploiting the unit-cost RAM model, it is possible to do better. On one hand, one can apply a four-Russian technique: All the possible blocks (submatrices of C) of size $t \times t$, for $t = O(\log_\sigma n)$, are precomputed, and matrix C is computed block-wise [9]. On the other hand, one can represent each cell in matrix C using a constant number of bits (as it can differ from neighboring cells by ± 1) so as to store and process several cells at once in a single machine word [10]. This latter technique is called *bit-parallelism* and assumes a machine word of $\Theta(\log n)$ bits.

Theorem 3 (Masek and Paterson 1980 [9]; Myers 1999 [10]) *There exist $O(n + mn/(\log_\sigma n)^2)$ and $O(n + mn/\log n)$ worst-case time solutions to the ASM problem under edit distance.*

Both complexities are retained for indel distance, yet not for Hamming distance.

For unit-cost edit distances, the complexity can depend on k rather than on m, as $k < m$ for the problem to be nontrivial, and usually k is a small fraction of m (or even $k = o(m)$). A classic technique [8] computes matrix C by processing in constant time diagonals $C[i + d, j + d], 0 \le d \le s$, along which cell values do not change. This is possible by preprocessing the suffix trees of T and P for lowest common ancestor queries.

Theorem 4 (Landau and Vishkin 1989 [8]) *There exists an $O(kn)$ worst-case time solution to the ASM problem under unit-cost edit distances.*

Other solutions exist which are better for small k, achieving time $O(n(1 + k^4/m))$ [4]. For the case of Hamming distance, one can achieve improved results using convolutions [1].

Theorem 5 (Amir et al. 2004 [1]) *There exist $O(n \sqrt{k \log k})$ and $O(n(1 + k^3/m) \log k)$ worst-case time solution to the ASM problem under Hamming distance.*

The last result for edit distance [4] achieves $O(n)$ time if k is small enough ($k = O(m^{1/4})$). It is also possible to achieve $O(n)$ time on unit-cost edit distances at the expense of an exponential additive term on m or k: The number of different columns in C is independent of n, so the transition from every possible column to the next can be precomputed as a finite-state machine.

Theorem 6 (Ukkonen 1985 [14]) *There exists an $O(n + m \min(3^m, m(2m\sigma)^k))$ worst-case time solution to the ASM problem under edit distance.*

Similar results apply for Hamming and indel distance, where the exponential term reduces slightly according to the particularities of the distances.

The worst-case complexity of the ASM problem is of course $\Omega(n)$, but it is not known if this can be attained for any m and k. Yet, the average-case complexity of the problem is known.

Theorem 7 (Chang and Marr 1994 [3]) *The average-case complexity of the ASM problem is $\Theta(n(k + \log_\sigma m)/m)$ under unit-cost edit distances.*

It is not hard to prove the lower bound as an extension to Yao's bound for exact string matching [15]. The lower bound was reached in the same paper [3], for $k/m < 1/3 - O\left(1/\sqrt{\sigma}\right)$. This was improved later to $k/m < 1/2 - O\left(1/\sqrt{\sigma}\right)$ [6] using a slightly different idea. The approach is to precompute the minimum distance to match every possible text substring (block) of length $O(\log_\sigma m)$ inside P. Then, a text window is scanned backwards, block-wise, adding up those minimum precomputed distances. If they exceed k before scanning all the window, then no occurrence of P with k errors can contain the scanned blocks, and the window can be safely slid over the scanned blocks, advancing in T. This is an example of a *filtration* algorithm, which discards most text areas and applies an ASM algorithm only over those areas that cannot be discarded.

Theorem 8 (Fredriksson and Navarro 2004 [6]) *There exists an optimal-on-average solution to the ASM problem under edit distance, for any $k/m \le \frac{1 - e/\sqrt{\sigma}}{2 - e/\sqrt{\sigma}} = 1/2 - O\left(1/\sqrt{\sigma}\right)$.*

The result applies verbatim to indel distance. The same complexity is achieved for Hamming distance, yet the limit on k/m improves to $1 - 1/\sigma$. Note that, when the limit k/m is reached, the average complexity is already $\Theta(n)$. It is not clear up to which k/m limit could one achieve linear time on average.

Applications

The problem has many applications in computational biology (to compare DNA and protein sequences, recovering from experimental errors, so as to spot mutations or predict similarity of structure or function), text retrieval (to recover from spelling, typing, or automatic recognition errors), signal processing (to recover from transmission and distortion errors), and several others. See a survey [11] for a more detailed discussion.

Many extensions of the ASM problem exist, particularly in computational biology. For example, it is possible to substitute whole substrings by others (called *generalized edit distance*), swap characters in the strings (*string matching with swaps or transpositions*), reverse substrings (*reversal distance*), have variable costs for insertions/deletions when they are grouped (*similarity with gap penalties*), and look for any pair of substrings of both strings that are sufficiently similar (*local similarity*). See, for example, Gusfield's book [7], where many related problems are discussed.

Open Problems

The worst-case complexity of the problem is not fully understood. For unit-cost edit distances, it is $\Theta(n)$ if $m = O(\min(\log n, (\log_\sigma n)^2))$ or $k = O(\min(m^{1/4}, \log_{m\sigma} n))$. For weighted edit distance, the complexity is $\Theta(n)$ if $m = O(\log_\sigma n)$. It is also unknown up to which k/m value can one achieve $O(n)$ average time; up to now this has been achieved up to $k/m = 1/2 - O\left(1/\sqrt{\sigma}\right)$.

Experimental Results

A thorough survey on the subject [11] presents extensive experiments. Nowadays, the fastest algorithms for edit distance are in practice filtration algorithms [6, 12] combined with bit-parallel algorithms to verify the candidate areas [2, 10]. Those filtration algorithms work well for small enough k/m; otherwise, the bit-parallel algorithms should be used stand-alone. Filtration al-

gorithms are easily extended to handle multiple patterns searched simultaneously.

URL to Code

Well-known packages offering efficient ASM are *agrep* (https://github.com/Wikinaut/agrep) and *nrgrep* (http://www.dcc.uchile.cl/~gnavarro/software).

Cross-References

▶ Approximate Regular Expression Matching is the more complex case where P can be a regular expression
▶ Indexed Approximate String Matching refers to the case where the text can be preprocessed
▶ String Matching is the simplified version where no errors are permitted

Recommended Reading

1. Amir A, Lewenstein M, Porat E (2004) Faster algorithms for string matching with k mismatches. J Algorithms 50(2):257–275
2. Baeza-Yates R, Navarro G (1999) Faster approximate string matching. Algorithmica 23(2):127–158
3. Chang W, Marr T (1994) Approximate string matching and local similarity. In: Proceedings of the 5th annual symposium on combinatorial pattern matching (CPM'94), Asilomar. LNCS, vol 807. Springer, Berlin, pp 259–273
4. Cole R, Hariharan R (2002) Approximate string matching: a simpler faster algorithm. SIAM J Comput 31(6):1761–1782
5. Crochemore M, Landau G, Ziv-Ukelson M (2003) A subquadratic sequence alignment algorithm for unrestricted scoring matrices. SIAM J Comput 32(6):1654–1673
6. Fredriksson K, Navarro G (2004) Average-optimal single and multiple approximate string matching. ACM J Exp Algorithms 9(1.4)
7. Gusfield D (1997) Algorithms on strings, trees and sequences. Cambridge University Press, Cambridge
8. Landau G, Vishkin U (1989) Fast parallel and serial approximate string matching. J Algorithms 10:157–169
9. Masek W, Paterson M (1980) A faster algorithm for computing string edit distances. J Comput Syst Sci 20:18–31

10. Myers G (1999) A fast bit-vector algorithm for approximate string matching based on dynamic programming. J ACM 46(3):395–415
11. Navarro G (2001) A guided tour to approximate string matching. ACM Comput Surv 33(1):31–88
12. Navarro G, Baeza-Yates R (1999) Very fast and simple approximate string matching. Inf Proc Lett 72:65–70
13. Sellers P (1980) The theory and computation of evolutionary distances: pattern recognition. J Algorithms 1:359–373
14. Ukkonen E (1985) Finding approximate patterns in strings. J Algorithms 6:132–137
15. Yao A (1979) The complexity of pattern matching for a random string. SIAM J Comput 8:368–387

Approximate Tandem Repeats

Gregory Kucherov[1] and Dina Sokol[2]
[1]CNRS/LIGM, Université Paris-Est, Marne-la-Vallée, France
[2]Department of Computer and Information Science, Brooklyn College of CUNY, Brooklyn, NY, USA

Keywords

Approximate periodicities; Approximate repetitions

Years and Authors of Summarized Original Work

2001; Landau, Schmidt, Sokol
2003; Kolpakov, Kucherov

Problem Definition

Identification of periodic structures in words (variants of which are known as *tandem repeats, repetitions, powers*, or *runs*) is a fundamental algorithmic task (see entry ▶ Squares and Repetitions). In many practical applications, such as DNA sequence analysis, considered repetitions admit a certain variation between copies of the repeated pattern. In other words, repetitions under interest are *approximate tandem repeats* and not necessarily exact repeats only.

The simplest instance of an approximate tandem repeat is an *approximate square*. An approximate square in a word w is a subword uv, where u and v are within a given distance k according to some distance measure between words, such as Hamming distance or edit (also called Levenshtein) distance. There are several ways to define approximate tandem repeats as successions of approximate squares, i.e., to generalize to the approximate case the notion of arbitrary periodicity (see entry ▶ Squares and Repetitions). In this entry, we discuss three different definitions of approximate tandem repeats. The first two are built upon the Hamming distance measure, and the third one is built upon the edit distance.

Let $h(\cdot,\cdot)$ denote the Hamming distance between two words of equal length.

Definition 1 A word $r[1\ldots n]$ is called a K-*repetition* of period p, $p \leq n/2$, iff $h(r[1\ldots n - p], r[p + 1\ldots n]) \leq K$.

Equivalently, a word $r[1\ldots n]$ is a K-repetition of period p, if the number of mismatches, i.e., the number of i such that $r[i] \neq r[i + p]$, is at most K. For example, *ataa atta ctta ct* is a 2-repetition of period 4. *atc atc atc atg atg atg atg atg* is a 1-repetition of period 3, but *atc atc atc att atc atc atc att* is not.

Definition 2 A word $r[1\ldots n]$ is called a K-*run*, of period p, $p \leq n/2$, iff for every $i \in [1\ldots n - 2p + 1]$, we have $h(r[i\ldots i + p - 1], r[i + p, i + 2p - 1]) \leq K$.

A K-run can be seen as a sequence of approximate squares uv such that $|u| = |v| = p$ and u and v differ by at most K mismatches. The total number of mismatches in a K-run is not bounded.

Let $ed(\cdot,\cdot)$ denote the edit distance between two strings.

Definition 3 A word r is a K-*edit* repeat if it can be partitioned into consecutive subwords, $r = v'w_1 w_2 \ldots w_\ell v''$, $\ell \geq 2$, such that

$$ed(v', w_1') + \sum_{i=1}^{\ell-1} ed(w_i, w_{i+1}) + ed(w_\ell'', v'') \leq K,$$

where w_1' is some suffix of w_1 and w_ℓ'' is some prefix of w_ℓ.

A K-edit repeat is a sequence of "evolving" copies of a pattern such that there are at most K insertions, deletions, and mismatches, overall, between all consecutive copies of the repeat. For example, the word $r = caagct\ cagct\ ccgct$ is a 2-edit repeat.

When looking for tandem repeats occurring in a word, it is natural to consider *maximal* repeats. Those are the repeats extended to the right and left as much as possible provided that the corresponding definition is still verified. Note that the notion of maximality applies to K-repetitions, to K-runs, and to K-edit repeats.

Under the Hamming distance, K-runs provide the weakest "reasonable" definition of approximate tandem repeats, since it requires that every square it contains cannot contain more than K mismatch errors, which seems to be a minimal reasonable requirement. On the other hand, K-repetition is the strongest such notion as it limits by K the *total* number of mismatches. This provides an additional justification that finding these two types of repeats is important as they "embrace" other intermediate types of repeats. Several intermediate definitions have been discussed in [9, Section 5].

In general, each K-repetition is a part of a K-run of the same period, and every K-run is the union of all K-repetitions it contains. Observe that a K-run can contain as many as a linear number of K-repetitions with the same period. For example, the word $(000\ 100)^n$ of length $6n$ is a 1-run of period 3, which contains $(2n - 1)$ 1-repetitions. In general, a K-run r contains $(s - K + 1)$ K-repetitions of the same period, where s is the number of mismatches in r.

Example 1 The following Fibonacci word contains three 3-runs of period 6. They are shown in regular font, in positions aligned with their occurrences. Two of them are identical and contain each four 3-repetitions, shown in italic for the first run only. The third run is a 3-repetition in itself.

010010   100100   101001   010010   010100   1001

10010   100100   101001
10010   *100100*   *10*
0010   *100100*   *101*
10   *100100*   *10100*
0   *100100*   *101001*
                      1001   010010   010100   1
                      10   010100   1001

Key Results

Given a word w of length n and an integer K, it is possible to find all K-runs, K-repetitions, and K-edit repeats within w in the following time and space bounds:

K**-runs** can be found in time $O(nK \log K + S)$ (S the output size) and working space $O(n)$ [9].

K**-repetitions** can be found in time $O(nK \log K + S)$ and working space $O(n)$ [9].

K**-edit repeats** can be found in time $O(nK \log K \log(n/K) + S)$ and working space $O(n + K^2)$ [14, 19].

All three algorithms are based on similar algorithmic tools that generalize corresponding techniques for the exact case [4, 15, 16] (see [10] for a systematic presentation). The first basic tool is a generalization of the *longest extension functions* [16] that, in the case of Hamming distance, can be exemplified as follows. Given a word w, we want to compute, for each position p and each $k \leq K$, the quantity $\max\{j \,|\, h(w[1 \ldots j], w[p \ldots p + j - 1]) \leq k\}$. Computing all those values can be done in time $O(nK)$ using a method based on the suffix tree and the computation of the *lowest common ancestor* described in [7].

The second tool is the Lempel-Ziv factorization used in the well-known compression method. Different variants of the Lempel-Ziv factorization of a word can be computed in linear time [7, 18].

The algorithm for computing K-repetitions from [9] can be seen as a direct generalization of the algorithm for computing maximal repetitions (runs) in the exact case [8, 15]. Although based on the same basic tools and ideas, the algorithm [9] for computing K-runs is much more involved and uses a complex "bootstrapping" technique for assembling runs from smaller parts.

The algorithm for finding the K-edit repeats uses both the recursive framework and the idea of the *longest extension functions* of [16]. The longest common extensions, in this case, allow up to K edit operations. Efficient methods for computing these extensions are based upon a combination of the results of [12] and [13]. The K-edit repeats are derived by combining the longest common extensions computed in the forward direction with those computed in the reverse direction.

Applications

Tandemly repeated patterns in DNA sequences are involved in various biological functions and are used in different practical applications.

Tandem repeats are known to be involved in regulatory mechanisms, e.g., to act as binding sites for regulatory proteins. Tandem repeats have been shown to be associated with recombination hotspots in higher organisms. In bacteria, a correlation has been observed between certain tandem repeats and virulence and pathogenicity genes.

Tandem repeats are responsible for a number of inherited diseases, especially those involving the central nervous system. Fragile X syndrome, Kennedy disease, myotonic dystrophy, and Huntington's disease are among the diseases that have been associated with triplet repeats.

Examples of different genetic studies illustrating abovementioned biological roles of tandem repeats can be found in introductive sections of [1, 6, 11]. Even more than just genomic elements associated with various biological functions, tandem repeats have been established to be a fundamental mutational mechanism in genome evolution [17].

A major practical application of short tandem repeats is based on the interindividual variability in copy number of certain repeats occurring at a single locus. This feature makes tandem repeats a convenient tool for genetic profiling of individuals. The latter, in turn, is applied to pedigree analysis and establishing phylogenetic relationships between species, as well as to forensic medicine [3].

Open Problems

The definition of K-edit repeats is similar to that of K-repetitions (for the Hamming distance case). It would be interesting to consider other definitions of maximal repeats over the edit distance. For example, a definition similar to the K-run would allow up to K edits between each pair of neighboring periods in the repeat. Other possible definitions would allow K errors between *any* pair of copies of a repeat, or between *all pairs* of copies, or between some *consensus* and each copy.

In general, a *weighted* edit distance scheme is necessary for biological applications. Known algorithms for tandem repeats based on a weighted edit distance scheme are not feasible, and thus, only heuristics are currently used.

URL to Code

The algorithms described in this entry have been implemented for DNA sequences and are publicly available. The Hamming distance algorithms (K-runs and K-repetitions) are part of the *mreps* software package, available at http://mreps.univ-mlv.fr/ [11]. The K-edit repeat software, *TRED*, is available at http://tandem.sci.brooklyn.cuny.edu/ [19]. The implementations of the algorithms are coupled with postprocessing filters, necessary due to the nature of biological sequences.

In practice, software based on heuristic and statistical methods is largely used. Among them, TRF (http://tandem.bu.edu/trf/trf.html) [1] is the most popular program used by the bioinformatics community. Other programs include ATRHunter (http://bioinfo.cs.technion.ac.il/atrhunter/) [20]

and TandemSWAN (http://favorov.bioinfolab.net/swan/) [2]. STAR (http://atgc.lirmm.fr/star/) [5] is another software, based on an information-theoretic approach, for computing approximate tandem repeats of a prespecified pattern.

Cross-References

▶ Squares and Repetitions

Acknowledgments This work was supported in part by the National Science Foundation Grant DB&I 0542751.

Recommended Reading

1. Benson G (1999) Tandem repeats finder: a program to analyze DNA sequences. Nucleic Acids Res 27:573–580
2. Boeva VA, Régnier M, Makeev VJ (2004) SWAN: searching for highly divergent tandem repeats in DNA sequences with the evaluation of their statistical significance. In: Proceedings of JOBIM 2004, Montreal, p 40
3. Butler JM (2001) Forensic DNA typing: biology and technology behind STR markers. Academic Press, San Diego
4. Crochemore M (1983) Recherche linéaire d'un carré dans un mot. C R Acad Sci Paris Sér I Math 296:781–784
5. Delgrange O, Rivals E (2004) STAR – an algorithm to search for tandem approximate repeats. Bioinformatics 20:2812–2820
6. Gelfand Y, Rodriguez A, Benson G (2007) TRDB – the tandem repeats database. Nucleic Acids Res 35(suppl. 1):D80–D87
7. Gusfield D (1997) Algorithms on strings, trees, and sequences. Cambridge University Press, Cambridge/New York
8. Kolpakov R, Kucherov G (1999) Finding maximal repetitions in a word in linear time. In: 40th symposium foundations of computer science (FOCS), New York, pp 596–604. IEEE Computer Society Press
9. Kolpakov R, Kucherov G (2003) Finding approximate repetitions under Hamming distance. Theor Comput Sci 33(1):135–156
10. Kolpakov R, Kucherov G (2005) Identification of periodic structures in words. In: Berstel J, Perrin D (eds) Applied combinatorics on words. Encyclopedia of mathematics and its applications. Lothaire books, vol 104, pp 430–477. Cambridge University Press, Cambridge
11. Kolpakov R, Bana G, Kucherov G (2003) *mreps*: efficient and flexible detection of tandem repeats in DNA. Nucleic Acids Res 31(13):3672–3678
12. Landau GM, Vishkin U (1988) Fast string matching with k differences. J Comput Syst Sci 37(1):63–78
13. Landau GM, Myers EW, Schmidt JP (1998) Incremental string comparison. SIAM J Comput 27(2):557–582
14. Landau GM, Schmidt JP, Sokol D (2001) An algorithm for approximate tandem repeats. J Comput Biol 8:1–18
15. Main M (1989) Detecting leftmost maximal periodicities. Discret Appl Math 25:145–153
16. Main M, Lorentz R (1984) An $O(n \log n)$ algorithm for finding all repetitions in a string. J Algorithms 5(3):422–432
17. Messer PW, Arndt PF (2007) The majority of recent short DNA insertions in the human genome are tandem duplications. Mol Biol Evol 24(5):1190–1197
18. Rodeh M, Pratt V, Even S (1981) Linear algorithm for data compression via string matching. J Assoc Comput Mach 28(1):16–24
19. Sokol D, Benson G, Tojeira J (2006) Tandem repeats over the edit distance. Bioinformatics 23(2):e30–e35
20. Wexler Y, Yakhini Z, Kashi Y, Geiger D (2005) Finding approximate tandem repeats in genomic sequences. J Comput Biol 12(7):928–942

Approximating Fixation Probabilities in the Generalized Moran Process

George B. Mertzios
School of Engineering and Computing Sciences, Durham University, Durham, UK

Keywords

Approximation algorithm; Evolutionary dynamics; Fixation probability; Markov-chain Monte Carlo; Moran process

Years and Authors of Summarized Original Work

2014; Diaz, Goldberg, Mertzios, Richerby, Serna, Spirakis

Problem Definition

Population and evolutionary dynamics have been extensively studied, usually with the assumption that the evolving population has no spatial structure. One of the main models in this area is the Moran process [17]. The initial population contains a single "mutant" with fitness $r > 0$, with all other individuals having fitness 1. At each step of the process, an individual is chosen at random, with probability proportional to its fitness. This individual reproduces, replacing a second individual, chosen uniformly at random, with a copy of itself.

Lieberman, Hauert, and Nowak introduced a generalization of the Moran process, where the members of the population are placed on the vertices of a connected graph which is, in general, directed [13, 19]. In this model, the initial population again consists of a single mutant of fitness $r > 0$ placed on a vertex chosen uniformly at random, with each other vertex occupied by a nonmutant with fitness 1. The individual that will reproduce is chosen as before, but now one of its neighbors is randomly selected for replacement, either uniformly or according to a weighting of the edges. The original Moran process can be recovered by taking the graph to be an unweighted clique.

Several similar models describing particle interactions have been studied previously, including the SIR and SIS epidemic models [8, Chapter 21], the voter model, the antivoter model, and the exclusion process [1, 7, 14]. Related models, such as the decreasing cascade model [12, 18], have been studied in the context of influence propagation in social networks and other models have been considered for dynamic monopolies [2]. However, these models do not consider different fitnesses for the individuals.

In general, the Moran process on a finite, connected, directed graph may end with all vertices occupied by mutants or with no vertex occupied by a mutant – these cases are referred to as *fixation* and *extinction*, respectively – or the process may continue forever. However, for undirected graphs and strongly connected digraphs, the process terminates almost surely, either at fixation or extinction. At the other extreme, in a directed graph with two sources, neither fixation nor extinction is possible. In this work we consider finite undirected graphs. The *fixation probability* for a mutant of fitness r in a graph G is the probability that fixation is reached and is denoted $f_{G,r}$.

Key Results

The fixation probability can be determined by standard Markov chain techniques. However, doing so for a general graph on n vertices requires solving a set of 2^n linear equations, which is not computationally feasible, even numerically. As a result, most prior work on computing fixation probabilities in the generalized Moran process has either been restricted to small graphs [6] or graph classes where a high degree of symmetry reduces the size of the set of equations – for example, paths, cycles, stars, and complete graphs [3–5] – or has concentrated on finding graph classes that either encourage or suppress the spread of the mutants [13, 16].

Because of the apparent intractability of exact computation, we turn to approximation. Using a potential function argument, we show that, with high probability, the Moran process on an undirected graph of order n reaches absorption (either fixation or extinction) within $\mathcal{O}(n^6)$ steps if $r = 1$ and $\mathcal{O}(n^4)$ and $\mathcal{O}(n^3)$ steps when $r > 1$ and $r < 1$, respectively. Taylor et al. [20] studied absorption times for variants of the generalized Moran process, but, in our setting, their results only apply to the process on regular graphs, where it is equivalent to a biased random walk on a line with absorbing barriers. The absorption time analysis of Broom et al. [3] is also restricted to cliques, cycles, and stars. In contrast to this earlier work, our results apply to all connected undirected graphs.

Our bound on the absorption time, along with polynomial upper and lower bounds for the fixation probability, allows the estimation of the fixation and extinction probabilities by Monte Carlo techniques. Specifically, we give a *fully polynomial randomized approximation scheme* (FPRAS) for these quantities. An FPRAS for a function $f(X)$ is a polynomial-time randomized algorithm g that, given input X and an error bound ε, satisfies $(1 - \varepsilon)f(X) \leqslant g(X) \leqslant (1 + \varepsilon)f(X)$ with probability at least $\frac{3}{4}$ and runs in time polynomial in the length of X and $\frac{1}{\varepsilon}$ [11].

For the case $r < 1$, there is no polynomial lower bound on the fixation probability so only the extinction probability can be approximated by this technique. Note that, when $f \ll 1$, computing $1 - f$ to within a factor of $1 \pm \varepsilon$ does not imply computing f to within the same factor.

Bounding the Fixation Probability

In the next two lemmas, we provide polynomial upper and lower bounds for the fixation probability of an arbitrary undirected graph G. Note that the lower bound of Lemma 1 holds only for $r \geqslant 1$. Indeed, for example, the fixation probability of the complete graph K_n is given by $f_{K_n,r} = (1 - \frac{1}{r})/(1 - \frac{1}{r^n})$ [13, 19], which is exponentially small for any $r < 1$.

Lemma 1 *Let* $G = (V, E)$ *be an undirected graph with n vertices. Then $f_{G,r} \geqslant \frac{1}{n}$ for any $r \geqslant 1$.*

Lemma 2 *Let* $G = (V, E)$ *be an undirected graph with n vertices. Then $f_{G,r} \leqslant 1 - \frac{1}{n+r}$ for any $r > 0$.*

Bounding the Absorption Time

In this section, we show that the Moran process on a connected graph G of order n is expected to reach absorption in a polynomial number of steps. To do this, we use the potential function given by

$$\phi(S) = \sum_{x \in S} \frac{1}{\deg x}$$

for any state $S \subseteq V(G)$ and we write $\phi(G)$ for $\phi(V(G))$. Note that $1 < \phi(G) < n$ and that $\phi(\{x\}) = 1/\deg x \leqslant 1$ for any vertex $x \in V$.

First, we show that the potential strictly increases in expectation when $r > 1$ and strictly decreases in expectation when $r < 1$.

Lemma 3 *Let* $(X_i)_{i \geqslant 0}$ *be a Moran process on a graph $G = (V, E)$ and let $\emptyset \subset S \subset V$. If $r \geqslant 1$, then*

$$\mathbb{E}[\phi(X_{i+1}) - \phi(X_i) \mid X_i = S] \geqslant \left(1 - \frac{1}{r}\right) \cdot \frac{1}{n^3},$$

with equality if and only if $r = 1$. For $r < 1$,

$$\mathbb{E}[\phi(X_{i+1}) - \phi(X_i) \mid X_i = S] < \frac{r - 1}{n^3}.$$

To bound the expected absorption time, we use martingale techniques. It is well known how to bound the expected absorption time using a potential function that decreases in expectation until absorption. This has been made explicit by Hajek [9] and we use the following formulation based on that of He and Yao [10]. The proof is essentially theirs but is modified to give a slightly stronger result.

Theorem 1 *Let* $(Y_i)_{i \geqslant 0}$ *be a Markov chain with state space Ω, where Y_0 is chosen from some set $I \subseteq \Omega$. If there are constants $k_1, k_2 > 0$ and a nonnegative function $\psi: \Omega \to \mathbb{R}$ such that*

- $\psi(S) = 0$ *for some $S \in \Omega$,*
- $\psi(S) \leqslant k_1$ *for all $S \in I$ and*
- $\mathbb{E}[\psi(Y_i) - \psi(Y_{i+1}) \mid Y_i = S] \geqslant k_2$ *for all $i \geqslant 0$ and all S with $\psi(S) > 0$,*

then $\mathbb{E}[\tau] \leqslant k_1/k_2$, where $\tau = \min\{i : \psi(Y_i) = 0\}$.

Using Theorem 1, we can prove the following upper bounds for the absorption time τ in the cases where $r < 1$ and $r > 1$, respectively.

Theorem 2 *Let* $G = (V, E)$ *be a graph of order n. For $r < 1$ and any $S \subseteq V$, the absorption time τ of the Moran process on G satisfies*

$$\mathbb{E}[\tau \mid X_0 = S] \leqslant \frac{1}{1 - r} n^3 \phi(S).$$

Theorem 3 *Let* $G = (V, E)$ *be a graph of order n. For $r > 1$ and any $S \subseteq V$, the absorption time τ of the Moran process on G satisfies*

$$\mathbb{E}[\tau \mid X_0 = S] \leqslant \frac{r}{r - 1} n^3 (\phi(G) - \phi(S))$$

$$\leqslant \frac{r}{r - 1} n^4.$$

The case $r = 1$ is more complicated as Lemma 3 shows that the expectation is constant. However, this allows us to use standard martingale techniques and the proof of the following is partly adapted from the proof of Lemma 3.4 in [15].

Theorem 4 *The expected absorption time for the Moran process* $(X_i)_{i \geqslant 0}$ *with* $r = 1$ *on a graph* $G = (V, E)$ *is at most* $n^4(\phi(G)^2 - \mathbb{E}[\phi(X_0)^2])$.

Approximation Algorithms

We now have all the components needed to present our fully polynomial randomized approximation schemes (FPRAS) for the problem of computing the fixation probability of a graph, where $r \geqslant 1$, and for computing the extinction probability for all $r > 0$. In the following two theorems, we give algorithms whose running times are polynomial in n, r, and $\frac{1}{\varepsilon}$. For the algorithms to run in time polynomial in the length of the input and thus meet the definition of FPRAS, r must be encoded in unary.

Theorem 5 *There is an FPRAS for* MORAN FIX-ATION, *for* $r \geqslant 1$.

Proof (sketch) The algorithm is as follows. If $r = 1$ then we return $\frac{1}{n}$. Otherwise, we simulate the Moran process on G for $T = \lceil \frac{8r}{r-1} N n^4 \rceil$ steps, $N = \lceil \frac{1}{2}\varepsilon^{-2} n^2 \ln 16 \rceil$ times and compute the proportion of simulations that reached fixation. If any simulation has not reached absorption (fixation or extinction) after T steps, we abort and immediately return an error value.

Note that each transition of the Moran process can be simulated in $\mathcal{O}(1)$ time. Maintaining arrays of the mutant and nonmutant vertices allows the reproducing vertex to be chosen in constant time, and storing a list of each vertex's neighbors allows the same for the vertex where the offspring is sent. Therefore, the total running time is $\mathcal{O}(NT)$ steps, which is polynomial in n and $\frac{1}{\varepsilon}$, as required.

For $i \in \{1, \ldots, N\}$, let $X_i = 1$ if the ith simulation of the Moran process reaches fixation and $X_i = 0$ otherwise. Assuming all simulation runs reach absorption, the output of the algorithm is $p = \frac{1}{N}\sum_i X_i$. □

Note that this technique fails for disadvantageous mutants ($r < 1$) because there is no analogue of Lemma 1 giving a polynomial lower bound on $f_{G,r}$. As such, an exponential number of simulations may be required to achieve the desired error probability. However, we can give an FPRAS for the extinction probability for all

$r > 0$. Although the extinction probability is just $1 - f_{G,r}$, there is no contradiction because a small relative error in $1 - f_{G,r}$ does not translate into a small relative error in $f_{G,r}$ when $f_{G,r}$ is, itself, small.

Theorem 6 *There is an FPRAS for* MORAN EX-TINCTION *for all* $r > 0$.

Proof (sketch) The algorithm and its correctness proof are essential as in the previous theorem. If $r = 1$, we return $1 - \frac{1}{n}$. Otherwise, we run $N = \lceil \frac{1}{2}\varepsilon^{-2}(r + n)^2 \ln 16 \rceil$ simulations of the Moran process on G for $T(r)$ steps each, where

$$T(r) = \begin{cases} \lceil \frac{8r}{r-1} N n^4 \rceil & \text{if } r > 1 \\ \lceil \frac{8}{1-r} N n^3 \rceil & \text{if } r < 1. \end{cases}$$

If any simulation has not reached absorption within $T(r)$ steps, we return an error value; otherwise, we return the proportion p of simulations that reached extinction. □

It remains open whether other techniques could lead to an FPRAS for MORAN FIXATION when $r < 1$.

Recommended Reading

1. Aldous DJ, Fill JA (2002) Reversible Markov chains and random walks on graphs. Monograph in preparation. Available at http://www.stat.berkeley.edu/aldous/RWG/book.html
2. Berger E (2001) Dynamic monopolies of constant size. J Comb Theory Ser B 83:191–200
3. Broom M, Hadjichrysanthou C, Rychtář J (2010) Evolutionary games on graphs and the speed of the evolutionary process. Proc R Soc A 466(2117):1327–1346
4. Broom M, Hadjichrysanthou C, Rychtář J (2010) Two results on evolutionary processes on general non-directed graphs. Proc R Soc A 466(2121):2795–2798
5. Broom M, Rychtář J (2008) An analysis of the fixation probability of a mutant on special classes of non-directed graphs. Proc R Soc A 464(2098):2609–2627
6. Broom M, Rychtář J, Stadler B (2009) Evolutionary dynamics on small order graphs. J Interdiscip Math 12:129–140
7. Durrett R (1988) Lecture notes on particle systems and percolation. Wadsworth Publishing Company, Pacific Grove
8. Easley D, Kleinberg J (2010) Networks, crowds, and markets: reasoning about a highly connected world. Cambridge University Press, New York

9. Hajek B (1982) Hitting-time and occupation-time bounds implied by drift analysis with applications. Adv Appl Probab 14(3):502–525
10. He J, Yao X (2001) Drift analysis and average time complexity of evolutionary algorithms. Artif Intell 127:57–85
11. Karp RM, Luby M (1983) Monte-Carlo algorithms for enumeration and reliability problems. In: Proceedings of 24th annual IEEE symposium on foundations of computer science (FOCS), Tucson, pp 56–64
12. Kempel D, Kleinberg J, Tardos E (2005) Influential nodes in a diffusion model for social networks. In: Proceedings of the 32nd international colloquium on automata, languages and programming (ICALP), Lisbon. Lecture notes in computer science, vol 3580, pp 1127–1138. Springer
13. Lieberman E, Hauert C, Nowak MA (2005) Evolutionary dynamics on graphs. Nature 433:312–316
14. Liggett TM (1985) Interacting particle systems. Springer, New York
15. Luby M, Randall D, Sinclair A (2001) Markov chain algorithms for planar lattice structures. SIAM J Comput 31(1):167–192
16. Mertzios GB, Nikoletseas S, Raptopoulos C, Spirakis PG (2013) Natural models for evolution on networks. Theor Comput Sci 477:76–95
17. Moran PAP (1958) Random processes in genetics. Proc Camb Philos Soc 54(1):60–71
18. Mossel E, Roch S (2007) On the submodularity of influence in social networks. In: Proceedings of the 39th annual ACM symposium on theory of computing (STOC), San Diego, pp 128–134
19. Nowak MA (2006) Evolutionary dynamics: exploring the equations of life. Harvard University Press, Cambridge
20. Taylor C, Iwasa Y, Nowak MA (2006) A symmetry of fixation times in evolutionary dynamics. J Theor Biol 243(2):245–251

Years and Authors of Summarized Original Work

1996; Bartal, Fakcharoenphol, Rao, Talwar
2004; Bartal, Fakcharoenphol, Rao, Talwar

Problem Definition

This problem is to construct a random tree metric that probabilistically approximates a given arbitrary metric well. A solution to this problem is useful as the first step for numerous approximation algorithms because usually solving problems on trees is easier than on general graphs. It also finds applications in on-line and distributed computation.

It is known that tree metrics approximate general metrics badly, e.g., given a cycle C_n with n nodes, any tree metric approximating this graph metric has distortion $\Omega(n)$ [17]. However, Karp [15] noticed that a random spanning tree of C_n approximates the distances between any two nodes in C_n well in expectation. Alon, Karp, Peleg, and West [1] then proved a bound of $\exp(O(\sqrt{\log n \log \log n}))$ on an average distortion for approximating any graph metric with its spanning tree.

Bartal [2] formally defined the notion of probabilistic approximation.

Approximating Metric Spaces by Tree Metrics

Jittat Fakcharoenphol[1], Satish Rao[2], and Kunal Talwar[3]
[1]Department of Computer Engineering, Kasetsart University, Bangkok, Thailand
[2]Department of Computer Science, University of California, Berkeley, CA, USA
[3]Microsoft Research, Silicon Valley Campus, Mountain View, CA, USA

Keywords

Embedding general metrics into tree metrics

Notations

A graph $G = (V, E)$ with an assignment of non-negative weights to the edges of G defines a metric space (V, d_G) where for each pair $u, v \in V$, $d_G(u, v)$ is the shortest path distance between u and v in G. A metric (V, d) is a *tree metric* if there exists some tree $T = (V', E')$ such that $V \subseteq V'$ and for all $u, v \in V$, $d_T(u, v) = d(u, v)$. The metric (V, d) is also called a metric induced by T.

Given a metric (V, d), a distribution $\mathcal{D}$ over tree metrics over V α-*probabilistically approximates* d if every tree metric $d_T \in \mathcal{D}$, $d_T(u, v) \geq d(u, v)$ and $\mathrm{E}_{d_T \in \mathcal{D}}[d_T(u, v)] \leq \alpha \cdot d(u, v)$, for every $u, v \in V$. The quantity α is referred to as the *distortion* of the approximation.

Although the definition of probabilistic approximation uses a distribution $\mathcal{D}$ over tree metrics, one is interested in a procedure that constructs a random tree metric distributed according to $\mathcal{D}$, i.e., an algorithm that produces a random tree metric that probabilistically approximates a given metric. The problem can be formally stated as follows.

Problem (APPROX-TREE)

INPUT: a metric (V, d)

OUTPUT: a tree metric (V, d_T) sampled from a distribution $\mathcal{D}$ over tree metrics that α-probabilistically approximates (V, d).

Bartal then defined a class of tree metrics, called hierarchically well-separated trees (HST), as follows. A *k-hierarchically well-separated tree* (*k*-HST) is a rooted weighted tree satisfying two properties: the edge weight from any node to each of its children is the same, and the edge weights along any path from the root to a leaf are decreasing by a factor of at least k. These properties are important to many approximation algorithms.

Bartal showed that any metric on n points can be probabilistically approximated by a set of k-HST's with $O(\log^2 n)$ distortion, an improvement from $\exp(O(\sqrt{\log n \log \log n}))$ in [1]. Later Bartal [3], following the same approach as in Seymour's analysis on the Feedback Arc Set problem [18], improved the distortion down to $O(\log n \log \log n)$. Using a rounding procedure of Calinescu, Karloff, and Rabani [5], Fakcharoenphol, Rao, and Talwar [9] devised an algorithm that, in expectation, produces a tree with $O(\log n)$ distortion. This bound is tight up to a constant factor.

Key Results

A tree metric is closely related to graph decomposition. The randomized rounding procedure of Calinescu, Karloff, and Rabani [5] for the 0-extension problem decomposes a graph into pieces with bounded diameter, cutting each edge with probability proportional to its length and a ratio between the numbers of nodes at certain distances. Fakcharoenphol, Rao, and Talwar [9] used the CKR rounding procedure to decompose the graph recursively and obtained the following theorem.

Theorem 1 *Given an n-point metric (V, d), there exists a randomized algorithm, which runs in time $O(n^2)$, that samples a tree metric from the distribution $\mathcal{D}$ over tree metrics that $O(\log n)$-probabilistically approximates (V, d). The tree is also a 2-HST.*

The bound in Theorem 1 is tight, as Alon et al. [1] proved the bound of an $\Omega(\log n)$ distortion when (V, d) is induced by a grid graph. Also note that it is known (as folklore) that even embedding a line metric onto a 2-HST requires distortion $\Omega(\log n)$.

If the tree is required to be a k-HST, one can apply the result of Bartal, Charikar, and Raz [4] which states that any 2-HST can be $O(k / \log k)$-probabilistically approximated by k-HST, to obtain an expected distortion of $O(k \log n / \log k)$.

Finding a distribution of tree metrics that probabilistically approximates a given metric has a dual problem that is to find a single tree T with small average weighted stretch. More specifically, given weight c_{uv} on edges, find a tree metric d_T such that for all $u, v \in V d_T(u, v) \geq d(u, v)$ and $\sum_{u,v \in V} c_{uv} \cdot d_T(u, v) \leq \alpha \sum_{u,v \in V} c_{uv} \cdot d(u, v)$.

Charikar, Chekuri, Goel, Guha, and Plotkin [6] showed how to find a distribution of $O(n \log n)$ tree metrics that α-probabilistically approximates a given metric, provided that one can solve the dual problem. The algorithm in Theorem 1 can be derandomized by the method of conditional expectation to find the required tree metric with $\alpha = O(\log n)$. Another algorithm based on modified region growing techniques is presented in [9], and independently by Bartal.

Theorem 2 *Given an n-point metric (V, d), there exists a polynomial-time deterministic algorithm that finds a distribution $\mathcal{D}$ over $O(n \log n)$ tree metrics that $O(\log n)$-probabilistically approximates (V, d).*

Note that the tree output by the algorithm contains Steiner nodes, however Gupta [10] showed how to find another tree metric without Steiner nodes while preserving all distances within a constant factor.

Applications

Metric approximation by random trees has applications in on-line and distributed computation, since randomization works well against oblivious adversaries, and trees are easy to work with and maintain. Alon et al. [1] first used tree embedding to give a competitive algorithm for the k-server problem. Bartal [3] noted a few problems in his paper: metrical task system, distributed paging, distributed k-server problem, distributed queuing, and mobile user.

After the paper by Bartal in 1996, numerous applications in approximation algorithms have been found. Many approximation algorithms work for problems on tree metrics or HST metrics. By approximating general metrics with these metrics, one can turn them into algorithms for general metrics, while, usually, losing only a factor of $O(\log n)$ in the approximation factors. Sample problems are metric labeling, buy-at-bulk network design, and group Steiner trees. Recent applications include an approximation algorithm to the Unique Games [12], information network design [13], and oblivious network design [11].

The SIGACT News article [8] is a review of the metric approximation by tree metrics with more detailed discussion on developments and techniques. See also [3, 9], for other applications.

Open Problems

Given a metric induced by a graph, some application, e.g., solving a certain class of linear systems, does not only require a tree metric, but a tree metric induced by a spanning tree of the graph. Elkin, Emek, Spielman, and Teng [7] gave an algorithm for finding a spanning tree with average distortion of $O(\log^2 n \log \log n)$. It remains open if this bound is tight.

Cross-References

▶ Metrical Task Systems
▶ Sparse Graph Spanners

Recommended Reading

1. Alon N, Karp RM, Peleg D, West D (1995) A graph-theoretic game and its application to the k-server problem. SIAM J Comput 24:78–100
2. Bartal Y (1996) Probabilistic approximation of metric spaces and its algorithmic applications. In: FOCS '96: proceedings of the 37th annual symposium on foundations of computer science, Washington, DC. IEEE Computer Society, pp 184–193
3. Bartal Y (1998) On approximating arbitrary metrics by tree metrics. In: STOC '98: proceedings of the thirtieth annual ACM symposium on theory of computing. ACM Press, New York, pp 161–168
4. Bartal Y, Charikar M, Raz D (2001) Approximating min-sum k-clustering in metric spaces. In: STOC '01: proceedings of the thirtythird annual ACM symposium on theory of computing. ACM Press, New York, pp 11–20
5. Calinescu G, Karloff H, Rabani Y (2001) Approximation algorithms for the 0-extension problem. In: SODA '01: proceedings of the twelfth annual ACM-SIAM symposium on Discrete algorithms. Society for Industrial and Applied Mathematics, Philadelphia, pp 8–16
6. Charikar M, Chekuri C, Goel A, Guha S (1998) Rounding via trees: deterministic approximation algorithms for group Steiner trees and k-median. In: STOC '98: proceedings of the thirtieth annual ACM symposium on theory of computing. ACM Press, New York, pp 114–123
7. Elkin M, Emek Y, Spielman DA, Teng S-H (2005) Lower-stretch spanning trees. In: STOC '05: proceedings of the thirty-seventh annual ACM symposium on theory of computing. ACM Press, New York, pp 494–503
8. Fakcharoenphol J, Rao S, Talwar K (2004) Approximating metrics by tree metrics. SIGACT News 35:60–70
9. Fakcharoenphol J, Rao S, Talwar K (2004) A tight bound on approximating arbitrary metrics by tree metrics. J Comput Syst Sci 69:485–497
10. Gupta A (2001) Steiner points in tree metrics don't (really) help. In: SODA '01: proceedings of the twelfth annual ACM-SIAM symposium on discrete algorithms. Society for Industrial and Applied Mathematics, Philadelphia, pp 220–227
11. Gupta A, Hajiaghayi MT, Räcke H (2006) Oblivious network design. In: SODA '06: proceedings of the seventeenth annual ACM-SIAM symposium on discrete algorithm. ACM Press, New York, pp 970–979

12. Gupta A, Talwar K (2006) Approximating unique games. In: SODA '06: proceedings of the seventeenth annual ACM-SIAM symposium on discrete algorithm, New York. ACM Press, New York, pp 99–106
13. Hayrapetyan A, Swamy C, Tardos É (2005) Network design for information networks. In: SODA '05: proceedings of the sixteenth annual ACM-SIAM symposium on discrete algorithms. Society for Industrial and Applied Mathematics, Philadelphia, pp 933–942
14. Indyk P, Matousek J (2004) Low-distortion embeddings of finite metric spaces. In: Goodman JE, O'Rourke J (eds) Handbook of discrete and computational geometry. Chapman&Hall/CRC, Boca Raton, chap. 8
15. Karp R (1989) A 2k-competitive algorithm for the circle. Manuscript
16. Matousek J (2002) Lectures on discrete geometry. Springer, New York
17. Rabinovich Y, Raz R (1998) Lower bounds on the distortion of embedding finite metric spaces in graphs. Discret Comput Geom 19:79–94
18. Seymour PD (1995) Packing directed circuits fractionally. Combinatorica 15:281–288

Approximating the Diameter

Liam Roditty
Department of Computer Science, Bar-Ilan University, Ramat-Gan, Israel

Keywords

Diameter; Graph algorithms; Shortest paths

Years and Authors of Summarized Original Work

1999; Aingworth, Chekuri, Indyk, Motwani
2013; Roditty, Vassilevska Williams
2014; Chechik, Larkin, Roditty, Schoenebeck, Tarjan, Vassilevska Williams

Problem Definition

The diameter of a graph is the largest distance between its vertices. Closely related to the diameter is the radius of the graph. The center of a graph is a vertex that minimizes the maximum distance to all other nodes, and the radius is the distance from the center to the node furthest from it. Being able to compute the diameter, center, and radius of a graph efficiently has become an increasingly important problem in the analysis of large networks [11]. For general weighted graphs the only known way to compute the exact diameter and radius is by solving the all-pairs shortest paths problem (APSP). Therefore, a natural question is whether it is possible to get faster diameter and radius algorithms by settling for an approximation. For a graph G with diameter D, a c-approximation of D is a value $\hat{D}$ such that $\hat{D} \in [D/c, D]$. The question is whether a c-approximation can be computed in sub-cubic time.

Key Results

For *sparse* directed or undirected unweighted graphs, the best-known algorithm (ignoring poly-logarithmic factors) for APSP, diameter, and radius does breadth-first search (BFS) from every node and hence runs in $O(mn)$ time, where m is the number of edges in the graph. For dense directed *unweighted* graphs, it is possible to compute both the diameter and the radius using fast matrix multiplication (this is folklore; for a recent simple algorithm, see [5]), thus obtaining $\tilde{O}(n^{\omega})$ time algorithms, where $\omega < 2.38$ is the matrix multiplication exponent [4, 9, 10] and n is the number of nodes in the graph.

A 2-approximation for both the diameter and the radius of an undirected graph can be obtained in $O(m + n)$ time using BFS from an arbitrary node. For APSP, Dor et al. [6] show that any $(2 - \epsilon)$-approximation algorithm in unweighted undirected graphs running in $T(n)$ time would imply an $O(T(n))$ time algorithm for Boolean matrix multiplication (BMM). Hence a priori it could be that $(2 - \epsilon)$-approximating the diameter and radius of a graph may also require solving BMM.

Aingworth et al. [1] showed that this is not the case by presenting a sub-cubic $(2 - \epsilon)$-approximation algorithm for the diameter in both directed and undirected graphs that does not use fast matrix multiplication. Their algorithm computes in $\tilde{O}(m\sqrt{n} + n^2)$ time an estimate $\hat{D}$ such that $\hat{D} \in [\lfloor 2D/3 \rfloor, D]$. Berman and Kasiviswanathan [2] showed that for the radius problem the approach of Aingworth et al. can be used to obtain in $\tilde{O}(m\sqrt{n} + n^2)$ time an estimate $\hat{r}$ that satisfies $r \in [\hat{r}, 3/2r]$, where r is the radius of the graph. For weighted graphs the algorithm of Aingworth et al. [1] guarantees that the estimate $\hat{D}$ satisfies $\hat{D} \in [\lfloor \frac{2}{3} \cdot D \rfloor - (M-1), D]$, where M is the maximum edge weight in the graph.

Roditty and Vassilevska Williams [8] gave a Las Vegas algorithm running in expected $\tilde{O}(m\sqrt{n})$ time that has the same approximation guarantee as Aingworth et al. for the diameter and the radius. They also showed that obtaining a $(\frac{3}{2} - \epsilon)$-approximation algorithm running in $O(n^{2-\delta})$ time in sparse undirected and unweighted graphs for constant $\epsilon, \delta > 0$ would be difficult, as it would imply a fast algorithm for CNF Satisfiability, violating the widely believed Strong Exponential Time Hypothesis of Impagliazzo, Paturi, and Zane [7].

Chechik et al. [3] showed that it is possible to remove the additive error while still keeping the running time (in terms of n) subquadratic for sparse graphs. They present two *deterministic* algorithms with $\frac{3}{2}$-approximation for the diameter, one running in $\tilde{O}(m^{\frac{3}{2}})$ time and one running in $\tilde{O}(mn^{\frac{3}{2}})$ time.

Open Problems

The main open problem is to understand the relation between the diameter computation and the APSP problem. Is there a truly sub-cubic time algorithm for computing the exact diameter or can we show sub-cubic equivalence between the exact diameter computation and APSP problem?

Another important open problem is to find an algorithm that distinguishes between graphs of diameter two to graphs of diameter three in sub-cubic time. Alternatively, can we show that it is sub-cubic equivalent to the problem of exact diameter?

Recommended Reading

1. Aingworth D, Chekuri C, Indyk P, Motwani R (1999) Fast estimation of diameter and shortest paths (without matrix multiplication). SIAM J Comput 28(4):1167–1181
2. Berman P, Kasiviswanathan SP (2007) Faster approximation of distances in graphs. In: Proceedings of the WADS, Halifax, pp 541–552
3. Chechik S, Larkin D, Roditty L, Schoenebeck G, Tarjan RE, Williams VV (2014) Better approximation algorithms for the graph diameter. In: SODA, Portland, pp 1041–1052
4. Coppersmith D, Winograd S (1990) Matrix multiplication via arithmetic progressions. J Symb Comput 9(3):251–280
5. Cygan M, Gabow HN, Sankowski P (2012) Algorithmic applications of Baur-strassen's theorem: shortest cycles, diameter and matchings. In: Proceedings of the FOCS, New Brunswick
6. Dor D, Halperin S, Zwick U (2000) All-pairs almost shortest paths. SIAM J Comput 29(5):1740–1759
7. Impagliazzo R, Paturi R, Zane F (2001) Which problems have strongly exponential complexity? J Comput Syst Sci 63(4):512–530
8. Roditty L, Vassilevska Williams V (2013) Fast approximation algorithms for the diameter and radius of sparse graphs. In: Proceedings of the 45th annual ACM symposium on theory of computing, STOC '13, Palo Alto. ACM, New York, pp 515–524. doi:10.1145/2488608. 2488673, http://doi.acm.org/10.1145/2488608.2488673
9. Stothers A (2010) On the complexity of matrix multiplication. PhD thesis, University of Edinburgh
10. Vassilevska Williams V (2012, to appear) Multiplying matrices faster than Coppersmith-Winograd. In: Proceedings of the STOC, New York
11. Watts DJ, Strogatz SH (1998) Collective dynamics of 'small-world' networks. Nature 393:440–442

Approximating the Partition Function of Two-Spin Systems

Pinyan Lu[1] and Yitong Yin[2]
[1] Microsoft Research Asia, Shanghai, China
[2] Nanjing University, Jiangsu, Nanjing, Gulou, China

Keywords

Approximate counting; Partition function; Two-state spin systems

Years and Authors of Summarized Original Work

1993; Jerrum, Sinclair
2003; Goldberg, Jerrum, Paterson
2006; Weitz
2012; Sinclair, Srivastava, Thurley
2013; Li, Lu, Yin
2015; Sinclair, Srivastava, Štefankovič, Yin

Problem Definition

Spin systems are well-studied objects in statistical physics and applied probability. An instance of a spin system is an undirected graph $G = (V, E)$ of n vertices. A *configuration* of a two-state spin system, or simply just *two-spin system* on G, is an assignment $\sigma : V \to \{0, 1\}$ of two *spin states* "0" and "1" (sometimes called "−" and "+" or seen as two colors) to the vertices of G. Let $\mathbf{A} = \begin{bmatrix} A_{0,0} & A_{0,1} \\ A_{1,0} & A_{1,1} \end{bmatrix}$ be a nonnegative symmetric matrix which specifies the local interactions between adjacent vertices and $\mathbf{b} = \begin{bmatrix} b_0 \\ b_1 \end{bmatrix}$ a nonnegative vector which specifies preferences of individual vertices over the two spin states. For each configuration $\sigma \in \{0, 1\}^V$, its weight is then given by the following product:

$$w(\sigma) = \prod_{\{u,v\} \in E} A_{\sigma(u),\sigma(v)} \prod_{v \in V} b_{\sigma(v)}.$$

The *partition function* $Z_{\mathbf{A},\mathbf{b}}(G)$ of a two-spin system on G is defined to be the following exponential summation over all possible configurations:

$$Z_{\mathbf{A},\mathbf{b}}(G) = \sum_{\sigma \in \{0,1\}^V} w(\sigma).$$

Up to normalization, $\mathbf{A}$ and $\mathbf{b}$ can be described by three parameters, so that one can assume that $\mathbf{A} = \begin{bmatrix} \beta & 1 \\ 1 & \gamma \end{bmatrix}$ and $\mathbf{b} = \begin{bmatrix} \lambda \\ 1 \end{bmatrix}$, where $\beta, \gamma \geq 0$ are the *edge activities* and $\lambda > 0$ is the *external field*. Since the roles of the two spin states are symmetric, it can be further assumed that $\beta \leq$ γ without loss of generality. Therefore, a two-spin system is completely specified by the three parameters (β, γ, λ) where it holds that $0 \leq \beta \leq \gamma$ and $\lambda > 0$. The resulting partition function is written as $Z_{(\beta,\gamma,\lambda)}(G) = Z_{\mathbf{A},\mathbf{b}}(G)$ and as $Z(G)$ for short if the parameters are clear from the context.

The two-spin systems are classified according to their parameters into two families with distinct physical and computational properties: the *ferromagnetic* two-spin systems ($\beta\gamma > 1$) in which neighbors favor agreeing spin states and the *antiferromagnetic* two-spin systems ($\beta\gamma < 1$) in which neighbors favor disagreeing spin states. Two-spin systems with $\beta\gamma = 1$ are trivial in both physical and computational senses and thus are usually not considered. The model of two-spin systems covers some of the most extensively studied statistical physics models as special cases, as well as being accepted in computer science as a framework for counting problems, for examples:

- When $\beta = 0$, $\gamma = 1$, and $\lambda = 1$, the $Z_{(\beta,\gamma,\lambda)}(G)$ gives the number of independent sets (or vertex covers) of G.
- When $\beta = 0$ and $\gamma = 1$, the $Z_{(\beta,\gamma,\lambda)}(G)$ is the partition function of the hardcore model with fugacity λ on G.
- When $\beta = \gamma$, the $Z_{(\beta,\gamma,\lambda)}(G)$ is the partition function of the Ising model with edge activity β and external field λ on G.

Given a set of parameters (β, γ, λ), the computational problem TWO-SPIN(β, γ, λ) is the problem of computing the value of the partition function $Z_{(\beta,\gamma,\lambda)}(G)$ when the graph G is given as input. This problem is known to be #P-hard except for the trivial cases where $\beta\gamma = 1$ or $\beta = \gamma = 0$ [1]. Therefore, the main focus here is the efficient approximation algorithms for TWO-SPIN(β, γ, λ). Formally, a fully polynomial-time approximation scheme (FPTAS) is an algorithm which takes G and any $\varepsilon > 0$ as input and outputs a number $\hat{Z}$ satisfying $Z(G)\exp(-\varepsilon) \leq \hat{Z} \leq Z(G)\exp(\varepsilon)$ within time polynomial in n and $1/\varepsilon$; and a fully polynomial-time randomized approximation scheme (FPRAS) is its randomized relaxation in which randomness is allowed and the above accuracy of approximation is required to be satisfied with high probability.

For many important two-spin systems (e.g., independent sets, antiferromagnetic Ising model), it is NP-hard to approximate the partition function on graphs of unbounded degrees. In these cases, the problem is further refined to consider the approximation algorithms for TWO-SPIN(β, γ, λ) on graphs with *bounded maximum degree*. In addition, in order to study the approximation algorithms on graphs which has bounded average degree or on special classes of lattice graphs, the approximation of partition function is studied on classes of graphs with bounded *connective constant*, a natural and well-studied notion of average degree originated from statistical physics.

Therefore, the main problem of interest is to characterize the regimes of parameters (β, γ, λ) for which there exist efficient approximation algorithms for TWO-SPIN(β, γ, λ) on classes of graphs with bounded maximum degree $\Delta_{\max}$, or on classes of graphs with bounded connective constant Δ, or on all graphs.

Key Results

Given a two-spin system on graph $G = (V, E)$, a natural probability distribution μ over all configurations $\sigma \in \{0, 1\}^V$, called the *Gibbs measure*, can be defined by $\mu(\sigma) = \frac{w(\sigma)}{Z(G)}$, where $w(\sigma) = \prod_{\{u,v\} \in E} A_{\sigma_u, \sigma_v} \prod_{v \in V} b_{\sigma_v}$ is the weight of σ and the normalizing factor $Z(G)$ is the partition function.

The Gibbs measure defines a marginal distribution at each vertex. Suppose that a configuration σ is sampled according to the Gibbs measure μ. Let p_v denote the probability of vertex v having spin state "0" in σ; and for a fixed configuration $\tau_\Lambda \in \{0, 1\}^\Lambda$ partially specified over vertices in $\Lambda \subset V$, let $p_v^{\tau_\Lambda}$ denote the probability of vertex v having spin state "0" conditioning on that the configuration of vertices in Λ in σ is as specified by τ_Λ.

The marginal probability plays a key role in computing the partition function. Indeed, the marginal probability $p_v^{\tau_\Lambda}$ itself is a quantity of main interest in many applications such as probabilistic inference. In addition, due to the standard procedure of self-reduction, an FPTAS for the partition function $Z(G)$ can be obtained if the value of $p_v^{\tau_\Lambda}$ can be approximately computed with an additive error ε in time polynomial in both n and $1/\varepsilon$. This reduces the problem of approximating the partition function (with multiplicative errors) to approximating the marginal probability (with additive errors), which is achieved either by rapidly mixing random walks or by recursions exhibiting a decay of correlation.

Ferromagnetic Two-Spin Systems

For the ferromagnetic case, the problem TWO-SPIN(β, γ, λ) is considered for $\beta\gamma > 1$ and without loss of generality for $\beta \leq \gamma$.

In a seminal work [3], Jerrum and Sinclair gave an FPRAS for approximately computing the partition function of the ferromagnetic Ising model, which is the TWO-SPIN(β, γ, λ) problem with $\beta = \gamma > 1$.

The algorithm uses the Markov chain Monte Carlo (MCMC) method; however very interestingly, it does not directly apply the random walk over configurations of two-spin system since such random walk might have a slow mixing time. Instead, it first transforms the two-spin system into configurations of the so-called "subgraphs world": each such configuration is a subgraph of G. A random walk over the subgraph configurations is applied and proved to be rapidly mixing for computing the new partition function defined over subgraphs, which is shown to be equal to the partition function $Z(G)$ of the two-spin system. This equivalence is due to that this transformation between the "spins world" and the "subgraphs world" is actually a holographic transformation, which is guaranteed to preserve the value of the partition function.

The result of [3] can be stated as the following theorem.

Theorem 1 *If $\beta = \gamma > 1$ and $\lambda > 0$, then there is an FPRAS for* TWO-SPIN(β, γ, λ).

The algorithm actually works for a stronger setting where the external fields are local (vertices have different external fields) as long as the external fields are homogeneous (all have the same preference over spin states).

For the two-spin system with general β and γ, one can translate it to the Ising model where $\beta = \gamma$ by delegating the effect of the general β, γ to the degree-dependent effective external fields. This extends the FPRAS for the ferromagnetic Ising model to certain regime of ferromagnetic two-spin systems, stated as follows.

Theorem 2 ([2, 6]) *If $\beta < \gamma$, $\beta\gamma > 1$, and $\lambda \leq \gamma/\beta$, then there is an FPRAS for* TWO-SPIN(β, γ, λ).

If one is restricted to the deterministic algorithms for approximating the partition function, then a deterministic FPTAS is known for a strictly smaller regime, implicitly stated in the following theorem.

Theorem 3 ([7]) *There is a continuous monotonically increasing function $\Gamma(\gamma)$ defined on $[1, +\infty)$ satisfying (1) $\Gamma(1) = 1$, (2) $1 < \Gamma(\gamma) < \gamma$ for all $\gamma > 1$, and (3) $\lim_{\gamma \to +\infty} \frac{\Gamma(\gamma)}{\gamma} = 1$, such that there is an FPTAS for* TWO-SPIN(β, γ, λ) *if $\beta\gamma > 1$, $\beta \leq \Gamma(\gamma)$, and $\lambda \leq 1$.*

This deterministic FPTAS uses the same holographic transformation from two-spin systems to the "subgraphs world" as in [3], and it approximately computes the marginal probability defined in the subgraphs world by a recursion. The accuracy of the approximation is guaranteed by the decay of correlation. This technique is more extensively and successfully used for the antiferromagnetic two-spin systems.

On the other hand, assuming certain complexity assumptions, it is unlikely that for every ferromagnetic two-spin system its partition function is easy to approximate.

Theorem 4 ([6]) *For any $\beta < \gamma$ with $\beta\gamma > 1$, there is a λ_0 such that* TWO-SPIN(β, γ, λ) *is #BIS-hard for all $\lambda \geq \lambda_0$.*

Antiferromagnetic Two-Spin Systems

For the antiferromagnetic case, the problem TWO-SPIN(β, γ, λ) is considered for $\beta\gamma < 1$ and without loss of generality for $\beta \leq \gamma$.

In [2], a heatbath random walk over spin configurations is applied to obtain an FPRAS for TWO-SPIN(β, γ, λ) for a regime of antiferromagnetic two-spin systems.

The regime of antiferromagnetic two-spin systems whose partition function is efficiently approximable is characterized by the *uniqueness condition*.

Given parameters (β, γ, λ) and $d \geq 1$, the *tree recursion $f(x)$* is given by

$$f(x) = \lambda \left(\frac{\beta x + 1}{x + \gamma} \right)^d. \tag{1}$$

For antiferromagnetic (β, γ, λ), the function $f(x)$ is decreasing in x; thus, there is a unique positive fixed point $\hat{x}$ satisfying $\hat{x} = f(\hat{x})$. Consider the absolute derivative of $f(x)$ at the fixed point:

$$\left| f'(\hat{x}) \right| = \frac{d(1 - \beta\gamma)\hat{x}}{(\beta\hat{x} + 1)(\hat{x} + \gamma)}.$$

Definition 1 Let $0 \leq \beta \leq \gamma$, $\beta\gamma < 1$, and $d \geq 1$. The *uniqueness condition* UNIQUE$(\beta, \gamma, \lambda, d)$ is satisfied if $|f'(\hat{x})| < 1$; and the condition NON-UNIQUE$(\beta, \gamma, \lambda, d)$ is satisfied if $|f'(\hat{x})| > 1$.

The condition UNIQUE$(\beta, \gamma, \lambda, d)$ holds if and only if the dynamical system (1) converges to its unique fixed point $\hat{x}$ at an exponential rate. The name uniqueness condition is due to that UNIQUE$(\beta, \gamma, \lambda, d)$ implies the uniqueness of the Gibbs measure of two-spin system of parameters (β, γ, λ) on the Bethe lattice (i.e., the infinite d-regular tree) and NON-UNIQUE$(\beta, \gamma, \lambda, d)$ implies that there are more than one such measures (Fig. 1).

Efficient approximation algorithms for TWO-SPIN(β, γ, λ) are discovered for special cases of antiferromagnetic two-spin systems within the uniqueness regime, including the hardcore model [12], the antiferromagnetic Ising model [8], and the antiferromagnetic two-spin

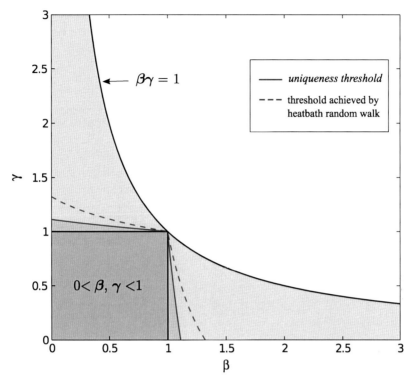

Approximating the Partition Function of Two-Spin Systems, Fig. 1 The regime of (β, γ) for which the uniqueness condition $\text{UNIQUE}(\beta, \gamma, \lambda, d)$ holds for $\lambda = 1$ and for all integer $d \geq 1$

systems without external field [4], and finally for all antiferromagnetic two-spin systems within the uniqueness regime [5].

Theorem 5 ([5]) *For* $0 \leq \beta \leq \gamma$ *and* $\beta\gamma < 1$, *there is an FPTAS for* $\text{TWO-SPIN}(\beta, \gamma, \lambda)$ *on graphs of maximum degree at most* $\Delta_{\max}$ *if* $\text{UNIQUE}(\beta, \gamma, \lambda, d)$ *holds for all integer* $1 \leq d \leq \Delta_{\max} - 1$.

This algorithmic result for graphs of bounded maximum degree can be extended to graphs of unbounded degrees.

Theorem 6 ([4, 5]) *For* $0 \leq \beta \leq \gamma$ *and* $\beta\gamma < 1$, *there is an FPTAS for* $\text{TWO-SPIN}(\beta, \gamma, \lambda)$ *if* $\text{UNIQUE}(\beta, \gamma, \lambda, d)$ *holds for all integer* $d \geq 1$.

All these algorithms follow the framework introduced by Weitz in his seminal work [12]. In this framework, the marginal probability $p_v^{\tau_\Lambda}$ is computed by applying the tree recursion (1) on the *tree of self-avoiding walks*, (In fact, (1) is the recursion for the ratio $p_v^{\tau_\Lambda}/(1 - p_v^{\tau_\Lambda})$ of

marginal probabilities.) which enumerates all paths originated from vertex v. Then, a decay of correlation, also called the *spatial mixing* property, is verified, so that a truncated recursion tree of polynomial size is sufficient to provide the required accuracy for the estimation of the marginal probability. For graphs of unbounded degrees, a stronger notion of decay of correlation, called the *computationally efficient correlation decay* [4], is verified to enforce the same cost and accuracy even when the branching number of the recursion tree is unbounded.

On the other hand, for antiferromagnetic two-spin systems in the nonuniqueness regime, the partition function is hard to approximate.

Theorem 7 ([11]) *Let* $0 \leq \beta \leq \gamma$ *and* $\beta\gamma < 1$. *For any* $\Delta_{\max} \geq 3$, *unless* $NP = RP$, *there does not exist an FPRAS for* $\text{TWO-SPIN}(\beta, \gamma, \lambda)$ *on graphs of maximum degree at most* $\Delta_{\max}$ *if* $\text{NON-UNIQUE}(\beta, \gamma, \lambda, d)$ *holds for some integer* $1 \leq d \leq \Delta_{\max} - 1$.

Altogether, this gives a complete classification of the approximability of partition function of antiferromagnetic two-spin systems except for the uniqueness threshold.

Algorithms for Graphs with Bounded Connective Constant

The *connective constant* is a natural and well-studied notion of the average degree of a graph, which, roughly speaking, measures the growth rate of the number of self-avoiding walks in the graph as their length grows. As a quantity originated from statistical physics, the connective constant has been especially well studied for various infinite regular lattices. In order to suit the algorithmic applications, the definition of connective constant was extended in [9] to families of finite graphs.

Given a vertex v in a graph G, let $N(v, l)$ denote the number of self-avoiding walks of length ℓ in G which start at v.

Definition 2 ([9]) Let $\mathcal{G}$ be a family of finite graphs. The connective constant of $\mathcal{G}$ is at most Δ if there exist constants a and c such that for any graph $G = (V, E)$ in $\mathcal{G}$ and any vertex v in G, it holds that $\sum_{i=1}^{\ell} N(v, i) \leq c\Delta^{\ell}$ for all $\ell \geq a \log |V|$.

The connective constant has a natural interpretation as the "average arity" of the tree of self-avoiding walks.

For certain antiferromagnetic two-spin systems, it is possible to establish the desirable decay of correlation on the tree of self-avoiding walks with bounded average arity instead of maximum arity, and hence the arity d in the uniqueness condition $\text{UNIQUE}(\beta, \gamma, \lambda, d)$ can be replaced with the connective constant Δ. The algorithmic implication of this is stated as the following theorem.

Theorem 8 ([10]) *For the following two cases:*

- *(The hardcore model) $\beta = 0$ and $\gamma = 1$;*
- *(The antiferromagnetic Ising model with zero field) $\beta = \gamma < 1$ and $\lambda = 1$;*

there exists an FPTAS for $\text{TWO-SPIN}(\beta, \gamma, \lambda)$ *on graphs of connective constant at most Δ if* $\text{UNIQUE}(\beta, \gamma, \lambda, \Delta)$ *holds.*

For the two-spin systems considered by this theorem, it holds that $\text{UNIQUE}(\beta, \gamma, \lambda, \Delta)$ implies $\text{UNIQUE}(\beta, \gamma, \lambda, d)$ for all $1 \leq d \leq \Delta$.

The connective constant of a graph of maximum degree $\Delta_{\max}$ is at most $\Delta_{\max} - 1$, but the connective constant of a family of graphs can be much smaller than this crude bound. For example, though the maximum degree of a graph drawn from the Erdös-Rényi model $G(n, d/n)$ is $\Theta(\log n / \log \log n)$ with high probability, the connective constant of such a graph is at most $d(1 + \varepsilon)$ with high probability for any fixed $\varepsilon > 0$. Therefore, for the considered two-spin systems, the algorithm in Theorem 8 works on strictly more general families of graphs than that of Theorem 5.

Cross-References

▶ Complexity Dichotomies for Counting Graph Homomorphisms

Recommended Reading

1. Bulatov AA, Grohe M (2005) The complexity of partition functions. Theor Comput Sci 348(2–3):148–186
2. Goldberg LA, Jerrum M, Paterson M (2003) The computational complexity of two-state spin systems. Random Struct Algorithms 23(2):133–154
3. Jerrum M, Sinclair A (1993) Polynomial-time approximation algorithms for the ising model. SIAM J Comput 22(5):1087–1116
4. Li L, Lu P, Yin Y (2012) Approximate counting via correlation decay in spin systems. In: Proceedings of SODA, Kyoto, pp 922–940
5. Li L, Lu P, Yin Y (2013) Correlation decay up to uniqueness in spin systems. In: Proceedings of SODA, New Orleans, pp 67–84
6. Liu J, Lu P, Zhang C (2014) The complexity of ferromagnetic two-spin systems with external fields. In: Proceedings of RANDOM, Barcelona. To appear
7. Lu P, Wang M, Zhang C (2014) Fptas for weighted fibonacci gates and its applications. In: Esparza J, Fraigniaud P, Husfeldt T, Koutsoupias E (eds) ICALP (1), Copenhagen. Lecture Notes in Computer Science, vol 8572. Springer, pp 787–799

8. Sinclair A, Srivastava P, Thurley M (2012) Approximation algorithms for two-state anti-ferromagnetic spin systems on bounded degree graphs. In: Proceedings of SODA, Kyoto, pp 941–953
9. Sinclair A, Srivastava P, Yin Y (2013) Spatial mixing and approximation algorithms for graphs with bounded connective constant. In: Proceedings of FOCS, Berkeley, pp 300–309
10. Sinclair A, Srivastava P, Štefankovič D, Yin Y (2015) Spatial mixing and the connective constant: Optimal bounds. In: Proceedings of SODA, San Diego. To appear
11. Sly A, Sun N (2012) The computational hardness of counting in two-spin models on d-regular graphs. In: Proceedings of FOCS, New Brunswick, pp 361–369
12. Weitz D (2006) Counting independent sets up to the tree threshold. In: Proceedings of STOC, Seattle, pp 140–149

Approximation Schemes for Bin Packing

Nikhil Bansal
Eindhoven University of Technology, Eindhoven, The Netherlands

Keywords

Cutting-stock problem

Years and Authors of Summarized Original Work

1982; Karmarker, Karp

Problem Definition

In the bin-packing problem, the input consists of a collection of items specified by their sizes. There are also identical bins, which without loss of generality can be assumed to be of size 1, and the goal is to pack these items using the minimum possible number of bins.

Bin packing is a classic optimization problem, and hundreds of its variants have been defined and studied under various settings such as average case analysis, worst-case off-line analysis, and worst-case online analysis. This note considers the most basic variant mentioned above under the off line model where all the items are given in advance. The problem is easily seen to be NP-hard by a reduction from the partition problem. In fact, this reduction implies that unless P = NP, it is impossible to determine in polynomial time whether the items can be packed into two bins or whether they need three bins.

Notations

The input to the bin-packing problem is a set of n items I specified by their sizes $s_1, \ldots, s_n$, where each s_i is a real number in the range $(0,1]$. A subset of items $S \subseteq I$ can be packed feasibly in a bin if the total size of items in S is at most 1. The goal is to pack all items in I into the minimum number of bins. Let $\text{OPT}(I)$ denote the value of the optimum solution and $\text{Size}(I)$ the total size of all items in I. Clearly, $\text{OPT}(I) \geq \lceil \text{Size}(I) \rceil$.

Strictly speaking, the problem does not admit a polynomial-time algorithm with an approximation guarantee better than 3/2. Interestingly, however, this does not rule out an algorithm that requires, say, $\text{OPT}(I) + 1$ bins (unlike other optimization problems, making several copies of a small hard instance to obtain a larger hard instance does not work for bin packing). It is more meaningful to consider approximation guarantees in an asymptotic sense. An algorithm is called an asymptotic ρ approximation if the number of bins required by it is $\rho \cdot \text{OPT}(I) + O(1)$.

Key Results

During the 1960s and 1970s, several algorithms with constant factor asymptotic and absolute approximation guarantees and very efficient running times were designed (see [1] for a survey). A breakthrough was achieved in 1981 by de la Vega and Lueker [3], who gave the first polynomial-time asymptotic approximation scheme.

Theorem 1 ([3]) *Given any arbitrary parameter $\epsilon > 0$, there is an algorithm that uses $(1 + \epsilon)\text{OPT}(I) + O(1)$ bins to pack I. The running* *time of this algorithm is* $O(n \log n) + (1 + \epsilon)^{O(1/\epsilon)}$.

The main insight of de la Vega and Lueker [3] was to give a technique for approximating the original instance by a simpler instance where large items have only $O(1)$ distinct sizes. Their idea was simple. First, it suffices to restrict attention to large items, say, with size greater than ε. These can be called I_b. Given an (almost) optimum packing of I_b, consider the solution obtained by greedily filling up the bins with remaining small items, opening new bins only if needed. Indeed, if no new bins are needed, then the solution is still almost optimum since the packing for I_b was almost optimum. If additional bins are needed, then each bin, except possibly one, must be filled to an extent $(1 - \epsilon)$, which gives a packing using $\text{Size}(I)/(1 - \epsilon) + 1 \leq \text{OPT}(I)/(1 - \epsilon) + 1$ bins. So it suffices to focus on solving I_b almost optimally. To do this, the authors show how to obtain another instance I' with the following properties. First, I' has only $O(1/\epsilon^2)$ distinct sizes, and second, I' is an approximation of I_b in the sense that $\text{OPT}(I_b) \geq \text{OPT}(I')$, and moreover, any solution of I' implies another solution of I_b using $O(\epsilon \cdot \text{OPT}(I))$ additional bins. As I' has only $1/\epsilon^2$ distinct item sizes, and any bin can obtain at most $1/\epsilon$ such items, there are at most $O\left(1/\epsilon^2\right)^{1/\epsilon}$ ways to pack a bin. Thus, I' can be solved optimally by exhaustive enumeration (or more efficiently using an integer programming formulation described below).

Later, Karmarkar, and Karp [4] proved a substantially stronger guarantee.

Theorem 2 ([4]) *Given an instance I, there is an algorithm that produces a packing of I using* $\text{OPT}(I) + O(\log^2 \text{OPT}(I))$ *bins. The running time of this algorithm is* $O(n^8)$.

Observe that this guarantee is significantly stronger than that of [3] as the additive term is $O(\log^2 \text{OPT})$ as opposed to $o(\epsilon \cdot \text{OPT})$. Their algorithm also uses the ideas of reducing the number of distinct item sizes and ignoring small items, but in a much more refined way. In particular, instead of obtaining a rounded instance

in a single step, their algorithm consists of a logarithmic number of steps where in each step they round the instance "mildly" and then solve it partially.

The starting point is an exponentially large linear programming (LP) relaxation of the problem commonly referred to as the configuration LP. Here, there is a variable x_S corresponding to each subset of items S that can be packed feasibly in a bin. The objective is to minimize $\sum_S x_S$ subject to the constraint that for each item i, the sum of x_S over all subsets S that contain i is at least 1. Clearly, this is a relaxation as setting $x_S = 1$ for each set S corresponding to a bin in the optimum solution is a feasible integral solution to the LP. Even though this formulation has exponential size, the separation problem for the dual is a knapsack problem, and hence the LP can be solved in polynomial time to any accuracy (in particular within an accuracy of 1) using the ellipsoid method. Such a solution is called a fractional packing. Observe that if there are n_i items each of size exactly s_i, then the constraints corresponding to i can be "combined" to obtain the following LP:

$$\min \sum_S x_S$$
$$\text{s.t.} \quad \sum_S a_{S,i} x_S \geq n_i \quad \forall \text{item sizes } i$$
$$x_S \geq 0 \quad \forall \text{ feasible sets } S.$$

Here, $a_{S,i}$ is the number of items of size s_i in the feasible S. Let $q(I)$ denote the number of distinct sizes in I. The number of nontrivial constraints in LP is equal to $q(I)$, which implies that there is a basic optimal solution to this LP that has only $q(I)$ variables set nonintegrally. Karmarkar and Karp exploit this observation in a very clever way. The following lemma describes the main idea.

Lemma 1 *Given any instance J, suppose there is an algorithmic rounding procedure to obtain another instance J' such that J' has $\text{Size}(J)/2$ distinct item sizes and J and J' are related in the following sense: given any fractional packing of J using ℓ bins gives a fractional packing of J' with at most ℓ bins, and given any packing of J' using ℓ' bins gives a packing of J using $\ell' + c$*

bins, where c is some fixed parameter. Then, J can be packed using $\text{OPT}(J) + c \cdot \log(\text{OPT}(J))$ *bins.*

Proof Let $I_0 = I$ and let I_1 be the instance obtained by applying the rounding procedure to I_0. By the property of the rounding procedure, $\text{OPT}(I) \leq \text{OPT}(I_1) + c$ and $\text{LP}(I_1) \leq \text{LP}(I)$. As I_1 has $\text{Size}(I_0)/2$ distinct sizes, the LP solution for I_1 has at most $\text{Size}(I_0)/2$ fractionally set variables. Remove the items packed integrally in the LP solution, and consider the residual instance I_1'. Note that $\text{Size}(I_1') \leq \text{Size}(I_0)/2$. Now, again apply the rounding procedure to I_1' to obtain I_2 and solve the LP for I_2. Again, this solution has at most $\text{Size}(I_1')/2 \leq \text{Size}(I_0)/4$ fractionally set variables and $\text{OPT}(I_1') \leq \text{OPT}(I_2) + c$ and $\text{LP}(I_2) \leq \text{LP}(I_1')$. The above process is repeated for a few steps. At each step, the size of the residual instance decreases by a factor of at least two, and the number of bins required to pack I_0 increases by additive c. After $\log(\text{Size}(I_0))(\approx \log(\text{OPT}(I)))$ steps, the residual instance has size $O(1)$ and can be packed into $O(1)$ additional bins. □

It remains to describe the rounding procedure. Consider the items in nondecreasing order $s_1 \geq s_2 \geq \cdots \geq s_n$ and group them as follows. Add items to current group until its size first exceeds 2. At this point, close the group and start a new group. Let $G_1, \ldots, G_k$ denote the groups formed and let $n_i = |G_i|$, setting $n_0 = 0$ for convenience. Define I' as the instance obtained by rounding the size of n_{i-1} largest items in G_i to the size of the largest item in G_i for $i = 1, \ldots, k$. The procedure satisfies the properties of Lemma 1 with $c = O(\log n_k)$ (left as an exercise to the reader). To prove Theorem 2, it suffices to show that $n_k = O(\text{Size}(I))$. This is done easily by ignoring all items smaller than $1/\text{Size}(I)$ and filling them in only in the end (as in the algorithm of de la Vega and Lueker).

In the case when the item sizes are not too small, the following corollary is obtained.

Corollary 1 *If all the item sizes are at least* δ, *it is easily seen that* $c = O(\log 1/\delta)$, *and the above algorithm implies a guarantee*

of $\text{OPT} + O(\log(1/\delta) \cdot \log \text{OPT})$, *which is* $\text{OPT} + O(\log \text{OPT})$ *if δ is a constant.*

Recently, Rothvoss gave the first improvement to the result of Karmarkar and Karp and improve their additive guarantee from $O(\log^2 \text{Opt})$ to $O(\log \text{Opt} \log \log \text{Opt})$. His algorithm also uses the configuration LP solution and is based on several new ideas and recent developments. First is the connection of bin packing to a problem in discrepancy theory known as the k-permutation problem. Second are the recently developed algorithmic approaches for addressing discrepancy minimization problems.

In addition to these, a key idea in Rothvoss' algorithm is to glue several small items contained in a configuration into a new large item. For more details, we refer the reader to [5].

Applications

The bin-packing problem is directly motivated from practice and has many natural applications such as packing items into boxes subject to weight constraints, packing files into CDs, packing television commercials into station breaks, and so on. It is widely studied in operations research and computer science. Other applications include the so-called cutting-stock problems where some material such as cloth or lumber is given in blocks of standard size from which items of certain specified size must be cut. Several variations of bin packing, such as generalizations to higher dimensions, imposing additional constraints on the algorithm and different optimization criteria, have also been extensively studied. The reader is referred to [1,2] for excellent surveys.

Open Problems

Except for the NP-hardness, no other hardness results are known, and it is possible that a polynomial-time algorithm with guarantee $\text{OPT} + 1$ exists for the problem. Resolving this is a key open question. A promising approach seems to be via the configuration LP (considered above).

In fact, no instance is known for which the additive gap between the optimum configuration LP solution and the optimum integral solution is more than 1. It would be very interesting to design an instance that has an additive integrality gap of two or more.

Cross-References

▶ Bin Packing
▶ Knapsack

Recommended Reading

1. Coffman EG, Garey MR, Johnson DS (1996) Approximation algorithms for bin packing: a survey. In: Hochbaum D (ed) Approximation algorithms for NP-hard problems. PWS, Boston, pp 46–93
2. Csirik J, Woeginger G (1998) On-line packing and covering problems. In: Fiat A, Woeginger G (eds) Online algorithms: the state of the art. LNCS, vol 1442. Springer, Berlin, pp 147–177
3. Fernandez de la Vega W, Lueker G (1981) Bin packing can be solved within $1 + \varepsilon$ in linear time. Combinatorica 1:349–355
4. Karmarkar N, Karp RM (1982) An efficient approximation scheme for the one-dimensional bin-packing problem. In: Proceedings of the 23rd IEEE symposium on foundations of computer science (FOCS), Chicago, pp 312–320
5. Rothvoss T (2013) Approximating bin packing withing O(log OPT log log OPT) bins. In: Proceedings of the 54th IEEE symposium on foundations of computer science (FOCS), Berkeley, pp 20–29

Approximation Schemes for Geometric Network Optimization Problems

Joseph S.B. Mitchell
Department of Applied Mathematics and Statistics, Stony Brook University, Stony Brook, NY, USA

Keywords

Approximation algorithms; Computational geometry; Geometric networks; Optimization; Polynomial-time approximation scheme; Traveling salesperson problem

Years and Authors of Summarized Original Work

1998; Arora
1999: Mitchell
1998; Rao, Smith

Problem Definition

Geometric network optimization is the problem of computing a network in a geometric space (e.g., the Euclidean plane), based on an input of geometric data (e.g., points, disks, polygons/polyhedra) that is optimal according to an objective function that typically involves geometric measures, such as Euclidean length, perhaps in addition to combinatorial metrics, such as the number of edges in the network. The desired network is required to have certain properties, such as being connected (or k-connected), having a specific topology (e.g., forming a path/cycle), spanning at least a certain number of input objects, etc.

One of the most widely studied optimization problems is the traveling salesperson problem (TSP): given a set S of n sites (e.g., cities), and distances between each pair of sites, determine a route or *tour* of minimum length that visits every member of S. The (symmetric) TSP is often formulated in terms of a graph optimization problem on an edge-weighted complete graph K_n, and the goal is to determine a Hamiltonian cycle (a cycle visiting each vertex exactly once), or a *tour*, of minimum total weight. In geometric settings, the sites are often points in the plane with distances measured according to the Euclidean metric.

The TSP is known to be NP-complete in graphs and NP-hard in the Euclidean plane. Many methods of combinatorial optimization, as well as heuristics, have been developed and applied successfully to solving to optimality instances of TSP; see Cook [7]. Our focus here is on provable approximation algorithms.

In the context of the TSP, a minimization problem, a c-approximation algorithm is an algorithm guaranteed to yield a solution whose objective function value (length) is guaranteed to be at most c times that of an optimal solution. A *polynomial-time approximation scheme* (PTAS) is a family of c-approximation algorithms, with $c = 1 + \varepsilon$, that runs in polynomial (in input size) time for any fixed $\varepsilon > 0$. A *quasi-polynomial-time approximation scheme* (QPTAS) is an approximation scheme, with factor $c = 1 + \varepsilon$ for any fixed $\varepsilon > 0$, whose running time is *quasi-polynomial*, $2^{O((\log n)^C)}$, for some C.

In the Euclidean *Steiner minimum spanning tree* (SMST) problem, the objective is to compute a minimum total length tree that spans all of the input points S, allowing nodes of the tree to be at points of the Euclidean space other than S (such points are known as *Steiner points*). The Euclidean SMST is known to be NP-hard, even in the plane.

Key Results

A simple 2-approximation algorithm for TSP follows from a "doubling" of a minimum spanning tree, assuming that the distances obey the triangle inequality. By augmenting the minimum spanning tree with a minimum-weight matching on the odd-degree nodes of the tree, Christofides obtained a 1.5-approximation for TSP with triangle inequality. This is the currently best-known approximation for general metric spaces; an outstanding open conjecture is that a 4/3-approximation (or better) may be possible. It is known that the TSP in a general metric space is APX-complete, implying that, unless P = NP, no PTAS exists, in general.

Research has shown that "geometry helps" in network optimization problems. Geometric structure has played a key role in solving combinatorial optimization problems. There are problems that are NP-hard in their abstract generality, yet are solvable exactly in polynomial time in geometric settings (e.g., maximum TSP in polyhedral metrics), and there are problems for which we have substantially better, or more efficient, approximation algorithms in geometric settings (e.g., TSP).

As shown by Arora [1] and Mitchell [10] in papers originally appearing in 1996, geometric instances of TSP and SMST have special structure that allows for the existence of a PTAS. Arora [1] gives a randomized algorithm that, with probability 1/2, yields a $(1 + \varepsilon)$-approximate tour in time $n(\log n)^{O(\sqrt{d}/\varepsilon)^{d-1}}$ in Euclidean d-space. Rao and Smith [14] obtain a deterministic algorithm with running time $2^{(d/\varepsilon)^{O(d)}} n + (d/\varepsilon)^{O(d)} n \log n$. This $O(n \log n)$ bound (for fixed d, ε) matches the $\Omega(n \log n)$ lower bound in the decision tree bound. In the real RAM model, with atomic floor or mod function, Bartal and Gottlieb [3] give a randomized linear-time PTAS that, with probability $1 - e^{-O(n^{1/3d})}$, computes a $(1 + \varepsilon)$-approximation to an optimal tour in time $2^{(d/\varepsilon)^{O(d)}} n$. The exponential dependence on d in the PTAS bounds is essentially best possible, since Trevisan has shown that if $d \geq \log n$, it is NP-hard to obtain a $(1 + \varepsilon)$-approximation.

A key insight of Rao and Smith is the application of the concept of "spanners" to the approximation schemes. A connected subgraph G of the complete Euclidean graph, joining every pair of points in S (within Euclidean d-dimensional space), is said to be a *t-spanner* for S if all points of S are nodes of G and, for any points $u, v \in S$, the length of a shortest path in G from u to v is at most t times the Euclidean distance, $d_2(u, v)$. It is known that for any point set S and $t > 1$, t-spanners exist and can be calculated in time $O(n \log n)$, with the property that the t-spanner is *lightweight*, meaning that the sum of its edge lengths is at most a constant factor (depending on d and t) greater than the Euclidean length of a minimum spanning tree on S.

Overview of Methods

The PTAS techniques are based on structure theorems showing that an optimal solution can be "rounded" to a "nearby" solution, of length at most a factor $(1 + \varepsilon)$ longer, that falls within a special class of recursively "nice" solutions for which optimization via dynamic programming can be done efficiently, because the interface between adjacent subproblems is "small" combinatorially. Arora's algorithm [1] is randomized, as is that of Rao and Smith [14]; both can be derandomized. The m-guillotine method (Mitchell [10]) is directly a deterministic method; however, the proof of its structure theorem is effectively an averaging argument.

Arora's Dissection Method

Arora [1, 2] gives a method based on geometric dissection using a quadtree (or its octtree analogue in d dimensions). On the boundary of each quadtree square are m equally spaced points ("portals"); a *portal-respecting tour* is one that crosses the boundaries of squares only at portals. Using an averaging argument based on a randomly shifted quadtree that contains the bounding square of S, Arora proves structure theorems, the simplest of which shows that, when $m > (\log n)/\epsilon$, the expected length of a shortest portal-respecting tour, T, is at most $(1 + \epsilon)$ times the length of an optimal tour. Within a quadtree square, T consists of at most m disjoint paths that together visit all sites within the square. Since the interface data specifying a subproblem has size $2^{O(m)}$, dynamic programming computes a shortest portal-respecting tour in time $2^{(O(m)}$ per quadtree square, for overall time $2^{O((\log n)/\epsilon)} = n^{O(1/\epsilon)}$. An improved, near-linear (randomized) running time is obtained via a stronger structure theorem, based on "(m, k)-light" tours, which are portal respecting and enter/leave each square at most k times (with $k = O(1/\varepsilon)$). Rao and Smith's improvement uses the observation that it suffices to restrict the algorithm to use edges of a $(1 + \varepsilon)$-spanner.

The m-Guillotine Method

The m-guillotine method of Mitchell [10] is based on the notion of an *m-guillotine structure*. A geometric graph G in the plane has the m-guillotine structure if the following holds: either (1) G has $O(m)$ edges or (2) there exists a *cut* by an axis-parallel line ℓ such that the intersection of ℓ with the edge set of G has $O(m)$ connected components and the subgraphs of G on each side of the cut recursively also have the m-guillotine

structure. The m-guillotine structure is defined in dimensions $d > 2$ as well, using hyperplane cuts orthogonal to the coordinate axes.

The m-guillotine structure theorem in 2 dimensions states that, for any positive integer m, a set E of straight line segments in the plane is either m-guillotine already or is "close" to being m-guillotine, in that there exists a superset, $E_m \supset E$ that has m-guillotine structure, where E_m is obtained from E by adding a set of axis-parallel segments (*bridges*, or *m-spans*) of total length at most $O(\varepsilon|E|)$. The proof uses a simple charging scheme.

The m-guillotine method originally (1996) yielded a PTAS for TSP and related problems in the plane, with running time $n^{O(1/\epsilon)}$; this was improved (1997) to $n^{O(1)}$. With the injection of the idea of Rao and Smith [14] to employ spanners, the m-guillotine method yields a simple, deterministic $O(n \log n)$ time PTAS for TSP and related problems in fixed dimension $d \geq 2$. The steps are the following: (a) construct (in $O(n \log n)$ time) a spanner, T; (b) compute its m-guillotine superset, T_m, by standard sweep techniques (in time $O(n \log n)$); and (c) use dynamic programming (time $O(n)$) applied to the recursive tree associated with T_m, to optimize over spanning subgraphs of T_m.

Generalizations to Other Metrics

The PTAS techniques described above have been significantly extended to variants of the Euclidean TSP. While we do not expect that a PTAS exists for general metric spaces (because of APX-hardness), the methods can be extended to a very broad class of "geometric" metrics known as *doubling metrics*, or metric spaces of bounded *doubling dimension*. A metric space $\mathcal{X}$ is said to have *doubling constant* c_d if any ball of radius r can be covered by c_d balls of radius $r/2$; the logarithm of c_d is the *doubling dimension* of $\mathcal{X}$. Euclidean d-space has doubling dimension $O(d)$. Bartal, Gottlieb, and Krauthgamer [4] have given a PTAS for TSP in doubling metrics, improving on a prior QPTAS.

For the discrete metric space induced by an edge-weighted planar graph, the TSP has a linear-time PTAS. The *subset TSP* for edge-weighted planar graphs, in which there is a subset $S \subseteq V$ of the vertex set V that must be visited, also has an efficient ($O(n \log n)$ time) PTAS; this implies a PTAS for the *geodesic metric* for TSP on a set S of sites in a polygonal domain in the plane, with distances given by the (Euclidean) lengths of geodesic shortest paths between pairs of sites.

Applications to Network Optimization

The approximation schemes we describe above have been applied to numerous geometric network optimization problems, including the list below. We do not give references for most of the results summarized below; see the surveys [2, 11, 12] and Har-Peled [9].

1. A PTAS for the Euclidean Steiner minimum spanning tree (SMST) problem.
2. A PTAS for the Euclidean minimum Steiner forest problem, in which one is to compute a minimum-weight forest whose connected components (Steiner trees) span given (disjoint) subsets $S_1, \ldots, S_K \subset S$ of the sites, allowing Steiner points.
3. A PTAS for computing a minimum-weight k-connected spanning graph of S in Euclidean d-space.
4. A PTAS for the *k-median problem*, in which one is to determine k *centers*, among S, in order to minimize the sum of the distances from the sites S to their nearest center points.
5. A PTAS for the *minimum latency problem* (MLP), also known as the *deliveryman problem* or the *traveling repairman problem*, in which one is to compute a tour on S that minimizes the sum of the "latencies" of all points, where the *latency* of a point p is the length of the tour from a given starting point to the point p. The PTAS of Sitters [15] runs in time $n^{O(1/\epsilon)}$, improving the prior QPTAS.
6. A PTAS for the k-TSP (and k-MST), in which one is to compute a shortest tour (tree) spanning at least k of the n sites of S.
7. A QPTAS for degree-bounded spanning trees in the plane.

8. A QPTAS for the capacitated vehicle routing problem (VRP) [8], in which one is to compute a minimum-length collection of tours, each visiting at most k sites of S. A PTAS is known for some values of k.
9. A PTAS for the *orienteering problem*, in which the goal is to maximize the number of sites visited by a length-bounded tour [6].
10. A PTAS for *TSP with Neighborhoods* (TSPN), in which each site of the input set S is a connected region of d-space (rather than a point), and the goal is to compute a tour that visits each region. The TSPN has a PTAS for regions in the plane that are "fat" and are weakly disjoint (no point lies in more than a constant number of regions) [13]. Chan and Elbassioni [5] give a QPTAS for fat, weakly disjoint regions in doubling metrics. For TSPN with disconnected regions, the problem is that of the "group TSP" (also known as "generalized TSP" or "one-of-a-set TSP"), which, in general metrics, is much harder than TSP; even in the Euclidean plane, the problem is NP-hard to approximate to within any constant factor for finite point sets and is NP-hard to approximate better than a fixed constant for visiting point pairs.
11. A PTAS for the *milling* and *lawnmowing* problems, in which one is to compute a shortest path/tour for a specified cutter so that all points of a given region R in the plane is swept over by the cutter head while keeping the cutter fully within the region R (milling), or allowing the cutter to sweep over points outside of region R (lawnmowing).
12. A PTAS for computing a minimum-length cycle that separates a given set of "red" points from a given set of "blue" points in the Euclidean plane.
13. A QPTAS for the *minimum-weight triangulation* (MWT) problem of computing a triangulation of the planar point set S in order to minimize the sum of the edge lengths. The MWT has been shown to be NP-hard.

14. A PTAS for the minimum-weight Steiner convex partition problem in the plane, in which one is to compute an embedded planar straight-line graph with convex faces whose vertex set contains the input set S.

Open Problems

A prominent open problem in approximation algorithms for network optimization is to determine if approximations better than factor 3/2 can be achieved for the TSP in general metric spaces.

Specific open problems for geometric network optimization problems include:

1. Is there a PTAS for minimum-weight triangulation (MWT) in the plane? (A QPTAS is known.)
2. Is there a PTAS for capacitated vehicle routing, for all k?
3. Is there a PTAS for Euclidean minimum spanning trees of bounded degree (3 or 4)? (A QPTAS is known for degree-3 trees.)
4. Is there a PTAS for TSP with Neighborhoods (TSPN) for connected disjoint regions in the plane?
5. Is there a PTAS for computing a minimum-weight t-spanner of a set of points in a Euclidean space?

Finally, can PTAS techniques be implemented to be competitive with other practical methods for TSP or related network optimization problems?

Cross-References

▶ Applications of Geometric Spanner Networks
▶ Euclidean Traveling Salesman Problem
▶ Metric TSP
▶ Minimum Geometric Spanning Trees
▶ Minimum k-Connected Geometric Networks
▶ Steiner Trees

Recommended Reading

1. Arora S (1998) Polynomial time approximation schemes for Euclidean traveling salesman and other geometric problems. J ACM 45(5):753–782
2. Arora S (2007) Approximation algorithms for geometric TSP. In: Gutin G, Punnen AP (eds) The traveling salesman problem and its variations. Combinatorial Optimization, vol 12. Springer, New York/Berlin, pp 207–221
3. Bartal Y, Gottlieb LA (2013) A linear time approximation scheme for Euclidean TSP. In: Proceedings of the 54th IEEE Foundations of Computer Science (FOCS). IEEE, Piscataway, pp 698–706
4. Bartal Y, Gottlieb LA, Krauthgamer R (2012) The traveling salesman problem: low-dimensionality implies a polynomial time approximation scheme. In: Proceedings of the 44th ACM Symposium on Theory of Computing. ACM, New York, pp 663–672
5. Chan TH, Elbassioni KM (2011) A QPTAS for TSP with fat weakly disjoint neighborhoods in doubling metrics. Discret Comput Geom 46(4):704–723
6. Chen K, Har-Peled S (2008) The Euclidean orienteering problem revisited. SIAM J Comput 38(1):385–397
7. Cook W (2011) In pursuit of the travelling salesman: mathematics at the limits of computation. Princeton University Press, Princeton
8. Das A, Mathieu C (2015) A quasipolynomial time approximation scheme for Euclidean capacitated vehicle routing. Algorithmica 73(1):115–142
9. Har-Peled S (2011) Geometric approximation algorithms, vol 173. American Mathematical Society, Providence
10. Mitchell JSB (1999) Guillotine subdivisions approximate polygonal subdivisions: a simple polynomial-time approximation scheme for geometric tsp, k-mst, and related problems. SIAM J Comput 28(4):1298–1309
11. Mitchell JSB (2000) Geometric shortest paths and network optimization. In: Sack JR, Urrutia J (eds) Handbook of Computational Geometry. Elsevier Science, North-Holland/Amsterdam, pp 633–701
12. Mitchell JSB (2004) Shortest paths and networks. In: Goodman JE, O'Rourke J (eds) Handbook of Discrete and Computational Geometry, 2nd edn. Chapman & Hall/CRC, Boca Raton, chap 27, pp 607–641
13. Mitchell JSB (2007) A PTAS for TSP with neighborhoods among fat regions in the plane. In: Proceedings of the 18th ACM-SIAM Symposium on Discrete Algorithms, New Orleans, pp 11–18
14. Rao SB, Smith WD (1998) Approximating geometrical graphs via spanners and banyans. In: Proceedings of the 30th ACM Symposium on Theory of Computing. ACM, New York, pp 540–550
15. Sitters R (2014) Polynomial time approximation schemes for the traveling repairman and other minimum latency problems. In: Proceedings of the 25th ACM-SIAM Symposium on Discrete Algorithms, Portland, pp 604–616

Approximation Schemes for Makespan Minimization

Asaf Levin
Faculty of Industrial Engineering and
Management, The Technion, Haifa, Israel

Keywords

Approximation scheme; Load balancing; Parallel machine scheduling

Years and Authors of Summarized Original Work

1987, 1988; Hochbaum, Shmoys

Problem Definition

Non-preemptive makespan minimization on m uniformly related machines is defined as follows. We are given a set $M = \{1, 2, \ldots, m\}$ of m machines where each machine i has a speed s_i such that $s_i > 0$. In addition we are given a set of jobs $J = \{1, 2, \ldots, n\}$, where each job j has a positive size p_j and all jobs are available for processing at time 0. The jobs need to be partitioned into m subsets $S_1, \ldots, S_m$, with S_i being the subset of jobs assigned to machine i, and each such (ordered) partition is a feasible solution to the problem. Processing job j on machine i takes $\frac{p_j}{s_i}$ time units. For such a solution (also known as a schedule), we let $L_i = (\sum_{j \in S_i} p_j)/s_i$ be the *completion time* or *load* of machine i. The *work* of machine i is $W_i = \sum_{j \in S_i} p_j = L_i \cdot s_i$, that is, the total size of the jobs assigned to i. The *makespan* of the schedule is defined as $\max\{L_1, L_2, \ldots, L_m\}$, and the goal is to find a schedule that minimizes the makespan. We also consider the problem on identical machines, that is, the special case of the above problem in which $s_i = 1$ for all i (in this special case, the work and the load of a given machine are always the same).

Key Results

A PTAS (polynomial-time approximation scheme) is a family of polynomial-time algorithms such that for all $\epsilon > 0$, the family has an algorithm such that for every instance of the makespan minimization problem, it returns a feasible solution whose makespan is at most $1 + \epsilon$ times the makespan of an optimal solution to the same instance. Without loss of generality, we can assume that $\epsilon < \frac{1}{5}$.

The Dual Approximation Framework and Common Preprocessing Steps

Using a guessing step of the optimal makespan, and scaling the sizes of all jobs by the value of the optimal makespan, we can assume that the optimal makespan is in the interval $[1, 1 + \epsilon)$ and it suffices to construct a feasible solution whose makespan is at most $1 + c\epsilon$ for a constant c (then scaling ϵ before applying the algorithm will give the claimed result). This assumption can be made since we can find in polynomial time two values LB and UB such that the optimal makespan is in the interval $[LB, UB]$ and $\frac{UB}{LB}$ is at most some constant (or even at most an exponential function of the length of the binary encoding of the input), then using a constant (or polynomial) number of iterations, we can find the minimum integer power of $1 + \epsilon$ for which the algorithm below will succeed to find a schedule with makespan at most $1 + c\epsilon$ times the optimal makespan. This approach is referred to as the *dual approximation method* [7, 8].

From now on, we assume that the optimal makespan is in the interval $[1, 1 + \epsilon)$. The next step is to round up the size of each job to the next integer power of $1 + \epsilon$ and to round down the speed of each machine to the next integer power of $1 + \epsilon$. That is, the rounded size of job j is $p'_j = (1 + \epsilon)^{\lceil \log_{1+\epsilon} p_j \rceil}$ and the rounded speed of machine i is $s'_i = (1 + \epsilon)^{\lfloor \log_{1+\epsilon} s_i \rfloor}$. Note that this rounding does not decrease the makespan of any feasible solution and increase the optimal makespan by a multiplicative factor of at most $(1 + \epsilon)^2$. Thus, in the new instance that we call *the rounded instance*, the makespan of an optimal solution is in the interval $[1, (1 + \epsilon)^3)$. We observe that if the original instance to the makespan minimization problem was for the special case of identical machines, so does the rounded instance. The next steps differ between the PTAS for identical machines and its generalization for related machines.

The Case of Identical Machines

We define a job to be *small* if its rounded size is at most ϵ, and otherwise it is *large*. The large jobs instance is the instance we obtain from the rounded instance by removing all small jobs. The first observation is that it is sufficient to design an algorithm for finding a feasible solution to the large jobs instance whose makespan is at most $1 + c\epsilon$ where $c \geq 5$ is some constant. This is so, because we can apply this algorithm on the large jobs instance and obtain a schedule of these large jobs. Later, we add to the schedule the small jobs one by one using the list scheduling heuristic [5]. In the analysis, there are two cases. In the first one, adding the small jobs did not increase the makespan of the resulting schedule, and in this case our claim regarding the makespan of the output of the algorithm clearly holds. In the second case, the makespan increased by adding the small jobs, and we consider the last iteration in which such increase happened. In that last iteration, the load of one machine was increased by the size of the job assigned in this iteration, that is by at most ϵ, and before this iteration its load was smaller than $\frac{\sum_j p'_j}{m} \leq (1 + \epsilon)^3$, where the inequality holds because the makespan of a feasible solution cannot be smaller than the average load of the machines. The claim now follows using $(1 + \epsilon)^3 + \epsilon \leq 1 + 5\epsilon$ as $\epsilon < \frac{1}{5}$.

The large jobs instance has a compact representation. There are m identical machines and jobs of at most $O(\log_{1+\epsilon} \frac{1}{\epsilon})$ distinct sizes. Note that each machine has at most $\frac{2}{\epsilon}$ large jobs assigned to it (in any solution with makespan smaller than 2), and thus there are a constant number of different schedules of one machine when we consider jobs of the same size as identical jobs. A schedule of one machine in a solution to the large jobs instance is called the *configuration of the machine*. Now, we can either perform a dynamic programming algorithm that assigns large jobs to one machine after the other while recalling in each step the number of jobs of each size that still need to be assigned (as done in [7]) or use an integer program of fixed dimension [9] to solve the problem of assigning all large jobs to configurations of machines while having at most m machines in the solution and allowing only configurations corresponding to machines with load at most $(1 + \epsilon)^3$ as suggested by Shmoys (see [6]).

We refer to [1, 2], and [10] for PTASs for other load balancing problems on identical machines.

The Case of Related Machines

Here, we still would like to consider separately the large jobs and the small jobs; however, a given job j can be large for one machine and small for another machine (it may even be too large for other machines, that is, processing it on such machine may take a period of time that is longer than the makespan of an optimal solution). Thus, for a given job j, we say that it is *huge* for machine i if $\frac{p_j}{s_i} > (1+\epsilon)^3$, it is *large* for machine i if $\epsilon < \frac{p_j}{s_i} \leq (1 + \epsilon)^3$, and otherwise it is *small* for machine i. A configuration of machine i is the number of large jobs of each rounded size that are assigned to machine i (observe that similarly to the case of identical machines, the number of sizes of large jobs for a given machine is a constant) as well as approximate value of the total size of small jobs assigned to machine i, that is a value Δ_i such that the total size of small jobs assigned to machine i is in the interval $\left((\Delta_i - 1)\epsilon \cdot \frac{1}{s_i}, \Delta_i \epsilon \cdot \frac{1}{s_i} \right]$. Note that the vector of configurations of machines defines some information about the schedule, but it does not give a one-to-one assignment of small jobs to the machines.

Once again, [8] suggested to use dynamic programming for assigning jobs to the machines by traversing the machines from the slowest one to the fastest one and, for each machine, decide the number of large jobs of each size as well as an approximate value of the total size of small jobs (for that machine) assigned to it. That is, the dynamic programming decides the configuration of each machine. To do so, it needs to recall the number of large jobs (with respect to the current machine) that are still not assigned, as well as the total size of small jobs (with respect to the current machine) that are still not assigned (this total size is a rounded value). At a postprocessing step, the jobs assigned as large jobs by the solution for the dynamic programming are scheduled accordingly, while the other jobs are assigned as small jobs as follows.

We assign the small jobs to the machines while traversing the machines from slowest to fastest and assigning the small jobs from the smallest to largest. At each time we consider the current machine i and the prefix of unassigned small jobs that are small with respect to the current machine. Denote by Δ_i the value of this parameter according to the solution of the dynamic programming. Due to the successive rounding of the total size of unassigned small jobs, we will allow assignments of a slightly larger total size of small jobs to machine i. So we will assign the small jobs one by one as long as their total size is at most $\frac{(\Delta_i+4)\epsilon}{s_i}$. If at some point there are no further unassigned small jobs that are small for machine i, we move to the next machine, otherwise we assign machine i small jobs (for machine i) of total size of at least $\frac{(\Delta_i+3)\epsilon}{s_i}$. This suffices to guarantee the feasibility of the resulting solution (i.e., all jobs are assigned) while increasing the makespan only by a small amount.

We refer to [3] and [4] for PTASs for other load balancing problems on related machines.

Cross-References

▶ Robust Scheduling Algorithms
▶ Vector Scheduling Problems

Recommended Reading

1. Alon N, Azar Y, Woeginger GJ, Yadid T (1997) Approximation schemes for scheduling. In: Proceedings of the 8th symposium on discrete algorithms (SODA), New Orleans, USA pp 493–500
2. Alon N, Azar Y, Woeginger GJ, Yadid T (1998) Approximation schemes for scheduling on parallel machines. J Sched 1(1):55–66
3. Azar Y, Epstein L (1998) Approximation schemes for covering and scheduling on related machines. In: Proceedings of the 1st international workshop on approximation algorithms for combinatorial optimization (APPROX), Aalborg, Denmark pp 39–47
4. Epstein L, Sgall J (2004) Approximation schemes for scheduling on uniformly related and identical parallel machines. Algorithmica 39(1):43–57
5. Graham RL (1966) Bounds for certain multiprocessing anomalies. Bell Syst Tech J 45(9):1563–1581
6. Hochbaum DS (1997) Various notions of approximations: Good, better, best and more. In: Hochbaum DS (ed) Approximation algorithms, PWS Publishing Company, Boston
7. Hochbaum DS, Shmoys DB (1987) Using dual approximation algorithms for scheduling problems: theoretical and practical results. J ACM 34(1):144–162
8. Hochbaum DS, Shmoys DB (1988) A polynomial approximation scheme for scheduling on uniform processors: using the dual approximation approach. SIAM J Comput 17(3):539–551
9. Lenstra HW Jr (1983) Integer programming with a fixed number of variables. Math Oper Res 8(4):538–548
10. Woeginger GJ (1997) A polynomial-time approximation scheme for maximizing the minimum machine completion time. Oper Res Lett 20(4):149–154

Approximation Schemes for Planar Graph Problems

Mohammad Taghi Hajiaghayi[1] and
Erik D. Demaine[3]
[1] Department of Computer Science, University of Maryland, College Park, MD, USA
[2] MIT Computer Science and Artificial Intelligence Laboratory, Cambridge, MA, USA

Keywords

Approximation algorithms in planar graphs; Baker's approach; Lipton-Tarjan approach

Years and Authors of Summarized Original Work

1983; Baker
1994; Baker

Problem Definition

Many NP-hard graph problems become easier to approximate on planar graphs and their generalizations. (A graph is *planar* if it can be drawn in the plane (or the sphere) without crossings. For definitions of other related graph classes, see the entry on ▶ Bidimensionality (2004; Demaine, Fomin, Hajiaghayi, Thilikos).) For example, a *maximum independent set* asks to find a maximum subset of vertices in a graph that induce no edges. This problem is inapproximable in general graphs within a factor of $n^{1-\epsilon}$ for any $\epsilon > 0$ unless NP = ZPP (and inapproximable within $n^{1/2-\epsilon}$ unless P = NP), while for planar graphs, there is a 4-approximation (or simple 5-approximation) by taking the largest color class in a vertex 4-coloring (or 5-coloring). Another is *minimum dominating set*, where the goal is to find a minimum subset of vertices such that every vertex is either in or adjacent to the subset. This problem is inapproximable in general graphs within $\epsilon \log n$ for some $\epsilon > 0$ unless P = NP, but as we will see, for planar graphs, the problem admits a *polynomial-time approximation scheme* (PTAS): a collection of $(1 + \epsilon)$-approximation algorithms for all $\epsilon > 0$.

There are two main general approaches to designing PTASs for problems on planar graphs and their generalizations: the separator approach and the Baker approach.

Lipton and Tarjan [15, 16] introduced the first approach, which is based on planar separators. The first step in this approach is to find a separator of $O(\sqrt{n})$ vertices or edges, where n is the size of the graph, whose removal splits the graph into two or more pieces each of which is a constant fraction smaller than the original graph. Then, recurse in each piece, building a recursion tree of separators, and stop when the pieces have some constant size such as $1/\epsilon$. The problem can be solved on these pieces by brute force, and then it remains to combine the solutions up the recursion tree. The induced error can often be bounded in terms of the total size of all separators, which in turn can be bounded by ϵn. If the optimal solution is at least some constant factor times n, this approach often leads to a PTAS.

There are two limitations to this planar-separator approach. First, it requires that the optimal solution be at least some constant factor times n; otherwise, the cost incurred by the separators can be far larger than the desired optimal solution. Such a bound is possible in some problems after some graph pruning (linear kernelization), e.g., independent set, vertex cover, and forms of the traveling salesman problem. But, for example, Grohe [12] states that the dominating set is a problem "to which the technique based on the separator theorem does not apply." Second, the approximation algorithms resulting from planar separators are often impractical because of large constant factors. For example, to achieve an approximation ratio of just 2, the base case requires exhaustive solution of graphs of up to $2^{2,400}$ vertices.

Baker [1] introduced her approach to address the second limitation, but it also addresses the first limitation to a certain extent. This approach is based on decomposition into overlapping subgraphs of bounded outerplanarity, as described in the next section.

Key Results

Baker's original result [1] is a PTAS for a maximum independent set (as defined above) on planar graphs, as well as the following list of problems on planar graphs: maximum tile salvage, partition into triangles, maximum H-matching, minimum vertex cover, minimum dominating set, and minimum edge-dominating set.

Baker's approach starts with a planar embedding of the planar graph. Then it divides vertices into *layers* by iteratively removing vertices on the outer face of the graph: layer j consists of the vertices removed at the jth iteration. If one now removes the layers congruent to i modulo k, for any choice of i, the graph separates into connected components each with at most k consecutive layers, and hence the graph becomes k-outerplanar. Many NP-complete problems can be solved on k-outerplanar graphs for fixed k using dynamic programming (in particular, such graphs have bounded treewidth). Baker's approximation algorithm computes these optimal solutions for each choice i of the congruence class of layers to remove and returns the best solution among these k solutions. The key argument for maximization problems considers the optimal solution to the full graph and argues that the removal of one of the k congruence classes of layers must remove at most a $1/k$ fraction of the optimal solution, so the returned solution must be within a $1 + 1/k$ factor of optimal. A more delicate argument handles minimization problems as well. For many problems, such as maximum independent set, minimum dominating set, and minimum vertex cover, Baker's approach obtains a $(1 + \epsilon)$-approximation algorithms with a running time of $2^{O(1/\epsilon)} n^{O(1)}$ on planar graphs.

Eppstein [10] generalized Baker's approach to a broader class of graphs called graphs of *bounded local treewidth*, i.e., where the treewidth of the subgraph induced by the set of vertices at a distance of at most r from any vertex is bounded above by some function $f(r)$ independent of n. The main differences in Eppstein's approach are replacing the concept of bounded outerplanarity with the concept of bounded treewidth, where dynamic programming can still solve many problems, and labeling layers according to a simple breadth-first search. This approach has led to PTASs for hereditary maximization problems such as maximum independent set and maximum clique, maximum triangle matching, maximum H-matching, maximum tile salvage, minimum vertex cover, minimum dominating set, minimum edge-dominating set, minimum color sum, and subgraph isomorphism for a fixed pattern [6, 8, 10]. Frick and Grohe [11] also developed a general framework for deciding any property expressible in first-order logic in graphs of bounded local treewidth.

The foundation of these results is Eppstein's characterization of minor-closed families of graphs with bounded local treewidth [10]. Specifically, he showed that a minor-closed family has bounded local treewidth if and only if it excludes some *apex graph*, a graph with a vertex whose removal leaves a planar graph. Unfortunately, the initial proof of this result

brought Eppstein's approach back to the realm of impracticality, because his bound on local treewidth in a general apex-minor-free graph is doubly exponential in r : $2^{2^{O(r)}}$. Fortunately, this bound could be improved to $2^{O(r)}$ [3] and even the optimal $O(r)$ [4]. The latter bound restores Baker's $2^{O(1/\epsilon)} n^{O(1)}$ running time for $(1 + \epsilon)$-approximation algorithms, now for all apex-minor-free graphs.

Another way to view the necessary decomposition of Baker's and Eppstein's approaches is that the vertices or edges of the graph can be split into any number k of pieces such that deleting any one of the pieces results in a graph of bounded treewidth (where the bound depends on k). Such decompositions in fact exist for arbitrary graphs excluding any fixed minor H [9], and they can be found in polynomial time [6]. This approach generalizes the Baker-Eppstein PTASs described above to handle general H-minor-free graphs.

This decomposition approach is effectively limited to *deletion-closed* problems, whose optimal solution only improves when deleting edges or vertices from the graph. Another decomposition approach targets *contraction-closed* problems, whose optimal solution only improves when contracting edges. These problems include classic problems such as dominating set and its variations, the traveling salesman problem, subset TSP, minimum Steiner tree, and minimum-weight c-edge-connected submultigraph. PTASs have been obtained for these problems in planar graphs [2, 13, 14] and in bounded-genus graphs [7] by showing that the edges can be decomposed into any number k of pieces such that contracting any one piece results in a bounded-treewidth graph (where the bound depends on k).

Applications

Most applications of Baker's approach have been limited to optimization problems arising from "local" properties (such as those definable in first-order logic). Intuitively, such local properties can be decided by locally checking every constant-size neighborhood. In [5], Baker's approach is generalized to obtain PTASs for nonlocal problems, in particular, connected dominating set. This generalization requires the use of two different techniques. The first technique is to use an ε-fraction of a constant-factor (or even logarithmic-factor) approximation to the problem as a "backbone" for achieving the needed nonlocal property. The second technique is to use subproblems that overlap by $\Theta(\log n)$ layers instead of the usual $\Theta(1)$ in Baker's approach.

Despite this advance in applying Baker's approach to more general problems, the planar-separator approach can still handle some different problems. Recall, though, that the planar-separator approach was limited to problems in which the optimal solution is at least some constant factor times n. This limitation has been overcome for a wide range of problems [5], in particular obtaining a PTAS for feedback vertex set, to which neither Baker's approach nor the planar-separator approach could previously apply. This result is based on evenly dividing the optimum solution instead of the whole graph, using a relation between treewidth and the optimal solution value to bound the treewidth of the graph, thus obtaining an $O(\sqrt{\text{OPT}})$ separator instead of an $O(\sqrt{n})$ separator. The $O(\sqrt{\text{OPT}})$ bound on treewidth follows from the bidimensionality theory described in the entry on ▶ Bidimensionality (2004; Demaine, Fomin, Hajiaghayi, Thilikos). We can divide the optimum solution into roughly even pieces, without knowing the optimum solution, by using existing constant-factor (or even logarithmic-factor) approximations for the problem. At the base of the recursion, pieces no longer have bounded size but do have bounded treewidth, so fast fixed-parameter algorithms can be used to construct optimal solutions.

Open Problems

An intriguing direction for future research is to build a general theory for PTASs of subset problems. Although PTASs for subset TSP and Steiner tree have recently been obtained for planar graphs [2, 14], there remain several open problems of this kind, such as subset feedback vertex set, group Steiner tree, and directed Steiner tree.

Another instructive problem is to understand the extent to which Baker's approach can be applied to nonlocal problems. Again there is an example of how to modify the approach to handle the nonlocal problem of connected dominating set [5], but, for example, the only known PTAS for feedback vertex set in planar graphs follows the separator approach.

Cross-References

▶ Bidimensionality
▶ Separators in Graphs
▶ Treewidth of Graphs

Recommended Reading

1. Baker BS (1994) Approximation algorithms for NP-complete problems on planar graphs. J Assoc Comput Mach 41(1):153–180
2. Borradaile G, Kenyon-Mathieu C, Klein PN (2007) A polynomial-time approximation scheme for Steiner tree in planar graphs. In: Proceedings of the 18th annual ACM-SIAM symposium on discrete algorithms (SODA'07), New Orleans
3. Demaine ED, Hajiaghayi M (2004) Diameter and treewidth in minor-closed graph families, revisited. Algorithmica 40(3):211–215
4. Demaine ED, Hajiaghayi M (2004) Equivalence of local treewidth and linear local treewidth and its algorithmic applications. In: Proceedings of the 15th ACM-SIAM symposium on discrete algorithms (SODA'04), New Orleans, pp 833–842
5. Demaine ED, Hajiaghayi M (2005) Bidimensionality: new connections between FPT algorithms and PTASs. In: Proceedings of the 16th annual ACM-SIAM symposium on discrete algorithms (SODA'05), Vancouver, pp 590–601
6. Demaine ED, Hajiaghayi M, Kawarabayashi K-I (2005) Algorithmic graph minor theory: decomposition, approximation, and coloring. In: Proceedings of the 46th annual IEEE symposium on foundations of computer science, Pittsburgh, pp 637–646
7. Demaine ED, Hajiaghayi M, Mohar B (2007) Approximation algorithms via contraction decomposition. In: Proceedings of the 18th annual ACM-SIAM symposium on discrete algorithms (SODA'07), New Orleans, 7–9 Jan 2007, pp 278–287

8. Demaine ED, Hajiaghayi M, Nishimura N, Ragde P, Thilikos DM (2004) Approximation algorithms for classes of graphs excluding single-crossing graphs as minors. J Comput Syst Sci 69(2):166–195
9. DeVos M, Ding G, Oporowski B, Sanders DP, Reed B, Seymour P, Vertigan D (2004) Excluding any graph as a minor allows a low tree-width 2coloring. J Comb Theory Ser B 91(1):25–41
10. Eppstein D (2000) Diameter and treewidth in minor-closed graph families. Algorithmica 27(3–4):275–291
11. Frick M, Grohe M (2001) Deciding first-order properties of locally tree-decomposable structures. J ACM 48(6):1184–1206
12. Grohe M (2003) Local tree-width, excluded minors, and approximation algorithms. Combinatorica 23(4):613–632
13. Klein PN (2005) A linear-time approximation scheme for TSP for planar weighted graphs. In: Proceedings of the 46th IEEE symposium on foundations of computer science, Pittsburgh, pp 146–155
14. Klein PN (2006) A subset spanner for planar graphs, with application to subset TSP. In: Proceedings of the 38th ACM symposium on theory of computing, Seattle, pp 749–756
15. Lipton RJ, Tarjan RE (1979) A separator theorem for planar graphs. SIAM J Appl Math 36(2):177–189
16. Lipton RJ, Tarjan RE (1980) Applications of a planar separator theorem. SIAM J Comput 9(3):615–627

Approximations of Bimatrix Nash Equilibria

Paul (Pavlos) Spirakis
Computer Engineering and Informatics, Research and Academic Computer Technology Institute, Patras University, Patras, Greece
Computer Science, University of Liverpool, Liverpool, UK
Computer Technology Institute (CTI), Patras, Greece

Keywords

ϵ-Nash equilibria; ϵ-Well-supported Nash equilibria

Years and Authors of Summarized Original Work

2003; Lipton, Markakis, Mehta

2006; Daskalaskis, Mehta, Papadimitriou
2006; Kontogiannis, Panagopoulou, Spirakis

Problem Definition

Nash [15] introduced the concept of Nash equilibria in noncooperative games and proved that any game possesses at least one such equilibrium. A well-known algorithm for computing a Nash equilibrium of a 2-player game is the Lemke-Howson algorithm [13]; however, it has exponential worst-case running time in the number of available pure strategies [18].

Daskalakis et al. [5] showed that the problem of computing a Nash equilibrium in a game with 4 or more players is *PPAD*-complete; this result was later extended to games with 3 players [8]. Eventually, Chen and Deng [3] proved that the problem is *PPAD*-complete for 2-player games as well.

This fact emerged the computation of *approximate* Nash equilibria. There are several versions of approximate Nash equilibria that have been defined in the literature; however, the focus of this entry is on the notions of ϵ-Nash equilibrium and ϵ-well-supported Nash equilibrium. An ϵ-Nash equilibrium is a strategy profile such that no deviating player could achieve a payoff higher than the one that the specific profile gives her, plus ϵ. A stronger notion of approximate Nash equilibria is the ϵ-*well-supported Nash equilibria*; these are strategy profiles such that each player plays only approximately best-response pure strategies with nonzero probability. These are *additive* notions of approximate equilibria; the problem of computing approximate equilibria of bimatrix games using a relative notion of approximation is known to be *PPAD*-hard even for constant approximations.

Notation

For a $n \times 1$ vector x, denote by $x_1, \ldots, x_n$ the components of x and by x^T the transpose of x. Denote by e_i the column vector with 1 at the ith coordinate and 0 elsewhere. For an $n \times m$ matrix A, denote a_{ij} the element in the i-th row and j-th column of A. Let $\mathbb{P}^n$ be the set of all probability vectors in n dimensions: $\mathbb{P}^n = \left\{ z \in \mathbb{R}^n_{\geq 0} : \sum_{i=1}^{n} z_i = 1 \right\}.$

Bimatrix Games

Bimatrix games [16] are a special case of 2-player games such that the payoff functions can be described by two real $n \times m$ matrices A and B. The n rows of A, B represent the *action set* of the first player (the *row player*), and the m columns represent the action set of the second player (the *column player*). Then, when the row player chooses action i and the column player chooses action j, the former gets payoff a_{ij}, while the latter gets payoff b_{ij}. Based on this, bimatrix games are denoted by $\Gamma = \langle A, B \rangle$.

A *strategy* for a player is any probability distribution on his/her set of actions. Therefore, a strategy for the row player can be expressed as a probability vector $x \in \mathbb{P}^n$, while a strategy for the column player can be expressed as a probability vector $y \in \mathbb{P}^m$. Each extreme point $e_i \in \mathbb{P}^n (e_j \in \mathbb{P}^m)$ that corresponds to the strategy assigning probability 1 to the i-th row (j-th column) is called a *pure strategy* for the row (column) player. A *strategy profile* (x, y) is a combination of (mixed in general) strategies, one for each player. In a given strategy profile (x, y), the players get *expected payoffs* $x^T A y$ (row player) and $x^T B y$ (column player).

If both payoff matrices belong to $[0, 1]^{m \times n}$, then the game is called a [0,1]-bimatrix (or else, *positively normalized*) game. The special case of bimatrix games in which all elements of the matrices belong to $\{0, 1\}$ is called a $\{0, 1\}$-bimatrix (or else, *win-lose*) game. A bimatrix game $\langle A, B \rangle$ is called *zero sum* if $B = -A$.

Approximate Nash Equilibria

Definition 1 (ϵ-Nash equilibrium) For any $\epsilon > 0$, a strategy profile (x, y) is an ϵ-*Nash equilibrium* for the $n \times m$ bimatrix game $\Gamma = \langle A, B \rangle$ if

1. For all pure strategies $i \in \{1, \ldots, n\}$ of the row player, $e_i^T A y \leq x^T A y + \epsilon$.

2. For all pure strategies $j \in \{1, \ldots, m\}$ of the column player, $x^T B e_j \leq x^T B y + \epsilon$.

Definition 2 (ϵ-well-supported Nash equilibrium) For any $\epsilon > 0$, a strategy profile (x, y) is an ϵ-*well-supported Nash equilibrium* for the $n \times m$ bimatrix game $\Gamma = \langle A, B \rangle$ if

1. For all pure strategies $i \in \{1, \ldots, n\}$ of the row player,

$$x_i > 0 \Rightarrow e_i^T A y \geq e_k^T A y - \epsilon \ \forall k \in \{1, \ldots, n\}$$

2. For all pure strategies $j \in \{1, \ldots, m\}$ of the column player,

$$y_i > 0 \Rightarrow x^T B e_j \geq x^T B e_k$$
$$- \epsilon \ \forall k \in \{1, \ldots, m\}.$$

Note that both notions of approximate equilibria are defined with respect to an additive error term ϵ. Although (exact) Nash equilibria are known not to be affected by any positive scaling, it is important to mention that approximate notions of Nash equilibria are indeed affected. Therefore, the commonly used assumption in the literature when referring to approximate Nash equilibria is that the bimatrix game is positively normalized, and this assumption is adopted in the present entry.

Key Results

The work of Althöfer [1] shows that, for *any* probability vector p, there exists a probability vector $\hat{P}$ with logarithmic supports, so that for a fixed matrix C, $\max_j |p^T C e_j - \hat{p}^T C e_j| \leq \epsilon$, for any constant $\epsilon > 0$. Exploiting this fact, the work of Lipton, Markakis, and Mehta [14] shows that, for any bimatrix game and for any *constant* $\epsilon > 0$, there exists an ϵ-Nash equilibrium with only logarithmic support (in the number n of available pure strategies). Consider a bimatrix game $\Gamma = \langle A, B \rangle$, and let (x, y) be a Nash equilibrium for Γ. Fix a positive integer k and form a multiset S_1 by sampling k times from the set of pure strategies of the row player, independently at random according to the distribution x. Similarly, form a multiset S_2 by sampling k times from set of pure strategies of the column player according to y. Let $\hat{x}$ be the mixed strategy for the row player that assigns probability $1/k$ to each member of S_1 and 0 to all other pure strategies, and let $\hat{y}$ be the mixed strategy for the column player that assigns probability $1/k$ to each member of S_2 and 0 to all other pure strategies. Then, $\hat{x}$ and $\hat{y}$ are called k-uniform [14], and the following holds:

Theorem 1 ([14]) *For any Nash equilibrium (x, y) of a $n \times n$ bimatrix game and for every $\epsilon > 0$, there exists, for every $k \geq (12 \ln n)/\epsilon^2$, a pair of k-uniform strategies $\hat{x}, \hat{y}$ such that $(\hat{x}, \hat{y})$ is an ϵ-Nash equilibrium.*

This result directly yields a quasi-polynomial $(n^{O(\ln n)})$ algorithm for computing such an approximate equilibrium. Moreover, as pointed out in [1], no algorithm that examines supports smaller than about $\ln n$ can achieve an approximation better than $1/4$.

Theorem 2 ([4]) *The problem of computing a $1/n^{\Theta(1)}$-Nash equilibrium of a $n \times n$ bimatrix game is PPAD-complete.*

Theorem 2 asserts that, unless $PPAD \subseteq P$, there exists no fully polynomial time approximation scheme for computing equilibria in bimatrix games. However, this does not rule out the existence of a polynomial approximation scheme for computing an ϵ-Nash equilibrium when ϵ is an absolute constant, or even when $\epsilon = \Theta(1 - /\text{poly}(\ln n))$. Furthermore, as observed in [4], if the problem of finding an ϵ-Nash equilibrium were *PPAD*-complete when ϵ is an absolute constant, then, due to Theorem 1, all *PPAD* problems would be solved in quasi-polynomial time, which is unlikely to be the case.

Two concurrent and independent works [6, 11] were the first to make progress in providing ϵ-Nash equilibria and ϵ-well-supported Nash equilibria for bimatrix games and some *constant* $0 < \epsilon < 1$. In particular, the work of Kontogiannis, Panagopoulou, and Spirakis [11] proposes a simple linear-time algorithm for computing a $3/4$-Nash equilibrium for any bimatrix game:

Theorem 3 ([11]) *Consider any $n \times m$ bimatrix game $\Gamma = \langle A, B \rangle$, and let $a_{i1,j1} = \max_{i,j} a_{ij}$ and $b_{i2,j2} = \max_{i,j} b_{ij}$. Then the pair of strategies $(\hat{x}, \hat{y})$ where $\hat{x}_{i_1} = \hat{x}_{i_2} = \hat{y}_{j_1} = \hat{y}_{j_2} = 1/2$ is a $3/4$-Nash equilibrium for Γ.*

The above technique can be extended so as to obtain a parametrized, stronger approximation:

Theorem 4 ([11]) *Consider a $n \times m$ bimatrix game $\Gamma = \langle A, B \rangle$. Let $\lambda_1{}^*(\lambda_2{}^*)$ be the minimum, among all Nash equilibria of Γ, expected payoff for the row (column) player, and let $\lambda = \max\{\lambda_1{}^*, \lambda_2{}^*\}$. Then, there exists a $(2 + \lambda)/4$-Nash equilibrium that can be computed in time polynomial in n and m.*

The work of Daskalakis, Mehta, and Papadimitriou [6] provides a simple algorithm for computing a $1/2$-Nash equilibrium: Pick an arbitrary row for the row player, say row i. Let $j = \arg \max{}'_j b'_{ij}$. Let $k = \arg \max{}'_k a'_{kj}$. Thus, j is a best-response column for the column player to the row i, and k is a best-response row for the row player to the column j. Let $\hat{x} = 1/2e_i + 1/2e_k$ and $\hat{y} = e_j$, i.e., the row player plays row i or row k with probability $1/2$ each, while the column player plays column j with probability 1. Then:

Theorem 5 ([6]) *The strategy profile $(\hat{x}, \hat{y})$ is a $1/2$-Nash equilibrium.*

A polynomial construction (based on linear programming) of a 0.38-Nash equilibrium is presented in [7].

For the more demanding notion of well-supported approximate Nash equilibrium, Daskalakis, Mehta, and Papadimitriou [6] propose an algorithm, which, under a quite interesting and plausible graph theoretic conjecture, constructs in polynomial time a $5/6$-well-supported Nash equilibrium. However, the status of this conjecture is still unknown. In [6], it is also shown how to transform a [0,1]-bimatrix game to a $\{0, 1\}$-bimatrix game of the same size, so that each ϵ-well-supported Nash equilibrium of the resulting game is $(1 + \epsilon)/2$-well-supported Nash equilibrium of the original game.

An algorithm given by Kontogiannis and Spirakis computes a 2/3-well-supported Nash equilibrium in polynomial time [12]. Their methodology for attacking the problem is based on the solvability of zero-sum bimatrix games (via its connection to linear programming) and provides a 0.5-well-supported Nash equilibrium for win-lose games and a 2/3-well-supported Nash equilibrium for normalized games. In [9], a polynomial-time algorithm computes an ε-well-supported Nash equilibrium with $\varepsilon < 2/3$, by extending the 2/3-algorithm of Kontogiannis and Spirakis. In particular, it is shown that either the strategies generated by their algorithm can be tweaked to improve the approximation or that we can find a sub-game that resembles matching pennies, which again leads to a better approximation. This allows to construct a (2/3-0.004735)-well-supported Nash equilibrium in polynomial time.

Two new results improved the approximation status of ϵ-Nash equilibria:

Theorem 6 ([2]) *There is a polynomial time algorithm, based on linear programming, that provides an 0.36392-Nash equilibrium.*

The second result below, due to Tsaknakis and Spirakis, is the best till now. Based on local search, it establishes that any local minimum of a very natural map in the space of pairs of mixed strategies or its dual point in a certain minimax problem used for finding the local minimum constitutes a 0.3393-Nash equilibrium.

Theorem 7 ([19]) *There exists a polynomial time algorithm, based on the stationary points of a natural optimization problem, that provides an 0.3393-Nash Equilibrium.*

In [20], it is shown that the problem of computing a Nash equilibrium for 2-person games can be polynomially reduced to an indefinite quadratic programming problem involving the spectrum of the adjacency matrix of a strongly connected directed graph on n vertices, where n is the total number of players' strategies. Based on that, a new method is presented for computing approximate equilibria, and it is shown that its complexity is a function of the average

spectral energy of the underlying graph. The implications of the strong connectedness properties on the energy and on the complexity of the method are discussed, and certain classes of graphs are described for which the method is a polynomial time approximation scheme (PTAS). The worst- case complexity is bounded by a subexponential function in the total number of strategies n.

Kannan and Theobald [10] investigate a hierarchy of bimatrix games $\langle A, B \rangle$ which results from restricting the rank of the matrix $A + B$ to be of fixed rank at most k. They propose a new model of ϵ-approximation for games of rank k and, using results from quadratic optimization, show that approximate Nash equilibria of constant rank games can be computed deterministically in time polynomial in $1/\epsilon$. Moreover, [10] provides a randomized approximation algorithm for certain quadratic optimization problems, which yields a randomized approximation algorithm for the Nash equilibrium problem. This randomized algorithm has similar time complexity as the deterministic one, but it has the possibility of finding an exact solution in polynomial time if a conjecture is valid. Finally, they present a polynomial time algorithm for *relative approximation* (with respect to the payoffs in an equilibrium) provided that the matrix $A + B$ has a nonnegative decomposition.

Applications

Noncooperative game theory and its main solution concept, i.e., the Nash equilibrium, have been extensively used to understand the phenomena observed when decision-makers interact and have been applied in many diverse academic fields, such as biology, economics, sociology, and artificial intelligence. Since however the computation of a Nash equilibrium is in general *PPAD*-complete, it is important to provide efficient algorithms for approximating a Nash equilibrium; the algorithms discussed in this entry are a first step towards this direction.

Recommended Reading

1. Althöfer I (1994) On sparse approximations to randomized strategies and convex combinations. Linear Algebr Appl 199:339–355
2. Bosse H, Byrka J, Markakis E (2007) New algorithms for approximate Nash equilibria in bimatrix games. In: Proceedings of the 3rd international workshop on internet and network economics (WINE 2007), San Diego, 12–14 Dec 2007. Lecture notes in computer science
3. Chen X, Deng X (2005) Settling the complexity of 2-player Nash-equilibrium. In: Proceedings of the 47th annual IEEE symposium on foundations of computer science (FOCS'06), Berkeley, 21–24 Oct 2005
4. Chen X, Deng X, Teng S-H (2006) Computing Nash equilibria: approximation and smoothed complexity. In: Proceedings of the 47th annual IEEE symposium on foundations of computer science (FOCS'06), Berkeley, 21–24 Oct 2006
5. Daskalakis C, Goldberg P, Papadimitriou C (2006) The complexity of computing a Nash equilibrium. In: Proceedings of the 38th annual ACM symposium on theory of computing (STOC'06), Seattle, 21–23 May 2006, pp 71–78
6. Daskalakis C, Mehta A, Papadimitriou C (2006) A note on approximate Nash equilibria. In: Proceedings of the 2nd workshop on internet and network economics (WINE'06), Patras, 15–17 Dec 2006, pp 297–306
7. Daskalakis C, Mehta A, Papadimitriou C (2007) Progress in approximate Nash equilibrium. In: Proceedings of the 8th ACM conference on electronic commerce (EC07), San Diego, 11–15 June 2007
8. Daskalakis C, Papadimitriou C (2005) Three-player games are hard. In: Electronic colloquium on computational complexity (ECCC TR 05-139)
9. Fearnley J, Goldberg PW, Savani R, Bjerre Sørensen T (2012) Approximate well-supported Nash equilibria below two-thirds. In: SAGT 2012, Barcelona, pp 108–119
10. Kannan R, Theobald T (2007) Games of fixed rank: a hierarchy of bimatrix games. In: Proceedings of the ACM-SIAM symposium on discrete algorithms, New Orleans, 7–9 Jan 2007
11. Kontogiannis S, Panagopoulou PN, Spirakis PG (2006) Polynomial algorithms for approximating Nash equilibria of bimatrix games. In: Proceedings of the 2nd workshop on internet and network economics (WINE'06), Patras, 15–17 Dec 2006, pp 286–296
12. Kontogiannis S, Spirakis PG (2010) Well supported approximate equilibria in bimatrix games. Algorithmica 57(4):653–667
13. Lemke CE, Howson JT (1964) Equilibrium points of bimatrix games. J Soc Indust Appl Math 12:413–423
14. Lipton RJ, Markakis E, Mehta A (2003) Playing large games using simple startegies. In: Proceedings of the 4th ACM conference on electronic commerce (EC'03), San Diego, 9–13 June 2003, pp 36–41
15. Nash J (1951) Noncooperative games. Ann Math 54:289–295
16. von Neumann J, Morgenstern O (1944) Theory of games and economic behavior. Princeton University Press, Princeton
17. Papadimitriou CH (1991) On inefficient proofs of existence and complexity classes. In: Proceedings of the 4th Czechoslovakian symposium on combinatorics 1990, Prachatice
18. Savani R, von Stengel B (2004) Exponentially many steps for finding a Nash equilibrium in a bimatrix game. In: Proceedings of the 45th annual IEEE symposium on foundations of computer science (FOCS'04), Rome, 17–19 Oct 2004, pp 258–267
19. Tsaknakis H, Spirakis P (2007) An optimization approach for approximate Nash equilibria. In: Proceedings of the 3rd international workshop on internet and network economics (WINE 2007). Lecture notes in computer science. Also in J Internet Math 5(4):365–382 (2008)
20. Tsaknakis H, Spirakis PG (2010) Practical and efficient approximations of Nash equilibria for win-lose games based on graph spectra. In: WINE 2010, Stanford, pp 378–390

Arbitrage in Frictional Foreign Exchange Market

Mao-cheng Cai[1] and Xiaotie Deng[2,3]
[1]Chinese Academy of Sciences, Institute of Systems Science, Beijing, China
[2]AIMS Laboratory (Algorithms-Agents-Data on Internet, Market, and Social Networks), Department of Computer Science and Engineering, Shanghai Jiao Tong University, Shanghai, China
[3]Department of Computer Science, City University of Hong Kong, Hong Kong, China

Keywords

Arbitrage; Complexity; Foreign exchange; Market

Years and Authors of Summarized Original Work

2003; Cai, Deng

Problem Definition

The simultaneous purchase and sale of the same securities, commodities, or foreign exchange in order to profit from a differential in the price. This usually takes place on different exchanges or marketplaces and is also known as a "riskless profit."

Arbitrage is, arguably, the most fundamental concept in finance. It is a state of the variables of financial instruments such that a riskless profit can be made, which is generally believed not in existence. The economist's argument for its nonexistence is that active investment agents will exploit any arbitrage opportunity in a financial market and thus will deplete it as soon as it may arise. Naturally, the speed at which such an arbitrage opportunity can be located and be taken advantage of is important for the profit-seeking investigators, which falls in the realm of analysis of algorithms and computational complexity.

The identification of arbitrage states is, at frictionless foreign exchange market (a theoretical trading environment where all costs and restraints associated with transactions are nonexistent), not difficult at all and can be reduced to existence of arbitrage on three currencies (see [11]). In reality, friction does exist. Because of friction, it is possible that there exist arbitrage opportunities in the market but difficult to find it and to exploit it to eliminate it. Experimental results in foreign exchange markets showed that arbitrage does exist in reality. Examination of data from 10 markets over a 12-day period by Mavrides [11] revealed that a significant arbitrage opportunity exists. Some opportunities were observed to be persistent for a long time. The problem becomes worse at forward and future markets (in which future contracts in commodities are traded) coupled with covered interest rates, as observed by Abeysekera and Turtle [1] and Clinton [4]. An obvious interpretation is that the arbitrage opportunity was not immediately identified because of information asymmetry in the market. However, that is not the only factor. Both the time necessary to collect the market information (so that an arbitrage opportunity would be identified) and the time people (or computer programs) need to find the arbitrage transactions are important factors for eliminating arbitrage opportunities.

The computational complexity in identifying arbitrage, the level in difficulty measured by arithmetic operations, is different in different models of exchange systems. Therefore, to approximate an ideal exchange market, models with lower complexities should be preferred to those with higher complexities.

To model an exchange system, consider n foreign currencies: $N = \{1, 2, \ldots, n\}$. For each ordered pair (i, j), one may change one unit of currency i to r_{ij} units of currency j. Rate r_{ij} is the *exchange rate* from i to j. In an ideal market, the exchange rate holds for any amount that is exchanged. An *arbitrage opportunity* is a set of exchanges between pairs of currencies such that the net balance for each involved currency is nonnegative and there is at least one currency for which the net balance is positive. Under ideal market conditions, there is no arbitrage if and only if there is no arbitrage among any three currencies (see [11]).

Various types of *friction* can be easily modeled in such a system. Bid-offer spread may be expressed in the present mathematical format as $r_{ij} r_{ji} < 1$ for some $i, j \in N$. In addition, usually the traded amount is required to be in multiples of a fixed integer amount, hundreds, thousands, or millions. Moreover, different traders may bid or offer at different rates, and each for a limited amount. A more general model to describe these market *imperfections* will include, for pairs $i \neq j \in N$, l_{ij} different rates r_{ij}^k of exchanges from currency i to j up to b_{ij}^k units of currency i, $k = 1, \ldots, l_{ij}$, where l_{ij} is the number of different exchange rates from currency i to j.

A currency exchange market can be represented by a digraph $G = (V, E)$ with vertex set V and arc set E such that each vertex $i \in V$ represents currency i and each arc $a_{ij}^k \in E$ represents the currency exchange relation from i to j with rate r_{ij}^k and bound b_{ij}^k. Note that parallel arcs may occur for different exchange rates. Such a digraph is called an exchange digraph. Let $x = (x_{ij}^k)$ denote a currency exchange vector (Fig. 1).

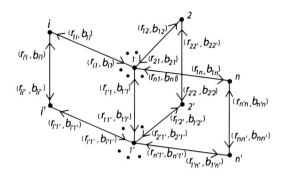

Arbitrage in Frictional Foreign Exchange Market, Fig. 1 Digraph G_1

Problem 1 The existence of arbitrage in a frictional exchange market can be formulated as follows:

$$\sum_{j\neq i}\sum_{k=1}^{l_{ji}}\lfloor r_{ji}^k x_{ji}^k\rfloor - \sum_{j\neq i}\sum_{k=1}^{l_{ij}} x_{ij}^k \geq 0,\ i=1,\ldots,n, \tag{1}$$

at least one strict inequality holds

$$0 \leq x_{ij}^k \leq b_{ij}^k,\ 1\leq k \leq l_{ij},\ 1\leq i\neq j \leq n, \tag{2}$$

$$x_{ij}^k \text{ is integer},\ 1\leq k \leq l_{ij},\ 1\leq i\neq j \leq n. \tag{3}$$

Note that the first term in the right-hand side of (1) is the revenue at currency i by selling other currencies and the second term is the expense at currency i by buying other currencies.

The corresponding optimization problem is

Problem 2 The maximum arbitrage problem in a frictional foreign exchange market with bid-ask spreads, bound, and integrality constraints is the following integer linear programming (P):

$$\text{maximize} \sum_{i=1}^n w_i \sum_{j\neq i}\left(\sum_{k=1}^{l_{ji}}\lfloor r_{ji}^k x_{ji}^k\rfloor - \sum_{k=1}^{l_{ij}} x_{ij}^k\right)$$

subject to

$$\sum_{j\neq i}\left(\sum_{k=1}^{l_{ji}}\lfloor r_{ji}^k x_{ji}^k\rfloor - \sum_{k=1}^{l_{ij}} x_{ij}^k\right) \geq 0,\ i=1,\ldots,n, \tag{4}$$

$$0 \leq x_{ij}^k \leq b_{ij}^k,\ 1\leq k \leq l_{ij},\ 1\leq i\neq j \leq n, \tag{5}$$

$$x_{ij}^k \text{ is integer},\ 1\leq k \leq l_{ij},\ 1\leq i\neq j \leq n, \tag{6}$$

where $w_i \geq 0$ is a given weight for currency $i, i=1,2,\ldots,n$, with at least one $w_i > 0$.

Finally, consider another

Problem 3 In order to eliminate arbitrage, how many transactions and arcs in a exchange digraph have to be used for the currency exchange system?

Key Results

A decision problem is called nondeterministic polynomial (*NP* for short) if its solution (if one exists) can be guessed and verified in polynomial time; nondeterministic means that no particular rule is followed to make the guess. If a problem is *NP* and all other *NP* problems are polynomial-time reducible to it, the problem is *NP*-complete. And a problem is called *NP*-hard if every other problem in *NP* is polynomial-time reducible to it (Fig. 2).

Theorem 1 *It is NP-complete to determine whether there exists arbitrage in a frictional foreign exchange market with bid-ask spreads, bound, and integrality constraints even if all $l_{ij} = 1$.*

Then, a further inapproximability result is obtained.

Theorem 2 *There exists fixed $\epsilon > 0$ such that approximating (P) within a factor of n^ϵ is NP-hard even for any of the following two special cases:*

(P_1) *all $l_{ij} = 1$ and $w_i = 1$.*
(P_2) *all $l_{ij} = 1$ and all but one $w_i = 0$.*

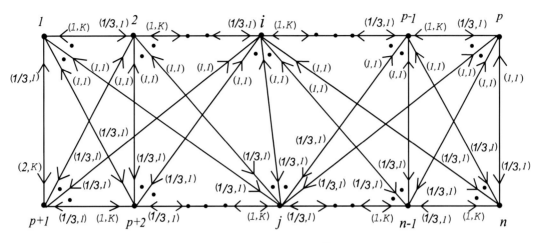

Arbitrage in Frictional Foreign Exchange Market, Fig. 2   Digraph G_2

Now, consider two polynomially solvable special cases when the number of currencies is constant or the exchange digraph is star shaped (a digraph is star shaped if all arcs have a common vertex).

Theorem 3 *There are polynomial time algorithms for* (P) *when the number of currencies is constant.*

Theorem 4 *It is polynomially solvable to find the maximum revenue at the center currency of arbitrage in a frictional foreign exchange market with bid-ask spread, bound, and integrality constraints when the exchange digraph is star shaped.*

However, if the exchange digraph is the coalescence of a star-shaped exchange digraph and its copy, shown by Digraph G_1, then the problem becomes *NP*-complete.

Theorem 5 *It is NP-complete to decide whether there exists arbitrage in a frictional foreign exchange market with bid-ask spreads, bound, and integrality constraints even if its exchange digraph is coalescent.*

Finally, an answer to Problem 3 is as follows

Theorem 6 *There is an exchange digraph of order* n *such that at least* $\lfloor n/2 \rfloor \lceil n/2 \rceil - 1$ *transactions and at least* $n^2/4 + n - 3$ *arcs are in need to bring the system back to non-arbitrage states.*

For instance, consider the currency exchange market corresponding to digraph $G_2 = (V, E)$, where the number of currencies is $n = |V|$, $p = \lfloor n/ \rfloor$, and $K = n^2$.

Set

$$C = \{a_{ij} \in E | 1 \leq i \leq p, p + 1 \leq j \leq n\}$$

$$\cup \{a_{1(p+1)}\}\backslash\{a_{(p+1)1}\} \cup \{a_{i(i-1)}| 2 \leq i \leq p\}$$

$$\cup \{a_{i(i+1)}| p + 1 \leq i \leq n - 1\}.$$

Then, $|C| = \lfloor n/2 \rfloor \lceil n/2 \rceil + n - 2 = |E|/ > n^2/4 + n - 3$. It follows easily from the rates and bounds that each arc in C has to be used to eliminate arbitrage. And $\lfloor n/2 \rfloor \lceil n/2 \rceil - 1$ transactions corresponding to $\{a_{ij} \in E | 1 \leq i \leq p, p + 1 \leq j \leq n\}\backslash\{a_{(p+1)1}\}$ are in need to bring the system back to non-arbitrage states.

Applications

The present results show that different foreign exchange systems exhibit quite different computational complexities. They may shed new light on how monetary system models are adopted and evolved in reality. In addition, it provides with a computational complexity point of view to the understanding of the now fast growing Internet electronic exchange markets.

Open Problems

The dynamic models involving in both spot markets (in which goods are sold for cash and delivered immediately) and futures markets are the most interesting ones. To develop good approximation algorithms for such general models would be important. In addition, it is also important to identify special market models for which polynomial time algorithms are possible even with future markets. Another interesting paradox in this line of study is why friction constraints that make arbitrage difficult are not always eliminated in reality.

Recommended Reading

1. Abeysekera SP, Turtle HJ (1995) Long-run relations in exchange markets: a test of covered interest parity. J Financ Res 18(4):431–447
2. Ausiello G, Crescenzi P, Gambosi G, Kann V, Marchetti-Spaccamela A, Protasi M (1999) Complexity and approximation: combinatorial optimization problems and their approximability properties. Springer, Berlin
3. Cai M, Deng X (2003) Approximation and computation of arbitrage in frictional foreign exchange market. Electron Notes Theor Comput Sci 78:1–10
4. Clinton K (1988) Transactions costs and covered interest arbitrage: theory and evidence. J Polit Econ 96(2):358–370
5. Deng X, Papadimitriou C (1994) On the complexity of cooperative game solution concepts. Math Oper Res 19(2):257–266
6. Deng X, Li ZF, Wang S (2002) Computational complexity of arbitrage in frictional security market. Int J Found Comput Sci 13(5):681–684
7. Deng X, Papadimitriou C, Safra S (2003) On the complexity of price equilibria. J Comput Syst Sci 67(2):311–324
8. Garey MR, Johnson DS (1979) Computers and intractability: a guide of the theory of NP-completeness. Freeman, San Francisco
9. Jones CK (2001) A network model for foreign exchange arbitrage, hedging and speculation. Int J Theor Appl Finance 4(6):837–852
10. Lenstra HW Jr (1983) Integer programming with a fixed number of variables. Math Oper Res 8(4):538–548
11. Mavrides M (1992) Triangular arbitrage in the foreign exchange market – inefficiencies, technology and investment opportunities. Quorum Books, London
12. Megiddo N (1978) Computational complexity and the game theory approach to cost allocation for a tree. Math Oper Res 3:189–196
13. Mundell RA (1981) Gold would serve into the 21st century. Wall Str J, Sep 30, p 33
14. Mundell RA (2000) Currency areas, exchange rate systems, and international monetary reform, paper delivered at Universidad del CEMA, Buenos Aires. http://www.robertmundell.net/pdf/Currency. Accessed 17 Apr 2000
15. Zhang S, Xu C, Deng X (2002) Dynamic arbitrage-free asset pricing with proportional transaction costs. Math Finance 12(1):89–97

Arithmetic Coding for Data Compression

Paul G. Howard[1] and Jeffrey Scott Vitter[2]
[1] Akamai Technologies, Cambridge, MA, USA
[2] University of Kansas, Lawrence, KS, USA

Keywords

Entropy coding; Statistical data compression

Years and Authors of Summarized Original Work

1994; Howard, Vitter

Problem Definition

Often it is desirable to encode a sequence of data efficiently to minimize the number of bits required to transmit or store the sequence. The sequence may be a file or message consisting of *symbols* (or *letters* or *characters*) taken from a fixed input alphabet, but more generally the sequence can be thought of as consisting of *events*, each taken from its own input set. Statistical data compression is concerned with encoding the data in a way that makes use of probability estimates of the events. Lossless compression has the property that the input sequence can be reconstructed exactly from the encoded sequence. Arithmetic

Arithmetic Coding for Data Compression, Table 1 Comparison of codes for Huffman coding, Hu-Tucker coding, and arithmetic coding for a sample 5-symbol alphabet

Symbol e_k	Prob.		Huffman		Hu-Tucker		Arithmetic
	p_k	$-\log_2 p_k$	Code	Length	Code	Length	Length
a	0.04	4.644	**1111**	4	**000**	3	4.644
b	0.18	2.474	**110**	3	**001**	3	2.474
c	0.43	1.218	**0**	1	**01**	2	1.218
d	0.15	2.737	**1110**	4	**10**	2	2.737
e	0.20	2.322	**10**	2	**11**	2	2.322

coding is a nearly optimal statistical coding technique that can produce a lossless encoding.

Problem (statistical data compression) INPUT: A sequence of m events $a_1, a_2, \ldots, a_m$. The ith event a_i is taken from a set of n distinct possible events $e_{i,1}, e_{i,2}, \ldots, e_{i,n}$, with an accurate assessment of the probability distribution P_i of the events. The distributions P_i need not be the same for each event a_i.

OUTPUT: A succinct encoding of the events that can be decoded to recover exactly the original sequence of events.

The goal is to achieve optimal or near-optimal encoding length. Shannon [10] proved that the smallest possible expected number of bits needed to encode the ith event is the *entropy* of P_i, denoted by

$$H(P_i) = \sum_{k=1}^{n} -p_{i,k} \log_2 p_{i,k}$$

where $p_{i,k}$ is the probability that e_k occurs as the ith event. An optimal code outputs $-\log_2 p$ bits to encode an event whose probability of occurrence is p.

The well-known Huffman codes [6] are optimal only among *prefix* (or *instantaneous*) codes, that is, those in which the encoding of one event can be decoded before encoding has begun for the next event. Hu-Tucker codes are prefix codes similar to Huffman codes and are derived using a similar algorithm, with the added constraint that coded messages preserve the ordering of original messages.

When an instantaneous code is not needed, as is often the case, arithmetic coding provides

a number of benefits, primarily by relaxing the constraint that the code lengths must be integers: (1) The code length is optimal ($-\log_2 p$ bits for an event with probability p), even when probabilities are not integer powers of $\frac{1}{2}$. (2) There is no loss of coding efficiency even for events with probability close to 1. (3) It is trivial to handle probability distributions that change from event to event. (4) The input message to output message ordering correspondence of Hu-Tucker coding can be obtained with minimal extra effort.

As an example, consider a 5-symbol input alphabet. Symbol probabilities, codes, and code lengths are given in Table 1.

The average code length is 2.13 bits per input symbol for the Huffman code, 2.22 bits per symbol for the Hu-Tucker code, and 2.03 bits per symbol for arithmetic coding.

Key Results

In theory, arithmetic codes assign one "codeword" to each possible input sequence. The codewords consist of half-open subintervals of the half-open unit interval $[0,1)$ and are expressed by specifying enough bits to distinguish the subinterval corresponding to the actual sequence from all other possible subintervals. Shorter codes correspond to larger subintervals and thus more probable input sequences. In practice, the subinterval is refined incrementally using the probabilities of the individual events, with bits being output as soon as they are known. Arithmetic codes almost always give better compression than prefix codes, but they lack the direct correspondence between the events in the input sequence and bits or groups of bits in the coded output file.

The algorithm for encoding a file using arithmetic coding works conceptually as follows:

1. The "current interval" $[L, H)$ is initialized to $[0,1)$.
2. For each event in the file, two steps are performed:
   (a) Subdivide the current interval into subintervals, one for each possible event. The size of an event's subinterval is proportional to the estimated probability that the event will be the next event in the file, according to the model of the input.
   (b) Select the subinterval corresponding to the event that actually occurs next and make it the new current interval.
3. Output enough bits to distinguish the final current interval from all other possible final intervals.

The length of the final subinterval is clearly equal to the product of the probabilities of the individual events, which is the probability p of the particular overall sequence of events. It can be shown that $\lfloor -\log_2 p \rfloor + 2$ bits are enough to distinguish the file from all other possible files.

For finite-length files, it is necessary to indicate the end of the file. In arithmetic coding, this can be done easily by introducing a special low-probability event that can be injected into the input stream at any point. This adds only $O(\log m)$ bits to the encoded length of an m-symbol file.

In step 2, one needs to compute only the subinterval corresponding to the event a_i that actually occurs. To do this, it is convenient to use two "cumulative" probabilities: the cumulative probability $P_C = \sum_{k=1}^{i-1} p_k$ and the next-cumulative probability $P_N = P_C + p_i = \sum_{k=1}^{i-1} p_k$. The new subinterval is $[L + P_C(H - L), L + P_N(H - L))$. The need to maintain and supply cumulative probabilities requires the model to have a sophisticated data structure, such

as that of Moffat [7], especially when many more than two events are possible.

Modeling

The goal of modeling for statistical data compression is to provide probability information to the coder. The modeling process consists of structural and probability estimation components; each may be adaptive (starting from a neutral model, gradually build up the structure and probabilities based on the events encountered), semi-adaptive (specify an initial model that describes the events to be encountered in the data and then modify the model during coding so that it describes only the events yet to be coded), or static (specify an initial model and use it without modification during coding).

In addition there are two strategies for probability estimation. The first is to estimate each event's probability individually based on its frequency within the input sequence. The second is to estimate the probabilities collectively, assuming a probability distribution of a particular form and estimating the parameters of the distribution, either directly or indirectly. For direct estimation, the data can yield an estimate of the parameter (the variance, for instance). For indirect estimation [4], one can start with a small number of possible distributions and compute the code length that would be obtained with each; the one with the smallest code length is selected. This method is very general and can be used even for distributions from different families, without common parameters.

Arithmetic coding is often applied to text compression. The events are the symbols in the text file, and the model consists of the probabilities of the symbols considered in some context. The simplest model uses the overall frequencies of the symbols in the file as the probabilities; this is a zero-order Markov model, and its entropy is denoted H_0. The probabilities can be estimated adaptively starting with counts of 1 for all symbols and incrementing after each symbol is coded, or the symbol counts can be coded before coding the file itself and either modified during coding (a decrementing semi-adaptive code) or left unchanged (a static code). In all cases, the code length is independent of the order of the symbols in the file.

Theorem 1 *For all input files, the code length L_A of an adaptive code with initial 1-weights is the same as the code length L_{SD} of the semiadaptive decrementing code plus the code length L_M of the input model encoded assuming that all symbol distributions are equally likely. This code length is less than $L_S = mH_0 + L_M$, the code length of a static code with the same input model. In other words, $L_A = L_{SD} + L_M < mH_0 + L_M = L_S$.*

It is possible to obtain considerably better text compression by using higher-order Markov models. Cleary and Witten [2] were the first to do this with their PPM method. PPM requires adaptive modeling and coding of probabilities close to 1 and makes heavy use of arithmetic coding.

Implementation Issues

Incremental Output
The basic implementation of arithmetic coding described above has two major difficulties: the shrinking current interval requires the use of high-precision arithmetic, and no output is produced until the entire file has been read. The most straightforward solution to both of these problems is to output each leading bit as soon as it is known and then to double the length of the current interval so that it reflects only the unknown part of the final interval. Witten, Neal, and Cleary [11] add a clever mechanism for preventing the current interval from shrinking too much when the endpoints are close to $\frac{1}{2}$ but straddle $\frac{1}{2}$. In that case, one does not yet know the next output bit, but whatever it is, the *following* bit will have the opposite value; one can merely keep track of that fact and expand the current interval symmetrically about $\frac{1}{2}$. This follow-on procedure may be repeated any number of times, so the current interval size is always strictly longer than $\frac{1}{4}$.

Before [11] other mechanisms for incremental transmission and fixed precision arithmetic were developed through the years by a number of researchers beginning with Pasco [8]. The bit-stuffing idea of Langdon and others at IBM [9] that limits the propagation of carries in the additions serves a function similar to that of the follow-on procedure described above.

Use of Integer Arithmetic
In practice, the arithmetic can be done by storing the endpoints of the current interval as sufficiently large integers rather than in floating point or exact rational numbers. Instead of starting with the real interval [0,1), start with the integer interval [0,N), N invariably being a power of 2. The subdivision process involves selecting nonoverlapping integer intervals (of length at least 1) with lengths approximately proportional to the counts.

Limited-Precision Arithmetic Coding
Arithmetic coding as it is usually implemented is slow because of the multiplications (and in some implementations, divisions) required in subdividing the current interval according to the probability information. Since small errors in probability estimates cause very small increases in code length, introducing approximations into the arithmetic coding process in a controlled way can improve coding speed without significantly degrading compression performance. In the Q-Coder work at IBM [9], the time-consuming multiplications are replaced by additions and shifts, and low-order bits are ignored.

Howard and Vitter [3] describe a different approach to approximate arithmetic coding. The fractional bits characteristic of arithmetic coding are stored as state information in the coder. The idea, called *quasi-arithmetic coding*, is to reduce the number of possible states and replace arithmetic operations by table lookups; the lookup tables can be precomputed.

The number of possible states (after applying the interval expansion procedure) of an arithmetic coder using the integer interval [0,N) is $3N^2/16$. The obvious way to reduce the number of states in order to make lookup tables practicable is to reduce N. Binary quasi-arithmetic coding causes an insignificant increase in the code length compared with pure arithmetic coding.

Theorem 2 *In a quasi-arithmetic coder based on full interval $[0, N)$, using correct probability estimates, and excluding very large and very small probabilities, the number of bits per input event by which the average code length obtained by the quasi-arithmetic coder exceeds that of an exact arithmetic coder is at most*

$$\frac{4}{\ln 2} \left(\log_2 \frac{2}{e \ln 2} \right) \frac{1}{N} + O\left(\frac{1}{N^2}\right)$$

$$\approx \frac{0.497}{N} + O\left(\frac{1}{N^2}\right),$$

and the fraction by which the average code length obtained by the quasi-arithmetic coder exceeds that of an exact arithmetic coder is at most

$$\left(\log_2 \frac{2}{e \ln 2}\right) \frac{1}{\log_2 N} + O\left(\frac{1}{(\log N)^2}\right)$$

$$\approx \frac{0.0861}{\log_2 N} + O\left(\frac{1}{(\log N)^2}\right).$$

General-purpose algorithms for parallel encoding and decoding using both Huffman and quasi-arithmetic coding are given in [5].

Applications

Arithmetic coding can be used in most applications of data compression. Its main usefulness is in obtaining maximum compression in conjunction with an adaptive model or when the probability of one event is close to 1. Arithmetic coding has been used heavily in text compression. It has also been used in image compression in the JPEG international standards for image compression and is an essential part of the JBIG international standards for bilevel image compression. Many fast implementations of arithmetic coding, especially for a two-symbol alphabet, are covered by patents; considerable effort has been expended in adjusting the basic algorithm to avoid infringing those patents.

Open Problems

The technical problems with arithmetic coding itself have been completely solved. The remaining unresolved issues are concerned with modeling, in which the issue is how to decompose an input data set into a sequence of events, so that the set of events possible at each point in the data set can be described by a probability distribution suitable for input into the coder. The modeling issues are entirely application-specific.

Experimental Results

Some experimental results for the Calgary and Canterbury corpora are summarized in a report by Arnold and Bell [1].

Data Sets

Among the most widely used data sets suitable for research in arithmetic coding are the Calgary Corpus and Canterbury Corpus (corpus. canterbury.ac.nz) and the Pizza&Chili Corpus (pizzachili.dcc.uchile.cl or http://pizzachili.di. unipi.it).

URL to Code

A number of implementations of arithmetic coding are available on The Data Compression Resource on the Internet, www.data-compression. info/Algorithms/AC/.

Cross-References

▶ Burrows-Wheeler Transform
▶ Huffman Coding

Recommended Reading

1. Arnold R, Bell T (1997) A corpus for the evaluation of lossless compression algorithms. In: Proceedings of the IEEE data compression conference, Snowbird, Mar 1997, pp 201–210

2. Cleary JG, Witten IH (1984) Data compression using adaptive coding and partial string matching. IEEE Trans Commun COM-32:396–402
3. Howard PG, Vitter JS (1992) Practical implementations of arithmetic coding. In: Storer JA (ed) Images and text compression. Kluwer Academic, Norwell
4. Howard PG, Vitter JS (1993) Fast and efficient lossless image compression. In: Proceedings of the IEEE data compression conference, Snowbird, Mar 1993, pp 351–360
5. Howard PG, Vitter JS (1996) Parallel lossless image compression using Huffman and arithmetic coding. Inf Process Lett 59:65–73
6. Huffman DA (1952) A method for the construction of minimum redundancy codes. Proc Inst Radio Eng 40:1098–1101
7. Moffat A (1999) An improved data structure for cumulative probability tables. Softw Pract Exp 29:647–659
8. Pasco R (1976) Source coding algorithms for fast data compression. Ph.D. thesis, Stanford University
9. Pennebaker WB, Mitchell JL, Langdon GG, Arps RB (1988) An overview of the basic principles of the Q-coder adaptive binary arithmetic coder. IBM J Res Dev 32:717–726
10. Shannon CE (1948) A mathematical theory of communication. Bell Syst Tech J 27:398–403
11. Witten IH, Neal RM, Cleary JG (1987) Arithmetic coding for data compression. Commun ACM 30:520–540

Assignment Problem

Samir Khuller
Computer Science Department, University of Maryland, College Park, MD, USA

Keywords

Weighted bipartite matching

Years and Authors of Summarized Original Work

1955; Kuhn
1957; Munkres

Problem Definition

Assume that a complete bipartite graph, $G(X, Y, X \times Y)$, with weights $w(x, y)$ assigned to every edge (x, y) is given. A matching M is a subset of edges so that no two edges in M have a common vertex. A perfect matching is one in which all the nodes are matched. Assume that $|X| = |Y| = n$. The **weighted matching problem** is to find a matching with the greatest total weight, where $w(M) = \sum_{e \in M} w(e)$. Since G is a complete bipartite graph, it has a perfect matching. An algorithm that solves the weighted matching problem is due to Kuhn [4] and Munkres [6]. Assume that all edge weights are non-negative.

Key Results

Define a *feasible vertex labeling* ℓ as a mapping from the set of vertices in G to the reals, where

$$\ell(x) + \ell(y) \geq w(x, y).$$

Call $\ell(x)$ the label of vertex x. It is easy to compute a feasible vertex labeling as follows:

$$\forall y \in Y \, \ell(y) = 0$$

and

$$\forall x \in X \quad \ell(x) = \max_{y \in Y} w(x, y).$$

Define the **equality subgraph**, G_ℓ, to be the spanning subgraph of G, which includes all vertices of G but only those edges (x, y) that have weights such that

$$w(x, y) = \ell(x) + \ell(y).$$

The connection between equality subgraphs and maximum-weighted matchings is provided by the following theorem:

Theorem 1 *If the equality subgraph, G_ℓ, has a perfect matching, M^*, then M^* is a maximum-weighted matching in G.*

In fact, note that the sum of the labels is an upper bound on the weight of the maximum-weighted perfect matching. The algorithm eventually finds a matching and a feasible labeling such that the weight of the matching is equal to the sum of all the labels.

High-Level Description

The above theorem is the basis of an algorithm for finding a maximum-weighted matching in a complete bipartite graph. Starting with a feasible labeling, compute the equality subgraph, and then find a maximum matching in this subgraph (here, one can ignore weights on edges). If the matching found is perfect, the process is done. If it is not perfect, more edges are added to the equality subgraph by revising the vertex labels. After adding edges to the equality subgraph, either the size of the matching goes up (an augmenting path is found) or the Hungarian tree continues to grow. (This is the structure of explored edges when one starts BFS simultaneously from all free nodes in S. When one reaches a matched node in T, one only explores the matched edge; however, all edges incident to nodes in S are explored.) In the former case, the phase terminates, and a new phase starts (since the matching size has gone up). In the latter case, the Hungarian tree grows by adding new nodes to it, and clearly, this cannot happen more than n times.

Let S be the set of free nodes in X. Grow Hungarian trees from each node in S. Let T be the nodes in Y encountered in the search for an augmenting path from nodes in S. Add all nodes from X that are encountered in the search to S.

Note the following about this algorithm:

$$\overline{S} = X \setminus S.$$

$$\overline{T} = Y \setminus T.$$

$$|S| > |T|.$$

There are no edges from S to $\overline{T}$ since this would imply that one did not grow the Hungarian trees completely. As the Hungarian trees are grown in G_ℓ, alternate nodes in the search are placed into S and T. To revise the labels, take the labels in S, and start decreasing them uniformly (say, by λ), and at the same time, increase the labels in T by λ. This ensures that the edges from S to T do not leave the equality subgraph (Fig. 1).

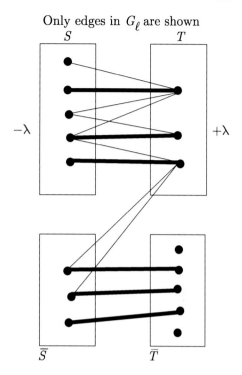

Only edges in G_ℓ are shown

Assignment Problem, Fig. 1 Sets S and T as maintained by the algorithm

As the labels in S are decreased, edges (in G) from S to $\overline{T}$ will potentially enter the equality subgraph, G_ℓ. As we increase λ, at some point in time, an edge enters the equality subgraph. This is when one stops and updates the Hungarian tree. If the node from $\overline{T}$ added to T is matched to a node in $\overline{S}$, both these nodes are moved to S and T, which yields a larger Hungarian tree. If the node from $\overline{T}$ is free, an augmenting path is found, and the phase is complete. One phase consists of those steps taken between increases in the size of the matching. There are at most n phases, where n is the number of vertices in G (since in each phase the size of the matching increases by 1). Within each phase, the size of the Hungarian tree is increased at most n times. It is clear that in $O(n^2)$ time, one can figure out which edge from S to $\overline{T}$ is the first to enter the equality subgraph (one simply scans all the edges). This yields an $O(n^4)$ bound on the total running time. How to implement it in $O(n^3)$ time is now shown.

More Efficient Implementation

Define the slack of an edge as follows:

$$slack\,(x, y) = \ell\,(x) + \ell\,(y) - w\,(x, y)\,.$$

Then,

$$\lambda = \min_{x \in S, y \in \overline{T}} slack\,(x, y)\,.$$

Naively, the calculation of λ requires $O(n^2)$ time. For every vertex $y \in \overline{T}$, keep track of the edge with the smallest slack, i.e.,

$$slack\,[y] = \min_{x \in S} slack\,(x, y)\,.$$

The computation of $slack[y]$ (for all $y \in \overline{T}$) requires $O(n^2)$ time at the start of a phase. As the phase progresses, it is easy to update all the *slack* values in $O(n)$ time since all of them change by the same amount (the labels of the vertices in S are going down uniformly). Whenever a node u is moved from $\overline{S}$ to S, one must recompute the slacks of the nodes in $\overline{T}$, requiring $O(n)$ time. But a node can be moved from $\overline{S}$ to S at most n times.

Thus, each phase can be implemented in $O(n^2)$ time. Since there are n phases, this gives a running time of $O(n^3)$. For sparse graphs, there is a way to implement the algorithm in $O(n(m + n \log n))$ time using min cost flows [1], where m is the number of edges.

Applications

There are numerous applications of biparitite matching, for example, scheduling unit-length jobs with integer release times and deadlines even with time-dependent penalties.

Open Problems

Obtaining a linear, or close to linear, time algorithm.

Recommended Reading

Several books on combinatorial optimization describe algorithms for weighted bipartite matching (see [2,5]). See also Gabow's paper [3].

1. Ahuja R, Magnanti T, Orlin J (1993) Network flows: theory, algorithms and applications. Prentice Hall, Englewood Cliffs
2. Cook W, Cunningham W, Pulleyblank W, Schrijver A (1998) Combinatorial Optimization. Wiley, New York
3. Gabow H (1990) Data structures for weighted matching and nearest common ancestors with linking. In: Symposium on discrete algorithms, San Francisco, pp 434–443
4. Kuhn H (1955) The Hungarian method for the assignment problem. Naval Res Logist Q 2:83–97
5. Lawler E (1976) Combinatorial optimization: networks and matroids. Holt, Rinehart and Winston, New York
6. Munkres J (1957) Algorithms for the assignment and transportation problems. J Soc Ind Appl Math 5:32–38

Asynchronous Consensus Impossibility

Maurice Herlihy
Department of Computer Science,
Brown University, Providence, RI, USA

Keywords

Agreement; Wait-free consensus

Years and Authors of Summarized Original Work

1985; Fischer, Lynch, Paterson

Problem Definition

Consider a distributed system consisting of a set of *processes* that communicate by sending and receiving messages. The network is a multiset of messages, where each message is addressed to some process. A process is a state machine that can take three kinds of *steps*.

- In a *send* step, a process places a message in the network.
- In a *receive* step, a process A either reads and removes from the network a message addressed to A, or it reads a distinguished *null* value, leaving the network unchanged. If a message addressed to A is placed in the network, and if A subsequently performs an infinite number of receive steps, then A will eventually receive that message.
- In a *computation* state, a process changes state without communicating with any other process.

Processes are *asynchronous*: there is no bound on their relative speeds. Processes can *crash*: they can simply halt and take no more steps. This article considers executions in which at most one process crashes.

In the *consensus* problem, each process starts with a private *input* value, communicates with the others, and then halts with a *decision* value. These values must satisfy the following properties:

- *Agreement:* all processes' decision values must agree.
- *Validity:* every decision value must be some process' input.
- *Termination:* every non-fault process must decide in a finite number of steps.

Fischer, Lynch, and Paterson showed that there is no protocol that solves consensus in any asynchronous message-passing system where even a single process can fail. This result is one of the most influential results in Distributed Computing, laying the foundations for a number of subsequent research efforts.

Terminology

Without loss of generality, one can restrict attention to *binary* consensus, where the inputs are 0 or 1. A *protocol state* consists of the states of the processes and the multiset of messages in transit in the network. An *initial state* is a protocol state before any process has moved, and a *final state* is a protocol state after all processes have finished.

The *decision value* of any final state is the value decided by all processes in that state.

Any terminating protocol's set of possible states forms a tree, where each node represents a possible protocol state, and each edge represents a possible step by some process. Because the protocol must terminate, the tree is finite. Each leaf node represents a final protocol state with decision value either 0 or 1.

A *bivalent* protocol state is one in which the eventual decision value is not yet fixed. From any bivalent state, there is an execution in which the eventual decision value is 0, and another in which it is 1. A *univalent* protocol state is one in which the outcome is fixed. Every execution starting from a univalent state decides the same value. A *1-valent* protocol state is univalent with eventual decision value 1, and similarly for a *0-valent* state.

A protocol state is *critical* if

- it is bivalent, and
- if any process takes a step, the protocol state becomes univalent.

Key Results

Lemma 1 *Every consensus protocol has a bivalent initial state.*

Proof Assume, by way of contradiction, that there exists a consensus protocol for $(n + 1)$ threads $A_0, \cdots, A_n$ in which every initial state is univalent. Let s_i be the initial state where processes $A_i, \cdots, A_n$ have input 0 and $A_0, \ldots, A_{i-1}$ have input 1. Clearly, s_0 is 0-valent: all processes have input 0, so all must decide 0 by the validity condition. If s_i is 0-valent, so is s_{i+1}. These states differ only in the input to process $A_i : 0$ in s_i, and 1 in s_{i+1}. Any execution starting from s_i in which A_i halts before taking any steps is indistinguishable from an execution starting from s_{i+1} in which A_i halts before taking any steps. Since processes must decide 0 in the first execution, they must decide 1 in the second. Since there is one execution starting from s_{i+1} that decides 0, and since s_{i+1} is univalent by hypothesis, s_{i+1} is 0-valent. It follows that the state s_{n+1}, in which all processes start with input 1, is 0-valent, a contradiction.   □

Lemma 2 *Every consensus protocol has a critical state.*

Proof by contradiction. By Lemma 1, the protocol has a bivalent initial state. Start the protocol in this state. Repeatedly choose a process whose next step leaves the protocol in a bivalent state, and let that process take a step. Either the protocol runs forever, violating the termination condition, or the protocol eventually enters a critical state.□

Theorem 1 *There is no consensus protocol for an asynchronous message-passing system where a single process can crash.*

Proof Assume by way of contradiction that such a protocol exists. Run the protocol until it reaches a critical state s. There must be two processes A and B such that A's next step carries the protocol to a 0-valent state, and B's next step carries the protocol to a 1-valent state.

Starting from s, let s_A be the state reached if A takes the first step, s_B if B takes the first step, s_{AB} if A takes a step followed by B, and so on. States s_A and s_{AB} are 0-valent, while s_B and s_{BA} are 1-valent. The rest is a case analysis.

Of all the possible pairs of steps A and B could be about to execute, most of them *commute*: states s_{AB} and s_{BA} are identical, which is a contradiction because they have different valences.

The only pair of steps that do not commute occurs when A is about to send a message to B (or vice versa). Let s_{AB} be the state resulting if A sends a message to B and B then receives it, and let s_{BA} be the state resulting if B receives a different message (or *null*) and then A sends its message to B. Note that every process other than B has the same local state in s_{AB} and s_{BA}. Consider an execution starting from s_{AB} in which every process other than B takes steps in round-robin order. Because s_{AB} is 0-valent, they will eventually decide 0. Next, consider an execution

starting from s_{BA} in which every process other than B takes steps in round-robin order. Because s_{BA} is 1-valent, they will eventually decide 1. But all processes other than B have the same local states at the end of each execution, so they cannot decide different values, a contradiction.   □

In the proof of this theorem, and in the proofs of the preceding lemmas, we construct scenarios where at most a single process is delayed. As a result, this impossibility result holds for any system where a single process can fail undetectably.

Applications

The consensus problem is a key tool for understanding the power of various asynchronous models of computation.

Open Problems

There are many open problems concerning the solvability of consensus in other models, or with restrictions on inputs.

Related Work

The original paper by Fischer, Lynch, and Paterson [8] is still a model of clarity.

Many researchers have examined alternative models of computation in which consensus can be solved. Dolev, Dwork, and Stockmeyer [5] examine a variety of alternative message-passing models, identifying the precise assumptions needed to make consensus possible. Dwork, Lynch, and Stockmeyer [6] derive upper and lower bounds for a semi-synchronous model where there is an upper and lower bound on message delivery time. Ben-Or [1] showed that introducing randomization makes consensus possible in an asynchronous message-passing system. Chandra and Toueg [3] showed that consensus becomes possible if in the presence of an oracle that can (unreliably) detect when a process has crashed. Each of the papers cited here has inspired many follow-up papers. A good place to start is the excellent survey by Fich and Ruppert [7].

A protocol is *wait-free* if it tolerates failures by all but one of the participants. A concurrent object implementation is *linearizable* if each method call seems to take effect instantaneously at some point between the method's invocation and response. Herlihy [9] showed that shared-memory objects can each be assigned a *consensus number*, which is the maximum number of processes for which there exists a wait-free consensus protocol using a combination of read-write memory and the objects in question. Consensus numbers induce an infinite hierarchy on objects, where (simplifying somewhat) higher objects are more powerful than lower objects. In a system of n or more concurrent processes, it is impossible to construct a lock-free implementation of an object with consensus number n from an object with a lower consensus number. On the other hand, any object with consensus number n is *universal* in a system of n or fewer processes: it can be used to construct a wait-free linearizable implementation of any object.

In 1990, Chaudhuri [4] introduced the *k-set agreement* problem (sometimes called *k-set consensus*, which generalizes consensus by allowing k or fewer distinct decision values to be chosen. In particular, 1-set agreement is consensus. The question whether k-set agreement can be solved in asynchronous message-passing models was open for several years, until three independent groups [2, 10, 11] showed that no protocol exists.

Cross-References

▶ Linearizability
▶ Topology Approach in Distributed Computing

Recommended Reading

1. Ben-Or M (1983) Another advantage of free choice (extended abstract): completely asynchronous agreement protocols. In: PODC '83: proceedings of the second annual ACM symposium on principles of distributed computing. ACM Press, New York, pp 27–30
2. Borowsky E, Gafni E (1993) Generalized FLP impossibility result for t-resilient asynchronous computations. In: Proceedings of the 1993 ACM symposium on theory of computing, May 1993, pp 206–215
3. Chandra TD, Toueg S (1996) Unreliable failure detectors for reliable distributed systems. J ACM 43(2):225–267
4. Chaudhuri S (1990) Agreement is harder than consensus: set consensus problems in totally asynchronous systems. In: Proceedings of the ninth annual ACM symposium on principles of distributed computing, Aug 1990, pp 311–324
5. Chandhuri S (1993) More choices allow more faults: set consensus problems in totally asynchronous systems. Inf Comput 105(1):132–158
6. Dwork C, Lynch N, Stockmeyer L (1988) Consensus in the presence of partial synchrony. J ACM 35(2):288–323
7. Fich F, Ruppert E (2003) Hundreds of impossibility results for distributed computing. Distrib Comput 16(2–3):121–163
8. Fischer M, Lynch N, Paterson M (1985) Impossibility of distributed consensus with one faulty process. J ACM 32(2):374–382
9. Herlihy M (1991) Wait-free synchronization. ACM Trans Program Lang Syst (TOPLAS) 13(1):124–149
10. Herlihy M, Shavit N (1999) The topological structure of asynchronous computability. J ACM 46(6):858–923
11. Saks ME, Zaharoglou F (2000) Wait-free k-set agreement is impossible: the topology of public knowledge. SIAM J Comput 29(5):1449–1483

Atomic Broadcast

Xavier Défago
School of Information Science, Japan Advanced Institute of Science and Technology (JAIST), Ishikawa, Japan

Keywords

Atomic multicast; Total order broadcast; Total order multicast

Years and Authors of Summarized Original Work

1995; Cristian, Aghili, Strong, Dolev

Problem Definition

The problem is concerned with allowing a set of processes to concurrently broadcast messages while ensuring that all destinations consistently deliver them in the exact same sequence, in spite of the possible presence of a number of faulty processes.

The work of Cristian, Aghili, Strong, and Dolev [7] considers the problem of atomic broadcast in a system with approximately synchronized clocks and bounded transmission and processing delays. They present successive extensions of an algorithm to tolerate a bounded number of omission, timing, or Byzantine failures, respectively.

Related Work

The work presented in this entry originally appeared as a widely distributed conference contribution [6], over a decade before being published in a journal [7], at which time the work was well-known in the research community. Since there was no significant change in the algorithms, the historical context considered here is hence with respect to the earlier version.

Lamport [11] proposed one of the first published algorithms to solve the problem of ordering broadcast messages in a distributed systems. That algorithm, presented as the core of a mutual exclusion algorithm, operates in a fully asynchronous system (i.e., a system in which there are no bounds on processor speed or communication delays), but does not tolerate failures. Although the algorithms presented here rely on physical clocks rather than Lamport's logical clocks, the principle used for ordering messages is essentially the same: message carry a timestamp of their sending time; messages are delivered in increasing order of the timestamp, using the sending processor name for messages with equal timestamps.

At roughly the same period as the initial publication of the work of Cristian et al. [6], Chang and Maxemchuck [3] proposed an atomic broadcast protocol based on a token passing protocol, and tolerant to crash failures of processors. Also,

Carr [1] proposed the Tandem global update protocol, tolerant to crash failures of processors.

Cristian [5] later proposed an extension to the omission-tolerant algorithm presented here, under the assumption that the communication system consists of $f + 1$ independent broadcast channels (where f is the maximal number of faulty processors). Compared with the more general protocol presented here, its extension generates considerably fewer messages.

Since the work of Cristian, Aghili, Strong, and Dolev [7], much has been published on the problem of atomic broadcast (and its numerous variants). For further reading, Défago, Schiper, and Urbán [8] surveyed more than sixty different algorithms to solve the problem, classifying them into five different classes and twelve variants. That survey also reviews many alternative definitions and references about two hundred articles related to this subject. This is still a very active research area, with many new results being published each year.

Hadzilacos and Toueg [10] provide a systematic classification of specifications for variants of atomic broadcast as well as other broadcast problems, such as reliable broadcast, FIFO broadcast, or causal broadcast.

Chandra and Toueg [2] proved the equivalence between atomic broadcast and the *consensus* problem. Thus, any application solved by a consensus can also be solved by atomic broadcast and vice-versa. Similarly, impossibility results apply equally to both problems. For instance, it is well-known that consensus, thus atomic broadcast, cannot be solved deterministically in an asynchronous system with the presence of a faulty process [9].

Notations and Assumptions

The system G consists of n distributed processors and m point-to-point communication links. A link does not necessarily exists between every pair of processors, but it is assumed that the communication network remains connected even in the face of faults (whether processors or links). All processors have distinct names and there exists a total order on them (e.g., lexicographic order).

A component (link or processor) is said to be *correct* if its behavior is consistent with its specification, and *faulty* otherwise. The paper considers three classes of component failures, namely, omission, timing, and Byzantine failures.

- An *omission* failure occurs when the faulty component fails to provide the specified output (e.g., loss of a message).
- A *timing* failure occurs when the faulty component omits a specified output, or provides it either too early or too late.
- A *Byzantine* failure [12] occurs when the component does not behave according to its specification, for instance, by providing output different from the one specified. In particular, the paper considers authentication-detectable Byzantine failures, that is, ones that are detectable using a message authentication protocol, such as error correction codes or digital signatures.

Each processor p has access to a local clock C_p with the properties that (1) two separate clock readings yield different values, and (2) clocks are ε-synchronized, meaning that, at any real time t, the deviation in readings of the clocks of any two processors p and q is at most ε.

In addition, transmission and processing delays, as measured on the clock of a correct processor, are bounded by a known constant δ. This bound accounts not only for delays in transmission and processing, but also for delays due to scheduling, overload, clock drift or adjustments. This is called a synchronous system model.

The diffusion time $d\delta$ is the time necessary to propagate information to all correct processes, in a surviving network of diameter d with the presence of a most π processor failures and λ link failures.

Problem Definition

The problem of atomic broadcast is defined in a synchronous system model as a broadcast primitive which satisfies the following three properties: atomicity, order, and termination.

Problem 1 (Atomic broadcast)
INPUT: A stream of messages broadcast by n concurrent processors, some of which may be faulty.
OUTPUT: The messages delivered in sequence, with the following properties:

1. *Atomicity*: if any correct processor delivers an update at time U on its clock, then that update was initiated by some processor and is delivered by each correct processor at time U on its clock.
2. *Order*: all updates delivered by correct processors are delivered in the same order by each correct processor.
3. *Termination*: every update whose broadcast is initiated by a correct processor at time T on its clock is delivered at all correct processors at time $T + \Delta$ on their clock.

Nowadays, problem definitions for atomic broadcast that do not explicitly refer to physical time are often preferred. Many variants of time-free definitions are reviewed by Hadzilacos and Toueg [10] and Défago et al. [8]. One such alternate definition is presented below, with the terminology adapted to the context of this entry.

Problem 2 (Total order broadcast)
INPUT: A stream of messages broadcast by n concurrent processors, some of which may be faulty.
OUTPUT: The messages delivered in sequence, with the following properties:

1. *Validity*: if a correct processor broadcasts a message m, then it eventually delivers m.
2. *Uniform agreement*: if a processor delivers a message m, then all correct processors eventually deliver m.
3. *Uniform integrity*: for any message m, every processor delivers m at most once, and only if m was previously broadcast by its sending processor.
4. *Gap-free uniform total order*: if some processor delivers message m' after message m, then a processor delivers m' only after it has delivered m.

Key Results

The paper presents three algorithms for solving the problem of atomic broadcast, each under an increasingly demanding failure model, namely, omission, timing, and Byzantine failures. Each protocol is actually an extension of the previous one.

All three protocols are based on a classical flooding, or information diffusion, algorithm [14]. Every message carries its initiation timestamp T, the name of the initiating processor s, and an update σ. A message is then uniquely identified by (s, T). Then, the basic protocol is simple. Each processor logs every message it receives until it is delivered. When it receives a message that was never seen before, it forwards that message to all other neighbor processors.

Atomic Broadcast for Omission Failures

The first atomic broadcast protocol, supporting omission failures, considers a termination time Δ_o as follows.

$$\Delta_o = \pi\delta + d\delta + \varepsilon. \tag{1}$$

The delivery deadline $T + \Delta_o$ is the time by which a processor can be sure that it has received copies of every message with timestamp T (or earlier) that could have been received by some correct process.

The protocol then works as follows. When a processor initiates an atomic broadcast, it propagates that message, similar to the diffusion algorithm described above. The main exception is that every message received after the local clock exceeds the delivery deadline of that message, is discarded. Then, at local time $T + \Delta_o$, a processor delivers all messages timestamped with T, in order of the name of the sending processor. Finally, it discards all copies of the messages from its logs.

Atomic Broadcast for Timing Failures

The second protocol extends the first one by introducing a hop count (i.e., a counter incremented each time a message is relayed) to the messages.

With this information, each relaying processor can determine when a message is timely, that is, if a message timestamped T with hop count h is received at time U then the following condition must hold.

$$T - h\varepsilon < U < T + h(\delta + \varepsilon). \tag{2}$$

Before relaying a message, each processor checks the acceptance test above and discard the message if it does not satisfy it. The termination time Δ_t of the protocol for timing failures is as follows.

$$\Delta_t = \pi(\delta + \varepsilon) + d\delta + \varepsilon. \tag{3}$$

The authors point out that discarding early messages is not necessary for correctness, but ensures that correct processors keep messages in their log for a bounded amount of time.

Atomic Broadcast for Byzantine Failures

Given some text, every processor is assumed to be able to generate a signature for it, that cannot be faked by other processors. Furthermore, every processor knows the name of every other processors in the network, and has the ability to verify the authenticity of their signature.

Under the above assumptions, the third protocol extends the second one by adding signatures to the messages. To prevent a Byzantine processor (or link) from tampering with the hop count, a message is co-signed by every processor that relays it. For instance, a message signed by k processors $p_1, \ldots, p_k$ is as follows.

$(relayed, \ldots (relayed, (first, T, \sigma, p_1, s_1), p_2, s_2),$

$\ldots p_k, s_k)$

Where σ is the update, T the timestamp, p_1 the message source, and s_i the signature generated by processor p_i. Any message for which one of the signature cannot be authenticated is simply discarded. Also, if several updates initiated by the same processor p carry the same timestamp, this indicates that p is faulty and the corresponding updates are discarded. The remainder of the protocol is the same as the second one, where the number of hops is given by the number of signatures. The termination time Δ_b is also as follows.

$$\Delta_b = \dot{\pi}(\delta + \varepsilon) + d\delta + \varepsilon. \qquad (4)$$

The authors insist however that, in this case, the transmission time δ must be considerably larger than in the previous case, since it must account for the time spent in generating and verifying the digital signatures; usually a costly operation.

Bounds

In addition to the three protocols presented above and their correctness, Cristian et al. [7] prove the following two lower bounds on the termination time of atomic broadcast protocols.

Theorem 1 *If the communication network G requires x steps, then any atomic broadcast protocol tolerant of up to π processor and λ link omission failures has a termination time of at least $x\delta + \varepsilon$.*

Theorem 2 *Any atomic broadcast protocol for a Hamiltonian network with n processors that tolerate $n - 2$ authentication-detectable Byzantine processor failures cannot have a termination time smaller than $(n - 1)(\delta + \varepsilon)$.*

Applications

The main motivation for considering this problem is its use as the cornerstone for ensuring fault-tolerance through process replication. In particular, the authors consider a *synchronous replicated storage*, which they define as a distributed and resilient storage system that displays the same content at every correct physical processor at any clock time. Using atomic broadcast to deliver updates ensures that all updates are applied at all correct processors in the same order. Thus, provided that the replicas are initially consistent, they will remain consistent. This technique, called *state-machine replication* [11, 13] or also *active replication*, is widely used in practice as

a means of supporting fault-tolerance in distributed systems.

In contrast, Cristian et al. [7] consider atomic broadcast in a *synchronous* system with bounded transmission and processing delays. Their work was motivated by the implementation of a highly-available replicated storage system, with tightly coupled processors running a real-time operating system.

Atomic broadcast has been used as a support for the replication of running processes in real-time systems or, with the problem reformulated to isolate explicit timing requirements, has also been used as a support for fault-tolerance and replication in many group communication toolkits (see survey of Chockler et al. [4]).

In addition, atomic broadcast has been used for the replication of database systems, as a means to reduce the synchronization between the replicas. Wiesmann and Schiper [15] have compared different database replication and transaction processing approaches based on atomic broadcast, showing interesting performance gains.

Cross-References

▶ Asynchronous Consensus Impossibility
▶ Causal Order, Logical Clocks, State Machine Replication
▶ Clock Synchronization
▶ Failure Detectors

Recommended Reading

1. Carr R (1985) The Tandem global update protocol. Tandem Syst Rev 1:74–85
2. Chandra TD, Toueg S (1996) Unreliable failure detectors for reliable distributed systems. J ACM 43:225–267
3. Chang J-M, Maxemchuk NF (1984) Reliable broadcast protocols. ACM Trans Comput Syst 2:251–273
4. Chockler G, Keidar I, Vitenberg R (2001) Group communication specifications: a comprehensive study. ACM Comput Surv 33:427–469
5. Cristian F (1990) Synchronous atomic broadcast for redundant broadcast channels. Real-Time Syst 2:195–212

6. Cristian F, Aghili H, Strong R, Dolev D (1985) Atomic broadcast: from simple message diffusion to Byzantine agreement. In: Proceedings of the 15th international symposium on fault-tolerant computing (FTCS-15), Ann Arbor, June 1985. IEEE Computer Society Press, pp 200—206

7. Cristian F, Aghili H, Strong R, Dolev D (1995) Atomic broadcast: from simple message diffusion to Byzantine agreement. Inform Comput 118:158–179

8. Défago X, Schiper A, Urbán P (2004) Total order broadcast and multicast algorithms: taxonomy and survey. ACM Comput Surv 36:372–421

9. Fischer MJ, Lynch NA, Paterson MS (1985) Impossibility of distributed consensus with one faulty process. J ACM 32:374–382

10. Hadzilacos V, Toueg S (1993) Fault-tolerant broadcasts and related problems. In: Mullender S (ed) Distributed systems, 2nd edn. ACM Press Books/Addison-Wesley, pp 97–146, Extended version appeared as Cornell Univ. TR 94-1425

11. Lamport L (1978) Time, clocks, and the ordering of events in a distributed system. Commun ACM 21:558–565

12. Lamport L, Shostak R, Pease M (1982) The Byzantine generals problem. ACM Trans Prog Lang Syst 4:382–401

13. Schneider FB (1990) Implementing fault-tolerant services using the state machine approach: a tutorial. ACM Comput Surv 22:299–319

14. Segall A (1983) Distributed network protocols. IEEE Trans Inf Theory 29:23–35

15. Wiesmann M, Schiper A (2005) Comparison of database replication techniques based on total order broadcast. IEEE Trans Knowl Data Eng 17:551–566

Attribute-Efficient Learning

Jyrki Kivinen
Department of Computer Science, University of Helsinki, Helsinki, Finland

Keywords

Learning with irrelevant attributes

Years and Authors of Summarized Original Work

1987; Littlestone

Problem Definition

Given here is a basic formulation using the *online mistake-bound* model, which was used by Littlestone [9] in his seminal work.

Fix a class C of Boolean functions over n variables. To start a learning scenario, a *target function* $f_* \in C$ is chosen but not revealed to the *learning algorithm*. Learning then proceeds in a sequence of *trials*. At trial t, an input $x_t \in \{0, 1\}^n$ is first given to the learning algorithm. The learning algorithm then produces its *prediction* $\hat{y}_t$, which is its guess as to the unknown value $f_*(x_t)$. The correct value $y_t = f_*(x_t)$ is then revealed to the learner. If $y_t \neq \hat{y}_t$, the learning algorithm made a *mistake*. The learning algorithm learns C with mistake-bound m, if the number of mistakes never exceeds m, no matter how many trials are made and how f_* and $x_1, x_2, \ldots$ are chosen.

Variable (or attribute) X_i is *relevant* for function $f : \{0, 1\}^n \rightarrow \{0, 1\}$ if $f(x_1, \ldots, x_i, \ldots, x_n) \neq f(x_1, \ldots, 1 - x_i, \ldots, x_n)$ holds for some $\vec{x} \in \{0, 1\}^n$. Suppose now that for some $k \leq n$, every function $f \in C$ has at most k relevant variables. It is said that a learning algorithm learns class C *attribute-efficiently*, if it learns C with a mistake-bound polynomial in k and $\log n$. Additionally, the computation time for each trial is usually required to be polynomial in n.

Key Results

The main part of current research of attribute-efficient learning stems from Littlestone's Winnow algorithm [9]. The basic version of Winnow maintains a weight vector $w_t = (w_{t,1}, \ldots, w_{t,n}) \in \mathbb{R}^n$. The prediction for input $x_t \in \{0, 1\}^n$ is given by

$$\hat{y}_t = \text{sign}\left(\sum_{i=1}^n w_{t,i} x_{t,i} - \theta\right)$$

where θ is a parameter of the algorithm. Initially $w_1 = (1, \ldots, 1)$, and after trial t, each component $w_{t,i}$ is updated according to

$$w_{t+1,i} = \begin{cases} \alpha w_{t,i} & \text{if } y_t = 1, \ \hat{y}_t = 0 \text{ and } x_{t,i} = 1 \\ w_{t,i}/\alpha & \text{if } y_t = 0, \ \hat{y}_t = 1 \text{ and } x_{t,i} = 1 \\ w_{t,i} & \text{otherwise} \end{cases}$$

$$(1)$$

where $\alpha > 1$ is a learning rate parameter.

Littlestone's basic result is that with a suitable choice of θ and α, Winnow learns the class of monotone k-literal disjunctions with mistake-bound $O(k \log n)$. Since the algorithm changes its weights only when a mistake occurs, this bound also guarantees that the weights remain small enough for computation times to remain polynomial in n. With simple transformations, Winnow also yields attribute-efficient learning algorithms for general disjunctions and conjunctions. Various subclasses of DNF formulas and decision lists [8] can be learned, too.

Winnow is quite robust against noise, i.e., errors in input data. This is extremely important for practical applications. Remove now the assumption about a target function $f_* \in C$ satisfying $y_t = f_*(x_t)$ for all t. Define *attribute error* of a pair (x, y) with respect to a function f as the minimum Hamming distance between x and x' such that $f(x') = y$. The attribute error of a sequence of trials with respect to f is the sum of attribute errors of the individual pairs (x_t, y_t). Assuming the sequence of trials has attribute error at most A with respect to some k-literal disjunction, Auer and Warmuth [1] show that Winnow makes $O(A + k \log n)$ mistakes. The noisy scenario can also be analyzed in terms of *hinge loss* [5].

The update rule (1) has served as a model for a whole family of *multiplicative update algorithms*. For example, Kivinen and Warmuth [7] introduce the exponentiated gradient algorithm, which is essentially Winnow modified for continuous-valued prediction, and show how it can be motivated by a relative entropy minimization principle.

Consider a function class C where each function can be encoded using $O(p(k) \log n)$ bits for some polynomial p. An example would be Boolean formulas with k relevant variables, when the size of the formula is restricted to $p(k)$

ignoring the size taken by the variables. The cardinality of C is then $|C| = 2^{O(p(k) \log n)}$. The classical halving algorithm (see [9] for discussion and references) learns any class consisting of m Boolean functions with mistake-bound $\log_2 m$ and would thus provide an attribute-efficient algorithm for such a class C. However, the running time would not be polynomial. Another serious drawback would be that the halving algorithm does not tolerate any noise. Interestingly, a multiplicative update similar to (1) has been used in Littlestone and Warmuth's weighted majority algorithm [10], and also Vovk's aggregating algorithm [14], to produce a noise-tolerant generalization of the halving algorithm.

Attribute-efficient learning has also been studied in other learning models than the mistake-bound model, such as Probably Approximately Correct learning [4], learning with uniform distribution [12], and learning with membership queries [3]. The idea has been further developed into learning with a potentially infinite number of attributes [2].

Applications

Attribute-efficient algorithms for simple function classes have a potentially interesting application as a component in learning more complex function classes. For example, any monotone k-term DNF formula over variables $x_1, \ldots, x_n$ can be represented as a monotone k-literal disjunction over 2^n variables z_A, where z_A is defined as $z_A = \prod_{i \in A} x_i$ for $A \subseteq \{1, \ldots, n\}$. Running Winnow with the transformed inputs $z \in \{0, 1\}^{2^n}$ would give a mistake-bound $O(k \log 2^n) = O(kn)$. Unfortunately the running time would be linear in 2^n, at least for a naive implementation. Khardon et al. [6] provide discouraging computational hardness results for this potential application.

Online learning algorithms have a natural application domain in signal processing. In this setting, the sender emits a true signal y_t at time t, for $t = 1, 2, 3, \ldots$. At some later time $(t + d)$, a receiver receives a signal z_t, which is a sum of the original signal y_t and various echoes of earlier signals $y_{t'}$, $t' < t$, all distorted by random noise. The task is to recover the true signal y_t based on received signals $z_t, z_{t-1}, \ldots, z_{t-l}$ over some time window l. Currently attribute-efficient algorithms are not used for such tasks, but see [11] for preliminary results.

Attribute-efficient learning algorithms are similar in spirit to statistical methods that find sparse models. In particular, statistical algorithms that use L_1 regularization are closely related to multiplicative algorithms such as winnow and exponentiated gradient. In contrast, more classical L_2 regularization leads to algorithms that are not attribute-efficient [13].

Cross-References

▶ Learning DNF Formulas

Recommended Reading

1. Auer P, Warmuth MK (1998) Tracking the best disjunction. Mach Learn 32(2):127–150
2. Blum A, Hellerstein L, Littlestone N (1995) Learning in the presence of finitely or infinitely many irrelevant attributes. J Comput Syst Sci 50(1):32–40
3. Bshouty N, Hellerstein L (1998) Attribute-efficient learning in query and mistake-bound models. J Comput Syst Sci 56(3):310–319
4. Dhagat A, Hellerstein L (1994) PAC learning with irrelevant attributes. In: Proceedings of the 35th Annual Symposium on Foundations of Computer Science, Santa Fe. IEEE Computer Society, Los Alamitos, pp 64–74
5. Gentile C, Warmuth MK (1999) Linear hinge loss and average margin. In: Kearns MJ, Solla SA, Cohn DA (eds) Advances in Neural Information Processing Systems, vol 11. MIT, Cambridge, pp 225–231
6. Khardon R, Roth D, Servedio RA (2005) Efficiency versus convergence of boolean kernels for on-line learning algorithms. J Artif Intell Res 24:341–356
7. Kivinen J, Warmuth MK (1997) Exponentiated gradient versus gradient descent for linear predictors. Inf Comput 132(1):1–64
8. Klivans AR, Servedio RA (2006) Toward attribute efficient learning of decision lists and parities. J Mach Learn Res 7:587–602
9. Littlestone N (1988) Learning quickly when irrelevant attributes abound: a new linear threshold algorithm. Mach Learn 2(4):285–318
10. Littlestone N, Warmuth MK (1994) The weighted majority algorithm. Inf Comput 108(2):212–261
11. Martin RK, Sethares WA, Williamson RC, Johnson CR Jr (2002) Exploiting sparsity in adaptive filters. IEEE Trans Signal Process 50(8):1883–1894
12. Mossel E, O'Donnell R, Servedio RA (2004) Learning functions of k relevant variables. J Comput Syst Sci 69(3):421–434
13. Ng AY (2004) Feature selection, L_1 vs. L_2 regularization, and rotational invariance. In: Greiner R, Schuurmans D (eds) Proceedings of the 21st International Conference on Machine Learning, Banff. The International Machine Learning Society, Princeton, pp 615–622
14. Vovk V (1990) Aggregating strategies. In: Fulk M, Case J (eds) Proceedings of the 3rd Annual Workshop on Computational Learning Theory, Rochester. Morgan Kaufmann, San Mateo, pp 371–383

Automated Search Tree Generation

Falk Hüffner
Department of Math and Computer Science, University of Jena, Jena, Germany

Keywords

Automated proofs of upper bounds on the running time of splitting algorithms

Years and Authors of Summarized Original Work

2004; Gramm, Guo, Hüffner, Niedermeier

Problem Definition

This problem is concerned with the automated development and analysis of search tree algorithms. Search tree algorithms are a popular way to find optimal solutions to NP-complete problems. (For ease of presentation, only decision problems are considered; adaption to optimization problems is straightforward.) The idea is to recursively solve several smaller instances in such a way that at least one branch is a yes-instance if and only if the original instance is. Typically, this is done by trying all possibilities to contribute to a solution certificate for a small part of the input, yielding a small local modification of the instance in each branch.

For example, consider the NP-complete CLUSTER EDITING problem: can a given graph be modified by adding or deleting up to k edges such that the resulting graph is a *cluster graph*, that is, a graph that is a disjoint union of cliques? To give a search tree algorithm for CLUSTER EDITING, one can use the fact that cluster graphs are exactly the graphs that do not contain a P_3 (a path of 3 vertices) as an induced subgraph. One can thus solve CLUSTER EDITING by finding a P_3 and splitting it into 3 branches: delete the first edge, delete the second edge, or add the missing edge. By this characterization, whenever there is no P_3 found, one already has a cluster graph. The original instance has a solution with k modifications if and only if at least one of the branches has a solution with $k - 1$ modifications.

Analysis

For NP-complete problems, the running time of a search tree algorithm only depends on the size of the search tree up to a polynomial factor , which depends on the number of branches and the reduction in size of each branch. If the algorithm solves a problem of size s and calls itself recursively for problems of sizes $s - d_1, \ldots, s - d_i$, then $(d_1, \ldots, d_i)$ is called the *branching vector* of this recursion. It is known that the size of the search tree is then $O(\alpha^s)$, where the *branching number* α is the only positive real root of the *characteristic polynomial*

$$z^d - z^{d-d_1} - \cdots - z^{d-d_i} , \qquad (1)$$

where $d = \max\{d_1, \ldots, d_i\}$. For the simple CLUSTER EDITING search tree algorithm and the size measure k, the branching vector is $(1, 1, 1)$ and the branching number is 3, meaning that the running time is up to a polynomial factor $O(3^k)$.

Case Distinction

Often, one can obtain better running times by distinguishing a number of cases of instances, and giving a specialized branching for each case. The overall running time is then determined by the branching number of the worst case. Several publications obtain such algorithms by hand (e.g., a search tree of size $O(2.27^k)$ for CLUSTER EDITING [4]); the topic of this work is how to automate this. That is, the problem is the following:

Problem 1 (Fast Search Tree Algorithm)
INPUT: An NP-hard problem $\mathcal{P}$ and a size measure $s(I)$ of an instance I of $\mathcal{P}$ where instances I with $s(I) = 0$ can be solved in polynomial time.
OUTPUT: A partition of the instance set of $\mathcal{P}$ into *cases*, and for each case a branching such that the maximum branching number over all branchings is as small as possible.

Note that this problem definition is somewhat vague; in particular, to be useful, the case an instance belongs to must be recognizable quickly. It is also not clear whether an optimal search tree algorithm exists; conceivably, the branching number can be continuously reduced by increasingly complicated case distinctions.

Key Results

Gramm et al. [3] describe a method to obtain fast search tree algorithms for CLUSTER EDITING and related problems, where the size measure is the number of editing operations k. To get a case distinction, a number of subgraphs are enumerated such that each instance is known to contain at least one of these subgraphs. It is next described how to obtain a branching for a particular case.

A standard way of systematically obtaining specialized branchings for instance cases is to use a combination of *basic branching* and *data reduction rules*. Basic branching is typically a very simple branching technique, and data reduction rules replace an instance with a smaller, solution-equivalent instance in polynomial time. Applying this to CLUSTER EDITING first requires a small modification of the problem: one considers an *annotated* version, where an edge can be marked as *permanent* and a non-edge can be marked as *forbidden*. Any such annotated vertex pair cannot

Automated Search Tree Generation, Fig. 1
Branching for a CLUSTER EDITING case using only basic branching on vertex pairs (*double circles*), and applications of the reduction rules (*asterisks*). Permanent edges are marked *bold*, forbidden edges *dashed*. The *numbers* next to the subgraphs state the change of the problem size k. The branching vector is $(1, 2, 3, 3, 2)$, corresponding to a search tree size of $O(2.27^k)$

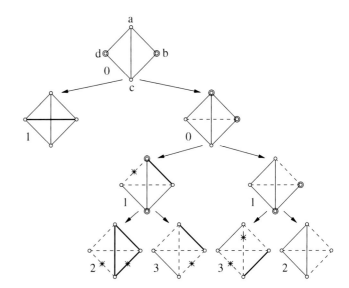

be edited anymore. For a pair of vertices, the basic branching then branches into two cases: permanent or forbidden (one of these options will require an editing operation). The reduction rules are: if two permanent edges are adjacent, the third edge of the triangle they induce must also be permanent; and if a permanent and a forbidden edge are adjacent, the third edge of the triangle they induce must be forbidden.

Figure 1 shows an example branching derived in this way.

Using a refined method of searching the space for all possible cases and to distinguish all branchings for a case, Gramm et al. [3] derive a number of search tree algorithms for graph modification problems.

Applications

Gramm et al. [3] apply the automated generation of search tree algorithms to several graph modification problems (see also Table 1). Further, Hüffner [5] demonstrates an application of DOMINATING SET on graphs with maximum degree 4, where the size measure is the size of the dominating set.

Fedin and Kulikov [2] examine variants of SAT; however, their framework is limited in that

it only proves upper bounds for a fixed algorithm instead of generating algorithms.

Skjernaa [6] also presents results on variants of SAT. His framework does not require user-provided data reduction rules, but determines reductions automatically.

Open Problems

The analysis of search tree algorithms can be much improved by describing the "size" of an instance by more than one variable, resulting in multivariate recurrences [1]. It is open to introduce this technique into an automation framework.

It has frequently been reported that better running time bounds obtained by distinguishing a large number of cases do not necessarily speed up, but in fact can slow down, a program. A careful investigation of the tradeoffs involved and a corresponding adaption of the automation frameworks is an open task.

Experimental Results

Gramm et al. [3] and Hüffner [5] report search tree sizes for several NP-complete problems. Further, Fedin and Kulikov [2] and Skjernaa [6]

Automated Search Tree Generation, Table 1
Summary of search tree sizes where automation gave
improvements. "Known" is the size of the best previously
published "hand-made" search tree. For the satisfiability
problems, m is the number of clauses and l is the length
of the formula

Problem	Trivial	Known	New
CLUSTER EDITING	3	2.27	1.92 [3]
CLUSTER DELETION	2	1.77	1.53 [3]
CLUSTER VERTEX DELETION	3	2.27	2.26 [3]
BOUNDED DEGREE DOMINATING SET	4		3.71 [5]
X3SAT, size measure m	3	1.1939	1.1586 [6]
$(n, 3)$-MAXSAT, size measure m	2	1.341	1.2366 [2]
$(n, 3)$-MAXSAT, size measure l	2	1.1058	1.0983 [2]

report on variants of satisfiability. Table 1 summarizes the results.

Cross-References

► Vertex Cover Search Trees

Acknowledgments Partially supported by the Deutsche
Forschungsgemeinschaft, Emmy Noether research group
PIAF (fixed-parameter algorithms), NI 369/4.

A

Recommended Reading

1. Eppstein D (2004) Quasiconvex analysis of backtracking algorithms. In: Proceedings of the 15th SODA.
ACM/SIAM, pp 788–797
2. Fedin SS, Kulikov AS (2006) Automated proofs of upper bounds on the running time of splitting algorithms.
J Math Sci 134:2383–2391. Improved results at http://
logic.pdmi.ras.ru/~kulikov/autoproofs.html
3. Gramm J, Guo J, Hüffner F, Niedermeier R (2004)
Automated generation of search tree algorithms
for hard graph modification problems. Algorithmica
39:321–347
4. Gramm J, Guo J, Hüffner F, Niedermeier R (2005)
Graph-modeled data clustering: exact algorithms
for clique generation. Theor Comput Syst 38:373–
392
5. Hüffner F (2003) Graph modification problems
and automated search tree generation. Diplomarbeit,
Wilhelm-Schickard-Institut für Informatik, Universität
Tübingen
6. Skjernaa B (2004) Exact algorithms for variants of
satisfiability and colouring problems. PhD thesis,
Department of Computer Science, University of
Aarhus

B

Backdoors to SAT

Serge Gaspers
Optimisation Research Group, National ICT
Australia (NICTA), Sydney, NSW, Australia
School of Computer Science and Engineering,
University of New SouthWales (UNSW),
Sydney, NSW, Australia

Keywords

Islands of tractability; Parameterized complexity;
Satisfiability

Years and Authors of Summarized Original Work

2013; Gaspers, Szeider

Problem Definition

In the satisfiability problem (SAT), the input is
a Boolean formula in conjunctive normal form
(CNF), and the question is whether the formula
is satisfiable, that is, whether there exists an
assignment of truth values to the variables such
that the formula evaluates to true. For example,
the formula

$$(x \vee \neg y) \wedge (\neg x \vee y \vee z) \wedge (x \vee \neg z)$$
$$\wedge (\neg x \vee y \vee z)$$

is satisfiable since it evaluates to true if we set x,
y, and z to true.

Several classes of CNF formulas are known
for which SAT can be solved in polynomial
time – so-called islands of tractability. For a given
island of tractability C, a C-backdoor is a set of
variables of a formula such that assigning truth
values to these variables gives a formula in C.

Formally, let C be a class of formulas for
which the recognition problem and the satisfiability problem can be solved in polynomial time.
For a subset of variables $X \subseteq \text{var}(F)$ of a
CNF formula F, and an assignment $\alpha : X \to$
$\{\text{true}, \text{false}\}$ of truth values to these variables, the
reduced formula $F[\alpha]$ is obtained from F by
removing all the clauses containing a true literal
under α and removing all false literals from the
remaining clauses. The notion of backdoors was
introduced by Williams et al. [15], and they come
in two variants:

Definition 1 ([15]) A *weak C-backdoor* of a
CNF formula F is a set of variables $X \subseteq \text{var}(F)$
such that there exists an assignment α to X such
that $F[\alpha] \in C$ and $F[\alpha]$ is satisfiable.

Definition 2 ([15]) A *strong C-backdoor* of a
CNF formula F is a set of variables $X \subseteq \text{var}(F)$
such that for each assignment α to X, we have
that $F[\alpha] \in C$.

There are two main computational problems
associated with backdoors. In the *detection* problem, the input is a CNF formula F and an integer k, and the question is whether F has a

© Springer Science+Business Media New York 2016
M.-Y. Kao (ed.), *Encyclopedia of Algorithms*,
DOI 10.1007/978-1-4939-2864-4

weak/strong C-backdoor of size k. In the *evaluation* problem, the input is a CNF formula F and a weak/strong C-backdoor X, and the question is whether F is satisfiable. (In the case of weak C-backdoors, one usually requires to find a satisfying assignment since every formula that has a weak C-backdoor is satisfiable.)

The size of a smallest weak/strong C-backdoor of a CNF formula F naturally defines the distance of F to C. The size of the backdoor then becomes a very relevant parameter in studying the parameterized complexity [1] of backdoor detection and backdoor evaluation.

For a base class C where #SAT (determine the number of satisfying assignments) or Max-SAT (find an assignment that maximizes the number or weight of satisfied clauses) can be solved in polynomial time, strong C-backdoors can also be used to solve these generalizations of SAT.

Key Results

While backdoor evaluation problems are fixed-parameter tractable for SAT, the parameterized complexity of backdoor detection depends on the particular island of tractability C that is considered. If weak (resp., strong) C-backdoor detection is fixed-parameter tractable parameterized by backdoor size, SAT is fixed-parameter tractable parameterized by the size of the smallest weak (resp., strong) C-backdoor. A sample of results for the parameterized complexity of backdoor detection is presented in Table 1. The considered islands of tractability are defined in Table 2.

It can be observed that restricting the input formulas to have bounded clause length can make backdoor detection more tractable. Also, weak backdoor detection is often no more tractable than strong backdoor detection; the outlier here is FOREST-backdoor detection for general CNF formulas, where the weak version is known to be W[2]-hard but the parameterized complexity of the strong variant is still open. A CNF formula belongs to the island of tractability Forest if its incidence graph is acyclic. Here, the *incidence graph* of a CNF formula F is the bipartite graph on the variables and clauses of F where a clause is incident to the variables it contains.

Backdoors to SAT, Table 1 The parameterized complexity of finding weak and strong backdoors of CNF formulas and r-CNF formulas, where $r \geq 3$ is a fixed integer

	Weak		Strong	
Island	CNF	r-CNF	CNF	r-CNF
HORN	W[2]-h [10]	FPT [7]	FPT [10]	FPT [10]
2CNF	W[2]-h [10]	FPT [7]	FPT [10]	FPT [10]
UP	W[P]-c [14]	W[P]-c [14]	W[P]-c [14]	W[P]-c [14]
FOREST	W[2]-h [6]	FPT [6]	Open	Open
RHORN	W[2]-h [7]	W[2]-h [7]	W[2]-h [7]	Open
CLU	W[2]-h [11]	FPT [7]	W[2]-h [11]	FPT [11]

Backdoors to SAT, Table 2 Some islands of tractability

Island	Description
HORN	Horn formulas, i.e., CNF formulas where each clause contains at most one positive literal
2CNF	Krom formulas, i.e., CNF formulas where each clause contains at most two literals
UP	CNF formulas from which the empty formula or an empty clause can be derived by unit propagation (setting the literals in unit length clauses to true)
FOREST	Acyclic formulas, i.e., CNF formulas whose incidence graphs are forests
RHORN	Renamable Horn formulas, i.e., CNF formulas that can be made Horn by flipping literals
CLU	Cluster formulas, i.e., CNF formulas that are variable disjoint unions of hitting formulas. A formula is *hitting* if every two of its clauses have at least one variable occurring positively in one clause and negatively in the other

The width of graph decompositions constitutes another measure for the tractability of CNF formulas that is orthogonal to backdoors. For example, Fischer et al. [2] and Samer and Szeider [12] give linear-time algorithms solving #SAT for CNF formulas in $\mathcal{W}_{\leq t}$.

Definition 3 For every integer $t \geq 0$, $\mathcal{W}_{\leq t}$ is the class of CNF formulas whose incidence graph has treewidth at most t.

Combining backdoor and graph decomposition methods, let us now consider backdoors to $\mathcal{W}_{\leq t}$. Since FOREST $= \mathcal{W}_{\leq 1}$, weak $\mathcal{W}_{\leq t}$-backdoor detection is already W[2]-hard for $t = 1$. Fomin et al. [3] give parameterized algorithms for weak $\mathcal{W}_{\leq t}$-backdoor detection when the input formula has bounded clause length. Concerning strong $\mathcal{W}_{\leq t}$-backdoor detection for formulas with bounded clause length, Fomin et al. [3] sidestep the issue of computing a backdoor by giving a fixed-parameter algorithm, where the parameter is the size of the smallest $\mathcal{W}_{\leq t}$-backdoor, that directly solves r-SAT. The parameterized complexity of strong $\mathcal{W}_{\leq t}$-backdoor detection remains open, even for $t = 1$. However, a fixed-parameter approximation algorithm was designed by Gaspers and Szeider.

Theorem 1 ([8]) *There is a cubic-time algorithm that, given a CNF formula F and two constants $k, t \geq 0$, either finds a strong $\mathcal{W}_{\leq t}$-backdoor of size at most 2^k or concludes that F has no strong $\mathcal{W}_{\leq t}$-backdoor set of size at most k.*

Using one of the #SAT algorithms for $\mathcal{W}_{\leq t}$ [2, 12], one can use Theorem 1 to obtain a fixed-parameter algorithm for #SAT parameterized by the size of the smallest strong backdoor to $\mathcal{W}_{\leq t}$.

Corollary 1 ([8]) *There is a cubic-time algorithm that, given a CNF formula F, computes the number of satisfying assignments of F or concludes that F has no strong $\mathcal{W}_{\leq t}$-backdoor of size k for any pair of constants $k, t \geq 0$.*

In general, a fixed-parameter approximation algorithm for weak/strong C-backdoor detection is sufficient to make SAT fixed-parameter tractable parameterized by the size of a smallest weak/strong C-backdoor.

Backdoors for SAT have been considered for combinations of base classes [4, 9], and the notion of backdoors has been extended to other computational reasoning problems such as constraint satisfaction, quantified Boolean formulas, planning, abstract argumentation, and nonmonotonic reasoning; see [7]. Other variants of the notion of backdoors include deletion backdoors where variables are deleted instead of instantiated, backdoors that are sensitive to clause-learning, pseudo-backdoors that relax the requirement that the satisfiability problem for an island of tractability be solved in polynomial time, and backdoor trees; see [5].

Applications

SAT is an NP-complete problem, but modern SAT solvers perform extremely well, especially on structured and industrial instances [13].

The study of backdoors, and especially the parameterized complexity of backdoor detection problems, is one nascent approach to try and explain the empirically observed running times of SAT solvers.

Open Problems

Major open problems in the area include to determine whether

- strong FOREST-backdoor detection is fixed-parameter tractable, and whether
- strong RHORN-backdoor detection is fixed-parameter tractable for 3-CNF formulas.

Experimental Results

Experimental results evaluate running times of algorithms to find backdoors in benchmark instances, evaluate the size of backdoors of known SAT benchmark instances, compare backdoor sizes for various islands of tractability, compare backdoor sizes for various notions of backdoors, evaluate what effect preprocessing has on backdoor size, compare how backdoor sizes of random instances compare to backdoor sizes of real-world industrial instances, and evaluate how SAT solver running times change if we force the solver to branch only on the variables of a given backdoor. The main messages are that the islands of tractability with the smallest backdoors are also those for which the backdoor detection problems are the most intractable and that existing SAT solvers can be significantly sped up on many real-world SAT instances if we feed them small backdoors. The issue is, of course, to compute these backdoors, and knowledge of the application domain, or specific SAT translations might help significantly with this task in practice. See [5] for a survey.

Cross-References

▶ Backtracking Based k-SAT Algorithms
▶ Exact Algorithms for General CNF SAT
▶ Parameterized SAT
▶ Treewidth of Graphs

Recommended Reading

1. Downey RG, Fellows MR (2013) Fundamentals of parameterized complexity. Springer, London
2. Fischer E, Makowsky JA, Ravve ER (2008) Counting truth assignments of formulas of bounded tree-width or clique-width. Discret Appl Math 156(4):511–529
3. Fomin FV, Lokshtanov D, Misra N, Ramanujan MS, Saurabh S (2015) Solving d-SAT via backdoors to small treewidth. In: Proceedings of the 26th annual ACM-SIAM symposium on discrete algorithms, SODA 2015, San Diego, 4–6 Jan 2015. SIAM, pp 630–641
4. Ganian R, Ramanujan M, Szeider S (2014) Discovering archipelagos of tractability: split-backdoors to constraint satisfaction. Presented at PCCR 2014 – the 2nd workshop on the parameterized complexity of computational reasoning, Vienna
5. Gario M (2011) Backdoors for SAT. Master's thesis, Dresden University of Technology. http://marco.gario.org/work/master/
6. Gaspers S, Szeider S (2012) Backdoors to acyclic SAT. In: Proceedings of the 39th international colloquium on automata, languages, and programming (ICALP 2012), Warwick. Lecture notes in computer science, vol 7391. Springer, pp 363–374
7. Gaspers S, Szeider S (2012) Backdoors to satisfaction. In: Bodlaender HL, Downey R, Fomin FV, Marx D (eds) The multivariate algorithmic revolution and beyond – essays dedicated to Michael R. Fellows on the occasion of his 60th birthday. Lecture notes in computer science, vol 7370. Springer, New York, pp 287–317
8. Gaspers S, Szeider S (2013) Strong backdoors to bounded treewidth SAT. In: 54th annual IEEE symposium on foundations of computer science, FOCS 2013, Berkeley, 26–29 Oct 2013. IEEE Computer Society, pp 489–498
9. Gaspers S, Misra N, Ordyniak S, Szeider S, Zivny S (2014) Backdoors into heterogeneous classes of SAT and CSP. In: Proceedings of the 28th AAAI conference on artificial intelligence (AAAI 2014), Québec City. AAAI Press, pp 2652–2658
10. Nishimura N, Ragde P, Szeider S (2004) Detecting backdoor sets with respect to Horn and binary clauses. In: Proceedings of the 7th international conference on theory and applications of satisfiability testing (SAT 2004), Vancouver, pp 96–103
11. Nishimura N, Ragde P, Szeider S (2007) Solving #SAT using vertex covers. Acta Informatica 44(7–8):509–523
12. Samer M, Szeider S (2010) Algorithms for propositional model counting. J Discret Algorithms 8(1):50–64
13. SAT competition (2002–) The international SAT competitions web page. http://www.satcompetition.org
14. Szeider S (2005) Backdoor sets for DLL subsolvers. J Autom Reason 35(1–3):73–88
15. Williams R, Gomes C, Selman B (2003) Backdoors to typical case complexity. In: Proceedings of the eighteenth international joint conference on artificial intelligence, IJCAI 2003. Morgan Kaufmann, San Francisco, pp 1173–1178

Backtracking Based k-SAT Algorithms

Ramamohan Paturi[1], Pavel Pudlák[2], Michael Saks[3], and Francis Zane[4]
[1]Department of Computer Science and Engineering, University of California at San Diego, San Diego, CA, USA
[2]Academy of Science of the Czech Republic, Mathematical Institute, Prague, Czech Republic
[3]Department of Mathematics, Rutgers, State University of New Jersey, Piscataway, NJ, USA
[4]Lucent Technologies, Bell Laboratories, Murray Hill, NJ, USA

Keywords

Boolean formulas; Conjunctive normal form satisfiability; Exponential time algorithms; Resolution

Years and Authors of Summarized Original Work

2005; Paturi, Pudlák, Saks, Zane

Problem Definition

Determination of the complexity of k-CNF satisfiability is a celebrated open problem: given a Boolean formula in conjunctive normal form with at most k literals per clause, find an assignment to the variables that satisfies each of the clauses or declare none exists. It is well known that the decision problem of k-CNF satisfiability is NP-complete for $k \geq 3$. This entry is concerned with algorithms that significantly improve the worst-case running time of the naive exhaustive search algorithm, which is $\text{poly}(n)2^n$ for a formula on n variables. Monien and Speckenmeyer [8] gave the first real improvement by giving a simple algorithm whose running time is $O(2_k^{(1-\varepsilon)n})$, with $\varepsilon_k > 0$ for all k. In a sequence of results [1, 3, 5–7, 9–12], algorithms with increasingly better running times (larger values of ε_k) have been proposed and analyzed.

These algorithms usually follow one of two lines of attack to find a satisfying solution. Backtrack search algorithms make up one class of algorithms. These algorithms were originally proposed by Davis, Logemann, and Loveland [4] and are sometimes called Davis-Putnam procedures. Such algorithms search for a satisfying assignment by assigning values to variables one by one (in some order), backtracking if a clause is made false. The other class of algorithms is based on local searches (the first guaranteed performance results were obtained by Schöning [12]). One starts with a randomly (or strategically) selected assignment and searches locally for a satisfying assignment guided by the unsatisfied clauses.

This entry presents **ResolveSat**, a randomized algorithm for k-CNF satisfiability which achieves some of the best known upper bounds. **ResolveSat** is based on an earlier algorithm of Paturi, Pudlák, and Zane [10], which is essentially a backtrack search algorithm where the variables are examined in a randomly chosen order. An analysis of the algorithm is based on the observation that as long as the formula has a satisfying assignment which is isolated from other satisfying assignments, a third of the variables are expected to occur as unit clauses as the variables are assigned in a random order. Thus, the algorithm needs to correctly guess the values of at most 2/3 of the variables. This analysis is extended to the general case by observing that either there exists an isolated satisfying assignment or there are many solutions, so the probability of guessing one correctly is sufficiently high.

ResolveSat combines these ideas with resolution to obtain significantly improved bounds [9]. In fact, **ResolveSat** obtains the best known upper bounds for k-CNF satisfiability for all $k \geq 5$. For $k = 3$ and 4, Iwama and Takami [6] obtained the best known upper bound with their randomized algorithm which combines the ideas from Schöning's local search algorithm and **ResolveSat**. Furthermore, for the promise problem of unique k-CNF satisfiability whose instances are conjectured to be among the hardest instances of k-CNF satisfiability [2], **ResolveSat** holds the best record for all $k \geq 3$. Bounds obtained by **ResolveSat** for unique k-SAT and k-SAT for $k = 3, 4, 5, 6$ are shown in Table 1. Here, these bounds are compared with those of Schöning [12], subsequently improved results based on local search [1, 5, 11], and the most recent improvements due to Iwama and Takami [6]. The upper bounds obtained by these algorithms are expressed in the form $2^{cn-o(n)}$ and the numbers in the table represent the exponent c. This comparison focuses only on the best bounds irrespective of the type of the algorithm (randomized versus deterministic).

Backtracking Based *k*-SAT Algorithms, Table 1 This table shows the exponent c in the bound $2^{cn-o(n)}$ for the unique k-SAT and k-SAT from the **ResolveSat** algorithm, the bounds for k-SAT from Schöning's algorithm [12], its improved versions for 3-SAT [1, 5, 11], and the hybrid version of [6]

	unique			*k*-SAT	
k	*k*-**SAT**[9]	*k*-**SAT**[9]	*k*-**SAT**[12]	[1,5,11]	*k*-**SAT**[6]
3	0.386 ...	0.521 ...	0.415 ...	0.409 ...	0.404 ...
4	0.554 ...	0.562 ...	0.584 ...		0.559 ...
5	0.650 ...		0.678 ...		
6	0.711 ...		0.736 ...		

Notation

In this entry, a CNF Boolean formula $F(x_1, x_2, \ldots, x_n)$ is viewed as both a Boolean function and a set of clauses. A Boolean formula F is a k-CNF if all the clauses have size at most k. For a clause C, write var(C) for the set of variables appearing in C. If $v \in$ var(C), the *orientation* of v is positive if the literal v is in C and is negative if $\bar{v}$ is in C. Recall that if F is a CNF Boolean formula on variables $(x_1, x_2, \ldots, x_n)$ and a is a partial assignment of the variables, the *restriction* of F by a is defined to be the formula $F' = F\lceil_a$ on the set of variables that are not set by a, obtained by treating each clause C of F as follows: if C is set to 1 by a, then delete C and otherwise replace C by the clause C' obtained by deleting any literals of C that are set to 0 by a. Finally, a *unit clause* is a clause that contains exactly one literal.

Key Results

ResolveSat Algorithm

The **ResolveSat** algorithm is very simple. Given a k-CNF formula, it first generates clauses that can be obtained by resolution without exceeding a certain clause length. Then it takes a random order of variables and gradually assigns values to them in this order. If the currently considered variable occurs in a unit clause, it is assigned as the only value that satisfies the clause. If it occurs in contradictory unit clauses, the algorithm starts over. At each step, the algorithm also checks if the formula is satisfied. If the formula is satisfied, then the input is accepted. This subroutine is repeated until either a satisfying assignment is found or a given time limit is exceeded.

The **ResolveSat** algorithm uses the following subroutine, which takes an arbitrary assignment y, a CNF formula F, and a permutation π as input, and produces an assignment u. The assignment u is obtained by considering the variables of y in the order given by π and modifying their values in an attempt to satisfy F.

Function **Modify**(CNF formula $G(x_1, x_2, \ldots, x_n)$, permutation π of $\{1, 2, \ldots, n\}$, assignment y) $\rightarrow$ (assignment u) $G_0 = G$.**for** $i = 1$ **to** n **if** G_{i-1} contains the unit clause $x_{\pi(i)}$ **then** $u_{\pi(i)} = 1$ **else if** G_{i-1} contains the unit clause $\bar{x}_{\pi(i)}$ **then** $u_{\pi(i)} = 0$ **else** $u_{\pi(i)} = y_{\pi(i)}$ $G_i = G_{i-1}\lceil_{x_{\pi(i)}=u_{\pi(i)}}$ **end** /* end for loop */**return** u;

The algorithm **Search** is obtained by running **Modify** (G, π, y) on many pairs (π, y), where π is a random permutation and y is a random assignment.

Search(CNF-formula F, integer I)**repeat** I times π = uniformly random permutation of $1, \ldots, n$ y = uniformly random vector $\in \{0, 1\}^n$ u = **Modify** (F, π, y); **if** u satisfies F **then** output(u); **exit**;**end**/* end repeat loop */output('Unsatisfiable');

The **ResolveSat** algorithm is obtained by combining **Search** with a preprocessing step consisting of *bounded resolution*. For the clauses C_1 and C_2, C_1 and C_2 *conflict* on variable v if one of them contains v and the other contains $\bar{v}$. C_1 and C_2 is a *resolvable pair* if they conflict on exactly one variable v. For such a pair, their *resolvent*, denoted $R(C_1, C_2)$, is the clause $C = D_1 \vee D_2$ where D_1 and D_2 are obtained by deleting v and $\bar{v}$ from C_1 and C_2. It is easy to see that any assignment satisfying C_1 and C_2 also satisfies C. Hence, if F is a satisfiable CNF formula containing the resolvable pair C_1, C_2 then the formula $F' = F \wedge R(C_1, C_2)$ has the same satisfying assignments as F. The resolvable pair C_1, C_2 is *s-bounded* if $|C_1|, |C_2| \leq s$ and $|R(C_1, C_2)| \leq s$. The following subroutine extends a formula F to a formula F_s by applying as many steps of s-bounded resolution as possible.

Resolve(CNF Formula F, integer s)$F_s = F$.**while** F_s has an s-bounded resolvable pair C_1, C_2 with $R(C_1, C_2) \notin F_s$ $F_s = F_s \wedge R(C_1, C_2)$.**return** ($F_s$).

The algorithm for k-SAT is the following simple combination of **Resolve** and **Search**:

ResolveSat(CNF-formula F, integer s, positive integer I)$F_s = $ **Resolve** (F, s).**Search** (F_s, I).

Analysis of ResolveSat

The running time of **ResolveSat** (F, s, I) can be bounded as follows. **Resolve** (F, s) adds at most $O(n^s)$ clauses to F by comparing pairs of clauses, so a naive implementation runs in time n^{3s}poly(n) (this time bound can be improved, but this will not affect the asymptotics of the main results). **Search** (F_s, I) runs in time $I(|F| + n^s)$poly(n). Hence, the overall running time of **ResolveSat** (F, s, I) is crudely bounded from above by $(n^{3s} + I(|F| + n^s))$poly(n). If $s = O(n/\log n)$, the overall running time can be bounded by $I|F|2^{O(n)}$ since $n^s = 2^{O(n)}$. It will be sufficient to choose s either to be some large constant or to be a *slowly growing* function of n. That is, $s(n)$ tends to infinity with n but is $O(\log n)$.

The algorithm **Search** (F, I) always answers "unsatisfiable" if F is unsatisfiable. Thus, the only problem is to place an upper bound on the error probability in the case that F is satisfiable. Define $\tau(F)$ to be the probability that **Modify** (F, π, y) finds some satisfying assignment. Then for a satisfiable F, the error probability of **Search** (F, I) is equal to $(1 - \tau(F))^I \le e^{-I\tau(F)}$, which is at most e^{-n} provided that $I \ge n/\tau(F)$. Hence, it suffices to give good upper bounds on $\tau(F)$.

Complexity analysis of **ResolveSat** requires certain constants μ_k for $k \ge 2$:

$$\mu_k = \sum_{j=1}^{\infty} \frac{1}{j\left(j + \frac{1}{k-1}\right)}.$$

It is straightforward to show that $\mu_3 = 4 - 4\ln 2 > 1.226$ using Taylor's series expansion of $\ln 2$. Using standard facts, it is easy to show that μ_k is an increasing function of k with the limit

$$\sum_{j=1}^{\infty}(1/j^2) = (\pi^2/6) = 1.644\ldots$$

The results on the algorithm **ResolveSat** are summarized in the following three theorems.

Theorem 1

(i) *Let $k \ge 5$, and let $s(n)$ be a function going to infinity. Then for any satisfiable k-CNF formula F on n variables,*

$$\tau(F_s) \le 2^{-\left(1 - \frac{\mu_k}{k-1}\right)n - o(n)}.$$

*Hence, **ResolveSat** (F, s, I) with $I = 2^{(1 - \mu_k/(k-1))n + O(n)}$ has error probability $O(1)$ and running time $2^{(1 - \mu_k/(k-1))n + O(n)}$ on any satisfiable k-CNF formula, provided that $s(n)$ goes to infinity sufficiently slowly.*

(ii) *For $k \ge 3$, the same bounds are obtained provided that F is uniquely satisfiable.*

Theorem 1 is proved by first considering the uniquely satisfiable case and then relating the general case to the uniquely satisfiable case. When $k \ge 5$, the analysis reveals that the asymptotics of the general case is no worse than that of the uniquely satisfiable case. When $k = 3$ or $k = 4$, it gives somewhat worse bounds for the general case than for the uniquely satisfiable case.

Theorem 2 *Let $s = s(n)$ be a slowly growing function. For any satisfiable n-variable 3-CNF formula, $\tau(F_s) \ge 2^{-0.521n}$, and so **ResolveSat** (F, s, I) with $I = n2^{0.521n}$ has error probability $O(1)$ and running time $2^{0.521n + O(n)}$.*

Theorem 3 *Let $s = s(n)$ be a slowly growing function. For any satisfiable n-variable 4-CNF formula, $\tau(F_s) \ge 2^{-0.5625n}$, and so **ResolveSat** (F, s, I) with $I = n2^{0.5625n}$ has error probability $O(1)$ and running time $2^{0.5625n + O(n)}$.*

Applications

Various heuristics have been employed to produce implementations of 3-CNF satisfiability algorithms which are considerably more efficient than exhaustive search algorithms. The **ResolveSat** algorithm and its analysis provide a rigorous explanation for this efficiency and identify the structural parameters (e.g., the width of clauses and the number of solutions), influencing the complexity.

Open Problems

The gap between the bounds for the general case and the uniquely satisfiable case when $k \in \{3, 4\}$ is due to a weakness in analysis, and it is conjectured that the asymptotic bounds for the uniquely satisfiable case hold in general for all k. If true, the conjecture would imply that **ResolveSat** is also faster than any other known algorithm in the $k = 3$ case.

Another interesting problem is to better understand the connection between the number of satisfying assignments and the complexity of finding a satisfying assignment [2]. A strong conjecture is that satisfiability for formulas with many satisfying assignments is strictly easier than for formulas with fewer solutions.

Finally, an important open problem is to design an improved k-SAT algorithm which runs faster than the bounds presented in here for the unique k-SAT case.

Cross-References

▶ Exact Algorithms for k SAT Based on Local Search
▶ Exact Algorithms for Maximum Two-Satisfiability
▶ Parameterized SAT
▶ Thresholds of Random k-SAT

Recommended Reading

1. Baumer S, Schuler R (2003) Improving a probabilistic 3-SAT algorithm by dynamic search and independent clause pairs. In: SAT, Santa Margherita Ligure, pp 150–161
2. Calabro C, Impagliazzo R, Kabanets V, Paturi R (2003) The complexity of unique k-SAT: an isolation lemma for k-CNFs. In: Proceedings of the eighteenth IEEE conference on computational complexity, Aarhus
3. Dantsin E, Goerdt A, Hirsch EA, Kannan R, Kleinberg J, Papadimitriou C, Raghavan P, Schöning U (2002) A deterministic $(2 - \frac{2}{k+1})^n$ algorithm for k-SAT based on local search. Theor Comp Sci 289(1):69–83
4. Davis M, Logemann G, Loveland D (1962) A machine program for theorem proving. Commun ACM 5:394–397
5. Hofmeister T, Schöning U, Schuler R, Watanabe O (2002) A probabilistic 3-SAT algorithm further improved. In: STACS, Antibes Juan-les-Pins. LNCS, vol 2285, Springer, Berlin, pp 192–202
6. Iwama K, Tamaki S (2004) Improved upper bounds for 3-SAT. In: Proceedings of the fifteenth annual ACM-SIAM symposium on discrete algorithms, New Orleans, pp 328–329
7. Kullmann O (1999) New methods for 3-SAT decision and worst-case analysis. Theor Comp Sci 223(1–2):1–72
8. Monien B, Speckenmeyer E (1985) Solving satisfiability in less than 2^n steps. Discret Appl Math 10:287–295
9. Paturi R, Pudlák P, Saks M, Zane F (2005) An improved exponential-time algorithm for k-SAT. J ACM 52(3):337–364. (An earlier version presented in Proceedings of the 39th annual IEEE symposium on foundations of computer science, 1998, pp 628–637)
10. Paturi R, Pudlák P, Zane F (1999) Satisfiability Coding Lemma. In: Proceedings of the 38th annual IEEE symposium on foundations of computer science, Miami Beach, 1997, pp. 566–574. Chicago J Theor Comput Sci. http://cjtcs.cs.uchicago.edu/
11. Rolf D (2003) 3-SAT $\in RTIME (1.32971^n)$. In: ECCC TR03-054
12. Schöning U (2002) A probabilistic algorithm for k-SAT based on limited local search and restart. Algorithmica 32:615–623. (An earlier version appeared in 40th annual symposium on foundations of computer science (FOCS '99), pp 410–414)

Bargaining Networks

Mohammad Taghi Hajiaghayi[1] and Hamid Mahini[1]
Department of Computer Science, University of Maryland, College Park, MD, USA

Keywords

Balanced; Bargaining; Cooperative game theory; Core; Kernel; Networks; Stable

Years and Authors of Summarized Original Work

2008; Kleinberg, Tardös
2010; Bateni, Hajiaghayi, Immorlica, Mahini
2013; Farczadi, Georgiou, Köenemann

Problem Definition

A network bargaining game can be represented by a graph $G = (V, E)$ along with a set of node capacities $\{c_i | i \in V\}$ and a set of edge weights

$\{w_{ij}|(i, j) \in E\}$, where V is a set of n agents, E is the set of all possible contracts, each agent $i \in V$ has a *capacity* c_i which the maximum number of contracts in which agent i may participate, and each edge $(i, j) \in E$ has a *weight* w_{ij} which represents the surplus of a possible contract between agent i and agent j which should be divided between agents i and j upon an agreement. The main goal is to find the outcome of bargaining among agents which is a set of contracts $M \subseteq E$ and the division of surplus $\{z_{ij}\}$ for all contracts in M.

Problem 1 (Computing the Final Outcome)

INPUT: *A network bargaining game $G = (V, E)$ along with capacities $\{c_i | i \in V\}$ and weights $\{w_{ij}|(i, j) \in E\}$.*

OUTPUT: *The final outcome of bargaining among agents.*

Solution Concept

Feasible Solution

The final outcome of the bargaining process might have many properties. The main one is the feasibility. A solution $(M, \{z_{ij}\})$ is *feasible* if and only if it has the following properties:

- The degree of each node i should be at most c_i in set M.
- For each edge $(i, j) \in M$, we should have $z_{ij} + z_{ji} = w_{ij}$. This means if there is a contract between agents i and j, the surplus should be divided between these two agents.
- For each edge $(i, j) \notin M$, we should have $z_{ij} = z_{ji} = 0$.

Outside Option

Given a feasible solution $(M, \{z_{ij}\})$, the *outside* option of agent is the best deal she can make outside of set M. For each edge $(i, k) \in E - M$, agent i has an outside option by offering agent k her current worst offer. In particular, if k has less than c_k contracts in M, agent i can offer agent k exactly 0, and thus the outside option of agent i regarding agent k would be w_{ik}. On the other hand, if k has exactly c_k contracts in M, agent i may offer agent k the minimum of

z_{kj} for all $(k, j) \in M$, and thus the outside option of agent i regarding agent k would be $w_{ik} - \min_{(k,j)\in M}\{z_{kj}\}$. Therefore, the outside option α_i of agent i regarding solution $(M, \{z_{ij}\})$ is defined as $\alpha_i = \max_{(i,k)\in E-M}\{w_{ik} - \eta_{ik}\}$ where

$$\eta_{ik} = \begin{cases} 0 & \text{if } k \text{ has less than} \\ & c_i \text{ contracts in } M \\ \min_{(k,j)\in M}\{z_{kj}\} & \text{if } k \text{ has exactly} \\ & c_i \text{ contracts in } M \end{cases}$$

Stable Solution

A solution $(M, \{z_{ij}\})$ is *stable* if for each contract $(i, j) \in M$, we have $z_{ij} \geq \alpha_i$ and for each agent i with less than c_i contracts in M, we have $\alpha_i = 0$. Otherwise, agent i has an incentive to deviate from M and makes a contract with agent k such that $(i, k) \notin M$.

Balanced Solution

John Nash [6] proposed a solution for the outcome of bargaining process between two agents. In his solution, known as the *Nash bargaining solution*, both agents will enjoy their outside options and then divide the surplus equally. One can leverage the intuition behind the Nash bargaining solution and defines the balanced solution in the network bargaining game. A feasible solution $(M, \{z_{ij}\})$ is *balanced* if for each contract (i, j) in M, the participants divide the net surplus equally, i.e., $z_{ik} = \alpha_i + \frac{w_{ik}-\alpha_i-\alpha_j}{2}$.

Problem 2 (Computing a Stable Solution)

INPUT: *A network bargaining game $G = (V, E)$ along with capacities $\{c_i | i \in V\}$ and weights $\{w_{ij}|(i, j) \in E\}$.*

OUTPUT: *A stable solution.*

Problem 3 (Computing a Balanced Solution)

INPUT: *A network bargaining game $G = (V, E)$ along with capacities $\{c_i | i \in V\}$ and weights $\{w_{ij}|(i, j) \in E\}$.*

OUTPUT: *A balanced solution.*

Key Results

The main goal of studying the network bargaining games is to find the right outcome of the game. Stable and balanced solutions are known to be good candidates. However, they might be too large, and moreover, some network bargaining games do not have stable and balanced solutions.

Existence of Stable and Balanced Solutions

It has been proved that a network bargaining game $G = (V, E)$ with set of weights $\{w_{ij} | (i, j) \in E\}$ has at least one stable solution if and only if the following linear program for the corresponding maximum weighted matching problem has an integral optimum solution [4, 5]:

$$
\begin{aligned}
\text{maximize} \quad & \textstyle\sum_{(i,j) \in E} x_{ij} w_{ij} \\
\text{subject to} \quad & \textstyle\sum_{(i,j) \in E} x_{ij} \le c_i, \quad \forall i \in V \\
& x_{ij} \le 1, \qquad\qquad \forall (i, j) \in E
\end{aligned}
$$
(LP1)

Kleinberg and Tardös [5] also study network bargaining games with unit capacities, i.e., $c_i = 1$ for each agent i, and show these games have at least one balanced solution if and only if they have a stable solution. Farczadi et al. [3] generalize this result and prove the same result for network bargaining games with general capacities.

Cooperative Game Theory Perspective

One can study network bargaining games from cooperative game theory perspective. A cooperative game is defined by a set of agents V and a value function $v : 2^V \to \mathcal{R}$, where $v(S)$ represents the surplus that all agents in S alone can generate. In order to consider our bargaining game as a cooperative game, we should first define a value function for our bargaining game. The value function $v(S)$ can be defined as the size of the maximum weighted c-matching of S.

Core

An outcome $\{x_i | i \in V\}$ is in core if for each subset of agents S, we have $\sum_{i \in S} x_i \ge v(S)$ and for set V we have $\sum_{i \in V} x_i = v(V)$. This means the agents should divide the total surplus $v(V)$

such that each subset of agents earns at least as much as they alone can generate.

Prekernel

Consider an outcome $\{x_i | i \in V\}$. The *power* of agent i over agent j regarding outcome $\{x_i | i \in V\}$ is defined as $s_{ij}(x) = \max\{v(S) - \sum_{i \in S} x_i | S \subseteq V, i \in S, j \in V - S\}$. An outcome $\{x_i | i \in V\}$ is in prekernel if for every two agents i and j, we have $s_{ij}(x) = s_{ji}(x)$.

Nucleolus

Consider an outcome $\{x_i | i \in V\}$. The *excess* of set S is defined as $\epsilon(S) = v(S) - \sum_{i \in S} x_i$. Let ϵ be the vector of all possible $2^{|V|}$ excesses which are sorted in nondecreasing order. The *nucleolus* is the outcome which lexicographically maximizes vector ϵ.

There is a nice connection between stable and balanced solutions in network bargaining games and core and prekernel outcomes in cooperative games [1, 2]. Bateni et al. [2] prove in a bipartite network where all nodes on one side have unit capacity, the set of stable solutions and the core coincide. Moreover, they map the set of balanced solutions to the prekernel in the same setting. Note that it is shown that this equivalence cannot be extended to a general bipartite network where nodes on both sides have general capacities [2, 3].

The set of stable and balanced solutions are quite large for many instances and thus may not be used for predicting the outcome of the game. Both Azar et al. [1] and Bateni et al. [2] leverage the connection between network bargaining games and cooperative games and suggest the nucleolus as a symmetric and unique solution for the outcome of a network bargaining game [1, 2]. Bateni et al. [2] also propose a polynomial-time algorithm for finding nucleolus in bipartite networks with unit capacities on one side.

Finding Stable and Balanced Solutions

Designing a polynomial-time algorithm for finding stable and balanced solutions of a network bargaining game is a well-known problem. Kleinberg and Tardös [5] were the first who studied this problem and proposed a polynomial-time algorithm which characterizes stable and balanced solutions when all agents have unit capacities. Their solution draws connection to the structure of matchings and the Edmonds-Gallai decomposition. Bateni et al. [2] generalize this results and design a polynomial-time algorithm for bipartite graphs where all agents on one side have general capacities and the other ones have unit capacities. They leverage the correspondence between the set of balanced solutions and the intersection of the core and prekernel and use known algorithms for finding a point in prekernel to solve the problem. Last but not least, Farczadi et al. [3] propose an algorithm for computing a balanced solution for general capacities. The main idea of their solution is to reduce an instance with general capacities to a network bargaining game with unit capacities.

Open Problems

- What is the right outcome of a network bargaining game on a general graph?
- How can we compute a proper outcome of a network bargaining game on a general graph in a polynomial time?

Cross-References

▶ Market Games and Content Distribution
▶ Matching Market Equilibrium Algorithms

Recommended Reading

1. Azar Y, Devanur NR, Jain K, Rabani Y (2010) Monotonicity in bargaining networks. In: SODA, Austin, Texas, USA, pp 817–826
2. Bateni MH, Hajiaghayi MT, Immorlica N, Mahini H (2010) The cooperative game theory foundations of network bargaining games. In: ICALP 2010, Bordeaux, France, pp 67–78
3. Farczadi L, Georgiou K, Köenemann J (2013) Network bargaining with general capacities. In: ESA, Sophia Antipolis, France, pp 433–444
4. Bayati M, Borgs C, Chayes J, Kanoria Y, Montanari A (2014) Bargaining dynamics in exchange networks. Journal of Economic Theory 156:417–454
5. Kleinberg J, Tardos E (2008) Balanced outcomes in social exchange networks. In: STOC, Victoria, BC, Canada, pp 295–304
6. Nash JF (1950) The bargaining problem. Econometrica 18:155–162

Bend Minimization for Orthogonal Drawings of Plane Graphs

Maurizio Patrignani
Engineering Department, Roma Tre University, Rome, Italy

Keywords

Bends; Grid embedding; Network flow; Orthogonal drawings; Planar embedding

Years and Authors of Summarized Original Work

1987; Tamassia

Problem Definition

A *drawing* of a graph $G = (V, E)$ maps each vertex $v \in V$ to a distinct point of the plane and each edge $e \in E$ to a simple open Jordan curve joining its end vertices. A drawing is *planar* if the edges do not intersect. A graph is *planar* if it admits a planar drawing. A planar drawing of a graph partitions the plane into connected regions called *faces*. The unbounded face is called *external face*. Two drawings of G are *equivalent* if they induce the same circular order of the edges incident to the vertices. A *planar embedding* of G is an equivalence class of such drawings. A *plane graph* is a planar graph together with a planar embedding and the specification of the external face.

A drawing of a graph is *orthogonal* if each edge is a sequence of alternate horizontal and vertical segments. Only planar graphs of maximum degree four admit planar orthogonal drawings.

Bend Minimization for Orthogonal Drawings of Plane Graphs, Fig. 1 (**a**) An orthogonal drawing of a graph. (**b**) An orthogonal drawing of the same graph with the minimum number of bends

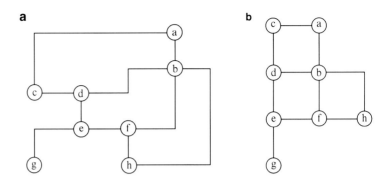

The points in common between two subsequent segments of the same edge are called *bends*. Figure 1 shows two orthogonal drawings of the same plane graph with seven bends and one bend, respectively.

Bend Minimization Problem

Formally, the main research problem can be defined as follows.

INPUT: *A plane graph* $G = (V, E)$ *of maximum degree four.*

OUTPUT: *An orthogonal drawing of* G *with the minimum number of bends.*

Since, given the shape of the faces, an orthogonal drawing of G with integer coordinates for vertices and bends can be computed in linear time, this problem may be alternatively viewed as that of embedding a 4-plane graph in the orthogonal grid with the minimum number of bends. Observe that if the planar embedding of the graph is not fixed, the problem of finding a minimum-bend orthogonal drawing is known to be NP-complete [9], unless the input graph has maximum degree three [5].

Key Results

The bend minimization problem can be solved in polynomial time by reducing it to that of finding a minimum-cost integer flow of a suitable network. Here, rather than describing the original model of [11], we describe the more intuitive model of [6].

Any orthogonal drawing of a maximum degree four plane graph $G = (V, E)$ corresponds to an integer flow in a network $\mathcal{N}(G)$ with value $4 \times n$, with $n = |V|$, where:

1. For each vertex $v \in V$, $\mathcal{N}(G)$ has a node n_v which is a source of 4 units of flow.
2. For each face f of G, $\mathcal{N}(G)$ has a node n_f which is a sink of $2\deg(f) - 4$ units if f is an internal face, or $2\deg(f) + 4$, otherwise, where $\deg(f)$ is the number of vertices encountered while traversing the boundary of face f (the same vertex may be counted multiple times).
3. For each edge $e \in E$, with adjacent faces f and g, $\mathcal{N}(G)$ has two arcs (n_f, n_g) and (n_g, n_f), both with cost 1 and lower bound 0.
4. For each vertex $v \in V$ incident to a face f of G, $\mathcal{N}(G)$ has an arc (n_v, n_f) with cost 0 and lower bound 1. Multiple incidences of the same vertex to the same face yield multiple arcs of $\mathcal{N}(G)$.

Figure 2 shows the two flows of cost 7 and 1, respectively, corresponding to the orthogonal drawings of Fig. 1. Intuitively, a flow of $\mathcal{N}(G)$ describes how 90° angles are distributed in the orthogonal drawing of G. Namely, each vertex has four 90° angles around it, hence "producing" four units of flow. The number of 90° angles needed to close a face f is given by the formula in [12], that is, $2\deg(f) - 4$ units if f is an internal face, and $2\deg(f) + 4$, otherwise. Finally, the flows traversing the edges account for their bends, where each bend allows a face

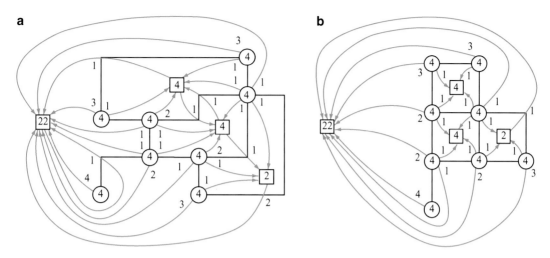

Bend Minimization for Orthogonal Drawings of Plane Graphs, Fig. 2 (**a**) The flow associated with the drawing of Fig. 1a has cost 7. (**b**) The flow associated with the drawing of Fig. 1b has cost 1

to "lose" a 90° angle and the adjacent face to "gain" it. More formally, we have the following theorem.

Theorem 1 *Let* $G = (V, E)$ *be a four-plane graph. For each orthogonal drawing of G with b bends, there exists an integer flow in $\mathcal{N}(G)$ whose value is $4 \times |V|$ and whose cost is b.*

Although several orthogonal drawings of G (e.g., with the order of the bends along edges permuted) may correspond to the same flow of $\mathcal{N}(G)$, starting from any flow, one of such drawings may be obtained in linear time. Namely, once the orthogonal shape of each face is fixed, it is possible to greedily add as many dummy edges and nodes as are needed to split the face into rectangular faces (the external face may require the addition of dummy vertices in the corners). Integer edge lengths can be consistently assigned to the sides of these rectangular faces, obtaining a grid embedding (a linear-time algorithm for doing this is described in [6]). The removal of dummy nodes and edges yields the desired orthogonal drawing. Hence, we have the following theorem.

Theorem 2 *Let* $G = (V, E)$ *be a four-plane graph. Given an integer flow in $\mathcal{N}(G)$ whose value is $4 \times |V|$ and whose cost is b, an orthogonal*

drawing of G with b bends can be found in linear time.

Since each bend of the drawing corresponds to a unit of cost for the flow, when the total cost of the flow is minimum, any orthogonal drawing that can be obtained from it has the minimum number of bends [11].

Hence, given a plane graph $G = (V, E)$ of maximum degree four, an orthogonal drawing of G with the minimum number of bends can be computed with the same asymptotic complexity of finding a minimum-cost integer flow of $\mathcal{N}(G)$. The solution to this problem proposed in [11] is based on an iterative augmentation algorithm. Namely, starting from the initial zero flow, the final $4 \times n$ flow is computed by augmenting the flow at each of the $O(n)$ steps along a minimum-cost path. Such a path can be found with the $O(n \log n)$-implementation of Dijkstra's algorithm that exploits a priority queue. The overall $O(n^2 \log n)$ time complexity was lowered first to $O(n^{7/4}\sqrt{\log n})$ [8] and then to $O(n^{3/2})$, exploiting the planarity of the flow network [4]. However, the latter time bound is increased by an additional logarithmic factor if some edges have constraints on the number of allowed bends [4] or if the Dijkstra's algorithm for the shortest path computation is preferred with respect to the rather theoretical algorithm in [10].

Bend Minimization for Orthogonal Drawings of Plane Graphs, Fig. 3 (**a**) A drawing on the hexagonal grid. (**b**) A drawing of the same graph in the Kandinsky model

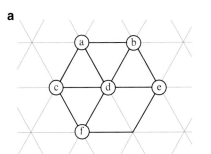

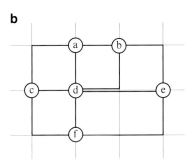

Applications

Orthogonal drawings with the minimum number of bends are of interest to VLSI circuit design, architectural floor plan layout, and aesthetic layout of diagrams used in information systems design. In particular, the orthogonal drawing standard is adopted for a wide range of diagrams, including entity-relationship diagrams, relational schemes, data-flow diagrams, flow charts, UML class diagrams, etc.

Open Problems

Several generalizations of the model have been proposed in order to deal with graphs of degree greater than four. The hexagonal grid, for example, would allow for vertices of maximum degree six (see Fig. 3a. Although the bend minimization problem is polynomial on such a grid [11], deciding edge lengths becomes NP-hard [2].

One of the most popular generalizations is the *Kandinsky* orthogonal drawing standard [7] where vertices of arbitrary degree are represented as small squares or circles of the same dimensions, while the first segments of the edges that leave a vertex in the same direction run very close together (see, e.g., Fig. 3b). Although the bend minimization problem in the Kandinsky orthogonal drawing standard has been shown to be NP-hard [3], this model is of great interest for applications. An extension of the flow model that makes it possible to solve this problem in polynomial time for a meaningful subfamily of Kandinsky orthogonal drawings has been proposed in [1]. Namely, in addition to the drawing

conventions of the Kandinsky model, each vertex with degree greater than four has at least one incident edge on each side, and each edge leaving a vertex either has no bend or has its first bend on the right.

Cross-References

▶ Planarisation and Crossing Minimisation
▶ Single-Source Shortest Paths

Recommended Reading

1. Bertolazzi P, Di Battista G, Didimo W (2000) Computing orthogonal drawings with the minimum number of bends. IEEE Trans Comput 49(8):826–840
2. Biedl TC (2007) Realizations of hexagonal graph representations. In: Bose P (ed) CCCG. Carleton University, Ottawa, pp 89–92
3. Bläsius T, Brückner G, Rutter I (2014) Complexity of higher-degree orthogonal graph embedding in the Kandinsky model. In: Schulz AS, Wagner D (eds) Algorithms – ESA 2014, Wroclaw. Lecture notes in computer science, vol 8737. Springer, Berlin/Heidelberg, pp 161–172
4. Cornelsen S, Karrenbauer A (2012) Accelerated bend minimization. J Graph Algorithms Appl 16(3):635–650
5. Di Battista G, Liotta G, Vargiu F (1998) Spirality and optimal orthogonal drawings. SIAM J Comput 27(6):1764–1811
6. Di Battista G, Eades P, Tamassia R, Tollis IG (1999) Graph drawing: algorithms for the visualization of graphs. Prentice-Hall, Upper Saddle River
7. Fößmeier U, Kaufmann M (1996) Drawing high degree graphs with low bend numbers. In: Brandenburg FJ (ed) Graph drawing. Lecture notes in computer science, vol 1027. Springer, Berlin/Heidelberg, pp 254–266

8. Garg A, Tamassia R (1996) A new minimum cost flow algorithm with applications to graph drawing. In: North SC (ed). Graph Drawing, Lecture notes in computer science, vol 1190. Springer, Berlin/Heidelberg, pp 201–216
9. Garg A, Tamassia R (2001) On the computational complexity of upward and rectilinear planarity testing. SIAM J Comput 31(2):601–625
10. Henzinger MR, Klein PN, Rao S, Subramanian S (1997) Faster shortest-path algorithms for planar graphs. J Comput Syst Sci 55(1):3–23
11. Tamassia R (1987) On embedding a graph in the grid with the minimum number of bends. SIAM J Comput 16(3):421–444
12. Vijayan G, Wigderson A (1985) Rectilinear graphs and their embeddings. SIAM J Comput 14(2):355–372

Best Response Algorithms for Selfish Routing

Paul (Pavlos) Spirakis
Computer Engineering and Informatics,
Research and Academic Computer Technology
Institute, Patras University, Patras, Greece
Computer Science, University of Liverpool,
Liverpool, UK
Computer Technology Institute (CTI), Patras,
Greece

Keywords

Atomic selfish flows

Years and Authors of Summarized Original Work

2005; Fotakis, Kontogiannis, Spirakis

Problem Definition

A setting is assumed in which n selfish users compete for routing their loads in a network. The network is an $s - t$ directed graph with a single source vertex s and a single destination vertex t. The users are ordered sequentially. It is assumed that each user plays after the user before her in the ordering, and the desired end result is a Pure Nash Equilibrium (PNE for short). It is assumed that, when a user plays (i.e., when she selects an $s - t$ path to route her load), the play is a best response (i.e., minimum delay), given the paths and loads of users currently in the net. The problem then is to find the class of directed graphs for which such an ordering exists so that the implied sequence of best responses leads indeed to a Pure Nash Equilibrium.

The Model

A *network congestion game* is a tuple $((w_i)_{i \in N}, G, (d_e)_{e \in E})$ where $N = \{1, \ldots, n\}$ is the set of users where user i controls w_i units of traffic demand. In *unweighted* congestion games $w_i = 1$ for $i = 1, \ldots, n$. $G(V,E)$ is a directed graph representing the communications network and d_e is the latency function associated with edge $e \in E$. It is assumed that the d_e's are non-negative and non-decreasing functions of the edge loads. The edges are called *identical* if $d_e(x) = x$, $\forall e \in E$. The model is further restricted to single-commodity network congestion games, where G has a single source s and destination t and the set of users' strategies is the set of $s - t$ paths, denoted P. Without loss of generality it is assumed that G is connected and that every vertex of G lies on a directed $s - t$ path.

A vector $P = (p_1, \ldots, p_n)$ consisting of an $s - t$ path p_i for each user i is a *pure strategies profile*. Let $l_e(P) = \sum_{i : e \in p_i} w_i$ be the load of edge e in P. The authors define *the cost* $\lambda_p^i(P)$ for user i routing her demand on path p in the profile P to be

$$\lambda_p^i(P) = \sum_{e \in p \cap p_i} d_e\left(l_e(P)\right) + \sum_{e \in p \smallsetminus p_i} d_e\left(l_e(P) + w_i\right).$$

The cost $\lambda^i(P)$ of user i in P is just $\lambda_{p_i}^i(P)$, i.e., the total delay along her path.

A pure strategies profile P is a Pure Nash Equilibrium (PNE) iff no user can reduce her total delay by *unilaterally deviating* i.e., by selecting another $s - t$ path for her load, while all other users keep their paths.

Best Response

Let p_i be the path of user i and $P^i = (p_1, \ldots, p_i)$ be the pure strategies profile for users $1, \ldots, i$. Then the *best response* of user $i + 1$ is a path p_{i+1} so that

$$p_{i+1} = avg \min_{p \in P^i} \left\{ \sum_{e \in p} \left(d_e \left(l_e \left(P^i \right) + w_{i+1} \right) \right) \right\}.$$

Flows and Common Best Response

A (feasible) flow on the set P of $s - t$ paths of G is a function $f : P \to \Re_{\geq 0}$ so that

$$\sum_{p \in P} f_p = \sum_{i=1}^{n} w_i.$$

The single-commodity network congestion game $((w_i)_{i \in N}, G, (d_e)_{e \in E})$ has the Common Best Response property if for every initial flow f (not necessarily feasible), all users have the same set of best responses with respect to f. That is, if a path p is a best response with respect to f for some user, then for all users j and all paths p'

$$\sum_{e \in p'} d_e \left(f_e + w_j \right) \geq \sum_{e \in p} d_e \left(f_e + w_j \right).$$

Furthermore, every segment π of a best response path p is a best response for routing the demand of any user between π's endpoints. It is allowed here that some users may already have contributed to the initial flow f.

Layered and Series-Parallel Graphs

A directed (multi)graph $G(V, E)$ with a distinguished source s and destination t is *layered* iff all directed $s - t$ paths have exactly the same length and each vertex lies on some directed $s - t$ path.

A multigraph is *series-parallel* with *terminals* (s, t) if

1. it is a single edge (s, t) or
2. it is obtained from two series-parallel graphs G_1, G_2 with terminals (s_1, t_1) and (s_2, t_2) by connecting them either in *series* or in *parallel*. In a series connection, t_1 is identified with s_2 and s_1 becomes s and t_2 becomes t. In a parallel connection, $s_1 = s_2 = s$ and $t_1 = t_2 = t$.

Key Results

The Greedy Best Response Algorithm (GBR)

GBR considers the users one-by-one in *non-increasing* order of weight (i.e., $w_1 \geq w_2 \geq \cdots \geq w_n$). Each user adopts her best response strategy on the set of (already adopted in the net) best responses of previous users. No user can change her strategy in the future. Formally, GBR *succeeds* if the eventual profile P is a Pure Nash Equilibrium (PNE).

The Characterization

In [3] it is shown:

Theorem 1 *If G is an $(s - t)$ series-parallel graph and the game $((w_i)_{i \in N}, G, (d_e)_{e \in E})$ has the common best response property, then GBR succeeds.*

Theorem 2 *A weighted single-commodity network congestion game in a layered network with identical edges has the common best response property for any set of user weights.*

Theorem 3 *For any single-commodity network congestion game in series-parallel networks, GBR succeeds if*

1. *The users are identical (if $w_i = 1$ for all i) and the edge-delays are arbitrary but non-decreasing or*
2. *The graph is layered and the edges are identical (for arbitrary user weights)*

Theorem 4 *If the network consists of bunches of parallel-links connected in series, then a PNE is obtained by applying GBR to each bunch.*

Theorem 5

1. *If the network is not series-parallel then there exist games where GBR fails, even for 2 identical users and identical edges.*
2. *If the network does not have the common best response property (and is not a sequence of parallel links graphs connected in series) then there exist games where GBR fails, even for 2-layered series-parallel graphs.*

Examples of such games are provided in [3].

Applications

GBR has a natural distributed implementation based on a leader election algorithm. Each player is now represented by a process. It is assumed that processes know the network and the edge latency functions. The existence of a message passing subsystem and an underlying synchronization mechanism (e.g., logical timestamps) is assumed, that allows a distributed protocol to proceed in logical rounds.

Initially all processes are active. In each round they run a leader election algorithm and determine the process of largest weight (among the active ones). This process routes its demand on its best response path, announces its strategy to all active processes, and becomes passive. Notice that each process can compute its best response locally.

Open Problems

What is the class of networks where (identical) users can achieve a PNE by a k-round repetition of a best responses sequence? What happens to weighted users? In general, how the network topology affects best response sequences? Such open problems are a subject of current research.

Cross-References

▶ General Equilibrium

Recommended Reading

1. Awerbuch B, Azar Y, Epstein A (2005) The price of routing unsplittable flows. In: Proceedings of the ACM symposium on theory of computing (STOC) 2005. ACM, New York, pp 57–66
2. Duffin RJ (1965) Topology of series-parallel networks. J Math Anal Appl 10:303–318
3. Fotakis D, Kontogiannis S, Spirakis P (2006) Symmetry in network congestion games: pure equilibria and anarchy cost. In: Proceedings of the 3rd workshop of approximate and on-line algorithms (WAOA 2005). Lecture notes in computer science (LNCS), vol 3879. Springer, Berlin/Heidelberg, pp 161–175
4. Fotakis D, Kontogiannis S, Spirakis P (2005) Selfish unsplittable flows. J Theor Comput Sci 348:226–239
5. Libman L, Orda A (2001) Atomic resource sharing in noncooperative networks. Telecommun Syst 17(4):385–409

Beyond Evolutionary Trees

Riccardo Dondi[1] and Yuri Pirola[2]
[1] Università degli Studi di Bergamo, Bergamo, Italy
[2] Università degli Studi di Milano-Bicocca, Milan, Italy

Keywords

Consensus networks; Lateral gene transfer; Phylogenetic networks

Years and Authors of Summarized Original Work

2005; Choy, Jansson, Sadakane, Sung
2010; Della Vedova, Dondi, Jiang, Pavesi, Pirola, Wang
2011; Tofigh, Hallett, Lagergren
2011; van Iersel, Kelk
2014; Kelk, Scornavacca

Problem Definition

Recent developments in phylogenetics have provided evidences that evolutionary histories cannot always be represented as a single tree; thus, more sophisticated representations are needed. *Phylogenetic networks* are natural extensions of phylogenetic trees that recently gathered general consensus in literature. Let Λ be a finite set of labels, representing a set of extant species (taxa). A *rooted phylogenetic N over Λ* (or, simply, *phylogenetic network* or *network*) is a directed acyclic connected graph $N = \langle V(N), A(N) \rangle$ containing a unique vertex with no incoming arcs, called *root* of N, and a labeling function from the set $L(N)$ of leaves of N to the set of labels Λ. The set of labels associated with the leaves of N is denoted by $\Lambda(N)$, and phylogenetic networks whose leaves are in bijection with the set of labels are called *uniquely labeled*. The undirected edges underlying the set $A(N)$ are denoted with $E(N)$.

We will discuss two important families of problems where phylogenetic networks have been introduced: consensus network computation and tree reconciliation. Other models (and the related problems) for representing and reconstructing non-treelike evolutionary scenarios are presented in [7]. The family of the consensus network computation problems asks for a single phylogenetic network (called consensus network) that best summarizes all the information provided by a collection of "structures" representing the evolutionary relationships among the set of taxa Λ. The family of tree reconciliation problems, instead, analyzes the evolution of a gene in order to either reconstruct the evolution of a set of species or infer the evolutionary scenario of the considered gene.

Observe that it is possible to topologically sort the vertices of a phylogenetic network, so that each vertex always appears after all its predecessors; hence, it is possible to define the *children*, *parents*, *ancestors*, and *descendants* of a given vertex, as usual for trees. Furthermore, as for trees, we can define the *least common ancestor* of a set of nodes. Given a subset A of nodes of a phylogenetic network N, then the least common ancestor (or, shortly, lca) of A in N is a node x of N that is an ancestor of each node in A and that is the furthest such node from the root.

Consensus Network Reconstruction

The aim of consensus network reconstruction problems is computing a unique phylogenetic network (called *consensus network*) that best summarizes all the information provided by a collection of "structures" representing the evolutionary relationships among the set of taxa Λ. Different specific computational problems have been defined in the literature depending (i) on the kind of input structures considered and (ii) on the definition of the optimality criterion used to choose the best consensus network. Simple formulations (such as *maximum agreement subtree* (MAST), *maximum compatible tree* (MCT), *maximum agreement supertree* (MASP)) compute a consensus network which is actually a phylogenetic tree. However, trees are not always sufficient for describing conflicting information and real evolutionary scenarios; hence, formulations which attempt to reconstruct phylogenetic networks have been proposed.

In terms of optimality criterion, two aims can be pursued: either finding the largest set of taxa that "share" (as defined below) a common substructure or finding the "simplest" network that represents all taxa. For measuring the complexity of a network, two natural parameters are considered: the *reticulation number* and the *level* of the network. *Reticulation* (or *hybrid*) nodes are nodes of the network with more than one parent. The *reticulation number* of a network N is defined as $|E(N)| - |V(N)| + 1$ and represents how "far" the network is from a phylogenetic tree (which has reticulation number 0). If all reticulation nodes have indegree 2 (which is often assumed in the literature), then the reticulation number is equal to the number of reticulation nodes. The *level* of a network N is defined as the maximum number of reticulation nodes in a biconnected component of $\langle V(N), E(N) \rangle$ (i.e., the undirected graph obtained from the network) [6]. Phylogenetic trees are level-0 networks, while level-1 networks are often called *galled trees* [13].

In terms of kinds of input structures, several options have been studied, and, among them, the most important ones are phylogenetic networks, triplets/quartets, and clusters. When input structures are phylogenetic networks, the usual aim is to reconstruct their *maximum agreement subnetwork* (MASN) [6], that is, a level-k phylogenetic network A (for some fixed k) uniquely labeled with a set $\Lambda' \subseteq \Lambda$ of maximum cardinality such that A is a subgraph of the topological restriction of each input network w.r.t. Λ'. The topological restriction of a uniquely labeled network N to a subset $\Lambda' \subseteq \Lambda$ of labels is defined as the network obtained by first deleting all nodes which are not on any directed path from the root to one of the leaves labeled in Λ' along with their incident edges, and then, for every node with outdegree 1 and indegree less than 2, contracting its outgoing edges. (Notice that the MAST problem is a special case of the MASN problem when $k = 0$.)

Triplets are rooted binary phylogenetic trees on exactly three species/leaves. They are generally represented as $xy|z$, indicating that the parent of x and y is a child of the parent of z. The consensus network reconstruction problem from triplets is the problem of finding, if possible, a phylogenetic network N *consistent* with each triplet given as input (or with the maximum number of them). A phylogenetic network N is said to be consistent with a triplet $xy|z$ if N contains two distinct vertices u, v and the four pairwise internally vertex-disjoint paths $u \rightarrow x$, $u \rightarrow y$, $v \rightarrow u$, $v \rightarrow z$. The resulting phylogenetic network is required to either have minimum level or have fixed level (possibly with minimum reticulation number) [18, 26, 27]. The related problem on *quartets* (i.e., unrooted phylogenetic trees on four species) has been also proposed [8].

Clusters are (strict) subsets of Λ. The consensus network reconstruction problem from clusters is the problems of finding a phylogenetic network N that *represents* all clusters given as input. There exist two main definitions for "represents," commonly referred to as the "hardwired" and the "softwired" definitions. A network N represents a cluster $\Lambda' \subset \Lambda$ in the *hardwired* sense if there exists an arc $(u, v) \in A(N)$ such that Λ' is exactly the set of labels associated with the leaves of a subnetwork rooted in v [16]. Instead, a network N represents a cluster $\Lambda' \subset \Lambda$ in the *softwired* sense if there exists an arc $(u, v) \in A(N)$ such that Λ' is the set of labels associated with the leaves of a subtree rooted in v obtained by removing, for each reticulation

node, all edges but one directed to that node [15]. In both cases, the computational problems focus on reconstructing networks with minimum level [29].

Reconciliation of Gene Trees and Species Trees

The evolution of a family of homologous genes in a given set of species is usually represented as a phylogenetic tree, called a *gene tree*, where each label can be associated with more than one leaf, while the evolution of the considered set of species is called a *species tree*, which is a uniquely labeled tree. Due to different evolutionary events that affect gene evolution (*duplications, losses, lateral gene transfer*), the evolution represented by a gene tree and a species tree (or by two different gene trees) can be different.

Two fundamental combinatorial problems have been studied in this field: the *reconstruction* of the species tree associated with the homologous genes considered [5, 12, 21] and the *reconciliation* of a gene tree with a given species tree [3, 4, 9, 21, 23], whose goal is the inference of the evolutionary events that occurred in the genes evolution. Here, we consider only the latter problem; hence, we assume that a gene tree (or a set of gene trees) and a (correct) species tree are given.

Given a set S of taxa, a species tree T_S and a gene tree T_G are two rooted binary trees, leaf-labeled by S, with $\Lambda(T_G) \subseteq \Lambda(T_S)$. Two nodes of a tree T are comparable when one is an ancestor of the other. T_G and T_S are compared introducing a mapping $\lambda : V(T_G) \rightarrow V(T_S)$, which usually corresponds to the *least common ancestor mapping*. Three biological events are considered for gene families' evolution: *duplications, losses,* and *lateral gene transfers*.

A *duplication* is a copy of a given gene, after which the two copies evolve independently. A duplication occurs in an internal node g of T_G if and only if $\lambda(f(g)) = \lambda(g)$, for a child $f(g)$ of g. A *loss* of a gene in some species consists in a copy of a gene disappearing during the evolution of a given gene family. The losses can be computed from the mapping λ between T_G and T_S [21].

When duplications and losses are the only evolutionary events considered, a gene tree and a species tree are compared with a *reconciled tree* $R(T_G, T_S)$ [2,5,9–11,21,23]. The reconciled tree is a binary tree that represents an embedding of a gene tree inside a species tree, and it allows to identify when duplications and losses occur. However, when considering also *lateral gene transfers* (also called horizontal gene transfers), the scenario changes, and the evolutionary history of a gene family must be represented by a phylogenetic network. A lateral gene transfer occurs when some genetic material is transferred from a taxon to another taxon which is not a descendant of the first taxon.

In order to represent a reconciliation of a gene tree T_G and a species tree T_S considering as evolutionary events, duplications, losses, and lateral gene transfers, the definition of *duplication-transfer-loss scenario* (DTL-scenario) has been introduced [25]. Notice that other models of reconciliation have been proposed, notably [10].

Definition 1 A DTL-scenario is a tuple $S = (T_S, T_G, \sigma, \lambda^*, \Sigma, \Delta, \Gamma, \Theta)$ where T_S is a species tree; T_G is gene tree; σ maps each leaf of T_G with the corresponding leaf of T_S; λ^* maps each node of T_G in a node of T_S (λ^* can be considered as a generalization of the least common ancestor mapping λ); $\{\Sigma, \Delta, \Gamma\}$ is a tripartition of the internal nodes of T_G in speciation nodes, duplication nodes, transfer nodes, respectively, while Θ is a subset of the edges of T_G; and the following properties hold:

1. For each leaf of T_G, $\lambda^*(u) = \sigma(u)$.
2. Consider a node x with children x_l and x_r, then $\lambda^*(x)$ is not a proper descendant of one of $\lambda^*(x_r)$, $\lambda^*(x_l)$, and one of $\lambda^*(x_r)$, $\lambda^*(x_l)$ is a descendant of $\lambda^*(x)$.
3. Given an edge (x, y) of T_G, then $(x, y) \in \Theta$ if and only if x is not comparable with y.
4. Given nodes x of T_G with children x_r and x_l, then:
   (a) $x \in \Gamma$ if and only if $(x, x_l) \in \Theta$ or $(x, x_r) \in \Theta$.
   (b) $x \in \Sigma$ only if $\lambda^*(x)$ is the least common ancestor of $\lambda^*(x_l)$ and $\lambda^*(x_r)$, and $\lambda^*(x_l)$ and $\lambda^*(x_r)$ are not comparable.

(c) $x \in \Delta$ only if $\lambda^*(x)$ is an ancestor of the least common ancestor of $\lambda^*(x_l)$ and $\lambda^*(x_r)$, and $\lambda^*(x_l)$ and $\lambda^*(x_r)$ are comparable.
5. Consider two edges $(x, x') \in \Theta$ and $(y, y') \in \Theta$, with x' an ancestor of y', then $\lambda^*(x')$ is an ancestor of $\lambda^*(y)$.

The number of losses is directly inferred from a given scenario S [1,25].

Now, we briefly discuss the biological motivations for the conditions introduced. Condition 1 guarantees the correspondence of each leaf T_G with the corresponding species (leaf) of T_S. Condition 2 guarantees that the order on the nodes of T_G is preserved by the mapping λ^*. Condition 3 defines the edges associated with a lateral gene transfer. Condition 4 establishes that the nodes of T_G can be either associated with a lateral gene transfer (condition 4.a), with a speciation (condition 4.b, then each node x and its two children must be mapped in different nodes of T_S), or with a duplication (condition 4.c, then each node x and at least one of its two children must be mapped in the same node of T_S). The last condition (condition 5) is introduced to ensure that different lateral gene transfers are biologically meaningful, that is, those events relate coexisting species, and that if (x, x') is lateral gene transfer, then there is no lateral gene transfer (y, y'), where y is a proper ancestor of x and y' is a proper descendant of x'.

Key Results

Consensus Networks

The maximum agreement subnetwork (MASN) problem is NP-hard even if the input is composed of a binary tree and an unbounded-level network [17]. If the input is composed of two level-1 networks, the problem can be solved in time $O(n^2)$, and it is fixed-parameter tractable if the input is composed of two level-k networks (where k is the parameter) [6].

Given a set of triplets, constructing a level-k phylogenetic network consistent with all of them is NP-hard for all $k \geq 1$, while it is NP-hard for all $k \geq 0$ if we want to construct a level-k network consistent with the maximum number of them [28]. If the input set $\mathcal{T}$ of triplets is *dense* (i.e., it contains a triplet for each cardinality three subset of taxa), then a level-k network consistent with all triplets can be found (if any) in time $O(\mathcal{T})$ for $k = 1$ [18], in time $O(\mathcal{T}^3)$ for $k = 2$ [27], and in polynomial time for any fixed k [24]. Recently, these results have been extended in order to minimize the reticulation number of the computed network [14, 26].

The reconstruction of a consensus network from a set of clusters in the hardwired sense has been tackled in [16], where an algorithm for reconstructing a phylogenetic network that represents a set $\mathcal{C}$ of clusters (and only $\mathcal{C}$) with $O(|\mathcal{C}|)$ nodes and $O(|\mathcal{C}|^2)$ edges is presented. An algorithm for reconstructing a level-1 network from clusters in the softwired sense has been shown in [15], which has been later extended in order to compute in polynomial time a level-k network or a network with reticulation number k for every fixed k [19]. An efficient algorithm for computing a level-k network (with $k \in \{1, 2\}$) representing a set of cluster in the softwired sense which also attempts to minimize the reticulation number has also been presented [29].

Reconciliation

The main combinatorial problem related to reconciliation is, given a species tree and a gene tree, the computation of a DTL-scenario of minimum cost, for some function that assigns positive cost to duplications, losses, and lateral gene transfers. The problem is known to be NP-hard [22, 25]. However, two tractable variants of the problem have been considered.

A first variant, called *cyclic DTL-scenario*, does not consider condition 5 of Definition 1. Computing a cyclic DTL-scenario of minimum cost is polynomial time solvable. First, an algorithm for cost function that assigns positive cost only to duplications and lateral gene transfers was presented [25]. Later a linear time algorithm for a general cost function (hence losses are assigned a positive cost) was given [1].

A second variant considered is the *dated version*, where nodes of the species are associated with labels that represent the divergence time and a lateral gene transfer is possible only between coexisting species. In this case, computing an acyclic DTL-scenario of minimum cost is polynomial time solvable [20].

Open Problems

The fixed parameter tractability of computing a minimum reticulation-number phylogenetic network that combines a set of nonbinary trees has not yet be assessed, and only a few attempts focus on approximate solutions. Answers to both questions could be of interest.

An interesting open problem related to the computation of a DTL-scenario of minimum cost is the investigation of the parameterized complexity for acyclic DTL-scenarios, when parameterized by the cost of the solution or by the number of lateral gene transfers. Another interesting direction for this problem is to investigate its approximation complexity.

Cross-References

▶ Algorithms for Combining Rooted Triplets into a Galled Phylogenetic Network
▶ Maximum Agreement Subtree (of 2 Binary Trees)
▶ Maximum Agreement Subtree (of 3 or More Trees)
▶ Maximum Agreement Supertree
▶ Maximum Compatible Tree

Recommended Reading

1. Bansal MS, Alm EJ, Kellis M (2012) Efficient algorithms for the reconciliation problem with gene duplication, horizontal transfer and loss. Bioinformatics 28(12):283–291. doi:10.1093/bioinformatics/bts225
2. Bonizzoni P, Della Vedova G, Dondi R (2005) Reconciling a gene tree to a species tree under the duplication cost model. Theor Comput Sci 347(1–2):36–53. doi:10.1016/j.tcs.2005.05.016

3. Burleigh JG, Bansal MS, Wehe A, Eulenstein O (2008) Locating multiple gene duplications through reconciled trees. In: Proceedings of the 12th annual international conference on research in computational molecular biology, RECOMB 2008, Singapore. LNCS, vol 4955. Springer, pp 273–284. doi:10.1007/978-3-540-78839-3_24

4. Chang WC, Eulenstein O (2006) Reconciling gene trees with apparent polytomies. In: Proceedings of the 12th annual international conference on computing and combinatorics, COCOON 2006, Taipei. LNCS, vol 4112. Springer, pp 235–244. doi:10.1007/11809678_26

5. Chauve C, El-Mabrouk N (2009) New perspectives on gene family evolution: losses in reconciliation and a link with supertrees. In: Proceedings of the 13th annual international conference on research in computational molecular biology, RECOMB 2009, Tucson. LNCS, vol 5541. Springer, pp 46–58. doi:10.1007/978-3-642-02008-7_4

6. Choy C, Jansson J, Sadakane K, Sung WK (2005) Computing the maximum agreement of phylogenetic networks. Theor Comput Sci 335(1):93–107. doi:10.1016/j.tcs.2004.12.012

7. Della Vedova G, Dondi R, Jiang T, Pavesi G, Pirola Y, Wang L (2010) Beyond evolutionary trees. Nat Comput 9(2):421–435. doi:10.1007/s11047-009-9156-6

8. Gambette P, Berry V, Paul C (2012) Quartets and unrooted phylogenetic networks. J Bioinform Comput Biol 10(4). doi:10.1142/S0219720012500047

9. Goodman M, Czelusniak J, Moore GW, Romero-Herrera AE, Matsuda G (1979) Fitting the gene lineage into its species lineage, a parsimony strategy illustrated by cladograms constructed from globin sequences. Syst Zool 28(2):132–163. doi:10.1093/sysbio/28.2.132

10. Gòrecki P (2004) Reconciliation problems for duplication, loss and horizontal gene transfer. In: Proceedings of the 8th annual international conference on computational molecular biology, RECOMB 2004, San Diego. ACM, pp 316–325. doi:10.1145/974614.974656

11. Gòrecki P, Tiuryn J (2006) DLS-trees: a model of evolutionary scenarios. Theor Comput Sci 359(1–3):378–399. doi:10.1016/j.tcs.2006.05.019

12. Guigò R, Muchnik I, Smith T (1996) Reconstruction of ancient molecular phylogeny. Mol Phylogenet Evol 6(2):189–213. doi:10.1006/mpev.1996.0071

13. Gusfield D, Eddhu S, Langley CH (2004) Optimal, efficient reconstruction of phylogenetic networks with constrained recombination. J Bioinform Comput Biol 2(1):173–214. doi:10.1142/S0219720004000521

14. Habib M, To TH (2012) Constructing a minimum phylogenetic network from a dense triplet set. J Bioinform Comput Biol 10(5). doi:10.1142/S0219720012500138

15. Huson DH, Klöpper TH (2007) Beyond galled trees – decomposition and computation of galled networks. In: Proceedings of the 11th annual international conference on research in computational molecular biology, RECOMB 2007, Oakland. LNCS, vol 4453. Springer, pp 211–225. doi:10.1007/978-3-540-71681-5_15

16. Huson DH, Rupp R (2008) Summarizing multiple gene trees using cluster networks. In: Proceedings of the 8th international workshop on algorithms in bioinformatics, WABI 2008, Karlsruhe. LNCS, vol 5251. Springer, pp 296–305. doi:10.1007/978-3-540-87361-7_25

17. Jansson J, Sung WK (2004) The maximum agreement of two nested phylogenetic networks. In: Proceedings of the 15th international symposium on algorithms and computation, ISAAC 2004, Hong Kong. LNCS, vol 3341. Springer, pp 581–593. doi:10.1007/978-3-540-30551-4_51

18. Jansson J, Nguyen NB, Sung WK (2006) Algorithms for combining rooted triplets into a galled phylogenetic network. SIAM J Comput 35(5):1098–1121. doi:10.1137/S0097539704446529

19. Kelk S, Scornavacca C (2014) Constructing minimal phylogenetic networks from softwired clusters is fixed parameter tractable. Algorithmica 68(4):886–915. doi:10.1007/s00453-012-9708-5

20. Libeskind-Hadas R, Charleston MA (2009) On the computational complexity of the reticulate cophylogeny reconstruction problem. J Comput Biol 16(1):105–117. doi:10.1089/ cmb.2008.0084

21. Ma B, Li M, Zhang L (2000) From gene trees to species trees. SIAM J Comput 30(3):729–752. doi:10.1137/S0097539798343362

22. Ovadia Y, Fielder D, Conow C, Libeskind-Hadas R (2011) The cophylogeny reconstruction problem is NP-complete. J Comput Biol 18(1):59–65. doi:10.1089/cmb.2009.0240

23. Page R (1994) Maps between trees and cladistic analysis of historical associations among genes. Syst Biol 43(1):58–77. doi:10.1093/sysbio/43.1.58

24. To TH, Habib M (2009) Level-k phylogenetic networks are constructable from a dense triplet set in polynomial time. In: Proceedings of the 20th annual symposium on combinatorial pattern matching, CPM 2009, Lille. LNCS, vol 5577. Springer, pp 275–288. doi:10.1007/978-3-642-02441-2_25

25. Tofigh A, Hallett MT, Lagergren J (2011) Simultaneous identification of duplications and lateral gene transfers. IEEE/ACM Trans Comput Biol Bioinform 8(2):517–535. doi:10.1109/TCBB.2010.14

26. van Iersel L, Kelk S (2011) Constructing the simplest possible phylogenetic network from triplets. Algorithmica 60(2):207–235. doi:10.1007/s00453-009-9333-0

27. van Iersel L, Keijsper J, Kelk S, Stougie L, Hagen F, Boekhout T (2009) Constructing level-2 phylogenetic networks from triplets. IEEE/ACM Trans Comput Biol Bioinform 6(4):667–681. doi:10.1145/1671403.1671415

28. van Iersel L, Kelk S, Mnich M (2009) Uniqueness, intractability and exact algorithms: reflections on level-

k phylogenetic networks. J Bioinform Comput Biol 7(4):597–623. doi:10.1142/S0219720009004308

29. van Iersel L, Kelk S, Rupp R, Huson D (2010) Phylogenetic networks do not need to be complex: using fewer reticulations to represent conflicting clusters. Bioinformatics 26(12):i124–i131. doi:10.1093/bioinformatics/btq202

Beyond Hypergraph Dualization

Lhouari Nourine[1] and Jean-Marc Petit[2]
[1]Clermont-Université, Université Blaise Pascal, LIMOS, CNRS, Aubière, France
[2]Université de Lyon, CNRS, INSA Lyon, LIRIS, Lyon, France

Keywords

Dualization; Enumeration; Hypergraph transversal; Partially ordered set

Years and Authors of Summarized Original Work

2012; Nourine, Petit

Problem Definition

This problem concerns hypergraph dualization and generalization to poset dualization.

A *hypergraph* $\mathcal{H} = (V, \mathcal{E})$ consists of a finite collection $\mathcal{E}$ of sets over a finite set V, i.e., $\mathcal{E} \subseteq \mathcal{P}(V)$ (the powerset of V). The elements of $\mathcal{E}$ are called *hyperedges*, or simply *edges*. A hypergraph is said simple if none of its edges is contained within another. A *transversal* (or *hitting set*) of $\mathcal{H}$ is a set $T \subseteq V$ that intersects every edge of $\mathcal{E}$. A transversal is *minimal* if it does not contain any other transversal as a subset. The set of all minimal transversal of $\mathcal{H}$ is denoted by $Tr(\mathcal{H})$. The hypergraph $(V, Tr(\mathcal{H}))$ is called the *transversal hypergraph* of $\mathcal{H}$. Given a simple hypergraph $\mathcal{H}$, the hypergraph dualization problem (TRANS-ENUM for short) concerns the enumeration without repetitions of $Tr(\mathcal{H})$.

The TRANS-ENUM problem can also be formulated as a dualization problem in posets. Let $(P, \leq)$ be a poset (i.e., $\leq$ is a reflexive, antisymmetric, and transitive relation on the set P). For $A \subseteq P$, $\downarrow A$ (resp. $\uparrow A$) is the downward (resp. upward) closure of A under the relation $\leq$ (i.e., $\downarrow A$ is an ideal and $\uparrow A$ a filter of $(P, \leq)$). Two antichains $(\mathcal{B}^+, \mathcal{B}^-)$ of P are said to be dual if $\downarrow \mathcal{B}^+ \cup \uparrow \mathcal{B}^- = P$ and $\downarrow \mathcal{B}^+ \cap \uparrow \mathcal{B}^- = \emptyset$. Given an implicit description of a poset P and an antichain $\mathcal{B}^+$ (resp. $\mathcal{B}^-$) of P, the poset dualization problem (DUAL-ENUM for short) enumerates the set $\mathcal{B}^-$ (resp. $\mathcal{B}^+$), denoted by $\mathrm{Dual}(\mathcal{B}^+) = \mathcal{B}^-$ (resp. $\mathrm{Dual}(\mathcal{B}^-) = \mathcal{B}^+$). Notice that the function dual is self-dual or idempotent, i.e., $\mathrm{Dual}(\mathrm{Dual}(\mathcal{B})) = \mathcal{B}$.

TRANS-ENUM is a particular case of DUAL-ENUM. Indeed, consider P the poset $(\mathcal{P}(V), \subseteq)$ for some set V. Then for every dual set $(\mathcal{B}^+, \mathcal{B}^-)$ of P, we have $\mathcal{B}^- = Tr(\overline{\mathcal{B}^+}) = \mathrm{Dual}(\mathcal{B}^+)$, or equivalently $\mathcal{B}^+ = \overline{Tr(\mathcal{B}^-)} = \mathrm{Dual}(\mathcal{B}^-)$ with $\overline{\mathcal{E}} = \{V \setminus E \mid E \in \mathcal{E}\}$ where $\mathcal{E} \subseteq \mathcal{P}(V)$.

Now we ask the following question: Which posets DUAL-ENUM can be reduced to TRANS-ENUM? To do so, we introduce the notions of duality gap, convex embedding, and poset reflexion.

Let $(P, \leq P)$ and $(Q, \leq_Q)$ be two posets and $f : P \rightarrow Q$ an *injective reflection*, i.e., for all $x, y \in P$, $f(x) \leq_Q f(y)$ implies $x \leq_P y$. Notice that the reflection f preserves incomparability, i.e., if x and y are incomparable in P, then $f(x)$ and $f(y)$ are incomparable in Q. Therefore, for every dual set $(\mathcal{B}^+, \mathcal{B}^-)$ of P, $\mathrm{Dual}(f(\mathcal{B}^+))$ contains $f(\mathcal{B}^-)$. The difference between the size of $\mathrm{Dual}(f(\mathcal{B}^+))$ and the size of $f(\mathcal{B}^-)$ is a positive integer, called the *duality gap*. We speak about *weak duality* when the gap is strictly positive, *strong duality* otherwise.

Duality gaps are important in enumeration problems because they provide an upper bound on the difference between the number of enumerated solutions and the number of solutions of the original problem.

Key Results

TRANS-ENUM has been intensively studied in the last two decades, and several results show that it is equivalent to many problems in computer science area (see the paper by Eiter and Gottlob [3]). The question whether TRANS-ENUM admits an output-polynomial time algorithm is still open. In fact, despite the number of papers on TRANS-ENUM, the best known algorithm is the one by Fredman and Khachiyan [8] which runs in time $O(n^{\log(n)})$ where n is the size of the hypergraph plus the number of minimal transversals. Other results on complexity can be found in [5, 6, 11, 12, 14]. For general posets, it is shown in [7] that the dualization over the products of some posets can be done with the same complexity as TRANS-ENUM. Recently, Nourine and Petit [16] have investigated dualization problems in general posets for which the duality gap is bounded by a polynomial.

Strong Duality

The following characterization theorem of the zero gap is a reformulation of a known result in [10, 15], where the poset Q is the powerset for some set.

Theorem 1 *Let $(P, \leq_P)$ and $(Q, \leq_Q)$ be two posets. Then the duality gap is zero iff there exists a map $f : P \rightarrow Q$ such that f is a bijective embedding, i.e., for all $x, y \in P$ $f(x) \leq_Q f(y)$ iff $x \leq_P y$.*

Many instances of problems have such a property, for example, frequent itemsets, monotone Boolean functions, minimal keys, inclusion dependencies, or minimal dominating sets [10, 13, 15]. Nevertheless, the bijective embedding between two posets does not always exist. In the following we give a relaxation of the bijection embedding in order to capture some polynomial reductions between enumeration problems.

Weak Duality

Let $(P, \leq_P)$ and $(Q, \leq_Q)$ be posets. A function $f : P \rightarrow Q$ is a *convex embedding* if for all $x, y \in P$ and $z \in Q$, $x \leq_P y$ iff $f(x) \leq_Q f(y)$ and $f(x) \leq_Q z \leq_Q f(y)$ implies there exists $t \in P$ such that $f(t) = z$.

The following result can be seen as a relaxation of the bijective embedding given in Theorem 1.

Proposition 1 *Let $(P, \leq_P)$ and $(Q, \leq_Q)$ be two posets and $f : P \rightarrow Q$ a convex embedding. Then there exist two antichains $\mathcal{B}_0^+$, $\mathcal{B}_0^-$ of Q such that $P \setminus \{\perp_P\}$ is isomorphic to $Q \setminus (\downarrow \mathcal{B}_0^+ \cup \uparrow \mathcal{B}_0^-)$, where $\perp_P$ is the bottom of P if it exists. Furthermore, the duality gap is bounded by $\mid \mathcal{B}_0^+ \mid + \mid \mathcal{B}_0^- \mid$.*

Complexity

For strong duality, [10, 15] points out how the result of Fredman and Khachiyan [8] can be reused to devise an incremental quasi-polynomial time algorithm, called `Dualize and Advance`, for some pattern mining problems. For weak duality, whenever the duality gap remains polynomial in the size of the problem and $(Q, \leq_Q)$ isomorphic to $(\mathcal{P}(E), \subseteq)$ for some set E, the `Dualize and Advance` algorithm can be reused with the same complexity if the following assumptions hold:

1. The reflexion f of $(P, \leq)$ to $(\mathcal{P}(E), \subseteq)$ and its inverse is computable in polynomial time.
2. Given two elements $x, y \in P$, checking $x \leq y$ is polynomial time.

Applications

The hypergraph dualization is a crucial step in many applications in logics, databases, artificial intelligence, and pattern mining [3, 4, 8, 11, 15], especially for hypergraphs, i.e., Boolean lattices. The main application domain concerns pattern mining problems, i.e., the identification of maximal interesting patterns in database by asking membership queries (predicate) to a database. In the rest of this section, we give two examples of pattern mining problems related to DUAL-ENUM and weak duality.

Frequent Conjunctive Queries

We consider the problem statement defined in [9]. Let $\mathbf{R} = \{R_1, \ldots, R_n\}$ be a database schema, $\mathcal{D}$ the domain of $\mathbf{R}$ and $sch(\mathbf{R}) = \{R_i.A \mid R_i \in \mathbf{R}, A \in R_i\}$. A (simple) conjunctive query Q

over $\mathbf{R}$ is of the form $\pi_X(\sigma_F(R_1 \times \ldots \times R_n))$ ($\pi_X(\sigma_F)$ for short) where $X \subseteq \text{sch}(\mathbf{R})$ and F a conjunction of equalities of the form $R_i.A = R_j.B$ or $R_i.A = c$ with $R_i.A, R_j.B \in \text{sch}(\mathbf{R})$ and $c \in \mathcal{D}$. Let $\mathcal{Q}_r$ be the set of all possible conjunctive queries over $\mathbf{R}$. For a given database d over $\mathbf{R}$, we note $\text{Adom}(d) \subseteq \mathcal{D}$ is the active domain of d and $Q(d)$ the result of the evaluation of Q against d. We note $\mathcal{F}$ is the finite set of all possible selection formula over $\mathbf{R}$ and $\text{Adom}(d)$, i.e., $\mathcal{F} = \{\{A, B\} \mid A \neq B, A \in \mathbf{R}, B \in \mathbf{R} \cup \text{Adom}(d)\}$.

Let Q_1, Q_2 be two conjunctive queries over $\mathbf{R}$. Q_1 is *contained* in Q_2, denoted by $Q_1 \subseteq Q_2$, if for every database d over $\mathbf{R}$, $Q_1(d) \subseteq Q_2(d)$. Q_1 is *diagonally contained* in Q_2, denoted $Q_1 \subseteq^\Delta Q_2$, if Q_1 is contained in a projection of Q_2, i.e., $Q_1 \subseteq \pi_X(Q_2)$. The frequency of $\pi_X(\sigma_F)$ in d is defined by $|\pi_X(\sigma_F)(d)|$. A query $\pi_X(\sigma_F)$ is *frequent* in d with respect to a given threshold ϵ if $|\pi_X(\sigma_F)(d)| \geq \epsilon$. The frequency is anti-monotonic with respect to $\subseteq^\Delta$ [9].

Proposition 2 *Let $Q_1 = \pi_{X_1}(\sigma_{F_1})$ and $Q_2 = \pi_{X_2}(\sigma_{F_2})$ be two queries of $\mathcal{Q}_r$. Then $Q_1 \subseteq^\Delta Q_2$ iff $X_1 \subseteq X_2$ and $F_2 \subseteq F_1$. Equivalently, $Q_1 \subseteq^\Delta Q_2$ iff $X_1 \cup (\mathcal{F} \setminus F_1) \subseteq X_2 \cup (\mathcal{F} \setminus F_2)$.*

From Proposition 2, $f : \mathcal{Q}_r \to \mathcal{P}(\mathbf{R} \cup \mathcal{F})$ with $f(\pi_X(\sigma_F)) = X \cup (\mathcal{F} \setminus F)$ is a bijective embedding. Thus $\mathcal{Q}_r$ ordered under $\subseteq^\Delta$ is a Boolean lattice and Theorem 1 can be applied. It is interesting to consider the subclass of $\mathcal{Q}_r$ restricted to *consistent queries*, i.e., queries for which there exists at least one database such that their evaluations return values different from zero. For instance, $\sigma(B = 1 \wedge B = 2)$ and $\sigma(A = B \wedge A = 1 \wedge B = 2)$ are not consistent. Let us consider the set $\mathcal{Q}_C \subset \mathcal{Q}_r$ of all consistent queries.

Lemma 1 *Let $Q_1 = \pi_{X_1}(\sigma_{F_1})$ and $Q_2 = \pi_{X_2}(\sigma_{F_2})$ be two queries of $\mathcal{Q}_r$ such that $Q_1 \subseteq^\Delta Q_2$. Then if Q_2 is consistent, it implies Q_1 is consistent.*

Notice that the restriction of f to $\mathcal{Q}_C$ is still a convex embedding, but no longer bijective. More interestingly, the associated duality gap is not polynomial. Indeed, $\mathcal{B}_0^+ = \emptyset$ but $\mathcal{B}_0^-$ has a size

exponential in the size of $\mathbf{R} \cup \text{Adom}(d)$ since the number of selections of the form $\sigma(A_1 = A_2 \wedge \ldots \wedge A_{n-1} = A_n \wedge A_1 = v \wedge A_n = v')$ is exponential in the number of attributes.

Rigid Sequences

Let us consider sequences with or without wild-card (denoted $\star$); see, e.g., [1]. Let Σ be an alphabet and $\star \notin \Sigma$. A rigid sequence $s[n]$ is a word of size n of $(\Sigma \cup \{\star\})^*$ such that $s[1] \neq \star$ and $s[n] \neq \star$. The set of all rigid sequences of size at most n are denoted by Σ_R^n and the empty sequence by ϵ. Let $s[l], t[k] \in \Sigma_R^n$. We consider the following classical (prefix and factor) partial orders on rigid sequences:

- $s \sqsubseteq_f t$, if there exists $j \in [1 \ldots k]$ such that for every $i \in [1 \ldots l]$, either $s[i] = t[j + i - 1]$ or $s[i] = \star$ (factor).
- $s \sqsubseteq_p t$, if for every $i \in [1 \ldots l]$, either $s[i] = t[i]$ or $s[i] = \star$ (prefix).

The following theorem shows that the duality gap between the dualization in prefix posets of rigid sequences and TRANS-ENUM is bounded by a polynomial in n and $|\Sigma|$.

Theorem 2 ([16]) *Let $f : (\Sigma_R^n \setminus \{\epsilon\}, \sqsubseteq_p) \to (\mathcal{P}(\{1, \ldots, n\} \times \Sigma), \subseteq)$ be a function defined by $f(s) = \{(i, s[i]) \mid s[i] \neq \star, i \leq n\}$. Then f is a convex embedding with $\mathcal{B}_0^+ = \{\{(i, x) \mid x \in \Sigma, i \in [2 \ldots n]\}\}$ and $\mathcal{B}_0^- = \{\{(1, x), (1, y)\} \mid x, y \in \Sigma, x \neq y\} \cup \{\{(1, x), (i, y), (i, z)\} \mid x, y, z \in \Sigma, y \neq z, i \in [2 \ldots n]\}$.*

Proposition 3 ([16]) *There is a poset reflection $f : (\Sigma_R^n, \sqsubseteq_f) \to (\Sigma_R^n, \sqsubseteq_p)$ with a duality gap bounded by a polynomial in n.*

Using Theorem 2 and Proposition 3, we conclude that the duality gap between the dualization in factor posets of rigid sequences and TRANS-ENUM is bounded by a polynomial the size of Σ and n [16].

Open Problems

1. The challenging question is to find an output-polynomial time algorithm for TRANS-ENUM.

2. Lattices are a particular class of posets. For example, the dualization over product of chains can be done with the same complexity as TRANS-ENUM which is equivalent to dualization in Boolean lattices. For distributive lattices class which contains Boolean lattice and the product of chains, the dualization is open.

3. Many connections have to be done between TRANS-ENUM and graph theory problems, such as minimal dominating sets [13].

4. Many problems in data mining can be formulated as dualization in posets, e.g., frequent subgraphs or frequent subtrees. An interesting direction is to identify posets for which the dualization is equivalent to TRANS-ENUM.

URLs to Code and Data Sets

Program Codes and Instances for Hypergraph Dualization can be found on the Takeaki Uno's webpage at http://research.nii.ac.jp/~uno/dualization.html. Some pattern mining problems, reducible to TRANS-ENUM with strong duality, can be found on the iZi webpage at http://liris.cnrs.fr/izi/.

Cross-References

▶ Efficient Polynomial Time Approximation Scheme for Scheduling Jobs on Uniform Processors
▶ Minimal Dominating Set Enumeration

Recommended Reading

1. Arimura H, Uno T (2009) Polynomial-delay and polynomial-space algorithms for mining closed sequences, graphs, and pictures in accessible set systems. In: SDM, Sparks, pp 1087–1098
2. Boros E, Makino K (2009) A fast and simple parallel algorithm for the monotone duality problem. In: Albers S, Marchetti-Spaccamela A, Matias Y, Nikoletseas SE, Thomas W (eds) ICALP (Part I), Rhodes. LNCS, vol 5555. Springer, pp 183–194
3. Eiter T, Gottlob G (1995) Identifying the minimal transversals of a hypergraph and related problems. SIAM J Comput 24(6):1278–1304
4. Eiter T, Gottlob G, Makino K (2003) New results on monotone dualization and generating hypergraph transversals. SIAM J Comput 32:514–537
5. Eiter T, Gottlob G, Makino K (2003) New results on monotone dualization and generating hypergraph transversals. SIAM J Comput 32(2):514–537
6. Elbassioni KM (2008) On the complexity of monotone dualization and generating minimal hypergraph transversals. Discret Appl Math 156(11): 2109–2123
7. Elbassioni KM (2009) Algorithms for dualization over products of partially ordered sets. SIAM J Discret Math 23(1):487–510
8. Fredman ML, Khachiyan L (1996) On the complexity of dualization of monotone disjunctive normal forms. J Algorithms 21(3):618–628
9. Goethals B, Van den Bussche J (2002) Relational association rules: getting warmer. In: Pattern detection and discovery, London, pp 125–139
10. Gunopulos D, Khardon R, Mannila H, Saluja S, Toivonen H, Sharm RS (2003) Discovering all most specific sentences. ACM Trans Database Syst 28(2):140–174
11. Gottlob G (2013) Deciding monotone duality and identifying frequent itemsets in quadratic logspace. In: PODS, NewYork, pp 25–36
12. Hagen M (2008) Algorithmic and computational complexity issues of MONET. Cuvillier Verlag ISBN 978-3-86727-826-3, 141 pages
13. Kanté MM, Limouzy V, Mary A, Nourine L (2014) On the enumeration of minimal dominating sets and related notions To appear in SIAM on Discrete Math.
14. Kavvadias DJ, Stavropoulos EC (2003) Monotone boolean dualization is in co-NP[log2n]. Inf Process Lett 85:1–6
15. Mannila H, Toivonen H (1997) Levelwise search and borders of theories in knowledge discovery. Data Min Knowl Discov 1(3):241–258
16. Nourine L, Petit J-M. (2012) Extending set-based dualization: application to pattern mining. In: ECAI Montpellier, pp 630–635

Beyond Worst Case Sensitivity in Private Data Analysis

Abhradeep Thakurta
Department of Computer Science, Stanford University, Stanford, CA, USA
Microsoft Research, CA, USA

Keywords

Algorithmic stability; Differential privacy; Robust estimators

Years and Authors of Summarized Original Work

2006; Cynthia Dwork, Frank McSherry, Kobbi
   Nissim, Adam Smith
2007; Kobbi Nissim, Sofya Raskhodnikova,
   Adam Smith
2009; Cynthia Dwork, Jing Lei
2013; Adam Smith, Abhradeep Thakurta

Problem Definition

Over the last few years, differential privacy [5,6]
has emerged as one of the most accepted notions of statistical data privacy. At a high level
differential privacy ensures that from the output
of an algorithm executed on a data set of potentially sensitive records, an adversary learns
"almost" the same thing about an individual irrespective of his presence or absence in the data set.
Formally, differential privacy is defined below
(Definition 1). Setting the privacy parameters
$\epsilon < 1$ and $\delta \ll \frac{1}{n^2}$ ensures semantically
meaningful privacy guarantees. For a detailed
survey on the semantics of differential privacy,
see [2,3,8,9].

Definition 1 ($((\epsilon, \delta)$-differential privacy [5, 6])
We call two data sets D and D' (with n records
from a fixed domain τ) neighboring if they differ
in exactly one entry, i.e., $|D \triangle D'| = 2$. An
algorithm $\mathcal{A}$ is (ϵ, δ)-differentially private if, for
all neighboring data sets D and D' and for all
measurable events S in the range space of $\mathcal{A}$, we
have

$$\Pr[\mathcal{A}(D) \in S] \leq e^{\epsilon} \Pr[\mathcal{A}(D') \in S] + \delta.$$

Initial efforts towards designing differentially
private algorithms have concentrated on settings
where the algorithms enjoy the same utility guarantees for *any* data set from the domain τ^n. (See
[3] for a survey on these efforts.) However, due to
the pessimistic nature of these algorithms, some
perform poorly in non-worst-case scenarios. With
the seminal work of [11], and followed by a
series of results [4, 7, 10, 12, 13], the community started focusing on designing differentially

private algorithms which are more useful in non-
worst-case settings, but in pessimistic scenarios
may only perform as poorly as the worst case
algorithms. In this entry, we provide an overview
of some of the recent efforts in this line of
research.

Computing the Median: A Motivating Example

To provide a flavor of the nature of these algorithms, we start with the following simple
example: Given a data set $D = \{d_1, \ldots, d_n\}$ of
n real numbers in $[0, R]$ (with $R \in \mathbb{R}^+$ and n
being odd), the task is to compute the median of
D while preserving differential privacy. Notice
that in the worst case, changing one entry in D
can change the median by R. So intuitively, any
algorithm that does not distinguish between worst
case and non-worst-case scenario will introduce
an error $\Omega(R)$ in the output.

Without loss of generality, assume that $d_1 \leq
\cdots \leq d_n$ and let $m = \frac{n+1}{2}$. Now it is not
hard to observe that by changing one data entry
in D, the median d_m can change by at most
$\max\{d_m - d_{m-1}, d_{m+1} - d_m\}$, which can be
potentially much smaller than R. An algorithm
that takes advantage of this observation can be
much more useful in non-worst-case settings as
compared to an algorithm that always introduces
an error of $\Theta(R)$. With this example in mind,
in the following section, we define some of the
notions in the differential privacy literature that
capture the non-worst-case change in the output
of a given computation task on neighboring data
sets.

Although the median might seem to be a very
simple example, but interestingly, this intuition
of capturing non-worst-case change extends to
a large class of problems. Especially in many
machine learning settings, where even for the
non-private algorithms the error guarantees are
over distributional assumptions on the data [1],
this intuition is very helpful in designing effective
differentially private variants.

N.B. Notice that in the case of computing the
average, there is no distinction between worst
case and non-worst change. Hence it is not a good
example for the scenarios we address in this entry.

Notions of Sensitivity

We describe some of the concepts which help us capture non-worst-case changes in the output of a given function $f : \tau^n \to \mathbb{R}$ on pairs of neighboring data sets D and D'. Later we use them to design differentially private algorithms which capture non-worst-case behavior of the data sets.

Global Sensitivity [6]

This notion of sensitivity refers to the maximum change the function f can have on any pair of neighboring data sets from the domain. Formally,

$$\mathsf{GS}(f) = \max_{D, D' \in \tau^n, |D \triangle D'| = 2} |f(D) - f(D')|.$$
(1)

The following algorithm in (2) is $(\epsilon, 0)$-differentially private. In the literature this is also called the *Laplace mechanism* [6]. Here $\mathsf{Lap}(\lambda)$ refers to the Laplace distribution with standard deviation $\sqrt{2}\lambda$.

$$\textbf{Output}: f(D) + \mathsf{Lap}\left(\frac{\mathsf{GS}(f)}{\epsilon}\right).$$
(2)

Notice that in (2) the distribution on the noise that is added is the same for all data sets. In general these style of algorithms that introduce data independent randomness have weaker utility guarantees in non-worst-case scenarios.

Local Sensitivity

While global sensitivity captures the maximum change in the output of f for any pairs of neighboring data sets, local sensitivity relaxes this notion to capture the maximum change in the output of f for any neighboring data set of a given data set D. Formally,

$$\mathsf{LS}(f, D) = \max_{D' \in \tau^n, |D \triangle D'| = 2} |f(D) - f(D')|.$$
(3)

With the similarity between the definitions of local sensitivity and global sensitivity, it might be tempting to use the same algorithm

as (2), with the GS replaced by LS. A careful observation indicates that this algorithm cannot be (ϵ, δ)-differentially private for any non-trivial choices of ϵ and δ. Consider the following setting where the data domain is $\{0, 1\}^n$, and the function f is the median value of D. Let D be a data set with $\lfloor \frac{n}{2} - 1 \rfloor$ entries as zero, and the rest as one. When n is odd, clearly $\mathsf{LS}(D)$ equals zero. But for a data set D' formed by changing one of the zeros in D to one, $\mathsf{LS}(D')$ equals one. So, if we replace GS by LS in (2), then for D there will be zero noise added, and for D' the noise will be $\Omega(1/\epsilon)$. Differential privacy prohibits this.

The counterexample above might give an impression that local sensitivity may not be a useful concept. However, in the following and in Algorithm 2, we show that one can use local sensitivity to obtain effective differentially private algorithms which are more useful in non-worst-case scenarios.

Smooth Sensitivity [11]

In the above example, we noticed that a direct use of local sensitivity in noise addition can result in trouble. However, using a related notion called smooth sensitivity, one can obtain a variant of the algorithm in (2) which is both differentially private and respects local properties of a given data set. At a high level, smooth sensitivity is an envelope over the local sensitivity which helps avoid abrupt change in the variance of the noise on neighboring data sets. Formally,

$$\mathsf{SS}(f, D, \beta) = \max_{D' \in \tau^n} \mathsf{LS}(f, D) e^{-\beta \cdot \mathsf{dist}(D, D')/2}.$$
(4)

Here dist is the symmetric difference between the two data sets D and D' and $\beta > 0$ is the smoothness parameter. Following observations on smooth sensitivity will be useful for designing differentially private algorithms.

1. **Observation 1 (Envelope on LS):** $\forall D, \beta > 0, \mathsf{SS}(f, D, \beta) \geq \mathsf{LS}(f, D)$.
2. **Observation 2 (Smaller than GS):** $\forall \beta > 0, D \in \tau^n, \mathsf{SS}(f, D, \beta) \leq \mathsf{GS}(f)$.
3. **Observation 3 (Smoothness):** For all neighbors $D, D', \mathsf{SS}(f, D, \beta) \leq e^{\beta} \mathsf{SS}(f, D', \beta)$.

Key Results

Using the concepts of local sensitivity and smooth sensitivity defined in the previous section, we provide two differentially private algorithmic frameworks which respect local (non-worst-case) properties of a given data set. Later in Applications 1 and 2, we instantiate them with specific problems.

Algorithm 1: Smooth Sensitivity Based

In order to use the notion of smooth sensitivity to design a differentially private algorithm analogous to (2), we need the following properties from the noise distribution to be added to $f(D)$. Let us define the following notation: For a subset S of $\mathbb{R}$, the set $S + \Delta$ defines $\{z + \Delta : z \in S\}$, and the set $e^\lambda \cdot S$ defines the set $\{e^\lambda \cdot z : z \in S\}$.

Definition 2 (Admissible noise distribution [11]) A probability distribution h on $\mathbb{R}$ is (α, β)-admissible if, for $\alpha = \alpha(\epsilon, \delta), \beta = \beta(\epsilon, \delta)$, the following conditions hold for all $|\Delta| \leq \alpha$ and $|\lambda| \leq \beta$, and for all subsets $S \subseteq \mathbb{R}$.

1. **Sliding property:** $\Pr_{Z \sim h}[Z \in S] \leq$
   $e^{\epsilon/2} \Pr_{Z \sim h}[Z \in S + \Delta] + \frac{\delta}{2}$.
2. **Dialation property:** $\Pr_{Z \sim h}[Z \in S] \leq$
   $e^{\epsilon/2} \Pr_{Z \sim h}\left[Z \in e^\lambda \cdot S\right] + \frac{\delta}{2}$.

With Definition 2 in hand, now we can define an algorithm which is analogous to (2). Let h be an (α, β)-admissible noise distribution and let Z be an independent sample from h. For a given data set D, the algorithm is as follows:

$$\textbf{Output: } f(D) + \frac{\text{SS}(f, D, \beta)}{\alpha} \cdot Z. \quad (5)$$

One can show that the above algorithm is (ϵ, δ)-differentially private [11]. An immediate question that arises: *Which natural distributions satisfy this property?* [11] showed that Laplace distribution $\frac{1}{2}e^{-|z|}$ is $\left(\epsilon/2, \frac{\epsilon}{2\ln(1/\delta)}\right)$-admissible, and $\mathcal{N}(0, 1)$ is $\left(\epsilon/\sqrt{\ln(1/\delta)}, \frac{\epsilon}{2\ln(1/\delta)}\right)$-admissible. Later we will see a concrete instantiation of (5) for the median problem.

Algorithm 2: Propose-Test-Release (PTR) Framework

In the previous section we saw a "noise-addition"-based algorithm that exploited the smooth upper bound on the local sensitivity to ensure differentially privacy. In this section, instead of obtaining a smooth bound on the local sensitivity, we seek an answer to the following question: *Given a proposed upper bound Λ on the local sensitivity of $f(D)$, how many data points (k) in D need to be changed to increase the local sensitivity beyond Λ?* If k is sufficiently large, then the algorithm uses the proposed bound Λ in (2) instead of $\text{GS}(f)$; otherwise the algorithm outputs a $\bot$ and fails. Once formalized, this algorithmic paradigm can be shown to be (ϵ, δ)-differentially private. A major component in the design of algorithms using this paradigm is to come up with tight upper bounds on the local sensitivity. In Applications 1 and 2 we state two approaches for getting such bounds.

In the following, we formally introduce the propose-test-release framework. The version in Algorithm 1 is a variant of the ones appeared in [4] and [13].

Algorithm 1 Propose-test-release (PTR) framework

Input: Data set: $D \in \tau^n$, function $f : \tau^n \to \mathbb{R}$, proposed local sensitivity bound: Λ, privacy parameters: (ϵ, δ).

1: **Distance to instability:** dist $\leftarrow$ Minimum $k \in [n]$ for which $\left[\max_{D', D \triangle D' = 2k} \text{LS}(f, D')\right] > \Lambda$.

2: **Noisy distance to instability:** $\widetilde{\text{dist}} \leftarrow$ dist $+$ Lap$\left(\frac{1}{\epsilon}\right)$.

3: **Test and release:** If $\widetilde{\text{dist}} > \frac{1}{\epsilon} \log(1/\delta)$, then output Λ, else output $\bot$.

One can show that Algorithm 1 is (ϵ, δ)-differentially private. (See [13] for more details.) Additionally, by the tail properties of Laplace distribution, it is not hard to show that if dist $> \frac{2}{\epsilon} \log(1/\delta)$, then with probability at least $1 - \delta$, the algorithm outputs Λ. In (6) we fit Algorithm 1 with the Laplace mechanism from (2) to obtain a differentially private estimate of $f(D)$. By the composition property of differential privacy [4, 5], the algorithm

(PTR+Laplace mechanism) in (6) is $(2\epsilon, \delta)$-differentially private.

If $\text{PTR}(f, D, \Lambda, \epsilon, \delta) \neq \bot$, then **output**

$$f(D) + \text{Lap}\left(\frac{\Lambda}{\epsilon}\right), \text{ else } \textbf{fail}. \tag{6}$$

One can notice that if the proposed bound Λ is much smaller than $\text{GS}(f)$, then whenever the algorithm succeeds, it would add much lesser noise to $f(D)$ as compared to (2). In Application 1 we will do a comparison between the global sensitivity based, the smooth sensitivity based, and the PTR-based algorithm for the problem of computing the median.

Application 1: Computing the Median

With the smooth sensitivity-based and the PTR-based algorithmic frameworks from the previous section in hand, we revisit the problem of computing the median. Let the data set $D = \{d_1, \ldots, d_n\} \in [0, R]^n$ with n being odd and $R \in \mathbb{R}$ being the range. W.l.o.g., assume that the entries in D are sorted in *ascending order*.

Smooth sensitivity-based algorithm for median computation. In order to use (5), we need to be able to efficiently compute the smooth sensitivity (4) of the median function for D with a given smoothness parameter β. The following theorem implies an $O(n \log n)$ algorithm for computing the smooth sensitivity.

Theorem 1 ([11]) *Let $m = \frac{n+1}{2}$. The smooth sensitivity of the median function with the smoothness parameter β is given by the following.*

$$\text{SS}(\text{Median}, D, \beta) = \max_{k=0,\ldots,n}$$

$$\times \left(e^{-k\beta} \max_{t=0,\ldots,k+1} (d_{m+t} - d_{m+t-k-1}) \right).$$

It can be computed in time $O(n \log n)$.

Once the smooth sensitivity bound is obtained, one can use it in (5) to obtain a differentially private approximation to the median of D. If Laplace distribution ($\frac{1}{2}e^{-|z|}$) is used as the noise, then set the admissible parameters $\alpha = \epsilon/2$ and $\beta = \frac{\epsilon}{2\ln(1/\delta)}$. An immediate question that arises is how does the noise added by the smooth sensitivity-based algorithm compare to the global sensitivity-based algorithm in (1). First notice that since global sensitivity is always an upper bound on smooth sensitivity, the noise added via smooth sensitivity can never be more than that via global sensitivity. Next we present a setting of the data set D where in fact the smooth sensitivity-based algorithm adds much lesser noise (and hence more accurate).

Consider the data set D where each $d_i = \frac{R \cdot i}{n}$ for all $i \in [n]$. In this case the term $A_k = \max_{t=0,\ldots,k+1} (d_{m+t} - d_{m+t-k-1}) = \frac{(k+1)R}{n}$. Thus the term $e^{-k\beta} A_k$ is maximized when $k = \frac{1}{\beta} - 1$. Assuming $\beta < 1$, the smooth sensitivity $\text{SS}(\text{Median}, D, \beta)$ is bounded by $\frac{R}{\beta n}$. If we use Laplace distribution in (5) to ensure (ϵ, δ)-differential privacy, then the noise that gets added to $\text{Median}(D)$ is $O\left(\frac{R \log(1/\delta)}{\epsilon^2 n}\right)$. In comparison, the global sensitivity-based algorithm in (2) will add noise $O\left(\frac{R}{\epsilon}\right)$, which is much higher.

One might argue that the global sensitivity-based algorithm guarantees stronger differential privacy ($(\epsilon, 0)$ as opposed to (ϵ, δ)) and hence it is not a fair comparison. Even when one uses a more concentrated noise distribution like Gaussian distribution (which ensures (ϵ, δ)-differential privacy) instead of Laplace distribution in (2), the error still remains the same.

PTR-based algorithm for computing the median. We now instantiate the PTR-based algorithm for the same problem of computing the median. In order to do so, we first partition the real line $\mathbb{R}$ into bins of width $h = \frac{R}{n^{1/3}}$ (or any width $\frac{R}{n^{1/2+\gamma}}$ for any $\gamma > 0$). Call this set of bins $\mathcal{B}$. Additionally consider the set $\mathcal{B}_{(+h/2)}$, which is the set of bins shifted by $h/2$.

In Algorithm 1 we set the proposed sensitivity bound $\Lambda = h$. We compute the *distance to instability* in Line 1 of Algorithm 1 by the following technique. Let k_1 be the minimum number of entries in D that needs to be changed to move the median from its bin in the set $\mathcal{B}$, and let k_2 be the corresponding minimum number for the set $\mathcal{B}_{(+h/2)}$. The distance to instability is $\mathsf{dist} \leftarrow \max\{k_1, k_2\}$. Now the rest of the algorithm follows as described for the PTR framework. The two sets of shifted bins $\mathcal{B}$ and $\mathcal{B}_{(+h/2)}$ were needed because the median might fall at the partition boundary of the bins. Notice that computing dist takes $O(n)$ time.

In terms of the utility guarantee for this algorithm, we have the following:

Theorem 2 ([4]) *Let the data set D be drawn i.i.d. from some fixed distribution $\mathcal{P}$, where the cumulative distribution function of $\mathcal{P}$ is differentiable with positive derivative at the median. Assuming the privacy parameter $\delta = 1/\mathsf{poly}(n)$, we have the following utility guarantees for the PTR-based median computation.*

$$\Pr[\mathsf{PTR}(D) = \bot] = O(e^{-\epsilon \log n})$$

and $\mathsf{PTR}(D)$ *converges in probability to the median of* $\mathcal{P}$ *as* $n \to \infty$.

Application 2: Selection from a Discrete Set

In this section we will see another application of the PTR framework. Although the exposition is fairly abstract, we will see that this tool is useful for a variety of machine learning problems, where we assume "very little" about the underlying learning algorithm. Some of the examples being sparse estimation, parameter tuning, and non-convex learning.

Given a data set $D \in \tau^n$, and a choice function $f : \tau^n \to \{S_1, \ldots, S_k\}$, the objective is to compute a differentially private approximation to $f(D)$. Here $\{S_1, \ldots, S_k\}$ form a discrete set of choices. In order to design the private algorithm, we instantiate the PTR framework in Algorithm 1, with $\Lambda = 0$ and f being the choice

function. (Notice that $\Lambda = 0$ means that the output of the function does not change *at all* by changing any one entry in the data set.) If the output of the PTR framework is *not equal to* $\bot$, then output $f(D)$ exactly, and output $\bot$ otherwise. From the privacy property of the PTR framework, it follows that the above algorithm is (ϵ, δ)-differentially private. In terms of utility one can show the following.

Theorem 3 ([13]) *If the distance to instability of the choice function (Line 1 of Algorithm 1) is at least $2\log(1/\delta)$, then with probability at least $1 - \delta$, the above PTR instantiation outputs $f(D)$ exactly.*

At a high level, Theorem 3 says that if one needs to change sufficient number of entries $(2\log(1/\delta))$ in the data set D to change $f(D)$, then with high probability the PTR framework will output $f(D)$ exactly. One issue with the current instantiation of the PTR framework is that it is not clear a priori how to efficiently compute the distance to instability in Line 1 of Algorithm 1. In the following we circumvent this problem by instantiating the PTR framework with a proxy function $\hat{f}$ instead of f, for which the distance to instability is always efficiently computable. Moreover, if on a (sufficiently large) random subset D_{sub} of D, with probability at least $3/4$ one can guarantee $f(D_{\mathsf{sub}}) = f(D)$, then the PTR framework outputs $f(D)$ exactly with high probability.

Subsample and aggregate framework. The basic idea of subsample aggregate framework first appeared in [11] and the current version is from [13]. Here we use a variant of that framework for instantiating the proxy function $\hat{f}$ corresponding to f.

Let $q = \frac{\epsilon}{32\log(1/\delta)}$ and $m = \frac{\log(n/\delta)}{q^2}$. Sample data sets $D_1, \ldots, D_m$, where each D_i is generated from D by sampling each entry in D with probability q, and D_i's are i.i.d. Let S_{first} be the choice which appears maximum number of times in $\mathcal{F} = \{f(D_1), \ldots, f(D_m)\}$, and let S_{second} be the corresponding second. Let the proxy function

$\hat{f}(D)$ equal S_{first}. Let count(S) be the number of times the choice S appears in $\mathcal{F}$. One can show that with probability at least $1 - \delta$, the distance to instability of $\hat{f}(D)$ equals dist $\leftarrow \frac{\text{count}(S_{\text{first}})-\text{count}(S_{\text{second}})}{4mq} - 1$. From this, one can conclude that using $\hat{f}$ as the proxy for the choice function f in the PTR framework ensures $(\epsilon, 2\delta)$-differential privacy. In terms of utility, for this "proxy" instantiation of the PTR one can show the following. Notice that in both Theorems 3 and 4, there is no dependence on the number of possible choices (k).

Theorem 4 ([13]) *If for each D_i (defined above) $f(D_i) = f(D)$ with probability at least $3/4$, then with probability at least $1 - 2\delta$, the above instantiation of the PTR framework outputs $f(D)$ exactly.*

One of the classic setting where the above algorithms can be used in the case of model or feature selection in machine learning. A specific example is the LASSO estimator, where the PTR-based algorithm achieves the *optimal* sample complexity even under the constraint of differential privacy. (See [13] for details.) Another example is in finding the best regularization parameter for a given regression problem (Dwork and Thakurta, Differentially private parameter tuning using subsample and aggregate framework. Personal communication, 2014). Let $\Lambda = \{\lambda_1, \ldots, \lambda_k\}$ be a candidate set of regularization parameters, with each $\lambda_i \in \mathbb{R}$. The idea is to estimate the best regularization parameter from the set Λ for each of the sampled data sets $D_1, \ldots, D_m$, and use the estimation algorithm itself as the choice function f in the PTR framework.

Notice that we almost assumed nothing about the regularization parameter selection algorithm, apart from the fact that on random subsamples of the original data set D, the algorithm selects the same regularization parameter most of the times. The subsampling-based algorithm can also be used in the context of learning non-convex models while preserving differential privacy (Bilenko et al., Private and robust non-convex learning. Personal communication, 2014). For

the purpose of brevity, we defer the exposition for differentially private learning with non-convex models.

Reference Notes

The global sensitivity-based and the smooth sensitivity-based algorithmic framework are due to [6] and [11] respectively. The propose-test-release (PTR) framework is initially due to [4], but the exposition in this note is from [4, 13]. The smooth sensitivity-based private median algorithm is due to [11], and the one based on PTR framework is due to [4]. The algorithm for privately selecting from a discrete set is from [13].

Recommended Reading

1. Anthony M, Bartlett PL (2009) Neural network learning: theoretical foundations. Cambridge University Press, Cambridge
2. Dwork C (2006) Differential privacy. In: 33rd international colloquium on automata, languages and programming, Venice, LNCS pp 1–12
3. Dwork C (2011) A firm foundation for private data analysis. Commun ACM 54(1):86–95
4. Dwork C, Lei J (2009) Differential privacy and robust statistics. In: Symposium on theory of computing (STOC), Bethesda, pp 371–380
5. Dwork C, Kenthapadi K, Mcsherry F, Mironov I, Naor M (2006) Our data, ourselves: privacy via distributed noise generation. In: EUROCRYPT. Springer, pp 486–503
6. Dwork C, McSherry F, Nissim K, Smith A (2006) Calibrating noise to sensitivity in private data analysis. In: Theory of cryptography conference. Springer, New York, pp 265–284
7. Hardt M, Roth A (2013) Beyond worst-case analysis in private singular vector computation. In: STOC, Palo Alto
8. Kasiviswanathan SP, Smith A (2008) A note on differential privacy: defining resistance to arbitrary side information. CoRR arXiv:0803.39461 [cs.CR]
9. Kifer D, Machanavajjhala A (2012) A rigorous and customizable framework for privacy. In: Principles of database systems (PODS 2012), Scottsdale
10. Kifer D, Smith A, Thakurta A (2012) Private convex empirical risk minimization and high-dimensional regression. In: Conference on learning theory, Edinburgh, pp 25.1–25.40

11. Nissim K, Raskhodnikova S, Smith A (2007) Smooth sensitivity and sampling in private data analysis. In: Symposium on theory of computing (STOC), San Diego, ACM, pp 75–84. Full paper: http://www.cse.psu.edu/~asmith/pubs/NRS07

12. Smith A (2011) Privacy-preserving statistical estimation with optimal convergence rates. In: Proceedings of the forty-third annual ACM symposium on theory of computing, San Jose, pp 813–822

13. Smith AD, Thakurta A (2013) Differentially private model selection via stability arguments and the robustness of the lasso. J Mach Learn Res Proc Track 30:819–850

BG Distributed Simulation Algorithm

Matthieu Roy
Laboratory of Analysis and Architecture of Systems (LAAS), Centre National de la Recherche Scientifique (CNRS), Université Toulouse, Toulouse, France

Keywords

Computability; Distributed tasks; Read/write shared memory; Reduction

Years and Authors of Summarized Original Work

1993; Borowsky, Gafni
2001; Borowsky, Gafni, Lynch, Rajsbaum

Problem Definition

How to effectively translate an algorithm from a distributed system model to another one?

Distributed systems come in diverse settings that are modeled by different assumptions (1) on the way processes communicate, e.g., using shared memory or messages, (2) on the fault model, (3) on synchrony assumptions, etc. Each of these parameters has a dramatic impact on the computing power of the model, and in practice, an algorithm or an impossibility result is usually tailored to a particular model and cannot be directly reused in another model.

This wide variety of models has given rise to many different impossibility theorems and numerous algorithms for many of the possible combinations of parameters that characterize them. Thus, a crucial question is the following: are there bridges between some models, i.e., is it possible to transfer an impossibility result or an algorithm from one model to another?

The Borowsky-Gafni simulation algorithm, or BG simulation, is one of the first steps toward direct translations of algorithms or impossibility results from one model to another. The BG simulation considers distributed systems made of asynchronous processes that communicate using a shared memory array. In a nutshell, this simulation allows a set of $t + 1$ asynchronous sequential processes, where up to t of them can stop during their execution, to simulate any set of $n \geq t + 1$ processes executing an algorithm that is designed to tolerate to up to t fail-stop failures.

The BG simulation has been used to prove solvability and unsolvability results for crash-prone asynchronous shared memory systems, paving the way for a more generic formal theory of reduction between problems in different models of distributed computing.

The BG-simulation algorithm is named after its authors, Elizabeth Borowsky and Eli Gafni, that introduced it as a side tool [3] in order to generalize the impossibility result of solving a weakened version of consensus, namely, k-set agreement [6]. It has been later on formalized and proven correct [4, 18] using the I/O automata formalism [17].

System Model

Processes

The simulation considers a system made of up to n asynchronous sequential processes that execute a distributed algorithm to solve a given colorless decision task, as defined below.

Failure Model

Processes may fail by stopping (crash failure). The simulation assumes that up to t processes can stop during the execution; $t < n$ is known before the execution, but the identity of processes that may crash is unknown to the simulation. This model of computation is referred to as the t-resilient model. A corner case of this model is the wait-free model where $t + 1$ processes execute concurrently and at most t of them may crash.

Communication

Processes communicate and coordinate using a reliable shared memory composed of n multiple-reader single-writer registers. Each process has the exclusive write access to one of these n registers, and processes can read all entries by invoking a *snapshot* operation, with the semantics that write and snapshot operations appear as if they are executed atomically. While using the snapshot abstraction eases the presentation of the algorithm, it has no impact on the power of the underlying computing model, since the snapshot/write model can be implemented wait-free using read/write registers [1].

Tasks

A colorless task is a distributed coordination problem in which every process p_i starts with a value, communicates with other processes, and has to decide eventually on a output value. Colorless tasks, or convergence tasks [12], are a restricted version of tasks in which a deciding process may adopt the decision value of any process, i.e., two participating processes may decide the same value. For more formal definitions of tasks using tools from algebraic topology, the reader should refer to [11].

Simulation

The simulation proceeds by executing concurrently, using $t + 1$ simulators $s_1, \ldots, s_{t+1}$, the code of $n > t$ processes that collaboratively solve a distributed colorless task. Hence, each simulator s_i is given the code of all simulated processes and handles the execution of n threads.

Key Results

Simulation of Memory

Each one of the $t + 1$ simulators s_i executes the sequential code of the n simulated processes p_j in parallel. By assumption, every simulated code is a sequence of instructions that are either (1) local processing, (2) a write operation into memory, or (3) a snapshot of the shared memory.

Every simulator s_i maintains its local view of the simulated memory for all simulated threads. These local views are synchronized between simulators by writing and reading (using snapshots) in a shared memory matrix array MEM that has one column per simulated thread and one row per *snapshot* instance.

To ensure global consistency between simulators that simulate concurrently all threads, operations on the memory must be coordinated between different simulators. This is achieved by ensuring that, for a given simulated thread, the sequence of snapshots of the memory as computed by all simulators is identical. As consensus cannot be implemented wait-free, the simulation coordinates snapshots using of a weaker form of agreement, the *safe agreement*.

The Safe-Agreement Object

Safe agreement is the most important building block of the simulation. First introduced as the *non-blocking busy-wait* agreement protocol [3], it has been further refined as safe agreement, with several blocking or non-blocking/wait-free implementations [2, 11, 14].

This weak form of agreement provides two methods to processes: *propose(v)* and *resolve()*. A participating process that proposes a value v first calls *propose(v)* once and is then allowed to make calls to *resolve()* that may return $\perp$ if safe agreement is not resolved yet or a value. In this later case, safe agreement is said to be resolved and the value returned is the decided value by the process. Formally, safe agreement is defined by three properties:

Termination: If no process crashes during the execution of *propose()*, then all processes decide, i.e., eventually all calls to *resolve()* return a non-$\perp$ value,

Validity: All processes that decide must decide a proposed value,
Agreement: All processes that decide must decide the same value.

The specification is almost identical to the one of consensus, apart from the weakened termination property. Safe agreement is wait-free solvable and thus solvable in t-resilient systems.

The crucial point of the BG simulation lies in the termination property of safe agreement: if a safe-agreement protocol cannot be resolved, i.e., if no process decides, then at least one process crashed during the call to *propose*(). Thus, a given safe-agreement instance can "capture" a calling process that crashed during the *propose* invocation.

Overview of the Simulation

The current state of the simulation and its history is thus represented by two twin data structures: (1) the shared memory matrix *MEM* that contains the consecutive memory status of all simulated threads, as seen by simulators, and (2) a matrix of safe-agreement objects SafeAgreement[$0 \ldots$][$1 \ldots n$] with n columns, each column representing the execution advancement of one of the simulated processes, as shown in Fig. 1.

In this view, the entry at row ℓ and column i corresponds to the state of the ℓth snapshot for simulated process p_j. Hence, the "program counter" of a simulated thread p_i is the greatest row of column i that is either unresolved or resolved. In this example, simulations of threads p_2, p_4, and p_6 are stopped with unresolved safe agreement that are due to (at least) one simulator stuck in the associated *propose*() methods. The program counter of all other threads is 9.

Each simulator s_i is given the code of the n threads it has to simulate, as well as an input value of one of the threads. Conceptually, the algorithm run by simulator s_i is as follows:

In the simulation, each *snapshot* invocation is mediated through a SafeAgreement object, lines 6 and 14. The only reason that could block the simulation of a given thread p_i is when the call to *resolve*, line 6, always returns $\bot$. By definition of the safe-agreement object, this situation can happen only when a simulator crashed during the call to *propose*() on the same safe-agreement instance: *the crash of a simulator can block the simulation of at most one simulated thread.*

Applications

The BG-simulation algorithm has been primarily used to reduce t-resilient solvability to wait-free

---

Algorithm 1 BG-simulation: code for a simulator s_j starting with input v

```
 1: procedure BG-SIMULATION(v)
 2:     ∀i = 1 . . . n, SafeAgreement[0][i].propose(v)
                                                              ▷ Initialization
 3:     loop
 4:         for i ← 1, n do                                    ▷ Simulate threads in round-robin
 5:             ℓ ← current program counter of pᵢ
 6:             snap ← SafeAgreement[ℓ][i].resolve()
 7:             if snap ≠ ⊥ then                               ▷ safe agreement is resolved
 8:                 perform local computation using snap, write operations in local memory
 9:                 execute write on behalf of pᵢ in MEM[ℓ][i]
10:                 if thread pᵢ is terminated then
11:                     return value and stop its simulation
12:                 else if at least (n − t) threads have program counter ≥ ℓ then
13:                     snap ← snapshot(MEM[ℓ])
14:                     SafeAgreement[ℓ + 1][i].propose(snap)
15:                 end if
16:             end if
17:         end for
18:     end loop
19: end procedure
```

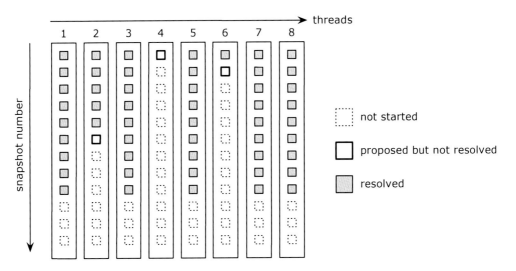

BG Distributed Simulation Algorithm, Fig. 1 Conceptual view of advancement for snapshots of all simulated process with $n = 8$ and $t = 3$

solvability for colorless tasks, that is, tasks that are agnostic on process identities. The initial application has been made to the k-set agreement problem, in which all processes have to agree on a final set of values of size at most k. If k-set agreement was solvable in a k-resilient system of $n > k + 1$ processes, then the BG simulation of this algorithm with $k + 1$ simulators would produce a wait-free solution to k-set agreement. Since k-set agreement is not wait-free solvable for $k + 1$ processes [13, 19], it follows a contradiction.

The BG simulation presented here only applies to *colorless tasks*. Gafni [8] extended further to more general classes of tasks and provided the general characterization of t-resilient solvable tasks, similarly to the Herlihy-Shavit conditions for wait-free computability [13]. This extension has been also studied in [14, 16].

In order to study the relationship between wait-freedom and t-resilience, [5] uses objects of type $\mathcal{S}$ in addition to read/write registers and shows that for any $t < k$, t-resilient k-process consensus can be implemented with objects of type $\mathcal{S}$ and registers if and only if wait-free $(t + 1)$-process consensus can be implemented with objects of type $\mathcal{S}$ and registers.

Imbs and Raynal [15] consider models equipped with registers and consensus objects and extend the results provided by BG simulation, showing equivalences between models based on the ratio between the maximum number of failures and the consensus number of consensus objects.

Chaudhuri and Reiners [7] use BG simulation to provide a characterization of the set consensus partial order, a refinement of Herlihy's consensus-based wait-free hierarchy [10]; a formal definition of *set consensus number* and a study of associated respective computing power have been later provided in [9].

Recommended Reading

1. Afek Y, Attiya H, Dolev D, Gafni E, Merritt M, Shavit N (1993) Atomic snapshots of shared memory. J ACM 40(4):873–890
2. Attiya H (2006) Adapting to point contention with long-lived safe agreement. In: Proceedings of the 13th international conference on structural information and communication complexity, SIROCCO'06, Chester. Springer, Berlin/Heidelberg, pp 10–23
3. Borowsky E, Gafni E (1993) Generalized FLP impossibility result for t-resilient asynchronous computations. In: STOC '93: proceedings of the twenty-fifth annual ACM symposium on theory

of computing, San Diego. ACM, New York, pp 91–100

4. Borowsky E, Gafni E, Lynch N, Rajsbaum S (2001) The BG distributed simulation algorithm. Distrib Comput 14(3):127–146

5. Chandra T, Hadzilacos V, Jayanti P, Toueg S (1994) Wait-freedom vs. *t*-resiliency and the robustness of wait-free hierarchies (extended abstract). In: PODC '94: proceedings of the thirteenth annual ACM symposium on principles of distributed computing, Los Angeles. ACM, New York, pp 334–343

6. Chaudhuri S (1993) More choices allow more faults: set consensus problems in totally asynchronous systems. Inf Comput 105(1):132–158

7. Chaudhuri S, Reiners P (1996) Understanding the set consensus partial order using the Borowsky-Gafni simulation (extended abstract). In: Proceedings of the 10th international workshop on distributed algorithms, Bologna. Springer, London, pp 362–379

8. Gafni E (2009) The extended BG-simulation and the characterization of t-resiliency. In: Proceedings of the 41st annual ACM symposium on theory of computing, STOC '09, Bethesda. ACM, New York, pp 85–92

9. Gafni E, Kuznetsov P (2011) On set consensus numbers. Distrib Comput 23(3–4):149–163

10. Herlihy M (1991) Wait-free synchronization. ACM Trans Program Lang Syst 13(1):124–149

11. Herlihy M, Kozlov D, Rajsbaum S (2013) Distributed Computing Through Combinatorial Topology. Morgan Kaufmann, Amsterdam

12. Herlihy M, Rajsbaum S (1997) The decidability of distributed decision tasks (extended abstract). In: Proceedings of the twenty-ninth annual ACM symposium on theory of computing, STOC '97, El Paso. ACM, New York, pp 589–598

13. Herlihy M, Shavit N (1999) The topological structure of asynchronous computability. J ACM 46(6):858–923

14. Imbs D, Raynal M (2009) Visiting gafni's reduction land: from the BG simulation to the extended BG simulation. In: SSS, pp 369–383

15. Imbs D, Raynal M (2010) The multiplicative power of consensus numbers. In: Proceedings of the 29th ACM SIGACT-SIGOPS symposium on principles of distributed computing, PODC '10, Zurich. ACM, New York, pp 26–35

16. Kuznetsov P (2013) Universal model simulation: BG and extended BG as examples. In: SSS, pp 17–31

17. Lynch NA (1996) Distributed algorithms. Morgan Kaufmann Publishers Inc., San Francisco

18. Lynch N, Rajsbaum S (1996) On the Borowsky-Gafni simulation algorithm. In: Proceedings of the fourth Israel symposium on theory of computing and systems, ISTCS '96, Jerusalem. IEEE Computer Society, pp 4–15

19. Saks M, Zaharoglou F (2000) Wait-free k-set agreement is impossible: the topology of public knowledge. SIAM J Comput 29(5):1449–1483

Bidimensionality

Fedor V. Fomin[1], Erik D. Demaine[2], Mohammad Taghi Hajiaghayi[3], and Dimitrios Thilikos[4,5]

[1] Department of Informatics, University of Bergen, Bergen, Norway
[2] MIT Computer Science and Artificial Intelligence Laboratory, Cambridge, MA, USA
[3] Department of Computer Science, University of Maryland, College Park, MD, USA
[4] AlGCo Project-Team, CNRS, LIRMM, France
[5] Department of Mathematics, National and Kapodistrian University of Athens, Athens, Greece

Keywords

EPTAS; Graph minors; Kernelization; Parametrized algorithms; Subexponential algorithms; Treewidth

Years and Authors of Summarized Original Work

2004; Demaine, Fomin, Hajiaghayi; Thilikos
2004; Demaine, Hajiaghayi
2005; Demaine, Fomin, Hajiaghayi, Thilikos
2005; Demaine, Hajiaghayi
2006; Demaine, Hajiaghayi, Thilikos
2008; Demaine, Hajiaghayi
2008; Dorn, Fomin, Thilikos
2009; Fomin, Golovach, Thilikos
2010; Demaine
2010; Fomin, Lokshtanov, Saurabh, Thilikos
2011; Fomin, Lokshtanov, Raman, Saurabh
2011; Fomin, Golovach, Thilikos
2012; Fomin, Lokshtanov, Saurabh
2013; Giannopoulou, Thilikos

2013; Demaine, Fomin, Hajiaghayi, Thilikos
2014; Grigoriev, Koutsonas, Thilikos

Problem Definition

The theory of bidimensionality provides general techniques for designing efficient fixed-parameter algorithms and approximation algorithms for a broad range of NP-hard graph problems in a broad range of graphs. This theory applies to graph problems that are "bidimensional" in the sense that (1) the solution value for the $k \times k$ grid graph and similar graphs grows with k, typically as $\Omega(k^2)$, and (2) the solution value goes down when contracting edges and optionally when deleting edges in the graph. Many problems are bidimensional; a few classic examples are vertex cover, dominating set, and feedback vertex set.

Graph Classes

Results about bidimensional problems have been developed for increasingly general families of graphs, all generalizing planar graphs.

The first two classes of graphs relate to embeddings on surfaces. A graph is *planar* if it can be drawn in the plane (or the sphere) without crossings. A graph has *(Euler) genus* at most g if it can be drawn in a surface of Euler characteristic g. A class of graphs has *bounded genus* if every graph in the class has genus at most g for a fixed g.

The next three classes of graphs relate to excluding minors. Given an edge $e = \{v, w\}$ in a graph G, the *contraction* of e in G is the result of identifying vertices v and w in G and removing all loops and duplicate edges. A graph H obtained by a sequence of such edge contractions starting from G is said to be a *contraction* of G. A graph H is a *minor* of G if H is a subgraph of some contraction of G. A graph class C is *minor closed* if any minor of any graph in C is also a member of C. A minor-closed graph class C is *H-minor-free* if $H \notin C$. More generally, the term "*H*-minor-free" refers to any minor-closed graph class that excludes some fixed graph H. A *single-crossing graph* is a minor of a graph that can be drawn in the plane with at most one pair

of edges crossing. A minor-closed graph class is *single-crossing-minor-free* if it excludes a fixed single-crossing graph. An *apex graph* is a graph in which the removal of some vertex leaves a planar graph. A graph class is *apex-minor-free* if it excludes some fixed apex graph.

Bidimensional Parameters

Although implicitly hinted at in [2, 5, 10, 11], the first use of the term "bidimensional" was in [3].

First, "parameters" are an alternative view on optimization problems. A *parameter P* is a function mapping graphs to nonnegative integers. The *decision problem associated with P* asks, for a given graph G and nonnegative integer k, whether $P(G) \leq k$. Many optimization problems can be phrased as such a decision problem about a graph parameter P.

Now, a parameter is $g(r)$-*bidimensional* (or just *bidimensional*) if it is at least $g(r)$ in an $r \times r$ "grid-like graph" and if the parameter does not increase when taking either minors ($g(r)$-*minor-bidimensional*) or contractions ($g(r)$-*contraction-bidimensional*). The exact definition of "grid-like graph" depends on the class of graphs allowed and whether one considers minor or contraction bidimensionality. For minor bidimensionality and for any H-minor-free graph class, the notion of a "grid-like graph" is defined to be the $r \times r$grid, i.e., the planar graph with r^2 vertices arranged on a square grid and with edges connecting horizontally and vertically adjacent vertices. For contraction bidimensionality, the notion of a "grid-like graph" is as follows:

1. For planar graphs and single-crossing-minor-free graphs, a "grid-like graph" is an $r \times r$ grid partially triangulated by additional edges that preserve planarity.
2. For bounded-genus graphs, a "grid-like graph" is such a partially triangulated $r \times r$ grid with up to genus (G) additional edges ("handles").
3. For apex-minor-free graphs, a "grid-like graph" is an $r \times r$ grid augmented with additional edges such that each vertex is incident to $O(1)$ edges to nonboundary vertices of the grid. (Here $O(1)$ depends on the excluded apex graph.)

Contraction bidimensionality is so far undefined for H-minor-free graphs (or general graphs).

Examples of bidimensional parameters include the number of vertices, the diameter, and the size of various structures such as feedback vertex set, vertex cover, minimum maximal matching, face cover, a series of vertex-removal parameters, dominating set, edge dominating set, R-dominating set, connected dominating set, connected edge dominating set, connected R-dominating set, unweighted TSP tour (a walk in the graph visiting all vertices), and chordal completion (fill-in). For example, feedback vertex set is $\Omega(r^2)$-minor-bidimensional (and thus also contraction-bidimensional) because (1) deleting or contracting an edge preserves existing feedback vertex sets and (2) any vertex in the feedback vertex set destroys at most four squares in the $r \times r$ grid, and there are $(r-1)^2$ squares, so any feedback vertex set must have $\Omega(r^2)$ vertices. See [1, 3] for arguments of either contraction or minor bidimensionality for the other parameters.

Key Results

Bidimensionality builds on the seminal graph minor theory of Robertson and Seymour, by extending some mathematical results and building new algorithmic tools. The foundation for several results in bidimensionality is the following two combinatorial results. The first relates any bidimensional parameter to treewidth, while the second relates treewidth to grid minors.

Theorem 1 ([1, 8]) *If the parameter P is $g(r)$-bidimensional, then for every graph G in the family associated with the parameter P, $\mathrm{tw}(G) = O(g^{-1}(P(G)))$. In particular, if $g(r) = \Theta(r^2)$, then the bound becomes $\mathrm{tw}(G) = O(\sqrt{P(G)})$.*

Theorem 2 ([8]) *For any fixed graph H, every H-minor-free graph of treewidth w has an $\Omega(w) \times \Omega(w)$ grid as a minor.*

The two major algorithmic results in bidimensionality are general subexponential fixed-parameter algorithm and general polynomial-time approximation scheme (PTASs).

Theorem 3 ([1, 8]) *Consider a $g(r)$-bidimensional parameter P that can be computed on a graph G in $h(w)n^{O(1)}$ time given a tree decomposition of G of width at most w. Then there is an algorithm computing P on any graph G in P's corresponding graph class, with running time $\left[h(O(g^{-1}(k))) + 2^{O(g-1(k))}\right]n^{O(1)}$. In particular, if $g(r) = \Theta(r^2)$ and $h(w) = 2^{O(w2)}$, then this running time is subexponential in k.*

Theorem 4 ([7]) *Consider a bidimensional problem satisfying the "separation property" defined in [4, 7].*

Suppose that the problem can be solved on a graph G with n vertices in $f(n, \mathrm{tw}(G))$ time. Suppose also that the problem can be approximated within a factor of α in $g(n)$ time. For contraction-bidimensional problems, suppose further that both of these algorithms also apply to the "generalized form" of the problem defined in [4, 7]. Then there is a $(1+^2)$-approximation algorithm whose running time is $O(nf(n, O(O^2/^2)) + n^3 g(n))$ for the corresponding graph class of the bidimensional problem.

Applications

The theorems above have many combinatorial and algorithmic applications.

Applying the parameter-treewidth bound of Theorem 1 to the parameter of the number of vertices in the graph proves that every H-minor-free graph on n vertices has treewidth $O(\sqrt{n})$, thus (re)proving the separator theorem for H-minor-free graphs. Applying the parameter-treewidth bound of Theorem 1 to the parameter of the diameter of the graph proves a stronger form of Eppstein's diameter-treewidth relation for apex-minor-free graphs. (Further work shows how to further strengthen the diameter-treewidth relation to linear [6].) The treewidth-grid relation of Theorem 2 can be used to bound the gap between half-integral multicommodity flow and fractional multicommodity flow in H-minor-free graphs. It also yields an $O(1)$-approximation for treewidth in H-minor-free graphs. The subexponential fixed-parameter algorithms of Theorem 3 subsume or strengthen all previous such results. These results can also be generalized to obtain fixed-parameter algorithms in arbitrary graphs. The PTASs of Theorem 4 in particular establish the first PTASs for connected dominating set and feedback vertex set even for planar graphs. For details of all of these results, see [4].

Open Problems

Several combinatorial and algorithmic open problems remain in the theory of bidimensionality and related concepts.

Can the grid-minor theorem for H-minor-free graphs, Theorem 2, be generalized to arbitrary graphs with a polynomial relation between treewidth and the largest grid minor? (The best relation so far is exponential.) Such polynomial generalizations have been obtained for the cases of "map graphs" and "power graphs" [9]. Good grid-treewidth bounds have applications to minor-bidimensional problems.

Can the algorithmic results (Theorems 3 and 4) be generalized to solve contraction-bidimensional problems beyond apex-minor-free graphs? It is known that the basis for these results, Theorem 1, does not generalize [1]. Nonetheless, Theorem 3 has been generalized for one specific contraction-bidimensional problem, dominating set [3].

Can the polynomial-time approximation schemes of Theorem 4 be generalized to more general algorithmic problems that do not correspond directly to bidimensional parameters? One general family of such problems arises when adding weights to vertices and/or edges, and the goal is, e.g., to find the minimum-weight dominating set. Another family of such problems arises when placing constraints (e.g., on coverage or domination) only on subsets of vertices and/or edges. Examples of such problems include Steiner tree and subset feedback vertex set.

For additional open problems and details about the problems above, see [4].

Cross-References

▶ Approximation Schemes for Planar Graph Problems
▶ Branchwidth of Graphs
▶ Treewidth of Graphs

Recommended Reading

1. Demaine ED, Fomin FV, Hajiaghayi M, Thilikos DM (2004) Bidimensional parameters and local treewidth. SIAM J Discret Math 18(3):501–511
2. Demaine ED, Fomin FV, Hajiaghayi M, Thilikos DM (2005) Fixed-parameter algorithms for (k, r)-center in planar graphs and map graphs. ACM Trans Algorithms 1(1):33–47
3. Demaine ED, Fomin FV, Hajiaghayi M, Thilikos DM (2005) Subexponential parameterized algorithms on graphs of bounded genus and H-minor-free graphs. J ACM 52(6):866–893
4. Demaine ED, Hajiaghayi M (to appear) The bidimensionality theory and its algorithmic applications. Comput J
5. Demaine ED, Hajiaghayi M (2004) Diameter and treewidth in minor-closed graph families, revisited. Algorithmica 40(3):211–215
6. Demaine ED, Hajiaghayi M (2004) Equivalence of local treewidth and linear local treewidth and its algorithmic applications. In: Proceedings of the 15th ACM-SIAM symposium on discrete algorithms (SODA'04), New Orleans, Jan 2004, pp 833–842
7. Demaine ED, Hajiaghayi M (2005) Bidimensionality: new connections between FPT algorithms and PTASs. In: Proceedings of the 16th annual ACM-SIAM symposium on discrete algorithms (SODA 2005), Vancouver, Jan 2005, pp 590–601
8. Demaine ED, Hajiaghayi M (2005) Graphs excluding a fixed minor have grids as large as treewidth, with combinatorial and algorithmic applications through bidimensionality. In: Proceedings of the 16th annual ACM-SIAM symposium on discrete algorithms (SODA 2005), Vancouver, Jan 2005, pp 682–689
9. Demaine ED, Hajiaghayi M, Kawarabayashi K (2006) Algorithmic graph minor theory: improved grid minor bounds and Wagner's contraction. In: Proceedings of the 17th annual international symposium on algorithms and computation, Calcutta, Dec 2006. Lecture notes in computer science, vol 4288, pp 3–15
10. Demaine ED, Hajiaghayi M, Nishimura N, Ragde P, Thilikos DM (2004) Approximation algorithms for classes of graphs excluding single-crossing graphs as minors. J Comput Syst Sci 69(2):166–195

11. Demaine ED, Hajiaghayi M, Thilikos DM (2005)
    Exponential speedup of fixed-parameter algorithms
    for classes of graphs excluding single-crossing graphs
    as minors. Algorithmica 41(4):245–267
12. Demaine ED, Hajiaghayi M, Thilikos DM (2006)
    The bidimensional theory of bounded-genus graphs.
    SIAM J Discret Math 20(2):357–371

Bin Packing

David S. Johnson
Department of Computer Science, Columbia
University, New York, NJ, USA
AT&T Laboratories, Algorithms and
Optimization Research Department, Florham
Park, NJ, USA

Keywords

Cutting stock problem

Years and Authors of Summarized Original Work

1997; Coffman, Garay, Johnson

Problem Definition

In the one-dimensional bin packing problem, one is given a list $L = (a_1, a_2, \ldots, a_n)$ of items, each item a_i having a size $s(a_i) \in (0, 1]$. The goal is to pack the items into a minimum number of unit-capacity bins, that is, to partition the items into a minimum number of sets, each having total size of at most 1. This problem is NP-hard, and so much of the research on it has concerned the design and analysis of approximation algorithms, which will be the subject of this article.

Although bin packing has many applications, it is perhaps most important for the role it has played as a proving ground for new algorithmic and analytical techniques. Some of the first worst- and average-case results for approximation algorithms were proved in this domain, as well as the first lower bounds on the competitive ratios of online algorithms. Readers interested in a more detailed coverage than is possible here are directed to two relatively recent surveys [4, 11].

Key Results

Worst-Case Behavior

Asymptotic Worst-Case Ratios
For most minimization problems, the standard worst-case metric for an approximation algorithm A is the maximum, over all instances I, of the ratio $A(I)/OPT(I)$, where $A(I)$ is the value of the solution generated by A and $OPT(I)$ is the optimal solution value. In the case of bin packing, however, there are limitations to this "absolute worst-case ratio" metric. Here it is already NP-hard to determine whether $OPT(I) = 2$, and hence no polynomial-time approximation algorithm can have an absolute worst-case ratio better than 1.5 unless P = NP. To better understand the behavior of bin packing algorithms in the typical situation where the given list L requires a large number of bins, researchers thus use a more refined metric for bin packing, the *asymptotic worst-case ratio* R_A^∞. This is defined in two steps as follows.

$$R_A^n = \max \{A(L)/OPT(L):$$
$$L \text{ is a list with } OPT(L) = n\}$$
$$R_A^\infty = \limsup_{n \to \infty} R_A^n$$

The first algorithm whose behavior was analyzed in these terms was *First Fit* (FF). This algorithm envisions an infinite sequence of empty bins $B_1, B_2, \ldots$ and, starting with the first item in the input list L, places each item in turn into the first bin which still has room for it. In a technical report from 1971 which was one of the very first papers in which worst-case performance ratios were studied, Ullman [22] proved the following.

Theorem 1 ([22]) $R_{FF}^\infty = 17/10$.

In addition to FF, five other simple heuristics received early study and have served as the inspiration for later research. *Best Fit* (BF) is the variant of FF in which each item is placed in the bin into which it will fit with the least space left over, with ties broken in favor of the earliest such bin. Both FF and BF can be implemented to run in time $O(n \log n)$ [12]. *Next Fit* (NF) is a still simpler and linear-time algorithm in which the first item is placed in the first bin, and thereafter each item is placed in the last nonempty bin if it will fit, otherwise a new bin is started. *First Fit Decreasing* (FFD) and *Best Fit Decreasing* (BFD) are the variants of those algorithms in which the input list is first sorted into nonincreasing order by size and then the corresponding packing rule is applied. The results for these algorithms are as follows.

Theorem 2 ([12]) $R_{NF}^{\infty} = 2$.

Theorem 3 ([13]) $R_{BF}^{\infty} = 17/10$.

Theorem 4 ([12, 13]) $R_{FFD}^{\infty} = R_{BFD}^{\infty} = 11/9 = 1.222 \ldots$

The above mentioned algorithms are relatively simple and intuitive. If one is willing to consider more complicated algorithms, one can do substantially better. The current best polynomial-time bin packing algorithm is very good indeed. This is the 1982 algorithm of Karmarkar and Karp [15], denoted here as "KK." It exploits the ellipsoid algorithm, approximation algorithms for the knapsack problem, and a clever rounding scheme to obtain the following guarantees.

Theorem 5 ([15]) $R_{KK}^{\infty} = 1$ *and there is a constant c such that for all lists L,*

$$KK(L) \leq OPT(L) + c \log^2(OPT(L)).$$

Unfortunately, the running time for KK appears to be worse than $O(n^8)$, and BFD and FFD remain much more practical alternatives.

Online Algorithms
Three of the abovementioned algorithms (FF, BF, and NF) are *online* algorithms, in that they pack items in the order given, without reference to the

sizes or number of later items. As was subsequently observed in many contexts, the online restriction can seriously limit the ability of an algorithm to produce good solutions. Perhaps the first limitation of this type to be proved was Yao's theorem [24] that no online algorithm A for bin packing can have $R_A^{\infty} < 1.5$. The bound has since been improved to the following.

Theorem 6 ([23]) *If A is an online algorithm for bin packing, then* $R_A^{\infty} \geq 1.540 \ldots$

Here the exact value of the lower bound is the solution to a complicated linear program.

Yao's paper also presented an online algorithm *Revised First Fit* (RFF) that had $R_{RFF}^{\infty} = 5/3 = 1.666 \ldots$ and hence got closer to this lower bound than FF and BF. This algorithm worked by dividing the items into four classes based on size and index, and then using different packing rules (and packings) for each class. Subsequent algorithms improved on this by going to more and more classes. The current champion is the online *Harmonic++* algorithm (H++) of [21]:

Theorem 7 ([21]) $R_{H++}^{\infty} \leq 1.58889$.

Bounded-Space Algorithms
The NF algorithm, in addition to being online, has another property worth noting: no more than a constant number of partially filled bins remain open to receive additional items at any given time. In the case of NF, the constant is 1 – only the last partially filled bin can receive additional items. Bounding the number of open bins may be necessary in some applications, such as packing trucks on loading docks. The bounded-space constraint imposes additional limits on algorithmic behavior however.

Theorem 8 ([17]) *For any online bounded-space algorithm A,* $R_A^{\infty} \geq 1.691 \ldots$.

The constant $1.691 \ldots$ arises in many other bin packing contexts. It is commonly denoted by h_{∞} and equals $\sum_{i=1}^{\infty}(1/t_i)$, where $t_1 = 1$ and, for $i > 1, t_i = t_{i-1}(t_{i-1} + 1)$.

The lower bound in Theorem 8 is tight, owing to the existence of the *Harmonic$_k$* algorithms (H_k) of [17]. H_k is a class-based algorithm in which the items are divided into classes C_h, $1 \leq h \leq k$, with C_k consisting of all items with size $1/k$ or smaller, and C_h, $1 \leq h < k$, consisting of all a_i with $1/(h+1) < s(a_i) \leq 1/h$. The items in each class are then packed by NF into a separate packing devoted just to that class. Thus, at most k bins are open at any time. In [17] it was shown that $\lim_{k \to \infty} R^\infty_{H_k} = h_\infty = 1.691\ldots$. This is even better than the asymptotic worst-case ratio of 1.7 for the unbounded-space algorithms FF and BF, although it should be noted that the bounded-space variant of BF in which all but the two most-full bins are closed also has $R^\infty_A = 1.7$ [8].

Average-Case Behavior

Continuous Distributions

Bin packing also served as an early test bed for studying the average-case behavior of approximation algorithms. Suppose F is a distribution on $(0, 1]$ and L_n is a list of n items with item sizes chosen independently according to F. For any list L, let $s(L)$ denote the lower bound on $OPT(L)$ obtained by summing the sizes of all the items in L. Then define

$$ER^n_A(F) = E\left[A(L_n)/OPT(L_n)\right],$$
$$ER^\infty_A(F) = \limsup_{n \to \infty} ER^n_A$$
$$EW^n_A(F) = E\left[A(L_n) - s(L_n)\right]$$

The last definition is included since $ER^\infty_A(F) = 1$ occurs frequently enough that finer distinctions are meaningful. For example, in the early 1980s, it was observed that for the distribution $F = U(0, 1]$ in which item sizes are uniformly distributed on the interval $(0, 1]$, $ER^\infty_{FFD}(F) = ER^\infty_{BFD}(F) = 1$, as a consequence of the following more-detailed results.

Theorem 9 ([16, 20]) *For $A \in \{FFD, BFD, OPT\}$,*
$EW^n_A(U(0, 1]) = \Theta(\sqrt{n})$.

Somewhat surprisingly, it was later discovered that the online FF and BF algorithms also had sublinear expected waste, and hence $ER^\infty_A(U(0, 1]) = 1$.

Theorem 10 ([5, 19])

$$EW^n_{FF}(U(0, 1]) = \Theta(n^{2/3})$$
$$EW^n_{BF}(U(0, 1]) = \Theta(n^{1/2} \log^{3/4} n)$$

This good behavior does not, however, extend to the bounded-space algorithms NF and H_k:

Theorem 11 ([6, 18])

$$ER^\infty_{NF}(U(0, 1]) = 4/3 = 1.333\ldots$$
$$\lim_{k \to \infty} ER_{H_k}(U(0, 1]) = \pi^2/3 - 2 = 1.2899\ldots$$

All the above results except the last two exploit the fact that the distribution $U(0, 1]$ is symmetric about $1/2$, and hence an optimal packing consists primarily of two-item bins, with items of size $s > 1/2$ matched with smaller items of size very close to $1 - s$. The proofs essentially show that the algorithms in question do good jobs of constructing such matchings. In practice, however, there will clearly be situations where more than matching is required. To model such situations, researchers first turned to the distributions $U(0, b]$, $0 < b < 1$, where item sizes are chosen uniformly from the interval $(0, b]$. Simulations suggest that such distributions make things worse for the online algorithms FF and BF, which appear to have $ER^\infty_A(U(0, b]) > 1$ for all $b \in (0, 1)$. Surprisingly, they make things better for FFD and BFD (and the optimal packing).

Theorem 12 ([2, 14])

1. *For $0 < b \leq 1/2$ and $A \in \{FFD, BFD\}$, $EW^n_A(U(0, b]) = O(1)$.*
2. *For $1/2 < b < 1$ and $A \in \{FFD, BFD\}$, $EW^n_A(U(0, b]) = \Theta(n^{1/3})$.*
3. *For $0 < b < 1$, $EW^n_{OPT}(U(0, b]) = O(1)$.*

Discrete Distributions

In many applications, the item sizes come from a finite set, rather than a continuous distribution like those discussed above. Thus, recently the study of average-case behavior for bin packing has turned to *discrete distributions*. Such a distribution is specified by a finite list $s_1, s_2, \ldots, s_d$ of rational sizes and for each s_i a corresponding rational probability p_i. A remarkable result of Courcoubetis and Weber [7] says the following.

Theorem 13 ([7]) *For any discrete distribution F, $EW_{OPT}^n(F)$ is either $\Theta(n)$, $\Theta(\sqrt{n})$, or $O(1)$.*

The discrete analogue of the continuous distribution $U(0, b]$ is the distribution $U\{j, k\}$, where the sizes are $1/k, 2/k, \ldots, j/k$ and all the probabilities equal $1/j$. Simulations suggest that the behavior of FF and BF in the discrete case are qualitatively similar to the behavior in the continuous case, whereas the behavior of FFD and BFD is considerably more bizarre [3]. Of particular note is the distribution $F = U\{6, 13\}$, for which $ER_{FFD}^\infty(F)$ is strictly greater than $ER_{FF}^\infty(F)$, in contrast to all the previously implied comparisons between the two algorithms.

For discrete distributions, however, the standard algorithms are all dominated by a new online algorithm called the *Sum-of-Squares* (SS) algorithm. Note that since the item sizes are all rational, they can be scaled so that they (and the bin size B) are all integral. Then at any given point in the operation of an online algorithm, the current packing can be summarized by giving, for each h, $1 \leq h \leq B$, the number n_h of bins containing items of total size h. In SS, one packs each item so as to minimize $\sum_{h=1}^{B-1} n_h^2$.

Theorem 14 ([9]) *For any discrete distribution F, the following hold.*

1. *If $EW_{OPT}^n(F) = \Theta(\sqrt{n})$, then $EW_{SS}^n(F) = \Theta(\sqrt{n})$.*
2. *If $EW_{OPT}^n(F) = O(1)$, then $EW_{SS}^n(F) \in \{O(1), \Theta(\log n)\}$.*

In addition, a simple modification to SS can eliminate the $\Theta(\log n)$ case of condition 2.

Applications

There are many potential applications of one-dimensional bin packing, from packing bandwidth requests into fixed-capacity channels to packing commercials into station breaks. In practice, simple heuristics like FFD and BFD are commonly used.

Open Problems

Perhaps the most fundamental open problem related to bin packing is the following. As observed above, there is a polynomial-time algorithm (KK) whose packings are within $O(\log^2(OPT))$ bins of optimal. Is it possible to do better? As far as is currently known, there could still be a polynomial-time algorithm that always gets within one bin of optimal, even if P $\neq$ NP.

Experimental Results

Bin packing has been a fertile ground for experimental analysis, and many of the theorems mentioned above were first conjectured on the basis of experimental results. For example, the experiments reported in [1] inspired Theorem 10 and 12, and the experiments in [10] inspired Theorem 14.

Cross-References

▶ Approximation Schemes for Bin Packing

Recommended Reading

1. Bentley JL, Johnson DS, Leighton FT, McGeoch CC (1983) An experimental study of bin packing. In: Proceedings of the 21st annual Allerton conference on communication, control, and computing, Urbana, University of Illinois, 1983, pp 51–60

2. Bentley JL, Johnson DS, Leighton FT, McGeoch CC, McGeoch LA (1984) Some unexpected expected behavior results for bin packing. In: Proceedings of the 16th annual ACM symposium on theory of computing. ACM, New York, pp 279–288

3. Coffman EG Jr, Courcoubetis C, Garey MR, Johnson DS, McGeoch LA, Shor PW, Weber RR, Yannakakis M (1991) Fundamental discrepancies between average-case analyses under discrete and continuous distributions. In: Proceedings of the 23rd annual ACM symposium on theory of computing, New York, 1991. ACM Press, New York, pp 230–240

4. Coffman EG Jr, Garey MR, Johnson DS (1997) Approximation algorithms for bin-packing: a survey. In: Hochbaum D (ed) Approximation algorithms for NP-hard problems. PWS Publishing, Boston, pp 46–93

5. Coffman EG Jr, Johnson DS, Shor PW, Weber RR (1997) Bin packing with discrete item sizes, part II: tight bounds on first fit. Random Struct Algorithms 10:69–101

6. Coffman EG Jr, So K, Hofri M, Yao AC (1980) A stochastic model of bin-packing. Inf Control 44:105–115

7. Courcoubetis C, Weber RR (1986) Necessary and sufficient conditions for stability of a bin packing system. J Appl Probab 23:989–999

8. Csirik J, Johnson DS (2001) Bounded space on-line bin packing: best is better than first. Algorithmica 31:115–138

9. Csirik J, Johnson DS, Kenyon C, Orlin JB, Shor PW, Weber RR (2006) On the sum-of-squares algorithm for bin packing. J ACM 53:1–65

10. Csirik J, Johnson DS, Kenyon C, Shor PW, Weber RR (1999) A self organizing bin packing heuristic. In: Proceedings of the 1999 workshop on algorithm engineering and experimentation. LNCS, vol 1619. Springer, Berlin, pp 246—265

11. Galambos G, Woeginger GJ (1995) Online bin packing – a restricted survey. ZOR Math Methods Oper Res 42:25–45

12. Johnson DS (1973) Near-optimal bin packing algorithms. PhD thesis, Massachusetts Institute of Technology, Department of Mathematics, Cambridge

13. Johnson DS, Demers A, Ullman JD, Garey MR, Graham RL (1974) Worst-case performance bounds for simple one-dimensional packing algorithms. SIAM J Comput 3:299–325

14. Johnson DS, Leighton FT, Shor PW, Weber RR.: The expected behavior of FFD, BFD, and optimal bin packing under $U(0, \alpha]$) distributions (in preparation)

15. Karmarkar N, Karp RM (1982) An efficient approximation scheme for the one-dimensional bin packing problem. In: Proceedings of the 23rd annual symposium on foundations of computer science. IEEE Computer Society, Los Alamitos, pp 312–320

16. Knödel W (1981) A bin packing algorithm with complexity O(n log n) in the stochastic limit. In: Proceedings of the 10th symposium on mathematical foundations of computer science. LNCS, vol 118. Springer, Berlin, pp 369–378

17. Lee CC, Lee DT (1985) A simple on-line packing algorithm. J ACM 32:562–572

18. Lee CC, Lee DT (1987) Robust on-line bin packing algorithms. Technical report, Department of Electrical Engineering and Computer Science, Northwestern University, Evanston

19. Leighton T, Shor P (1989) Tight bounds for minimax gridmatching with applications to the average case analysis of algorithms. Combinatorica 9:161–187

20. Lueker GS (1982) An average-case analysis of bin packing with uniformly distributed item sizes. Technical report, Report No 181, Department of Information and Computer Science, University of California, Irvine

21. Seiden SS (2002) On the online bin packing problem. J ACM 49:640–671

22. Ullman JD (1971) The performance of a memory allocation algorithm. Tech. Rep. 100, Princeton University, Princeton

23. van Vliet A (1992) An improved lower bound for online bin packing algorithms. Inf Process Lett 43:277–284

24. Yao AC (1980) New algorithms for bin packing. J ACM 27:207–227

Bin Packing with Cardinality Constraints

Hiroshi Fujiwara[1] and Koji M. Kobayashi[2]
[1]Shinshu University, Nagano, Japan
[2]National Institute of Informatics, Chiyoda-ku, Tokyo, Japan

Keywords

Approximation algorithm; Bin packing problem; Cardinality constraint; Competitive analysis; Online algorithm

Years and Authors of Summarized Original Work

1975; Krause, Shen, Schwetman
2003; Babel, Chen, Kellerer, Kotov
2006; Epstein
2010; Epstein, Levin
2013; Fujiwara, Kobayashi
2014; Dósa, Epstein

Problem Definition

In the *bin packing problem*, one is given a sequence of *items*, each of *size* in the range $(0, 1]$, and an infinite number of *bins*. The goal is to pack each item into some bin using as few bins as possible, under the constraint that the sum of sizes of items in each bin is at most one. In the *bin packing problem with cardinality constraints*, an additional constraint is imposed that each bin can contain at most k items.

This problem for $k = 2$ is solvable in polynomial time by reducing it to the *cardinality matching problem*. Nevertheless, this problem for $k \geq 3$ is NP-hard, since one can reduce *3-PARTITION* to it. Therefore, much work has been done on *approximation algorithms*. We remark that, in particular, it has also been of interest to design *online algorithms* that pack each item upon its arrival.

The standard performance measure of an approximation algorithm for this problem is the *asymptotic performance ratio*. For a sequence of items L and an approximation algorithm A, let $A(L)$ denote the value of the solution generated by A for L, and let $OPT(L)$ denote the value of the optimal solution for L. The asymptotic performance ratio of A is defined as

$$R_A^\infty = \limsup_{n \to \infty} \sup_L \left\{ \frac{A(L)}{OPT(L)} \mid OPT(L) = n \right\}.$$

The bin packing problem with cardinality constraints is formally defined as follows:

Problem 1 (Bin Packing with Cardinality Constraints)

Input: *A sequence* $L = (a_1, a_2, \ldots, a_n) \in (0, 1]^n$ *and an integer* $k \geq 2$. Output: *An integer* $m \geq 1$ *and a partition of* $\{1, 2, \ldots, n\}$ *into disjoint subsets* $S_1, S_2, \ldots, S_m$ *such that (1) m is minimum, (2) $\sum_{i \in S_j} a_i \leq 1$ for all $1 \leq j \leq m$, and (3) $|S_j| \leq k$ for all $1 \leq j \leq m$.*

Key Results

Approximation Algorithms

Krause et al. [9, 10] gave approximation algorithms whose asymptotic performance ratios are all two. Kellerer and Pferschy [8] presented an improved approximation algorithm with asymptotic performance ratio $\frac{3}{2}$. Caprara et al. [3] provided an APTAS (asymptotic polynomial-time approximation scheme): a collection of approximation algorithms that, for any parameter $\varepsilon > 0$, guarantees an asymptotic performance ratio of $1 + \varepsilon$. Finally, a better polynomial-time scheme was developed:

Theorem 1 ([6]) *There exists an AFPTAS (asymptotic fully polynomial-time approximation scheme) for the bin packing problem with cardinality constraints, that is, an APTAS whose running time is polynomial in the input size and $\frac{1}{\varepsilon}$.*

Online Algorithms

An *online algorithm* is an approximation algorithm which, for each $i = 1, 2, \ldots, n$, decides into which bin to place the ith item without information on the sizes of later items or the value of n. The *First-Fit, Best-Fit,* and *Next-Fit* algorithms may be the most common online algorithms for the bin packing problem without cardinality constraints.

Krause et al. [9, 10] adapted the *First-Fit* algorithm to the problem with cardinality constraints and showed that its asymptotic performance ratio is at most $2.7 - \frac{12}{5k}$. The result was later improved. Some work was done for individual values of k. We thus summarize best known upper and lower bounds on the asymptotic performance ratio for each $2 \leq k \leq 6$ in Table 1. We say here that u is an *upper bound* on the asymptotic performance ratio if there exists an online algorithm A such that $R_A^\infty = u$. On the other hand, we say that l is a *lower bound* on the asymptotic performance ratio if $R_A^\infty \geq l$ holds for any online algorithm A.

Babel et al. [1] designed an online algorithm, denoted here by $BCKK$, which guarantees an asymptotic performance ratio regardless of the value of k. For $k \geq 7$, $BCKK$ is the best so far.

Theorem 2 ([1]) *For any k, $R^\infty_{BCKK} = 2$.*

Recently, Dósa and Epstein [4] showed a lower bound on the asymptotic performance ratio for each $7 \leq k \leq 11$. Fujiwara and Kobayashi [7] established a lower bound for each $12 \leq k \leq 41$. Some results on the bin packing problem without cardinality constraints can be interpreted as lower bounds on the asymptotic performance ratio for large k: a lower bound of $\frac{217}{141}(\approx 1.53900)$ for $42 \leq k \leq 293$ [11] and $\frac{10,633}{6,903}(\approx 1.54034)$ for $294 \leq k \leq 2,057$ [2].

Bounded-Space Online Algorithms

A *bounded-space online algorithm* is an online algorithm which has only a constant number of bins available to accept given items at any time point. For example, the *Next-Fit* algorithm is a bounded-space online algorithm for the bin packing problem without cardinality constraints, since for the arrival of each new item, the algorithm always keeps a single bin which contains some item(s). All algorithms that appeared in the previous section, except *Next-Fit*, do not satisfy this property; such algorithms are sometimes called *unbounded-space online algorithms*.

For the bin packing problem with cardinality constraints, a bounded-space online algorithm called CCH_k [5] is known to be optimal, which is based on the Harmonic algorithm. Its asymptotic performance ratio is $\mathcal{R}_k = $ $\sum_{i=1}^{k} \max\left\{\frac{1}{t_i-1}, \frac{1}{k}\right\}$, where t_i is the sequence defined by $t_1 = 2$, $t_{i+1} = t_i(t_i - 1) + 1$ for $i \geq 1$. For example, we have $\mathcal{R}_2 = \frac{3}{2} = 1.5$, $\mathcal{R}_3 = \frac{11}{6} \approx 1.83333$, $\mathcal{R}_4 = 2$, $\mathcal{R}_5 = 2.1$, and $\mathcal{R}_6 = \frac{13}{6} \approx 2.16667$. The value of $\mathcal{R}_k$ increases as k grows and approaches $1 + \sum_{i=1}^{\infty} \frac{1}{t_i-1} \approx 2.69103$.

Theorem 3 ([5]) *For every k, $R^\infty_{CCH_k} = \mathcal{R}_k$. Besides, $R^\infty_A \geq \mathcal{R}_k$ holds for any bounded-space online algorithm A.*

Applications

In the paper by Krause et al. [9, 10], the aim was to analyze task scheduling algorithms for multiprocessor systems. Not only this but a constraint on the number of objects in a container is important in application, such as a limit to the number of files on a hard disk drive or a limit to the number of requests assigned to each node in a distributed system.

Open Problems

Many problems concerning (unbounded-space) online algorithms remain open. Even for small values of k, an optimal online algorithm has yet to be found. It is also interesting whether, for general k, there is an online algorithm whose asymptotic performance ratio is strictly smaller than two.

Cross-References

▶ Approximation Schemes for Bin Packing
▶ Bin Packing

Acknowledgments This work was supported by KAKENHI (23700014, 23500014, 26330010, and 26730008).

Recommended Reading

1. Babel L, Chen B, Kellerer H, Kotov V (2004) Algorithms for on-line bin-packing problems with cardinality constraints. Discret Appl Math 143(1-3):238–251

Bin Packing with Cardinality Constraints, Table 1 Best known upper and lower bounds on the asymptotic performance ratio of online algorithms for $2 \leq k \leq 6$

k	Upper bound	Lower bound
2	$1 + \frac{\sqrt{5}}{5}(\approx 1.44721)$ [1]	1.42764 [7]
3	1.75 [5]	1.5 [1]
4	$\frac{71}{38}(\approx 1.86843)$ [5]	1.5 [7]
5	$\frac{771}{398}(\approx 1.93719)$ [5]	1.5 [4]
6	$\frac{287}{144}(\approx 1.99306)$ [5]	1.5 [12]

2. Balogh J, Békési J, Galambos G (2012) New lower bounds for certain classes of bin packing algorithms. Theor Comput Sci 440–441:1–13
3. Caprara A, Kellerer H, Pferschy U (2003) Approximation schemes for ordered vector packing problems. Naval Res Logist 50(1):58–69
4. Dósa G, Epstein L (2014) Online bin packing with cardinality constraints revisited. CoRR abs/1404.1056
5. Epstein L (2006) Online bin packing with cardinality constraints. SIAM J Discret Math 20(4):1015–1030
6. Epstein L, Levin A (2010) AFPTAS results for common variants of bin packing: a new method for handling the small items. SIAM J Optim 20(6):3121–3145
7. Fujiwara H, Kobayashi K (2013) Improved lower bounds for the online bin packing problem with cardinality constraints. J Comb Optim 1–21. e10.1007/s10878-013-9679-8
8. Kellerer H, Pferschy U (1999) Cardinality constrained bin-packing problems. Ann Oper Res 92(1):335–348
9. Krause KL, Shen VY, Schwetman HD (1975) Analysis of several task-scheduling algorithms for a model of multiprogramming computer systems. J ACM 22(4):522–550
10. Krause KL, Shen VY, Schwetman HD (1977) Errata: "Analysis of several task-scheduling algorithms for a model of multiprogramming computer systems". J ACM 24(3):527
11. van Vliet A (1992) An improved lower bound for on-line bin packing algorithms. Inf Process Lett 43(5):277–284
12. Yao AC (1980) New algorithms for bin packing. J ACM 27(2):207–227

Bin Packing, Variants

Leah Epstein
Department of Mathematics, University of Haifa, Haifa, Israel

Keywords

Approximation schemes; Bin packing; Concave costs

Years and Authors of Summarized Original Work

1994; Anily, Bramel, Simchi-Levi
2012; Epstein, Levin

Problem Definition

The well-known bin packing problem [3, 8] has numerous variants [4]. Here, we consider one natural variant, called the bin packing problem with general cost structures (GCBP) [1, 2, 6]. In this problem, the action of an algorithm remains as in standard bin packing. We are given n items of rational sizes in $(0, 1]$. These items are to be assigned into unit size bins. Each bin may contain items of total size at most 1. While in the standard problem the goal is to minimize the number of used bins, the goal in GCBP is different; the cost of a bin is not 1, but it depends on the number of items actually packed into this bin. This last function is a concave function of the number of packed items, where the cost of an empty bin is zero. More precisely, the input consists of n items $I = \{1, 2, \ldots, n\}$ with sizes $1 \geq s_1 \geq s_2 \geq \cdots \geq s_n \geq 0$ and a function $f : \{0, 1, 2, \ldots, n\} \to \mathbb{R}_0^+$, where f is a monotonically nondecreasing concave function, satisfying $f(0) = 0$. The goal is to partition I into some number of sets $S_1, \ldots, S_\mu$, called bins, such that $\sum_{j \in S_i} s_j \leq 1$ for any $1 \leq i \leq \mu$, and so that $\sum_{i=1}^{\mu} f(|S_i|)$ is minimized (where $|S_i|$ denotes the cardinality of the set S_i). An instance of GCBP is defined not only by its input item sizes but also using the function f. It can be assumed that $f(1) = 1$ (by possible scaling of the cost function f). The problem is strongly NP-hard for multiple functions f, and as standard bin packing, it was studied using the asymptotic approximation ratio.

Key Results

There are two kinds of results for the problem. The first kind of results is algorithms that do not take f into account. The second kind is those that base their action on the values of f.

A class of (concave and monotonically nondecreasing) functions $\{f_q\}_{q \in \mathbb{N}}$, which was considered in [1], is the following. These are functions that grow linearly (with a slope of 1) up to an integer point q, and then, they are constant (starting from that point). Specifically, $f_q(t) = t$ for $t \leq q$ and $f_q(t) = q$ for $t > q$. It was shown in [1] that focusing on such functions is sufficient when computing upper bounds on algorithms that act independently of the cost function. Note that $f_1 \equiv 1$, and thus GCBP with the cost function f_1 is equivalent to standard bin packing.

Before describing the results, we present a simple example showing the crucial differences between GCBP and standard bin packing. Consider the function $f = f_3$ (where $f(1) = 1$, $f(2) = 2$, and $f(k) = 3$ for $k \geq 3$). Given an integer $N \geq 1$, consider an input consisting of $3N$ items, each of size $\frac{2}{3}$, called large items, and $6N$ items, each of size $\frac{1}{6}$, called small items. An optimal solution for this input with respect to standard bin packing uses $3N$ bins, each containing one large item and two small items. This is the unique optimal solution (up to swapping positions of identical items). The cost of this solution for GCBP with the function $f = f_3$ is $9N$. Consider a solution that uses $4N$ bins, the first N bins receive six small items each, and each additional bin receives one large item. This solution is not optimal for standard bin packing, but its cost for GCBP with $f = f_3$ is $6N$.

Anily, Bramel, and Simchi-Levi [1] analyzed the worst-case performance of some natural bin packing heuristics [8], when they are applied to GCBP. They showed that many common heuristics for bin packing, such as First Fit, Best Fit, and Next Fit, do not have a finite asymptotic approximation ratio. Moreover, running the modifications of the first two heuristics after sorting the lists of items (in a nonincreasing order), i.e., applying the algorithms First Fit Decreasing and Best Fit Decreasing, leads to similar results. However, Next Fit Decreasing was shown to have an asymptotic approximation ratio of exactly 2. The algorithm Next Fit packs items into its last bin as long as this is possible and opens a new bin when necessary. Sorting the items in nondecreasing order gives a better asymptotic approximation ratio of approximately 1.691 (in this case, the three algorithms, First Fit Increasing, Best Fit Increasing, and Next Fit Increasing, are the same algorithm). It is stated in [1] that any heuristic that is independent of f has an asymptotic approximation ratio of at least $\frac{4}{3}$. An improved approximation algorithm, called MatchHalf (MH), was developed in [6]. The asymptotic approximation ratio of this algorithm does not exceed 1.5. The idea of MH is to create bins containing pairs of items. The candidate items to be packed into those bins are half of the items of size above $\frac{1}{2}$ (large items), but they can only be packed with smaller items. Naturally, the smallest large items are selected, and the algorithm tries to match them with smaller items. The remaining items and unmatched items are packed using Next Fit Increasing. Interestingly, it was shown [6] that matching a larger fraction of large items can harm the asymptotic approximation ratio.

A fully polynomial approximation scheme (asymptotic FPTAS or AFPTAS) for GCBP was given in [6]. This is a family of approximation algorithms that contains, for any $\varepsilon > 0$, an approximation algorithm whose asymptotic approximation ratio is at most $1 + \varepsilon$. The running time must be polynomial in the input and in $\frac{1}{\varepsilon}$. An AFPTAS for GCBP must use the function f in its calculations (this can be shown using the example above and similar examples and can also be deduced from the lower bound of $\frac{4}{3}$ on the asymptotic approximation ratio of an algorithm that is oblivious of f [1]). An AFPTAS for GCBP is presented in [6]. One difficulty in designing such a scheme is that the nature of packing of small items is important, unlike approximation schemes for standard bin packing, where small items can be added greedily [7,9]. While in our problem we can impose cardinality constraints on bins (upper bounds on numbers of packed items) as in [5], still the cost function introduces major difficulties. Another ingredient of the scheme is preprocessing where some very small items are packed into relatively full bins. It is impossible to do this for all very small items as bins consisting of only such items will have a relatively large cost (as each such bin will contain a very large number of items). This AFPTAS and those of [5] require column generation as in [9] but require fairly complicated configuration linear programs.

Cross-References

▶ Harmonic Algorithm for Online Bin Packing

Recommended Reading

1. Anily S, Bramel J, Simchi-Levi D (1994) Worst-case analysis of heuristics for the bin packing problem with general cost structures. Oper Res 42(2):287–298
2. Bramel J, Rhee WT, Simchi-Levi D (1998) Average-case analysis of the bin packing problem with general cost structures. Naval Res Logist 44(7):673–686
3. Coffman E Jr, Csirik J (2007) Performance guarantees for one-dimensional bin packing. In: Gonzalez TF (ed) Handbook of approximation algorithms and metaheuristics, chap 32. Chapman & Hall/CRC, Boca Raton, pp (32–1)–(32–18)
4. Coffman E Jr, Csirik J (2007) Variants of classical one-dimensional bin packing. In: Gonzalez TF (ed) Handbook of approximation algorithms and metaheuristics, chap 33. Chapman & Hall/CRC, Boca Raton, pp (33–1)–(33–14)
5. Epstein L, Levin A (2010) AFPTAS results for common variants of bin packing: a new method for handling the small items. SIAM J Optim 20(6):3121–3145
6. Epstein L, Levin A (2012) Bin packing with general cost structures. Math Program 132(1–2): 355–391
7. Fernandez de la Vega W, Lueker GS (1981) Bin packing can be solved within $1 + \varepsilon$ in linear time. Combinatorica 1(4):349–355
8. Johnson DS, Demers AJ, Ullman JD, Garey MR, Graham RL (1974) Worst-case performance bounds for simple one-dimensional packing algorithms. SIAM J Comput 3(4):299–325
9. Karmarkar N, Karp RM (1982) An efficient approximation scheme for the one-dimensional bin packing problem. In: Proceedings of the 23rd annual symposium on foundations of computer science (FOCS1982), Chicago, Illinois, USA, pp 312–320

Binary Decision Graph

Adnan Aziz[1] and Amit Prakash[2]
[1]Department of Electrical and Computer Engineering, University of Texas, Austin, TX, USA
[2]Microsoft, MSN, Redmond, WA, USA

Keywords

BDDs; Binary decision diagrams

Years and Authors of Summarized Original Work

1986; Bryant

Problem Definition

Boolean Functions

The concept of a *Boolean function* – a function whose domain is $\{0, 1\}^n$ and range is $\{0, 1\}$ – is central to computing. Boolean functions are used in foundational studies of complexity [7, 9] as well as the design and analysis of logic circuits [4, 13], A Boolean function can be represented using a *truth table* – an enumeration of the values taken by the function on each element of $\{0, 1\}^n$. Since the truth table representation requires memory exponential in n, it is impractical for most applications. Consequently, there is a need for data structures and associated algorithms for efficiently representing and manipulating Boolean functions.

Boolean Circuits

Boolean functions can be represented in many ways. One natural representation is a *Boolean combinational circuit*, or circuit for short [6, Chapter 34]. A circuit consists of *Boolean combinational elements* connected by *wires*. The Boolean combinational elements are *gates* and *primary inputs*. Gates come in three types: NOT, AND, and OR. The NOT gate functions as follows: it takes a single Boolean-valued *input* and produces a single Boolean-valued *output* which takes value 0 if the input is 1, and 1 if the input is 0. The AND gate takes two Boolean-valued inputs and produces a single output; the output is 1 if both inputs are 1, and 0 otherwise. The OR gate is similar to AND, except that its output is 1 if one or both inputs are 1, and 0 otherwise.

Circuits are required to be acyclic. The absence of cycles implies that a Boolean assignment to the primary inputs can be unambiguously propagated through the gates in topological order. It follows that a circuit on n ordered primary inputs with a designated gate called the *primary output* corresponds to a Boolean function on $\{0, 1\}^n$. Every Boolean function can be represented by a circuit, e.g., by building a circuit that mimics the truth table.

The circuit representation is very general – any decision problem that is computable in polynomial time on a Turing machine can be computed by circuit polynomial in the instance size, and the circuits can be constructed efficiently from the Turing machine program [15]. However, the key analysis problems on circuits, namely, satisfiability and equivalence, are NP-hard [7].

Boolean Formulas

A *Boolean formula* is defined recursively: a *Boolean variable* x_i is a *Boolean formula*, and if φ and ψ are Boolean formulas, then so are $(\neg\varphi), (\varphi \wedge \psi), (\varphi \vee \psi), (\varphi \to \psi)$, and $(\varphi \leftrightarrow \psi)$. The operators $\neg, \vee, \wedge, \to, \leftrightarrow$ are referred to as *connectives*; parentheses are often dropped for notational convenience. Boolean formulas also can be used to represent arbitrary Boolean functions; however, formula satisfiability and equivalence are also NP-hard. Boolean formulas are not as succinct as Boolean circuits: for example, the parity function has linear sized circuits, but formula representations of parity are super-polynomial. More precisely, $XOR_n :$ $\{0, 1\}^n \to \{0, 1\}$ is defined to take the value 1 on exactly those elements of $\{0, 1\}^n$ which contain an odd number of 1s. Define the *size* of a formula to be the number of connectives appearing in it. Then for any sequence of formulas $\theta_1, \theta_2, \ldots$ such that θ_k represents XOR_k, the size of θ_k is $\Omega(k^c)$ for all $c \in Z^+$ [14, Chapters 11, 12].

A *disjunct* is a Boolean formula in which $\wedge$ and $\neg$ are the only connectives, and $\neg$ is applied only to variables; for example, $x_1 \wedge \neg x_3 \wedge \neg x_5$ is a disjunct. A Boolean formula is said to be in *Disjunctive Normal Form* (DNF) if it is of the form $D_0 \vee D_1 \vee \cdots \vee D_{k-1}$, where each D_i is a disjunct. DNF formulas can represent arbitrary Boolean functions, e.g., by identifying each input on which the formula takes the value 1 with a disjunct. DNF formulas are useful in logic design, because it can be translated directly into a PLA implementation [4]. While satisfiability of DNF formulas is trivial, equivalence is

NP-hard. In addition, given DNF formulas φ and ψ, the formulas $\neg\varphi$ and $\varphi \wedge \psi$ are not DNF formulas, and the translation of these formulas to DNF formulas representing the same function can lead to exponential growth in the size of the formula.

Shannon Trees

Let f be a Boolean function on domain $\{0, 1\}^n$. Associate the n dimensions with variables $x_0, \ldots, x_{n-1}$. Then the *positive cofactor* of f with respect to x_i, denoted by f_{x_i}, is the function on domain $\{0, 1\}^n$, which is defined by

$$f_{x_i}(\alpha_0, \ldots, \alpha_{i-1}, a_i, \alpha_{i+1}, \ldots, \alpha_{n-1})$$
$$= f(\alpha_0, \ldots, \alpha_{i-1}, 1, \alpha_{i+1}, \ldots, \alpha_{n-1}).$$

The *negative cofactor* of f with respect to x_i, denoted by $f_{x_i'}$, is defined similarly, with 0 taking the place of 1 in the right-hand side.

Every Boolean function can be decomposed using Shannon's expansion theorem:

$$f(x_1, \ldots, x_n) = x_i \cdot f_{x_i} + x_i' \cdot f_{x_i'}.$$

This observation can be used to represent f by a *Shannon tree* – a kill binary tree [6, Appendix B.5] of height n, where each path to a leaf node defines a complete assignment to the n variables that f is defined over, and the leaf node holds a 0 or a 1, based on the value f takes for the assignment.

The Shannon tree is not a particularly useful representation, since the height of the tree representing every Boolean function on $\{0, 1\}^n$ is n, and the tree has 2^n leaves. The Shannon tree can be made smaller by merging isomorphic subtrees and bypassing nodes which have identical children. At first glance the reduced Shannon tree representation is not particularly useful, since it entails creating the full binary tree in the first place. Furthermore, it is not clear how to efficiently perform computations on the reduced Shannon tree representation, such as equivalence checking or computing the conjunction of functions presented as reduced Shannon trees.

Bryant [5] recognized that adding the restriction that variables appear in fixed order from root to leaves greatly reduced the complexity of manipulating reduced Shannon trees. He referred to this representation as a binary decision diagram (BDD).

Key Results

Definitions

Technically, a BDD is a directed acyclic graph (DAG), with a designated root, and at most two sinks – one labeled 0, the other labeled 1. Nonsink nodes are labeled with a variable. Each nonsink node has two outgoing edges – one labeled with a 1 leading to the *1-child*, the other is a 0, leading to the *0-child*. Variables must be ordered – that is, if the variable label x_i appears before the label x_j on some path from the root to a sink, then the label x_j is precluded from appearing before x_i on any path from the root to a sink. Two nodes are *isomorphic* if both are equilabeled sinks, or they are both nonsink nodes, with the same variable label, and their 0- and 1-children are isomorphic. For the DAG to be a valid BDD, it is required that there are no isomorphic nodes, and for no nodes are its 0- and 1-children the same.

A key result in the theory of BDDs is that given a fixed variable ordering, the representation is unique up to isomorphism, i.e., if F and G are both BDDs representing $f : \{0, 1\}^n \rightarrow \{0, 1\}$ under the variable ordering $x_1 \prec x_2 \prec \ldots x_n$, then F and G are isomorphic.

The definition of isomorphism directly yields a recursive algorithm for checking isomorphism. However, the resulting complexity is exponential in the number of nodes – this is illustrated, for example, by checking the isomorphism of the BDD for the parity function against itself on inspection, the exponential complexity arises from repeated checking of isomorphism between pairs of nodes – this naturally suggest dynamic programming. Caching isomorphism checks reduces the complexity of isomorphism checking to $O(|F| \cdot |G|)$, where $|B|$ denotes the number of nodes in the BDD B.

BDD Operations

Many logical operations can be implemented in polynomial time using BDDs: *bdd_and* which computes a BDD representing the logical AND of the functions represented by two BDDs, *bdd_or* and *bdd_not* which are defined similarly, and *bdd_compose* which takes a BDD representing a function f, a variable v, and a BDD representing a function g and returns the BDD for f where v is substituted by g are examples.

The example of *bdd_and* is instructive – it is based on the identity $f \cdot g = x \cdot (f_x \cdot g_x) + x' \cdot (f_{x'} \cdot g_{x'})$. The recursion can be implemented directly: the base cases are when either f or g are 0 and when one or both are 1. The recursion chooses the variable v labeling either the root of the BDD for f or g, depending on which is earlier in the variable ordering, and recursively computes BDDs for $f_v \cdot g_v$ and $f_{v'} \cdot g_{v'}$; these are merged if isomorphic. Given a BDD F for f, if v is the variable labeling the root of F, the BDDs for $f_{v'}$ and f_v, respectively, are simply the 0-child and 1-child of F's root.

The implementation of *bdd_and* as described has exponential complexity because of repeated subproblems arising. Dynamic programming again provides a solution – caching the intermediate results of *bdd_and* reduced complexity to $O(|F| \cdot |G|)$.

Variable Ordering

All symmetric functions on $\{0, 1\}^n$ have a BDD that is polynomial in n, independent of the variable ordering. Other useful functions such as comparators, multiplexers, adders, and subtracters can also be efficiently represented, if the variable ordering is selected correctly. Heuristics for ordering selection are presented in [1, 2, 11]. There are functions which do not have a polynomial-sized BDD under any variable ordering – the Unction representing the n-th bit of the output of a multiplier taking two n-bit unsigned integer inputs is an example [5]. Wegener [17] presents many more examples of the impact of variable ordering.

Applications

BDDs have been most commonly applied in the context of formal verification of digital hardware [8]. Digital hardware extends the notion of circuit described above by adding *state elements* which hold a Boolean value between updates and are updated on a *clock* signal.

The gates comprising a design are often updated based on performance requirements; these changes typically are not supposed to change the logical functionality of the design. BDD-based approaches have been used for checking the equivalence of digital hardware designs [10].

BDDs have also been used for checking properties of digital hardware. A typical formulation is that a set of "good" states and a set of "initial" states are specified using Boolean formulas over the state elements; the property holds iff there is no sequence of inputs which leads a state in the initial state to a state not in the set of good states. Given a design with n registers, a set of states A in the design can be characterized by a formula φ_A over n Boolean variables: φ_A evaluates to true on an assignment to the variables iff the corresponding state is in A. The formula φ_A represents a Boolean function, and so BDDs can be used to represent sets of states. The key operation of computing the *image* of a set of states A, i.e., the set of states that can be reached on application of a single input from states in A, can also be implemented using BDDs [12].

BDDs have been used for *test generation*. One approach to test generation is to specify legal inputs using constraints, in essence Boolean formulas over the primary input and state variables. Yuan et al. [18] have demonstrated that BDDs can be used to solve these constraints very efficiently.

Logic synthesis is the discipline of realizing hardware designs specified as logic equations using gates. Mapping equations to gates is straightforward; however, in practice a direct mapping leads to implementations that are not acceptable from a performance perspective, where performance is measured by gate area or timing delay.

Manipulating logic equations in order to reduce area (e.g., through constant propagation, identifying common sub-expressions, etc.), and delay (e.g., through propagating late arriving signals closer to the outputs), is conveniently done using BDDs.

Experimental Results

Bryant reported results on verifying two qualitatively distinct circuits for addition. He was able to verify on a VAX 11/780 (a 1 MIP machine) that two 64-bit adders were equivalent in 95.8 min. He used an ordering that he derived manually.

Normalizing for technology, modern BDD packages are two orders of magnitude faster than Bryant's original implementation. A large source the improvement comes from the use of the *strong canonical form*, wherein a global database of BDD nodes is maintained, and no new node is added without checking to see if a node with the same label and 0- and 1-children exists in the database [3]. (For this approach to work, it is also required that the children of any node being added be in strong canonical form.) Other improvements stem from the use of complement pointers (if a pointer has its least-significant bit set, it refers to the complement of the function), better memory management (garbage collection based on reference counts, keeping nodes that are commonly accessed together close in memory), better hash functions, and better organization of the computed table (which keeps track of subproblems that have already been encountered) [16].

Data Sets

The SIS (http://embedded.eecs.berkeley.edu/pubs/downloads/sis/) system from UC Berkeley is used for logic synthesis. It comes with a number of combinational and sequential circuits that have been used for benchmarking BDD packages.

The VIS (http://embedded.eecs.berkeley.edu/pubs/downloads/vis) system from UC Berkeley and UC Boulder is used for design verification; it uses BDDs to perform checks. The distribution includes a large collection of verification problems, ranging from simple hardware circuits to complex multiprocessor cache systems.

URL to Code

A number of BDD packages exist today, but the package of choice is CUDD (http://vlsi.colorado.edu/~fabio/CUDD/). CUDD implements all the core features for manipulating BDDs, as well as variants. It is written in C++ and has extensive user and programmer documentation.

Cross-References

▶ Symbolic Model Checking

Recommended Reading

1. Aziz A, Tasiran S, Brayton R (1994) BDD variable ordering for interacting finite state machines. In: ACM design automation conference, San Diego, pp 283–288
2. Berman CL (1989) Ordered binary decision diagrams and circuit structure. In: IEEE international conference on computer design, Cambridge
3. Brace K, Rudell R, Bryant R (1990) Efficient implementation of a BDD package. In: ACM design automation conference, Orlando
4. Brayton R, Hachtel G, McMullen C, Sangiovanni-Vincentelli A (1984) Logic minimization algorithms for VLSI synthesis. Kluwer Academic, Boston
5. Bryant R (1986) Graph-based algorithms for Boolean function manipulation. IEEE Trans Comput C-35:677–691
6. Cormen TH, Leiserson CE, Rivest RH, Stein C (2001) Introduction to algorithms. MIT, Cambridge
7. Garey MR, Johnson DS (1979) Computers and intractability. W.H. Freeman and Co, New York
8. Gupta A (1993) Formal hardware verification methods: a survey. Formal Method Syst Des 1:151–238
9. Karchmer M (1989) Communication complexity: a new approach to circuit depth. MIT, Cambridge
10. Kuehlmann A, Krohm F (1997) Equivalence checking using cuts and heaps. In: ACM design automation conference, Anaheim
11. Malik S, Wang AR, Brayton RK, Sangiovanni-Vincentelli A (1988) Logic verification using binary decision diagrams in a logic synthesis environment. In: IEEE international conference on computer-aided design, Santa Clara, pp 6–9
12. McMillan KL (1993) Symbolic model checking. Kluwer Academic, Boston
13. De Micheli G (1994) Synthesis and optimization of digital circuits. McGraw Hill, New York
14. Schoning U, Pruim R (1998) Gems of theoretical computer science. Springer, Berlin/New York
15. Sipser M (2005) Introduction to the theory of computation, 2nd edn. Course Technology, Boston
16. Somenzi F (2012) Colorado University Decision Diagram package. http://vlsi.colorado.edu/~fabio/CUDD
17. Wegener I (2000) Branching programs and binary decision diagrams. SIAM, Philadelphia
18. Yuan J, Pixley C, Aziz A (2006) Constraint-based verification. Springer, New York

Binary Space Partitions

Adrian Dumitrescu[1] and Csaba D. Tóth[2,3]
[1]Computer Science, University of Wisconsin–Milwaukee, Milwaukee, WI, USA
[2]Department of Computer Science, Tufts University, Medford, MA, USA
[3]Department of Mathematics, California State University Northridge, Los Angeles, CA, USA

Keywords

BSP tree; Computational geometry; Convex decomposition; Recursive partition

Years and Authors of Summarized Original Work

1990; Paterson, Yao
1992; D'Amore, Franciosa
1992; Paterson, Yao
2002; Berman, DasGupta, Muthukrishnan
2003; Tóth
2004; Dumitrescu, Mitchell, Sharir
2005; Hershberger, Suri, Tóth
2011; Tóth

Problem Definition

The *binary space partition* (for short, *BSP*) is a scheme for subdividing the ambient space $\mathbb{R}^d$

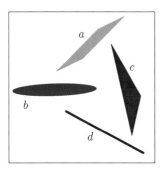 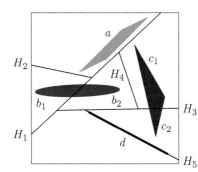 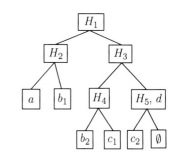

Binary Space Partitions, Fig. 1 Three 2-dimensional convex objects and a line segment (*left*), a binary space partition with five partition lines $H_1, \ldots, H_5$ (*center*), and the corresponding BSP tree (*right*)

into open convex sets (called *cells*) by hyperplanes in a recursive fashion. Each subdivision step for a cell results in two cells, in which the process may continue, independently of other cells, until a stopping criterion is met. The binary recursion tree, also called *BSP-tree*, is traditionally used as a data structure in computer graphics for efficient rendering of polyhedral scenes. Each node v of the BSP-tree, except for the leaves, corresponds to a cell $C_v \subseteq \mathbb{R}^d$ and a partitioning hyperplane H_v. The cell of the root r is $C_r = \mathbb{R}^d$, and the two children of a node v correspond to $C_v \cap H_v^-$ and $C_v \cap H_v^+$, where H_v^- and H_v^+ denote the open half-spaces bounded by H_v. Refer to Fig. 1.

A binary space partition *for* a set of n pairwise disjoint (typically polyhedral) objects in $\mathbb{R}^d$ is a BSP where the space is recursively partitioned until each cell intersects at most one object. When the BSP-tree is used as a data structure, every leaf v stores the fragment of at most one object clipped in the cell C_v, and every interior node v stores the fragments of any lower-dimensional objects that lie in $C_v \cap H_v$.

A BSP for a set of objects has two parameters of interest: the *size* and the *height* of the corresponding BSP-tree. Ideally, a BSP partitions space so that each object lies entirely in a single cell or in a cutting hyperplane, yielding a so-called *perfect* BSP [4]. However, in most cases this is impossible, and the hyperplanes H_v partition some of the input objects into *fragments*. Assuming that the input objects are k-dimensional, for some $k \leq d$, the BSP typically stores only k-dimensional fragments, i.e., object

parts clipped in leaf cells C_v or in $C_v \cap H_v$ at interior nodes.

The size of the BSP-tree is typically proportional to the number of k-dimensional fragments that the input objects are partitioned into, or the number of nodes in the tree. Given a set S of objects in $\mathbb{R}^d$, one would like to find a BSP for S with small size and/or height. The *partition complexity* of a set of objects S is defined as the minimum size of a BSP for S.

Glossary

- **Autopartition**: a class of BSPs obtained by imposing the constraint that each cut is along a hyperplane containing a facet of one of the input objects.
- **Axis-aligned BSP**: a class of BSPs obtained by imposing the constraint that each cut is orthogonal to a coordinate axis.
- **Round-robin BSP:** An axis-aligned BSP in $\mathbb{R}^d$ where any d consecutive recursive cuts are along hyperplanes orthogonal to the d coordinate axes.
- **Tiling** in $\mathbb{R}^d$: a set of interior-disjoint polyhedra that partition $\mathbb{R}^d$.
- **Axis-aligned tiling**: a set of full-dimensional boxes that partition $\mathbb{R}^d$.
- d-**dimensional box**: the cross product of d real-valued intervals.

Key Results

The theoretical study of BSPs was initiated by Paterson and Yao [10, 11].

Line Segments in the Plane

A classical result of Paterson and Yao [10] is a simple and elegant randomized algorithm, which, given n disjoint segments, produces a BSP whose expected size is $O(n \log n)$; see also [3, Ch. 12]. It was widely believed for decades that every set of n disjoint line segments in the plane admits a BSP of size $O(n)$; see e.g., [10, p. 502]; this was until Tóth proved a tight super-linear bound for this problem, first by constructing a set of segments for which any BSP must have size $\Omega(n \log n / \log \log n)$ and later by matching this bound algorithmically:

Theorem 1 ([12, 15]) *Every set of n disjoint line segments in the plane admits a BSP of size $O(n \log n / \log \log n)$. This bound is the best possible, and a BSP of this size can be computed in $O(n \log^2 n)$ time.*

Simplices in $\mathbb{R}^d$

The randomized partition technique of Paterson and Yao generalizes to higher dimensions yielding the following.

Theorem 2 ([10]) *Every set of n $(d - 1)$-dimensional simplices in $\mathbb{R}^d$, where $d \geq 3$ admits a BSP of size $O(n^{d-1})$.*

While there exist n disjoint triangles in $\mathbb{R}^3$ that require a BSP of size $\Omega(n^2)$, no super-quadratic lower bound is known in any dimension d. Near-linear upper bounds are known for "realistic" input models in $\mathbb{R}^3$ such as uncluttered scenes [5] or fat axis-aligned rectangles [14].

Axis-Parallel Segments, Rectangles, and Hyperrectangles

Theorem 3 ([1, 7, 10]) *Every set of n pairwise disjoint axis-parallel line segments in the plane admits an auto-partition of size at most $2n - 1$. Such a BSP can be computed using $O(n \log n)$ time and space and has the additional property that no input segment is cut more than once. The upper bound on the size is the best possible apart from lower-order terms.*

Theorem 4 ([7, 11]) *Let Γ be a collection of n line segments in $\mathbb{R}^d$, where $d \geq 3$, consisting of n_i segments parallel to the x_i-axis, for $i = 1, \ldots, d$. Then Γ admits a BSP of size at most*

$$4^{1/(d-1)}(d-1)(n_1 n_2 \ldots n_d)^{1/(d-1)} + 2n.$$

Theorem 5 ([7]) *For constants $1 \leq k \leq d - 1$, every set of n axis-parallel k-rectangles in d-space admits an axis-aligned BSP of size $O(n^{d/(d-k)})$. This bound is the best possible for $k < d/2$ apart from the constant factor.*

For $k \geq d/2$, the best known upper and lower bounds do not match. No super-quadratic lower bound is known in any dimension d. In $\mathbb{R}^4$, Dumitrescu et al. [7] constructed n 2-dimensional disjoint rectangles whose partition complexity is $\Omega(n^{5/3})$.

Tilings

Already in the plane, the worst-case partition complexity of axis-aligned tilings is smaller than that for disjoint boxes. Berman, DasGupta, and Muthukrishnan [6] showed that every axis-aligned tiling of size n admits an axis-aligned BSP of size at most $2n$; apart from lower-order terms, this bound is the best possible. For higher dimensions, Hershberger, Suri, and Tóth obtained the following result.

Theorem 6 ([9]) *Every axis-aligned tiling of size n in $\mathbb{R}^d$, where $d \geq 2$, admits a round-robin BSP of size $O(n^{(d+1)/3})$. On the other hand, there exist tilings of size n in $\mathbb{R}^d$ for which every BSP has size $\Omega(n^{\beta(d)})$, where $\beta(3) = 4/3$, and $\lim_{d \to \infty} \beta(d) = (1 + \sqrt{5})/2 \approx 1.618$.*

In dimensions $d = 3$, the partition complexity of axis-aligned tilings of size n is $O(n^{4/3})$, which is tight by a construction of Hershberger and Suri [8].

Applications

The initial and most prominent applications are in computer graphics: BSPs support fast *hidden-surface removal* and *ray tracing* for moving viewpoints [10]. Rendering is used for visualizing spatial opaque surfaces on the screen. A common and efficient rendering technique is the so-called *painter's algorithm*. Every object is drawn sequentially according to the *back-to-front* order, starting with the deepest object and continuing with the objects closer to the viewpoint. When all the objects have been drawn, every pixel represents the color of the object closest to the viewpoint. Further computer graphics applications include *constructive solid geometry* and *shadow generation*. Other applications of BSP trees include *range counting, point location, collision detection, robotics, graph drawing*, and *network design*; see, for instance, [13] and the references therein.

In the original setting, the input objects of the BSP were assumed to be static. Recent research on BSPs for moving objects can be seen in the context of *kinetic data structures* (KDS) of Basch, Guibas, and Hershberger [2]. In this model, objects move continuously along a given trajectory (flight plan), typically along a line or a low-degree algebraic curve. The splitting hyperplanes are defined by faces of the input objects, and so they move continuously, too. The BSP is updated only at discrete *events*, though, when the combinatorial structure of the BSP changes.

Open Problems

- What is the maximum partition complexity of n disjoint $(d-1)$-dimensional simplices in $\mathbb{R}^d$ for $d \geq 3$?
- What is the maximum partition complexity of n disjoint (axis-aligned) boxes in $\mathbb{R}^d$ for $d \geq 3$?
- What is the maximum (axis-aligned) partition complexity of a tiling of n axis-aligned boxes in $\mathbb{R}^d$ for $d \geq 4$?
- Are there families of n disjoint objects in $\mathbb{R}^d$ whose partition complexity is super-quadratic in n?
- How many combinatorial changes can occur in the kinetic BSP of n points moving with constant velocities in the plane?

In all five open problems, the dimension $d \in \mathbb{N}$ of the ambient space $\mathbb{R}^d$ is constant, and asymptotically tight bounds in terms of n are sought.

Recommended Reading

1. d'Amore F, Franciosa PG (1992) On the optimal binary plane partition for sets of isothetic rectangles. Inf Process Lett 44:255–259
2. Basch J, Guibas LJ, Hershberger J (1999) Data structures for mobile data. J Algorithms 31(1):1–28
3. de Berg M, Cheong O, van Kreveld M, Overmars M (2008) Computational geometry, 3rd edn. Springer, Berlin
4. de Berg M, de Groot M, Overmars MH (1997) Perfect binary space partitions. Comput Geom Theory Appl 7:81–91
5. de Berg M, Katz MJ, van der Stappen AF, Vleugels J (2002) Realistic input models for geometric algorithms. Algorithmica 34:81–97
6. Berman P, DasGupta B, Muthukrishnan S (2002) On the exact size of the binary space partitioning of sets of isothetic rectangles with applications. SIAM J Discret Math 15(2):252–267
7. Dumitrescu A, Mitchell JSB, Sharir M (2004) Binary space partitions for axis-parallel segments, rectangles, and hyperrectangles. Discret Comput Geom 31(2):207–227
8. Hershberger J, Suri S (2003) Binary space partitions for 3D subdivisions. In: Proceedings of the 14th ACM-SIAM symposium on discrete algorithms, Baltimore. ACM, pp 100–108
9. Hershberger J, Suri S, Tóth CsD (2005) Binary space partitions of orthogonal subdivisions. SIAM J Comput 34(6):1380–1397
10. Paterson MS, Yao FF (1990) Efficient binary space partitions for hidden-surface removal and solid modeling. Discret Comput Geom 5:485–503
11. Paterson MS, Yao FF (1992) Optimal binary space partitions for orthogonal objects. J Algorithms 13:99–113
12. Tóth CsD (2003) A note on binary plane partitions. Discret Comput Geom 30:3–16
13. Tóth CsD (2005) Binary space partitions: recent developments. In: Goodman JE, Pach J, Welzl E (eds) Combinatorial and Computational Geometry. Volume 52 of MSRI Publications, Cambridge University Press, Cambridge, pp 529–556
14. Tóth CsD (2008) Binary space partition for axis-aligned fat rectangles. SIAM J Comput 38(1):429–447
15. Tóth CsD (2011) Binary plane partitions for disjoint line segments. Discret Comput Geom 45(4):617–646

Block Shaping in Floorplan

Chris Chu
Department of Electrical and Computer Engineering, Iowa State University, Ames, IA, USA

Keywords

Block shaping; Fixed-outline floorplanning; Floorplanning; VLSI physical design

Years and Authors of Summarized Original Work

2012; Yan, Chu
2013; Yan, Chu

Problem Definition

Floorplanning is an early stage of the very-large-scale integration (VLSI) design process in which a coarse layout of a set of rectangular circuit blocks is determined. A floorplan enables designers to quickly estimate circuit performance, routing congestion, etc., of the circuit. In modern VLSI design flow, a fixed outline of the floorplanning region is given. On the other hand, many circuit blocks do not have a fixed shape during floorplanning as their internal circuitries have not yet been laid out. Those blocks are called soft blocks. Others blocks with predetermined shapes are called hard blocks. Given the geometric (i.e., left-right and above-below) relationship among the blocks, the block shaping problem is to determine the shapes of the soft blocks such that all blocks can be packed without overlap into the fixed-outline region.

To handle the block shaping problem in fixed-outline floorplanning, Yan and Chu [1, 2] provide a problem formulation in which the floorplan height is minimized, while the width is upper bounded. The formulation is described below. Let W be the upper bound on the width of the floorplanning region. Given a set of n blocks, each block i has area A_i, width w_i, and height h_i. A_i is fixed, while w_i and h_i may vary as long as they satisfy $w_i \times h_i = A_i$, $W_i^{min} \leq w_i \leq W_i^{max}$, and $H_i^{min} \leq h_i \leq H_i^{max}$. If $W_i^{min} = W_i^{max}$ and $H_i^{min} = H_i^{max}$, then block i is a hard block. Two constraint graphs G_H and G_V [3, Chapter 10] are given to specify the geometric relationship among the blocks. G_H and G_V consist of $n + 2$ vertices. Vertices 1 to n represent the n blocks. In addition, dummy vertices 0 (called source) and $n + 1$ (called sink) are added. In G_H, vertices 0 and $n + 1$ represent the leftmost and rightmost boundaries of the floorplanning region, respectively. In G_V, vertices 0 and $n + 1$ represent

the bottommost and topmost boundaries of the floorplanning region, respectively. A_0, w_0, h_0, A_{n+1}, w_{n+1}, and h_{n+1} are all set to 0. If block i is on the left of block j, $(i, j) \in G_H$. If block i is below block j, $(i, j) \in G_V$.

Let x_i and y_i be the x- and y-coordinates of the bottom-left corner of block i in the floorplan. Then, the block shaping problem formulation in [1, 2] can be written as the following geometric program:

Minimize y_{n+1}
subject to $x_{n+1} \leq W$
$$x_i + w_i \leq w_j \qquad \forall (i, j) \in G_H$$
$$y_i + h_i \leq y_j \qquad \forall (i, j) \in G_V$$
$$w_i \times h_i = A_i \qquad 1 \leq i \leq n$$
$$W_i^{min} \leq w_i \leq W_i^{max} \quad 1 \leq i \leq n$$
$$H_i^{min} \leq h_i \leq H_i^{max} \quad 1 \leq i \leq n$$
$$x_0 = y_0 = 0$$

To solve the original problem of packing all blocks into a fixed-outline region, we can take any feasible solution of the geometric program in which y_{n+1} is less than or equal to the height of the region.

Key Results

Almost all previous works target the classical floorplanning formulation, which minimizes the floorplan area. Such a formulation is not compatible with modern design methodologies [4], but those works may be modified to help fixed-outline floorplanning to various extents. For the special case of slicing floorplan [5], the block shaping problem can be solved by the elegant shape curve idea [6]. For a general floorplan which may not have a slicing structure, various heuristics have been proposed [7–9]. Moh et al. [10] formulated the shaping problem as a geometric program and optimally solved it using standard convex optimization. Young et al. [11] solved the geometric program formulation by Lagrangian relaxation. Lin et al. [12] minimized the floorplan area indirectly by minimizing its perimeter optimally using min-cost flow and trust region method. For previous works which directly tackled the block shaping problem in fixed-outline floorplanning, Adya and Markov [13] proposed a simple greedy heuristic, and Lin and Hung [14] used second-order cone programming. Previous works are either non-optimal or time-consuming.

Yan and Chu [1, 2] presented a simple and optimal algorithm called slack-driven shaping (SDS). SDS iteratively shapes the soft blocks to reduce the floorplan height while not exceeding the floorplan width bound. We first present a simplified version called basic slack-driven shaping (basic SDS), which almost always produces an optimal solution. Then, we present its extension to the SDS algorithm.

Given some initial block shapes, the blocks can be packed to the four boundaries of the floorplanning region. For block i ($1 \leq i \leq n$), let Δ_{x_i} be the difference in x_i between the two layouts generated by packing all blocks to $x = W$ and to $x = 0$, respectively. Similarly, let Δ_{y_i} be the difference in y_i between the two layouts generated by packing all blocks to $y = y_{n+1}$ and to $y = 0$, respectively. The horizontal slack s_i^H and vertical slack s_i^V are defined as follows:

$$s_i^H = \max(0, \Delta_{x_i}), \; s_i^V = \max(0, \Delta_{y_i}).$$

Horizontal critical path (HCP) is defined as a path in G_H from source to sink such that all blocks along the path have zero horizontal slack. Vertical critical path (VCP) is similarly defined. We also define two subsets of blocks:

$$SH = \{i \text{ is soft}\} \cap \{s_i^H > 0, s_i^V = 0\}$$
$$\cap \{w_i < W_i^{\max}\}$$
$$SV = \{i \text{ is soft}\} \cap \{s_i^H = 0, s_i^V > 0\}$$
$$\cap \{h_i < H_i^{\max}\}$$

Note that y_{n+1} can be reduced by decreasing the height (i.e., increasing the width) of the blocks in SH, and x_{n+1} can be reduced by decreasing the width (i.e., increasing the height) of the blocks in SV. We call the blocks in the sets SH and SV target soft blocks. In each iteration of basic SDS, we would like to increase the width w_i of each block $i \in$ SH by δ_i^H and the height h_i of each block $i \in$ SV by δ_i^V. The basic SDS algorithm is shown below:

Basic Slack-Driven Shaping Algorithm
Input: A set of n blocks, upper-bound width W, G_H and G_V.
Output: Optimized y_{n+1}, w_i and h_i for all i.
Begin
1. Set w_i to $W_i^{\min}$ for all i.
2. Pack blocks to $x = 0$ and compute x_{n+1}.
3. If $x_{n+1} > W$,
4.   Return no feasible solution.
5. Else,
6.   Repeat
7.   Pack blocks to $y = 0$, $y = y_{n+1}$, $x = 0$, and $x = W$.
8.   Calculate s_i^H and s_i^V for all i.
9.   Identify target soft blocks in SH and SV.
10. $\forall i \in$ SH, increase w_i by $\delta_i^H = \dfrac{(W_i^{\max} - w_i)}{\text{MAX}_{p \in P_i^H} \left(\sum_{k \in p}(W_k^{\max} - w_k)\right)} s_i^H$,
           where P_i^H is the set of paths in G_H passing through block i.
11. $\forall i \in$ SV, increase h_i by $\delta_i^V = \beta \times \dfrac{(H_i^{\max} - h_i)}{\text{MAX}_{p \in P_i^V} \left(\sum_{k \in p}(H_k^{\max} - h_k)\right)} s_i^V$,
           where P_i^V is the set of paths in G_V passing through block i.
12.   Until there is no target soft block.
End

Note that all δ_i^H and δ_i^V in Lines 10 and 11 can be computed using dynamic programming in linear time. Packing of blocks can also be done in linear time by longest path algorithm on a directed acyclic graph. Hence, each iteration of basic SDS takes linear time. The way δ_i^H and δ_i^V are set in Lines 10 and 11 is the key to the convergence of the algorithm.

Lemma 1 *For any path p from source to sink in G_H, we have $\sum_{i \in p} \delta_i^H \leq s_{\max\_H}^p$, where $s_{\max\_H}^p$ is the maximum horizontal slack over all blocks along p.*

Basically, $s_{\max\_H}^p$ gives us a budget on the total amount of increase in the block width along path p. Hence, Lemma 1 implies that the width of the floorplan will not be more than W after shaping of the blocks in SH at Line 10. The shaping of blocks in SV is done similarly, but a factor β is introduced in Line 11.

Lemma 2 *For any path p from source to sink in G_V, we have $\sum_{i \in p} \delta_i^V \leq \beta \times s_{\max\_V}^p$, where $s_{\max\_V}^p$ is the maximum vertical slack over all blocks along p.*

Lemma 2 guarantees that by setting $\beta \leq 1$, y_{n+1} will not increase after each iteration. In other words, the height of the floorplan will monotonically decrease during the whole shaping process. β is almost always set to 1. However, if $\beta = 1$, it is possible that the floorplan height may remain the same after one iteration even when the solution is not yet optimal. To avoid getting stuck at a local minimum, if the floorplan height does

not decrease for two consecutive iterations, β is set to 0.9 for the next iteration.

Consider a shaping solution L generated by basic SDS (i.e., without any target soft block). Blocks at the intersection of some HCP and some VCP are called intersection blocks. The following optimality conditions were derived in [1,2].

Lemma 3 *If L contains one VCP in which all intersection blocks are hard, then L is optimal.*

Lemma 4 *If L contains at most one HCP in which some intersection blocks are soft, then L is optimal.*

Lemma 5 *If L contains at most one VCP in which some intersection blocks are soft, then L is optimal.*

In practice, it is very rare for a shaping solution generated by basic SDS to satisfy none of the three optimality conditions. According to the experiments in [1,2], all solutions by basic SDS satisfy at least one of the optimality conditions, i.e., are optimal. However, [1,2] showed that it is possible for basic SDS to converge to non-optimal solutions. If a non-optimal solution is produced by basic SDS, it can be used as a starting solution to the geometric program above and then be improved by a single step of any descent-based optimization technique (e.g., deepest descent). This perturbed and improved solution can be fed to basic SDS again to be further improved. The resulting SDS algorithm, which is guaranteed optimal, is shown below:

Slack-Driven Shaping Algorithm

Input: A set of n blocks, upper-bound width W, G_H and G_V.

Output: Optimal y_{n+1}, w_i and h_i for all i.

Begin

1. Run basic SDS to generate shaping solution L.

2. If Lemma 3 or Lemma 4 or Lemma 5 is satisfied,

3. L is optimal. Exit.

4. Else,

5.   Improve L by a single step of geometric programming.

6.   Go to Line 1.

End

Applications

Floorplanning is a very important step in modern VLSI design. It enables designers to explore different alternatives in the design space and make critical decisions early in the design process. Typically, a huge number of alternatives need to be evaluated during the floorplanning stage. Hence, an efficient block shaping algorithm is a crucial component of a floorplanning tool. SDS is tens to hundreds of times faster than previous algorithms in practice. It also directly handles a fixed-outline floorplanning formulation, which is the standard in modern design methodologies. Hence, SDS should be able to improve the quality while also reduce the design time of VLSI circuits.

Open Problems

An interesting open problem is to derive a theoretical bound on the number of iterations for SDS to converge to an optimal solution. Although experimental results have shown that the number of iterations is small in practice, no theoretical bound is known.

Another interesting problem is to design an algorithm to achieve optimal block shaping entirely by simple slack-driven operations without resorting to geometric programming.

Besides, because of the similarity of the concept of slack in floorplanning and in circuit timing analysis, it would be interesting to see if a slack-driven approach similar to that in SDS can be applied to buffer and wire sizing for timing optimization.

Cross-References

▶ Floorplan and Placement
▶ Gate Sizing

▶ Slicing Floorplan Orientation
▶ Wire Sizing

Recommended Reading

1. Yan JZ, Chu C (2012) Optimal slack-driven block shaping algorithm in fixed-outline floorplanning. In: Proceedings of international symposium on physical design, Napa, pp 179–186
2. Yan JZ, Chu C (2013) SDS: an optimal slack-driven block shaping algorithm for fixed-outline floorplanning. IEEE Trans Comput Aided Design 32(2): 175–188
3. Wang L-T, Chang Y-W, Cheng K-T (eds) (2009) Electronic design automation: synthesis, verification, and test. Morgan Kaufmann, Burlington
4. Kahng A (2000) Classical floorplanning harmful? In: Proceedings of international symposium on physical design, San Diego, pp 207–213
5. Otten RHJM (1982) Automatic floorplan design. In: Proceedings of ACM/IEEE design automation conference, Las Vegas, pp 261–267
6. Stockmeyer L (1983) Optimal orientations of cells in slicing floorplan designs. Inf Control 57: 91–101
7. Wang TC, Wong DF (1992) Optimal floorplan area optimization. IEEE Trans Comput Aided Design 11(8):992–1001
8. Pan P, Liu CL (1995) Area minimization for floorplans. IEEE Trans Comput Aided Design 14(1): 129–132
9. Kang M, Dai WWM (1997) General floorplanning with L-shaped, T-shaped and soft blocks based on bounded slicing grid structure. In: Proceedings of ASP-DAC, Chiba, pp 265–270
10. Moh TS, Chang TS, Hakimi SL (1996) Globally optimal floorplanning for a layout problem. IEEE Trans Circuits Syst I 43:713–720
11. Young FY, Chu CCN, Luk WS, Wong YC (2001) Handling soft modules in general non-slicing floorplan using Lagrangian relaxation. IEEE Trans Comput Aided Design 20(5):687–692
12. Lin C, Zhou H, Chu C (2006) A revisit to floorplan optimization by Lagrangian relaxation. In: Proceedings of ICCAD, San Jose, pp 164–171
13. Adya SN, Markov IL (2003) Fixed-outline floorplanning: enabling hierarchical design. IEEE Trans VLSI Syst 11(6):1120–1135
14. Lin J-M, Hung Z-X (2011) UFO: unified convex optimization algorithms for fixed-outline floorplanning considering pre-placed modules. IEEE Trans Comput Aided Design 30(7):1034–1044

Boosting Textual Compression

Paolo Ferragina[1] and Giovanni Manzini[2,3]
[1]Department of Computer Science, University of Pisa, Pisa, Italy
[2]Department of Computer Science, University of Eastern Piedmont, Alessandria, Italy
[3]Department of Science and Technological Innovation, University of Piemonte Orientale, Alessandria, Italy

Keywords

Context-aware compression; High-order compression models

Years and Authors of Summarized Original Work

2005; Ferragina, Giancarlo, Manzini, Sciortino

Problem Definition

Informally, a boosting technique is a method that, when applied to a particular class of algorithms, yields improved algorithms. The improvement must be provable and well defined in terms of one or more of the parameters characterizing the algorithmic performance. Examples of boosters can be found in the context of randomized algorithms (here, a booster allows one to turn a *BPP* algorithm into an *RP* one [6]) and computational learning theory (here, a booster allows one to improve the prediction accuracy of a weak learning algorithm [10]). The problem of compression boosting consists of designing a technique that improves the compression performance of a wide class of algorithms. In particular, the results of Ferragina et al. provide a general technique for turning a compressor that uses no context information into one that always uses the best possible context.

The classic Huffman and arithmetic coding algorithms [1] are examples of *statistical* compressors which typically encode an input symbol according to its *overall* frequency in the data to be compressed. (In their dynamic versions these algorithms consider the frequency of a symbol in the already scanned portion of the input.) This approach is efficient and easy to implement but achieves poor compression. The compression performance of statistical compressors can be improved by adopting *higher*-order models that obtain better estimates for the frequencies of the input symbols. The PPM compressor [9] implements this idea by collecting (the frequency of) all symbols which follow *any* k-long context and by compressing them via arithmetic coding. The length k of the context is a parameter of the algorithm that depends on the data to be compressed: it is different if one is compressing English text, a DNA sequence, or an XML document. There exist other examples of sophisticated compressors that use context information in an *implicit* way, such as Lempel-Ziv and Burrows-Wheeler compressors [9]. All these context-aware algorithms are effective in terms of compression performance, but are usually rather complex to implement and difficult to analyze.

Applying the boosting technique of Ferragina et al. to Huffman or arithmetic coding yields a new compression algorithm with the following features: (i) the new algorithm uses the boosted compressor as a black box; (ii) the new algorithm compresses in a PPM-like style, automatically choosing the *optimal* value of k; and (iii) the new algorithm has essentially the same time/space asymptotic performance of the boosted compressor. The following sections give a precise and formal treatment of the three properties (i)–(iii) outlined above.

Key Results

Notation: The Empirical Entropy

Let s be a string over the alphabet $\Sigma = \{a_1, \ldots, a_h\}$, and for each $a_i \in \Sigma$, let n_i be the number of occurrences of a_i in s. The *0th order empirical entropy* of the string s is defined as $H_0(s) = -\sum_{i=1}^{h} (n_i/|s|) \log(n_i/|s|)$, where it is assumed that all logarithms are taken to the base 2 and $0 \log 0 = 0$. It is well known that H_0 is the maximum compression one can achieve using a uniquely decodable code in which a fixed codeword is assigned to each alphabet symbol. Greater compression is achievable if the codeword of a symbol depends on the k symbols following it (namely, its *context*). (In data compression it is customary to define the context looking at the symbols *preceding* the one to be encoded. The present entry uses the nonstandard "forward" contexts to simplify the notation of the following sections. Note that working with "forward" contexts is equivalent to working with the traditional "backward" contexts on the string s reversed (see [3] for details).) Let us define w_s as the string of single symbols immediately preceding the occurrences of w in s. For example, for $s = \mathtt{bcabcabdca}$, it is $\mathtt{ca}_s = \mathtt{bbd}$. The value

$$H_k(s) = \frac{1}{|s|} \sum_{w \in \Sigma^k} |w_s| H_0(w_s) \qquad (1)$$

is the k-th order empirical entropy of s and is a lower bound to the compression one can achieve using codewords which only depend on the k symbols immediately following the one to be encoded.

Example 1 Let $s = \mathtt{mississippi}$. For $k = 1$, it is $\mathtt{i}_s = \mathtt{mssp}$, $\mathtt{s}_s = \mathtt{isis}$, $\mathtt{p}_s = \mathtt{ip}$. Hence,

$$H_1(s) = \frac{4}{11} H_0(\mathtt{mssp}) + \frac{4}{11} H_0(\mathtt{isis})$$

$$+ \frac{2}{11} H_0(\mathtt{ip})$$

$$= \frac{6}{11} + \frac{4}{11} + \frac{2}{11} = \frac{12}{11}.$$

Note that the empirical entropy is defined for any string and can be used to measure the performance of compression algorithms without any assumption on the input source. Unfortunately, for some (highly compressible) strings, the empirical entropy provides a lower bound that is too conservative. For example, for $s = \mathtt{a}^n$, it is $|s| H_k(s) = 0$ for any $k \geq 0$. To better deal with

highly compressible strings, [7] introduced the notion of 0th *order modified empirical* entropy $H_0^*(s)$ whose property is that $|s| H_0^*(s)$ is at least equal to the number of bits needed to write down the length of s in binary. The kth *order modified empirical entropy* H_k^* is then defined in terms of H_0^* as the maximum compression one can achieve by looking at *no more than* k symbols following the one to be encoded.

The Burrows-Wheeler Transform

Given a string s, the Burrows-Wheeler transform [2] (*bwt*) consists of three basic steps: (1) append to the end of s a special symbol $\$$ smaller than any other symbol in Σ; (2) form a *conceptual* matrix $\mathcal{M}$ whose rows are the cyclic shifts of the string $s\$$, sorted in lexicographic order; and (3) construct the transformed text $\hat{s} = \mathtt{bwt}(s)$ by taking the last column of $\mathcal{M}$ (see Fig. 1). In [2] Burrows and Wheeler proved that $\hat{s}$ is a permutation of s, and that from $\hat{s}$ it is possible to recover s in $O(|s|)$ time.

To see the power of the *bwt*, the reader should reason in terms of empirical entropy. Fix a positive integer k. The first k columns of the *bwt* matrix contain, lexicographically ordered, all length-k substrings of s (and k substrings containing the symbol $\$$). For any length-k substring w of s, the symbols immediately preceding every occurrence of w in s are grouped together in a set of consecutive positions of $\hat{s}$ since they are the last symbols of the rows of $\mathcal{M}$ prefixed by w. Using the notation introduced for defining H_k, it is possible to rephrase this property by saying that the symbols of w_s are consecutive within $\hat{s}$ or, equivalently, that $\hat{s}$ contains, as a substring, a permutation $\pi_w(w_s)$ of the string w_s.

Example 2 Let $s = \mathtt{mississippi}$ and $k = 1$. Figure 1 shows that $\hat{s}[1, 4] = \mathtt{pssm}$ is a permutation of $\mathtt{i}_s = \mathtt{mssp}$. In addition, $\hat{s}[6, 7] = \mathtt{pi}$ is a permutation of $\mathtt{p}_s = \mathtt{ip}$, and $\hat{s}[8, 11] = \mathtt{ssii}$ is a permutation of $\mathtt{s}_s = \mathtt{isis}$.

Since permuting a string does not change its (modified) 0th order empirical entropy (that is, $H_0(\pi_w(w_s)) = H_0(w_s)$), the Burrows-Wheeler transform can be seen as a tool for reducing the

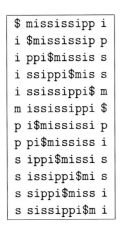

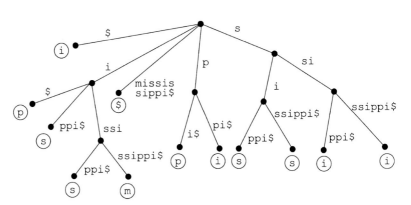

Boosting Textual Compression, Fig. 1 The *bwt* matrix (*left*) and the suffix tree (*right*) for the string $s =$ mississippi$. The output of the *bwt* is the last column of the *bwt* matrix, i.e., $\hat{s} = \text{bwt}(s) = $ ipssm$pissii

problem of compressing s up to its kth order entropy to the problem of compressing *distinct portions* of $\hat{s}$ up to their 0*th order* entropy. To see this, assume partitioning of $\hat{s}$ into the substrings $\pi_w(w_s)$ by varying w over Σ^k. It follows that $\hat{s} = \bigsqcup\limits_{w \in \Sigma^k} \pi_w(w_s)$ where $\bigsqcup$ denotes the concatenation operator among strings. (In addition to $\bigsqcup_{w \in \Sigma^k} \pi_w(w_s)$, the string $\hat{s}$ also contains the last k symbols of s (which do not belong to any w_s) and the special symbol $. For simplicity these symbols will be ignored in the following part of the entry.) By (1) it follows that

$$\sum_{w \in \Sigma^k} |\pi_w(w_s)| H_0(\pi_w(w_s))$$

$$= \sum_{w \in \Sigma^k} |w_s| H_0(w_s) = |s| H_k(s).$$

Hence, to compress s up to $|s| H_k(s)$, it suffices to compress each substring $\pi_w(w_s)$ up to its 0th order empirical entropy. Note, however, that in the above scheme the parameter k must be chosen in advance. Moreover, a similar scheme cannot be applied to H_k^* which is defined in terms of contexts of length *at most* k. As a result, no efficient procedure is known for computing the partition of $\hat{s}$ corresponding to $H_k^*(s)$. The compression booster [3] is a natural complement to the *bwt* and allows one to compress any string

s up to $H_k(s)$ (or $H_k^*(s)$) simultaneously for all $k \geq 0$.

The Compression Boosting Algorithm

A crucial ingredient of compression boosting is the relationship between the *bwt* matrix and the suffix tree data structure. Let $\mathcal{T}$ denote the suffix tree of the string s. $\mathcal{T}$ has $|s| + 1$ leaves, one per suffix of s, and edges labeled with substrings of s (see Fig. 1). Any node u of $\mathcal{T}$ has *implicitly associated* a substring of s, given by the concatenation of the edge labels on the downward path from the root of $\mathcal{T}$ to u. In that implicit association, the leaves of $\mathcal{T}$ correspond to the suffixes of s. Assume that the suffix tree edges are sorted lexicographically. Since each row of the *bwt* matrix is prefixed by one suffix of s and rows are lexicographically sorted, the ith leaf (counting from the left) of the suffix tree corresponds to the ith row of the *bwt* matrix. Associate to the ith leaf of $\mathcal{T}$ the ith symbol of $\hat{s} = \text{bwt}(s)$. In Fig. 1 these symbols are represented inside circles.

For any suffix tree node u, let $\hat{s}\langle u \rangle$ denote the substring of $\hat{s}$ obtained by concatenating, from left to right, the symbols associated to the leaves descending from node u. Of course $\hat{s}\langle root(\mathcal{T}) \rangle = \hat{s}$. A subset $\mathcal{L}$ of $\mathcal{T}$'s nodes is called a *leaf cover* if every leaf of the suffix tree has a *unique* ancestor in $\mathcal{T}$. Any leaf cover $\mathcal{L} = \{u_1, \ldots, u_p\}$ naturally induces a partition of the leaves of $\mathcal{T}$.

Because of the relationship between $\mathcal{T}$ and the *bwt* matrix, this is also a partition of $\hat{s}$, namely, $\{\hat{s}\langle u_1\rangle, \ldots, \hat{s}\langle u_p\rangle\}$.

Example 3 Consider the suffix tree in Fig. 1. A leaf cover consists of all nodes of depth one. The partition of $\hat{s}$ induced by this leaf cover is $\{\text{i}, \text{pssm}, \$, \text{pi}, \text{ssii}\}$.

Let C denote a function that associates to every string x over $\Sigma \cup \{\$\}$ a positive real value $C(x)$. For any leaf cover $\mathcal{L}$, define its cost as $C(\mathcal{L}) = \sum_{u \in \mathcal{L}} C(\hat{s}\langle u\rangle)$. In other words, the cost of the leaf cover $\mathcal{L}$ is equal to the sum of the costs of the strings in the partition induced by $\mathcal{L}$. A leaf cover $\mathcal{L}_{\min}$ is called *optimal* with respect to C if $C(\mathcal{L}_{\min}) \le C(\mathcal{L})$, for any leaf cover $\mathcal{L}$.

Let A be a compressor such that, for any string x, its output size is bounded by $|x| H_0(x) + \eta|x| + \mu$ bits, where η and μ are constants. Define the cost function $C_A(x) = |x| H_0(x) + \eta|x| + \mu$. In [3] Ferragina et al. exhibit a linear-time greedy algorithm that computes the optimal leaf cover $\mathcal{L}_{\min}$ with respect to C_A. The authors of [3] also show that, for any $k \ge 0$, there exists a leaf cover $\mathcal{L}_k$ of cost $C_A(\mathcal{L}_k) = |s| H_k(s) + \eta|s| + O(|\Sigma|^k)$. These two crucial observations show that, if one uses A to compress each substring in the partition induced by the optimal leaf cover $\mathcal{L}_{\min}$, the total output size is bounded in terms of $|s| H_k(s)$, for any $k \ge 0$. In fact,

$$\sum_{u \in \mathcal{L}_{\min}} C_A(\hat{s}\langle u\rangle) = C_A(\mathcal{L}_{\min}) \le C_A(\mathcal{L}_k)$$

$$= |s| H_k(s) + \eta|s| + O(|\Sigma|^k)$$

In summary, boosting the compressor A over the string s consists of three main steps:

1. Compute $\hat{s} = \text{bwt}(s)$.
2. Compute the optimal leaf cover $\mathcal{L}_{\min}$ with respect to C_A and partition $\hat{s}$ according to $\mathcal{L}_{\min}$.
3. Compress each substring of the partition using the algorithm A.

So the boosting paradigm reduces the design of effective compressors that use context information, to the (usually easier) design of 0th order compressors. The performance of this paradigm is summarized by the following theorem.

Theorem 1 ([3]) *Let A be a compressor that squeezes any string x in at most $|x| H_0(x) + \eta|x| + \mu$ bits. The compression booster applied to A produces an output whose size is bounded by $|s| H_k(s) + \log|s| + \eta|s| + O(|\Sigma|^k)$ bits simultaneously for all $k \ge 0$. With respect to A, the booster introduces a space overhead of $O(|s| \log|s|)$ bits and no asymptotic time overhead in the compression process.* □

A similar result holds for the modified entropy H_k^* as well (but it is much harder to prove): given a compressor A that squeezes any string x in at most $\lambda|x| H_0^*(x) + \mu$ bits, the compression booster produces an output whose size is bounded by $\lambda|s| H_k^*(s) + \log|s| + O(|\Sigma|^k)$ bits, simultaneously for all $k \ge 0$. In [3] the authors also show that no compression algorithm, satisfying some mild assumptions on its inner working, can achieve a similar bound in which both the multiplicative factor λ and the additive logarithmic term are dropped simultaneously. Furthermore [3] proposes an instantiation of the booster which compresses any string s in at most $2.5 |s| H_k^*(s) + \log|s| + O(|\Sigma|^k)$ bits. This bound is analytically superior to the bounds proven for best existing compressors including Lempel-Ziv, Burrows-Wheeler, and PPM compressors.

Applications

Apart from the natural application in data compression, compressor boosting has been used also to design compressed full-text indexes [8].

Open Problems

The boosting paradigm may be generalized as follows: given a compressor A, find a permutation $\mathcal{P}$ for the symbols of the string s *and* a partitioning strategy such that the boosting approach, applied to them, minimizes the output size. These pages have provided convincing evidence that the Burrows-Wheeler transform is an elegant and efficient permutation $\mathcal{P}$. Surprisingly enough, other classic data compression problems fall into this framework: shortest common superstring (which is MAX-SNP hard), run length encoding for a set of strings (which is polynomially solvable), *LZ77*, and minimum number of phrases (which is MAX-SNP hard). Therefore, the boosting approach is general enough to deserve further theoretical and practical attention [5].

Experimental Results

An investigation of several compression algorithms based on boosting and a comparison with other state-of-the-art compressors are presented in [4]. The experiments show that the boosting technique is more robust than other *bwt*-based approaches and works well even with less effective 0th order compressors. However, these positive features are achieved using more (time and space) resources.

Data Sets

The data sets used in [4] are available from http://people.unipmn.it/manzini/boosting. Other data sets for compression and indexing are available at the Pizza&Chili site http://pizzachili.di.unipi.it/.

URL to Code

The compression boosting page (http://people.unipmn.it/manzini/boosting) contains the source code of all the algorithms tested in [4]. The code is organized in a highly modular library that can be used to boost any compressor even without knowing the *bwt* or the boosting procedure.

Cross-References

▶ Arithmetic Coding for Data Compression
▶ Burrows-Wheeler Transform
▶ Compressing and Indexing Structured Text
▶ Suffix Array Construction
▶ Suffix Tree Construction
▶ Suffix Tree Construction in Hierarchical Memory
▶ Table Compression

Recommended Reading

1. Bell TC, Cleary JG, Witten IH (1990) Text compression. Prentice Hall, Englewood
2. Burrows M, Wheeler D (1994) A block sorting lossless data compression algorithm. Technical report 124, Digital Equipment Corporation
3. Ferragina P, Giancarlo R, Manzini G, Sciortino M (2005) Boosting textual compression in optimal linear time. J ACM 52:688–713
4. Ferragina P, Giancarlo R, Manzini G (2006) The engineering of a compression boosting library: theory vs practice in bwt compression. In: Proceedings of the 14th European symposium on algorithms (ESA), Zurich. LNCS, vol 4168. Springer, Berlin, pp 756–767
5. Giancarlo R, Restivo A, Sciortino M (2007) From first principles to the Burrows and Wheeler transform and beyond, via combinatorial optimization. Theor Comput Sci 387(3):236–248
6. Karp R, Pippenger N, Sipser M (1985) A time-randomness tradeoff. In: Proceedings of the conference on probabilistic computational complexity, Santa Barbara. AMS, pp 150–159
7. Manzini G (2001) An analysis of the Burrows-Wheeler transform. J ACM 48:407–430
8. Navarro G, Mäkinen V (2007) Compressed full text indexes. ACM Comput Surv 39(1):Article No. 2
9. Salomon D (2007) Data compression: the complete reference, 4th edn. Springer, New York
10. Schapire RE (1990) The strength of weak learnability. Mach Learn 2:197–227

Branchwidth of Graphs

Fedor V. Fomin[1] and Dimitrios Thilikos[2,3]
[1]Department of Informatics, University of Bergen, Bergen, Norway
[2]AlGCo Project Team, CNRS, LIRMM, France
[3]Department of Mathematics, National and Kapodistrian University of Athens, Athens, Greece

Keywords

Tangle Number

Years and Authors of Summarized Original Work

2003; Fomin, Thilikos

Problem Definition

Branchwidth, along with its better-known counterpart, treewidth, are measures of the "global connectivity" of a graph.

Definition

Let G be a graph on n vertices. A *branch decomposition* of G is a pair (T, τ), where T is a tree with vertices of degree 1 or 3 and τ is a bijection from the set of leaves of T to the edges of G. The *order*, we denote it as $\alpha(e)$, of an edge e in T is the number of vertices v of G such that there are leaves t_1, t_2 in T in different components of $T(V(T), E(T) - e)$ with $\tau(t_1)$ and $\tau(t_2)$ both containing v as an endpoint.

The *width* of (T, τ) is equal to $\max_{e \in E(T)} \{\alpha(e)\}$, i.e., is the maximum order over all edges of T. The *branchwidth* of G is the minimum width over all the branch decompositions of G (in the case where $|E(G)| \leq 1$, then we define the branchwidth to be 0; if $|E(G)| = 0$, then G has no branch decomposition; if $|E(G)| = 1$, then G has a branch decomposition consisting of a tree with one vertex – the width of this branch decomposition is considered to be 0).

The above definition can be directly extended to hypergraphs where τ is a bijection from the leaves of T to the hyperedges of G. The same definition can easily be extended to matroids.

Branchwidth was first defined by Robertson and Seymour in [25] and served as a main tool for their proof of Wagner's Conjecture in their Graph Minors series of papers. There, branchwidth was used as an alternative to the parameter of treewidth as it appeared easier to handle for the purposes of the proof. The relation between branchwidth and treewidth is given by the following result.

Theorem 1 ([25]) *If G is a graph, then* branchwidth$(G) \leq$ treewidth$(G) + 1 \leq \lfloor 3/2$ branchwidth$(G) \rfloor$.

The algorithmic problems related to branchwidth are of two kinds: first find fast algorithms computing its value and, second, use it in order to design fast dynamic programming algorithms for other problems.

Key Results

Algorithms for Branchwidth

Computing branchwidth is an NP-hard problem ([29]). Moreover, the problem remains NP-hard even if we restrict its input graphs to the class of split graphs or bipartite graphs [20].

On the positive side, branchwidth is computable in polynomial time on interval graphs [20, 24], and circular arc graphs [21]. Perhaps the most celebrated positive result on branchwidth is an $O(n^2)$ algorithm for the branchwidth of planar graphs, given by Seymour and Thomas in [29]. In the same paper they also give an $O(n^4)$ algorithm to compute an optimal branch decomposition. (The running time of this algorithm has been improved to $O(n^3)$ in [18].) The algorithm in [29] is basically an algorithm for a parameter called carving width, related to telephone routing and the result for branchwidth follows from the fact that the branch width of a planar graph is half of the carving-width of its medial graph.

The algorithm for planar graphs [29] can be used to construct an approximation algorithm for branchwidth of some non-planar graphs. On graph classes excluding a single crossing graph as a minor branchwidth can be approximated within a factor of 2.25 [7] (a graph H is a *minor* of a graph G if H can be obtained by a subgraph of G after applying edge contractions). Finally, it follows from [13] that for every minor closed graph class, branchwidth can be approximated by a constant factor.

Branchwidth cannot increase when applying edge contractions or removals. According to the Graph Minors theory, this implies that, for any fixed k, there is a finite number of minor minimal graphs of branchwidth more than k and we denote this set of graphs by $\mathcal{B}_k$. Checking whether a graph G contains a fixed graph as a minor can be done in polynomial time [27].

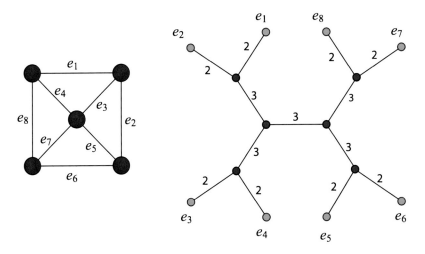

Branchwidth of Graphs, Fig. 1 Example of a graph and its branch decomposition of width 3

Therefore, the knowledge of $\mathcal{B}_k$ implies the construction of a polynomial time algorithm for deciding whether branchwidth$(G) \leq k$, for any fixed k. Unfortunately $\mathcal{B}_k$ is known only for small values of k. In particular, $\mathcal{B}_0 = \{P_2\}$, $\mathcal{B}_1 = \{P_4, K_3\}$, $\mathcal{B}_2 = \{K_4\}$ and $\mathcal{B}_3 = \{K_5, V_8, M_6, Q_3\}$ (here K_r is a clique on r vertices, P_r is a path on r edges, V_8 is the graph obtained by a cycle on 8 vertices if we connect all pairs of vertices with cyclic distance 4, M_6 is the octahedron, and Q_3 is the 3-dimensional cube). However, for any fixed k, one can construct a linear, on $n = |V(G)|$, algorithm that decides whether an input graph G has branchwidth $\leq k$ and, if so, outputs the corresponding branch decomposition (see [3]). In technical terms, this implies that the problem of asking, for a given graph G, whether branchwidth$(G) \leq k$, parameterized by k is fixed parameter tractable (i.e., belongs in the parameterized complexity class **FPT**). (See [12] for further references on parameterized algorithms and complexity.) The algorithm in [3] is complicated and uses the technique of characteristic sequences, which was also used in order to prove the analogous result for treewidth. For the particular cases where $k \leq 3$, simpler algorithms exist that use the "reduction rule" technique (see [4]). We stress that $\mathcal{B}_4$ remains unknown while several elements of it have been detected so far (including the

dodecahedron and the icosahedron graphs). There is a number of algorithms that for a given k in time $2^{O(k)} \cdot n^{O(1)}$ either decide that the branchwidth of a given graph is at least k, or construct a branch decomposition of width $O(k)$ (see [26]). These results can be generalized to compute the branchwidth of matroids and even more general parameters.

An exact algorithm for branchwidth appeared in [14]. Its complexity is $O((2 \cdot \sqrt{3})^n \cdot n^{O(1)})$. The algorithm exploits special properties of branchwidth (see also [24]).

In contrast to treewidth, edge maximal graphs of given branchwidth are not so easy to characterize (for treewidth there are just k-trees, i.e., chordal graphs with all maximal cliques of size $k + 1$). An algorithm for generating such graphs has been given in [23] and reveals several structural issues on this parameter.

It is known that a large number of graph theoretical problems can be solved in linear time when their inputs are restricted to graphs of small (i.e., fixed) treewidth or branchwidth (see [2]).

Branchwidth appeared to be a useful tool in the design of exact subexponential algorithms on planar graphs and their generalizations. The basic idea behind this approach is very simple: Let $\mathcal{P}$ be a problem on graphs and G be a class of graphs such that

- for every graph $G \in \mathcal{G}$ of branchwidth at most ℓ, the problem $\mathcal{P}$ can be solved in time $2^{c \cdot \ell} \cdot n^{O(1)}$, where c is a constant, and;
- for every graph $G \in \mathcal{G}$ on n vertices a branch decomposition (not necessarily optimal) of G of width at most $h(n)$ can be constructed in polynomial time, where $h(n)$ is a function.

Then for every graph $G \in \mathcal{G}$, the problem $\mathcal{P}$ can be solved in time $2^{c \cdot h(n)} \cdot n^{O(1)}$. Thus, everything boils down to computations of constants c and functions $h(n)$. These computations can be quite involved. For example, as was shown in [17], for every planar graph G on n vertices, the branchwidth of G is at most $\sqrt{4.5n} < 2.1214\sqrt{n}$. For extensions of this bound to graphs embeddable on a surface of genus g, see [15].

Dorn [9] used fast matrix multiplication in dynamic programming to estimate the constants c for a number of problems. For example, for the MAXIMUM INDEPENDENT SET problem, $c \leq \omega/2$, where $\omega < 2.376$ is the matrix product exponent over a ring, which implies that the INDEPENDENT SET problem on planar graphs is solvable in time $O(2^{2.52\sqrt{n}})$. For the MINIMUM DOMINATING SET problem, $c \leq 4$, thus implying that the branch decomposition method runs in time $O(2^{3.99\sqrt{n}})$. It appears that algorithms of running time $2^{O(\sqrt{n})}$ can be designed even for some of the "non-local" problems, such as the HAMILTONIAN CYCLE, CONNECTED DOMINATING SET, and STEINER TREE, for which no time $2^{O(\ell)} \cdot n^{O(1)}$ algorithm on general graphs of branchwidth ℓ is known [11]. Here one needs special properties of some optimal planar branch decompositions, roughly speaking that every edge of T corresponds to a disk on a plane such that all edges of G corresponding to one component of $T - e$ are inside the disk and all other edges are outside. Some of the subexponential algorithms on planar graphs can be generalized for graphs embedded on surfaces [10] and, more generally, to graph classes that are closed under taking of minors [8].

A similar approach can be used for parameterized problems on planar graphs. For example,

a parameterized algorithm that finds a dominating set of size $\leq k$ (or reports that no such set exists) in time $2^{O(\sqrt{k})}n^{O(1)}$ can be obtained based on the following observations: there is a constant c such that every planar graph of branchwidth at least $c\sqrt{k}$ does not contain a dominating set of size at most k. Then for a given k the algorithm computes an optimal branch decomposition of a palanar graph G and if its width is more than $c\sqrt{k}$ concludes that G has no dominating set of size k. Otherwise, find an optimal dominating set by performing dynamic programming in time $2^{O(\sqrt{k})}n^{O(1)}$. There are several ways of bounding a parameter of a planar graph in terms of its branchwidth or treewidth including techniques similar to Baker's approach from approximation algorithms [1], the use of separators, or by some combinatorial arguments, as shown in [16]. Another general approach of bounding the branchwidth of a planar graph by parameters, is based on the results of Robertson et al. [28] regarding quickly excluding a planar graph. This brings us to the notion of *bidimensionality* [6]. Parameterized algorithms based on branch decompositions can be generalized from planar graphs to graphs embedded on surfaces and to graphs excluding a fixed graph as a minor.

Applications

See [5] for using branchwidth for solving TSP.

Open Problems

1. It is known that any planar graph G has branchwidth at most $\sqrt{4.5} \cdot \sqrt{|V(G)|}$ (or at most $\frac{3}{2} \cdot \sqrt{|E(G)|} + 2$) [17]. Is it possible to improver this upper bound? Any possible improvement would accelerate many of the known exact or parameterized algorithms on planar graphs that use dynamic programming on branch decompositions.

2. In contrast to treewidth, very few graph classes are known where branchwidth is computable in polynomial time. Find graphs

classes where branchwidth can be computed or approximated in polynomial time.

3. Find $\mathcal{B}_k$ for values of k bigger than 3. The only structural result on $\mathcal{B}_k$ is that its planar elements will be either self-dual or pairwise-dual. This follows from the fact that dual planar graphs have the same branchwidth [29, 16].

4. Find an exact algorithm for branchwidth of complexity $O^*(2^n)$ (the notation $O^*()$ assumes that we drop the non-exponential terms in the classic $O()$ notation).

5. The dependence on k of the linear time algorithm for branchwidth in [3] is huge. Find an $2^{O(k)} \cdot n^{O(1)}$ step algorithm, deciding whether the branchwidth of an n-vertex input graph is at most k.

Cross-References

▶ Bidimensionality
▶ Treewidth of Graphs

Recommended Reading

1. Alber J, Bodlaender HL, Fernau H, Kloks T, Niedermeier R (2002) Fixed parameter algorithms for dominating set and related problems on planar graphs. Algorithmica 33:461–493
2. Arnborg S (1985) Efficient algorithms for combinatorial problems on graphs with bounded decomposability – a survey. BIT 25:2–23
3. Bodlaender HL, Thilikos DM (1997) Constructive linear time algorithms for branchwidth. In: Automata, languages and programming (Bologna, 1997). Lecture notes in computer science, vol 1256. Springer, Berlin, pp 627–637
4. Bodlaender HL, Thilikos DM (1999) Graphs with branchwidth at most three. J Algoritm 32:167–194
5. Cook W, Seymour PD (2003) Tour merging via branch-decomposition. INFORMS J Comput 15:233–248
6. Demaine ED, Fomin FV, Hajiaghayi M, Thilikos DM (2004) Bidimensional parameters and local treewidth. SIAM J Discret Math 18:501–511
7. Demaine ED, Hajiaghayi MT, Nishimura N, Ragde P, Thilikos DM (2004) Approximation algorithms for classes of graphs excluding single-crossing graphs as minors. J Comput Syst Sci 69:166–195
8. Dorn F, Fomin FV, Thilikos DM (2006) Subexponential algorithms for non-local problems on H-minor-free graphs. In: Proceedings of the nineteenth annual ACM-SIAM symposium on discrete algorithms (SODA 2008). Society for Industrial and Applied Mathematics, Philadelphia, pp 631–640
9. Dorn F (2006) Dynamic programming and fast matrix multiplication. In: Proceedings of the 14th annual European symposium on algorithms (ESA 2006). Lecture notes in computer science, vol 4168. Springer, Berlin, pp 280–291
10. Dorn F, Fomin FV, Thilikos DM (2005) Fast subexponential algorithm for non-local problems on graphs of bounded genus. In: Proceedings of the 10th Scandinavian workshop on algorithm theory (SWAT 2006). Lecture notes in computer science. Springer, Berlin
11. Dorn F, Penninkx E, Bodlaender H, Fomin FV (2005) Efficient exact algorithms on planar graphs: exploiting sphere cut branch decompositions. In: Proceedings of the 13th annual European symposium on algorithms (ESA 2005). Lecture notes in computer science, vol 3669. Springer, Berlin, pp 95–106
12. Downey RG, Fellows MR (1999) Parameterized complexity. In: Monographs in computer science. Springer, New York
13. Feige U, Hajiaghayi M, Lee JR (2005) Improved approximation algorithms for minimum-weight vertex separators. In: Proceedings of the 37th annual ACM symposium on theory of computing (STOC 2005). ACM Press, New York, pp 563–572
14. Fomin FV, Mazoit F, Todinca I (2005) Computing branchwidth via efficient triangulations and blocks. In: Proceedings of the 31st workshop on graph theoretic concepts in computer science (WG 2005). Lecture notes computer science, vol 3787. Springer, Berlin, pp 374–384
15. Fomin FV, Thilikos DM (2004) Fast parameterized algorithms for graphs on surfaces: linear kernel and exponential speed-up. In: Proceedings of the 31st international colloquium on automata, languages and programming (ICALP 2004). Lecture notes computer science, vol 3142. Springer, Berlin, pp 581–592
16. Fomin FV, Thilikos DM (2006) Dominating sets in planar graphs: branch-width and exponential speed-up. SIAM J Comput 36:281–309
17. Fomin FV, Thilikos DM (2006) New upper bounds on the decomposability of planar graphs. J Graph Theory 51:53–81
18. Gu QP, Tamaki H (2005) Optimal branch-decomposition of planar graphs in O(n3) time. In: Proceedings of the 32nd international colloquium on automata, languages and programming (ICALP 2005). Lecture notes computer science, vol 3580. Springer, Berlin, pp 373–384
19. Gu QP, Tamaki H (2006) Branch-width, parse trees, and monadic second-order logic for matroids. J Comb Theory Ser B 96:325–351
20. Kloks T, Kratochvíl J, Müller H (2005) Computing the branchwidth of interval graphs. Discret Appl Math 145:266–275

21. Mazoit F (2006) The branch-width of circular-arc graphs. In: 7th Latin American symposium on theoretical informatics (LATIN 2006)
22. Oum SI, Seymour P (2006) Approximating clique-width and branch-width. J Comb Theory Ser B 96:514–528
23. Paul C, Proskurowski A, Telle JA (2006) Generating graphs of bounded branchwidth. In: Proceedings of the 32nd workshop on graph theoretic concepts in computer science (WG 2006). Lecture notes computer science, vol 4271. Springer, Berlin, pp 205–216
24. Paul C, Telle JA (2005) New tools and simpler algorithms for branchwidth. In: Proceedings of the 13th annual European symposium on algorithms (ESA 2005)
25. Robertson N, Seymour PD (1991) Graph minors. X. Obstructions to tree-decomposition. J Comb Theory Ser B 52:153–190
26. Robertson N, Seymour PD (1995) Graph minors. XII. Distance on a surface. J Comb Theory Ser B 64:240–272
27. Robertson N, Seymour PD (1995) Graph minors. XIII. The disjoint paths problem. J Comb Theory Ser B 63:65–110
28. Robertson N, Seymour PD, Thomas R (1994) Quickly excluding a planar graph. J Comb Theory Ser B 62:323–348
29. Seymour PD, Thomas R (1994) Call routing and the ratcatcher. Combinatorica 14:217–241

Broadcast Scheduling – Minimizing Average Response Time

Ravishankar Krishnaswamy
Computer Science Department, Princeton University, Princeton, NJ, USA

Keywords

Broadcast scheduling; Combinatorial optimization; Discrepancy; Linear programming; Rounding; Scheduling

Years and Authors of Summarized Original Work

2005; Bansal, Charikar, Khanna, Naor
2008; Bansal, Coppersmith, Sviridenko
2014; Bansal, Charikar, Krishnaswamy, Li

Problem Definition

In this entry, we consider the classical broadcast scheduling problem and discuss some recent advances on this problem. The problem is formalized as follows: there is a server which has a collection of unit-sized *pages* $P = \{1, \ldots, n\}$. The server can broadcast pages in integer time slots in response to *requests*, which are given as the following sequence: at time t, the server receives $w_p(t) \in \mathbb{Z}_{\geq 0}$ requests for each page $p \in P$. We say that a request ρ for page p that arrives at time t is satisfied at time $c_p(t)$ if $c_p(t)$ is the first time after t by which the server has completely transmitted page p. The response time of the request ρ is defined to be $c_p(t) - t$, i.e., the time that elapses from its arrival till the time it is satisfied. Notice that by definition, the response time for any request is at least 1. The goal is to find a schedule for broadcasting pages to minimize the average response time, i.e., $(\sum_{t,p} w_p(t)(c_p(t)-t))/\sum_{t,p} w_p(t)$. Recall that the problem we discuss here is an *offline problem*, where the entire request sequence is specified as part of the input. There has also been much research on the online version of the problem, and we briefly discuss this toward the end of the entry.

Key Results

Erlebach and Hall [5] were the first to show complexity theoretic hardness for this problem by showing that it is NP complete. The techniques we describe below were introduced in [1, 3]. By fine-tuning these ideas, [2] shows the following result on the approximability of the offline problem, which will be the main result we will build toward this entry.

Theorem 1 ([2]) *Let $\gamma > 0$ be any arbitrary parameter. There is a polynomial time algorithm that finds a schedule with average response time $(2+\gamma) \cdot OPT + O((\sqrt{\log_{1+\gamma} n \cdot \log \log n}) \log n)$, where OPT denotes the value of the average response time in the optimum solution.*

By setting $\gamma = \Theta(\log n)$ above, we can get an approximation guarantee of $O(\log^{1.5} n)$. Also note that the $O(\log^{1.5} n)$ term in Theorem 1 is additive. As a result, for instance, where OPT is large (say $\Omega(\log^{1.5+\epsilon} n)$ for some $\epsilon > 0$), we can set γ arbitrarily small to get an approximation ratio arbitrarily close to 2.

Linear Programming Formulation

All of the algorithmic results in [1–3] are based on rounding the following natural LP relaxation for the problem. For each page $p \in [n]$ and each time t, there is a variable y_{pt} which indicates whether page p was transmitted at time t. We have another set of variables $x_{ptt'}$ s.t $t' > t$, which indicates whether a request for page p which arrives at time t is satisfied at t'. Let $w_p(t)$ denote the total weight of requests for page p that arrive at time t.

$$\min \sum_{p,t,t'>t} (t' - t) \cdot \omega_p(t) \cdot x_{ptt'} \qquad (1)$$

$$\text{s.t.} \sum_{p} y_{pt} \leq 1 \qquad \forall t \qquad (2)$$

$$\sum_{t'>t} x_{ptt'} \geq 1 \qquad \forall p,t \qquad (3)$$

$$x_{ptt'} \leq y_{pt'} \qquad \forall p,t,t' \geq t \qquad (4)$$

$$x_{ptt'}, y_{pt'} \in [0,1] \qquad \forall p,t,t' \qquad (5)$$

Constraint (2) ensures that only one page is transmitted in each time, (3) ensures that each request must be satisfied, and (4) ensures that a request for page p can be satisfied at time t only if p is transmitted at time t. Finally, a request arriving at time t that is satisfied at time t' contributes $(t' - t)$ to the objective. Now consider the linear program obtained by relaxing the integrality constraints on $x_{ptt'}$ and y_{pt}.

Rounding Techniques

The following points illustrate the main ideas that form the building blocks of the rounding algorithms in [1–3].

The Half-Integrality Assumption

In what follows, we discuss the techniques in the special case that the LP solution is *half-integral*, i.e., where all the $x_{ptt'} \in \{0, \frac{1}{2}\}$. The general case essentially builds upon this intuition, and all main technical ingredients are contained in this special case.

Viewing the LP Solution as a Convex Combination of Blocks

In half-integral solutions, note that every request is satisfied by the two earliest half broadcasts of the corresponding page. For any page p, let $\tau_p = \{t_{p,1}, t_{p,2}, \ldots\}$ denote the times when the fractional solution broadcasts $\frac{1}{2}$ units of page p. Notice that the fractional solution can be entirely characterized by these sets τ_p for all pages p. The main intuition now is to view the fractional broadcast of each page as a *convex combination* of two different solutions, one which broadcasts the page p integrally at the odd times $\{t_{p,1}, t_{p,3}, \ldots\}$ and another which broadcasts the page p integrally at the even times $\{t_{p,2}, t_{p,4}, \ldots\}$. We call these the *odd schedule* and even *schedule* for page p.

Rounding the Solution to Minimize Backlog: Attempt 1 [1]

Our first and most natural rounding idea is to round the convex combination for each page into one of the odd or even schedules, each with probability $1/2$. Let us call this the *tentative schedule*. Note that on average, the tentative schedule broadcasts one page per time slot, and moreover, the expected response time of any request is equal to its fractional response time. The only issue, however, is that different pages may broadcast at the same time slots. Indeed, there could some time interval $[t_1, t_2)$ where the tentative schedule makes many more than $t_2 - t_1$ broadcasts! A natural manner to resolve this issue is to broadcast conflicting pages in a first-come first-serve manner, breaking ties arbitrarily. Now, the typical request waits for at most its fractional cost (on average), plus the *backlog* due to conflicting broadcasts. Formally, the backlog is defined as $\max_{t_1,t_2>t_1} N_A(t_1,t_2)-(t_2-t_1)$, where $N_A(t_1,t_2)$ is the number of broadcasts made in the interval $[t_1, t_2)$ by the tentative schedule. For this simple randomized algorithm, note that the backlog of any interval $[t_1, t_2)$ is at most $\tilde{O}(\sqrt{n})$ w.h.p by a standard concentration bound. This can be formalized to give us the $\tilde{O}(\sqrt{n})$ approximation algorithm of [1].

Rounding the Solution to Minimize Backlog: Attempt 2 [3]

Our next attempt involves controlling the backlog by explicitly enforcing constraints which periodically reset the backlog to 0. For this, we write an auxiliary LP as follows: we divide the set τ_p for each page p into blocks of size $B = \Theta(\log n)$ units each and have a variable for choosing the odd schedule or even schedule within each block. (If the first block chooses an odd schedule and the second block chooses an even schedule, then the requests which arrived at the boundary may incur greater costs, but [3] argues that these can be bounded for $B \geq \Omega(\log n)$.) Since each block has $B/2$ units of fractional transmission (recall that the LP is half-integral), the total number of blocks is at most $2T/B$, where T is the time horizon of all broadcasts made by the LP solution. Therefore, the total number of variables is at most $4T/B$ (each block chooses either an odd schedule or even schedule). Now, instead of asking for the LP to choose schedules such that each time has at most one transmission, suppose we *group the time slots into intervals of size B and ask for the LP to choose schedules such that each interval has at most B transmissions.* Now, there are T/B such constraints, and there are $2T/B$ constraints which enforce that we pick one schedule for each block.

Therefore, in this relaxed LP, we start with a solution that has $4T/B$ variables each set to $1/2$ and a total of $3T/B$ constraints. But now, we can convert this into an *optimal basic feasible solution* (where the number of nonzero variables is at most the number of constraints). This implies that at least a constant fraction of the blocks chooses either the odd or even schedules integrally. It is easy to see that the backlog incurred in any time interval $[t_1, t_2)$ is at most $O(B)$ since we

explicitly enforce 0 backlog for consecutive intervals of size B. Therefore, by repeating this process $O(\log T)$ time, we get a fully integral schedule with backlog $O(\log TB) = O(\log^2 n)$. This then gives us the $2 \cdot \text{OPT} + O(\log^2 n)$ approximation guarantee of [3].

Rounding the Solution to Minimize Backlog: Attempt 3 [2]

Our final attempt involves combining the ideas of attempts 1 and 2. Indeed, the main issue with approach 2 is that, when we solve for a basic feasible solution, we lose all control over how the solution looks! Therefore, we would ideally like for a rounding which enforces the constraints on time intervals of size B, but still *randomly selects the schedules within each block.* This way, we'll be able to argue that within each time interval of size B, the maximum backlog is $O(\sqrt{B})$. Moreover, if we look at a larger time interval I, we can decompose this into intervals of size B for which we have constraints in the LP, and a prefix and suffix of size at most B. Therefore, the backlog is constrained to be 0 by the LP for all intermediate intervals except the prefix and suffix which can have a backlog of $O(\sqrt{B})$. This will immediately give us the $O(\log^{1.5} n)$ approximation of [2].

Indeed, the main tool which lets us achieve this comes from a recent rounding technique of Lovett and Meka [7]. They prove the following result which they used as a subroutine for minimizing discrepancy of set systems, but it turns out to be a general result applicable in our setting as well [2].

Theorem 2 (Constructive partial coloring theorem [7]) *Let* $\mathbf{y} \in [0, 1]^m$ *be any starting point,* $\delta > 0$ *be an arbitrary error parameter,* $v_1, \ldots, v_n \in \mathbf{R}^n$ *vectors, and* $\lambda_1, \ldots, \lambda_n \geq 0$ *parameters with*

$$\sum_{i=1}^{n} e^{-\lambda_i^2/16} \leq \frac{m}{16}. \tag{6}$$

Then, there is a randomized $\tilde{O}((m+n)^3/\delta^2)$*-time algorithm to compute a vector* $\mathbf{z} \in [0, 1]^m$ *with* *(i)* $z_j \in [0, \delta] \cup [1 - \delta, 1]$ *for at least* $m/32$ *of the indices* $j \in [m]$.

(ii) $|v_i \cdot \mathbf{z} - v_i \cdot \mathbf{y}| \leq \lambda_i \|v_i\|_2$, *for each* $i \in [n]$.

Hardness Results

The authors [2] also complement the above algorithmic result with the following negative results.

Theorem 3 *The natural LP relaxation for the broadcast problem has an integrality gap of* $\Omega(\log n)$.

Interestingly, Theorem 3 is based on establishing a new connection with the problem of minimizing the discrepancy of 3 permutations. In the 3-permutation problem, we are given 3 permutations π_1, π_2, π_3 of $[n]$. The goal of the problem is to find a coloring χ that minimizes the discrepancy. The discrepancy of $\Pi = (\pi_1, \pi_2, \pi_3)$ w.r.t a ± 1 coloring χ is the worst case discrepancy of all prefixes. That is, $\max_{i=1}^{3} \max_{k=1}^{n} \left| \sum_{j=1}^{k} \chi(\pi_{i,j}) \right|$, where $\pi_{i,j}$ is the jth element in π_i. Newman and Nikolov [8] showed a tight $\Omega(\log n)$ lower bound on the discrepancy of 3 permutations, resolving a long-standing conjecture. The authors [2] note that this can be used to give an integrality gap for the broadcast scheduling problem as well. Then, by generalizing the connection to the discrepancy of ℓ permutations, [2] shows the following hardness results (prior to this, only NP hardness was known).

Theorem 4 *There is no* $O(\log^{1/2 - \epsilon} n)$ *approximation algorithm for the problem of minimizing average response time, for any* $\epsilon > 0$, *unless* $\mathsf{NP} \subseteq \cup_{t>0} \mathsf{BPTIME}(2^{\log^t n})$. *Moreover, for any sufficiently large* ℓ, *there is no* $O(\ell^{1/2})$ *approximation algorithm for the* ℓ-*permutation problem, unless* $\mathsf{NP} = \mathsf{RP}$.

Online Broadcast Scheduling

The broadcast scheduling problem has also been studied in the *online scheduling model*, where the algorithm is made aware of requests *only when*

they arrive, and it has to make the broadcast choices without knowledge of the future requests. Naturally, the performance of our algorithms degrade when compared to the offline model, but remarkably, we can get nontrivial algorithms even in the online model! The only additional assumption we need in the online model is that our scheduling algorithm may broadcast two pages (instead of one) every $1/\epsilon$ time slots, in order to get approximation ratios that depend on $1/\epsilon$. In particular, several $(1 + \epsilon)$-speed, $O(\text{poly}(1/\epsilon))$ are now known [4, 6], and it is also known that extra speed is necessary to obtain $n^{o(1)}$ competitive algorithms.

Recommended Reading

1. Bansal N, Charikar M, Khanna S, Naor J (2005) Approximating the average response time in broadcast scheduling. In: Proceedings of the 16th annual ACM-SIAM symposium on discrete algorithms, Vancouver
2. Bansal N, Charikar M, Krishnaswamy R, Li S (2014) Better algorithms and hardness for broadcast scheduling via a discrepancy approach. In: Proceedings of the 25th annual ACM-SIAM symposium on discrete algorithms, Portland
3. Bansal N, Coppersmith D, Sviridenko M (2008) Improved approximation algorithms for broadcast scheduling. SIAM J Comput 38(3): 1157–1174
4. Bansal N, Krishnaswamy R, Nagarajan V (2010) Better scalable algorithms for broadcast scheduling. In: Automata, languages and programming (ICALP), Bordeaux, pp 324–335
5. Erlebach T, Hall A (2002) NP-hardness of broadcast scheduling and inapproximability of singlesource unsplittable min-cost flow. In: Proceedings of the 13th ACM-SIAM symposium on discrete algorithms, San Francisco, pp 194–202
6. Im S, Moseley B (2012) An online scalable algorithm for average flow time in broadcast scheduling. ACM Trans Algorithms (TALG) 8(4):39
7. Lovett S, Meka R (2012) Constructive discrepancy minimization by walking on the edges. In: 53rd annual IEEE symposium on foundations of computer science (FOCS), New Brunswick, pp 61–67
8. Newman A, Neiman O, Nikolov A (2012) Beck's three permutations conjecture: a counterexample and some consequences. In: 2012 IEEE 53rd annual symposium on foundations of computer science (FOCS), New Brunswick. IEEE, APA

Broadcasting in Geometric Radio Networks

Andrzej Pelc
Department of Computer Science, University of
Québec-Ottawa, Gatineau, QC, Canada

Keywords

Broadcasting; Collision detection; Deterministic
algorithm; Geometric; Knowledge radius; Radio
network

Synonyms

Wireless information dissemination in geometric
networks

Years and Authors of Summarized Original Work

2001; Dessmark, Pelc

Problem Definition

The Model Overview
Consider a set of stations (nodes) modeled as
points in the plane, labeled by natural numbers,
and equipped with transmitting and receiving capabilities. Every node u has a *range* r_u depending
on the power of its transmitter, and it can reach
all nodes at distance at most r_u from it. The
collection of nodes equipped with ranges determines a directed graph on the set of nodes, called
a *geometric radio network* (GRN), in which a
directed edge (uv) exists if node v can be reached
from u. In this case u is called a *neighbor* of v. If
the power of all transmitters is the same, then all
ranges are equal and the corresponding GRN is
symmetric.

Research partially supported by NSERC discovery grant
and by the Research Chair in Distributed Computing at
the Université du Québec en Outaouais.

Nodes send messages in synchronous *rounds*.
In every round, every node acts either as a *transmitter* or as a *receiver*. A node gets a message in
a given round, if and only if it acts as a receiver
and exactly one of its neighbors transmits in this
round. The message received in this case is the
one that was transmitted. If at least two neighbors
of a receiving node u transmit simultaneously in
a given round, none of the messages is received
by u in this round. In this case, it is said that a
collision occurred at u.

The Problem
Broadcasting is one of the fundamental network
communication primitives. One node of the network, called the *source*, has to transmit a message
to all other nodes. Remote nodes are informed via
intermediate nodes, along directed paths in the
network. One of the basic performance measures
of a broadcasting scheme is the total time, i.e., the
number of rounds it uses to inform all the nodes
of the network.

For a fixed real $s \geq 0$, called the *knowledge
radius*, it is assumed that each node knows the
part of the network within the circle of radius s
centered at it, i.e., it knows the positions, labels,
and ranges of all nodes at distance at most s. The
following problem is considered:

How does the size of the knowledge radius influence deterministic broadcasting time in GRN?

Terminology and Notation
Fix a finite set $R = \{r_1, \ldots, r_\rho\}$ of positive reals
such that $r_1 < \cdots < r_\rho$. Reals r_i are called
ranges. A *node* v is a triple $[l, (x, y), r_i]$, where l
is a binary sequence called the *label* of v; (x, y)
are coordinates of a point in the plane, called the
position of v; and $r_i \in R$ is called the *range*
of v. It is assumed that labels are consecutive
integers 1 to n, where n is the number of nodes,
but all the results hold if labels are integers in the
set $\{1, \ldots, M\}$, where $M \in O(n)$. Moreover, it
is assumed that all nodes know an upper bound
Γ on n, where Γ is polynomial in n. One of
the nodes is distinguished and called the *source*.
Any set of nodes C with a distinguished source,

such that positions and labels of distinct nodes are different, is called a *configuration*.

With any configuration C, the following directed graph $\mathcal{G}(C)$ is associated. Nodes of the graph are nodes of the configuration and a directed edge (uv) exists in the graph, if and only if the distance between u and v does not exceed the range of u. (The word "distance" always means the geometric distance in the plane and not the distance in a graph.) In this case u is called a neighbor of v. Graphs of the form $\mathcal{G}(C)$ for some configuration C are called *geometric radio networks* (GRN). In what follows, only configurations C such that in $\mathcal{G}(C)$ there exists a directed path from the source to any other node are considered. If the size of the set R of ranges is ρ, a resulting configuration and the corresponding GRN are called a ρ-configuration and ρ-GRN, respectively. Clearly, all 1-GRN are symmetric graphs. D denotes the *eccentricity* of the source in a GRN, i.e., the maximum length of all shortest paths in this graph from the source to all other nodes. D is of order of the diameter if the graph is symmetric but may be much smaller in general. $\Omega(D)$ is an obvious lower bound on broadcasting time.

Given any configuration, fix a nonnegative real s, called the *knowledge radius*, and assume that every node of C has initial input consisting of all nodes whose positions are at distance at most s from its own. Thus, it is assumed that every node knows a priori labels, positions, and ranges of all nodes within a circle of radius s centered at it. All nodes also know the set R of available ranges.

It is not assumed that nodes know any global parameters of the network, such as its size or diameter. The only global information that nodes have about the network is a polynomial upper bound on its size. Consequently, the broadcast process may be finished but no node needs to be aware of this fact. Hence, the adopted definition of broadcasting time is the same as in [3]. An algorithm accomplishes broadcasting in t rounds, if all nodes know the source message after round t, and no messages are sent after round t.

Only deterministic algorithms are considered. Nodes can transmit messages even before getting the source message, which enables preprocessing in some cases. The algorithms are *adaptive*, i.e.,

nodes can schedule their actions based on their local history. A node can obviously gain knowledge from previously obtained messages. There is, however, another potential way of acquiring information during the communication process. The availability of this method depends on what happens during a collision, i.e., when u acts as a receiver and two or more neighbors of u transmit simultaneously. As mentioned above, u does not get any of the messages in this case. However, two scenarios are possible. Node u may either hear nothing (except for the background noise), or it may receive *interference noise* different from any message received properly but also different from background noise. In the first case, it is said that there is no *collision detection*, and in the second case – that collision detection is available (cf., e.g., [1]). A discussion justifying both scenarios can be found in [1, 7].

Related Work

Broadcasting in geometric radio networks and some of their variations was considered, e.g., in [6, 8, 9, 11, 12]. In [12] the authors proved that scheduling optimal broadcasting is NP hard even when restricted to such graphs and gave an $O(n \log n)$ algorithm to schedule an optimal broadcast when nodes are situated on a line. In [11] broadcasting was considered in networks with nodes randomly placed on a line. In [9] the authors discussed fault-tolerant broadcasting in radio networks arising from regular locations of nodes on the line and in the plane, with reachability regions being squares and hexagons, rather than circles. Finally, in [6] broadcasting with restricted knowledge was considered but the authors studied only the special case of nodes situated on the line.

Key Results

The results summarized below are based on the paper [5], of which [4] is a preliminary version.

Arbitrary GRN in the Model Without Collision Detection

Clearly all upper bounds and algorithms are valid in the model with collision detection as well.

Large Knowledge Radius

Theorem 1 *The minimum time to perform broadcasting in an arbitrary GRN with source eccentricity D and knowledge radius $s > r_\rho$ (or with global knowledge of the network) is $\Theta(D)$.*

This result yields a centralized $O(D)$ broadcasting algorithm when global knowledge of the GRN is available. This is in sharp contrast with broadcasting in arbitrary graphs, as witnessed by the graph from [10] which has bounded diameter but requires time $\Omega(\log n)$ for broadcasting.

Knowledge Radius Zero

Next consider the case when knowledge radius $s = 0$, i.e., when every node knows only its own label, position, and range. In this case, it is possible to broadcast in time $O(n)$ for arbitrary GRN. It should be stressed that this upper bound is valid for arbitrary GRN, not only symmetric, unlike the algorithm from [3] designed for arbitrary symmetric graphs.

Theorem 2 *It is possible to broadcast in arbitrary n-node GRN with knowledge radius zero in time $O(n)$.*

The above upper bound for GRN should be contrasted with the lower bound from [2,3] showing that some graphs require broadcasting time $\Omega(n \log n)$. Indeed, the graphs constructed in [2,3] and witnessing to this lower bound are not GRN. Surprisingly, this sharper lower bound does not require very unusual graphs. While counterexamples from [2, 3] are not GRN, it turns out that the reason for a longer broadcasting time is really not the topology of the graph but the difference in knowledge available to nodes. Recall that in GRN with knowledge radius 0, it is assumed that each node knows its own position (apart from its label and range): the upper bound $O(n)$ uses this geometric information extensively.

If this knowledge is not available to nodes (i.e., each node knows only its label and range), then there exists a family of GRN requiring broadcasting time $\Omega(n \log n)$. Moreover, it is possible to show such GRN resulting from configurations with only 2 distinct ranges. (Obviously for 1 configurations, this lower bound does not hold, as

these configurations yield symmetric GRN, and in [3], the authors showed an $O(n)$ algorithm working for arbitrary symmetric graphs).

Theorem 3 *If every node knows only its own label and range (and does not know its position), then there exist n-node GRN requiring broadcasting time $\Omega(n \log n)$.*

Symmetric GRN

The Model with Collision Detection

In the model with collision detection and knowledge radius zero, optimal broadcast time is established by the following pair of results.

Theorem 4 *In the model with collision detection and knowledge radius zero, it is possible to broadcast in any n-node symmetric GRN of diameter D in time $O(D + \log n)$.*

The next result is the lower bound $\Omega(\log n)$ for broadcasting time, holding for some GRN of diameter 2. Together with the obvious bound, $\Omega(D)$ this matches the upper bound from Theorem 4.

Theorem 5 *For any broadcasting algorithm with collision detection and knowledge radius zero, there exist n-node symmetric GRN of diameter 2 for which this algorithm requires time $\Omega(\log n)$.*

The Model Without Collision Detection

For the model without collision detection, it is possible to maintain complexity $O(D + \log n)$ of broadcasting. However, we need a stronger assumption concerning knowledge radius: it is no longer 0, but positive, although arbitrarily small.

Theorem 6 *In the model without collision detection, it is possible to broadcast in any n-node symmetric GRN of diameter D in time $O(D + \log n)$, for any positive knowledge radius.*

Applications

The radio network model is applicable to wireless networks using a single frequency. The specific model of geometric radio networks described in section "Problem Definition" is applicable to wireless networks where stations are located in a relatively flat region without large obstacles (natural or human made), e.g., in the sea or a desert, as opposed to a large city or a mountain region. In such a terrain, the signal of a transmitter reaches receivers at the same distance in all directions, i.e., the set of potential receivers of a transmitter is a disc.

Open Problems

1. Is it possible to broadcast in time $o(n)$ in arbitrary n-node GRN with eccentricity D sublinear in n for knowledge radius zero?
   Note: In view of Theorem 2, it is possible to broadcast in time $O(n)$.
2. Is it possible to broadcast in time $O(D + \log n)$ in all symmetric n-node GRN with eccentricity D, without collision detection, when knowledge radius is zero?
   Note: In view of Theorems 4 and 6, the answer is positive if either collision detection or a positive (even arbitrarily small) knowledge radius is assumed.

Cross-References

▶ Deterministic Broadcasting in Radio Networks
▶ Randomized Broadcasting in Radio Networks
▶ Randomized Gossiping in Radio Networks
▶ Routing in Geometric Networks

Recommended Reading

1. Bar-Yehuda R, Goldreich O, Itai A (1992) On the time complexity of broadcast in radio networks: an exponential gap between determinism and randomization. J Comput Syst Sci 45:104–126
2. Bruschi D, Del Pinto M (1997) Lower bounds for the broadcast problem in mobile radio networks. Distrib Comput 10:129–135
3. Chlebus BS, Gasieniec L, Gibbons A, Pelc A, Rytter W (2002) Deterministic broadcasting in ad hoc radio networks. Distrib Comput 15:27–38
4. Dessmark A, Pelc A (2001) Tradeoffs between knowledge and time of communication in geometric radio networks. In: Proceedings of 13th annual ACM symposium on parallel algorithms and architectures (SPAA 2001), Heraklion, pp 59–66
5. Dessmark A, Pelc A (2007) Broadcasting in geometric radio networks. J Discret Algorithms 5:187–201
6. Diks K, Kranakis E, Krizanc D, Pelc A (2002) The impact of knowledge on broadcasting time in linear radio networks. Theor Comput Sci 287:449–471
7. Gallager R (1985) A perspective on multiaccess channels. IEEE Trans Inf Theory 31:124–142
8. Gasieniec L, Kowalski DR, Lingas A, Wahlen M (2008) Efficient broadcasting in known geometric radio networks with non-uniform ranges. In: Proceedings of 22nd international symposium on distributed computing (DISC 2008), Arcachon, pp 274–288
9. Kranakis E, Krizanc D, Pelc A (2001) Fault-tolerant broadcasting in radio networks. J Algorithms 39:47–67
10. Pelc A, Peleg D (2007) Feasibility and complexity of broadcasting with random transmission failures. Theor Comput Sci 370:279–292
11. Ravishankar K, Singh S (1994) Broadcasting on $[0, L]$. Discret Appl Math 53:299–319
12. Sen A, Huson ML (1996) A new model for scheduling packet radio networks. In: Proceedings of 15th annual joint conference of the IEEE computer and communication societies (IEEE INFOCOM'96), San Francisco, pp 1116–1124

B-trees

Jan Vahrenhold
Department of Computer Science, Westfälische
Wilhelms-Universität Münster, Münster,
Germany

Keywords

Data structures; Dictionary; External memory; Indexing; Searching

Years and Authors of Summarized Original Work

1972; Bayer, McCreight

Problem Definition

This problem is concerned with storing a linearly ordered set of elements such that the DICTIO-

NARY operations FIND, INSERT, and DELETE can be performed efficiently.

In 1972, Bayer and McCreight introduced the class of B-trees as a possible way of implementing an "index for a dynamically changing random access file" [7, p. 173]. B-trees have received considerable attention both in the database and in the algorithms community ever since; a prominent witness to their immediate and widespread acceptance is the fact that the authoritative survey on B-trees authored by Comer [10] appeared as soon as 1979 and, already at that time, referred to the B-tree data structure as the "ubiquitous B-tree."

Notations

A *B-tree* is a multiway search tree defined as follows (the definition of Bayer and McCreight [7] is restated according to Knuth [19, Sec. 6.2.4] and Cormen et al. [11, Ch. 18.1]):

Definition 1 Let $m \geq 3$ be a positive integer. A tree T is a *B-tree* of degree m if it is either empty or fulfills the following properties:

1. All leaves of T appear on the same level of T.
2. Every node of T has at most m children.
3. Every node of T, except for the root and the leaves, has at least $m/2$ children.
4. The root of T is either a leaf or has at least two children.
5. An internal node with k children $c_1[v], \ldots, c_k[v]$ stores $k - 1$ keys, and a leaf stores between $m/2 - 1$ and $m - 1$ keys. The keys $\text{key}_i[v]$, $1 \leq i \leq k - 1$, of a node $v \in T$ are maintained in sorted order, i.e., $\text{key}_1[v] \leq \ldots \leq \text{key}_{k-1}[v]$.
6. If v is an internal node of T with k children $c_1[v], \ldots, c_k[v]$, the $k - 1$ keys $\text{key}_1[v], \ldots, \text{key}_{k-1}[v]$ of v separate the range of keys stored in the subtrees rooted at the children of v. If x_i is any key stored in the subtree rooted at $c_i[v]$, the following holds:

$$x_1 \leq \text{key}_1[v] \leq x_2 \leq \text{key}_2[v] \leq \cdots \leq x_{k-1}$$
$$\leq \text{key}_{k-1}[v] \leq x_k$$

To search a B-tree for a given key x, the algorithm starts with the root of the tree being the current node. If x matches one of the current node's keys, the search terminates successfully. Otherwise, if the current node is a leaf, the search terminates unsuccessfully. If the current node's key does not contain x and if the current node is not a leaf, the algorithm identifies the unique subtree rooted at the child of the current node that may contain x and recurses on this subtree. Since the keys of a node guide the search process, they are also referred to as *routing elements*.

Variants and Extensions

Knuth [19] defines a B^**-tree* to be a B-tree where Property 3 in Definition 1 is modified such that every node (except for the root) contains at least $2m/3$ keys.

A B^+*-tree* is a leaf-oriented B-tree, i.e., a B-tree that stores the keys in the leaves only. Additionally, the leaves are linked in left-to-right order to allow for fast sequential traversal of the keys stored in the tree. In a leaf-oriented tree, the routing elements usually are copies of certain keys stored in the leaves ($\text{key}_i[v]$ can be set to be the largest key stored in the subtree rooted at $c_i[v]$), but any set of routing elements that fulfills Properties 5 and 6 of Definition 1 can do as well.

Huddleston and Mehlhorn [16] extended Definition 1 to describe a more general class of multiway search trees that includes the class of B-trees as a special case. Their class of so-called (a, b)-trees is parameterized by two integers a and b with $a \geq 2$ and $2a - 1 \leq b$. Property 2 of Definition 1 is modified to allow each node to have up to b children, and Property 3 is modified to require that, except for the root and the leaves, every node of an (a, b)-tree has at least a children. All other properties of Definition 1 remain unchanged for (a, b)-trees. Usually, (a, b)-trees are implemented as leaf-oriented trees.

By the above definitions, a B-tree is a $(b/2, b)$-tree (if b is even) or an $(a, 2a - 1)$-tree (if b is odd). The subtle difference between even and odd maximum degree becomes relevant in an important amortization argument of Huddleston and Mehlhorn (see below) where the inequality

$b \geq 2a$ is required to hold. This amortization argument actually caused (a, b)-trees with $b \geq 2a$ to be given a special name: *weak* B-trees [16].

Update Operations

An INSERT operation on an (a, b)-tree first tries to locate the key x to be inserted. After an unsuccessful search that stops at some leaf ℓ, x is inserted into ℓ's set of keys. If ℓ becomes too full, i.e., contains more than b elements, two approaches are possible to resolve this *overflow* situation: (1) the node ℓ can be split around its median key into two nodes with at least a keys each, or (2) the node ℓ can have some of its keys be distributed to its left or right siblings (if this sibling has enough space to accommodate the new keys). In the first case, a new routing element separating the keys in the two new subtrees of ℓ's parent μ has to be inserted into the key set of μ, and in the second case, the routing element in μ separating the keys in the subtree rooted at ℓ from the keys rooted at ℓ's relevant sibling needs to be updated. If ℓ was split, the node μ needs to be checked for a potential overflow due to the insertion of a new routing element, and the split may propagate all the way up to the root.

A DELETE operation also first locates the key x to be deleted. If (in a non-leaf-oriented tree) x resides in an internal node, x is replaced by the largest key in the left subtree of x (or the smallest key in the right subtree of x) which resides in a leaf and is deleted from there. In a leaf-oriented tree, keys are deleted from leaves only (the correctness of a routing element on a higher levels is not affected by this deletion). In any case, a DELETE operation may result in a leaf node ℓ containing less than a elements. Again, there are two approaches to resolve this *underflow* situation: (1) the node ℓ is merged with its left or right sibling node or (2) keys from ℓ's left or right sibling node are moved to ℓ (unless the sibling node would underflow as a result of this). Both underflow handling strategies require updating the routing information stored in the parent of ℓ which (in the case of merging)

may underflow itself. As with overflow handling, this process may propagate up to the root of the tree.

Note that the root of the tree can be split as a result of an INSERT operation and that it may disappear if the only two children of the root are merged to form the new root. This implies that B-trees grow and shrink at the top, and thus all leaves are guaranteed to appear on the same level of the tree (Property 1 of Definition 1).

Key Results

Since B-trees are a premier index structure for external storage, the results given in this section are stated not only in the RAM-model of computation but also in the I/O-model of computation introduced by Aggarwal and Vitter [2]. In the I/O-model, not only the number N of elements in the problem instance but also the number M of elements that simultaneously can be kept in main memory and the number B of elements that fit into one disk block are (nonconstant) parameters, and the complexity measure is the number of I/O-operations needed to solve a given problem instance. If B-trees are used in an external memory setting, the degree m of the B-tree is usually chosen such that one node fits into one disk block, i.e., $m \in \Theta(B)$, and this is assumed implicitly whenever the I/O-complexity of B-trees is discussed.

Theorem 1 *The height of an N-key B-tree of degree $m \geq 3$ is bounded by $\log_{\lceil m/2 \rceil}(N + 1)/2$.*

Theorem 2 ([22]) *The storage utilization for B-trees of high order under random insertions and deletions is approximately $\ln 2 \approx 69\%$.*

Theorem 3 *A B-tree may be used to implement the abstract data type* DICTIONARY *such that the operations* FIND, INSERT, *and* DELETE *on a set of N elements from a linearly ordered domain can be performed in $\mathcal{O}(\log N)$ time (with $\mathcal{O}(\log_B N)$ I/O-operations) in the worst case.*

Remark 1 By threading the nodes of a B-tree, i.e., by linking the nodes according to their in-order traversal number, the operations PREV and

NEXT can be performed in constant time (with a constant number of I/O-operations).

A (one-dimensional) *range query* asks for all keys that fall within a given query range (interval).

Lemma 1 *A B-tree supports (one-dimensional) range queries with $\mathcal{O}(\log N + K)$ time complexity ($\mathcal{O}(\log_B N + K/B)$ I/O-complexity) in the worst case where K is the number of keys reported.*

Under the convention that each update to a B-tree results in a new "version" of the B-tree, a *multiversion* B-tree is a B-tree that allows for updates of the current version but also supports queries in earlier versions.

Theorem 4 ([9]) *A multiversion B-tree can be constructed from a B-tree such that it is optimal with respect to the worst-case complexity of the* FIND, INSERT, *and* DELETE *operations as well as to the worst-case complexity of answering range queries.*

Applications

Databases
One of the main reasons for the success of the B-tree lies in its close connection to databases: any implementation of Codd's relational data model (introduced incidentally in the same year as B-trees were invented) requires an efficient indexing mechanism to search and traverse relations that are kept on secondary storage. If this index is realized as a B^+-tree, all keys are stored in a linked list of leaves which is indexed by the top levels of the B^+-tree, and thus both efficient logarithmic searching and sequential scanning of the set of keys is possible.

Due to the importance of this indexing mechanism, a wide number of results on how to incorporate B-trees and their variants into database systems and how to formulate algorithms using these structures have been published in the database community. Comer [10] and Graefe [14] summarize early and recent results, but due to the bulk of results, even these summaries cannot be fully comprehensive. Also, B-trees have been shown to work well in the presence of concurrent operations [8], and Mehlhorn [20, p. 212] notes that they perform especially well if a top-down splitting approach is used. The details of this splitting approach may be found, e.g., in the textbook of Cormen et al. [11, Ch. 18.2].

Priority Queues
A B-tree may be used to serve as an implementation of the abstract data type PRIORITYQUEUE since the smallest key always resides in the first slot of the leftmost leaf.

Lemma 2 *An implementation of a priority queue that uses a B-tree supports the* MIN *operation in $\mathcal{O}(1)$ time (with $\mathcal{O}(1)$ I/O-operations). All other operations (including* DECREASEKEY*) have a time complexity of $\mathcal{O}(\log N)$ (an I/O-complexity of $\mathcal{O}(\log_B N)$) in the worst case.*

Mehlhorn [20, Sec. III, 5.3.1] examined B-trees (and, more general, (a, b)-trees with $a \geq 2$ and $b \geq 2a - 1$) in the context of *mergeable* priority queues. *Mergeable priority queues* are priority queues that additionally allow for concatenating and splitting priority queues. Concatenating priority queues for a set $S_1 \neq \emptyset$ and a set $S_2 \neq \emptyset$ is only defined if $\max\{x \mid x \in S_1\} < \min\{x \mid x \in S_2\}$ and results in a single priority queue for $S_1 \cup S_2$. Splitting a priority queue for a set $S_3 \neq \emptyset$ according to some $y \in \mathrm{dom}(S_3)$ results in a priority queue for the set $S_4 := \{x \in S_3 \mid x \leq y\}$ and a priority queue for the set $S_5 := \{x \in S_3 \mid x > y\}$ (one of these sets may be empty). Mehlhorn's result restated in the context of B-trees is as follows:

Theorem 5 (Theorem 6 in [20, Sec. III, 5.3.1]) *If sets $S_1 \neq \emptyset$ and $S_2 \neq \emptyset$ are represented by a B-tree each, then operation* CONCATENATE(S_1, S_2) *takes time $\mathcal{O}(\log \max\{|S_1|, |S_2|\})$ (has an I/O-complexity of $\mathcal{O}(\log_B \max\{|S_1|, |S_2|\})$) and operation* SPLIT$(S_1, y)$ *takes time $\mathcal{O}(\log |S_1|)$ (has an I/O-complexity of $\mathcal{O}(\log_B |S_1|)$). All bounds hold in the worst case.*

Buffered Data Structures

Many applications (including sorting) that involve massive data sets allow for batched data processing. A variant of B-trees that exploits this relaxed problem setting is the so-called *buffer tree* proposed by Arge [4]. A *buffer tree* is a B-trees of degree $m \in \Theta(M/B)$ (instead of $m \in \Theta(B)$) where each node is assigned a buffer of size $\Theta(M)$. These buffers are used to collect updates and query requests that are passed further down the tree only if the buffer gets full enough to allow for cost amortization.

Theorem 6 (Theorem 1 in [4]) *The total cost of an arbitrary sequence of N intermixed* INSERT *and* DELETE *operations on an initially empty buffer tree is* $\mathcal{O}(N/B \log_{M/B} N/B)$ *I/O operations, that is, the amortized I/O-cost of an operation is* $\mathcal{O}(1/B \log_{M/B} N/B)$.

As a consequence, N elements can be sorted spending an optimal number of $\mathcal{O}(N/B \log_{M/B} N/B)$ I/O-operations by inserting them into a (leaf-oriented) buffer tree in a batched manner and then traversing the leaves. By the preceding discussion, buffer trees can also be used to implement (batched) priority queues in the external memory setting. Arge [4] extended his analysis of buffer trees to show that they also support DELETEMIN operations with an amortized I/O-cost of $\mathcal{O}(1/B \log_{M/B} N/B)$.

Since the degree of a buffer tree is too large to allow for efficient Single shot, i.e., non-batched operations, Arge et al. [6] discussed how buffers can be attached to (and later detached from) a multiway tree while at the same time keeping the degree of the base structure in $\Theta(B)$. Their discussion uses the R-tree index structure as a running example; the techniques presented, however, carry over to the B-tree. The resulting data structure is accessed through standard methods and additionally allows for batched update operations, e.g., *bulk loading*, and queries. The amortized I/O-complexity of all operations is analogous to the complexity of the buffer tree operations.

Using this buffering technique along with weight balancing [5], Achakeev and Seeger [1]

showed how to efficiently bulk load and bulk update partially persistent data structures such as the multiversion B-tree.

Variants of the B-tree base structure that support modern architectures such as many-core processors and that can be updated efficiently have also been proposed by Sewall et al. [21], Graefe et al. [15], and Erb et al. [12].

B-trees as Base Structures

Several external memory data structures are derived from B-trees or use a B-tree as their base structure – see the survey by Arge [3] for a detailed discussion. One of these structures, the so-called *weight-balanced* B-tree is particularly useful as a base tree for building dynamic external data structures that have secondary structures attached to all (or some) of their nodes. The weight-balanced B-tree, developed by Arge and Vitter [5], is a variant of the B-tree that requires all subtrees of a node to have approximately, i.e., up to a small constant factor, the same number of leaves. Weight-balanced B-trees can be shown to have the following property:

Theorem 7 ([5]) *In a weight-balanced B-tree, rebalancing after an update operation is performed by splitting or merging nodes. When a rebalancing operation involves a node v that is the root of a subtree with $w(v)$ leaves, at least $\Theta(w(v))$ update operations involving leaves below v have to be performed before v itself has to be rebalanced again.*

Using the above theorem, amortized bounds for maintaining secondary data structures attached to nodes of the base tree can be obtained – as long as each such structure can be updated with an I/O-complexity linear in the number of elements stored below the node it is attached to [3, 5].

Amortized Analysis

Most of the amortization arguments used for (a, b)-trees, buffer trees, and their relatives are based upon a theorem due to Huddleston and Mehlhorn [16, Theorem 3]. This theorem states that the total number of rebalancing operations in any sequence of N intermixed insert and delete operations performed on an initially empty *weak* B-tree, i.e., an (a, b)-tree with $b \geq 2a$, is at most linear in N. This result carries over to buffer trees since they are $(M/4B, M/B)$-trees. Since B-trees are (a, b)-trees with $b = 2a - 1$ (if b is odd), the result in its full generality is not valid for B-trees, and Huddleston and Mehlhorn present a simple counterexample for $(2, 3)$-trees.

A crucial fact used in the proof of the above amortization argument is that the sequence of operations to be analyzed is performed on an initially *empty* data structure. Jacobsen et al. [17] proved the existence of *non-extreme* (a, b)-trees, i.e., (a, b)-trees where only few nodes have a degree of a or b. Based upon this, they re-established the above result that the rebalancing cost in a sequence of operations is amortized constant (and thus the related result for buffer trees) also for operations on initially nonempty data structures.

In connection with concurrent operations in database systems, it should be noted that the analysis of Huddleston and Mehlhorn actually requires $b \geq 2a + 2$ if a top-down splitting approach is used. In can be shown, though, that even in the general case, few node splits (in an amortized sense) happen close to the root.

URLs to Code and Data Sets

There is a variety of (commercial and free) implementations of B-trees and (a, b)-trees available for download. Representatives are the C++-based implementations that are part of the LEDA-library (http://www.algorithmic-solutions.com), the STXXL-library (http://stxxl.sourceforge.net), and the TPIE-library (http://www.madalgo.au.dk/tpie/) as well as the Java-based implementation that is part of the javaxxl-library (http://xxl.googlecode.com). Furthermore, (pseudo-code) implementations can be found in almost every textbook on database systems or on algorithms and data structures – see, e.g., [11, 13]. Since textbooks almost always leave developing the implementation details of the DELETE operation as an exercise to the reader, the discussion by Jannink [18] is especially helpful.

Cross-References

▶ Cache-Oblivious B-Tree
▶ I/O-Model
▶ R-Trees

Recommended Reading

1. Achakeev D, Seeger B (2013) Efficient bulk updates on multiversion B-trees. PVLDB 14(6):1834–1845
2. Aggarwal A, Vitter JS (1988) The input/output complexity of sorting and related problems. Commun ACM 31(9):1116–1127
3. Arge LA (2002) External memory data structures. In: Abello J, Pardalos PM, Resende MGC (eds) Handbook of massive data sets. Kluwer, Dordrecht, pp 313–357
4. Arge LA (2003) The buffer tree: a technique for designing batched external data structures. Algorithmica 37(1):1–24
5. Arge LA, Vitter JS (2003) Optimal external interval management. SIAM J Comput 32(6):1488–1508
6. Arge LA, Hinrichs KH, Vahrenhold J, Vitter JS (2002) Efficient bulk operations on dynamic R-trees. Algorithmica 33(1):104–128
7. Bayer R, McCreight EM (1972) Organization and maintenance of large ordered indexes. Acta Inform 1(3):173–189
8. Bayer R, Schkolnick M (1977) Concurrency of operations on B-trees. Acta Inform 9(1):1–21
9. Becker B, Gschwind S, Ohler T, Seeger B, Widmayer P (1996) An asymptotically optimal multiversion B-tree. VLDB J 5(4):264–275
10. Comer DE (1979) The ubiquitous B-tree. ACM Comput Surv 11(2):121–137
11. Cormen TH, Leiserson CE, Rivest RL, Stein C (2009) Introduction to algorithms. The MIT electrical engineering and computer science series, 3rd edn. MIT, Cambridge
12. Erb S, Kobitzsch M, Sanders P (2014) Parallel bi-objective shortest paths using weight-balanced B-trees with bulk updates. In: Gudmundsson J, Katajainen J (eds) Proceedings of the 13th international symposium on experimental algorithms (SEA 2014). Lecture notes in computer science, vol 8504. Springer, Berlin, pp 111–122
13. Garcia-Molina H, Ullman JD, Widom J (2009) Database systems: the complete book, 2nd edn. Prentice Hall, Upper Saddle River
14. Graefe G (2006) B-tree indexes for high update rates. SIGMOD RECORD 35(1):39–44
15. Graefe G, Kimura H, Kuno H (2012) Foster B-trees. ACM Trans Database Syst 37(3):Article 17, 29 Pages
16. Huddleston S, Mehlhorn K (1982) A new data structure for representing sorted lists. Acta Inform 17(2):157–184

17. Jacobsen L, Larsen KS, Nielsen MN (2002) On the existence of non-extreme (a, b)-trees. Inf Process Lett 84(2):69–73
18. Jannink J (1995) Implementing deletions in B$^+$-trees. SIGMOD RECORD 24(1):33–38
19. Knuth DE (1998) Sorting and searching, the art of computer programming, vol 3, 2nd edn. Addison-Wesley, Reading
20. Mehlhorn K (1984) Data structures and algorithms 1: sorting and searching. EATCS monographs on theoretical computer science, vol 1. Springer, Berlin
21. Sewall J, Chhugani J, Kim C, Satish N, Dubey P (2011) PALM: parallel architecture-friendly latch-free modifications to B+-trees on many-core processors. PVLDB 11(4):795–806
22. Yao ACC (1978) On random 2–3 trees. Acta Inform 9:159–170

Burrows-Wheeler Transform

Paolo Ferragina[1] and Giovanni Manzini[2,3]
[1]Department of Computer Science, University of Pisa, Pisa, Italy
[2]Department of Computer Science, University of Eastern Piedmont, Alessandria, Italy
[3]Department of Science and Technological Innovation, University of Piemonte Orientale, Alessandria, Italy

Keywords

Block-sorting data compression

Years and Authors of Summarized Original Work

1994; Burrows, Wheeler

Problem Definition

The Burrows-Wheeler transform is a technique used for the lossless compression of data. It is the algorithmic core of the tool *bzip2* which has become a standard for the creation and distribution of compressed archives.

Before the introduction of the Burrows-Wheeler transform, the field of lossless data compression was dominated by two approaches

(see [2, 21] for comprehensive surveys). The first approach dates back to the pioneering works of Shannon and Huffman, and it is based on the idea of using shorter codewords for the more frequent symbols. This idea has originated the techniques of Huffman and arithmetic coding and, more recently, the PPM (prediction by partial matching) family of compression algorithms. The second approach originated from the works of Lempel and Ziv and is based on the idea of adaptively building a dictionary and representing the input string as a concatenation of dictionary words. The best-known compressors based on this approach form the so-called ZIP-family; they have been the standard for several years and are available on essentially any computing platform (e.g., *gzip*, *zip*, *winzip*, just to cite a few).

The Burrows-Wheeler transform introduced a completely new approach to lossless data compression based on the idea of *transforming* the input to make it easier to compress. In the authors' words: "(this) technique [...] works by applying a reversible transformation to a block of text to make redundancy in the input more accessible to simple coding schemes" [5, Sect. 7]. Not only has this technique produced some state-of-the-art compressors, but it also originated the field of compressed indexes [20] and it has been successfully extended to compress (and index) structured data such as XML files [11] and tables [22].

Key Results

Notation
Let s be a string of length n drawn from an alphabet Σ. For $i = 0, \ldots, n - 1$, $s[i]$ denotes the i-th character of s and $s[i, n - 1]$ denotes the suffix of s starting at position i (i.e., starting with the character $s[i]$). Given two strings s and t, the notation $s \prec t$ is used to denote that s lexicographically precedes t.

The Burrows-Wheeler Transform
In [5] Burrows and Wheeler introduced a new compression algorithm based on a reversible

transformation, now called the *Burrows-Wheeler transform* (*bwt*). Given a string s, the computation of $bwt(s)$ consists of three basic steps (see Fig. 1):

1. Append to the end of s a special symbol $ smaller than any other symbol in Σ.
2. Form a *conceptual* matrix $\mathcal{M}$ whose rows are the cyclic shifts of the string $s$$ sorted in lexicographic order.
3. Construct the transformed text $\hat{s} = bwt(s)$ by taking the last column of $\mathcal{M}$.

Notice that every column of $\mathcal{M}$, hence also the transformed text $\hat{s}$, is a permutation of $s$$. As an example F, the first column of the *bwt* matrix $\mathcal{M}$ consists of all characters of s alphabetically sorted. In Fig. 1 it is $F = $iiiimppssss$.

Although it is not obvious from its definition, the *bwt* is an invertible transformation, and both the *bwt* and its inverse can be computed in $O(n)$ optimal time. To be consistent with the more recent literature, the following notation and proof techniques will be slightly different from the ones in [5].

Definition 1 For $1 \le i \le n$, let $s[k_i, n-1]$ denote the suffix of s prefixing row i of $\mathcal{M}$, and define $\Psi(i)$ as the index of the row prefixed by $s[k_i + 1, n - 1]$.

For example, in Fig. 1 it is $\Psi(2) = 7$ since row 2 of $\mathcal{M}$ is prefixed by *ippi* and row 7 is prefixed by *ppi*. Note that $\Psi(i)$ is not defined for $i = 0$ since row 0 is not prefixed by a proper suffix of s. (In [5] instead of Ψ the authors make use of a map which is essentially the inverse of Ψ. The use of Ψ has been introduced in the literature of compressed indexes where Ψ and its inverse play an important role (see [20]).)

Lemma 1 *For* $i = 1, \ldots, n$, *it is* $F[i] = \hat{s}[\Psi(i)]$.

Proof Since each row contains a cyclic shift of $s$$, the last character of the row prefixed by $s[k_i + 1, n - 1]$ is $s[k_i]$. Definition 1 then implies $\hat{s}[\Psi(i)] = s[k_i] = F[i]$ as claimed. □

Lemma 2 *If* $1 \le i < j \le n$ *and* $F[i] = F[j]$, *then* $\Psi(i) < \Psi(j)$.

Proof Let $s[k_i, n - 1]$ (resp. $s[k_j, n - 1]$) denote the suffix of s prefixing row i (resp. row j). The hypothesis $i < j$ implies that $s[k_i, n - 1] \prec s[k_j, n - 1]$. The hypothesis $F[i] = F[j]$ implies $s[k_i] = s[k_j]$; hence, it must be $s[k_i + 1, n - 1] \prec s[k_j + 1, n - 1]$. The thesis follows since by construction $\Psi(i)$ (resp. $\Psi(j)$) is the lexicographic position of the row prefixed by $s[k_i + 1, n - 1]$ (resp. $s[k_j + 1, n - 1]$). □

Lemma 3 *For any character* $c \in \Sigma$, *if* $F[j]$ *is the ℓ-th occurrence of c in F, then $\hat{s}[\Psi(j)]$ is the ℓ-th occurrence of c in $\hat{s}$.*

Proof Take an index h such that $h < j$ and $F[h] = F[j] = c$(the case $h > j$ is symmetric).

Burrows-Wheeler Transform, Fig. 1
Example of Burrows-Wheeler transform for the string $s = mississippi$. The matrix on the right has the rows sorted in lexicographic order. The output of the *bwt* is the last column of the sorted matrix; in this example, the output is $\hat{s} = bwt(s) = ipssm$pissii$

```
mississippi$              $ mississipp i
ississippi$m              i $mississip p
ssissippi$mi              i ppi$missis s
sissippi$mis              i ssippi$mis s
issippi$miss              i ssissippi$ m
ssippi$missi      ⟹      m ississippi $
sippi$missis              p i$mississi p
ippi$mississ              p pi$mississ i
ppi$mississi              s ippi$missi s
pi$mississip              s issippi$mi s
i$mississipp              s sippi$miss i
$mississippi              s sissippi$m i
```

Lemma 2 implies $\Psi(h) < \Psi(j)$ and Lemma 1 implies $\hat{s}[\Psi(h)] = \hat{s}[\Psi(j)] = c$. Consequently, the number of c's preceding (resp. following) $F[j]$ in F coincides with the number of c's preceding (resp. following) $\hat{s}[\Psi(j)]$ in $\hat{s}$ and the lemma follows. □

In Fig. 1 it is $\Psi(2) = 7$ and both $F[2]$ and $\hat{s}[7]$ are the second i in their respective strings. This property is usually expressed by saying that corresponding characters maintain the *same relative order* in both strings F and $\hat{s}$.

Lemma 4 *For any i, $\Psi(i)$ can be computed from $\hat{s} = bwt(s)$.*

Proof Retrieve F simply by sorting alphabetically the symbols of $\hat{s}$. Then compute $\Psi(i)$ as follows: (1) set $c = F(i)$, (2) compute ℓ such that $F[i]$ is the ℓ-th occurrence of c in F, and (3) return the index of the ℓ-th occurrence of c in $\hat{s}$. □

Referring again to Fig. 1, to compute $\Psi(10)$ it suffices to set $c = F[10] = s$ and observe that $F[10]$ is the second s in F. Then it suffices to locate the index j of the second s in $\hat{s}$, namely, $j = 4$. Hence, $\Psi(10) = 4$, and in fact row 10 is prefixed by *sissippi* and row 4 is prefixed by *issippi*.

Theorem 1 *The original string s can be recovered from $bwt(s)$.*

Proof Lemma 4 implies that the column F and the map Ψ can be retrieved from $bwt(s)$. Let j_0 denote the index of the special character \$ in $\hat{s}$. By construction, the row j_0 of the bwt matrix is prefixed by $s[0, n-1]$; hence, $s[0] = F[j_0]$. Let $j_1 = \Psi(j_0)$. By Definition 1 row j_1 is prefixed by $s[1, n-1]$; hence, $s[1] = F[j_1]$. Continuing in this way, it is straightforward to prove by induction that $s[i] = F[\Psi^i(j_0)]$, for $i = 1, \ldots, n-1$. □

Algorithmic Issues

A remarkable property of the bwt is that both the direct and the inverse transform admit efficient algorithms that are extremely simple and elegant.

Theorem 2 *Let $s[1,n]$ be a string over a constant size alphabet Σ. String $\hat{s} = bwt(s)$ can be computed in $O(n)$ time using $O(n\log n)$ bits of working space.*

Proof The suffix array of s can be computed in $O(n)$ time and $O(n\log n)$ bits of working space by using, for example, the algorithm in [17]. The suffix array is an array of integers $sa[1, n]$ such that for $i = 1, \ldots, n$, $s[sa[i], n-1]$ is the i-th suffix of s in the lexicographic order. Since each row of $\mathcal{M}$ is prefixed by a unique suffix of s followed by the special symbol \$, the suffix array provides the ordering of the rows in $\mathcal{M}$. Consequently, $bwt(s)$ can be computed from sa in linear time using the procedure *sa2bwt* of Fig. 2. □

Theorem 3 *Let $s[1,n]$ be a string over a constant size alphabet Σ. Given $bwt(s)$, the string s can be retrieved in $O(n)$ time using $O(n\log n)$ bits of working space.*

Proof The algorithm for retrieving s follows almost verbatim the procedure outlined in the proof of Theorem 1. The only difference is that, for efficiency reasons, all the values of the map Ψ are computed in one shot. This is done by the procedure *bwt2psi* in Fig. 2. In *bwt2psi* instead of working with the column F, it uses the array *count* which is a "compact" representation of F. At the beginning of the procedure, for any character $c \in \Sigma$, $count[c]$ provides the index of the first row of $\mathcal{M}$ prefixed by c. For example, in Fig. 1 $count[i] = 1$, $count[m] = 5$, and so on. In the main *for* loop of *bwt2psi*, the array *bwt* is scanned and $count[c]$ is increased every time an occurrence of character c is encountered (line 6). Line 6 also assigns to h the index of the ℓ-th occurrence of c in F. By Lemma 3, line 7 stores correctly in $psi[h]$ the value $i = \Psi(h)$. After the computation of array psi, s is retrieved by using the procedure *psi2text* of Fig. 2, whose correctness immediately follows from Theorem 1. Clearly, the procedures *bwt2psi* and *psi2text* in Fig. 2 run in $O(n)$ time. Their working space is dominated by the cost of storing the array psi which takes $O(n\log n)$ bits. □

```
Procedure sa2bwt          Procedure bwt2psi          Procedure psi2text
1. bwt[0]=s[n-1];         71. for(i=0;i<=n;i++)      891. k = j0; i=0;
2. for(i=1;i<=n;i++)      2.  c = bwt[i];            2.  do
3.  if(sa[i] == 1)        3.  if(c == '$')           3.   k = psi[k];
4.  bwt[i]='$';           4.   j0 = i;               4.   s[i++] = bwt[k];
5.  else                  5.  else                        while(i<n);
6.  bwt[i]=s[sa[i]-1];    6.   h = count[c]++;
                          7.   psi[h]=i;
```

Burrows-Wheeler Transform, Fig. 2 Algorithms for computing and inverting the Burrows-Wheeler transform. Procedure *sa2bwt* computes *bwt(s)* given *s* and its suffix array *sa*. Procedure *bwt2psi* takes *bwt(s)* as input and computes the Ψ map storing it in the array *psi*. *bwt2psi* also stores in j_0 the index of the row prefixed

by $s[0, n-1]$. *bwt2psi* uses the auxiliary array *count* [1, $|\Sigma|$] which initially contains in *count[i]* the number of occurrences in *bwt(s)* of the symbols $1, \ldots, i-1$. Finally, procedure *psi2text* recovers the string *s* given *bwt(s)*, the array *psi*, and the value j_0

The Burrows-Wheeler Compression Algorithm

The rationale for using the *bwt* for data compression is the following. Consider a string *w* that appears *k* times within *s*. In the *bwt* matrix of *s*, there will be *k* consecutive rows prefixed by *w*, say rows $r_w + 1, r_w + 2, \ldots, r_w + k$. Hence, the positions $r_w + 1, \ldots, r_w + k$ of $\hat{s} = bwt(s)s$ will contain precisely the symbols that immediately precede *w* in *s*. If in *s* certain patterns are more frequent than others, then for many substrings *w*, the corresponding positions $r_w + 1, \ldots, r_w + k$ of $\hat{s}$ will contain only a few distinct symbols. For example, if *s* is an English text and *w* is the string *his*, the corresponding portion of $\hat{s}$ will likely contain many *t*'s and blanks and only a few other symbols. Hence, $\hat{s}$ is a permutation of *s* that is usually *locally homogeneous*, in that its "short" substrings usually contain only a few distinct symbols. (Obviously this is true only if *s* has some regularity: if *s* is a random string $\hat{s}$ will be random as well!)

To take advantage of this property, Burrows and Wheeler proposed to process the string $\hat{s}$ using move-to-front encoding [4] (*mtf*). *mtf* encodes each symbol with the number of distinct symbols encountered since its previous occurrence. To this end, *mtf* maintains a list of the symbols ordered by recency of occurrence; when the next symbol arrives, the encoder outputs its current rank and moves it to the front of the list. Note that *mtf* produces a string which has the same length as

$\hat{s}$ and, if $\hat{s}$ is locally homogeneous, the string $mtf(\hat{s})$ will mainly consist of small integers. (If *s* is an English text, $mtf(\hat{s})$ usually contains more that 50 % zeroes.) Given this skewed distribution, $mtf(\hat{s})$ can be easily compressed: Burrows and Wheeler proposed to compress it using Huffman or Arithmetic coding, possibly preceded by the run-length encoding of runs of equal integers.

Burrows and Wheeler were mainly interested in proposing an algorithm with good practical performance. Indeed their simple implementation outperformed, in terms of compression ratio, the tool *gzip* that was the current standard for lossless compression. A few years after the introduction of the *bwt*, [14, 18] have shown that the compression ratio of the Burrows-Wheeler compression algorithm can be bounded in terms of the *k*-th order empirical entropy of the input string for any $k \geq 0$. For example, Kaplan et al. [14] showed that for any input string *s* and real $\mu > 1$, the length of the compressed string is bounded by $\mu n H_k(s) + n \log(\zeta(\mu)) + \mu g_k + O(\log n)$ bits, where $\zeta(\mu)$ is the standard Zeta function and g_k is a function depending only on *k* and the size of Σ. This bound holds *pointwise* for *any* string *s*, *simultaneously* for any $k \geq 0$ and $\mu > 1$, and it is remarkable since similar bounds have not been proven for any other known compressor. The theoretical study on the performance of *bwt*-based compressors is an active area of research. For more recent results, see [6, 12].

Applications

After the seminal paper of Burrows and Wheeler, many researchers have proposed compression algorithms based on the *bwt* (see [8, 9] and references therein). Of particular theoretical interest are the results in [10] showing that the *bwt* can be used to design a "compression booster," that is, a tool for improving the performance of other compressors in a well-defined and measurable way.

Today the main area of application of the *bwt* is the design of Compressed Full-text Indexes [20]. These indexes take advantage of the relationship between the *bwt* and the suffix array to provide a compressed representation of a string supporting the efficient search and retrieval of the occurrences of an arbitrary pattern.

Open Problems

In addition to the investigation on the performance of *bwt*-based compressors, an open problem of great practical significance is the space-efficient computation of the *bwt*. Given a string *s* of length *n* over an alphabet Σ, both *s* and $\hat{s} = bwt(s)$ take $O(n\log |\Sigma|)$ bits. Unfortunately, the linear time algorithms shown in Fig. 2 make use of auxiliary arrays (i.e., *sa* and Ψ) whose storage takes $\Theta(n\log n)$ bits. This poses a serious limitation to the size of the largest *bwt* that can be computed in main memory. The problem of space- and time-efficient computation of the *bwt* is still open, even if interesting results are reported in [1, 3, 7, 13, 15, 19]. The problem of designing space-efficient algorithms for inverting the *bwt* is also open; see [7, 16, 20] and references therein for further details.

Experimental Results

An experimental study of the performance of several compression algorithms based on the *bwt*

and a comparison with other state-of-the-art compressors is presented in [8].

Data Sets

The data sets used in [8] are available from http:// people.unipmn.it/manzini/boosting. Other data sets relevant for compression and compressed indexing are available at the Pizza&Chili site http://pizzachili.di.unipi.it/.

URL to Code

The compression boosting page (http://people. unipmn.it/manzini/boosting) contains the source code of the algorithms tested in [8]. An extremely efficient code for the computation of the suffix array and the *bwt* (without compression) is available at http://code.google. com/p/libdivsufsort. The code of *bzip2* is available at http://www.bzip.org.

Cross-References

▶ Arithmetic Coding for Data Compression
▶ Compressing and Indexing Structured Text
▶ Compressed Suffix Array
▶ Suffix Array Construction
▶ Table Compression

Recommended Reading

1. Bauer MJ, Cox AJ, Rosone G (2013) Lightweight algorithms for constructing and inverting the BWT of string collections. Theor Comput Sci 483: 134–148
2. Bell TC, Cleary JG, Witten IH (1990) Text compression. Prentice Hall, Englewood Cliffs
3. Beller T, Zwerger M, Gog S, Ohlebusch E (2013) Space-efficient construction of the Burrows-Wheeler transform. In: Proceedings of the 20th international symposium on string processing and information retrieval (SPIRE), Jerusalem. LNCS, vol 8214. Springer, Berlin, pp 5–16

4. Bentley J, Sleator D, Tarjan R, Wei V (1986) A locally adaptive compression scheme. Commun ACM 29:320–330
5. Burrows M, Wheeler D (1994) A block sorting lossless data compression algorithm. Technical report 124, Digital Equipment Corporation
6. Culpepper JS, Petri M, Puglisi SJ (2012) Revisiting bounded context block-sorting transformations. Softw Pract Exp 42:1037–1054
7. Ferragina P, Gagie T, Manzini G (2012) Lightweight data indexing and compression in external memory. Algorithmica 63:707–730
8. Ferragina P, Giancarlo R, Manzini G (2006) The engineering of a compression boosting library: theory vs practice in BWT compression. In: Proceedings of the 14th European symposium on algorithms (ESA), Zurich. LNCS, vol 4168. Springer, Berlin, pp 756–767
9. Ferragina P, Giancarlo R, Manzini G (2009) The myriad virtues of wavelet trees. Inf Comput 207:849–866
10. Ferragina P, Giancarlo R, Manzini G, Sciortino M (2005) Boosting textual compression in optimal linear time. J ACM 52:688–713
11. Ferragina P, Luccio F, Manzini G, Muthukrishnan S (2009) Compressing and indexing labeled trees, with applications. J ACM 57
12. Gagie T, Manzini G (2010) Move-to-front, distance coding, and inversion frequencies revisited. Theor Comput Sci 411:2925–2944
13. Hon W, Sadakane K, Sung W (2009) Breaking a time-and-space barrier in constructing full-text indices. SIAM J Comput 38:2162–2178
14. Kaplan, H., Landau, S., Verbin, E.: A simpler analysis of Burrows-Wheeler-based compression. Theoretical Computer Science **387**, 220–235 (2007)
15. Kärkkäinen J (2007) Fast BWT in small space by blockwise suffix sorting. Theor Comput Sci 387:249–257
16. Kärkkäinen J, Kempa D, Puglisi SJ (2012) Slashing the time for BWT inversion. In: Proceedings of the IEEE data compression conference (DCC), Snowbird. IEEE Computer Society, pp 99–108
17. Kärkkäinen J, Sanders P, Burkhardt S (2006) Linear work suffix array construction. J ACM 53(6):918–936
18. Manzini G (2001) An analysis of the Burrows-Wheeler transform. J ACM 48:407–430
19. Na JC, Park K (2007) Alphabet-independent linear-time construction of compressed suffix arrays using o(n logn)-bit working space. Theor Comput Sci 385:127–136
20. Navarro G, Mäkinen V (2007) Compressed full text indexes. ACM Comput Surv 39(1):2
21. Salomon D (2007) Data compression: the complete reference, 4th edn. Springer, New York
22. Vo BD, Vo KP (2007) Compressing table data with column dependency. Theor Comput Sci 387(3):273–283

Byzantine Agreement

Michael Okun
Weizmann Institute of Science, Rehovot, Israel

Keywords

Byzantine generals; Consensus; Interactive consistency

Years and Authors of Summarized Original Work

1980; Pease, Shostak, Lamport

Problem Definition

The study of Pease, Shostak and Lamport was among the first to consider the problem of achieving a coordinated behavior between processors of a distributed system in the presence of failures [21]. Since the paper was published, this subject has grown into an extensive research area. Below is a presentation of the main findings regarding the specific questions addressed in their paper. In some cases this entry uses the currently accepted terminology in this subject, rather than the original terminology used by the authors.

System Model

A distributed system is considered to have n independent processors, $p_1, \ldots, p_n$, each modeled as a (possibly infinite) state machine. The processors are linked by a communication network that supports direct communication between every pair of processors. The processors can communicate only by exchanging messages, where the sender of every message can be identified by the receiver. While the processors may fail, it is assumed that the communication subsystem is fail-safe. It is not known in advance which processors will not fail (remain correct) and which ones will fail. The types of processor failures are classified according to the following hierarchy.

Crash failure A crash failure means that the processor no longer operates (ad infinitum, starting from the failure point). In particular, other processors will not receive messages from a faulty processor after it crashes.

Omission failure A processor fails to send and receive an arbitrary subset of its messages.

Byzantine failure A faulty processor behaves arbitrarily.

The Byzantine failure is further subdivided into two cases, according to the ability of the processors to create unfalsifiable signatures for their messages. In the *authenticated Byzantine failure* model it is assumed that each message is *signed* by its sender and that no other processor can fake a signature of a correct processor. Thus, even if such a message is forwarded by other processors, its authenticity can be verified. If the processors represent malevolent (human) users of a distributed system, a Public Key Infrastructure (PKI) is typically used to sign the messages (which involves cryptography related issues [17], not discussed here). Practically, in systems where processors are just "processors", a simple signature, such as CRC (cyclic redundancy check), might be sufficient [13]. In the *unauthenticated Byzantine failure* model there are no message signatures.

Definition of the Byzantine Agreement Problem

In the beginning, each processor p_i has an externally provided input value v_i, from some set V (of at least size 2). In the *Byzantine Agreement* (BA) problem, every correct processor p_i is required to decide on an output value $d_i \in V$ such that the following conditions hold:

Termination Eventually, p_i decides, i.e., the algorithm cannot run indefinitely.

Validity If the input value of all the processors is v, then the correct processors decide v.

Agreement All the correct processors decide on the same value.

For crash failures and omission failures there exists a stronger agreement condition:

Uniform Agreement No two processors (either correct or faulty) decide differently.

The termination condition has the following stronger version.

Simultaneous Termination All the correct processors decide in the *same round* (see definition below).

Timing Model

The BA problem was originally defined for *synchronous* distributed systems [18, 21]. In this timing model the processors are assumed to operate in lockstep, which allows to partition the execution of a protocol to rounds. Each round consists of a send phase, during which a processor can send a (different) message to each processor directly connected to it, followed by a receive phase, in which it receives messages sent by these processors in the current round. Unlimited local computations (state transitions) are allowed in both phases, which models the typical situation in real distributed systems, where computation steps are faster than the communication steps by several orders of magnitude.

Overview

This entry deals only with *deterministic* algorithms for the BA problem in the synchronous model. For algorithms involving randomization see the ▶ Optimal Probabilistic Synchronous Byzantine Agreement entry in this volume. For results on BA in other models of synchrony, see ▶ Asynchronous Consensus Impossibility, ▶ Failure Detectors, ▶ Consensus with Partial Synchrony entries in this volume.

Key Results

The maximum possible number of faulty processors is assumed to be bounded by an a priori specified number t (e.g., estimated from the failure probability of individual processor and the requirements on the failure probability of the system as a whole). The number of processors that actually become faulty in a given execution is denoted by f, where $f \leq t$.

The complexity of synchronous distributed algorithms is measured by three complementary parameters. The first is the *round complexity*, which measures the number of rounds required by the algorithm. The second is the *message complexity*, i.e., the total number of messages (and sometimes also their size in bits) sent by all the processors (in case of Byzantine failures, only messages sent by correct processors are counted). The third complexity parameter measures the number of local operations, as in sequential algorithms.

All the algorithms presented bellow are efficient, i.e., the number of rounds, the number of messages and their size, and the local operations performed by each processor are polynomial in n. In most of the algorithms, both the exchanged messages and the local computations involve only the basic data structures (e.g., arrays, lists, queues). Thus, the discussion is restricted only to the round and the message complexities of the algorithms.

The network is assumed to be fully connected, unless explicitly stated otherwise.

Crash Failures

A simple BA algorithm which runs in $t + 1$ rounds and sends $O(n^2)$ messages, together with a proof that this number of rounds is optimal, can be found in textbooks on distributed computing [19]. Algorithms for deciding in $f + 1$ rounds, which is the best possible, are presented in [7, 23] (one additional round is necessary before the processors can stop [11]). Simultaneous termination requires $t + 1$ rounds, even if no failures actually occur [11], however there exists an algorithm that in any given execution stops in the earliest possible round [14]. For uniform agreement, decision can be made in $\min(f + 2, t + 1)$ rounds, which is tight [7].

In case of crash failures it is possible to solve the BA problem with $O(n)$ messages, which is also the lower bound. However, all known message-optimal BA algorithms require a superlinear time. An algorithm that runs in

$O(f + 1)$ rounds and uses only $O(n$ polylog $n)$ messages, is presented in [8], along with an overview of other results on BA message complexity.

Omission Failures

The basic algorithm used to solve the crash failure BA problem works for omission failures as well, which allows to solve the problem in $t + 1$ rounds [23]. An algorithm which terminates in $\min(f + 2, t + 1)$ rounds was presented in [22]. Uniform agreement is impossible for $t \geq n/2$ [23]. For $t < n/2$, there is an algorithm that achieves uniform agreement in $\min(f + 2, t + 1)$ rounds (and $O(n^2 f)$ message complexity) [20].

Byzantine Failures with Authentication

A $(t + 1)$-round BA algorithm is presented in [12]. An algorithm which terminates in $\min(f + 2, t + 1)$ rounds can be found in [24]. The message complexity of the problem is analyzed in [10], where it is shown that the number of signatures and the number of messages in any authenticated BA algorithm are $\Omega(nt)$ and $\Omega(n + t^2)$, respectively. In addition, it is shown that $\Omega(nt)$ is the bound on the number of messages for the unauthenticated BA.

Byzantine Failures Without Authentication

In the unauthenticated case, the BA problem can be solved if and only if $n > 3t$. The proof can be found in [1, 19]. An algorithm that decides in $\min(f + 3, t + 1)$ rounds (it might require two additional rounds to stop) is presented in [16]. Unfortunately, this algorithm is complicated. Simpler algorithms, that run in $\min(2f + 4, 2t + 1)$ and $3 \min(f + 2, t + 1)$ rounds, are presented in [24] and [5], respectively. In these algorithms the number of sent messages is $O(n^3)$, moreover, in the latter algorithm the messages are of constant size (2 bits). Both algorithms assume $V = \{0, 1\}$. To solve the BA problem for a larger V, several instances of a binary algorithm can be run in parallel. Alternatively, there exists a simple 2-round protocol that reduces a BA problem with arbitrary initial values to the binary case, e.g., see Sect. 6.3.3 in [19]. For algorithms with optimal $O(nt)$ message complexity and $t + o(t)$ round complexity see [4, 9].

Arbitrary Network Topologies

When the network is not fully connected, BA can be solved for crash, omission and authenticated Byzantine failures if and only if it is $(t + 1)$-connected [12]. In case of Byzantine failures without authentication, BA has a solution if and only if the network is $(2t + 1)$-connected and $n > 3t$ [19]. In both cases the BA problem can be solved by simulating the algorithms for the fully connected network, using the fact that the number of disjoint communication paths between any pair of non-adjacent processors exceeds the number of faulty nodes by an amount that is sufficient for reliable communication.

Interactive Consistency and Byzantine Generals

The BA (consensus) problem can be stated in several similar ways. Two widely used variants are the *Byzantine Generals* (BG) problem and the *Interactive Consistency (*IC) problem. In the BG case there is a designated processor, say p_1, which is the only one to have an input value. The termination and agreement requirements of the BG problem are exactly as in BA, while the validity condition requires that if the input value of p_1 is v and p_1 is correct, then the correct processors decide v. The IC problem is an extension of BG, where every processor is "designated", so that each processor has to decide on a vector of n values, where the conditions for the i-th entry are as in BG, with p_i as the designated processor. For deterministic synchronous algorithms BA, BG and IC problems are essentially equivalent, e.g., see the discussion in [15].

Firing Squad

The above algorithms assume that the processors share a "global time", i.e., all the processors start in the same (first) round, so that their round counters are equal throughout the execution of the algorithm. However, there are cases in which the processors run in a synchronous network, yet each processor has its own notion of time (e.g., when each processor starts on its own, the round counter values are distinct among the processors). In these cases, it is desirable to have a protocol that allows the processors to agree on some specific round, thus creating a common round which synchronizes all the correct processors. This synchronization task, known as the *Byzantine firing squad* problem [6], is tightly realted to BA.

General Translation Techniques

One particular direction that was pursued as part of the research on the BA problem is the development of methods that automatically translate any protocol that tolerates a more benign failure type into one which tolerates more severe failures [24]. Efficient translations spanning the entire failure hierarchy, starting from crash failures all the way to unauthenticated Byzantine failures, can be found in [3] and in Ch. 12 of [1].

Applications

Due to the very tight synchronization assumptions made in the algorithms presented above, they are used mainly in real-time, safety-critical systems, e.g., aircraft control [13]. In fact, the original interest of Pease, Shostak and Lamport in this problem was raised by such an application [21]. In addition, BA protocols for the Byzantine failure case serve as a basic building block in many cryptographic protocols, e.g., secure multi-party computation [17], by providing a broadcast channel on top of pairwise communication channels.

Cross-References

▶ Asynchronous Consensus Impossibility
▶ Atomic Broadcast
▶ Consensus with Partial Synchrony
▶ Failure Detectors
▶ Optimal Probabilistic Synchronous Byzantine Agreement
▶ Randomization in Distributed Computing

▶ Renaming
▶ Set Agreement

Recommended Reading

1. Attiya H, Welch JL (1998) Distributed computing: fundamentals, simulations and advanced topics. McGraw-Hill, London
2. Barborak M, Dahbura A, Malek M (1993) The consensus problem in fault-tolerant computing. ACM Comput Surv 25(2):171–220
3. Bazzi RA, Neiger G (2001) Simplifying fault-tolerance: providing the abstraction of crash failures. J ACM 48(3):499–554
4. Berman P, Garay JA, Perry KJ (1992) Bit optimal distributed consensus. In: Yaeza-Bates R, Manber U (eds) Computer science research. Plenum Publishing, New York, pp 313–322
5. Berman P, Garay JA, Perry KJ (1992) Optimal early stopping in distributed consensus. In: Proceedings of the 6th international workshop on distributed algorithms (WDAG), Haifa, Nov 1992, pp 221–237
6. Burns JE, Lynch NA (1987) The Byzantine firing squad problem. Adv Comput Res 4:147–161
7. Charron-Bost B, Schiper A (2004) Uniform consensus is harder than consensus. J Algorithms 51(1):15–37
8. Chlebus BS, Kowalski DR (2006) Time and communication efficient consensus for crash failures. In: Proceedings of the 20th international international symposium on distributed computing (DISC), Stockholm, Sept 2006, pp 314–328
9. Coan BA, Welch JL (1992) Modular construction of a Byzantine agreement protocol with optimal message bit complexity. Inf Comput 97(1):61–85
10. Dolev D, Reischuk R (1985) Bounds on information exchange for Byzantine agreement. J ACM 32(1):191–204
11. Dolev D, Reischuk R, Strong HR (1990) Early stopping in Byzantine agreement. J ACM 37(4):720–741
12. Dolev D, Strong HR (1983) Authenticated algorithms for Byzantine agreement. SIAM J Comput 12(4):656–666
13. Driscoll K, Hall B, Sivencrona H, Zumsteg P (2003) Byzantine fault tolerance, from theory to reality. In: Proceedings of the 22nd international conference on computer safety, reliability, and security (SAFECOMP), Edinburgh, Sept 2003, pp 235–248
14. Dwork C, Moses Y (1990) Knowledge and common knowledge in a Byzantine environment: crash failures. Inf Comput 88(2):156–186
15. Fischer MJ (1983) The consensus problem in unreliable distributed systems (a brief survey). Research report, YALEU/DCS/RR-273, Yale University, New Heaven
16. Garay JA, Moses Y (1998) Fully polynomial Byzantine agreement for n > 3t processors in t + 1 rounds. SIAM J Comput 27(1):247–290
17. Goldreich O (2004) Foundations of cryptography, vols. 1-2. Cambridge University Press, Cambridge (2001)
18. Lamport L, Shostak RE, Pease MC (1982) The Byzantine generals problem. ACM Trans Program Lang Syst 4(3):382–401
19. Lynch NA (1996) Distributed algorithms. Morgan Kaufmann, San Francisco
20. Parvédy PR, Raynal M (2004) Optimal early stopping uniform consensus in synchronous systems with process omission failures. In: Proceedings of the 16th annual ACM symposium on parallel algorithms (SPAA), Barcelona, June 2004, pp 302–310
21. Pease MC, Shostak RE, Lamport L (1980) Reaching agreement in the presence of faults. J ACM 27(2):228–234
22. Perry KJ, Toueg S (1986) Distributed agreement in the presence of processor and communication faults. IEEE Trans Softw Eng 12(3):477–482
23. Raynal M (2002) Consensus in synchronous systems: a concise guided tour. In: Proceedings of the 9th Pacific rim international symposium on dependable computing (PRDC), Tsukuba-City, Dec 2002, pp 221–228
24. Toueg S, Perry KJ, Srikanth TK (1987) Fast distributed agreement. SIAM J Comput 16(3):445–457

C

Cache-Oblivious B-Tree

Rolf Fagerberg
Department of Mathematics and Computer
Science, University of Southern Denmark,
Odense, Denmark

Keywords

Cache-oblivious; Dictionary; External memory;
Search tree

Years and Authors of Summarized Original Work

2005; Bender, Demaine, Farach-Colton

Problem Definition

Computers contain a hierarchy of memory levels,
with vastly differing access times. Hence, the
time for a memory access depends strongly on
what is the innermost level containing the data
accessed. In analysis of algorithms, the standard
RAM (or von Neumann) model cannot capture
this effect, and external memory models were
introduced to better model the situation. The most
widely used of these models is the two-level
I/O-model [4], also called the external memory
model or the disk access model. The I/O-model
approximates the memory hierarchy by modeling
two levels, with the inner level having size M,
the outer level having infinite size, and transfers
between the levels taking place in blocks of
B consecutive elements. The cost of an algorithm
is the number of such transfers it makes.

The cache-oblivious model, introduced by
Frigo et al. [26], elegantly generalizes the I/O-
model to a multilevel memory model by a simple
measure: the algorithm is not allowed to know
the value of B and M. More precisely, a cache-
oblivious algorithm is an algorithm formulated
in the RAM model, but analyzed in the I/O-
model, with an analysis valid for *any* value of
B and M. Cache replacement is assumed to
take place automatically by an optimal off-line
cache replacement strategy. Since the analysis
holds for any B and M, it holds for all levels
simultaneously (for a detailed version of this
statement, see [26]).

The subject of the present chapter is that
of efficient dictionary structures for the cache-
oblivious model.

Key Results

The first cache-oblivious dictionary was given
by Prokop [32], who showed how to lay
out a static binary tree in memory such that
searches take $O(\log_B n)$ memory transfers.
This layout, often called the *van Emde Boas*

Research supported by Danish Council for Independent
Research, Natural Sciences.

layout because it is reminiscent of the classic van Emde Boas data structure, also ensures that range searches take $O(\log_B n + k/B)$ memory transfers [8], where k is the size of the output. Both bounds are optimal for comparison-based searching.

The first dynamic, cache-oblivious dictionary was given by Bender et al. [13]. Making use of a variant of the van Emde Boas layout, a density maintenance algorithm of the type invented by Itai et al. [28], and weight-balanced B-trees [5], they arrived at the following results:

Theorem 1 ([13]) *There is a cache-oblivious dictionary structure supporting searches in $O(\log_B n)$ memory transfers and insertions and deletions in amortized $O(\log_B n)$ memory transfers.*

Theorem 2 ([13]) *There is a cache-oblivious dictionary structure supporting searches in $O(\log_B n)$ memory transfers, insertions and deletions in amortized $O(\log_B n + (\log^2 n)/B)$ memory transfers, and range searches in $O(\log_B n + k/B)$ memory transfers, where k is the size of the output.*

Later, Bender et al. [10] developed a cache-oblivious structure for maintaining linked lists which supports insertion and deletion of elements in $O(1)$ memory transfers and scanning of k consecutive elements in amortized $O(k/B)$ memory transfers. Combining this structure with the structure of the first theorem above, the following result can be achieved.

Theorem 3 ([10,13]) *There is a cache-oblivious dictionary structure supporting searches in $O(\log_B n)$ memory transfers, insertions and deletions in amortized $O(\log_B n)$ memory transfers, and range searches in amortized $O(\log_B n + k/B)$ memory transfers, where k is the size of the output.*

A long list of extensions of these basic cache-oblivious dictionary results has been given. We now survey these.

Bender et al. [12] and Brodal et al. [20] gave very similar proposals for reproducing the result of Theorem 2 but with simpler structures (avoiding the use of weight-balanced B-trees). Based on exponential trees, Bender et al. [11] gave a proposal with $O(\log_B n)$ worst-case queries and updates. They also gave a solution with partial persistence, where searches (in all versions of the structure) and updates (in latest version of the structure) require amortized $O(\log_B(m + n))$ memory transfers, where m is the number of versions and n is the number of elements in the version operated on. Bender et al. [14] extended the cache-oblivious model to a concurrent setting and gave three proposals for cache-oblivious B-trees in this setting. Bender et al. [16] presented cache-oblivious dictionary structures exploring trade-offs between faster insertion costs and slower search cost, and Brodal et al. [21] later gave improved structures meeting lower bounds. Franceschini and Grossi [25] showed how to achieve $O(\log_B n)$ worst-case queries and updates while using $O(1)$ space besides the space for the n elements stored. Brodal and Kejlberg-Rasmussen [19] extended this to structures adaptive to the working set bound and allowing predecessor queries. Cache-oblivious dictionaries for other data types such as strings [15, 18, 22–24, 27] and geometric data [1, 2, 6, 7, 9] have been given. The expected number of I/Os for hashing was studied in the cache-oblivious model in [31].

It has been shown [17] that the best possible multiplicative constant in the $\Theta(\log_B n)$ search bound for comparison-based searching is different in the I/O-model and in the cache-oblivious model. It has also been shown [1,3] that for three-sided range reporting in 2D, the best possible space bound for structures with worst-case optimal query times is different in the two models. The latter result implies that linear space cache-oblivious persistent B-trees with optimal worst-case bounds for (1D) range reporting are not possible.

Applications

Dictionaries solve a fundamental data structuring problem which is part of solutions for a very high number of computational problems. Dictionaries for external memory are useful in settings where memory accesses are dominating the running time, and cache-oblivious dictionaries in particular stand out by their ability to optimize the access to all levels of an unknown memory hierarchy. This is an asset, e.g., when developing programs to be run on diverse or unknown architectures (such as software libraries or programs for heterogeneous distributed computing like grid computing and projects such as SETI@home). Even on a single, known architecture, the memory parameters available to a computational process may be nonconstant if several processes compete for the same memory resources. Since cache-oblivious algorithms are optimized for all parameter values, they have the potential to adapt more graceful to these changes and also to varying input sizes forcing different memory levels to be in use.

Open Problems

It is an open problem to find a data structure achieving worst-case versions of all of the bounds in Theorem 3.

Experimental Results

Cache-oblivious dictionaries have been studied empirically in [12, 15, 20, 29, 30, 33]. The overall conclusion of these investigations is that cache-oblivious methods easily can outperform RAM algorithms, although sometimes not as much as algorithms tuned to the specific memory hierarchy and problem size at hand. On the other hand, cache-oblivious algorithms seem to perform well on all levels of the memory hierarchy and to be more robust to changing problem sizes.

Cross-References

▶ B-trees
▶ Cache-Oblivious Model
▶ Cache-Oblivious Sorting

Recommended Reading

1. Afshani P, Zeh N (2011) Improved space bounds for cache-oblivious range reporting. In: Proceedings of the 22nd annual ACM-SIAM symposium on discrete algorithms, San Francisco, pp 1745–1758
2. Afshani P, Hamilton CH, Zeh N (2010) A general approach for cache-oblivious range reporting and approximate range counting. Comput Geom 43(8):700–712. Conference version appeared at SoCG 2009
3. Afshani P, Hamilton CH, Zeh N (2011) Cache-oblivious range reporting with optimal queries requires superlinear space. Discret Comput Geom 45(4):824–850. Conference version appeared at SoCG 2009
4. Aggarwal A, Vitter JS (1988) The input/output complexity of sorting and related problems. Commun ACM 31(9):1116–1127
5. Arge L, Vitter JS (2003) Optimal external memory interval management. SIAM J Comput 32(6):1488–1508
6. Arge L, Zeh N (2006) Simple and semi-dynamic structures for cache-oblivious planar orthogonal range searching. In: Proceedings of the 22nd ACM symposium on computational geometry, Sedona, pp 158–166
7. Arge L, de Berg M, Haverkort HJ (2005) Cache-oblivious R-trees. In: Proceedings of the 21st ACM symposium on computational geometry, Pisa, pp 170–179
8. Arge L, Brodal GS, Fagerberg R (2005) Cache-oblivious data structures. In: Mehta D, Sahni S (eds) Handbook on data structures and applications. CRC, Boca Raton
9. Arge L, Brodal GS, Fagerberg R, Laustsen M (2005) Cache-oblivious planar orthogonal range searching and counting. In: Proceedings of the 21st ACM symposium on computational geometry, Pisa, pp 160–169
10. Bender M, Cole R, Demaine E, Farach-Colton M (2002) Scanning and traversing: maintaining data for traversals in a memory hierarchy. In: Proceedings of the 10th annual European symposium on algorithms, Rome. LNCS, vol 2461, pp 139–151
11. Bender M, Cole R, Raman R (2002) Exponential structures for cache-oblivious algorithms. In: Proceedings of the 29th international colloquium on automata, languages, and programming, Málaga. LNCS, vol 2380, pp 195–207
12. Bender MA, Duan Z, Iacono J, Wu J (2004) A locality-preserving cache-oblivious dynamic dictionary. J Algorithms 53(2):115–136. Conference version appeared at SODA 2002
13. Bender MA, Demaine ED, Farach-Colton M (2005) Cache-oblivious B-trees. SIAM J Comput 35(2):341–358. Conference version appeared at FOCS 2000
14. Bender MA, Fineman JT, Gilbert S, Kuszmaul BC (2005) Concurrent cache-oblivious B-trees. In: Proceedings of the 17th annual ACM symposium

on parallelism in algorithms and architectures, Las Vegas, pp 228–237

15. Bender MA, Farach-Colton M, Kuszmaul BC (2006) Cache-oblivious string B-trees. In: Proceedings of the 25th ACM SIGACT-SIGMOD-SIGART symposium on principles of database systems, Chicago, pp 233–242

16. Bender MA, Farach-Colton M, Fineman JT, Fogel YR, Kuszmaul BC, Nelson J (2007) Cache-oblivious streaming B-trees. In: Proceedings of the 19th annual ACM symposium on parallelism in algorithms and architectures, San Diego, pp 81–92

17. Bender MA, Brodal GS, Fagerberg R, Ge D, He S, Hu H, Iacono J, López-Ortiz A (2011) The cost of cache-oblivious searching. Algorithmica 61(2):463–505. Conference version appeared at FOCS 2003

18. Brodal GS, Fagerberg R (2006) Cache-oblivious string dictionaries. In: Proceedings of the 17th annual ACM-SIAM symposium on discrete algorithms, Miami, pp 581–590

19. Brodal GS, Kejlberg-Rasmussen C (2012) Cache-oblivious implicit predecessor dictionaries with the working set property. In: Proceedings of the 29th annual symposium on theoretical aspects of computer science, Paris. Leibniz international proceedings in informatics, vol 14, pp 112–123

20. Brodal GS, Fagerberg R, Jacob R (2002) Cache-oblivious search trees via binary trees of small height. In: Proceedings of the 13th annual ACM-SIAM symposium on discrete algorithms, San Francisco, pp 39–48

21. Brodal GS, Demaine ED, Fineman JT, Iacono J, Langerman S, Munro JI (2010) Cache-oblivious dynamic dictionaries with update/query tradeoffs. In: Proceedings of the 21st annual ACM-SIAM symposium on discrete algorithms, Austin, pp 1448–1456

22. Ferragina P (2013) On the weak prefix-search problem. Theor Comput Sci 483:75–84. Conference version appeared at CPM 2011

23. Ferragina P, Venturini R (2013) Compressed cache-oblivious string B-tree. In: Proceedings of the algorithms – ESA 2013 – 21st annual European symposium, Sophia Antipolis. LNCS, vol 8125, pp 469–480

24. Ferragina P, Grossi R, Gupta A, Shah R, Vitter JS (2008) On searching compressed string collections cache-obliviously. In: Proceedings of the 27th ACM SIGMOD-SIGACT-SIGART symposium on principles of database systems, Vancouver, pp 181–190

25. Franceschini G, Grossi R (2003) Optimal worst-case operations for implicit cache-oblivious search trees. In: Proceedings of the 8th international workshop on algorithms and data structures (WADS), Ottawa. LNCS, vol 2748, pp 114–126

26. Frigo M, Leiserson CE, Prokop H, Ramachandran S (2012) Cache-oblivious algorithms. ACM Trans Algorithms 8(1):4. Conference version appeared at FOCS 1999

27. Hon W, Lam TW, Shah R, Tam S, Vitter JS (2011) Cache-oblivious index for approximate string matching. Theor Comput Sci 412(29):3579–3588. Conference version appeared at CPM 2007

28. Itai A, Konheim AG, Rodeh M (1981) A sparse table implementation of priority queues. In: 8th International colloquium on Automata, languages and programming, Acre (Akko). LNCS, vol 115, pp 417–431

29. Ladner RE, Fortna R, Nguyen BH (2002) A comparison of cache aware and cache oblivious static search trees using program instrumentation. In: Experimental algorithmics. LNCS, Springer-Verlag, Berlin/Heidelberg, vol 2547, pp 78–92

30. Lindstrom P, Rajan D (2014) Optimal hierarchical layouts for cache-oblivious search trees. In: IEEE 30th international conference on data engineering, Chicago, pp 616–627

31. Pagh R, Wei Z, Yi K, Zhang Q (2014) Cache-oblivious hashing. Algorithmica 69(4):864–883. Conference version appeared at PODS 2010

32. Prokop H (1999) Cache-oblivious algorithms. Master's thesis, Massachusetts Institute of Technology

33. Rahman N, Cole R, Raman R (2001) Optimised predecessor data structures for internal memory. In: Proceedings of the algorithm engineering, 5th international workshop (WAE), Aarhus. LNCS, vol 2141, pp 67–78

Cache-Oblivious Model

Rolf Fagerberg
Department of Mathematics and Computer Science, University of Southern Denmark, Odense, Denmark

Keywords

Cache-oblivious; Computational models; External memory

Years and Authors of Summarized Original Work

1999; Frigo, Leiserson, Prokop, Ramachandran

Problem Definition

The memory system of contemporary computers consists of a hierarchy of memory levels, with

Research supported by Danish Council for Independent Research, Natural Sciences.

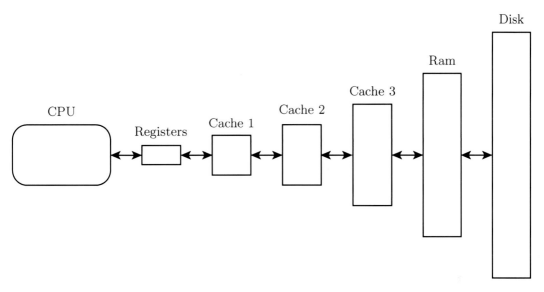

Cache-Oblivious Model, Fig. 1 The memory hierarchy

each level acting as a cache for the next; a typical hierarchy consists of registers, level 1 cache, level 2 cache, level 3 cache, main memory, and disk (Fig. 1). One characteristics of the hierarchy is that the memory levels get larger and slower the further they get from the processor, with the access time increasing most dramatically between RAM memory and disk. Another characteristics is that data is moved between levels in blocks of consecutive elements.

As a consequence of the differences in access time between the levels, the cost of a memory access depends highly on what is the current lowest memory level holding the element accessed. Hence, the memory access pattern of an algorithm has a major influence on its practical running time. Unfortunately, the RAM model (Fig. 2) traditionally used to design and analyze algorithms is not capable of capturing this, as it assumes that all memory accesses take equal time.

To better account for the effects of the memory hierarchy, a number of computational models have been proposed. The simplest and most successful is the two-level I/O-model introduced by Aggarwal and Vitter [3] (Fig. 3). In this model a two-level memory hierarchy is assumed, consisting of a fast memory of size M and a slower memory of infinite size, with

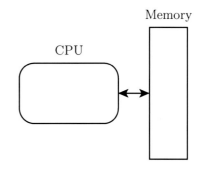

Cache-Oblivious Model, Fig. 2 The RAM model

data transferred between the levels in blocks of B consecutive elements. Computation can only be performed on data in the fast memory, and algorithms are assumed to have complete control over transfers of blocks between the two levels. Such a block transfer is denoted a *memory transfer*. The complexity measure is the number of memory transfers performed. The strength of the I/O-model is that it captures part of the memory hierarchy while being sufficiently simple to make design and analysis of algorithms feasible. Over the last two decades, a large body of results for the I/O-model has been produced, covering most areas of algorithmics. For an overview, see the surveys [5, 32, 34–36].

Cache-Oblivious Model,
Fig. 3 The I/O-model

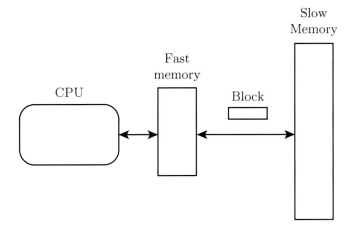

Cache-Oblivious Model,
Fig. 4 The
cache-oblivious model

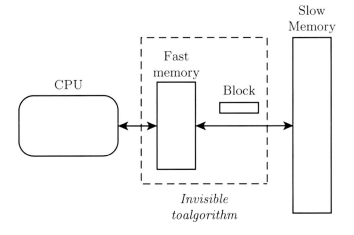

More elaborate models of multilevel memory have been proposed (see e.g., [34] for an overview) but these models have been less successful than the I/O-model, mainly because of their complexity which makes analysis of algorithms harder. All these models, including the I/O-model, assume that the characteristics of the memory hierarchy (the level and block sizes) are known.

In 1999 the *cache-oblivious model* (Fig. 4) was introduced by Frigo et al. [30]. In short, a cache-oblivious algorithm is an algorithm formulated in the RAM model but analyzed in the I/O-model, with the analysis required to hold for *any* block size B and memory size M. Memory transfers are assumed to take place automatically by an optimal off-line cache replacement strategy.

The crux of the cache-oblivious model is that because the I/O-model analysis holds for any block and memory size, it holds for all levels of a multilevel memory hierarchy (see [30] for a detailed version of this statement). Put differently, by optimizing an algorithm to one unknown level of the memory hierarchy, it is optimized to all levels simultaneously. Thus, the cache-oblivious model elegantly generalizes the I/O-model to a multilevel memory model by one simple measure: the algorithm is not allowed to know the value of B and M. The challenge, of course, is to develop algorithms having good memory transfer analyses under these conditions.

Besides capturing the entire memory hierarchy in a conceptually simple way, the cache-oblivious model has other benefits: Algorithms developed in the model do not rely on knowing the parameters of the memory hierarchy, which is an asset when developing programs to be run on diverse or unknown architectures (e.g., software libraries or programs for heterogeneous distributed computing such as grid computing and projects like SETI@home). Even on a single, known architecture, the memory parameters available to a computational process may be nonconstant if several processes compete for the same memory resources. Since cache-oblivious algorithms are optimized for all parameter values, they have the potential to adapt more graceful to these changes. Also, the same code will adapt to varying input sizes forcing different memory levels to be in use. Finally, cache-oblivious algorithms automatically are optimizing the use of translation lookaside buffers (a cache holding recently accessed parts of the page table used for virtual memory) of the CPU, which may be seen as second memory hierarchy parallel to the one mentioned in the introduction.

Possible weak points of the cache-oblivious model are the assumption of optimal off-line cache replacement and the lack of modeling of the limited associativity of many of the levels of the hierarchy. The first point is mitigated by the fact that normally, the provided analysis of a proposed cache-oblivious algorithm will work just as well assuming a least recently used cache replacement policy, which is closer to actual replacement strategies of computers. The second point is also a weak point of most other memory models.

Key Results

This section surveys a number of the known results in the cache-oblivious model. Other surveys available include [6, 15, 26, 32].

First of all, note that scanning an array of N elements takes $O(N/B)$ memory transfers for any values of B and M and hence is an optimal cache-oblivious algorithm. Thus, standard RAM algorithms based on scanning may already posses good analysis in the cache-oblivious model – for instance, the classic deterministic linear time selection algorithm has complexity $O(N/B)$ [26].

For sorting, a fundamental fact in the I/O-model is that comparison-based sorting of N elements takes $\Theta(\mathrm{Sort}(N))$ memory transfers [3],

where $\mathrm{Sort}(N) = \frac{N}{B} \log_{M/B} \frac{N}{M}$. Also in the cache-oblivious model, sorting can be carried out in $\Theta(\mathrm{Sort}(N))$ memory transfer, if one makes the so-called *tall cache* assumption $M \geq B^{1+\varepsilon}$ [16, 30]. Such an assumption has been shown to be necessary [18], which proves a separation in power between cache-oblivious algorithms and algorithms in the I/O-model (where this assumption is not needed for the sorting bound).

For searching, B-trees have cost $O(\log_B N)$, which is optimal in the I/O-model for comparison-based searching. This cost is also attainable in the cache-oblivious model, as shown for the static case in [33] and for the dynamic case in [12]. Also for searching, a separation between cache-oblivious algorithms and algorithms in the I/O-model has been shown [13] in the sense that the constants attainable in the $O(\log_B N)$ bound are provably different.

By now, a large number of cache-oblivious algorithms and data structures in many areas have been given. These include priority queues [7, 17]; many dictionaries for standard data, string data, and geometric data (see survey in section on cache-oblivious B-trees); and algorithms for other problems in computational geometry [8, 16, 22], for graph problems [4, 7, 19, 23, 31], for scanning dynamic sets [9], for layout of static trees [11], for search problems on multi-sets [28], for dynamic programming [14, 24], for adaptive sorting [20], for inplace sorting [29], for sorting of strings [27], for partial persistence [10], for matrix operations [30], and for the fast Fourier transform [30].

In the negative direction, a few further separations in power between cache-oblivious algorithms and algorithms in the I/O-model are known. Permuting in the I/O-model has complexity $\Theta(\min\{\mathrm{Sort}(N), N\})$, assuming that elements are indivisible [3]. It has been proven [18] that this asymptotic complexity cannot be attained in the cache-oblivious model. A separation with respect to space complexity has been proven [2] for three-sided range reporting in 2D where the best possible space bound for structures with worst-case optimal query times is different in the two models. This result also implies that linear space cache-oblivious persistent B-trees with optimal worst-case bounds for (1D) range reporting are not possible.

Applications

The cache-oblivious model is a means for design and analysis of algorithms that use the memory hierarchy of computers efficiently.

Experimental Results

Cache-oblivious algorithms have been evaluated empirically in a number of areas, including sorting [21], searching (see survey in section on cache-oblivious B-trees), matrix algorithms [1, 30, 37], and dynamic programming [24, 25].

The overall conclusion of these investigations is that cache-oblivious methods often outperform RAM algorithms but not always exactly as much as do algorithms tuned to the specific memory hierarchy and problem size. On the other hand, cache-oblivious algorithms seem to perform well on all levels of the memory hierarchy and to be more robust to changing problem sizes than cache-aware algorithms.

Cross-References

▶ Cache-Oblivious B-Tree
▶ Cache-Oblivious Sorting
▶ I/O-Model

Recommended Reading

1. Adams MD, Wise DS (2006) Seven at one stroke: results from a cache-oblivious paradigm for scalable matrix algorithms. In: Proceedings of the 2006 workshop on memory system performance and correctness, San Jose, pp 41–50
2. Afshani P, Zeh N (2011) Improved space bounds for cache-oblivious range reporting. In: Proceedings of the 22nd annual ACM-SIAM symposium on discrete algorithms, San Francisco, pp 1745–1758
3. Aggarwal A, Vitter JS (1988) The Input/Output complexity of sorting and related problems. Commun ACM 31(9):1116–1127
4. Allulli L, Lichodzijewski P, Zeh N (2007) A faster cache-oblivious shortest-path algorithm for undirected graphs with bounded edge lengths. In: Proceedings of the 18th annual ACM-SIAM symposium on discrete algorithms, New Orleans, pp 910–919
5. Arge L (2002) External memory data structures. In: Abello J, Pardalos PM, Resende MGC (eds) Handbook of massive data sets. Kluwer Academic, Dordrecht/Boston/London, pp 313–358
6. Arge L, Brodal GS, Fagerberg R (2005) Cache-oblivious data structures. In: Mehta D, Sahni S (eds) Handbook on data structures and applications. Chapman & Hall/CRC, Boca Raton/London/New York/Washington, D.C
7. Arge L, Bender MA, Demaine ED, Holland-Minkley B, Munro JI (2007) An optimal cache-oblivious priority queue and its application to graph algorithms. SIAM J Comput 36(6):1672–1695. Conference version appeared at STOC 2002
8. Arge L, Mølhave T, Zeh N (2008) Cache-oblivious red-blue line segment intersection. In: Proceedings of the 16th annual European symposium on algorithms, Karlsruhe. LNCS, vol 5193, pp 88–99
9. Bender M, Cole R, Demaine E, Farach-Colton M (2002) Scanning and traversing: maintaining data for traversals in a memory hierarchy. In: Proceedings of the 10th annual European symposium on algorithms, Rome. LNCS, vol 2461, pp 139–151
10. Bender M, Cole R, Raman R (2002) Exponential structures for cache-oblivious algorithms. In: Proceedings of the 29th international colloquium on automata, languages, and programming, Málaga. LNCS, vol 2380, pp 195–207
11. Bender M, Demaine E, Farach-Colton M (2002) Efficient tree layout in a multilevel memory hierarchy. In: Proceedings of the 10th annual European symposium on algorithms, Rome. LNCS, vol 2461, pp 165–173, corrected full version at http://arxiv.org/abs/cs/0211010
12. Bender MA, Demaine ED, Farach-Colton M (2005) Cache-oblivious B-trees. SIAM J Comput 35(2):341–358. Conference version appeared at FOCS 2000
13. Bender MA, Brodal GS, Fagerberg R, Ge D, He S, Hu H, Iacono J, López-Ortiz A (2011) The cost of cache-oblivious searching. Algorithmica 61(2):463–505. Conference version appeared at FOCS 2003
14. Bille P, Stöckel M (2012) Fast and cache-oblivious dynamic programming with local dependencies. In: Proceedings of the 6th international conference on language and automata theory and applications, A Coruña. LNCS, vol 7183, pp 131–142
15. Brodal GS (2004) Cache-oblivious algorithms and data structures. In: Proceedings of the 9th Scandinavian workshop on algorithm theory, Humlebæk. LNCS, vol 3111, pp 3–13
16. Brodal GS, Fagerberg R (2002) Cache oblivious distribution sweeping. In: Proceedings of the 29th international colloquium on automata, languages, and programming, Málaga. LNCS, vol 2380, pp 426–438

17. Brodal GS, Fagerberg R (2002) Funnel heap – a cache oblivious priority queue. In: Proceedings of the 13th international symposium on algorithms and computation, Vancouver. LNCS, vol 2518, pp 219–228

18. Brodal GS, Fagerberg R (2003) On the limits of cache-obliviousness. In: Proceedings of the 35th annual ACM symposium on theory of computing, San Diego, pp 307–315

19. Brodal GS, Fagerberg R, Meyer U, Zeh N (2004) Cache-oblivious data structures and algorithms for undirected breadth-first search and shortest paths. In: Proceedings of the 9th Scandinavian workshop on algorithm theory, Humlebæk. LNCS, vol 3111, pp 480–492

20. Brodal GS, Fagerberg R, Moruz G (2005) Cache-aware and cache-oblivious adaptive sorting. In: Proceedings of the 32nd international colloquium on automata, languages and programming, Lisbon. LNCS, vol 3580, pp 576–588

21. Brodal GS, Fagerberg R, Vinther K (2007) Engineering a cache-oblivious sorting algorithm. ACM J Exp Algorithmics 12:Article 2.2. Conference version appeared at ALENEX 2004

22. Chan TM, Chen EY (2010) Optimal in-place and cache-oblivious algorithms for 3-D convex hulls and 2-D segment intersection. Comput Geom 43(8):636–646

23. Chowdhury RA, Ramachandran V (2004) Cache-oblivious shortest paths in graphs using buffer heap. In: Proceedings of the 16th annual ACM symposium on parallelism in algorithms and architectures, Barcelona

24. Chowdhury RA, Ramachandran V (2006) Cache-oblivious dynamic programming. In: Proceedings of the 17th annual ACM-SIAM symposium on discrete algorithms, Miami, pp 591–600

25. Chowdhury RA, Le HS, Ramachandran V (2010) Cache-oblivious dynamic programming for bioinformatics. IEEE/ACM Trans Comput Biol Bioinf 7(3):495–510

26. Demaine ED (2002) Cache-oblivious algorithms and data structures. Lecture notes from the EEF summer school on massive data sets. Online version at http://theory.csail.mit.edu/~edemaine/papers/BRICS2002/

27. Fagerberg R, Pagh A, Pagh R (2006) External string sorting: faster and cache-oblivious. In: Proceedings of the 23rd annual symposium on theoretical aspects of computer science, Marseille. LNCS, vol 3884, pp 68–79

28. Farzan A, Ferragina P, Franceschini G, Munro JI (2005) Cache-oblivious comparison-based algorithms on multisets. In: Proceedings of the 13th annual European symposium on algorithms, Palma de Mallorca. LNCS, vol 3669, pp 305–316

29. Franceschini G (2004) Proximity mergesort: optimal in-place sorting in the cache-oblivious model. In: Proceedings of the 15th annual ACM-SIAM symposium on discrete algorithms, New Orleans, pp 291–299

30. Frigo M, Leiserson CE, Prokop H, Ramachandran S (2012) Cache-oblivious algorithms. ACM Trans Algorithms 8(1):4. Conference version appeared at FOCS 1999

31. Jampala H, Zeh N (2005) Cache-oblivious planar shortest paths. In: Proceedings of the 32nd international colloquium on automata, languages, and programming, Lisbon. LNCS, vol 3580, pp 563–575

32. Meyer U, Sanders P, Sibeyn JF (eds) (2003) Algorithms for memory hierarchies. LNCS, vol 2625. Springer, Berlin/Heidelberg/New York

33. Prokop H (1999) Cache-oblivious algorithms. Master's thesis, Department of Electrical Engineering and Computer Science, Massachusetts Institute of Technology

34. Vitter JS (2001) External memory algorithms and data structures: dealing with MASSIVE data. ACM Comput Surv 33(2):209–271

35. Vitter JS (2005) Geometric and spatial data structures in external memory. In: Mehta D, Sahni S (eds) Handbook on data structures and applications. Chapman & Hall/CRC, Boca Raton/London/New York/Washington, D.C

36. Vitter JS (2008) Algorithms and data structures for external memory. Found Trends Theor Comput Sci 2(4):305–474

37. Yotov K, Roeder T, Pingali K, Gunnels JA, Gustavson FG (2007) An experimental comparison of cache-oblivious and cache-conscious programs. In: Proceedings of the 19th annual ACM symposium on parallelism in algorithms and architectures, San Diego, pp 93–104

Cache-Oblivious Sorting

Gerth Stølting
Department of Computer Science, University of Aarhus, Århus, Denmark

Keywords

Funnelsort

Problem Definition

Sorting a set of elements is one of the most well-studied computational problems. In the cache-oblivious setting, the first study of sorting was presented in 1999 in the seminal paper by Frigo et al. [8] that introduced the cache-oblivious framework for developing algorithms aimed at machines with (unknown) hierarchical memory.

Model

In the cache-oblivious setting, the computational model is a machine with two levels of memory: a cache of limited capacity and a secondary memory of infinite capacity. The capacity of the cache is assumed to be M elements, and data is moved between the two levels of memory in blocks of B consecutive elements. Computations can only be performed on elements stored in cache, i.e., elements from secondary memory need to be moved to the cache before operations can access the elements. Programs are written as acting directly on one unbounded memory, i.e., programs are like standard RAM programs. The necessary block transfers between cache and secondary memory are handled automatically by the model, assuming an optimal offline cache replacement strategy. The core assumption of the cache-oblivious model is that M and B are *unknown to the algorithm*, whereas in the related I/O model introduced by Aggarwal and Vitter [1], the algorithms know M and B, and the algorithms perform the block transfers explicitly. A thorough discussion of the cache-oblivious model and its relation to multilevel memory hierarchies is given in [8].

Sorting

For the sorting problem, the input is an array of N elements residing in secondary memory, and the output is required to be an array in secondary memory, storing the input elements in sorted order.

Key Results

In the I/O model, tight upper and lower bounds were proved for the sorting problem and the problem of permuting an array [1]. In particular it was proved that sorting requires $\Omega\left(\frac{N}{B}\log_{M/B}\frac{N}{B}\right)$ block transfers and permuting an array requires $\Omega\left(\min\left\{N,\frac{N}{B}\log_{M/B}\frac{N}{B}\right\}\right)$ block transfers. Since lower bounds for the

I/O model also hold for the cache-oblivious model, the lower bounds from [1] immediately give a lower bound of $\Omega\left(\frac{N}{B}\log_{M/B}\frac{N}{B}\right)$ block transfers for cache-oblivious sorting and $\Omega\left(\min\left\{N,\frac{N}{B}\log_{M/B}\frac{N}{B}\right\}\right)$ block transfers for cache-oblivious permuting. The upper bounds from [1] cannot be applied to the cache-oblivious setting since these algorithms make explicit use of B and M.

Binary mergesort performs $O(N\log_2 N)$ comparisons, but analyzed in the cache-oblivious model, it performs $O\left(\frac{N}{B}\log_2\frac{N}{M}\right)$ block transfers which is a factor $\Theta\left(\log\frac{M}{B}\right)$ from the lower bound (assuming a recursive implementation of binary mergesort, in order to get M in the denominator in the $\log N/M$ part of the bound on the block transfers). Another comparison-based sorting algorithm is the classical quicksort sorting algorithm from 1962 by Hoare [9] that performs expected $O(N\log_2 N)$ comparisons and expected $O\left(\frac{N}{B}\log_2\frac{N}{M}\right)$ block transfers. Both these algorithms achieve their relatively good performance for the number of block transfers from the fact that they are based on repeated scanning of arrays – a property not shared with, e.g., heapsort [10] that has a very poor performance of $\Theta\left(\frac{N}{B}\log_{M/B}\frac{N}{B}\right)$ block transfers. In the I/O model, the optimal performance of $O\left(\frac{M}{B}\log_{M/B}\frac{N}{B}\right)$ is achieved by generalizing binary mergesort to $\Theta\left(\frac{M}{B}\right)$-way mergesort [1].

Frigo et al. in [8] presented two cache-oblivious sorting algorithms (which can also be used to permute an array of elements). The first algorithm [8, Section 4] is denoted as *funnelsort* and is a reminiscent of classical binary mergesort, whereas the second algorithm [8, Section 5] is a distribution-based sorting algorithm. Both algorithms perform *optimal* $O\left(\frac{N}{B}\log_{M/B}\frac{N}{B}\right)$ block transfers – provided a *tall cache assumption* $M=\Omega(B^2)$ is satisfied.

Funnelsort

The basic idea of funnelsort is to rearrange the sorting process performed by binary mergesort, such that the processed data is stored "locally." This is achieved by two basic ideas: (1) a

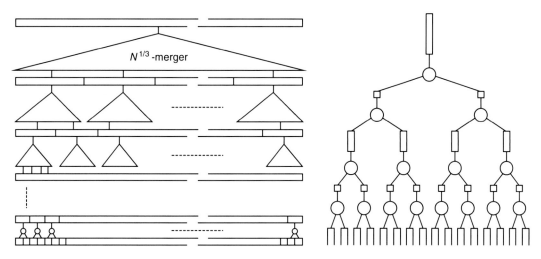

Cache-Oblivious Sorting, Fig. 1 The overall recursion of funnelsort (*left*) and a 16-merger (*right*)

top-level recursion that partitions the input into $N^{1/3}$ sequences of size $N^{2/3}$ funnelsorts these sequences recursively and merges the resulting sorted subsequences using an $N^{1/3}$-*merger*. (2) A k-merger is recursively defined to perform binary merging of k input sequences in a clever schedule with an appropriate recursive layout of data in memory using buffers to hold suspended merging processes (see Fig. 1). Subsequently two simplifications were made, without sacrificing the asymptotic number of block transfers performed. In [3], it was proved that the binary merging could be performed lazily, simplifying the scheduling of merging. In [5], it was further observed that the recursive layout of k-mergers is not necessary. It is sufficient that a k-merger is stored in a consecutive array, i.e., the buffers can be laid out in an arbitrary order which simplifies the construction algorithm for the k-merger.

Implicit Cache-Oblivious Sorting

Franceschini in [7] showed how to perform optimal cache-oblivious sorting implicitly using only $O(1)$ space, i.e., all data is stored in the input array except for $O(1)$ additional words of information. In particular the output array is just a permutation of the input array.

The Role of the Tall Cache Assumption

The role of the tall cache assumption on cache-oblivious sorting was studied by Brodal and

Fagerberg in [4]. If no tall cache assumption is made, they proved the following theorem:

Theorem 1 ([4], Corollary 3) *Let* $B_1 = 1$ *and* $B_2 = M/2$. *For any cache-oblivious comparison-based sorting algorithm, let* t_1 *and* t_2 *be upper bounds on the number of I/Os performed for block sizes* B_1 *and* B_2. *If for a real number* $d \geq 0$, *it is satisfied that* $t_2 = d \cdot \frac{N}{B_2} \log_{M/B_2} \frac{N}{B_2}$ *then* $t_1 > 1/8 \cdot N \log_2 N/M$.

The theorem shows that cache-oblivious comparison-based sorting without a tall cache assumption cannot match the performance of algorithms in the I/O model where M and B are known to the algorithm. It also has the natural interpretation that if a cache-oblivious algorithm is required to be I/O optimal for the case $B = M/2$, then binary mergesort is best possible – any other algorithm will be the same factor of $\Theta(\log M)$ worse than the optimal block transfer bound for the case $M >> D$.

For the related problem of permuting an array, the following theorem states that for all possible tall cache assumptions $B \leq M^\delta$, no cache-oblivious permuting algorithm exists with a block transfer bound (even only in the average case sense), matching the worst case bound in the I/O model.

Theorem 2 ([4], Theorem 2) *For all* $\delta > 0$, *there exists no cache-oblivious algorithm for*

permuting that for all $M \geq 2B$ *and* $1 \leq B \leq M^{\delta}$ *achieves* $O\left(\min\left\{N, \frac{N}{B} \log_{M/B} \frac{N}{B}\right\}\right)$ *I/Os averaged over all possible permutations of size* N.

Applications

Many problems can be reduced to cache-oblivious sorting. In particular Arge et al. [2] developed a cache-oblivious priority queue based on a reduction to sorting. They furthermore showed how a cache-oblivious priority queue can be applied to solve a sequence of graph problems, including list ranking, BFS, DFS, and minimum spanning trees.

Brodal and Fagerberg in [3] showed how to modify the cache-oblivious lazy funnelsort algorithm to solve several problems within computational geometry, including orthogonal line segment intersection reporting, all the nearest neighbors, 3D maxima problem, and batched orthogonal range queries. All these problems can be solved by a computation process very similarly to binary mergesort with an additional problem-dependent twist. This general framework to solve computational geometry problems is denoted as *distribution sweeping*.

Open Problems

Since the seminal paper by Frigo et al. [8] introducing the cache-oblivious framework, there has been a lot of work on developing algorithms with a good theoretical performance, but only a limited amount of work has been done on implementing these algorithms. An important issue for future work is to get further experimental results consolidating the cache-oblivious model as a relevant model for dealing efficiently with hierarchical memory.

Experimental Results

A detailed experimental study of the cache-oblivious sorting algorithm funnelsort was performed in [5]. The main result of [5] is that a carefully implemented cache-oblivious sorting algorithm can be faster than a tuned implementation of quicksort already for input sizes well within the limits of RAM. The implementation is also at least as fast as the recent cache-aware implementations included in the test. On disk, the difference is even more pronounced regarding quicksort and the cache-aware algorithms, whereas the algorithm is slower than a careful implementation of multiway mergesort optimized for external memory such as in TPIE [6].

URL to Code

http://kristoffer.vinther.name/projects/funnelsort/

Cross-References

▶ Cache-Oblivious Model
▶ External Sorting and Permuting
▶ I/O-Model

Recommended Reading

1. Aggarwal A, Vitter JS (1988) The input/output complexity of sorting and related problems. Commun ACM 31(9):1116–1127
2. Arge L, Bender MA, Demaine ED, Holland-Minkley B, Munro JI (2002) Cache-oblivious priority queue and graph algorithm applications. In: Proceedings of the 34th annual ACM symposium on theory of computing. ACM, New York, pp 268–276
3. Brodal GS, Fagerberg R (2002) Cache oblivious distribution sweeping. In: Proceedings of the 29th international colloquium on automata, languages, and programming. Lecture notes in computer science, vol 2380, pp 426–438. Springer, Berlin
4. Brodal GS, Fagerberg R (2003) On the limits of cache-obliviousness. In: Proceedings of the 35th annual ACM symposium on theory of computing. ACM, New York, pp 307–315
5. Brodal GS, Fagerberg R, Vinther K (2007) Engineering a cache-oblivious sorting algorithm. ACM J Exp Algoritmics (Special Issue of ALENEX 2004) 12(2.2):23
6. Department of Computer Science, Duke University. TPIE: a transparent parallel I/O environment. http://www.cs.duke.edu/TPIE/. Accessed 2002

7. Franceschini G (2004) Proximity mergesort: optimal in-place sorting in the cache-oblivious model. In: Proceedings of the 15th annual ACM-SIAM symposium on discrete algorithms (SODA). SIAM, Philadelphia, p 291

8. Frigo M, Leiserson CE, Prokop H, Ramachandran S (1999) Cache-oblivious algorithms. In: Proceedings of the 40th annual symposium on foundations of computer science. IEEE Computer Society Press, Los Alamitos, pp 285–297

9. Hoare CAR (1962) Quicksort. Comput J 5(1):10–15

10. Williams JWJ (1964) Algorithm 232: Heapsort. Commun ACM 7(6):347–348

Cache-Oblivious Spacetree Traversals

Michael Bader[1] and Tobias Weinzierl[2]
[1]Department of Informatics, Technical University of Munich, Garching, Germany
[2]School of Engineering and Computing Sciences, Durham University, Durham, UK

Keywords

Cache-oblivious algorithms; Grid traversals; Octree; Quadtree; Space-filling curves; Spacetree; Tree-structured grids

Years and Authors of Summarized Original Work

2009; Weinzierl
2013; Bader

Background

In scientific computing and related fields, mathematical functions are often approximated on meshes where each mesh cell contains a local approximation (e.g., using polynomials) of the represented quantity (density functions, physical quantities such as temperature or pressure, etc.). The grid cells may adaptively refine within areas of high interest or where the applied numerical algorithms demand improved resolution.

The resolution even may dynamically change throughout the computation.

In this context, we consider *tree-structured* adaptive meshes, i.e., meshes that result from a recursive subdivision of grid cells. They can be represented via trees – quadtrees or octrees being the most prominent examples. In typical problem settings, quantities are stored on entities (vertices, edges, faces, cells) of the grid. The computation of these variables is usually characterized by local interaction rules and involves variables of adjacent grid cells only. Hence, efficient algorithms are required for the (parallel) traversal of such tree-structured grids and their associated variables.

Problem Definition

Consider a hierarchical mesh of grid cells (triangles, squares, tetrahedra, cubes, etc.), in which all grid cells result from recursively splitting an existing grid cell into a fixed number k of geometrically similar child cells. The resulting grid is equivalent to a tree with uniform degree k. We refer to it as a *spacetree*. Special cases are quadtrees (based upon squares, i.e., dimension $d = 2$, and $k = 4$) and octrees (cubes, $d = 3$, and $k = 8$). Depending on the problem, only the mesh defined by the leaves of the tree may be of interest, or a *multilevel* grid may be considered. The latter also includes all cells corresponding to interior nodes of the tree. Also, meshes resulting from a collection of spacetrees may be considered. Such a generalized data structure is called a *forest* of spacetrees. A mathematical function shall be defined on such a mesh via coefficients that are associated with entities (vertices, edges, faces, cells) of the grid cells. Each coefficient contributes to the representation of the mathematical function in the grid cells adjacent to its entity. For typical computations, we then require efficient algorithms for *mesh traversals* processing all unknowns on entities.

Mesh traversal (definition): Run through all leaf, i.e., unrefined, grid cells and process all function coefficients associated to each cell or to entities adjacent to it.

Multiscale traversal (definition): Perform one mesh traversal including all (or a certain subset of the) grid cells (tree-interior and leaf cells), thus processing the coarse-grid cells of the grid hierarchy as well.

Mesh traversals may be used to define a sequential order (linearization) on the mesh cells or coefficients. Sequential orders that preserve locality may be used to define partitions for parallel processing and load balancing. Of special interest are algorithms that minimize the memory accesses during traversals as well as the memory required to store the tree-structured grids.

Key Results

Space-Filling Curve Orders on Tree-Structured Grids

Space-filling curves (SFCs) [1, 5] are continuous surjective mappings from a one-dimensional interval to a higher-dimensional target domain (typically squares, cubes, etc.). They are constructed

via an "infinite" recursion process analogous to the generation of tree-structured meshes. Space-filling curves thus induce a sequential order on a corresponding tree-structured grid (an example is given in Fig. 1).

The construction of the curve may be described via a grammar, in which the nonterminals reflect the local orientation of the curve within a grid cell (e.g., Fig. 2). Terminals are used to indicate transfers between grid cells or levels.

Together with this grammar, a bitstream encoding of the refinement information (as in Fig. 1) provides a minimal-memory data structure to encode a given tree-structured grid. Using Hilbert or Lebesgue (Morton order) SFCs, for example, respective algorithms can be formulated for quadtrees and octrees. Peano curves lead to traversals for (hyper)cube-based spacetrees with 3-refinement along each dimension.

Space-Filling-Curve Traversals

Depth-first traversals of the SFC/bitstream-encoded tree visit all leaf cells of the tree-structured grid in space-filling order (*SFC*

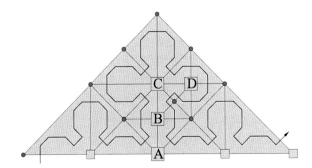

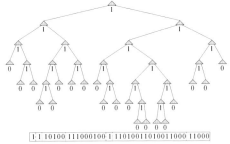

Cache-Oblivious Spacetree Traversals, Fig. 1 Recursively structured triangular mesh and corresponding binary tree with bitstream encoding. The illustrated iteration of a Sierpinski SFC defines a sequential order

on the leaf cells (equivalent to a depth-first traversal of the tree) and classifies vertices into groups left (●) and right (□) of the curve. Vertices *A*, *B*, *C*, and *D* are visited in last-in-first-out order during the traversal

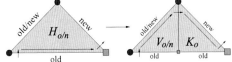

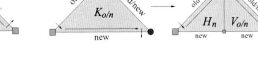

Cache-Oblivious Spacetree Traversals, Fig. 2 Replacement rules of a grammar (with six non-terminals $H_{o/n}, K_{o/n}, V_{o/n}$; illustration for $V_{o/n}$ is skipped) to construct the Sierpinski curve. The nonterminals classify

for each vertex and edge of a cell, whether it is located left (●) or right (□) of the curve. The *old/new* labels indicate whether the grid cell adjacent to this edge occurs earlier/later in the Sierpinski order *traversal*) sequentially. Cell-local data can be held in a stream. All other entities (in 2D: vertices and edges) are to be processed by all adjacent grid cells, i.e., are processed multiple times. For them, a storage scheme for intermediate values or repeated access is required. Figure 1 illustrates that the SFC induces a left/right classification of these entities. During SFC traversals, these entities are accessed in a LIFO fashion, such that intermediate values can be stored on two stacks (left vs. right). Local access rules may be inferred from an augmented grammar (as in Fig. 2). While the left/right classification determines the involved stack, the old/new classification determines whether entities are accessed for the first time during traversal (*first touch*) or have been processed by all adjacent cells (*last touch*). First and last touch trigger loading and storing the respective variables from/onto data streams. These *stack properties* hold for several space-filling curves.

SFC Traversals in the Cache-Oblivious Model

Memory access in SFC traversals is restricted to stack and stream accesses. Random access to memory is entirely avoided. Thus, the number of cache and memory accesses can be accurately described by the I/O or cache-oblivious model (see Cross-References). For the 2D case, assume that for a subtree with K grid cells, the number of edges on the boundary of the respective subgrid is $O(\sqrt{K})$ – which is always satisfied by regularly refined meshes. It can be shown that if we choose K such that all boundary elements fit into cache (size: M words), only boundary edges will cause non-compulsory cache misses [2]. For an entire SFC traversal, the number of cache misses is $O\left(\frac{N}{MB}\right)$, which is asymptotically optimal (B is the number of words per cache line). For adaptively refined meshes, it is an open question what kind of restrictions are posed on the mesh by the $O(\sqrt{K})$ criterion. While it is easy to construct degenerate grids that violate the condition, it is interesting whether grids that result from useful refinement processes (with physically motivated refinement criteria) tend to satisfy the $O(\sqrt{K})$ requirement.

Multiscale Depth-First and Breadth-First Traversals

Multiscale traversals find applications in multiphysics simulations, where different physical models are used on the different levels, as well as in (additive) multigrid methods or data analysis, where the different grid levels hold data in different resolutions. If algorithms compute not only results on one level, it is typically sufficient to have two levels available throughout the traversal at one time. Multiscale algorithms then can be constructed recursively.

As variables exist on multiple levels, their (intermediate) access scheme is more elaborate than for a pure SFC traversal. A stack-based management is trivial if we apply one set of stacks per resolution level. Statements on cache obliviousness then have to be weakened, as the maximum number of stacks is not resolution independent anymore. They depend on the number of refinement levels. For depth-first and Peano, $2d + 2$ stacks have been proven to be sufficient (d the spatial dimension). Such a multiscale scheme remains cache oblivious independent of the refinement. It is unknown though doubtful whether schemes for other curves and depth-first traversal exist that allow accesses to unknowns using a resolution-independent number of stacks for arbitrary d.

Toward Parallel Tree Traversals

Data decomposition is the predominant parallelization paradigm in scientific computing: operations are executed on different sets of data in parallel. For distinct sets, the parallelization does not require any synchronization mechanism. For spacetree data structures, distinct pieces of data are given by spacetree cells that do not share a vertex. A parallel data traversal then can be rewritten as a mesh traversal where (in the parallel traversal phases) succeeding cells along the traversal do not share grid entities – a contradiction to connected space-filling curves. For three-partitioning in 2D, such a reordering allows a maximum concurrency level of four (see Fig. 3).

For breadth-first traversals, a reordering is trivial if we drop the space-filling curve ordering and instead reorder all leaves of one level to ob-

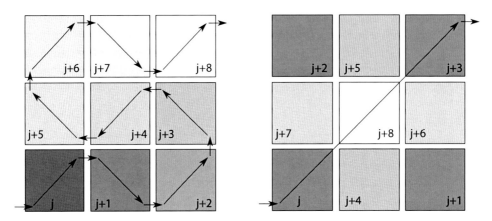

Cache-Oblivious Spacetree Traversals, Fig. 3 Peano space-filling curve with numbering on a level $n+1$ (*left*). The ordering then is rearranged to have a concurrency level of four (*right*) illustrated via different shades of gray

tain the highest concurrency level. For depth-first traversals, in contrast, the maximum concurrency level is strictly bounded, even if we drop the SFC paradigm. It remains one for all bipartitioning schemes, is at most 2^d for three-partitioning, and is at most $(\lfloor k/2 \rfloor)^d$ for k-partitioning.

Recursion Unrolling and Parallelism

As concurrency on the cell level is important for many applications, *recursion unrolling* becomes an important technique: (Regular) Subtrees or, more general, fragments along the space-filling curve's depth-first ordering are identified and locally replaced by a breadth-first traversal. This can be done without modifying any data access order on the surfaces of the cut-out curve fragment if the data load and store sequence is preserved along the fragment throughout the unrolling while only computations are reordered. Recursion unrolling then has an impact on the execution overhead as it eliminates recursive function calls and it improves the concurrency of the computations. It can be controlled by an on-the-fly analysis of the tree structure and thus seamlessly integrates into changing grids.

Other Tree-Structured Meshes and Space-Filling Curves

Traversals and stacks can team up only for certain space-filling curves and dimensions:

- In 2D, the stack property is apparently satisfied by all *connected* space-filling curves (for connected SFCs, two contiguous subdomains share an edge). SFC traversals are induced by a grammar that allows a left/right classification. However, no formal proof for this claim has been given yet.
- In 3D and all higher dimensions, the Peano curve is the only connected curve that has been found to satisfy all required properties [6].
- For octree meshes (in 3D), it is an open problem whether SFC traversals exist that can exploit stack properties. Hilbert curves and Lebesgue curves yield data access patterns with spatial and temporal locality but do not provide a stack property.

Applications

Practical applications comprise (parallel) numerical simulations on spacetree meshes that require adaptive refinement and coarsening in each time step or after each iteration [3,4,6]. SFC traversals on spacetrees induce sequential orders that may be exploited to create balanced partitions with favorable quality (due to SFCs being Hölder continuous). Using spacetrees as helper data structures, respective SFC orders can be defined also for entirely unstructured meshes or particle sets.

Space-filling curves are thus a frequently used tool to define parallel partitions.

Cross-References

▶ Cache-Oblivious Model
▶ I/O-Model

Recommended Reading

1. Bader M (2013) Space-filling curves – an introduction with applications in scientific computing. Texts in computational science and engineering, vol 9. Springer, Heidelberg/New York http://link.springer.com/book/10.1007/978-3-642-31046-1/page/1
2. Bader M, Rahnema K, Vigh CA (2012) Memory-efficient Sierpinski-order traversals on dynamically adaptive, recursively structured triangular grids. In: Jonasson K (ed) Applied parallel and scientific computing – 10th international conference, PARA 2010. Lecture notes in computer science, vol 7134. Springer, Berlin/New York, pp 302–311
3. Burstedde C, Wilcox LC, Ghattas O (2011) p4est: scalable algorithms for parallel adaptive mesh refinement on forests of octrees. SIAM J Sci Comput 33(3):1103–1133
4. Mitchell WF (2007) A refinement-tree based partitioning method for dynamic load balancing with adaptively refined grids. J Parallel Distrib Comput 67(4):417–429
5. Sagan H (1994) Space-filling curves. Universitext. Springer, New York
6. Weinzierl T (2009) A framework for parallel PDE solvers on multiscale adaptive Cartesian grids. Dissertation, Institut für Informatik, Technische Universität München, München, http://www.dr.hut-verlag.de/978-3-86853-146-6.html

Canonical Orders and Schnyder Realizers

Stephen Kobourov
Department of Computer Science, University of Arizona, Tucson, AZ, USA

Keywords

Canonical order; Planar graph drawing algorithms; Schnyder realizer

Years and Authors of Summarized Original Work

1990; de Fraysseix, Pach, Pollack
1990; Schnyder

Problem Definition

Every planar graph has a crossings-free drawing in the plane. Formally, a *straight-line drawing* of a planar graph G is one where each vertex is placed at a point in the plane and each edge is represented by a straight-line segment between the two corresponding points such that no two edges cross each other, except possibly at their common end points. A *straight-line grid drawing* of G is a straight-line drawing of G where each vertex of G is placed on an integer grid point. The area for such a drawing is defined by the minimum-area axis-aligned rectangle, or *bounding box*, that contains the drawing.

Wagner in 1936 [12], Fáry in 1948 [5], and Stein in 1951 [10] proved independently that every planar graph has a straight-line drawing. It was not until 1990 that the first algorithms for drawing a planar graph on a grid of polynomial area were developed. The concepts of canonical orders [4] and Schnyder realizers [9] were independently introduced for the purpose of efficiently computing straight-line grid drawings on the $O(n) \times O(n)$ grid. These two seemingly very different combinatorial structures turn out to be closely related and have since been used in many different problems and in many applications.

Key Results

We first describe canonical orders for planar graphs and a linear time procedure to construct them. Then we describe Schnyder realizers and a linear time procedure to compute them. Finally, we show how they can be used to compute straight-line grid drawings for planar graph.

Canonical Order

A planar graph G along with a planar embedding (a cyclic order of the neighbors for each vertex) is called a *plane graph*. Given a graph G, testing for planarity and computing a planar embedding can be done in linear time [6]. Let G be a maximal plane graph with outer vertices u, v, w in counterclockwise order. Then a *canonical order* or *shelling order* of G is a total order of the vertices $v_1 = u, v_2 = v, v_3, \ldots, v_n = w$ that meets the following criteria for every $4 \leq i \leq n$:

(a) The subgraph $G_{i-1} \subseteq G$ induced by $v_1, v_2, \ldots, v_{i-1}$ is 2-connected, and the boundary of its outerface is a cycle C_{i-1} containing the edge (v_1, v_2).
(b) The vertex v_i is in the outerface of G_{i-1}, and its neighbors in G_{i-1} form a subinterval of the path $C_{i-1} - (u, v)$ with at least two vertices; see Fig. 1a–1b.

Every maximal plane graph G admits a canonical order; see Fig. 1a–1b. Moreover, computing such an order can be done in $O(n)$ time where n is the number of vertices in G. Before proving these claims, we need a simple lemma.

Lemma 1 *Let G be a maximal plane graph with canonical order $v_1, v_2 = v, v_3, \ldots, v_n$. Then for $i \in \{3, \ldots, n-1\}$, any separating pair $\{x, y\}$ of G_i is a chord of C_i.*

The proof of the lemma is simple. Recall that a *chord* of a cycle C is an edge between nonadjacent vertices of C. Since G is a maximal plane graph and since each vertex $v_j, j \in \{n, \ldots, i+1\}$ is in the outerface of G_j, all the internal faces of G_i are triangles. Adding

a dummy vertex d along with edges from d to each vertex on the outerface of G_i yields a maximal plane graph G'. Then G' is 3-connected, and for each separation pair $\{x, y\}$ of G, the set $T = \{x, y, d\}$ is a separating set of G' since $G = G' \setminus d$. The set T is a separating triangle in G' [1], and therefore, the edge (x, y) is a chord on C_i.

Theorem 1 *A canonical order of a maximal plane graph G can be computed in linear time.*

This is also easy to prove. If the number of vertices n in G is 3, then the canonical ordering of G is trivially defined. Let $n > 3$ and choose the vertices $v_n = w, v_{n-1}, \ldots, v_3$ in this order so that conditions (a)–(b) of the definition are satisfied. Since G is a maximal plane graph, G is 3-connected, and hence, $G_{n-1} = G \setminus w$ is 2-connected. Furthermore, the set of vertices adjacent to $v_n = w$ forms a cycle C_{n-1}, which is the boundary of the outerface of G_{n-1}. Thus, conditions (a)–(b) hold for $k = n$.

Assume by induction hypothesis that the vertices $v_n, v_{n-1}, \ldots, v_{i+1}, i \geq 3$ have been appropriately chosen. We now find the next vertex v_i. If we can find a vertex x on C_i, which is not an end vertex of a chord, then we can choose $v_k = x$. Indeed, if deleting x from G_i violated the 2-connectivity, then the cut vertex y of $G_i - x$, together with x, would form a separating pair for G_i, and hence, (x, y) would be a chord in G_i (from the lemma above). We now show that we can find a vertex v_i on C_i which is not an end vertex of a chord.

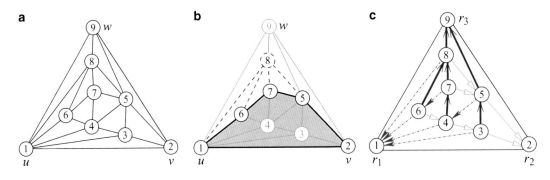

Canonical Orders and Schnyder Realizers, Fig. 1 (a) A canonical order of vertices for a maximal plane graph G, (b) insertion of v_8 in G_7, (c) a Schnyder realizer for G

If there is no chord of C_i, then we can choose any vertex of C_i other than u and v as v_k. Otherwise, label the vertices of $C_i - \{(u, v)\}$ by $p_1 = u, p_2, \ldots, p_t = v$ consecutively from u to v. By definition, any chord $(p_k, p_l), k < l$ must have $k < l - 1$. We say that a chord $(p_k, p_l), k < l$, includes another chord $(p_{k'}, p_{l'}), k' < l'$, if $k \leq k' < l' \leq l$. Then take an inclusion-minimal chord (p_k, p_l) and any vertex p_j for $k < j < l$ can be chosen as v_i. Since v_i is not an end vertex of a chord for $C_{i-1}, G_{i-1} = G_i \setminus v_i$ remains 2-connected. Furthermore, due to the maximal planarity of G, the neighborhood of v_i on C_{i-1} forms a subinterval for $C_{i-1} - (u, v)$.

The algorithm, implicit in the above argument, can be implemented to run in linear time by keeping a variable for each vertex x on C_i, counting the number of chords x is incident to. After each vertex v_i is chosen, the variables for all its neighbors can be updated in $O(\deg(v_i))$ time. Summing over all vertices in the graph leads to an overall linear running time [2], and this concludes the proof of the theorem.

Schnyder Realizer

Let G be a maximal plane graph. A *Schnyder realizer* S of G is a partition of the internal edges of G into three sets T_1, T_2, and T_3 of directed edges, so that for each interior vertex v, the following conditions hold:

(a) v has outdegree exactly one in each of T_1, T_2, and T_3.
(b) The clockwise order of edges incident to v is outgoing T_1, incoming T_2, outgoing T_3, incoming T_1, outgoing T_2, and incoming T_3; see Fig. 1c.

Since a maximal plane graph has exactly $n - 3$ internal vertices and exactly $3n - 9$ internal edges, the three outgoing edges for each internal vertex imply that all the edges incident to the outer vertices are incoming. In fact, these two conditions imply that for each outer vertex $r_i, i = 1, 2, 3$, the incident edges belong to the same set, say T_i, where r_1, r_2, r_3 are in counterclockwise order around the outerface and each set of edges T_i forms a directed tree, spanning all the internal

vertices and one external vertex r_i, oriented towards r_i [3]; see Fig. 1c. Call r_i the root of T_i for $i = 1, 2, 3$.

Note that the existence of a decomposition of a maximal planar graph G into three trees was proved earlier by Nash-Williams [8] and by Tutte [11]. Kampen [7] showed that these three trees can be oriented towards any three specified root vertices r_1, r_2, r_3 of G so that each vertex other than these roots has exactly one outgoing edges in each tree. Schnyder [9] proved the existence of the special decomposition defined above, along with a linear time algorithm to compute it. Before we describe the algorithm, we need to define the operation of *edge contraction*. Let G be a graph and $e = (x, y)$ be an edge of G. Then we denote by G/e, the simple graph obtained by deleting x, y and all their incident edges from G, adding a new vertex z and inserting an edge (z, v) for each vertex v that is adjacent to either x or y in G. Note that for a maximal plane graph G, contracting an edge $e = (x, y)$ yields a maximal plane graph if and only if there are exactly two common neighbors of x and y. Two end vertices of an edge $e = (x, y)$ have exactly two common neighbors if and only if the edge e is not on the boundary of a separating triangle.

Lemma 2 *Let G be a maximal plane graph with at least 4 vertices, where u is an outer vertex of G. Then there exists an internal vertex v of G such that (u, v) is an edge in G and vertices u and v have exactly two common neighbors.*

This is easy to prove. If G has exactly 4 vertices, then it is K_4 and the internal vertex of G has exactly two common neighbors. Consider graph G with more than 4 vertices. If u is not on the boundary of any separating triangle, then taking any neighbor of u as v is sufficient. Else, if u is on the boundary of a separating triangle Δ, we can find a desired vertex v by induction on the subgraph of G inside Δ.

Theorem 2 *A Schnyder realizer of a maximal plane graph G can be computed in linear time.*

The proof of the theorem is by induction. If G has exactly 3 vertices, its Schnyder realizer is computed trivially. Consider graph G with

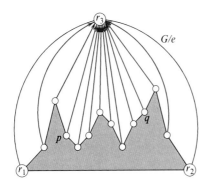
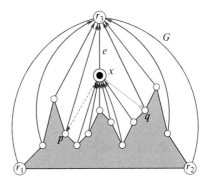

Canonical Orders and Schnyder Realizers, Fig. 2 Computing a Schnyder realizer of a maximal plane graph G from that of G/e

more than 3 vertices. Let r_1, r_2, and r_3 be the three outer vertices in counterclockwise order. Then by the above lemma, there is an internal vertex x in G and edge $e = (r_3, x)$ so that r_3 and x have exactly two common neighbors. Let $G' = G/e$. Then by the induction hypothesis, G' has a Schnyder realizer with the three trees T_1, T_2, and T_3, rooted at r_1, r_2, and r_3. We now modify this to find a Schnyder realizer for G. The orientation and partitioning of all the edges not incident to x remain unchanged from G/e. Among the edges incident to x, we add e to T_3, oriented towards r_3. We add the two edges that are just counterclockwise of e and just clockwise of e in the ordering around x, to T_1 and T_2, respectively, both oriented away from x. Finally we put all the remaining edges in T_3, oriented towards x; see Fig. 2. It is now straightforward to check that these assignment of edges to the trees satisfy the two conditions. The algorithm implicit in the proof can be implemented in linear time, given the edge contraction sequence. The edge contraction sequence itself can be computed in linear time by taking the reverse order of a canonical order of the vertices and in every step contracting the edge between r_3 and the current vertex in this order.

Drawing Planar Graphs

We now show how canonical orders and Schnyder realizers can be used to compute straight-line grid drawings of maximal plane graphs.

Theorem 3 *Let G be a maximal plane graph with n vertices. A straight-line grid drawing of G on the $(2n-4) \times (n-2)$ grid can be computed in linear time.*

This is a constructive proof. Let $\mathcal{O} = v_1, \ldots, v_n$ be a canonical order of G, G_i, the subgraph of G induced by the vertices $v_1, \ldots, v_i$, and C_i the boundary of the outerface of $G_i, i = 3, 4, \ldots, n$. We incrementally obtain straight-line drawing Γ_i of G_i for $i = 3, 4, \ldots, n$. We also maintain the following invariants for Γ_i:

(i) The x-coordinates of the vertices on the path $C_i \setminus \{(v_1, v_2)\}$ are monotonically increasing as we go from v_1 to v_2.
(ii) Each edge of the path $C_i - \{(v_1, v_2)\}$ is drawn with slope 1 or -1.

We begin with G_3, giving v_1, v_2, and v_3 coordinates $(0, 0), (2, 0)$, and $(1, 1)$; the drawing Γ_3 satisfies conditions (i)–(ii); see Fig. 3a. Suppose the drawing for Γ_{i-1} for some $i > 3$ has already been computed; we now show how to obtain Γ_i. We need to add vertex v_i and its incident edges in G_i to Γ_{i-1}. Let $w_1 = v_1, \ldots, w_t = v_2$ be the vertices on $C_{i-1} \setminus \{(v_1, v_2)\}$ in this order from v_1 to v_2. By the property of canonical orders, v_i is adjacent to a subinterval of this path. Let $w_l, \ldots, w_r, 1 \leq l < r \leq t$, be the vertices adjacent to v_i, in this order. We want to place v_i at the intersection point p between the straight line from w_l with slope 1 and the straight line from w_r with slope -1. Note that by condition (ii),

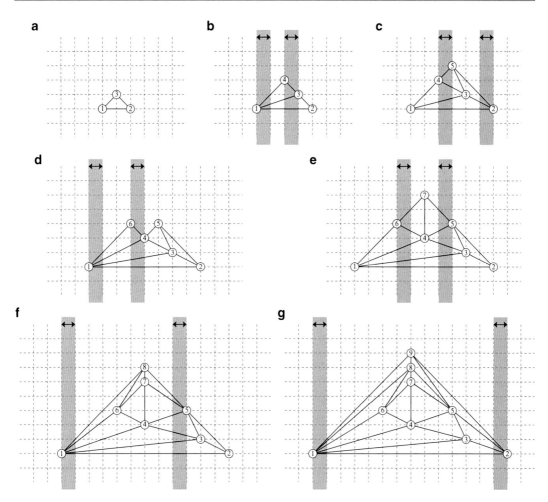

Canonical Orders and Schnyder Realizers, Fig. 3 Illustration for the straight-line drawing algorithm using canonical order

the two vertices w_l and w_r are at even Manhattan distance in Γ, and hence, point p is a grid point. However, if we place v_i at p, then the edges $v_i w_l$ and $v_i w_r$ might overlap with the edges $w_l w_{l+1}$ and $w_{r-1} w_r$, since they are drawn with slopes 1 or -1. We thus shift all the vertices to the left of w_l in Γ_{i-1} (including w_l) one unit to the left at all the vertices to the right of w_r in Γ_{i-1} (including w_r) one unit to the right; see Fig. 3.

Consider a rooted tree T, spanning all the internal vertex of G along with one external vertex v_n, where v_n is the root of T and for each internal vertex x, the parent of x is the highest numbered successor in G. (Later we see that T can be one of the three trees in a Schnyder realizer of G.)

For any internal vertex x of G, denote by $U(x)$ the set of vertices that are in the subtree of T rooted at v (including v itself). Then the shifting of the vertices above can be obtained by shifting the vertices in $U(w_i), i = 1, \ldots, l$ one unit to the left and the vertices in $U(w_i), i = r, \ldots, t$ one unit to the right. After these shifts, v_i can be safely placed at the intersection of the line with slope 1 from w_l and the line with slope -1 from w_r. Note that this drawing satisfies conditions (i)–(ii). This algorithm can be implemented in linear time, even though the efficient vertex shifting requires careful relative offset computation [2].

The next theorem shows how Schnyder realizers can be used to compute a straight-line grid

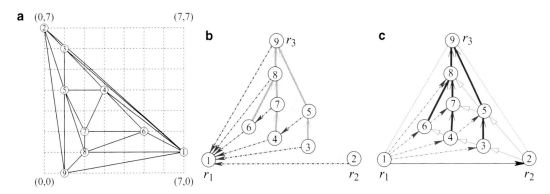

Canonical Orders and Schnyder Realizers, Fig. 4 (a) A straight-line drawing for the graph in Fig. 1 using Schnyder realizer, (b)–(c) computation of a canonical order from a Schnyder realizer

drawing of a plane graph. Let T_1, T_2, and T_3 be the trees in a Schnyder realizer of G, rooted at outer vertices r_1, r_2, and r_3. Since each internal vertex v of G has exactly one outgoing edge in each of the trees, there is a directed path $P_i(v)$ in each of the three trees T_i from v to r_i. These three paths $P_1(v)$, $P_2(v)$, and $P_3(v)$ are vertex disjoint except for v, and they define three regions $R_1(v), R_2(v)$, and $R_3(v)$ for v. Here $R_i(v)$ is the region between the two paths $P_{i-1}(v)$ and $P_{i+1}(v)$, where the addition and subtraction are modulo 3. Let $\eta_i(v)$ denote the number of vertices in $R_i(v) \setminus P_{i-1}(v)$, where $i = 1, 2, 3$ and the subtraction is modulo 3. Extend these definitions to the outer vertices as follows: $\eta_i(r_i) = n - 2, \eta_{i+1}(r_i) = 1, \eta_{i-1}(r_i) = 0$.

Theorem 4 *The coordinates $((\eta_1(v), \eta_2(v)))$ for each vertex v in G give a straight-line drawing Γ of G on a grid of size $(n-2) \times (n-2)$.*

Place each vertex v at the point with coordinates $(\eta_1(v), \eta_2(v)$, and $\eta_3(v))$. Since $\eta(v) = \eta_1(v) + \eta_2(v) + \eta_3(v)$ counts the number of vertices in all the three regions of v, each vertex of G except v is counted exactly once in $\eta(v)$. Thus, $\eta_1(v) + \eta_2(v) + \eta_3(v) = n - 1$ for each vertex v. Thus, the drawing Γ' obtained by these coordinates $(\eta_1(v), \eta_2(v), \eta_3(v))$ lies on the plane $x + y + z = n - 1$. Furthermore, the drawing does not induce any edge crossings; see [9]. Thus, Γ' is a straight-line drawing of G on the plane $x + y + z = n - 1$. Then Γ is just a projection of Γ' on the plane $z = 0$ and hence is planar. Since each coordinate in the drawing is

bounded between 0 and $n - 1$, the area is at most $(n - 2) \times (n - 2)$; see Fig. 4a.

Equivalency of Canonical Orders and Schnyder Realizers

Here we show that canonical orders and Schnyder realizers are in fact equivalent in the sense that a canonical order of a graph defines a Schnyder realizer and vice versa [3].

Lemma 3 *A canonical order for a maximal plane graph G defines a unique Schnyder realizer where the three parents for each vertex v of G are its leftmost predecessor, its rightmost predecessor, and its highest-labeled successor.*

See Fig. 1a, 1c for a canonical order $\mathcal{O}$ and the corresponding Schnyder realizer $\mathcal{S}$ defined by $\mathcal{O}$ for a maximal plane graph. One can easily verify that this definition of $\mathcal{S}$ satisfies the two conditions for each internal vertex for a maximal plane graph. A canonical order $\mathcal{O}$ and the Schnyder realizer $\mathcal{S}$ obtained from $\mathcal{O}$ for a maximal plane graph are said to be *compatible*.

We now describe two ways to obtain a canonical order from a Schnyder realizer $\mathcal{S}$. In both cases, we obtain a canonical order which is compatible with $\mathcal{S}$.

Lemma 4 *Let G be a maximal plane graph with outer vertices r_1, r_2, and r_3 in counterclockwise order and let T_1, T_2, and T_3 be the three trees in a Schnyder realizer of G rooted at r_1, r_2, and r_3. Then a compatible canonical order of G can be obtained as follows:*

1. By taking the counterclockwise depth-first traversal order of the vertices in the graph $T_1 \cup \{(r_2, r_1), (r_3, r_1)\}$; see Fig. 4b.

2. By taking the topological order of the directed acyclic graph $T_1^{-1} \cup T_2^{-1} \cup T_3$, where T_i^{-1}, $i = 1, 2$ is the Schnyder tree T_i with reversed edge directions; see Fig. 4c.

It is not difficult to show that the directed graph $T_1^{-1} \cup T_2^{-1} \cup T_3$ is in fact acyclic [3, 9]. Then it is easy to verify that the canonical orders obtained from a Schnyder realizer S are compatible with S; i.e., defining a Schnyder realizer from them by Lemma 3 produces the original Schnyder realizer S.

Cross-References

▸ Bend Minimization for Orthogonal Drawings of Plane Graphs
▸ Convex Graph Drawing
▸ Force-Directed Graph Drawing
▸ Planarity Testing

Recommended Reading

1. Baybars I (1982) On k-path hamiltonian maximal planar graphs. Discret Math 40(1):119–121
2. Chrobak M, Payne TH (1995) A linear-time algorithm for drawing a planar graph on a grid. Inf Process Lett 54(4):241–246
3. de Fraysseix H, de Mendez PO (2001) On topological aspects of orientations. Discret Math 229(1–3):57–72
4. de Fraysseix H, Pach J, Pollack R (1990) How to draw a planar graph on a grid. Combinatorica 10(1):41–51
5. Fáry I (1948) On straight lines representation of planar graphs. Acta Scientiarum Mathematicarum 11:229–233
6. Hopcroft JE, Tarjan RE (1974) Efficient planarity testing. J ACM 21(4):549–568
7. Kampen G (1976) Orienting planar graphs. Discret Math 14(4):337–341
8. Nash-Williams CSJA (1961) Edge-disjoint spanning trees of finite graphs. J Lond Math Soc 36:445–450
9. Schnyder W (1990) Embedding planar graphs on the grid. In: ACM-SIAM symposium on discrete algorithms (SODA 1990), San Francisco, pp 138–148
10. Stein SK (1951) Convex maps. Am Math Soc 2(3):464–466
11. Tutte WT (1961) On the problem of decomposing a graph into n connected factors. J Lond Math Soc 36:221–230
12. Wagner K (1936) Bemerkungen zum Vierfarbenproblem. Jahresbericht der Deutschen Mathematiker-Vereinigung 46:26–32

Causal Order, Logical Clocks, State Machine Replication

Xavier Défago
School of Information Science, Japan Advanced Institute of Science and Technology (JAIST), Ishikawa, Japan

Keywords

State-machine replication: active replication

Years and Authors of Summarized Original Work

1978; Lamport

Problem Definition

This entry covers several problems, related with each other. The first problem is concerned with maintaining the causal relationship between events in a distributed system. The motivation is to allow distributed systems to reason about time with no explicit access to a physical clock. Lamport [5] defines a notion of logical clocks that can be used to generate timestamps that are consistent with causal relationships (in a conservative sense). He illustrates logical clocks (also called Lamport clocks) with a distributed mutual exclusion algorithm. The algorithm turns out to be an illustration of state-machine replication. Basically, the algorithm generates a total ordering of the events that is consistent across processes. With all processes starting in the same state, they evolve consistently with no need for further synchronization.

System Model

The system consists of a collection of processes. Each process consists of a sequence of events.

Processes have no shared memory and communicate only by exchanging messages. The exact definition of an event depends on the system actually considered and the abstraction level at which it is considered. One distinguishes between three kinds of events: internal (affects only the process executing it), send, and receive events.

Causal Order

Causal order is concerned with the problem that the occurrence of some events may affect other events in the future, while other events may not influence each other. With processes that do not measure time, the notion of simultaneity must be redefined in such a way that simultaneous events are those that cannot possibly affect each other. For this reason, it is necessary to define what it means for an event to happen before another event.

The following "happened before" relation is defined as an irreflexive partial ordering on the set of all events in the system [5].

Definition 1 The relation "$\rightarrow$" on the set of events of a system is the smallest relation satisfying the following three conditions:

1. If a and b are events in the same process, and a comes before b, then $a \rightarrow b$.
2. If a is the sending of a message by one process and b is the receipt of the same message by another process, then $a \rightarrow b$.
3. If $a \rightarrow b$ and $b \rightarrow c$ then $a \rightarrow c$.

Definition 2 Two distinct events a and b are said to be *concurrent* if $a \nrightarrow b$ and $b \nrightarrow a$.

Logical Clocks

Lamport also defines clocks in a generic way, as follows.

Definition 3 A clock C_i for a process p_i is a function which assigns a number $C_i \langle a \rangle$ to any event a on that process. The entire system of clocks is represented by the function C which assigns to any event b the number $C \langle b \rangle$, where $C \langle b \rangle = C_j \langle b \rangle$ if b is an event in process p_j. The system of clocks must meet the following *clock condition*.

- For any events a and b, if $a \rightarrow b$ then $C \langle a \rangle < C \langle b \rangle$.

Assuming that there is some arbitrary total ordering $\prec$ of the processes (e.g., unique names ordered lexicographically), Lamport extends the "happened before" relation and defines a relation "$\Rightarrow$" as a total ordering on the set of all events in the system.

Definition 4 The total order relation $\Rightarrow$ is defined as follows. If a is an event in process p_i and b is an event in process p_j, then $a \Rightarrow b$ if and only if either one of the following conditions is satisfied.

1. $C_i \langle a \rangle < C_j \langle b \rangle$
2. $C_i \langle a \rangle = C_j \langle b \rangle$ and $p_i \prec p_j$.

In fact, Lamport [5] also discusses an adaptation of these conditions to physical clocks, and provides a simple clock synchronization algorithm. This is however not discussed further here.

State Machine Replication

The problem of state-machine replication was originally presented by Lamport [4, 5]. In a later review of the problem, Schneider [8] defines the problem as follows (formulation adapted to the context of the entry).

Problem 1 (State-machine replication)

INPUT: A set of concurrent requests.
OUTPUT: A sequence of the requests processed at each process, such that:

1. *Replica coordination*: all replicas receive and process the same sequence of requests.
2. *Agreement*: every non-faulty state-machine replica receives every request.
3. *Order*: every non-faulty state-machine replica processes the requests it receives in the same relative order.

In his paper on logical time [5] and discussed in this entry, Lamport does not consider failures. He does however consider them in another paper on state-machine replication for fault-tolerance [4], which he published the same year.

Key Results

Lamport [5] proposed many key results related to the problems described above.

Logical Clocks

Lamport [5] defines an elegant system of logical clocks that meets the clock condition presented in Definition 3. The clock of a process p_i is represented by a register C_i, such that $C_i \langle a \rangle$ is the value held by C_i when a occurs. Each message m carries a timestamp T_m, which equals the time at which m was sent. The clock system can be described in terms of the following rules.

1. Each process p_i increments C_i between any two successive events.
2. If event a is the sending of a message m by process p_i, then the message m contains a timestamp $T_m = C_i \langle a \rangle$.
3. Upon receiving a message m, process p_j sets C_j to $\max(C_j, T_m + 1)$ (before actually executing the receive event).

State Machine Replication

As an illustration for the use of logical clocks, Lamport [5] describes a mutual exclusion algorithm. He also mentions that the approach is more general and discusses the concept of state-machine replication that he refines in a different paper [4].

The mutual exclusion algorithm is based on the idea that every process maintains a copy of a request queue, and the algorithm ensures that the copies remain consistent across the processes. This is done by generating a total ordering of the request messages, according to timestamps obtained from the logical clocks of the sending processes.

The algorithm described works under the following simplifying assumptions:

- Every message that is sent is eventually received.

- For any processes p_i and p_j, messages from p_i to p_j are received in the same order as they are sent.
- A process can send messages directly to every other processes.

The algorithm requires that each process maintains its own request queue, and ensures that the request queues of different processes always remain consistent. Initially, request queues contain a single message $(T_0, p_0, request)$, where p_0 is the process that holds the resource and the timestamp T_0 is smaller than the initial value of every clock. Then, the algorithm works as follows.

1. When a process p_i requests the resource, it sends a request message $(T_m, p_i, request)$ to all other processes and puts the message in its request queue.
2. When a process p_j receives a message $(T_m, p_i, request)$, it puts that message in its request queue and sends an acknowledgment $(T_{m'}, p_j, ack)$ to p_i.
3. When a process p_i releases the resource, it removes all instances of messages $(-, p_i, request)$ from its queue, and sends a message $(T_{m'}, p_i, release)$ to all other processes.
4. When a process p_j receives a release message from process p_i, it removes all instances of messages $(-, p_i, request)$ from its queue, and sends a timestamped acknowledgment to p_i.
5. Messages in the queue are sorted according to the total order relation $\Rightarrow$ of Definition 4. A process p_i can use the resource when (a) a message $(T_m, p_i, request)$ appears first in the queue, and (b) p_i has received from all other processes a message with a timestamp greater than T_m (or equal from any process p_j where $p_i \prec p_j$).

Applications

A brief overview of some applications of the concepts presented in this entry has been provided.

First of all, the notion of causality in distributed systems (or lack thereof) leads to a famous problem in which a user may potentially see an answer before she can see the relevant question. The time-independent characterization of causality of Lamport lead to the development of efficient solutions to enforce causal order in communication. In his later work, Lamport [3] gives a more general definition to the "happened before" relation, so that a system can be characterized at various abstraction levels.

About a decade after Lamport's work on logical clock, Fidge [2] and Mattern [6] have developed the notion of vector clocks, with the advantage of a complete characterization of causal order. Indeed, the clock condition enforced by Lamport's logical clocks is only a one-way implication (see Definition 3). In contrast, vector clocks extend Lamport clocks by ensuring that, for any events a and b, if $C\langle a \rangle < C\langle b \rangle$, then $a \rightarrow b$. This is for instance useful for choosing a set of checkpoints after recovery of a distributed system, for distributed debugging, or for deadlock detection. Other extensions of logical time have been proposed, that have been surveyed by Raynal and Singhal [7].

The state-machine replication also has many applications. In particular, it is often used for replicating a distributed service over several processors, so that the service can continue to operate even in spite of the failure of some of the processors. State-machine replication ensures that the different replicas remain consistent.

The mutual exclusion algorithm proposed by Lamport [5] and described in this entry is actually one of the first known solution to the *atomic broadcast* problem (see relevant entry). Briefly, in a system with several processes that broadcast messages concurrently, the problem requires that all processes deliver (and process) all message in the same order. Nowadays, there exist several approaches to solving the problem. Surveying many algorithms, Défago et al. [1] have classified Lamport's algorithm as *communication history* algorithms, because of the way the ordering is generated.

Cross-References

► Atomic Broadcast
► Clock Synchronization
► Concurrent Programming, Mutual Exclusion
► Linearizability
► Quorums

Recommended Reading

1. Défago X, Schiper A, Urbán P (2004) Total order broadcast and multicast algorithms: taxonomy and survey. ACM Comput Surv 36:372–421
2. Fidge CJ (1991) Logical time in distributed computing systems. IEEE Comput 24:28–33
3. Lamport L (1986) On interprocess communication. Part I: basic formalism. Distrib Comput 1:77–85
4. Lamport L (1978) The implementation of reliable distributed multiprocess systems. Comput Netw 2:95–114
5. Lamport L (1978) Time, clocks, and the ordering of events in a distributed system. Commun ACM 21:558–565
6. Mattern F (1989) Virtual time and global states of distributed systems. In: Cosnard M, Quinton P (eds) Parallel and distributed algorithms. North-Holland, Amsterdam, pp 215–226
7. Raynal M, Singhal M (1996) Capturing causality in distributed systems. IEEE Comput 29:49–56
8. Schneider FB (1990) Implementing fault-tolerant services using the state machine approach: a tutorial. ACM Comput Surv 22:299–319

Certificate Complexity and Exact Learning

Lisa Hellerstein
Department of Computer Science and Engineering, NYU Polytechnic School of Engineering, Brooklyn, NY, USA

Keywords

Boolean formulae; Cerfiticates; Computational learning; Exact learning; Query complexity

Years and Authors of Summarized Original Work

1995; Hellerstein, Pilliapakkamnatt, Raghavan, Wilkins

Problem Definition

This problem concerns the query complexity of proper learning in a widely studied learning model: exact learning with membership and equivalence queries. Hellerstein et al. [10] showed that the number of (polynomially sized) queries required to learn a concept class in this model is closely related to the size of certain certificates associated with that class. This relationship gives a combinatorial characterization of the concept classes that can be learned with polynomial query complexity. Similar results were shown by Hegedüs based on the work of Moshkov [8, 13].

The Exact Learning Model

Concepts are functions $f : X \to \{0, 1\}$ where X is an arbitrary domain. In exact learning, there is a hidden concept f from a known class of concepts C, and the problem is to exactly identify the concept f.

Algorithms in the exact learning model obtain information about f, the target concept, by querying two oracles, a membership oracle and an equivalence oracle. A membership oracle for f answers membership queries (i.e., point evaluation queries), which are of the form "What is $f(x)$?" where $x \in X$. The membership oracle responds with the value $f(x)$. An equivalence oracle for f answers equivalence queries, which are of the form "Is $h \equiv f$?" where h is a representation of a concept defined on the domain X. Representation h is called a hypothesis. The equivalence oracle responds "yes" if $h(x) = f(x)$ for all $x \in X$. Otherwise, it returns a counterexample, a value $x \in X$ such that $f(x) \neq h(x)$.

The exact learning model is due to Angluin [2]. Angluin viewed the combination of membership and equivalence oracles as constituting a "minimally adequate teacher." Equivalence queries can be simulated both in Valiant's well-known PAC model, and in the online mistake-bound learning model.

Let R be a set of representations of concepts, and let C_R be the associated set of concepts. For example, if R were a set of DNF formulas, then C_R would be the set of Boolean functions (concepts) represented by those formulas. An exact learning algorithm is said to learn R if, given access to membership and equivalence oracles for any f in C_R, it ends by outputting a hypothesis h that is a representation of f.

Query Complexity of Exact Learning

There are two aspects to the complexity of exact learning: query complexity and computational complexity. The results of Hellerstein et al. concern query complexity.

The query complexity of an exact learning algorithm measures the number of queries it asks and the size of the hypotheses it uses in those queries (and as the final output). We assume that each representation class R has an associated size function that assigns a nonnegative number to each $r \in R$. The size of a concept c with respect to R, denoted by $|c|_R$, is the size of the smallest representation of c in R; if $c \notin C_R$, $|c|_R = \infty$. Ideally, the query complexity of an exact learning algorithm will be polynomial in the size of the target and other relevant parameters of the problem.

Many exact learning results concern learning classes of representations of Boolean functions. Algorithms for learning such classes R are said to have polynomial query complexity if the number of hypotheses used, and the size of those hypotheses, is bounded by some polynomial $p(m, n)$, where n is the number of variables on which the target f is defined, and $m = |f|_R$. We assume that algorithms for learning Boolean representation classes are given the value of n as input.

Since the number and size of queries used by an algorithm are a lower bound on the time taken by that algorithm, query complexity lower bounds imply computational complexity lower bounds.

Improper Learning and the Halving Algorithm

An algorithm for learning a representation class R is said to be proper if all hypotheses used in its equivalence queries are from R, and it outputs a representation from R. Otherwise, the algorithm is said to be improper.

When C_R is a finite concept class, defined on a finite domain X, a simple, generic algorithm called the halving algorithm can be used to exactly learn R using $\log |C_R|$ equivalence queries and no membership queries. The halving algorithm is based on the following idea. For any $V \subseteq C_R$, define the majority hypothesis MAJ_V to be the concept defined on X such that for all $x \in X$, $MAJ_V(x) = 1$ if $g(x) = 1$ for more than half the concepts g in V, and $MAJ_V(x) = 0$ otherwise. The halving algorithm begins by setting $V = C_R$. It then repeats the following:

1. Ask an equivalence query with the hypothesis MAJ_V.
2. If the answer is yes, then output MAJ_V.
3. Otherwise, the answer is a counterexample x. Remove from V all g such that $g(x) = MAJ_V(x)$.

Each counterexample eliminates the majority of the elements currently in V, so the size of V is reduced by a factor of at least 2 with each equivalence query. It follows that the algorithm cannot ask more than $\log_2 |C_R|$ queries.

The halving algorithm cannot necessarily be implemented as a proper algorithm, since the majority of hypotheses may not be representable in C_R. Even when they are representable in C_R, the representations may be exponentially larger than the target concept.

Proper Learning and Certificates

In the exact model, the query complexity of proper learning is closely related to the size of certain certificates.

For any concept f defined on a domain X, a certificate that f has property P is a subset $S \subseteq X$ such that for all concepts g defined on X, if $g(x) = f(x)$ for all $x \in S$, then g has property P. The size of the certificate S is $|S|$, the number of elements in it.

We are interested in properties of the form "g is not a member of the concept class C." To take a simple example, let D be the class of constant-valued n-variable Boolean functions, i.e., D consists of the two functions $f_1(x_1, \ldots, x_n) = 1$ and $f_2(x_1, \ldots, x_n) = 0$. Then if g is an n-variable Boolean function that is not a member of D, a certificate that g is not in C could be just a pair $a \in \{0, 1\}^n$ and $b \in \{0, 1\}^n$ such that $g(a) = 1$ and $g(b) = 0$.

For C a class of concepts defined on X define the exclusion dimension of C to be the maximum, over all concepts g not in C, of the size of the smallest certificate that g is not in C. Let $XD(C)$ denote the exclusion dimension of C. In the above example, $XD(C) = 2$.

Key Results

Theorem 1 *Let R be a finite class of representations. Then there exists a proper learning algorithm in the exact model that learns C using at most $XD(C) \log |C|$ queries. Further, any such algorithm for C must make at least $XD(C)$ queries.*

Independently, Hegedüs proved a theorem that is essentially identical to the above theorem. The algorithm in the theorem is a variant of the ordinary halving algorithm. As noted by Hegedüs, a similar result to Theorem 1 was proved earlier by Moshkov, and Moshkov's techniques can be used to improve the upper bound by a factor of $\dfrac{2}{\log_2 XD(C)}$.

An extension of the above result characterizes the representation classes that have polynomial query complexity. The following theorem presents the extended result as it applies to representation classes of Boolean functions.

Theorem 2 *Let R be a class of representations of Boolean functions. Then there exists a proper* *learning algorithm in the exact model that learns R with polynomial query complexity iff there exists a polynomial $p(m, n)$ such that for all $m, n > 0$, and all n-variable Boolean functions g, if $|g|_R > m$, then there exists a certificate of size at most $p(m, n)$ proving that $|g|_R > m$.*

A concept class having certificates of the type specified in this theorem is said to have polynomial certificates.

The algorithm in the above theorem does not run in polynomial time. Hellerstein et al. give a more complex algorithm that runs in polynomial time using a Σ_4^P oracle, provided R satisfies certain technical conditions. Köbler and Lindner subsequently gave an algorithm using a Σ_2^P oracle [12].

Theorem 2 and its generalization give a technique for proving bounds on proper learning in the exact model. Proving upper bounds on the size of the appropriate certificates yields upper bounds on query complexity. Proving lower bounds on the size of appropriate certificates yields lower bounds on query complexity and hence also on time complexity. Moreover, unlike many computational hardness results in learning, computational hardness results achieved in this way do not rely on any unproven complexity theoretic or cryptographic hardness assumptions.

One of the most widely studied problems in computational learning theory has been the question of whether DNF formulas can be learned in polynomial time in common learning models. The following result on learning DNF formulas was proved using Theorem 2, by bounding the size of the relevant certificates.

Theorem 3 *There is a proper algorithm that learns DNF formulas in the exact model with query complexity bounded above by a polynomial $p(m, r, n)$, where m is the size of the smallest DNF representing the target function f, n is the number of variables on which f is defined, and r is the size of the smallest CNF representing f.*

The size of a DNF is the number of its terms; the size of a CNF is the number of its clauses. The above theorem does not imply polynomial-time learnability of arbitrary DNF formulas, since the running time of the algorithm depends not just on the size of the smallest DNF representing the target but also on the size of the smallest CNF.

Building on results of Alekhnovich et al., Feldman showed that if NP $\neq$ RP, DNF formulas cannot be properly learned in polynomial time in the PAC model augmented with membership queries. The same negative result then follows immediately for the exact model [1, 7]. Hellerstein and Raghavan used certificate size lower bounds and Theorem 1 to prove that DNF formulas cannot be learned by a proper exact algorithm with polynomial query complexity, if the algorithm is restricted to using DNF hypotheses that are only slightly larger than the target [9].

The main results of Hellerstein et al. apply to learning with membership and equivalence queries. Hellerstein et al. also considered the model of exact learning with membership queries alone and showed that in this model, a projection-closed Boolean function class is polynomial query learnable iff it has polynomial teaching dimension. Teaching dimension was previously defined by Goldman and Kearns. Hegedüs defined the extended teaching dimension and showed that all classes are polynomially query learnable with membership queries alone iff they have polynomial extended teaching dimension.

Balcázar et al. introduced the general dimension, which generalizes the combinatorial dimensions discussed above [5]. It can be used to characterize polynomial query learnability for a wide range of different queries. Balcan and Hanneke have investigated related combinatorial dimensions in the active learning setting [4].

Open Problems

It remains open whether DNF formulas can be learned in polynomial time in the exact model, using hypotheses that are not DNF formulas.

Feldman's results show the computational hardness of proper learning of DNF in the exact learning model based on complexity-theoretic assumptions. However, it is unclear whether query complexity is also a barrier to efficient learning of DNF formulas. It is still open whether the class of DNF formulas has polynomial certificates; showing they do not have polynomial certificates would give a hardness result for proper learning of DNF based only on query complexity, with no complexity-theoretic assumptions (and without the hypothesis-size restrictions used by Hellerstein and Raghavan). DNF formulas do have certain sub-exponential certificates [11].

It is open whether decision trees have polynomial certificates.

Certificate techniques are used to prove lower bounds on learning when we restrict the type of hypotheses used by the learning algorithm. These types of results are called representation dependent, since they depend on the restriction of the representations used as hypotheses. Although there are some techniques for proving representation-independent hardness results, there is a need for more powerful techniques.

Cross-References

▶ Cryptographic Hardness of Learning
▶ Hardness of Proper Learning
▶ Learning DNF Formulas
▶ Learning with the Aid of an Oracle

Recommended Reading

1. Alekhnovich M, Braverman M, Feldman V, Klivans AR, Pitassi T (2004) Learnability and automatizability. In: Proceedings of the 45th annual IEEE symposium on foundations of computer science (FOCS '04), Rome. IEEE Computer Society, Washington, DC, pp 621–630
2. Angluin D (1987) Queries and concept learning. Mach Learn 2(4):319–342
3. Angluin D (2004) Queries revisited. Theor Comput Sci 313(2):175–194
4. Balcan N, Hanneke S (2012) Robust interactive learning. In: Proceedings of the twenty fifth annual conference on learning theory (COLT '12), Edinburgh, pp 20.1–20.34
5. Balcázar JL, Castro J, Guijarro D, Köbler J, Lindner W (2007) A general dimension for query learning. J Comput Syst Sci 73(6):924–940
6. Balcázar JL, Castro J, Guijarro D, Simon H-U (2002) The consistency dimension and distribution-dependent learning from queries. Theor Comput Sci 288(2):197–215
7. Feldman V (2006) Hardness of approximate two-level logic minimization and PAC learning with membership queries. In: Proceedings of the 38th annual ACM symposium on the theory of computing (STOC '06), Seattle. ACM, New York, pp 363–372
8. Hegedüs T (1995) Generalized teaching dimensions and the query complexity of learning. In: Proceedings of the 8th annual conference on computational learning theory (COLT '95), Santa Cruz, pp 108–117
9. Hellerstein L, Raghavan V (2005) Exact learning of DNF formulas using DNF hypotheses. J Comput Syst Sci 70(4):435–470
10. Hellerstein L, Pillaipakkamnatt K, Raghavan V, Wilkins D (1996) How many queries are needed to learn? J ACM 43(5):840–862
11. Hellerstein L, Kletenik D, Sellie L, Servedio R (2012) Tight bounds on proper equivalence query learning of DNF. In: Proceedings of the 25th annual conference on learning theory (COLT '12), Edinburgh, pp 31.1–31.18
12. Köbler J, Lindner W (2000) Oracles in Σ^p_2 are sufficient for exact learning. Int J Found Comput Sci 11(4):615–632
13. Moshkov MY (1983) Conditional tests. Probl Kibern (in Russian) 40:131–170

Channel Assignment and Routing in Multi-Radio Wireless Mesh Networks

Mansoor Alicherry, Randeep Bhatia, and Li (Erran) Li
Bell Laboratories, Alcatel-Lucent, Murray Hill, NJ, USA

Keywords

Ad-hoc networks; Graph coloring

Years and Authors of Summarized Original Work

2005; Alicherry, Bhatia, Li

Problem Definition

One of the major problems facing wireless networks is the capacity reduction due to interference among multiple simultaneous transmissions. In wireless mesh networks providing mesh routers with multiple-radios can greatly alleviate this problem. With multiple-radios, nodes can transmit and receive simultaneously or can transmit on multiple channels simultaneously. However, due to the limited number of channels available the interference cannot be completely eliminated and in addition careful channel assignment must be carried out to mitigate the effects of interference. Channel assignment and routing are inter-dependent. This is because channel assignments have an impact on link bandwidths and the extent to which link transmissions interfere. This clearly impacts the routing used to satisfy traffic demands. In the same way traffic routing determines the traffic flows for each link which certainly affects channel assignments. Channel assignments need to be done in a way such that the communication requirements for the links can be met. Thus, the problem of throughput maximization of wireless mesh networks must be solved through channel assignment, routing, and scheduling.

Formally, given a wireless mesh backbone network modeled as a graph (V, E): The node $t \in V$ represents the wired network. An edge $e = (u, v)$ exists in E iff u and v are within *communication range* R_T. The set $V_G \subseteq V$ represents the set of gateway nodes. The system has a total of K channels. Each node $u \in V$ has $I(u)$ network interface cards, and has an aggregated demand $l(u)$ from its associated users. For each edge e the set $I(e) \subset E$ denotes the set of edges that it interferes with. A pair of nodes that use the same channel and are within *interference range* $R_I x$ may interfere with each other's communication, even if they cannot directly communicate. Node pairs using different channels can transmit packets simultaneously without interference. The problem is to maximize λ where at least $\lambda l(u)$ amount of throughput can be routed from each node u to the Internet (represented by a node t). The $\lambda l(u)$ throughput for each node u is achieved by computing $g(1)$ a network flow that associates with each edge $e = (u, v)$ values $f(e(i)), 1 \leq i \leq K$ where $f(e(i))$ is the rate at which traffic is transmitted by node u for node v on channel i; (2) a feasible channel assignment $F(u)$ ($F(u)$

is an ordered set where the ith interface of u operates on the ith channel in $F(u)$) such that, whenever $f(e(i)) > 0$, $i \in F(u) \cap F(v)$; (3) a *feasible* schedule S that decides the set of edge channel pair (e, i) (edge e using channel, i.e., $f(e(i)) > 0$ scheduled at time slot τ, for $\tau = 1, 2, \ldots, T$ where T is the period of the schedule. A schedule is feasible if the edges of no two edge pairs $(e_1, i), (e_2, i)$ scheduled in the same time slot for a common channel i interfere with each other ($e_1 \notin I(e_2)$ and $e_2 \notin I(e_1)$). Thus, a feasible schedule is also referred to as an interference free edge schedule. An indicator variable $X_{e,i,\tau}, e \in E, i \in F(e), \tau \geq 1$ is used. It is assigned 1 if and only if link e is active in slot τ on channel i. Note that $1/T \sum_{1 \leq \tau \leq T} X_{e,i,\tau} c(e) = f(e(i))$. This is because communication at rate $c(e)$ happens in every slot that link e is active on channel i and since $f(e(i))$ is the average rate attained on link e for channel i. This implies $1/T \sum_{1 \leq \tau \leq T} X_{e,i,\tau} = \frac{f(e(i))}{c(e)}$.

Joint Routing, Channel Assignment, and Link Scheduling Algorithm

Even the interference free edge scheduling subproblem given the edge flows is NP-hard [5]. An approximation algorithm called RCL for the joint routing, channel assignment, and link scheduling problem has been developed. The algorithm performs the following steps in the given order:

1. **Solve LP:** First optimally solve a LP relaxation of the problem. This results in a flow on the flow graph along with a not necessarily feasible channel assignment for the node radios. Specifically, a node may be assigned more channels than the number of its radios. However, this channel assignment is "optimal" in terms of ensuring that the interference for each channel is minimum. This step also yields a lower bound on the λ value which is used in establishing the worst case performance guarantee of the overall algorithm.

2. **Channel Assignment:** This step presents a channel assignment algorithm which is used to adjust the flow on the flow graph (routing changes) to ensure a feasible channel assignment. This flow adjustment also strives to keep the increase in interference for each channel to a minimum.

3. **Interference Free Link Scheduling:** This step obtains an interference free link schedule for the edge flows corresponding to the flow on the flow graph.

Each of these steps is described in the following subsections.

A Linear Programming-Based Routing Algorithm

A linear program LP (1) to find a flow that maximizes λ is given below:

$$\max \lambda \qquad (1)$$

Subject to

$$\lambda l(v) + \sum_{e=(u,v)\in E} \sum_{i=1}^{K} f(e(i))$$

$$= \sum_{e=(v,u)\in E} \sum_{i=1}^{K} f(e(i)), \forall v \in V - V_G \qquad (2)$$

$$f(e(i)) \le c(e), \quad \forall e \in E \qquad (3)$$

$$\sum_{1\le i \le K} \left(\sum_{e=(u,v)\in E} \frac{f(e(i))}{c(e)} + \sum_{e=(v,u)\in E} \frac{f(e(i))}{c(e)} \right)$$

$$\le I(v), v \in V \qquad (4)$$

$$\frac{f(e(i))}{c(e)} + \sum_{e'\in I(e)} \frac{f(e'(i))}{c(e')} \le c(q),$$

$$\forall e \in E, \ 1 \le i \le K. \qquad (5)$$

The first two constraints are *flow constraints*. The first one is the flow conservation constraint; the second one ensures no link capacity is violated. The third constraint is the *node radio constraints*. Recall that a IWMN node $v \in V$ has $I(v)$ radios and hence can be assigned at most $I(v)$ channels from $1 \le i \le K$. One way to model this constraint is to observe that due to interference constraints v can be involved in at most $I(v)$ simultaneous communications (with different one hop neighbors). In other words this constraint follows from $\sum_{1\le i \le K} \sum_{e=(u,v)\in E} X_{e,i,\tau} + \sum_{1\le i \le K} \sum_{e=(v,u)\in E} X_{e,i,\tau} \le I(v)$. The fourth constraint is the *link congestion constraints* which are discussed in detail in section "Link Flow Scheduling". Note that all the constraints listed above are necessary conditions for any feasible solution. However, these constraints are not necessarily sufficient. Hence if a solution is found that satisfies these constraints it may not be a feasible solution. The approach is to start with a "good" but not necessarily feasible solution that satisfies all of these constraints and use it to construct a feasible solution without impacting the quality of the solution.

A solution to this LP can be viewed as a flow on a *flow graph* $H = (V, E^H)$ where $E^H = \{e(i)|\forall e \in E, 1 \le i \le K\}$. Although the optimal solution to this LP yields the best possible λ (say λ^*) from a practical point of view more improvements may be possible:

- The flow may have directed cycles. This may be the case since the LP does not try to minimize the amount of interference directly. By removing the flow on the directed cycle (equal amount off each edge) flow conservation is maintained and in addition since there are fewer transmissions the amount of interference is reduced.

- The flow may be using a long path when shorter paths are available. Note that longer paths imply more link transmissions. In this case it is often the case that by moving the flow to shorter paths, system interference may be reduced.

The above arguments suggest that it would be practical to find among all solutions that attain the optimal λ value of λ^* the one for which the total value of the following quantity is minimized:

$$\sum_{1 \leq i \leq K} \sum_{e=(v,u) \in E} \frac{f(e(i))}{c(e)} .$$

The LP is then re-solved with this objective function and with λ fixed at λ^*.

Channel Assignment

The solution to the LP (1) is a set of flow values $f(e(i))$ for edge e and channel i that maximize the value λ. Let λ^* denote the optimal value of λ. The flow $f(e(i))$ implies a channel assignment where the two end nodes of edge e are both assigned channel i if and only if $f(e(i)) > 0$. Note that for the flow $f(e(i))$ the implied channel assignment may not be feasible (it may require more than $I(v)$ channels at node v). The channel assignment algorithm transforms the given flow to fix this infeasibility. Below only a sketch of the algorithm is given. More details can be found in [1].

First observe that in an idle scenario, where all nodes v have the same number of interfaces I (i.e., $I = I(v)$) and where the number of available channels K is also I, the channel assignment implied by the LP (1) is feasible. This is because even the trivial channel assignment where all nodes are assigned all the channels 1 to I is feasible. The main idea behind the algorithm is to first transform the LP (1) solution to a new flow in which every edge e has flow $f(e(i)) > 0$ only for the channels $1 \leq i \leq I$. The basic operation that the algorithm uses for this is to equally distribute, for every edge e, the flow $f(e(i))$, for $I < i \leq K$ to the edges $e(j)$, for $1 \leq i \leq I$. This ensures that all $f(e(i)) = 0$, for $I < i \leq K$ after the operation. This operation is called Phase I of the Algorithm. Note that the Phase I operation does not violate the flow conservation constraints or the node radio constraints (5) in the LP (1). It can be shown that in the resulting solution the flow $f(e(i))$ may exceed the capacity of edge e by at most a factor $\phi = K/I$. This is called the "inflation factor" of Phase I. Likewise in the new flow, the *link congestion constraints* (5) may also be violated for edge e and channel i by no more

than the inflation factor φ. In other words in the resulting flow

$$\frac{f(e(i))}{c(e)} + \sum_{e' \in I(e)} \frac{f(e'(i))}{c(e')} \leq \phi c(q) .$$

This implies that if the new flow is scaled by a fraction $1/\phi$ than it is feasible for the LP (1). Note that the implied channel assignment (assign channels 1 to I to every node) is also feasible. Thus, the above algorithm finds a feasible channel assignment with a λ value of at least λ^*/ϕ.

One shortcoming of the channel assignment algorithm (Phase I) described so far is that it only uses I of the K available channels. By using more channels the interference may be further reduced thus allowing for more flow to be pushed in the system. The channel assignment algorithm uses an additional heuristic for this improvement. This is called Phase II of the algorithm.

Now define an operation called "channel switch operation." Let A be a maximal connected component (the vertices in A are not connected to vertices outside A) in the graph formed by the edges e for a given channel i for which $f(e(i)) > 0$. The main observation to use is that for a given channel j, the operation of completely moving flow $f(e(i))$ to flow $f(e(j))$ for every edge e in A, does not impact the feasibility of the implied channel assignment. This is because there is no increase in the number of channels assigned per node after the flow transformation: the end nodes of edges e in A which were earlier assigned channel i are now assigned channel j instead. Thus, the transformation is equivalent to switching the channel assignment of nodes in A so that channel i is discarded and channel j is gained if not already assigned.

The Phase II heuristic attempts to re-transform the unscaled Phase I flows $f(e(i))$ so that there are multiple connected components in the graphs $G(e, i)$ formed by the edges e for each channel $1 \leq i \leq I$. This re-transformation is done so that the LP constraints are kept satisfied with an inflation factor of at most φ, as is the case for the unscaled flow after Phase I of the algorithm.

Next in Phase III of the algorithm the connected components within each graph $G(e, i)$ are grouped such that there are as close to K (but no more than) groups overall and such that the maximum interference within each group is minimized. Next the nodes within the lth group are assigned channel l, by using the channel switch operation to do the corresponding flow transformation. It can be shown that the channel assignment implied by the flow in Phase III is feasible. In addition the underlying flows $f(e(i))$ satisfy the LP (1) constraints with an inflation factor of at most $\phi = K/I$.

Next the algorithm scales the flow by the largest possible fraction (at least $1/\phi$) such that the resulting flow is a feasible solution to the LP (1) and also implies a feasible channel assignment solution to the channel assignment. Thus, the overall algorithm finds a feasible channel assignment (by not necessarily restricting to channels 1 to I only) with a λ value of at least λ^*/ϕ.

Link Flow Scheduling

The results in this section are obtained by extending those of [4] for the single channel case and for the Protocol Model of interference [2]. Recall that the time slotted schedule S is assumed to be periodic (with period T) where the indicator variable $X_{e,i,\tau}, e \in E, i \in F(e), \tau \geq 1$ is 1 if and only if link e is active in slot τ on channel i and i is a channel in common among the set of channels assigned to the end-nodes of edge e.

Directly applying the result (Claim 2) in [4] it follows that a necessary condition for interference free link scheduling is that for every $e \in E, i \in F(e), \tau \geq 1: X_{e,i,\tau} + \sum_{e' \in I(e)} X_{e',i,\tau} \leq c(q)$. Here $c(q)$ is a constant that only depends on the interference model. In the interference model this constant is a function of the fixed value q, the ratio of the interference range R_I to the transmission range R_T, and an intuition for its derivation for a particular value $q = 2$ is given below.

Lemma 1 $c(q) = 8$ for $q = 2$.

Proof Recall that an edge $e' \in I(e)$ if there exist two nodes $x, y \in V$ which are at most $2R_T$ apart

and such that edge e is incident on node x and edge e' is incident on node y. Let $e = (u, v)$. Note that u and v are at most R_T apart. Consider the region C formed by the union of two circles C_u and C_v of radius $2R_T$ each, centered at node u and node v, respectively. Then $e' = (u', v') \in I(e)$ if an only if at least one of the two nodes u', v' is in C; Denote such a node by $C(e')$. Given two edges $e_1, e_2 \in I(e)$ that do not interfere with each other it must be the case that the nodes $C(e_1)$ and $C(e_2)$ are at least $2R_T$ apart. Thus, an upper bound on how many edges in $I(e)$ do not pair-wise interfere with each other can be obtained by computing how may nodes can be put in C that are pair-wise at least $2R_T$ apart. It can be shown [1] that this number is at most 8. Thus, in schedule S in a given slot only one of the two possibilities exist: either edge e is scheduled or an "independent" set of edges in $I(e)$ of size at most 8 is scheduled implying the claimed bound.   □

A necessary condition: (*Link Congestion Constraint*) Recall that $\frac{1}{T} \sum_{1 \leq \tau \leq T} X_{e,i,\tau} = \frac{f(e(i))}{c(e)}$. Thus: Any valid "interference free" edge flows must satisfy for every link e and every channel i the Link Congestion Constraint:

$$\frac{f(e(i))}{c(e)} + \sum_{e' \in I(e)} \frac{f(e'(i))}{c(e')} \leq c(q). \quad (6)$$

A matching sufficient condition can also established [1].

A sufficient condition: (*Link Congestion Constraint*) If the edge flows satisfy for every link e and every channel i the following *Link Schedulability Constraint* than an interference free edge communication schedule can be found using an algorithm given in [1].

$$\frac{f(e(i))}{c(e)} + \sum_{e' \in I(e)} \frac{f(e'(i))}{c(e')} \leq 1. \quad (7)$$

The above implies that if a flow $f(e(i))$ satisfies the *Link Congestion Constraint* then by scaling the flow by a fraction $1/c(q)$ it can be scheduled free of interference.

Key Results

Theorem *The RCL algorithm is a $Kc(q)/I$ approximation algorithm for the Joint Routing and Channel Assignment with Interference Free Edge Scheduling problem.*

Proof Note that the flow $f(e(i))$ returned by the channel assignment algorithm in Sect. "Channel Assignment" satisfies the *Link Congestion Constraint*. Thus, from the result of Sect. "Link Flow Scheduling" it follows that by scaling the flow by an additional factor of $1/c(q)$ the flow can be realized by an interference free link schedule. This implies a feasible solution to the joint routing, channel assignment and scheduling problem with a λ value of at least $\lambda^*/\phi c(q)$. Thus, the RCL algorithm is a $\phi c(q) = Kc(q)/I$ approximation algorithm. □

Applications

Infrastructure mesh networks are increasingly been deployed for commercial use and law enforcement. These deployment settings place stringent requirements on the performance of the underlying IWMNs. Bandwidth guarantee is one of the most important requirements of applications in these settings. For these IWMNs, topology change is infrequent and the variability of aggregate traffic demand from each mesh router (client traffic aggregation point) is small. These characteristics admit periodic optimization of the network which may be done by a system management software based on traffic demand estimation. This work can be directly applied to IWMNs. It can also be used as a benchmark to compare against heuristic algorithms in multi-hop wireless networks.

Open Problems

For future work, it will be interesting to investigate the problem when routing solutions can be enforced by changing link weights of a distributed routing protocol such as OSPF. Also, can the worst case bounds of the algorithm be improved (e.g., a constant factor independent of K and I)?

Cross-References

▶ Graph Coloring
▶ Stochastic Scheduling

Recommended Reading

1. Alicherry M, Bhatia R, Li LE (2005) Joint channel assignment and routing for throughput optimization in multi-radio wireless mesh networks. In: Proceedings of the ACM MOBICOM, pp 58–72
2. Gupta P, Kumar PR (2000) The capacity of wireless networks. IEEE Trans Inf Theory IT-46(2):388–404
3. Jain K, Padhye J, Padmanabhan VN, Qiu L (2003) Impact of interference on multi-hop wireless network performance. In: Proceedings of the ACM MOBICOM, pp 66–80
4. Kumar VSA, Marathe MV, Parthasarathy S, Srinivasan A (2004) End-to-end packet-scheduling in wireless ad-hoc networks. In: Proceedings of the ACM-SIAM symposium on discrete algorithms, pp 1021–1030
5. Kumar VSA, Marathe MV, Parthasarathy S, Srinivasan A (2005) Algorithmic aspects of capacity in wireless networks. In: Proceedings of the ACM SIGMETRICS, pp 133–144
6. Kyasanur P, Vaidya N (2005) Capacity of multi-channel wireless networks: impact of number of channels and interfaces. In: Proceedings of the ACM MOBICOM, pp 43–57

Circuit Partitioning: A Network-Flow-Based Balanced Min-Cut Approach

Martin D.F. Wong[1] and Honghua Hannah Yang[2]
[1]Department of Electrical and Computer Engineering, University of Illinois at Urbana-Champaign, Urbana, IL, USA
[2]Strategic CAD Laboratories, Intel Corporation, Hillsboro, OR, USA

Keywords

Hypergraph partitioning; Netlist partitioning

Years and Authors of Summarized Original Work

1994; Yang, Wong

Problem Definition

Circuit partitioning is a fundamental problem in many areas of VLSI layout and design. *Min-cut balanced bipartition* is the problem of partitioning a circuit into two disjoint components with equal weights such that the number of nets connecting the two components is minimized. The min-cut balanced bipartition problem was shown to be NP-complete [5]. The problem has been solved by heuristic algorithms, e.g., Kernighan and Lin type (K&L) iterative improvement methods [4, 11], simulated annealing approaches [10], and analytical methods for the ratio-cut objective [2, 7, 13, 15]. Although it is a natural method for finding a min-cut, the network max-flow min-cut technique [6, 8] has been overlooked as a viable approach for circuit partitioning. In [16], a method was proposed for exactly modeling a circuit netlist (or, equivalently, a hypergraph) by a flow network, and an algorithm for balanced bipartition based on repeated applications of the max-flow min-cut technique

was proposed as well. Our algorithm has the same asymptotic time complexity as one max-flow computation.

A *circuit netlist* is defined as a digraph $N = (V, E)$, where V is a set of nodes representing logic gates and registers and E is a set of edges representing wires between gates and registers. Each node $v \in V$ has a weight $w(v) \in R^+$. The total weight of a subset $U \subseteq V$ is denoted by $w(U) = \Sigma_{v \in U} w(v)$. $W = w(V)$ denotes the total weight of the circuit. A *net* $n = (v; v_1, \ldots, v_l)$ is a set of outgoing edges from node v in N. Given two nodes s and t in N, an $s - t$ *cut* (or *cut* for short) $(X, \bar{X})$ of N is a bipartition of the nodes in V such that $s \in X$ and $t \in \bar{X}$. The *net-cut net* $(X, \bar{X})$ of the cut is the set of nets in N that are incident to nodes in both X and $\bar{X}$. A cut $(X, \bar{X})$ is a *min-net-cut* if $|net(X, \bar{X})|$ is minimum among all $s - t$ cuts of N. In Fig. 1, net $a = [r_1; g_1, g_2)$, net cuts $net(X, \bar{X}) = \{b, e\}$ and $net(Y, \bar{Y}) = \{c, a, b, e\}$, and $(X, \bar{X})$ is a min-net-cut.

Formally, given an aspect ratio r and a deviation factor ϵ, *min-cut r-balanced bipartition* is the problem of finding a bipartition $(X, \bar{X})$ of the netlist N such that (1) $(1 - \epsilon)rW \leq W(X) \leq (1 + \epsilon)rW$ and (2) the size of the cut $net(X, \bar{X})$ is minimum among all bipartitions satisfying (1). When $r = 1/2$, this becomes a min-cut balanced-bipartition problem.

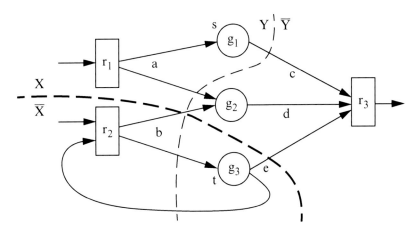

Circuit Partitioning: A Network-Flow-Based Balanced Min-Cut Approach, Fig. 1 A circuit netlist with two net-cuts

Circuit Partitioning: A Network-Flow-Based Balanced Min-Cut Approach, Fig. 2 Modeling a net in N in the flow network N'

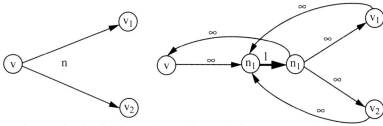

A net n in circuit N   The nodes and edges correspond to net n in N'

Circuit Partitioning: A Network-Flow-Based Balanced Min-Cut Approach, Fig. 3 The flow network for Fig. 1

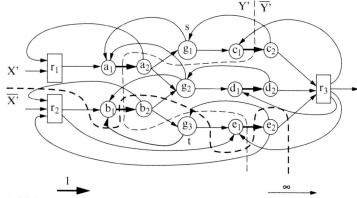

A bridging edge with unit capacity   An ordinary edge with infinite capacity

Key Results

Optimal-Network-Flow-Based Min-Net-Cut Bipartition

The problem of finding a min-net-cut in $N = (V, E)$ is reduced to the problem of finding a cut of minimum capacity. Then the latter problem is solved using the max-flow min-cut technique. A flow network $N' = (V', E')$ is constructed from $N = (V, E)$ as follows (see Figs. 2 and 3):

1. V' contains all nodes in V.
2. For each net $n = (v; v_1, \ldots, v_l)$ in N, add two nodes n_1 and n_2 in V' and a *bridging edge* $bridge(n) = (n_1, n_2)$ in E'.
3. For each node $u \in \{v, v_1, \ldots, v_l\}$ incident on net n, add two edges (u, n_1) and (n_2, u) in E'.
4. Let s be the source of N' and t the sink of N'.
5. Assign unit capacity to all bridging edges and infinite capacity to all other edges in E'.
6. For a node $v \in V'$ corresponding to a node in V, $w(v)$ is the weight of v in N. For a node $u \in V'$ split from a net, $w(u) = 0$.

Note that all nodes incident on net n are connected to n_1 and are connected from n_2 in N'. Hence, the flow network construction is symmetric with respect to all nodes incident on a net. This construction also works when the netlist is represented as a hypergraph.

It is clear that N' is a strongly connected digraph. This property is the key to reducing the bidirectional min-net-cut problem to a minimum-capacity cut problem that counts the capacity of the forward edges only.

Theorem 1 *N has a cut of net-cut size at most C if and only if N' has a cut of capacity at most C.*

Corollary 1 *Let $\left(X', \bar{X}'\right)$ be a cut of minimum capacity C in N'. Let $N_{cut} = \{n \mid bridge(n) \in \left(X', \bar{X}'\right)\}$. Then $N_{cut} = \left(X, \bar{X}\right)$ is a min-net-cut in N and $|N_{cut}| = C$.*

Corollary 2 *A min-net-cut in a circuit $N = (V, E)$ can be found in $O(|V||E|)$ time.*

Min-Cut Balanced-Bipartition Heuristic

First, a repeated max-flow min-cut heuristic algorithm, flow-balanced bipartition (FBB), is developed for finding an r-balanced bipartition that minimizes the number of crossing nets. Then, an efficient implementation of FBB is developed that has the same asymptotic time complexity as one max-flow computation. For ease of presentation, the FBB algorithm is described on the original circuit rather than the flow network constructed from the circuit. The heuristic algorithm is described in Fig. 4. Figure 5 shows an example.

Table 1 compares the best bipartition net-cut sizes of FBB with those produced by the analytical-method-based partitioners EIG1 [7] and PARABOLI (PB) [13]. The results produced by PARABOLI were the best previously known results reported on the benchmark circuits. The results for FBB were the best of ten runs. On average, FBB outperformed EIG1 and PARABOLI by 58.1 and 11.3 %, respectively. For circuit S38417, the suboptimal result from FBB can be improved by (1) running more times and (2) applying clustering techniques to the circuit based on connectivity before partitioning.

In the FBB algorithm, the node-collapsing method is chosen instead of a more gradual method (e.g., [9]) to ensure that the capacity of a cut always reflects the real net-cut size. To pick a node at steps 4.2 and 5.2, a threshold R is given for the number of nodes in the uncollapsed subcircuit. A node is randomly picked if the number of nodes is larger than R. Otherwise, all nodes adjacent to C are tried and the one whose collapse induces a min-net-cut with the smallest size is picked. A naive implementation of step 2 by computing the max-flow from the zero flow would incur a high time complexity. Instead, the flow value in the flow network is retained, and additional flow is explored to saturate the bridging edges of the min-net-cut from one iteration to the next. The procedure is shown in Fig. 6. Initially, the flow network retains the flow function computed in the previous iteration. Since the max-flow computation using the augmenting-path method is insensitive to the initial flow values in the flow network and the order in which the augmenting paths are found, the above procedure correctly finds a max-flow with the same flow value as a max-flow computed in the collapsed flow network from the zero flow.

Circuit Partitioning: A Network-Flow-Based Balanced Min-Cut Approach, Fig. 4 FBB algorithm

Algorithm: Flow-Balanced-Bipartition (FBB)
1. Pick a pair of nodes s and t in N;
2. Find a min-net-cut C in N;
   Let X be the subcircuit reachable from s through augmenting paths in the flow network, and $\overline{X}$ the rest;
3. **if** $(1-\epsilon)rW \le w(X) \le (1+\epsilon)rW$
      return C as the answer;
4. **if** $w(X) < (1-\epsilon)rW$
   4.1. Collapse all nodes in X to s;
   4.2. Pick a node $v \in \overline{X}$ adjacent to C and collapse it to s;
   4.3. Goto 1;
5. **if** $w(X) > (1+\epsilon)rW$
   5.1. Collapse all nodes in $\overline{X}$ to t;
   5.2. Pick a node $v \in X$ adjacent to C and collapse it to t;
   5.3. Goto 1;

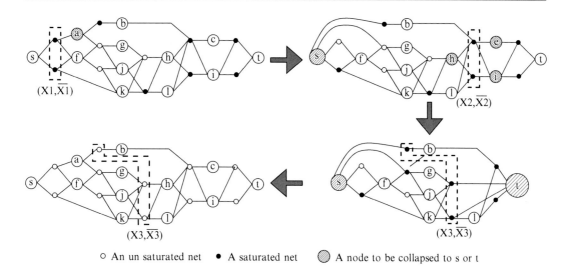

○ An un saturated net   ● A saturated net   ⬤ A node to be collapsed to s or t

Circuit Partitioning: A Network-Flow-Based Balanced Min-Cut Approach, Fig. 5 FBB on the example in Fig. 3 for $r = 1/2, \epsilon = 0.15$ and unit weight for each node. The algorithm terminates after finding cut $(X_2, \bar{X}_2)$. A small solid node indicates that the bridging edge corresponding to the net is saturated with flow

Circuit Partitioning: A Network-Flow-Based Balanced Min-Cut Approach, Table 1 Comparison of EIG1, PB, and FBB ($r = 1/2, \epsilon = 0.1$). All allow $\leq 10\%$ deviation

Circuit				Best net-cut size			Improve. % over		
Name	Gates and latches	Nets	Avg. deg	EIG1	PB	FBB	EIG1	PB	FBB elaps. sec.
S1423	731	743	2.7	23	16	13	43.5	18.8	1.7
S9234	5,808	5,805	2.4	227	74	70	69.2	5.4	55.7
S13207	8,696	8,606	2.4	241	91	74	69.3	18.9	100.0
S15850	10,310	10,310	2.4	215	91	67	68.8	26.4	96.5
S35932	18,081	17,796	2.7	105	62	49	53.3	21.0	2,808
S38584	20,859	20,593	2.7	76	55	47	38.2	14.5	1,130
S38417	24,033	23,955	2.4	121	49	58	52.1	−18.4	2,736
Average							58.5	11.3	

Circuit Partitioning: A Network-Flow-Based Balanced Min-Cut Approach, Fig. 6 Incremental max-flow computation

Procedure: Incremental Flow Computation

1. **while** $\exists$ an additional augmenting path from s to t
   increase flow value along the augmenting path;

2. Mark all nodes u s.t. $\exists$ an augmenting path from s to u;

3. Let C' be the set of bridging edges whose starting nodes are marked and ending nodes are not marked;

4. Return the nets corresponding to the bridging edges in C' as the min-net-cut C, and the marked nodes as X.

Theorem 2 *FBB has time complexity $O(|V||E|)$ for a connected circuit $N = (V, E)$.*

Theorem 3 *The number of iterations and the final net-cut size are nonincreasing functions of $\in$.*

In practice, FBB terminates much faster than this worst-case time complexity as shown in the section "Experimental Results." Theorem 3 allows us to improve the efficiency of FBB and the partition quality for a larger $\in$. This is not true for other partitioning approaches such as the K&L heuristics.

Applications

Circuit partitioning is a fundamental problem in many areas of VLSI layout and design automation. The FBB algorithm provides the first efficient predictable solution to the min-cut balanced-circuit-partitioning problem. It directly relates the efficiency and the quality of the solution produced by the algorithm to the deviation factor $\in$. The algorithm can be easily extended to handle nets with different weights by simply assigning the weight of a net to its bridging edge in the flow network. K-way min-cut partitioning for $K > 2$ can be accomplished by recursively applying FBB or by setting $r = 1/K$ and then using FBB to find one partition at a time. A flow-based method for directly solving the problem can be found in [12]. Prepartitioning circuit clustering according to the connectivity or the timing information of the circuit can be easily incorporated into FBB

by treating a cluster as a node. Heuristic solutions based on K&L heuristics or simulated annealing with low temperature can be used to further fine-tune the solution.

Experimental Results

The FBB algorithm was implemented in SIS/MISII [1] and tested on a set of large ISCAS and MCNC benchmark circuits on a SPARC 10 workstation with 36-MHz CPU and 32 MB memory.

Table 2 compares the average bipartition results of FBB with those reported by Dasdan and Aykanat in [3]. SN is based on the K&L heuristic algorithm in Sanchis [14]. PFM3 is based on the K&L heuristic with free moves as described in [3]. For each circuit, SN was run 20 times and PFM3 10 times from different randomly generated initial partitions. FBB was run 10 times from different randomly selected s and t. With only one exception, FBB outperformed both SN and PFM3 on the five circuits. On average, FBB found a bipartition with 24.5 and 19.0 % fewer crossing nets than SN and PFM3, respectively. The runtimes of SN, PFM3, and FBB were not compared since they were run on different workstations.

Cross-References

▶ Circuit Placement
▶ Circuit Retiming

Circuit Partitioning: A Network-Flow-Based Balanced Min-Cut Approach, Table 2 Comparison of SN. PFM3. and FBB ($r = 1/2, \epsilon = 0.1$)

Circuit				Avg. net-cut size			FBB bipart. ratio	Improve. %	
Name	Gates and latches	Nets	Avg. deg	SN	PFM3	FBB		Over SN	Over PFM3
C1355	514	523	3.0	38.9	29.1	26.0	1:1.08	33.2	10.7
C2670	1,161	1,254	2.6	51.9	46.0	37.1	1:1.15	28.5	19.3
C3540	1,667	1,695	2.7	90.3	71.0	79.8	1:1.11	11.6	–12.4
C7552	3,466	3,565	2.7	44.3	81.8	42.9	1:1.08	3.2	47.6
S838	478	511	2.6	27.1	21.0	14.7	1:1.04	45.8	30.0
Ave							1:1.10	24.5	19.0

Recommended Reading

1. Brayton RK, Rudell R, Sangiovanni-Vincentelli AL (1987) MIS: a multiple-level logic optimization. IEEE Trans CAD 6(6):1061–1081
2. Cong J, Hagen L, Kahng A (1992) Net partitions yield better module partitions. In: Proceedings of the 29th ACM/IEEE design automation conference, Anaheim, pp 47–52
3. Dasdan A, Aykanat C (1994) Improved multiple-way circuit partitioning algorithms. In: International ACM/SIGDA workshop on field programmable gate arrays, Berkeley
4. Fiduccia CM, Mattheyses RM (1982) A linear time heuristic for improving network partitions. In: Proceedings of the ACM/IEEE design automation conference, Las Vegas, pp 175–181
5. Garey M, Johnson DS (1979) Computers and intractability: a guide to the theory of NP-completeness. Freeman, Gordonsville
6. Goldberg AW, Tarjan RE (1988) A new approach to the maximum flow problem. J SIAM 35: 921–940
7. Hagen L, Kahng AB (1991) Fast spectral methods for ratio cut partitioning and clustering. In: Proceedings of the IEEE international conference on computer-aided design, Santa Clara, pp 10–13
8. Hu TC, Moerder K (1985) Multiterminal flows in a hypergraph. In: Hu TC, Kuh ES (eds) VLSI circuit layout: theory and design. IEEE, New York, pp 87–93
9. Iman S, Pedram M, Fabian C, Cong J (1993) Finding uni-directional cuts based on physical partitioning and logic restructuring. In: 4th ACM/SIGDA physical design workshop, Lake Arrowhead
10. Kirkpatrick S, Gelatt CD, Vecchi MP (1983) Optimization by simulated annealing. Science 4598:671–680
11. Kernighan B, Lin S (1970) An efficient heuristic procedure for partitioning of electrical circuits. Bell Syst Tech J 49:291–307
12. Liu H, Wong DF (1998) Network-flow-based multiway partitioning with area and pin constraints. IEEE Trans CAD Integr Circuits Syst 17(1): 50–59
13. Riess BM, Doll K, Frank MJ (1994) Partitioning very large circuits using analytical placement techniques. In: Proceedings of the 31th ACM/IEEE design automation conference, San Diego, pp 646–651
14. Sanchis LA (1989) Multiway network partitioning. IEEE Trans Comput 38(1):62–81
15. Wei YC, Cheng CK (1989) Towards efficient hierarchical designs by ratio cut partitioning. In: Proceedings of the IEEE international conference on computer-aided design, Santa Clara, pp 298–301
16. Yang H, Wong DF (1994) Efficient network flow based min-cut balanced partitioning. In: Proceedings of the IEEE international conference on computer-aided design, San Jose, pp 50–55

Circuit Placement

Andrew A. Kennings[1] and Igor L. Markov[2]
[1] Department of Electrical and Computer Engineering, University of Waterloo, Waterloo, ON, Canada
[2] Department of Electrical Engineering and Computer Science, University of Michigan, Ann Arbor, MI, USA

Keywords

Algorithm; Circuit; Combinatorial optimization; Hypergraph; Large-scale optimization; Linear programming; Network flow; Nonlinear optimization; Partitioning; Physical design; Placement; VLSI CAD

Synonyms

Analytical placement; EDA; Layout; Mathematical programming; Min-cost max-flow; Min-cut placement; Netlist

Years and Authors of Summarized Original Work

2000; Caldwell, Kahng, Markov
2006; Kennings, Vorwerk
2012; Kim, Lee, Markov

Problem Definition

This problem is concerned with determining constrained positions of objects while minimizing a measure of interconnect between the objects, as in physical layout of integrated circuits, commonly done in 2 dimensions. While most formulations are NP-hard, modern circuits are so large that practical placement algorithms must have near-linear run time and memory requirements, but not necessarily produce optimal solutions. Research in placement algorithms has identified scalable techniques which are now being adopted in the electronic design automation industry.

One models a circuit by a hypergraph $G_h(V_h, E_h)$ with (i) vertices $V_h = \{v_1, \ldots, v_n\}$ representing logic gates, standard cells, larger modules, or fixed I/O pads and (ii) hyperedges $E_h = \{e_1, \ldots, e_m\}$ representing connections between modules. Vertices and hyperedges connect through *pins* for a total of P pins in the hypergraph. Each vertex $v_i \in V_h$ has width w_i, height h_i, and area A_i. Hyperedges may also be weighted. Circuit placement seeks center positions (x_i, y_i) for vertices that optimize a hypergraph-based objective subject to constraints (see below). A placement is captured by $\mathbf{x} = (x_1, \cdots, x_n)$ and $\mathbf{y} = (y_1, \cdots, y_n)$.

Objective: Let C_k be the index set of the hypergraph vertices incident to hyperedge e_k. The total half-perimeter wire length (HPWL) of the circuit hypergraph is given by $\text{HPWL}(G_h) = \sum_{e_k \in E_h} \text{HPWL}(e_k) = \sum_{e_k \in E_h} \left[\max_{i,j \in C_k} |x_i - x_j| + \max_{i,j \in C_k} |y_i - y_j| \right]$. HPWL is piecewise linear, separable in the x and y directions, convex, but not strictly convex. Among many objectives for circuit placement, it is the simplest and most common.

Constraints:

1. **No overlap.** The area occupied by any two vertices cannot overlap; i.e., either $|x_i - x_j| \geq \frac{1}{2}(w_i + w_j)$ or $|y_i - y_j| \geq \frac{1}{2}(h_i + h_j)$, $\forall v_i, v_j \in V_h$.
2. **Fixed outline.** Each vertex $v_i \in V_h$ must be placed entirely within a specified rectangular region bounded by $x_{\min}$ ($y_{\min}$) and $x_{\max}$ ($y_{\max}$) which denote the left (bottom) and right (top) boundaries of the specified region.
3. **Discrete slots.** There is only a finite number of discrete positions, typically on a grid. However, in large-scale circuit layout, slot constraints are often ignored during *global placement* and enforced only during *legalization* and *detail placement*.

Other constraints may include alignment, minimum and maximum spacing, etc. Many placement techniques temporarily relax overlap constraints into density constraints to avoid vertices clustered in small regions. A $m \times n$ regular bin structure B is superimposed over the fixed outline and vertex area is assigned to bins based on the positions of vertices. Let D_{ij} denote the density of bin $B_{ij} \in B$, defined as the total cell area assigned to bin B_{ij} divided by its capacity. Vertex overlap is limited implicitly by satisfying $D_{ij} \leq K$, $\forall B_{ij} \in B$, for some $K \leq 1$ (density target).

Problem 1 (Circuit Placement) INPUT: Circuit hypergraph $G_h(V_h, E_h)$ and a fixed outline for the placement area.
OUTPUT: Positions for each vertex $v_i \in V_h$ such that (1) wire length is minimized and (2) the area-density constraints $D_{ij} \leq K$ are satisfied for all $B_{ij} \in B$.

Key Results

An unconstrained optimal position of a single placeable vertex connected to fixed vertices can be found in linear time as the median of adjacent positions [7]. Unconstrained HPWL minimization for multiple placeable vertices can be formulated as a linear program [6, 11]. For each $e_k \in E_h$, upper and lower bound variables U_k and L_k are added. The cost of e_k (x-direction only) is the difference between U_k and L_k. Each $U_k (L_k)$ comes with p_k inequality constraints that restrict its value to be larger (smaller) than the position of every vertex $i \in C_k$.

Linear programming has poor scalability and integrating constraint-tracking into optimization is difficult. Other approaches include nonlinear optimization and partitioning-based methods.

Combinatorial Techniques for Wire Length Minimization
The no-overlap constraints are not convex and cannot be directly added to the linear program for HPWL minimization. Vertices often cluster in small regions of high density. One can lower bound the distance between closely placed vertices with a single linear constraint that depends on the relative placement of these vertices [11]. The resulting optimization problem is incrementally resolved, and the process repeats until the desired density is achieved.

The *min-cut placement* technique is based on balanced min-cut partitioning of hypergraphs and is more focused on density constraints [12]. Vertices of the initial hypergraph are first partitioned in two similar-sized groups. One of them is assigned to the left half of the placement region, and the other one to the right half. Partitioning is performed by the Multilevel Fiduccia-Mattheyses (MLFM) heuristic [10] to minimize connections between the two groups of vertices (the net-cut objective). Each half is partitioned again but takes into account the connections to the other half [12]. At the large scale, ensuring the similar sizes of bipartitions corresponds to density constraints, and cut minimization corresponds to HPWL minimization. When regions become small and contain <10 vertices, optimal positions can be found with respect to discrete slot constraints by branch-and-bound [2]. Balanced hypergraph partitioning is NP-hard [4], but the MLFM heuristic takes $O((V + E) \log V)$ time. The entire min-cut placement procedure takes $O((V + E)(\log V)^2)$ time and can process hypergraphs with millions of vertices in several hours.

A special case of interest is that of one-dimensional placement. When all vertices have identical width and none of them are fixed, one obtains the NP-hard MINIMUM LINEAR ARRANGEMENT problem [4] which can be approximated in polynomial time within $O(\log V)$ and solved exactly for trees in $O(V^3)$ time as shown by Yannakakis. The min-cut technique described above also works well for the related NP-hard MINIMUM-CUT LINEAR ARRANGEMENT problem [4].

Quadratic and Nonlinear Wire Length Approximations

Quadratic and generic nonlinear optimization may be faster than linear programming while reasonably approximating the original formulation.

Quadratic, Linearized Quadratic, and Bound-to-Bound Placement

The hypergraph is represented by a weighted graph where w_{ij} represents the weight on the 2-pin edge connecting vertices v_i and v_j in the weighted graph. When an edge is absent, $w_{ij} = 0$, and in general $w_{ii} = -\Sigma_{i \neq j} w_{ij}$. A quadratic placement (x-direction only) is given by

$$\Phi(x) = \sum_{i,j} w_{ij} \left[(x_i - x_j)^2 \right]$$

$$= \frac{1}{2} \mathbf{x}^T \mathbf{Q} \mathbf{x} + \mathbf{c}^T \mathbf{x} + \text{const.} \quad (1)$$

The global minimum of $\Phi(x)$ is found by solving $\mathbf{Q}\mathbf{x} + \mathbf{c} = \mathbf{0}$ which is a sparse, symmetric positive definite system of linear equations (assuming ≥ 1 fixed vertex), efficiently solved using any number of iterative solvers. Quadratic placement may have different optima depending on the model (clique or star) used to represent hyperedges. However, for a k-pin hyperedge, if $w_{ij} = W_c$ in a clique model and $w_{ij} = k W_c$ is a star model, then the models are equivalent in quadratic placement [6].

Quadratic placement can produce lower quality placements. To approximate a linear objective, one can iteratively solve Eq. 1 with $w_{ij} = 1/|x_i - x_j|$ computed at every iteration. Alternatively, one can solve a single β-regularized optimization problem given by $\Phi^\beta(\mathbf{x}) = \min_x \sum_{i,j} w_{ij} \sqrt{(x_i - x_j)^2 + \beta}$, $\beta > 0$, e.g., using a Primal-Dual Newton method [1].

In bound-to-bound placement, instead of a clique or star model, hyperedges are decomposed based on the relative placement of vertices. For a k-pin hyperedge, the extreme vertices (min and max) are connected to each other and to each internal vertex with weights $w_{ij} = 1/(k - 1)|x_i - x_j|$. With these weights, the quadratic objective captures HPWL exactly, but only for the given placement. As placement changes, updates to the quadratic placement objective are required to reduce discrepancies [8].

Half-perimeter Wire Length Placement:
HPWL can be provably approximated by strictly convex and differentiable functions. For 2-pin hyperedges, β-regularization can be used [1]. For an k-pin hyperedge ($k \geq 3$), one can rewrite HPWL as the maximum (l_∞-norm) of

all $k(k-1)/2$ pairwise distances $|x_i - x_j|$ and approximate the l_∞-norm by the l_p-norm. This removes all non-differentiabilities except at $\mathbf{0}$ which is then removed with β-regularization. The resulting HPWL approximation is given by

$$\text{HPWL}_{\text{REG}}(G_h) = \sum_{e_k \in E_h} \left(\sum_{i,j \in C_k} |x_i - x_j|^p + \beta \right)^{1/p} \tag{2}$$

which overestimates HPWL with arbitrarily small relative error as $p \to \infty$ and $\beta \to 0$ [6]. Alternatively, HPWL can be approximated via the log-sum-exp (LSE) formula given by

$$\text{HPWL}_{\text{LSE}}(G_h) = \alpha \sum_{e_k \in E_h} \left[\ln \left(\sum_{i \in C_k} \exp \left(\frac{x_i}{\alpha} \right) \right) \right. $$
$$\left. + \ln \left(\sum_{v_i \in C_k} \exp \left(\frac{-x_i}{\alpha} \right) \right) \right] \tag{3}$$

where $\alpha > 0$ is a smoothing parameter [5]. Both approximations can be optimized using conjugate gradient methods. Other convex and differentiable HPWL approximations exist.

Analytic Techniques for Target Density Constraints

The target density constraints are non-differentiable and are typically handled by approximation.

Force-Based Spreading

The key idea is to add constant forces $\mathbf{f}$ that pull vertices always from overlaps, and recompute the forces over multiple iterations to reflect changes in vertex distribution. For quadratic placement, the new optimality conditions are $\mathbf{Qx} + \mathbf{c} + \mathbf{f} = \mathbf{0}$ [7]. The constant force can perturb a placement in any number of ways to satisfy the target density

constraints. The force $\mathbf{f}$ is computed using a discrete version of Poisson's equation.

Fixed-Point Spreading

A fixed point f is a pseudo-vertex with zero area, fixed at (x_f, y_f), and connected to one vertex $H(f)$ in the hypergraph through the use of a pseudo-edge with weight $w_{f,H(f)}$. Each fixed point introduces a single quadratic term into the objective function; quadratic placement with fixed points is given by $\Phi(x) = \sum_{i,j} w_{i,j}(x_i - x_j)^2 + \sum_f w_{f,H(f)}(x_{H(f)} - x_f)^2$. By manipulating the positions of fixed points, one can perturb a placement to satisfy the target density constraints. Fixed points improve the controllability and stability of placement iterations, in particular by improving the conditioning number of resulting numerical problem instances. A particularly effective approach to find fixed points is through the use of fast *LookAhead Legalization* (LAL) [8, 9]. Given locations found by quadratic placement, LAL gradually modifies them into a relatively overlap-free placement that satisfies density constraints and seeks to preserve the ordering of x and y positions, while avoiding unnecessary movement. The resulting locations are used as fixed target points. LAL can be performed by top-down geometric partitioning with nonlinear scaling between partitions. As described in [8, 9], this approach is particularly effective at handling rectilinear obstacles. Subsequent work developed extensions to account for routing congestion and other considerations arising in global placement. At the most recent (ISPD 2014) placement contest, the contestants ranked in top three used the framework outlined in [8].

Generalized Force-Directed Spreading

The Helmholtz equation models a diffusion process and makes it ideal for spreading vertices [3]. The Helmholtz equation is given by

$$\frac{\partial^2 \phi(x,y)}{\partial x^2} + \frac{\partial^2 \phi(x,y)}{\partial y^2} - \epsilon \phi(x,y) = D(x,y), \quad (x,y) \in \text{R}$$
$$\frac{\partial \phi}{\partial v} = 0, \quad\quad\quad (x,y) \text{ on the boundary of } R \tag{4}$$

where $\epsilon > 0$, v is an outer unit normal, R represents the fixed outline, and $D(x, y)$ represents the continuous density function. The boundary conditions, $\frac{\partial \phi}{\partial v} = 0$, specify that forces pointing outside of the fixed outline be set to zero – this is a key difference with the Poisson method which assumes that forces become zero at infinity. The value ϕ_{ij} at the center of each bin B_{ij} is found by discretization of Eq. 4 using finite differences. The density constraints are replaced by $\phi_{ij} = \hat{K}, \forall B_{ij} \in B$ where $\hat{K}$ is a scaled representative of the density target K. Wire length minimization subject to the smoothed density constraints can be solved via Uzawa's algorithm. For quadratic wire length, this algorithm is a generalization of force-based spreading.

Potential Function Spreading

Target density constraints can also be satisfied via a penalty function. The area assigned to bin B_{ij} by vertex v_i is represented by Potential(v_i, B_{ij}) which is a bell-shaped function. The use of piecewise quadratic functions makes the potential function non-convex but smooth and differentiable [5]. The wire length approximation can be combined together with a penalty term given by Penalty $= \sum_{B_{ij} \in B} \left(\sum_{v_i \in V_h} \text{Potential}(v_i, B_{ij}) - K \right)^2$ to arrive at an unconstrained optimization problem which is solved using a conjugate gradient method [5].

Applications

Practical applications involve more sophisticated interconnect objectives, such as circuit delay, routing congestion, power dissipation, power density, and maximum thermal gradient. The above techniques are adapted to handle multiobjective optimization. Many such extensions are based on heuristic assignment of net weights that encourage the shortening of some (e.g., timing critical and frequently switching) connections at the expense of other connections. To moderate routing congestion, predictive congestion maps are used to decrease the maximal density constraint for placement in congested regions.

Another application is in physical synthesis, where incremental placement is used to evaluate changes in circuit topology.

Experimental Results and Data Sets

Circuit placement has been actively studied for the past 30 years, and a wealth of experimental results have been reported. A 2003 result showed that placement tools could produce results as much as 1.41× to 2.09× known optimal wire lengths on average. In a 2006 placement contest, academic software for placement produced results that differed by as much as 1.39× on average when the objective was the simultaneous minimization of wire length, routability, and run time. Placement run times for instances with 2M movable objects ranged into hours. More recently, the gap in wire length between different tools has decreased, and run times have improved, in part due to the use of multicore CPUs and vectorized arithmetics. Over the last 10 years, wire length has improved by 20–25 % and run time by 15–20 times [8, 9]. More recent work in circuit placement has focused on other objectives such as routability in addition to wire length minimization.

Modern benchmark suites include the ISPD05, ISPD06, ISPD11, and ISPD14 suites (http://www.ispd.cc). Additional benchmark suites include the ICCAD12 (http://cad_contest.cs.nctu.edu.tw/CAD-contest-at-ICCAD2012), ICCAD13 (http://cad_contest.cs.nctu.edu.tw/CAD-contest-at-ICCAD2013), and ICCAD14 (http://cad_contest.ee.ncu.edu.tw/CAD-contest-at-ICCAD2014) suites. Instances in these benchmark suites contain between several hundred thousand to several million placeable objects. Additional benchmark suites also exist.

Cross-References

▶ Circuit Partitioning: A Network-Flow-Based Balanced Min-Cut Approach
▶ Floorplan and Placement
▶ Performance-Driven Clustering

Recommended Reading

1. Alpert CJ, Chan T, Kahng AB, Markov IL, Mulet P (1998) Faster minimization of linear wirelength for global placement. IEEE Trans CAD 17(1):3–13
2. Caldwell AE, Kahng AB, Markov IL (2000) Optimal partitioners and end-case placers for standard-cell layout IEEE Trans CAD 19(11):1304–1314
3. Chan T, Cong J, Sze K (2005) Multilevel generalized force-directed method for circuit placement. In: Proceedings of international symposium on physical design, San Francisco, pp 185–192
4. Crescenzi P, Kann V (1998) A compendium of NP optimization problems. Springer, Berlin/ Heidelberg
5. Kahng AB, Wang Q (2005) Implementation and extensibility of an analytic placer. IEEE Trans CAD 24(5):734–747
6. Kennings A, Markov IL (2002) Smoothing max-terms and analytical minimization of half-perimeter wirelength. VLSI Design 14(3):229–237
7. Kennings A, Vorwerk K (2006) Force-directed methods for generic placement. IEEE Trans CAD 25(10):2076–2087
8. Kim M-C, Lee D, Markov IL (2012) SimPL: an effective placement algorithm. IEEE Trans CAD 31(1):50–60
9. Lin T, Chu C, Shinnerl JR, Bustany I, Nedelchev I (2013) POLAR: placement based on novel rough legalization and refinement. In: International conference on computer-aided design, San Jose, pp 357–362
10. Papa DA, Markov IL (2007) Hypergraph partitioning and clustering. In: Gonzalez T (ed) Approximation algorithms and metaheuristics. Chapman & Hall/CRC computer and information science series. Chapman & Hall/CRC, Florida
11. Reda S, Chowdhary A (2006) Effective linear programming based placement methods. In: International symposium on physical design, San Jose, pp 186–191
12. Roy JA, Adya SN, Papa DA, Markov IL (2006) Min-cut floorplacement. IEEE Trans CAD 25(7):1313–1326

Circuit Retiming

Hai Zhou
Electrical Engineering and Computer Science (EECS) Department, Northwestern University, Evanston, IL, USA

Keywords

Min-area retiming; Min-period retiming

Years and Authors of Summarized Original Work

1991; Leiserson, Saxe

Problem Definition

Circuit retiming is one of the most effective structural optimization techniques for sequential circuits. It moves the registers within a circuit without changing its function. Besides clock period, retiming can be used to minimize the number of registers in the circuit. It is also called minimum area retiming problem Leiserson and Saxe [3] started the research on retiming and proposed algorithms for both minimum period and minimum area retiming. Both their algorithms for minimum area and minimum period will be presented here.

The problems can be formally described as follows. Given a directed graph $G = (V, E)$ representing a circuit – each node $v \in V$ represents a gate and each edge $e \in E$ represents a signal passing from one gate to another – with gate delays $d : V \to \mathbb{R}^+$ and register numbers $w : E \to \mathbb{N}$, the minimum area problem asks for a relocation of registers $w' : E \to \mathbb{N}$ such that the number of registers in the circuit is minimum under a given clock period φ. The minimum period problem asks for a solution with the minimum clock period.

Notations

To guarantee that the new registers are actually a relocation of the old ones, a label $r : V \to \mathbb{Z}$ is used to represent how many registers are moved from the outgoing edges to the incoming edges of each node. Using this notation, the new number of registers on an edge (u, v) can be computed as

$$w'[u, v] = w[u, v] + r[v] - r[u].$$

The same notation can be extended from edges to paths. However, between any two nodes u and v, there may be more than one path. Among these paths, the ones with the minimum number of registers will decide how many registers can be moved outside of u and v. The number is denoted by $W[u, v]$ for any $u, v \in V$, that is,

$$W[u, v] \triangleq \min_{p:u \rightsquigarrow v} \sum_{(x,y) \in p} w[x, y]$$

The maximal delay among all the paths from u to v with the minimum number of registers is also denoted by $D[u, v]$, that is,

$$D[u, v] \triangleq \max_{w[p:u \rightsquigarrow v]=W[u,v]} \sum_{x \in p} d[x]$$

Constraints

Based on the notations, a valid retiming r should not have any negative number of registers on any edge. Such a validity condition is given as

$$P0(r) \triangleq \forall (u, v) \in E : w[u, v] + r[v] - r[u] \geq 0$$

On the other hand, given a retiming r, the minimum number of registers between any two nodes u and v is $W[u, v] - r[u] + r[v]$. This number will not be negative because of the previous constraint. However, when it is zero, there will be a path of delay $D[u, v]$ without any register on it. Therefore, to have a retimed circuit working for clock period φ, the following constraint must be satisfied.

$$P1(r) \triangleq \forall u, v \in V : D[u, v] > \phi$$

$$\Rightarrow W[u, v] + r[v] - r[u] \geq 1$$

Key Results

The object of the minimum area retiming is to minimize the total number of registers in the circuit, which is given by $\sum_{(u,v) \in E} w'[u, v]$. Expressing $w'[u, v]$ in terms of r, the objective becomes

$$\sum_{v \in V} (\text{in}[v] - \text{out}[v]) * r[v] + \sum_{(u,v) \in E} w[u, v]$$

where in$[v]$ is the in-degree and out$[v]$ is the out-degree of node v. Since the second term is a constant, the problem can be formulated as the following integer linear program.

$$\text{Minimize} \sum_{v \in V} (\text{in}[v] - \text{out}[v]) * r[v]$$

s.t. $w[u, v] + r[v] - r[u] \geq 0 \quad \forall (u, v) \in E$

$W[u, v] + r[v] - r[u] \geq 1 \quad \forall u, v \in V : D[u, v] > \phi$

$r[v] \in \mathbb{Z} \quad \forall v \in V$

Since the constraints have only difference inequalities with integer-constant terms, solving the relaxed linear program (without the integer constraint) will only give integer solutions. Even better, it can be shown that the problem is the dual of a minimum cost network flow problem and, thus, can be solved efficiently.

Theorem 1 *The integer linear program for the minimum area retiming problem is the dual of the following minimum cost network flow problem.*

$$\text{Minimize} \sum_{(u,v) \in E} w[u, v] * f[u, v]$$

$$+ \sum_{D[u,v]>\phi} (W[u, v] - 1) * f[u, v]$$

s.t. in$[v] + \sum_{(v,w) \in E \vee D[v,w]>\phi} f[v, w] = \text{out}[v]$

$$+ \sum_{(u,v) \in ED[u,v]>\phi} f[u, v] \quad \forall v \in V$$

$f[u, v] \geq 0 \quad \forall (u, v) \in ED[u, v] > \phi$

From the theorem, it can be seen that the network graph is a dense graph where a new edge (u, v) needs to be introduced for any node pair u, v such that $D[u, v] > \phi$. There may be redundant constraints in the system.

For example, if $W[u, w] = W[u, v] + w[v, w]$ and $D[u, v] > \phi$ then the constraint $W[u, w] + r[w] - r[u] \geq 1$ is redundant, since there are already $W[u, v] + r[v] - r[u] \geq 1$ and $w[v, w] + r[w] - r[v] \geq 0$. However, it may not be easy to check and remove all redundancy in the constraints.

In order to build the minimum cost flow network, it is needed to first compute both matrices W and D. Since $W[u, v]$ is the shortest path from u to v in terms of w, the computation of W can be done by an all-pair shortest paths algorithm such as Floyd-Warshall's algorithm [1]. Furthermore, if the ordered pair $(w[x, y], -d[x])$ is used as the edge weight for each $(x, y) \in E$, an all-pair shortest paths algorithm can also be used to compute both W and D. The algorithm will add weights by component-wise addition and will compare weights by lexicographic ordering.

Leiserson and Saxe's [3] first algorithm for the minimum period retiming was also based on the matrices W and D. The idea was that the constraints in the integer linear program for the minimum area retiming can be checked efficiently by Bellman-Ford's shortest paths algorithm [1], since they are just difference inequalities. This gives a feasibility checking for any given clock period φ. Then the optimal clock period can be found by a binary search on a range of possible periods. The feasibility checking can be done in $O(|V|^3)$ time, thus the runtime of such an algorithm is $O(|V|^3 \log |V|)$.

Their second algorithm got rid of the construction of the matrices W and D. It still used a clock period feasibility checking within a binary search. However, the feasibility checking was done by incremental retiming. It works as follows: Starting with $r = 0$, the algorithm computes the arrival time of each node by the longest paths computation on a DAG (Directed Acyclic Graph). For each node v with an arrival time larger than the given period φ, the $r[v]$ will be increased by one. The process of the arrival time computation and r increasing will be repeated $|V| - 1$ times. After that, if there is still arrival time that is larger than φ, then the period is infeasible. Since the feasibility checking is done in $O(|V||E|)$ time, the runtime for the minimum period retiming is $O(|V||E| \log |V|)$.

retiming algorithms with some efficiency improvements. For minimum period retiming, they implemented the second algorithm and, in order to find out infeasibility earlier, they introduced a pointer from one node to another where at least one register is required between them. A cycle formed by the pointers indicates the feasibility of the given period. For minimum area retiming, they removed some of the redundancy in the constraints and used the cost-scaling algorithm of Goldberg and Tarjan [2] for the minimum cost flow computation.

As can be seen from the second minimum period retiming algorithm here and Zhou's algorithm [9] in another entry (▶ Circuit Retiming: An Incremental Approach), incremental computation of the longest combinational paths (i.e., those without register on them) is more efficient than constructing the dense graph (via matrices W and D). However, the minimum area retiming algorithm is still based on a minimum cost network flow on the dense graph. A more efficient algorithm based on incremental retiming has recently been designed for the minimum area problem by Wang and Zhou [8].

Experimental Results

Sapatnekar and Deokar [6] and Pan [5] proposed continuous retiming as an efficient approximation for minimum period retiming and reported the experimental results. Maheshwari and Sapatnekar [4] also proposed some efficiency improvements to the minimum area retiming algorithm and reported their experimental results.

Cross-References

▶ Circuit Retiming: An Incremental Approach

Applications

Shenoy and Rudell [7] implemented Leiserson and Saxe's minimum period and minimum area

Recommended Reading

1. Cormen TH, Leiserson CE, Rivest RL, Stein C (2001) Introduction to algorithms, 2nd edn. MIT, Cambridge

2. Goldberg AV, Taijan RE (1987) Solving minimum cost flow problem by successive approximation. In: Proceedings of ACM symposium on the theory of computing, New York, pp 7–18. Full paper in: Math Oper Res 15:430–466 (1990)
3. Leiserson CE, Saxe JB (1991) Retiming synchronous circuitry. Algorithmica 6:5–35
4. Maheshwari N, Sapatnekar SS (1998) Efficient retiming of large circuits. IEEE Trans Very Large-Scale Integr Syst 6:74–83
5. Pan P (1997) Continuous retiming: algorithms and applications. In: Proceedings of international conference on computer design, Austin. IEEE, Los Alamitos pp 116–121
6. Sapatnekar SS, Deokar RB (1996) Utilizing the retiming- skew equivalence in a practical algorithm for retiming large circuits. IEEE Trans Comput Aided Des 15:1237–1248
7. Shenoy N, Rudell R (1994) Efficient implementation of retiming. In: Proceedings of international conference on computer-aided design, San Jose. IEEE, Los Alamitos, pp 226–233
8. Wang J, Zhou H (2008) An efficient incremental algorithm for min-area retiming. In: Proceedings of design automation conference, Anaheim, CA, pp 528–533
9. Zhou H (2005) Deriving a new efficient algorithm for min-period retiming. In: Asia and South Pacific design automation conference, Shanghai, Jan 2005. ACM, New York

Circuit Retiming: An Incremental Approach

Hai Zhou
Electrical Engineering and Computer Science (EECS) Department, Northwestern University, Evanston, IL, USA

Keywords

Minimum period retiming; Min-period retiming

Years and Authors of Summarized Original Work

2005; Zhou

Problem Definition

Circuit retiming is one of the most effective structural optimization techniques for sequential circuits. It moves the registers within a circuit without changing its function. The minimal period retiming problem needs to minimize the longest delay between any two consecutive registers, which decides the clock period.

The problem can be formally described as follows. Given a directed graph $G = (V, E)$ representing a circuit – each node $v \in V$ represents a gate and each edge $e \in E$ represents a signal passing from one gate to another – with gate delays $d : V \rightarrow \mathbb{R}^+$ and register numbers $w : E \rightarrow \mathbb{N}$, it asks for a relocation of registers $w' : E \rightarrow \mathbb{N}$ such that the maximal delay between two consecutive registers is minimized.

Notations To guarantee that the new registers are actually a relocation of the old ones, a label $r : V \rightarrow \mathbb{Z}$ is used to represent how many registers are moved from the outgoing edges to the incoming edges of each node. Using this notation, the new number of registers on an edge (u, v) can be computed as

$$w'[u, v] = w[u, v] + r[v] - r[u].$$

Furthermore, to avoid explicitly enumerating the paths in finding the longest path, another label $t : V \rightarrow \mathbb{R}^+$ is introduced to represent the output arrival time of each gate, that is, the maximal delay of a gate from any preceding register. The condition for t to be at least the combinational delays is

$$\forall [u, v] \in E : w'[u, v] = 0 \Rightarrow t[v] \geq t[u] + d[u].$$

Constraints and Objective Based on the notations, a valid retiming r should not have any negative number of registers on any edge. Such a validity condition is given as

$$P0(r) \stackrel{\Delta}{=} \forall (u, v) \in E : w[u, v] + r[v] - r[u] \geq 0.$$

As already stated, the conditions for t to be valid arrival time is given by the following two predicates:

$$P1(t) \overset{\Delta}{=} \forall v \in V : t[v] \geq d[v]$$
$$P2(r,t) \overset{\Delta}{=} \forall (u,v) \in E : r[u] - r[v] = w[u,v]$$
$$\Rightarrow t[v] - t[u] \geq d[v].$$

A predicate P is used to denote the conjunction of the above conditions:

$$P(r,t) \overset{\Delta}{=} P0(r) \wedge P1(t) \wedge P2(r,t).$$

A minimal period retiming is a solution r,t satisfying the following optimality condition:

$$P3 \overset{\Delta}{=} \forall r',t' : P(r',t') \Rightarrow \max(t) \leq \max(t')$$

where
$$\max(t) \overset{\Delta}{=} \max_{v \in V} [v].$$

Since only a valid retiming (r',t') will be discussed in the sequel, to simplify the presentation, the range condition $P(r',t')$ will often be omitted; the meaning shall be clear from the context.

Key Results

This section will show how an efficient algorithm is designed for the minimal period retiming problem. Contrary to the usual way of only presenting the final product, i.e., the algorithm, but not the ideas on its design, a step-by-step design process will be shown to finally arrive at the algorithm.

To design an algorithm is to construct a procedure such that it will terminate in finite steps and will satisfy a given predicate when it terminates. In the minimal period retiming problem, the predicate to be satisfied is $P0 \wedge P1 \wedge P2 \wedge P3$. The predicate is also called the *post-condition*. It can be argued that any nontrivial algorithm will have at least one loop; otherwise, the processing length is only proportional to the text length. Therefore, some part of the post-condition will be iteratively satisfied by the loop, while the remaining part will be initially satisfied by an initialization and made invariant during the loop.

The first decision needed to make is to partition the post-condition into possible invariant and loop goal. Among the four conjuncts, the

predicate $P3$ gives the optimality condition and is the most complex one. Therefore, it will be used as a loop goal. On the other hand, the predicates $P0$ and $P1$ can be easily satisfied by the following simple initialization:

$$r,t := 0,d.$$

Based on these, the plan is to design an algorithm with the following scheme:

$$r,t := 0,d$$
$$\text{do}\{P0 \wedge P1\}$$
$$\quad \neg P2 \rightarrow \text{update } t$$
$$\quad \neg P3 \rightarrow \text{update } r$$
$$\text{od}\{P0 \wedge P1 \wedge P2 \wedge P3\}.$$

The first command in the loop can be refined as

$$\exists (u,v) \in E : r[u] - r[v] = w[u,v] \wedge t[v]$$
$$- t[u] < d[v] \rightarrow t[v]$$
$$:= t[u] + d[v].$$

This is simply the Bellman-Ford relaxations for computing the longest paths.

The second command is more difficult to refine. If $\neg P3$, that is, there exists another valid retiming r',t' such that $\max(t) > \max(t')$, then on any node v such that $t[v] = \max(t)$ it must have $t'[v] < t[v]$. One property known on these nodes is

$$\forall v \in V : t'[v] < t[v]$$
$$\Rightarrow (\exists u \in V : r[u] - r[v] > r'[u] - r'[v]),$$

which means that if the arrival time of v is smaller in another retiming r',t', then there must be a node u such that r' gives more registers between u and v. In fact, one such a u is the starting node of the longest combinational path to v that gives the delay of $t[v]$.

To reduce the clock period, the variable r needs to be updated to make it closer to r'. It should be noted that it is not the absolute values of r but their differences that are relevant in the retiming. If r,t is a solution to a retiming problem, then $r+c, t$, where $c \in \mathbb{Z}$ is an arbitrary constant, is also a solution. Therefore r can be made "closer" to r' by allocating more registers between u and v, that is, by either decreasing $r[u]$ or increasing $r[v]$. Notice that v can be easily identified by $t[v] = \max(t)$. No matter whether $r[v]$ or $r[u]$ is selected to change, the amount of change should be only one since r should not be overadjusted. Thus, after the adjustment, it is still true that $r[v] - r[u] \le r'[v] - r'[u]$ or equivalently $r[v] - r'[v] \le r[u] - r'[u]$. Since v is easy to identify, $r[v]$ is selected to increase. The arrival time $t[v]$ can be immediately reduced to $d[v]$. This gives a refinement of the second command:

$$\neg P3 \wedge P2 \wedge \exists v \in V : t[v] = \max(t)$$
$$\rightarrow r[v], t[v] := r[v] + 1, d[v].$$

Since registers are moved in the above operation, the predicate $P2$ may be violated. However, the first command will take care of it. That command will increase t on some nodes; some may even become larger than $\max(t)$ before the register move. The same reasoning using r', t' shows that their r values shall be increased, too. Therefore, to implement this as-soon-as-possible (ASAP) increase of r, a snapshot of $\max(t)$ needs to be taken when $P2$ is valid. Physically, such a snapshot records one feasible clock period ϕ and can be implemented by adding one more command in the loop:

$$P2 \wedge \phi > \max(t) \rightarrow \phi := \max(t).$$

However, such an ASAP operation may increase $r[u]$ even when $w[u, v] - r[u] + r[v] = 0$ for an edge (u, v). It means that $P0$ may no longer be an invariant. But moving $P0$ from invariant to loop goal will not cause a problem since one more command can be added in the loop to take care of it:

$$\exists (u, v) \in E : r[u] - r[v] > w[u, v]$$
$$\rightarrow r[v] := r[u] - w[u, v].$$

Putting all things together, the algorithm now has the following form:

$r, t, \phi := 0, d, \infty;$
$\text{do}\{P1\}$
$\quad \exists (u, v) \in E : r[u] - r[v] = w[u, v]$
$\quad \wedge t[v] - t[u] < d[v] \rightarrow t[v] := t[u] + d[v]$
$\quad \neg P3 \wedge \exists v \in V : t[v] \ge \phi$
$\quad \rightarrow r[v], t[v] := r[v] + 1, d[v]$
$\quad P0 \wedge P2 \wedge \phi > \max(t) \rightarrow \phi := \max(t)$
$\quad \exists (u, v) \in E : r[u] - r[v] > w[u, v]$
$\quad \rightarrow r[v] := r[u] - w[u, v]$
$\text{od}\{P0 \wedge P1 \wedge P2 \wedge P3\}.$

The remaining task to complete the algorithm is how to check $\neg P3$. From previous discussion, it is already known that $\neg P3$ implies that there is a node u such that $r[u] - r'[u] \ge r'[v] - r'[v]$ every time after $r[v]$ is increased. This means that $\max_{v \in V} r[v] - r'[v]$ will not increase. In other words, there is at least one node v whose $r[v]$ will not change. Before $r[v]$ is increased, it also has $w_{u \leadsto v} - r[u] + r[v] \le 0$, where $w_{u \leadsto v} \ge 0$ is the original number of registers on one path from u to v, which gives $r[v] - r[u] \le 1$ even after the increase of $r[v]$. This implies that there will be at least $i + 1$ nodes whose r is at most i for $0 \le i < |V|$. In other words, the algorithm can keep increasing r and when there is any r reaching $|V|$ it shows that $P3$ is satisfied. Therefore, the complete algorithm will have the following form:

$r, t, \phi := 0, d, \infty;$
$\text{do}\{P1\}$
$\quad \exists (u, v) \in E : r[u] - r[v] = w[u, v]$
$\quad \wedge t[v] - t[u] < d[v] \rightarrow t[v] := t[u] + d[v]$
$\quad (\forall v \in V : r[v] < |V|)$
$\quad \wedge \exists v \in V : t[v] \ge \phi \rightarrow r[v], t[v] := r[v] + 1, d[v]$
$\quad (\exists v \in V : r[v] \ge |V|)$
$\quad \wedge \exists v \in V : t[v] \ge \phi \rightarrow r[v], t[v] := r[v] + 1, d[v]$
$\quad P0 \wedge P2 \wedge \phi > \max(t) \rightarrow \phi := \max(t)$
$\quad \exists (u, v) \in E : r[u] - r[v] > w[u, v]$
$\quad \rightarrow r[v] := r[u] - w[u, v]$
$\text{od}\{P0 \wedge P1 \wedge P2 \wedge P3\}.$

The correctness of the algorithm can be proved easily by showing that the invariant $P1$ is maintained and the negation of the guards implies

$P0 \wedge P2 \wedge P3$. The termination is guaranteed by the monotonic increase of r and an upper bound on it. In fact, the following theorem gives its worst-case runtime.

Theorem 1 *The worst-case running time of the given retiming algorithm is upper bounded by* $O(|V|^2|E|)$.

The runtime bound of the retiming algorithm is got under the worst-case assumption that each increase on r will trigger a timing propagation on the whole circuit ($|E|$ edges). This is only true when the r increase moves all registers in the circuit. However, in such a case, the r is upper bounded by 1, thus the running time is not larger than $O(|V||E|)$. On the other hand, when the r value is large, the circuit is partitioned by the registers into many small parts, thus the timing propagation triggered by one r increase is limited within a small tree.

Applications

In the basic algorithm, the optimality $P3$ is verified by an $r[v] \geq |V|$. However, in most cases, the optimality condition can be discovered much earlier. Since each time $r[v]$ is increased, there must be a "safeguard" node u such that $r[u] - r'[u] \geq r[v] - r'[v]$ after the operation. Therefore, if a pointer is introduced from v to u when $r[v]$ is increased, the pointers cannot form a cycle under $\neg P3$. In fact, the pointers will form a forest

where the roots have $r = 0$ and a child can have an r at most one larger than its parent. Using a cycle by the pointers as an indication of $P3$, instead of an $r[v] \geq |V|$, the algorithm can have much better practical performance.

Retiming is usually used to optimize either the clock period or the number of registers in the circuit. The discussed algorithm solves only the minimal period retiming problem. The retiming problem for minimizing the number of registers under a given period has been solved by Leiserson and Saxe [1] and is presented in another entry in this encyclopedia. Their algorithm reduces the problem to the dual of a minimal cost network problem on a denser graph. An efficient iterative algorithm similar to Zhou's algorithm has been designed for the minimal register problem recently [3].

Experimental Results

Experimental results are reported by Zhou [4] which compared the runtime of the algorithm with an efficient heuristic called ASTRA [2]. The results on the ISCAS89 benchmarks are reproduced here in Table 1 from [4], where columns A and B are the running time of the two stages in ASTRA.

Cross-References

▶ Circuit Retiming

Circuit Retiming: An Incremental Approach, Table 1 Experimental results

Name	#gates	Clock period		$\sum r$	#updates	Time(s)	ASTRA	
		Before	After				A(s)	B(s)
s1423	490	166	127	808	7,619	0.02	0.03	0.02
s1494	558	89	88	628	7,765	0.02	0.01	0.01
s9234	2,027	89	81	2,215	76,943	0.12	0.11	0.09
s9234.1	2,027	89	81	2,164	77,644	0.16	0.11	0.10
s13207	2,573	143	82	4,086	28,395	0.12	0.38	0.12
s15850	3,448	186	77	12,038	99,314	0.36	0.43	0.17
s35932	12,204	109	100	16,373	108,459	0.28	0.24	0.65
s38417	8,709	110	56	9,834	155,489	0.58	0.89	0.64
s38584	11,448	191	163	19,692	155,637	0.41	0.50	0.67
s38584.1	11,448	191	183	9,416	114,940	0.48	0.55	0.78

Recommended Reading

1. Leiserson CE, Saxe JB (1991) Retiming synchronous circuitry. Algorithmica 6:5–35
2. Sapatnekar SS, Deokar RB (1996) Utilizing the retiming-skew equivalence in a practical algorithm for retiming large circuits. IEEE Trans Comput Aided Des 15:1237–1248
3. Wang J, Zhou H (2008) An efficient incremental algorithm for min-area retiming. In: Proceedings of the design automation conference, Anaheim, pp 528–533
4. Zhou H (2005) Deriving a new efficient algorithm for min-period retiming. In: Asia and South Pacific design automation conference, Shanghai, Jan 2005

Clique Enumeration

Etsuji Tomita
The Advanced Algorithms Research Laboratory,
The University of Electro-Communications,
Chofu, Tokyo, Japan

Keywords

Enumeration; Maximal clique; Maximal independent set; Time complexity

Years and Authors of Summarized Original Work

1977; Tsukiyama, Ide, Ariyoshi, Shirakawa
2004; Makino, Uno
2006; Tomita, Tanaka, Takahashi

Problem Definition

We discuss a simple undirected and connected graph $G = (V, E)$ with a finite set V of vertices and a finite set $E \subseteq V \times V$ of edges. A pair of vertices v and w is said to be adjacent if $(v, w) \in E$. For a subset $R \subseteq V$ of vertices, $G(R) = (R, E \cap (R \times R))$ is an induced subgraph. An induced subgraph $G(Q)$ is said to be a *clique* if $(v, w) \in E$ for all $v, w \in Q \subseteq V$ with $v \neq w$. In this case, we may simply state that Q is a clique. In particular, a clique that is not

properly contained in any other clique is called *maximal*. An induced subgraph $G(S)$ is said to be an *independent set* if $(v, w) \notin E$ for all $v, w \in S \subseteq V$. For a vertex $v \in V$, let $\Gamma(v) = \{w \in V | (v, w) \in E\}$. We call $|\Gamma(v)|$ the degree of v.

The problem is to enumerate all maximal cliques of the given graph $G = (V, E)$. It is equivalent to enumerate all maximal independent sets of the *complementary graph* $\bar{G} = (V, \bar{E})$, where $\bar{E} = \{(v, w) \in V \times V | (v, w) \notin E, v \neq w\}$.

Key Results

Efficient Algorithms for Clique Enumeration

Efficient algorithms to solve the problem can be found in the following approaches (1) and (2).

(1) Clique Enumeration by Depth-First Search with Pivoting Strategy

The basis of the first approach is a simple depth-first search. It begins with a clique of size 0 and continues with finding all of the progressively larger cliques until they can be verified as maximal. Formally, this approach maintains a global variable $Q = \{p_1, p_2, \ldots, p_d\}$ that consists of vertices of a current clique found so far. Let

$$SUBG = V \cap \Gamma(p_1) \cap \Gamma(p_2) \cap \cdots \cap \Gamma(p_d).$$

We begin the algorithm by letting $Q = \varnothing$ and $SUBG := V$ (the set of all vertices). We select a certain vertex p from $SUBG$ and add p to $Q (Q := Q \cup \{p\})$. Then we compute $SUBG_p := SUBG \cap \Gamma(p)$ as the new set of vertices in question. In particular, the first selected vertex $u \in SUBG$ is called a *pivot*. This procedure (EXPAND()) is applied recursively while $SUBG_p \neq \varnothing$.

When $SUBG_p = \varnothing$ is reached, Q constitutes a *maximal* clique. We then backtrack by removing the lastly inserted vertex from Q and $SUBG$. We select a new vertex p from the resulting $SUBG$ and continue the same procedure until $SUBG = \varnothing$. This process can be represented by a depth-first *search forest*. See Fig. 2b as an

```
procedure CLIQUES(G)
begin                                                              /* Q := ∅ */
 1 : EXPAND(V,V)
end of CLIQUES

      procedure EXPAND(SUBG, CAND)
      begin
 2 :     if SUBG = ∅
 3 :       then print("clique,")                                   /* Q is a maximal clique */
 4 :       else u := a vertex u in SUBG that maximizes | CAND ∩ Γ(u) |;    /* pivot */
 5 :            while CAND − Γ(u) ≠ ∅
 6 :              do q := a vertex in (CAND − Γ(u));
 7 :                 print (q,",");                                /* Q := Q ∪{ q } */
 8 :                 SUBG_q := SUBG ∩ Γ(q);
 9 :                 CAND_q := CAND ∩ Γ(q);
10:                  EXPAND(SUBG_q, CAND_q);
11:                  CAND := CAND −{ q };
12:                  print ("back,")                               /* Q := Q −{ q } */
              od
       fi
      end of EXPAND
```

Clique Enumeration, Fig. 1 Algorithm CLIQUES

example of an essential part of a search forest. It clearly generates all maximal cliques.

The above-generated maximal cliques, however, could contain duplications or nonmaximal ones, so we *prune* unnecessary parts of the search forest as in the Bron-Kerbosch algorithm [3].

First, let *FINI* be a subset of vertices of *SUBG* that have already been processed by the algorithm. (*FINI* is short for *finished*.) Then we denote by *CAND* the set of remaining candidates for expansion: $CAND := SUBG - FINI$, where for two sets X and Y, $X - Y = \{v | v \in X \text{ and } v \notin Y\}$. At the beginning, $FINI := \varnothing$ and $CAND := SUBG$. In the subgraph $G(SUBG_q)$ with $SUBG_q : = SUBG \cap \Gamma(q)$, let

$$FINI_q := SUBG_q \cap FINI,$$
$$CAND_q := SUBG_q - FINI_q.$$

Then only the vertices in $CAND_q$ can be candidates for expanding the clique $Q \cup \{q\}$ to find *new* larger cliques.

Second, for the first selected pivot u in $SUBG$, any maximal clique R in $G(SUBG \cap \Gamma(u))$ is not maximal in $G(SUBG)$, since $R \cup \{u\}$ is a larger clique in $G(SUBG)$. Therefore, searching for maximal cliques from $SUBG \cap \Gamma(u)$ should be excluded.

When the previously described pruning method is also taken into consideration, we find that the only search subtrees to be expanded are from the vertices in $(SUBG - SUBG \cap \Gamma(u)) - FINI = CAND - \Gamma(u)$. Here, in order to minimize $|CAND - \Gamma(u)|$, we choose the pivot $u \in SUBG$ to be the one that *maximizes* $|CAND \cap \Gamma(u)|$. This is *crucial* to establish the *optimality* of the worst-case time complexity of the algorithm. This kind of pivoting strategy was proposed by Tomita et al. [11]. (Recommended Reading [11] was reviewed by Pardalos and Xue [10] and Bomze et al. [2].)

The algorithm CLIQUES by Tomita et al. [12] is shown in Fig. 1. It enumerates all maximal cliques based upon the above approach, but all maximal cliques enumerated are presented in a tree-like form. Here, if Q is a *maximal* clique that is found at statement 2, then the algorithm only prints out a string of characters *clique* instead of Q itself at statement 3. Otherwise, it is impossible to achieve the optimal worst-case running time. Instead, in addition to printing *clique* at statement 3, we print out q followed by a *comma* at statement 7 every time q is picked out as a new element of a larger clique, and we print out a string of characters *back* at statement 12 after q is moved from *CAND* to *FINI* at statement 11. We can easily obtain a tree representation of all

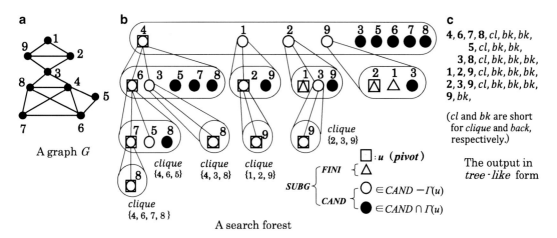

Clique Enumeration, Fig. 2 An example run of CLIQUES [12]. (**a**) A graph G. (**b**) A search forest. (**c**) The output in *tree-like* form

the maximal cliques from the sequence printed by statements 3, 7, and 12. The output in a tree-like format is also important *practically* since it saves space in the output file. An example run of CLIQUES to Fig. 2a is shown in Fig. 2b, c with appropriate indentations.

The worst-case time complexity of CLIQUES was proved to be $O(3^{n/3})$ for an n-vertex graph [11, 12]. This is *optimal* as a function of n since there exist up to $3^{n/3}$ cliques in an n-vertex graph [9].

Eppstein et al. [5] used this approach and proposed an algorithm for enumerating all maximal cliques that runs in time $O(dn3^{d/3})$ for an n-vertex graph G, where d is the *degeneracy* of G that is defined to be the smallest number such that every subgraph of G contains a vertex of degree at most d. If graph G is *sparse*, d can be much smaller than n and hence $O(dn3^{d/3})$ can be much smaller than $O(3^{n/3})$.

(2) Clique Enumeration by Reverse Search

The second approach is regarded to be based upon the *reverse search* that was introduced by Avis and Fukuda [1] to solve enumeration problems efficiently.

Given the graph $G = (V, E)$ with $V = \{v_1, v_2, \ldots, v_n\}$ where $n = |V|$, let $V_i = \{v_1, v_2, \ldots, v_i\}$. Then $\{v_1\}$ is simply a maximal clique in $G(V_1)$. All the maximal cliques in $G(V_i)$ are enumerated based on those in $G(V_{i-1})$

step by step for $i = 2, 3, \ldots, n$. The process forms an *enumeration tree* whose root is $\{v_1\}$, where the root is considered at depth 1 of the enumeration tree for the sake of simplicity. The children at depth i are all the maximal cliques in $G(V_i)$ for $i = 2, 3, \ldots, n$. For two subsets $X, Y \subseteq V$, we say that X precedes Y in lexicographic order if for $v_i \in (X - Y) \cup (Y - X)$ with the minimum index i it holds that $v_i \in X$.

Let Q be a maximal clique in $G(V_{i-1})$. If v_i is adjacent to all vertices in Q, then $Q \cup \{v_i\}$ is the only child of Q at depth i. Otherwise, Q itself is the first child of Q at depth i. In addition, if $Q \cap \Gamma(v_i) \cup \{v_i\}$ is a *maximal* clique in $G(V_i)$, then it is a candidate for the second child of Q at depth i. The unique parent of $Q \cap \Gamma(v_i) \cup \{v_i\}$ is defined to be the *lexicographically first* maximal clique in $G(V_{i-1})$ that contains $Q \cap \Gamma(v_i)$. (In general, there exist multiple numbers of distinct maximal cliques that contain $Q \cap \Gamma(v_i)$ at depth $i - 1$.)

The algorithm of Tsukiyama et al. [14] traverses the above enumeration tree in a depth-first way. Such a traversal is considered to be reverse search [8]. To be more precise, the algorithm MIS in [14] is to enumerate all maximal independent sets, and we are concerned here with its complementary algorithm in [8] that enumerates all maximal cliques, which we call here $\overline{\text{MIS}}$. An example run of $\overline{\text{MIS}}$ to Fig. 2a is shown in Fig. 3a. Algorithms MIS and $\overline{\text{MIS}}$ run in time $O(m'n)$ and

a

$G(V_1)$ $G(V_2)$ $G(V_3)$ $G(V_4)$ $G(V_5)$ $G(V_6)$ $G(V_7)$ $G(V_8)$ $G(V_9)$

$\{1\} \rightarrow \{1,2\} \rightarrow \{1,2\} \rightarrow \{1,2\} \rightarrow \{1,2\} \rightarrow \{1,2\} \rightarrow \{1,2\} \rightarrow \{1,2\} \rightarrow \{1,2,\mathbf{9}\}$

$\{2,3\} \rightarrow \{2,3\} \rightarrow \{2,3\} \rightarrow \{2,3\} \rightarrow \{2,3\} \rightarrow \{2,3\} \rightarrow \{2,3,\mathbf{9}\}$

$\{3,4\} \rightarrow \{3,4\} \rightarrow \{3,4\} \rightarrow \{3,4\} \rightarrow \{3,4,\mathbf{8}\} \rightarrow \{3,4,8\}$

$V_i = \{1,2,\dots,i\}$

$\{4,5\} \rightarrow \{4,5,6\} \rightarrow \{4,5,6\} \rightarrow \{4,5,6\} \rightarrow \{4,5,6\}$

$\{4,6,7\} \rightarrow \{4,6,7,\mathbf{8}\} \rightarrow \{4,6,7,8\}$

By $\overline{\text{MIS}}$ [14,8]

b

$\{1,2,9\}$
$\downarrow$
$\{2,\mathbf{3},9\}$
$\downarrow$
$\{3,\mathbf{4},8\}$
$\downarrow$
$\{4,\mathbf{5},6\}$
$\downarrow$
$\{4,6,7,\mathbf{8}\}$

By AMC [8]

Clique Enumeration, Fig. 3 Enumeration trees for Fig. 2a in reverse search. (**a**) By $\overline{\text{MIS}}$ [14,8]. (**b**) By AMC [8]

$O(mn)$ per maximal clique, respectively, where $n = |V|, m = |E|$, and $m' = |\bar{E}|$ [8,14].

Chiba and Nishizeki [4] reduced the time complexity of $\overline{\text{MIS}}$ to $O(a(G)m)$ per maximal clique, where $a(G)$ is the *arboricity* of G with $m/(n-1) \le a(G) \le O(m^{1/2})$ for a connected graph G. Johnson et al. [7] presented an algorithm that enumerates all maximal cliques in lexicographic order in time $O(mn)$ per maximal clique [7,8].

Makino and Uno [8] proposed new algorithms that are based on the algorithm of Tsukiyama et al. [14]. Let $C(Q)$ denote the lexicographically first maximal clique containing Q in graph G. The root of their enumeration tree is the lexicographically first maximal clique Q_0 in G. For a maximal clique $Q \ne Q_0$ in the enumeration tree, define the parent of Q to be $C(Q \cap V_i)$ where i is the maximum index such that $C(Q \cap V_i) \ne Q$. Such a parent uniquely exists for every $Q \ne Q_0$. In the enumeration tree, $Q' = C(Q \cap V_j \cap \Gamma(v_j) \cup \{v_j\})$ is a child of Q if and only if Q is a parent of Q'. (In general, a parent has at most $|V|$ children.) This concludes the description of the enumeration tree of ALLMAXCLIQUES (AMC for short) in [8]. An example run to Fig. 2a is shown in Fig. 3b, where the bold-faced vertex is the minimum i such that $Q \cap V_i = Q$. Algorithm AMC runs in time $O(M(n))$ per maximal clique, where $M(n)$ denotes the time required to multiply two $n \times n$ matrices. Another algorithm in [8] runs in time $O(\Delta^4)$ per maximal clique, where Δ is the maximum degree of G. Here, if G is sparse, then Δ can be small. In addition, they presented an algorithm that enumerates all

maximal bipartite cliques in a bipartite graph in time $O(\Delta^3)$ per maximal bipartite clique.

Applications

Clique enumeration has diverse applications in clustering, data mining, information retrieval, bioinformatics, computer vision, wireless networking, computational topology, and many other areas. Here, one of Makino and Uno's algorithms [8] was successfully applied for enumerating *frequent closed itemsets* [16]. See Recommended Reading [2, 5, 6, 8, 10, 12, 13, 16] for details. For practical applications, enumeration of *pseudo cliques* is sometimes more important [15].

Experimental Results

Experimental Results are shown in Recommended Reading [12, 6, 14, 8]. Tomita et al.'s algorithm CLIQUES [12] is fast especially for graphs with high and medium density. Eppstein et al.'s algorithm [5] is effective for very large and sparse graphs [6]. Makino and Uno's algorithms [8] can be fast for sparse graphs especially when they have a small number of maximal cliques.

Cross-References

▶ Reverse Search; Enumeration Algorithms

Recommended Reading

1. Avis D, Fukuda K (1996) Reverse search for enumeration. Discret Appl Math 65:21–46
2. Bomze IM, Budinich M, Pardalos PM, Pelillo M (1999) The maximum clique problem. In: Du D-Z, Pardalos PM (eds) Handbook of combinatorial optimization, supplement, vol A. Kluwer Academic, Dordrecht, pp 1–74
3. Bron C, Kerbosch J (1973) Algorithm 457, finding all cliques of an undirected graph. Commun ACM 16:575–577
4. Chiba N, Nishizeki T (1985) Arboricity and subgraph listing algorithms. SIAM J Comput 14: 210–223
5. Eppstein D, Löffler M, Strash D (2010) Listing all maximal cliques in sparse graphs in near-optimal time. In: ISAAC 2010, Jeju Island. Lecture notes in computer science, vol 6506, pp 403–414
6. Eppstein D, Strash D (2011) Listing all maximal cliques in large sparse real-world graphs. In: SEA 2011, Chania. Lecture notes in computer science, vol 6630, pp 364–375
7. Johnson DS, Yanakakis M, Papadimitriou CH (1998) On generating all maximal independent sets. Inf Process Lett 27:119–123
8. Makino K, Uno T (2004) New algorithms for enumerating all maximal cliques. In: SWAT 2004, Humlebaek. Lecture notes in computer science, vol 3111, pp 260–272
9. Moon JW, Moser L (1965) On cliques in graphs. Isr J Math 3:23–28
10. Pardalos PM, Xue J (1994) The maximum clique problem. J Glob Optim 4:301–328
11. Tomita E, Tanaka A, Takahashi H (1988) The worst-case time complexity for finding all maximal cliques. Technical Report of the University of Electro-Communications, UEC-TR-C5(2)
12. Tomita E, Tanaka A, Takahashi H (2006) The worst-case time complexity for generating all maximal cliques and computational experiments. Theor Comput Sci 363 (Special issue on COCOON 2004, Jeju Island. Lecture notes in computer science, 3106): 28–42
13. Tomita E, Akutsu T, Matsunaga T (2011) Efficient algorithms for finding maximum and maximal cliques: Effective tools for bioinformatics. In: Laskovski AN (ed) Biomedical engineering, trends in electronics, communications and software. InTech, Rijeka, pp 625–640. Available from: http://www.intechopen.com/articles/show/title/effici ent-algorithms-for-finding-maximum-and-maximal-cliques-effective-tools-for-bioinformatics
14. Tsukiyama S, Ide M, Ariyoshi H, Shirakawa I (1977) A new algorithm for generating all the maximal independent sets. SIAM J Comput 6:505–517
15. Uno T (2010) An efficient algorithm for solving pseudo clique enumeration problem. Algorithmica 56:3–16
16. Uno T, Asai T, Arimura H, Uchida Y (2003) LCM: an efficient algorithm for enumerating frequent closed item sets. In: Workshop on frequent itemset mining implementations (FIMI), Melbourne

Clock Synchronization

Boaz Patt-Shamir
Department of Electrical Engineering, Tel-Aviv University, Tel-Aviv, Israel

Years and Authors of Summarized Original Work

1994; Patt-Shamir, Rajsbaum

Problem Definition

Background and Overview

Coordinating processors located in different places is one of the fundamental problems in distributed computing. In his seminal work, Lamport [4, 5] studied the model where the only source of coordination is message exchange between the processors; the time that elapses between successive steps at the same processor, as well as the time spent by a message in transit, may be arbitrarily large or small. Lamport observed that in this model, called the *asynchronous model*, temporal concepts such as "past" and "future" are derivatives of causal dependence, a notion with a simple algorithmic interpretation. The work of Patt-Shamir and Rajsbaum [10] can be viewed as extending Lamport's qualitative treatment with quantitative concepts. For example, a statement like "event *a* happened before event *b*" may be refined to a statement like "event *a* happened at least 2 time units and at most 5 time units before event *b*". This is in contrast to most previous theoretical work, which focused on the linear-programming aspects of clock synchronization (see below).

The basic idea in [10] is as follows. First, the framework is extended to allow for upper and lower bounds on the time that elapses between pairs of events, using the system's *real-time specification*. The notion of real-time specification is a very natural one. For example, most processors have local clocks, whose rate of progress is typically bounded with respect to real time (these bounds are usually referred to as the clock's "drift bounds"). Another example is send and receive events of a given message: It is always true that the receive event occurs before the send event, and in many cases, tighter lower and upper bounds are available. Having defined real-time specification, [10] proceeds to show how to combine these local bounds global bounds in the best possible way using simple graph-theoretic concepts. This allows one to derive optimal protocols that say, for example, what is the current reading of a remote clock. If that remote clock is the standard clock, then the result is optimal clock synchronization in the common sense (this concept is called "external synchronization" below).

Formal Model

The system consists of a fixed set of interconnected *processors*. Each processor has a *local clock*. An *execution* of the system is a sequence of events, where each event is either a *send* event, a *receive* event, or an *internal* event. Regarding communication, it is only assumed that each receive event of a message m has a unique corresponding send event of m. This means that messages may be arbitrarily lost, duplicated or reordered, but not corrupted. Each event e occurs at a single specified processor, and has two real numbers associated with it: its *local time*, denoted $LT(e)$, and its *real time*, denoted $RT(e)$. The local time of an event models the reading of the local clock when that event occurs, and the local processor may use this value, e.g., for calculations, or by sending it over to another processor. By contrast, the real time of an event is not observable by processors: it is an abstract concept that exists only in the analysis.

Finally, the real-time properties of the system are modeled by a pair of functions that map each pair of events to $\mathbb{R} \cup \{-\infty, \infty\}$: given two events e and e', $L(e, e') = \ell$ means that

$RT(e') - RT(e) \geq \ell$, and $H(e, e') = h$ means that $RT(e') - RT(e) \leq h$, i.e., that the number of (real) time units since the occurrence of event e until the occurrence of e' is at least ℓ and at most h. Without loss of generality, it is assumed that $L(e, e') = -H(e', e)$ for all events e, e' (just use the smaller of them). Henceforth, only the upper bounds function H is used to represent the real-time specification.

Some special cases of real time properties are particularly important. In a completely asynchronous system, $H(e', e) = 0$ if either e occurs before e' in the same processor, or if e and e' are the send and receive events, respectively, of the same message. (For simplicity, it is assumed that two ordered events may have the same real time of occurrence.) In all other cases $H(e, e') = \infty$. On the other extreme of the model spectrum, there is the *drift-free* clocks model, where all local clocks run at exactly the rate of real time. Formally, in this case $H(e, e') = LT(e') - LT(e)$ for any two events e and e' occurring at the same processor. Obviously, it may be the case that only some of the clocks in the system are drift-free.

Algorithms

In this work, message generation and delivery is completely decoupled from message information. Formally, messages are assumed to be generated by some "send module", and delivered by the "communication system". The task of algorithms is to add contents in messages and state variables in each node. (The idea of decoupling synchronization information from message generation was introduced in [1].) The algorithm only has local information, i.e., contents of the local state variables and the local clock, as well as the contents of the incoming message, if we are dealing with a receive event. It is also assumed that the real time specification is known to the algorithm. The conjunction of the events, their and their local times (but not their real times) is called as the *view* of the given execution. Algorithms, therefore, can only use as input the view of an execution and its real time specification.

Problem Statement

The simplest variant of clock synchronization is *external synchronization*, where one of the processors, called the source, has a drift-free clock, and the task of all processors is to maintain the tightest possible estimate on the current reading of the source clock. This formulation corresponds to the Newtonian model, where the processors reside in a well-defined time coordinate system, and the source clock is reading the standard time. Formally, in external synchronization each processor v has two output variables Δ_v and ε_v; the estimate of v of the source time at a given state is $LT_v + \Delta_v$, where LT_v is the current local time at v. The algorithm is required to guarantee that the difference between the source time and it estimate is at most ε_v (note that Δ_v, as well as ε_v, may change dynamically during the execution). The performance of the algorithm is judged by the value of the ε_v variables: the smaller, the better.

In another variant of the problem, called *internal synchronization*, there is no distinguished processor, and the requirement is essentially that all clocks will have values which are close to each other. Defining this variant is not as straightforward, because trivial solutions (e.g., "set all clocks to 0 all the time") must be disqualified.

Key Results

The key construct used in [10] is the *synchronization graph* of an execution, defined by combining the concepts of local times and real-time specification as follows.

Definition 1 Let β be a view of an execution of the system, and let H be a real time specification for β. The *synchronization graph* generated by β and H is a directed weighted graph $\Gamma_{\beta H} = (V, E, w)$, where V is the set of events in β, and for each ordered pair of events p q in β such that $H(p, q) < \infty$, there is a directed edge $(p, q) \in E$. The *weight* of an edge (p, q) is $w(p, q) \overset{\text{def}}{=} H(p, q) - LT(p) + LT(q)$.

The natural concept of *distance* from an event p to an event q in a synchronization graph Γ, denoted $d_\Gamma(p, q)$, is defined by the length of the shortest weight path from p to q, or infinity if q is not reachable from p. Since weights may be negative, one has to prove that the concept is well defined: indeed, it is shown that if $\Gamma_{\beta H}$ is derived from an execution with view β that satisfies real time specification H, then $\Gamma_{\beta H}$ does not contain directed cycles of negative weight.

The main algorithmic result concerning synchronization graphs is summarized in the following theorem.

Theorem 1 *Let α be an execution with view β. Then α satisfies the real time specification H if and only if* $\mathrm{RT}(p) - \mathrm{RT}(q) \le d_\Gamma(p, q) + \mathrm{LT}(p) - \mathrm{LT}(q)$ *for any two events p and q in $\Gamma_{\beta H}$.*

Note that all quantities in the r.h.s. of the inequality are available to the synchronization algorithm, which can therefore determine upper bounds on the real time that elapses between events. Moreover, these bounds are the best possible, as implied by the next theorem.

Theorem 2 *Let $\Gamma_{\beta H} = (V, E, w)$ be a synchronization graph obtained from a view β satisfying real time specification H. Then for any given event $p_0 \in V$, and for any finite number $N > 0$, there exist executions α_0 and α_1 with view β, both satisfying H, and such that the following real time assignments hold.*

- *In α_0, for all $q \in V$ with $d_\Gamma(q, p_0) < \infty$, $\mathrm{RT}_{\alpha_0}(q) = \mathrm{LT}(q) + d_\Gamma(q, p_0)$, and for all $q \in V$ with $d_\Gamma(q, p_0) = \infty$, $\mathrm{RT}_{\alpha_0}(q) > \mathrm{LT}(q) + N$.*
- *In α_1, for all $q \in V$ with $d_\Gamma(p_0, q) < \infty$, $\mathrm{RT}_{\alpha_1}(q) = \mathrm{LT}(q) - d_\Gamma(p_0, q)$, and for all $q \in V$ with $d_\Gamma(p_0, q) = \infty$, $\mathrm{RT}_{\alpha_1}(q) < \mathrm{LT}(q) - N$.*

From the algorithmic viewpoint, one important drawback of results of Theorems 1 and 2 is that they depend on the view of an execution, which may grow without bound. Is it really necessary?

The last general result in [10] answers this question in the affirmative. Specifically, it is shown that in some variant of the *branching program* computational model, the space complexity of any synchronization algorithm that works with arbitrary real time specifications cannot be bounded by a function of the system size. The result is proved by considering multiple scenarios on a simple system of four processors on a line.

Later Developments

Based on the concept of synchronization graph, Ostrovsky and Patt-Shamir present a refined general optimal algorithm for clock synchronization [9]. The idea in [9] is to discard parts of the synchronization graphs that are no longer relevant. Roughly speaking, the complexity of the algorithm is bounded by a polynomial in the system size and the ratio of processors speeds.

Much theoretical work was invested in the internal synchronization variant of the problem. For example, Lundelius and Lynch [7] proved that in a system of n processors with full connectivity, if message delays can take arbitrary values in $[0, 1]$ and local clocks are drift-free, then the best synchronization that can be guaranteed is $1 - \frac{1}{n}$. Helpern et al. [3] extended their result to general graphs using linear-programming techniques. This work, in turn, was extended by Attiya et al. [1] to analyze any given execution (rather than only the worst case for a given topology), but the analysis is performed off-line and in a centralized fashion. The work of Patt-Shamir and Rajsbaum [10] extended the "per execution" viewpoint to on-line distributed algorithms, and shifted the focus of the problem to external synchronization.

Recently, Fan and Lynch [2] proved that in a line of n processors whose clocks may drift, no algorithm can guarantee that the difference between the clock readings of all pairs of neighbors is $o(\log n / \log\log n)$.

Clock synchronization is very useful in practice. See, for example, Liskov [6] for some motivation. It is worth noting that the Internet provides a protocol for external clock synchronization called NTP [8].

Applications

Theorem 1 immediately gives rise to an algorithm for clock synchronization: every processor maintains a representation of the synchronization graph portion known to it. This can be done using a full information protocol: In each outgoing message this graph is sent, and whenever a message arrives, the graph is extended to include the new information from the graph in the arriving message. By Theorem 2, the synchronization graph obtained this way represents at any point in time all information available required for optimal synchronization. For example, consider external synchronization. Directly from definitions it follows that all events associated with a drift-free clock (such as events in the source node) are at distance 0 from each other in the synchronization graph, and can therefore be considered, for distance computations, as a single node s. Now, assuming that the source clock actually shows real time, it is easy to see that for any event p,

$$\mathrm{RT}(p) \in [\mathrm{LT}(p) - d(s, p), \mathrm{LT}(p) + d(p, s)],$$

and furthermore, no better bounds can be obtained by any correct algorithm.

The general algorithm described above (maintaining the complete synchronization graph) can be used also to obtain optimal results for internal synchronization; details are omitted.

An interesting special case is where all clocks are drift free. In this case, the size of the synchronization graph remains fixed: similarly to a source node in external synchronization, all events occurring at the same processor can be mapped to a single node; parallel edges can be replaced by a single new edge whose weight is minimal among all old edges. This way one can obtain a particularly efficient distributed algorithm solving external clock synchronization, based on the distributed Bellman–Ford algorithm for distance computation.

Finally, note that the asynchronous model may also be viewed as a special case of this general theory, where an event p "happens before" an event q if and only if $d(p, q) \leq 0$.

Open Problems

One central issue in clock synchronization is faulty executions, where the real time specification is violated. Synchronization graphs detect any detectable error: views which do not have an execution that conforms with the real time specification will result in synchronization graphs with negative cycles. However, it is desirable to overcome such faults, say by removing from the synchronization graph some edges so as to break all negative-weight cycles. The natural objective in this case is to remove the least number of edges. This problem is APX-hard as it generalizes the Feedback Arc Set problem. Unfortunately, no non-trivial approximation algorithms for it are known.

Cross-References

▶ Causal Order, Logical Clocks, State Machine Replication

Recommended Reading

1. Attiya H, Herzberg A, Rajsbaum S (1996) Optimal clock synchronization under different delay assumptions. SIAM J Comput 25(2):369–389
2. Fan R, Lynch NA (2006) Gradient clock synchronization. Distrib Comput 18(4):255–266
3. Halpern JY, Megiddo N, Munshi AA (1985) Optimal precision in the presence of uncertainty. J Complex 1:170–196
4. Lamport L (1978) Time, clocks, and the ordering of events in a distributed system. Commun ACM 21(7):558–565
5. Lamport L (1986) The mutual exclusion problem. Part I: a theory of interprocess communication. J ACM 33(2):313–326
6. Liskov B (1993) Practical uses of synchronized clocks in distributed systems. Distrib Comput 6:211–219. Invited talk at the 9th annual ACM symposium on principles of distributed computing, Quebec City, 22–24 Aug 1990
7. Lundelius J, Lynch N (1988) A new fault-tolerant algorithm for clock synchronization. Inf Comput 77:1–36
8. Mills DL (2006) Computer network time synchronization: the network time protocol. CRC, Boca Raton
9. Ostrovsky R, Patt-Shamir B (1999) Optimal and efficient clock synchronization under drifting clocks. In: Proceedings of the 18th annual symposium on principles of distributed computing, Atlanta, May 1999, pp 3–12
10. Patt-Shamir B, Rajsbaum S (1994) A theory of clock synchronization. In: Proceedings of the 26th annual ACM symposium on theory of computing, Montreal, 23–25 May 1994, pp 810–819

Closest String and Substring Problems

Lusheng Wang[1], Ming Li[2], and Bin Ma[2,3]
[1] Department of Computer Science, City University of Hong Kong, Hong Kong, Hong Kong
[2] David R. Cheriton School of Computer Science, University of Waterloo, Waterloo, ON, Canada
[3] Department of Computer Science, University of Western Ontario, London, ON, Canada

Keywords

Approximation algorithm; Fixed-parameter algorithms

Years and Authors of Summarized Original Work

2000; Li, Ma, Wang
2003; Deng, et al.
2008; Marx
2009; Ma, Sun
2011; Chen, Wang
2012; Chen, Ma, Wang

Problem Definition

The problem of finding a center string that is "close" to every given string arises and has applications in computational molecular biology [4,5,9–11,18,19] and coding theory [1,6,7].

This problem has two versions: The first problem comes from coding theory when we are looking for a code not too far away from a given set of codes.

Problem 1 (The closest string problem) Input: a set of strings $\mathcal{S} = \{s_1, s_2, \ldots, s_n\}$, each of length m.

Output: the smallest d and a string s of length m which is within Hamming distance d to each $s_i \in \mathcal{S}$.

The second problem is much more elusive than the closest string problem. The problem is formulated from applications in finding conserved regions, genetic drug target identification, and genetic probes in molecular biology.

Problem 2 (The closest substring problem) Input: an integer L and a set of strings $\mathcal{S} = \{s_1, s_2, \ldots, s_n\}$, each of length m.

Output: the smallest d and a string s, of length L, which is within Hamming distance d away from a length L substring t_i of s_i for $i = 1, 2, \ldots, n$.

The following results on approximation algorithms are from [12–15].

Theorem 1 *There is a polynomial time approximation scheme for the closest string problem.*

Theorem 2 *There is a polynomial time approximation scheme for the closest substring problem.*

A faster approximation algorithm for the closest string problem was given in [16].

Lots of results have been obtained in terms of parameterized complexity and fixed-parameter algorithms. In 2005, Marx showed that the closest substring problem is W[1]-hard even if both d and n are parameters [17]. Two algorithms for the closest substring problem have been developed [17] for the cases where d and n are small. The running times for

the two algorithms are $f(d) \cdot m^{O(\log d)}$ and $g(d, n) \cdot n^{O(\log \log n)}$ for some functions f and g, respectively. The first fixed-parameter algorithm for closest string problem has a running time complexity $O(nd^{d+1})$ [8]. Ma and Sun designed a fixed-parameter algorithm with running time $O(nm + nd \cdot (16|\Sigma|)^d)$ for the closest string problem [16]. Extending the algorithm for the closest string problem, an $O(nL + nd \times 2^{4d}|\Sigma|^d \times m^{\lceil \log d \rceil + 1})$ time algorithm was given for the closest substring problem [16].

Since then, a series of improved algorithms have been obtained. Wang and Zhu gave an $O(nL + nd \cdot (2^{3.25}(|\Sigma| - 1))^d)$ algorithm for the closest string problem [20]. Chen and Wang gave an algorithm with running times $O(nL + nd \cdot 47.21^d)$ for protein with $|\Sigma| = 20$ and $O(nL + nd \cdot 13.92^d)$ for DNA with $|\Sigma| = 4$, respectively [2]. They also developed a software package for the (L, d) motif model. Currently the fastest fixed-parameter algorithm for the closest string problem was given by Chen, Ma, and Wang. They developed a three-string approach and the running time of the algorithm is $O(nL + nd^3 \cdot d^6.731)$ for binary strings [3].

Results for other measures with applications in computational biology can be found in [5, 9, 18, 19].

Applications

Many problems in molecular biology involve finding similar regions common to each sequence in a given set of DNA, RNA, or protein sequences. These problems find applications in locating binding sites and finding conserved regions in unaligned sequences [5, 9, 18, 19], genetic drug target identification [10], designing genetic probes [10], universal PCR primer design [5, 10], and, outside computational biology, in coding theory [1, 6, 7]. Such problems may be considered to be various generalizations of the common substring problem, allowing errors. Many measures have been proposed for finding such regions common to every given string. A popular and one of the most fundamental measures is the Hamming distance. Moreover, two popular objective functions are used in these areas. One is the total sum of distances between the center string (common substring) and each of the given strings. The other is the maximum distance between the center string and a given string. For more details, see [10].

A More General Problem

The *distinguishing substring selection* problem has as input two sets of strings, $\mathcal{B}$ and $\mathcal{G}$. It is required to find a substring of unspecified length (denoted by L) such that it is, informally, close to a substring of every string in $\mathcal{B}$ and far away from every length L substring of strings in $\mathcal{G}$. However, we can go through all the possible length L substrings of strings in $\mathcal{G}$, and we may assume that every string in $\mathcal{G}$ has the same length L since $\mathcal{G}$ can be reconstructed to contain all substrings of length L in each of the good strings.

The problem is formally defined as follows: Given a set $\mathcal{B} = \{s_1, s_2, \ldots, s_n\}$ of n_1 (bad) strings of length at least L, and a set $\mathcal{G} = \{g_1, g_2, \ldots g_{n_2}\}$ of n_2 (good) strings of length exactly L, as well as two integers d_b and d_g ($d_b \leq d_g$), the distinguishing substring selection problem (DSSP) is to find a string s such that for each string, there exists a length L substring t_i of s_i with $d(s, t_i) \leq d_b$ and for any string $g_i \in \mathcal{G}$, $d(s, g_i) \geq d_g$. Here $d(,)$ represents the Hamming distance between two strings. If all strings in $\mathcal{B}$ are also of the same length L, the problem is called the distinguishing string problem (DSP).

The distinguishing string problem was first proposed in [10] for generic drug target design. The following results are from [4].

Theorem 3 *There is a polynomial time approximation scheme for the distinguishing substring selection problem. That is, for any constant $\epsilon > 0$, the algorithm finds a string s of length L such that for every $s_i \in \mathcal{B}$, there is a length L substring t_i of s_i with $d(t_i, s) \leq (1 + \epsilon)d_b$ and for every substring u_i of length L of every $g_i \in \mathcal{G}$,*

$d(u_i, s) \geq (1 - \epsilon)d_g$, *if a solution to the original pair ($d_b \leq d_g$) exists. Since there are a polynomial number of such pairs (db,dg), we can exhaust all the possibilities in polynomial time to find a good approximation required by the corresponding application problems.*

Open Problems

The PTASs designed here use linear programming and randomized rounding technique to solve some cases for the problem. Thus, the running time complexity of the algorithms for both the closest string and closest substring is very high. An interesting open problem is to design more efficient PTASs for both problems.

Recommended Reading

1. Ben-Dor A, Lancia G, Perone J, Ravi R (1997) Banishing bias from consensus sequences. In: Proceedings of the 8th annual symposium on combinatorial pattern matching conference, Aarhus, pp 247–261
2. Chen Z, Wang L (2011) Fast exact algorithms for the closest string and substring problems with application to the planted (L, d)-motif model. IEEE/ACM Trans Comput Biol Bioinform 8(5):1400–1410
3. Chen Z-Z, Ma B, Wang L (2012) A three-string approach to the closest string problem. J Comput Syst Sci 78(1):164–178
4. Deng X, Li G, Li Z, Ma B, Wang L (2003) Genetic design of drugs without side-effects. SIAM J Comput 32(4):1073–1090
5. Dopazo J, Rodríguez A, Sáiz JC, Sobrino F (1993) Design of primers for PCR amplification of highly variable genomes. CABIOS 9:123–125
6. Frances M, Litman A (1997) On covering problems of codes. Theor Comput Syst 30:113–119
7. Gasieniec L, Jansson J, Lingas A (1999) Efficient approximation algorithms for the hamming center problem. In: Proceedings of the 10th ACM-SIAM symposium on discrete algorithms, Baltimore, pp 135–S906
8. Gramm J, Niedermeier R, Rossmanith P 2003 Fixed-parameter algorithms for closest string and related problems. Algorithmica 37(1):25–42
9. Hertz G, Stormo G (1995) Identification of consensus patterns in unaligned DNA and protein sequences: a large-deviation statistical basis for penalizing gaps. In: Proceedings of the 3rd international conference on bioinformatics and genome research, Tallahassee, pp 201–216

10. Lanctot K, Li M, Ma B, Wang S, Zhang L (1999) Distinguishing string selection problems. In: Proceedings of the 10th ACM-SIAM symposium on discrete algorithms, Baltimore, pp 633–642

11. Lawrence C, Reilly A (1990) An expectation maximization (EM) algorithm for the identification and characterization of common sites in unaligned biopolymer sequences. Proteins 7:41–51

12. Li M, Ma B, Wang L (2002) Finding similar regions in many sequences. J Comput Syst Sci 65(1):73–96

13. Li M, Ma B, Wang L (1999) Finding similar regions in many strings. In: Proceedings of the thirty-first annual ACM symposium on theory of computing, Atlanta, pp 473–482

14. Li M, Ma B, Wang L (2002) On the closest string and substring problems. J ACM 49(2):157–171

15. Ma B (2000) A polynomial time approximation scheme for the closest substring problem. In: Proceedings of the 11th annual symposium on combinatorial pattern matching, Montreal, pp 99–107

16. Ma B, Sun X (2009) More efficient algorithms for closest string and substring problems. SIAM J Comput 39(4):1432–1443

17. Marx D (2008) Closest substring problems with small distances. SIAM J Comput 38(4):1382–1410

18. Stormo G (1990) Consensus patterns in DNA. In: Doolittle RF (ed) Molecular evolution: computer analysis of protein and nucleic acid sequences. Methods Enzymol 183:211–221

19. Stormo G, Hartzell GW III (1991) Identifying protein-binding sites from unaligned DNA fragments. Proc Natl Acad Sci USA 88:5699–5703

20. Wang L, Zhu B (2009) Efficient algorithms for the closest string and distinguishing string selection problems. In: Proceedings of 3rd international workshop on frontiers in algorithms, Hefei. Lecture notes in computer science, vol 5598, pp 261–270

Closest Substring

Jens Gramm
WSI Institute of Theoretical Computer Science,
Tübingen University, Tübingen, Germany

Keywords

Common approximate substring

Years and Authors of Summarized Original Work

2005; Marx

Problem Definition

CLOSEST SUBSTRING is a core problem in the field of consensus string analysis with, in particular, applications in computational biology. Its decision version is defined as follows.

CLOSEST SUBSTRING

Input: k strings $s_1, s_2, \ldots, s_k$ over alphabet Σ and non-negative integers d and L.

Question: Is there a string s of length L and, for all $i = 1, \ldots, k$, a length-L substring s_i' of s_i such that $d_H(s, s_i') \leq d$?

Here $d_H(s, s_i')$ denotes the Hamming distance between s and s_i', i.e., the number of positions in which s and s_i' differ. Following the notation used in [7], m is used to denote the average length of the input strings and n to denote the total size of the problem input.

The optimization version of CLOSEST SUBSTRING asks for the minimum value of the distance parameter d for which the input strings still allow a solution.

Key Results

The classical complexity of CLOSEST SUBSTRING is given by

Theorem 1 ([4, 5]) CLOSEST SUBSTRING is NP-complete, and remains so for the special case of the CLOSEST STRING problem, where the requested solution string s has to be of same length as the input strings. CLOSEST STRING is NP-complete even for the further restriction to a binary alphabet.

The following theorem gives the central statement concerning the problem's approximability:

Theorem 2 ([6]) CLOSEST SUBSTRING (as well as CLOSEST STRING) admit polynomial time approximation schemes (PTAS's), where the objective function is the minimum Hamming distance d.

In its randomized version, the PTAS cited by Theorem 2 computes, with high probability, a solution with Hamming distance $(1 + \epsilon)d_{\mathrm{opt}}$ for an optimum value d_{opt} in $(k^2m)^{O(\log|\Sigma|/\epsilon^4)}$ running time. With additional overhead, this randomized PTAS can be derandomized. A straightforward and efficient factor-2 approximation for CLOSEST STRING is obtained by trying all length-L substrings of one of the input strings.

The following two statements address the problem's parametrized complexity, with respect to both obvious problem parameters d and k:

Theorem 3 ([3]) CLOSEST SUBSTRING *is W[1]-hard with respect to the parameter k, even for binary alphabet.*

Theorem 4 ([7]) CLOSEST SUBSTRING *is W[1]-hard with respect to the parameter d, even for binary alphabet.*

For non-binary alphabet the statement of Theorem 3 has been shown independently by Evans et al. [2]. Theorem 3 and Theorem 4 show that an exact algorithm for CLOSEST SUBSTRING with polynomial running time is unlikely for a constant value of d as well as for a constant value of k, i.e., such an algorithm does not exist unless 3-SAT can be solved in subexponential time.

Theorem 4 also allows additional insights into the problem's approximability: In the PTAS for CLOSEST SUBSTRING, the exponent of the polynomial bounding the running time depends on the approximation factor. These are not "efficient" PTAS's (EPTAS's), i.e., PTAS's with a $f(\epsilon) \cdot n^c$ running time for some function f and some constant c, and therefore are probably not useful in practice. Theorem 4 implies that most likely the PTAS with the $n^{O(1/\epsilon^4)}$ running time presented in [6] cannot be improved to an EPTAS. More precisely, there is no $f(\epsilon) \cdot n^{o(\log 1/\epsilon)}$ time PTAS for CLOSEST SUBSTRING unless 3-SAT can be solved in subexponential time. Moreover, the proof of Theorem 4 also yields.

Theorem 5 ([7]) *There are no $f(d, k) \cdot n^{o(\log d)}$ time and no $g(d, k) \cdot n^{o(\log \log k)}$ exact algorithms solving CLOSEST SUBSTRING for some functions f and g unless 3-SAT can be solved in subexponential time.*

For unbounded alphabet the bounds have been strengthened by showing that Closest Substring has no PTAS with running time $f(\epsilon) \cdot n^{o(1/\epsilon)}$ for any function f unless 3-SAT can be solved in subexponential time [10]. The following statements provide exact algorithms for CLOSEST SUBSTRING with small fixed values of d and k, matching the bounds given in Theorem 5:

Theorem 6 ([7]) CLOSEST SUBSTRING *can be solved in time $f(d) \cdot n^{O(\log d)}$ for some function f, where, more precisely, $f(d) = |\Sigma|^{d(\log d + 2)}$.*

Theorem 7 ([7]) CLOSEST SUBSTRING *can be solved in time $g(d, k) \cdot n^{O(\log \log k)}$ for some function g, where, more precisely, $g(d, k) = (|\Sigma|d)^{O(kd)}$.*

With regard to problem parameter L, CLOSEST SUBSTRING can be trivially solved in $O(|\Sigma|^L \cdot n)$ time by trying all possible strings over alphabet Σ.

Applications

An application of CLOSEST SUBSTRING lies in the analysis of biological sequences. In motif discovery, a goal is to search "signals" common to a set of selected strings representing DNA or protein sequences. One way to represent these signals are approximately preserved substrings occurring in each of the input strings. Employing Hamming distance as a biologically meaningful distance measure results in the problem formulation of CLOSEST SUBSTRING.

For example, Sagot [9] studies motif discovery by solving CLOSEST SUBSTRING (and generalizations thereof) using suffix trees; this approach has a worst-case running time of $O(k^2m \cdot L^d \cdot |\Sigma|^d)$. In the context of motif discovery, also heuristics applicable to CLOSEST SUBSTRING were proposed, e.g., Pevzner and Sze [8] present an algorithm called WINNOWER and Buhler and Tompa [1] use a technique called random projections.

Open Problems

It is open [7] whether the $n^{O(1/\epsilon^4)}$ running time of the approximation scheme presented in [6] can be improved to $n^{O(\log 1/\epsilon)}$, matching the bound derived from Theorem 4.

Cross-References

The following problems are close relatives of CLOSEST SUBSTRING:

- ▶ Closest String and Substring Problems is the special case of CLOSEST SUBSTRING, where the requested solution string s has to be of same length as the input strings.
- Distinguishing Substring Selection is the generalization of CLOSEST SUBSTRING, where a second set of input strings and an additional integer d' are given and where the requested solution string s has – in addition to the requirements posed by CLOSEST SUBSTRING – Hamming distance at least d' with every length-L substring from the second set of strings.
- Consensus Patterns is the problem obtained by replacing, in the definition of CLOSEST SUBSTRING, the maximum of Hamming distances by the sum of Hamming distances. The resulting modified question of CONSENSUS PATTERNS is: Is there a string s of length L with

$$\sum_{i=1,\dots,m} d_H(s, s'_i) \le d?$$

CONSENSUS PATTERNS is the special case of SUBSTRING PARSIMONY in which the phylogenetic tree provided in the definition of SUBSTRING PARSIMONY is a star phylogeny.

Recommended Reading

1. Buhler J, Tompa M (2002) Finding motifs using random projections. J Comput Biol 9(2):225–242
2. Evans PA, Smith AD, Wareham HT (2003) On the complexity of finding common approximate substrings. Theor Comput Sci 306(1–3):407–430
3. Fellows MR, Gramm J, Niedermeier R (2006) On the parameterized intractability of motif search problems. Combinatorica 26(2):141–167
4. Frances M, Litman A (1997) On covering problems of codes. Theor Comput Syst 30:113–119
5. Lanctot JK, Li M, Ma B, Wang S, Zhang L (2003) Distinguishing string search problems. Inf Comput 185:41–55
6. Li M, Ma B, Wang L (2002) On the closest string and substring problems. J ACM 49(2):157–171
7. Marx D (2005) The closest substring problem with small distances. In: Proceedings of the 46th FOCS. IEEE Press, pp 63–72
8. Pevzner PA, Sze SH (2000) Combinatorial approaches to finding subtle signals in DNA sequences. In: Proceedings of 8th ISMB. AAAI Press, pp 269–278
9. Sagot MF (1998) Spelling approximate repeated or common motifs using a suffix tree. In: Proceedings of the 3rd LATIN. LNCS, vol 1380. Springer, pp 111–127
10. Wang J, Huang M, Cheng J (2007) A lower bound on approximation algorithms for the closest substring problem. In: Proceedings of the COCOA 2007. LNCS, vol 4616, pp 291–300

Clustered Graph Drawing

Fabrizio Frati
School of Information Technologies, The University of Sydney, Sydney, NSW, Australia
Engineering Department, Roma Tre University, Rome, Italy

Keywords

Clustered graph; Convex drawings; Graph drawing; Planarity testing; Straight-line drawings

Years and Authors of Summarized Original Work

1995; Feng, Cohen, Eades

Problem Definition

A *clustered graph* $C(G, T)$ consists of a graph G, called *underlying graph*, and of a rooted tree T, called *inclusion tree*. The leaves of T are

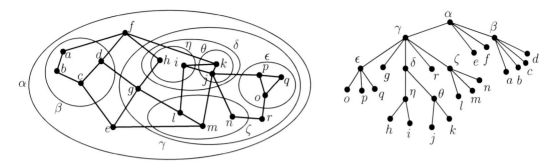

Clustered Graph Drawing, Fig. 1 A clustered graph $C(G, T)$ (*left*) and its inclusion tree (*right*)

Clustered Graph Drawing, Fig. 2 Types of crossings in a drawing of a clustered graph

the vertices of G; each internal node μ of T represents a *cluster*, that is, the set of vertices of G that are the leaves of the subtree of T rooted at μ. Figure 1 shows a clustered graph and its inclusion tree.

Clustered graphs are widely used in applications where it is needed at the same time to represent relationships between entities and to group entities with semantic affinities. For example, in the Internet network, the routers and the links between them are the vertices and edges of a graph, respectively; geographically close routers are grouped into areas that are hence associated with clusters of vertices. In turn, areas are grouped into autonomous systems that are hence associated with clusters of vertices.

Visualizing clustered graphs is a difficult problem, due to the simultaneous need for a readable drawing of the underlying structure and of the clustering relationship. As for the visualization of graphs, the most important aesthetic criterion for the readability of a drawing of a clustered graph is the *planarity*, whose definition needs a refinement in the context of clustered graphs, in order to deal with the clustering structure.

In a *drawing* of a clustered graph $C(G, T)$, vertices and edges of G are drawn as points and open curves, respectively, and each cluster μ is represented by a simple closed region R_μ containing all and only the vertices of μ. A drawing

of C can have three types of crossings. *Edge-edge crossings* are crossings between edges of G (see Fig. 2, left). Consider an edge e of G and a cluster μ in T. If e intersects the boundary of R_μ more than once, we have an *edge-region crossing* (see Fig. 2, middle). Finally, consider two clusters μ and ν in T; if the boundary of R_μ intersects the boundary of R_ν, we have a *region-region crossing* (see Fig. 2, right). A drawing of a clustered graph is *c-planar* (short for *clustered planar*) if it does not have any edge-edge, edge-region, or region-region crossing. A clustered graph is *c-planar* if it admits a c-planar drawing. A drawing of a clustered graph is *straight line* if each edge is represented by a straight-line segment; also, it is *convex* if each cluster is represented by a convex region.

The notion of c-planarity was first introduced by Feng, Cohen, and Eades in 1995 [10, 11]. The graph drawing community has subsequently adopted this definition as a standard, and the topological and geometric properties of c-planar drawings of clustered graphs have been investigated in tens of papers. The two main questions raised by Feng, Cohen, and Eades were the following.

Problem 1 (C-Planarity Testing)

QUESTION: What's the time complexity of testing the c-planarity of a clustered graph?

Problem 2 (Straight-Line Convex C-Planar Drawability)

QUESTION: Does every c-planar-clustered graph admit a straight-line convex c-planar drawing?

Key Results

Almost 20 years after the publication of the seminal papers by Feng et al. [10, 11], a solution for Problem 1 remains an elusive goal, arguably the most intriguing and well-studied algorithmic problem in the graph drawing research area (see, e.g., [3, 4, 7, 12, 13, 15–17]).

Polynomial-time algorithms have been presented to test the c-planarity of a large number of classes of clustered graphs. A particular attention has been devoted to *c-connected* clustered graphs that are clustered graphs $C(G, T)$ such that each cluster $\mu \in T$ induces a connected component G_μ of G. The following theorem reveals the importance of c-connected clustered graphs.

Theorem 1 (Feng, Cohen, and Eades [11]) *A clustered graph is c-planar if and only if it is a subgraph of a c-planar c-connected clustered graph.*

Feng, Cohen, and Eades provided in [11] a nice and simple quadratic-time testing algorithm, which is described in the following.

Theorem 2 (Feng, Cohen, and Eades [11]) *The c-planarity of an n-vertex c-connected clustered graph can be tested in $O(n^2)$ time.*

The starting point of Feng et al. result is a characterization of c-planar drawings.

Theorem 3 (Feng, Cohen, and Eades [11]) *A drawing of a c-connected clustered graph $C(G, T)$ is c-planar if and only if it is planar, and, for each cluster μ, all the vertices and edges of $G - G_\mu$ are in the outer face of the drawing of G_μ.*

The algorithm of Feng et al. [11] performs a bottom-up traversal of T.

When a node $\mu \in T$ is considered, the algorithm tests whether a drawing of G_μ exists such that (P1) for each descendant v of μ, all the vertices and edges of $G_\mu - G_v$ are in the outer face of the drawing of G_v in Γ_μ and (P2) all the vertices of G_μ having neighbors in $G - G_\mu$ are incident to the outer face of G_μ in Γ_μ. Feng et al. show how a PQ-tree P_μ [2] can be used to efficiently represent all the (possibly exponentially many) orderings in which the edges incident to μ can cross the boundary of R_μ in any planar drawing Γ_μ of G_μ satisfying properties P1 and P2 (see Fig. 3, left and middle).

PQ-tree P_μ can be easily computed for each leaf $\mu \in T$. Consider an internal node $\mu \in T$ and assume that PQ-trees $P_{\mu_1}, \ldots, P_{\mu_k}$ have been associated to the children $\mu_1, \ldots, \mu_k$ of μ. *Representative graphs* $H_{\mu_1}, \ldots, H_{\mu_k}$ are constructed from $P_{\mu_1}, \ldots, P_{\mu_k}$; the embeddings of H_{μ_i} are in bijection with the embeddings of G_{μ_i} satisfying properties P1 and P2 (see Fig. 3, left and right). Then, a graph G'_μ is constructed composed of $H_{\mu_1}, \ldots, H_{\mu_k}$, of a dummy vertex v_μ, and of length-2 paths connecting v_μ with every vertex of $H_{\mu_1}, \ldots, H_{\mu_k}$ that has a neighbor in $G - G_\mu$. Feng et al. argue that the embeddings of G_μ satisfying properties P1 and P2 are in bijection with the embeddings of G'_μ in which v_μ is incident to the outer face. Hence, a planarity testing for G'_μ is performed. This allows to determine P_μ, thus allowing the visit of T to go on.

If no planarity test fails, the algorithm completes the visit of T. Top-down traversing T and fixing an embedding for the PQ-tree associated to each node of T determines a c-planar drawing of $C(G, T)$.

Involved linear-time algorithms to test the c-planarity of c-connected clustered graphs are known nowadays [5, 6]. The algorithm in [5] relies on a structural characterization of the c-planarity of a c-connected clustered graph $C(G, T)$ based on the decomposition of G into triconnected components. The characterization allows one to test in linear time the c-planarity of $C(G, T)$ via a bottom-up visit of the SPQR-tree [8] of G, which is a data structure efficiently representing the planar embeddings of G.

Problem 1 is fundamental for the graph drawing research area. However, no less importance has to be attributed to the task of designing

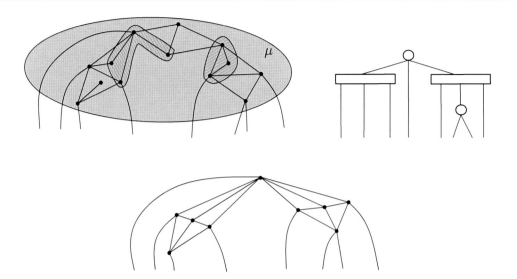

Clustered Graph Drawing, Fig. 3 *Left*: Graph G_μ and the edges incident to μ. *Middle*: The PQ-tree representing the possible orderings in which the edges incident to μ can cross the boundary of R_μ in a planar drawing of G_μ satisfying properties P1 and P2. *Right*: The representative graph H_μ for G_μ

algorithms for constructing *geometric representations* of clustered graphs. The milestones in this research direction have been established by Feng et al., who provided in [10] a full answer to Problem 2.

Theorem 4 (Feng, Cohen, and Eades [10]) *Every c-planar clustered graph admits a straight-line convex c-planar drawing.*

The proof of Theorem 4 relies on a positive answer for the following question: Does every *planar hierarchical graph* admit a *planar straight-line hierarchical drawing*? A *hierarchical graph* is a graph with an assignment of its vertices to k layers $l_1, \ldots, l_k$. A hierarchical drawing maps each vertex assigned to layer l_i to a point on the horizontal line $y = i$ and each edge to a y-monotone curve between the corresponding endpoints. A hierarchical graph is *planar* if it admits a planar hierarchical drawing. Feng et al. [10] showed an algorithm to construct a planar straight-line hierarchical drawing of any planar hierarchical graph H. Their algorithm splits H into some subgraphs, inductively constructs planar straight-line hierarchical drawings of such subgraphs, and glues these drawings together to obtain a planar straight-line hierarchical drawing of H.

Feng et al. also showed how the result on hierarchical graphs leads to a proof of Theorem 4, namely:

1. Starting from any c-planar clustered graph $C(G, T)$, construct a hierarchical graph H by assigning the vertices of G to n layers, in the same order as in an st-numbering of G in which vertices of the same cluster are numbered consecutively.

2. Construct a planar straight-line hierarchical drawing Γ_H of H.

3. Construct a straight-line convex c-planar drawing of $C(G, T)$ starting from Γ_H by drawing each cluster as a convex region slightly surrounding the convex hull of its vertices.

Angelini, Frati, and Kaufmann [1] recently strengthened Theorem 4 by proving that every c-planar clustered graph admits a straight-line c-planar drawing in which each cluster is represented by a scaled copy of an arbitrary convex shape.

Hong and Nagamochi [14] studied straight-line convex c-planar drawings in which the faces of the underlying graph are delimited by convex polygons. They proved that a c-connected clustered graph admits such a drawing if and only if it is c-planar, completely connected (i.e., for every cluster μ, both G_μ and $G - G_\mu$ are connected), and internally triconnected (i.e., for every separation pair $\{u, v\}$, vertices u and v are incident to the outer face of G and each connected component of $G - \{u, v\}$ contains vertices incident to the outer face of G).

The drawings constructed by the algorithm in [10] use real coordinates. Hence, when displaying these drawings on a screen with a finite resolution rule, exponential area might be required for the visualization. This drawback is however unavoidable. Namely, Feng et al. proved in [10] that there exist clustered graphs requiring exponential area in *any* straight-line convex c-planar drawing with a finite resolution rule. This harshly differentiates c-planar clustered graphs from planar graphs, as straight-line convex planar drawings of planar graphs can be constructed in polynomial area.

Theorem 5 (Feng, Cohen, and Eades [10])
There exist n-vertex c-planar clustered graphs requiring $2^{\Omega(n)}$ area in any straight-line convex c-planar drawing.

The proof of Theorem 5 adapts techniques introduced by Di Battista et al. [9] to prove area lower bounds for *straight-line upward planar drawings* of *directed graphs*.

Open Problems

After almost 20 years since it was first posed by Feng, Cohen, and Eades, Problem 1 still represents a terrific challenge for researchers working in graph drawing.

A key result of Feng, Cohen, and Eades [11] shows that testing the c-planarity of a clustered graph $C(G, T)$ is a polynomial-time solvable problem if $C(G, T)$ is c-connected – see Theorem 2. Moreover, a clustered graph is c-planar if and only if it is a subgraph of a c-planar c-connected clustered graph – see Theorem 1. Hence, the core of testing the c-planarity of a non-c-connected clustered graph $C(G, T)$ is an

augmentation problem, asking whether $C(G, T)$ can be augmented to a c-connected c-planar clustered graph $C'(G', T)$ by inserting edges in G.

This augmentation problem seems far from being solved. A particular attention [3,4,7,15,16] has been devoted to the case in which a planar embedding for G is prescribed as part of the input. In this case, edges might only be inserted inside faces of G, in order to guarantee the planarity of G'. Thus, the problem becomes equivalent to the one selecting a set S of edges into a set M of topological embedded multigraphs, where each cluster μ defines a multigraph M_μ in M consisting of all the edges that can be inserted inside faces of G in order to connect distinct connected components of G_μ. Then the edges in S are those that are selected to augment G to G' – hence no two edges in S are allowed to cross. Even in this prescribed-embedding version, only partial results are known. For example, polynomial-time algorithms to test the c-planarity of $C(G, T)$ are known if the faces of G have at most five incident vertices [7], or if each cluster induces at most two connected components [15], or if each cluster has at most two incident vertices on each face of G [3].

Cross-References

► Convex Graph Drawing
► Upward Graph Drawing

Recommended Reading

1. Angelini P, Frati F, Kaufmann M (2011) Straight-line rectangular drawings of clustered graphs. Discret Comput Geom 45(1):88–140
2. Booth KS, Lueker GS (1976) Testing for the consecutive ones property, interval graphs, and graph planarity using PQ-tree algorithms. J Comput Syst Sci 13(3):335–379
3. Chimani M, Di Battista G, Frati F, Klein K (2014) Advances on testing c-planarity of embedded flat clustered graphs. In: Graph drawing (GD '14), Würzburg, pp 416–427
4. Cortese PF, Di Battista G, Patrignani M, Pizzonia M (2005) Clustering cycles into cycles of

clusters. J Graph Algorithms Appl 9(3):391–413. doi:10.7155/jgaa.00115

5. Cortese PF, Di Battista G, Frati F, Patrignani M, Pizzonia M (2008) C-planarity of c-connected clustered graphs. J Graph Algorithms Appl 12(2):225–262

6. Dahlhaus E (1998) A linear time algorithm to recognize clustered graphs and its parallelization. In: Lucchesi CL, Moura AV (eds) Latin American theoretical informatics (LATIN '98), Campinas. LNCS, vol 1380. Springer, pp 239–248

7. Di Battista G, Frati F (2009) Efficient c-planarity testing for embedded flat clustered graphs with small faces. J Graph Algorithms Appl 13(3):349–378. Special issue from GD '07

8. Di Battista G, Tamassia R (1996) On-line planarity testing. SIAM J Comput 25:956–997

9. Di Battista G, Tamassia R, Tollis IG (1992) Area requirement and symmetry display of planar upward drawings. Discret Comput Geom 7: 381–401

10. Feng Q, Cohen RF, Eades P (1995) How to draw a planar clustered graph. In: Du D, Li M (eds) Computing and combinatorics conference (COCOON '95), Xi'an. LNCS, vol 959. Springer, pp 21–30

11. Feng Q, Cohen RF, Eades P (1995) Planarity for clustered graphs. In: Spirakis P (ed) European symposium on algorithms (ESA '95), Corfu. LNCS, vol 979. Springer, pp 213–226

12. Goodrich MT, Lueker GS, Sun JZ (2006) C-planarity of extrovert clustered graphs. In: Healy P, Nikolov N (eds) International symposium on graph drawing (GD '05), Limerick. LNCS, vol 3843. Springer, pp 211–222

13. Gutwenger C, Jünger M, Leipert S, Mutzel P, Percan M, Weiskircher R (2002) Advances in c-planarity testing of clustered graphs. In: Goodrich MT, Kobourov SG (eds) International symposium on graph drawing (GD '02), Irvine. LNCS, vol 2528. Springer, pp 220–235

14. Hong SH, Nagamochi H (2010) Convex drawings of hierarchical planar graphs and clustered planar graphs. J Discret Algorithms 8(3): 282–295

15. Jelínek V, Jelínková E, Kratochvíl J, Lidický B (2009) Clustered planarity: embedded clustered graphs with two-component clusters. In: Tollis IG, Patrignani M (eds) Graph drawing (GD '08), Heraklion. LNCS, vol 5417, pp 121–132. doi:10.1007/978-3-642-00219-9_13

16. Jelínková E, Kára J, Kratochvíl J, Pergel M, Suchý O, Vyskocil T (2009) Clustered planarity: small clusters in cycles and Eulerian graphs. J Graph Algorithms Appl 13(3):379–422

17. Schaefer M (2013) Toward a theory of planarity: Hanani-Tutte and planarity variants. J Graph Algorithms Appl 17(4):367–440

Clustering Under Stability Assumptions

Pranjal Awasthi
Department of Computer Science, Princeton University, Princeton, NJ, USA
Department of Electrical Engineering, Indian Institute of Technology Madras, Chennai, Tamilnadu, India

Keywords

Approximation stability; Clustering; k-median; k-means; Non-worst-case analysis; Separability

Years and Authors of Summarized Original Work

2006; Ostrovsky, Rabani, Schulman, Swami
2009; Balcan, Blum, Gupta
2010; Bilu, Linial
2010; Awasthi, Blum, Sheffet
2010; Kumar, Kannan
2011; Awasthi, Blum, Sheffet
2012; Balcan, Liang

Problem Definition

The problem of clustering consists of partitioning a set of objects such as images, text documents, etc. into groups of related items. The information available to the clustering algorithm consists of pairwise similarity information between objects. One of the most popular approaches to clustering is to map the objects into data points in a metric space, define an objective function over the data points, and find a partitioning which achieves the optimal solution, or an approximately optimal solution to the given objective function. In this entry, we will focus on two of the most widely studied objective functions for clustering: the k-median objective and the k-means objective.

In k-median clustering, the input is a set P of n points in a metric space (X, d), where $d(\cdot)$ is the distance function. The objective is to find k center points $c_1, c_2, \cdots c_k$. The clustering is then formed by assigning each data point to the closest center point. If a point x is assigned to center $c(x)$, then the cost incurred is $d(x, c(x))$. The goal is to find center points and a partitioning of the data so as to minimize the total cost $\Phi = \min_{c_1, c_2, \dots c_k} \sum_x \min_i d(x, c_i)$. This objective is closely related to the well-studied facility location problem [1, 9] in theoretical computer science.

Similarly, for the k-means objective, the goal is to find k center points. However, the cost incurred by a point x is the distance squared to its center. Hence, the goal is to minimize $\Phi = \min_{c_1, c_2, \dots c_k} \sum_x \min_i d^2(x, c_i)$. A special case of the k-means objective which is of particular interest is the Euclidean k-means problem where the points are in $\Re^m$ and the distance function is the squared Euclidean distance. Again, the goal is to choose k center points and assign each point to the closest center while minimizing the total cost. However, unlike k-median and k-means in metric spaces, the center points do not necessarily have to belong to the data set P and can be arbitrarily chosen from $\Re^m$. Unfortunately, optimizing both these objectives turns out to be NP-hard. Hence, a lot of the work in the theoretical computer science community focuses on designing good approximation algorithms for these problems[1, 8–10, 12] with formal guarantees on worst-case instances.

However, in most practical scenarios, the clustering instances which one encounters are not worst case but instead have additional structure/stability associated with them. In such cases, it is natural to ask if one can abstract out this structure in the form of a stability notion, formally study it, and exploit this additional structure in order to obtain optimal or nearly optimal solutions and bypass NP-hardness which only applies to worst-case instances. This modern take on clustering research has, in recent years, produced new insights and deeper understanding of what we know about clustering. In this entry, we will survey some key results on clustering under stability assumptions.

Key Results

ϵ-separability:

This notion of stability was proposed by Ostrovsky et al. [15]. The motivation comes from the fact that in practice, when solving a clustering instance, one typically has to decide how many clusters to partition the data into, i.e., the value of k. If the k-means objective is the underlying criteria being used to judge the quality of a clustering, and the optimal $(k-1)$-means clustering is comparable to the optimal k-means clustering; then, one can in principle also use $(k-1)$ clusters to describe the data set. Hence, the particular clustering instance is not well behaved or not stable. In fact this particular method is a very popular heuristic to find out the number of hidden clusters in the data set suggesting that real-world instances have this property.

Definition 1 (ϵ-Separability) Given an instance of Euclidean k-means clustering, let $\text{OPT}(k)$ denote the cost of the optimal k-means solution. We can also decompose $\text{OPT}(k)$ as $\text{OPT} = \sum_{i=1}^{k} \text{OPT}_i$, where OPT_i denotes the 1-means cost of cluster C_i, i.e., $\sum_{x \in C_i} d(x, c_i)^2$. Such an instance is called ϵ-separable if it satisfies $\text{OPT}(k-1) > \frac{1}{\epsilon^2} \text{OPT}(k)$.

It was shown by Ostrovsky et al. [15] that one can design much better approximation algorithms for such instances. The algorithm is based on over sampling $O(k)$ candidate centers using a distance weighted sampling scheme, followed by a greedy deletion strategy to reduce the k centers without incurring too much increase in the k-means cost.

Theorem 1 ([15]) *There is a polynomial time algorithm which given any ϵ-separable Euclidean k-means instance, outputs a clustering of cost at most $\frac{\text{OPT}}{1-\rho}$ with probability $1 - O((\rho)^{1/4})$ where $\rho = \Theta(\epsilon^2)$.*

$(1 + \alpha, \epsilon)$-Approximation-Stability:

Balcan et al. [5] introduced and analyzed a class of approximation stable instances for which one can find near optimal clusterings in polynomial time. The motivation comes from the fact that for many problems of interest to machine learning, there is an unknown underlying correct "target" clustering. In such cases, the implicit hope when pursuing an objective-based clustering approach (k-means or k-median) is that approximately optimizing the objective function will in fact produce a clustering of low clustering error, i.e., a clustering that is point wise close to the target clustering. Balcan et al. showed that by making this implicit assumption explicit, one can efficiently compute a low-error clustering even in cases when the approximation problem of the objective function is NP-hard!

Definition 2 (($1 + \alpha, \epsilon$)-approximation-stability) Let P be a set of n points residing in a metric space (M, d). Given an objective function Φ (such as k-median, k-means), we say that instance (M, P) *satisfies* $(1 + \alpha, \epsilon)$-*approximation-stability for* Φ if all clusterings C with $\Phi(C) \leq (1 + \alpha) \cdot OPT(k)$ are point-wise ϵ-close to the target clustering $\mathcal{T}$ for (M, P).

Here, the term "target" clustering refers to the ground truth clustering of P which one is trying to approximate. The distance between any two k clusterings C and C^* of n points is measured as $\text{dist}(C, C^*) = \min_{\sigma \in S_k} \frac{1}{n} \sum_{i=1}^{k} |C_i \setminus C^*_{\sigma(i)}|$. Interestingly, this approximation stability condition implies a lot of structure about the problem instance which could be exploited algorithmically. For example, one can show the following:

Theorem 2 ([5]) *If the given instance (M, P) satisfies $(1 + \alpha, \epsilon)$-approximation-stability for the k-median or the k-means objective, then we can efficiently produce a clustering that is $O(\epsilon + \epsilon/\alpha)$-close to the target clustering $\mathcal{T}$.*

As mentioned above, this theorem is valid even for values of α for which getting a $(1 + \alpha)$-approximation to k-median and k-means is NP-hard! The algorithm first creates a graph over data points by connecting points which are within a certain distance threshold. The next step involves iteratively peeling off connected components in the graph and simultaneously de-noising the instance.

Related Notions
The notion of ϵ-separability and $(1 + \alpha, \epsilon)$-approximation-stability are related to each other. For example, Theorem 5.1 in [15] shows that ϵ-separability implies that any near-optimal solution to k-means is $O(\epsilon^2)$-close to the k-means optimal clustering. However, the converse is not necessarily the case; an instance could satisfy approximation-stability without being ϵ-separated. In [6], Balcan et al. present a specific example of points in Euclidean space with $\alpha = 1$. In fact, when k is much larger than $1/\epsilon$, the difference between the two properties can be more substantial.

The notion of separability and approximation stability was generalized in [2] where the authors study a notion of stability called α-*weak deletion stability*. A clustering instance is stable under this notion if in the optimal clustering merging any two clusters into one increases the cost by a multiplicative factor of $(1 + \alpha)$. This definition captures both ϵ-separability and approximation stability in the case of large cluster sizes. Remarkably, [2] show that for such instances of k-median and Euclidean k-means, one can design a $(1 + \epsilon)$ approximation algorithm for any constant $\epsilon > 0$. This leads to immediate improvements over the works of [5] (for the case of large clusters) and of [15]. However, the run time of the resulting algorithm depends polynomially in n and k and exponentially in the parameters $1/\alpha$ and $1/\epsilon$, so the simpler algorithms of [2] and [5] are more suitable for scenarios where one expects the stronger properties to hold.

Kumar and Kannan [11] study a separation condition motivated by the k-means objective and the kind of instances produced by Gaussian and related mixture models. They consider the setting of points in Euclidean space and show that if the projection of any data point onto the line joining the mean of its cluster in the target clustering to the mean of any other cluster of the target is $\Omega(k)$ standard deviations closer to its own mean than the other mean, then they can recover the target clusters in polynomial time. This condition was further analyzed and reduced by work of [3].

Bilu and Linial [7] study clustering instances which are perturbation resilient. An instance is c-perturbation resilient if it has the property that the optimal solution to the objective remains optimal even after bounded perturbations (up to factor c) to the input weight matrix. They give an algorithm for maxcut (which can be viewed as a 2-clustering problem) under the assumption that the optimal solution is stable to (roughly) $O(n^{2/3})$-factor multiplicative perturbations to the edge weights. This was improved by [14]. Awasthi et al. [3] study perturbation resilience for center-based clustering objectives such as k-median and k-means and give an algorithm that finds the optimal solution when the input is stable to only factor-3 perturbations. This factor is improved to $1 + \sqrt{2}$ by [4], who also design algorithms under a relaxed (c, ϵ)-stability to perturbation condition in which the optimal solution need not be identical on the c-perturbed instance, but may change on an ϵ fraction of the points (in this case, the algorithms require $c = 2 + \sqrt{3}$).

For the k-median objective, (c, ϵ)-approximation-stability with respect to $\mathcal{C}^*$ implies (c, ϵ)-stability to perturbations because an optimal solution in a c-perturbed instance is guaranteed to be a c-approximation on the original instance. Similarly, for k-means, (c, ϵ)-stability to perturbations is implied by (c^2, ϵ)-approximation-stability. However, as noted above, the values of c known to lead to efficient clustering in the case of stability to perturbations are larger than for approximation-stability, where any constant $c > 1$ suffices.

In a different direction, one can also consider relaxations of the perturbation-resilience condition. For example, Balcan et al. [4] also consider instances that are "mostly resilient" to c-perturbations; under any c-perturbation of the underlying metric, no more than an ϵ-fraction of the points gets mislabeled under the optimal solution. For sufficiently large constant c and sufficiently small constant ϵ, they present algorithms that get good approximations to the objective under this condition. A different kind of relaxation would be to consider a notion of *resilience to perturbations on average*: a clustering instance whose optimal clustering is likely not to change, assuming the perturbation is *random* from a suitable distribution. Can this weaker notion be used to still achieve positive guarantees?

Finally, the notion of stability can also shed light on practically interesting instances of many other important problems. Can stability assumptions, preferably ones of a mild nature, allow us to bypass NP-hardness results of other problems? One particularly intriguing direction is the problem of Sparsest-Cut, for which no PTAS or constant-approximation algorithm is known, yet a powerful heuristics based on spectral techniques work remarkably well in practice.

Open Problems

The algorithm proposed in [15] for ϵ-separability is a variant of the popular Lloyd's heuristic for k-means [13]. Hence, the result can also be viewed as a characterization of when such heuristics work in practice. It would be interesting to establish weaker sufficient conditions for such heuristics. For instance, is it possible that weak-deletion stability is sufficient for a version of the Lloyd's heuristic to converge to the optimal clustering? Another open direction concerns the notion of perturbation resilience. Can one reduce the perturbation factor c needed for efficient clustering? Alternatively, if one cannot find the *optimal* clustering for small values of c, can one still find a near-optimal clustering, of approximation ratio better than what is possible on worst-case instances?

Recommended Reading

1. Arya V, Garg N, Khandekar R, Meyerson A, Munagala K, Pandit V (2004) Local search heuristics for k-median and facility location problems. SIAM J Comput 33(3):544–562
2. Awasthi P, Blum A, Sheffet O (2010) Stability yields a PTAS for k-median and k-means clustering. In: Proceedings of the 2010 IEEE 51st annual symposium on foundations of computer science, Las Vegas
3. Awasthi P, Blum A, Sheffet O (2012) Center-based clustering under perturbation stability. Inf Process Lett 112(1–2):49–54
4. Balcan M-F, Liang Y (2012) Clustering under perturbation resilience. In: Proceedings of the 39th international colloquium on automata, languages and programming, Warwick
5. Balcan M-F, Blum A, Gupta A (2009) Approximate clustering without the approximation. In: Proceedings of the ACM-SIAM symposium on discrete algorithms, New York
6. Balcan M-F, Blum A, Gupta A (2013) Clustering under approximation stability. J ACM 60:1–34

7. Bilu Y, Linial N (2010) Are stable instances easy? In: Proceedings of the first symposium on innovations in computer science, Beijing

8. Charikar M, Guha S, Tardos E, Shmoy DB (1999) A constant-factor approximation algorithm for the k-median problem. In: Proceedings of the thirty-first annual ACM symposium on theory of computing, Atlanta

9. Jain K, Mahdian M, Saberi A (2002) A new greedy approach for facility location problems. In: Proceedings of the 34th annual ACM symposium on theory of computing, Montreal

10. Kanungo T, Mount DM, Netanyahu NS, Piatko CD, Silverman R, Wu AY (2002) A local search approximation algorithm for k-means clustering. In: Proceedings of the eighteenth annual symposium on computational geometry. ACM, New York

11. Kumar A, Kannan R (2010) Clustering with spectral norm and the k-means algorithm. In: Proceedings of the 51st annual IEEE symposium on foundations of computer science, Las Vegas

12. Kumar A, Sabharwal Y, Sen S (2004) A simple linear time $(1 + \epsilon)$-approximation algorithm for k-means clustering in any dimensions. In: Proceedings of the 45th annual IEEE symposium on foundations of computer science, Washington, DC

13. Lloyd SP (1982) Least squares quantization in PCM. IEEE Trans Inf Theory 28(2):129–137

14. Makarychev K, Makarychev Y, Vijayaraghavan A (2014) Bilu-linial stable instances of max cut and minimum multiway cut. In: SODA, Portland

15. Ostrovsky R, Rabani Y, Schulman L, Swamy C (2006) The effectiveness of lloyd-type methods for the k-means problem. In: Proceedings of the 47th annual IEEE symposium on foundations of computer science, Berkeley

Color Coding

Noga Alon[1], Raphael Yuster[2], and Uri Zwick[3]
[1]Department of Mathematics and Computer Science, Tel-Aviv University, Tel-Aviv, Israel
[2]Department of Mathematics, University of Haifa, Haifa, Israel
[3]Department of Mathematics and Computer Science, Tel-Aviv University, Tel-Aviv, Israel

Keywords

Finding small subgraphs within large graphs

Years and Authors of Summarized Original Work

1995; Alon, Yuster, Zwick

Problem Definition

Color coding [2] is a novel method used for solving, in polynomial time, various subcases of the generally NP-Hard *subgraph isomorphism* problem. The input for the subgraph isomorphism problem is an ordered pair of (possibly directed) graphs (G, H). The output is either a mapping showing that H is isomorphic to a (possibly induced) subgraph of G, or **false** if no such subgraph exists. The subgraph isomorphism problem includes, as special cases, the HAMILTON-PATH, CLIQUE, and INDEPENDENT SET problems, as well as many others. The problem is also interesting when H is *fixed*. The goal, in this case, is to design algorithms whose running times are significantly better than the running time of the naïve algorithm.

Method Description

The color coding method is a randomized method. The vertices of the graph $G = (V, E)$ in which a subgraph isomorphic to $H = (V_H, E_H)$ is sought are randomly colored by $k = |V_H|$ colors. If $|V_H| = O(\log |V|)$, then with a small probability, but only polynomially small (i.e., one over a polynomial), all the vertices of a subgraph of G which is isomorphic to H, if there is such a subgraph, will be colored by distinct colors. Such a subgraph is called *color coded*. The color coding method exploits the fact that, in many cases, it is easier to detect color coded subgraphs than uncolored ones.

Perhaps the simplest interesting subcases of the subgraph isomorphism problem are the following: Given a directed or undirected graph $G = (V, E)$ and a number k, does G contain a simple (directed) path of length k? Does G contain a simple (directed) cycle of length *exactly* k? The following describes a $2^{O(k)} \cdot |E|$

time algorithm that receives as input the graph $G = (V, E)$, a coloring $c: V \rightarrow \{1, \ldots, k\}$ and a vertex $s \in V$, and finds a colorful path of length $k - 1$ that starts at s, if one exists. To find a colorful path of length $k - 1$ in G that starts somewhere, just add a new vertex s' to V, color it with a new color 0 and connect it with edges to all the vertices of V. Now look for a colorful path of length k that starts at s'.

A colorful path of length $k - 1$ that starts at some specified vertex s is found using a dynamic programming approach. Suppose one is already given, for each vertex $v \in V$, the possible sets of colors on colorful paths of length i that connect s and v. Note that there is no need to record all colorful paths connecting s and v. Instead, record the color sets appearing on such paths. For each vertex v there is a collection of at most $\binom{k}{i}$ color sets. Now, inspect every subset C that belongs to the collection of v, and every edge $(v, u) \in E$. If $c(u) \notin C$, add the set $C \cup \{c(u)\}$ to the collection of u that corresponds to colorful paths of length $i + 1$. The graph G contains a colorful path of length $k - 1$ with respect to the coloring c if and only if the final collection, that corresponding to paths of length $k - 1$, of at least one vertex is non-empty. The number of operations performed by the algorithm outlined is at most $O(\sum_{i=0}^{k} i \binom{k}{i} \cdot |E|)$ which is clearly $O(k 2^k \cdot |E|)$.

Derandomization

The randomized algorithms obtained using the color coding method are derandomized with only a small loss in efficiency. All that is needed to derandomize them is a family of colorings of $G = (V, E)$ so that every subset of k vertices of G is assigned distinct colors by at least one of these colorings. Such a family is also called a family of *perfect hash functions* from $\{1, 2, \ldots, |V|\}$ to $\{1, 2, \ldots, k\}$. Such a family is explicitly constructed by combining the methods of [1, 9, 12, 16]. For a derandomization technique yielding a constant factor improvement see [5].

Key Results

Lemma 1 *Let $G = (V, E)$ be a directed or undirected graph and let $c: V \rightarrow \{1, \ldots, k\}$ be a coloring of its vertices with k colors. A colorful path of length $k - 1$ in G, if one exists, can be found in $2^{O(k)} \cdot |E|$ worst-case time.*

Lemma 2 *Let $G = (V, E)$ be a directed or undirected graph and let $c: V \rightarrow \{1, \ldots, k\}$ be a coloring of its vertices with k colors. All pairs of vertices connected by colorful paths of length $k - 1$ in G can be found in either $2^{O(k)} \cdot |V||E|$ or $2^{O(k)} \cdot |V|^{\omega}$ worst-case time (here $\omega < 2.376$ denotes the matrix multiplication exponent).*

Using the above lemmata the following results are obtained.

Theorem 3 *A simple directed or undirected path of length $k - 1$ in a (directed or undirected) graph $G = (V, E)$ that contains such a path can be found in $2^{O(k)} \cdot |V|$ expected time in the undirected case and in $2^{O(k)} \cdot |E|$ expected time in the directed case.*

Theorem 4 *A simple directed or undirected cycle of size k in a (directed or undirected) graph $G = (V, E)$ that contains such a cycle can be found in either $2^{O(k)} \cdot |V||E|$ or $2^{O(k)} \cdot |V|^{\omega}$ expected time.*

A cycle of length k in minor-closed families of graphs can be found, using color coding, even faster (for planar graphs, a slightly faster algorithm appears in [6]).

Theorem 5 *Let C be a non-trivial minor-closed family of graphs and let $k \geq 3$ be a fixed integer. Then, there exists a randomized algorithm that given a graph $G = (V, E)$ from C, finds a C_k (a simple cycle of size k) in G, if one exists, in $O(|V|)$ expected time.*

As mentioned above, all these theorems can be derandomized at the price of a $\log |V|$ factor. The algorithms are also easily to parallelize.

Applications

The initial goal was to obtain efficient algorithms for finding simple paths and cycles in graphs. The color coding method turned out, however, to have a much wider range of applicability. The linear time (i.e., $2^{O(k)} \cdot |E|$ for directed graphs and $2^{O(k)} \cdot |V|$ for undirected graphs) bounds for simple paths apply in fact to any *forest* on k vertices. The $2^{O(k)} \cdot |V|^{\omega}$ bound for simple cycles applies in fact to any *series-parallel* graph on k vertices. More generally, if $G = (V, E)$ contains a subgraph isomorphic to a graph $H = (V_H, E_H)$ whose *tree-width* is at most t, then such a subgraph can be found in $2^{O(k)} \cdot |V|^{t+1}$ expected time, where $k = |V_H|$. This improves an algorithm of Plehn and Voigt [14] that has a running time of $k^{O(k)} \cdot |V|^{t+1}$. As a very special case, it follows that the LOG PATH problem is in P. This resolves in the affirmative a conjecture of Papadimitriou and Yannakakis [13]. The exponential dependence on k in the above bounds is probably unavoidable as the problem is NP-complete if k is part of the input.

The color coding method has been a fruitful method in the study of parametrized algorithms and parametrized complexity [7, 8]. Recently, the method has found interesting applications in computational biology, specifically in detecting signaling pathways within protein interaction networks, see [10, 17, 18, 19].

Open Problems

Several problems, listed below, remain open.

- Is there a polynomial time (deterministic or randomized) algorithm for deciding if a given graph $G = (V, E)$ contains a path of length, say, $\log^2 |V|$? (This is unlikely, as it will imply the existence of an algorithm that decides in time $2^{O(\sqrt{n})}$ whether a given graph on n vertices is Hamiltonian.)

- Can the $\log |V|$ factor appearing in the derandomization be omitted?
- Is the problem of deciding whether a given graph $G = (V, E)$ contains a triangle as difficult as the Boolean multiplication of two $|V| \times |V|$ matrices?

Experimental Results

Results of running the basic algorithm on biological data have been reported in [17, 19].

Cross-References

▶ Approximation Schemes for Planar Graph Problems
▶ Graph Isomorphism
▶ Treewidth of Graphs

Recommended Reading

1. Alon N, Goldreich O, Håstad J, Peralta R (1992) Simple constructions of almost k-wise independent random variables. Random Struct Algorithms 3(3):289–304
2. Alon N, Yuster R, Zwick U (1995) Color coding. J ACM 42:844–856
3. Alon N, Yuster R, Zwick U (1997) Finding and counting given length cycles. Algorithmica 17(3):209–223
4. Björklund A, Husfeldt T (2003) Finding a path of superlogarithmic length. SIAM J Comput 32(6):1395–1402
5. Chen J, Lu S, Sze S, Zhang F (2007) Improved algorithms for path, matching, and packing problems. In: Proceedings of the 18th ACM-SIAM symposium on discrete algorithms (SODA), pp 298–307
6. Eppstein D (1999) Subgraph isomorphism in planar graphs and related problems. J Graph Algorithms Appl 3(3):1–27
7. Fellows MR (2003) New directions and new challenges in algorithm design and complexity, parameterized. In: Lecture notes in computer science, vol 2748, pp 505–519
8. Flum J, Grohe M (2004) The parameterized complexity of counting problems. SIAM J Comput 33(4):892–922
9. Fredman ML, Komlós J, Szemerédi E (1984) Storing a sparse table with $O(1)$ worst case access time. J ACM 31:538–544

10. Hüffner F, Wernicke S, Zichner T (2007) Algorithm engineering for color coding to facilitate signaling pathway detection. In: Proceedings of the 5th Asia-Pacific bioinformatics conference (APBC), pp 277–286

11. Monien B (1985) How to find long paths efficiently. Ann Discret Math 25:239–254

12. Naor J, Naor M (1993) Small-bias probability spaces: efficient constructions and applications. SIAM J Comput 22(4):838–856

13. Papadimitriou CH, Yannakakis M (1996) On limited nondeterminism and the complexity of the V-C dimension. J Comput Syst Sci 53(2):161–170

14. Plehn J, Voigt B (1990) Finding minimally weighted subgraphs. Lect Notes Comput Sci 484:18–29

15. Robertson N, Seymour P (1986) Graph minors. II. Algorithmic aspects of tree-width. J Algorithms 7:309–322

16. Schmidt JP, Siegel A (1990) The spatial complexity of oblivious k-probe hash functions. SIAM J Comput 19(5):775–786

17. Scott J, Ideker T, Karp RM, Sharan R (2006) Efficient algorithms for detecting signaling pathways in protein interaction networks. J Comput Biol 13(2):133–144

18. Sharan R, Ideker T (2006) Modeling cellular machinery through biological network comparison. Nat Biotechnol 24:427–433

19. Shlomi T, Segal D, Ruppin E, Sharan R (2006) QPath: a method for querying pathways in a protein-protein interaction network. BMC Bioinform 7:199

Colouring Non-sparse Random Intersection Graphs

Christoforos L. Raptopoulos
Computer Science Department, University of Geneva, Geneva, Switzerland
Computer Technology Institute and Press "Diophantus", Patras, Greece
Research Academic Computer Technology Institute, Greece and Computer Engineering and Informatics Department, University of Patras, Patras, Greece

Keywords

Martingales; Probabilistic method; Proper coloring; Random intersection graphs; Random hypergraphs

Years and Authors of Summarized Original Work

2009; Nikoletseas, Raptopoulos, Spirakis

Problem Definition

A *proper coloring* of a graph $G = (V, E)$ is an assignment of colors to all vertices in V in such a way that no two adjacent vertices have the same color. A k-*coloring* of G is a coloring that uses k colors. The minimum number of colors that can be used to properly color G is the (*vertex*) *chromatic number* of G and is denoted by $\chi(G)$.

Deciding whether a given graph admits a k-coloring for a given $k \geq 3$ is well known to be NP complete. In particular, it is NP hard to compute the chromatic number [5]. The best known approximation algorithm computes a coloring of size at most within a factor $O\left(\frac{n(\log\log n)^2}{(\log n)^3}\right)$ of the chromatic number [6]. Furthermore, for any constant $\epsilon > 0$, it is NP hard to approximate the chromatic number within a factor $n^{1-\epsilon}$ [14].

The intractability of the vertex coloring problem for arbitrary graphs leads researchers to the study of the problem for appropriately generated random graphs. In the current entry, we consider coloring random instances of the *random intersection graphs model*, which is defined as follows:

Definition 1 (Random Intersection Graph – $\mathcal{G}_{n,m,p}$ [9, 13]) Consider a universe $\mathcal{M} = \{1, 2, \ldots, m\}$ of elements and a set of n vertices V. Assign independently to each vertex $v \in V$ a subset S_v of $\mathcal{M}$, choosing each element $i \in \mathcal{M}$ independently with probability p, and draw an edge between two vertices $v \neq u$ if and only if $S_v \cap S_u \neq \emptyset$. The resulting graph is an instance $G_{n,m,p}$ of the random intersection graphs model.

We will say that a property holds in $\mathcal{G}_{n,m,p}$ *with high probability (whp)* if the probability that a random instance of the $\mathcal{G}_{n,m,p}$ model has the property is at least $1 - o(1)$.

In this model, we will refer to the elements in the universe $\mathcal{M}$ as *labels*. We also denote by L_i the set of vertices that have chosen label $i \in M$.

Given $G_{n,m,p}$, we will refer to $\{L_i, i \in \mathcal{M}\}$ as its *label representation*. Consider the bipartite graph with vertex set $V \cup \mathcal{M}$ and edge set $\{(v,i) : i \in S_v\} = \{(v,i) : v \in L_i\}$. We will refer to this graph as the *bipartite random graph* $B_{n,m,p}$ *associated to* $G_{n,m,p}$. Notice that the associated bipartite graph is uniquely defined by the label representation.

It follows from the definition of the model that the edges in $G_{n,m,p}$ are not independent. This dependence becomes stronger as the number of labels decreases. In fact, the authors in [3] prove the equivalence (measured in terms of total variation distance) of the random intersection graphs model $\mathcal{G}_{n,m,p}$ and the Erdős-Rényi random graphs model $\mathcal{G}_{n,\hat{p}}$, for $\hat{p} = 1 - (1 - p^2)^m$, when $m = n^\alpha, \alpha > 6$. This bound on the number of labels was improved in [12], by showing equivalence of sharp threshold functions among the two models for $\alpha \geq 3$. We note that $1 - (1 - p^2)^m$ is in fact the (unconditioned) probability that a specific edge exists in $G_{n,m,p}$. In view of this equivalence, in this entry, we consider the interesting range of values $m = n^\alpha, \alpha < 1$, where random intersection graphs seem to differ the most from Erdős-Rényi random graphs.

In [1] the authors propose algorithms that whp probability color sparse instances of $G_{n,m,p}$. In particular, for $m = n^\alpha, \alpha > 0$ and $p = o\left(\sqrt{\frac{1}{nm}}\right)$, they show that $G_{n,m,p}$ can be colored optimally. Also, in the case where $m = n^\alpha, \alpha < 1$ and $p = o\left(\frac{1}{m \ln n}\right)$, they show that $\chi(G_{n,m,p}) \sim np$ whp. To do this, they prove that $G_{n,m,p}$ is chordal whp (or equivalently, the associated bipartite graph does not contain cycles), and so a perfect elimination scheme can be used to find a coloring in polynomial time. The range of values we consider here is different than the one needed for the algorithms in [1] to work. In particular, we study coloring $G_{n,m,p}$ for the wider range $mp \leq (1 - \alpha) \ln n$, as well as the denser range $mp \geq \ln^2 n$. We have to note also that the proof techniques used in [1] cannot be used in the range we consider, since the properties that they examine do not hold in our case. Therefore, a completely different approach is required.

Key Results

In this entry, we initially considered the problem of properly coloring almost all vertices in $G_{n,m,p}$. In particular, we proved the following:

Theorem 1 *Let* $m = n^\alpha, \alpha < 1$ *and* $mp \leq \beta \ln n$, *for any constant* $\beta < 1 - \alpha$. *Then a random instance of the random intersection graphs model* $G_{n,m,p}$ *contains a subset of at least* $n - o(n)$ *vertices that can be colored using* np *colors, with probability at least* $1 - e^{-n^{0.99}}$.

Note that the range of values of m, p considered in the above Theorem is quite wider than the one studied in [1]. For the proof, we combine ideas from [4] (see also [7]) and [10]. In particular, we define a Doob martingale as follows: Let $v_1, v_2, \ldots, v_n$ be an arbitrary ordering of the vertices of $G_{n,m,p}$. For $i = 1, 2, \ldots, n$, let B_i be the subgraph of the associated bipartite graph for $G_{n,m,p}$ (namely, $B_{n,m,p}$) induced by $\cup_{j=1}^{i} v_j \bigcup \mathcal{M}$. We denote by H_i the intersection graph whose bipartite graph has vertex set $V \bigcup \mathcal{M}$ and edge set that is exactly as B_i between $\cup_{j=1}^{i} v_j$ and $\mathcal{M}$, whereas every other edge (i.e., the ones between $\cup_{j=i+1}^{n} v_j$ and $\mathcal{M}$) appears independently with probability p.

Let also X denote the size of the largest np-colorable subset of vertices in $G_{n,m,p}$, and let X_i denote the expectation of the largest np-colorable subset in H_i. Notice that X_i is a random variable depending on the overlap between $G_{n,m,p}$ and H_i. Obviously, $X = X_n$ and setting $X_0 = E[X]$, we have $|X_i - X_{i+1}| \leq 1$, for all $i = 1, 2, \ldots, n$. It is straightforward to verify that the sequence $X_0, X_1, \ldots, X_n$ is a Doob martingale, and thus we can use Azuma's inequality to prove concentration of X_n around its mean value. However, the exact value of $E[X]$ is unknown. Nevertheless, we could prove a lower bound on $E[X]$ by providing a lower bound on the probability that X takes sufficiently large values. In particular, we showed that, for any positive constant $\epsilon > 0$, the probability that X takes values at least $(1 - \epsilon)n$ is larger than the upper bound given by Azuma's inequality, implying that $E[X] \geq n - o(n)$.

It is worth noting here that the proof of Theorem 1 can also be used to prove that $\Theta(np)$

colors are enough to color $n - o(n)$ vertices, even in the case where $mp = \beta \ln n$, for any constant $\beta > 0$. However, finding the exact constant multiplying np is technically more difficult. Finally, note that Theorem 1 does not provide any direct information for the chromatic number of $G_{n,m,p}$, because the vertices that remain uncolored could induce a clique in $G_{n,m,p}$ in the worst case.

An Efficient Algorithm

Following our existential result of Theorem 1, we also proposed and analyzed an algorithm CliqueColor for finding a proper coloring of a random instance of $\mathcal{G}_{n,m,p}$, for any $mp \geq \ln^2 n$, where $m = n^\alpha, \alpha < 1$. The algorithm uses information of the label sets assigned to the vertices of $G_{n,m,p}$, and it runs in $O\left(\frac{n^2 m p^2}{\ln n}\right)$ time whp (i.e., polynomial in n and m). In the algorithm, every vertex initially chooses independently uniformly at random a preference in colors from a set C, denoted by $shade(\cdot)$, and every label l chooses a preference in the colors of the vertices in L_l, denoted by $c_l(\cdot)$. Subsequently, the algorithm visits every label clique and fixes the color (according to preference lists) for as many vertices as possible without causing collisions with already-colored vertices. Finally, it finds a proper coloring to the remaining uncolored vertices, using a new set of colors C'. Algorithm CliqueColor is described below:

It is evident that algorithm CliqueColor always finds a proper coloring of $G_{n,m,p}$, but its efficiency depends on the number of colors included in C and C'. The main idea is that if we have enough colors in the initial color set C, then the subgraph $\mathcal{H}$ containing uncolored vertices will have sufficiently small maximum degree $\Delta(\mathcal{H})$, so that we can easily color it using $\Delta(\mathcal{H})$ extra colors. More specifically, we prove the following:

Theorem 2 (Efficiency) *Let $m = n^\alpha, \alpha < 1$ and $mp \geq \ln^2 n, p = o\left(\frac{1}{\sqrt{m}}\right)$. Then algorithm CliqueColor succeeds in finding a proper $\Theta\left(\frac{nmp^2}{\ln n}\right)$-coloring using for $G_{n,m,p}$ in polynomial time whp.*

Algorithm CliqueColor:

Input: An instance $G_{n,m,p}$ of $\mathcal{G}_{n,m,p}$ and its associated bipartite $B_{n,m,p}$.
Output: A proper coloring of $G_{n,m,p}$.

1. for every $v \in V$ choose a color denoted by $shade(v)$ independently, uniformly at random among those in C;
2. for every $l \in \mathcal{M}$, choose a coloring of the vertices in L_l such that, for every color in $\{c \in C : \exists v \in L_l$ with $shade(v) = c\}$, there is exactly one vertex in the set $\{u \in L_l : shade(u) = c\}$ having $c_l(u) = c$, while the rest remain uncolored;
3. set $U = \emptyset$ and $C = \emptyset$;
4. **for** $l = 1$ to m do {
5. color every vertex in $L_l \backslash (U \cup C)$ according to $c_l(\cdot)$ iff there is no collision with the color of a vertex in $L_l \cap C$;
6. include every vertex in L_l colored that way in C and the rest in U; } **enddo**
7. let $\mathcal{H}$ denote the (intersection) subgraph of $G_{n,m,p}$ induced by the vertices in U and let $\Delta(\mathcal{H})$ be its maximum degree;
8. give a proper $\Delta(\mathcal{H})$-coloring of $\mathcal{H}$ using a new set of colors C' of cardinality $\Delta(\mathcal{H})$;
9. **output** a coloring of $G_{n,m,p}$ using $|C \cup C'|$ colors;

It is worth noting that the number of colors used by algorithm CliqueColor in the case $mp \geq \ln^2 n, p = O\left(\frac{1}{\sqrt[4]{m}}\right)$ and $m = n^\alpha, \alpha < 1$ is of the correct order of magnitude (i.e., it is optimal up to constant factors). Indeed, by the concentration of the values of $|S_v|$ around mp for any vertex v with high probability, we can use the results of [11] on the independence number of the *uniform random intersection graphs model* $\mathcal{G}_{n,m,\lambda}$, with $\lambda \sim mp$, to provide a lower bound on the chromatic number. Indeed, it can be easily verified that the independence number of $G_{n,m,\lambda}$ for $\lambda = mp \geq \ln^2 n$ is at most $\Theta\left(\frac{\ln n}{mp^2}\right)$, which implies that the chromatic number of $G_{n,m,\lambda}$ (hence also of the $G_{n,m,p}$ because of the concentration of the values of $|S_v|$) is at least $\Omega\left(\frac{nmp^2}{\ln n}\right)$.

Coloring Random Hypergraphs

The model of random intersection graphs $\mathcal{G}_{n,m,p}$ could also be thought of as generating random hypergraphs. The hypergraphs generated have vertex set V and edge set $\mathcal{M}$. There is a huge amount of literature concerning coloring hypergraphs. However, the question about coloring there seems to be different from the one we considered here. More specifically, a proper coloring of a hypergraph is any assignment of colors to the vertices, so that no monochromatic edge exists. This of course implies that fewer colors than the chromatic number (studied in this entry) are needed in order to achieve this goal.

In $G_{n,m,p}$, the problem of finding a coloring such that no label is monochromatic seems to be quite easier when p is not too small. The proof of the following Theorem is based on the method of conditional expectations (see [2, 8]).

Theorem 3 *Let $G_{n,m,p}$ be a random instance of the model $G_{n,m,p}$, for $p = \omega\left(\frac{\ln m}{n}\right)$ and $m = n^\alpha$, for any fixed $\alpha > 0$. Then with high probability, there is a polynomial time algorithm that finds a k-coloring of the vertices such that no label is monochromatic, for any fixed integer $k \geq 2$.*

Applications

Graph coloring enjoys many practical applications as well as theoretical challenges. Beside the classical types of problems, different limitations can also be set on the graph, or on the way a color is assigned, or even on the color itself. Some of the many applications of graph coloring include modeling scheduling problems, register allocation, pattern matching, etc.

Random intersection graphs are relevant to and capture quite nicely social networking. Indeed, a social network is a structure made of nodes (individuals or organizations) tied by one or more specific types of interdependency, such as values, visions, financial exchange, friends, conflicts, web links, etc. Social network analysis views social relationships in terms of nodes and ties. Nodes are the individual actors within the networks and ties are the relationships between the actors. Other applications include oblivious resource sharing in a (general) distributed setting, efficient and secure communication in sensor networks, interactions of mobile agents traversing the web, etc. Even epidemiological phenomena (like spread of disease) tend to be more accurately

captured by this "interaction-sensitive" random graphs model.

Open Problems

In [1], the authors present (among other results) an algorithm for coloring $G_{n,m,p}$ in the case where $m = n^\alpha, \alpha < 1$ and $mp = o\left(\frac{1}{\log n}\right)$. In contrast, we presented algorithm CliqueColor, which finds a proper coloring of $G_{n,m,p}$ using $\Theta(\chi(G_{n,m,p}))$ whp, in the case $m = n^\alpha, \alpha < 1$ and $mp \geq \ln^2 n$. It remains open whether we can construct efficient algorithms (both in terms of the running time and the number of colors used) for finding proper colorings of $G_{n,m,p}$ for the range of values $\Omega\left(\frac{1}{\log n}\right) \leq mp \leq \ln^2 n$.

Recommended Reading

1. Behrisch M, Taraz A, Ueckerdt M (2009) Colouring random intersection graphs and complex networks. SIAM J Discret Math 23(1):288–299
2. Erdős P, Selfridge J (1973) On a combinatorial game. J Comb Theory (A) 14:298–301
3. Fill JA, Sheinerman ER, Singer-Cohen KB (2000) Random intersection graphs when $m = \omega(n)$: an equivalence theorem relating the evolution of the $G(n, m, p)$ and $G(n, p)$ models. Random Struct Algorithms 16(2):156–176
4. Frieze A (1990) On the independence number of random graphs. Discret Math 81:171–175
5. Garey MR, Johnson DS, Stockmeyer L (1974) Some simplified NP-complete problems. In: Proceedings of the sixth annual ACM symposium on theory of computing, Seattle, pp 47–63. doi:10.1145/800119.803884
6. Halldórsson MM (1993) A still better performance guarantee for approximate graph coloring. Inf Process Lett 45:19–23. doi:10.1016/0020-0190(93)90246-6
7. Łuczak L (2005) The chromatic number of random graphs. Combinatorica 11(1):45–54
8. Molloy M, Reed B (2002) Graph colouring and the probabilistic method. Springer, Berlin/Heidelberg
9. Karoński M, Scheinerman ER, Singer-Cohen KB (1999) On random intersection graphs: the subgraph problem. Comb Probab Comput J 8:131–159
10. Nikoletseas SE, Raptopoulos CL, Spirakis PG (2008) Large independent sets in general random intersection graphs. Theor Comput Sci (TCS) J Spec Issue Glob Comput 406(3):215–224

11. Nikoletseas SE, Raptopoulos CL, Spirakis PG (2009) Combinatorial properties for efficient communication in distributed networks with local interactions. In: Proceedings of the 23rd IEEE international parallel and distributed processing symposium (IPDPS), Rome, pp 1–11

12. Rybarczyk K (2011) Equivalence of a random intersection graph and $G(n, p)$. Random Struct Algorithms 38(1–2):205–234

13. Singer-Cohen KB (1995) Random intersection graphs. PhD thesis, John Hopkins University

14. Zuckerman D (2007) Linear degree extractors and the inapproximability of max clique and chromatic number. Theory Comput 3:103–128. doi:10.4086/toc.2007.v003a006

Combinatorial Gray Code

Frank Ruskey
Department of Computer Science, University of Victoria, Victoria, BC, Canada

Keywords

Combinatorial object; Gray code; Hamilton cycle; Loopless algorithm

Years and Authors of Summarized Original Work

1989; Savage
1993; Pruesse, Ruskey
2012; Ruskey, Sawada, Williams
2012; Sawada, Williams

Problem Definition

In the field of *combinatorial generation*, the goal is to have fast elegant algorithms and code for exhaustively listing the elements of various combinatorial classes such as permutations, combinations, partitions, trees, graphs, and so on. Often it is desirable that successive objects in the listing satisfy some closeness condition, particularly conditions where successive objects differ only by a constant amount. Such listings are called *combinatorial Gray codes*; thus the study of combinatorial Gray codes is an important subfield of combinatorial generation.

There are a variety of applications of combinatorial objects where there is an inherent *closeness operation* that takes one object to another object and vice versa. There is a natural *closeness graph* $G = (V, E)$ that is associated with this setting. The vertices, V, of this graph are the combinatorial objects and edges, E, are between objects that are close. A combinatorial Gray code then becomes a Hamilton path in G and a *Gray cycle* is a Hamilton cycle in G.

The term *Combinatorial Gray Code* seems to have first appeared in print in Joichi, White, and Williamson [4]. An excellent survey up to 1997 was provided by Savage [10], and many examples of combinatorial Gray codes may be found in Knuth [5]. There are literally thousands of papers and patents on Gray codes in general, and although fewer about combinatorial Gray codes, this article can only scratch the surface and will focus on fundamental concepts, generalized settings, and recent results.

The Binary Reflected Gray Codes

The binary reflected Gray code (BRGC) is a well-known circular listing of the bit strings of a fixed length in which successive bit strings differ by a single bit. (The word *bit string* is used in this article instead of "binary string.") Let B_n be the Gray code list of all bit strings of length n. The list is defined by the following simple recursion:

$$B_0 = \varepsilon, \quad \text{and for } n > 0 \quad B_{n+1} = 0B_n, 1B_n^R. \tag{1}$$

In this definition, ε is the empty string, the comma represents concatenation of lists, the symbol x preceding a list indicates that an x is to be prepended to every string in list, and the superscript R indicates that the list is reversed. For example, $B_1 = 0, 1$, $B_1 = 00, 01, 11, 10$, and $B_3 = 000, 001, 011, 010, 110, 111, 101, 100$.

Figure 1 illustrates one of the most useful applications of the BRGC. Imagine a rotating "shaft," like a photocopier drum or a shutoff valve

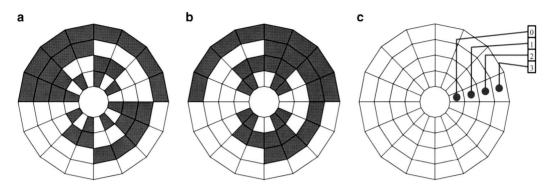

Combinatorial Gray Code, Fig. 1 Applications of the binary reflected Gray code (**a**, **b**, **c**: see explanation in the text)

on a pipeline, in which the amount of rotation is to be determined by reading n "bits" from a sensor (c). If the sensor position is between two adjacent binary strings, then the result is indeterminate; so, for example, if the sensor fell between 0110 and 1100, then it could return either of these, but also 0100 or 1110 could be returned. If the bits are encoded around the shaft as a sequence of bit strings in lexicographic order, as shown in (a), and the sensor falls between 0000 and 1111, then any bit string might be returned! This problem is entirely mitigated by using a Gray code, such as the BRGC (b), since then only one of the two bit strings that lay under the sensor will be returned.

The term "Gray" comes from Frank Gray and engineer at Bell Labs who was issued a patent that uses the BRGC (Pulse code communication, March 17, 1953. U.S. patent no. 2,632,058). However, the BRGC was known much earlier; in particular it occurs in solutions to the Chinese rings puzzle and was used to list diagrams in the I Ching. See Knuth [5] for further historical background.

Surprisingly, new things are yet being proved about the BRGC; e.g., Williams [15] shows that the BRGC is generated by the following iterative "greedy" rule: Starting with the all 0s bit string, flip the rightmost bit that yields a bit string that has not already been generated. There are many other useful Gray code listings of bit strings known, for example, where the bit flips are equally distributed over the indexing positions or where all bit strings of density k (i.e., having k 1s) are listed before any of those of density $k + 2$.

Combinations as Represented by Bit Strings

Many other simple Gray codes can be described using rules in the style of (1). For example, $B_{n,k} = 0B_{n-1,k}, 1B_{n-1,k-1}^R$ gives a Gray code of all bit strings of length n with exactly k 1s. It has the property that successive bit strings differ by the transposition of two (possibly nonadjacent) bits. Observe also that this is the list obtained from the BRGC by deleting all bit strings that do not contain k 1s.

A more restrictive Gray code for combinations is that of Eades and McKay [1]. Here the recursive construction ($1 < k < n$) is $E_{n,k} = 0E_{n-1,k-1}, 10E_{n-2,k-1}^R, 11E_{n-2,k-2}$. This Gray code has the nice property that successive bit strings differ by a transposition of two bits *and* only 0s lie between the two transposed bits. It is worth noting that there is no Gray code for combinations by transpositions in which the transposed bits are always adjacent (unless n is even and k is odd).

Generating Permutations via Plain Changes

The second most well-known combinatorial Gray code lists all $n!$ permutations of $\{1, 2, \ldots, n\}$ in one-line notation in such a way that successive permutations differ by the transposition of two adjacent elements. In its algorithmic form it is usually attributed to Johnson [3] and Trotter [13] although it has been used for centuries by campanologists [5], who refer to it as *plain changes*. Given such a list L_{n-1} for $\{1, 2, \ldots, n - 1\}$, the list L_n can be created by successively sweeping the n back-and-forth through each permutation in L_{n-1}. For example, if L_3 is 123, 132, 312, 321, 231, 213, then the first 8 permutations of L_4 are 1234, 1243, 1423, 4123, 4132, 1432, 1342, 1324. The "weave" below illustrates plane changes for $n = 5$, where the permutations occur as columns and the leftmost column corresponds to the permutation 12345, read top to bottom.

Hamiltonicity

In our BRGC example above, the closeness operation is the flipping of a bit, and the closeness graph is the hypercube. In a Gray code for permutations, a natural closeness operation is the transposing of two adjacent elements; in this case the closeness graph is sometimes called the permutohedron.

Sometimes it happens that the closeness graph G has no Hamilton path. If it is not connected, then there is no hope of finding a Gray code, but if it is connected, then there are several approaches that have proved successful in finding a Gray code in a related graph; the result is a weaker Gray code that permits 2 or 3 applications of the closeness operation. One approach is to consider the prism $G \times e$ [7]. If G is bipartite and Hamiltonian, then the Gray code consists of every other vertex along the Hamilton cycle. As a last resort the result of Sekanina below [12] can be used. An implementation of Sekanina's proof, called prepostorder, is given by Knuth [5].

Theorem 1 (Sekanina) *If G is connected, then G^3 is Hamiltonian.*

At this point the reader may be wondering: "What distinguishes the study of Gray codes from the study of Hamiltonicity of graphs?" The underlying motivation for most combinatorial Gray codes is algorithmic; proving the existence of the Hamilton cycle is only the first step; one then tries to get an efficient implementation according to the criteria discussed in the next section (e.g., the recent "middle-level" result gives an existence proof via a complex construction – but it lacks the nice construction that one would strive for when thinking about a Gray code). Another difference is that the graphs studied usually have a very specific underlying structure; often this structure is recursive and is exploited in the development of efficient algorithms.

The Representation Issue

Many combinatorial structures exhibit a chameleonic nature. The canonical example of this is the objects counted by the Catalan numbers, C_n, of which there are hundreds. For example, they count the well-formed parentheses strings with n left and n right parentheses but also count the binary trees with n nodes. The most natural ways of representing these are very different; well-formed parentheses are naturally represented by bit strings of length $2n$, whereas binary trees are naturally represented using a linked structure with two pointers per cell. Furthermore, the natural closeness conditions are also different; for parentheses, a swap of two parentheses is natural, and for binary trees, the class rotation operation is most natural. When discussing Gray codes it is imperative to know precisely what representation is being used (and then what closeness operation is being used).

Algorithmic Issues

For the vast majority of combinatorial Gray codes, *space* is the main enemy. The first task of the algorithm designer is to make sure that their algorithm uses an amount of space that is a small polynomial of the input size; algorithms that rely on sublists of the objects being listed are doomed even though many Hamiltonicity proofs naïvely lead to such algorithms. For example, an efficient algorithm for generating the BRGC cannot directly use (1) since that would require exponential space.

CAT Algorithms, Loopless Algorithms

From the global point of view, the best possible algorithms are those that output the objects (V)

in time that is proportional to the number of objects ($|V|$). Such algorithms are said to be CAT, standing for constant amortized time.

From a more local point of view, the best possible algorithms are those that output successive objects so that the amount of work between successive objects is constant. Such algorithms are said to be *loopless*, a term introduced by Ehrlich [2]. Both the BRGC and the plain changes algorithm for permutations mentioned above can be implemented as loopless algorithms.

Note that in both of these definitions we ignore the time that it would take to actually output the objects; what is being measured is the amount of data structure change that is occurring. This measure is the correct one to use because in many applications it is only the part of the data structure that changes that is really needed by the application (e.g., to update an objective function).

Key Results

Below are listed some combinatorial Gray codes, focusing on those that are representative, are very general, or are recent breakthroughs.

Numerical Partitions [5, 9]

Objects: All numerical partitions of an integer n.

Representation: Sequence of positive integers $a_1 \geq a_2 \geq \cdots$ such that $a_1 + a_2 \cdots = n$.

Closeness operation: Two partitions a and a' are close if there are two indices i and j such that $a'_i = a_i + 1$ and $a'_j = a_j - 1$.

Efficiency: CAT.

Comments: Results have been extended to the case where all parts are distinct and where the number of parts or the largest part is fixed.

Spanning Trees

Objects: Spanning trees of connected unlabeled graphs.

Representation: List of edges in the graph.

Closeness operation: Successive trees differ by a single edge replacement.

Efficiency: CAT.

References: University of Victoria MSc thesis of Malcolm Smith. Knuth 4A, [5], pages 468–469. Knuth has an implementation of the Smith algorithm on his website mentioned below; see the programs *SPAN.

Basic Words of Antimatroids [7]

Objects: Let $\mathcal{A}$ be a set system over $[n] := \{1, 2, \ldots, n\}$ that (a) is closed under union ($S \cup T \in \mathcal{A}$ for all $S, T \in \mathcal{A}$) and (b) is accessible (for all $S \in \mathcal{A}$ with $S \neq \emptyset$, there is an $x \in S$ such that $S \setminus \{x\} \in \mathcal{A}$). Such a set system $\mathcal{A}$ is an *antimatroid*. Repeated application of (b) starting with the set $[n]$ and ending with $\emptyset$ gives a permutation of $[n]$ called a *basic word*.

Representation: A permutation of $[n]$ in one-line notation.

Closeness operation: Successive permutations differ by the transposition of one or two adjacent elements.

Efficiency: CAT if there is an $O(1)$ "oracle" for determining whether the closeness operation applied to a basic word gives another basic word of $\mathcal{A}$; loopless if the antimatroid is the set of ideals of a poset.

Important special cases: Linear extensions of posets (partially ordered sets), convex shellings of finite point sets, and perfect elimination orderings of chordal graphs. Linear extensions have as special cases permutations (of a multiset), k-ary trees, standard young tableau, alternating permutations, etc.

Additional notes: If G is the cover graph of an antimatroid $\mathcal{A}$ where the sets are ordered by set inclusion, then the prism of G is Hamiltonian. Thus there is a Gray code for the elements of $\mathcal{A}$. No CAT algorithm is known for this Gray code, even in the case where the antimatroid consists of the ideals of a poset.

Words in a Bubble Language [8, 11]

Objects: A language L over alphabet $\{0, 1\}$ is a *bubble language* if it is closed under the operation of changing the rightmost 01 to a 10.

Representation: Bit strings of length n (i.e., elements of $\{0, 1\}^n$).

Closeness operation: Two 01 $\leftrightarrow$ 10 swaps, or (equivalently) one rotation of a prefix.

Important special cases: Combinations, well-formed parentheses, necklaces, and prefix normal words.

Efficiency: CAT if there is an $O(1)$ oracle for determining membership under the operation.

Permutations via σ and τ [16]

Objects: Permutations of $[n]$.

Representation: One-line notation.

Closeness operation: Successive permutations differ by the rotation $\sigma = (1\ 2\ \cdots\ n)$ or the transposition $\tau = (1\ 2)$ as applied to the indices of the representation.

Efficiency: CAT, but has a loopless implementation if only the successive σ or τ generators are output or if the permutation is represented using linked lists.

Comments: This is known as Wilf's (directed) $\sigma - \tau$ problem.

Middle Two Levels of the Boolean Lattice [6]

Objects: All subsets S of $\mathcal{B}_{2n+1}$ of density n or $n + 1$.

Representation: Characteristic bit strings ($b_i = 1$ if and only if $i \in S$).

Closeness operation: Bit strings differ by a transposition of adjacent bits.

Efficiency: Unknown. A good open problem.

Comments: This is famously known as the "middle two-level problem."

URLs to Code and Data Sets

Don Knuth maintains a web page which contains some implementations of combinatorial Gray codes at http://www-cs-faculty.stanford.edu/~uno/programs.html. See, in particular, the programs GRAYSPAN, SPSPAN, GRAYSPSPAN, KODA-RUSKEY, and SPIDERS.

Jeorg Arndt maintains a website (http://www.jjj.de/fxt/) and book with many C programs for generating combinatorial objects, some of which are combinatorial Gray codes. The book may be freely downloaded. Chapter 14, entitled "Gray codes for strings with restrictions", is devoted to combinatorial Gray codes, but they can also be found in other chapters; e.g., 15.2 and 15.3 both contain Gray codes for well-formed parentheses strings.

The "Combinatorial Object Server" at http://www.theory.cs.uvic.ca/~cos/ allows you to produce small lists of combinatorial objects, often in various Gray code orders.

Cross-References

Entries relevant to combinatorial generation (not necessarily Gray codes):

- ▶ Enumeration of Non-crossing Geometric Graphs
- ▶ Enumeration of Paths, Cycles, and Spanning Trees
- ▶ Geometric Object Enumeration
- ▶ Permutation Enumeration
- ▶ Tree Enumeration

Recommended Reading

1. Eades P, McKay B (1984) An algorithm for generating subsets of fixed size with a strong minimal change property. Inf Process Lett 19:131–133
2. Ehrlich G (1973) Loopless algorithms for generating permutations, combinations, and other combinatorial configurations. J ACM 20:500–513
3. Johnson S (1963) Generation of permutations by adjacent transposition. Math Comput 17:282–285
4. Joichi JT, White DE, Williamson SG (1980) Combinatorial Gray codes. SIAM J Comput 9(1):130–141
5. Knuth DE (2011) The art of computer programming. Combinatorial algorithms, vol 4A, Part 1. Addison-Wesley, Upper Saddle River, xvi+883pp. ISBN 0-201-03804-8
6. Mütze T (2014) Proof of the middle levels conjecture. arXiv:1404.4442
7. Pruesse G, Ruskey F (1993) Gray codes from antimatroids. Order 10:239–252
8. Ruskey F, Sawada J, Williams A (2012) Binary bubble languages. J Comb Theory Ser A 119(1):155–169
9. Savage C (1989) Gray code sequences of partitions. J Algorithms 10:577–595
10. Savage C (1997) A survey of combinatorial Gray codes. SIAM Rev 39:605–629
11. Sawada J, Williams A (2012) Efficient oracles for generating binary bubble languages. Electron J Comb 19:Paper 42
12. Sekanina M (1960) Spisy Přírodovědecké. Fakulty University v Brně 412:137–140

13. Trotter H (1962) Algorithm 115: Perm. Commun ACM 5:434–435
14. Wilf HS (1989) Combinatorial algorithms: an update. In: CBMS-NSF regional conference series in applied mathematics. SIAM. http://epubs.siam.org/doi/book/10.1137/1.9781611970166
15. Williams A (2013) The greedy Gray code algorithm. In: Algorithms and data structures symposium (WADS 2013), London, Canada. LNCS, vol 8037, pp 525–536
16. Williams A (2014) Hamiltonicity of the Cayley digraph on the symmetric group generated by $\sigma = (1\ 2\ \cdots\ n)$ and $\tau = (1\ 2)$. arXiv:1307.2549

Combinatorial Optimization and Verification in Self-Assembly

Robert Schweller
Department of Computer Science, University of Texas Rio Grande Valley, Edinburg, TX, USA

Keywords

Algorithmic self-assembly; Tile Assembly Model; Tile complexity; Two-Handed Assembly; Self-assembly

Years and Authors of Summarized Original Work

2002; Adleman, Cheng, Goel, Huang, Kempe, Moisset, Rothemund
2013; Cannon, Demaine, Demaine, Eisenstat, Patitz, Schweller, Summers, Winslow

Problem Definition

Tile Assembly Models

Two of the most studied tile self-assembly models in the literature are the abstract Tile Assembly Model (aTAM) [7] and the Two-Handed Tile Assembly Model (2HAM) [4]. Both models constitute a mathematical model of self-assembly in which system components are four-sided Wang tiles with glue types assigned to each tile edge. Any pair of glue types are assigned some nonneg-

ative interaction strength denoting how strongly the pair of glues bind. The models differ in their rules for growth in that the aTAM allows singleton tiles to attach one at a time to a growing seed, whereas the 2HAM permits any two previously built assemblies to combine given enough affinity for attachment.

In more detail, an aTAM system is an ordered triplet (T, τ, σ) consisting of a set of tiles T, a positive integer threshold parameter τ called the system's *temperature*, and a special tile $\sigma \in T$ denoted as the *seed* tile. Assembly proceeds by attaching copies of tiles from T to a growing seed assembly whenever the placement of a tile on the 2D grid achieves a total strength of attachment from abutting edges, determined by the sum of pairwise glue interactions, that meets or exceeds the temperature parameter τ. An additional twist that is often considered is the ability to specify a relative concentration distribution on the tiles in T. The growth from the initial seed then proceeds randomly with higher concentrated tile types attaching more quickly than lower concentrated types. Even when the final assembly is deterministic, adjusting concentration profiles may substantially alter the expected time to reach the unique terminal state.

The Two-Handed Tile Assembly Model (2HAM) [4] is similar to the aTAM, but removes the concept of a seed tile. Instead, a 2HAM system (T, τ) *produces* a new assembly whenever any two previously built (and potentially large) assemblies may translate together into a new stable assembly based on glue interactions and temperature. The distinction between the 2HAM and the aTAM is that the 2HAM allows large assemblies to grow independently and attach as large, pre-built assemblies, while the aTAM grows through the step-by-step attachment of singleton tiles to a growing seed.

A typical goal in tile self-assembly is to design an *efficient* tile system that uniquely assembles a target shape. Two primary efficiency metrics are (1) the number of distinct tile types used to self-assemble the target shape and (2) the expected time the system takes to self-assemble the target shape. Toward minimizing the number of tiles used to build a shape, the Minimum Tile Set

Problem is considered. Toward the goal of minimizing assembly time, the problem of selecting an optimal concentration distribution over the tiles of a given set is considered in the Tile Concentration Problem. Finally, the computational problem of simply verifying whether a given system correctly and uniquely self-assembles a target shape is considered in the Unique Assembly Verification Problem. Formally, the problems are as follows:

Problem 1 (The Minimum Tile Set Problem [2]) Given a shape, find the tile system with the minimum number of tile types that uniquely self-assembles into this shape.

Problem 2 (The Tile Concentration Problem [2]) Given a shape and a tile system that uniquely produces the given shape, assign concentrations to each tile type so that the expected assembly time for the shape is minimized.

Problem 3 (The Unique Assembly Verification Problem [2, 4]) Given a tile system and an assembly, determine if the tile system uniquely self-assembles into the assembly.

Key Results

Minimum Tile Set Problem

The NP-completeness of the Minimum Tile Set Problem within the aTAM is proven in [2] by a reduction from 3CNF-SAT. The proof is notable in that the polynomial time reduction relies on the polynomial time solution of the Minimum Tile Set Problem for tree shapes, which the authors show is polynomial time solvable. The authors also show that the Minimum Tile Set Problem is polynomial time solvable for $n \times n$ squares by noting that since the optimal solution has at most $O(\log n)$ tile types [7], a brute force search of candidate tile sets finishes in polynomial time as long as the temperatures of the systems under consideration are all a fixed constant. Extending the polynomial time solution to find the minimum tile system over any temperature is achieved in [5].

Theorem 1 *The Minimum Tile Set Problem is NP-complete within the aTAM. For the restricted classes of shapes consisting of squares and trees, the Minimum Tile Set Problem is polynomial time solvable.*

Concentration Optimization

The next result provides an approximation algorithm for the Tile Concentration Problem for a restricted class of aTAM tile system called *partial order* systems. Partial order systems are systems in which a unique assembly is constructed, and for any pair of adjacent tiles in the final assembly which have positive bonding strength, there is a strict order in which the two tiles are placed with respect to each other for all possible assembly sequences. For such systems, a $O(\log n)$-approximation algorithm is presented [2].

Theorem 2 *For any partial order aTAM system (T, τ, σ) that uniquely self-assembles a size-n assembly, there exists a polynomial time $O(\log n)$-approximation algorithm for the Tile Concentration Problem.*

Assembly Verification

The next result provides an important distinction in verification complexity between the aTAM and the 2HAM. In [2] a straightforward quadratic time algorithm for assembly verification is presented. In contrast, the problem is shown to be co-NP-complete in [4] through a reduction from 3CNF-SAT. The hardness holds for a 3D generalization of the 2HAM, but requires only 1 step into the third dimension. To achieve this reduction, the exponentially many candidate 3CNF-SAT solutions are engineered into the order in which the system might grow while maintaining that these candidate paths all collapse into a single polynomial-sized final assembly in the case that no satisfying solution exists. This reduction fundamentally relies on the third dimension and thus leaves open the complexity of 2D verification in the 2HAM.

Theorem 3 *The Unique Assembly Verification Problem is co-NP-complete for the 3D 2HAM and solvable in polynomial time $O(|A|^2 + |A||T|)$ in the aTAM.*

Open Problems

A few open problems in this area are as follows. The Minimum Tile Set Problem has an efficient solution for squares which stems from a logarithmic upper bound on the complexity of assembling such shapes. This holds more generally for thick rectangles, but this ceases to be true when the width of the rectangle becomes sufficiently thin [3]. The complexity of the Minimum Tile Set Problem is open for this class of simple geometric shapes. For the Tile Concentration Problem, an exact solution is conjectured to be #P-hard for partial order systems [2], but this has not been proven. More generally, little is known about the Tile Concentration Problem for non-partial order systems. Another direction within the scope of minimizing assembly time is to consider optimizing over the tiles used, as well as the concentration distribution over the tile set. Some work along these lines has been done with respect to the fast assembly of $n \times n$ squares [1] and the fast implementation of basic arithmetic primitives in self-assembly [6]. In the case of the Unique Assembly Verification Problem, the complexity of the problem for the 2HAM in 2D is still unknown. For the aTAM, it is an open question as to whether the quadratic run time of verification can be improved.

Cross-References

▶ Active Self-Assembly and Molecular Robotics with Nubots
▶ Combinatorial Optimization and Verification in Self-Assembly
▶ Intrinsic Universality in Self-Assembly
▶ Patterned Self-Assembly Tile Set Synthesis
▶ Randomized Self-Assembly
▶ Robustness in Self-Assembly
▶ Self-Assembly at Temperature 1
▶ Self-Assembly of Fractals
▶ Self-Assembly of Squares and Scaled Shapes
▶ Self-Assembly with General Shaped Tiles
▶ Temperature Programming in Self-Assembly

Recommended Reading

1. Adleman L, Cheng Q, Goel A, Huang M-D (2001) Running time and program size for self-assembled squares. In: Proceedings of the thirty-third annual ACM symposium on theory of computing. ACM, New York, pp 740–748
2. Adleman LM, Cheng Q, Goel A, Huang M-DA, Kempe D, de Espanés PM, Rothemund PWK (2002) Combinatorial optimization problems in self-assembly. In: Proceedings of the thirty-fourth annual ACM symposium on theory of computing (STOC '02), Montreal. ACM, New York, pp 23–32
3. Aggarwal G, Cheng Q, Goldwasser MH, Kao M-Y, de Espanés PM, Schweller RT (2005) Complexities for generalized models of self-assembly. SIAM J Comput 34:1493–1515
4. Cannon S, Demaine ED, Demaine ML, Eisenstat S, Patitz MJ, Schweller R, Summers SM, Winslow A (2013) Two hands are better than one (up to constant factors): self-assembly in the 2HAM vs. aTAM. In: Portier N, Wilke T (eds) STACS. LIPIcs, vol 20, Schloss Dagstuhl – Leibniz-Zentrum fuer Informatik, Wadern, pp 172–184
5. Chen H-L, Doty D, Seki S (2011) Program size and temperature in self-assembly. In: ISAAC 2011: proceedings of the 22nd international symposium on algorithms and computation. Lecture notes in computer science, vol 7074. Springer, Berlin/New York, pp 445–453
6. Keenan A, Schweller R, Sherman M, Zhong X (2014) Fast arithmetic in algorithmic self-assembly. In: Unconventional computation and natural computation. Lecture notes in computer science. Springer, Cham, pp 242–253
7. Rothemund PWK, Winfree E (2000) The program-size complexity of self-assembled squares (extended abstract). In: Proceedings of the 32nd ACM symposium on theory of computing, STOC'00. ACM, New York, pp 459–468

Communication Complexity

Amit Chakrabarti
Department of Computer Science, Dartmouth College, Hanover, NH, USA

Keywords

Circuits; Communication; Compression; Data streams; Data structures; Direct sum; Fingerprinting; Information theory; Lower bounds; Randomization

Years and Authors of Summarized Original Work

Problem Definition

Two players – Alice and Bob – are playing a game in which their shared goal is to compute a function $f : \mathcal{X} \times \mathcal{Y} \to \mathcal{Z}$ efficiently. The game starts with Alice holding a value $x \in \mathcal{X}$ and Bob holding $y \in \mathcal{Y}$. They then communicate by sending each other messages according to a predetermined *protocol*, at the end of which they must both arrive at some output $z \in \mathcal{Z}$. The protocol is deemed *correct* if $z = f(x, y)$ for all inputs (x, y). Each message from Alice (resp. Bob) is an arbitrary binary-string-valued function of x (resp. y) and all previous messages received during the protocol's execution. The *cost* of the protocol is the maximum total length of all such messages, over all possible inputs, and is the basic measure of efficiency of the protocol. The central goals in communication complexity [23] are (i) to design protocols with low cost for given problems of interest, and (ii) to prove lower bounds on the cost that must be paid to solve a

given problem. The minimum possible such cost is a natural measure of complexity of the function f and is denoted $D(f)$.

Notably, the "message functions" in the above definition are not required to be efficiently computable. Thus, communication complexity focuses on certain basic information theoretic aspects of computation, abstracting away messier and potentially unmanageable lower-level details. Arguably, it is this aspect of communication complexity that has made it such a successful paradigm for proving lower bounds in a wide range of areas in computer science.

Most work in communication complexity focuses on *randomized* protocols, wherein random coin tosses (equivalently, a single random binary string) may be used to determine the messages sent. These coin tosses may be performed either in private by each player or in public: the resulting protocols are called private coin and public coin, respectively. A randomized protocol is said to compute f with error bounded by $\varepsilon \geq 0$ if, for all inputs (x, y), its output on (x, y) differs from $f(x, y)$ with probability at most ε. With this notion in place, one can then define the ε-error randomized communication complexity of f, denoted $R_\varepsilon(f)$, analogously to the deterministic one. By convention, this notation assumes private coins; the analogous public-coin variant is denoted $R_\varepsilon^{\mathrm{pub}}(f)$. Further, when f has Boolean output, it is convenient to put $R(f) = R_{1/3}(f)$. Clearly, one always has $R^{\mathrm{pub}}(f) \leq R(f) \leq D(f)$.

Consider a probability distribution μ on the input domain $\mathcal{X} \times \mathcal{Y}$. A protocol's error under μ is the probability that it errs when given a random input $(X, Y) \sim \mu$. The ε-error μ-*distributional complexity* of f, denoted $D_\varepsilon^\mu(f)$, is then the minimum cost of a deterministic protocol for f whose error, under μ, is at most ε; an easy averaging argument shows that the restriction of determinism incurs no loss of generality. The fundamental minimax principle of Yao [22] says that $R_\varepsilon(f) = \max_\mu D_\varepsilon^\mu(f)$. In particular, exhibiting a lower bound on $D_\varepsilon^\mu(f)$ for a wisely chosen μ lower bounds $R_\varepsilon(f)$; this is a key lower-bounding technique in the area.

Let Π be a protocol that uses a public random string R as well as private random strings: R_A for Alice, R_B for Bob. Let $\Pi(x, y, R, R_A, R_B)$ denote the *transcript* of conversation between Alice and Bob, on input (x, y). The internal and external information costs of Π with respect to the distribution μ are then defined as follows:

$$\text{icost}(\Pi) = I(\Pi(X, Y, R, R_A, R_B) : X \mid Y, R)$$
$$+ I(\Pi(X, Y, R, R_A, R_B):Y \mid X, R),$$
$$\text{icost}^{\text{ext}}(\Pi) = I(\Pi(X, Y, R, R_A, R_B) : X, Y \mid R),$$

where $(X, Y) \sim \mu$ and I denotes mutual information. These definitions capture the amount of *information* learned by each player about the other's input (in the internal case) and by an external observer about the total input (in the external case) when Π is run on a random μ-distributed input. It is elementary to show that these quantities lower bound the actual communication cost of the protocol. Therefore, the corresponding *information complexity* [3,9] measures – denoted $\text{IC}_\varepsilon^\mu(f)$ and $\text{IC}_\varepsilon^{\mu,\text{ext}}(f)$ – defined as the infima of these costs over all worst-case ε-error protocols for f, naturally lower bound $R_\varepsilon(f)$. This is another important lower-bounding technique.

Key Results

In a number of basic communication problems, Alice's and Bob's inputs are n-bit strings, denoted x and y, respectively, and the goal is to compute a Boolean function $f(x, y)$. We shall denote the ith bit of x as x_i. The bound $D(f) \leq n + 1$ is then trivial, because Alice can always just sent Bob her input, for a cost of n.

The textbook by Kushilevitz and Nisan [14] gives a thorough introduction to the subject and contains detailed proofs of several of the results summarized below. We first present results about specific communication problems, then move on to more abstract results about general problems, and close with a few applications of these results and ideas.

Problem-Specific Results

Equality Testing

This problem is defined by the *equality* function, given by $\text{EQ}_n(x, y) = 1$ if $x = y$ and $\text{EQ}_n(x, y) = 0$ otherwise. This can be solved nontrivially by a randomized protocol wherein Alice sends Bob a *fingerprint* of x, which Bob can compare with the corresponding fingerprint of y generated using the same random seed. Using public coins, a random n-bit string r can be used as a seed to generate the fingerprint $\langle x, r \rangle = \sum_{i=1}^{n} x_i r_i \bmod 2$. One can readily check that this yields $\text{R}^{\text{pub}}(\text{EQ}_n) = O(1)$ and, more generally, $\text{R}_\varepsilon^{\text{pub}}(\text{EQ}_n) = O(\log(1/\varepsilon))$. In the private coin setting, one can use a different kind of fingerprinting, e.g., by treating the bits of x as the coefficients of a degree-n polynomial and evaluating it at a random element of $\mathbb{F}_q$ (for a large enough prime q) to obtain a fingerprint. This idea leads to the bound $\text{R}(\text{EQ}_n) = O(\log n)$.

Randomization is essential for the above results: it can be shown that $\text{D}(\text{EQ}_n) \geq n$. The argument relies on the fundamental *rectangle property* of deterministic protocols, which states that the set of inputs (x, y) that lead to the same transcript must form a combinatorial rectangle inside the input space $\mathcal{X} \times \mathcal{Y}$. This can be proved by induction on the length of the transcript. This rectangle property then implies that if $u \neq v \in \{0, 1\}^n$, then a correct protocol for EQ_n cannot have the same transcript on inputs (u, u) and (v, v); otherwise, it would have the same transcript of (u, v) as well, and therefore err, because $\text{EQ}_n(u, u) \neq \text{EQ}_n(u, v)$. It follows that the protocol must have at least 2^n distinct transcripts, whence one of them must have length at least n.

It can also be shown that the upper bounds above are optimal [14]. The lower bound $\text{R}(\text{EQ}_n) = \Omega(\log n)$ is a consequence of the more general result that $\text{D}(f) = 2^{O(\text{R}(f))}$ for every Boolean function f. The lower bound $\text{R}_\varepsilon^{\text{pub}}(\text{EQ}_n) = \Omega(\log(1/\varepsilon))$ follows from Yao's minimax principle and a version of the above rectangle-property argument.

More refined results can be obtained by considering the expected (rather than worst case) cost of an r-round protocol, i.e., a protocol in which a total of r are sent: in this case, $\text{R}^{\text{pub},(r)}_{\varepsilon}(\text{EQ}_n) = \Theta(\log \log \cdots \log(\min\{n, \log(1/\varepsilon)\}))$ with the outer chain of logarithms iterated $(r-1)$ times. This is tight [7]. Another, incomparable, result is that EQ_n has a zero-error randomized protocol with information cost only $O(1)$, regardless of the joint distribution from which the inputs are drawn [5].

Comparison

This problem is defined by the *greater-than* function, given by $\text{GT}_n(x,y) = 1$ if $x > y$ and $\text{GT}_n(x,y) = 0$ otherwise, where we treat x and y as integers written in binary. Like EQ_n, it has no nontrivial deterministic protocol, for much the same reason. As before, this implies $\text{R}(\text{GT}_n) = \Omega(\log n)$.

In fact the tight bound $\text{R}(\text{GT}_n) = \Theta(\log n)$ holds, but the proof requires a subtle argument. Binary search based on equality testing on substrings of x and y allows one to zoom in, in $O(\log n)$ rounds, on the most significant bit position where x differs from y. If each equality test is allowed $O(1/\log n)$ probability of error, a straightforward union bound gives an $O(1)$ overall error rate, but this uses $\Theta(\log \log n)$ communication per round. The improvement to an overall $O(\log n)$ bound is obtained by preceding each binary search step with an extra "sanity check" equality test on prefixes of x and y and backtracking to the previous level of the binary search if the check fails: this allows one to use only $O(1)$ communication per round.

The bounded-round complexity of GT_n is also fairly well understood. Replacing the binary search above with an $n^{1/r}$-ary search leads to the r-round bound $\text{R}^{(r)}_{\varepsilon}(\text{GT}_n) = O(n^{1/r} \log n)$. A lower bound of $\Omega(n^{1/r}/r^2)$ can be proven by carefully analyzing information cost.

Indexing and Bipartite Pointer Jumping

The *indexing* or *index* problem is defined by the Boolean function $\text{INDEX}_n(x,k) = x_k$, where $x \in \{0,1\}^n$ as usual, but Bob's input $k \in [n]$, where $[n] = \{1, 2, \ldots, n\}$. Straightforward information-theoretic arguments show that the one-round complexity $\text{R}^{(1)}(\text{INDEX}_n) = \Omega(n)$,

where the single message must go from Alice to Bob. Without this restriction, clearly $\text{R}(\text{INDEX}_n) = O(\log n)$. A more delicate result [17] is that in a $(1/3)$-error protocol for INDEX_n, for any $b \in [1, \log n]$, either Bob must send b bits or Alice must send $n/2^{O(b)}$ bits; an easy 2-round protocol shows that this trade-off is optimal. Even more delicate results, involving information cost, are known, and these are useful in certain applications (see below).

The indexing problem illustrates that interaction can improve communication cost exponentially. This can be generalized to show that $r + 1$ rounds can be exponentially more powerful than r rounds. For this, one considers the *bipartite pointer jumping* problem, where Alice and Bob receive functions $f, g : [n] \to [n]$, respectively, and Bob also receives $y \in [n]$. Their goal is to compute $\text{PJ}_{r,n}(f,g,y) = h_r(\cdots h_2(h_1(y)) \cdots) \bmod 2$, where $h_i = f$ for odd i and $h_i = g$ for even i. Notice that $\text{PJ}_{1,n}$ is essentially the same as INDEX_n and that $\text{R}^{(r+1)}(\text{PJ}_{r,n}) = O(r \log n)$. Suitably generalizing the information-theoretic arguments for INDEX_n shows that $\text{R}^{(r)}(\text{PJ}_{r,n}) = \Omega(n/r^2)$.

Inner Product Parity

The Boolean function $\text{IP}_n(x,y) = \langle x, y \rangle$, which is the parity of the inner product $\sum_{i=1}^{n} x_i y_i$, is the most basic *very* hard communication problem: solving it to error $\frac{1}{2} - \delta$ (for constant δ) requires $n - O(\log(1/\delta))$ communication. This is proved by considering the distributional complexity $\text{D}^{\mu}_{\varepsilon}(\text{IP}_n)$, where μ is the uniform distribution and lower bounding it using the *discrepancy method*. Observe that a deterministic protocol Π with cost C induces a partition of the input domain into 2^C combinatorial rectangles, on each of which Π has the same transcript, hence the same output. If Π has error at most $\frac{1}{2} - \delta$ under μ, then the μ-discrepancies of these rectangles – i.e., the differences between the μ-measures of the 0s and 1s within them – must sum up to at least 2δ. Letting $\text{disc}^{\mu}(f)$ denote the maximum over all rectangles R in $\mathcal{X} \times \mathcal{Y}$ of the μ-discrepancy of R, we then obtain $2^C \text{disc}^{\mu}(f) \geq 2\delta$, which allows us to lower bound C if we can upper bound the discrepancy $\text{disc}^{\mu}(f)$.

For the function IP, the matrix $(\mathrm{IP}(x, y))_{x \in \{0,1\}^n, y \in \{0,1\}^n}$ is easily seen to be a Hadamard matrix, whose spectrum is well understood. With a bit of matrix analysis, this enables the discrepancy of IP under a uniform μ to be computed very accurately. This in turn yields the claimed communication complexity lower bound.

Set Disjointness

The problem of determining whether Alice's set $x \subseteq [n]$ is disjoint from Bob's set $y \subseteq [n]$, denoted $\mathrm{DISJ}_n(x, y)$, is, along with its natural generalizations, the most studied and widely useful problem in communication complexity. It is easy to prove, from first principles, the strong lower bound $\mathrm{D}(\mathrm{DISJ}_n(x, y)) = n - o(n)$. Obtaining a similarly strong lower bound for randomized complexity turns out to be quite a challenge, one that has driven a number of theoretical innovations.

The discrepancy method outlined above is provably very weak at lower bounding $\mathrm{R}(\mathrm{DISJ}_n)$. Instead, one considers a refinement called the *corruption* technique: it consists of showing that "large" rectangles in the matrix for DISJ_n cannot come close to consisting purely of 1-inputs (i.e., disjoint pairs (x, y)) but must be corrupted by a "significant" fraction of 0-inputs. On the other hand, a sufficiently low-cost communication protocol for DISJ_n would imply that at least one such large rectangle must exist. The tension between these two facts gives rise to a lower bound on $\mathrm{D}_\varepsilon^\mu(\mathrm{DISJ}_n)$, where μ and ε figure in the quantification of "large" and "significant" above. Following this outline, Babai et al. [2] proved an $\Omega(\sqrt{n})$ lower bound using the uniform distribution. This was then improved by using a certain non-product input distribution, i.e., one where the inputs x and y are correlated – a provably necessary complication – to the optimal $\Omega(n)$, initially via a complicated Kolmogorov complexity technique, but eventually via elementary (though ingenious) combinatorics by Razborov [20]. Subsequently, Bar-Yossef et al. [4] re-proved the $\Omega(n)$ bound via a novel notion of *conditional information complexity*, a proof that has since been reworked

to use the more natural internal information complexity [5].

Disjointness is also interesting and natural as a multiparty problem, where each of t players holds a subset of $[n]$ and they wish to determine if these are disjoint. An important result with many applications (see below) is that under a promise that the sets are pairwise disjoint except perhaps at one element, this requires $\Omega(n/t)$ communication, even if the players communicate via broadcast; this bound is essentially tight. Without this promise, and with only private message channels between the t players, disjointness requires $\Omega(tn)$ communication.

Gap Hamming Distance

This problem is defined by the partial Boolean function on $\{0, 1\}^n \times \{0, 1\}^n$ given by $\mathrm{GHD}_n(x, y) = 0$ if $\|x - y\|_1 \leq \frac{1}{2}n - \sqrt{n}$; $\mathrm{GHD}_n(x, y) = 1$ if $\|x - y\| \geq \frac{1}{2}n + \sqrt{n}$; and $\mathrm{GHD}_n(x, y) = \star$ otherwise. Correctness and error probability for protocols for GHD_n are based only on inputs not mapped to $\star$. After several efforts giving special-case lower bounds, it was eventually proved [8] that $\mathrm{R}(\mathrm{GHD}_n) = \Omega(n)$ and in particular $\mathrm{D}_\varepsilon^\mu(\mathrm{GHD}_n) = \Omega(n)$ with μ being uniform and $\varepsilon = \Theta(1)$ being sufficiently small. This bound provably does not follow from the corruption method because of the presence of large barely-corrupted rectangles in the matrix for GHD_n; instead it was proved using the so-called *smooth corruption* technique [12].

General Complexity-Theoretic Results

There is a vast literature on general results connecting various notions of complexity for communication problems. As before, we survey some highlights. Throughout, we consider a general function $f : \{0, 1\}^n \times \{0, 1\}^n \to \{0, 1\}$.

Determinism vs. Public vs. Private Randomness

A private-coin protocol can be deterministically simulated by direct estimation of the probability of generating each possible transcript. This leads to the relation $\mathrm{R}(f) = \Omega(\log \mathrm{D}(f))$. This separation is the best possible, as witnessed by EQ_n. Further, a public-coin protocol can be restricted to draw its random string – no matter how long – from a fixed set S consisting of $O(n)$ strings, for a constant additive increase in error probability [18]. This implies that it can be simulated using a private coin (which is used only to draw a random element of S) at an additional communication cost of $\log n + O(1)$. Therefore, $R(f) \leq R^{\text{pub}}(f) + \log n + O(1)$. Again the EQ_n function shows that this separation is the best possible.

The Log-Rank Conjecture and Further Matrix Analysis

The rectangle property of communication protocols readily implies that $D(f) \geq \log_2 \text{rk} f$, where $\text{rk} f$ is the rank of the matrix $(f(x, y))_{\{0,1\}^n \times \{0,1\}^n}$. It has long been conjectured that $D(f) = \text{poly}(\log \text{rk} f)$. This famous conjecture remains wide open; the best-known relevant upper bound is $D(f) \leq O(\sqrt{\text{rk} f} \log \text{rk} f)$ [16].

Other, more sophisticated, matrix analysis techniques can be used to establish lower bounds on $R(f)$ by going through the approximate rank and factorization norms. The survey by Lee and Shraibman [15] has a strong focus on such techniques and provides a comprehensive coverage of results.

Direct Sum, Direct Product, and Amortization

Does the complexity of f grow n-fold, or even $\Omega(n)$-fold, if we have to compute f on n independent instances? This is called a direct sum question. Attempts to answer it and its variants, for general, as well as specific functions f have spurred a number of developments. Let f^k denote the (non-Boolean) function that, on input $((x^{(1)}, \ldots, x^{(k)}), (y^{(1)}, \ldots, y^{(k)}))$, outputs $(f(x^{(1)}, y^{(1)}), \ldots, f(x^{(k)}, y^{(k)}))$. Let $R_\varepsilon^k(f^k)$ denote the cost of the best randomized protocol that computes each entry of the k-tuple of of values of f^k up to error ε. This is in contrast to the usual $R_\varepsilon(f^k)$, which is concerned with getting the entire k-tuple correct except with probability ε. An alternate way of posing the direct sum question is to ask how the *amortized* randomized complexity $\overline{R}_\varepsilon(f) = \lim_{k \to \infty} R_\varepsilon^k(f^k)/k$ compares with $R_\varepsilon(f)$.

An early result along these lines shows that $\overline{R}(EQ_n) = O(1)$ [10]. Recalling that $R(EQ_n) = \Omega(\log n)$, this shows that (in the private-coin setting) EQ_n exhibits economy of scale; we say that it does not satisfy the direct sum property. It had long been conjectured that no such economy of scale is possible in a public-coin setting; in fact information complexity rose to prominence as a technique precisely because in the *informational* setting, the direct sum property is easy to prove [9]. Thus, $IC^\mu(f)$, for every distribution μ, lower bounds not just $R(f)$ but also $\overline{R}(f)$. More interestingly, the opposite inequality also holds, so that information and amortized randomized complexities are in fact equal; the proof uses a sophisticated *interactive protocol compression* technique [6] that should be seen as an analog of classical information theoretic results about single-message compression, e.g., via Huffman coding.

Thus, proving a general direct sum theorem for randomized communication is equivalent to compressing a protocol with information cost I down to $O(I)$ bits of communication. However, the best-known compression uses $2^{O(I)}$ bits [5], and this is optimal [11]. The proof of optimality – despite showing that a fully general direct sum theorem is impossible – is such that a slightly weakened direct sum result, such as $R^k(f^k) = \Omega(k) R(f) - O(\log n)$, remains possible and open. Meanwhile, fully general and strong direct sum theorems can be proven by restricting the model to bounded-round communication, or restricting the function to those whose complexity is captured by the smooth corruption bound.

Round Elimination

For problems that are hard only under a limitation on the number of rounds – e.g., GT_n, discussed above – strong bounded-round lower bounds are proved using the round elimination technique. Here, one shows that if an r-round protocol for f starts with a short enough first message, then this message can be eliminated altogether, and the resulting $(r - 1)$-round protocol will solve a "subproblem" of f. Repeating this operation r times results in a 0-round protocol that, hopefully, solves a nontrivial subproblem, giving a contradiction.

To make this useful, one then has to identify a reasonable notion of subproblem. It is especially useful to have this subproblem be a smaller instance of f itself. This does happen in several cases and can be illustrated by looking at GT_n: restricting n-length strings to indices in $[\ell, h]$ and forcing agreement at indices in $[1, \ell - 1]$ shows that GT_n contains $GT_{h-\ell}$ as a subproblem. The proof of the aforementioned lower bound $R^{(r)}(GT_n) = \Omega(n^{1/r}/r^2)$ uses exactly this observation. The pointer jumping lower bound, also mentioned before, proceeds along similar lines.

Applications

Data Stream Algorithms

Consider a data stream algorithm using s bits of working memory and p passes over its input. Splitting the stream into two pieces, giving the first to Alice and the second to Bob, creates a communication problem and a $(2p - 1)$-round protocol for it, using s bits of communication per round. A generalization to multiplayer communication is immediate. These observations [1] allow us to infer several space (or pass/space trade-off) lower bounds for data stream algorithms from suitable communication lower bounds, often after a suitable reduction.

For instance, the problem of approximating the number of distinct items in a stream and its generalization to the problem of approximating frequency moments are almost fully understood, based on lower bounds for EQ_n, GHD_n, $DISJ_n$, and its generalization to multiple players, under the unique intersection restriction noted above. A large number of graph-theoretic problems can be shown to require $\Omega(n)$ space, n being the number of vertices, based on lower bounds for $INDEX_n$, PJ_n and variants, and again $DISJ_n$. For several other data streaming problems – e.g., approximating ℓ_∞ and cascaded norms and approximating a maximum cut or maximum matching – a reduction using an off-the-shelf communication lower bound is not known, but one can still obtain strong space lower bounds by considering a tailor-made communication problem in each case and applying the familiar lower-bounding techniques outlined above.

Data Structures

The cell-probe model of Yao is designed to capture all conceivable data structures on a modern computer: it models the query/update process as a sequence of probes into the entries (memory words) of a table containing the data structure. Focusing on static data structures for the moment, note that a t-probe algorithm using an s-word table with w-bit words directly leads to a $2t$-round communication protocol in which Alice (the querier) sends $(\log_2 s)$-bit messages and Bob (the table holder) sends w-bit messages. Lower bounds trading off t against w and s are therefore implied by suitable *asymmetric* communication lower bounds, where Alice's messages need to be much shorter than Bob's and Alice also has a correspondingly smaller input.

The study of these kinds of lower bounds was systematized by Miltersen et al. [17], who used round elimination as well as corruption-style techniques to obtain cell-probe lower bounds for set membership, predecessor search, range query, and further static data structure problems. Pătraşcu [19] derived an impressive number of cell-probe lower bounds – for problems ranging from union-find to dynamic stabbing and range reporting (in low dimension) to approximate near neighbor searching – by a tree of reductions starting from the *lopsided set disjointness* problem. This latter problem, denoted $LSD_{k,n}$, gives Alice a set $x \subseteq [kn]$ with $|x| \leq k$ and Bob a set $y \subseteq [kn]$. Using information complexity techniques and a direct sum result for the basic $INDEX$ problem, one can use the Alice/Bob trade-off result for $INDEX$ discussed earlier to establish the nearly optimal trade-off that, for each $\delta > 0$, solving $LSD_{k,n}$ to $\frac{1}{3}$ error (say) requires either Alice to send at least $\delta n \log k$ bits or Bob to send $nk^{1-O(\delta)}$ bits.

Circuit Complexity

Early work in circuit complexity had identified certain conjectured communication complexity lower bounds as a route towards strong lower bounds for circuit size and depth and related complexity measures for Boolean formulas and branching programs. Several of these conjectures remain unproven, especially ones involving the *number-on-the-forehead* (NOF) communication model, where the input is "written on the fore-heads" of a large number, t, of players. The resulting high degree of input sharing allows for some rather novel nontrivial protocols, making lower bounds very hard to prove. Nevertheless, the discrepancy technique has been extended to NOF communication, and some of the technically hardest work in communication complexity has gone towards using it effectively for concrete problems, such as set disjointness [21]. While NOF lower bounds strong enough to imply circuit lower bounds remain elusive, certain other communication lower bounds, such as two-party bounds for computing *relations*, have had more success. In particular, monotone circuits for directed and undirected graph connectivity have been shown to require super-logarithmic depth, via the influential idea of Karchmer-Wigderson games [13].

Further Applications

We note in passing that there are plenty more applications of communication complexity than are possible to even outline in this short article. These touch upon such diverse areas as proof complexity; extension complexity for linear and semidefinite programming; AT^2 lower bounds in VLSI design; query complexity in the classical and quantum models; and time complexity on classical Turing machines. Kushilevitz and Nisan [14] remains the best starting point for further reading about these and more applications.

Recommended Reading

1. Alon N, Matias Y, Szegedy M (1999) The space complexity of approximating the frequency moments. J Comput Syst Sci 58(1):137–147. Preliminary version in Proceedings of the 28th annual ACM symposium on the theory of computing, 1996, pp 20–29
2. Babai L, Frankl P, Simon J (1986) Complexity classes in communication complexity theory. In: Proceedings of the 27th annual IEEE symposium on foundations of computer science, Toronto, pp 337–347
3. Barak B, Braverman M, Chen X, Rao A (2010) How to compress interactive communication. In: Proceedings of the 41st annual ACM symposium on the theory of computing, Cambridge, pp 67–76
4. Bar-Yossef Z, Jayram TS, Kumar R, Sivakumar D (2004) An information statistics approach to data stream and communication complexity. J Comput Syst Sci 68(4):702–732
5. Braverman M (2012) Interactive information complexity. In: Proceedings of the 44th annual ACM symposium on the theory of computing, New York, pp 505–524
6. Braverman M, Rao A (2011) Information equals amortized communication. In: Proceedings of the 52nd annual IEEE symposium on foundations of computer science, Palm Springs, pp 748–757
7. Brody J, Chakrabarti A, Kondapally R, Woodruff DP, Yaroslavtsev G (2014) Certifying equality with limited interaction. In: Proceedings of the 18th international workshop on randomization and approximation techniques in computer science, Barcelona, pp 545–581
8. Chakrabarti A, Regev O (2012) An optimal lower bound on the communication complexity of GAP-HAMMING-DISTANCE. SIAM J Comput 41(5):1299–1317. Preliminary version in Proceedings of the 43rd annual ACM symposium on the theory of computing, 2011, pp 51–60
9. Chakrabarti A, Shi Y, Wirth A, Yao AC (2001) Informational complexity and the direct sum problem for simultaneous message complexity. In: Proceedings of the 42nd annual IEEE symposium on foundations of computer science, Las Vegas, pp 270–278
10. Feder T, Kushilevitz E, Naor M, Nisan N (1995) Amortized communication complexity. SIAM J Comput 24(4):736–750. Preliminary version in Proceedings of the 32nd annual IEEE symposium on foundations of computer science, 1991, pp 239–248
11. Ganor A, Kol G, Raz R (2014) Exponential separation of information and communication for boolean functions. Technical report TR14-113, ECCC
12. Jain R, Klauck H (2010) The partition bound for classical communication complexity and query complexity. In: Proceedings of the 25th annual IEEE conference on computational complexity, Cambridge, pp 247–258
13. Karchmer M, Wigderson A (1990) Monotone circuits for connectivity require super-logarithmic depth. SIAM J Discrete Math 3(2):255–265. Preliminary version in Proceedings of the 20th annual ACM symposium on the theory of computing, 1988, pp 539–550
14. Kushilevitz E, Nisan N (1997) Communication complexity. Cambridge University Press, Cambridge

15. Lee T, Shraibman A (2009) Lower bounds in communication complexity. Found Trends Theor Comput Sci 3(4):263–399

16. Lovett S (2014) Communication is bounded by root of rank. In: Proceedings of the 46th annual ACM symposium on the theory of computing, New York, pp 842–846

17. Miltersen PB, Nisan N, Safra S, Wigderson A (1998) On data structures and asymmetric communication complexity. J Comput Syst Sci 57(1):37–49. Preliminary version in Proceedings of the 27th annual ACM symposium on the theory of computing, 1995, pp 103–111

18. Newman I (1991) Private vs. common random bits in communication complexity. Inf Process Lett 39(2):67–71

19. Pătraşcu M (2011) Unifying the landscape of cell-probe lower bounds. SIAM J Comput 40(3):827–847

20. Razborov A (1992) On the distributional complexity of disjointness. Theor Comput Sci 106(2):385–390. Preliminary version in Proceedings of the 17th international colloquium on automata, languages and programming, 1990, pp 249–253

21. Sherstov AA (2014) Communication lower bounds using directional derivatives. J. ACM 61:1–71. Preliminary version in Proceedings of the 45th annual ACM symposium on the theory of computing, 2013, pp 921–930

22. Yao AC (1977) Probabilistic computations: towards a unified measure of complexity. In: Proceedings of the 18th annual IEEE symposium on foundations of computer science, Providence, pp 222–227

23. Yao AC (1979) Some complexity questions related to distributive computing. In: Proceedings of the 11th annual ACM symposium on the theory of computing, Atlanta, pp 209–213

Communication in Ad Hoc Mobile Networks Using Random Walks

Ioannis Chatzigiannakis
Department of Computer Engineering and Informatics, University of Patras and Computer Technology Institute, Patras, Greece

Keywords

Data mules; Delay-tolerant networks; Disconnected ad hoc networks; Message ferrying; Message relays; Sink mobility

Years and Authors of Summarized Original Work

2003; Chatzigiannakis, Nikoletseas, Spirakis

Problem Definition

A mobile ad hoc network is a temporary dynamic interconnection network of wireless mobile nodes without any established infrastructure or centralized administration. A *basic communication problem*, in ad hoc mobile networks, is to send information from a *sender* node, A, to another designated *receiver* node, B. If mobile nodes A and B come within wireless range of each other, then they are able to communicate. However, if they do not, they can communicate if other network nodes of the network are willing to forward their packets. One way to solve this problem is the protocol of notifying every node that the sender A meets and provide it with *all the information* hoping that some of them will eventually meet the receiver B.

> Is there a more efficient technique (other than notifying every node that the sender meets, in the hope that some of them will then eventually meet the receiver) that will effectively solve the communication establishment problem without flooding the network and exhausting the battery and computational power of the nodes?

The problem of communication among mobile nodes is one of the most fundamental problems in ad hoc mobile networks and is at the core of many algorithms, such as for counting the number of nodes, electing a leader, data processing etc. For an exposition of several important problems in ad hoc mobile networks see [13]. The work of Chatzigiannakis, Nikoletseas and Spirakis [5] focuses on wireless mobile networks that are subject to highly dynamic structural changes created by mobility, channel fluctuations and device failures. These changes affect topological connectivity, occur with high frequency and may not be predictable in advance. Therefore, the environment where the nodes move (in three-dimensional space with possible obstacles) as well as the motion that the nodes perform are *input* to any distributed algorithm.

The Motion Space

The space of possible motions of the mobile nodes is combinatorially abstracted by a *motion-graph*, i.e., the detailed geometric characteristics of the motion are neglected. Each mobile node is assumed to have a transmission range represented by a sphere *tr* centered by itself. Any other node inside *tr* can receive any message broadcast by this node. This sphere is approximated by a cube *tc* with volume $V(tc)$, where $V(tc) < V(tr)$. The size of *tc* can be chosen in such a way that its volume $V(tc)$ is the maximum that preserves $V(tc) < V(tr)$, and if a mobile node inside *tc* broadcasts a message, this message is received by any other node in *tc*. Given that the mobile nodes are moving in the space S, S is divided into consecutive cubes of volume $V(tc)$.

Definition 1 The motion graph $G(V, E)$, ($|V| = n, |E| = m$), which corresponds to a quantization of S is constructed in the following way: a vertex $u \in G$ represents a cube of volume $V(tc)$ and an edge $(u, v) \in G$ exists if the corresponding cubes are adjacent.

The number of vertices *n*, actually approximates the ratio between the volume $V(S)$ of space S, and the space occupied by the transmission range of a mobile node $V(tr)$. In the extreme case where $V(S) \approx V(tr)$, the transmission range of the nodes approximates the space where they are moving and $n = 1$. Given the transmission range *tr*, *n* depends linearly on the volume of space S regardless of the choice of *tc*, and $n = O(V(S)/V(tr))$. The ratio $V(S)/V(tr)$ is the *relative motion space size* and is denoted by ρ. Since the edges of *G* represent neighboring polyhedra each vertex is connected with a constant number of neighbors, which yields that $m = \Theta(n)$. In this example where *tc* is a cube, *G* has maximum degree of six and $m \leq 6n$. Thus *motion graph G* is (usually) a *bounded degree graph* as it is derived from a regular graph of small degree by deleting parts of it corresponding

to motion or communication obstacles. Let Δ be the maximum vertex degree of *G*.

The Motion of the Nodes-Adversaries

In the general case, the motions of the nodes are decided by an *oblivious adversary*: The adversary determines motion patterns in any possible way but independently of the distributed algorithm. In other words, the case where some of the nodes are deliberately trying to *maliciously affect* the protocol, e.g., avoid certain nodes, are excluded. This is a pragmatic assumption usually followed by applications. Such kind of motion adversaries are called *restricted motion adversaries*.

For purposes of studying efficiency of distributed algorithms for ad hoc networks *on the average*, the motions of the nodes are modeled by *concurrent and independent random walks*. The assumption that the mobile nodes move randomly, either according to uniformly distributed changes in their directions and velocities or according to the random waypoint mobility model by picking random destinations, has been used extensively by other research.

Key Results

The key idea is to take advantage of the mobile nodes natural movement by exchanging information whenever mobile nodes meet incidentally. It is evident, however, that if the nodes are spread in remote areas and they do not move beyond these areas, there is no way for information to reach them, unless the protocol takes special care of such situations. The work of Chatzigiannakis, Nikoletseas and Spirakis [5] proposes the idea of forcing only a small subset of the deployed nodes to move as per the needs of the protocol; they call this subset of nodes the *support* of the network. Assuming the availability of such nodes, they are used to provide a simple, correct and efficient strategy for communication between any pair of nodes of the network that avoids message flooding.

Let *k* nodes be a predefined set of nodes that become the nodes of the support. These

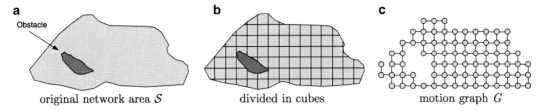

Communication in Ad Hoc Mobile Networks Using Random Walks, Fig. 1 The original network area S (**a**), how it is divided in consecutive cubes of volume $V(tc)$ (**b**) and the resulting motion graph G (**c**)

nodes move randomly and fast enough so that they visit in sufficiently short time the entire motion graph. When some node of the support is within transmission range of a sender, it notifies the sender that it may send its message(s). The messages are then stored "somewhere within the support structure". When a receiver comes within transmission range of a node of the support, the receiver is notified that a message is "waiting" for him and the message is then forwarded to the receiver.

Protocol 1 (The "Snake" Support Motion Co-ordination Protocol) Let $S_0, S_1, \ldots, S_{k-1}$ be the members of the support and let S_0 denote the leader node (possibly elected). The protocol forces S_0 to perform a random walk on the motion graph and each of the other nodes S_i execute the simple protocol "move where S_{i-1} was before". When S_0 is about to move, it sends a message to S_1 that states the new direction of movement. S_1 will change its direction as per instructions of S_0 and will propagate the message to S_2. In analogy, S_i will follow the orders of S_{i-1} after transmitting the new directions to S_{i+1}. Movement orders received by S_i are positioned in a queue Q_i for sequential processing. The very first move of S_i, $\forall i \in \{1, 2, \ldots, k-1\}$ is delayed by a δ period of time.

The purpose of the random walk of the head S_0 is to ensure a *cover*, within some finite time, of the whole graph G without knowledge and memory, other than local, of topology details. This memoryless motion also ensures fairness, low-overhead and inherent robustness to structural changes.

Consider the case where any sender or receiver is allowed a general, unknown motion strategy, but its strategy is provided by a restricted motion adversary. This means that each node not in the support either (a) executes a deterministic motion which either stops at a vertex or cycles forever after some initial part or (b) it executes a stochastic strategy which however is *independent* of the motion of the support. The authors in [5] prove the following correctness and efficiency results. The reader can refer to the excellent book by Aldous and Fill [1] for a nice introduction on Makrov Chains and Random Walks.

Theorem 1 *The support and the "snake" motion coordination protocol guarantee reliable communication between any sender-receiver (A, B) pair in finite time, whose expected value is bounded only by a function of the relative motion space size ρ and does not depend on the number of nodes, and is also independent of how MH_S, MH_R move, provided that the mobile nodes not in the support do not deliberately try to avoid the support.*

Theorem 2 *The expected communication time of the support and the "snake" motion coordination protocol is bounded above by $\Theta(\sqrt{mc})$ when the (optimal) support size $k = \sqrt{2mc}$ and c is $e/(e-1)u$, u being the "separation threshold time" of the random walk on G.*

Theorem 3 *By having the support's head move on a regular spanning subgraph of G, there is an absolute constant $\gamma > 0$ such that the expected meeting time of A (or B) and the support is bounded above by $\gamma n^2/k$. Thus the protocol* *guarantees a total expected communication time of $\Theta(\rho)$, independent of the total number of mobile nodes, and their movement.*

The analysis assumes that the head S_0 moves according to a continuous time random walk of total rate 1 (rate of exit out of a node of G). If S_0 moves ψ *times faster* than the rest of the nodes, all the estimated times, except the inter-support time, will be divided by ψ. Thus the expected total communication time can be made to be as small as $\Theta(\gamma\rho/\sqrt{\psi})$ where γ is an absolute constant. In cases where S_0 can take advantage of the network topology, all the estimated times, except the inter-support time are improved:

Theorem 4 *When the support's head moves on a regular spanning subgraph of G the expected meeting time of A (or B) and the support cannot be less than $(n-1)^2/2m$. Since $m = \Theta(n)$, the lower bound for the expected communication time is $\Theta(n)$. In this sense, the "snake" protocol's expected communication time is optimal, for a support size which is $\Theta(n)$.*

The "on-the-average" analysis of the time-efficiency of the protocol assumes that the motion of the mobile nodes not in the support *is a random walk on the motion graph G*. The random walk of each mobile node is performed independently of the other nodes.

Theorem 5 *The expected communication time of the support and the "snake" motion coordination protocol is bounded above by the formula*

$$E(T) \leq \frac{2}{\lambda_2(G)}\Theta\left(\frac{n}{k}\right) + \Theta(k) \ .$$

The upper bound is minimized when $k = \sqrt{2n/\lambda_2(G)}$, where λ_2 is the second eigenvalue of the motion graph's adjacency matrix.

The way the support nodes move and communicate is robust, in the sense that it can tolerate failures of the support nodes. The types of failures of nodes considered are permanent, i.e., stop failures. Once such a fault happens, the support node of the fault does not participate in the ad

hoc mobile network anymore. A communication protocol is β-*faults tolerant*, if it still allows the members of the network to communicate correctly, under the presence of at most β permanent faults of the nodes in the support ($\beta \geq 1$). [5] shows that:

Theorem 6 *The support and the "snake" motion coordination protocol is 1-fault tolerant.*

Applications

Ad hoc mobile networks are rapidly deployable and self-configuring networks that have important applications in many critical areas such as disaster relief, ambient intelligence, wide area sensing and surveillance. The ability to network *anywhere*, *anytime* enables teleconferencing, home networking, sensor networks, personal area networks, and embedded computing applications [13].

Related Work

The most common way to establish communication is to form paths of intermediate nodes that lie within one another's transmission range and can directly communicate with each other. The mobile nodes act as hosts and routers at the same time in order to propagate packets along these paths. This approach of maintaining a global structure with respect to the temporary network is a difficult problem. Since nodes are moving, the underlying communication graph is changing, and the nodes have to adapt quickly to such changes and reestablish their routes. Busch and Tirthapura [2] provide the first analysis of the performance of some characteristic protocols [8, 13] and show that in some cases they require $\Omega(u^2)$ time, where u is the number of nodes, to stabilize, i.e., be able to provide communication.

The work of Chatzigiannakis, Nikoletseas and Spirakis [5] focuses on networks where topological connectivity is subject to frequent, unpredictable change and studies the problem of efficient data delivery in sparse networks where network partitions can last for a significant period of time. In such cases, it is possible to have a small team of fast moving and versatile vehicles, to implement the support. These vehicles can be cars, motorcycles, helicopters or a collection of independently controlled mobile modules, i.e., robots. This specific approach is inspired by the work of Walter, Welch and Amato [14] that study the problem of motion coordination in distributed systems consisting of such robots, which can connect, disconnect and move around.

The use of mobility to improve performance in ad hoc mobile networks has been considered in different contexts in [6, 9, 11, 15]. The primary objective has been to provide intermittent connectivity in a disconnected ad hoc network. Each solution achieves certain properties of end-to-end connectivity, such as delay and message loss among the nodes of the network. Some of them require long-range wireless transmission, other require that all nodes move pro-actively under the control of the protocol and collaborate so that they meet more often. The *key idea* of forcing only a subset of the nodes to facilitate communication is used in a similar way in [10, 15]. However, [15] focuses in cases where only one node is available. Recently, the application of mobility to the domain of wireless sensor networks has been addressed in [3, 10, 12].

Open Problems

A number of problems related to the work of Chatzigiannakis, Nikoletseas and Spirakis [5] remain open. It is clear that the size of the support, *k*, the shape and the way the support moves affects the performance of end-to-end connectivity. An open issue is to investigate alternative structures for the support, different motion coordination strategies and comparatively study the corresponding effects on communication times. To this end, the support idea is extended to hierarchical and highly changing motion graphs in [4]. The idea of cooperative routing based on the existence of support nodes may also improve security and trust.

An important issue for the case where the network is sparsely populated or where the rate of motion is too high is to study the performance of path construction and maintenance protocols. Some work has be done in this direction in [2] that can be also used to investigate the end-to-end communication in wireless sensor networks. It is still unknown if there exist impossibility results for distributed algorithms that attempt to maintain structural information of the implied fragile network of virtual links.

Another open research area is to analyze the properties of end-to-end communication given certain support motion strategies. There are cases where the mobile nodes interactions may behave in a similar way to the Physics paradigm of *interacting particles* and their modeling. Studies of interaction times and propagation times in various graphs are reported in [7] and are still important to further research in this direction.

Experimental Results

In [5] an experimental evaluation is conducted via simulation in order to model the different possible situations regarding the geographical area covered by an ad-hoc mobile network. A number of experiments were carried out for grid-graphs (2D, 3D), random graphs ($G_{n,p}$ model), bipartite multi-stage graphs and two-level motion graphs. All results verify the theoretical analysis and provide useful insight on how to further exploit the support idea. In [4] the model of hierarchical and highly changing ad-hoc networks is investigated. The experiments indicate that, the pattern of the "snake" algorithm's performance remains the same even in such type of networks.

URL to Code

http://ru1.cti.gr

Cross-References

▶ Mobile Agents and Exploration

Recommended Reading

1. Aldous D, Fill J (1999) Reversible markov chains and random walks on graphs. http://stat-www.berkeley.edu/users/aldous/book.html. Accessed 1999
2. Busch C, Tirthapura S (2005) Analysis of link reversal routing algorithms. SIAM J Comput 35(2):305–326
3. Chatzigiannakis I, Kinalis A, Nikoletseas S (2006) Sink mobility protocols for data collection in wireless sensor networks. In: Zomaya AY, Bononi L (eds) 4th international mobility and wireless access workshop (MOBIWAC 2006), Terromolinos, pp 52–59
4. Chatzigiannakis I, Nikoletseas S (2004) Design and analysis of an efficient communication strategy for hierarchical and highly changing ad-hoc mobile networks. J Mobile Netw Appl 9(4):319–332. Special issue on parallel processing issues in mobile computing
5. Chatzigiannakis I, Nikoletseas S, Spirakis P (2003) Distributed communication algorithms for ad hoc mobile networks. J Parallel Distrib Comput 63(1):58–74. Special issue on wireless and mobile ad-hoc networking and computing, edited by Boukerche A
6. Diggavi SN, Grossglauser M, Tse DNC (2005) Even one-dimensional mobility increases the capacity of wireless networks. IEEE Trans Inf Theory 51(11):3947–3954
7. Dimitriou T, Nikoletseas SE, Spirakis PG (2004) Analysis of the information propagation time among mobile hosts. In: Nikolaidis I, Barbeau M, Kranakis E (eds) 3rd international conference on ad-hoc, mobile, and wireless networks (ADHOCNOW 2004). Lecture notes in computer science (LNCS), vol 3158. Springer, Berlin, pp 122–134
8. Gafni E, Bertsekas DP (1981) Distributed algorithms for generating loop-free routes in networks with frequently changing topology. IEEE Trans Commun 29(1):11–18
9. Grossglauser M, Tse DNC (2002) Mobility increases the capacity of ad hoc wireless networks. IEEE/ACM Trans Netw 10(4):477–486
10. Jain S, Shah R, Brunette W, Borriello G, Roy S (2006) Exploiting mobility for energy efficient data collection in wireless sensor networks. J Mobile Netw Appl 11(3):327–339
11. Li Q, Rus D (2003) Communication in disconnected ad hoc networks using message relay. J Parallel Distrib Comput 63(1):75–86. Special issue on wireless and mobile ad-hoc networking and computing, edited by A Boukerche
12. Luo J, Panchard J, Piórkowski M, Grossglauser M, Hubaux JP (2006) Mobiroute: routing towards a mobile sink for improving lifetime in sensor networks. In: Gibbons PB, Abdelzaher T, Aspnes J, Rao R (eds) 2nd IEEE/ACM international conference on distributed computing in sensor systems (DCOSS 2005). Lecture notes in computer science (LNCS), vol 4026. Springer, Berlin, pp 480–497
13. Perkins CE (2001) Ad hoc networking. Addison-Wesley, Boston
14. Walter JE, Welch JL, Amato NM (2004) Distributed reconfiguration of metamorphic robot chains. J Distrib Comput 17(2):171–189
15. Zhao W, Ammar M, Zegura E (2004) A message ferrying approach for data delivery in sparse mobile ad hoc networks. In: Murai J, Perkins C, Tassiulas L (eds) 5th ACM international symposium on mobile ad hoc networking and computing (MobiHoc 2004). ACM, Roppongi Hills, pp 187–198

Compact Routing Schemes

Shiri Chechik
Department of Computer Science, Tel Aviv University, Tel Aviv, Israel

Keywords

Approximate shortest paths; Compact routing; Stretch factor

Years and Authors of Summarized Original Work

2013; Chechik

Problem Definition

Routing is a distributed mechanism that allows sending messages between any pair of nodes of the network. As in all distributed algorithms, a routing scheme runs locally on every processor/node of the network. Each node/processor of the network has a routing daemon running on it whose responsibility is to forward arriving messages while utilizing local information that is stored at the node itself. This local information is usually referred to as the routing table of the node.

A routing scheme involves two phases, a preprocessing phase and a routing phase. In the preprocessing phase, the algorithm assigns every node of the network a routing table and a small-size label. The label is used as the address of the node and therefore is usually expected to be of small size – poly-logarithmic in the size of the network.

In the routing phase, some node of the networks wishes to send a message to some other nodes of the network in a distributed manner. During the routing phase, each node of the network may receive this message, and it has to decide whether this message reached its final destination, and if not, the node needs to decide to which of its neighbors this message should be forwarded next. In order to make these decisions, the node may use its own routing table and the header of the message that usually contains the label of the final destination and perhaps some additional information.

The stretch of a routing scheme is defined as the worst case ratio between the length of the path obtained by the routing scheme and the length of the shortest path between the source and the destination. There are two main objectives in designing the routing scheme. The first is to minimize the stretch of the routing scheme, and the second is to minimize the size of the routing tables. Much of the work on designing routing schemes focuses on the trade-off between these two objectives.

One extreme case is when it is allowed to use linear-size routing tables. In this case, one can store a complete routing table at all nodes, i.e., for every source node s and every potential destination node t, store at s the port number of the neighbor of s on the shortest path from s to t. In this case, the stretch is 1, i.e., the algorithm can route on exact shortest paths. However, a clear drawback is that the size of the routing tables is large, linear in the size of the network.

One may wish to use smaller routing tables at the price of a larger stretch. A routing scheme is considered to be compact if the size of the routing tables is sublinear in the number of nodes.

Key Results

This section presents a survey on compact routing schemes and a highlight of some recent new developments.

Many papers focus on the trade-off between the size of the routing tables and the stretch (e.g., [1, 2, 4, 5, 7–9]). The first trade-off was obtained by Peleg and Upfal [9]. Their scheme considered unweighted graph and achieved a bound on the total size of the routing tables.

Later, Awerbuch et al. [1] considered weighted graphs and achieved a routing scheme with a guarantee on the maximum table size. Their routing scheme uses table size of $\tilde{O}(n^{1/k})$ and was with $O(k^2 9^k)$ stretch. A better trade-off was later achieved by Awerbuch and Peleg [2].

Until very recently, the best-known trade-off was due to Thorup and Zwick [10]. They presented a routing scheme that uses routing tables of size $\tilde{O}(n^{1/k})$, a stretch of $4k - 5$, and label size of $O(k \log^2 n)$. Moreover, they showed that if a handshaking is allowed, namely, if the source node and the destination are allowed to exchange an information of size $O(\log^2 n)$ bits, then the stretch can be improved to $2k - 1$. Clearly, in many cases, it would be desirable to avoid the use of handshaking, as the overhead of establishing a handshake can be as high as sending the original message itself.

A natural question is, what is the best trade-off between routing table size and stretch one can hope for with or without a handshake? In fact, assuming the girth conjecture of Erdős [6], one can show that with table size of $O(n^{1/k})$, the best stretch possible is $2k - 1$ with or without a handshake. Hence, in the case of a handshake, Thorup and Zwick's scheme [10] is essentially optimal. However, in the case of no handshake, there is still a gap between the lower and upper bound. A main open problem in the area of compact routing schemes is on the gap between the stretch $4k - 5$ and $2k - 1$.

Recently, Chechik [3] gave the first evidence that the asymptotically optimal stretch is less than $4k$. Chechik [3] presented the first improvement to the stretch-space trade-off of compact routing scheme since the result of Thorup and Zwick [10]. More specifically, [3] presented a compact routing scheme for weighted general undirected graphs that uses tables of size $\tilde{O}(n^{1/k})$ and has stretch $c \cdot k$ for some absolute constant $c < 4$.

Open Problems

The main question that still remains unresolved is to prove or disprove the existence of a compact routing scheme that utilizes tables of size $\tilde{O}\left(n^{1/k}\right)$ and has stretch of $2k$ without the use of a handshake.

Recommended Reading

1. Awerbuch B, Bar-Noy A, Linial N, Peleg D (1990) Improved routing strategies with succinct tables. J Algorithms 11(3):307–341
2. Awerbuch B, Peleg D (1990) Sparse partitions. In: Proceedings of 31st IEEE symposium on foundations of computer science (FOCS), St. Louis, pp 503–513
3. Chechik S (2013) Compact routing schemes with improved stretch. In: 32nd ACM symposium on principles of distributed computing (PODC), Montreal, pp 33–41
4. Cowen LJ (2001) Compact routing with minimum stretch. J Algorithms 38:170–183
5. Eilam T, Gavoille C, Peleg D (2003) Compact routing schemes with low stretch factor. J Algorithms 46:97–114
6. Erdős P (1964) Extremal problems in graph theory. In: Theory of graphs and its applications. Methuen, London, pp 29–36
7. Gavoille C, Peleg D (2003) Compact and localized distributed data structures. Distrib Comput 16:111–120
8. Peleg D (2000) Distributed computing: a locality-sensitive approach. SIAM, Philadelphia
9. Peleg D, Upfal E (1989) A trade-off between space and efficiency for routing tables. J ACM 36(3):510–530
10. Thorup M, Zwick U (2001) Compact routing schemes. In: Proceedings of 13th ACM symposium on parallel algorithms and architectures (SPAA), Heraklion, pp 1–10

Competitive Auction

Tian-Ming Bu
Software Engineering Institute, East China Normal University, Shanghai, China

Keywords

Auction design; Optimal mechanism design

Years and Authors of Summarized Original Work

2001; Goldberg, Hartline, Wright
2002; Fiat, Goldberg, Hartline, Karlin

Problem Definition

This problem studies the *one round, sealed-bid* auction model where an auctioneer would like to sell an idiosyncratic commodity with unlimited copies to n bidders and each bidder $i \in \{1, \ldots, n\}$ will get at most one item.

First, for any i, bidder i bids a value b_i representing the price he is willing to pay for the item. They submit the bids simultaneously. After receiving the bidding vector $\mathbf{b} = (b_1, \ldots, b_n)$, the auctioneer computes and outputs the allocation vector $\mathbf{x} = (x_1, \ldots, x_n) \in \{0, 1\}^n$ and the price vector $\mathbf{p} = (p_1, \ldots, p_n)$. If for any i, $x_i = 1$, then bidder i gets the item and pays p_i for it. Otherwise, bidder i loses and pays nothing. In the auction, the auctioneer's revenue is $\sum_{i=1}^{n} \mathbf{x}\mathbf{p}^T$.

Definition 1 (Optimal Single Price Omniscient Auction $\mathcal{F}$) Given a bidding vector $\mathbf{b}$ sorted in decreasing order,

$$\mathcal{F}(\mathbf{b}) = \max_{1 \leq i \leq n} i \cdot b_i$$

Further,

$$\mathcal{F}^{(m)}(\mathbf{b}) = \max_{m \leq i \leq n} i \cdot b_i$$

Obviously, $\mathcal{F}$ maximizes the auctioneer's revenue if only uniform price is allowed.

However, in this problem, each bidder i is associated with a private value v_i representing the item's value in his opinion. So if bidder i gets the item, his payoff should be $v_i - p_i$. Otherwise, his payoff is 0. So for any bidder i, his payoff function can be formulated as $(v_i - p_i)x_i$. Furthermore, free will is allowed in the model. In other words, each bidder would bid some b_i different from his true value v_i, to maximize his payoff.

The objective of the problem is to design a *truthful* auction which could still maximize the auctioneer's revenue. An auction is *truthful* if for every bidder i, bidding his true value would maximize his payoff, regardless of the bids submitted by the other bidders [12, 13].

Definition 2 (Competitive Auctions)

INPUT: the submitted bidding vector $\mathbf{b}$.
OUTPUT: the allocation vector $\mathbf{x}$ and the price vector $\mathbf{p}$.
CONSTRAINTS:

(a) Truthful;
(b) The auctioneer's revenue is within a constant factor of the optimal single pricing for all inputs.

Key Results

Let $\mathbf{b}_{-i} = (b_1, \ldots, b_{i-1}, b_{i+1}, \ldots, b_n)$. f is any function from $\mathbf{b}_{-i}$ to the price.

Algorithm 1 Bid-independent auction: $\mathcal{A}_f(\mathbf{b})$

1: **for** $i = 1$ to n **do**
2:    **if** $f(\mathbf{b}_{-i}) \leq b_i$ **then**
3:       $x_i = 1$ and $p_i = f(\mathbf{b}_i)$
4:    **else**
5:       $x_i = 0$
6:    **end if**
7: **end for**

Theorem 1 ([6]) *An auction is truthful if and only if it is equivalent to a bid-independent auction.*

Definition 3 A truthful auction $\mathcal{A}$ is β-competitive against $\mathcal{F}^{(m)}$ if for all bidding vectors $\mathbf{b}$, the expected profit of $\mathcal{A}$ on $\mathbf{b}$ satisfies

$$\mathbf{E}(\mathcal{A}(\mathbf{b})) \geq \frac{\mathcal{F}^{(m)}(\mathbf{b})}{\beta}$$

Definition 4 (CostShare$_C$ [11]) Given bids $\mathbf{b}$, this mechanism finds the largest k such that the highest k bidders' biddings are at least $\frac{C}{k}$. Charge each of such k bidders $\frac{C}{k}$.

Algorithm 2 Sampling cost-sharing auction (SCS)

1: Partition bidding vector $\mathbf{b}$ uniformly at random into two sets $\mathbf{b}'$ and $\mathbf{b}''$.
2: Computer $\mathcal{F}' = \mathcal{F}(\mathbf{b}')$ and $\mathcal{F}'' = \mathcal{F}(\mathbf{b}'')$.
3: Running CostShare$_{\mathcal{F}''}$ on $\mathbf{b}'$ and CostShare$_{\mathcal{F}'}$ on $\mathbf{b}''$.

Theorem 2 ([6]) *SCS is 4-competitive against $\mathcal{F}^{(2)}$, and the bound is tight.*

SCS could be extended for partitioning into k parts for any k. In fact, $k = 3$ is the optimal partition.

Theorem 3 ([10]) *The random three partitioning cost sharing auction is 3.25-competitive.*

Theorem 4 ([9]) *Let $\mathcal{A}$ be any truthful randomized auction. There exists an input bidding vector $\mathbf{b}$ on which $E(\mathcal{A}(\mathbf{b})) \leq \frac{\mathcal{F}^{(2)}(\mathbf{b})}{2.42}$.*

Applications

As the Internet becomes more popular, more and more auctions are beginning to appear. Further, the items on sale in the auctions vary from antiques, paintings, and digital goods, for example, mp3, licenses, network resources, and so on. Truthful auctions can reduce the bidders' cost of investigating the competitors' strategies, since truthful auctions encourage bidders to bid their true values. On the other hand, competitive auctions can also guarantee the auctioneer's profit. So this problem is very practical and significant. These years, designing and analyzing competitive auctions under various auction models has become a hot topic [1–5, 7, 8].

Cross-References

▶ CPU Time Pricing
▶ Multiple Unit Auctions with Budget Constraint

Recommended Reading

1. Abrams Z (2006) Revenue maximization when bidders have budgets. In: Proceedings of the seventeenth annual ACM-SIAM symposium on discrete algorithms (SODA-06), Miami, pp 1074–1082
2. Bar-Yossef Z, Hildrum K, Wu F (2002) Incentive-compatible online auctions for digital goods. In: Proceedings of the 13th annual ACM-SIAM symposium on discrete mathematics (SODA-02), New York, pp 964–970
3. Borgs C, Chayes JT, Immorlica N, Mahdian M, Saberi A (2005) Multi-unit auctions with budget-constrained bidders. In: ACM conference on electronic commerce (EC-05), Vancouver, pp 44–51
4. Bu TM, Qi Q, Sun AW (2008) Unconditional competitive auctions with copy and budget constraints. Theor Comput Sci 393(1–3):1–13
5. Deshmukh K, Goldberg AV, Hartline JD, Karlin AR (2002) Truthful and competitive double auctions. In: Möhring RH, Raman R (eds) Algorithms – ESA 2002, 10th annual European symposium, Rome. Lecture notes in computer science, vol 2461. Springer, pp 361–373
6. Fiat A, Goldberg AV, Hartline JD, Karlin AR (2002) Competitive generalized auctions. In: Proceedings of the 34th annual ACM symposium on theory of computing (STOC-02), New York, pp 72–81
7. Goldberg AV, Hartline JD (2001) Competitive auctions for multiple digital goods. In: auf der Heide FM (ed) Algorithms – ESA 2001, 9th annual european symposium, Aarhus. Lecture notes in computer science, vol 2161. Springer, pp 416–427
8. Goldberg AV, Hartline JD (2003) Envy-free auctions for digital goods. In: Proceedings of the 4th ACM conference on electronic commerce (EC-03), New York, pp 29–35
9. Goldberg AV, Hartline JD, Wright A (2001) Competitive auctions and digital goods. In: Proceedings of the twelfth annual ACM-SIAM symposium on discrete algorithms (SODA-01), New York, pp 735–744
10. Hartline JD, McGrew R (2005) From optimal limited to unlimited supply auctions. In: Proceedings of the 6th ACM conference on electronic commerce (EC-05), Vancouver, pp 175–182
11. Moulin H (1999) Incremental cost sharing: characterization by coalition strategy-proofness. Soc Choice Welf 16:279–320
12. Nisan N, Ronen A (1999) Algorithmic mechanism design. In: Proceedings of the thirty-first annual ACM symposium on theory of computing (STOC-99), New York, pp 129–140
13. Parkes DC (2004) Chapter 2: iterative combinatorial auctions. PhD thesis, University of Pennsylvania

Complexity Dichotomies for Counting Graph Homomorphisms

Jin-Yi Cai[1,2], Xi Chen[3,4], and Pinyan Lu[5]
[1]Beijing University, Beijing, China
[2]Computer Sciences Department, University of Wisconsin–Madison, Madison, WI, USA
[3]Computer Science Department, Columbia University, New York, NY, USA
[4]Computer Science and Technology, Tsinghua University, Beijing, China
[5]Microsoft Research Asia, Shanghai, China

Keywords

Computational complexity; Counting complexity; Graph homomorphisms; Partition functions

Years and Authors of Summarized Original Work

2000; Dyer, Greenhill
2005; Bulatov, Grohe
2010; Goldberg, Grohe, Jerrum, Thurley
2013; Cai, Chen, Lu

Problem Definition

It is well known that if NP $\neq$ P, there is an infinite hierarchy of complexity classes between them [10]. However, for some broad classes of problems, a *complexity dichotomy* exists: every problem in the class is either in polynomial time or NP-hard. Such results include Schaefer's theorem [13], the dichotomy of Hell and Nešetřil for H-coloring [9], and some subclasses of the general constraint satisfaction problem [4]. These developments lead to the following questions: How far can we push the envelope and show dichotomies for even broader classes of problems? Given a class of problems, what is the criterion that distinguishes the tractable problems from the intractable ones? How does it help in solving the tractable problems efficiently? Now replacing NP with #P [15], all the questions above can be asked for *counting* problems.

One family of counting problem concerns graph homomorphisms. Given two undirected graphs G and H, a graph homomorphism from G to H is a map ξ from the vertex set $V(G)$ to $V(H)$ such that (u, v) is an edge in G if and only if $(\xi(u), \xi(v))$ is an edge in H. The counting problem for graph homomorphism is to compute the number of homomorphisms from G to H. For a fixed graph H, this problem is also known as the #H-coloring problem. In addition to #H-coloring, a more general family of problems that has been studied intensively over the years is to count graph homomorphisms with weights. Formally, we use $\mathbf{A}$ to denote an $m \times m$ symmetric matrix with entries $(A_{i,j})$, $i, j \in [m] = \{1, \ldots, m\}$. Given any undirected graph $G = (V, E)$, we define the graph homomorphism function $Z_\mathbf{A}(G)$ as follows:

$$Z_\mathbf{A}(G) = \sum_{\xi: V \to [m]} \prod_{(u,v) \in E} A_{\xi(u), \xi(v)}. \quad (1)$$

This is also called the *partition function* from statistical physics. It is clear from the definition that $Z_\mathbf{A}(G)$ is exactly the number of homomorphisms from G to H, when $\mathbf{A}$ is the $\{0, 1\}$ adjacency matrix of H.

Graph homomorphism can express many natural graph properties. For example, if one takes H to be the graph over two vertices $\{0, 1\}$ with an edge $(0, 1)$ and a self-loop at 1, then the set of vertices mapped to 1 in a graph homomorphism from G to H corresponds to a VERTEX COVER of G, and the counting problem simply counts the number of vertex covers. As another example, if H is the complete graph over k vertices (without self-loops), then the problem is exactly the k-COLORING problem for G. Many additional graph invariants can be expressed as $Z_\mathbf{A}(G)$ for appropriate $\mathbf{A}$. Consider the Hadamard matrix

$$\mathbf{H} = \begin{pmatrix} 1 & 1 \\ 1 & -1 \end{pmatrix}, \quad (2)$$

where we index the rows and columns by $\{0, 1\}$. In $Z_\mathbf{H}(G)$, every product

$$\prod_{(u,v) \in E} H_{\xi(u), \xi(v)} \in \{1, -1\}$$

and is -1 precisely when the induced subgraph of G on $\xi^{-1}(1)$ has an odd number of edges. Thus, $(2^n - Z_\mathbf{H}(G))/2$ is the number of induced subgraphs of G with an odd number of edges. Also expressible as $Z_\mathbf{A}(\cdot)$ are S-flows where S is a subset of a finite Abelian group closed under inversion [6], and a scaled version of the Tutte polynomial $\hat{T}(x, y)$ where $(x - 1)(y - 1)$ is a positive integer. In [6], Freedman, Lovász, and Schrijver characterized the graph functions that can be expressed as $Z_\mathbf{A}(\cdot)$.

Key Results

In [5], Dyer and Greenhill first prove a complexity dichotomy theorem for all undirected graphs H. To state it formally, we give the following definition of *block-rank-1* matrices:

Definition 1 (Block-rank-1 matrices) A nonnegative *(but not necessarily symmetric)* matrix $\mathbf{A} \in \mathbb{C}^{m \times n}$ is said to be *block-rank-1* if after separate appropriate permutations of its rows and columns, $\mathbf{A}$ becomes a block diagonal matrix and every block is of rank 1.

It is clear that a nonnegative matrix $\mathbf{A}$ is block-rank-1 iff every 2×2 submatrix of $\mathbf{A}$ with at least three positive entries is of rank 1. Here is the dichotomy theorem of Dyer and Greenhill [5]:

Theorem 1 ([5]) *Given any undirected graph H, the #H-coloring problem is in polynomial time if its adjacency matrix is block-rank-1 and is #P-hard otherwise.*

For the special case when H has two vertices, the dichotomy above states that #H-coloring is in polynomial time if the number of 1s in its adjacency matrix is $0, 1, 2,$ or 4 and is #P-hard otherwise. For the latter case, one of the diagonal entries is 0 (as H is undirected), and #H-coloring is indeed the problem of counting independent sets [16]. However, proving a dichotomy theorem for H of arbitrary size is much more challenging. Besides counting independent sets, the other starting point used in [5] is the problem of counting proper q-colorings [12]. To show that there is a reduction from one of these two problems whenever H violates the block-rank-1 criterion, Dyer and Greenhill need to define a more general counting problem with vertex weights and employ the technique of interpolation [14, 16], as well as two tools often used with interpolation, stretching, and thickening.

Later in [1], Bulatov and Grohe give a sweeping complexity dichotomy theorem that generalizes the result of Dyer and Greenhill to nonnegative symmetric matrices:

Theorem 2 ([1]) *Given any symmetric and nonnegative algebraic matrix* $\mathbf{A}$, *computing* $Z_{\mathbf{A}}(\cdot)$ *is in polynomial time if* $\mathbf{A}$ *is block-rank-1 and is #P-hard otherwise.*

This dichotomy theorem has since played an important role in many of the new developments in the study of counting graph homomorphisms as well as counting constraint satisfaction problem because of its enormous applicability. Many #P-hardness results are built on top of this dichotomy. A proof of the dichotomy theorem with a few shortcuts can also be found in [8].

Recently in a paper with both exceptional depth and conceptual vision, Goldberg, Jerrum, Grohe, and Thurley [7] proved a complexity dichotomy for all real-valued symmetric matrices:

Theorem 3 ([7]) *Given any symmetric and real algebraic matrix* $\mathbf{A}$, *the problem of computing* $Z_{\mathbf{A}}(\cdot)$ *is either in polynomial time or #P-hard.*

The exact tractability criterion in the dichotomy above, however, is much more technical and involved. Roughly speaking, the proof of the theorem proceeds by establishing a sequence of successively more stringent properties that a tractable $\mathbf{A}$ must satisfy. Ultimately, it arrives at a point where the satisfaction of these properties together implies

that the computation of $Z_{\mathbf{A}}(G)$ can be reduced to the following sum:

$$\sum_{x_1,\ldots,x_n \in \mathbb{Z}_2} (-1)^{f_G(x_1,\ldots,x_n)} \qquad (3)$$

where f_G is a quadratic polynomial over $\mathbb{Z}_2$ constructed from the input graph G efficiently. This sum is known to be computable in polynomial time in n, the number of variables (e.g., see [3] and [11, Theorem 6.30]). In particular, the latter immediately implies that the following two Hadamard matrices

$$\mathbf{H}_2 = \begin{pmatrix} 1 & 1 \\ 1 & -1 \end{pmatrix} \quad \text{and} \quad \mathbf{H}_4 = \begin{pmatrix} 1 & 1 & 1 & 1 \\ 1 & 1 & -1 & -1 \\ 1 & -1 & 1 & -1 \\ 1 & -1 & -1 & 1 \end{pmatrix}$$

are both tractable. This can be seen from the following polynomial view of these two matrices. If we index the rows and columns of $\mathbf{H}_2$ by $\mathbb{Z}_2$ and index the rows and columns of $\mathbf{H}_4$ by $(\mathbb{Z}_2)^2$, then their (x, y)th entry and $((x_1, x_2), (y_1, y_2))$th entry are

$$(-1)^{xy} \quad \text{and} \quad (-1)^{x_1 y_2 + x_2 y_1},$$

respectively. From here, it is easy to reduce $Z_{\mathbf{H}_2}(\cdot)$ and $Z_{\mathbf{H}_4}(\cdot)$ to (3).

Compared with the nonnegative domain [1, 5], there are a lot more interesting tractable cases over the real numbers, e.g., the two Hadamard matrices above as well as their arbitrary tensor products. It is not surprising that the potential cancelations in the sum $Z_{\mathbf{A}}(\cdot)$ may in fact be the source of efficient algorithms for computing $Z_{\mathbf{A}}(\cdot)$ itself. This motivates Cai, Chen, and Lu to continue to investigate the computational complexity of $Z_{\mathbf{A}}(\cdot)$ with $\mathbf{A}$ being a symmetric complex matrix [2], because over the complex domain, there is a significantly richer variety of possible cancelations with the roots of unit, and more interesting tractable cases are expected. This turns out to be the case, and they prove the following complexity dichotomy:

Theorem 4 ([2]) *given any symmetric and algebraic complex matrix* $\mathbf{A} \in \mathbb{C}^{m \times m}$, *the problem of* *computing $Z_{\mathbf{A}}(\cdot)$ is either in polynomial time or #P-hard.*

Applications

None is reported.

Open Problems

The efficient approximation of $Z_{\mathbf{A}}(\cdot)$ remains widely open even for small nonnegative matrices. See the entry "Approximating the Partition Function of Two-Spin Systems" for the current state of the art on this. Two families of counting problems that generalize $Z_{\mathbf{A}}(\cdot)$ are counting constraint satisfaction and Holant problems. Open problems in these two areas can be found in the two entries "Complexity Dichotomies for the Counting Constraint Satisfaction Problem" and "Holant Problems."

Experimental Results

None is reported.

URLs to Code and Data Sets

None is reported.

Cross-References

▶ Approximating the Partition Function of Two-Spin Systems
▶ Holant Problems
▶ Holographic Algorithms

Recommended Reading

1. Bulatov A, Grohe M (2005) The complexity of partition functions. Theor Comput Sci 348(2):148–186
2. Cai JY, Chen X, Lu P (2013) Graph homomorphisms with complex values: a dichotomy theorem. SIAM J Comput 42(3):924–1029
3. Carlitz L (1969) Kloosterman sums and finite field extensions. Acta Arith 16:179–193
4. Creignou N, Khanna S, Sudan M (2001) Complexity classifications of boolean constraint satisfaction problems. SIAM monographs on discrete mathematics and applications. Society for Industrial and Applied Mathematics, Philadelphia
5. Dyer M, Greenhill C (2000) The complexity of counting graph homomorphisms. Random Struct Algorithms 17(3–4):260–289
6. Freedman M, Lovász L, Schrijver A (2007) Reflection positivity, rank connectivity, and homomorphism of graphs. J Am Math Soc 20:37–51
7. Goldberg L, Grohe M, Jerrum M, Thurley M (2010) A complexity dichotomy for partition functions with mixed signs. SIAM J Comput 39(7):3336–3402
8. Grohe M, Thurley M (2011) Counting homomorphisms and partition functions. In: Grohe M, Makowsky J (eds) Model theoretic methods in finite combinatorics. Contemporary mathematics, vol 558. American Mathematical Society, Providence
9. Hell P, Nešetřil J (1990) On the complexity of H-coloring. J Comb Theory Ser B 48(1):92–110
10. Ladner R (1975) On the structure of polynomial time reducibility. J ACM 22(1):155–171
11. Lidl R, Niederreiter H (1997) Finite fields. Encyclopedia of mathematics and its applications, vol 20. Cambridge University Press, Cambridge
12. Linial N (1986) Hard enumeration problems in geometry and combinatorics. SIAM J Algebraic Discret Methods 7:331–335
13. Schaefer T (1978) The complexity of satisfiability problems. In: Proceedings of the 10th annual ACM symposium on theory of computing, San Diego, California, pp 216–226
14. Vadhan S (2002) The complexity of counting in sparse, regular, and planar graphs. SIAM J Comput 31:398–427
15. Valiant L (1979) The complexity of computing the permanent. Theor Comput Sci 8:189–201
16. Valiant L (1979) The complexity of enumeration and reliability problems. SIAM J Comput 8:410–421

Complexity of Bimatrix Nash Equilibria

Xi Chen
Computer Science Department, Columbia University, New York, NY, USA
Computer Science and Technology, Tsinghua University, Beijing, China

Keywords

Bimatrix games; Nash Equilibria; Two-player games

Years and Authors of Summarized Original Work

2006; Chen, Deng

Problem Definition

In the middle of the last century, Nash [8] studied general noncooperative games and proved that there exists a set of mixed strategies, now commonly referred to as a Nash equilibrium, one for each player, such that no player can benefit if he/she changes his/her own strategy unilaterally. Since the development of Nash's theorem, researchers have worked on how to compute Nash equilibria efficiently. Despite much effort in the last half century, no significant progress has been made on characterizing its algorithmic complexity, though both hardness results and algorithms have been developed for various modified versions.

An exciting breakthrough, which shows that computing Nash equilibria is possibly hard, was made by Daskalakis, Goldberg, and Papadimitriou [5], for games among four players or more. The problem was proven to be complete in **PPAD** (polynomial parity argument, directed version), a complexity class introduced by Papadimitriou in [9]. The work of [5] is based on the techniques developed in [6]. This hardness result was then improved to the three-player case by Chen and Deng [1] and Daskalakis and Papadimitriou [4], independently and with different proofs. Finally, Chen and Deng [2] proved that NASH, the problem of finding a Nash equilibrium in a bimatrix game (or two-player game), is **PPAD**-complete.

A bimatrix game is a noncooperative game between two players in which the players have m and n choices of actions (or pure strategies), respectively. Such a game can be specified by two $m \times n$ matrices $\mathbf{A} = (a_{i,j})$ and $\mathbf{B} = (b_{i,j})$. If the first player chooses action i and the second player chooses action j, then their payoffs are $a_{i,j}$ and $b_{i,j}$, respectively. A mixed strategy of a player is a probability distribution over his/her choices. Let $\mathbb{P}^n$ denote the set of all probability

vectors in $\mathbb{R}^n$, i.e., nonnegative vectors whose entries sum to 1. The Nash equilibrium theorem on noncooperative games, when specialized to bimatrix games, states that for every bimatrix game $\mathcal{G} = (\mathbf{A}, \mathbf{B})$, there exists a pair of mixed strategies $(\mathbf{x}^* \in \mathbb{P}^m, \mathbf{y}^* \in \mathbb{P}^n)$, called a Nash equilibrium, such that for all $\mathbf{x} \in \mathbb{P}^m$ and $\mathbf{y} \in \mathbb{P}^n$, $(\mathbf{x}^*)^\mathsf{T} \mathbf{A} \mathbf{y}^* \geq \mathbf{x}^\mathsf{T} \mathbf{A} \mathbf{y}^*$ and $(\mathbf{X}^*)^\mathsf{T} \mathbf{B} \mathbf{y}^* \geq (\mathbf{x}^*)^\mathsf{T} \mathbf{B} \mathbf{y}$.

Computationally, one might settle with an approximate Nash equilibrium. Let $\mathbf{A}_i$ denote the ith row vector of $\mathbf{A}$ and $\mathbf{B}_i$ denote the ith column vector of $\mathbf{B}$. An ε-well-supported Nash equilibrium of game $(\mathbf{A},\mathbf{B})$ is a pair of mixed strategies $(\mathbf{x}^*, \mathbf{y}^*)$ such that

$$\mathbf{A}_i \mathbf{y}^* > \mathbf{A}_j \mathbf{y}^* + \epsilon \Rightarrow x_j^* = 0, \, \forall \, i, j : 1 \leq i, j \leq m;$$

$$(\mathbf{x}^*)^\mathsf{T} \mathbf{B}_i > (\mathbf{x}^*)^\mathsf{T} \mathbf{B}_j + \epsilon \Rightarrow y_j^* = 0, \, \forall \, i, j : 1 \leq i, j \leq n.$$

Definition 1 (2-NASH and NASH) The input instance of problem 2-NASH is a pair $(\mathcal{G}, 0^k)$ where $\mathcal{G}$ is a bimatrix game and the output is a 2^{-k}-well-supported Nash equilibrium of $\mathcal{G}$. The input of problem NASH is a bimatrix game $\mathcal{G}$ and the output is an exact Nash equilibrium of $\mathcal{G}$.

Key Results

A binary relation $R \subset \{0, 1\}^* \times \{0, 1\}^*$ is *polynomially balanced* if there exists a polynomial p such that for all pairs $(x, y) \in R$, $|y| \leq p(|x|)$. It is a *polynomial-time computable relation* if for each pair (x, y), one can decide whether or not $(x, y) \in R$ in polynomial time in $|x| + |y|$. The **NP** search problem Q_R specified by R is defined as follows: given $x \in \{0, 1\}^*$, if there exists y such that $(x, y) \in R$, return y; otherwise, return a special string "no."

Relation R is *total* if for every $x \in \{0, 1\}^*$, there exists a y such that $(x, y) \in R$. Following [7], let **TFNP** denote the class of all **NP** search problems specified by total relations. A search problem $Q_{R_1} \in$ **TFNP** is *polynomial-time reducible* to problem $Q_{R_2} \in$ **TFNP** if there exists a pair of polynomial-time computable functions (f, g) such that for every x of R_1, if y satisfies that $(f(x), y) \in R_2$, then $(x, g(y)) \in R_1$.

Furthermore, Q_{R1} and Q_{R2} are polynomial-time equivalent if Q_{R2} is also reducible to Q_{R1}.

The complexity class **PPAD** is a subclass of **TFNP**, containing all the search problems which are polynomial-time reducible to:

Definition 2 (Problem LEAFD) The input instance of LEAFD is a pair $(M, 0^n)$, where M defines a polynomial-time Turing machine satisfying:

1. For every $v \in \{0, 1\}^n$, $M(v)$ is an ordered pair (u_1, u_2) with $u_1, u_2 \in \{0, 1\}^n \cup \{\text{"no"}\}$.
2. $M(0^n) = (\text{"no,"} 1^n)$ and the first component of $M(1^n)$ is 0^n.

This instance defines a directed graph $G = (V, E)$ with $V = \{0, 1\}^n$. Edge $(u, v) \in E$ iff v is the second component of $M(u)$ and u is the first component of $M(v)$.

The output of problem LEAFD is a directed leaf of G other than 0^n. Here a vertex is called a *directed leaf* if its out-degree plus in-degree equals one.

A search problem in **PPAD** is said to be *complete* in **PPAD** (or **PPAD**-complete) if there exists a polynomial-time reduction from LEAFD to it.

Theorem ([2]) *2-Nash* and *Nash* are **PPAD**-*complete*.

Applications

The concept of Nash equilibria has traditionally been one of the most influential tools in the study of many disciplines involved with strategies, such as political science and economic theory. The rise of the Internet and the study of its anarchical environment have made the Nash equilibrium an indispensable part of computer science. Over the past decades, the computer science community has contributed a lot to the design of efficient algorithms for related problems. This sequence of results [1–6], for the first time, provides *some evidence* that the problem of finding a Nash equilibrium is possibly hard for **P**. These results are very important to the emerging discipline, algorithmic game theory.

Open Problems

This sequence of works shows that $(r + 1)$-player games are polynomial-time reducible to r-player games for every $r \geq 2$, but the reduction is carried out by first reducing $(r + 1)$-player games to a fixed-point problem and then further to r-player games. Is there a natural reduction that goes directly from $(r + 1)$-player games to r-player games? Such a reduction could provide a better understanding for the behavior of multiplayer games. Although many people believe that **PPAD** is hard for **P**, there is no strong evidence for this belief or intuition. The natural open problem is: can one rigorously prove that class **PPAD** is hard, under one of those generally believed assumptions in theoretical computer science, like "**NP** is not in **P**" or "one-way function exists"? Such a result would be extremely important to both computational complexity theory and algorithmic game theory.

Cross-References

▶ Matching Market Equilibrium Algorithms

Recommended Reading

1. Chen X, Deng X (2005) 3-Nash is ppad-complete. ECCC, TR05-134
2. Chen X, Deng X (2006) Settling the complexity of two-player Nash-equilibrium. In: Proceedings of the 47th annual IEEE symposium on foundations of computer science (FOCS'06), Berkeley, pp 261–272
3. Chen X, Deng X, Teng SH (2006) Computing Nash equilibria: approximation and smoothed complexity. In: Proceedings of the 47th annual IEEE symposium on foundations of computer science (FOCS'06), Berkeley, pp 603–612
4. Daskalakis C, Papadimitriou CH (2005) Three-player games are hard. ECCC, TR05-139
5. Daskalakis C, Goldberg PW, Papadimitriou CH (2006) The complexity of computing a Nash equilibrium. In: Proceedings of the 38th ACM symposium on theory of computing (STOC'06), Seattle, pp 71–78

6. Goldberg PW, Papadimitriou CH (2006) Reducibility among equilibrium problems. In: Proceedings of the 38th ACM symposium on theory of computing (STOC'06), Seattle, pp 61–70
7. Megiddo N, Papadimitriou CH (1991) On total functions, existence theorems and computational complexity. Theor Comput Sci 81:317–324
8. Nash JF (1950) Equilibrium point in n-person games. Proc Natl Acad USA 36(1):48–49
9. Papadimitriou CH (1994) On the complexity of the parity argument and other inefficient proofs of existence. J Comput Syst Sci 48: 498–532

Complexity of Core

Qizhi Fang
School of Mathematical Sciences, Ocean University of China, Qingdao, Shandong Province, China

Keywords

Balanced; Least core

Years and Authors of Summarized Original Work

2001; Fang, Zhu, Cai, Deng

Problem Definition

The core is one of the most important solution concepts in cooperative game, which is based on the coalition rationality condition: no subgroup of the players will do better if they break away from the joint decision of all players to form their own coalition. The principle behind this condition can be seen as an extension to that of the Nash Equilibrium in noncooperative games. The work of Fang, Zhu, Cai, and Deng [4] discusses the computational complexity problems related to the cores of some cooperative game models arising from combinatorial optimization problems, such as flow games and Steiner tree games.

A cooperative game with side payments is given by the pair (N, v), where $N = \{1, 2, \ldots, n\}$ is the player set and $v : 2^N \to R$

is the characteristic function. For each coalition $S \subseteq N$, the value $v(S)$ is interpreted as the profit or cost achieved by the collective action of players in S without any assistance of players in $N \setminus S$. A game is called a profit (cost) game if the characteristic function values measure the profit (cost) achieved by the coalitions. Here, the definitions are only given for profit games, symmetric statements hold for cost games.

A vector $x = \{x_1, x_2, \ldots, x_n\}$ is called an imputation if it satisfies $\sum_{i \in N} x_i = v(N)$ and $\forall i \in N : x_i \geq v(\{i\})$. The core of the game (N, v) is defined as:

$$\mathcal{C}(v) = \{x \in R^n : x(N) = v(N)$$
$$\text{and } x(S) \geq v(S), \ \forall S \subseteq N\},$$

where $x(S) = \sum_{i \in S} x_i$ for $S \subseteq N$. A game is called *balanced*, if its core is nonempty, and *totally balanced*, if every subgame (i.e., the game obtained by restricting the player set to a coalition and the characteristic function to the power set of that coalition) is balanced.

It is a challenge for the algorithmic study of the core, since there are an exponential number of constraints imposed on its definition. The following computational complexity questions have attracted much attention from researchers:

1. *Testing balancedness:* Can it be tested in polynomial time whether a given instance of the game has a nonempty core?
2. *Checking membership:* Can it be checked in polynomial time whether a given imputation belongs to the core?
3. *Finding a core member:* Is it possible to find an imputation in the core in polynomial time?

In reality, however, there is an important case in which the characteristic function value of a coalition can be evaluated via a combinatorial optimization problem, subject to constraints of resources controlled by the players of this coalition. In such circumstances, the input size of a game is the same as that of the related optimization problem, which is usually polynomial in the number of players. Therefore, this class of games, called combinatorial optimization games, fits well into the framework of algorithm and complexity analysis. Flow games and Steiner tree games discussed in Fang et al. [4] fall within this scope.

Flow Game Let $D = (V, E; \omega; s, t)$ be a directed flow network, where V is the vertex set, E is the arc set, $\omega: E \to R^+$ is the arc capacity function, and s and t are the source and the sink of the network, respectively. Assume that each player controls one arc in the network. The value of a maximum flow can be viewed as the profit achieved by the players in cooperation. The flow game $\Gamma_f = (E, \nu)$ associated with the network D is defined as follows:

(i) The player set is E.
(ii) $\forall S \subseteq E$, $\nu(S)$ is the value of a maximum flow from s to t in the subnetwork of D consisting only of arcs belonging to S.

In Kailai and Zemel [6] and Deng et al. [1], it was shown that the flow game is totally balanced and finding a core member can be done in polynomial time.

Problem 1 (Checking membership for flow game) INSTANCE: A flow network $D = (V, E; \omega; s, t)$ and $x : E \to R^+$.

QUESTION: Is it true that $x(E) = \nu(E)$ and $x(S) \geq \nu(S)$ for all subsets $S \subset E$?

Steiner Tree Game Let $G = (V, E; \omega)$ be an edge-weighted graph with $V = \{v_0\} \cup N \cup M$, where $N, M \subseteq V \setminus \{v_0\}$ are disjoint. v_0 represents a central supplier, N represents the consumer set, M represents the switch set, and $\omega(e)$ denotes the cost of connecting the two endpoints of edge e directly. It is required to connect all the consumers in N to the central supplier v_0. The connection is not limited to using direct links between two consumers or a consumer and the central supplier; it may pass through some switches in M. The aim is to construct the cheapest connection and distribute the connection cost among the consumers fairly. Then, the associated Steiner tree game $\Gamma_s = (N, \gamma)$ is defined as follows:

(i) The player set is N.
(ii) $\forall S \subseteq N$, $\gamma(S)$ is the weight of a minimum Steiner tree on G w.r.t. the set $S \cup \{v_0\}$, that is, $\gamma(S) = \min\{\sum_{e \in E_S} \omega(e) : T_S = (V_S, E_S)$ is a subtree of G with $V_S \supseteq S \cup \{v_0\}\}$.

Different from flow games, the core of a Steiner tree game may be empty. An example with an empty core was given in Megiddo [9].

Problem 2 (Testing balancedness for a Steiner tree game) INSTANCE: An edge-weighted graph $G = (V, E; \omega)$ with $V = \{v_0\} \cup N \cup M$.

QUESTION: Does there exist a vector $x : N \to R^+$ such that $x(N) = \gamma(N)$ and $x(S) \leq \gamma(S)$ for all subsets $S \subset N$?

Problem 3 (Checking membership for a Steiner tree game) INSTANCE: An edge-weighted graph $G = (V, E; \omega)$ with $V = \{v_0\} \cup N \cup M$ and $x : N \to R^+$.

QUESTION: Is it true that $x(N) = \gamma(N)$ and $x(S) \leq \gamma(S)$ for all subsets $S \subset N$?

Key Results

Theorem 1 *It is $\mathcal{NP}$-complete to show that given a flow game $\Gamma_f = (E, \nu)$ defined on network $D = (V, E; \omega; s, t)$ and a vector $x : E \to R^+$ with $x(E) = \nu(E)$, whether there exists a coalition $S \subset N$ such that $x(S) < \nu(S)$. That is, checking membership of the core for flow games is co-$\mathcal{NP}$-complete.*

The proof of Theorem 1 yields directly the same conclusion for linear production games. In Owen's linear production game [10], each player j ($j \in N$) is in possession of an individual resource vector b^j. For a coalition S of players, the profit obtained by S is the optimum value of the following linear program:

$$\max \left\{ c^t y : Ay \leq \sum_{j \in S} b^j, \ y \geq 0 \right\}.$$

That is, the characteristic function value is what the coalition can achieve in the linear production model with the resources under the control of its players. Owen showed that one imputation in the core can also be constructed through an optimal dual solution to the linear program which determines the value of N. However, in general, there exist some imputations in the core which cannot be obtained in this way.

Theorem 2 *Checking membership of the core for linear production games is co-$\mathcal{NP}$-complete.*

The problem of finding a minimum Steiner tree in a network is $\mathcal{NP}$-hard; therefore, in a Steiner tree game, the value $\gamma(S)$ of each coalition S may not be obtained in polynomial time. It implies that the complement problem of checking membership of the core for Steiner tree games may not be in $\mathcal{NP}$.

Theorem 3 *It is $\mathcal{NP}$-hard to show that given a Steiner tree game $\Gamma_s = (N, \gamma)$ defined on network $G = (V, E; \omega)$ and a vector $x : N \to R^+$ with $x(N) = \gamma(N)$, whether there exists a coalition $S \subset N$ such that $x(S) > \gamma(S)$. That is, checking membership of the core for Steiner tree games is $\mathcal{NP}$-hard.*

Theorem 4 *Testing balancedness for Steiner tree games is $\mathcal{NP}$-hard.*

Given a Steiner tree game $\Gamma_s = (N, \gamma)$ defined on network $G = (V, E; \omega)$ and a subset $S \subseteq N$, in the subgame (S, γ_S), the value $\gamma(S')$ ($S' \subseteq S$) is the weight of a minimum Steiner tree of G w.r.t. the subset $S' \cup \{v_0\}$, where all the vertices in $N \setminus S$ are treated as switches but not consumers. It is further proved in Fang et al. [4] that determining whether a Steiner tree game is totally balanced is also $\mathcal{NP}$-hard. This is the first example of $\mathcal{NP}$-hardness for the totally balanced condition.

Theorem 5 *Testing total balancedness for Steiner tree games is $\mathcal{NP}$-hard.*

Applications

The computational complexity results on the cores of combinatorial optimization games have been as diverse as the corresponding combinatorial optimization problems. For example:

1. In matching games [2], testing balancedness, checking membership, and finding a core member can all be done in polynomial time.
2. In both flow games and minimum-cost spanning tree games [3, 4], although their cores are always nonempty and a core member can be found in polynomial time, the problem of checking membership is co-$\mathcal{NP}$-complete.
3. In facility location games [5], the problem of testing balancedness is in general $\mathcal{NP}$-hard; however, given the information that the core is nonempty, both finding a core member and checking membership can be solved efficiently.
4. In a game of sum of edge weight defined on a graph [1], all the problems of testing balancedness, checking membership, and finding a core member are $\mathcal{NP}$-hard.

Based on the concept of bounded rationality [3, 8], it is suggested that computational complexity be taken as an important factor in considering rationality and fairness of a solution concept. That is, the players are not willing to spend super-polynomial time to search for the most suitable solution. In the case when the solutions of a game do not exist or are difficult to compute or to check, it may not be simple to dismiss the problem as hopeless, especially when the game arises from important applications. Hence, various conceptual approaches are proposed to resolve this problem.

When the core of a game is empty, it motivates conditions ensuring nonemptiness of approximate cores. A natural way to approximate the core is the *least core*. Let (N, v) be a profit cooperative game. Given a real number ε, the ε-core is defined to contain the imputations such that $x(S) \geq v(S) - \varepsilon$ for each nonempty proper subset S of N. The *least core* is the intersection of all nonempty ε-cores. Let ε^* be the minimum value of ε such that the ε-core is empty and then the least core is the same as the ε^*-core.

The concept of the least core poses new challenges in regard to algorithmic issues. The most natural problem is how to efficiently compute the value ε^* for a given cooperative game. The catch is that the computation of ε^* requires solving of a linear program with an exponential number of constrains. Though there are cases where this value can be computed in polynomial time [7], it is in general very hard. If the value of ε^* is considered to represent some subsidies given by the central authority to ensure the existence of the cooperation, then it is significant to give the approximate value of it even when its computation is $\mathcal{NP}$-hard.

Another possible approach is to interpret approximation as bounded rationality. For example, it would be interesting to know if there is any game with a property that for any $\varepsilon > 0$, checking membership in the ε-core can be done in polynomial time, but it is $\mathcal{NP}$-hard to tell if an imputation is in the core. In such cases, the restoration of cooperation would be a result of bounded rationality. That is to say, the players would not care an extra gain or loss of ε as the expense of another order of degree of computational resources. This methodology may be further applied to other solution concepts.

Cross-References

▶ General Equilibrium
▶ Nucleolus

Recommended Reading

1. Deng X, Papadimitriou C (1994) On the complexity of cooperative game solution concepts. Math Oper Res 19:257–266
2. Deng X, Ibaraki T, Nagamochi H (1999) Algorithmic aspects of the core of combinatorial optimization games. Math Oper Res 24:751–766
3. Faigle U, Fekete S, Hochstättler W, Kern W (1997) On the complexity of testing membership in the core of min-cost spanning tree games. Int J Game Theory 26:361–366
4. Fang Q, Zhu S, Cai M, Deng X (2001) Membership for core of LP games and other games. In: COCOON 2001. Lecture notes in computer science, vol 2108. Springer, Berlin/Heidelberg, pp 247–246
5. Goemans MX, Skutella M (2004) Cooperative facility location games. J Algorithms 50:194–214
6. Kalai E, Zemel E (1982) Generalized network problems yielding totally balanced games. Oper Res 30:998–1008
7. Kern W, Paulusma D (2003) Matching games: the least core and the nucleolus. Math Oper Res 28:294–308
8. Megiddo N (1978) Computational complexity and the game theory approach to cost allocation for a tree. Math Oper Res 3:189–196
9. Megiddo N (1978) Cost allocation for steiner trees. Networks 8:1–6
10. Owen G (1975) On the core of linear production games. Math Program 9:358–370

Compressed Document Retrieval on String Collections

Sharma V. Thankachan
School of CSE, Georgia Institute of Technology, Atlanta, USA

Keywords

Compressed data structures; Document retrieval; String algorithms; Top-k

Years and Authors of Summarized Original Work

2009; Hon, Shah, Vitter
2013; Belazzougui, Navarro, Valenzuela
2013; Tsur
2014; Hon, Shah, Thankachan, Vitter
2014; Navarro, Thankachan

Problem Definition

We face the following problem.

Problem 1 (Top-k document retrieval) *Let* $\mathcal{D} = \{T_1, T_2, \ldots, T_D\}$ *be a collection of* D *documents of* n *characters in total, drawn from an alphabet set* $\Sigma = [\sigma]$. *The relevance of a document* T_d *with respect to a pattern* P, *denoted by* $w(P, d)$ *is a function of the set of occurrences of* P *in* T_d. *Our task is to index* $\mathcal{D}$, *such that whenever a pattern* $P[1, p]$ *and a parameter* k *comes as a query, the* k *documents with the highest* $w(P, \cdot)$ *values can be reported efficiently.*

Compressed Document Retrieval on String Collections, Table 1 Indexes of space $2|\text{CSA}| + D\log(n/D) + O(D) + o(n)$ bits

Source	Report time per document
Hon et al. [3]	$O(t_{\text{SA}} \log^{3+\epsilon} n)$
Gagie et al. [2]	$O(t_{\text{SA}} \log D \log(D/k) \log^{1+\epsilon} n)$
Belazzougui et al. [1]	$O(t_{\text{SA}} \log k \log(D/k) \log^{\epsilon} n)$
Hon et al. [4]	$O(t_{\text{SA}} \log k \log^{\epsilon} n)$

Compressed Document Retrieval on String Collections, Table 2 Indexes of space $|\text{CSA}| + D\log(n/D) + O(D) + o(n)$ bits

Source	Report time per document
Tsur [12]	$O(t_{\text{SA}} \log k \log^{1+\epsilon} n)$
Navarro and Thankachan [8]	$O(t_{\text{SA}} \log^2 k \log^{\epsilon} n)$

Traditionally, inverted indexes are employed for this task in Information Retrieval. However, they are not powerful enough to handle scenarios where the documents need to be treated as general strings over an arbitrary alphabet set (e.g., genome sequences in bioinformatics, text in many East-Asian languages) [5]. Hon et al. [3] proposed the first solution for Problem 1, requiring $O(n \log n)$ bits of space and $O(p + k \log k)$ query time. Later, optimal $O(p + k)$ query time indexes were proposed by Navarro and Nekrich [7], and also by Shah et al. [11]. There also exist compressed/compact space solutions, tailored to specific relevance functions (mostly term-frequency or PageRank). In this article, we briefly survey the compressed space indexes for Problem 1 for the case where the relevance function is term-frequency (i.e., $w(P, d)$ is the number of occurrences of P in T_d), which we call the *Compressed Top-k Frequent Document Retrieval* (CTFDR) problem.

Key Results

First we introduce some notations. For convenience, we append a special character $ to every document. Then, $\text{T} = \text{T}_1 \circ \text{T}_2 \circ \cdots \circ \text{T}_D$ is the concatenation of all documents. GST, SA, and CSA are the suffix tree, suffix array, and a compressed suffix array of T, respectively. Notice that both GST and SA take $O(n \log n)$ bits of space, whereas the space of CSA ($|\text{CSA}|$ bits)

can be made as close as the minimum space for maintaining $\mathcal{D}$ (which is not more than $n \log \sigma$ bits) by choosing an appropriate version of CSA [6]. Using CSA, the suffix range $[sp, ep]$ of $P[1, p]$, as well as any $\text{SA}[\cdot]$, can be computed in times $\text{search}(p)$ and t_{SA}, respectively. Hon et al. [3] gave the first solution for the CTFDR Problem, requiring roughly $2|\text{CSA}|$ bits of space, whereas the first space-optimal index was given by Tsur [12]. Various improvements on both results have been proposed and are summarized in Tables 1 and 2. Notice that the total query time is $\text{search}(p)$ plus k times per reported document.

Notations and Basic Framework

The suffix tree GST of T can be considered as a generalized suffix tree of $\mathcal{D}$, where

- ℓ_i is the ith leftmost leaf in GST
- $\text{doc}(\ell_i)$ is the document to which the suffix corresponding to ℓ_i belongs
- $\text{Leaf}(u)$ is the set of leaves in the subtree of node u
- $\text{tf}(u, d)$ is the number of leaves in $\text{Leaf}(u)$ with $\text{doc}(\cdot) = d$
- $\text{Top}(u, k)$ is the set k of document identifiers with the highest $\text{tf}(u, \cdot)$ value

From now onwards, we assume all solutions consists of a fully compressed representation of GST in $|\text{CSA}| + o(n)$ bits [10], and a bitmap

$B[1, n]$, where $B[i] = 1$ iff $T[i] = \$$. We use a $D \log(n/D) + O(D) + o(n)$ bits representation of B with constant time rank/select query support [9]. Therefore, $\mathrm{doc}(\ell_i)$ can be computed as 1 plus the number of 1's in $B[1, \mathsf{SA}[i] - 1]$ in time $t_{\mathsf{SA}} + O(1)$. Observe that a CTFDL query (P, k) essentially asks to return the set $\mathsf{Top}(u_P, k)$, where u_P is the *locus node* of P in GST. Any superset of $\mathsf{Top}(u_P, k)$ be called as a candidate set of (P, k). The following lemma is crucial.

Lemma 1 *The set* $\mathsf{Top}(w, k) \cup \{\mathrm{doc}(\ell_i) | \ell_i \in \mathsf{Leaf}(u_P) \backslash \mathsf{Leaf}(w)\}$ *is a candidate set of* (P, k), *where* w *is any node in the subtree of* u_P.

All query processing algorithms consist of the following two steps: (i) Generate a candidate set $\mathcal{C}$ of size as close to k as possible. (ii) Compute $\mathrm{tf}(u_P, d)$ of all $d \in \mathcal{C}$ and report those k document identifiers with the highest $\mathrm{tf}(u_P, \cdot)$ values as output.

An Index of Size $\approx 2|\mathsf{CSA}|$ Bits

Queries are categorized into $O(\log D)$ different types.

Definition 1 A query (P, k) is of type x if $\lceil \log k \rceil = x$.

We start with the description of a structure DS_x (of size $|\mathsf{DS}_x|$ bits) that along with GST and B can generate a candidate set of size proportional to k for any type-x query. The first step is to identify a set Mark_g of nodes in GST using the scheme described in Lemma 2 (parameter g will be fixed later). Then maintain $\mathsf{Top}(u', 2^x)$ for all $u' \in \mathsf{Mark}_g$.

Lemma 2 ([3]) *There exists a scheme to identify a set* Mark_g *of nodes in* GST *(called marked nodes) based on a grouping factor* g, *where the following conditions are satisfied: (i)* $|\mathsf{Mark}_g| = O(n/g)$, *(ii) if it exists, the highest marked node* $u' \in \mathsf{Mark}_g$ *in the subtree of any node* u *is unique, and* $\mathsf{Leaf}(u) \backslash \mathsf{Leaf}(u') \le 2g$. *For example,* Mark_g *can be the set of all lowest common ancestor (LCA) nodes of* ℓ_i *and* ℓ_{i+g}, *where* i *is an integer multiple of* g.

Using DS_x, any type-x query (P, k) can be processed as follows: find the highest node u'_P in the subtree of u_P and generate the following candidate set.

$$\mathsf{Top}(u'_P, 2^x) \cup \{\mathrm{doc}(\ell_i) | \ell_i \in \mathsf{Leaf}(u_P) \backslash \mathsf{Leaf}(u'_P)\}$$

From the properties described in Lemma 2, the cardinality of this set is $O(2^x + g) = O(k + g)$ and the size of DS_x is $O((n/g)2^x \log D)$ bits. By fixing $g = 2^x \log^{2+\epsilon} n$ [3], we can bound the set cardinality by $O(k \log^{2+\epsilon} n)$ and $|\mathsf{DS}_x|$ by $O(n/\log^{1+\epsilon} n)$ bits. Therefore, we maintain DS_x for $x = 1, 2, 3, \ldots, \log D$ in $o(n)$ bits overall, and whenever a query comes, generate a candidate set of size $O(k \log^{2+\epsilon} n)$ using appropriate structures. The observation by Belazzougui et al. [1] is that $g = x2^x \log^{1+\epsilon} n$ in the above analysis yields a candidate set of even lower cardinality $O(k \log k \log^{1+\epsilon} n)$, without blowing up the space.

Later, Hon et al. [4] came up with another strategy for generating a candidate set of even smaller size, $O(k \log k \log^\epsilon n)$. They associate another structure DS_x^* (of space $|\mathsf{DS}_x^*|$ bits) with each DS_x. Essentially, DS_x^* maintains $\mathsf{Top}(u'', 2^x)$ of every $u'' \in \mathsf{Mark}_h$ with $h = x2^x \log^\epsilon n$ in an *encoded* form. Now, whenever a type-x query (P, k) comes, we first find the highest node u''_P in the subtree of u_P that belongs to Mark_h and generate the candidate set $\mathsf{Top}(u''_P, 2^x) \cup \{\mathrm{doc}(\ell_i) | \ell_i \in \mathsf{Leaf}(u_P) \backslash \mathsf{Leaf}(u''_P)\}$, whose cardinality is $O(2^x + h) = O(k \log k \log^\epsilon n)$.

We now describe the scheme for encoding a particular $\mathsf{Top}(u'', 2^x)$. Let u' be the highest node in the subtree of u'', that belongs to Mark_g. Then, $\mathsf{Top}(u'', 2^x) \subseteq \mathsf{Top}(u', 2^x) \cup \{\mathrm{doc}(\ell_i) | \ell_i \in \mathsf{Leaf}(u'') \backslash \mathsf{Leaf}(u')\}$. Notice that $\mathsf{Top}(u'', 2^x)$ is stored in DS_x and any $\mathrm{doc}(\ell_i)$ can be decoded in $O(t_{\mathsf{SA}})$ time. Therefore, instead of explicitly storing an entry d within $\mathsf{Top}(u'', 2^x)$ in $\log D$ bits, we can refer to the position of d in $\mathsf{Top}(u', 2^x)$ if $d \in \mathsf{Top}(u', 2^x)$, else refer to the relative position of a leaf node ℓ_i in $\mathsf{Leaf}(u'') \backslash \mathsf{Leaf}(u')$ with $\mathrm{doc}(\ell_i) = d$. Therefore, maintaining the following two bitmaps is sufficient.

- $F[1, 2^x]$, where $F[i] = 1$ iff ith entry in $\text{Top}(u', 2^x)$ is present in $\text{Top}(u'', 2^x)$
- $F'[1, |\text{Leaf}(u'') \backslash \text{Leaf}(u')|]$, $F'[i] = 1$ iff $\text{doc}(\cdot)$ of ith leaf node in $\text{Leaf}(u'') \backslash \text{Leaf}(u')$ is present in $\text{Top}(u'', 2^x)$, but not in $\text{Top}(u', 2^x)$.

As the total length and the number of 1's over F and F' is $O(g + 2^x)$ and $O(2^x)$, respectively, we can encode them in $O(2^x \log(g/2^x)) = O(2^x \log \log n)$ bits. Therefore, $|\text{DS}_x^*| = O((n/(x2^x \log^\epsilon n))2^x \log \log n)$ bits and $\sum_{x=1}^{\log D} |\text{DS}_x| = o(n)$ bits.

Lemma 3 *By maintaining a $|\text{CSA}| + o(n) + D \log(n/D) + O(D)$ bits space structure (which includes the space of CSA and B), a candidate set of size $O(k \log k \log^\epsilon n)$ can be generated for any query (P, k) in time $O(t_{\text{SA}} \cdot k \log k \log^\epsilon n)$.*

We now turn our attention to Step 2 of the query algorithm. Let $[sp, ep]$ be the suffix range of P in CSA and $[sp_d, ep_d]$ be the suffix range of P in CSA_d, the compressed suffix array of T_d. Hon et al. [3] showed that by additionally maintaining all CSA_d's (in space roughly $\approx |\text{CSA}|$ bits), any $[sp_d, ep_d]$ can be computed in time $O(t_{\text{SA}} \log n)$ (and thus $\text{tf}(P, d) = ep_d - sp_d + 1$). Belazzougui et al. [1] improved this time to $O(t_{\text{SA}} \log \log n)$ using $o(n)$ extra bits. Combined with Lemma 3, this gives the following.

Theorem 1 ([4]) *Using a $2|\text{CSA}| + o(n) + D \log(n/D) + O(D)$ bits space index, top-k frequent document retrieval queries can be answered in $O(\text{search}(p) + k \cdot t_{\text{SA}} \log k \log^\epsilon n)$ time.*

Space-Optimal Index

Space-optimal indexes essentially circumvent the need of CSA_d's. We first present a simplified version of Tsur's index (with slightly worse query time). To handle type-x queries, first identify the set of nodes Mark_g based on a grouping factor g (to be fixed later). Tsur proved that each node $u' \in \text{Mark}_g$ can be associated with a set $\text{Set}(u')$ of $O(2^x + \sqrt{2^x g})$ document identifiers, such

that $\text{Set}(u'_p)$ represents a candidate set, where u'_p is the highest node in the subtree of u_P that belongs to Mark_g. Therefore, we can store $S(u')$ for every $u' \in \text{Mark}_g$ along with $\text{tf}(u', d)$ of every $d \in S(u')$ in $O((n/g)(2^x + \sqrt{2^x g}) \log n)$ bits. Now a type-x query can be processed as follows:

1. Find u'_p, the highest node in the subtree of u_P that belongs to Mark_g.
2. Extract $\text{Set}(u'_p)$ and $\text{tf}(u'_p, d)$ of all $d \in \text{Set}(u'_p)$.
3. Scan the leaves in $\text{Leaf}(u_P) \backslash \text{Leaf}(u'_p)$, decode the corresponding $\text{doc}(\cdot)$ values and compute $\text{tf}(u_P, d) - \text{tf}(u'_p, d)$ for all $d \in \text{Set}(u'_p)$.
4. Then obtain $\text{tf}(u_P, d) = \text{tf}(u'_p, d) + (\text{tf}(u_P, d) - \text{tf}(u'_p, d))$ for all $d \in \text{Set}(u'_p)$.
5. Report k documents within $\text{Set}(u'_p)$ with highest $\text{tf}(u_P, \cdot)$ values as output.

In summary, an $O((n/g)(2^x + \sqrt{2^x g}) \log n)$-bit structure (along with GST and B) can answer any type-x query in $O(\text{search}(p) + (|\text{Set}(u'_p)| + |\text{Leaf}(u_P) \backslash \text{Leaf}(u'_p)|) \cdot t_{\text{SA}}) = O(\text{search}(p) + (2^x + \sqrt{2^x g} + g) \cdot t_{\text{SA}}) = O(\text{search}(p) + (g + 2^x) \cdot t_{\text{SA}})$ time. By fixing $g = x^2 2^x \log^{2+\epsilon} n$, the query time can be bounded by $O(\text{search}(p) + k \cdot t_{\text{SA}} \log^2 k \log^{2+\epsilon} n)$ and the overall space corresponding to $x = 1, 2, 3, \ldots, \log D$ is $o(n)$ bits. We remark that the index originally proposed by Tsur is even faster.

Navarro and Thankachan [8] observed that each document identifier and the associated $\text{tf}(\cdot, \cdot)$ value can be compressed into $O(\log \log n)$ bits. For compressing document identifiers, ideas from Hon et al. [4] were borrowed. For compressing $\text{tf}(\cdot, \cdot)$ values, they introduced an $o(n)$-bit structure, called *sampled document array* that can compute an approximate value of any $\text{tf}(\cdot, \cdot)$ (denoted by $\text{tf}^*(\cdot, \cdot)$) in time $O(\log \log n)$ within an additive error of at most $\log^2 n$. This means that, instead of storing $\text{tf}(\cdot, \cdot)$, storing $\text{tf}(\cdot, \cdot) - \text{tf}^*(\cdot, \cdot)$ (in just $O(\log \log n)$ bits) is sufficient. In summary, by maintaining an $O((n/g)(2^x + \sqrt{2^x g}) \log \log n)$-bit structure (along with GST, B and the sampled document array), any type-x query can be answered in $O((g + 2^x) \cdot t_{\text{SA}})$ time.

A similar analysis with $g = x^2 2^x \log^\epsilon n$ gives the following result.

Theorem 2 ([8]) *Top-k frequent document retrieval queries can be answered in* $O(\mathsf{search}(p) + k \cdot t_{\mathsf{SA}} \log^2 k \log^\epsilon n)$ *time using a* $|\mathsf{CSA}| + o(n) + D \log(n/D) + O(D)$-*bit index.*

Cross-References

▶ Compressed Suffix Array
▶ Compressed Suffix Trees
▶ Document Retrieval on String Collections
▶ Rank and Select Operations on Bit Strings
▶ Suffix Trees and Arrays

Recommended Reading

1. Belazzougui D, Navarro G, Valenzuela D (2013) Improved compressed indexes for full-text document retrieval. J Discret Algorithms 18:3–13
2. Gagie T, Kärkkäinen J, Navarro G, Puglisi SJ (2013) Colored range queries and document retrieval. Theor Comput Sci 483:36–50
3. Hon WK, Shah R, Vitter JS (2009) Space-efficient framework for top-k string retrieval problems. In: FOCS, Atlanta, pp 713–722
4. Hon WK, Shah R, Thankachan SV, Vitter JS (2014) Space-efficient frameworks for top-k string retrieval. J ACM 61(2):9
5. Navarro G (2014) Spaces, trees, and colors: the algorithmic landscape of document retrieval on sequences. ACM Comput Surv 46(4):52
6. Navarro G, Mäkinen V (2007) Compressed full-text indexes. ACM Comput Surv 39(1):2
7. Navarro G, Nekrich Y (2012) Top-k document retrieval in optimal time and linear space. In: SODA, Kyoto, pp 1066–1077
8. Navarro G, Thankachan SV (2014) New space/time tradeoffs for top-k document retrieval on sequences. Theor Comput Sci 542:83–97
9. Raman R, Raman V, Satti SR (2007) Succinct indexable dictionaries with applications to encoding k-ary trees, prefix sums and multisets. ACM Trans Algorithms 3(4):43
10. Russo L, Navarro G, Oliveira AL (2011) Fully compressed suffix trees. ACM Trans Algorithms 7(4):53
11. Shah R, Sheng C, Thankachan SV, Vitter JS (2013) Top-k document retrieval in external memory. In: ESA, Sophia Antipolis, pp 803–814
12. Tsur D (2013) Top-k document retrieval in optimal space. Inf Process Lett 113(12): 440–443

Compressed Range Minimum Queries

Johannes Fischer
Technical University Dortmund, Dortmund, Germany

Keywords

Compressed data structures; Succinct data structures

Years and Authors of Summarized Original Work

2011; Fischer, Heun

Problem Definition

Given a static array A of n totally ordered objects, the range minimum query problem (RMQ problem) is to build a data structure $\mathcal{D}$ on A that allows us to answer efficiently subsequent online queries of the form "what is the position of a minimum element in the subarray ranging from i to j?" (We consider the *minimum*; all results hold for *maximum* as well.) Such queries are denoted by $\mathsf{RMQ}_A(i, j)$ and are formally defined by $\mathsf{RMQ}_A(i, j) = \operatorname{argmin}_{i \le k \le j}\{A[k]\}$ for an array $A[1, n]$ and indices $1 \le i \le j \le n$. In the succinct or compressed setting, the goal is to use as few *bits* as possible for $\mathcal{D}$, hopefully sublinear in the space needed for storing A itself. The space for A is denoted by $|A|$ and is $|A| = \Theta(n \log n)$ bits if A stores numbers from a universe of size $n^{\Theta(1)}$.

Indexing Versus Encoding Model

There are two variations of the problem, depending on whether the input array A is available at query time (*indexing model*) or not (*encoding model*). In the indexing model, some space for the data structure $\mathcal{D}$ can in principle be saved, as the query algorithm can substitute the "missing information" by consulting A when answering the queries, and this is indeed what all indexing data structures make heavy use of. However, due to the need to access A at query time, the *total* space (which is $|A| + |\mathcal{D}|$ bits) will never be sublinear in the space needed for storing the array A itself.

This is different in the *encoding model*, where the data structure $\mathcal{D}$ must be built in a way such that the query algorithm can derive its answers *without* consulting A. Such encoding data structures are important when only the *positions* of the minima matter (and not the actual *values*) or when the access to A itself is slow.

Any encoding data structure $\mathcal{D}_E$ is automatically also an indexing data structure; conversely, an indexing data structure $\mathcal{D}_I$ can always be "converted" to an encoding data structure by storing $\mathcal{D}_I$ plus (a copy of) A. Hence, differentiating between the two concepts only makes sense if there are indexing data structures that use less space than the best encoding data structures and if there exist encoding data structures which use space sublinear in $|A|$. Interestingly, for range minimum queries, exactly this is the case.

Model of Computation

All results assume the usual word RAM model of computation with word size $\Omega(\log n)$ bits.

Key Results

Table 1 summarizes the key results from [12] by showing the sizes of data structures for range minimum queries (left column). The first data structure is in the indexing model, and the last two are encoding data structures. The leading terms ($2n/c(n)$ bits with $O(c(n))$ query time in the indexing model and $2n$ bits for arbitrary query time in the encoding model) are optimal: in the encoding model, this is rather easy to see by establishing a bijection between the class

of binary trees and the class of arrays with different answers for at least one RMQ [12], and in the indexing model, Brodal et al. [4] prove the lower bound. Particular emphasis is placed on the additional space needed for constructing the data structure (middle column), where it is important to use asymptotically less space than the final data structure.

Extensions

Surpassing the Lower Bound

Attempts have been made to break the lower bound for special cases. If A is compressible under the order-k empirical entropy measure $H_k(A)$, then also the leading term $2n/c(n)$ of the indexing data structure can be compressed to $nH_k(A)$ [12]. Other results (in both the indexing and the encoding model) exist for compressibility measures based on the number of *runs* in A [2]. Davoodi et al. [8] show that *random* input arrays can be encoded in expected $1.919n + o(n)$ bits for RMQs. All of the above results retain constant query times.

Top-k Queries

A natural generalization of RMQs is listing the k smallest (or largest) values in a query range (k needs only be specified at query time). In the indexing model, any RMQ structure with constant query time can be used to answer top-k queries in $O(k)$ time [16].

Recently, an increased interest in encoding data structures for top-k queries can be observed. For general k, Grossi et al. [15] show that $\Omega(n \log k)$ bits are needed for answering top-k queries. Therefore, interesting encodings can only exist if an upper limit κ on k is given at construction time. This lower bound is matched

Compressed Range Minimum Queries, Table 1 Data structures [12] for range minimum queries, where $|A|$ denotes the space of the (read-only) input array A. Construction space is *in addition* to the final space of the data structure. All data structures can be constructed in $O(n)$ time, and query time is $O(1)$ unless noted otherwise

Final space (bits)	Construction space	Comment
$|A| + \frac{2n}{c(n)} - \Theta\left(\frac{n \log \log n}{c(n) \log n}\right)$	$O(\log^3 n)$	Query time $O(c(n))$ for $c(n) = O(n^\varepsilon), 0 < \varepsilon < 1$
$2n + O(n \log \log n / \log n)$	$|A| + n + O\left(\frac{n \log \log n}{\log n}\right)$	Construction space improved to $|A| + o(n)$ [8]
$2n + O(n/\text{polylog } n)$	$|A| + O(n)$	Using succinct data structures from [23] and [21]

asymptotically by an encoding data structure using $O(n \log \kappa)$ bits and $O(k)$ query time by Navarro et al. [22]. For the specific case $\kappa = 2$, Davoodi et al. [8] provide a lower bound of $2.656n - O(\log n)$ bits (using computer-assisted search) and also give an encoding data structure using at most $3.272n + o(n)$ bits supporting top-2 queries in $O(1)$ time.

Range Selection

Another generalization are queries asking for the k-th smallest (or largest) value in the query range (k is again part of the query). This problem is harder for nonconstant k, as Jørgensen and Larsen prove a lower bound of $\Omega(\log k / \log \log n)$ on the query time when using $O(n \text{polylog} n)$ words of space [17]. Also, the abovementioned space lower bounds on encoding top-k queries also apply to range selection. Again, Navarro et al. [22] give a matching upper bound: $O(n \log \kappa)$ bits suffice to answer queries asking for the k-largest element in a query range ($k \leq \kappa$) in $O(\log k / \log \log n)$ time. Note that this includes queries asking for the *median* in a query range.

Higher Dimensions

Indexing and encoding data structures for range minima also exist for higher-dimensional arrays (matrices) of total size N. Atallah and Yuan [25] show an (uncompressed) indexing data structure of size $O(2^d d! N)$ words with $O(3^d)$ query time, where d is the dimension of the underlying matrix A.

Tighter results exist in the two-dimensional case [1], where A is an ($m \times n$) matrix consisting of $N = m \cdot n$ elements (w.l.o.g. assume $m \leq n$). In the indexing model, the currently best solution is a data structure of size $|A| + O(N/c)$ bits ($1 \leq c \leq n$) that answers queries in $O(c \log c \log^2 \log c)$ time [3], still leaving a gap between the highest lower bounds of $\Omega(c)$ query time for $O(N/c)$ bits of space [4].

In the encoding model, a lower bound of $\Omega(N \log m)$ bits exists [4], but the best data structure with constant query time achieves only $O(N \min\{m, \log n\})$ bits of space [4], which still leaves a gap unless $m = n^{\Omega(1)}$. However,

Brodal et al. [5] *do* show an encoding using only $O(N \log m)$ bits, but nothing better than the trivial $O(N)$ can be said about its query time. Special cases for small (constant) values of m (e.g., optimal $5n + o(n)$ bits for $m = 2$) and also for random input arrays are considered by Golin et al. [14].

Further Extensions

The indexing technique [12] has been generalized such that a specific minimum (e.g., the position of the median of the minima) can be returned if the minimum in the query range is not unique [11]. Further generalizations include functions other than the "minimum" on the query range, e.g., median [17], mode [6], etc. RMQs have also been generalized to edge-weighted trees [9], where now a query specifies two nodes v and w, and a minimum-weight edge on the path from v to w is sought.

Applications

Data structures for RMQs have many applications. Most notably, the problem of preprocessing a tree for *lowest common ancestor* (LCA) queries is equivalent to the RMQ problem. In succinctly encoded trees (using balanced parentheses), RMQs can also be used to answer LCA queries in constant time; in this case, the RMQ structure is built on the virtual excess sequence of the parentheses and uses only $o(n)$ bits in addition to the parenthesis sequence [24]. Other applications of RMQs include document retrieval [20], succinct trees [21], compressed suffix trees [13], text-index construction [10], Lempel-Ziv text compression [e.g., 18], orthogonal range searching [19], and other kinds of range queries [7].

Cross-References

Recommended Reading

1. Amir A, Fischer J, Lewenstein M (2007) Two-dimensional range minimum queries. In: Proceedings of the CPM, London, LNCS, vol 4580. Springer, pp 286–294
2. Barbay J, Fischer J, Navarro G (2012) LRM-trees: compressed indices, adaptive sorting, and compressed permutation. Theor Comput Sci 459:26–41
3. Brodal GS, Davoodi P, Lewenstein M, Raman R, Rao SS (2012) Two dimensional range minimum queries and Fibonacci lattices. In: Proceedings of the ESA, Ljubljana, LNCS, vol 7501. Springer, pp 217–228
4. Brodal GS, Davoodi P, Rao SS (2012) On space efficient two dimensional range minimum data structures. Algorithmica 63(4):815–830
5. Brodal GS, Brodnik A, Davoodi P (2013) The encoding complexity of two dimensional range minimum data structures. In: Proceedings of the ESA, Sophia Antipolis, LNCS, vol 8125. Springer, pp 229–240
6. Chan TM, Durocher S, Larsen KG, Morrison J, Wilkinson BT (2014) Linear-space data structures for range mode query in arrays. Theory Comput Syst 55(4):719–741
7. Chen KY, Chao KM (2004) On the range maximum-sum segment query problem. In: Proceedings of the ISAAC, Hong Kong, LNCS, vol 3341. Springer, pp 294–305
8. Davoodi P, Navarro G, Raman R, Rao SS (2014) Encoding range minima and range top-2 queries. Philos Trans R Soc A 372:20130,131
9. Demaine ED, Landau GM, Weimann O (2014) On Cartesian trees and range minimum queries. Algorithmica 68(3):610–625
10. Fischer J (2011) Inducing the LCP-array. In: Proceedings of the WADS, New York, LNCS, vol 6844. Springer, pp 374–385
11. Fischer J, Heun V (2010) Finding range minima in the middle: approximations and applications. Math Comput Sci 3(1):17–30
12. Fischer J, Heun V (2011) Space efficient preprocessing schemes for range minimum queries on static arrays. SIAM J Comput 40(2):465–492
13. Fischer J, Mäkinen V, Navarro G (2009) Faster entropy-bounded compressed suffix trees. Theor Comput Sci 410(51):5354–5364
14. Golin M, Iacono J, Krizanc D, Raman R, Rao SS (2011) Encoding 2d range maximum queries. In: Proceedings of the ISAAC, Yokohama, LNCS, vol 7074. Springer, pp 180–189
15. Grossi R, Iacono J, Navarro G, Raman R, Rao SS (2013) Encodings for range selection and top-k queries. In: Proceedings of the ESA, Sophia Antipolis, LNCS, vol 8125. Springer, pp 553–564
16. Hon WK, Shah R, Thankachan SV, Vitter JS (2014) Space-efficient frameworks for top-k string retrieval. J ACM 61(2):Article No. 9
17. Jørgensen AG, Larsen KG (2011) Range selection and median: tight cell probe lower bounds and adaptive data structures. In: Proceedings of the SODA, San Francisco. ACM/SIAM, pp 805–813
18. Kärkkäinen J, Kempa D, Puglisi SJ (2013) Lightweight Lempel-Ziv parsing. In: Proceedings of the SEA, Rome, LNCS, vol 7933. Springer, pp 139–150
19. Lewenstein M (2013) Orthogonal range searching for text indexing. In: Space-efficient data structures, streams, and algorithms, LNCS, vol 8066. Springer, Heidelberg, pp 267–302
20. Navarro G (2014) Spaces, trees, and colors: the algorithmic landscape of document retrieval on sequences. ACM Comput Surv 46(4):Article No. 52
21. Navarro G, Sadakane K (2014) Fully functional static and dynamic succinct trees. ACM Trans Algorithms 10(3):Article No. 16
22. Navarro G, Raman R, Satti SR (2014) Asymptotically optimal encodings for range selection. In: Proceedings of the FSTTCS, New Delhi. IBFI Schloss Dagstuhl, paper to be published
23. Pătraşcu M (2008) Succincter. In: Proceedings of the FOCS, Washington, DC. IEEE Computer Society, pp 305–313
24. Sadakane K (2007) Compressed suffix trees with full functionality. Theory Comput Syst 41(4):589–607
25. Yuan H, Atallah MJ (2010) Data structures for range minimum queries in multidimensional arrays. In: Proceedings of the SODA, Austin. ACM/SIAM, pp 150–160

Compressed Representations of Graphs

J. Ian Munro[1] and Patrick K. Nicholson[2]
[1]David R. Cheriton School of Computer Science, University of Waterloo, Waterloo, ON, Canada
[2]Department D1: Algorithms and Complexity, Max Planck Institut für Informatik, Saarbrücken, Germany

Keywords

Digraphs; Graph representations; k-page graphs; Partial orders; Planar graphs; Succinct data structures

Years and Authors of Summarized Original Work

1989; Jacobson

Problem Definition

The problem is to represent a graph of a given *size* and *type* (e.g., an n node planar graph) in a compressed form while still supporting efficient navigation operations. More formally, if $G = (V, E)$ is a graph of a given type χ, with n nodes and m edges, then represent G using $\lg |\chi| + o(\lg |\chi|)$ bits of space, and support a set of appropriate operations on the graph in constant time (assuming we can access $\Theta(\lg n)$ consecutive bits in one operation). This may not be possible; if not, then explore what trade-offs are possible. Data structures that achieve this space bound are called *succinct* [13]. To simplify the statement of results, we assume the graph G in question contains no self-loops, and we also restrict ourselves to the static case.

Key Results

Outerplanar, Planar, and k-Page Graphs

The area of succinct data structures was initiated by Jacobson [13], who presented a succinct representation of planar graphs. His approach was to decompose the planar graph into at most *four* one-page (or outerplanar) graphs by applying a theorem of Yannikakis [20]. Each one-page graph is then represented as a sequence of *balanced parentheses*: this representation extends naturally to k-pages for $k \geq 1$. Using this representation, it is straightforward to support the following operations efficiently:

- Adjacent(x, y): report whether there is an edge $(x, y) \in E$.
- Neighbors(x): report all vertices that are adjacent to vertex x.
- Degree(x): report the degree of vertex x.

Munro and Raman [16] improved Jacobson's balanced parenthesis representation, thereby im-

proving the constant factor in the space bound for representing both k-page and planar graphs. We present Table 1 which compares a representative selection of the various succinct data structures for representing planar graphs. Subsequent simplifications to the representation of balanced parentheses have been presented; cf., [11, 17]. Barbay et al. [1] present results for larger values of k, as well as the case where the edges or vertices of the graph have labels.

The decomposition into four one-page graphs is not the only approach to representing static planar graphs. Chuang et al. [7] presented another encoding based on *canonical orderings* of a planar graph and represented the graph using a *multiple parentheses sequence* (a sequence of balanced parentheses of more than one type). Later, Chiang et al. [6] generalized the notion of canonical orderings to *orderly spanning trees*, yielding improved constant factors in terms of n and m. Gavoille and Hanusse [10] presented an alternate encoding scheme for k-page graphs that yields a trade-off based on the number of isolated nodes (connected components with one vertex). Further improvements have been presented by Chuang et al. [7], as well as Castelli Aleardi et al. [5], for the special case of planar triangulations.

Blandford et al. [2] considered unlabelled *separable graphs*. A separable graph is one that admits an $O(n^c)$ separator for $c < 1$. Their structure occupies $O(n)$ bits and performs all three query types optimally. Subsequently, Blelloch and Farzan [3] made the construction of Blandford et al. [2] succinct in the sense that, given a graph G from a separable class of graphs χ (e.g., the class of arbitrary planar graphs), their data structure represents G using $\lg |\chi| + o(n)$ bits. Interestingly, we need not even *know* the value of $\lg |\chi|$ in order to use this representation.

Arbitrary Directed Graphs, DAGs, Undirected Graphs, and Posets

We consider the problem of designing succinct data structures for *arbitrary* digraphs. In a directed graph, we refer to the set of vertices $\{y : (x, y) \in E\}$ as the *successors* of x and

Compressed Representations of Graphs, Table 1
Comparison of various succinct planar graph representations. The second column indicates whether the representation supports multigraphs. For the entries marked with a †, the query cost is measured in *bit* accesses. Notation: n is the number of vertices in G, m is the number of edges in G, i is the number of isolated vertices in G, ε is an arbitrary positive constant, $\tau = \min\{\lg k / \lg \lg m, \lg \lg k\}$, and H is the information theoretic lower space bound for storing a graph G drawn from a class of separable graphs

Type	Multi	Ref.	Space in bits	Adjacent(x, y)	Neighbors(x)	Degree(x)
k-page	N	[13]	$O(kn)$	$O(\lg n + k)$†	$O(\deg(x) \lg n + k)$†	$O(\lg n)$†
	N	[10]	$(2(m+i) + o(m+i)) \lg k$	$O(\tau \lg k)$	$O(\deg(x)\tau)$	$O(1)$
	N	[1]	$2m \lg k + n + o(m \lg k)$	$O(\lg k \lg \lg k)$	$O(\deg(x) \lg \lg k)$	$O(1)$
	N	[1]	$(2 + \varepsilon)m \lg k + n + o(m \lg k)$	$O(\lg k)$	$O(\deg(x))$	$O(1)$
	Y	[16]	$2m + 2kn + o(kn)$	$O(k)$	$O(\deg(x) + k)$	$O(1)$
Planar	N	[13]	$O(n)$	$O(\lg n)$†	$O(\deg(x) \lg n)$†	$O(\lg n)$†
	N	[10]	$12n + 4i + o(n)$	$O(1)$	$O(\deg(x))$	$O(1)$
	N	[7]	$\frac{5}{3}m + (5 + \varepsilon)n + o(n)$	$O(1)$	$O(\deg(x))$	$O(1)$
	N	[6]	$2m + 2n + o(m + n)$	$O(1)$	$O(\deg(x))$	$O(1)$
	Y	[16]	$2m + 8n + o(n)$	$O(1)$	$O(\deg(x))$	$O(1)$
	Y	[7]	$2m + (5 + \varepsilon)n + o(n)$	$O(1)$	$O(\deg(x))$	$O(1)$
	Y	[6]	$2m + 3n + o(m + n)$	$O(1)$	$O(\deg(x))$	$O(1)$
Triangulation	N	[7]	$2m + n + o(n)$	$O(1)$	$O(\deg(x))$	$O(1)$
	Y	[7]	$2m + 2n + o(n)$	$O(1)$	$O(\deg(x))$	$O(1)$
Separable	N	[2]	$O(n)$	$O(1)$	$O(\deg(x))$	$O(1)$
	N	[3]	$H + o(n)$	$O(1)$	$O(\deg(x))$	$O(1)$

the set of vertices $\{y : (y, x) \in E\}$ as the *predecessors* of x.

It is well known that an arbitrary directed graph can be represented using $n \times n$ bits by storing its adjacency matrix. This representation supports the Adjacency(x, y) operation by probing a single bit in the table in constant time. On the other hand, we can represent the graph using $\Theta(m \lg n)$ bits using an adjacency list representation, such that the following operations can be supporting in constant time:

- Successor(x, i): list the ith successor of vertex x
- Predecessor(x, i): list the ith predecessor of vertex x

The information theoretic lower bound dictates that essentially $\lg \binom{n^2}{m}$ bits are necessary for representing an arbitrary digraph. By representing each row (resp. column) of the adjacency matrix using an *indexable dictionary* [19], we get a data structure that supports Adjacency and Successor (resp. Predecessor) queries in constant time and occupies $\lg \binom{n^2}{m} + o\left(\lg \binom{n^2}{m}\right)$

bits of space. Note that we can only support two of the three operations with this approach. Farzan and Munro [9] showed that if $\Theta(n^\varepsilon) \leq m \leq \Theta\left(n^{2-\varepsilon}\right)$ for some constant $\varepsilon > 0$, then $(1 + \varepsilon')\binom{n^2}{m}$ bits are sufficient and required to support all three operations. In a more general setting, Golynski [12] had proven that the difficulty in supporting both Successor and Predecessor queries simultaneously using succinct space relates to the fact that they have the so-called *reciprocal property*. In other words, if the graph is not extremely sparse or dense, then it is impossible to support all three operations succinctly. On the other hand, if $m = o(n^\varepsilon)$ or $m = \Omega(n^2/\lg^{1-\varepsilon} n)$, for some constant $\varepsilon > 0$, then Farzan and Munro [9] showed that succinctness can be achieved while supporting the three operations.

Suppose G is a directed acyclic graph instead of an arbitrary digraph. In this case, one can exploit the fact that the graph is acyclic by ordering the vertices topologically. This ordering induces an adjacency matrix which is upper triangular. Exploiting this fact, when $\Theta(n^\varepsilon) \leq$

$m \leq \Theta(n^{2-\varepsilon})$, the data structure of Farzan and Munro can support all three operations using $(1 + \varepsilon) \lg \binom{\binom{n}{2}}{m}$ bits, for an arbitrary positive constant ε, which is optimal. If $m = o(n^{\varepsilon})$ or $m = \Omega(n^2/\lg^{1-\varepsilon} n)$, for some constant $\varepsilon > 0$, then they achieve $\lg \binom{\binom{n}{2}}{m}(1 + o(1))$ bits. By orienting the edges in an *undirected* graph so that they are directed toward the vertex with the with the larger label, this representation can also be used to support the operations `Adjacency` and `Neighbors` on an arbitrary undirected graph.

A partial order or *poset* is a directed acyclic graph with the additional transitivity constraint on the set of edges E: if (x, y) and (y, z) are present in E, then it is implied that $(x, z) \in E$. Farzan and Fischer [8] showed that a poset can be stored using $2nw(1 + o(1)) + (1 + \varepsilon)n \lg n$ bits of space, where w is the *width* of the poset – i.e., the length of the maximum antichain – and ε is an arbitrary positive constant. Their data structure supports `Adjacency` queries in constant time, as well as many other operations in time proportional to w. This matches a lower bound of Brightwell and Goodall [4] up to the additive $\varepsilon n \lg n$ term when n is sufficiently large relative to w. For an arbitrary poset, Kleitman and Rothschild showed that $n^2/4 + O(n)$ bits are sufficient and necessary by a constructive counting argument [14]. Munro and Nicholson [15, 18] showed that there is a data structure that occupies $n^2/4 + o(n^2)$ bits, such that `Adjacency`, `Predecessor`, and `Successor` queries can be supported in constant time.

Cross-References

▶ Compressed Tree Representations
▶ Rank and Select Operations on Bit Strings
▶ Recursive Separator Decompositions for Planar Graphs

Recommended Reading

1. Barbay J, Castelli Aleardi L, He M, Munro JI (2012) Succinct representation of labeled graphs. Algorithmica 62(1–2):224–257

2. Blandford DK, Blelloch GE, Kash IA (2003) Compact representations of separable graphs. In: SODA, ACM/SIAM, pp 679–688

3. Blelloch GE, Farzan A (2010) Succinct representations of separable graphs. In: Amir A, Parida L (eds) CPM. Lecture notes in computer science, vol 6129. Springer, Berlin/New York, pp 138–150

4. Brightwell G, Goodall S (1996) The number of partial orders of fixed width. Order 13(4):315–337. doi:10.1007/BF00405592, http://dx.doi.org/10.1007/BF00405592

5. Castelli Aleardi L, Devillers O, Schaeffer G (2005) Succinct representation of triangulations with a boundary. In: Dehne FKHA, López-Ortiz A, Sack JR (eds) WADS. Lecture notes in computer science, vol 3608. Springer, Berlin/New York, pp 134–145

6. Chiang YT, Lin CC, Lu HI (2005) Orderly spanning trees with applications. SIAM J Comput 34(4):924–945

7. Chuang RCN, Garg A, He X, Kao MY, Lu HI (1998) Compact encodings of planar graphs via canonical orderings and multiple parentheses. In: Larsen KG, Skyum S, Winskel G (eds) ICALP. Lecture notes in computer science, vol 1443. Springer, Berlin/New York, pp 118–129

8. Farzan A, Fischer J (2011) Compact representation of posets. In: Asano T, Nakano SI, Okamoto Y, Watanabe O (eds) ISAAC. Lecture notes in computer science, vol 7074. Springer, Berlin/New York, pp 302–311

9. Farzan A, Munro JI (2013) Succinct encoding of arbitrary graphs. Theor Comput Sci 513:38–52

10. Gavoille C, Hanusse N (2008) On compact encoding of pagenumber. Discret Math Theor Comput Sci 10(3)

11. Geary RF, Rahman N, Raman R, Raman V (2006) A simple optimal representation for balanced parentheses. Theor Comput Sci 368(3):231–246

12. Golynski A (2009) Cell probe lower bounds for succinct data structures. In: Proceedings of the 20th annual ACM-SIAM symposium on discrete algorithms (SODA). Society for Industrial and Applied Mathematics, pp 625–634

13. Jacobson G (1989) Space-efficient static trees and graphs. 30th annual IEEE symposium on foundations of computer science, pp 549–554

14. Kleitman DJ, Rothschild BL (1975) Asymptotic enumeration of partial orders on a finite set. Trans Am Math Soc 205:205–220

15. Munro JI, Nicholson PK (2012) Succinct posets. In: Epstein L, Ferragina P (eds) ESA. Lecture notes in computer science, vol 7501. Springer, Berlin/New York, pp 743–754

16. Munro JI, Raman V (2001) Succinct representation of balanced parentheses and static trees. SIAM J Comput 31(3):762–776

17. Navarro G, Sadakane K (2014) Fully-functional static and dynamic succinct trees. ACM Trans Algorithms 10(3):article 16

18. Nicholson PK (2013) Space efficient data structures in the word-RAM and bitprobe models. PhD thesis, University of Waterloo

19. Raman R, Raman V, Rao SS (2007) Succinct indexable dictionaries with applications to encoding *k*-ary trees, prefix sums and multisets. ACM Trans Algorithms 3(4)

20. Yannakakis M (1989) Embedding planar graphs in four pages. J Comput Syst Sci 38(1):36–67

Compressed Suffix Array

Djamal Belazzougui, Veli Mäkinen, and Daniel Valenzuela
Department of Computer Science, Helsinki Institute for Information Technology (HIIT), University of Helsinki, Helsinki, Finland

Keywords

Burrows-wheeler transform; FM-index; Self-indexing

Years and Authors of Summarized Original Work

2000, 2003; Sadakane
2000, 2005; Grossi, Vitter
2000, 2005; Ferragina, Manzini

Problem Definition

Given a *text string* $T = t_1 t_2 \ldots t_n$ over an alphabet Σ of size σ, the suffix array $A[1, n]$ is a permutation of the interval $[1, n]$ that sorts the suffixes of T. More precisely, it satisfies $T[A[i], n] < T[A[i + 1], n]$ for all $1 \leq i < n$, where "$<$" between strings is the lexicographical order. The *suffix array* is the canonical full-text index that allows to efficiently compute basic string matching queries on T.

The *compressed suffix array (CSA)* problem asks to replace A with a space-efficient data structure that is capable of efficiently computing $A[i]$.

If a CSA does not require T to operate, and is capable of efficiently answering substring queries on T, it is called a *self-index*, as it can be seen as a replacement of T itself. Typical queries required from such an index are the following:

- count(P): count how many times a given *pattern string* $P = p_1 p_2 \ldots p_m$ occurs in T.
- locate(P): return the locations where P occurs in T.
- display(i, j): return $T[i, j]$.

Key Results

Ψ-Based CSAs

The first solution to the problem is by Grossi and Vitter [8], who exploit the regularities of the suffix array via the Ψ-function:

Definition 1 Given suffix array $A[1, n]$, *function* $\Psi : [1, n] \to [1, n]$ is defined so that, for all $1 \leq i \leq n$, $A[\Psi(i)] = A[i] + 1$. The exception is $A[1] = n$, in which case the requirement is that $A[\Psi(1)] = 1$ so that Ψ is a permutation.

The following lemma shows that Ψ is appealing to compression:

Lemma 1 *Given a text $T[1, n]$, its suffix array $A[1, n]$, and the corresponding function Ψ, it holds $\Psi(i) < \Psi(i + 1)$ whenever $T_{A[i]} = T_{A[i+1]}$.*

Grossi and Vitter used a hierarchical decomposition of Ψ into $h = \lceil \log \log n \rceil$ levels. The piecewise increasing property of Ψ can be used to represent each level of Ψ in $\frac{1}{2} n \log \sigma$ bits [8]. By storing some sampled values of A in the bottom level, any $A[i]$ can be computed by traversing the hierarchical structure. Other tradeoffs are possible using different amount of levels. The following one involves the use of a constant number of levels:

Theorem 1 (inspired from [8]) *The Compressed Suffix Array of Grossi and Vitter supports retrieving $A[i]$ in $O(\log^\epsilon n)$ time using $(\frac{1}{\epsilon} n) \log \sigma + O(n \log \log \sigma)$ bits of space, for any $0 < \epsilon < 1$.*

As a consequence, simulating the classical binary searches [13] to find the range of suffix array containing all the occurrences of a pattern $P[1, m]$ in $T[1, n]$, can then be done in $O(m \log^{1+\epsilon} n)$ time.

Sadakane [16] shows how the above compressed suffix array can be converted into a self-index, and at the same time optimized it in several ways.

Sadakane represents both A and T using the full function Ψ, and a few extra structures. Imagine one wishes to compare P against $T[A[i], n]$. For the binary search, one needs to extract enough characters from $T[A[i], n]$ so that its lexicographical relation to P is clear. Retrieving character $T[A[i]]$, given i, is easy. Use a bit vector $F[1, n]$ marking the suffixes of $A[i]$ where the first character changes from that of $A[i-1]$. After preprocessing F for $rank$-queries, computing $j = \text{rank}_1(F, i)$ tells us that $T[A[i]] = c_j$, where c_j is the j-th smallest alphabet character. Once $T[A[i]] = c_j$ is determined this way, one needs to obtain the next character, $T[A[i] + 1]$. But $T[A[i]+1] = T[A[\Psi(i)]]$, so one can simply move to $i' = \Psi(i)$ and keep extracting characters with the same method, as long as necessary. Note that at most $|P| = m$ characters suffice to decide a comparison with P. Thus the binary search is simulated in $O(m \log n)$ time.

Up to now, the space used is $n + o(n) + \sigma \log \sigma$ bits for F and Σ. Sadakane [16] gives an improved representation for Ψ using $O(nH_0 + n \log \log \sigma)$ bits, where H_0 is the *zeroth order entropy* of T.

Sadakane also shows how $A[i]$ can be retrieved, by plugging in the hierarchical scheme of Grossi and Vitter. He adds to the scheme the retrieval of the inverse $A^{-1}[j]$. This is used in order to retrieve arbitrary text substrings $T[p, r]$, by first applying $i = A^{-1}[p]$ and then continuing as before to retrieve $r - p + 1$ first characters of suffix $T[A[i], n]$. This capability turns the compressed suffix array into self-index. The following bound is a modified version of Sadakane's CSA taken from [15]:

Theorem 2 *The* Compressed Suffix Array *of Sadakane is a self-index occupying* $\frac{1}{\epsilon} n H_0 +$ $O(n \log \log \sigma)$ *bits, and supporting retrieval of values* $A[i]$ *and* $A^{-1}[j]$ *in* $O(\log^\epsilon n)$ *time, counting of pattern occurrences in* $O(m \log n)$ *time, and displaying any substring of T of length* ℓ *in* $O(\ell + \log^\epsilon n)$ *time. Here* $0 < \epsilon \le 1$ *is an arbitrary constant.*

Grossi, Gupta, Vitter, and Foschini [6, 9] have improved the space requirement of compressed suffix arrays to depend on the *k-th order entropy* H_k of T. The idea behind this improvement is a more careful analysis of regularities captured by the Ψ-function when combined with the indexing capabilities of their new elegant data structure, *wavelet tree*. They obtain, among other results, the following tradeoff:

Theorem 3 (Grossi, Gupta, and Vitter [9]) *The* Compressed Suffix Array *of Grossi, Gupta, and Vitter is a self-index of size* $\frac{1}{\epsilon} n H_k + o(n \log \sigma)$ *bits, that supports* $A[i]$ *and* $A^{-1}[j]$ *in* $O(\log^{1+\epsilon} n/\epsilon)$ *time,* count(P) *in* $O(m \log \sigma + \log^{2+\epsilon} n/\epsilon)$ *time, and* display(i, j) *in* $O((j-i)/\log_\sigma n + \log^{1+\epsilon} n/\epsilon)$ *time. Here* $0 < \epsilon \le 1$ *is an arbitrary constant,* $k \le \alpha \log_\sigma n$ *for some constant* $0 < \alpha < 1$.

They also obtain an interesting special case:

Theorem 4 (Grossi, Gupta, and Vitter [9]) *The space optimized* Compressed Suffix Array *of Grossi, Gupta, and Vitter is a self-index of size* $n H_k + o(n \log \sigma)$ *bits, that supports* $A[i]$ *and* $A^{-1}[j]$ *in* $O(\log^2 n/\log \log n)$ *time,* count(P) *in* $O(m \log n \log \sigma + \log^3 n/\log \log n)$ *time, and* display(i, j) *in* $O((j - i)/\log \sigma + \log^2 n/\log \log n)$ *time. Here* $k \le \alpha \log_\sigma n$ *for some constant* $0 < \alpha < 1$.

In the above results, value k must be fixed before building the indexes. Later, they notice that a simple coding of Ψ-values yields an nH_k-dependent bound without the need of fixing k beforehand [6].

FM-Index

A different solution to the problem (at least on the surface) is obtained by exploiting the connection of *Burrows-Wheeler Transform (BWT)* [2] and *Suffix Array* data structure

[13]. The BWT is formed by a permutation T^{bwt} of T defined as $T^{\text{bwt}}[i] = T[A[i] - 1]$ for $A[i] > 1$ and $T^{\text{bwt}}[i] = T[n]$ for $A[i] = 1$. Without lack of generality, one can assume that T ends with $T[n] = \$$ with $\$$ being distinct symbol smaller than other symbols in T. Then $T^{\text{bwt}}[1] = T[n-1]$.

A property of the BWT is that symbols having the same context (i.e., string following them in T) are consecutive in T^{bwt}. This makes it easy to compress T^{bwt} achieving space close to high-order empirical entropies [14].

Ferragina and Manzini [3] discovered a way to combine the compressibility of the BWT and the indexing properties of the suffix array. The structure is essentially a compressed representation of the BWT plus some small additional structures to make it searchable.

To retrieve the whole text from the structure (that is, to support display$(1, n)$), it is enough to invert the BWT. For this purpose, let us consider a table $LF[1, n]$ defined such that if $T[i]$ is permuted to $T^{\text{bwt}}[j]$ and $T[i-1]$ to $T^{\text{bwt}}[j']$ then $LF[j] = j'$. It is then immediate that T can be retrieved *backwards* by printing $\$ \cdot T^{\text{bwt}}[1] \cdot T^{\text{bwt}}[LF[1]] \cdot T^{\text{bwt}}[LF[LF[1]]] \ldots$.

To represent array LF space-efficiently, Ferragina and Manzini noticed that each $LF[i]$ can be expressed as follows:

Lemma 2 (Ferragina and Manzini [3])
$LF[i] = C(c) + \text{rank}_c(i)$, *where* $c = T^{\text{bwt}}[i]$, $C(c)$ *tells how many times symbols smaller than* c *appear in* T^{bwt} *and* $\text{rank}_c(i)$ *tells how many times symbol* c *appears in* $T^{\text{bwt}}[1, i]$.

It was later observed that LF is in fact the inverse of Ψ.

It also happens that the very same two-part expression of $LF[i]$ enables efficient count(P) queries. The idea is that if one knows the range of the suffix array, say $A[sp_i, ep_i]$, such that the suffixes $T[A[sp_i], n], T[A[sp_i + 1], n], \ldots, T[A[ep_i], n]$ are the only ones containing $P[i, m]$ as a prefix, then one can compute the new range $A[sp_{i-1}, ep_{i-1}]$ where the suffixes contain $P[i-1, m]$ as a prefix, as follows:

$sp_{i-1} = C(P[i-1]) + \text{rank}_{P[i-1]}(sp_i - 1) + 1$ and $ep_{i-1} = C(P[i-1]) + \text{rank}_{P[i-1]}(ep_i)$. It is then enough to scan the pattern *backwards* and compute values $C()$ and $\text{rank}_c()$ $2m$ times to find out the (possibly empty) range of the suffix array where all the suffixes start with the complete P. Returning $ep_1 - sp_1 + 1$ solves the count(P) query without the need of having the suffix array available at all.

For locating each such occurrence $A[i]$, $sp_1 \leq i \leq ep_1$, one can compute the sequence i, $LF[i]$, $LF[LF[i]]$, …, until $LF^k[i]$ is a *sampled suffix array position*; sampled positions can be marked in a bit vector B such that $B[LF^k[i]] = 1$ indicates that $\text{samples}[\text{rank}_1(B, LF^k[i])] = A[LF^k[i]]$, where samples is a compact array storing the sampled suffix array values. Then $A[i] = A[LF^k[i]] + k = \text{samples}[\text{rank}_1(B, LF^k[i])] + k$. A similar structure can be used to support display(i, j).

Values $C()$ can be stored trivially in a table of $\sigma \log_2 n$ bits. $T^{\text{bwt}}[i]$ can be computed in $O(\sigma)$ time by checking for which c is $\text{rank}_c(i) \neq \text{rank}_c(i-1)$. The suffix array sampling rate can be chosen as $s = \Theta(\log^{1+\epsilon} n)$ so that the samples require $o(n)$ bits. The real challenge is to preprocess the text for $\text{rank}_c()$ queries. The original proposal builds several small partial sum data structures on top of the compressed BWT, and achieves the following result:

Theorem 5 (Ferragina and Manzini [3])
The FM-Index (FMI) *is a self-index of size* $5nH_k + o(n \log \sigma)$ *bits that supports* count(P) *in* $O(m)$ *time,* locate(P) *in* $O(\sigma \log^{1+\epsilon} n)$ *time per occurrence, and* display(i, j) *in* $O(\sigma(j - i + \log^{1+\epsilon} n))$ *time. Here* $\sigma = o(\log n / \log \log n)$, $k \leq \log_\sigma(n / \log n) - \omega(1)$, *and* $\epsilon > 0$ *is an arbitrary constant.*

The original FM-Index has a severe restriction on the alphabet size. This has been removed in follow-up works. Conceptually, the easiest way to achieve a more alphabet-friendly instance of the FM-index is to build a wavelet tree [9] on T^{bwt}. It allows one to simulate a single $\text{rank}_c()$ query or to obtain $T^{\text{bwt}}[i]$ in $O(\log \sigma)$ time. Some later enhancements have improved the time requirement, so as to obtain, for example, the following result:

Theorem 6 (Mäkinen and Navarro [11]) *The CSA problem can be solved using a so-called* Succinct Suffix Array (SSA), *of size* $nH_0 + o(n \log \sigma)$ *bits that supports* count(P) *in* $O(m(1 + \log \sigma / \log \log n))$ *time,* locate(P) *in* $O(\log^{1+\epsilon} n \, (1 + \log \sigma / \log \log n))$ *time per occurrence, and* display(i, j) *in* $O((j - i + \log^{1+\epsilon} n)(1 + \log \sigma / \log \log n))$ *time. Here* $\sigma = o(n)$ *and* $\epsilon > 0$ *is an arbitrary constant.*

Ferragina et al. [4] developed a technique called *compression boosting* that finds an optimal partitioning of T^{bwt} such that, when one compresses each piece separately using its zero-order model, the result is proportional to the k-th order entropy. It was observed in [10] that a fixed block partitioning achieves the same result.

Compression boosting can be combined with the idea of SSA by building a wavelet tree separately for each piece and some additional structures in order to solve global rank$_c$() queries from the individual wavelet trees:

Theorem 7 (Ferragina et al. [5]) *The CSA problem can be solved using a so-called* Alphabet-Friendly FM-Index (AF-FMI), *of size* $nH_k + o(n \log \sigma)$ *bits, with the same time complexities and restrictions of SSA with* $k \leq \alpha \log_\sigma n$, *for any constant* $0 < \alpha < 1$.

A careful analysis [12] reveals that the space of the plain SSA is bounded by the same $nH_k + o$ $(n \log \sigma)$ bits, making the boosting approach to achieve the same result unnecessary in theory. By plugging a better wavelet tree implementation [7], the space of Theorem 7 can be improved to $nH_k + o(n)$ bits.

The wavelet tree is space-efficient, but it cannot operate in time better than $O(1 + \frac{\log \sigma}{\log \log n})$. To achieve better performance, some other techniques must be used. One example, is the following fastest FM-index with dominant term nH_k in the space.

Theorem 8 (Belazzougui and Navarro [1]) *The CSA problem can be solved using an index of size* $nH_k + o(n \log \sigma)$, *that supports* count(P) *in* $O(m)$ *time,* locate(P) *in* $O(\log_\sigma n \log \log n)$ *time per occurrence, and* display(i, j) *in* $O((j - i) + \log_\sigma n \log \log n)$ *with* $k \leq \alpha \log_\sigma n$, $\sigma = O(n)$, *and* α *is any constant such that* $0 < \alpha < 1$.

Cross-References

▶ Burrows-Wheeler Transform
▶ Rank and Select Operations on Bit Strings
▶ Rank and Select Operations on Sequences
▶ Suffix Trees and Arrays
▶ Wavelet Trees

Recommended Reading

1. Belazzougui D, Navarro G (2011) Alphabet-independent compressed text indexing. In: ESA, Saarbrücken, pp 748–759
2. Burrows M, Wheeler D (1994) A block sorting lossless data compression algorithm. Technical report 124, Digital Equipment Corporation
3. Ferragina P, Manzini G (2005) Indexing compressed texts. J ACM 52(4):552–581
4. Ferragina P, Giancarlo R, Manzini G, Sciortino M (2005) Boosting textual compression in optimal linear time. J ACM 52(4):688–713
5. Ferragina P, Manzini G, Mäkinen V, Navarro G (2007) Compressed representations of sequences and full-text indexes. ACM Trans Algorithms 3(2):20
6. Foschini L, Grossi R, Gupta A, Vitter JS (2006) When indexing equals compression: experiments with compressing suffix arrays and applications. ACM Trans Algorithms 2(4):611–639
7. Golynski A, Raman R, Srinivasa Rao S (2008) On the redundancy of succinct data structures. In: SWAT, Gothenburg, pp 148–159
8. Grossi R, Vitter J (2006) Compressed suffix arrays and suffix trees with applications to text indexing and string matching. SIAM J Comput 35(2):378–407
9. Grossi R, Gupta A, Vitter J (2003) High-order entropy-compressed text indexes. In: Proceedings of the 14th annual ACM-SIAM symposium on discrete algorithms (SODA), Baltimore, pp 841–850
10. Kärkkäinen J, Puglisi SJ (2011) Fixed block compression boosting in fm-indexes. In: SPIRE, Pisa, pp 174–184
11. Mäkinen V, Navarro G (2005) Succinct suffix arrays based on run-length encoding. Nord J Comput 12(1):40–66
12. Mäkinen V, Navarro G (2008) Dynamic entropy-compressed sequences and full-text indexes. ACM Trans Algorithms 4(3):32

13. Manber U, Myers G (1993) Suffix arrays: a new method for on-line string searches. SIAM J Comput 22(5):935–948
14. Manzini G (2001) An analysis of the Burrows-Wheeler transform. J ACM 48(3):407–430
15. Navarro G, Mäkinen V (2007) Compressed full-text indexes. ACM Comput Surv 39(1): Article 2
16. Sadakane K (2003) New text indexing functionalities of the compressed suffix arrays. J Algorithms 48(2):294–313

Compressed Suffix Trees

Luís M.S. Russo
Departamento de Informática, Instituto Superior Técnico, Universidade de Lisboa, Lisboa, Portugal
INESC-ID, Lisboa, Portugal

Keywords

Compressed index; Data compression; Enhanced suffix array; Longest common prefix; Range minimum query; Succinct data structure; Suffix link

Years and Authors of Summarized Original Work

2007; Sadakane
2009; Fischer, Mäkinen, Navarro
2010; Ohlebusch, Fischer, Gog
2011; Russo, Navarro, Oliveira

Problem Definition

The problem consists in representing suffix trees in main memory. The representation needs to support operations efficiently, using a reasonable amount of space.

Suffix trees were proposed by Weiner in 1973 [16]. Donald Knuth called them the "Algorithm of the Year." Their ubiquitous nature was quickly perceived and used to solve a myriad of string processing problems. The downside of this flexibility was the notorious amount of space necessary to keep it in main memory. A direct implementation is several times larger than the file it is indexing. Initial research into this matter discovered smaller data structures, sometimes by sacrificing functionality, namely, suffix arrays [6], directed acyclic word graphs [5], or engineered solutions.

A suffix tree is obtained from a sequence of characters T by considering all its suffixes. The suffixes are collated together by their common prefixes into a labeled tree. This means that suffixes that share a common prefix are united by that prefix and split only when the common prefix ends, i.e., in the first letter where they mismatch. A special terminator character \$ is placed at the end to force these mismatches for the small suffixes of T. The string depth of a node is the number of letters between the node and the root. Figure 1 shows the suffix tree of the string *abbbab*.

A viable representation needs to support several operations: tree navigation, such as finding a parent node, a child node, or a sibling node; labeling, such as reading the letters along a branch or using a letter to choose a child node; and indexing operations, such as determining a leaf's index or a node's string depth.

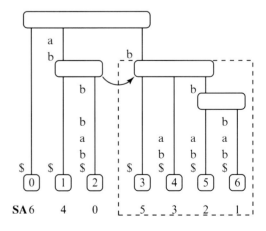

Compressed Suffix Trees, Fig. 1 Suffix tree of string *abbbab*, with the leaves numbered. The *arrow* shows the SLINK between nodes *ab* and *b*. Below we show the suffix array. The portion of the tree corresponding to node *b* and respective leaves interval is within a *dashed box*

Besides the usual tree topology, suffix trees contain suffix links. For a given node, the suffix link points to a second node. The string from the root to the second node can be obtained by removing the first letter of the string from the root to the first node. For example, the suffix link of node ab is node b. See Fig. 1.

An acceptable lower bound to represent a DNA sequence is $2n$ bits, i.e., $n \log \sigma$ bits (we use logarithms in base 2) for a text of size n with an alphabet of size σ. The size of the suffix array for such a sequence, on a 32-bit machine, is 16 times bigger than this bound. A space-engineered suffix tree would be 40 times larger. Succinct data structures are functional representations whose space requirements are close to the space needed to represent the data (i.e., close to $2n$ bits, in our example). If we consider the order-k statistical compressibility of the text, then the information-theory lower bound is even smaller, $nH_k + o(n)$ bits, where H_k is the empirical entropy of the text. Such representations require data compression techniques, which need to be made functional. The prospect of representing suffix trees within this space was significant.

Objective

Obtain a representation of a suffix tree that requires $n \log \sigma + o(n)$ bits, or even $nH_k + o(n)$ bits, or close, and that supports all operations efficiently.

Key Results

Classical Results

Suffix arrays [6] are a common alternative to suffix trees. They do not provide the same set of operations or the same time bounds, and still they require only 5 bytes per text character as opposed to the >10 bytes of suffix trees.

The suffix array stores the lexicographical order of the suffixes of T. Figure 1 shows the suffix array SA of our running example, i.e., the suffixes in the suffix tree are lexicographically ordered. Suffix arrays lack node information. Still they can represent nodes as an interval of suffixes, for example, node b corresponds to the interval $[3, 6]$.

This mapping is injective, i.e., no two nodes can map to the same interval. Hence, a given interval corresponds to no more than one node. Some intervals do not correspond to any node, for example, $[4, 6]$ does not correspond to a node on the suffix tree. To determine which intervals are invalid and speed up navigation operations, the suffix array can be augmented with longest common prefix (LCP) information. It is enough to store the length of the LCP between consecutive suffixes. For an arbitrary pair of suffixes, the LCP value can be computed as a range minimum query (RMQ) on the corresponding leaf interval. For example, $\text{LCP}(3, 5)$ can be computed as the minimum of $1, 1, 2$. The suffix array enhanced with LCP information [2] can now be used to emulate several algorithms that required suffix tree navigation.

Another approach is to reduce suffix tree redundancy, by factoring repeated structures. The following lemma is used to build directed acyclic word graphs.

Lemma 1 *If the sub-trees rooted at nodes v and v', of the suffix tree of T, have the same number of leaves and v' is the suffix link of v, then the sub-trees are isomorphic.*

Succinct Results

A fundamental component of compressed suffix trees is the underlying compressed index, which provides the suffix array information. Additionally these indexes provide support for ψ and LF, which are used to compute suffix links and backward search. In fact, ψ is the equivalent to suffix links over the suffix array, i.e., $\text{SA}[\psi(i)] = \text{SA}[i] + 1$. Moreover, LF is the inverse of ψ, i.e., $\text{LF}(\psi(i)) = \psi(\text{LF}(i)) = i$. There is a wide variety of these indexes, depending on the underlying data compression technique, namely, the Burrows-Wheeler transform, δ-coding, or Lempel-Ziv. For a complete survey on these indexes, consult the survey by Navarro and Mäkinen [7]. For our purposes, we consider the following index.

Theorem 1 *For a string T, over an alphabet of size* $\text{polylog} \, n$, *there exists a suffix array* *representation that requires $nH_k + o(n)$ bits and computes ψ, LF and retrieves letters $T[SA[i]]$ in $O(1)$ time, while it obtains values $SA[i]$ in $O(\log n \log \log n)$ time.*

Sadakane was the first to combine a compressed suffix array with a succinct tree and a succinct representation of string depth information. The combination of these three ingredients leads to the first succinct representation of suffix trees, which set the basic structure for later developments.

Theorem 2 ([14]) *There is a compressed suffix tree representation that requires $nH_k + 6n + o(n)$ bits and supports child in $O(\log^2 n \log \log n)$, string depth and edge letters in $O(\log n \log \log n)$ time, and the remaining operations in $O(1)$ time.*

The $6n$ term is composed of $4n$ for the succinct tree and $2n$ to store LCP values. A tree can be represented succinctly as a sequence of balanced parentheses. A suffix tree contains at most $2n - 1$ nodes; since each node requires exactly 2 parentheses, this accounts for the $4n$ parcel. For example, the parentheses representation of the tree in Fig. 1 is ((0) ((1) (2)) ((3) (4) ((5) (6)))); the numbers correspond to the leaf indexes and are not part of the representation; only the parentheses are necessary. These parentheses are encoded with bits, set to 0 or 1, respectively.

We refer to the position of T in the suffix array as t, i.e., $SA[t] = 0$, in our running example $t = 2$. A technique used to store the SA values is to use the relation $SA[\psi(i)] = SA[i] + 1$. This means that, if ψ is supported, we can store the $SA[\psi^{il}(t)]$ values, with $l = \log n \log \log n$, and obtain the missing values in at most l steps. The resulting SA values require only $n/\log \log n$ bits to store. To encode the LCP of internal nodes, Sadakane used that $LCP(\psi(i), \psi(i)+1) \geq LCP(i, i+1) - 1$. Hence $LCP(\psi^k(i), \psi^k(i) + 1) + k$ forms an increasing sequence of numbers, which can be encoded in at most $2n$ bits such that any element can be accessed with a "select" operation on the bits

(find the position of the jth 1, which can be solved in constant time with an $o(n)$-bit extra index). Hence computing LCP requires determining k and subtracting it from the number in the sequence; this can be achieved with SA. Subsequent research focused on eliminating the $6n$ term. Fischer, Mäkinen, and Navarro obtained a smaller representation.

Theorem 3 ([4]) *There is a compressed suffix tree representation which, for any constant $\epsilon > 0$, requires $nH_k(2\log(1/H_k) + (1/\epsilon) + O(1)) + o(n \log \sigma)$ bits and supports all operations in $O(\log^\epsilon n)$ time, except level ancestor queries (LAQ) which need $O(\log^{1+\epsilon} n)$ time.*

This bound is obtained by compressing the differential LCP values in Sadakane's representation and discarding the parentheses representation. Instead it relies exclusively on range minimum queries over the LCP values. The next smaller value (NSV) and previous smaller value (PSV) operations are presented to replace the need to find matching closing or opening parentheses. Later, Fischer [3] further improved the speed at which ϵ vanishes.

Russo, Navarro, and Oliveira [13] obtained the smallest representation by showing that $\text{LCA}(\text{SLINK}(v), \text{SLINK}(v')) = \text{SLINK}(\text{LCA}(v, v'))$ holds for any nodes v and v'. $\text{LCA}(v, v')$ means the lowest common ancestor of nodes v and v', and $\text{SLINK}(v)$ means the suffix link of node v. This relation generalizes Lemma 1.

Theorem 4 ([13]) *There is a compressed suffix tree representation that requires $nH_k + o(n)$ bits and supports child in $O(\log n (\log \log n)^2)$ and the other operations in time $O(\log n \log \log n)$.*

The reduced space requirements are obtained by storing information about a few sampled nodes. Information for the remaining nodes is computed with the property above. Only the sampled nodes are stored in a parentheses representation and moreover string depth is stored only for these nodes. Although Theorem 4 obtains optimal space, the logarithmic time is significant, in theory and in practice. This limitation was recently improved by Navarro and Russo [9] with a new approach to compare sampled nodes. They obtain $O(\text{polyloglog } n)$ time for all operations within the same space, except for child, which retains the previous time bound.

Applications

There is a myriad of practical applications for suffix trees. An extensive description can be found in the book by Gusfield [5]. Nowadays, most applications are related to bioinformatics, due to the string-like nature of DNA sequences. Most of these problems can be solved with compressed suffix trees and in fact can only be computed in reasonable time if the ever-increasing DNA database can be kept in main memory with suffix tree functionality.

Experimental Results

The results we have described provided a firm ground for representing suffix trees efficiently in compressed space. Still the goal was not a theoretical endeavor, since these data structures play a center role in the analysis of genomic data, among others. In practice, several aspects of computer architecture come into play, which can significantly impact the resulting performance. Different approaches can sacrifice space optimality to be orders of magnitude faster than the smaller variants.

Abeliuk, Cánovas, and Navarro [1] presented an exhaustive experimental analysis of existing CSTs. They obtained practical CSTs by implementing the PSV and NSV operations of Fischer, Mäkinen, and Navarro [4], with a range min-max tree [10]. Their CSTs covered a wide range in the space and time spectrum. Roughly, they need 8–12 bits per character (bpc) and perform the operations in microseconds. Further, practical variants considered reducing the $6n$ term of Sadakane's representation by using a single data structure that simultaneously provides RMQ/PSV/NSV. Ohlebusch and Gog

[11] used only $2n + o(n)$ bits, obtaining around 10–12 bpc and operations in microseconds. Ohlebusch, Fischer, and Gog [12] also used this approach to obtain the same time performance of Theorem 2, within $3n$ extra bits instead of $6n$. This yields around 16 bpc and operations running in micro- to nanoseconds. An implementation of Theorem 4 [13] needed around 4 bpc but queries required milliseconds, although better performance is expected for the new version [9].

The proposal in [1] is also designed to adapt efficiently to highly repetitive sequence collections, obtaining 1–2 bpc and operations in milliseconds. Also for repetitive texts, Navarro and Ordóñez [8] obtained a speedup to microsecond operations, with a slight space increase, 1–3 bpc, by representing the parenthesis topology explicitly but in grammar-compressed form.

URLs to Code and Data Sets

An implementation of Sadakane's compressed suffix tree [14] is available from the SuDS group at http://www.cs.helsinki.fi/group/suds/cst/. Implementation details and engineering decisions are described by Välimäki, Gerlach, Dixit, and Mäkinen [15]. The compressed suffix tree of Abeliuk, Cánovas, and Navarro [1] is available in the libcds library at https://github.com/fclaude/libcds.

An alternative implementation of Sadakane's CST is available in the Succinct Data Structure Library at https://github.com/simongog/sdsl-lite. The SDSL also contains an implementation of the CST++ [12].

The Pizza and Chili site contains a large and varied dataset to test compressed indexes, http://pizzachili.dcc.uchile.cl.

Cross-References

▶ Burrows-Wheeler Transform
▶ Compressed Range Minimum Queries
▶ Compressed Suffix Array
▶ Compressed Tree Representations

▶ Grammar Compression
▶ Rank and Select Operations on Bit Strings
▶ String Matching
▶ Suffix Trees and Arrays

Recommended Reading

1. Abeliuk A, Cánovas R, Navarro G (2013) Practical compressed suffix trees. Algorithms 6(2):319–351
2. Abouelhoda MI, Kurtz S, Ohlebusch E (2004) Replacing suffix trees with enhanced suffix arrays. J Discr Algorithms 2(1):53–86
3. Fischer J (2010) Wee LCP. Inf Process Lett 110(8–9):317–320
4. Fischer J, Mäkinen V, Navarro G (2009) Faster entropy-bounded compressed suffix trees. Theor Comput Sci 410(51):5354–5364
5. Gusfield D (1997) Algorithms on strings, trees and sequences computer science and computational biology. Cambridge University Press, Cambridge/New York
6. Manber U, Myers G (1993) Suffix arrays: a new method for on-line string searches. SIAM J Comput 22(5):935–948
7. Navarro G, Mäkinen V (2007) Compressed full-text indexes. ACM Comput Surv 39(1): article 2
8. Navarro G, Ordóñez A (2014) Faster compressed suffix trees for repetitive text collections. In: Proceedings of the 13th international symposium on experimental algorithms (SEA), Copenhagen. LNCS 8504, pp 424–435
9. Navarro G, Russo L (2014) Fast fully-compressed suffix trees. In: Proceedings of the 24th data compression conference (DCC), Snowbird, pp 283–291
10. Navarro G, Sadakane K (2014) Fully-functional static and dynamic succinct trees. ACM Trans Algorithms 10(3):article 16
11. Ohlebusch E, Gog S (2009) A compressed enhanced suffix array supporting fast string matching. In: String processing and information retrieval, Saariselkä. Springer, pp 51–62
12. Ohlebusch E, Fischer J, Gog S (2010) CST++. In: String processing and information retrieval, Los Cabos. Springer, pp 322–333
13. Russo L, Navarro G, Oliveira A (2011) Fully-compressed suffix trees. ACM Trans Algorithms (TALG) 7(4):article 53, 35p
14. Sadakane K (2007) Compressed suffix trees with full functionality. Theory Comput Syst 41(4):589–607
15. Välimäki N, Gerlach W, Dixit K, Mäkinen V (2007) Engineering a compressed suffix tree implementation. In: Experimental algorithms, Rome. Springer, pp 217–228
16. Weiner P (1973) Linear pattern matching algorithms. In: IEEE conference record of 14th annual symposium on switching and automata theory, 1973. SWAT'08. IEEE, pp 1–11. http://ieeexplore.ieee.org/xpl/mostRecentIssue.jsp?punumber=4569717

Compressed Text Indexing

Veli Mäkinen[1] and Gonzalo Navarro[2]
[1]Department of Computer Science, Helsinki Institute for Information Technology (HIIT), University of Helsinki, Helsinki, Finland
[2]Department of Computer Science, University of Chile, Santiago, Chile

Keywords

Compressed full-text indexing; Self-indexing; Space-efficient text indexing

Years and Authors of Summarized Original Work

2005; Ferragina, Manzini

Problem Definition

Given a *text string* $T = t_1 t_2 \ldots t_n$ over an alphabet Σ of size σ, the *compressed text indexing (CTI)* problem asks to *replace* T with a space-efficient data structure capable of efficiently answering basic string matching and substring queries on T. Typical queries required from such an index are the following:

- *count(P)*: count how many times a given *pattern string* $P = p_1 p_2 \ldots p_m$ occurs in T.
- *locate(P)*: return the locations where P occurs in T.
- *display(i, j)*: return $T[i, j]$.

Key Results

An elegant solution to the problem is obtained by exploiting the connection of *Burrows-Wheeler Transform (BWT)* [1] and *Suffix Array* data structure [9]. The suffix array $SA[1, n]$ of T is the permutation of text positions $(1 \ldots n)$ listing the *suffixes* $T[i, n]$ in lexicographic order. That is, $T[SA[i], n]$ is the ith smallest suffix. The BWT is formed by (1) a permutation T^{bwt} of T defined as $T^{bwt}[i] = T[SA[i] - 1]$, where $T[0] = T[n]$, and (2) the number $i^* = SA^{-1}[1]$.

A property of the BWT is that symbols having the same context (i.e., string following them in T) are consecutive in T^{bwt}. This makes it easy to compress T^{bwt} achieving space close to high-order empirical entropies [10]. On the other hand, the suffix array is a versatile text index, allowing for example $O(m \log n)$ time counting queries (using two binary searches on SA) after which one can locate the occurrences in optimal time.

Ferragina and Manzini [3] discovered a way to combine the compressibility of the BWT and the indexing properties of the suffix array. The structure is essentially a compressed representation of the BWT plus some small additional structures to make it searchable.

We first focus on retrieving arbitrary substrings from this compressed text representation, and later consider searching capabilities. To retrieve the whole text from the structure (that is, to support $display(1, n)$), it is enough to invert the BWT. For this purpose, let us consider a table $LF[1, n]$ defined such that if $T[i]$ is permuted to $T^{bwt}[j]$ and $T[i - 1]$ to $T^{bwt}[j']$ then $LF[j] = j'$. It is then immediate that T can be retrieved *backwards* by printing $T^{bwt}[i^*] \cdot T^{bwt}[LF[i^*]] \cdot T^{bwt}[LF[LF[i^*]]] \ldots$ (by definition $T^{bwt}[i^*]$ corresponds to $T[n]$).

To represent array LF space-efficiently, Ferragina and Manzini noticed that each $LF[i]$ can be expressed as follows:

Lemma 1 (Ferragina and Manzini [3]) $LF[i] = C(c) + rank_c(i)$, where $c = T^{bwt}[i]$, $C(c)$ *tells how many times symbols smaller than* c *appear in* T^{bwt} *and* $rank_c(i)$ *tells how many times symbol* c *appears in* $T^{bwt}[1, i]$.

General $display(i, j)$ queries rely on a regular sampling of the text. Every text position of the form $j' \cdot s$, being s the sampling rate, is stored together with $SA^{-1}[j' \cdot s]$, the suffix array position pointing to it. To solve $display(i, j)$ we start from the smallest sampled text position $j' \cdot s > j$ and apply the BWT inversion procedure starting with $SA^{-1}[j' \cdot s]$ instead of i^*. This gives the

characters in reverse order from $j' \cdot s - 1$ to i, requiring at most $j - i + s$ steps.

It also happens that the very same two-part expression of $LF[i]$ enables efficient $count(P)$ queries. The idea is that if one knows the range of the suffix array, say $SA[sp_i, ep_i]$, such that the suffixes $T[SA[sp_i], n], T[SA[sp_i + 1], n], \ldots, T[SA[ep_i], n]$ are the only ones containing $P[i, m]$ as a prefix, then one can compute the new range $SA[sp_{i-1}, ep_{i-1}]$ where the suffixes contain $P[i - 1, m]$ as a prefix, as follows: $sp_{i-1} = C(P[i - 1]) + rank_{P[i-1]}(sp_i - 1) + 1$ and $ep_{i-1} = C(P[i - 1]) + rank_{P[i-1]}(ep_i)$. It is then enough to scan the pattern *backwards* and compute values $C()$ and $rank_c()$ $2m$ times to find out the (possibly empty) range of the suffix array where all the suffixes start with the complete P. Returning $ep_1 - sp_1 + 1$ solves the $count(P)$ query without the need of having the suffix array available at all.

For locating each such occurrence $SA[i]$, $sp_1 \leq i \leq ep_1$, one can compute the sequence $i, LF[i], LF[LF[i]], \ldots,$ until $LF^k[i]$ is a sampled suffix array position and thus it is explicitly stored in the sampling structure designed for $display(i, j)$ queries. Then $SA[i] = SA[LF^k[i]] + k$. As we are virtually moving sequentially on the text, we cannot do more than s steps in this process.

Now consider the space requirement. Values $C()$ can be stored trivially in a table of $\sigma \log_2 n$ bits. $T^{bwt}[i]$ can be computed in $O(\sigma)$ time by checking for which c is $rank_c(i) \neq rank_c(i - 1)$. The sampling rate can be chosen as $s = \Theta(\log^{1+\epsilon} n)$ so that the samples require $o(n)$ bits. The only real challenge is to preprocess the text for $rank_c()$ queries. This has been a subject of intensive research in recent years and many solutions have been proposed. The original proposal builds several small partial sum data structures on top of the compressed BWT, and achieves the following result:

Theorem 2 (Ferragina and Manzini [3]) *The CTI problem can be solved using a so-called FM-Index (FMI), of size* $5nH_k + o(n \log \sigma)$ *bits, that supports* $count(P)$ *in* $O(m)$ *time,* $locate(P)$ *in* $O(\sigma \log^{1+\epsilon} n)$ *time per occurrence, and*

$display(i, j)$ in $O(\sigma(j - i + \log^{1+\epsilon} n))$ time. Here H_k is the k th order empirical entropy of T, $\sigma = o(\log n / \log \log n)$, $k \leq \log_\sigma (n / \log n)$, $-\omega(1)$, and $\epsilon > 0$ is an arbitrary constant.

The original FM-Index has a severe restriction on the alphabet size. This has been removed in follow-up works. Conceptually, the easiest way to achieve a more alphabet-friendly instance of the FM-index is to build a *wavelet tree* [5] on T^{bwt}. This is a binary tree on Σ such that each node v handles a subset $S(v)$ of the alphabet, which is split among its children. The root handles Σ and each leaf handles a single symbol. Each node v encodes those positions i so that $T^{\mathrm{bwt}}[i] \in S(v)$. For those positions, node v only stores a bit vector telling which go to the left, which to the right. The node bit vectors are preprocessed for constant time $rank_1()$ queries using $o(n)$-bit data structures [6, 12]. Grossi et al. [4] show that the wavelet tree built using the encoding of [12] occupies $nH_0 + o(n \log \sigma)$ bits. It is then easy to simulate a single $rank_c()$ query by $\log_2 \sigma$ $rank_1()$ queries. With the same cost one can obtain $T^{\mathrm{bwt}}[i]$. Some later enhancements have improved the time requirement, so as to obtain, for example, the following result:

Theorem 3 (Mäkinen and Navarro [7]) *The CTI problem can be solved using a so-called Succinct Suffix Array (SSA), of size $nH_0 + o(n \log \sigma)$ bits, that supports $count(P)$ in $O(m(1 + \log \sigma / \log \log n))$ time, $locate(P)$ in $O(\log^{1+\epsilon} n \log \sigma / \log \log n)$ time per occurrence, and $display(i, j)$ in $O((j - i + \log^{1+\epsilon} n) \log \sigma / \log \log n)$ time. Here H_0 is the zero-order entropy of T, $\sigma = o(n)$, and $\epsilon > 0$ is an arbitrary constant.*

Ferragina et al. [2] developed a technique called *compression boosting* that finds an optimal partitioning of T^{bwt} such that, when one compresses each piece separately using its zero-order model, the result is proportional to the kth order entropy. This can be combined with the idea of SSA by building a wavelet tree separately for each piece and some additional structures in order to solve global $rank_c()$ queries from the individual wavelet trees:

Theorem 4 (Ferragina et al. [4]) *The CTI problem can be solved using a so-called Alphabet-Friendly FM-Index (AF-FMI), of size $nH_k + o(n \log \sigma)$ bits, with the same time complexities and restrictions of SSA with $k \leq \alpha \log_\sigma n$, for any constant $0 < \alpha < 1$.*

A very recent analysis [8] reveals that the space of the plain SSA is bounded by the same $nH_k + o(n \log \sigma)$ bits, making the boosting approach to achieve the same result unnecessary in theory. In practice, implementations of [4, 7] are superior by far to those building directly on this simplifying idea.

Applications

Sequence analysis in Bioinformatics, search and retrieval on oriental and agglutinating languages, multimedia streams, and even structured and traditional database scenarios.

URL to Code and Data Sets

Site Pizza-Chili http://pizzachili.dcc.uchile.cl or http://pizzachili.di.unipi.it contains a collection of standardized library implementations as well as data sets and experimental comparisons.

Cross-References

▶ Burrows-Wheeler Transform
▶ Compressed Suffix Array
▶ Text Indexing

Recommended Reading

1. Burrows M, Wheeler D (1994) A block sorting lossless data compression algorithm. Technical report 124, Digital Equipment Corporation
2. Ferragina P, Giancarlo R, Manzini G, Sciortino M (2005) Boosting textual compression in optimal linear time. J ACM 52(4):688–713
3. Ferragina P, Manzini G (2005) Indexing compressed texts. J ACM 52(4):552–581
4. Ferragina P, Manzini G, Mäkinen V, Navarro G (2007) Compressed representation of sequences and

full-text indexes. ACM Trans Algorithms 3(2):Article 20

5. Grossi R, Gupta A, Vitter J (2003) High-order entropy-compressed text indexes. In: Proceedings of the 14th annual ACM-SIAM symposium on discrete algorithms (SODA), pp 841–850

6. Jacobson G (1989) Space-efficient static trees and graphs. In: Proceedings of the 30th IEEE symposium on foundations of computer science (FOCS), pp 549–554

7. Mäkinen V, Navarro G (2005) Succinct suffix arrays based on run-length encoding. Nord J Comput 12(1):40–66

8. Mäkinen V, Navarro G (2006) Dynamic entropy-compressed sequences and full-text indexes. In: Proceedings of the 17th annual symposium on combinatorial pattern matching (CPM). LNCS, vol 4009. Extended version as TR/DCC-2006-10, Department of Computer Science, University of Chile, July 2006, pp 307–318

9. Manber U, Myers G (1993) Suffix arrays: a new method for on-line string searches. SIAM J Comput 22(5):935–948

10. Manzini G (2001) An analysis of the Burrows-Wheeler transform. J ACM 48(3):407–430

11. Navarro G, Mäkinen V (2007) Compressed full-text indexes. ACM Comput Surv 39(1):Article 2

12. Raman R, Raman V, Rao S (2002) Succinct indexable dictionaries with applications to encoding k-ary trees and multisets. In: Proceedings of the 13th annual ACM-SIAM symposium on discrete algorithms (SODA), pp 233–242

Compressed Tree Representations

Gonzalo Navarro[1] and Kunihiko Sadakane[2]
[1]Department of Computer Science, University of Chile, Santiago, Chile
[2]Graduate School of Information Science and Technology, The University of Tokyo, Tokyo, Japan

Keywords

Balanced parentheses; Ordered tree; Succinct data structure

Years and Authors of Summarized Original Work

1989; Jacobson
2001; Munro, Raman

2005; Benoit, Demaine, Munro, Raman, S. Rao
2014; Navarro, Sadakane

Problem Definition

The problem is, given a tree, to encode it compactly so that basic operations on the tree are done quickly, preferably in constant time for static trees. Here, we consider the most basic class of trees: rooted ordered unlabeled trees. The information-theoretic lower bound for representing an n-node ordered tree is $2n - o(n)$ bits because there are $\binom{2n-2}{n-1}/n$ different trees. Therefore, the aim is to encode an ordered tree in $2n + o(n)$ bits including auxiliary data structures so that basic operations are done quickly. We assume that the computation model is the $\Theta(\log n)$-bit word RAM, that is, memory access for consecutive $\Theta(\log n)$ bits and arithmetic and logical operations on two $\Theta(\log n)$-bit integers are done in constant time.

Preliminaries

Let X be a string on alphabet $\mathcal{A}$. The number of occurrences of $c \in \mathcal{A}$ in $X[1 \ldots i]$ is denoted by $rank_c(X, i)$, and the position of j-th c from the left is denoted by $select_c(j)$ (with $select_c(0) = 0$). For binary strings ($|\mathcal{A}| = 2$) of length n, $rank$ and $select$ are computed in constant time using an $n + o(n)$-bit data structure [4]. Let us define for simplicity $prev_c(i) = select_c(rank_c(i - 1))$ and $next_c(i) = select_c(rank_c(i) + 1)$ the position of the c preceding and following, respectively, position i in X.

Key Results

Basically, there are three representations of ordered trees: LOUDS (Level-Order Unary Degree Sequence) [11], DFUDS (Depth-First Unary Degree Sequence) [2], and BP (Balanced Parenthesis sequence) [16]. An example is shown in Fig. 1. All these representations are succinct, that is, of $2n + o(n)$ bits. However, their functionality is slightly different.

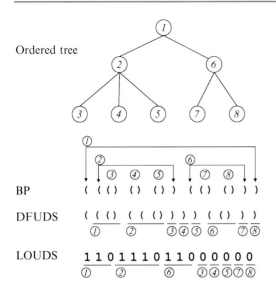

Ordered tree

BP

DFUDS

LOUDS

Compressed Tree Representations, Fig. 1 Succinct representations of trees

LOUDS

In LOUDS representation, the degree of each node is encoded by a unary representation, that is, a node with d children is encoded in d ones, followed by a zero. Codes for the nodes are stored in level order: the root node is encoded first, then its children are encoded from left to right, all the nodes at depth 2 are encoded next, and so on. Let L be the LOUDS representation of a tree. Then L is a 0,1-string of length $2n - 1$. Tree navigation operations are expressed by *rank* and *select* operations. The i-th node in level order is represented by $select_0(i - 1) + 1$. The operations $isleaf(i)$, $parent(i)$, $first\_child(i)$, $last\_child(i)$, $next\_sibling(i)$, $prev\_sibling(i)$, $degree(i)$, $child(i, q)$, and $child\_rank(i)$ (see Table 1) are computed in constant time using *rank* and *select* operations, for example, $degree(i) = next_0(i - 1) - i$, $child(i, q) = select_0(rank_1(i - 1 + q)) + 1$, and $parent(i) = prev_0(select_1(rank_0(i - 1))) + 1$. However, because nodes are stored in level order, other operations cannot be done efficiently, such as $depth(i)$, $subtree\_size(i)$, $lca(i, j)$, etc. A merit of the LOUDS representation is that in practice it is fast and simple to implement because all the operations are done by only *rank* and *select* operations.

BP Representation

In BP representation, the tree is traversed in depth-first order, appending an open parenthesis "(" to the sequence when we reach a node and a closing parenthesis ")" when we leave it. These parentheses are represented with the bits 1 and 0, respectively. The result is a sequence of $2n$ parentheses that is balanced: for any open (resp. close) parenthesis, there is a matching close (resp. open) parenthesis to the right (resp. left), so that the areas between two pairs of matching parentheses either nest or are disjoint. Each node is identified with the position of its open parenthesis.

Munro and Raman [16] showed how to implement operations $findclose(i)$, $findopen(i)$, and $enclose(i)$ in constant time and $2n + o(n)$ bits in total. Later, Geary et al. [8] considerably simplified the solutions. With those operations and *rank* and *select* support, many operations in Table 1 are possible. For example, $pre\_rank(i) = rank_1(i)$, $pre\_select(j) = select_1(j)$, $isleaf(i)$ iff there is a 0 at position $i + 1$, $isancestor(i, j)$ if $i \leq j \leq findclose(i)$, $depth(i) = rank_1(i) - rank_0(i)$, $parent(i) = enclose(i)$, $first\_child(i) = i + 1$, $next\_sibling(i) = findclose(i) + 1$, $subtree\_size(i) = (findclose(i) - i + 1)/2$, etc. Operations lca, $height(i)$, and $deepest\_node(i)$ could be added with additional structures for range minimum queries, $rmqi(i, j)$, in $o(n)$ further bits [20]. Some other operations can be also supported in constant time by adding different additional structures. Lu and Yeh [14] gave $o(n)$-bit data structures for $degree(i)$, $child(i, q)$, and $child\_rank(i)$. Geary et al. [9] gave $o(n)$-bit data structures for $LA(i, d)$. However, these extra structures are complicated and add considerably extra space in practice.

DFUDS

In DFUDS representation, nodes are also encoded by a unary representation of their degrees, but stored in depth-first order. The bits 1 and 0 are interpreted as open parenthesis "(" and close parenthesis ")", respectively. By adding a dummy open parenthesis at the beginning, the DFUDS sequence becomes balanced. Each node is identified with the first position of its unary description.

Compressed Tree Representations, Table 1
Operations supported by the data structure of [19]. The time complexities are for the dynamic case; in the static case, all operations take $\mathcal{O}(1)$ time. The first group, of basic operations, is used to implement the others, but could have other uses

Operation	Description	Time complexity	
		Variant 1	Variant 2
$inspect(i)$	$P[i]$	$\mathcal{O}(\frac{\log n}{\log\log n})$	
$findclose(i)$ / $findopen(i)$	Position of parenthesis matching $P[i]$		
$enclose(i)$	Position of tightest open parent. enclosing i		
$rmqi(i,j)$ / $RMQi(i,j)$	Position of min/max excess value in range $[i,j]$		
$pre\_rank(i)$ / $post\_rank(i)$	Preorder/postorder rank of node i	$\mathcal{O}(\frac{\log n}{\log\log n})$	
$pre\_select(i)$ / $post\_select(i)$	The node with preorder/postorder i		
$isleaf(i)$	Whether $P[i]$ is a leaf		
$isancestor(i,j)$	Whether i is an ancestor of j		
$depth(i)$	Depth of node i		
$parent(i)$	Parent of node i		
$first\_child(i)$ / $last\_child(i)$	First/last child of node i		
$next\_sibling(i)$ / $prev\_sibling(i)$	Next/previous sibling of node i		
$subtree\_size(i)$	Number of nodes in the subtree of node i		
$lca(i,j)$	The lowest common ancestor of two nodes i,j		
$deepest\_node(i)$	The (first) deepest node in the subtree of i		
$height(i)$	The height of i (distance to its deepest node)		
$in\_rank(i)$	Inorder of node i		
$in\_select(i)$	Node with inorder i		
$leaf\_rank(i)$	Number of leaves to the left of leaf i		
$leaf\_select(i)$	i-th leaf		
$lmost\_leaf(i)$ / $rmost\_leaf(i)$	Leftmost/rightmost leaf of node i		
$LA(i,d)$	Ancestor j of i s.t. $depth(j) = depth(i) - d$	$\mathcal{O}(\log n)$	
$level\_next(i)$ / $level\_prev(i)$	Next/previous node of i in BFS order		
$level\_lmost(d)$ / $level\_rmost(d)$	Leftmost/rightmost node with depth d		
$degree(i)$	q = number of children of node i	$\mathcal{O}(\frac{q\log n}{\log\log n})$	$\mathcal{O}(\log n)$
$child(i,q)$	q-th child of node i		
$child\_rank(i)$	q = number of siblings to the left of node i		
$insert(i,j)$	Insert node given by matching parent at i and j	$\mathcal{O}(\frac{\log n}{\log\log n})$	$\mathcal{O}(\log n)$
$delete(i)$	Delete node i		

A merit of using DFUDS is that it retains many operations supported in BP, but $degree(i)$, $child(i,q)$, and $child\_rank(i)$ are done in constant time by using only the basic operations supported by Munro and Raman [16]. For example, $degree(i) = next_0(i-1) - i$, $child(i,q) = findclose(i + degree(i) - q) + 1$, and $parent(i) = prev_0(findopen(i-1)) + 1$. Operation $lca(i,j)$ can also be computed using $rmqi(i,j)$ [12].

Another important feature of DFUDS is that some trees can be represented in less than $2n$ bits. The number of ordered trees having n_i nodes with degree i ($i = 0, 1, \ldots, n-1$) is $\frac{1}{n}\binom{n}{n_0 n_1 \cdots n_{n-1}}$ if $\sum_{i\geq0} n_i(i-1) = -1$ and 0 otherwise (i.e., there are no trees satisfying the condition). Jansson et al. [12] proposed a compression algorithm for DFUDS sequences that achieves the lower bound $\lg\left(\frac{1}{n}\binom{n}{n_0 n_1 \cdots n_{n-1}}\right) +$

$o(n)$ bits. For example, a full binary tree, that is, a tree in which all internal nodes have exactly two children, is encoded in n bits.

A demerit of DUFDS is that computing $depth(i)$ and $LA(i, d)$ is complicated, though it is possible in constant time [12].

Fully Functional BP Representation

Navarro and Sadakane [19] proposed a data structure for ordered trees using the BP representation [16]. The data structure is called *range min-max tree*. Let P be the BP sequence. For each entry $P[i]$ of the sequence, we define its *excess value* as the number of open parentheses minus the number of close parentheses in $P[1, i]$. If $P[i]$ is an open parenthesis, its excess value is equal to $depth(i)$. Let $E[i]$ denote the excess value for $P[i]$. With the range min-max tree and small additional data structures, they support in constant time the operations $fwd\_search(i, d) = \min\{\{j > i | E[j] = E[i] + d\} \cup \{n + 1\}\}$, $bwd\_search(i, d) = \max\{\{j < i | E[j] = E[i] + d\} \cup \{0\}\}$, and $rmqi(i, j) = \mathrm{argmin}_{i \le k \le j} E[k]$. Then, the basic operations can be expressed as $findclose(i) = fwd\_search(i, -1)$, $findopen(i) = bwd\_search(i, 0) + 1$, and $enclose(i) = bwd\_search(i, -2) + 1$. In addition to those operations and $rank/select$, other operations can be computed, such as $LA(i, d) = bwd\_search(i, -d - 1) + 1$. Operation $child(i, q)$ and related ones are also done in constant time using small additional structures. The results are summarized as follows:

Theorem 1 ([19]) *For any ordinal tree with n nodes, all operations in Table 1 except insert and delete are carried out in constant time $\mathcal{O}(c)$ with a data structure using $2n + \mathcal{O}(n/\log^c n)$ bits of space on a $\Theta(\log n)$-bit word RAM, for any constant $c > 0$. The data structure can be constructed from the balanced parenthesis sequence of the tree, in $\mathcal{O}(n)$ time using $\mathcal{O}(n)$ bits of space.*

Another merit of using the range min-max tree is that it is easy to dynamize. By using a balanced tree, we obtain the following:

Theorem 2 ([19]) *On a $\Theta(\log n)$-bit word RAM, all operations on a dynamic ordinal tree with n nodes can be carried out within the worst-case complexities given in Table 1, using a data structure that requires $2n + \mathcal{O}(n \log \log n / \log n)$ bits. Alternatively, all the operations of the table can be carried out in $\mathcal{O}(\log n)$ time using $2n + \mathcal{O}(n/\log n)$ bits of space.*

The time complexity $\mathcal{O}(\log n / \log \log n)$ is optimal [3].

Other Representations

There are other representations of ordered trees and other types of trees [6, 7, 10]. The idea is that the entire tree is partitioned into mini-trees using the tree cover technique [10], and mini-trees are again partitioned into micro-trees.

Applications

There are many applications of succinct ordered trees because trees are fundamental data structures. A typical application is compressed suffix trees [20]. Balanced parentheses have also been used to represent planar and k-page graphs [16]. We also have other applications to encoding permutations and functions [17], grammar compression [15], compressing BDDs/ZDDs (binary decision diagrams/zero-suppressed BDDs) [5], etc.

Open Problems

An open problem is to give a dynamic data structure supporting all operations in the optimal $\mathcal{O}(\log n / \log \log n)$ time.

Experimental Results

Arroyuelo et al. [1] implemented LOUDS and the static version of the fully functional BP representation. LOUDS uses little space (as little as $2.1n$ bits) and solves its operations in half a microsecond or less, but its functionality is limited. The range min-max tree [19] requires about $2.4n$ bits and can be used to represent both BP and DFUDS. It solves all the operations within 1–2 microseconds. Previous implementations [8, 18] require more space and are generally slower.

Joannou and Raman [13] proposed an efficient implementation of the dynamic version using splay trees to represent the range min-max tree.

URLs to Code and Data Sets

An implementation of the BP representation using the range min-max tree by the original authors is available at https://github.com/fclaude/libcds. Another implementation by Simon Gog is at https://github.com/simongog/sdsl-lite.

Cross-References

▶ Compressed Range Minimum Queries
▶ Compressed Representations of Graphs
▶ Compressed Suffix Trees
▶ Grammar Compression
▶ Lowest Common Ancestors in Trees
▶ Rank and Select Operations on Bit Strings
▶ Succinct and Compressed Data Structures for Permutations and Integer Functions

Recommended Reading

1. Arroyuelo D, Cánovas R, Navarro G, Sadakane K (2010) Succinct trees in practice. In: Proceedings of 11th workshop on algorithm engineering and experiments (ALENEX), Austin. SIAM Press, pp 84–97
2. Benoit D, Demaine ED, Munro JI, Raman R, Raman V, Rao SS (2005) Representing trees of higher degree. Algorithmica 43(4):275–292
3. Chan HL, Hon WK, Lam TW, Sadakane K (2007) Compressed indexes for dynamic text collections. ACM Trans Algorithms 3(2):article 21
4. Clark DR (1996) Compact pat trees. PhD thesis, University of Waterloo
5. Denzumi S, Kawahara J, Tsuda K, Arimura H, Minato Si, Sadakane K (2014) DenseZDD: a compact and fast index for families of sets. In: Experimental algorithms (SEA), Copenhagen. LNCS 8503, pp 187–198
6. Farzan A, Munro JI (2014) A uniform paradigm to succinctly encode various families of trees. Algorithmica 68(1):16–40
7. Farzan A, Raman R, Rao SS (2009) Universal succinct representations of trees? In: Automata, languages and programming, 36th international colloquium (ICALP), Rhodes, pp 451–462
8. Geary RF, Rahman N, Raman R, Raman V (2006a) A simple optimal representation for balanced parentheses. Theor Comput Sci 368:231–246
9. Geary RF, Raman R, Raman V (2006b) Succinct ordinal trees with level-ancestor queries. ACM Trans Algorithms 2:510–534
10. He M, Munro JI, Satti SR (2012) Succinct ordinal trees based on tree covering. ACM Trans Algorithms 8(4):42
11. Jacobson G (1989) Space-efficient static trees and graphs. In: Proceedings of IEEE FOCS, Research Triangle Park, pp 549–554
12. Jansson J, Sadakane K, Sung WK (2012) Ultra-succinct representation of ordered trees with applications. J Comput Syst Sci 78(2):619–631
13. Joannou S, Raman R (2012) Dynamizing succinct tree representations. In: 11th international symposium experimental algorithms (SEA) 2012, Bordeaux, pp 224–235
14. Lu HI, Yeh CC (2008) Balanced parentheses strike back. ACM Trans Algorithms (TALG) 4(3):article 28
15. Maruyama S, Tabei Y, Sakamoto H, Sadakane K (2013) Fully-online grammar compression. In: Proceedings of string processing and information retrieval (SPIRE), Jerusalem. LNCS 8214, pp 218–229
16. Munro JI, Raman V (2001) Succinct representation of balanced parentheses and static trees. SIAM J Comput 31(3):762–776
17. Munro JI, Raman R, Raman V, Rao SS (2012) Succinct representations of permutations and functions. Theor Comput Sci 438:74–88
18. Navarro G (2009) Implementing the LZ-index: theory versus practice. J Exp Algorithmics 13:article 2
19. Navarro G, Sadakane K (2014) Fully-functional static and dynamic succinct trees. ACM Trans Algorithms 10(3):article 16
20. Sadakane K (2007) Compressed suffix trees with full functionality. Theory Comput Syst 41(4):589–607

Compressing and Indexing Structured Text

Paolo Ferragina[1] and Srinivasa Rao Satti[2]
[1] Department of Computer Science, University of Pisa, Pisa, Italy
[2] Department of Computer Science and Engineering, Seoul National University, Seoul, South Korea

Keywords

Tree compression; Tree indexing; (labeled) Tree succinct representations; XML compression and indexing

Years and Authors of Summarized Original Work

2005; Ferragina, Luccio, Manzini, Muthukrishnan

Problem Definition

Trees are a fundamental structure in computing. They are used in almost every aspect of modeling and representation for computations like searching for keys, maintaining directories, and representations of parsing or execution traces, to name just a few. One of the latest uses of trees is *XML*, the de facto format for data storage, integration, and exchange over the Internet (see http://www.w3.org/XML/). Explicit storage of trees, with one pointer per child as well as other auxiliary information (e.g., label), is often taken as given but can account for the dominant storage cost. Just to have an idea, a simple tree encoding needs at least 16 bytes per tree node: one pointer to the auxiliary information (e.g., node label) plus three node pointers to the parent, the first child, and the next sibling. This large space occupancy may even prevent the processing of medium-sized trees, e.g., *XML* documents. This entry surveys the best-known storage solutions for unlabeled and labeled trees that are space efficient and support fast navigational and search operations over the tree structure. In the literature, they are referred to as *succinct/compressed tree-indexing* solutions.

Notation and Basic Facts

The information-theoretic storage cost for any item of a universe U can be derived via a simple *counting argument*: at least $\log |U|$ bits are needed to distinguish any two items of U. (Throughout the entry, all logarithms are taken to the base 2, and it is assumed $0 \log 0 = 0$.) Now, let $\mathcal{T}$ be a rooted tree of arbitrary degree and shape, and consider the following three main classes of trees:

Ordinal trees $\mathcal{T}$ is unlabeled and its children are left-to-right *ordered*. The number of ordinal trees on t nodes is $C_t = \binom{2t}{t} / (t+1)$ which induces a lower bound of $2t - \Theta(\log t)$ bits

Cardinal k-ary Trees $\mathcal{T}$ is labeled on its *edges* with symbols drawn from the alphabet $\Sigma = \{1, \ldots, k\}$. Any node has degree at most k because the edges outgoing from each node have *distinct* labels. Typical examples of cardinal trees are the binary tree ($k = 2$), the (uncompacted) trie, and the *Patricia tree*. The number of k-ary cardinal trees on t nodes is $C_t^k = \binom{kt+1}{t} / (kt+1)$ which induces a lower bound of $t(\log k + \log e) - \Theta(\log kt)$ bits, when k is a slowly growing function of t

(Multi-)labeled trees $\mathcal{T}$ is an ordinal tree, labeled on its *nodes* with symbols drawn from the alphabet Σ. In the case of multi-labeled trees, every node has *at least* one symbol as its label. The *same* symbols may repeat among sibling nodes, so that the degree of each node is *unbounded*, and the same labeled subpath may occur many times in $\mathcal{T}$, anchored anywhere. The information-theoretic lower bound on the storage complexity of this class of trees on t nodes comes easily from the decoupling of the tree structure and the storage of tree labels. For labeled trees, it is $\log C_t + t \log |\Sigma| = t(\log |\Sigma| + 2) - \Theta(\log t)$ bits

The following query operations should be supported over $\mathcal{T}$:

Basic navigational queries They ask for the parent of a given node u, the ith child of u, and the degree of u. These operations may be restricted to some label $c \in \Sigma$, if $\mathcal{T}$ is labeled

Sophisticated navigational queries They ask for the jth level-ancestor of u, the depth of u, the subtree size of u, the lowest common ancestor of a pair of nodes, and the ith node according to some node ordering over $\mathcal{T}$, possibly restricted to some label $c \in \Sigma$ (if $\mathcal{T}$ is labeled). For even more operations, see [2, 14]

Subpath query Given a labeled subpath Π, it asks for the (number *occ* of) nodes of $\mathcal{T}$ that immediately descend from every occurrence of

Π in $\mathcal{T}$. Each subpath occurrence may be anchored anywhere in the tree (i.e., not necessarily in its root)

The elementary solution to the tree-indexing problem consists of encoding the tree $\mathcal{T}$ via a mixture of pointers and arrays, thus taking a total of $\Theta(t \log t)$ bits. This supports basic navigational queries in constant time, but it is not space efficient and requires visiting the whole tree to implement the subpath query or the more sophisticated navigational queries. Here, the goal is to design tree storage schemes that are either *succinct*, namely, "close to the information-theoretic lower bound" mentioned before or *compressed* in that they achieve "entropy-bounded storage." Furthermore, these storage schemes do *not* require the whole visit of the tree for most navigational operations. Thus, succinct/compressed tree-indexing solutions are distinct from simply compressing the input and then uncompressing it at query time.

In this entry, it is assumed that $t \geq |\Sigma|$. The model of computation used is the random access machine (RAM) with word size $\Theta(\log t)$, where one can perform various arithmetic and bit-wise Boolean operations on single words in constant time.

Key Results

The notion of *succinct* data structures was introduced by Jacobson [13] in a seminal work over 25 years ago. He presented a storage scheme for ordinal trees using $2t + o(t)$ bits that supports basic navigational queries in $O(\log \log t)$ time (i.e., parent, first child, and next sibling of a node). Later, Munro and Raman [16] closed the issue for ordinal trees on basic navigational queries and the subtree-size query by achieving constant query time and $2t + o(t)$ bits of storage. Their storage scheme is called *balanced parenthesis* (*BP*) representation (some papers [Chiang et al., ACM-SIAM SODA '01; Sadakane, ISAAC '01; Munro et al., J.ALG '01; Munro and Rao, ICALP '04] have extended *BP* to support in constant time other sophisticated navigational queries like *LCA*,

node degree, rank/select on leaves and number of leaves in a subtree, level-ancestor and level-successor) representation. Subsequently, Benoit et al. [3] proposed a storage scheme called *depth-first unary degree sequence* (shortly, *DFUDS*) that still uses $2t + o(t)$ bits but performs more navigational queries like ith child, child rank, and node degree in constant time. Geary et al. [10] gave another representation still taking optimal space that extends *DFUDS*'s operations with the level-ancestor query.

Although these three representations achieve the optimal space occupancy, none of them supports every existing operation in constant time: e.g., *BP* does not support ith child and child rank and *DFUDS* and Geary et al.'s representation do not support *LCA*. Later, Jansson et al. [14] extended the *DFUDS* storage scheme in two directions: (1) they showed how to implement in constant time all navigational queries above and (the *BP* representation and the one of Geary et al. [10] have been recently extended to support further operations-like depth/height of a node, next node in the same level, rank/select over various node orders-still in constant time and $2t + o(t)$ bits see [11] and references therein) (2) they showed how to compress the new tree storage scheme up to $H^*(\mathcal{T})$, which denotes the entropy of the distribution of node degrees in $\mathcal{T}$.

Theorem 1 (Jansson et al. [14]) *For any rooted (unlabeled) tree $\mathcal{T}$ with t nodes, there exists a tree-indexing scheme that uses $tH^*(\mathcal{T}) + O(t (\log \log t)^2 / \log t)$ bits and supports all navigational queries in constant time.*

This improves the standard tree pointer-based representation, since it needs no more than $H^*(\mathcal{T})$ bits per node and does not compromise the performance of sophisticated navigational queries. Since it is $H^*(\mathcal{T}) \leq 2$, this solution is also never worse than BP or DFUDS, but its improvement may be significant! This result can be extended to achieve the kth-order entropy of the *DFUDS* sequence, by adopting any compressed-storage scheme for strings (see, e.g., [7] and references therein).

Further work in the area of succinct ordinal tree representations came in the form of (i) a *uniform approach* to succinct tree representations [5] that simplified and extended the representation of Geary et al., (ii) a *universal representation* [6] that emulates all three representations mentioned above, and (iii) a *fully functional representation* [18] that obtains a simplified ordinal tree encoding with reduced space occupancy.

Benoit et al. [3] extended the use of *DFUDS* to cardinal trees and proposed a tree-indexing scheme whose space occupancy is close to the information-theoretic lower bound and supports various navigational queries in constant time. Raman et al. [19] improved the space by using a different approach (based on storing the tree as a set of edges), thus proving the following:

Theorem 2 (Raman et al. [19]) *For any k-ary cardinal tree $\mathcal{T}$ with t nodes, there exists a tree-indexing scheme that uses $\log C_t^k + o(t) + O(\log \log k)$ bits and supports in constant time the following operations: finding the parent, the degree, the ordinal position among its siblings, the child with label c, and the ith child of a node.*

The subtree-size operation cannot be supported efficiently using this representation, so [3] should be resorted to in case this operation is a primary concern.

Despite this flurry of activity, the fundamental problem of indexing *labeled* trees succinctly has remained mostly unsolved. In fact, the succinct encoding for ordered trees mentioned above might be replicated $|\Sigma|$ times (once for each symbol of Σ), and then the divide-and-conquer approach of [10] might be applied to reduce the final space occupancy. However, the final space bound would be $2t + t \log |\Sigma| + O(t|\Sigma|(\log \log \log t)/(\log \log t))$ bits, which is nonetheless far from the information-theoretic storage bound even for moderately large Σ. On the other hand, if subpath queries are of primary concern (e.g., *XML*), one can use the approach of [15] which consists of a variant of the suffix-tree data structure properly designed to index all $\mathcal{T}$'s labeled paths. Subpath queries can be supported in $O(|\Pi| \log |\Sigma| + occ)$ time, but the required space would still be $\Theta(t \log t)$ bits (with large hidden constants, due to the use of suffix trees). Subsequently, some papers [1, 2, 8, 12] addressed this problem in its whole generality by either dealing simultaneously with subpath and basic navigational queries [8] or by considering *multi-labeled* trees and a larger set of navigational operations [1, 2, 12].

In particular, [8] introduced a *transform* of the labeled tree $\mathcal{T}$, denoted $xbw[\mathcal{T}]$, which linearizes it into *two* coordinated arrays $\langle S_{\text{last}}, S_\alpha \rangle$: the former capturing the tree structure and the latter keeping a permutation of the labels of $\mathcal{T}$. $xbw[\mathcal{T}]$ has the optimal (up to lower-order terms) size of $2t + t \log |\Sigma|$ bits and can be built and inverted in optimal linear time. In designing the *XBW*-transform, the authors were inspired by the elegant Burrows-Wheeler transform for strings [4]. The power of $xbw[\mathcal{T}]$ relies on the fact that it allows one to transform compression and indexing problems on labeled trees into easier problems over strings. Namely, the following two string-search primitives are key tools for indexing $xbw[\mathcal{T}]$: $\text{rank}_c(S, i)$ returns the number of occurrences of the symbol c in the string prefix $S[1, i]$, and $\text{select}_c(S, j)$ returns the position of the jth occurrence of the symbol c in string S. The literature offers many time-/space-efficient solutions for these primitives that could be used as a *black box* for the compressed indexing of $xbw[\mathcal{T}]$ (see, e.g., [2, 17] and references therein).

Theorem 3 (Ferragina et al. [8]) *Consider a tree $\mathcal{T}$ consisting of t nodes labeled with symbols drawn from alphabet Σ. There exists a compressed tree-indexing scheme that achieves the following performance:*

- *If $|\Sigma| = O(polylog(t))$, the index takes at most $tH_0(S_\alpha) + 2t + o(t)$ bits and supports basic navigational queries in constant time and (counting) subpath queries in $O(|\Pi|)$ time.*
- *For any alphabet Σ, the index takes less than $tH_k(S_\alpha) + 2t + o(t \log |\sum|))$ bits, but label-based navigational queries and (counting) subpath queries are slowed down by a factor $o((\log \log |\Sigma|)3)$.*

Here, $H_k(s)$ is the kth-order empirical entropy of string s, with $H_k(s) \leq H_{k-1}(s)$ for any $k > 0$.

Since $H_k(S_\alpha) \leq H_0(S_\alpha) \log |\sum|$, the indexing of $xbw[\mathcal{T}]$ takes at most as much space as its plain representation, up to lower-order terms, but with the additional feature of being able to navigate and search $\mathcal{T}$ efficiently. This is indeed a sort of *pointerless representation* of the labeled tree $\mathcal{T}$ with additional search functionalities (see [8] for details).

If sophisticated navigational queries over labeled trees are a primary concern, and subpath queries are not necessary, then the approach of Barbay et al. [1, 2] should be followed. They proposed the novel concept of *succinct index*, which is different from the concept of *succinct/compressed encoding* implemented by all the above solutions. A succinct index does not *touch* the data to be indexed; it just accesses the data via basic operations offered by the underlying abstract data type (ADT) and requires asymptotically less space than the information-theoretic lower bound on the storage of the data itself. The authors reduce the problem of indexing labeled trees to the one of indexing ordinal trees and strings and the problem of indexing multi-labeled trees to the one of indexing ordinal trees and binary relations. Then, they provide succinct indexes for strings and binary relations. In order to present their result, the following definitions are needed. Let m be the total number of symbols in $\mathcal{T}$, let t_c be the number of nodes labeled c in $\mathcal{T}$, and let ρ_c be the maximum number of labels c in any rooted path of $\mathcal{T}$ (called the *recursivity* of c). Define ρ as the average recursivity of $\mathcal{T}$, namely, $\rho = (1/m) \sum_{c \in \sum} (t_c \rho_c)$.

Theorem 4 (Barbay et al. [1]) *Consider a tree $\mathcal{T}$ consisting of t nodes (multi-)labeled with possibly many symbols drawn from alphabet Σ. Let m be the total number of symbols in $\mathcal{T}$, and assume that the underlying ADT for $\mathcal{T}$ offers basic navigational queries in constant time and retrieves the ith label of a node in time f. There is a succinct index for $\mathcal{T}$ using $m(\log \rho + o(\log(|\Sigma|\rho)))$ bits that supports for a given node u the following operations (where $L = \log \log |\sum| \log \log \log |\sum|):$*

- *Every c-descendant or c-child of u can be retrieved in $O(L (f + \log \log |\sum|))$ time.*
- *The set A of c-ancestors of u can be retrieved in $O(L (f + \log \log |\sum|)) + |A|(\log \log \rho_c + \log \log \log |\sum|(f + \log \log |\sum|)))$ time.*

More recently, He et al. [12] obtained new representations that support a much broader collection of operations than the ones mentioned above.

Applications

As trees are ubiquitous in many applications, this section concentrates just on two examples that, in their simplicity, highlight the flexibility and power of succinct/compressed tree indexes.

The first example regards suffix trees, which are a crucial algorithmic block of many string processing applications – ranging from bioinformatics to data mining, from data compression to search engines. Standard implementations of suffix trees take at least 80 bits per node. The compressed suffix tree of a string $S[1,s]$ consists of three components: the tree topology, the string depths stored into the internal suffix-tree nodes, and the suffix pointers stored in the suffix-tree leaves (also called *suffix array* of S). The succinct tree representation of [14] can be used to encode the suffix-tree topology and the string depths taking $4s + o(s)$ bits (assuming w.l.o.g. that $|\Sigma| = 2$). The suffix array can be compressed up to the kth-order entropy of S via any solution surveyed in [17]. The overall result is never worse than 80 bits per node, but can be significantly better for highly compressible strings.

The second example refers to the *XML* format which is often modeled as a labeled tree. The succinct/compressed indexes in [1, 2, 8] are theoretical in flavor but turn out to be relevant for practical *XML* processing systems. As an example, [9] has published some encouraging experimental results that highlight the impact of the *XBW*-transform on real *XML* datasets. The authors show that a proper adaptation of the *XBW*-transform allows one to compress *XML* data up to state-of-the-art XML-conscious compressors and to provide access to its content, navigate up and down the *XML* tree structure, and search for simple path expressions and substrings in a few milliseconds over MBs of *XML* data, by uncompressing only a tiny fraction of them at each operation. Previous solutions took several seconds per operation!

Open Problems

For recent results, open problems, and further directions of research in the general area of succinct tree representation, the interested reader is referred to [2, 11, 14, 18] and references therein. Here, we describe two main problems, which naturally derive from the discussion above.

Motivated by *XML* applications, one may like to extend the subpath search operation to the *efficient* search for all leaves of $\mathcal{T}$ whose labels *contain* a substring β and that *descend from* a given subpath Π. The term "efficient" here means in time proportional to $|\Pi|$ and to the number of retrieved occurrences, but independent as much as possible of $\mathcal{T}$'s size in the worst case. Currently, this search operation is possible only for the leaves which are immediate descendants of Π, and even for this setting, the solution proposed in [9] is not optimal.

There are two main encodings for trees which lead to the results above: ordinal tree representation (*BP*, *DFUDS* or the representation of Geary et al. [10]) and *XBW*. The former is at the base of solutions for sophisticated navigational operations, and the latter is at the base of solutions for sophisticated subpath searches. Is it possible to devise *one unique* transform for the labeled tree $\mathcal{T}$ which combines the best of the two worlds and is still compressible?

Experimental Results

See http://mattmahoney.net/dc/text.html and at the paper [9] for numerous experiments on *XML* datasets.

Data Sets

See http://mattmahoney.net/dc/text.html and the references in [9].

URL to Code

Paper [9] contains a list of software tools for compression and indexing of *XML* data.

Cross-References

▶ Burrows-Wheeler Transform
▶ Compressed Suffix Array
▶ Compressed Tree Representations
▶ Rank and Select Operations on Bit Strings
▶ Rank and Select Operations on Sequences
▶ Suffix Trees and Arrays
▶ Table Compression

Recommended Reading

1. Barbay J, Golynski A, Munro JI, Rao SS (2007) Adaptive searching in succinctly encoded binary relations and tree-structured documents. Theor Comput Sci 387:284–297
2. Barbay J, He M, Munro JI, Rao SS (2011) Succinct indexes for strings, binary relations and multi-labeled trees. ACM Trans Algorithms 7(4):article 52
3. Benoit D, Demaine E, Munro JI, Raman R, Raman V, Rao SS (2005) Representing trees of higher degree. Algorithmica 43:275–292
4. Burrows M, Wheeler D (1994) A block sorting lossless data compression algorithm. Technical report 124, Digital Equipment Corporation
5. Farzan A, Munro JI (2014) A uniform paradigm to succinctly encode various families of trees. Algorithmica 68(1):16–40
6. Farzan A, Raman R, Rao SS (2009) Universal succinct representations of trees? In: Proceedings of the 36th international colloquium on automata, languages and programming (ICALP, Part I), Rhodes. Lecture notes in computer science, vol 5555. Springer, pp 451–462
7. Ferragina P, Venturini R (2007) A simple storage scheme for strings achieving entropy bounds. Theor Comput Sci 372(1):115–121
8. Ferragina P, Luccio F, Manzini G, Muthukrishnan S (2005) Structuring labeled trees for optimal succinctness, and beyond. In: Proceedings of the 46th

IEEE symposium on foundations of computer science (FOCS), Cambridge, pp 184–193. The journal version of this paper appear in J ACM 57(1) (2009)

9. Ferragina P, Luccio F, Manzini G, Muthukrishnan S (2006) Compressing and searching XML data via two zips. In: Proceedings of the 15th World Wide Web conference (WWW), Edingburg, pp 751–760
10. Geary R, Raman R, Raman V (2006) Succinct ordinal trees with level-ancestor queries. ACM Trans Algorithms 2:510–534
11. He M, Munro JI, Rao SS (2012) Succinct ordinal trees based on tree covering. ACM Trans Algorithms 8(4):article 42
12. He M, Munro JI, Zhou G (2012) A framework for succinct labeled ordinal trees over large alphabets. In: Proceedings of the 23rd international symposium on algorithms and computation (ISAAC), Taipei. Lecture notes in computer science, vol 7676. Springer, pp 537–547
13. Jacobson G (1989) Space-efficient static trees and graphs. In: Proceedings of the 30th IEEE symposium on foundations of computer science (FOCS), Triangle Park, pp 549–554
14. Jansson J, Sadakane K, Sung W Ultra-succinct representation of ordered trees. J Comput Syst Sci 78: 619–631 (2012)
15. Kosaraju SR (1989) Efficient tree pattern matching. In: Proceedings of the 20th IEEE foundations of computer science (FOCS), Triangle Park, pp 178–183
16. Munro JI, Raman V (2001) Succinct representation of balanced parentheses and static trees. SIAM J Comput 31(3):762–776
17. Navarro G, Mäkinen V (2007) Compressed full text indexes. ACM Comput Surv 39(1):article 2
18. Navarro G, Sadakane K (2014) Fully functional static and dynamic succinct trees. ACM Trans Algorithms 10(3):article 16
19. Raman R, Raman V, Rao SS (2007) Succinct indexable dictionaries with applications to encoding k-ary trees and multisets. ACM Trans Algorithms 3(4):article 43

Compressing Integer Sequences

Alistair Moffat
Department of Computing and Information Systems, The University of Melbourne, Melbourne, VIC, Australia

Keywords

Binary code; Byte-based code; Compressed set representations; Elias code; Elias-Fano code; Golomb code; Integer code; Interpolative code

Years and Authors of Summarized Original Work

1966; Golomb
1974; Elias
2000; Moffat, Stuiver

Problem Definition

An n-symbol message $M = \langle s_0, s_1, \ldots, s_{n-1} \rangle$ is given, where each symbol s_i is an integer in the range $0 \leq s_i < U$. If the s_is are strictly increasing, then M identifies an n-subset of $\{0, 1, \ldots, U - 1\}$.

Objective To economically encode M as a binary string over $\{0, 1\}$.

Constraints

1. **Short messages.** The message length n may be small relative to U.
2. **Monotonic equivalence.** Message M is converted to a strictly increasing message M' over the alphabet $U' \leq Un$ by taking prefix sums, $s_i' = i + \sum_{j=0}^{i} s_j$ and $U' = s_{n-1}' + 1$. The inverse is to "take gaps," $g_i = s_i - s_{i-1} - 1$, with $g_0 = s_0$.
3. **Combinatorial Limit.** If M is monotonic then $\left\lceil \log_2 \binom{U}{n} \right\rceil \leq U$ bits are required in the worst case. When $n \ll U$, $\log_2 \binom{U}{n} \approx n(\log_2(U/n) + \log_2 e)$.

Key Results

Monotonic sequences can be coded in $\min\{U, n(\log_2(U/n) + 2)\}$ bits. Non-monotonic sequences can be coded in $\min\{\sum_{i=0}^{n-1}(\log_2(1 + s_i) + o(\log s_i)), n(\log_2(U'/n) + 2)\}$ bits.

Unary and Binary Codes
The *unary code* represents symbol x as x 1-bits followed by a single 0-bit. The unary code for x is $1 + x$ bits long; hence, the corresponding ideal symbol probability distribution (for which this

pattern of codeword lengths yields the minimal message length) is given by $p_x = 2^{-(1+x)}$. Unary is an *infinite code*, for which knowledge of U is not required. But unless M is dominated by small integers, unary is expensive – the representation of a message $M = \langle s_0 \ldots s_{n-1} \rangle$ requires $n + \sum_i s_i = U' + 1$ bits.

If $U \leq 2^k$ for integer k, then s_i can be represented in k bits using the *binary code*. Binary is *finite*, with an ideal probability distribution given by $p_x = 2^{-k}$. When $U = 2^k$, the ideal probability $p_x = 2^{-\log_2 U} = 1/U$. When $2^{k-1} < U < 2^k$, then $2^k - U$ of the codewords can be shortened to $k - 1$ bits, in a *minimal binary code*. It is usual (but not necessary) to assign the short codewords to $0 \ldots 2^k - U - 1$, leaving the codewords for $2^k - U \ldots U - 1$ as k bits.

Elias Codes

Peter Elias described a suite of hybrids between unary and binary in work published in 1975 [7]. This family of codes are defined recursively, with unary being the base method. To code a value x, the "predecessor" Elias code is used to specify $x' = \lfloor \log_2(1 + x) \rfloor$, followed by an x'-bit binary code for $x - (2^{x'} - 1)$. The second member of the Elias family is C_γ and is a unary-binary code: unary for the prefix part and then binary to indicate the value of x within the range it specifies. The first few C_γ codewords are 0, 10 0, 10 1, 110 00, and so on, where spaces are used to illustrate the split between the components. The C_γ codeword for a value x requires $1 + \lfloor \log_2(1 + x) \rfloor$ bits for the unary part and a further $\lfloor \log_2(1+x) \rfloor$ bits for the binary part. The ideal probability distribution is thus given by $p_x \approx 1/(2(1 + x)^2)$. After C_γ, the next member of the Elias family is C_δ, in which C_γ is used to store x'. The first few codewords are 0, 100 0, 100 1, and 101 00; like unary and C_γ, C_δ is infinite, but now the codeword for x requires $1 + 2\lfloor \log_2(1+x') \rfloor + x'$ bits, where $x' = \lfloor \log_2(1 + x) \rfloor$. Further members of the family can be generated, but for most practical purposes, C_δ is the last useful one. To see why, note that $|C_\gamma(x')| \leq |C_\delta(x')|$ whenever $x' \leq 30$, meaning that the next Elias code is shorter than C_δ only for values $x' \geq 31$, that is, for $x \geq 2^{31} - 1$.

Fibonacci-Based Codes

Another infinite code arises from the Fibonacci sequence described (for this purpose) as $F_0 = 1$, $F_1 = 2$, $F_2 = 3$, $F_3 = 5$, $F_4 = 8$, and in general $F_i = F_{i-1} + F_{i-2}$. The *Zeckendorf* representation of a natural number is a list of Fibonacci values that add up to it, such that no two adjacent Fibonacci numbers are used. For example, 9 is the sum of $1 + 8 = F_0 + F_4$. The *Fibonacci code* for $x \geq 0$ is derived from the Zeckendorf representation of $x + 1$ and consists of a 1 bit in the ith position (counting from the left) if F_i appears in the sum and a 0 bit if not. Because it is not possible for both F_i and F_{i+1} to be part of the sum, the last 2 bits must be 01; hence, appending a further 1 bit provides a unique sentinel for the codeword. The code for $x = 0$ is 11, and the next few codewords are 011, 0011, 1011, 00011, and 10011. Because (for large i) $F_i \approx \phi^{i+2}/\sqrt{5}$, where $\phi = (1 + \sqrt{5})/2 \approx 1.62$, the codeword for x requires approximately $\lfloor \log_\phi \sqrt{5} + \log_\phi(1 + x)\sqrt{5} \rfloor \approx \lfloor 1.67 + 1.44 \log_2(1 + x) \rfloor$ bits and is longer than C_γ only for $x = 0$ and $x = 2$. The Fibonacci code is also as good as, or better than, C_δ between $x = 1$ and $F_{18} - 2 = 6,763$. Higher-order variants are also possible, with increased minimum codeword lengths and decreased coefficients on the logarithmic term. Fenwick [8] provides coverage of Fibonacci codes.

Byte-Aligned Codes

Extracting bits from bitstrings can slow down decoding rates, especially if each bit is then tested in a loop guard. Operations on larger units tend to be faster. The simplest *byte-aligned code* is an interleaved 8-bit analog of the Elias C_γ mechanism. The top bit in each byte is reserved for a flag that indicates (when 0) that this is the last byte of the codeword and (when 1) that the codeword continues; the other 7 bits in each byte are for data. A total of $8\lceil (\log_2 x)/7 \rceil$ bits are used, which makes it more effective asymptotically than the Elias C_γ code or the Fibonacci code. However, the minimum 8 bits means that byte-aligned codes are expensive on messages dominated by small values. A further advantage of byte codes is that compressed sequences can be searched, using a search pattern converted using the same code [5]. The zero top bit in all final bytes means that false matches can be easily eliminated.

An improvement to the simple byte-aligned coding mechanism arises from the observation that different values for the separating value between the *stopper* and *continuer* bytes lead to different trade-offs in overall codeword lengths [3]. In the (S, C)-*byte-aligned code*, values for $S + C = 256$ are chosen, and each codeword consists of a sequence of zero or more continuer bytes with values greater than or equal to S, followed by a stopper byte with a value less than S. Other variants include methods that use bytes as the coding units to form Huffman codes, either using 8-bit coding symbols or tagged 7-bit units [5] and methods that partially permute the alphabet, but avoid the need for a complete mapping [4]. Culpepper and Moffat [4] also describe a byte-aligned coding method that creates a set of byte-based codewords with the property that the first byte uniquely identifies the length of the codeword. Similarly, *nibble* codes can be designed as a 4-bit analog of the byte-aligned approach, with 1 bit reserved for a stopper-continuer flag, and 3 bits used for data.

Golomb Codes

In 1966 Solomon Golomb observed that the intervals between consecutive members of a random n-subset of the items $0 \ldots U-1$ could be modeled by a geometric probability distribution $p_x = p(1 - p)^{x-1}$, where $p = n/U$ [10]. This probability distribution implies that in a *Golomb code*, the representation for $x + b$ should be 1 bit longer than the representation for x when $(1 - p)^b \approx 0.5$, that is, when $b = \log 0.5 / \log(1 - p) \approx (\ln 2)/p \approx 0.69 U/n$. Assuming a monotonic message, each s_i is converted to a gap; then g_i div b is coded in unary; and finally g_i mod b is coded in minimal binary with respect to b. Like unary, the Golomb code is infinite. If the sequence is non-monotonic, then the values s_i are coded directly using parameter $b = 0.69 U'/n$. Each 1-bit that is part of a unary part spans b elements of U, meaning that there are at most $\lfloor U/b \rfloor$ of them in total; and there are exactly

n 0-bits in the unary parts. The minimal binary parts, one per symbol, take fewer than $n \lceil \log_2 b \rceil$ bits in total. Summing these components, and maximizing the cost over different values of n and U by assuming an adversary that forces the use of the first long minimal binary codeword whenever possible, yields a total Golomb code length for a monotonic sequence of at most $n(\log_2(U/n) + 2)$ bits. The variant in which $b = 2^k$ is used, $k = \lfloor \log_2(U/n) \rfloor$, is called a *Rice code*. Note that Golomb and Rice codes are infinite, but require that a parameter be set and that one way of estimating the parameter is based on knowing a value for U.

Other Static Codes

Elias codes and Golomb codes are examples of methods specified by a set of *buckets*, with symbol x coded in two parts: a bucket identifier, followed by an offset within the bucket, the latter usually using minimal binary. For example, the Elias C_γ code employs a vector of bucket sizes $\langle 2^0, 2^1, 2^2, 2^3, 2^4, \ldots \rangle$. Teuhola (see Moffat and Turpin [14]) proposed a hybrid in which a parameter k is chosen, and the vector of bucket sizes is given by $\langle 2^k, 2^{k+1}, 2^{k+2}, 2^{k+3}, \ldots \rangle$. One way of setting the parameter k is the length in bits of the median sequence value, so that the first bit of each codeword approximately halves the range of observed symbol values. Another variant is described by Boldi and Vigna [2], using vector $\langle 2^k-1, (2^k-1)2^k, (2^k-1)2^{2k}, (2^k-1)2^{3k}, \ldots \rangle$ to obtain a family of codes that are analytically and empirically well suited to power-law probability distributions, especially (taking k in the range 2–4) those associated with web-graph compression. Fraenkel and Klein [9] observed that the sequence of symbol *magnitudes* (i.e., the sequence of values $\lfloor \log_2(1 + s_i) \rfloor$) provides a denser range than the message itself and that it can be effective to use a principled code for them. For example, rather than using unary for the prefix part, a Huffman code can be used. Moffat and Anh [12] consider other ways in which the prefix part of each codeword can be reduced; and Fenwick [8] provides general coverage of other static coding methods.

Elias-Fano Codes

In 1974 Elias [6] presented another coding method, noting that it was described independently in 1971 by Robert Fano. The approach is now known as *Elias-Fano coding*. For monotonic sequence M, parameter $k = \lfloor \log_2(U/n) \rfloor$ is used to again break each codeword into quotient and remainder, *without* first taking gaps, with codewords formed relative to a sequence of buckets each of width $b = 2^k$. The number of symbols in the buckets is stored in a bitstring of $\lceil U/b \rceil < 2n$ unary codes. The n remainder parts $r_i = s_i \bmod b$ are stored as a sequence of k-bit binary codes. Each symbol contributes k bits as a binary part and adds 1 bit to one of the unary parts; plus there are at most $\lceil U/b \rceil$ 0-bits terminating the unary parts. Based on these relationships, the total length of an Elias-Fano code can be shown to be at most $n(\log_2(U/n) + 2)$ bits. Vigna [15] has deployed Elias-Fano codes to good effect.

Packed Codes

If each n-subset of $0 \ldots U - 1$ is equally likely, the Golomb code is effective in the average case; and the Elias-Fano code is effective in the worst case. But if the elements in the subset are *clustered*, then it is possible to obtain smaller representations, provided that groups of elements themselves can be employed as part of the process of determining the code. The *word-aligned* codes of Anh and Moffat [1] fit as many binary values into each output word as possible. For example, in their Simple-9 method, 32-bit output words are used, and the first 4 bits of each word contain a *selector* which specifies how to decode the next 28 bits: one 28-bit binary number, or two 14-bit binary numbers, or three 9-bit numbers, and so on. Other variants use 64-bit words [1]. In these codes, clusters of low s_i (or g_i) values can be represented more compactly than would occur with the Golomb code and an all-of-message b parameter; and decoding is fast because whole words are expanded without any need for conditionals or branching.

Zukowski et al. [16] describe a different approach, in which blocks of z values from M are coded in binary using k bits each, where k is chosen such that 2^k is larger than most, but not necessarily all, of the z elements in the block. Any s_i's in the block that are larger than $2^k - 1$ are noted as *exceptions* and handled separately; a variety of methods for coding the exceptions have been used in different forms of the *pfordelta code*. This mechanism is again fast when decoding, due to the absence of conditional bit evaluations, and, for typical values such as $z = 128$, also yields effective compression. Lemire and Boytsov have carried out detailed experimentation with packed codes [11].

Context-Sensitive Codes

If the objective is to create the smallest representation, rather than provide a balance between compression effectiveness and decoding speed, the nonsequential *binary interpolative code* of Moffat and Stuiver [13] can be used. As an example, consider message M, shown in Table 1, and suppose that the decoder is aware that $U' = 29$, that is, that $s_i' < 29$. Every item in M is greater than or equal to $lo = 0$ and less than $hi = 29$, and the mid-value in M, in this example $s_4 = 6$ (it doesn't matter which mid-value is chosen), can be transmitted to the decoder using $\lceil \log_2 29 \rceil = 5$ bits. Once that middle number is pinned, the remaining values can be coded recursively within more precise ranges and might require fewer than 5 bits each.

In fact, there are four distinct values in M that precede s_4' and another five that follow it, so a more restricted range for s_4' can be inferred: it must be greater than or equal to $lo' = lo + 4 = 4$ and less than $hi' = hi - 5 = 24$. That is, $s_4' = 6$ can be minimal binary coded as the value $6 - lo' = 2$ within the range $[0, 20)$ using just 4 bits.

It remains to transmit the left part, $\langle 0, 3, 4, 5 \rangle$, against the knowledge that every value is greater than or equal to $lo = 0$ and less than $hi = 6$, and the right part, $\langle 16, 24, 26, 27, 28 \rangle$, against the knowledge that every value is greater than or equal to $lo = 7$ and less than $hi = 29$. These two sublists are processed recursively in the order shown in the remainder of Table 1, with the tighter ranges $[lo', hi')$ also shown at

Compressing Integer Sequences, Table 1 Example encodings of message $M = \langle 0, 3, 4, 5, 6, 16, 24, 26, 27, 28 \rangle$ using the interpolative code. When a minimal binary code is applied to each value in its corresponding range, a total of 20 bits are required. No bits are output if $lo' = hi' - 1$

i	s_i	s_i'	lo	hi	lo'	hi'	$s_i - lo'$	$hi' - lo'$	Binary	MinBin
4	0	6	0	29	4	24	2	20	00010	0010
1	2	3	0	6	1	4	2	3	10	11
0	0	0	0	3	0	3	0	3	00	0
2	0	4	4	6	4	5	0	1	- -	- -
3	0	5	5	6	5	6	0	1	- -	- -
7	1	26	7	29	9	27	17	18	10001	11111
5	9	16	7	26	7	25	9	18	01001	1001
6	7	24	17	26	17	26	7	9	0111	1110
8	1	27	27	29	27	28	0	1	- -	- -
9	0	28	28	29	28	29	0	1	- -	- -

each step. One key feature of the interpolative code is that the when the range is just one, codewords of length zero are used. Four of the symbols in M benefit in this way. The interpolative code is particularly effective when the subset contains runs of consecutive items, or localized regions where there is a high density.

The final column of Table 1 uses minimal binary for each value within its bounded range. A refinement is to use a *centered minimal binary code*, so that the short codewords are assigned in the middle of the range rather than at the beginning, recognizing that the midpoint of a set is more likely to be near the middle of the range spanned by the set than it is to be near the ends of the range. Adding this enhancement tends to be beneficial. But improvement is not guaranteed, and on M it adds 1 bit to the compressed representation compared to what is shown in Table 1.

Applications

Messages in this form are often the output of a modeling step in a data compression system. Other examples include recording the hyperlinks in the graph representation of the World Wide Web and storing inverted indexes for large document collections.

Cross-References

▶ Arithmetic Coding for Data Compression
▶ Huffman Coding

Recommended Reading

1. Anh VN, Moffat A (2010) Index compression using 64-bit words. Softw Pract Exp 40(2):131–147
2. Boldi P, Vigna S (2005) Codes for the world-wide web. Internet Math 2(4):405–427
3. Brisaboa NR, Fariña A, Navarro G, Esteller MF (2003) (S, C)-dense coding: an optimized compression code for natural language text databases. In: Proceedings of the symposium on string processing and information retrieval, Manaus, pp 122–136
4. Culpepper JS, Moffat A (2005) Enhanced byte codes with restricted prefix properties. In: Proceedings of the symposium on string processing and information retrieval, Buenos Aires, pp 1–12
5. de Moura ES, Navarro G, Ziviani N, Baeza-Yates R (2000) Fast and flexible word searching on compressed text. ACM Trans Inf Syst 18(2):113–139
6. Elias P (1974) Efficient storage and retrieval by content and address of static files. J ACM 21(2):246–260
7. Elias P (1975) Universal codeword sets and representations of the integers. IEEE Trans Inf Theory IT-21(2):194–203
8. Fenwick P (2003) Universal codes. In: Sayood K (ed) Lossless compression handbook. Academic, Boston, pp 55–78
9. Fraenkel AS, Klein ST (1985) Novel compression of sparse bit-strings—preliminary report. In: Apostolico A, Galil Z (eds) Combinatorial algorithms on words. NATO ASI series F, vol 12. Springer, Berlin, pp 169–183

10. Golomb SW (1966) Run-length encodings. IEEE Trans Inf Theory IT–12(3):399–401

11. Lemire D, Boytsov L (2014, to appear) Decoding billions of integers per second through vectorization. Softw Pract Exp. http://dx.doi.org/10.1002/spe.2203

12. Moffat A, Anh VN (2006) Binary codes for locally homogeneous sequences. Inf Process Lett 99(5):75–80

13. Moffat A, Stuiver L (2000) Binary interpolative coding for effective index compression. Inf Retr 3(1):25–47

14. Moffat A, Turpin A (2002) Compression and coding algorithms. Kluwer Academic, Boston

15. Vigna S (2013) Quasi-succinct indices. In: Proceedings of the international conference on web search and data mining, Rome, pp 83–92

16. Zukowski M, Héman S, Nes N, Boncz P (2006) Super-scalar RAM-CPU cache compression. In: Proceedings of the international conference on data engineering, Atlanta. IEEE Computer Society, Washington, DC, paper 59

Computing Cutwidth and Pathwidth of Semi-complete Digraphs

Michał Pilipczuk
Institute of Informatics, University of Bergen, Bergen, Norway
Institute of Informatics, University of Warsaw, Warsaw, Poland

Keywords

Cutwidth; Minor; Pathwidth; Semi-complete digraph; Tournament

Years and Authors of Summarized Original Work

2012; Chudnovsky, Fradkin, Seymour
2013; Fradkin, Seymour
2013; Fomin, Pilipczuk
2013; Pilipczuk
2013; Fomin, Pilipczuk

Problem Definition

Recall that a simple digraph T is a tournament if for every two vertices $u, v \in V(T)$, exactly one

of the arcs (u, v) and (v, u) exists in T. If we relax this condition by allowing both these arcs to exist at the same time, then we obtain the definition of a *semi-complete* digraph. We say that a digraph T contains a digraph H as a *topological minor* if one can map vertices of H to different vertices of T, and arcs of H to directed paths connecting respective images of the endpoints that are internally vertex disjoint. By relaxing vertex disjointness to arc disjointness, we obtain the definition of an *immersion*. (For simplicity, we neglect here the difference between weak immersions and strong immersions, and we work with weak immersions only.) Finally, we say that T contains H as a *minor* if vertices of H can be mapped to vertex disjoint strongly connected subgraphs of T in such a manner that for every arc (u, v) of H, there exists a corresponding arc of T going from a vertex belonging to the image of u to a vertex belonging to the image of v.

The topological minor, immersion, and minor relations form fundamental containment orderings on the class of digraphs. Mirroring the achievements of the graph minors program of Robertson and Seymour, it is natural to ask what is the complexity of testing these relations when the pattern graph H is assumed to be small. For general digraphs, even very basic problems of this nature are NP-complete [5]; however, the structure of semi-complete digraphs allow us to design efficient algorithms.

On semi-complete digraphs, the considered containment relations are tightly connected to digraph parameters *cutwidth* and *pathwidth*. For a digraph T and an ordering $(v_1, v_2, \ldots, v_n)$ of $V(T)$, by *width* of this ordering, we mean the maximum over $1 \leq t \leq n - 1$ of the number of arcs going from $\{v_1, v_2, \ldots, v_t\}$ to $\{v_{t+1}, v_{t+2}, \ldots, v_n\}$ in T. The cutwidth of a digraph T, denoted by $\mathbf{ctw}(T)$, is the minimum width of an ordering of $V(T)$. A *path decomposition* of a digraph T is a sequence $\mathcal{P} = (W_1, W_2, \ldots, W_r)$ of subsets of vertices, called bags, such that (i) $\bigcup_{i=1}^{r} W_i = V(T)$, (ii) $W_j \supseteq W_i \cap W_k$ for all $1 \leq i < j < k \leq r$, and (iii) whenever (u, v) is an edge of T, then u and v appear together in some bag of $\mathcal{P}$ or all the bags in which u appears are placed after all the bags in which v appears. The *width* of $\mathcal{P}$ is equal to $\max_{1 \leq i \leq r} |W_i| - 1$. The pathwidth of T, denoted by $\mathbf{pw}(T)$, is the minimum width of a path decomposition of T.

It appears that if a semi-complete digraph T excludes some digraph H as an immersion, then its cutwidth is bounded by a constant c_H depending on H only. Similarly, if T excludes H as a minor or as a topological minor, then its pathwidth is bounded by a constant p_H depending on H only. These Erdős-Pósa-style results were proven by Chudnovsky et al. [2] and Fradkin and Seymour [7], respectively. Based on this understanding of the links between containment relations and width parameters, it has been shown that immersion and minor relations are well-quasi-orderings of the class of semi-complete digraphs [1,6].

The aforementioned theorems give also raise to natural algorithms for testing the containment relations. We try to approximate the appropriate width measure: If we obtain a guarantee that it is larger than the respective constant c_H or p_H, then we can conclude that H is contained in T for sure. Otherwise, we can construct a decomposition of T of small width on which a dynamic programming algorithm can be employed. In fact, the proofs of Chudnovsky et al. [2] and Fradkin and Seymour [7] can be turned into (some) approximation algorithms for cutwidth and pathwidth on semi-complete digraphs. Therefore, it is natural to look for more efficient such algorithms, both in terms of the running time and the approximation ratio. The efficiency of an approximation subroutine is, namely, the crucial ingredient of the overall running time for testing containment relations.

Key Results

As a by-product of their proof, Chudnovsky et al. [2] obtained an algorithm that, given an n-vertex semi-complete digraph T and an integer k, either finds an ordering of $V(T)$ of width $\mathcal{O}(k^2)$ or concludes that $\mathbf{ctw}(T) > k$ by finding an appropriate combinatorial obstacle

embedded in T. The running time is $\mathcal{O}(n^3)$. Similarly, a by-product of the proof of Fradkin and Seymour [7] is an algorithm that, for the same input, either finds a path decomposition of T of width $\mathcal{O}(k^2)$ or concludes that $\mathbf{pw}(T) > k$, again certifying this by providing an appropriate obstacle. Unfortunately, here the running time is $\mathcal{O}(n^{\mathcal{O}(k)})$; in other words, the exponent of the polynomial grows with k.

The proofs of the Erdős-Pósa statements proceed as follows: One shows that if the found combinatorial obstacle is large enough, i.e., it certifies that $\mathbf{ctw}(T) > k$ or $\mathbf{pw}(T) > k$ for large enough k, then already inside this obstacle one can find an embedding of every digraph H of a fixed size. Of course, the final values of constants c_H and p_H depend on how efficiently we can extract a model of H from an obstacle and, more precisely, how large k must be in terms of $|H|$ in order to guarantee that an embedding of H can be found. (We denote $|H| = |V(H)| + |E(H)|$.) Unfortunately, in the proofs of Chudnovsky et al. [2] and Fradkin and Seymour [7], this dependency is exponential (even multiple exponential in the case of p_H) and so is the overall dependency of constants c_H and p_H on $|H|$. Using the framework presented before, one can obtain an $f(|H|) \cdot n^3$-time algorithm for testing whether H can be immersed into an n-vertex semi-complete digraph T, as well as similar tests for the (topological) minor relations with running time $n^{g(|H|)}$. Here, f and g are some multiple-exponential functions.

The running time of the immersion testing algorithm is fixed-parameter tractable (FPT), while the running time for (topological) minor testing is only XP. It is natural to ask for an FPT algorithm also for the latter problem. Fomin and the current author [3,4,8] approached the issue from a different angle, which resulted in reproving the previous results with better constants, refined running times, and also in giving FPT algorithms for all the containment tests. Notably, the framework seems to be simpler and more uniform compared to the previous work. We now state the results of [3,4,8] formally, since they constitute the best known so far algorithms for topological problems in semi-complete digraphs.

Theorem 1 ([8]) *There exists an algorithm that, given an n-vertex semi-complete digraph T, runs in time $\mathcal{O}(n^2)$ and returns an ordering of $V(T)$ of width at most $\mathcal{O}(\mathbf{ctw}(T)^2)$.*

Theorem 2 ([8, 9]) *There exists an algorithm that, given an n-vertex semi-complete digraph T and an integer k, runs in time $\mathcal{O}(k\, n^2)$ and either returns a path decomposition of $V(T)$ of width at most $6k$ or correctly concludes that $\mathbf{pw}(T) > k$.*

Theorem 3 ([4, 8, 9]) *There exist algorithms that, given an n-vertex semi-complete digraph T and an integer k, determine whether:*

- $\mathbf{ctw}(T) \leq k$ *in time* $2^{\mathcal{O}(\sqrt{k \log k})} \cdot n^2$.
- $\mathbf{pw}(T) \leq k$ *in time* $2^{\mathcal{O}(k \log k)} \cdot n^2$.

Theorem 4 ([8]) *There exist algorithms that, given a digraph H and an n-vertex semi-complete digraph T, determine whether:*

- H *can be immersed in T in time* $2^{\mathcal{O}(|H|^2 \log |H|)} \cdot n^2$.
- H *is a topological minor of T in time* $2^{\mathcal{O}(|H| \log |H|)} \cdot n^2$.
- H *is a minor of T in time* $2^{\mathcal{O}(|H| \log |H|)} \cdot n^2$.

Thus, Theorems 1 and 2 provide approximation algorithms for cutwidth and pathwidth, Theorem 3 provides FPT algorithms for computing the exact values of these parameters, and Theorem 4 utilizes the approximation algorithms to give efficient algorithms for containment testing. We remark that the exact algorithm for cutwidth (the first bullet of Theorem 3) is a combination of the results of [8] (which gives a $2^{\mathcal{O}(k)} \cdot n^2$-time algorithm) and of [4] (which gives a $2^{\mathcal{O}(\sqrt{k \log k})} \cdot n^{\mathcal{O}(1)}$-time algorithm). A full exposition of this algorithm can be found in the PhD thesis of the current author [9], which contains a compilation of [3, 4, 8]. Moreover, for Theorem 2 work [8] claims only a 7-approximation, which has been consequently improved to a 6-approximation in the aforementioned PhD thesis [9]. Finally, it also follows that in the Erdős-Pósa results, one can take $c_H = \mathcal{O}(|H|^2)$ and $p_H = \mathcal{O}(|H|)$;

this claim is not mentioned explicitly in [8], but follows easily from the results proven there.

To conclude, let us shortly deliberate on the approach that led to these results. The key to the understanding is the work [8]. The main observation there is that a large cluster of vertices with very similar outdegrees is already an obstacle for admitting a path decomposition of small width. More precisely, if one finds $4k + 2$ vertices whose outdegrees pairwise differ by at most k (a so-called $(4k + 2, k)$-*degree tangle*), then this certifies that $\mathbf{pw}(T) > k$; see Lemma 46 of [9]. As it always holds that $\mathbf{pw}(T) \leq 2\mathbf{ctw}(T)$, this conclusion also implies that $\mathbf{ctw}(T) > k/2$. Therefore, in semi-complete digraphs of small pathwidth or cutwidth, the outdegrees of vertices must be spread evenly; there is no "knot" with a larger density of vertices around some value of the outdegree. If we then order the vertices of T by their outdegrees, then this ordering should crudely resemble an ordering with the optimal width, as well as the order in which the vertices appear on an optimal path decomposition of T. Indeed, it can be shown that any ordering of $V(T)$ w.r.t. nondecreasing outdegrees has width at most $\mathcal{O}(\mathbf{ctw}(T)^2)$ [8]. Hence, the algorithm of Theorem 1 is, in fact, trivial: We just sort the vertices by their outdegrees. The pathwidth approximation algorithm (Theorem 2) is obtained by performing a left-to-right scan through the outdegree ordering that constructs a path decomposition in a greedy manner. For the exact algorithms (Theorem 3), in the scan we maintain a dynamic programming table of size exponential in k, whose entries correspond to possible endings of partial decompositions for prefixes of the ordering. The key to improving the running time for cutwidth to subexponential in terms of k (shown in [4]) is relating the states of the dynamic programming algorithm to *partition numbers*, a sequence whose subexponential asymptotics is well understood. Finally, the obstacles yielded by the approximation algorithms of Theorems 1 and 2 are more useful for finding embeddings of small digraphs H than the ones used in the previous works. This leads to a better dependence on $|H|$ of constants

c_H, p_H in the Erdős-Pósa results, as well as of the running times of the containment tests (Theorem 4).

Recommended Reading

1. Chudnovsky M, Seymour PD (2011) A well-quasi-order for tournaments. J Comb Theory Ser B 101(1):47–53
2. Chudnovsky M, Ovetsky Fradkin A, Seymour PD (2012) Tournament immersion and cutwidth. J Comb Theory Ser B 102(1):93–101
3. Fomin FV, Pilipczuk M (2013) Jungles, bundles, and fixed parameter tractability. In: Khanna S (ed) SODA, New Orleans. SIAM, pp 396–413
4. Fomin FV, Pilipczuk M (2013) Subexponential parameterized algorithm for computing the cutwidth of a semi-complete digraph. In: Bodlaender HL, Italiano GF (eds) ESA, Sophia Antipolis. Lecture notes in computer science, vol 8125. Springer, pp 505–516
5. Fortune S, Hopcroft JE, Wyllie J (1980) The directed subgraph homeomorphism problem. Theor Comput Sci 10:111–121
6. Kim I, Seymour PD (2012) Tournament minors. CoRR abs/1206.3135
7. Ovetsky Fradkin A, Seymour PD (2013) Tournament pathwidth and topological containment. J Comb Theory Ser B 103(3):374–384
8. Pilipczuk M (2013) Computing cutwidth and pathwidth of semi-complete digraphs via degree orderings. In: Portier N, Wilke T (eds) STACS, Schloss Dagstuhl – Leibniz-Zentrum fuer Informatik, LIPIcs, vol 20, pp 197–208
9. Pilipczuk M (2013) Tournaments and optimality: new results in parameterized complexity. PhD thesis, University of Bergen, Norway. Available at the webpage of the author

Computing Pure Equilibria in the Game of Parallel Links

Spyros Kontogiannis
Department of Computer Science, University of Ioannina, Ioannina, Greece

Keywords

Convergence of Nash dynamics; Incentive compatible algorithms; Load balancing game; Nashification

Years and Authors of Summarized Original Work

2002; Fotakis, Kontogiannis, Koutsoupias, Mavronicolas, Spirakis
2003; Even-Dar, Kesselman, Mansour
2003; Feldman, Gairing, Lücking, Monien, Rode

Problem Definition

This problem concerns the construction of pure Nash equilibria (PNE) in a special class of atomic congestion games, known as the Parallel Links Game (PLG). The purpose of this note is to gather recent advances in the *existence and tractability of PNE in PLG*.

THE PURE PARALLEL LINKS GAME. Let $N \equiv [n]$ ($\forall k \in \mathbb{N}$, $[k] \equiv \{1, 2, \ldots, k\}$.) be a set of (selfish) players, each of them willing to have her good served by a *unique* shared resource (link) of a system. Let $E = [m]$ be the set of all these links. For each link $e \in E$, and each player $i \in N$, let $D_{i,e}(\cdot) : \mathbb{R}_{\geq 0} \mapsto \mathbb{R}_{\geq 0}$ be the **charging mechanism** according to which link e charges player i for using it. Each player $i \in [n]$ comes with a service requirement (e.g., traffic demand, or processing time) $W[i, e] > 0$, if she is to be served by link $e \in E$. A service requirement $W[i, e]$ is allowed to get the value ∞, to denote the fact that player i would never want to be assigned to link e. The charging mechanisms are functions of each link's cumulative congestion.

Any element $\sigma \in E$ is called a **pure strategy** for a player. Then, this player is assumed to assign her own good to link e. A collection of pure strategies for all the players is called a **pure strategies profile**, or a **configuration** of the players, or a **state** of the game.

The **individual cost** of player i wrt the profile σ is: $\mathrm{IC}_i(\sigma) = D_{i,\sigma_i}(\sum_{j \in [n]: \sigma_j = \sigma_i} W[j, \sigma_j])$. Thus, the **Pure Parallel Links Game** (PLG) is the game in strategic form defined as $\Gamma = \langle N, (\Sigma_i = E)_{i \in N}, (\mathrm{IC}_i)_{i \in N} \rangle$, whose acceptable solutions are only PNE. Clearly, an arbitrary instance of PLG can be described by the tuple $\langle N, E, (W[i, e])_{i \in N, e \in E}, (D_{i,e}(\cdot))_{i \in N, e \in E} \rangle$.

DEALING WITH SELFISH BEHAVIOR. The dominant solution concept for finite games in strategic form, is the Nash Equlibrium [14]. The definition of pure Nash Equilibria for PLG is the following:

Definition 1 (Pure Nash Equilibrium) For any instance $\langle N, E, (W[i, e])_{i \in N, e \in E}, (D_{i,e}(\cdot))_{i \in N, e \in E} \rangle$ of PLG, a pure strategies profile $\sigma \in E^n$ is a **Pure Nash Equilibrium** (PNE in short), iff: $\forall i \in N, \forall e \in E, \mathrm{IC}_i(\sigma) = D_{i,\sigma_i} \left(\sum_{j \in [n]:\sigma_j = \sigma_i} W[j, \sigma_i] \right) \leq D_{i,e} \left(W[i, e] + \sum_{j \in [n] \setminus \{i\}:\sigma_j = e} W[j, e] \right)$.

A refinement of PNE are the **k- robust PNE**, for $n \geq k \geq 1$ [9]. These are pure profiles for which no subset of at most k players may concurrently change their strategies in such a way that the worst possible individual cost among the movers is *strictly decreased*.

QUALITY OF PURE EQUILIBRIA. In order to determine the quality of a PNE, a social cost function that measures it must be specified. The typical assumption in the literature of PLG, is that the social cost function is the worst individual cost paid by the players: $\forall \sigma \in E^n, \mathrm{SC}(\sigma) = \max_{i \in N} \{\mathrm{IC}_i(\sigma)\}$ and $\forall \mathbf{p} \in (\Delta_m)^n, \mathrm{SC}(\mathbf{p}) = \sum_{\sigma \in E^n} (\prod_{i \in N} p_i(\sigma_i)) \cdot \max_{i \in N} \{\mathrm{IC}_i(\sigma)\}$. Observe that, for mixed profiles, the social cost is the *expectation* of the maximum individual cost among the players.

The measure of the quality of an instance of PLG wrt PNE, is measured by the **Pure Price of Anarchy** (PPoA in short) [12]: $\mathrm{PPoA} = \max \{(\mathrm{SC}(\sigma))/\mathrm{OPT} : \sigma \in E^n \text{ is PNE}\}$ where $\mathrm{OPT} \equiv \min_{\sigma \in E^n} \{\mathrm{SC}(\sigma)\}$.

DISCRETE DYNAMICS. Crucial concepts of strategic games are the best and better responses. Given a configuration $\sigma \in E^n$, an **improvement step**, or **selfish step**, or **better response** of player $i \in N$ is the choice by i of a pure strategy $\alpha \in E \setminus \{\sigma_i\}$, so that player i would have a positive gain by this *unilateral* change (i.?e., provided that the other players maintain the

same strategies). That is, $\mathrm{IC}_i(\sigma) > \mathrm{IC}_i(\sigma \oplus_i \alpha)$ where, $\sigma \oplus_i \alpha \equiv (\sigma_1, \ldots, \sigma_{i-1}, \alpha, \sigma_{i+1}, \ldots, \sigma_n)$. A **best response**, or **greedy selfish step** of player i, is any change from the current link σ_i to an element $\alpha^* \in \arg\max_{a \in E} \{\mathrm{IC}_i(\sigma \oplus_i \alpha)\}$. An **improvement path** (aka a **sequence of selfish steps** [6], or an **elementary step system** [3]) is a sequence of configurations $\pi = \langle \sigma(1), \ldots, \sigma(k) \rangle$ such that

$$\forall 2 \leq r \leq k, \exists i_r \in N, \exists \alpha_r \in E :$$
$$[\sigma(r) = \sigma(r-1) \oplus_{i_r} \alpha_r] \wedge [\mathrm{IC}_{i_r}(\sigma(r))$$
$$< \mathrm{IC}_{i_r}(\sigma(r-1))] .$$

A game has the **Finite Improvement Property** (FIP) iff any improvement path has *finite* length. A game has the **Finite Best Response Property** (FBRP) iff any improvement path, each step of whose is a best response of some player, has *finite* length.

An alternative trend is to, rather than consider *sequential* improvement paths, let the players conduct selfish improvement steps *concurrently*. Nevertheless, the selfish decisions are no longer deterministic, but rather distributions over the links, in order to have some notion of a priori Nash property that justifies these moves. The selfish players try to minimize their *expected* individual costs this time. Rounds of concurrent moves occur until a posteriori Nash Property is achieved. This is called a **selfish rerouting** policy [4].

Subclasses of PLG

[PLG$_1$] Monotone PLG: The charging mechanism of each pair of a link and a player, is a *non–decreasing function* of the resource's cumulative congestion.

[PLG$_2$] Resource Specific Weights PLG (RSPLG): Each player may have a different service demand from every link.

[PLG$_3$] Player Specific Delays PLG (PSPLG): Each link may have a different charging mechanism for each player. Some special cases of PSPLG are the following:

[PLG$_{3.1}$] Linear Delays PSPLG: Every link has a (player specific) *affine* charging mechanism: $\forall i \in N, \forall e \in E, D_{i,e}(x) = a_{i,e}x + b_{i,e}$ for some $a_{i,e} > 0$ and $b_{i,e} \geq 0$.

[PLG$_{3.1.1}$] Related Delays PSPLG: Every link has a (player specific) *non–uniformly related* charging mechanism: $\forall i \in N, \forall e \in E, W[i,e] = w_i$ and $D_{i,e}(x) = a_{i,e}x$ for some $a_{i,e} > 0$.

[PLG$_4$] Resource Uniform Weights PLG (RUPLG): Each player has a unique service demand from all the resources. Ie, $\forall i \in N, \forall e \in E, W[i,e] = w_e > 0$. A special case of RUPLG is:

[PLG$_{4.1}$] Unweighted PLG: All the players have identical demands from all the links: $\forall i \in N, \forall e \in E, W[i,e] = 1$.

[PLG$_5$] Player Uniform Delays PLG (PU-PLG): Each resource adopts a unique charging mechanism, for all the players. That is, $\forall i \in N, \forall e \in E, D_{i,e}(x) = d_e(x)$.

[PLG$_{5.1}$] Unrelated Parallel Machines, or **Load Balancing PLG (LBPLG):** The links behave as parallel machines. That is, they charge each of the players for the cumulative load assigned to their hosts. One may think (wlog) that all the machines have as charging mechanisms the identity function. That is, $\forall i \in N, \forall e \in E, D_{i,e}(x) = x$.

[PLG$_{5.1.1}$] Uniformly Related Machines LBPLG: Each player has the same demand at every link, and each link serves players at a fixed rate. That is: $\forall i \in N, \forall e \in E, W[i,e] = w_i$ and $D_{i,e}(x) = \frac{x}{s_e}$. Equivalently, service demands proportional to the capacities of the machines are allowed, but the identity function is required as the charging mechanism: $\forall i \in N, \forall e \in E, W[i,e] = \frac{w_i}{s_e}$ and $D_{i,e}(x) = x$.

[PLG$_{5.1.1.1}$] Identical Machines LBPLG: Each player has the same demand at every link, and all the delay mechanisms are the identity function: $\forall i \in N, \forall e \in E, W[i,e] = w_i$ and $D_{i,e}(x) = x$.

[PLG$_{5.1.2}$] Restricted Assignment LBPLG: Each traffic demand is either of unit or infinite size. The machines are identical. Ie, $\forall i \in N, \forall e \in E, W[i,e] \in \{1, \infty\}$ and $D_{i,e}(x) = x$.

Algorithmic Questions Concerning PLG

The following algorithmic questions are considered:

Problem 1 (PNEExistsInPLG(E, N, W, D))
INPUT: An instance $\langle N, E, (W[i,e])_{i \in N, e \in E}, (D_{i,e}(\cdot))_{i \in N, e \in E} \rangle$ of PLG
OUTPUT: Is there a configuration $\sigma \in E^n$ of the players to the links, which is a PNE?

Problem 2 (PNEConstructionInPLG(E, N, W, D))
INPUT: An instance $\langle N, E, (W[i,e])_{i \in N, e \in E}, (D_{i,e}(\cdot))_{i \in N, e \in E} \rangle$ of PLG
OUTPUT: An assignment $\sigma \in E^n$ of the players to the links, which is a PNE.

Problem 3 (BestPNEInPLG(E, N, W, D))
INPUT: An instance $\langle N, E, (W[i,e])_{i \in N, e \in E}, (D_{i,e}(\cdot))_{i \in N, e \in E} \rangle$ of PLG. A **social cost** function $SC : (\mathbb{R}_{\geq 0})^m \mapsto \mathbb{R}_{\geq 0}$ that characterizes the quality of any configuration $\sigma \in E^N$.
OUTPUT: An assignment $\sigma \in E^n$ of the players to the links, which is a PNE and minimizes the value of the social cost, compared to other PNE of PLG.

Problem 4 (WorstPNEInPLG(E, N, W, D))
INPUT: An instance $\langle N, E, (W[i,e])_{i \in N, e \in E}, (D_{i,e}(\cdot))_{i \in N, e \in E} \rangle$ of PLG. A **social cost** function $SC : (\mathbb{R}_{\geq 0})^m \mapsto \mathbb{R}_{\geq 0}$ that characterizes the quality of any configuration $\sigma \in E^N$.
OUTPUT: An assignment $\sigma \in E^n$ of the players to the links, which is a PNE and *maximizes* the value of the social cost, compared to other PNE of PLG.

Problem 5 (DynamicsConvergeInPLG(E, N, W, D))
INPUT: An instance $\langle N, E, (W[i,e])_{i \in N, e \in E}, (D_{i,e}(\cdot))_{i \in N, e \in E} \rangle$ of PLG
OUTPUT: Does FIP (or FBRP) hold? How long does it take then to reach a PNE?

Problem 6 (ReroutingConvergeInPLG(E, N, W, D))
INPUT: An instance $\langle N, E, (W[i,e])_{i \in N, e \in E}, (D_{i,e}(\cdot))_{i \in N, e \in E} \rangle$ of PLG

OUTPUT: Compute (if any) a selfish rerouting policy that converges to a PNE.

Status of Problem 1

Player uniform, unweighted atomic congestion games always possess a PNE [15], with no assumption on monotonicity of the charging mechanisms. Thus, Problem 1 is already answered for all unweighted PUPLG. Nevertheless, this is not necessarily the case for weighted versions of PLG:

Theorem 1 ([13]) *There is an instance of (monotone) PSPLG with only three players and three strategies per player, possessing no PNE. On the other hand, any unweighted instance of monotone PSPLG possesses at least one PNE.*

Similar (positive) results were given for LBPLG. The key observation that lead to these results, is the fact that the lexicographically minimum vector of machine loads is always a PNE of the game.

Theorem 2 *There is always a PNE for any instance of Uniformly Related LB-PLG* [7], *and actually for any instance of LBPLG* [3]. *Indeed, there is a* $k-$*robust PNE for any instance of LBPLG, and any* $1 \leq k \leq n$ [9].

Status of Problem 2, 5 and 6

Milchtaich [13] gave a constructive proof of existence for PNE in unweighted, monotone PSPLG, and thus implies a path of length at most n that leads to a PNE. Although this is a very efficient construction of PNE, it is not necessarily an improvement path, when all players are considered to coexist all the time, and therefore there is no justification for the adoption of such a path by the players. Milchtaich [13] proved that from an arbitrary initial configuration and allowing only best reply defections, there is a best reply improvement path of length at most $m \cdot \binom{n+1}{2}$. Finally, [11] proved for unweighted, Related PSPLG that it possesses FIP. Nevertheless, the convergence time is poor.

For LBPLG, the implicit connection of PNE construction to classical scheduling problems, has lead to quite interesting results.

Theorem 3 ([7]) *The LPT algorithm of Graham, yields a PNE for the case of Uniformly Related LBPLG, in time* $\mathcal{O}(m \log m)$.

The drawback of the LPT algorithm is that it is centralized and not selfishly motivated. An alternative approach, called **Nashification**, is to start from an arbitrary initial configuration $\sigma \in E^n$ and then try to construct a PNE of at most the same maximum individual cost among the players.

Theorem 4 ([6]) *There is an* $\mathcal{O}(nm^2)$ *time Nashification algorithm for any instance of Uniformly Related PLG.*

An alternative style of Nashification, is to let the players follow an arbitrary improvement path. Nevertheless, it is not always the case that this leads to a polynomial time construction of a PNE, as the following theorem states:

Theorem 5 *For Identical Machines LBPLG:*

- *There exist best response improvement paths of length* $\Omega\left(\max\left\{2^{\sqrt{n}}, \left(\frac{n}{m^2}\right)^m\right\}\right)$ [3,6].
- *Any best response improvement path is of length* $\mathcal{O}(2^n)$ [6].
- *Any best response improvement path, which gives priority to players of maximum weight among those willing to defect in each improvement step, is of length at most n* [3].
- *If all the service demands are integers, then any improvement path which gives priority to unilateral improvement steps, and otherwise allows only selfish 2-flips (ie, swapping of hosting machines between two goods) converges to a 2-robust PNE in at most* $\frac{1}{2}(\sum_{i \in N} w_i)^2$ *steps* [9].

The following result concerns selfish rerouting policies:

Theorem 6 ([4])

- *For unweighted Identical Machines LBPLG, a simple policy (BALANCE) forcing all the* *players of overloaded links to migrate to a new (random) link with probability proportional to the load of the link, converges to a PNE in $O(\log \log n + \log m)$ rounds of concurrent moves. The same convergence time holds also for a simple Nash Rerouting Policy, in which each mover actually has an incentive to move.*

- *For unweighted Uniformly Related LBPLG, BALANCE has the same convergence time, but the Nash Rerouting Policy may converge in $\Omega(\sqrt{n})$ rounds.*

Finally, a generic result of [5] is mentioned, that computes a PNE for arbitrary *unweighted, player uniform* symmetric network congestion games in polynomial time, by a nice exploitation of Rosenthal's potential and the solution of a proper minimum cost flow problem. Therefore, for PLG the following result is implied:

Theorem 7 ([5]) *For unweighted, monotone PU-PLG, a PNE can be constructed in polynomial time.*

Of course, this result provides no answer, e.g., for Restricted Assignment LBPLG, for which it is still not known how to efficiently compute PNE.

Status of Problem 3 and 4

The proposed LPT algorithm of [7] for constructing PNE in Uniformly Related LBPLG, actually provides a solution which is at most $1.52 < \text{PPoA}(LPT) < 1.67$ times worse than the optimum PNE (which is indeed the allocation of the goods to the links that minimizes the make-span). The construction of the optimum, as well as the worst PNE are hard problems, which nevertheless admits a PTAS (in some cases):

Theorem 8 *For LBPLG with a social cost function as defined in the Quality of Pure Equilibria paragraph:*

- *For Identical Machines, constructing the optimum or the worst PNE is* **NP**−*hard* [7].
- *For Uniformly Related Machines, there is a PTAS for the optimum PNE* [6].

- *For Uniformly Related Machines, it holds that* $\text{PPoA} = \Theta\big(\min\big\{(\log m)/(\log \log m),$ $\log(s_{\max})/(s_{\min})\big\}\big)$ [2].
- *For the Restricted Assignments,* $\text{PPoA} = \Omega((\log m)/(\log \log m))$ [10].
- *For a generalization of the Restricted Assignments, where the players have goods of any positive, otherwise infinite service demands from the links (and not only elements of $\{1, \infty\}$), it holds that $m - 1 \leq \text{PPoA} < m$ [10].*

It is finally mentioned that a recent result [1] for unweighted, single commodity network congestion games with linear delays, is translated to the following result for PLG:

Theorem 9 ([1]) *For unweighted PUPLG with linear charging mechanisms for the links, the worst case PNE may be a factor of $\text{PPoA} = 5/2$ away from the optimum solution, wrt the social cost defined in the Quality of Pure Equilibria paragraph.*

Key Results

None

Applications

Congestion games in general have attracted much attention from many disciplines, partly because they capture a large class of routing and resource allocation scenarios.

PLG in particular, is the most elementary (non–trivial) atomic congestion game among a large number of players. Despite its simplicity, it was proved ([8] that it is asymptotically the worst case instance wrt the maximum individual cost measure, for a large class atomic congestion games involving the so called *layered networks*. Therefore, PLG is considered an excellent starting point for studying congestion games in large scale networks.

The importance of seeking for PNE, rather than arbitrary (mixed in general) NE, is quite obvious in sciences like the economics, ecology, and biology. It is also important for computer scientists, since it enforces deterministic costs to the players, and both the players and the network designer may feel safer in this case about what they will actually have to pay.

The question whether the Nash Dynamics converge to a PNE in a reasonable amount of time, is also quite important, since (in case of a positive answer) it justifies the selfish, decentralized, local dynamics that appear in large scale communications systems. Additionally, the selfish rerouting schemes are of great importance, since this is what should actually be expected from selfish, decentralized computing environments.

Open Problems

Open Question 1 *Determine the (in)existence of PNE for all the instances of PLG that do not belong in LBPLG, or in monotone PSPLG.*

Open Question 2 *Determine the (in)existence of k−robust PNE for all the instances of PLG that do not belong in LBPLG.*

Open Question 3 *Is there a polynomial time algorithm for constructing k−robust PNE, even for the Identical Machines LBPLG and k ≥ 1 being a constant?*

Open Question 4 *Do the improvement paths of instances of PLG other than PSPLG and LBPLG converge to a PNE?*

Open Question 5 *Are there selfish rerouting policies of instances of PLG other than Identical Machines LBPLG converge to a PNE? How long much time would they need, in case of a positive answer?*

Cross-References

▶ Best Response Algorithms for Selfish Routing
▶ Price of Anarchy
▶ Selfish Unsplittable Flows: Algorithms for Pure Equilibria

Recommended Reading

1. Christodoulou G, Koutsoupias E (2005) The price of anarchy of finite congestion games. In: Proceedings of the 37th ACM symposium on theory of computing (STOC '05). ACM, Baltimore, pp 67–73
2. Czumaj A, Vöcking B (2002) Tight bounds for worst-case equilibria. In: Proceedings of the 13th ACM-SIAM symposium on discrete algorithms (SODA '02). SIAM, San Francisco, pp 413–420
3. Even-Dar E, Kesselman A, Mansour Y (2003) Convergence time to nash equilibria. In: Proceedings of the 30th international colloquium on automata, languages and programming (ICALP '03). LNCS. Springer, Eindhoven, pp 502–513
4. Even-Dar E, Mansour Y (2005) Fast convergence of selfish rerouting. In: Proceedings of the 16th ACM-SIAM symposium on discrete algorithms (SODA '05). SIAM, Vancouver, pp 772–781
5. Fabrikant A, Papadimitriou C, Talwar K (2004) The complexity of pure nash equilibria. In: Proceedings of the 36th ACM symposium on theory of computing (STOC '04). ACM, Chicago
6. Feldmann R, Gairing M, Lücking T, Monien B, Rode M (2003) Nashification and the coordination ratio for a selfish routing game. In: Proceedings of the 30th international colloquium on automata, languages and programming (ICALP '03). LNCS. Springer, Eindhoven, pp 514–526
7. Fotakis D, Kontogiannis S, Koutsoupias E, Mavronicolas M, Spirakis P (2002) The structure and complexity of nash equilibria for a selfish routing game. In: Proceedings of the 29th international colloquium on automata, languages and programming (ICALP '02). LNCS. Springer, Málaga, pp 123–134
8. Fotakis D, Kontogiannis S, Spirakis P (2005) Selfish unsplittable flows. Theor Comput Sci 348:226–239. Special Issue dedicated to ICALP (2004) (TRACK-A).
9. Fotakis D, Kontogiannis S, Spirakis P (2006) Atomic congestion games among coalitions. In: Proceedings of the 33rd international colloquium on automata, languages and programming (ICALP '06). LNCS, vol 4051. Springer, Venice, pp 572–583
10. Gairing M, Luecking T, Mavronicolas M, Monien B (2006) The price of anarchy for restricted parallel links. Parallel Process Lett 16:117–131, Preliminary version appeared in STOC 2004
11. Gairing M, Monien B, Tiemann K (2006) Routing (un-)splittable flow in games with player specific linear latency functions. In: Proceedings of the 33rd international colloquium on automata, languages and programming (ICALP '06). LNCS. Springer, Venice, pp 501–512
12. Koutsoupias E, Papadimitriou C (1999) Worst-case equilibria. In: Proceedings of the 16th annual symposium on theoretical aspects of computer science (STACS '99). Springer, Trier, pp 404–413

13. Milchtaich I (1996) Congestion games with player-specific payoff functions. Games Econ Behav 13:111–124
14. Nash J (1951) Noncooperative games. Ann Math 54:289–295
15. Rosenthal R (1973) A class of games possessing pure-strategy nash equilibria. Int J Game Theory 2:65–67

Concurrent Programming, Mutual Exclusion

Gadi Taubenfeld
Department of Computer Science,
Interdiciplinary Center Herzlia, Herzliya, Israel

Keywords

Critical section problem

Years and Authors of Summarized Original Work

1965; Dijkstra

Problem Definition

Concurrency, Synchronization and Resource Allocation

A *concurrent* system is a collection of processors that communicate by reading and writing from a shared memory. A distributed system is a collection of processors that communicate by sending messages over a communication network. Such systems are used for various reasons: to allow a large number of processors to solve a problem together much faster than any processor can do alone, to allow the distribution of data in several locations, to allow different processors to share resources such as data items, printers or discs, or simply to enable users to send electronic mail.

A *process* corresponds to a given computation. That is, given some program, its execution is a process. Sometimes, it is convenient to refer to the program code itself as a process. A process runs on a *processor*, which is the physical hardware. Several processes can run on the same processor although in such a case only one of them may be active at any given time. Real concurrency is achieved when several processes are running simultaneously on several processors.

Processes in a concurrent system often need to synchronize their actions. *Synchronization* between processes is classified as either cooperation or contention. A typical example for cooperation is the case in which there are two sets of processes, called the producers and the consumers, where the producers produce data items which the consumers then consume.

Contention arises when several processes compete for exclusive use of shared resources, such as data items, files, discs, printers, etc. For example, the integrity of the data may be destroyed if two processes update a common file at the same time, and as a result, deposits and withdrawals could be lost, confirmed reservations might have disappeared, etc. In such cases it is sometimes essential to allow at most one process to use a given resource at any given time.

Resource allocation is about interactions between processes that involve contention. The problem is, how to resolve conflicts resulting when several processes are trying to use shared resources. Put another way, how to allocate shared resources to competing processes. A special case of a general resource allocation problem is the *mutual exclusion* problem where only a single resource is available.

The Mutual Exclusion Problem

The *mutual exclusion* problem, which was first introduced by Edsger W. Dijkstra in 1965, is the guarantee of mutually exclusive access to a single shared resource when there are several competing processes [6]. The problem arises in operating systems, database systems, parallel supercomputers, and computer networks, where it is necessary to resolve conflicts resulting when several processes are trying to use shared resources. The problem is of great significance, since it lies at the heart of many interprocess synchronization problems.

The problem is formally defined as follows: it is assumed that each process is executing a sequence of instructions in an infinite loop. The instructions are divided into four continuous sections of code: the *remainder, entry, critical section* and *exit*. Thus, the structure of a mutual exclusion solution looks as follows:

loop forever
  remainder code;
  entry code;
  critical section;
  exit code
end loop

A process starts by executing the remainder code. At some point the process might need to execute some code in its critical section. In order to access its critical section a process has to go through an entry code which guarantees that while it is executing its critical section, no other process is allowed to execute its critical section. In addition, once a process finishes its critical section, the process executes its exit code in which it notifies other processes that it is no longer in its critical section. After executing the exit code the process returns to the remainder.

The Mutual exclusion problem is to write the code for the *entry code* and the *exit code* in such a way that the following two basic requirements are satisfied.

Mutual exclusion: *No two processes are in their critical sections at the same time.*

Deadlock-freedom: *If a process is trying to enter its critical section, then some process, not necessarily the same one, eventually enters its critical section.*

The deadlock-freedom property guarantees that the system as a whole can always continue to make progress. However deadlock-freedom may still allow "starvation" of individual processes. That is, a process that is trying to enter its critical section, may never get to enter its critical section, and wait forever in its entry code. A stronger requirement, which does not allow starvation, is defined as follows.

Starvation-freedom: *If a process is trying to enter its critical section, then this process must eventually enter its critical section.*

Although starvation-freedom is strictly stronger than deadlock-freedom, it still allows processes to execute their critical sections arbitrarily many times before some trying process can execute its critical section. Such a behavior is prevented by the following fairness requirement.

First-in-first-out (FIFO): *No beginning process can enter its critical section before a process that is already waiting for its turn to enter its critical section.*

The first two properties, mutual exclusion and deadlock freedom, were required in the original statement of the problem by Dijkstra. They are the minimal requirements that one might want to impose. In solving the problem, it is assumed that once a process starts executing its critical section the process always finishes it regardless of the activity of the other processes. Of all the problems in interprocess synchronization, the mutual exclusion problem is the one studied most extensively. This is a deceptive problem, and at first glance it seems very simple to solve.

Key Results

Numerous solutions for the problem have been proposed since it was first introduced by Edsger W. Dijkstra in 1965 [6]. Because of its importance and as a result of new hardware and software developments, new solutions to the problem are still being designed. Before the results are discussed, few models for interprocess communication are mentioned.

Atomic Operations

Most concurrent solutions to the problem assumes an architecture in which n processes communicate asynchronously via a shared objects. All architectures support *atomic registers*, which are shared objects that support atomic reads and writes operations. A weaker notion than an atomic register, called a *safe* register, is also considered in the literature. In a safe register, a read not concurrent with any writes must obtain the correct value, however, a read that is concurrent with some write, may return an arbitrary value. Most modern architectures support also some form of atomicity which is stronger than simple reads and writes. Common atomic operations have special names. Few examples are,

- *Test-and-set*: takes a shared registers r and a value *val*. The value *val* is assigned to r, and the old value of r is returned.
- *Swap*: takes a shared registers r and a local register ℓ, and atomically exchange their values.
- *Fetch-and-increment*: takes a register r. The value of r is incremented by 1, and the old value of r is returned.
- *Compare-and-swap*: takes a register r, and two values: *new* and *old*. If the current value of the register r is equal to *old*, then the value of r is set to *new* and the value *true* is returned; otherwise r is left unchanged and the value *false* is returned.

Modern operating systems (such as Unix and Windows) implement synchronization mechanisms, such as semaphores, that simplify the implementation of mutual exclusion locks and hence the design of concurrent applications. Also, modern programming languages (such as Modula and Java) implement the monitor concept which is a program module that is used to ensure exclusive access to resources.

Algorithms and Lower Bounds

There are hundreds of beautiful algorithms for solving the problem some of which are also very efficient. Only few are mentioned below. First algorithms that use only atomic registers, or even safe registers, are discussed.

The Bakery Algorithm. The Bakery algorithm is one of the most known and elegant mutual exclusion algorithms using only safe registers [9]. The algorithm satisfies the FIFO requirement, however it uses unbounded size registers. A modified version, called the Black-White Bakery algorithm, satisfies FIFO and uses bounded number of bounded size atomic registers [14].

Lower bounds. A space lower bound for solving mutual exclusion using only atomic registers

is that: any deadlock-free mutual exclusion algorithm for n processes must use at least n shared registers [5]. It was also shown in [5] that this bound is tight. A time lower bound for any mutual exclusion algorithm using atomic registers is that: there is no a priori bound on the number of steps taken by a process in its entry code until it enters its critical section (counting steps only when no other process is in its critical section or exit code) [2]. Many other interesting lower bounds exist for solving mutual exclusion.

A Fast Algorithm. A *fast* mutual exclusion algorithm, is an algorithm in which in the absence of contention only a constant number of shared memory accesses to the shared registers are needed in order to enter and exit a critical section. In [10], a fast algorithm using atomic registers is described, however, in the presence of contention, the winning process may have to check the status of all other n processes before it is allowed to enter its critical section. A natural question to ask is whether this algorithm can be improved for the case where there is contention.

Adaptive Algorithms. Since the other contending processes are waiting for the winner, it is particularly important to speed their entry to the critical section, by the design of an *adaptive* mutual exclusion algorithm in which the time complexity is independent of the total number of processes and is governed only by the current degree of contention. Several (rather complex) adaptive algorithms using atomic registers are known [1, 3, 14]. (Notice that, the time lower bound mention earlier implies that no adaptive algorithm using only atomic registers exists when time is measured by counting *all* steps.)

Local-spinning Algorithms. Many algorithms include busy-waiting loops. The idea is that in order to wait, a process *spins* on a flag register, until some other process terminates the spin with a single write operation. Unfortunately, under contention, such spinning may generate lots of traffic on the interconnection network between the process and the memory. An algorithm satisfies local spinning if the only type of spinning required is local spinning. Local Spinning is the situation where a process is spinning on locally-accessible registers. Shared registers may be locally-accessible as a result of either coherent caching or when using distributed shared memory where shared memory is physically distributed among the processors.

Three local-spinning algorithms are presented in [4, 8, 11]. These algorithms use strong atomic operations (i.e., fetch-and-increment, swap, compare-and-swap), and are also called *scalable* algorithms since they are both local-spinning and adaptive. Performance studies done, have shown that these algorithms scale very well as contention increases. Local spinning algorithms using only atomic registers are presented in [1, 3, 14].

Only few representative results have been mentioned. There are dozens of other very interesting algorithms and lower bounds. All the results discussed above, and many more, are described details in [15]. There are also many results for solving mutual exclusion in distributed message passing systems [13].

Applications

Synchronization is a fundamental challenge in computer science. It is fast becoming a major performance and design issue for concurrent programming on modern architectures, and for the design of distributed and concurrent systems.

Concurrent access to resources shared among several processes must be synchronized in order to avoid interference between conflicting operations. Mutual exclusion *locks* (i.e., algorithms) are the de facto mechanism for concurrency control on concurrent applications: a process accesses the resource only inside a critical section code, within which the process is guaranteed exclusive access. The popularity of this approach is largely due the apparently simple programming model of such locks and the availability of implementations which are efficient and scalable. Essentially all concurrent programs (including operating systems) use various types of mutual exclusion locks for synchronization.

When using locks to protect access to a resource which is a large data structure (or a database), the *granularity* of synchronization is important. Using a single lock to protect the

whole data structure, allowing only one process at a time to access it, is an example of *coarse-grained* synchronization. In contrast, *fine-grained* synchronization enables to lock "small pieces" of a data structure, allowing several processes with non-interfering operations to access it concurrently. Coarse-grained synchronization is easier to program but is less efficient and is not fault-tolerant compared to fine-grained synchronization. Using locks may degrade performance as it enforces processes to wait for a lock to be released. In few cases of simple data structures, such as queues, stacks and counters, locking may be avoided by using lock-free data structures.

Cross-References

▶ Registers
▶ Self-Stabilization

Recommended Reading

In 1968, Edsger Wybe Dijkstra has published his famous paper "Co-operating sequential processes" [7], that originated the field of concurrent programming. The mutual exclusion problem was first stated and solved by Dijkstra in [6], where the first solution for two processes, due to Dekker, and the first solution for *n* processes, due to Dijkstra, have appeared. In [12], a collection of some early algorithms for mutual exclusion are described. In [15], dozens of algorithms for solving the mutual exclusion problems and wide variety of other synchronization problems are presented, and their performance is analyzed according to precise complexity measures.

1. Afek Y, Stupp G, Touitou D (2002) Long lived adaptive splitter and applications. Distrib Comput 30:67–86
2. Alur R, Taubenfeld G (1992) Results about fast mutual exclusion. In: Proceedings of the 13th IEEE real-time systems symposium, Dec 1992, pp 12–21
3. Anderson JH, Kim Y-J (2000) Adaptive mutual exclusion with local spinning. In: Proceedings of the 14th international symposium on distributed computing, Lecture notes in computer science, vol 1914, pp 29–43
4. Anderson TE (1990) The performance of spin lock alternatives for shared-memory multiprocessor. IEEE Trans Parallel Distrib Syst 1(1):6–16

5. Burns JN, Lynch NA (1993) Bounds on shared-memory for mutual exclusion. Inf Comput 107(2):171–184
6. Dijkstra EW (1965) Solution of a problem in concurrent programming control. Commun ACM 8(9):569
7. Dijkstra EW (1968) Co-operating sequential processes. In: Genuys F (ed) Programming languages. Academic, New York, pp 43–112. Reprinted from: technical report EWD-123, Technological University, Eindhoven (1965)
8. Graunke G, Thakkar S (1990) Synchronization algorithms for shared-memory multiprocessors. IEEE Comput 28(6):69–69
9. Lamport L (1974) A new solution of Dijkstra's concurrent programming problem. Commun ACM 17(8):453–455
10. Lamport L (1987) A fast mutual exclusion algorithm. ACM Trans Comput Syst 5(1):1–11
11. Mellor-Crummey JM, Scott ML (1991) Algorithms for scalable synchronization on shared-memory multiprocessors. ACM Trans Comput Syst 9(1):21–65
12. Raynal M (1986) Algorithms for mutual exclusion. MIT, Cambridge. Translation of: Algorithmique du parallélisme (1984)
13. Singhal M (1993) A taxonomy of distributed mutual exclusion. J Parallel Distrib Comput 18(1):94–101
14. Taubenfeld G (2004) The black-white bakery algorithm. In: 18th international symposium on distributed computing, Oct 2004. LNCS, vol 3274. Springer, Berlin, pp 56–70
15. Taubenfeld G (2006) Synchronization algorithms and concurrent programming. Pearson Education/Prentice-Hall, Upper Saddle River. ISBN:0131972596

Connected Dominating Set

Feng Wang[1], Ding-Zhu Du[2,4], and Xiuzhen Cheng[3]
[1]Mathematical Science and Applied Computing, Arizona State University at the West Campus, Phoenix, AZ, USA
[2]Computer Science, University of Minnesota, Minneapolis, MN, USA
[3]Department of Computer Science, George Washington University, Washington, DC, USA
[4]Department of Computer Science, The University of Texas at Dallas, Richardson, TX, USA

Keywords

Techniques for partition

Years and Authors of Summarized Original Work

2003; Cheng, Huang, Li, Wu, Du

Problem Definition

Consider a graph $G = (V, E)$. A subset C of V is called a *dominating set* if every vertex is either in C or adjacent to a vertex in C. If, furthermore, the subgraph induced by C is connected, then C is called a *connected dominating set*. A connected dominating set with a minimum cardinality is called a *minimum connected dominating set (MCDS)*. Computing an MCDS is an NP-hard problem and there is no polynomial-time approximation with performance ratio $\rho H(\Delta)$ for $\rho < 1$ unless $NP \subseteq DTIME(n^{O(\ln \ln n)})$ where H is the harmonic function and Δ is the maximum degree of the input graph [11].

A unit disk is a disk with radius one. A *unit disk graph (UDG)* is associated with a set of unit disks in the Euclidean plane. Each node is at the center of a unit disk. An edge exists between two nodes u and v if and only if $|uv| \le 1$ where $|uv|$ is the Euclidean distance between u and v. This means that two nodes u and v are connected with an edge if and only if u's disk covers v and v's disk covers u.

Computing an MCDS in a unit disk graph is still NP-hard. How hard is it to construct a good approximation for MCDS in unit disk graphs? Cheng et al. [5] answered this question by presenting a polynomial-time approximation scheme.

Historical Background

The connected dominating set problem has been studied in graph theory for many years [23]. However, recently it becomes a hot topic due to its application in wireless networks for virtual backbone construction [4]. Guha and Khuller [11] gave a two-stage greedy approximation for the minimum connected dominating set in general graphs and showed that its performance ratio is $3 + \ln \Delta$ where Δ is the maximum node degree in the graph. To design a one-step greedy approximation to reach a similar performance ratio, the difficulty is to find a submodular potential function. In [22], Ruan et al. successfully designed a one-step greedy approximation that reaches a better performance ratio $c + \ln \Delta$ for any $c > 2$. Du et al. [7] showed that there exists a polynomial-time approximation with a performance ratio $a(1 + \ln \Delta)$ for any $a > 1$. The importance of those works is that the potential functions used in their greedy algorithm are non-submodular and they managed to complete its theoretical performance evaluation with fresh ideas.

Guha and Khuller [11] also gave a negative result that there is no polynomial-time approximation with a performance ratio $\rho \ln \Delta$ for $\rho < 1$ unless $NP \subseteq DTIME(n^{O(\ln \ln n)})$. As indicated by [9], dominating sets cannot be approximated arbitrarily well, unless P almost equal to NP. These results move ones' attention from general graphs to unit disk graphs because the unit disk graph is the model for wireless sensor networks, and in unit disk graphs, MCDS has a polynomial-time approximation with a constant performance ratio. While this constant ratio is getting improved step-by-step [1, 2, 20, 25], Cheng et al. [5] closed this story by showing the existence of a polynomial-time approximation scheme (PTAS) for the MCDS in unit disk graphs. This means that theoretically, the performance ratio for polynomial-time approximation can be as small as $1 + \varepsilon$ for any positive number ε.

Dubhashi et al. [8] showed that once a dominating set is constructed, a connected dominating set can be easily computed in a distributed fashion. Most centralized results for dominating sets are available at [19]. In particular, a simple constant approximation for dominating sets in unit disk graphs was presented in [19]. Constant-factor approximation for minimum-weight (connected) dominating sets in UDGs was studied in [3]. A PTAS for the minimum dominating set problem in UDGs was proposed in [21]. Kuhn et al. [16] proved that a maximal independent set (MIS) (and hence also a dominating set) can be computed in asymptotically optimal time $O(\log n)$ in UDGs and a large class of bounded independence graphs. Luby [18] reported an elegant local $O(\log n)$ algorithm for

MIS on general graphs. Jia et al. [12] proposed a fast $O(\log n)$ distributed approximation for dominating set in general graphs. The first constant-time distributed algorithm for dominating sets that achieves a nontrivial approximation ratio for general graphs was reported in [13]. The matching $\Omega(\log n)$ lower bound is considered to be a classic result in distributed computing [17]. For UDGs a PTAS is achievable in a distributed fashion [15]. The fastest deterministic distributed algorithm for dominating sets in UDGs was reported in [14], and the fastest randomized distributed algorithm for dominating sets in UDGs was presented in [10].

Key Results

The construction of PTAS for MCDS is based on the fact that there is a polynomial-time approximation with a constant performance ratio. Actually, this fact is quite easy to see. First, note that a unit disk contains at most five independent vertices [2]. This implies that every maximal independent set has a size at most $1 + 4opt$ where opt is the size of an MCDS. Moreover, every maximal independent set is a dominating set and it is easy to construct a maximal independent set with a spanning tree of all edges with length two. All vertices in this spanning tree form a connected dominating set of a size at most $1 + 8opt$. By improving the upper bound for the size of a maximal independent set [26] and the way to interconnecting a maximal independent set [20], the constant ratio has been improved to 6.8 with a distributed implementation.

The basic techniques in this construction are nonadaptive partition and shifting. Its general picture is as follows: First, the square containing all vertices of the input unit disk graph is divided into a grid of small cells. Each small cell is further divided into two areas, the central area and the boundary area. The central area consists of points h distance away from the cell boundary. The boundary area consists of points within distance $h + 1$ from the boundary. Therefore, two areas are overlapping. Then a minimum union of connected dominating sets is computed in each cell for connected components of the central area of the cell. The key lemma is to prove that the union of all such minimum unions is no more than the minimum connected dominating set for the whole graph. For vertices not in central areas, just use the part of an 8-approximation lying in boundary areas to dominate them. This part together with the above union forms a connected dominating set for the whole input unit disk graph. By shifting the grid around to get partitions at different coordinates, a partition having the boundary part with a very small upper bound can be obtained.

The following details the construction.

Given an input connected unit disk graph $G = (V, E)$ residing in a square $Q = \{(x, y)|0 \leq x \leq q, 0 \leq y \leq q\}$ where $q \leq |V|$. To construct an approximation with a performance ratio $1 + \varepsilon$ for $\varepsilon > 0$, choose an integer $m = O((1/\varepsilon)\ln(1/\varepsilon))$. Let $p = \lfloor q/m \rfloor + 1$. Consider the square $\bar{Q}''$. Partition $\bar{Q}$ into $(p + 1) \times (p + 1)$ grids so that each cell is an $m \times m$ square excluding the top and the right boundaries, and hence, no two cells are overlapping each other. This partition of $\bar{Q}$s is denoted by $P(0)$ (Fig. 1). In general, the partition $P(a)$ is obtained from $P(0)$ by shifting the bottom-left corner of $\bar{Q}$ from $(-m, -m)$ to $(-m + a, -m + a)$. Note that shifting from $P(0)$ to $P(a)$ for $0 \leq a \leq m$ keeps Q covered by the partition.

For each cell e (an $m \times m$ square), $C_e(d)$ denotes the set of points in e away from the boundary by distance at least d, e.g., $C_e(0)$ is the cell e itself. Denote $B_e(d) = C_e(0) - C_e(d)$. Fix a positive integer $h = 7 + 3\lfloor \log_2(4m^2/\pi) \rfloor$. Call $C_e(h)$ the *central area* of e and $B_e(h + 1)$ the *boundary area* of e. Hence, the boundary area and the central area of each cell are overlapping with width one.

Central Area

Let $G_e(d)$ denote the part of input graph G lying in area $C_e(d)$. In particular, $G_e(h)$ is the part of graph G lying in the central area of e. $G_e(h)$ may consist of several connected components. Let K_e be a subset of vertices in $G_e(0)$ with a minimum cardinality such that for each connected component H of $G_e(h)$, K_e contains a connected component dominating H. In other words, K_e is a minimum union of connected dominating sets in $G(0)$ for the connected components of $G_e(h)$.

Now, denote by $K(a)$ the union of K_e for e over all cells in partition $P(a)$. $K(a)$ has two important properties:

Lemma 1 $K(a)$ can be computed in time $n^{O(m2)}$.

Lemma 2 $|K^a| \leq opt$ for $0 \leq a \leq m - 1$.

Lemma 1 is not hard to see. Note that in a square with edge length $\sqrt{2}/2$, all vertices induce a complete subgraph in which any vertex must dominate all other vertices. It follows that the minimum dominating set for the vertices of $G_e(0)$ has size at most $\left(\lfloor \sqrt{2}m \rfloor\right)^2$. Hence, the size of K_e is at most $3\left(\lfloor \sqrt{2}m \rfloor\right)^2$ because any dominating set in a connected graph has a spanning tree with an edge length at most three. Suppose cell $G_e(0)$ has n_e vertices. Then the number of candidates for K_e is at most

$$\sum_{k=0}^{3(\lceil \sqrt{2}m \rceil)^2} \binom{n_e}{k} = n_e^{O(m^2)}.$$

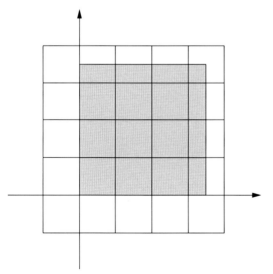

Connected Dominating Set, Fig. 1 Squares Q and $\bar{Q}$

Hence, computing $K(a)$ can be done in time

$$\sum_e n_e^{O(m^2)} \leq \left(\sum_e n_e\right)^{O(m^2)} = n^{O(m^2)}.$$

However, the proof of Lemma 2 is quite tedious. The reader who is interested in it may find it in [5].

Boundary Area
Let F be a connected dominating set of G satisfying $|F| \leq 8opt + 1$. Denote by $F(a)$ the subset of F lying in the boundary area $B_a(h+1)$. Since F is constructed in polynomial time, only the size of $F(a)$ needs to be studied.

Lemma 3 *Suppose* $h = 7 + 3\lfloor \log_2(4m^2/\pi) \rfloor$ *and* $\lfloor m/(h+1) \rfloor \geq 32/\varepsilon$. *Then there is at least half of* $i = 0, 1, \ldots, \lfloor m/(h+1) \rfloor - 1$ *such that* $|F(i(h+1))| \leq \varepsilon \cdot opt$.

Proof Let $F_H(a)$ $(F_V(a))$ denote the subset of vertices in $F(a)$ each with distance $< h+1$ from the horizontal (vertical) boundary of some cell in $P(a)$. Then $F(a) = F_H(a) \cup F_V(a)$. Moreover, all $F_H(i(h+1))$ for $i = 0, 1, \ldots, \lfloor m/(h+1) \rfloor - 1$ are disjoint. Hence,

$$\sum_{i=0}^{\lfloor m/(h+1) \rfloor - 1} |F_H(i(h+1))| \leq |F| \leq 8opt.$$

Similarly, all $F_V(i(h+1))$ for $i = 0, 1, \ldots, \lfloor m/(h+1) \rfloor - 1$ are disjoint and

$$\sum_{i=0}^{\lfloor m/(h+1) \rfloor - 1} |F_V(i(h+1))| \leq |F| \leq 8opt.$$

Thus,

$$\sum_{i=0}^{\lfloor m/(h+1) \rfloor - 1} |F(i(h+1))| \leq$$

$$\sum_{i=0}^{\lfloor m/(h+1) \rfloor - 1} (|F_H(i(h+1))| + |F_V(i(h+1))|)$$

$$\leq 16opt.$$

That is,

$$\frac{1}{\lfloor m/(h+1) \rfloor} \sum_{i=0}^{\lfloor m/(h+1) \rfloor - 1} |F(i(h+1))| \leq (\varepsilon/2)opt.$$

This means that there are at least half of $F(i(h+1))$ for $i = 0, 1, \lfloor m/(h+1) \rfloor - 1$ satisfying

$$|F(i(h+1))| \leq \varepsilon \cdot opt.$$

$\square$

Putting Together
Now put $K(a)$ and $F(a)$. By Lemmas 2 and 3, there exists $a \in \{0, h+1, \ldots, (\lfloor m/(h+1) \rfloor - 1)(h+1)\}$ such that

$$|K(a) \cup F(a)| \leq (1+\varepsilon)opt.$$

Lemma 4 *For* $0 \leq a \leq m - 1$, $K(a) \cup F(a)$ *is a connected dominating for input connected graph* G.

Proof $K(a) \cup F(a)$ is clearly a dominating set for input graph G. Its connectivity can be shown as follows. Note that the central area and the boundary area are overlapping with an area of width one. Thus, for any connected component H of the subgraph $G_e(h)$, $F(a)$ has a vertex in H. Hence, $F(a)$ must connect to any connected dominating set for H, especially, the one D_H in $K(a)$. This means that D_H is making up the connections of F lost from cutting a part in H. Therefore, the connectivity of $K(a) \cup F(a)$ follows from the connectivity of F. $\square$

By summarizing the above results, the following result is obtained:

Theorem 1 *There is a* $(1+\varepsilon)$-*approximation for MCDS in connected unit disk graphs, running in time* $n^{O((1/\varepsilon)\log(1/\varepsilon)2)}$.

Applications

An important application of connected dominating sets is to construct virtual backbones for wireless networks, especially, wireless sensor networks [4]. The topology of a wireless sensor network is often a unit disk graph.

Open Problems

In general, the topology of a wireless network is a disk graph, that is, each vertex is associated with a disk. Different disks may have different sizes. There is an edge from vertex u to vertex v if and only if the disk at u covers v. A virtual backbone in disk graphs is a subset of vertices, which induces a strongly connected subgraph, such that every vertex not in the subset has an in-edge coming from a vertex in the subset and also has an out-edge going into a vertex in the subset. Such a virtual backbone can be considered as a *connected dominating set* in disk graph. Is there a polynomial-time approximation with a constant performance ratio? It is still open right now [6]. Thai et al. [24] has made some effort towards this direction.

Cross-References

▶ Enumeration of Paths, Cycles, and Spanning Trees
▶ Exact Algorithms for Dominating Set
▶ Greedy Set-Cover Algorithms

Recommended Reading

1. Alzoubi KM, Wan P-J, Frieder O (2002) Message-optimal connected dominating sets in mobile ad hoc networks. In: ACM MOBIHOC, Lausanne, 09–11 June 2002
2. Alzoubi KM, Wan P-J, Frieder O (2002) New distributed algorithm for connected dominating set in wireless ad hoc networks. In: HICSS35, Hawaii
3. Ambuhl C, Erlebach T, Mihalak M, Nunkesser M (2006) Constant-factor approximation for minimum-weight (connected) dominating sets in unit disk graphs. In: Approximation, randomization, and combinatorial optimization. Algorithms and techniques. LNCS, vol 4110. Springer, Berlin, pp 3–14
4. Blum J, Ding M, Thaeler A, Cheng X (2004) Applications of connected dominating sets in wireless networks. In: Du D-Z, Pardalos P (eds) Handbook of combinatorial optimization. Kluwer, pp 329–369
5. Cheng X, Huang X, Li D, Wu W, Du D-Z (2003) A polynomial-time approximation scheme for minimum connected dominating set in ad hoc wireless networks. Networks 42:202–208
6. Du D-Z, Wan P-J (2012) Connected dominating set: theory and applications. Springer, New York
7. Du D-Z, Graham RL, Pardalos PM, Wan P-J, Wu W, Zhao W (2008) Analysis of greedy approximations with nonsubmodular potential functions. In: Proceedings of the 19th annual ACM-SIAM symposium on discrete algorithms (SODA), San Francisco, pp 167–175
8. Dubhashi D, Mei A, Panconesi A, Radhakrishnan J, Srinivasan A (2003) Fast distributed algorithms for (weakly) connected dominating sets and linear-size skeletons. In: SODA, Baltimore, pp 717–724
9. Feige U (1998) A threshold of ln n for approximating set cover. J ACM 45(4):634–652
10. Gfeller B, Vicari E (2007) A randomized distributed algorithm for the maximal independent set problem in growth-bounded graphs. In: PODC, Portland
11. Guha S, Khuller S (1998) Approximation algorithms for connected dominating sets. Algorithmica 20:374–387
12. Jia L, Rajaraman R, Suel R (2001) An efficient distributed algorithm for constructing small dominating sets. In: PODC, Newport
13. Kuhn F, Wattenhofer R (2003) Constant-time distributed dominating set approximation. In: PODC, Boston
14. Kuhn F, Moscibroda T, Nieberg T, Wattenhofer R (2005) Fast deterministic distributed maximal independent set computation on growth-bounded graphs. In: DISC, Cracow
15. Kuhn F, Moscibroda T, Nieberg T, Wattenhofer R (2005) Local approximation schemes for ad hoc and sensor networks. In: DIALM-POMC, Cologne
16. Kuhn F, Moscibroda T, Wattenhofer R (2005) On the locality of bounded growth. In: PODC, Las Vegas
17. Linial N (1992) Locality in distributed graph algorithms. SIAM J Comput 21(1):193–201
18. Luby M (1986) A simple parallel algorithm for the maximal independent set problem. SIAM J Comput 15:1036–1053
19. Marathe MV, Breu H, Hunt III HB, Ravi SS, Rosenkrantz DJ (1995) Simple heuristics for unit disk graphs. Networks 25:59–68
20. Min M, Du H, Jia X, Huang X, Huang C-H, Wu W (2006) Improving construction for connected dominating set with Steiner tree in wireless sensor networks. J Glob Optim 35:111–119
21. Nieberg T, Hurink JL (2006) A PTAS for the minimum dominating set problem in unit disk graphs. In: Approximation and online algorithms. LNCS, vol 3879. Springer, Berlin, pp 296–306
22. Ruan L, Du H, Jia X, Wu W, Li Y, Ko K-I (2004) A greedy approximation for minimum connected dominating set. Theor Comput Sci 329:325–330

23. Sampathkumar E, Walikar HB (1979) The connected domination number of a graph. J Math Phys Sci 13:607–613
24. Thai MT, Wang F, Liu D, Zhu S, Du D-Z (2007) Connected dominating sets in wireless networks with different transmission range. IEEE Trans Mob Comput 6(7):721–730
25. Wan P-J, Alzoubi KM, Frieder O (2002) Distributed construction of connected dominating set in wireless ad hoc networks. In: IEEE INFOCOM, New York
26. Wu W, Du H, Jia X, Li Y, Huang C-H (2006) Minimum connected dominating sets and maximal independent sets in unit disk graphs. Theor Comput Sci 352:1–7

Connected Set-Cover and Group Steiner Tree

Lidong Wu[1], Huijuan Wang[2], and Weili Wu[3,4,5]
[1]Department of Computer Science, The University of Texas, Tyler, TX, USA
[2]Shandong University, Jinan, China
[3]College of Computer Science and Technology, Taiyuan University of Technology, Taiyuan, Shanxi Province, China
[4]Department of Computer Science, California State University, Los Angeles, CA, USA
[5]Department of Computer Science, The University of Texas at Dallas, Richardson, TX, USA

Keywords

Approximation algorithms; Connected set-cover; Group Steiner tree; Performance ratio

Years and Authors of Summarized Original Work

2012; Zhang, Wu, Lee, Du
2012; Elbassion, Jelic, Matijevic
2013; Wu, Du, Wu, Li, Lv, Lee

Problem Definition

Given a collection C of subsets of a finite set X, find a minimum subcollection C' of C such that every element of X appears in some subset

in C'. This problem is called the *minimum set-cover* problem. Every feasible solution, i.e., a subcollection C' satisfying the required condition, is called a *set-cover*. The minimum set-cover problem is NP-hard, and the complexity of approximation for it is well solved. It is well known that (1) the minimum set-cover problem has a polynomial-time $(1 + \ln n)$-approximation where $n = |X|$ [2,7,8], and moreover (2) if the minimum set-cover problem has a polynomial-time $(\rho \ln n)$-approximation for any $0 < \rho < 1$, then $NP \subseteq DTIME(n^{O(\log \log n)})$ [4].

The *minimum connected set-cover* problem is closely related to the minimum set-cover problem, which can be described as follows: Given a collection C of subsets of a finite set X and a graph G with vertex set C, find a minimum set-cover $C' \subseteq C$ such that the subgraph induced by C' is connected. An issue of whether the minimum connected set-cover problem has a polynomial-time $O(\log n)$-approximation or not [1,9,11] was open for several years.

Key Results

Zhang et al. [12] solved this problem by discovering a relationship between the minimum connected set-cover problem and the group Steiner tree problem.

Given a graph $G = (V, E)$ with edge nonnegative weight $c : E \rightarrow N$ and k subsets (called groups) of vertices, $V_1, \ldots, V_k$, find the minimum edge-weight tree interconnecting those k vertex subsets, i.e., containing at least one vertex from each subset. This is called the *group Steiner tree* problem. It has another formulation as follows: Given a graph $G = (V, E)$ with edge nonnegative weight $c : E \rightarrow R^+$, a special vertex r, and k subsets of vertices, $V_1, \ldots, V_k$, find the minimum edge-weight tree with root r, interconnecting those k vertex subsets.

These two formulations are equivalent in the sense that one has a polynomial-time ρ-approximation and so does the other one. Actually, consider vertex r as a group with only one member. Then it is immediately known that if the first formulation has a polynomial-time ρ-

approximation, so does the second formulation. Next, assume the second formulation has a polynomial-time ρ-approximation. In the first formulation, fix a group V_1, for each vertex $v \in V_1$, and apply the polynomial-time ρ-approximation algorithm for the second formulation to the root $r = v$ and $k - 1$ groups $V_2, \ldots, V_k$. Choose the shortest one from $|V_1|$ obtained trees, which would be a polynomial-time ρ-approximation for the first formulation.

The following are well-known results for the group Steiner tree problem

Theorem 1 (Halperin and Krauthgamer [6]) *The group Steiner tree problem has no polynomial-time $O(\log^{2-\varepsilon} n)$-approximation for any $\varepsilon > 0$ unless NP has quasi-polynomial Las-Vega algorithm.*

Theorem 2 (Garg, Konjevod, Ravi [5]) *The group Steiner tree problem has a polynomial-time random $O(\log^2 n \log k)$-approximation where n is the number of nodes in the input graph and k is the number of groups.*

Zhang et al. [12] showed that if the minimum connected set-cover problem has a polynomial-time ρ-approximation, then for any $\varepsilon > 0$, there is a polynomial-time $(\rho + \varepsilon)$-approximation for the group Steiner tree problem. Therefore, by Theorem 1 they obtained the following result.

Theorem 3 (Zhang et al. [12]) *The connected set-cover problem has no polynomial-time $O(\log^{2-\varepsilon} n)$-approximation for any $\varepsilon > 0$ unless NP has quasi-polynomial Las-Vega algorithm.*

To obtain a good approximation for the minimum connected set-cover problem, Wu et al. [10] showed that if the group Steiner tree problem has a polynomial-time ρ-approximation, so does the minimum connected set-cover problem. Therefore, they obtained the following theorem.

Theorem 4 (Wu et al. [10]) *The connected set-cover problem has a polynomial-time random $O(\log^2 n \log k)$-approximation where $n = |C|$ and $k = |X|$.*

Combining what have been proved by Zhang et al. [12] and by Wu et al. [10], it is easy to know the following relation.

Theorem 5 *The connected set-cover problem has a polynomial-time $(\rho + \varepsilon)$-approximation for any $\varepsilon > 0$ if and only if the group Steiner tree problem has a polynomial-time $(\rho + \varepsilon)$-approximation.*

This equivalence is also independently discovered by [3]. Actually, this equivalence is similar to the one between the minimum set-cover problem and the *minimum hitting set* problem.

For each element $x \in X$, define a collection of subsets:

$$C_x = \{S \mid x \in S \in C\}.$$

Then, the minimum set-cover problem becomes the minimum hitting set problem as follows: Given a finite set C and a collection of subsets of C, $\{C_x \mid x \in\}$, find the minimum hitting set, i.e., a subset C' of C such that for every $x \in X$, $C' \cap C_x \neq \emptyset$.

Similarly, the minimum connected set-cover problem becomes the equivalent *connected hitting set* problem as follows: Given a finite set C, a graph G with vertex set C, and a collection of subsets of C, $\{C_x \mid x \in\}$, find the minimum connected hitting set where a connected hitting set is a hitting set C' such that C' induces a connected subgraph of G.

To see the equivalence between the minimum connected hitting set problem and the group Steiner tree problem, it is sufficient to note the following two facts.

First, the existence of a connected hitting set C' is equivalent to the existence of a tree with weight $|C'| - 1$, interconnecting groups C_x for all $x \in X$ when the graph G is given unit weight for each edge. This is because in the subgraph induced by C', we can construct a spanning tree with weight $|C'| - 1$.

Second, a graph with nonnegative integer edge weight can be turned into an equivalent graph with unit edge weight by adding some new vertices to cut each edge with weight bigger than one into several edges with unit weight.

Open Problems

It is an open problem whether there exists or not a polynomial-time approximation for the minimum connected set-cover problem with performance ratio $O(\log^\alpha n)$ for $2 < \alpha < 3$.

Cross-References

▶ Minimum Connected Sensor Cover
▶ Performance-Driven Clustering

Recommended Reading

1. Cerdeira JO, Pinto LS (2005) Requiring connectivity in the set covering problem. J Comb Optim 9:35–47
2. Chvátal V (1979) A greedy heuristic for the set-covering problem. Math Oper Res 4(3):233–235
3. Elbassion K, Jelic S, Matijevic D (2012) The relation of connected set cover and group Steiner tree. Theor Comput Sci 438:96–101
4. Feige U (1998) A threshold of ln n for approximating set-cover. J ACM 45(4):634–652
5. Garg N, Konjevod G, Ravi R (1998) A polylogarithmic approximation algorithm for the group Steiner tree problem. In: SODA, San Francisco
6. Halperin E, Krauthgamer R (2003) Polylogarithmic inapproximability. In: STOC, San Diego, pp 585–594
7. Johnson DS (1974) Approximation algorithms for combinatorial problems. J Comput Syst Sci 9(3):256–278
8. Lovász L (1975) On the ratio of optimal integral and fractional covers. Discret Math 13(4):383–390
9. Shuai T-P, Hu X (2006) Connected set cover problem and its applications. In: Proceedings of the AAIM 2006, Hong Kong. LNCS, vol 4041, pp 243–254
10. Wu L, Du H, Wu W, Li D, Lv J, Lee W (2013) Approximations for minimum connected sensor cover. In: INFOCOM, Turin
11. Zhang Z, Gao X, Wu W (2009) Algorithms for connected set cover problem and fault-tolerant connected set cover problem. Theor Comput Sci 410:812–817
12. Zhang W, Wu W, Lee W, Du D-Z (2012) Complexity and approximation of the connected set-cover problem. J Glob Optim 53(3):563–572

Connectivity and Fault Tolerance in Random Regular Graphs

Sotiris Nikoletseas
Computer Engineering and Informatics Department, University of Patras, Patras, Greece
Computer Technology Institute and Press "Diophantus", Patras, Greece

Keywords

Average case analysis of algorithms; Connectivity; Random graphs; Robustness

Years and Authors of Summarized Original Work

2000; Nikoletseas, Palem, Spirakis, Yung

Problem Definition

A new model of random graphs was introduced in [10], that of random regular graphs with edge faults (denoted hereafter by $G_{n,p}^r$), obtained by selecting the edges of a random member of the set of all regular graphs of degree r independently and with probability p. Such graphs can represent a communication network in which the links fail independently and with probability $f = 1 - p$. A formal definition of the probability space $G_{n,p}^r$ follows.

Definition 1 (The $G_{n,p}^r$ Probability Space) Let G_n^r be the probability space of all random regular graphs with n vertices where the degree of each vertex is r. The probability space $G_{n,p}^r$ of random regular graphs with edge faults is constructed by the following two subsequent random experiments: first, a random regular graph is chosen from the space G_n^r and, second, each edge is randomly and independently deleted from this graph with probability $f = 1 - p$.

Important connectivity properties of $G_{n,p}^r$ are investigated in this entry by estimating the ranges of r, f for which, with high probability, $G_{n,p}^r$

graphs (a) are highly connected (b) become disconnected and (c) admit a giant (i.e., of $\Theta(n)$ size) connected component of small diameter.

Notation. The terms "almost certainly" (a.c.) and "with high probability" (w.h.p.) will be frequently used with their standard meaning for random graph properties. A property defined in a random graph holds almost certainly when its probability tends to 1 as the independent variable (usually the number of vertices in the graph) tends to infinity. "With high probability" means that the probability of a property of the random graph (or the success probability of a randomized algorithm) is at least $1 - n^{-\alpha}$, where $\alpha > 0$ is a constant and n is the number of vertices in the graph.

The interested reader can further study [1] for an excellent exposition of the probabilistic method and its applications, [3] for a classic book on random graphs, as well as [6] for an excellent book on the design and analysis of randomized algorithms.

Key Results

Summary. This entry studies several important connectivity properties of random regular graphs with edge faults. In order to deal with the $G_{n,p}^r$ model, [10] first extends the notion of configurations and the translation lemma between configurations and random regular graphs provided by B. Bollobás [2, 3], by introducing the concept of *random configurations* to account for edge faults and by also providing an *extended translation lemma* between random configurations and random regular graphs with edge faults.

For this new model of random regular graphs with edge faults [10] shows that:

1. For all failure probabilities $f = 1 - p \leq n^{-\epsilon}$ ($\epsilon \geq \frac{3}{2r}$ fixed) and any $r \geq 3$ the biggest part of $G_{n,p}^r$ (i.e., the whole graph except of $O(1)$ vertices) remains connected and this connected part cannot be separated, almost certainly, unless more than r vertices are removed. Note interestingly that the situation for

this range of f and r is very similar, despite the faults, to the properties of G_n^r which is r-connected for $r \geq 3$.

2. $G_{n,p}^r$ is *disconnected* a.c. for constant f and any $r = o(\log n)$ but is *highly connected*, almost certainly, when $r \geq \alpha \log n$, where $\alpha > 0$ an appropriate constant.

3. Even when $G_{n,p}^r$ becomes disconnected, it still has *a giant component of small diameter*, even when $r = O(1)$. An $O(n \log n)$-time algorithm to construct a giant component is provided.

Configurations and Translation Lemmata

Note that it is not as easy (from the technical point of view) as in the $G_{n,p}$ case to argue about random regular graphs, because of the stochastic dependencies on the existence of the edges due to regularity. The following notion of *configurations* was introduced by B. Bollobás [2, 3] to translate statements for random regular graphs to statements for the corresponding configurations which avoid the edge dependencies due to regularity and thus are much easier to deal with:

Definition 2 (Bollobás [2]) Let $w = \cup_{j=1}^{n} w_j$ be a fixed set of $2m = \sum_{j=1}^{n} d_j$ labeled vertices where $|w_j| = d_j$. A configuration F is a partition of w into m pairs of vertices, called edges of F.

Given a configuration F, let $\theta(F)$ be the (multi)graph with vertex set V in which (i, j) is an edge if and only if F has a pair (edge) with one element in w_i and the other in w_j. Note that every regular graph $G \in G_n^r$ is of the form $\theta(F)$ for exactly $(r!)^n$ configurations. However not every configuration F with $d_j = r$ for all j corresponds to a $G \in G_n^r$ since F may have an edge entirely in some w_j or parallel edges joining w_i and w_j.

Let ϕ be the set of all configurations F and let G_n^r be the set of all regular graphs. Given a property (set) $Q \subseteq G_n^r$ let $Q^* \subseteq \phi$ such that $Q^* \cap \theta^{-1}(G_n^r) = \theta^{-1}(Q)$. By estimating the probability of possible cycles of length one (self-loops) and two (loops) among pairs w_i, w_j in $\theta(F)$, the following important lemma follows:

Lemma 1 (Bollobás [3]) *If $r \geq 2$ is fixed and property Q^* holds for a.e. configuration, then property Q holds for a.e. $r-$regular graph.*

The main importance of the above lemma is that when studying random regular graphs, instead of considering the set of all random regular graphs, one can study the (much more easier to deal with) set of configurations.

In order to deal with edge failures, [10] introduces here the following extension of the notion of configurations:

Definition 3 (Random Configurations) Let $w = \cup_{j=1}^{n} w_j$ be a fixed set of $2m = \sum_{j=1}^{n} d_j$ labeled "vertices" where $|w_j| = d_j$. Let F be any configuration of the set ϕ. For each edge of F, remove it with probability $1 - p$, independently. Let $\hat{\phi}$ be the new set of objects and $\hat{F}$ the outcome of the experiment. $\hat{F}$ is called a *random configuration.*

By introducing probability p in every edge, an extension of the proof of Lemma 1 leads (since in both $\bar{Q}$ and $\hat{Q}$ each edge has the same probability and independence to be deleted, thus the modified spaces follow the properties of Q and Q^*) to the following extension to random configurations.

Lemma 2 (Extended Translation Lemma) *Let $r \geq 2$ fixed and $\bar{Q}$ be a property for $G_{n,p}^r$ graphs. If $\hat{Q}$ holds for a.e. random configuration, then the corresponding property $\bar{Q}$ holds for a.e. graph in $G_{n,p}^r$.*

Multiconnectivity Properties of $G_{n,p}^r$

The case of constant link failure probability f is studied, which represents a worst case for connectivity preservation. Still, [10] shows that logarithmic degrees suffice to guarantee that $G_{n,p}^r$ remains w.h.p. highly connected, despite these constant edge failures. More specifically:

Theorem 1 *Let G be an instance of $G_{n,p}^r$ where $p = \Theta(1)$ and $r \geq \alpha \log n$, where $\alpha > 0$ an appropriate constant. Then G is almost certainly k-connected, where*

$$k = O\left(\frac{\log n}{\log \log n}\right)$$

The proof of the above theorem uses Chernoff bounds to estimate the vertex degrees in $G_{n,p}^r$ and "similarity" of $G_{n,p}^r$ and $G_{n,p'}$ (whose properties are known) for a suitably chosen p'.

Now the (more practical) case in which $f = 1 - p = o(1)$ is considered and it is proved that the desired connectivity properties of random regular graphs are almost preserved despite the link failures. More specifically:

Theorem 2 *Let $r \geq 3$ and $f = 1 - p = O(n^{-\epsilon})$ for $\epsilon \geq \frac{3}{2r}$. Then the biggest part of $G_{n,p}^r$ (i.e., the whole graph except of $O(1)$ vertices) remains connected and this connected part (excluding the vertices that were originally neighbors of the $O(1)$-sized disconnected set) cannot be separated unless more than r vertices are removed, with probability tending to 1 as n tends to $+\infty$.*

The proof is carefully extending, in the case of faults, a known technique for random regular graphs about not admitting small separators.

$G_{n,p}^r$ Becomes Disconnected

Next remark that a constant link failure probability dramatically alters the connectivity structure of the regular graph in the case of low degrees. In particular, by using the notion of random configurations, [10] proves the following theorem:

Theorem 3 *When $2 \leq r \leq \frac{\sqrt{\log n}}{2}$ and $p = \Theta(1)$ then $G_{n,p}^r$ has at least one isolated node with probability at least $1 - n^{-k}, k \geq 2$.*

The regime for disconnection is in fact larger, since [10] shows that $G_{n,p}^r$ is a.c. disconnected even for any $r = o(\log n)$ and constant f. The proof of this last claim is complicated by the fact that due to the range for r one has to avoid using the extended translation lemma.

Existence of a Giant Component in $G_{n,p}^r$

Since $G_{n,p}^r$ is a.c. disconnected for $r = o(\log n)$ and $1 - p = f = \Theta(1)$, it would be interesting to know whether at least a large part of the network represented by $G_{n,p}^r$ is still connected, i.e., whether the biggest connected component of $G_{n,p}^r$ is large. In particular, [10] shows that:

Theorem 4 *When $f < 1 - \frac{32}{r}$ then $G_{n,p}^r$ admits a giant (i.e., $\Theta(n)$-sized) connected component for any $r \geq 64$ with probability at least $1 - O\left(\frac{\log^2 n}{n^{\alpha/3}}\right)$, where $\alpha > 0$ a constant that can be selected.*

In fact, the proof of the existence of the component includes first proving the existence (w.h.p.) of a sufficiently long (of logarithmic size) path as a basis for a BFS process starting from the vertices of that path that creates the component. The proof is quite complex: occupancy arguments are used (bins correspond to the vertices of the graphs while balls correspond to its edges); however, the random variables involved are not independent, and in order to use Chernoff-Hoeffding bounds for concentration one must prove that these random variables, although not independent, are negatively associated. Furthermore, the evaluation of the success of the BFS process uses a careful, detailed average case analysis.

The path construction and the BFS process can be viewed as an algorithm that (in case of no failures) actually reveals a giant connected component. This algorithm is very efficient, as shown by the following result:

Theorem 5 *A giant component of $G_{n,p}^r$ can be constructed in $O(n \log n)$ time, with probability at least $1 - O\left(\frac{\log^2 n}{n^{\alpha/3}}\right)$, where $\alpha > 0$ a constant that can be selected.*

Applications

In recent years the development and use of distributed systems and communication networks has increased dramatically. In addition, state-of-the-art multiprocessor architectures compute over structured, regular interconnection networks. In such environments, several applications may share the same network while executing concurrently. This may lead to unavailability of certain network resources (e.g., links) for certain applications. Similarly, faults may cause unavailability of links or nodes. The aspect of *reliable distributed computing* (which means computing with the available resources and resisting faults) adds value to applications developed in such environments.

When computing in the presence of faults, one cannot assume that the actual structure of the computing environment is known. Faults may happen even in execution time. In addition, what is a "faulty" or "unavailable" link for one application may in fact be the de-allocation of that link because it is assigned (e.g., by the network operation system) to another application. The problem of analyzing allocated computation or communication in a network over a *randomly assigned subnetwork* and *in the presence of faults* has a nature different from fault analysis of special, well-structured networks (e.g., hypercube), which does not deal with network aspects. The work presented in this entry addresses this interesting issue, i.e., analyzing the average case taken over a set of possible topologies and focuses on multiconnectivity and existence of giant component properties required for reliable distributed computing in such randomly allocated unreliable environments.

The following important application of this work should be noted: multitasking in distributed memory multiprocessors is usually performed by assigning an arbitrary subnetwork (of the interconnection network) to each task (called the *computation graph*). Each parallel program may then be expressed as communicating processors over the computation graph. Note that a multiconnectivity value k of the computation graph means also that the execution of the application can tolerate up to $k - 1$ *online additional faults*.

Open Problems

The ideas presented in [10] inspired already further interesting research. Andreas Goerdt [4] continued the work presented in a preliminary version [8] of [10] and showed the following results: if the degree r is fixed then $p = \frac{1}{r-1}$ is a threshold probability for the existence of a linear-sized component in the faulty version of almost all random regular graphs. In fact, he further shows that if each edge of an *arbitrary* graph G with maximum degree bounded above by r is present with probability $p = \frac{\lambda}{r-1}$, when $\lambda < 1$, then the faulty version of G has only components whose size is at most logarithmic in the number of nodes, with high probability. His result implies some kind of optimality of random regular graphs with edge faults. Furthermore, [5, 7] investigates important expansion properties of random regular graphs with edge faults, as well as [9] does in the case of fat trees, a common type of interconnection networks. It would be also interesting to further pursue this line of research, by also investigating other combinatorial properties (and also provide efficient algorithms) for random regular graphs with edge faults.

Recommended Reading

1. Alon N, Spencer J (1992) The probabilistic method. Wiley, New York
2. Bollobás B (1980) A probabilistic proof of an asymptotic formula for the number of labeled regular graphs. Eur J Comb 1:311–316
3. Bollobás B (1985) Random graphs. Academic, London
4. Goerdt A (1997) The giant component threshold for random regular graphs with edge faults. In: Proceedings of mathematical foundations of computer science (MFCS'97), Bratislava, pp 279–288
5. Goerdt A (2001) Random regular graphs with edge faults: expansion through cores. Theor Comput Sci (TCS) J 264(1):91–125
6. Motwani R, Raghavan P (1995) Randomized algorithms. Cambridge University Press, Cambridge
7. Nikoletseas S, Spirakis P (1995) Expander properties in random regular graphs with edge faults. In: 12th annual symposium on theoretical aspects of computer science (STACS), München, pp 421–432
8. Nikoletseas S, Palem K, Spirakis P, Yung M (1994) Short vertex disjoint paths and multiconnectivity in random graphs: reliable network computing. In: 21st international colloquium on automata, languages and programming (ICALP), Jerusalem, pp 508–515
9. Nikoletseas S, Pantziou G, Psycharis P, Spirakis P (1997) On the reliability of fat-trees. In: 3rd international European conference on parallel processing (Euro-Par), Passau, pp 208–217
10. Nikoletseas S, Palem K, Spirakis P, Yung M (2000) Connectivity properties in random regular graphs

with edge faults. Spec Issue Random Comput Int J Found Comput Sci (IJFCS) 11(2):247–262. World Scientific Publishing Company

Consensus with Partial Synchrony

Bernadette Charron-Bost[1] and André Schiper[2]
[1]Laboratory for Informatics, The Polytechnic School, Palaiseau, France
[2]EPFL, Lausanne, Switzerland

Keywords

Agreement problem

Years and Authors of Summarized Original Work

1988; Dwork, Lynch, Stockmeyer

Problem Definition

Reaching agreement is one of the central issues in fault tolerant distributed computing. One version of this problem, called *Consensus*, is defined over a fixed set $\Pi = \{p_1, \ldots, p_n\}$ of n processes that communicate by exchanging messages along channels. Messages are correctly transmitted (no duplication, no corruption), but some of them may be lost. Processes may fail by prematurely stopping (crash), may omit to send or receive some messages (omission), or may compute erroneous values (Byzantine faults). Such processes are said to be *faulty*. Every process $p \in \Pi$ has an initial value v_p and non-faulty processes must decide irrevocably on a common value v. Moreover, if the initial values are all equal to the same value v, then the common decision value is v. The properties that define Consensus can be split into safety properties (processes decide on the same value; the decision value must be consistent with initial values) and a liveness property (processes must eventually decide).

Various Consensus algorithms have been described [6, 12] to cope with any type of process failures if there is a known (Intuitively, "known bound" means that the bound can be "built into" the algorithm. A formal definition is given in the next section.) bound on the transmission delay of messages (*communication is synchronous*) and a known bound on process relative speeds (*processes are synchronous*). In completely asynchronous systems, where there exists no bound on transmission delays and no bound on process relative speeds, Fischer, Lynch, and Paterson [8] have proved that there is no Consensus algorithm resilient to even one crash failure. The paper by Dwork, Lynch, and Stockmeyer [7] introduces the concept of *partial synchrony*, in the sense it lies between the completely synchronous and completely asynchronous cases, and shows that partial synchrony makes it possible to solve Consensus in the presence of process failures, whatever the type of failure is.

For this purpose, the paper examines the quite realistic case of asynchronous systems that behave synchronously during some "good" periods of time. Consensus algorithms designed for synchronous systems do not work in such systems since they may violate the safety properties of Consensus during a bad period, that is when the system behaves asynchronously. This leads to the following question: is it possible to design a Consensus algorithm that never violates safety conditions in an asynchronous system, while ensuring the liveness condition when some additional conditions are met?

Key Results

The paper has been the first to provide a positive and comprehensive answer to the above question. More precisely, the paper (1) defines various types of partial synchrony and introduces a new round based computational model for partially synchronous systems, (2) gives various Consensus algorithms according to the severity of failures (crash, omission, Byzantine faults with or without authentication), and (3) shows how to implement the round based computational model in each type of partial synchrony.

Partial Synchrony

Partial synchrony applies both to communications and to processes. Two definitions for *partially synchronous communications* are given: (1) for each run, there exists an upper bound Δ on communication delays, but Δ is *unknown* in the sense it depends on the run; (2) there exists an upper bound Δ on communication delays that is common for all runs (Δ is *known*), but holds only after some time T, called the *Global Stabilization Time* (*GST*) that may depend on the run (*GST* is *unknown*). Similarly, *partially synchronous processes* are defined by replacing "transmission delay of messages" by "relative process speeds" in (1) and (2) above. That is, the upper bound on relative process speed Φ is unknown, or Φ is known but holds only after some unknown time.

Basic Round Model

The paper considers a round based model: computation is divided into *rounds* of message exchange. Each round consists of a *send step*, a *receive step*, and then a *computation step*. In a send step, each process sends messages to any subset of processes. In a receive step, some subset of the messages sent to the process during the send step at the *same* round is received. In a computation step, each process executes a state transition based on its current state and the set of messages just received.

Some of the messages that are sent may not be received, i.e., some can be lost. However, the *basic round model* assumes that there is some round GSR, such that all messages sent from non faulty processes to non faulty processes at round GSR or afterward are received.

Consensus Algorithm for Benign Faults (Requires $f < n/2$)

In the paper, the algorithm is only described informally (textual form). A formal expression is given by Algorithm 1: the code of each process is given round by round, and each round is specified by the send and the computation steps (the receive step is implicit). The constant f denotes the maximum number of processes that may be faulty (crash or omission). The algorithm requires $f < n/2$.

Algorithm 1 Consensus algorithm in the basic round model for benign faults ($f < n/2$)

```
 1: Initialization:
 2:    Acceptable_p := {v_p}                                                                    {v_p is the initial value of p }
 3:    Proper_p := {v_p}                                      {All the lines for maintaining Proper_p are trivial to write, and so are omitted}
 4:    vote_p := ⊥
 5:    Lock_p := ∅

 6: Round r = 4k − 3 :
 7:    Send:
 8:        send ⟨Acceptable_p⟩ to coord_k

 9:    Compute:
10:        if p = coord_k and p receives at least ≥ n − f messages containing a common value then
11:            vote_p := select one of these common acceptable values

12: Round r = 4k − 2 :
13:    Send:
14:        if p = coord_k and vote_p ≠ ⊥ then
15:            send ⟨vote_p⟩ to all processes

16:    Compute:
17:        if received ⟨v⟩ from coord_k then
18:            Lock_p := Lock_p ∖ {v, −}; Lock_p := Lock_p ∪ {(v, k)};

19: Round r = 4k − 1 :
20:    Send:
21:        if ∃v s.t. (v , k) ∈ Lock_p then
22:            send ⟨ack⟩ to coord_k

23:    Compute:
24:        if p = coord_k then
25:            if received at least ≥ f + 1 ack messages then
26:                DECIDE(vote_p);
27:            vote_p := ⊥

28: Round r = 4k :
29:    Send:
30:        send ⟨Lock_p⟩ to all processes

31:    Compute:
32:        for all (v , θ) ∈ Lock_p do
33:            if received (w , θ̄) s.t. w ≠ v and θ̄ ≥ θ then                                            {release lock on v}
34:                Lock_p := Lock_p ∪ {(w, θ̄)} ∖ {(v, θ)};
35:        if |Lock_p| = 1 then
36:            Acceptable_p := v where (v, −) ∈ Lock_p
37:        else
38:            if Lock_p = ∅ then Acceptable_p := Proper_p else Acceptable_p := ∅
```

Rounds are grouped into phases, where each phase consists in four consecutive rounds. The algorithm includes the rotating coordinator strategy: each phase k is led by a unique coordinator – denoted by $coord_k$ – defined as process p_i for phase $k = i(mod\ n)$. Each process p maintains a set $Proper_p$ of values that p has heard of (*proper* values), initialized to $\{v_p\}$ where v_p is p's initial value. Process p attaches $Proper_p$ to each message it sends.

Process p may *lock* value v when p thinks that some process might decide v. Thus value v is an *acceptable* value to p if (1) v is a proper value to p, and (2) p does not have a lock on any value except possibly v (lines 35–38).

At the first round of phase k (round $4k − 3$), each process sends the list of its acceptable values to $coord_k$. If $coord_k$ receives at least $n − f$ sets of acceptable values that all contain some value

v, then $coord_k$ votes for v (line 11), and sends its vote to all at second round $4k − 2$. Upon receiving a vote for v, any process locks v in the current phase (line 18), releases any earlier lock on v, and sends an acknowledgment to $coord_k$ at the next round $4k − 1$. If the latter process receives acknowledgments from at least $f + 1$ processes, then it decides (line 26). Finally locks are released at round $4k$ – for any value v, only the lock from the most recent phase is kept, see line 34 – and the set of values *acceptable* to p is updated (lines 35–38).

Consensus Algorithm for Byzantine Faults (Requires $f < n/3$)

Two algorithms for Byzantine faults are given. The first algorithm assumes *signed messages*, which means that any process can verify the origin of all messages. This fault model is called *Byzantine faults with authentication*. The algorithm has the same phase structure as Algorithm 1. The difference is that (1) messages are signed, and (2) "proofs" are carried by some messages. A proof carried by message m sent by some process p_i in phase k consists of a set of signed messages $sgn_j(m', k)$, proving that p_i received message (m', k) in phase k from p_j before sending m. A proof is carried by the message send at line 16 and line 30 (Algorithm 1). Any process receiving a message carrying a proof accepts the message and behaves accordingly if – and only if the proof is found valid. The algorithm requires $f < n/3$ (less than a third of the processes are faulty).

The second algorithm does not assume a mechanism for signing messages. Compared to Algorithm 1, the structure of a phase is slightly changed. The problem is related to the vote sent by the coordinator (line 15). Can a Byzantine coordinator fool other processes by not sending the right vote? With signed messages, such a behavior can be detected thanks to the "proofs" carried by messages. A different mechanism is needed in the absence of signature.

The mechanism is a small variation of the *Consistent Broadcast* primitive introduced by Srikanth and Toueg [15]. The broadcast primitive ensures that (1) if a non faulty process broadcasts m, then every non faulty process delivers m, and (2) if some non faulty process delivers m, then all non faulty processes also eventually deliver m. The implementation of this broadcast primitive requires two rounds, which define a *superround*. A phase of the algorithm consists now of three superrounds. The superrounds $3k - 2$, $3k - 1$, $3k$ mimic rounds $4k - 3$, $4k - 2$, and $4k - 1$ of Algorithm 1, respectively. Lock-release of phase k occurs at the end of superround $3k$, i.e., does not require an additional round, as it does in the two previous algorithms. The algorithm also requires $f < n/3$.

The Special Case of Synchronous Communication

By strengthening the round based computational model, the authors show that synchronous communication allow higher resiliency. More pre-

cisely, the paper introduces the model called the *basic round model with signals*, in which upon receiving a signal at round r, every process knows that all the non faulty processes have received the messages that it has sent during round r. At each round after *GSR*, each non faulty process is guaranteed to receive a signal. In this computational model, the authors present three new algorithms tolerating less than n benign faults, $n/2$ Byzantine faults with authentication, and $n/3$ Byzantine faults respectively.

Implementation of the Basic Round Model

The last part of the paper consists of algorithms that simulate the basic round model under various synchrony assumption, for crash faults and Byzantine faults: first with partially synchronous communication and synchronous processes (case 1), second with partially synchronous communication and processes (case 2), and finally with partially synchronous processes and synchronous communication (case 3).

In case 1, the paper first assumes the basic case $\Phi = 1$, i.e., all non faulty process progress exactly at the same speed, which means that they have a common notion of time. Simulating the basic round model is simple in this case. In case 2 processes do not have a common notion of time. The authors handle this case by designing an algorithm for clock synchronization. Then each process uses its private clock to determine its current round. So processes alternate between steps of the clock synchronization algorithm and steps simulating rounds of the basic round model. With synchronous communication (case 3), the authors show that for any type of faults, the so-called basic round model with signals is implementable.

Note that, from the very definition of partial synchrony, the six algorithms share the fundamental property of tolerating message losses, provided they occur during a finite period of time.

Upper Bound for Resiliency

In parallel, the authors exhibit upper bounds for the resiliency degree of Consensus algorithms in each partially synchronous model, according to the type of faults. They show that their Consensus

Consensus with Partial Synchrony, Table 1 Tight resiliency upper bounds (P stands for "process", C for "communication"; 0 means "asynchronous", 1/2 means "partially synchronous", and 1 means "synchronous")

	$P = 0 \quad C = 0$	$P = 1/2 \quad C = 1/2$	$P = 1 \quad C = 1/2$	$P = 1/2 \quad C = 1$	$P = 1 \quad C = 1$
Benign	0	$\lceil (n-1)/2 \rceil$	$\lceil (n-1)/2 \rceil$	$n - 1$	$n - 1$
Authenticated Byzantine	0	$\lceil (n-1)/3 \rceil$	$\lceil (n-1)/3 \rceil$	$\lceil (n-1)/2 \rceil$	$n - 1$
Byzantine	0	$\lceil (n-1)/3 \rceil$	$\lceil (n-1)/3 \rceil$	$\lceil (n-1)/3 \rceil$	$\lceil (n-1)/3 \rceil$

algorithms achieve these upper bounds, and so are optimal with respect to their resiliency degree. These results are summarized in Table 1.

Applications

Availability is one of the key features of critical systems, and is defined as the ratio of the time the system is operational over the total elapsed time. Availability of a system can be increased by replicating its critical components. Two main classes of replication techniques have been considered: *active* replication and *passive* replication. The Consensus problem is at the heart of the implementation of these replication techniques. For example, active replication, also called *state machine replication* [10, 14], can be implemented using the group communication primitive called *Atomic Broadcast*, which can be reduced to Consensus [3].

Agreement needs also to be reached in the context of distributed transactions. Indeed, all participants of a distributed transaction need to agree on the output *commit* or *abort* of the transaction. This agreement problem, called *Atomic Commitment*, differs from Consensus in the validity property that connects decision values (*commit* or *abort*) to the initial values (favorable to commit, or demanding abort) [9]. In the case decisions are required in all executions, the problem can be reduced to Consensus if the abort decision is acceptable although all processes were favorable to commit, in some restricted failure cases.

Open Problems

A slight modification to each of the algorithms given in the paper is to force a process repeatedly to broadcast the message "Decide v" after it decides v. Then the resulting algorithms share the property that all non faulty processes definitely make a decision within $O(f)$ rounds after GSR, and the constant factor varies between 4 (benign faults) and 12 (Byzantine faults). A question raised by the authors at the end of the paper is whether this constant can be reduced. Interestingly, a positive answer has been given later, in the case of benign faults and $f < n/3$, with a constant factor of 2 instead of 4. This can be achieved with deterministic algorithms, see [4], based on the communication schema of the Rabin randomized Consensus algorithm [13].

The second problem left open is the generalization of this algorithmic approach – namely, the design of algorithms that are always safe and that terminate when a sufficiently long good period occurs – to other fault tolerant distributed problems in partially synchronous systems. The latter point has been addressed for the Atomic Commitment and Atomic Broadcast problems (see section "Applications").

Cross-References

▶ Asynchronous Consensus Impossibility
▶ Failure Detectors
▶ Randomization in Distributed Computing

Recommended Reading

1. Bar-Noy A, Dolev D, Dwork C, Strong HR (1987) Shifting gears: changing algorithms on the fly To expedite Byzantine agreement. In: PODC, pp 42–51
2. Chandra TD, Hadzilacos V, Toueg S (1996) The weakest failure detector for solving consensus. J ACM 43(4):685–722
3. Chandra TD, Toueg S (1996) Unreliable failure detectors for reliable distributed systems. J ACM 43(2):225–267
4. Charron-Bost B, Schiper A (2007) The "Heard-Of" model: computing in distributed systems with benign failures. Technical report, EPFL

5. Dolev D, Dwork C, Stockmeyer L (1987) On the minimal synchrony needed for distributed consensus. J ACM 34(1):77–97
6. Dolev D, Strong HR (1983) Authenticated algorithms for Byzantine agreement. SIAM J Comput 12(4):656–666
7. Dwork C, Lynch N, Stockmeyer L (1988) Consensus in the presence of partial synchrony. J ACM 35(2):288–323
8. Fischer M, Lynch N, Paterson M (1985) Impossibility of distributed consensus with one faulty process. J ACM 32:374–382
9. Gray J (1990) A comparison of the Byzantine agreement problem and the transaction commit problem. In: Fault-tolerant distributed computing [Asilomar-Workshop 1986], vol 448, LNCS. Springer, Berlin, pp 10–17
10. Lamport L (1978) Time, clocks, and the ordering of events in a distributed system. Commun ACM 21(7):558–565
11. Lamport L (1998) The part-time parliament. ACM Trans Comput Syst 16(2):133–169
12. Pease MC, Shostak RE, Lamport L (1980) Reaching agreement in the presence of faults. J ACM 27(2):228–234
13. Rabin M (1983) Randomized Byzantine generals. In: Proceedings of the 24th annual ACM symposium on foundations of computer science, pp 403–409
14. Schneider FB (1993) Replication management using the state-machine approach. In: Mullender S (ed) Distributed systems. ACM, pp 169–197
15. Srikanth TK, Toueg S (1987) Simulating authenticated broadcasts to derive simple fault-tolerant algorithms. Distrib Comput 2(2):80–94

Convex Graph Drawing

Md. Saidur Rahman
Department of Computer Science and Engineering, Bangladesh University of Engineering and Technology, Dhaka, Bangladesh

Keywords

Convex drawing; Linear-time algorithm; Planar graphs

Years and Authors of Summarized Original Work

1984; Chiba, Yamanouchi, Nishizeki

Problem Definition

A convex drawing of a planar graph G is a planar drawing of G where every vertex is drawn as a point, every edge is drawn as a straight line segment, and every face is drawn as a convex polygon. Not every planar graph has a convex drawing. The planar graph in Fig. 1a has a convex drawing as shown in Fig. 1b whereas the planar graph in Fig. 1d has no convex drawing. Tutte [11] showed that every 3-connected planar graph has a convex drawing, and obtained a necessary and sufficient condition for a planar graph to have a convex drawing with a prescribed outer polygon. Furthermore, he gave a "barycentric mapping" method for finding a convex drawing, which requires solving a system of $O(n)$ linear equations [12] and leads to an $O(n^{1.5})$ time convex drawing algorithm for a planar graph with a fixed embedding. Development of faster algorithms for determining whether a planar graph (where the embedding is not fixed) has a convex drawing and finding such a drawing if it exits is addressed in the paper of Chiba, Yamanouchi, and Nishizeki [2].

A Characterization for Convex Drawing A plane graph is a planar graph with a fixed embedding. In a convex drawing of a plane graph G, the outer cycle $C_o(G)$ is also drawn as a convex polygon. The polygonal drawing C_o^* of $C_o(G)$, called an *outer convex polygon*, plays a crucial role in finding a convex drawing of G. The plane graph G in Fig. 1a admits a convex drawing if an outer convex polygon C_o^* has all vertices 1, 2, 3, 4, and 5 of $C_o(G)$ as the *apices* (i.e., geometric vertices) of C_o^*, as illustrated in Fig. 1b. However, if C_o^* has only apices 1, 2, 3, and 4, then G does not admit a convex drawing as depicted in Fig. 1c. We say that an outer convex polygon C_o^* is *extendible* if there exists a convex drawing of G in which $C_o(G)$ is drawn as C_o^*. Thus, the outer convex polygon drawn by thick lines in Fig. 1b is extendible, while that in Fig. 1c is not. If the outer facial cycle C_o has an extendible outer convex polygon, we say that the facial cycle C_o is *extendible*.

Tutte established a necessary and sufficient condition for an outer convex polygon to be

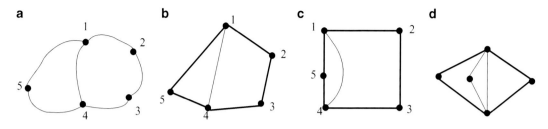

Convex Graph Drawing, Fig. 1 Plane graphs and drawings

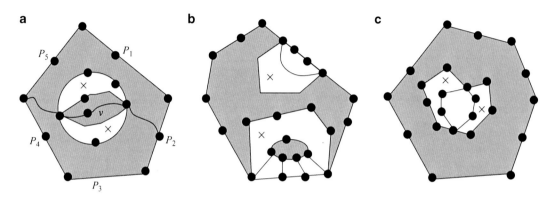

Convex Graph Drawing, Fig. 2 G and C_o^* violating Conditions (i)–(iii) in Theorem 1

extendible [11]. The following theorem obtained by Thomassen [9] is slightly more general than the result of Tutte.

Theorem 1 *Let G be a 2-connected plane graph, and let C_o^* be an outer convex polygon of G. Let C_o^* be a k-gon, $k \geq 3$, and let $P_1, P_2, \ldots, P_k$ be the paths in $C_o(G)$, each corresponding to a side of the polygon C_o^*, as illustrated in Fig. 2a. Then, C_o^* is extendible if and only if the following Conditions (i)–(iii) hold.*

(i) *For each inner vertex v with $d(v) \geq 3$, there exist three paths disjoint except v, each joining v and an outer vertex.*

(ii) *$G - V(C_o(G))$ has no connected component H such that all the outer vertices adjacent to vertices in H lie on a single path P_i, and no two outer vertices in each path P_i are joined by an inner edge.*

(iii) *Any cycle containing no outer edge has at least three vertices of degree ≥ 3.*

Figure 2a–c violate Conditions (i)–(iii) of Theorem 1, respectively, where each of the faces marked by $\times$ cannot be drawn as a convex polygon.

Key Results

Two linear algorithms for convex drawings are the key contribution of the paper of Chiba, Yamanouchi, and Nishizeki [2]. One algorithm is for finding a convex drawing of a plane graph if it exists, and the other algorithm is for testing whether there is a planar embedding of a given planar graph which has a convex drawing. Thus, the main result of the paper can be stated as in the following theorem.

Theorem 2 *Let G be a 2-connected planar graph. Then, one can determine whether G has a convex drawing in linear time and find such a drawing in linear time if it exists.*

Convex Drawing Algorithm

In this section, we describe the drawing algorithm of Chiba, Yamanouchi, and Nishizeki [2] which is based on Thomassen's short proof of Theorem 1. Suppose that a 2-connected plane graph G and an outer convex polygon C_o^* satisfy conditions in Theorem 1. The convex drawing algorithm extends C_o^* into a convex drawing of G in linear time. For simplicity, it is assumed that every inner vertex has degree three or more in G. Otherwise, replace each maximal induced path not on $C_o(G)$ by a single edge joining its ends (the resulting simple graph G' satisfies Conditions (i)–(iii) of Theorem 1); then, find a convex drawing of G'; and finally, subdivide each edge substituting a maximal induced path.

They reduce the convex drawing of G to those of several subgraphs of G as follows: delete from G an arbitrary apex v of the outer convex polygon C_o^* together with the edges incident to v; divide the resulting graph $G' = G - v$ into blocks $B_1, B_2, \ldots, B_p$, $p \geq 1$ as illustrated in Fig. 3; determine an outer convex polygon C_i^* of each block B_i so that B_i with C_i^* satisfies Conditions (i)–(iii) of Theorem 1; and recursively apply the algorithm to each block B_i with C_i^* to determine the position of inner vertices of B_i.

The recursive algorithm described above can be implemented in linear time by ensuring that only the edges which newly appear on the outer face are traversed in each recursive step.

Convex Testing Algorithm

In this section, we describe the convex testing algorithm of Chiba, Yamanouchi, and Nishizeki [2]

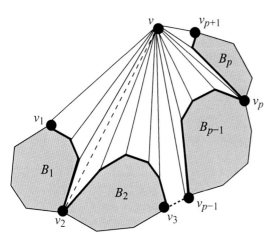

Convex Graph Drawing, Fig. 3 Reduction of the convex drawing of G into subproblems

which implies a constructive proof of Theorem 2. They have modified the conditions in Theorem 1 into a form suitable for the convex testing, which is represented in terms of 3-connected components. Using the form, they have shown that the convex testing of a planar graph G can be reduced to the planarity testing of a certain graph obtained from G.

To describe the convex testing algorithm, we need some definitions. A pair $\{x, y\}$ of vertices of a 2-connected graph $G = (V, E)$ is called a *separation pair* if there exists two subgraphs $G_1' = (V_1, E_1')$ and $G_2' = (V_2, E_2')$ satisfying the following conditions (a) and (b): (a) $V = V_1 \cup V_2, V_1 \cap V_2 = \{x, y\}$; and (b) $E = E_1' \cup E_2', E_1' \cap E_2' = \emptyset, |E_1'| \geq 2, |E_2'| \geq 2$. For a separation pair $\{x, y\}$ of G, $G_1 = (V_1, E_1' + (x, y))$ and $G_2 = (V_2, E_2' + (x, y))$ are called the *split graphs* of G. The new edges (x, y) added to G_1 and G_2 are called the *virtual edges*. Dividing a graph G into two split graphs G_1 and G_2 is called *splitting*. Reassembling the two split graphs G_1 and G_2 into G is called *merging*. Suppose that a graph G is split, the split graphs are split, and so on, until no more splits are possible. The graphs constructed in this way are called the *split components* of G. The split components are of three types: triple bonds (i.e., a set of three multiple edges), triangles, and 3-connected graphs. The *3-connected components* of G are obtained from the split components of G by merging triple bonds into a bond and triangles into a ring, as far as possible, where a *bond* is a set of multiple edges and a *ring* is a cycle. Note that the split components of G are not necessarily unique, but the 3-connected components of G are unique [5].

A separation pair $\{x, y\}$ is *prime* if x and y are the end vertices of a virtual edge contained in a 3-connected component. Suppose that $\{x, y\}$ is a prime separation pair of a graph G and that G is split at $\{x, y\}$, the split graphs are split, and so on, until no more splits are possible at $\{x, y\}$. A graph constructed in this way is called an $\{x, y\}$-*split component* of G if it has at least one real (i.e., non-virtual) edge.

In some cases, it can be easily known only from the $\{x, y\}$-split components for a single separation pair $\{x, y\}$ that a graph G has no convex drawing. A prime separation pair $\{x, y\}$ of G is called a *forbidden separation pair* if there are either (a) at least four $\{x, y\}$-split components or (b) exactly three $\{x, y\}$-split components, each of which is neither a ring nor a bond. Note that an $\{x, y\}$-split component corresponds to an edge (x, y) if it is a bond and to a subdivision of an edge (x, y) if it is a ring. One can easily know that if a planar graph G has a forbidden separation pair, then any plane embedding of G has no convex drawing, that is, G has no extendible facial cycle. On the other hand, the converse of the fact above is not true. A prime separation pair $\{x, y\}$ is called a *critical separation pair* if there are either (i) exactly three $\{x, y\}$-split components including a bond or a ring or (ii) exactly two $\{x, y\}$-split components each of which is neither a bond nor a ring. When a planar graph G has no forbidden separation pair, two cases occur: if G has no critical separation pair either, then G is a subdivision of a 3-connected graph, and so every facial cycle of G is extendible; otherwise, that is, if G has critical separation pairs, then a facial cycle F of G may or may not be extendible, depending on the interaction of F and critical separation pairs.

Using the concepts of forbidden separation pairs and critical separation pairs, Chiba et al. gave the following condition in Theorem 3 which is suitable for the testing algorithm. They proved that the condition in Theorem 3 is equivalent to the condition in Theorem 1 under a restriction that the outer convex polygon C_o^* is strict, that is, every vertex of $C_o(G)$ is an apex of C_o^* [2].

Theorem 3 *Let $G = (V, E)$ be a 2-connected plane graph with the outer facial cycle $F = C_o(G)$, and let C_o^* be an outer strict convex polygon of G. Then, C_o^* is extendible if and only if G and F satisfy the following conditions.*

(a) *G has no forbidden separation pair.*
(b) *For each critical separation pair $\{x, y\}$ of G, there is at most one $\{x, y\}$-split component having no edge of F, and, if any, it is either a bond if $(x, y) \in E$ or a ring, otherwise.*

The convex testing condition in Theorem 3 is given for a plane graph. Note that Condition (a) does not depend on a plane embedding. Thus, to test whether a planar graph G has a convex drawing, it is needed to test whether G satisfies Condition (a) or not and if G satisfies Condition (a) then test whether G has a plane embedding such that its outer face F satisfies Condition (b) in Theorem 3. With some simple observation, it is shown that every graph G having no forbidden separation pair has an embedding such that the outer face satisfies Condition (b) if G has at most one critical separation pair. Hence, every planar graph with no forbidden separation pair and at most one critical separation pair has a convex drawing.

The convex testing problem of G for the case where G has no forbidden separation pair and has two or more critical separation pairs is reduced to the planarity testing problem of a certain graph obtained from G. If G has a plane embedding which has a convex drawing, the outer face F of the embedding must satisfy Condition (b) of Theorem 3. Then, F contains every vertex of critical separation pairs and any split component which is neither a bond nor a ring must have an edge on the outer face. Observe that if a critical separation pair $\{x, y\}$ has exactly three $\{x, y\}$-split components, then two of them can have edges on F and one cannot have an edge on F; the $\{x, y\}$-split component which will not have an edge on F must be either a bond or a ring. Thus, to test whether G has such an embedding or not, a new graph from G is constructed as follows. For each critical separation pair $\{x, y\}$, if (x, y) is an edge of G, then delete the edge $\{x, y\}$ from G. If (x, y) is not an edge of G and exactly one $\{x, y\}$-split component is a ring, then delete the x-y path in the component from G. Let G_1 be the resulting graph, as illustrated in Fig. 4b. Let G_2 be the graph obtained from G_1 by adding a new vertex v and joining v to all vertices of critical separation pairs of G, as illustrated in Fig. 4c. If G_2 has a planar embedding Γ_2 such that v is embedded on the outer face of Γ_2 as illustrated in Fig. 4d, we get a planar embedding Γ_1 of G_1 from Γ_2 by deleting v from the embedding as illustrated in Fig. 4e. The outer facial cycle of Γ_1 will be the outer facial cycle F of a planar embedding of G as illustrated in Fig. 4f which satisfies the Condition (b) of Theorem 3. Thus,

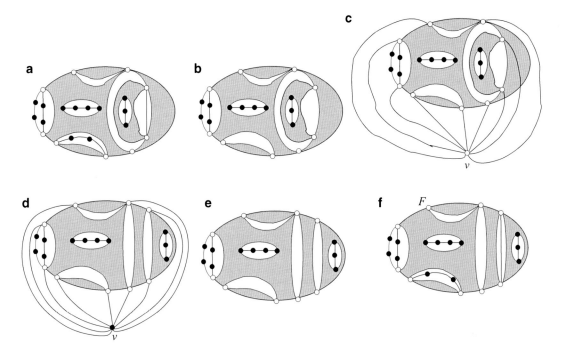

Convex Graph Drawing, Fig. 4 Illustration for convex testing; (**a**) G, (**b**) G_1, (**c**) G_2, (**d**) Γ_2, (**e**) Γ_1, and (**f**) Γ

the strict convex polygon F^* of F is extendible. Hence, G has a convex drawing if G_2 has a planar embedding with v on the outer face. It is not difficult to show the converse implication. Hence, Theorem 2 holds.

Observe that F may not be the only extendible facial cycle of G, that is, Γ may not be the only planar embedding of G which has a convex drawing. Chiba et al. [2] also gave a linear algorithm which finds all extendible facial cycles of G.

Applications

Thomassen [10] showed applications of convex representations in proving a conjecture of Grünbaum and Shephard on convex deformation of convex graphs and for giving a short proof of the result of Mani-Levistka, Guigas, and Klee on convex representation of infinite doubly periodic 3-connected planar graphs. The research on convex drawing of planar graphs was motivated by the desire of finding aesthetic drawings of graphs [3]. Arkin et al. [1] showed that there is a monotone path in some direction between every pair of vertices in any strictly convex drawing of a planar graph.

Open Problems

A convex drawing is called a *convex grid drawing* if each vertex is drawn at a grid point of an integer grid. Using canonical ordering and shift method, Chrobak and Kant [4] showed that every 3-connected plane graph has a convex grid drawing on an $(n-2) \times (n-2)$ grid and such a grid drawing can be found in linear time. However, the question of whether every planar graph which has a convex drawing admits a convex grid drawing on a grid of polynomial size remained as an open problem. Several research works are concentrated in this direction [6, 7, 13]. For example, Zhou and Nishizeki showed that every internally triconnected plane graph G whose decomposition tree $T(G)$ has exactly four leaves has a convex grid drawing on a $2n \times 4n = O(n^2)$ grid and presented a linear algorithm to find such a drawing [13].

Recommended Reading

1. Arkin EM, Connelly R, Mitchell JS (1989) On monotone paths among obstacles with applications to planning assemblies. In: Proceedings of the fifth annual symposium on computational geom-

etry (SCG '89), Saarbrücken. ACM, New York, pp 334–343

2. Chiba N, Yamanouchi T, Nishizeki T (1984) Linear algorithms for convex drawings of planar graphs. In: Bondy JA, Murty USR (eds) Progress in graph theory. Academic, Toronto, pp 153–173
3. Chiba N, Onoguchi K, Nishizeki T (1985) Drawing planar graphs nicely. Acta Inform 22:187–201
4. Chrobak M, Kant G (1997) Convex grid drawings of 3-connected planar graphs. Int J Comput Geom Appl 7:221–223
5. Hopcroft JE, Tarjan RE (1973) Dividing a graph into triconnected components. SIAM J Comput 2:135–158
6. Miura K, Azuma M, Nishizeki T (2006) Convex drawings of plane graphs of minimum outer apices. Int J Found Comput Sci 17:1115–1127
7. Miura K, Nakano S, Nishizeki T (2006) Convex grid drawings of four connected plane graphs. Int J Found Comput Sci 17:1032–1060
8. Nishizeki T, Rahman MS (2004) Planar graph drawing. Lecture notes series on computing, vol 12. World Scientific, Singapore
9. Thomassen C (1980) Planarity and duality of finite and infinite graphs. J Comb Theory 29:244–271
10. Thomassen C (1984) Plane representations of graphs. In: Bondy JA, Murty USR (eds) Progress in graph theory. Academic, Toronto, pp 43–69
11. Tutte WT (1960) Convex representations of graphs. Proc Lond Math Soc. 10:304–320
12. Tutte WT (1963) How to draw a graph. Proc Lond Math Soc. 13:743–768
13. Zhou X, Nishizeki T (2010) Convex drawings of internally triconnected plane graphs on $o(n^2)$ grids. Discret Math Algorithms Appl 2:347–362

Convex Hulls

Michael Hemmer[1] and Christiane Schmidt[2]
[1]Department of Computer Science, TU Braunschweig, Braunschweig, Germany
[2]The Selim and Rachel Benin School of Computer Science and Engineering, The Hebrew University of Jerusalem, Jerusalem, Israel

Keywords

Computational geometry; Point sets

Years and Authors of Summarized Original Work

1972; Graham
1973; Jarvis
1977; Preparata, Hong
1996; Chan

Problem Definition

The *convex hull* of a set P of n points in $\mathbb{R}^d$ is the intersection of all convex regions that contain P. While convex hulls are defined for arbitrary d, the focus here is on $d = 2$ (and $d = 3$). For a more general overview, we recommend reading [7,9] as well as [3].

A frequently used visual description for a convex hull in 2D is a rubber band: when we imagine the points in the plane to be nails and put a rubber band around them, the convex hull is exactly the structure we obtain by a tight rubber band; see Fig. 1.

The above definition, though intuitive, is hard to use for algorithms to compute the convex hull – one would have to intersect all convex supersets of P. However, one can show that there is an alternative definition of the convex hull of P: it is the set of all convex combinations of P.

Notation

For a point set $P = \{p_1, \dots, p_n\}$, a *convex combination* is of the form

$$\lambda_1 p_1 + \lambda_2 p_2 + \dots \lambda_n p_n \ \text{ with } \ \lambda_i \geq 0, \sum \lambda_i = 1. \tag{1}$$

The convex hull, $CH(P)$, of P is the polygon that consists of all convex combinations of P. The *ordered* convex hull gives the ordered sequence of vertices on the boundary of $CH(P)$, instead of only the set of vertices that constitute the hull.

Key Results

In the following, we present algorithms that compute the ordered convex hull of a given point set P in the plane. We start with a short proof for a lower bound of $\Omega(n \log n)$.

Lower Bound

Theorem 1 *Let P be a set of n points in the plane. An algorithm that computes the ordered convex hull is in $\Omega(n \log n)$.*

Convex Hulls, Fig. 1 The convex hull of a set of points in $\mathbb{R}^2$

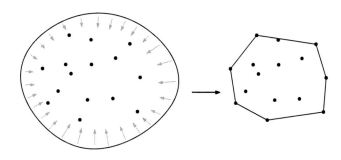

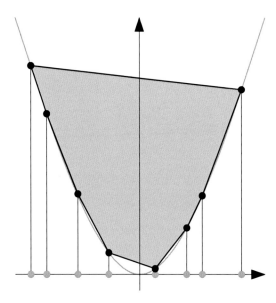

Convex Hulls, Fig. 2 Set of numbers X in *gray*, the point set P in *black* and the convex hull $CH(P)$ in *blue*

Proof Given an unsorted list of numbers $X = \{x_1, x_2, \ldots, x_n\}$, we can lift these to a parabola as depicted in Fig. 2. Computing the convex hull $CH(P)$ of the resulting set $P = \{(x_i, x_i^2) \mid x_i \in X\}$ allows to output the sorted numbers by reading off the x-values of the vertices on the lower chain of $CH(P)$ in $O(n)$ time. Thus, a computation of $CH(P)$ in $o(n \log n)$ time would contradict the lower bound $\Omega(n \log n)$ for sorting.

Divide and Conquer by Preparata and Hong [8]

In the first step, the elements of P are sorted, and then the algorithm recursively divides P into subsets A and B of similar size. This is done until at most three points are left in each set,

for which the convex hull is trivially computed. The sorting assures that the computed convex hulls are disjoint. Thus, in each step of the merge phase, the algorithm is given two ordered convex hulls CH(A) and CH(B), which are separated by a vertical line. To compute CH($A \cup B$), it needs to find the two tangents supporting CH(A) and CH(B) from above and below, respectively. The procedure, which is often referred to as a "wobbly stick", is exemplified in Fig. 3.

It is easy to see that each level of the merge phase requires $O(n)$ time, resulting in a total running time of $O(n \log n)$. A similar idea is also applicable in 3D.

Graham Scan [4]

The Graham Scan starts with a known point on the hull as an anchor, the bottommost (rightmost) point p_1. The remaining $n - 1$ points are sorted according to the angles they form with the negative x-axis through p_1, from the largest angle to the smallest angle. The points p_i are processed using this angular order. For the next point p_i, the algorithm determines whether the last constructed hull edge xy and p_i form a left or a right turn. In case of a right turn, y is not part of the hull and is discarded. The discarding is continued as long as the last three points form a right turn. Once xy and p_i form a left turn, the next point in the angular order is considered. Because of the initial sorting step, the total running time of the Graham Scan is $O(n \log n)$.

Jarvis's March or Gift Wrapping [5]

Jarvis's March is an output-sensitive algorithm, i.e., the running time does depend on the size of the output. Its total running time is $O(nh)$, where h is the number of points on CH(P).

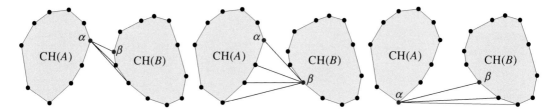

Convex Hulls, Fig. 3 Finding the tangent supporting CH(A) and CH(B) from below: considering CH(A) and CH(B) as obstacles, the algorithm iteratively tries to increment either α or β in clockwise (cw) and coun-

terclockwise (ccw) order, respectively, while maintaining visibility of α and β. That is, it stops as soon as α does not see the ccw neighbor of β and β does not see the cw neighbor of α

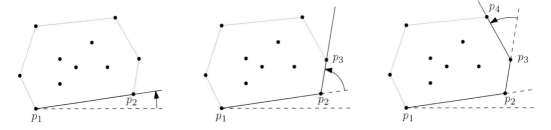

Convex Hulls, Fig. 4 The first three steps of the gift wrapping algorithm. Starting at p_1, the algorithm acts as if it was wrapping a string around the point set

The algorithm starts with a known point on the hull, i.e., the bottommost (rightmost) point p_1. Just like wrapping a string around the point set, it then computes the next point on CH(P) in counterclockwise order: compare angles to all other points and choose the point with the largest angle to the negative x-axis. In general, the wrapping step is as follows: let p_{k-1} and p_k be the two previously computed hull vertices, then p_{k+1} is set to the point $p \in P, p \neq p_k$, that maximizes the angle $\angle p_{k-1} p_k p_{k+1}$; see Fig. 4. Each steps takes $O(n)$ time and finds one point on CH(P). Thus, the total running time is $O(nh)$.

Chan's Algorithm [2]

In 1996, Chan presented an output-sensitive algorithm, with a worst-case optimal running time of $O(n \log h)$. This does not contradict the lower bound presented above, as it features n points on the hull. Let us for now assume that h is known, the number of points on the final convex hull. The algorithm runs in two phases.

Phase 1 splits P into $\lceil n/h \rceil$ groups of size at most h. Computing the convex hull for each set using, e.g., Graham Scan takes $O(h \log h)$.

This results in $\lceil n/h \rceil$ (potentially overlapping) convex hulls and takes $O(\lceil n/h \rceil \cdot h \log h) = O(n \log h)$; see Fig. 5.

Phase 2 essentially applies Jarvis's March. Starting at the lowest leftmost point, it wraps a string around the set of convex hulls, i.e., for each hull it computes the proper tangent to the current point and chooses the tangent with the best angle in order to obtain the next point on the final convex hull. Computing the tangent for a hull of size h takes $O(\log h)$, which must be done for each of the $\lceil n/h \rceil$ hulls in each of the h rounds. Thus, the running time is $O(h \cdot \lceil n/h \rceil \log h) = O(n \log h)$.

Because h is not known, the algorithm does several such rounds for increasing values of h, until h is determined. In the initial round, it starts with a very small $h_0 = 4 = 2^{2^0}$ and continues with $h_i = 2^{2^i}$ in round i. As long as $h_i < h$, phase 1 is very quick. The second phase stops with an incomplete hull, knowing that h_i is still too small. That is, round i costs at most $n \log(h_i) = n2^i$. The algorithm terminates as soon as $h_i > h$. Thus, in total we obtain $\lceil \log \log h \rceil$ rounds. Therefore, the total

Convex Hulls, Fig. 5
Initial step of Jarvis's
March in the first round of
Chan's algorithm
($h_0 = 4$). Starting at p_1,
the algorithm computes
tangents to each convex
hull (indicated in different
colors) and selects the first
tangent in
counterclockwise order

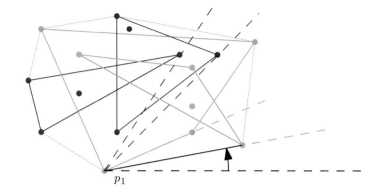

cost is:

$$O\left(\sum_{t=1}^{\lceil \log \log h \rceil} n2^t \right) = O(n2^{\lceil \log \log h \rceil + 1}) = O(n \log h).$$

(2)

Implementation

Like many geometric algorithms, the computation of the convex hull can be very sensitive to inconsistencies, due to rounding errors [6]. A well-maintained collection of exact implementations that eliminates problems due to rounding errors can be found in CGAL, the Computational Geometry Algorithms Library [1].

Cross-References

▶ Engineering Geometric Algorithms
▶ Robust Geometric Computation

Recommended Reading

1. CGAL (2014) Computational geometry algorithms library. http://www.cgal.org
2. Chan T (1996) Optimal output-sensitive convex hull algorithms in two and three dimensions. Discret Comput Geom 16(4):361–368. doi:10.1007/BF02712873
3. Devadoss SL, O'Rourke J (2011) Discrete and computational geometry. Princeton University Press, Princeton
4. Graham RL (1972) An efficient algorithm for determining the convex hull of a finite planar set. Inf Process Lett 1(4):132–133
5. Jarvis R (1973) On the identification of the convex hull of a finite set of points in the plane. Information Process Lett 2(1):18–21. doi:10.1016/0020-0190(73)90020-3
6. Kettner L, Mehlhorn K, Pion S, Schirra S, Yap CK (2004) Classroom examples of robustness problems in geometric computations. In: Proceedings of the 12th annual European symposium on algorithms, vol 3221, pp 702–713
7. O'Rourke J (1998) Computational geometry in C, 2nd edn. Cambridge University Press, New York
8. Preparata FP, Hong SJ (1977) Convex hulls of finite sets of points in two and three dimensions. Commun ACM 20(2):87–93
9. Sack JR, Urrutia J (eds) (2000) Handbook of computational geometry. North-Holland, Amsterdam

Coordinated Sampling

Edith Cohen
Tel Aviv University, Tel Aviv, Israel
Stanford University, Stanford, CA, USA

Keywords

Bottom-k sampling; Coordinated random sampling; Monotone estimation problems; Permanent random numbers

Years and Authors of Summarized Original Work

1972; Brewer, Early, Joice
1997; Cohen
1997; Broder

2013; Cohen, Kaplan
2014; Cohen

Problem Definition

Data is often sampled as a means of addressing resource constraints on storage, bandwidth, or processing – even when we have the resources to store the full data set, processing queries exactly over the data can be very expensive, and we therefore may opt for approximate fast answers obtained from the much smaller sample.

Our focus here is on data sets that have the form of a set of *keys* from some universe and multiple *instances*, which are assignments of nonnegative values to keys. We denote by v_{hi} the value of key h in instance i. Examples of data sets with this form include measurements of a set of parameters; snapshots of a state of a system; logs of requests, transactions, or activity; IP flow records in different time periods; and occurrences of terms in a set of documents. Typically, this matrix is very sparse – the vast majority of entries are $v_{ih} = 0$.

The sampling algorithms we apply can scan the full data set but retain the values v_{hi} of only a subset of the entries (h, i). From the sample, we would like to estimate functions (or statistics) specified over the original data. In particular, many common queries can be expressed as (or as a function of) the sum $\sum_{h \in H} f(v_{h\cdot})$ over selected keys $h \in H$ of a basic function f applied to the values of the key h in one or more instances. These include domain (subset) queries $\sum_{h \in H} v_{hi}$ which are the total weight of keys H in instance i, L_p distances between instances i, j which use $f(v_{h\cdot}) \equiv |v_{hi} - v_{hj}|^p$, one-sided distances which use $f(v_{h\cdot}) \equiv \max\{0, v_{hi} - v_{hj}\}^p$, and sums of quantiles (say maximum or median) of the tuple $(v_{h1}, \ldots, v_{hr})$.

The objective is to design a family of sampling scheme that is suitable for one or more query types. When the sampling scheme is specified, we are interested in designing estimators that use the information in the sample in the "best" way. The estimators are functions $\hat{f}$ that are applied to the sample and return an approximate answer to the query.

When sampling a single-instance i and aiming for approximation quality on domain queries and on sparse or skewed data, we use a *weighted sample*, meaning that the inclusion probability of an entry (h, i) in the sample depends (usually is increasing) with its weight v_{hi}. In particular, zero entries are never sampled. Two popular weighted sampling schemes are Poisson sampling, where entries are sampled independently, and bottom-k (order) sampling [12, 35, 36]. It is convenient for our purposes to specify these sampling schemes through a *rank function*, $r : [0, 1] \times V \to \mathbb{R}$, which maps seed-value pairs to a number $r(u, v)$ that is nonincreasing with u and nondecreasing with v. For each item h, we draw a *seed* $u(h) \sim U[0, 1]$ uniformly at random and compute the rank value $r(u(h), v_{hi})$. A Poisson sample includes a key $h \iff r(u(h), v_{hi}) \geq T_{hi}$, where T_{hi} are fixed thresholds. A bottom-k sample includes the k items with the highest ranks. (The term bottom-k is due to equivalently using the inverse rank function and lowest k ranks [12–14, 35, 36].)

Specifically, Poisson probability proportional to size (PPS) [28] samples include each key with probability proportion to its value. They are specified using the rank function $r(u, v) = v/u$ and a fixed $T_i \equiv T_{hi}$ across all keys in the instance. Priority (sequential Poisson) samples [22, 33, 38] are bottom-k samples utilizing the PPS ranks $r(u, v) = v/u$, and successive weighted samplings without replacement [12, 23, 35] are bottom-k samples with the rank function $r(u, v) = -v/\ln(1 - u)$. All these sampling algorithms can be easily implemented when the data is streamed or distributed.

Queries over a single instance can be estimated using inverse probabilities [29]. For each sampled key h, we can compute the probability $q(h)$ that it is included in the sample. With Poisson sampling, this probability is that of $u \sim U[0, 1]$ satisfying $r(u(h), v_{hi}) \geq T_{hi}$. With bottom-$k$ sampling, we use a conditioned version [13, 22, 38]: The probability $q(h)$ is that of $r(u(h), v_{hi})$ being larger than the kth largest value among $r(u(y), v_{yi})$, where y are all keys other than h and $u(y)$ is fixed to be as in the current sample. Note that this threshold value is available to us from the sample, and hence

keys:	1	2	3	4	5	6	7	8
Instance1:	1	3	2	0	1	4	0	1
Instance2:	0	1	3	2	0	1	2	3

PPS sampling probabilities for T=4 (sample of expected size 3):

Instance1:	0.25 0.75 0.50 0.00 0.25 1.00 0.00 0.25
Instance2:	0.00 0.25 0.75 0.50 0.00 0.25 0.50 0.75

Coordinated Sampling, Fig. 1 Two instances with 8 keys and respective PPS sampling probabilities for threshold value 4, so a key with value v is sampled with probability $\min\{1, v/4\}$. To obtain two coordinated PPS samples of the instances, we associate an independent $u(i) \sim U[0, 1]$ with each key $i \in [8]$. We then sample $i \in [8]$ in instance $h \in [2]$ if and only if $u(i) \le v_{hi}/4$, where v_{hi} is the value of i in instance h. When coordinating the samples this way, we make them as similar as possible. In the example, key 8 will always (for any drawing of seeds) be sampled in instance 2 if it is sampled in instance 1 and vice versa for key 2

$q(h)$ can be computed for each sampled key. We then estimate $\sum_{h \in H} f(v_{hi})$ using the sum over sampled keys that are in H of $f(v_{hi})/q(h)$. This estimator is a *sum* estimator in that it is applied separately for each key: $\hat{f}(h) = 0$ when the key is not sampled and is $\hat{f}(h) = f(v_{hi})/q(h)$ otherwise. When the sampling scheme is such that $q(h) > 0$ whenever $f(v_{hi}) > 0$, this estimator is unbiased. Moreover, this is also the optimal sum estimator, in terms of minimizing variance. We note that tighter estimators can be obtained when the total weight of the instance is available to the estimator [13].

functions $u(h)$, where the only requirement for our purposes is uniformity and pairwise independence.

Coordinated sampling has the property that the samples of different instances are more similar when the instances are more similar, a property also known as Locality Sensitive Hashing (LSH) [26, 30, 31]. Figure 1 contains an example data set of two instances and the PPS sampling probabilities of each item in each instance. When the samples are coordinated, a key sampled in one instance is always sampled in the instance with a higher inclusion probability.

What Is Sample Coordination?

When the data has multiple instances, we distinguish between a data source that is *dispersed*, meaning that different entries of each key occur in different times or locations and *co-located* if all entries occur together. These scenarios [17] impose different constraints on the sampling algorithm. In particular, with co-located data, it is easy to include the complete tuple $v_{h\cdot}$ of each key h that is sampled in at least one instance, whereas with dispersed data, we would want the sampling of one entry v_{hi} not to depend on the values v_{hj} in other instances j.

When sampling, we can redraw a fresh set of random seed values $u(h)$ for each instance, which results in the samples being independent. Samples of different instances are *coordinated* when the set of seeds $u(h)$ is common in all instances. Scalable sharing of seeds when data is dispersed is facilitated through random hash

Why Use Coordination?

Co-located Data
With co-located data, coordination allows us to minimize the sample size, which is the (expected) total number of included keys, while ensuring that the sample "contains" a desired Poisson or bottom-k sample of each instance.

For a key h, we can consider its respective inclusion probabilities in each of the instances (for bottom-k, inclusion may be conditioned on seeds of other keys). With coordination, the inclusion probability in the combined sample, $q(h)$, is always the maximum of these probabilities. Estimation with these samples is easy: We estimate $\sum_{h \in H} f(v_{h\cdot})$ using the Horvitz-Thompson estimator (inverse probabilities) [29]. The estimate is the sum $f(v_{h\cdot})/q(h)$ over sampled keys that are in H. Since the complete tuple $v_{h\cdot}$ is available for each sampled key, we can compute $f(v_{h.})$, $q(h)$, and thus the estimate.

When the query involves entries from a single-instance i, the variance of this estimate is at most that obtained from a respective sample of i. This is because the inclusion probability $q(h)$ of each key is at least as high as in the respective single-instance sample. Therefore, by coordinating the samples, we minimize the total number of keys included while ensuring estimation quality with respect to each instance.

Dispersed Data

With dispersed data, coordination is useful when we are interested in both domain queries over a single instance at a time and some queries that involve complex relation, such as similarity queries between multiple instances. Estimation of more complex relations, however, can be much more involved. Intuitively, this is because the sample can provide us with partial information on $f(v_{h.})$ that is short of the exact value but still lower bounds it by a positive amount. Therefore, in this case, inverse probability estimators may not be optimal or even well defined – there could be zero probability of knowing the exact value, but the function f can have a nonnegative unbiased estimator. In the sequel, we overview estimators that are able to optimally use the available information.

Implicit Data

Another setting where coordination arises is when the input is not explicit, for example, expressed as relations in a graph, and coordinated samples can be obtained much more efficiently than independent samples. In this case, we work with coordinated samples even when we are interested in queries that are approximated well with independent samples.

Key Results

We now overview results on estimators that are applied to coordinated samples of dispersed data.

We first observe that coordinated PPS and bottom-k samples are mergeable, meaning that a respective sample of the union, or more generally,

of a new instance whose weight of each key is the coordinate-wise maxima of several instances can be computed from the individual samples of each instance. This makes some estimation problems very simple. Even in these cases, however, better estimators of cardinalities of unions and intersections (key-wise maxima or minima) can be computed when we consider all sampled keys in the two sets rather than just the sample of the union. Such estimators for the 0/1 case are presented in [14] and for general nonnegative weights in [17].

The general question is to derive estimators for an arbitrary function $f \geq 0$ with respect to a coordinated sampling scheme. This problem can be formalized as a Monotone Estimation Problem (MEP) [9]: The smaller the seed $u(h) \in U[0, 1]$ is, the more information we have on the values $v_{h.}$ and therefore on $f(v_{h.})$. We are interested in deriving estimators $\hat{f}$ that are nonnegative; this is because we are interested in nonnegative functions and estimates should be from the same range. We also seek unbiasedness, because we are ultimately estimating a sum over many keys, and bias accumulates even when sampling of different keys are independent. Other desirable properties are finite variance (for any $v_{h.}$ in the domain) or bounded estimates. A complete characterization of functions f for which estimators with subsets of these properties exist is given in [15].

We are also interested in deriving estimators that are *admissible*. Admissibility is Pareto optimality, meaning that any other estimator with lower variance on some data would have higher variance on another. Derivations of admissible estimators (for any MEP for which an unbiased nonnegative estimator exists) are provided in [9]. Of particular interest is the L* estimator, which is the unique admissible monotone estimator. By monotone, we mean that when there is more information, that is, when $u(h)$ is higher, the estimate $\hat{f}$ is at least as high.

A definition of *variance competitiveness* of MEP estimators is provided in [15]. The competitive ratio of an estimator $\hat{f}$ is the maximum, over all possible inputs (data values) $v_{h.}$, of the ratio of the integral of the square of $\hat{f}$ to the minimum possible by a nonnegative unbiased estimator. It turns out that the L* has ratio of at most 4 on any

MEP for which a nonnegative unbiased estimator with finite variances exists [9]. Two interesting remaining open problems, partially addressed in [10], are to design an estimator with minimum ratio for a given MEP and also to bound the maximum minimum ratio over all MEPs that admit an unbiased nonnegative estimator with finite variance.

Applications

We briefly discuss the history and some applications of coordinated sampling.

Sample coordination was proposed in 1972 by Brewer, Early, and Joice [2], as a method to maximize overlap and therefore minimize overhead in repeated surveys [34, 36, 37]: The values of keys change, and therefore, there is a new set of PPS sampling probabilities. With coordination, the sample of the new instance is as similar as possible to the previous sample, and therefore, the number of keys that need to be surveyed again is minimized. The term permanent random numbers (PRN) is used in the statistics literature for sample coordination.

Coordination was subsequently used by computer scientists to facilitate efficient processing of large data sets, as estimates obtained over coordinated samples are much more accurate than possible with independent samples [1, 3–6, 8, 13, 14, 16, 17, 21, 24, 25, 27, 32].

In some applications, the representation of the data is not explicit, and coordinated samples are much easier to compute than independent samples. One such example is computing all-distance sketches, which are coordinated samples of (all) d-neighborhoods of all nodes in a graph [6, 7, 11–13, 32]. These sketches support centrality and similarity and influence queries useful in the analysis of massive graph data sets such as social networks or Web graphs [18–20].

Recommended Reading

1. Beyer KS, Haas PJ, Reinwald B, Sismanis Y, Gemulla R (2007) On synopses for distinct-value estimation under multiset operations. In: SIGMOD, Beijing. ACM, pp 199–210
2. Brewer KRW, Early LJ, Joyce SF (1972) Selecting several samples from a single population. Aust J Stat 14(3):231–239
3. Broder AZ (1997) On the resemblance and containment of documents. In: Proceedings of the compression and complexity of sequences, Salerno. IEEE, pp 21–29
4. Broder AZ (2000) Identifying and filtering near-duplicate documents. In: Proceedings of the 11th annual symposium on combinatorial pattern matching, Montreal. LNCS, vol 1848. Springer, pp 1–10
5. Byers JW, Considine J, Mitzenmacher M, Rost S (2004) Informed content delivery across adaptive overlay networks. IEEE/ACM Trans Netw 12(5):767–780
6. Cohen E (1997) Size-estimation framework with applications to transitive closure and reachability. J Comput Syst Sci 55:441–453
7. Cohen E (2013) All-distances sketches, revisited: HIP estimators for massive graphs analysis. Tech. Rep. cs.DS/1306.3284, arXiv http://arxiv.org/abs/1306.3284
8. Cohen E (2014) Distance queries from sampled data: accurate and efficient. In: ACM KDD, New York. Full version: http://arxiv.org/abs/1203.4903
9. Cohen E (2014) Estimation for monotone sampling: competitiveness and customization. In: ACM PODC, Paris. http://arxiv.org/abs/1212.0243, full version http://arxiv.org/abs/1212.0243
10. Cohen E (2014) Variance competitiveness for monotone estimation: tightening the bounds. Tech. Rep. cs.ST/1406.6490, arXiv http://arxiv.org/abs/1406.6490
11. Cohen E, Kaplan H (2007) Spatially-decaying aggregation over a network: model and algorithms. J Comput Syst Sci 73:265–288. Full version of a SIGMOD 2004 paper
12. Cohen E, Kaplan H (2007) Summarizing data using bottom-k sketches. In: ACM PODC, Portland
13. Cohen E, Kaplan H (2008) Tighter estimation using bottom-k sketches. In: Proceedings of the 34th international conference on very large data bases (VLDB), Auckland. http://arxiv.org/abs/0802.3448
14. Cohen E, Kaplan H (2009) Leveraging discarded samples for tighter estimation of multiple-set aggregates. In: ACM SIGMETRICS, Seattle
15. Cohen E, Kaplan H (2013) What you can do with coordinated samples. In: The 17th international workshop on randomization and computation (RANDOM), Berkeley. Full version: http://arxiv.org/abs/1206.5637
16. Cohen E, Wang YM, Suri G (1995) When piecewise determinism is almost true. In: Proceedings of the pacific rim international symposium on fault-tolerant systems, Newport Beach, pp 66–71
17. Cohen E, Kaplan H, Sen S (2009) Coordinated weighted sampling for estimating aggregates over multiple weight assignments. In: Proceedings of the VLDB endowment, Lyon, France, vol 2(1–2). Full version: http://arxiv.org/abs/0906.4560

18. Cohen E, Delling D, Fuchs F, Goldberg A, Goldszmidt M, Werneck R (2013) Scalable similarity
estimation in social networks: closeness, node labels,
and random edge lengths. In: ACM COSN, Boston

19. Cohen E, Delling D, Pajor T, Werneck RF (2014)
Sketch-based influence maximization and computation: scaling up with guarantees. In: ACM CIKM,
Shanghai. http://research.microsoft.com/apps/pubs/?
id=226623, full version http://research.microsoft.
com/apps/pubs/?id=226623

20. Cohen E, Delling D, Pajor T, Werneck RF (2014)
Timed influence: computation and maximization.
Tech. Rep. cs.SI/1410.6976, arXiv http://arxiv.org/
abs/1410.06976

21. Das A, Datar M, Garg A, Rajaram S (2007) Google
news personalization: scalable online collaborative
filtering. In: WWW, Banff, Alberta, Canada

22. Duffield N, Thorup M, Lund C (2007) Priority sampling for estimating arbitrary subset sums. J Assoc
Comput Mach 54(6)

23. Efraimidis PS, Spirakis PG (2006) Weighted random sampling with a reservoir. Inf Process Lett
97(5):181–185

24. Gibbons PB (2001) Distinct sampling for highly-
accurate answers to distinct values queries and event
reports. In: International conference on very large
databases (VLDB), Roma, pp 541–550

25. Gibbons P, Tirthapura S (2001) Estimating simple
functions on the union of data streams. In: Proceedings of the 13th annual ACM symposium on
parallel algorithms and architectures, Crete Island.
ACM

26. Gionis A, Indyk P, Motwani R (1999) Similarity
search in high dimensions via hashing. In: Proceedings of the 25th international conference on very large
data bases (VLDB'99), Edinburgh

27. Hadjieleftheriou M, Yu X, Koudas N, Srivastava D
(2008) Hashed samples: selectivity estimators for set
similarity selection queries. In: Proceedings of the
34th international conference on very large data bases
(VLDB), Auckland

28. Hájek J (1981) Sampling from a finite population.
Marcel Dekker, New York

29. Horvitz DG, Thompson DJ (1952) A generalization
of sampling without replacement from a finite universe. J Am Stat Assoc 47(260):663–685

30. Indyk P (2001) Stable distributions, pseudorandom
generators, embeddings and data stream computation.
In: Proceedings of the 41st IEEE annual symposium
on foundations of computer science, Redondo Beach.
IEEE, pp 189–197

31. Indyk P, Motwani R (1998) Approximate nearest
neighbors: towards removing the curse of dimen-
sionality. In: Proceedings of the 30th annual ACM
symposium on theory of computing, Texas. ACM, pp
604–613

32. Mosk-Aoyama D, Shah D (2006) Computing separable functions via gossip. In: ACM PODC, Denver

33. Ohlsson E (1998) Sequential poisson sampling. J Off
Stat 14(2):149–162

34. Ohlsson E (2000) Coordination of PPS samples over
time. In: The 2nd international conference on estab-
lishment surveys. American Statistical Association,
pp 255–264

35. Rosén B (1972) Asymptotic theory for successive
sampling with varying probabilities without replacement, I. Ann Math Stat 43(2):373–397. http://www.
jstor.org/stable/2239977

36. Rosén B (1997) Asymptotic theory for order sampling. J Stat Plan Inference 62(2):135–158

37. Saavedra PJ (1995) Fixed sample size PPS approximations with a permanent random number. In: Proceedings of the section on survey research methods, Alexandria. American Statistical Association,
pp 697–700

38. Szegedy M (2006) The DLT priority sampling is es-
sentially optimal. In: Proceedings of the 38th annual
ACM symposium on theory of computing, Seattle.
ACM

Counting by ZDD

Shin-ichi Minato
Graduate School of Information Science and
Technology, Hokkaido University, Sapporo,
Japan

Keywords

BDD; Binary decision diagram; Data compression; Dynamic programming; Enumeration;
Graph algorithm; Indexing; ZDD; Zero-
suppressed BDD

Years and Authors of Summarized Original Work

1986; R. E. Bryant
1993; S. Minato
2009; D. E. Knuth
2013; S. Minato
2013; T. Inoue, H. Iwashita, J. Kawahara, S.
Minato

Problem Definition

We consider a framework of the problems to
enumerate all the subsets of a given graph, each

subset of which satisfies a given constraint. For example, enumerating all Hamilton cycles, all spanning trees, all paths between two vertices, all independent sets of vertices, etc. When we assume a graph $G = (V, E)$ with the vertex set $V = \{v_1, v_2, \ldots, v_n\}$ and the edge set $E = \{e_1, e_2, \ldots, e_m\}$, a graph enumeration problem is to compute a subset of the power set 2^E (or 2^V), each element of which satisfies a given constraint. In this model, we can consider that each solution is a combination of edges (or vertices), and the problem is how to represent the set of solutions and how to generate it efficiently.

Constraints

Any kind of constraint for the graph edges and vertices can be considered. For example, we consider to enumerate all the simple (self-avoiding) paths connecting the two vertices s and t on the graph shown in Fig. 1. The constraint is described as:

1. At a terminal vertex (s and t), only one edge is selected and connected to the vertex.
2. At the other vertices, none or just two edges are selected and connected to the vertex, respectively.
3. The set of selected edges forms a connected component.

In this example, the set of solutions can be represented as a combination of the edges, $\{e_1 e_3 e_5, \ e_1 e_4, \ e_2 e_3 e_4, \ e_2 e_5\}$.

Key Results

A binary decision diagram (BDD) is a representation of a Boolean function, one of the most basic models of discrete structures. After the epoch-making paper [1] by Bryant in 1986, BDD-based

methods have attracted a great deal of attention. BDD was originally invented for the efficient Boolean function manipulation required in VLSI logic design, but Boolean functions are also used for modeling many kinds of combinatorial problems. Zero-suppressed BDD (ZDD) [8] is a variant of BDD, customized for manipulating "sets of combinations." ZDDs have been successfully applied not only for VLSI design but also for solving various combinatorial problems, such as constraint satisfaction, frequent pattern mining, and graph enumeration.

Recently, D. E. Knuth presented a surprisingly fast algorithm "Simpath" [7] to construct a ZDD which represents all the paths connecting two points in a given graph structure. This work is important because many kinds of practical problems are efficiently solved by some variations of this algorithm. We generically call such ZDD construction method "frontier-based methods."

BDDs/ZDDs for Graph Enumeration

A binary decision diagram (BDD) is a graph representation for a Boolean function, developed for VLSI design. A BDD is derived by reducing a binary decision tree, which represents a decision-making process by the input variables. If we fix the order of input variables and apply the following two reduction rules, then we have a compact canonical form for a given Boolean function:

1. Delete all redundant nodes whose two edges have the same destination.
2. Share all equivalent nodes having the same child nodes and the same variable.

The compression ratio achieved by using a BDD instead of a decision tree depends on the property of Boolean function to be represented, but it can be 10–100 times in some practical cases.

A zero-suppressed BDD (ZDD) is a variant of BDD, customized for manipulating sets of combinations. This data structure was first introduced by Minato [8]. ZDDs are based on the special reduction rules different from ordinary ones, as follows:

Counting by ZDD, Fig. 1
Example of a graph

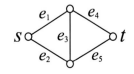

1. Delete all nodes whose 1-edge directly points to the 0-terminal node, but do not delete the nodes which were deleted in ordinary BDDs.

This new reduction rule is extremely effective, when it is applied to a set of sparse combinations. If each item appears in 1 % of combinations in average, ZDDs are possibly more compact than ordinary BDDs, by up to 100 times. Such situations often appear in real-life problems, for example, in a supermarket, the number of items in a customer's basket is usually much less than all the items displayed there. Because of such an advantage, ZDDs are now widely recognized as the most important variant of BDD.

ZDDs can be utilized for enumerating and indexing the solutions of a graph problem. For example, Fig. 2 shows the ZDD enumerating all the simple paths of the graph same as Fig. 1. The ZDD has four paths from the root node to the 1-terminal node, and each path corresponds to the solution of the problem, where $e_i = 1$ means to use the edge e_i and $e_i = 0$ means not to use e_i.

Frontier-Based Method

In 2009, Knuth published the surprisingly fast algorithm "Simpath" [7] (Vol. 4, Fascicle 1, p. 121, or p. 254 of Vol. 4A) to construct a ZDD which represents all the simple (or self-avoiding) paths connecting two points s and t in a given graph (not necessarily the shortest ones but

ones not passing through the same point twice). This work is important because many kinds of practical problems can be efficiently solved by some variations of this algorithm. Knuth provides his own C source codes on his web page for public access, and the program is surprisingly fast. For example, in a 14×14 grid graph (420 edges in total), the number of self-avoiding paths between opposite corners is exactly 227449714676812739631826459327989863387 613323440 ($\approx 2.27 \times 10^{47}$) ways. By applying the Simpath algorithm, the set of paths can be compressed into a ZDD with only 144759636 nodes, and the computation time is only a few minutes.

The Simpath algorithm is minutely written in Knuth's book, and his source codes are also provided, but it is not easy to read. The survey paper [9] will be helpful for understanding the basic mechanism of the Simpath algorithm.

The Simpath algorithm belongs to the method of dynamic programming, by scanning the given graph from left to right like a moving frontier line. If the frontier grows larger in the computation process, more intermediate states appear and more computation time is required. Thus, it is important to keep the frontier small. The maximum size of the frontier depends on the given graph structures and the order of the edges. Planar and narrow graphs tend to have small frontier.

Knuth described in his book [7] that the Simpath algorithm can easily be modified to generate not only $s - t$ paths but also Hamilton paths, directed paths, some kinds of cycles, and many other problems by slightly changing the mate data structure. We generically call such ZDD construction method "frontier-based methods."

Applications

Here we list graph problems which can be enumerated and indexed by a ZDD using a frontier-based method.

All $s - t$ paths, $s - t$ paths with length k, k-pairs of $s - t$ paths, all cycles, cycles with length k, Hamilton paths/cycles, directed

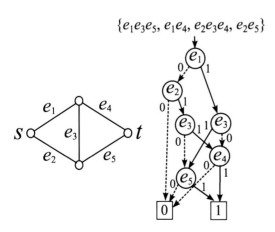

Counting by ZDD, Fig. 2 ZDD representing the paths from s to t

paths/cycles, all connected components, spanning trees/forests, Steiner trees, all cutsets, k-partitioning, calculating probability of connectivity, all cliques, all independent sets, graph colorings, tilings, and perfect/imperfect matching.

These problems are strongly related to many kinds of real-life problems. For example, path enumeration is of course important in geographic information systems and is also used for dependency analysis of a process flow chart, fault analysis of industrial systems, etc. Recently, Inoue et al. [5] discussed the design of electric power distribution systems. Such civil engineering systems are usually near to planar graphs, so the frontier-based method is very effective in many cases. They succeeded in generating a ZDD to enumerate all the possible switching patterns in a realistic benchmark of an electric power distribution system with 468 switches. The obtained ZDD represents as many as 10^{60} of the valid switching patterns, but the actual ZDD size is less than 100 MB, and computation time is around 30 min. After generating the ZDD, all valid switching patterns are compactly represented, and we can efficiently discover the switching patterns with maximum, minimum, and average cost. We can also efficiently apply additional constraints to the current solutions. In this way, frontier-based methods can be utilized for many kinds of real-life problems.

Open Problems

Frontier-based method is a general framework of the algorithm, and we have to develop particular algorithm for enumerating graphs to satisfy a given constraint. It is sometimes time consuming, and it is not clearly understood which kind of graphs can be generated easily and which are hard or impossible.

Experimental Results

It is an interesting problem how large n is possible to count the number of simple paths included in an $n \times n$ grid graph with s and t at the opposite corners. We have worked for this problems and succeeded in counting the total number of self-avoiding $s - t$ paths for the 26×26 grid graph. The number is exactly:

17369931586279272931175440421236498900372229588288140604663703720910342413276134762789218193498006107082296223143380491348290026721931129627708738890853908108906396.

This is the current world record and is officially registered in the On-Line Encyclopedia of Integer Sequences [10] in November 2013. The detailed techniques for solving larger problems are presented in the report by Iwashita et al. [6]

A Related YouTube Video

In 2012, Minato supervised a short educational animation video (Fig. 3). The video is mainly designed for junior high school to college students, to show the power of combinatorial explosion and the state-of-the-art techniques for solving such hard problems. This video uses the simple path enumeration problem for $n \times n$ grid graphs. The story is that the teacher counts the total number of paths for children starting from $n = 1$, but she will be faced with a difficult situation, since the number grows unbelievably fast. She would spend 250,000 years to count the paths for the 10×10 grid graph by using a supercomputer if she used a naive method. The story ends by telling that a state-of-the-art algorithm can finish the same problem in a few seconds. The video is now shown in YouTube [2] and received more than 1.5 million views, which is an extraordinary case in the scientific educational contents. We hear that Knuth also enjoyed this video and shared it to several of his friends.

Graphillion: Open Software Library

The above techniques of data structures and algorithms have been implemented and published as an open software library, named "Graphillion"

Counting by ZDD, Fig. 3
Screenshots of the
animation video [2]

[3, 4]. Graphillion is a library for manipulating very large sets of graphs, based on ZDDs and frontier-based method. Traditional graph libraries maintain each graph individually, which causes poor scalability, while Graphillion handles a set of graphs collectively without considering individual graph. Graphillion is implemented as a Python extension in C++, to encourage easy development of its applications without introducing significant performance overhead.

URLs to Code and Data Sets

The open software library "Graphillion" can be found on the web page at http://graphillion. org, the YouTube video http://www.youtube. com/watch?v=Q4gTV4r0zRs, and the On-Line Encyclopedia of Integer Sequences (OEIS) on the self-avoiding path enumeration problem https:// oeis.org/A007764.

Recommended Reading

1. Bryant RE (1986) Graph-based algorithms for Boolean function manipulation. IEEE Trans Comput C-35(8):677–691
2. Doi S et al (2012) Time with class! let's count! (the art of 10^{64} – understanding vastness –). YouTube video, http://www.youtube.com/watch?v=Q4gTV4r0zRs
3. Inoue T et al (2013) Graphillion. http://graphillion. org/
4. Inoue T, Iwashita H, Kawahara J, Minato S (2013) Graphillion: ZDD-based software library for very large sets of graphs. In: Proceedings of the workshop on synthesis and simulation meeting and international interchange (SASIMI-2013), R4-6, Sapporo, pp 237–242
5. Inoue T, Takano K, Watanabe T, Kawahara J, Yoshinaka R, Kishimoto A, Tsuda K, Minato S,

Hayashi Y (2014) Distribution loss minimization with guaranteed error bound. IEEE Trans Smart Grid 5(1):102–111
6. Iwashita H, Nakazawa Y, Kawahara J, Uno T, Minato S (2013) Efficient computation of the number of paths in a grid graph with minimal perfect hash functions. TCS technical reports TCS-TR-A-10-64, Division of Computer Science, Hokkaido University
7. Knuth DE (2009) Bitwise tricks & techniques; binary decision diagrams. The art of computer programming, vol 4, fascicle 1. Addison-Wesley, Upper Saddle River
8. Minato S (1993) Zero-suppressed BDDs for set manipulation in combinatorial problems. In: Proceedings of 30th ACM/IEEE design automation conference (DAC'93), Dallas, pp 272–277
9. Minato S (2013) Techniques of BDD/ZDD: brief history and recent activity. IEICE Trans Inf Syst E96-D(7):1419–1429
10. OEIS-simpath (2013) Number of nonintersecting (or self-avoiding) rook paths joining opposite corners of an $n \times n$ grid, the on-line encyclopedia of integer sequences. https://oeis.org/A007764

Counting Triangles in Graph Streams

Madhav Jha
Sandia National Laboratories, Livermore, CA, USA
Zenefits, San Francisco, CA, USA

Keywords

Counting triangles; Streaming algorithms

Years and Authors of Summarized Original Work

2013; Jha, Seshadhri, Pinar

Problem Definition

A *graph stream* is a sequence of unordered pairs of elements from a set V implicitly describing an underlying graph G on vertex set. The unordered pairs represent edges of the graph. A *triangle* is a triple of vertices $u, v, w \in V$ which form a 3-clique, that is, every unordered pair of vertices of the set $\{u, v, w\}$ is connected by an edge. In this article, we investigate the problem of counting the number of triangles in an input graph G given as a graph stream. Furthermore, we restrict our attention to algorithms which are severely limited in total space (in particular, they cannot store the entire stream) and are allowed only a single scan over the stream of edges.

Next we describe the streaming setting more formally. Consider a sequence of *distinct* unordered pairs or, equivalently, edges $e_1, e_2, \ldots, e_m$ on the set V. Let G_t be the graph formed by the first t edges of the stream where $t \in \{1, \ldots, m\}$. Denoting the empty graph by G_0, we see that graph G_t is obtained from G_{t-1} by *inserting* edge e_t for all $t \in \{1, \ldots, m\}$. An edge $\{u, v\}$ of the stream implicitly introduces vertex labels u and v. New vertices are therefore implicitly added as new labels. In this article, we do not consider edge or vertex deletions, nor do we allow repeated edges. The problem of counting triangles even in this simple setting has received a lot of attention [1, 2, 6, 12–15, 17].

A streaming algorithm has a small working memory M and gets to scan the input stream in a sequential fashion at most a few times. In this article, we only consider algorithms which make a single pass over the input stream. Thus, the algorithm proceeds by sequentially reading each edge e_t from the input graph stream and updating its data structures in M using e_t. The algorithm cannot read an edge that has already passed again. (It may remember it in M.) Since the size of M is much smaller than m, the algorithm must work with only a "sketch" of the input graph stored in M. The streaming algorithm has access to random coins and typically maintains a random sketch of the input graph (e.g., a random subsample of the input edge stream).

The aim is to output an accurate estimate of the number of triangles in graph G_m. In fact, we require that the algorithm output a running estimate of the number of triangles T_t in graph G_t seen so far as it is reading the edge stream. It is also of interest to output an estimate of another quantity, related to the number of triangles, called *transitivity*. The transitivity of a graph, denoted κ, is the fraction of length-2 paths which form triangles. A path of length 2 is also called a *wedge*. A wedge $\{\{u, v\}, \{u, w\}\}$ is *open* (respectively, *closed*) if the edge $\{v, w\}$ is absent (respectively, present) in the graph. Every closed wedge is part of a triangle, and every triangle has exactly three closed wedges. This immediately gives a formula for transitivity: $\kappa = 3T/W$, where T and W are the total number of triangles and wedges in the graph, respectively. We use the subscript t, as in T_t, W_t, and κ_t, to denote corresponding quantities for graph G_t. The key result described in this article is a small space single-pass streaming algorithm for maintaining a running estimate for each of T_t, W_t, and κ_t.

Outline of the Rest of the Article

The bulk of this article is devoted to the description of results and algorithms of [12]. This is complemented by a section on related work where we briefly describe some of the other algorithms for triangle counting. This is followed by applications of triangle counting, an open problem, a section describing experimental results, and, finally, references used in this article.

Key Results

The main result (from [12]) presented in this article is a single pass, $O(m/\sqrt{T})$-space streaming algorithm to estimate the transitivity and the number of triangles of the input graph with small additive error.

The Algorithm of [12] and Its Analysis

The starting point of the algorithm is the wedge sampling idea from [19]. The transitivity of a graph is precisely the probability that a uniformly random wedge from the graph is closed. Thus, estimating transitivity amounts to approximating the bias of a coin flip simulated by the following probabilistic experiment: sample a wedge $\{\{u, v\}, \{u, w\}\}$ uniformly from the set of wedges and output "Heads" if it is closed and "Tails" otherwise. One can check whether a wedge is closed by checking if $\{v, w\}$ is an edge in the graph. If, in addition, we have an accurate estimate of W, the triangle count can be estimated using $T = \kappa/3 \cdot W$.

If we adopt the described strategy, we need a way to sample a wedge uniformly from the *graph stream*. While this task by itself appears rather challenging, we note that sampling an edge uniformly from the graph stream can be done easily via an adaptation of reservoir sampling. (See below for details.) Can we use an edge sampling primitive to sample wedges uniformly from a graph stream? This is exactly what [12] achieves. Before we describe the algorithm of [12], we present a key primitive which is also used in other works on counting triangles in graph streams.

A key Algorithmic Tool: A Reservoir of Uniform Edges

This algorithmic tool allows one to maintain a set R_t of k edges while reading the edge stream with the following guarantee for every t: each of the k edges in R_t is selected uniformly from the edges of G_t and all edges are mutually independent. The idea is to adapt the classic reservoir sampling [21]. More precisely, at the beginning of the stream, R_0 consists of k empty slots. While reading the edge stream, on observing edge e_t, do the following probabilistic experiment independently for each of the k slots of R_{t-1} to yield R_t: (i) sample a random number x uniformly from $[0, 1]$, and (ii) if $x \leq 1/t$, replace the slot with e_t. Otherwise, keep the slot unchanged.

How does a reservoir of edges R_t help in sampling wedges from the graph stream? When the edge reservoir R_t is large enough, there are many pairs of edges in R_t sharing a common vertex, yielding many wedges. Further, if the transitivity of the input graph is high, many of these wedges

will in fact be closed wedges, aka, triangles. This is the idea behind the algorithm of [12]. The key theoretical insight is that there is a *birthday paradox*-like situation here: many uniform edges result in sufficiently many "collisions" to form wedges. Recall (e.g., from Chap.II.3 of [10]) that the classic birthday paradox states that if we choose 23 random people, the probability that 2 of them share a birthday is at least 1/2. In our setting, edges correspond to people and "sharing a birthday" corresponds to sharing a common vertex.

A Key Analytical Tool: The Birthday Paradox

Suppose $R_1, \ldots, R_k$ are i.i.d. samples from the uniform distribution on the edges of G. Then for distinct $i, j \in [k]$, $\Pr(\{R_i, R_j\}$ forms a wedge$) = 2W/m^2$. By linearity of expectation, the expected number of expected wedges is $\binom{k}{2}(2W/m^2)$. In particular, setting k to be c times a large constant multiple of $m/\sqrt{W}$ results in expectation of at least c^2 wedges. Even better is the fact that these wedges are uniformly (but not independently) distributed in the set of all wedges. Similarly, one can show that when k is $\Omega(m/\sqrt{T})$, one obtains many closed wedges. This is the heart of the argument in the analysis of the algorithm of [12].

While we do not wish to present all the technical details of the algorithm of [12] here, we do provide some high-level ideas from [12]. The algorithm maintains a reservoir of edges R_t as explained above. Further, it maintains a set C_t of wedges $\{e_a, e_b\}$ with $e_a, e_b \in R_t$ such that (i) $\{e_a, e_b\}$ is a closed wedge in G_t and (ii) the closing edge appears after edges e_a and e_b in the stream. In other words, C_t are the wedges in R_t which can be *detected to be closed*. The algorithm outputs a random bit b_t whose expectation is close to κ_t.

For each edge e_t of the graph stream:

(a) Update reservoir R_t of k edges as described above.
(b) Let W_t be the set of wedges in R_t. Let N_t be the set of wedges in R_t which have e_t as their closing edge.

(c) Update C_t to $C_{t-1} \cap W_t \cup N_t$.

(d) If there are no wedges in R_t, output $b_t = 0$. Otherwise, sample a uniform wedge in R_t and output $b_t = 1$ if this wedge is in C_t and $b_t = 0$ otherwise.

The next theorem shows that the expectation of b_m is close to $\kappa/3$. Moreover, it shows that $|W_m|$ can be used to estimate W. Getting a good estimate on $\mathbf{E}[b_m]$ and W allows one to estimate T by multiplying the two estimates together. We note that while the theorem is stated for the final index $t = m$, it holds for any large enough t.

Theorem 1 ([12]) *Assume $W \geq m$ and fix $\beta \in (0, 1)$. Suppose $k = |R_m|$ is $\Omega(m/(\beta^3 \sqrt{T}))$. Set $\hat{W} = m^2|W_m|/(k(k-1))$. Then $|\kappa/3 - \mathbf{E}[b_m]| < \beta$ and $\Pr[|W - \hat{W}| < \beta W]$ are at least $1 - \beta$.*

Related Work

The problem of counting triangles in graphs has been studied extensively in a variety of different settings: exact, sampling, streaming, and MapReduce. We refer the reader to references in [12] for a comprehensive list of these works. Here we focus on a narrow topic: single-pass streaming algorithms for estimating triangle counts. In particular, we do not discuss streaming algorithms that make multiple passes (e.g., [11,13]) or algorithms that compute triangles incident on every vertex (e.g., Becchetti et al. [4]).

Bar-Yossef et al. [3] were the first to study the problem of triangle counting in the streaming setting. Since then there have been a long line of work improving the guarantees of the algorithm in various ways [1, 2, 6, 12–15, 17]. Specifically, Buriol et al. [6] gave an $O(mn/T)$ space algorithm. The algorithm maintains samples of the form $((u, v), w)$ where (u, v) is a uniform edge in the stream and w is a uniform node label other than u and v. The algorithm checks for presence of edges (u, w) and (v, w) to detect triangle. They also give an implementation of their algorithm which is practical but relative error in triangle estimates can be high. The $O(m/\sqrt{T})$ algorithm described in this algorithm is from Jha et al. [12].

In parallel with [12], Pavan et al. [17] independently gave an $O(m\Delta/T)$ space algorithm where Δ is the maximum degree of the graph. Their algorithm is based on a sampling technique they introduced called *neighborhood sampling*. Neighborhood sampling maintains edge pair samples of the form (r_1, r_2). In each pair, edge r_1 is sampled uniformly from the edges observed so far. Edge r_2 is sampled uniformly from the set $N(r_1)$ where $N(r_1)$ consists of the edges adjacent to edge r_1 that appear after edge r_1 in the stream. When r_2 is nonempty, the pair (r_1, r_2) forms a wedge which can be monitored to see if it forms a triangle. Observe that a triangle formed by edges e_{t_1}, e_{t_2}, and e_{t_3} appearing in this order is detected as a closed triangle with probability $1/|N(e_{t_1})| \cdot m$. Accounting for this bias by keeping track of the quantity $N(r_1)$ for each sample (r_1, r_2), one gets an unbiased estimator for triangle counts. In a recent work, Ahmed et al. [1] give practical algorithms for triangle counting which seem to empirically improve on some of the previous results. In this work, the authors present a generic technique called *graph sample and hold* and use it for estimating triangle counts. At a high level, graph sample and hold associates a nonzero probability p_{e_t} to each edge e_t which corresponds to the probability with which edge e_t is independently sampled. Importantly, the probability p_{e_t} may depend on the graph sampled so far. But the actual probability with which the edge is sampled is *recorded*. Now estimates about the original graph can be obtained from the sampled graph using the selection estimator. For example, the number of triangles is estimated by summing $1/(p_{e_{t_1}} \cdot p_{e_{t_2}} \cdot p_{e_{t_3}})$ over all sampled triangles $\{e_{t_1}, e_{t_2}, e_{t_3}\}$. This can be seen as a generalization of neighborhood sampling.

On the lower bound side, Braverman et al. [5] show that any single-pass streaming algorithm which gives a good *multiplicative* approximation of triangle counts must use $\Omega(m)$ bits of storage even if the input graph has $\Omega(n)$ triangles. This improves lower bounds from [3, 13]. For algorithms making a constant c number of passes, for every constant c, the lower bound is shown to be $\Omega(m/T)$ in the same work.

Applications

The number of triangles in a graph and the transitivity of a graph are important measures used widely in the study of networks across many different domains. For example, these measures appear in social science research [7, 8, 18, 22], in data mining applications such as spam detection and finding common topics on the World Wide Web [4, 9], and in bioinformatics for motif counting [16]. For a more detailed list of applications,

we point the reader to introductory sections of references in the related work.

Open Problems

Give a tight lower bound on the space required by a single-pass streaming algorithm for estimating the number of triangles in graph stream with additive error. The lower bound for multiplicative approximation is known to be $\Omega(m)$ [3, 5].

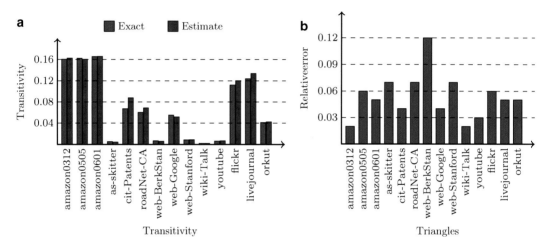

Counting Triangles in Graph Streams, Fig. 1 Output of STREAMING-TRIANGLES algorithm of [12] on a variety of real datasets while storing at most $40\,\mathrm{K}$ edges. (**a**) Gives the estimated transitivity values alongside their exact values. (**b**) Gives the relative error of STREAMING-TRIANGLES's estimates of triangles T. Observe that the relative error for T is mostly below 8 % and often below 4 %

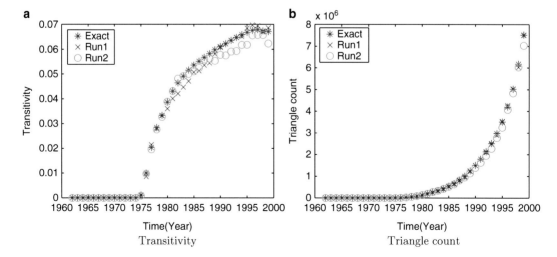

Counting Triangles in Graph Streams, Fig. 2 Real-time tracking of number of triangles and transitivities on cit-Patents (16 M edges), storing only 200 K edges (**a**) Transitivity. (**b**) Triangle count

Experimental Results

Figures 1 and 2 give a glimpse of experimental results from [12] on performance of algorithm STREAMING-TRIANGLES. Specifically, Fig. 1 shows the result of running the algorithm on a variety of graph datasets obtained from [20]. This includes run on graph datasets such as Orkut social network consisting of 200 M edges. The relative errors on κ and T are mostly less than 5 % (except for graphs with tiny κ). The storage used by the algorithm stated in terms of number of edges is at most 40 K. (The storage roughly corresponds to the size of edge reservoir used in the algorithm of [12] described in this article.)

An important aspect of the algorithm presented in [12] is that it can track the quantities κ_t and T_t for all values of t in real time. This is exhibited in Fig. 2.

Cross-References

▶ Data Stream Verification
▶ Graph Sketching
▶ Sliding Window Algorithms

Recommended Reading

1. Ahmed N, Duffield N, Neville J, Kompella R (2014) Graph sample and hold: a framework for big-graph analytics. In: The 20th ACM SIGKDD international conference on knowledge discovery and data mining (KDD '14), New York, 24–27 Aug 2014, pp 1446–1455. doi:10.1145/2623330.2623757. http://doi.acm.org/10.1145/2623330.2623757
2. Ahn KJ, Guha S, McGregor A (2012) Graph sketches: sparsification, spanners, and subgraphs. In: Proceedings of the 31st ACM SIGMOD-SIGACT-SIGART symposium on principles of database systems, PODS 2012, Scottsdale, 20–24 May 2012, pp 5–14. doi:10.1145/2213556.2213560. http://doi.acm.org/10.1145/2213556.2213560
3. Bar-Yossef Z, Kumar R, Sivakumar D (2002) Reductions in streaming algorithms, with an application to counting triangles in graphs. In: Proceedings of the thirteenth annual ACM-SIAM symposium on discrete algorithms, San Francisco, 6–8 Jan 2002, pp 623–632. http://dl.acm.org/citation.cfm?id=545381.545464
4. Becchetti L, Boldi P, Castillo C, Gionis A (2008) Efficient semi-streaming algorithms for local triangle counting in massive graphs. In: Proceedings of the 14th ACM SIGKDD international conference on knowledge discovery and data mining, Las Vegas, 24–27 Aug 2008, pp 16–24. doi:10.1145/1401890.1401898. http://doi.acm.org/10.1145/1401890.1401898
5. Braverman V, Ostrovsky R, Vilenchik D (2013) How hard is counting triangles in the streaming model? In: Proceedings of the 40th international colloquium automata, languages, and programming (ICALP 2013) Part I, Riga, 8–12 July 2013, pp 244–254. doi:10.1007/978-3-642-39206-1_21. http://dx.doi.org/10.1007/978-3-642-39206-1_21
6. Buriol LS, Frahling G, Leonardi S, Marchetti-Spaccamela A, Sohler C (2006) Counting triangles in data streams. In: Proceedings of the twenty-fifth ACM SIGACT-SIGMOD-SIGART symposium on principles of database systems, Chicago, 26–28 June 2006, pp 253–262. doi:10.1145/1142351.1142388. http://doi.acm.org/10.1145/1142351.1142388
7. Burt RS (2004) Structural holes and good ideas. Am J Soc 110(2):349–399. http://www.jstor.org/stable/10.1086/421787
8. Coleman JS (1988) Social capital in the creation of human capital. Am J Soc 94:S95–S120. http://www.jstor.org/stable/2780243
9. Eckmann JP, Moses E (2002) Curvature of co-links uncovers hidden thematic layers in the World Wide Web. Proc Nat Acad Sci (PNAS) 99(9):5825–5829. doi:10.1073/pnas.032093399
10. Feller W (1968) An Introduction to probability theory and applications, vol I, 3rd edn. John Wiley & Sons, New York
11. García-Soriano D, Kutzkov K (2014) Triangle counting in streamed graphs via small vertex covers. In: Proceedings of the 2014 SIAM international conference on data mining, Philadelphia, 24–26 Apr 2014, pp 352–360. doi:10.1137/1.9781611973440.40. http://dx.doi.org/10.1137/1.9781611973440.40
12. Jha M, Seshadri C, Pinar A (2013) A space efficient streaming algorithm for triangle counting using the birthday paradox. In: Proceedings of the 19th ACM SIGKDD international conference on knowledge discovery and data mining (KDD '13), Chicago. ACM, New York, pp 589–597. doi:10.1145/2487575.2487678. http://doi.acm.org/10.1145/2487575.2487678
13. Jowhari H, Ghodsi M (2005) New streaming algorithms for counting triangles in graphs. In: Proceedings of the 11th annual international conference computing and combinatorics (COCOON 2005), Kunming, 16–29 Aug 2005, pp 710–716. doi:10.1007/11533719_72. http://dx.doi.org/10.1007/11533719_72
14. Kane DM, Mehlhorn K, Sauerwald T, Sun H (2012) Counting arbitrary subgraphs in data streams. In: Proceedings of the 39th international colloquium

automata, languages, and programming (ICALP 2012), Warwick, 9–13 July 2012, pp 598–609. doi:10.1007/978-3-642-31585-5_53.   http://dx.doi.org/10.1007/978-3-642-31585-5_53

15. Kutzkov K, Pagh R (2014) Triangle counting in dynamic graph streams. In: Proceedings of the 14th Scandinavian symposium and workshops algorithm theory (SWAT 2014), Copenhagen, 2–4 July 2014, pp 306–318. doi:10.1007/978-3-319-08404-6_27. http://dx.doi.org/10.1007/978-3-319-08404-6_27

16. Milo R, Shen-Orr S, Itzkovitz S, Kashtan N, Chklovskii D, Alon U (2002) Network motifs: simple building blocks of complex networks. Science 298(5594):824–827

17. Pavan A, Tangwongsan K, Tirthapura S, Wu KL (2013) Counting and sampling triangles from a graph stream. PVLDB 6(14):1870–1881. http://www.vldb.org/pvldb/vol6/p1870-aduri.pdf

18. Portes A (1998) Social capital: its origins and applications in modern sociology. Ann Rev Soc 24(1):1–24. doi:10.1146/annurev.soc.24.1.1

19. Seshadhri C, Pinar A, Kolda TG (2014) Wedge sampling for computing clustering coecients and triangle counts on large graphs. Stat Anal Data Mining 7(4):294–307. doi:10.1002/sam.11224. http://dx.doi.org/10.1002/sam.11224

20. SNAP (2013) Stanford network analysis project. Available at http://snap.stanford.edu/

21. Vitter J (1985) Random sampling with a reservoir. ACM Trans Math Softw 11(1):37–57. doi:10.1145/3147.3165. http://doi.acm.org/10.1145/3147.3165

22. Welles BF, Devender AV, Contractor N (2010) Is a friend a friend? Investigating the structure of friendship networks in virtual worlds. In: Proceedings of the 28th international conference on human factors in computing systems (CHI 2010), extended abstracts volume, Atlanta, 10–15 April 2010, pp 4027–4032. doi:10.1145/1753846.1754097. http://doi.acm.org/10.1145/1753846.1754097

Count-Min Sketch

Graham Cormode
Department of Computer Science, University of Warwick, Coventry, UK

Keywords

Approximate counting; Frequent items; Sketch; Streaming algorithms

Years and Authors of Summarized Original Work

2004; Cormode, Muthukrishnan

Problem Definition

The problem of *sketching* a large mathematical object is to produce a compact data structure that approximately represents it. The Count-Min (CM) sketch is an example of a sketch that allows a number of related quantities to be estimated with accuracy guarantees, including point queries and dot product queries. Such queries are at the core of many computations, so the structure can be used in order to answer a variety of other queries, such as frequent items (heavy hitters), quantile finding, join size estimation, and more. Since the sketch can process updates in the form of additions or subtractions to dimensions of the vector (which may correspond to insertions or deletions or other transactions), it is capable of working over streams of updates, at high rates.

The data structure maintains the linear projection of the vector with a number of other random vectors. These vectors are defined implicitly by simple hash functions. Increasing the range of the hash functions increases the accuracy of the summary, and increasing the number of hash functions decreases the probability of a bad estimate. These trade-offs are quantified precisely below. Because of this linearity, CM sketches can be scaled, added, and subtracted, to produce summaries of the corresponding scaled and combined vectors.

Key Results

The Count-Min sketch was first proposed in 2003 [4], following several other sketch techniques, such as the Count sketch [2] and the AMS sketch [1]. The sketch is similar to a counting Bloom filter or multistage filter [7].

Count-Min Sketch, Fig. 1
Each item i is mapped to
one cell in each row of the
array of counts: when an
update of c to item i_t
arrives, c_t is added to each
of these cells

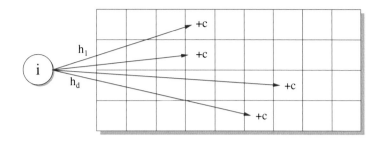

Data Structure Description

The CM sketch is simply an array of counters of width w and depth d, $CM[1,1] \ldots CM[d,w]$. Each entry of the array is initially zero. Additionally, d hash functions

$$h_1 \ldots h_d : \{1 \ldots n\} \to \{1 \ldots w\}$$

are chosen uniformly at random from a pairwise-independent family. Once w and d are chosen, the space required is fixed: the data structure is represented by wd counters and d hash functions (which can each be represented in $O(1)$ machine words [12]).

Update Procedure

A vector $\boldsymbol{a}$ of dimension n is described incrementally. Initially, $\boldsymbol{a}(0)$ is the zero vector, $\boldsymbol{0}$, so $a_i(0)$ is 0 for all i. Its state at time t is denoted $\boldsymbol{a}(t) = [a_1(t), \ldots a_i(t), \ldots a_n(t)]$. Updates to individual entries of the vector are presented as a stream of pairs. The tth update is (i_t, c_t), meaning that

$$a_{i_t}(t) = a_{i_t}(t-1) + c_t$$
$$a_{i'}(t) = a_{i'}(t-1) \qquad i' \neq i_t.$$

For convenience, the subscript t is dropped, and the current state of the vector simply referred to as $\boldsymbol{a}$. For simplicity of description, it is assumed here that although values of a_i increase and decrease with updates, each $a_i \geq 0$. However, the sketch can also be applied to the case where a_is can be less than zero with some increase in costs [4].

When an update (i_t, c_t) arrives, c_t is added to one count in each row of the Count-Min sketch; the counter is determined by h_j. Formally, given (i_t, c_t), the following modifications

are performed:

$$\forall 1 \leq j \leq d : CM[j, h_j(i_t)] \leftarrow CM[j, h_j(i_t)] + c_t.$$

This procedure is illustrated in Fig. 1. Because computing each hash function takes constant time, the total time to perform an update is $O(d)$, independent of w. Since d is typically small in practice (often less than 10), updates can be processed at high speed.

Point Queries

A *point query* is to estimate the value of an entry in the vector a_i. Given a query point i, an estimate is found from the sketch as $\hat{a}_i = \min_{1 \leq j \leq d} CM[j, h_j(i)]$. The approximation guarantee is that if $w = \lceil \frac{e}{\varepsilon} \rceil$ and $d = \lceil \ln \frac{1}{\delta} \rceil$, the estimate $\hat{a}_i$ obeys $a_i \leq \hat{a}_i$; and, with probability at least $1 - \delta$,

$$\hat{a}_i \leq a_i + \varepsilon \|\boldsymbol{a}\|_1.$$

Here, $\|\boldsymbol{a}\|_1$ is the L_1 norm of $\boldsymbol{a}$, i.e., the sum of the (absolute) values. The proof follows by using the Markov inequality to limit the error in each row and then using the independence of the hash functions to amplify the success probability.

This analysis makes no assumption about the distribution of values in $\boldsymbol{a}$. In many applications, there are Zipfian, or power law, distributions of item frequencies. Here, the (relative) frequency of the ith most frequent item is proportional to i^{-z}, for some parameter z, where z is typically in the range 1–3. Here, the skew in the distribution can be used to show a stronger space/accuracy trade-off: for a Zipf distribution with parameter z, the space required to answer point queries with error $\varepsilon \|a\|_1$ with probability at least $1 - \delta$ is given by $O(\varepsilon^{-\min\{1,1/z\}} \ln 1/\delta)$ [5].

Range, Heavy Hitter, and Quantile Queries

A *range query* is to estimate $\sum_{i=l}^{r} a_i$ for a range $[l \ldots r]$. For small ranges, the range sum can be estimated as a sum of point queries; however, as the range grows, the error in this approach also grows linearly. Instead, $\log n$ sketches can be kept, each of which summarizes a derived vector a^k where

$$a^k[j] = \sum_{i=j2^k}^{(j+1)2^k - 1} a_i$$

for $k = 1 \ldots \log n$. A range of the form $j2^k \ldots (j+1)2^k - 1$ is called a *dyadic range*, and any arbitrary range $[l \ldots r]$ can be partitioned into at most $2 \log n$ dyadic ranges. With appropriate rescaling of accuracy bounds, it follows that Count-Min sketches can be used to find an estimate $\hat{r}$ for a range query on $l \ldots r$ such that

$$\hat{r} - \varepsilon \|a\|_1 \leq \sum_{i=l}^{r} a_i \leq \hat{r}.$$

The right inequality holds with certainty, and the left inequality holds with probability at least $1 - \delta$. The total space required is $O(\frac{\log^2 n}{\varepsilon} \log \frac{1}{\delta})$ [4]. The closely related ϕ-*quantile query* is to find a point j such that

$$\sum_{i=1}^{j} a_i \leq \phi \|a\|_1 \leq \sum_{i=1}^{j+1} a_i.$$

Range queries can be used to (binary) search for a j which satisfies this requirement approximately (i.e., tolerates up to $\varepsilon \|a\|_1$ error in the above expression) given ϕ. The overall cost is space that depends on $1/\varepsilon$, with further log factors for the rescaling necessary to give the overall guarantee [4]. The time for each insert or delete operation and the time to find any quantile are logarithmic in n, the size of the domain.

Heavy hitters are those points i such that $a_i \geq \phi \|a\|_1$ for some specified ϕ. The range query

primitive based on Count-Min sketches can again be used to find heavy hitters, by recursively splitting dyadic ranges into two and querying each half to see if the range is still heavy, until a range of a single, heavy item is found. The cost of this is similar to that for quantiles, with space dependent on $1/\varepsilon$ and $\log n$. The time to update the data structure and to find approximate heavy hitters is also logarithmic in n. The guarantee is that every item with frequency at least $(\phi + \varepsilon) \|a\|_1$ is output, and with probability $1 - \delta$ no item whose frequency is less than $\phi \|a\|_1$ is output.

Inner Product Queries

The Count-Min sketch can also be used to estimate the inner product between two vectors. The inner product $a \cdot b$ can be estimated by treating the Count-Min sketch as a collection of d vectors of length w and finding the minimum inner product between corresponding rows of sketches of the two vectors. With probability $1 - \delta$, this estimate is at most an additive quantity $\varepsilon \|a\|_1 \|b\|_1$ above the true value of $a \cdot b$. This is to be compared with AMS sketches which guarantee $\varepsilon \|a\|_2 \|b\|_2$ additive error but require space proportional to $\frac{1}{\varepsilon^2}$ to make this guarantee.

Conservative Update

If only positive updates arrive, then the "conservative update" process (due to Estan and Varghese [7]) can be used. For an update (i, c), $\hat{a}_i$ is computed, and the counts are modified according to $\forall 1 \leq j \leq d : CM[j, h_j(i)] \leftarrow \max(CM[j, h_j(i)], \hat{a}_i + c)$. This procedure still ensures for point queries that $a_i \leq \hat{a}_i$, and that the error is no worse than in the normal update procedure; it has been observed that conservative update can improve accuracy "up to an order of magnitude" [7]. However, deletions or negative updates can no longer be processed, and the new update procedure is slower than the original one.

Applications

The Count-Min sketch has found a number of applications.

- Indyk [9] used the Count-Min sketch to estimate the residual mass after removing a set of items, that is, given a (small) set of indices I, to estimate $\sum_{i \notin I} a_i$. This supports clustering over streaming data.
- The *entropy* of a data stream is a function of the relative frequencies of each item or character within the stream. Using Count-Min sketches within a larger data structure based on additional hashing techniques, B. Lakshminath and Ganguly [8] showed how to estimate this entropy to within relative error.
- Sarlós et al. [14] gave approximate algorithms for personalized page rank computations which make use of Count-Min sketches to compactly represent web-sized graphs.
- In describing a system for building selectivity estimates for complex queries, Spiegel and Polyzotis [15] use Count-Min sketches in order to allow clustering over a high-dimensional space.
- Sketches that reduce the amount of information stored seem like a natural candidate to preserve privacy of information. However, proving privacy requires more care. Roughan and Zhang use the Count-Min sketch to allow private computation of a sketch of a vector [13]. Dwork et al. show that the Count-Min sketch can be made *pan-private*, meaning that information about individuals contributing to the data structure is held private.

Experimental Results

There have been a number of experimental studies of COUNT-MIN and related algorithms for a variety of computing models. These have shown that the algorithm is accurate and fast to execute [3, 11]. Implementations on desktop machines achieve many millions of updates per second, primarily limited by IO throughput. Other implementations have incorporated Count-Min sketch into high-speed streaming systems such as Gigascope [6] and tuned it to process packet streams of multi-gigabit speeds. Lai and Byrd report on an implementation of Count-Min sketches on a low-power stream processor [10] capable of processing 40 byte packets at a throughput rate of up to 13 Gbps. This is equivalent to about 44 million updates per second.

URLs to Code and Data Sets

Sample implementations are widely available in a variety of languages.

C code is given by the MassDal code bank: http://www.cs.rutgers.edu/~muthu/massdal-code-index.html.

C++ code by Marios Hadjieleftheriou is available from http://hadjieleftheriou.com/sketches/index.html.

The MADlib project has SQL implementations for Postgres/Greenplum http://madlib.net/.

OCaml implementation is available via https://github.com/ezyang/ocaml-cminsketch.

Cross-Reference

▶ AMS Sketch

Recommended Reading

1. Alon N, Matias Y, Szegedy M (1996) The space complexity of approximating the frequency moments. In: ACM symposium on theory of computing, Philadelphia, pp 20–29
2. Charikar M, Chen K, Farach-Colton M (2002) Finding frequent items in data streams. In: Procedings of the international colloquium on automata, languages and programming (ICALP), Málaga
3. Cormode G, Hadjieleftheriou M (2009) Finding the frequent items in streams of data. Commun ACM 52(10):97–105
4. Cormode G, Muthukrishnan S (2005) An improved data stream summary: the Count-Min sketch and its applications. J Algorithms 55(1):58–75
5. Cormode G, Muthukrishnan S (2005) Summarizing and mining skewed data streams. In: SIAM conference on data mining, Newport Beach
6. Cormode G, Korn F, Muthukrishnan S, Johnson T, Spatscheck O, Srivastava D (2004) Holistic UDAFs at streaming speeds. In: ACM SIGMOD international conference on management of data, Paris, pp 35–46
7. Estan C, Varghese G (2002) New directions in traffic measurement and accounting. In: Proceedings of

ACM SIGCOMM, computer communication review, vol 32, 4, Pittsburgh, PA, pp 323–338

8. Ganguly S, Lakshminath B (2006) Estimating entropy over data streams. In: European symposium on algorithms (ESA), Zurich
9. Indyk P (2003) Better algorithms for high-dimensional proximity problems via asymmetric embeddings. In: ACM-SIAM symposium on discrete algorithms, Baltimore
10. Lai YK, Byrd GT (2006) High-throughput sketch update on a low-power stream processor. In: Proceedings of the ACM/IEEE symposium on architecture for networking and communications systems, San Jose
11. Manerikar N, Palpanas T (2009) Frequent items in streaming data: an experimental evaluation of the state-of-the-art. Data Knowl Eng 68(4): 415–430
12. Motwani R, Raghavan P (1995) Randomized algorithms. Cambridge University Press, Cambridge/New York
13. Roughan M, Zhang Y (2006) Secure distributed data mining and its application in large-scale network measurements. In: ACM SIGCOMM computer communication review (CCR), Pisa
14. Sarlós T, Benzúr A, Csalogány K, Fogaras D, Rácz B (2006) To randomize or not to randomize: space optimal summaries for hyperlink analysis. In: International conference on World Wide Web (WWW), Edinburgh
15. Spiegel J, Polyzotis N (2006) Graph-based synopses for relational selectivity estimation. In: ACM SIGMOD international conference on management of data, Chicago

CPU Time Pricing

Li-Sha Huang
Department of Computer Science and
Technology, Tsinghua University, Beijing, China

Keywords

Competitive auction; Market equilibrium; Resource scheduling

Years and Authors of Summarized Original Work

2005; Deng, Huang, Li

Problem Definition

This problem is concerned with a Walrasian equilibrium model to determine the prices of CPU time. In a market model of a CPU job scheduling problem, the owner of the CPU processing time sells time slots to customers and the prices of each time slot depends on the seller's strategy and the customers' bids (valuation functions). In a Walrasian equilibrium, the market is clear and each customer is most satisfied according to its valuation function and current prices. The work of Deng, Huang, and Li [1] establishes the existence conditions of Walrasion equilibrium, and obtains complexity results to determine the existence of equilibrium. It also discusses the issues of excessive supply of CPU time and price dynamics.

Notations

Consider a combinatorial auction (Ω, I, V):

- Commodities: The seller sells m kinds of indivisible commodities. Let $\Omega = \{\omega_1 \times \delta_1, \ldots, \omega_m \times \delta_m\}$ denote the set of commodities, where δ_j is the available quantity of the item ω_j.
- Agents: There are n agents in the market acting as buyers, denoted by $I = \{1, 2, \ldots, n\}$.
- Valuation functions: Each buyer $i \in I$ has a valuation function $v_i : 2^{\Omega} \to \mathbb{R}^+$ to submit the maximum amount of money he is willing to pay for a certain bundle of items. Let $V = \{v_1, v_2, \ldots, v_n\}$.

An XOR combination of two valuation functions v_1 and v_2 is defined by:

$$(v_1 \text{ XOR } v_2)(S) = \max \{v_1(S), v_2(S)\}$$

An *atomic bid* is a valuation function v denoted by a pair (S, q), where $S \subset \Omega$ and $q \in \mathbb{R}^+$:

$$v(T) = \begin{cases} q, & \text{if } S \subset T \\ 0, & \text{otherwise} \end{cases}$$

Any valuation function v_i can be expressed by an XOR combination of atomic bids,

$$v_i = (S_{i1}, q_{i1}) \text{XOR} (S_{i2}, q_{i2}) \ldots \text{XOR} (S_{in}, q_{in})$$

Given (Ω, I, V) as the input, the seller will determine an *allocation* and a *price vector* as the output:

- An *allocation* $X = \{X_0, X_1, X_2, \ldots, X_n\}$ is a partition of Ω, in which X_i is the bundle of commodities assigned to buyer i and X_0 is the set of unallocated commodities.
- A *price* vector p is a non-negative vector in $\mathbb{R}^m$, whose jth entry is the price of good $\omega_j \in \Omega$.

For any subset $T = \{\omega_1 \times \sigma_1, \ldots, \omega_m \times \sigma_m\} \subset \Omega$, define $p(T)$ by $p(T) = \sum_{j=1}^{m} \sigma_j p_j$. If buyer i is assigned to a bundle X_i, his *utility* is $u_i(X_i, p) = v_i(X_i) - p(X_i)$.

Definition A *Walrasian equilibrium* for a combinatorial auction (Ω, I, V) is a tuple (X, p), where $X = \{X_0, X_1, \ldots, X_n\}$ is an allocation and p is a price vector, satisfying that:

(1) $p(X_0) = 0$;

(2) $u_i(X_i, p) \geq u_i(B, p), \quad \forall B \subset \Omega,$

$$\forall 1 \leq i \leq n$$

Such a price vector is also called a market clearing price, or Walrasian price, or equilibrium price.

The CPU Job-Scheduling Problem

There are two types of players in a market-driven CPU resource allocation model: a resource provider and n consumers. The provider sells to the consumers CPU time slots and the consumers each have a job that requires a fixed number of CPU time, and its valuation function depends on the time slots assigned to the job, usually the last assigned CPU time slot. Assume that all jobs are released at time $t = 0$ and the ith job needs

s_i time units. The jobs are interruptible without preemption cost, as is often modeled for CPU jobs.

Translating into the language of combinatorial auctions, there are m commodities (time units), $\Omega = \{\omega_1, \ldots, \omega_m\}$, and n buyers (jobs), $I = \{1, 2, \ldots, n\}$, in the market. Each buyer has a valuation function v_i, which only depends on the completion time. Moreover, if not explicitly mentioned, every job's valuation function is non-increasing w.r.t. the completion time.

Key Results

Consider the following linear programming problem:

$$\max \sum_{i=1}^{n} \sum_{j=1}^{k_i} q_{ij} x_{ij}$$

$$\text{s.t.} \sum_{i,j | \omega_k \in S_{ij}} x_{ij} \leq \delta_k, \quad \forall \omega_k \in \Omega$$

$$\sum_{j=1}^{r_i} x_{ij} \leq 1, \quad \forall 1 \leq i \leq n$$

$$0 \leq x_{ij} \leq 1, \quad \forall i, j$$

Denote the problem by **LPR** and its integer restriction by **IP**. The following theorem shows that a non-zero gap between the integer programming problem **IP** and its linear relaxation implies the non-existence of the Walrasian equilibrium.

Theorem 1 *In a combinatorial auction, the Walrasian equilibrium exists if and only if the optimum of **IP** equals the optimum of **LPR**. The size of the LP problem is linear to the total number of XOR bids.*

Theorem 2 *Determination of the existence of Walrasian equilibrium in a CPU job scheduling problem is strong NP-hard.*

Now consider a job scheduling problem in which the customers' valuation functions are all linear. Assume n jobs are released at the time $t = 0$ for a single machine, the jth job's time span is $s_j \in \mathbb{N}^+$ and weight $w_j \geq 0$. The goal of the scheduling is to minimize the weighted completion time: $\sum_{i=1}^{n} w_i t_i$, where t_i is the completion time of job i. Such a problem is called an MWCT (Minimal Weighted Completion Time) problem.

Theorem 3 *In a single-machine MWCT job scheduling problem, Walrasian equilibrium always exists when $m \geq EM + \Delta$, where m is the total number of processor time, $EM = \sum_{i=1}^{n} s_i$ and $\Delta = \max_k \{s_k\}$. The equilibrium can be computed in polynomial time.*

The following theorem shows the existence of a non-increasing price sequence if Walrasian equilibrium exists.

Theorem 4 *If there exists a Walrasian equilibrium in a job scheduling problem, it can be adjusted to an equilibrium with consecutive allocation and a non-increasing equilibrium price vector.*

Applications

Information technology has changed people's lifestyles with the creation of many digital goods, such as word processing software, computer games, search engines, and online communities. Such a new economy has already demanded many theoretical tools (new and old, of economics and other related disciplines) be applied to their development and production, marketing, and pricing. The lack of a full understanding of the new economy is mainly due to the fact that digital goods can often be re-produced at no additional cost, though multi-fold other factors could also be part of the difficulty. The work of Deng, Huang, and Li [1] focuses on CPU time as a product for sale in the market, through the Walrasian pricing model in economics. CPU time as a commercial product is extensively studied in grid computing. Singling out CPU time pricing will help us to set aside other complicated issues caused by secondary factors, and a complete understanding of this special digital product (or service) may shed some light on the study of other goods in the digital economy.

The utilization of CPU time by multiple customers has been a crucial issue in the development of operating system concept. The rise of grid computing proposes to fully utilize computational resources, e.g., CPU time, disk space, bandwidth. Market-oriented schemes have been proposed for efficient allocation of computational grid recourses, by [2, 5]. Later, various practical and simulation systems have emerged in grid resource management. Besides the resource allocation in grids, an economic mechanism has also been introduced to TCP congestion control problems, see Kelly [4].

Cross-References

▶ Adwords Pricing
▶ Competitive Auction
▶ Incentive Compatible Selection
▶ Price of Anarchy

Recommended Reading

1. Deng X, Huang L-S, Li M (2007) On Walrasian price of CPU time. In: Proceedings of COCOON'05, Knming, 16–19 Aug 2005, pp 586–595. Algorithmica 48(2):159–172
2. Ferguson D, Yemini Y, Nikolaou C (1988) Microeconomic algorithms for load balancing in distributed computer systems. In: Proceedings of DCS'88, San Jose, 13–17 June 1988, pp 419–499
3. Goldberg AV, Hartline JD, Wright A (2001) Competitive auctions and digital goods. In: Proceedings of SODA'01, Washington, DC, 7–9 Jan 2001, pp 735–744
4. Kelly FP (1997) Charging and rate control for elastic traffic. Eur Trans Telecommun 8:33–37
5. Kurose JF, Simha R (1989) A microeconomic approach to optimal resource allocation in distributed computer systems. IEEE Trans Comput 38(5):705–717
6. Nisan N (2000) Bidding and allocation in combinatorial auctions. In: Proceedings of EC'00, Minneapolis, 17–20 Oct 2000, pp 1–12

Critical Range for Wireless Networks

Chih-Wei Yi
Department of Computer Science, National
Chiao Tung University, Hsinchu City, Taiwan

Keywords

Connectivity; Delaunay triangulations; Gabriel graphs; Greedy forward routing; Isolated nodes; Monotonic properties; Random geometric graphs

Years and Authors of Summarized Original Work

2004; Wan, Yi

Problem Definition

Given a point set V, a graph of the vertex set V in which two vertices have an edge if and only if the distance between them is at most r for some positive real number r is called a r-disk graph over the vertex set V and denoted by $G_r(V)$. If $r_1 \leq r_2$, obviously $G_{r_1}(V) \subseteq G_{r_2}(V)$. A graph property is monotonic (increasing) if a graph is with the property, then every supergraph with the same vertex set also has the property. The critical-range problem (or critical-radius problem) is concerned with the minimal range r such that $G_r(V)$ is with some monotonic property. For example, graph connectivity is monotonic and crucial to many applications. It is interesting to know whether $G_r(V)$ is connected or not. Let $\rho_{\text{con}}(V)$ denote the minimal range r such that $G_r(V)$ is connected. Then, $G_r(V)$ is connected if $r \geq \rho_{\text{con}}(V)$, and otherwise not connected. Here $\rho_{\text{con}}(V)$ is called the critical range for connectivity of V. Formally, the critical-range problem is defined as follows.

Definition 1 The critical range for a monotonic graph property π over a point set V, denoted by $\rho_\pi(V)$, is the smallest range r such that $G_r(V)$ has property π.

From another aspect, for a given geometric property, a corresponding geometric structure is usually embedded. In many cases, the critical-range problem for graph properties is related or equivalent to the longest-edge problem of corresponding geometric structures. For example, if $G_r(V)$ is connected, it contains a Euclidean minimal spanning tree (EMST), and $\rho_{\text{con}}(V)$ is equal to the largest edge length of the EMST. So the critical range for connectivity problem is equivalent to the longest edge of the EMST problem, and the critical range for connectivity is the smallest r such that $G_r(V)$ contains the EMST.

In most cases, given an instance, the critical range can be calculated by polynomial time algorithms. So it is not a hard problem to decide the critical range. Researchers are interested in the probabilistic analysis of the critical range, especially asymptotic behaviors of r-disk graphs over random point sets. Random geometric graphs [8] is a general term for the theory about r-disk graphs over random point sets.

Key Results

In the following, problems are discussed in a 2D plane. Let $X_1, X_2, \cdots$ be independent and uniformly distributed random points on a bounded region A. Given a positive integer n, the point process $\{X_1, X_2, \ldots, X_n\}$ is referred to as the uniform n-point process on A, and is denoted by $X_n(A)$. Given a positive number λ, let $Po(\lambda)$ be a Poisson random variable with parameter λ, independent of $\{X_1, X_2, \ldots\}$. Then the point process $\{X_1, X_2, \ldots, X_{Po(n)}\}$ is referred to as the Poisson point process with mean n on A, and is denoted by $P_n(A)$. A is called a deployment region. An event is said to be asymptotic almost sure if it occurs with a probability that converges to 1 as $n \to \infty$.

In a graph, a node is "isolated" if it has no neighbor. If a graph is connected, there exists no isolated node in the graph. The asymptotic distribution of the number of isolated nodes is given by the following theorem [2, 6, 14].

Theorem 1 *Let* $r_n = \sqrt{\frac{\ln n + \xi}{\pi n}}$ *and* Ω *be a unit-area disk or square. The number of isolated nodes in* $G_r(X_n(\Omega))$ *or* $G_r(P_n(\Omega))$ *is asymptotically Poisson with mean* $e^{-\xi}$.

According to the theorem, the probability of the event that there is no isolated node is asymptotically equal to $\exp(-e^{-\xi})$. In the theory of random geometric graphs, if a graph has no isolated node, it is almost surely connected. Thus, the next theorem follows [6, 8, 9].

Theorem 2 *Let* $r_n = \sqrt{\frac{\ln n + \xi}{\pi n}}$ *and* Ω *be a unit-area disk or square. Then,* $\Pr[G_r(X_n(\Omega))$ *is connected]* $\to \exp(-e^{-\xi})$, *and*

$\Pr[G_r(P_n(\Omega))$ *is connected]* $\to \exp(-e^{-\xi})$.

In wireless sensor networks, the deployment region is k-covered if every point in the deployment region is within the coverage ranges of at least k sensors (vertices). Assume the coverage ranges are disks of radius r centered at the vertices. Let k be a fixed non-negative integer, and Ω be the unit-area square or disk centered at the origin o. For any real number t, let $t\Omega$ denote the set $\{tx : x \in \Omega\}$, i. e., the square or disk of area t^2 centered at the origin. Let $C_{n,r}$ (respectively, $C'_{n,r}$) denote the event that Ω is $(k+1)$-covered by the (open or closed) disks of radius r centered at the points in $P_n(\Omega)$ (respectively, $X_n(\Omega)$). Let $K_{s,n}$ (respectively, $K'_{s,n}$) denote the event that $\sqrt{s}\Omega$ is $(k+1)$-covered by the unit-area (closed or open) disks centered at the points in $P_n(\sqrt{s}\Omega)$ (respectively, $X_n(\sqrt{s}\Omega)$). To simplify the presentation, let η denote the peripheral of Ω, which is equal to 4 (respectively, $2\sqrt{\pi}$) if Ω is a square (respectively, disk). For any $\xi \in \mathbb{R}$, let

$$\alpha(\xi) = \begin{cases} \dfrac{\left(\frac{\sqrt{\pi}\eta}{2} + e^{-\frac{\xi}{2}}\right)^2}{16\left(2\sqrt{\pi}\eta + e^{-\frac{\xi}{2}}\right)} e^{-\frac{\xi}{2}}, & \text{if } k = 0; \\ \dfrac{\sqrt{\pi}\eta}{2^{k+6}(k+2)!} e^{-\frac{\xi}{2}}, & \text{if } k \geq 1. \end{cases}$$

and

$$\beta(\xi) = \begin{cases} 4e^{-\xi} + 2\left(\sqrt{\pi} + \frac{1}{\sqrt{\pi}}\right)\eta e^{-\frac{\xi}{2}}, & \text{if } k = 0; \\ \dfrac{\sqrt{\pi} + \frac{1}{\sqrt{\pi}}}{2^{k-1}k!}\eta e^{-\frac{\xi}{2}}, & \text{if } k \geq 1. \end{cases}$$

The asymptotics of $\Pr[C_{n,r}]$ and $\Pr[C'_{n,r}]$ as n approaches infinity, and the asymptotics of $\Pr[K_{s,n}]$ and $\Pr[K'_{s,n}]$ as s approaches infinity are given in the following two theorems [4, 10, 16].

Theorem 3 *Let* $r_n = \sqrt{\frac{\ln n + (2k+1)\ln\ln n + \xi_n}{\pi n}}$. *If* $\lim_{n\to\infty} \xi_n = \xi$ *for some* $\xi \in \mathbb{R}$, *then*

$$1 - \beta(\xi) \leq \lim_{n\to\infty} \Pr[C_{n,r_n}] \leq \frac{1}{1 + \alpha(\xi)}, \text{ and}$$

$$1 - \beta(\xi) \leq \lim_{n\to\infty} \Pr[C'_{n,r_n}] \leq \frac{1}{1 + \alpha(\xi)}.$$

If $\lim_{n\to\infty} \xi_n = \infty$, *then*

$$\lim_{n\to\infty} \Pr[C_{n,r_n}] = \lim_{n\to\infty} \Pr[C'_{n,r_n}] = 1.$$

If $\lim_{n\to\infty} \xi_n = -\infty$, *then*

$$\lim_{n\to\infty} \Pr[C_{n,r_n}] = \lim_{n\to\infty} \Pr[C'_{n,r_n}] = 0.$$

Theorem 4 *Let* $\mu(s) = \ln s + 2(k+1)\ln\ln s + \xi(s)$. *If* $\lim_{s\to\infty} \xi(s) = \xi$ *for some* $\xi \in \mathbb{R}$, *then*

$$1 - \beta(\xi) \leq \lim_{s\to\infty} \Pr[K_{s,\mu(s)s}] \leq \frac{1}{1 + \alpha(\xi)}, \text{ and}$$

$$1 - \beta(\xi) \leq \lim_{s\to\infty} \Pr[K'_{s,\mu(s)s}] \leq \frac{1}{1 + \alpha(\xi)}.$$

If $\lim_{s\to\infty} \xi(s) = \infty$, *then*

$$\lim_{s\to\infty} \Pr[K_{s,\mu(s)s}] = \lim_{s\to\infty} \Pr[K'_{s,\mu(s)s}] = 1.$$

If $\lim_{s\to\infty} \xi(s) = -\infty$, *then*

$$\lim_{s\to\infty} \Pr[K_{s,\mu(s)s}] = \lim_{s\to\infty} \Pr[K'_{s,\mu(s)s}] = 0.$$

In Gabriel graphs (GG), two nodes have an edge if and only if there is no other node in the disk using the segment of these two nodes as its diameter. If V is a point set and l is a positive real number, we use $\rho_{GG}(V)$ to denote the largest edge length of the GG over V, and $N(V, l)$ denotes the number of GG edges over V whose length is at least l. Wan and Yi (2007) [11] gave the following theorem.

Theorem 5 *Let Ω be a unit-area disk. For any constant ξ, $N\left(P_n(\Omega), 2\sqrt{\frac{\ln n+\xi}{\pi n}}\right)$ is asymptotically Poisson with mean $2e^{-\xi}$, and*

$$\lim_{n\to\infty} \Pr\left[\rho_{GG}(P_n(\Omega)) < 2\sqrt{\frac{\ln n+\xi}{\pi n}}\right]$$

$$= \exp\left(-2e^{-\xi}\right).$$

Let $\rho_{Del}(V)$ denote the largest edge length of the Delaunay triangulation over a point set V. The following theorem is given by Kozma et al. [3].

Theorem 6 *Let Ω be a unit-area disk. Then,*

$$\rho_{Del}(X_n(\Omega)) = O\left(\sqrt[3]{\frac{\ln n}{n}}\right).$$

In wireless networks with greedy forward routing (GFR), each node discards a packet if none of its neighbors is closer to the destination of the packet than itself, or otherwise forwards the packet to the neighbor that is the closest to the destination. Since each node only needs to maintain the locations of its one-hop neighbors and each packet should contain the location of the destination node, GFR can be implemented in a localized and memoryless manner. Because of the existence of local minima where none of the neighbors is closer to the destination than the current node, a packet may be discarded before it reaches its destination. To ensure that every packet can reach its destination, all nodes should have sufficiently large transmission radii to avoid the existence of local minima. Applying the r-disk model, we assume every node has the same transmission radius r, and each pair of nodes with distance at most r has a link. For a point set V, the *critical*

transmission radius for GFR is given by

$$\rho_{GFR}(V) = \max_{(u,v)\in V^2, u\neq v}\left(\min_{\|w-v\|<\|u-v\|} \|w-u\|\right).$$

In the definition, (u, v) is a source–destination pair and w is a node that is closer to v than u. If every node is with a transmission radius not less than $\rho_{GFR}(V)$, GFR can guarantee the deliverability between any source–destination pair [12].

Theorem 7 *Let Ω be a unit-area convex compact region with bounded curvature, and $\beta_0 = 1/\left(2/3 - \sqrt{3}/2\pi\right) \approx 1.6^2$. Suppose that $n\pi r_n^2 = (\beta + o(1))\ln n$ for some $\beta > 0$. Then,*

1. *If $\beta > \beta_0$, then $\rho_{GFR}(P_n(\Omega)) \leq r_n$ is asymptotically almost sure.*
2. *If $\beta < \beta_0$, then $\rho_{GFR}(P_n(\Omega)) > r_n$ is asymptotically almost sure.*

Applications

In the literature, r-disk graphs (or unit disk graphs by proper scaling) are widely used to model homogeneous wireless networks in which each node is equipped with an omnidirectional antenna. According to the path loss of radio frequency, the transmission ranges (radii) of wireless devices depend on transmission powers. For simplicity, the power assignment problem usually is modeled by a corresponding transmission range assignment problem. Recently, wireless ad-hoc networks have attracted attention from a lot of researchers because of various possible applications. In many of the possible applications, since wireless devices are powered by batteries, transmission range assignment has become one of the most important tools for prolonging system lifetime. By applying the theory of critical ranges, a randomly deployed wireless ad-hoc network may have good properties in high probability if the transmission range is larger than some critical value.

One application of critical ranges is to connectivity of networks. A network is k-vertex-connected if there exist k node-disjoint paths between any pair of nodes. With such a property, at least k distinct communication paths exist between any pair of nodes, and the network is connected even if $k - 1$ nodes fail. Thus, with a higher degree of connectivity, a network may have larger bandwidth and higher fault tolerance capacity. In addition, in [9, 14], and [15], networks with node or link failures were considered.

Another application is in topology control. To efficiently operate wireless ad-hoc networks, subsets of network topology will be constructed and maintained. The related topics are called topology control. A spanner is a subset of the network topology in which the minimal total cost of a path between any pair of nodes, e.g., distance or energy consumption, is only a constant fact larger than the minimal total cost in the original network topology. Hence spanners are good candidates for virtual backbones. Geometric structures, including Euclidean minimal spanning trees, relative neighbor graphs, Gabriel graphs, Delaunay triangulations, Yao's graphs, etc., are widely used ingredients to construct spanners [1, 5, 13]. By applying the knowledge of critical ranges, the complexity of algorithm design can be reduced, e.g., [3, 11].

Open Problems

A number of problems related to critical ranges remain open. Most problems discussed here apply 2-D plane geometry. In other words, the point set is in the plane. The first direction for future work is to study those problems in high-dimension spaces. Another open research area is on the longest-edge problems for other geometric structures, e.g., relative neighbor graphs and Yao's graphs. A third direction for future work involves considering relations between graph properties. A well-known result in random geometric graphs is that vanishment of isolated nodes asymptotically implies connectivity of networks. But for the wireless networks with unreliable links, this property is still open. In addition, in wireless sensor networks, the rela-

tions between connectivity and coverage are also interesting.

Cross-References

▶ Applications of Geometric Spanner Networks
▶ Connected Dominating Set
▶ Dilation of Geometric Networks
▶ Geometric Spanners
▶ Minimum Geometric Spanning Trees
▶ Minimum k-Connected Geometric Networks
▶ Randomized Broadcasting in Radio Networks
▶ Randomized Gossiping in Radio Networks

Recommended Reading

1. Cartigny J, Ingelrest F, Simplot-Ryl D, Stojmenovic I (2004) Localized LMST and RNG based minimum-energy broadcast protocols in ad hoc networks. Ad Hoc Netw 3(1):1–16
2. Dette H, Henze N (1989) The limit distribution of the largest nearest-neighbour link in the unit d-cube. J Appl Probab 26:67–80
3. Kozma G, Lotker Z, Sharir M, Stupp G (2004) Geometrically aware communication in random wireless networks. In: Proceedings of the twenty-third annual ACM symposium on principles of distributed computing, 25–28 July 2004, pp 310–319
4. Kumar S, Lai TH, Balogh J (2004) On k-coverage in a mostly sleeping sensor network. In: Proceedings of the 10th annual international conference on Mobile Computing and Networking (MobiCom'04), 26 Sept–1 Oct 2004
5. Li N, Hou JC, Sha L (2003) Design and analysis of a MST-based distributed topology control algorithm for wireless ad-hoc networks. In: 22nd annual joint conference of the IEEE computer and communications societies (INFOCOM 2003), vol 3, 1–3 Apr 2003, pp 1702–1712
6. Penrose M (1997) The longest edge of the random minimal spanning tree. Ann Appl Probab 7(2):340–361
7. Penrose M (1999) On k-connectivity for a geometric random graph. Random Struct Algorithms 15(2):145–164
8. Penrose M (2003) Random geometric graphs. Oxford University Press, Oxford
9. Wan P-J, Yi C-W (2005) Asymptotic critical transmission ranges for connectivity in wireless ad hoc networks with Bernoulli nodes. In: IEEE wireless communications and networking conference (WCNC 2005), 13–17 Mar 2005

10. Wan P-J, Yi C-W (2005) Coverage by randomly deployed wireless sensor networks. In: Proceedings of the 4th IEEE international symposium on Network Computing and Applications (NCA 2005), 27–29 July 2005

11. Wan P-J, Yi C-W (2007) On the longest edge of Gabriel graphs in wireless ad hoc networks. Trans Parallel Distrib Syst 18(1):1–16

12. Wan P-J, Yi C-W, Yao F, Jia X (2006) Asymptotic critical transmission radius for greedy forward routing in wireless ad hoc networks. In: Proceedings of the 7th ACM international symposium on mobile ad hoc networking and computing, 22–25 May 2006, pp 25–36

13. Wang Y, Li X-Y (2003) Localized construction of bounded degree and planar spanner for wireless ad hoc networks, In: Proceedings of the 2003 joint workshop on foundations of mobile computing (DIALM-POMC'03), 19 Sept 2003, pp 59–68

14. Yi C-W, Wan P-J, Li X-Y, Frieder O (2003) Asymptotic distribution of the number of isolated nodes in wireless ad hoc networks with Bernoulli nodes. In: IEEE wireless communications and networking conference (WCNC 2003), March 2003

15. Yi C-W, Wan P-J, Lin K-W, Huang C-H (2006) Asymptotic distribution of the number of isolated nodes in wireless ad hoc networks with unreliable nodes and links. In: The 49th annual IEEE GLOBECOMTechnical conference (GLOBECOM2006), 27 Nov–1 Dec 2006

16. Zhang H, Hou J (2004) On deriving the upper bound of α-lifetime for large sensor networks. In: Proceedings of the 5th ACM international symposium on mobile ad hoc networking & computing (MobiHoc 2004), 24–26 Mar 2004

Cryptographic Hardness of Learning

Adam Klivans
Department of Computer Science, University of Texas, Austin, TX, USA

Keywords

Representation-independent hardness for learning

Years and Authors of Summarized Original Work

1994; Kearns, Valiant

Problem Definition

This entry deals with proving negative results for distribution-free PAC learning. The crux of the problem is proving that a polynomial-time algorithm for learning various concept classes in the PAC model implies that several well-known cryptosystems are insecure. Thus, if we assume a particular cryptosystem is secure, we can conclude that it is impossible to efficiently learn a corresponding set of concept classes.

PAC Learning

We recall here the PAC learning model. Let C be a concept class (a set of functions over n variables), and let D be a distribution over the input space $\{0, 1\}^n$. With C we associate a size function $size$ that measures the complexity of each $c \in C$. For example, if C is a class of Boolean circuits, then $size(c)$ is equal to the number of gates in c. Let A be a randomized algorithm that has access to an oracle which returns labeled examples $(x, c(x))$ for some unknown $c \in C$; the examples x are drawn according to D. Algorithm A PAC learns concept class C by hypothesis class H if for any $c \in C$, for any distribution D over the input space, and any $\epsilon, \delta > 0$, A runs in time $poly(n, 1/\epsilon, 1/\delta, size(c))$ and produces a hypothesis $h \in H$ such that with probability at least $(1 - \delta)$, $Pr_D[c(x) \neq h(x)] < \epsilon$. This probability is taken over the random coin tosses of A as well as over the random labeled examples seen from distribution D. When $H = C$ (the hypothesis must be some concept in C), then A is a *proper* PAC learning algorithm. In this entry it is not assumed $H = C$, i.e., hardness results for *representation-independent* learning algorithms are discussed. The only assumption made on H is that for each $h \in H$, h can be evaluated in polynomial time for every input of length n.

Cryptographic Primitives

Also required is knowledge of various cryptographic primitives such as public-key cryptosystems, one-way functions, and one-way trapdoor functions. For a formal treatment of these primitives, refer to Goldreich [3].

Informally, a function f is *one way* if, after choosing a random x of length n and giving an adversary A only $f(x)$, it is computationally intractable for A to find y such that $f(y) = f(x)$. Furthermore, given x, $f(x)$ can be evaluated in polynomial time. That is, f is easy to compute one way, but there is no polynomial-time algorithm for finding pre-images of f on randomly chosen inputs. Say a function f is *trapdoor* if f is one way, but if an adversary A is given access to a secret "trapdoor" d, then A *can* efficiently find random pre-images of f.

Trapdoor functions that are permutations are closely related to *public-key cryptosystems*: imagine a person Alice who wants to allow others to secretly communicate with her. She publishes a one-way trapdoor permutation f so that it is publicly available to everyone, but keeps the "trapdoor" d to herself. Then Bob can send Alice a secret message x by sending her $f(x)$. Only Alice is able to invert f (recall f is a permutation) and recover x because only she knows d.

Key Results

The main insight in Kearns and Valiant's work is the following: if f is a trapdoor one-way function, and C is a circuit class containing the set of functions capable of inverting f given access to the trapdoor, then C is not efficiently PAC learnable, i. e., assuming the difficulty of inverting trap-door function f, there is a distribution on $\{0, 1\}^n$ where no learning algorithm can succeed in learning f's associated decryption function.

The following theorem is stated in the (closely related) language of public-key cryptosystems:

Theorem 1 (Cryptography and learning; cf. Kearns and Valiant [4]) *Consider a public-key cryptosystem for encrypting individual bits into n-bit strings. Let C be a concept class that contains all the decryption functions $\{0, 1\}^n \to \{0, 1\}$ of the cryptosystem. If C is PAC learnable in polynomial time, then there is a polynomial-time distinguisher between the encryptions of 0 and 1.*

The intuition behind the proof is as follows: fix an encryption function f, associated secret key d, and let C be a class of functions such that the problem of inverting $f(x)$ given d can be computed by an element c of C; notice that knowledge of d is not necessary to generate a polynomial-size sample of $(x, f(x))$ pairs.

If C is PAC learnable, then given a relatively small number of encrypted messages $(x, f(x))$, a learning algorithm A can find a hypothesis h that will approximate c and thus have a non-negligible advantage for decrypting future *randomly* encrypted messages. This violates the security properties of the cryptosystem.

A natural question follows: "what is the simplest concept class that can compute the decryption function for secure public-key cryptosystems?" For example, if a public-key cryptosystem is proven to be secure, and encrypted messages can be decrypted (given the secret key) by polynomial-size DNF formulas, then, by Theorem 1, one could conclude that polynomial-size DNF formulas cannot be learned in the PAC model.

Kearns and Valiant do not obtain such a hardness result for learning DNF formulas (it is still an outstanding open problem), but they do obtain a variety of hardness results assuming the security of various well-known public-key cryptosystems based on the hardness of number-theoretic problems such as factoring.

The following list summarizes their main results:

- Let C be the class of polynomial-size Boolean formulas (not necessarily DNF formulas) or polynomial-size circuits of logarithmic depth. Assuming that the RSA cryptosystem is secure, or recognizing quadratic residues is intractable, or factoring Blum integers is intractable, C is not PAC learnable.
- Let C be the class of polynomial-size deterministic finite automata. Under the same assumptions as above, C is not PAC learnable.
- Let C be the class of constant depth threshold circuits of polynomial size. Under the same assumptions as above, C is not PAC learnable.

The depth of the circuit class is not specified but it can be seen to be at most 4.

Kearns and Valiant also prove the intractability of finding optimal solutions related to coloring problems assuming the security of the above cryptographic primitives (e.g., breaking RSA).

Relationship to Hardness Results for Proper Learning

The key results above should not be confused with the extensive literature regarding hardness results for *properly* PAC learning concept classes. For example, it is known [1] that, unless RP = NP, it is impossible to properly PAC learn polynomial-size DNF formulas (i.e., require the learner to learn DNF formulas by outputting a DNF formula as its hypothesis). Such results are *incomparable* to the work of Kearns and Valiant, as they require something much stronger from the learner but take a much weaker assumption (RP $\neq$ NP is a weaker assumption than the assumption that RSA is secure).

Applications and Related Work

Valiant [10] was the first to observe that the existence of a particular cryptographic primitive (pseudorandom functions) implies hardness results for PAC learning concept classes. The work of Kearns and Valiant has subsequently found many applications. Klivans and Sherstov have recently shown [7] that the problem of PAC learning intersections of halfspaces (a very simple depth-2 threshold circuit) is intractable unless certain lattice-based cryptosystems due to Regev [9] are not secure. Their result makes use of the Kearns and Valiant approach. Angluin and Kharitonov [2] have extended the Kearns and Valiant paradigm to give cryptographic hardness results for learning concept classes even if the learner has *query access* to the unknown concept. Kharitonov [6] has given hardness results for learning polynomial-size, constant depth circuits that assumes the existence of secure pseudo-random generators rather than the existence of public-key cryptosystems.

Open Problems

The major open problem in this line of research is to prove a cryptographic hardness result for PAC learning polynomial-size DNF formulas. Currently, polynomial-size DNF formulas seem far too weak to compute cryptographic primitives such as the decryption function for a well-known cryptosystem. The fastest known algorithm for PAC learning DNF formulas runs in time $2^{\tilde{O}(n^{1/3})}$ [8].

Cross-References

▶ PAC Learning

Recommended Reading

1. Alekhnovich M, Braverman M, Feldman V, Klivans AR, Pitassi T (2004) Learnability and automatizability. In: Proceedings of the 45th symposium on foundations of computer science, Rome
2. Angluin D, Kharitonov M (1995) When won't membership queries help? J Comput Syst Sci 50(2):336-355
3. Goldreich O (2001) Foundations of cryptography: basic tools. Cambridge University Press, Cambridge/New York
4. Kearns M, Valiant L (1994) Cryptographic limitations on learning boolean formulae and finite automata. J ACM 41(1):67–95
5. Kearns M, Vazirani U (1994) An introduction to computational learning theory. MIT, Cambridge
6. Kharitonov M (1993) Cryptographic hardness of distribution-specific learning. In: Proceedings of the twenty-fifth annual symposium on theory of computing, San Diego, pp 372–381
7. Klivans A, Sherstov AA (2006) Cryptographic hardness for learning intersections of halfspaces. In: Proceedings of the 47th symposium on foundations of computer science, Berkeley
8. Klivans A, Servedio R (2001) Learning DNF in time $2^{\tilde{O}(n^{1/3})}$. In: Proceedings of the 33rd annual symposium on theory of computing, Heraklion
9. Regev O (2004) New lattice-based cryptographic constructions. J ACM 51:899–942
10. Valiant L (1984) A theory of the learnable. Commun ACM 27(11):1134–1142

Cuckoo Hashing

Rasmus Pagh
Theoretical Computer Science, IT University of
Copenhagen, Copenhagen, Denmark

Keywords

Dictionary; Hash table; Key-value store; Open
addressing

Years and Authors of Summarized Original Work

2001; Pagh, Rodler

Problem Definition

A *dictionary* (also known as an *associative array*)
is an abstract data structure capable of storing a
set S of elements, referred to as *keys*, and information associated with each key. The operations
supported by a dictionary are insertion of a key
(and associated information), deletion of a key,
and lookup of a key (retrieving the associated
information). In case a lookup is made on a key
that is not in S, this must be reported by the data
structure.

The *hash table* is a class of data structures
use to implement dictionaries in the RAM model
of computation. *Open addressing* hash tables are
a particularly simple type of hash table, where
the data structure is an array such that each
entry either contains a key of S or is marked
"empty." *Cuckoo hashing* addresses the problem
of implementing an open addressing hash table
with *worst-case* constant lookup time. Specifically, a constant number of entries in the hash
table should be associated with each key x, such
that x is present in one of these entries if $x \in S$.

In the following it is assumed that a key and
the information associated with a key are single
machine words. This is essentially without loss
of generality: If more associated data is wanted,
it can be referred to using a pointer. If keys
are longer than one machine word, they can be
mapped down to a single (or a few) machine

words using universal hashing [4] and the described method used on the hash values (which
are unique to each key with high probability).
The original key must then be stored as associated
data. Let n denote an upper bound on the size of
S. To allow the size of the set to grow beyond n,
global rebuilding can be used.

Key Results

Prehistory
It has been known since the advent of universal
hashing [4] that if the hash table has $r \geq n^2$
entries, a lookup can be implemented by retrieving just a single entry in the hash table. This is
done by storing a key x in entry $h(x)$ of the
hash table, where h is a function from the set of
machine words to $\{1, \dots, n^2\}$. If h is chosen at
random from a *universal family* of hash functions
[4], then with probability at least 1/2 every key in
S is assigned a unique entry. The same behavior
would be seen if h was a random function, but in
contrast to random functions, there are universal
families that allow efficient storage and evaluation of h (constant number of machine words and
constant evaluation time).

This overview concentrates on the case where
the space used by the open- addressing hash table
is linear, $r = O(n)$. It was shown by Azar et al.
[1] that it is possible to combine linear space
with worst-case constant lookup time. It was
not considered how to construct the data structure. Since randomization is used, all schemes
discussed have a probability of error. However,
this probability is small, $O(1/n)$ or less for
all schemes, and an error can be handled by
rehashing (changing the hash functions and rebuilding the hash table). The result of [1] was
shown under the assumption that the algorithm
is given free access to a number of truly random
hash functions. In many of the subsequent papers,
it is shown how to achieve the bounds using
explicitly defined hash functions. However, no
attempt is made here to cover these constructions.

In the following, let ε denote an arbitrary positive constant. Pagh [11] showed that retrieving
two entries from the hash table suffices when

$r \geq (2 + \varepsilon)n$. Specifically, lookup of a key x can be done by retrieving entries $h_1(x)$ and $h_2(x)$ of the hash table, where h_1 and h_2 are random hash functions mapping machine words to $\{1,\ldots,r\}$. The same result holds if h_1 has range $\{1,\ldots,r/2\}$ and h_2 has range $\{r/2 + 1,\ldots,r\}$, that is, if the two lookups are done in disjoint parts of memory.

It follows from [11] that it is not possible to perform lookup by retrieving a *single* entry in the worst case unless the hash table is of size $n^{2-o(1)}$.

Cuckoo Hashing

Pagh and Rodler [12] showed how to maintain the data structure of Pagh [11] under insertions. They considered the variant in which the lookups are done in disjoint parts of the hash table. It will be convenient to think of these as separate arrays, T_1 and T_2. Let $\perp$ denote the contents of an empty hash table entry, and let $x \leftrightarrow y$ express that the values of variables x and y are swapped. The proposed dynamic algorithm, called *cuckoo hashing*, performs insertions by the following procedure:

```
procedure insert(x)
  i := 1;
  repeat
    x ↔ Ti[hi(x)]; i := 3 − i;
  until x = ⊥
  end
```

At any time the variable x holds a key that needs to be inserted in the table, or $\perp$. The value of i changes between 1 and 2 in each iteration, so the algorithm is alternately exchanging the contents of x with a key from Table 1 and Table 2. Conceptually, what happens is that the algorithm moves a sequence of zero or more keys from one table to the other to make room for the new key. This is done in a greedy fashion, by kicking out any key that may be present in the location where a key is being moved. The similarity of the insertion procedure and the nesting habits of the European cuckoo is the reason for the name of the algorithm.

The pseudocode above is slightly simplified. In general the algorithm needs to make sure not to insert the same key twice and handle the possibility that the insertion may not succeed (by rehashing if the loop takes too long).

Theorem 1 *Assuming that $r \geq (2 + \varepsilon)n$, the expected time for the cuckoo hashing insertion procedure is $O(1)$.*

Generalizations of Cuckoo Hashing

Cuckoo hashing has been generalized in several directions. Kirsh et al. [8] showed that keeping a small *stash* of memory locations that are inspected at every lookup can significantly reduce the error probability of cuckoo hashing.

More generally the case of $k > 2$ hash functions has been considered. Also, the hash table may be divided into "buckets" of size b, such that the lookup procedure searches an entire bucket for each hash function. Let (k, b)-cuckoo denote a scheme with k hash functions and buckets of size b. What was described above is a $(2,1)$-cuckoo scheme. Already in 1999, $(4,1)$-cuckoo was described in a patent application by David A. Brown (US patent 6,775,281). Fotakis et al. described and analyzed a $(k, 1)$-cuckoo scheme in [7], and a $(2,b)$-cuckoo scheme was described and analyzed by Dietzfelbinger and Weidling [5]. In both cases, it was shown that space utilization arbitrarily close to 100% is possible and that the necessary fraction of unused space decreases exponentially with k and b. The insertion procedure considered in [5, 7] is a breadth-first search for the shortest sequence of key moves that can be made to accommodate the new key. Panigrahy [13] studied $(2,2)$-cuckoo schemes in detail, showing that a space utilization of 83% can be achieved dynamically, still supporting constant time insertions using breadth-first search. In a *static* setting with no updates, thresholds for general (k, b)-cuckoo hashing have been established (see LeLarge [10] and its references).

Applications

Dictionaries (sometimes referred to as key-value stores) have a wide range of uses in computer science and engineering. For example, dictionaries arise in many applications in string algorithms and data structures, database systems, data compression, and various information retrieval applications. Also, cuckoo hashing has been used in oblivious RAM simulations and other cryptographic constructions [2, 14].

Open Problems

The results above provide a good understanding of the properties of open-addressing schemes with worst-case constant lookup time. However, several aspects are still not understood satisfactorily.

First of all, there is no practical class of hash functions for which the above results can be shown. The only explicit classes of hash functions that are known to make the methods work either have evaluation time $\Theta(\log n)$ or use space $n^{\Omega(1)}$. It is an intriguing open problem to construct a class having constant evaluation time and space usage.

For the generalizations of cuckoo hashing, the use of breadth-first search is not so attractive in practice, due to the associated overhead in storage. A simpler approach that does not require any storage is to perform a random walk where keys are moved to a random, alternative position. (This generalizes the cuckoo hashing insertion procedure, where there is only one alternative position to choose.) Panigrahy [13] showed that this works for (2,2)-cuckoo when the space utilization is low. However, it is unknown whether this approach works well as the space utilization approaches 100 %.

Finally, many of the analyses that have been given are not tight. In contrast, most classical open addressing schemes have been analyzed very precisely. It seems likely that precise analysis of cuckoo hashing and its generalizations is possible using techniques from analysis of algorithms and tools from the theory of random graphs. In particular, the relationship between space utilization and insertion time is not well understood. A precise analysis of the probability that cuckoo hashing fails has been given by Kutzelnigg [9].

Experimental Results

All experiments on cuckoo hashing and its generalizations so far presented in the literature have been done using simple, heuristic hash functions. Pagh and Rodler [12] presented experiments showing that, for space utilization 1/3, cuckoo hashing is competitive with open addressing schemes that do not give a worst-case guarantee. Zukowski et al. [15] showed how to implement cuckoo hashing such that it runs very efficiently on pipelined processors with the capability of processing several instructions in parallel. For hash tables that are small enough to fit in cache, cuckoo hashing was 2 to 4 times faster than chained hashing in their experiments. Erlingsson et al. [6] considered (k, b)-cuckoo schemes for various combinations of small values of k and b, showing that very high space utilization is possible even for modestly small values of k and b. For example, a space utilization of 99.9 % is possible for $k = b = 4$. It was further found that the resulting algorithms were very robust. Experiments in [7] indicate that the random walk insertion procedure performs as well as one could hope for.

Cross-References

▶ Dictionary Matching
▶ Online Load Balancing of Temporary Tasks

Recommended Reading

1. Azar Y, Broder AZ, Karlin AR, Upfal E (1999) Balanced allocations. SIAM J Comput 29(1):180–200
2. Berman I, Haitner I, Komargodski I, Naor M (eds) (2013) Hardness preserving reductions via cuckoo hashing. In: Theory of cryptography. Springer, Berlin/New York, pp 40–59
3. Cain JA, Sanders P, Wormald N (2007) The random graph threshold for k-orientability and a fast algorithm for optimal multiple-choice allocation. In: Proceedings of the 18th annual ACM-SIAM symposium on discrete algorithms (SODA), New Orleans. ACM, pp 469–476
4. Carter JL, Wegman MN (1979) Universal classes of hash functions. J Comput Syst Sci 18(2):143–154

5. Dietzfelbinger M, Weidling C (2005) Balanced allocation and dictionaries with tightly packed constant size bins. In: ICALP, Lisbon. Lecture notes in computer science, vol 3580. Springer, pp 166–178
6. Erlingsson Ú, Manasse M, McSherry F (2006) A cool and practical alternative to traditional hash tables. In: Proceedings of the 7th workshop on distributed data and structures (WDAS), Santa Clara
7. Fotakis D, Pagh R, Sanders P, Spirakis PG (2005) Space efficient hash tables with worst case constant access time. Theory Comput Syst 38(2):229–248
8. Kirsch A, Mitzenmacher M, Wieder U (2009) More robust hashing: cuckoo hashing with a stash. SIAM J Comput 39(4):1543–1561
9. Kutzelnigg R (2006) Bipartite random graphs and cuckoo hashing. In: Proceedings of 4th colloquium on mathematics and computer science, Nancy
10. LeLarge M (2012) A new approach to the orientation of random hypergraphs. In: Proceedings of the 23rd annual ACM-SIAM symposium on discrete algorithms (SODA), Kyoto. ACM, pp 251–264
11. Pagh R (2001) On the cell probe complexity of membership and perfect hashing. In: Proceedings of the 33rd annual ACM symposium on theory of computing (STOC), Heraklion. ACM, pp 425–432
12. Pagh R, Rodler FF (2004) Cuckoo hashing. J Algorithms 51:122–144
13. Panigrahy R (2005) Efficient hashing with lookups in two memory accesses. In: Proceedings of the 16th annual ACM-SIAM symposium on discrete algorithms (SODA), Vancouver. SIAM, pp 830–839
14. Pinkas B, Reinman T (2010) Oblivious RAM revisited. In: Advances in cryptology–CRYPTO 2010, Santa Barbara. Springer, pp 502–519
15. Zukowski M, Heman S, Boncz PA (2006) Architecture-conscious hashing. In: Proceedings of the international workshop on data management on new hardware (DaMoN), Chicago. ACM, Article No 6

Current Champion for Online Bin Packing

Rob van Stee
University of Leicester, Leicester, UK

Keywords

Bin packing; Champion; Competitive analysis; Online algorithms

Years and Authors of Summarized Original Work

2002; Seiden

Problem Definition

In the online bin packing problem, a sequence of *items* with sizes in the interval $(0, 1]$ arrive one by one and need to be packed into *bins*, so that each bin contains items of total size at most 1. Each item must be irrevocably assigned to a bin before the next item becomes available. The algorithm has no knowledge about future items. There is an unlimited supply of bins available, and the goal is to minimize the total number of used bins (bins that receive at least one item).

The most common performance measure for online bin packing algorithms is the asymptotic performance ratio, or asymptotic competitive ratio, which is defined as

$$R_{\text{ASY}}(A) := \limsup_{n \to \infty} \left\{ \max_L \left\{ \frac{A(L)}{n} \middle| \text{OPT}(L) = n \right\} \right\}.$$

(1)

Hence, for any input L, the number of bins used by an online algorithm A is compared to the optimal number of bins needed to pack the same input. Note that calculating the optimal number of bins might take exponential time; moreover, it requires that the entire input is known in advance.

Key Results

This paper presents a new framework for analyzing online bin packing algorithms. It can be used to analyze all known versions of the well-known Harmonic algorithm, including a new version introduced in this paper.

The Harmonic algorithm [4] partitions the input into types depending on its size and packs each type separately. Harmonic-k has k types, and type i consists of items of size in the interval $(1/(i + 1), 1/i]$. Harmonic has k open bins at all times, one for each type, and packs i items of type i in one bin. Thus it achieves an asymptotic performance ratio of 1.691.

For some inputs, this algorithm wastes a lot of space in some bins, for instance, if many items of size $1/2 + \varepsilon$ arrive for some small $\varepsilon > 0$. Several authors improved on the basic Harmonic algorithm by combining some items of different types together into bins. Typically this is done by partitioning the intervals $(1/2, 1]$ and $(1/3, 1/2]$ further, guaranteeing that items can be combined. As a simple example, by introducing intervals $(1/2, 0.6]$ and $(1/3, 0.4]$, we can guarantee that items of these two new types can always be packed together in a single bin. Furthermore, the remaining intervals $(0.6, 1]$ and $(0.4, 0.5]$ now give better area guarantees than before: bins with items of these types will be at least 0.6 and at least 0.8 full, respectively.

Seiden builds on this idea and gives a new algorithm, Harmonic++, which beats all previously known algorithms and is still the best algorithm known. He used a computer-assisted search to set the many parameters of this algorithm. The algorithm partitions the intervals $(1/2, 1]$ and $(1/3, 1/2]$ in ten matching subintervals (in the sense described above) and also partitions intervals of several smaller types, using no less than 70 intervals in total. Also using a computer search, he proves that the asymptotic performance ratio of this algorithm is at most 1.58889.

Seiden also showed that the asymptotic performance ratio of a similar algorithm presented earlier, Harmonic+1, is at least 1.5972, disproving a claim by Richey [6] that Harmonic+1 is 1.58872-competitive.

The framework introduced by Seiden was used later in other contexts, for instance, for two-dimensional bin packing, where a set of rectangles needs to be packed into square bins. Han et al. [3] presented an algorithm with asymptotic performance ratio 2.5545. For the special case of packing squares, Han et al. [2] presented an algorithm with asymptotic performance ratio 2.1187.

Open Problems

The algorithm is very close to optimal for Harmonic-type algorithms, for which Ramanan et al. [5] showed a lower bound of 1.58333.... However, the general lower bound for this problem is only 1.54037 [1]. It is very difficult to see how either result can be improved, and this remains a challenging open problem which will require new ideas. There has been essentially no improvement in this area in over a decade.

Cross-References

▶ Harmonic Algorithm for Online Bin Packing
▶ Lower Bounds for Online Bin Packing

Recommended Reading

1. Balogh J, Békési J, Galambos G (2012) New lower bounds for certain bin packing algorithms. Theor Comput Sci 440–441:1–13
2. Han X, Ye D, Zhou Y (2010) A note on online hypercube packing. Cent Eur J Oper Res 18:221–239
3. Han X, Chin FYL, Ting H-F, Zhang G, Zhang Y (2011) A new upper bound 2.5545 on 2D online bin packing. ACM Trans Algorithms 7(4):50
4. Lee CC, Lee DT (1985) A simple online bin packing algorithm. J ACM 32:562–572
5. Ramanan P, Brown DJ, Lee CC, Lee DT (1989) Online bin packing in linear time. J Algorithms 10:305–326
6. Richey MB (1991) Improved bounds for harmonic-based bin packing algorithms. Discret Appl Math 34:203–227

Curve Reconstruction

Stefan Funke
Department of Computer Science, Universität Stuttgart, Stuttgart, Germany

Keywords

Computational geometry; Curve reconstruction

Years and Authors of Summarized Original Work

1998; Amenta, Bern, Eppstein
1999; Dey, Kumar
1999; Dey, Mehlhorn, Ramos
1999; Giesen
2001; Funke, Ramos
2003; Cheng, Funke, Golin, Kumar, Poon, Ramos

Problem Definition

Given a set S of sample points from a collection Γ of simple (nonintersecting) curves in the Euclidean plane, *curve reconstruction* is the problem of computing the graph $G(S, \Gamma)$, called the *correct reconstruction*, whose vertex set is S and that has an edge between two vertices if and only if the respective samples are adjacent on a curve in Γ; see Fig. 1.

Obviously, it is not possible to correctly reconstruct a given collection of curves from an arbitrary sample set from it. Therefore, some restriction on the sample set S – a so-called sampling condition – is required which specifies how dense a sampling has to be to guarantee a correct output of an algorithm. The difficulty for an algorithm to solve the curve reconstruction problem (and to come up with a suitable sampling condition) varies with the classes of allowed curves in Γ and whether the set S is actually sampled from the curves or noisy.

Key Results

If the curves are closed, smooth, and uniformly sampled – that is, with a uniform maximum distance between adjacent sample points – several methods for the curve reconstruction problem are known to work ranging over minimum spanning trees, α-shapes, β-skeletons [KR85], and r-regular shapes; see the survey by Edelsbrunner [7]. The focus of this section are approaches which can deal with *nonuniform sampling conditions*, that is, conditions which allow sparser sampling in areas of low detail and require higher sampling only in areas of high detail.

Closed Smooth Curves

Amenta, Bern, and Epstein [2] introduced the concept of the *local feature size* lfs(p) of a point $p \in \Gamma$ which is defined as the distance to the *medial axis* of Γ. The medial axis of a collection of curves Γ is defined as the set of points in the plane which have more than one closest point on a curve in Γ; see Fig. 2. Roughly speaking, a neighborhood of a point of size equal to its local feature size is intersected by the curves in a single piece that winds up only a small angle.

The introduction of the local feature size allowed for a very elegant sampling condition: A sample set S is called an ϵ-sampling for a collection of curves Γ, if for all $p \in \Gamma$, $\exists s \in S$ with $|ps| \leq \epsilon \text{lfs}(p)$. This condition naturally captures the intuition that "complicated" areas of the curve require higher sampling density than areas of low detail.

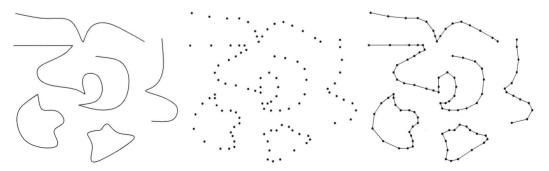

Curve Reconstruction, Fig. 1 A collection of curves Γ, a sample S from Γ, and the correct reconstruction G(S,Γ)

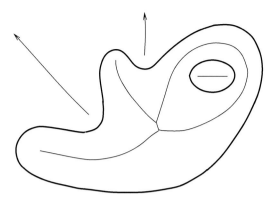

Curve Reconstruction, Fig. 2 The light curves are the medial axis of the heavy curves (Courtesy of N. Amenta, M. Bern, and D. Eppstein)

For small enough ϵ, the Voronoi nodes in the Voronoi diagram $VD(S)$ of an ϵ-sampling S for Γ approximate the medial axis of Γ. Based on that intuition, the *CRUST* algorithm in [1] outputs as correct reconstruction the edges of the Delaunay triangulation of S having a ball empty of Voronoi vertices in the Voronoi diagram $VD(S)$. For $\epsilon < 0.252$ CRUST provably outputs the correct reconstruction of an ϵ-sampling S with respect to a collection of closed smooth curves Γ.

In the same paper, the authors could also show that a known algorithm – the β-skeleton – for suitable choice of β also correctly reconstructs a collection of closed smooth curves for $\epsilon < 0.297$.

Later, Dey and Kumar [4] presented an extremely simple and straightforward algorithm connecting essentially the nearest neighbors on opposite sides. They could prove this algorithm to be correct under the local feature size sampling condition for $\epsilon \leq 1/3$. What is particularly interesting about this algorithm is the fact that decisions which points to connect are made based on a very local neighborhood of the respective points. This idea later nicely translated to algorithms for the 3- and higher-dimensional manifold reconstruction problem.

Open Smooth Curves

When considering the larger class of *open* and closed smooth curves, there is a little caveat. While one can guarantee for sufficiently dense samplings, i.e., small enough values of ϵ, that all edges of the correct reconstruction are present in the output of a reconstruction algorithm, one cannot always avoid the inclusion of additional edges in the output of an algorithm. Essentially, the problem is that a sample set, S set, might be an ϵ-sampling for two collections of curves Γ and Γ' with different correct reconstructions $G(S, \Gamma)$ and $G(S, \Gamma')$ irrespectively how small ϵ is chosen.

Dey, Mehlhorn, and Ramos in [5] introduced the concept of a *witness curve* Γ^*, proving the following guarantee for their *CONSERVATIVE CRUST* algorithm: If S is an ϵ-sampling for a collection of open and closed smooth curves Γ, their algorithm returns a reconstruction H such that $H \supseteq G(S, \Gamma)$, that is, H contains all edges of the correct reconstruction. Furthermore, the algorithm outputs a curve Γ^* such that S is an $\epsilon' \approx \epsilon$-sampling for Γ^* and $H = G(S, \Gamma^*)$. Their algorithm is similar to the CRUST algorithm in that it identifies a subcomplex of the Delaunay triangulation.

Closed Curves with Corners

Another natural extension of the allowed classes of curves in Γ is the inclusion of curves with corners, that is, points where left and right tangent do not coincide. Unfortunately, in this case, one cannot use the sampling condition based on the local feature size since the medial axis actually touches the corners, requiring an infinitely dense sampling near corners. The first algorithms to deal with a *single* closed curve possibly with corners were by Giesen [9] and Althaus/Mehlhorn [2] based on the construction of a travelling salesman tour. In their sampling condition, areas of the curve nearby a corner were exempt from the ϵ-sampling condition. [2] could even prove that the respective TSP instance can be solved in polynomial time for sufficiently dense sample sets. Dey and Wenger [6] proposed a non-TSP-based approach for collections of closed curves with corners.

Open and Closed Curves with Corners

Finally Funke/Ramos [8] considered the case of collections of open and closed curves. While their algorithm also comes with a guarantee for some variant of an ϵ-sampling condition (with special condition near corners like [9] and [2]), they also propose a sampling condition which is expressed with respect to the correct reconstruction $G(S, \Gamma)$. Not being based on a travelling salesman tour computation, their algorithm can also handle collections containing several open curves. As [5] the algorithm also produces a collection of witness curves Γ'.

Noisy Sample Sets

A generalization in a different direction is the consideration of sample sets S which do not consist of points exactly *on* the curves in Γ but – e.g., due to measurement errors – lie only "nearby." In [3] the authors considered such noisy sample sets from a collection of disjoint smooth closed curves and could prove for a perturbed locally uniform sample set that their algorithm computes as output a set of polygonal curves converging to Γ as the sampling density increases.

Cross-References

▶ Manifold Reconstruction

Recommended Reading

1. Amenta N, Bern M, Eppstein D (1998) The Crust and the beta-skeleton: combinatorial curve reconstruction. Graph Models Image Process 60/2:2:125–135
2. Althaus E, Mehlhorn K (2000) TSP-based curve reconstruction in polynomial time. In: Proceedings of the 11th annual ACM-SIAM symposium on discrete algorithms, San Francisco, pp 686–695
3. Cheng S-W, Funke S, Golin MJ, Kumar P, Poon S-H, Ramos E A (2003) Curve reconstruction from noisy samples. In: Proceedings of the 19th ACM symposium on computational geometry, San Diego, pp 302–311
4. Dey TK, Kumar P (1999) A simple provable algorithm for curve reconstruction. In: Proceedings of the 10th ACM-SIAM symposium on discrete algorithms, Baltimore, pp 893–894
5. Dey TK, Mehlhorn K, Ramos EA (1999) Curve reconstruction: connecting dots with good reason. In: Proceedings of the 15th annual acm symposium on computational geometry, Miami Beach, pp 197–206
6. Dey TK, Wenger R (2000) Reconstruction curves with sharp corners. In: Proceedings of the 16th annual ACM symposium on computational geometry, Hong Kong, pp 233–241
7. Edelsbrunner H (1998) Shape reconstruction with delaunay complex. In: Proceedings of the 2nd Latin American theoretical informatics symposium, Campinas. LNCS, vol 1380, pp 119–132
8. Funke S, Ramos EA (2001) Reconstructing a collection of curves with corners and endpoints. In: Proceedings of the 12th annual ACM-SIAM symposium on discrete algorithms, Washington, DC, pp 344–353
9. Giesen J (1999) Curve reconstruction, the traveling salesman problem and Menger's theorem on length. In: Proceedings of the 15th annual ACM symposium on computational geometry, Miami Beach, pp 207–216

D

Data Migration

Yoo-Ah Kim
Computer Science and Engineering Department,
University of Connecticut, Storrs, CT, USA

Keywords

Data movements; File transfers

Years and Authors of Summarized Original Work

2004; Khuller, Kim, Wan

Problem Definition

The problem is motivated by the need to manage data on a set of storage devices to handle dynamically changing demand. To maximize utilization, the data layout (i.e., a mapping that specifies the subset of data items stored on each disk) needs to be computed based on disk capacities as well as the demand for data. Over time as the demand for data changes, the system needs to create new data layout. The data migration problem is to compute an efficient schedule for the set of disks to convert an initial layout to a target layout.

The problem is defined as follows. Suppose that there are N disks and Δ data items, and an initial layout and a target layout are given (see Fig. 1a for an example). For each item i, source disks S_i is defined to be a subset of disks which have item i in the initial layout. Destination disks D_i is a subset of disks that want to receive item i. In other words, disks in D_i have to store item i in the target layout but do not have to store it in the initial layout. Figure 1b shows the corresponding S_i and D_i. It is assumed that $S_i \neq \emptyset$ and $D_i \neq \emptyset$ for each item i. Data migration is the transfer of data to have all D_i receive data item i residing in S_i initially, and the goal is to minimize the total amount of time required for the transfers.

Assume that the underlying network is fully connected and the data items are all the same size. In other words, it takes the same amount of time to migrate an item from one disk to another. Therefore, migrations are performed in rounds. Consider the half-duplex model, where each disk can participate in the transfer of only one item – either as a sender or as a receiver. The objective is to find a migration schedule using the minimum number of rounds. No bypass nodes (A bypass node is a node that is not the target of a move operation, but is used as an intermediate holding point for a data item.) can be used and therefore all data items are sent only to disks that desire them.

Key Results

Khuller et al. [11] developed a 9.5-approximation for the data migration problem, which was later improved to $6.5 + o(1)$. In the next subsection,

© Springer Science+Business Media New York 2016
M.-Y. Kao (ed.), *Encyclopedia of Algorithms*,
DOI 10.1007/978-1-4939-2864-4

Data Migration, Fig. 1
Left An example of initial
and target layout and
right their corresponding
S_i's and D_i's

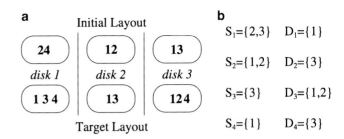

the lower bounds of the problem are first
examined.

Notations and Lower Bounds

1. **Maximum in-degree (β):** Let β_j be the
number of data items that a disk j has to
receive. In other words, $\beta_j = |\{i \mid j \in D_i\}|$.
Then $\beta = \max_j \beta_j$ is a lower bound on the
optimal as a disk can receive only one data
item in one round.
2. **Maximum number of items that a disk may
be a source or destination for (α):** For each
item i, at least one disk in S_i should be used
as a source for the item, and this disk is
called a *primary source*. A unique primary
source $s_i \in S_i$ for each item i that minimizes
$\alpha = \max_{j=1,...,N}(|\{i \mid j = s_i\}| + \beta_j)$ can be
found using a network flow. Note that $\alpha \geq \beta$,
and α is also a lower bound on the optimal
solution.
3. **Minimum time required for cloning (M):**
Let a disk j make a copy of item i at the kth
round. At the end of the mth round, the number
of copies that can be created from the copy
is at most 2^{m-k} as in each round the number
of copies can only be doubled. Also note that
each disk can make a copy of only one item
in one round. Since at least $|D_i|$ copies of
item i need to be created, the minimum m that
satisfies the following linear program gives
a lower bound on the optimal solution: $\mathbf{L(m)}$:

$$\sum_j \sum_{k=1}^{m} 2^{m-k} x_{ijk} \geq |D_i| \quad \text{for all } i \quad (1)$$

$$\sum_i x_{ijk} \leq 1 \quad \text{for all } j, k \quad (2)$$

$$0 \leq x_{ijk} \leq 1 \quad (3)$$

Data Migration Algorithm
A 9.5-approximation can be obtained as follows.
The algorithm first computes representative sets
for each item and sends the item to the representative sets first, which in turn send the item to the
remaining set. Representative sets are computed
differently depending on the size of D_i.

Representatives for Big Sets
For sets with size at least β, a *disjoint* collection of *representative sets* $R_i, i = 1 \ldots \Delta$ has to
satisfy the following properties: Each R_i should
be a subset of D_i and $|R_i| = \lfloor |D_i|/\beta \rfloor$. The
representative sets can be found using a network
flow.

Representatives for Small Sets
For each item i, let $\gamma_i = |D_i| \bmod k$. A *sec-
ondary representative* r_i in D_i for the items with
$\gamma_i \neq 0$ needs to be computed. A disk j can be
a secondary representative r_i for several items as
long as $\sum_{i \in I_j} \gamma_i \leq 2\beta - 1$, where I_j is a set of
items for which j is a secondary representative.
This can be done by applying the Shmoys–Tardos
algorithm [17] for the generalized assignment
problem.

Scheduling Migrations
Given representatives for all data items, migrations can be done in three steps as follows:

1. **Migration to R_i:** Each item i is first sent to
the set R_i. By converting a fractional solution given in $L(M)$, one can find a migration
schedule from s_i to R_i and it requires at most
$2M + \alpha$ rounds.
2. **Migration to r_i:** Item i is sent from primary
source s_i to r_i. The migrations can be done
in 1.5α rounds, using an algorithm for edge
coloring [16].

3. **Migration to the remaining disks:** A transfer graph from representatives to the remaining disks can now be created as follows. For each item i, add directed edges from disks in R_i to $(\beta - 1)\lfloor \frac{|D_i|}{\beta} \rfloor$ disks in $D_i \setminus R_i$ such that the out-degree of each node in R_i is at most $\beta - 1$ and the in-degree of each node in $D_i \setminus R_i$ from R_i is 1. A directed edge is also added from the secondary representative r_i of item i to the remaining disks in D_i which do not have an edge coming from R_i. It has been shown that the maximum degree of the transfer graph is at most $4\beta - 5$ and the multiplicity is $\beta + 2$. Therefore, migration for the transfer graph can be done in $5\beta - 3$ rounds using an algorithm for multigraph edge coloring [18].

Analysis
Note that the total number of rounds required in the algorithm described in "Data Migration Algorithm" is at most $2M + 2.5\alpha + 5\beta - 3$. As α, β and M are lower bounds on the optimal number of rounds, the abovementioned algorithm gives a 9.5-approximation.

Theorem 1 ([11]) *There is a 9.5-approximation algorithm for the data migration problem.*

Khuller et al. [10] later improved the algorithm and obtained a $(6.5 + o(1))$-approximation.

Theorem 2 ([10]) *There is a $(6.5 + o(1))$-approximation algorithm for the data migration problem.*

Applications

Data Migration in Storage Systems
Typically, a large storage server consists of several disks connected using a dedicated network, called a *storage area network*. To handle high demand, especially for multimedia data, a common approach is to replicate data objects within the storage system. Disks typically have constraints on storage as well as the number of clients that can access data from a single disk simultaneously. Approximation algorithms have been developed to map known demand for data to a specific data layout pattern to maximize utilization (The utilization is the total number of clients that can be assigned to a disk that contains the data they want.) [4, 8, 14, 15]. In the layout, they compute not only how many copies of each item need to be created, but also a layout pattern that specifies the precise subset of items on each disk. The problem is NP-hard, but there are polynomial-time approximation schemes [4, 8, 14]. Given the relative demand for data, the algorithm computes an almost optimal layout.

Over time as the demand for data changes, the system needs to create new data layouts. To handle high demand for popular objects, new copies may have to be dynamically created and stored on different disks. The data migration problem is to compute a specific schedule for the set of disks to convert an initial layout to a target layout. Migration should be done as quickly as possible since the performance of the system will be suboptimal during migration.

Gossiping and Broadcasting
The data migration problem can be considered as a generalization of gossiping and broadcasting. The problems of gossiping and broadcasting play an important role in the design of communication protocols in various kinds of networks and have been extensively studied (see for example [6, 7] and the references therein). The gossip problem is defined as follows. There are n individuals and each individual has an item of gossip that he/she wish to communicate to everyone else. Communication is typically done in rounds, where in each round an individual may communicate with at most one other individual. Some communication models allow for the full exchange of all items of gossip known to each individual in a single round. In addition, there may be a communication graph whose edge indicates which pairs of individuals are allowed to communicate directly in each round. In the broadcast problem, one individual needs to convey an item of gossip to every other individual. The data migration problem generalizes the gossiping and broadcasting in three ways: (1) each item of gossip needs to be communicated to only a subset of individuals; (2) several items of gossip may be known to an

individual; (3) a single item of gossip can initially be shared by several individuals.

Open Problems

The data migration problem is NP-hard by reduction from the edge coloring problem. However, no inapproximability results are known for the problem. As the current best approximation factor is relatively high $(6.5 + o(1))$, it is an interesting open problem to narrow the gap between the approximation guarantee and the inapproximability.

Another open problem is to combine data placement and migration problems. This question was studied by Khuller et al. [9]. Given the initial layout and the new demand pattern, their goal was to find a set of data migrations that can be performed in a specific number of rounds and gives the best possible layout to the current demand pattern. They showed that even one-round migration is NP-hard and presented a heuristic algorithm for the one-round migration problem. The experiments showed that performing a few rounds of one-round migration consecutively works well in practice. Obtaining nontrivial approximation algorithms for this problem would be interesting future work.

Data migration in a heterogeneous storage system is another interesting direction for future research. Most research on data migration has focused mainly on homogeneous storage systems, assuming that disks have the same fixed capabilities and the network connections are of the same fixed bandwidth. In practice, however, large-scale storage systems may be heterogenous. For instance, disks tend to have heterogeneous capabilities as they are added over time owing to increasing demand for storage capacity. Lu et al. [13] studied the case when disks have variable bandwidth owing to the loads on different disks. They used a control-theoretic approach to generate adaptive rates of data migrations which minimize the degradation of the quality of the service. The algorithm reduces the latency experienced by clients significantly compared with the previous schemes. However, no theoretical bounds on the efficiency of data

migrations were provided. Coffman et al. [2] studied the case when each disk i can handle p_i transfers simultaneously and provided approximation algorithms. Some papers [2, 12] considered the case when the lengths of data items are heterogenous (but the system is homogeneous), and present approximation algorithms for the problem.

Experimental Results

Golubchik et al. [3] conducted an extensive study of the performance of data migration algorithms under different changes in user-access patterns. They compared the 9.5-approximation [11] and several other heuristic algorithms. Some of these heuristic algorithms cannot provide constant approximation guarantees, while for some of the algorithms no approximation guarantees are known. Although the worst-case performance of the algorithm by Khuller et al. [11] is 9.5, in the experiments the number of rounds required was less than 3.25 times the lower bound.

They also introduced the *correspondence problem*, in which a matching between disks in the initial layout with disks in the target layout is computed so as to minimize changes. A good solution to the correspondence problem can improve the performance of the data migration algorithms by a factor of 4.4 in their experiments, relative to a bad solution.

URL to Code

http://www.cs.umd.edu/projects/smart/data-migration/

Cross-References

▶ Broadcasting in Geometric Radio Networks
▶ Deterministic Broadcasting in Radio Networks

Recommended Reading

A special case of the data migration problem was studied by Anderson et al. [1] and Hallet al. [5].

They assumed that a data transfer graph is given, in which a node corresponds to each disk and a directed edge corresponds to each move operation that is specified (the creation of new copies of data items is not allowed). Computing a data movement schedule is exactly the problem of edge-coloring the transfer graph. Algorithms for edge-coloring multigraphs can now be applied to produce a migration schedule since each color class represents a matching in the graph that can be scheduled simultaneously. Computing a solution with the minimum number of rounds is NP-hard, but several good approximation algorithms are available for edge coloring. With space constraints on the disk, the problem becomes more challenging. Hall et al. [5] showed that with the assumption that each disk has one spare unit of storage, very good constant factor approximations can be developed. The algorithms use at most $4\lceil \Delta/4 \rceil$ colors with at most $n/3$ bypass nodes, or at most $6\lceil \Delta/4 \rceil$ colors without bypass nodes.

Most of the results on the data migration problem deal with the half-duplex model. Another interesting communication model is the full-duplex model where each disk can act as a sender and a receiver in each round for a single item. There is a $(4 + o(1))$-approximation algorithm for the full-duplex model [10].

1. Anderson E, Hall J, Hartline J, Hobbes M, Karlin A, Saia J, Swaminathan R, Wilkes J (2001) An experimental study of data migration algorithms. In: Workshop on algorithm engineering
2. Coffman E, Garey M, Jr, Johnson D, Lapaugh A (1985) Scheduling file transfers. SIAM J Comput 14(3):744–780
3. Golubchik L, Khuller S, Kim Y, Shargorodskaya S, Wan Y (2004) Data migration on parallel disks. In: 12th annual european symposium on algorithms (ESA)
4. Golubchik L, Khanna S, Khuller S, Thurimella R, Zhu A (2000) Approximation algorithms for data placement on parallel disks. In: Symposium on discrete algorithms. Society for industrial and applied mathematics, Philadelphia, pp 223–232
5. Hall J, Hartline J, Karlin A, Saia J, Wilkes J (2001) On algorithms for efficient data migration. In: SODA. Society for industrial and applied mathematics, Philadelphia, pp 620–629
6. Hedetniemi SM, Hedetniemi ST, Liestman A (1988) A survey of gossiping and broadcasting in communication networks. Networks 18: 129–134
7. Hromkovic J, Klasing R, Monien B, Peine R (1996) Dissemination of information in interconnection networks (broadcasting and gossiping). In: Du DZ, Hsu F (eds) Combinatorial network theory. Kluwer Academic, Dordrecht, pp 125–212
8. Kashyap S, Khuller S (2003) Algorithms for non-uniform size data placement on parallel disks. In: Conference on FST&TCS conference. LNCS, vol 2914. Springer, Heidelberg, pp 265–276
9. Kashyap S, Khuller S, Wan Y-C, Golubchik L (2006) Fast reconfiguration of data placement in parallel disks. In: Workshop on algorithm engineering and experiments
10. Khuller S, Kim Y, Malekian A (2006) Improved algorithms for data migration. In: 9th international workshop on approximation algorithms for combinatorial optimization problems
11. Khuller S, Kim Y, Wan Y-C (2004) Algorithms for data migration with cloning. SIAM J Comput 33(2):448–461
12. Kim Y-A (2005) Data migration to minimize the average completion time. J Algorithms 55:42–57
13. Lu C, Alvarez GA, Wilkes J (2002) Aqueduct:online datamigration with performance guarantees. In: Proceedings of the conference on file and storage technologies
14. Shachnai H, Tamir T (2001) Polynomial time approximation schemes for class-constrained packing problems. J Sched 4(6):313–338
15. Shachnai H, Tamir T (2001) On two class-constrained versions of the multiple knapsack problem. Algorithmica 29(3):442–467
16. Shannon CE (1949) A theorem on colouring lines of a network. J Math Phys 28:148–151
17. Shmoys DB, Tardos E (1993) An approximation algorithm for the generalized assignment problem. Math Program 62(3):461–474
18. Vizing VG (1964) On an estimate of the chromatic class of a p-graph (Russian). Diskret Analiz 3:25–30

Data Reduction for Domination in Graphs

Rolf Niedermeier
Department of Mathematics and Computer Science, University of Jena, Jena, Germany
Institut für Softwaretechnik und Theoretische Informatik, Technische Universität Berlin, Berlin, Germany

Keywords

Dominating set; Kernelization; Reduction to a problem kernel

Years and Authors of Summarized Original Work

2004; Alber, Fellows, Niedermeier

Problem Definition

The NP-complete DOMINATING SET problem is a notoriously hard problem:

Problem 1 (Dominating Set)
INPUT: An undirected graph $G = (V, E)$ and an integer $k \geq 0$.
QUESTION: Is there an $S \subseteq V$ with $|S| \leq k$ such that every vertex $v \in V$ is contained in S or has at least one neighbor in S?

For instance, for an n-vertex graph its optimization version is known to be polynomial-time approximable only up to a factor of $\Theta(\log n)$ unless some standard complexity-theoretic assumptions fail [9]. In terms of parametrized complexity, the problem is shown to be W[2]-complete [8]. Although still NP-complete when restricted to planar graphs, the situation much improves here. In her seminal work, Baker showed that there is an efficient polynomial-time approximation scheme (PTAS) [6], and the problem also becomes fixed-parameter tractable [2, 4] when restricted to planar graphs. In particular, the problem becomes accessible to fairly effective data reduction rules and a kernelization result (see [16] for a general description of data reduction and kernelization) can be proven. This is the subject of this entry.

Key Results

The key idea behind the data reduction is preprocessing based on locally acting simplification rules. Exemplary, here we describe a rule where the local neighborhood of each graph vertex is considered. To this end, we need the following definitions.

We partition the neighborhood $N(v)$ of an arbitrary vertex $v \in V$ in the input graph into three disjoint sets $N_1(v)$, $N_2(v)$, and $N_3(v)$ depending on local neighborhood structure. More specifically, we define

- $N_1(v)$ to contain all neighbors of v that have edges to vertices that are not neighbors of v;
- $N_2(v)$ to contain all vertices from $N(v) \backslash N_1(v)$ that have edges to at least one vertex from $N_1(v)$;
- $N_3(v)$ to contain all neighbors of v that are neither in $N_1(v)$ nor in $N_2(v)$.

An example which illustrates such a partitioning is given in Fig. 1 (left-hand side). A helpful and intuitive interpretation of the partition is to see vertices in $N_1(v)$ as *exits* because they have direct connections to the world outside the closed neighborhood of v, vertices in $N_2(v)$ as *guards* because they have direct connections to exits, and vertices in $N_3(v)$ as *prisoners* because they do not see the world outside $\{v\} \cup N(v)$.

Now consider a vertex $w \in N_3(v)$. Such a vertex only has neighbors in $\{v\} \cup N_2(v) \cup N_3(v)$. Hence, to dominate w, at least one vertex of $\{v\} \cup N_2(v) \cup N_3(v)$ *must* be contained in a dominating set for the input graph. Since v can dominate all vertices that would be dominated by choosing a vertex from $N_2(v) \cup N_3(v)$ into the dominating set, we obtain the following data reduction rule.

If $N_3(v) \neq \emptyset$ for some vertex v, then remove

$$N_2(v) \text{ and } N_3(v) \text{ from } G$$

and add a new vertex v'

with the edge $\{v, v'\}$ to G.

Note that the new vertex v' can be considered as a "gadget vertex" that "enforces" v to be chosen into the dominating set. It is not hard to verify the correctness of this rule, that is, the original graph has a dominating set of size k iff the reduced graph has a dominating set of size k. Clearly, the data reduction can be executed in polynomial time [5]. Note, however, that there are particular "diamond" structures that are not amenable to this reduction rule. Hence, a second, somewhat more complicated rule based on considering the joint neighborhood of *two* vertices has been introduced [5].

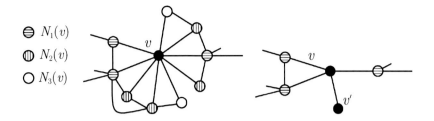

Data Reduction for Domination in Graphs, Fig. 1 The *left-hand side* shows the partitioning of the neighborhood of a single vertex v. The *right-hand side* shows the result of applying the presented data reduction rule to this particular (sub)graph

Altogether, the following core result could be shown [5].

Theorem 1 *A planar graph $G = (V, E)$ can be reduced in polynomial time to a planar graph $G' = (V', E')$ such that G has a dominating set of size k iff G' has a dominating set of size k and $|V'| = O(k)$.*

In other words, the theorem states that the DOM-INATING SET in planar graphs has a linear-size problem kernel. The upper bound on $|V'|$ was first shown to be $335k$ [5] and was then further improved to $67k$ [7]. Moreover, the results can be extended to graphs of bounded genus [10]. In addition, similar results (linear kernelization) have been recently obtained for the FULL-DEGREE SPANNING TREE problem in planar graphs [13]. Very recently, these results have been generalized into a methodological framework [12].

Applications

DOMINATING SET is considered to be one of the most central graph problems [14, 15]. Its applications range from facility location to bioinformatics.

Open Problems

The best lower bound for the size of a problem kernel for DOMINATING SET in planar graphs is $2k$ [7]. Thus, there is quite a gap between known upper and lower bounds. In addition, there have been some considerations concerning a generalization of the above-discussed data reduction rules [3]. To what extent such extensions are of practical use remains to be explored. Finally, a study of deeper connections between Baker's PTAS results [6] and linear kernelization results for DOMINATING SET in planar graphs seems to be worthwhile for future research. Links concerning the class of problems amenable to both approaches have been detected recently [12]. The research field of data reduction and problem kernelization as a whole together with its challenges is discussed in a recent survey [11].

Experimental Results

The above-described theoretical work has been accompanied by experimental investigations on synthetic as well as real-world data [1]. The results have been encouraging in general. However, note that grid structures seem to be a hard case where the data reduction rules remained largely ineffective.

Cross-References

▶ Connected Dominating Set

Recommended Reading

1. Alber J, Betzler N, Niedermeier R (2006) Experiments on data reduction for optimal domination in networks. Ann Oper Res 146(1):105–117
2. Alber J, Bodlaender HL, Fernau H, Kloks T, Niedermeier R (2002) Fixed parameter algorithms for Dominating Set and related problems on planar graphs. Algorithmica 33(4):461–493

3. Alber J, Dorn B, Niedermeier R (2006) A general data reduction scheme for domination in graphs. In: Proceedings of 32nd SOFSEM. LNCS, vol 3831. Springer, Berlin, pp 137–147
4. Alber J, Fan H, Fellows MR, Fernau H, Niedermeier R, Rosamond F, Stege U (2005) A refined search tree technique for dominating Set on planar graphs. J Comput Syst Sci 71(4):385–405
5. Alber J, Fellows MR, Niedermeier R (2004) Polynomial time data reduction for Dominating Set. J ACM 51(3):363–384
6. Baker BS (1994) Approximation algorithms for NP-complete problems on planar graphs. J ACM 41(1):153–180
7. Chen J, Fernau H, Kanj IA, Xia G (2007) Parametric duality and kernelization: lower bounds and upper bounds on kernel size. SIAM J Comput 37(4):1077–1106
8. Downey RG, Fellows MR (1999) Parameterized complexity. Springer, New York
9. Feige U (1998) A threshold of ln n for approximating set cover. J ACM 45(4):634–652
10. Fomin FV, Thilikos DM (2004) Fast parameterized algorithms for graphs on surfaces: linear kernel and exponential speed-up. In: Proceedings of 31st ICALP. LNCS, vol 3142. Springer, Berlin, pp 581–592
11. Guo J, Niedermeier R (2007) Invitation to data reduction and problemkernelization. ACM SIGACT News 38(1):31–45
12. Guo J, Niedermeier R (2007) Linear problem kernels for NPhard problems on planar graphs. In: Proceedings of 34th ICALP. LNCS, vol 4596. Springer, Berlin, pp 375–386
13. Guo J, Niedermeier R, Wernicke S (2006) Fixed-parameter tractability results for full-degree spanning tree and its dual. In: Proceedings of 2nd IWPEC. LNCS, vol 4196. Springer, Berlin, pp 203–214
14. Haynes TW, Hedetniemi ST, Slater PJ (1998) Domination in graphs: advanced topics. Pure and applied mathematics, vol 209. Marcel Dekker, New York
15. Haynes TW, Hedetniemi ST, Slater PJ (1998) Fundamentals of domination in graphs. Pure and applied mathematics, vol 208. Marcel Dekker, New York
16. Niedermeier R (2006) Invitation to fixed-parameter algorithms. Oxford University Press, New York

Data Stream Verification

Justin Thaler
Yahoo! Labs, New York, NY, USA

Keywords

Delegation; Interactive proofs; Streaming algorithms; Verification

Years and Authors of Summarized Original Work

2008; Goldwasser, Kalai, Rothblum
2011; Chung, Kalai, Liu, Raz
2012; Cormode, Thaler, Yi
2013; Gur, Raz
2014; Chakrabarti, Cormode, McGregor, Thaler

Problem Definition

The problem is concerned with the following setting. A computationally limited client wants to compute some property of a massive input, but lacks the resources to store even a small fraction of the input, and hence cannot perform the desired computation locally. The client therefore accesses a powerful but untrusted service provider (e.g., a commercial cloud computing service), who not only performs the requested computation but also proves that the answer is correct. An array of closely related models have been introduced to capture this scenario. The following section provides a unified presentation of these models, emphasizing their common features before delineating their differences.

Streaming Verification Model

Let $\sigma = \langle a_1, a_2, \ldots, a_m \rangle$ be a data stream, where each a_i comes from a data universe $\mathcal{U}$ of size n, and let F be a function mapping data streams to a finite range $\mathcal{R}$. A stream verification protocol for F involves two parties: a *prover* $\mathcal{P}$ and a (randomized) *verifier* $\mathcal{V}$. The protocol consists of two stages: a stream observation stage and a proof verification stage.

In the stream observation stage, $\mathcal{V}$ processes the stream σ, subject to the standard constraints of the data-stream model, i.e., sequential access to σ and limited memory. In the proof verification stage, $\mathcal{V}$ and $\mathcal{P}$ exchange a sequence of one or more messages, and afterward $\mathcal{V}$ outputs a value b. $\mathcal{V}$ is allowed to output a special symbol $\perp$ indicating a rejection of $\mathcal{P}$'s claims. Formally, $\mathcal{V}$ constitutes a stream verification protocol if the following two properties are satisfied:

- **Completeness**: There is some prover strategy $\mathcal{P}$ such that, for all streams σ, the probability that $\mathcal{V}$ outputs $F(\sigma)$ after interacting with $\mathcal{P}$ is at least $2/3$.
- **Soundness**: For all streams σ and all prover strategies $\mathcal{P}$, the probability that $\mathcal{V}$ outputs a value not in $\{F(x), \perp\}$ after interacting with $\mathcal{P}$ is at most $\epsilon \leq 1/3$.

Here, the probabilities are taken over $\mathcal{V}$'s internal randomness. The constants $2/3$ and $1/3$ are not essential and are chosen by convention. The parameter ϵ is referred to as the *soundness error* of the protocol.

Costs

There are five primary costs in any stream verification protocol: (1) $\mathcal{V}$'s space usage, (2) the total communication cost, (3) $\mathcal{V}$'s runtime, (4) $\mathcal{P}$'s runtime, and (5) the number of messages exchanged.

Differences Between Models

There are three primary differences between the various models of stream verification that have been put forth in the literature. The first is whether the soundness condition is required to hold against all cheating provers (such protocols are called *information-theoretically* or *statistically* sound), or only against cheating provers that run in polynomial time (such protocols are called *computationally* sound). The second is the amount and format of the interaction allowed between $\mathcal{P}$ and $\mathcal{V}$. The third is the temporal relationship between the stream observation and proof verification stage – in particular, several models permit $\mathcal{P}$ and $\mathcal{V}$ to exchange messages before and during the stream observation stage and sometimes permit the prover's messages to depend on parts of the data stream that $\mathcal{V}$ has not yet seen. In general, more permissive models allow a larger class of problems to be solved efficiently, but may yield protocols that are less realistic.

Summary of Models

The *annotated data streaming* (ADS) model [3] is noninteractive: $\mathcal{P}$ is permitted to send a single message to $\mathcal{V}$, with no communication allowed in the reverse direction. Technically, this model permits the contents of $\mathcal{P}$'s message to be interleaved with the stream, in which case each bit of $\mathcal{P}$'s message may be viewed as an "annotation" associated with a particular stream update. However, for most ADS protocols that have been developed, $\mathcal{P}$'s message can be sent after the stream observation phase. There are two kinds of ADS protocols: *prescient* protocols, in which the annotation sent at any given time can depend on parts of the data stream that $\mathcal{V}$ has not yet seen, and *online* protocols, which disallow this kind of dependence.

Streaming interactive proofs (SIPs) extend the ADS model to allow the prover and verifier to exchange many messages [6]. The *Arthur–Merlin streaming model* [10] is equivalent to a restricted class of SIPs, in which $\mathcal{V}$ is only allowed to send a single message to $\mathcal{P}$ (which must consist entirely of random coin tosses, in analogy with the classical complexity class AM), before receiving $\mathcal{P}$'s reply. The *streaming delegation model* [5] corresponds to SIPs that only satisfy computational, rather than information-theoretic, soundness.

Key Results

Obtaining exact answers even for basic problems in the standard data streaming model is impossible using $o(n)$ space. In contrast, stream verification protocols with $o(n)$ space and communication costs have been developed for (exactly solving) a wide variety of problems. Many of these protocols have adapted powerful algebraic techniques originally developed in the classical literature on interactive proofs, particularly the *sum-check* protocol of Lund et al. [14]. All of the protocols described here apply even to streams in the strict turnstile update model, where universe items can be deleted as well as inserted.

Annotated Data Streams

Chakrabarti et al. [3] showed that prescient ADS protocols can be exponentially more powerful than online ones for some problems. For example, there is a prescient ADS protocol with logarithmic space and communication costs for computing the median of a sequence of numbers: $\mathcal{P}$ sends $\mathcal{V}$ the claimed median τ at the start of the stream, and while observing the stream, $\mathcal{V}$ checks that $|\{j : a_j < \tau\}| \leq m/2$, and $|\{j : a_j > \tau\}| \leq m/2$, which can be done using an $O(\log m)$-bit counter. Meanwhile, [3] proved that any online protocol for MEDIAN with communication cost h and space cost v requires $h \cdot v = \Omega(n)$ and gave an online ADS protocol achieving these communication–space trade-offs up to logarithmic factors.

Chakrabarti et al. [3] also gave online ADS protocols achieving identical trade-offs between space and communication costs for problems including FREQUENCY MOMENTS and FREQUENT ITEMS and used a lower bound due to Klauck [11] on the *Merlin–Arthur communication complexity* of the SET-DISJOINTNESS function to show that these trade-offs are optimal for these problems even among prescient protocols. Subsequent work gave similarly optimal online ADS protocols for several more problems, including maximum matching and counting triangles in graphs and matrix-vector multiplication [8, 18]. Chakrabarti et al. [2] gave optimized protocols for streams whose length m is much smaller than the universe size n.

Streaming Interactive Proofs
Cormode, Thaler, and Yi [6] showed that several general protocols from the classical literature on interactive proofs can be simulated in the SIP model. In particular, this includes a powerful, general-purpose protocol due to Goldwasser, Kalai, and Rothblum [9] (henceforth, the GKR protocol). Given any problem computed by an arithmetic or Boolean circuit of polynomial size and polylogarithmic depth, the GKR protocol requires only polylogarithmic space and communication while using polylogarithmic rounds of verifier–prover interaction. This yields SIPs for exactly solving many basic streaming problems with polylogarithmic space and communication costs, including FREQUENCY MOMENTS, FREQUENT ITEMS, and GRAPH CONNECTIVITY.

Optimized protocols for specific problems, including FREQUENCY MOMENTS (see the detailed example below), were also presented.

Chakrabarti et al. [4] give *constant-round* SIPs with logarithmic space and communication costs for many problems, including INDEX, RANGE-COUNTING, and NEAREST-NEIGHBOR SEARCH. Gur and Raz [10] gave an Arthur–Merlin streaming protocol for the DISTINCT ELEMENTS problem with communication cost $\tilde{O}(h)$ space cost $\tilde{O}(v)$ for any h, v satisfying $h \cdot v \geq n$. Klauck and Prakash [13] extended this protocol to give an SIP for Distinct Elements with polylogarithmic space and communication costs and logarithmically many rounds of prover–verifier interaction.

Computationally Sound Protocols
Computationally sound protocols may achieve properties that are unattainable in the information-theoretic setting. For example, they typically achieve reusability, allowing the verifier to use the same randomness to answer many queries. In contrast, most SIPs only support "one-shot" queries, because they require the verifier to reveal secret randomness to the prover over the course of the protocol. Chung et al. [5] combined the GKR protocol with fully homomorphic encryption (FHE) to give reusable two-message protocols with polylogarithmic space and communication costs for any problem in the complexity class NC. They also gave reusable four-message protocols with polylogarithimic space and communication costs for any problem in the complexity class P. Papamanthou et al. [15] gave improved protocols for a class of low-complexity queries including point queries and range search: these protocols avoid the use of FHE and allow the prover to answer such queries in polylogarithmic time. (In contrast, protocols based on the GKR protocol [5, 6] require the prover to spend time quasilinear in the size of the data stream after receiving a query, even if the answer itself can be computed in sublinear time.)

Implementations
Implementations of the GKR protocol were provided in [7, 17]. Cormode, Mitzenmacher, and Thaler [7] also provided optimized

implementations of several ADS protocols from [3, 8]. Thaler et al. [19] provided parallelized implementations using Graphics Processing Units.

Detailed Example

The sum-check protocol can be directly applied to give an SIP for the kth frequency moment problem with $\log n$ rounds of prover–verifier iteration and $O(\log^2(n))$ space and communication costs. The sum-check protocol is described in Fig. 1.

Properties and Costs of the Sum-Check Protocol

The sum-check protocol satisfies perfect completeness and has soundness error $\epsilon \leq \deg(g)/|\mathbb{F}|$, where $\deg(g)$ denotes the total degree of g [14]. There is one round of prover–verifier interaction in the sum-check protocol for each of the v variables of g, and the total communication is $O(\deg(g))$ field elements.

Note that as described in Fig. 1, the sum-check protocol assumes that the verifier has oracle access to g. However, this will not be the case in applications, as g will ultimately be a polynomial that depends on the input data stream.

The SIP for Frequency Moments

In the kth frequency moment problem, the goal is to output $\sum_{i \in \mathcal{U}} f_i^k$, where f_i is the number of times item i appears in the data stream σ. For a vector $\mathbf{i} = (i_1, \ldots, i_{\log n}) \in \{0, 1\}^{\log n}$, let $\chi_{\mathbf{i}}(x_1, \ldots, x_{\log n}) = \prod_{k=1}^{\log n} \chi_{i_k}(x_k)$, where $\chi_0(x_k) = 1 - x_k$ and $\chi_1(x_k) = x_k$. $\chi_{\mathbf{i}}$ is the unique multilinear polynomial that maps $\mathbf{i} \in \{0, 1\}^{\log n}$ to 1 and all other values in $\{0, 1\}^{\log n}$ to 0, and it is referred to as the *multilinear extension* of $\mathbf{i}$.

For each $i \in \mathcal{U}$, associate i with a vector $\mathbf{i} \in \{0, 1\}^{\log n}$ in the natural way, and let $\mathbb{F}$ be a finite field with $n^k \leq |\mathbb{F}| \leq 4 \cdot n^k$. Define the polynomial $\hat{f} : \mathbb{F}^{\log n} \to \mathbb{F}$ via

---

Input: $\mathcal{V}$ is given oracle access to a v-variate polynomial g over finite field $\mathbb{F}$ and an $H \in \mathbb{F}$.
Goal: Determine whether $H = \sum_{(x_1, \ldots, x_v) \in \{0,1\}^v} g(x_1, \ldots, x_v)$.

- In the first round, $\mathcal{P}$ computes the univariate polynomial

$$g_1(X_1) := \sum_{x_2, \ldots, x_v \in \{0,1\}^{v-1}} g(X_1, x_2, \ldots, x_v),$$

and sends g_1 to $\mathcal{V}$. $\mathcal{V}$ checks that g_1 is a univariate polynomial of degree at most $\deg_1(g)$, and that $H = g_1(0) + g_1(1)$, rejecting if not.
- $\mathcal{V}$ chooses a random element $r_1 \in \mathbb{F}$, and sends r_1 to $\mathcal{P}$.
- In the jth round, for $1 < j < v$, $\mathcal{P}$ sends to $\mathcal{V}$ the univariate polynomial

$$g_j(X_j) = \sum_{(x_{j+1}, \ldots, x_v) \in \{0,1\}^{v-j}} g(r_1, \ldots, r_{j-1}, X_j, x_{j+1}, \ldots, x_v).$$

$\mathcal{V}$ checks that g_j is a univariate polynomial of degree at most $\deg_j(g)$, and that $g_{j-1}(r_{j-1}) = g_j(0) + g_j(1)$, rejecting if not.
- $\mathcal{V}$ chooses a random element $r_j \in \mathbb{F}$, and sends r_j to $\mathcal{P}$.
- In round v, $\mathcal{P}$ sends to $\mathcal{V}$ the univariate polynomial

$$g_v(X_v) = g(r_1, \ldots, r_{v-1}, X_v).$$

$\mathcal{V}$ checks that g_v is a univariate polynomial of degree at most $\deg_v(g)$, rejecting if not.
- $\mathcal{V}$ chooses a random element $r_v \in \mathbb{F}$ and evaluates $g(r_1, \ldots, r_v)$ with a single oracle query to g. $\mathcal{V}$ checks that $g_v(r_v) = g(r_1, \ldots, r_v)$, rejecting if not.
- If $\mathcal{V}$ has not yet rejected, $\mathcal{V}$ halts and accepts.

Data Stream Verification, Fig. 1 Description of the sum-check protocol. $\deg_i(g)$ denotes the degree of g in the ith variable

$$\hat{f} = \sum_{\mathbf{i} \in \{0,1\}^{\log n}} f_i \cdot \chi_{\mathbf{i}}. \qquad (1)$$

Note that $\hat{f}$ is the unique multilinear polynomial satisfying the property that $\hat{f}(\mathbf{i}) = f_i$ for all $\mathbf{i} \in \{0,1\}^{\log n}$.

The kth frequency moment of σ is equal to

$$\sum_{\mathbf{i} \in \{0,1\}^{\log n}} f_i^k = \sum_{\mathbf{i} \in \{0,1\}^{\log n}} (\hat{f}^k)(\mathbf{i}) .$$

Hence, in order to compute the kth frequency moment of σ, it suffices to apply the sum-check protocol to the polynomial $g = \hat{f}^k$. This requires $\log n$ rounds of prover–verifier interaction, and since the total degree of $\hat{f}^k$ is $k \cdot \log n$, the total communication cost is $O(k \log n)$ field elements, which require $O(k^2 \log^2 n)$ total bits to specify.

At the end of the sum-check protocol, $\mathcal{V}$ must compute

$$g(r_1, \ldots, r_{\log n}) = (\hat{f}^k)(r_1, \ldots, r_{\log n})$$

for randomly chosen $(r_1, \ldots, r_{\log n}) \in \mathbb{F}^{\log n}$. It suffices for $\mathcal{V}$ to evaluate $z := \hat{f}(r_1, \ldots, r_{\log n})$, since $(\hat{f}^k)(r_1, \ldots, r_{\log n}) = z^k$. The following lemma establishes that $\mathcal{V}$ can evaluate z with a single pass over σ, while storing $O(\log n)$ field elements.

Lemma 1 $\mathcal{V}$ *can compute* $z = \hat{f}(r_1, \ldots, r_{\log n})$ *with a single streaming pass over* σ*, while storing* $O(\log n)$ *field elements.*

Proof Given any stream update $a_j \in \mathcal{U}$, let $\mathbf{a_j}$ denote the binary vector associated with a_j. It follows from Eq. (1) that $\hat{f}(r_1, \ldots, r_{\log n}) = \sum_{j=1}^{m} \chi_{\mathbf{a_j}}(r_1, \ldots, r_{\log n})$. Thus, $\mathcal{V}$ can compute $\hat{f}(r_1, \ldots, r_{\log n})$ incrementally from the raw stream by initializing $\hat{f}(r_1, \ldots, r_{\log n}) \leftarrow 0$ and processing each update a_j via

$$\hat{f}(r_1, \ldots, r_{\log n}) \leftarrow \hat{f}(r_1, \ldots, r_{\log n})$$
$$+ \chi_{\mathbf{a_j}}(r_1, \ldots, r_{\log n}).$$

$\mathcal{V}$ only needs to store $(r_1, \ldots, r_{\log n})$ and $\hat{f}(r_1, \ldots, r_{\log n})$, which is $O(\log n)$ field elements in total.

Open Problems

- For several functions $F : \{0,1\}^n \rightarrow \{0,1\}$, it is known that any online ADS protocol for F with communication cost h and space cost v requires $h \cdot v = \Omega(n)$. This lower bound is tight in many cases, such as for the INDEX function [3]. However, it is open to exhibit a function that cannot be computed by any online ADS protocol with communication and space costs both bounded above by h, for some $h = \omega(n^{1/2})$.
- Two-message online SIP protocols with logarithmic space and communication costs are known for several functions, including the INDEX function [4]. It is also known that *existing techniques* cannot yield 2- or 3-message online SIPs of polylogarithmic cost for the MEDIAN or FREQUENCY MOMENT problems. However, it is open to exhibit a function $F : \{0,1\}^n \rightarrow \{0,1\}$ that cannot be computed by any online two-message SIP with communication and space costs both bounded above by h, for some $h = \omega(\log n)$.

Cross-References

▸ Communication Complexity

Recommended Reading

1. Aaronson S, Wigderson A (2009) Algebrization: a new barrier in complexity theory. ACM Trans Comput Theory 1(1):2:1–2:54
2. Chakrabarti A, Cormode G, Goyal N, Thaler J (2014) Annotations for sparse data streams. In: Chekuri C (ed) SODA, Portland. SIAM, Philadelphia, pp 687–706
3. Chakrabarti A, Cormode G, McGregor A, Thaler J (2014) Annotations in data streams. ACM Trans Algorithms 11(1):7. Preliminary version appeared in ICALP, 2009
4. Chakrabarti A, Cormode G, McGregor A, Thaler J, Venkatasubramanian S (2015) Verifiable stream computation and Arthur-Merlin communication. In: 30th conference on computational complexity (CCC), Portland, 17–19 June 2015, pp 217–243
5. Chung K-M, Kalai YT, Liu F-H, Raz R (2011) Memory delegation. In: Rogaway P (ed) CRYPTO, Santa Barbara. Volume 6841 of Lecture notes in computer science. Springer, Heidelberg/New York, pp 151–168

6. Cormode G, Thaler J, Yi K (2011) Verifying computations with streaming interactive proofs. PVLDB 5(1):25–36

7. Cormode G, Mitzenmacher M, Thaler J (2012) Practical verified computation with streaming interactive proofs. In: Goldwasser S (ed) ITCS, Cambridge, MA. ACM, pp 90–112

8. Cormode G, Mitzenmacher M, Thaler J (2013) Streaming graph computations with a helpful advisor. Algorithmica 65(2):409–442

9. Goldwasser S, Kalai YT, Rothblum GN (2008) Delegating computation: interactive proofs for muggles. In: Proceedings of the 40th annual ACM symposium on theory of computing (STOC'08), Victoria. ACM, New York, pp 113–122

10. Gur T, Raz R (2013) Arthur-Merlin streaming complexity. In: Proceedings of the 40th international colloquium on automata, languages and programming: Part I (ICALP'13). Springer, Berlin/Heidelberg

11. Klauck H (2003) Rectangle size bounds and threshold covers in communication complexity. In: IEEE conference on computational complexity, Aarhus, pp 118–134

12. Klauck H, Prakash V (2013) Streaming computations with a loquacious prover. In: Kleinberg RD (ed) ITCS, Berkeley. ACM, pp 305–320

13. Klauck H, Prakash V (2014) An improved interactive streaming algorithm for the distinct elements problem. In: Esparza J, Fraigniaud P, Husfeldt T, Koutsoupias E (eds) ICALP (1), Copenhagen. Volume 8572 of Lecture notes in computer science. Springer, Heidelberg, pp 919–930

14. Lund C, Fortnow L, Karloff H, Nisan N (1992) Algebraic methods for interactive proof systems. J ACM 39:859–868

15. Papamanthou C, Shi E, Tamassia R, Yi K (2013) Streaming authenticated data structures. In: Johansson T, Nguyen PQ (eds) EUROCRYPT, Athens. Volume 7881 of Lecture notes in computer science. Springer, Heidelberg/New York, pp 353–370

16. Schröder D, Schröder H (2012) Verifiable data streaming. In: Yu T, Danezis G, Gligor VD (eds) ACM conference on computer and communications security, Raleigh. ACM, New York, pp 953–964

17. Thaler J (2013) Time-optimal interactive proofs for circuit evaluation. In: Proceedings of the 33rd annual conference on advances in cryptology (CRYPTO'13), Santa Barbara. Springer, Berlin/Heidelberg

18. Thaler J (2014) Semi-streaming algorithms for annotated graph streams. Electron Colloq Comput Complex 21:90

19. Thaler J, Roberts M, Mitzenmacher M, Pfister H (2012) Verifying computations with massively parallel interactive proofs. In: HotCloud, Boston

20. Vu V, Setty S, Blumberg AJ, Walfish M (2013) A hybrid architecture for interactive verifiable computation. In: IEEE symposium on security and privacy, San Francisco

3D Conforming Delaunay Triangulation

Siu-Wing Cheng
Department of Computer Science and Engineering, Hong Kong University of Science and Technology, Hong Kong, China

Keywords

Delaunay refinement; Protecting ball; Radius-edge ratio; Weighted Delaunay triangulation

Years and Authors of Summarized Original Work

1998; Shewchuk
2004; Cohen-Steiner, de Verdière, Yvinec
2006; Cheng, Poon
2012; Cheng, Dey, Shewchuk

Problem Definition

A three-dimensional domain with piecewise linear boundary elements can be represented as a *piecewise linear complex* (PLC) of *linear cells* – vertices, edges, polygons, and polyhedra – that satisfy the following properties [4]. First, no vertex lies in the interior of an edge and every two edges are interior-disjoint. Second, the boundary of a polygon or polyhedra are union of cells in the PLC. Third, if two cells f and g intersect, the intersection is a union of cells in the PLC with dimensions lower than f or g. A triangulation of an input PLC is *conforming* if every edge and polygon appear as a union of segments and triangles in the triangulation. Additional Steiner vertices are often necessary. The 3D conforming Delaunay triangulation problem is to construct a triangulation of an input PLC that is both conforming and Delaunay. Figure 1a, b shows an input PLC and its conforming Delaunay triangulation. In many applications, it is often desired that the triangulation is not unnecessarily dense and the resulting tetrahedra are of *bounded*

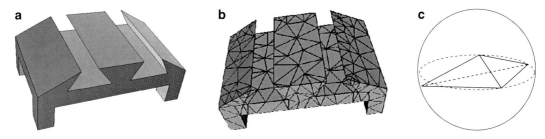

3D Conforming Delaunay Triangulation, Fig. 1 A PLC and its conforming Delaunay triangulation in (**a**) and (**b**). A sliver with negligible volume and edge lengths similar to its circumradius in (**c**).

aspect ratio. Another popular shape measure is the *radius-edge ratio*, which is the ratio of the circumradius of the tetrahedron to its shortest edge length. Tetrahedra with bounded radius-edge ratio may still have negligible volume, and they are known as *slivers.* Figure 1c shows a sliver.

Key Results

Since the Delaunayhood of edges and triangles are guaranteed by the emptiness of their circumspheres, one would imagine that a conforming Delaunay triangulation can be obtained by sprinkling Steiner vertices on the input edges and polygons. Indeed, Murphy, Mount, and Gable [6] showed a way to do this, but the resulting triangulation is very dense, and no shape guarantee is offered.

Shewchuk [9] gave the first algorithm that offers shape guarantee for PLCs in which two adjoining elements do not make an acute angle. (The exact requirement is more general and is called the projection condition [4, 9].) The algorithm is a generalization of Ruppert's Delaunay refinement algorithm in the plane [8]. An initial Delaunay triangulation is formed using the input vertices, and then Steiner vertices are added incrementally. Boundary conformity takes precedence. Therefore, whenever a segment on an edge or a triangle on a polygon has a nonempty diametric ball (the ball enclosed by the smallest circumsphere), that segment or triangle is split by inserting the center of its diametric ball. A Delaunay tetrahedron with radius-edge ratio larger than a prescribed constant $\rho > 2$ is split by inserting

its circumcenter. However, if this circumcenter lies inside the diametric ball of a segment or a triangle, then the insertion is aborted and that segment or triangle is split instead. Similarly, the insertion of a triangle's diametric ball center is aborted if it lies in the diametric ball of a segment, and that segment is split instead. The following theorem states the main result.

Theorem 1 ([4, 9]) *Let ρ be a constant greater than 2. Let P be a PLC with no acute input angle. A conforming Delaunay triangulation of P can be obtained by Delaunay refinement and all tetrahedra obtained have radius-edge ratio at most ρ.*

In the presence of acute angles, the splitting of segments and triangles may lead to an infinite loop as illustrated in Fig. 2. Notice that the Steiner vertices inserted are approaching the input vertex in Fig. 2. In the plane, Ruppert [8] proposed a fix: place some protecting circles centered at the input vertices, disallow the insertion of Steiner vertices inside these protecting circles, and triangulate the inside of these protecting circles using a separate mechanism. A key change is that if a Steiner vertex to be inserted is too close to an arc on a protecting circle, then the insertion of the Steiner vertex is aborted and the circular arc is split by inserting its midpoint. This is analogous to the splitting of segments.

Cohen-Steiner, de Verdière, and Yvinec [5] generalized this idea partly to three dimensions. They proposed to place protecting balls centered at the input vertices as well as at some appropriate points in the interior of input edges. These protecting balls cover all input vertices and edges. The intersection between a protecting ball

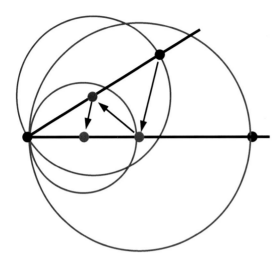

3D Conforming Delaunay Triangulation, Fig. 2 The midpoint of the segment with the largest diametric ball triggers the splitting of the segment with the second largest diametric ball, which in turn triggers the splitting of the segment with the smallest diametric ball. This may go on indefinitely

boundary and an input polygon is analogous to a protecting circle in 2D. Therefore, when we want to insert Steiner vertices in a polygon f to recover the Delaunay triangles on f, if such a Steiner vertex v is too close to an arc α at the intersection of f and some protecting ball boundary, the insertion of v is aborted and α is split instead. The portions of polygons inside the protecting balls are triangulated using a separate mechanism. If a tetrahedron τ has large radius-edge ratio but its circumcenter lies inside some protecting ball, then τ is just kept in the triangulation. As a result, no shape guarantee is offered.

Theorem 2 ([5]) *There is a Delaunay refinement algorithm that constructs a conforming Delaunay triangulation of any input PLC.*

Cheng and Poon [2] extended Delaunay refinement by observing that segments, circular arcs, triangles, and spherical triangles can all be handled in a uniform way.

Let $\mathcal{B}$ be the union of protecting balls with centers at the input vertices and interior of input edges. Let B be a protecting ball. Let ∂ denote the boundary operator. For every input polygon f, the Steiner vertices on $f \cap B \cap \partial B$ divide $f \cap B \cap \partial B$ into circular arcs. For every pro-

tecting B, the projection of the convex hull of the Steiner vertices on B onto ∂B divides ∂B into some spherical triangles. The diametric ball of a segment or triangle can be viewed as the circumscribing ball whose boundary intersects the affine hull of the segment or triangle at right angle. Analogously, the "diametric ball" of a circular arc in ∂B or a spherical triangle with vertices in ∂B can be defined as the circumscribing ball whose boundary intersects B at right angle. If a Steiner vertex to be inserted lies inside this "diametric ball," the insertion is aborted and the circular arc or spherical triangle is split instead. A circular arc is split by inserting its midpoint. A spherical triangle is split by inserting the intersection point between ∂B and the line segment joining the centers of B and the "diametric ball." The last ingredient is that for every pair of protecting balls whose centers are adjacent on an input edge, their boundaries should intersect at right angle. That is, the protecting ball B' adjacent to B serves as the "diametric ball" of the spherical triangles that have vertices on the circle $\partial B \cap \partial B'$.

Theorem 3 ([2]) *There exist a constant $\rho > 2$ and a Delaunay refinement algorithm that constructs a conforming Delaunay triangulation of any input PLC such that all resulting tetrahedra have radius-edge ratio at most ρ.*

In fact, all tetrahedra inside the union of protecting balls have aspect ratios that depend only on the smallest angle in the input PLC. Subsequently, different simplifications and algorithms with less expensive primitives have been proposed [3,7].

Placing Steiner vertices on the protecting balls and constructing the convex hull of the Steiner vertices on a protecting ball are fairly expensive. Even checking whether a Steiner vertex to be inserted lies inside any protecting ball is a burden. These expensive computations can be bypassed by switching to the *weighted Delaunay triangulation* – a more general variant of Delaunay triangulation.

Let B_x and B_y be two balls with centers x and y and radii $r_x \geq 0$ and $r_y \geq 0$. The *power distance* between B_x and B_y is defined to be

$$\pi(B_x, B_y) = d(x, y)^2 - r_x^2 - r_y^2.$$

This definition allows B_x or B_y to degenerate to a single point. As in the Euclidean case, the bisector between B_x and B_y is also a plane perpendicular to the line through x and y; however, the bisector plane may not pass through the midpoint of xy. Using the power distance, one can define a weighted version of the Voronoi diagram called the *power diagram*. The dual of the power diagram is known as the *weighted Delaunay triangulation*. For each segment, triangle or tetrahedron σ in the triangulation, there is a point z at equal and minimum power distances D from the vertices of σ. This point z is known as the *orthocenter* of σ. The ball centered at z with radius $\sqrt{D}$ is called the *orthoball* of σ, which is at zero power distances from the vertices of σ.

The key idea is to use a weighted Delaunay triangulation after placing the protecting balls. The Delaunay refinement strategy is then modified to insert orthocenters instead of centers of diametric balls. If the protecting balls are not too large, every triangle or tetrahedron σ in the initial weighted Delaunay triangulation involve a pair of nonoverlapping protecting ball, which must be a positive power distance apart. It follows that the orthocenter of σ lies outside all protecting balls. As the refinement progresses, an edge, triangle, or tetrahedron σ may involve Steiner vertices which can be viewed as balls of zero radii. Such a Steiner vertex must be at positive power distances from the other vertices of σ, so the orthocenter of σ also lies outside all protecting balls. In summary, the indefinite insertions of Steiner vertices at decreasing distances from the input vertices and edges as shown in Fig. 2 cannot happen. The following theorem summarizes the result.

Theorem 4 ([4]) *Let P be a PLC. Let ρ be any constant at least 2. There is an algorithm that constructs a conforming weighted Delaunay triangulation of P in which no tetrahedron has an orthoradius-edge ratio greater than ρ. Therefore, tetrahedra with no weighted vertices have circumradius-edge ratio at most ρ.*

There are known methods for eliminating slivers from a conforming weighted Delaunay triangulation of a PLC [1, 4].

Cross-References

▶ Voronoi Diagrams and Delaunay Triangulations
▶ Meshing Piecewise Smooth Complexes

Recommended Reading

1. Cheng S-W, Dey TK (2003) Quality meshing with weighted Delaunay refinement. SIAM J Comput 33(1):69–93
2. Cheng S-W, Poon S-H (2006) Three-dimensional Delaunay mesh generation. Discret Comput Geom 36(3):419–456; conference version in Proceedings of the ACM-SIAM symposium on discrete algorithms (2003)
3. Cheng S-W, Dey TK, Ramos EA, Ray T (2005) Quality meshing for polyhedra with small angles. Int J Comput Geom Appl 15(4):421–461; conference version in Proceedings of the annual symposium on computational geometry (2004)
4. Cheng S-W, Dey TK, Shewchuk JR (2012) Delaunay mesh generation. CRC, Boca Raton
5. Cohen-Steiner D, de Verdière ÉC, Yvinec M (2004) Conforming Delaunay triangulation in 3D. Comput Geom Theory Appl 28(2–3):217–233
6. Murphy M, Mount DM, Gable CW (2001) A point-placement strategy for conforming Delaunay tetrahedralization. Int J Comput Geom Appl 11(6):669–682
7. Pav SE, Walkington NJ (2004) Robust three-dimensional Delaunay refinement. In: Proceedings of the international meshing roundtable, Williamsburg, pp 145–156
8. Ruppert J (1995) A Delaunay refinement algorithm for quality 2-dimensional mesh generation. J. Algorithms 18(3):548–585
9. Shewchuk JR (1998) Tetrahedral mesh generation by Delaunay refinement. In: Proceedings of the annual symposium on computational geometry, Minneapolis, pp 86–95

Decoding Reed–Solomon Codes

Venkatesan Guruswami
Department of Computer Science and Engineering, University of Washington, Seattle, WA, USA

Keywords

Decoding; Error correction

Years and Authors of Summarized Original Work

1999; Guruswami, Sudan

Problem Definition

In order to ensure the integrity of data in the presence of errors, an *error-correcting code* is used to *encode* data into a redundant form (called a *codeword*). It is natural to view both the original data (or *message*) as well as the associated codeword as strings over a finite alphabet. Therefore, an error-correcting code C is defined by an injective encoding map $E: \Sigma^k \to \Sigma^n$, where k is called the *message length*, and n the *block length*. The codeword, being a redundant form of the message, will be longer than the message. The *rate* of an error-correcting code is defined as the ratio k/n of the length of the message to the length of the codeword. The rate is a quantity in the interval $(0, 1]$, and is a measure of the redundancy introduced by the code. Let $R(C)$ denote the rate of a code C.

The redundancy built into a codeword enables detection and hopefully also correction of any errors introduced, since only a small fraction of all possible strings will be legitimate codewords. Ideally, the codewords encoding different messages should be "far-off" from each other, so that one can recover the original codeword even when it is distorted by moderate levels of noise. A natural measure of distance between strings is the Hamming distance. The Hamming distance between strings $x, y \in \Sigma^*$ of the same length, denoted $\mathsf{dist}(x, y)$, is defined to be the number of positions i for which $x_i \neq y_i$.

The *minimum distance*, or simply *distance*, of an error-correcting code C, denoted $d(C)$, is defined to be the smallest Hamming distance between the encodings of two distinct messages. The *relative distance* of a code C of block length n, denoted $\delta(C)$, is the ratio between its distance and n. Note that arbitrary corruption of any $\lfloor (d(C) - 1)/2 \rfloor$ of locations of a codeword of C cannot take it closer (in Hamming distance) to

any other codeword of C. Thus in principle (i.e., efficiency considerations apart) error patterns of at most $\lfloor (d(C) - 1)/2 \rfloor$ errors can be corrected. This task is called *unique decoding* or *decoding up to half-the-distance*. Of course, it is also possible, and will often be the case, that error patterns with more than $d(C)/2$ errors can also be corrected by decoding the string to the closest codeword in Hamming distance. The latter task is called *Nearest-Codeword decoding* or *Maximum Likelihood Decoding (MLD)*.

One of the fundamental trade-offs in the theory of error-correcting codes, and in fact one could say all of combinatorics, is the one between rate $R(C)$ and distance $d(C)$ of a code. Naturally, as one increases the rate and thus number of codewords in a code, some two codewords must come closer together thereby lowering the distance. More qualitatively, this represents the tension between the redundancy of a code and its error-resilience. To correct more errors requires greater redundancy, and thus lower rate.

A code defined by encoding map $E : \Sigma^k \to \Sigma^n$ with minimum distance d is said to be an (n, k, d) code. Since there are $|\Sigma|^k$ codewords and only $|\Sigma^{k-1}|$ possible projections onto the first $k = 1$ coordinates, some two codewords must agree on the first $k - 1$ positions, implying that the distance d of the code must obey $d \leq n - k + 1$ (this is called the *Singleton bound*). Quite surprisingly, over large alphabets Σ there are well-known codes called Reed–Solomon codes which meet this bound exactly and have the optimal distance $d = n - k + 1$ for any given rate k/n. (In contrast, for small alphabets, such as $\Sigma = \{0, 1\}$, the optimal trade-off between rate and relative distance for an asymptotic family of codes is unknown and is a major open question in combinatorics.)

This article will describe the best known algorithmic results for error-correction of Reed–Solomon codes. These are of central theoretical and practical interest given the above-mentioned optimal trade-off achieved by Reed–Solomon codes, and their ubiquitous use in our everyday lives ranging from compact disc players to deep-space communication.

Reed–Solomon Codes

Definition 1 A *Reed–Solomon code* (or RS code), $\mathsf{RS}_{\mathbb{F},S}[n,k]$, is parametrized by integers n,k satisfying $1 \leq k \leq n$, a finite field $\mathbb{F}$ of size at least n, and a tuple $S = (\alpha_1, \alpha_2, \ldots, \alpha_n)$ of n *distinct* elements from $\mathbb{F}$. The code is described as a subset of $\mathbb{F}^n$ as:

$$\mathsf{RS}_{\mathbb{F},S}[n,k] = \{(p(\alpha_1), p(\alpha_2), \ldots, p(\alpha_n))$$

$$|p(X) \in \mathbb{F}[X] \text{ is a polynomial of degree} \leq k-1\}.$$

In other words, the message is viewed as a polynomial, and it is encoded by evaluating the polynomial at n distinct field elements $\alpha_1, \ldots, \alpha_n$. The resulting code is linear of dimension k, and its minimum distance equals $n - k + 1$, which matches the Singleton bound.

The distance property of RS codes follows from the fact that the evaluations of two distinct polynomials of degree less than k can agree on at most $k - 1$ field elements. Note that in the absence of errors, given a codeword $\mathbf{y} \in \mathbb{F}^n$, one can recover its corresponding message by polynomial interpolation on any k out of the n codeword positions. In fact, this also gives an *erasure decoding* algorithm when all but the information-theoretically bare minimum necessary k symbols are erased from the codeword (but the receiver *knows* which symbols have been erased and the correct values of the rest of the symbols). The RS decoding problem, therefore, amounts to a noisy polynomial interpolation problem when some of the evaluation values are incorrect.

The holy grail in decoding RS codes would be to find the polynomial $p(X)$ whose RS encoding is closest in Hamming distance to a noisy string $\mathbf{y} \in \mathbb{F}^n$. One could then decode $\mathbf{y}$ to this message $p(X)$ as the maximum likelihood choice. No efficient algorithm for such nearest-codeword decoding is known for RS codes (or for that matter any family of "good" or non-trivial codes), and it is believed that the problem is NP-hard. Guruswami and Vardy [6] proved the problem to NP-hard over exponentially large fields, but this is a weak negative result since normally one considers Reed–Solomon codes over fields of size at most $O(n)$.

Given the intractability of nearest-codeword decoding in its extreme generality, lot of attention has been devoted to the *bounded distance decoding* problem, where one assumes that the string $\mathbf{y} \in \mathbb{F}^n$ to be decoded has at most e errors, and the goal is to find the Reed–Solomon codeword(s) within Hamming distance e from $\mathbf{y}$.

When $e < (n - k)/2$, this corresponds to decoding up to half the distance. This is a classical problem for which a polynomial time algorithm was first given by Peterson [8]. (It is notable that this even before the notion of polynomial time was put forth as the metric of theoretical efficiency.) The focus of this article is on a *list decoding* algorithm for Reed–Solomon codes due to Guruswami and Sudan [5] that decode beyond half the minimum distance. The formal problem and the key results are stated next.

Key Results

In this section, the main result of focus concerning decoding Reed–Solomon codes is stated. Given the target of decoding errors beyond half-the-minimum distance, one needs to deal with inputs where there may be more than one codeword within the radius e specified in the bounded distance decoding problem. This is achieved by a relaxation of decoding called *list decoding* where the decoder outputs a list of all codewords (or the corresponding messages) within Hamming distance e from the received word. If one wishes, one can choose the closest codeword among the list as the "most likely" answer, but there are many applications of Reed–Solomon decoding, for example to decoding concatenated codes and several applications in complexity theory and cryptography, where having the entire list of codewords adds to the power of the decoding primitive. The main result of Guruswami and Sudan [5], building upon the work of Sudan [9], is the following:

Theorem 1 ([5]) *Let* $C = \mathsf{RS}_{\mathbb{F},S}[n,k]$ *be a Reed–Solomon code over a field* $\mathbb{F}$ *of size* $q \geq n$ *with* $S = (\alpha_1, \alpha_2, \ldots, \alpha_n)$. *There is a deterministic algorithm running in time polynomial in* q *that on input* $\mathbf{y} \in \mathbb{F}_q^n$ *outputs* *a list of all polynomials $p(X) \in \mathbb{F}[X]$ of degree less than k for which $p(\alpha_i) \neq y_i$ for less than $n - \sqrt{(k-1)n}$ positions $i \in \{1, 2, \ldots, n\}$. Further, at most $O(n^2)$ polynomials will be output by the algorithm in the worst-case.*

Alternatively, one can correct a RS code of block length n and rate $R = k/n$ up to $n - \sqrt{(k-1)}$ errors, or equivalently a fraction $1 - \sqrt{R}$ of errors.

The Reed–Solomon decoding algorithm is based on the solution to the following more general polynomial reconstruction problem which seems like a natural algebraic question in itself. (The problem is more general than RS decoding since the α_i's need not be distinct.)

Problem 1 (Polynomial Reconstruction)
Input: Integers $k, t \leq n$ and n distinct pairs $\{(\alpha_i, y_i)\}_{i=1}^{n}$ where $\alpha_i, y_i \in \mathbb{F}$.
Output: A list of all polynomials $p(X) \in \mathbb{F}[X]$ of degree less than k which satisfy $p(\alpha_i) = y_i$ for at least t values of $i \in [n]$.

Theorem 2 *The polynomial reconstruction problem can be solved in time polynomial in $n, |\mathbb{F}|$, provided $t > \sqrt{(k-1)n}$.*

The reader is referred to the original papers [5, 9], or a recent survey [1], for details on the above algorithm. A quick, high level peek into the main ideas is given below. The first step in the algorithm consists of an interpolation step where a nonzero bivariate polynomial $Q(X,Y)$ is "fit" through the n pairs (α_i, y_i), so that $Q(\alpha_i, y_i) = 0$ for every i. The key is to do this with relatively low degree; in particular one can find such a $Q(X,Y)$ with so-called $(1, k-1)$-weighted degree at most $D \approx \sqrt{2(k-1)n}$. This degree budget on Q implies that for any polynomial $p(X)$ of degree less than k, $Q(X, p(X))$ will have degree at most D. Now whenever $p(\alpha_i) = y_i$, $Q(\alpha_i, p(\alpha)i)) = Q(\alpha_i, y_i) = 0$. Therefore, if a polynomial $p(X)$ satisfies $p(\alpha_i) = y_i$ for at least t values of i, then $Q(X, p(X))$ has at least t roots. On the other hand the polynomial $Q(X, p(X))$ has degree at most D. Therefore, if $t > D$, one must have $Q(X, p(X)) = 0$, or in other words $Y - p(X)$ is a factor of $Q(X,Y)$. The second step of the algorithm factorized the

polynomial $Q(X,Y)$, and all polynomials $p(X)$ that must be output will be found as factors $Y - p(X)$ of $Q(X,Y)$.

Note that since $D \approx \sqrt{2(k-1)n}$ this gives an algorithm for polynomial reconstruction provided the agreement parameter t satisfies $t > \sqrt{2(k-1)n}$ [9]. To get an algorithm for $t > \sqrt{(k-1)n}$, and thus decode beyond half the minimum distance $(n-k)/2$ for all parameter choices for k, n, Guruswami and Sudan [5] use the crucial idea of allowing "multiple roots" in the interpolation step. Specifically, the polynomial Q is required to have $r \geq 1$ roots at each pair (α_i, y_i) for some integer multiplicity parameter r (the notion needs to be formalized properly, see [5] for details). This necessitates an increase in the $(1, k-1)$-weighted degree of a factor of about $r/\sqrt{2}$, but the gain is that one gets a factor r more roots for the polynomial $Q(X, p(X))$. These facts together lead to an algorithm that works as long as $t > \sqrt{(k-1)n}$.

There is an additional significant benefit offered by the multiplicity based decoder. The multiplicities of the interpolation points need not all be equal and they can picked in proportion to the reliability of different received symbols. This gives a powerful way to exploit "soft" information in the decoding stage, leading to impressive coding gains in practice. The reader is referred to the paper by Koetter and Vardy [7] for further details on using multiplicities to encode symbol level reliability information from the channel.

Applications

Reed–Solomon codes have been extensively studied and are widely used in practice. The above decoding algorithm corrects more errors beyond the traditional half the distance limit and therefore directly advances the state of the art on this important algorithmic task. The RS list decoding algorithm has also been the backbone for many further developments in algorithmic coding theory. In particular, using this algorithm in concatenation schemes leads to good binary list-decodable codes. A variant of

RS codes called folded RS codes have been used to achieve the optimal trade-off between error-correction radius and rate [3] (see the companion encyclopedia entry by Rudra on folded RS codes).

The RS list decoding algorithm has also found many surprising applications beyond coding theory. In particular, it plays a key role in several results in cryptography and complexity theory (such as constructions of randomness extractors and pseudorandom generators, hardness amplification, constructions to hardcore predicates, traitor tracing, reductions connecting worst-case hardness to average-case hardness, etc.); more information can be found, for instance, in [10] or Chap. 12 in [2].

Open Problems

The most natural open question is whether one can improve the algorithm further and correct more than a fraction $1 - \sqrt{R}$ of errors for RS codes of rate R. It is important to note that there is a combinatorial limitation to the number of errors one can list decode from. One can only list decode in polynomial time from a fraction ρ of errors if for every received word $\mathbf{y}$ the number of RS codewords within distance $e = \rho n$ of $\mathbf{y}$ is bounded by a polynomial function of the block length n. The largest ρ for which this holds as a function of the rate R is called the list decoding radius $\rho_{\mathsf{LD}} = \rho_{\mathsf{LD}}(R)$ of RS codes. The RS list decoding algorithm discussed here implies that $\rho_{\mathsf{LD}}(R) \geq 1 - \sqrt{R}$, and it is trivial to see than $\rho_{\mathsf{LD}}(R) \leq 1 - R$. Are there RS codes (perhaps based on specially structured evaluation points) for which $\rho_{\mathsf{LD}}(R) > 1 - \sqrt{R}$? Are there RS codes for which the $1 - \sqrt{R}$ radius (the so-called "Johnson bound") is actually tight for list decoding? For the more general polynomial reconstruction problem the $\sqrt{(k-1)n}$ agreement cannot be improved upon [4], but this is not known for RS list decoding.

Improving the NP-hardness result of [6] to hold for RS codes over polynomial sized fields and for smaller decoding radii remains an important challenge.

Cross-References

▶ Learning Heavy Fourier Coefficients of Boolean Functions
▶ List Decoding near Capacity: Folded RS Codes
▶ LP Decoding

Recommended Reading

1. Guruswami V (2007) Algorithmic results in list decoding. In: Foundations and trends in theoretical computer science, vol 2, issue 2. NOW Publishers, Hanover
2. Guruswami V (2004) List decoding of error-correcting codes. Lecture notes in computer science, vol 3282. Springer, Berlin
3. Guruswami V, Rudra A (2008) Explicit codes achieving list decoding capacity: Error-correction with optimal redundancy. IEEE Trans Inform Theory 54(1):135–150
4. Guruswami V, Rudra A (2006) Limits to list decoding Reed–Solomon codes. IEEE Trans Inf Theory 52(8):3642–3649
5. Guruswami V, Sudan M (1999) Improved decoding of Reed–Solomon and algebraic-geometric codes. IEEE Trans Inf Theory 45(6):1757–1767
6. Guruswami V, Vardy A (2005) Maximum likelihood decoding of Reed–Solomon codes is NP-hard. IEEE Trans Inf Theory 51(7):2249–2256
7. Koetter R, Vardy A (2003) Algebraic soft-decision decoding of Reed–Solomon codes. IEEE Trans Inf Theory 49(11):2809–2825
8. Peterson WW (1960) Encoding and error-correction procedures for Bose-Chaudhuri codes. IEEE Trans Inf Theory 6:459–470
9. Sudan M (1997) Decoding of Reed–Solomon codes beyond the error-correction bound. J Complex 13(1):180–193
10. Sudan M (2000) List decoding: algorithms and applications. SIGACT News 31(1):16–27

Decremental All-Pairs Shortest Paths

Camil Demetrescu and Giuseppe
F. Italiano
Department of Computer and Systems Science,
University of Rome, Rome, Italy
Department of Information and Computer
Systems, University of Rome, Rome, Italy

Keywords

Deletions-only dynamic all-pairs shortest paths

Years and Authors of Summarized Original Work

2004; Demetrescu, Italiano

Problem Definition

A dynamic graph algorithm maintains a given property P on a graph subject to dynamic changes, such as edge insertions, edge deletions and edge weight updates. A dynamic graph algorithm should process queries on property P quickly, and perform update operations faster than recomputing from scratch, as carried out by the fastest static algorithm. An algorithm is *fully dynamic* if it can handle both edge insertions and edge deletions. A *partially dynamic* algorithm can handle either edge insertions or edge deletions, but not both: it is *incremental* if it supports insertions only, and *decremental* if it supports deletions only.

This entry addressed the *decremental* version of the all-pairs shortest paths problem (APSP), which consists of maintaining a directed graph with real-valued edge weights under an inter-mixed sequence of the following operations:

delete(u, v): delete edge (u, v) from the graph.
distance(x, y): return the distance from vertex x to vertex y.
path(x, y): report a shortest path from vertex x to vertex y, if any.

A natural variant of this problem supports a generalized delete operation that removes a vertex and all edges incident to it. The algorithms addressed in this entry can deal with this generalized operation within the same bounds.

History of the Problem
A simple-minded solution to this problem would be to rebuild shortest paths from scratch after each deletion using the best static APSP algorithm so that distance and path queries can be reported in optimal time. The fastest known static APSP algorithm for arbitrary real weights has a running time of $O(mn + n^2 \log \log n)$, where m is the number of edges and n is the number of vertices in the graph [13]. This is $\Omega(n^3)$ in the worst case. Fredman [6] and later Takaoka [19] showed how to break this cubic barrier: the best asymptotic bound is by Takaoka, who showed how to solve APSP in $O(n^3 \sqrt{\log \log n / \log n})$ time.

Another simple-minded solution would be to answer queries by running a point-to-point shortest paths computation, without the need to update shortest paths at each deletion. This can be done with Dijkstra's algorithm [3] in $O(m + n \log n)$ time using the Fibonacci heaps of Fredman and Tarjan [5]. With this approach, queries are answered in $O(m + n \log n)$ worst-case time and updates require optimal time.

The dynamic maintenance of shortest paths has a long history, and the first papers date back to 1967 [11, 12, 17]. In 1985 Even and Gazit [4] presented algorithms for maintaining shortest paths on directed graphs with arbitrary real weights. The worst-case bounds of their algorithm for edge deletions were comparable to recomputing APSP from scratch. Also Ramalingam and Reps [15, 16] and Frigioni et al. [7, 8] considered dynamic shortest path algorithms with real weights, but in a different model. Namely, the running time of their algorithm is analyzed in terms of the output change rather than the input size (*output bounded complexity*). Again, in the worst case the running times of output-bounded dynamic algorithms are comparable to recomputing APSP from scratch.

The first decremental algorithm that was provably faster than recomputing from scratch was devised by King for the special case of graphs with integer edge weights less than C: her algorithm can update shortest paths in a graph subject to a sequence of $\Omega(n^2)$ deletions in $O(C \cdot n^2)$ amortized time per deletion [9]. Later, Demetrescu and Italiano showed how to deal with graphs with real non-negative edge weights in $O(n^2 \log n)$ amortized time per deletion [2] in a sequence of $\Omega(m/n)$ operations. Both algorithms work in the more general case where edges are not deleted from the graph, but their weight is increased at each update. Moreover, since they update shortest paths explicitly after each deletion, queries are answered in optimal time at any time during a sequence of operations.

Key Results

The decremental APSP algorithm by Demetrescu and Italiano hinges upon the notion of locally shortest paths [2].

Definition 1 A path is *locally shortest* in a graph if all of its proper subpaths are shortest paths.

Notice that by the optimal-substructure property, a shortest path is locally shortest. The main idea of the algorithm is to keep information about locally shortest paths in a graph subject to edge deletions. The following theorem derived from [2] bounds the number of changes in the set of locally shortest paths due to an edge deletion:

Theorem 1 *If shortest paths are unique in the graph, then the number of paths that start or stop being shortest at each deletion is $O(n^2)$ amortized over $\Omega(m/n)$ update operations.*

The result of Theorem 1 is purely combinatorial and assumes that shortest paths are unique in the graph. The latter can be easily achieved using any consistent tie-breaking strategy (see, e.g., [2]). It is possible to design a deletions-only algorithm that pays only $O(\log n)$ time per change in the set of locally shortest paths, using a simple modification of Dijkstra's algorithm [3]. Since by Theorem 1 the amortized number of changes is bounded by $O(n^2)$, this yields the following result:

Theorem 2 *Consider a graph with n vertices and an initial number of m edges subject to a sequence of $\Omega(m/n)$ edge deletions. If shortest paths are unique and edge weights are nonnegative, it is possible to support each delete operation in $O(n^2 \log n)$ amortized time, each distance query in O(1) worst-case time, and each path query in $O(\ell)$ worst-case time, where ℓ is the number of vertices in the reported shortest path. The space used is O(mn).*

Applications

Application scenarios of dynamic shortest paths include network optimization [1], document formatting [10], routing in communication systems, robotics, incremental compilation, traffic information systems [18], and dataflow analysis. A comprehensive review of real-world applications of dynamic shortest path problems appears in [14].

URL to Code

An efficient C language implementation of the decremental algorithm addressed in section "Key Results" is available at the URL: http://www.dis.uniroma1.it/~demetres/experim/dsp.

Cross-References

▶ All Pairs Shortest Paths in Sparse Graphs
▶ All Pairs Shortest Paths via Matrix Multiplication
▶ Fully Dynamic All Pairs Shortest Paths

Recommended Reading

1. Ahuja R, Magnanti T, Orlin J (1993) Network flows: theory, algorithms and applications. Prentice Hall, Englewood Cliffs
2. Demetrescu C, Italiano G (2004) A new approach to dynamic all pairs shortest paths. J Assoc Comput Mach 51:968–992
3. Dijkstra E (1959) A note on two problems in connexion with graphs. Numer Math 1:269–271
4. Even S, Gazit H (1985) Updating distances in dynamic graphs. Methods Oper Res 49:371–387
5. Fredman M, Tarjan R (1987) Fibonacci heaps and their use in improved network optimization algorithms. J ACM 34:596–615
6. Fredman ML (1976) New bounds on the complexity of the shortest path problems. SIAM J Comput 5(1):87–89
7. Frigioni D, Marchetti-Spaccamela A, Nanni U (1998) Semi-dynamic algorithms for maintaining single source shortest paths trees. Algorithmica 22: 250–274
8. Frigioni D, Marchetti-Spaccamela A, Nanni U (2000) Fully dynamic algorithms for maintaining shortest paths trees. J Algorithm 34:351–381

9. King V (1999) Fully dynamic algorithms for maintaining all-pairs shortest paths and transitive closure in digraphs. In: Proceedings of 40th IEEE symposium on foundations of computer science (FOCS'99). IEEE Computer Society, New York, pp 81–99

10. Knuth D, Plass M (1981) Breaking paragraphs into lines. Softw-Pract Exp 11:1119–1184

11. Loubal P (1967) A network evaluation procedure. Highw Res Rec 205:96–109

12. Murchland J (1967) The effect of increasing or decreasing the length of a single arc on all shortest distances in a graph. Technical report, LBS-TNT-26, London Business School. Transport Network Theory Unit, London

13. Pettie S (2003) A new approach to all-pairs shortest paths on realweighted graphs. Theor Comput Sci 312:47–74, Special issue of selected papers from ICALP (2002)

14. Ramalingam G (1996) Bounded incremental computation. Lect Note Comput Sci 1089

15. Ramalingam G, Reps T (1996) An incremental algorithm for a generalization of the shortest path problem. J Algorithm 21:267–305

16. Ramalingam G, Reps T (1996) On the computational complexity of dynamic graph problems. Theor Comput Sci 158:233–277

17. Rodionov V (1968) The parametric problem of shortest distances. USSR Comput Math Math Phys 8:336–343

18. Schulz F, Wagner D, Weihe K (1999) Dijkstra's algorithm on-line: an empirical case study frompublic railroad transport. In: Proceedings of 3rd workshop on algorithm engineering (WAE'99), London. Notes in computer science, vol 1668, pp 110–123

19. Takaoka T (1992) A new upper bound on the complexity of the all pairs shortest path problem. Inf Proc Lett 43:195–199

Decremental Approximate-APSP in Directed Graphs

Aaron Bernstein
Department of Computer Science, Columbia University, New York, NY, USA

Keywords

Directed graphs; Dynamic graph algorithms; Shortest paths

Years and Authors of Summarized Original Work

2013; Bernstein

Problem Definition

A dynamic graph algorithm maintains information about a graph that is changing over time. Given a property $\mathcal{P}$ of the graph (e.g., minimum spanning tree), the algorithm must support an online sequence of query and update operations, where an update operation changes the underlying graph, while a query operation asks for the state of $\mathcal{P}$ in the current graph. In the typical model studied, each update only affects a single edge. In a *fully dynamic* setting, an update can insert or delete an edge or change the weight of an existing edge; in a *decremental* setting an update can only delete an edge or increase a weight; in an *incremental* setting an update can insert an edge or decrease a weight.

This entry addresses the decremental $(1 + \epsilon)$-approximate all-pairs shortest path problem (APSP) in weighted directed graphs. The goal is to maintain a directed graph G with real-valued nonnegative edge weights under an online intermixed sequence of the following operations:

- **delete**(u, v) (update): remove edge (u, v) from G.
- **increase-weight**(u, v, w) (update): increase the weight of edge (u, v) to w.
- **distance**(u, v) (query): return a $(1 + \epsilon)$-approximation to the shortest $u - v$ distance in G.
- **path**(u, v) (query): return a $(1 + \epsilon)$-approximate shortest path from u to v.

A History of Decremental APSP

The naive approach to the decremental APSP problem is to recompute shortest paths from scratch after every update, allowing queries to be answered in optimal time. Letting n be the number of vertices and m the number of edges, computing APSP requires $O(mn + n^2 \log \log(n))$

time in sparse graphs [11] or slightly less than n^3 in dense graphs (see [13, 14]). Another simple-minded approach would be to not perform any computation during the updates and to simply compute the shortest $u - v$ path from scratch when a query arrived. This would lead to a constant update time and a query time of $O(m + n \log(n))$ using Dijkstra's algorithm with Fibonacci heaps [6].

One can improve significantly upon both the above approaches by reusing information between updates. Decremental shortest path algorithms have a long history, with the current state of the art for the general case of directed graphs with real-valued weights being an algorithm of Demetrescu and Italiano which achieves constant query time and update time $O(n^2 \log(n))$ [5]. Later papers improved upon $O(n^2)$ update time in restricted types of graph. In *unweighted* directed graphs, Baswana et al. achieve an amortized update of $O(n^3 \log^2(n)/m)$ for exact distances and $\tilde{O}(\epsilon^{-1} n^2/\sqrt{m})$ for $(1+\epsilon)$ approximate distances [1]. (We say that $f(n) = \tilde{O}(g(n))$ if $f(n) = O(g(n)\text{polylog}(n))$). Keeping the $(1 + \epsilon)$ approximation, Roditty and Zwick further reduced the amortized update time to $\tilde{O}(n/\epsilon)$ [12].

An amortized update time of $\tilde{O}(n)$ forms a natural barrier for decremental APSP because if edges are deleted from the graph one at a time, an $\tilde{O}(n)$ update time allows us to maintain APSP over the entire sequence of deletions in a *total* of $\tilde{O}(mn)$ time; excepting fast matrix multiplication in dense graphs, this $\tilde{O}(mn)$ matches the best known bound for the much simpler problem of computing APSP a single time in the *static* setting. Roddity and Zwick achieve this desired *total* update time of $\tilde{O}(mn)$ only for undirected, unweighted graphs; this entry focuses on a result of Bernstein that achieves the same $\tilde{O}(mn)$ for directed graphs with weights polynomial in n [3].

There have recently been several results on breaking through the $\tilde{O}(n)$ amortized update time barrier in undirected graphs by allowing a larger than $(1 + \epsilon)$ approximation (see [4, 7, 9]), as well in directed graphs for *single-source* shortest paths [8].

Key Results

Bernstein's result shows that in a directed graph with weights polynomial in n, we can maintain $(1 + \epsilon)$-approximate decremental APSP with constant query time and a *total* update time of $\tilde{O}(mn)$ over the entire sequence of deletions and weight increases (see Theorem 2 below). At a high level, Bernstein's result uses the framework from his earlier paper on fully dynamic APSP in undirected graphs [2], but all the details and techniques change significantly in the shift to directed graphs.

From Weighted Distances to Hop Distances

The majority of dynamic APSP algorithms use as a building block an algorithm of King for maintaining a *single-source* shortest path tree under deletions [10]. King's algorithm maintains a shortest path tree up to distance d (assuming integral weights) in the total update time $O(md)$ over all deletions (amortized $O(d)$ per update); hence, $O(mn)$ is the total update time in unweighted graphs where $d \le n$. This makes it an extremely efficient building block for small distances, but with two main drawbacks: it is inefficient at handling vertices that are far apart and it completely fails in graphs with large weights where d can be very big. Bernstein's algorithm overcomes the second of these problems by showing that if we allow a $(1 + \epsilon)$ approximation, then with a simple scaling approach, we can shift the dependency from the weighted distance d between two vertices to the unweighted *hop distance* between them.

Definition 1 The hop distance between two vertices x and y is the number of edges on the shortest $x - y$ path. The $(1 + \epsilon)$-approximate $x - y$ hop distance, denoted $h(x, y)$, is the minimum number of edges among any $(1 + \epsilon)$-approximate $x - y$ path.

Theorem 1 *Given a directed graph G with non-negative real weights and a source s and letting R be the ratio of the heaviest to the lightest nonzero edge weight in the graph, we can for any* *hop distance h decrementally maintain $(1 + \epsilon)$-shortest paths from s to all vertices v for which $h(s, v) \leq h$. The total update time over the whole sequence of deletions and weight increases is $O(nh \log(R)/\epsilon)$, which is $O(nh)$ if weights are polynomial in n.*

We refer to the above decremental SSSP algorithm as h-SSSP. In short, Theorem 1 tells us that with a $(1 + \epsilon)$ approximation, we can decrementally maintain a shortest path tree in time proportional not to the maximum distance of the tree ($O(nd)$) but to the maximum *hop distance* ($O(nh)$) of the tree. This is a big improvement in weighted graphs where $h \ll d$, but still inadequate as h can be $\Omega(n)$. Bernstein's key idea is that regardless of whether the original graph is weighted or not, we can add weighted edges that reduce hop distances in the graph and hence allow h-SSSP to run extremely efficiently.

Shortcut Edges

The algorithm of Bernstein works by adding many different (weighted) shortcut edges (x, y) to the original graph G, which are defined as edges that do not exist in G itself and have weight $w(x, y)$ satisfying $\delta(x, y) \leq w(x, y) \leq (1 + \epsilon)\delta(x, y)$, where $\delta(x, y)$ is the shortest $x-y$ distance. Note that as the graph changes, $\delta(x, y)$ will increase, and so the algorithm will have to increase $w(x, y)$ for the shortcut edge to remain valid; a shortcut edge (x, y) is not simply computed once, but must be maintained over the whole sequence of edge deletions and weight increases.

It is clear that because the weight of a shortcut edge (x, y) is tethered to $\delta(x, y)$, the shortcut edges do not change shortest distances in the graph. But they do drastically reduce hop distances. In an unweighted graph with $\delta(x, y) = 1,000$, a single $x - y$ edge of weight 1,000 (or slightly larger) decreases $h(x, y)$ from 1,000 to 1. Moreover, any path that goes through x and y can also use the (x, y) shortcut edge to reduce its hop distance by 999.

Bernstein's algorithm runs in phases, each of which adds more shortcut edges to successively decrease all hop distances in the graph by a factor of 2. It starts by defining a small set of pairs S_1 for which it maintains approximate shortest paths over the entire sequence of updates; this first step can easily be done in the desired $\tilde{O}(mn \log(R))$ total update time as instead of maintaining *all-pairs* shortest paths, the algorithm only has to maintain $|S_1|$ pairs. Now that the algorithm maintains $\delta(x, y)$ for every pair (x, y) in S_1, it can add shortcut edges (x, y) to the graph. These shortcut edges decrease hop distances in the graph, thus increasing the efficiency of the h-SSSP building block and allowing it to maintain approximate shortest distances within a slightly larger set of pairs S_2 in the same $\tilde{O}(mn \log(R))$ total update time. Since knowing the shortest distance between two vertices allows us to maintain a corresponding shortcut edge, maintaining the larger set of distances S_2 directly leads to a larger set of shortcut edges, which further reduce hop distances in the graph, allowing h-SSSP to efficiently maintain shortest distances for an even larger set of pairs S_3; this in turn leads to more shortcut edges, thus further reducing hop distances and allowing h-SSSP to maintain a larger distance set S_4, and so on. After $\log(n)$ such layers, there are enough shortcut edges to ensure that all hop distances are constant, so by Theorem 1 h-SPPP can decrementally maintain a shortest path tree in the graph in a total update time of only $\tilde{O}(n \log(R)/\epsilon)$; doing this from every vertex yields the desired bound of $\tilde{O}(mn \log(R)/\epsilon)$ for decremental APSP.

Theorem 2 *Let G be a graph with nonnegative real-valued edge weights, n vertices, and m initial edges subject to an arbitrary sequence of Σ edge deletions and weight increases. Let R be the ratio of the heaviest weight that ever appears in G to the lightest nonzero weight. It is possible to support the whole sequence of updates in a total time $\tilde{O}(mn \log(R)/\epsilon) + O(\Sigma)$ while answering queries with a single $O(1)$ time lookup.*

Cross-References

▶ Decremental All-Pairs Shortest Paths
▶ Dynamic Approximate All-Pairs Shortest Paths: Breaking the $O(mn)$ Barrier and Derandomization
▶ Dynamic Approximate-APSP
▶ Trade-Offs for Dynamic Graph Problems

Recommended Reading

1. Baswana S, Hariharan R, Sen S (2007) Improved decremental algorithms for maintaining transitive closure and all-pairs shortest paths. J Algorithms 62(2):74–92
2. Bernstein A (2009) Fully dynamic approximate all-pairs shortest paths with constant query and close to linear update time. In: Proceedings of the 50th FOCS, Atlanta, pp 50–60
3. Bernstein A (2013) Maintaining shortest paths under deletions in weighted directed graphs: [extended abstract]. In: STOC, Palo Alto, pp 725–734
4. Bernstein A, Roditty L (2011) Improved dynamic algorithms for maintaining approximate shortest paths under deletions. In: Proceedings of the 22nd SODA, San Francisco, pp 1355–1365
5. Demetrescu C, Italiano GF (2004) A new approach to dynamic all pairs shortest paths. J ACM 51(6): 968–992. doi:http://doi.acm.org/10.1145/1039488.1039492
6. Fredman ML, Tarjan RE (1987) Fibonacci heaps and their uses in improved network optimization algorithms. J ACM 34(3):596–615
7. Henzinger M, Krinninger S, Nanongkai D (2013) Dynamic approximate all-pairs shortest paths: breaking the $o(mn)$ barrier and derandomization. In: FOCS 2013, Berkeley, pp 538–547
8. Henzinger M, Krinninger S, Nanongkai D (2014) Sublinear-time decremental algorithms for single-source reachability and shortest paths on directed graphs. In: STOC, New York, pp 674–683
9. Henzinger M, Krinninger S, Nanongkai D (2014) A subquadratic-time algorithm for decremental single-source shortest paths. In: SODA 2014, Portland
10. King V (1999) Fully dynamic algorithms for maintaining all-pairs shortest paths and transitive closure in digraphs. In: FOCS, New York, pp 81–91
11. Pettie S (2004) A new approach to all-pairs shortest paths on real-weighted graphs. Theor Comput Sci 312(1):47–74. doi:10.1016/S0304-3975(03)00402-X, http://dx.doi.org/10.1016/S0304-3975(03)00402-X
12. Roditty L, Zwick U (2012) Dynamic approximate all-pairs shortest paths in undirected graphs. SIAM J Comput 41(3):670–683
13. Takaoka T (1992) A new upper bound on the complexity of the all pairs shortest path problem. Inf Process Lett 43(4):195–199. doi:10.1016/0020-0190(92)90200-F, http://dx.doi.org/10.1016/0020-0190(92)90200-F
14. Williams R (2014) Faster all-pairs shortest paths via circuit complexity. In: STOC, New York, pp 664–673

Degree-Bounded Planar Spanner with Low Weight

Wen-Zhan Song[1], Xiang-Yang Li[2], and Weizhao Wang[3]
[1]School of Engineering and Computer Science, Washington State University, Vancouver, WA, USA
[2]Department of Computer Science, Illinois Institute of Technology, Chicago, IL, USA
[3]Google Inc., Irvine, CA, USA

Keywords

Unified energy-efficient unicast and broadcast topology control

Years and Authors of Summarized Original Work

2005; Song, Li, Wang

Problem Definition

An important requirement of wireless ad hoc networks is that they should be self-organizing, and transmission ranges and data paths may need to be dynamically restructured with changing topology. Energy conservation and network performance are probably the most critical issues in wireless ad hoc networks, because wireless devices are usually powered by batteries only and have limited computing capability and memory. Hence, in such a dynamic and resource-limited environment, each wireless node needs to locally select communication neighbors and adjust its transmission power accordingly, such that all nodes together self-form a topology that is energy efficient for both unicast and broadcast communications.

To support energy-efficient unicast, the topology is preferred to have the following features in the literature:

1. POWER SPANNER: [1, 9, 13, 16, 17] Formally speaking, a subgraph H is called a *power spanner* of a graph G if there is a positive real constant ρ such that for any two nodes, the power consumption of the shortest path in H is at most ρ times of the power consumption of the shortest path in G. Here ρ is called the *power stretch factor* or *spanning ratio*.

2. DEGREE BOUNDED: [1, 9, 11, 13, 16, 17] It is also desirable that the logical node degree in the constructed topology is bounded from above by a small constant. Bounded logical degree structures find applications in Bluetooth wireless networks since a *master* node can have only seven active slaves simultaneously. A structure with small logical node degree will save the cost of updating the routing table when nodes are mobile. A structure with a small degree and using shorter links could improve the overall network throughout [6].

3. PLANAR: [1, 4, 13, 14, 16] A network topology is also preferred to be planar (no two edges crossing each other in the graph) to enable some localized routing algorithms to work correctly and efficiently, such as *Greedy Face Routing* (GFG) [2], *Greedy Perimeter Stateless Routing* (GPSR) [5], *Adaptive Face Routing* (AFR) [7], and *Greedy Other Adaptive Face Routing* (GOAFR) [8]. Notice that with planar network topology as the underlying routing structure, these localized routing protocols guarantee the message delivery without using a routing table: each intermediate node can decide which logical neighboring node to forward the packet to using only local information and the position of the source and the destination.

To support energy-efficient broadcast [15], the locally constructed topology is preferred to be *low-weighted* [10, 12]: the total link length of the final topology is within a constant factor of

that of *EMST*. Recently, several localized algorithms [10, 12] have been proposed to construct low-weighted structures, which indeed approximate the energy efficiency of EMST as the network density increases. However, none of them is power efficient for unicast routing.

Before this work, all known topology control algorithms could not support power efficient unicast and broadcast in the same structure. It is indeed challenging to design a unified topology, especially due to the trade off between spanner and low weight property. The main contribution of this algorithm is to address this issue.

Key Results

This algorithm is the *first* localized topology control algorithm for all nodes to maintain a *unified* energy-efficient topology for unicast and broadcast in wireless ad hoc/sensor networks. In one single structure, the following network properties are guaranteed:

1. **Power efficient unicast**: given any two nodes, there is a path connecting them in the structure with total power cost no more than $2\rho + 1$ times the power cost of any path connecting them in the original network. Here $\rho > 1$ is some constant that will be specified later in this algorithm. It assumes that each node u can adjust its power sufficiently to cover its next-hop v on any selected path for unicast.

2. **Power efficient broadcast**: the power consumption for broadcast is within a constant factor of the optimum among all *locally* constructed structures. As proved in [10], to prove this, it equals to prove that the structure is *low-weighted*. Here we called a structure low-weigthed, if its total edge length is within a constant factor of the total length of the Euclidean Minimum Spanning Tree (EMST). For broadcast or generally multicast, it assumes that each node u can adjust its power sufficiently to cover its farthest down-stream node on any selected structure (typically a tree) for multicast.

3. **Bounded logical node degree**: each node has to communicate with at most $k - 1$ logical

---

Algorithm 1 $S\Theta GG$: Power-Efficient Unicast Topology

1: First, each node self-constructs the Gabriel graph GG locally. The algorithm to construct GG locally is well-known, and a possible implementation may refer to [13]. Initially, all nodes mark themselves WHITE, i.e., *unprocessed*.

2: Once a WHITE node u has the smallest ID among all its WHITE neighbors in $N(u)$, it uses the following strategy to select neighbors:

    1. Node u first sorts all its BLACK neighbors (if available) in $N(u)$ in the distance-increasing order, then sorts all its WHITE neighbors (if available) in $N(u)$ similarly. The sorted results are then restored to $N(u)$, by first writing the sorted list of BLACK neighbors then appending the sorted list of WHITE neighbors.

    2. Node u scans the sorted list $N(u)$ from left to right. In each step, it keeps the current pointed neighbor w in the list, while deletes every *conflicted* node v in the remainder of the list. Here a node v is conflicted with w means that node v is in the θ-dominating region of node w. Here $\theta = 2\pi/k$ ($k \geq 9$) is an adjustable parameter.

    Node u then marks itself BLACK, i.e. *processed*, and notifies each deleted neighboring node v in $N(u)$ by a broadcasting message UPDATEN.

3: Once a node v receives the message UPDATEN from a neighbor u in $N(v)$, it checks whether itself is in the nodes set for deleting: if so, it deletes the sending node u from list $N(v)$, otherwise, marks u as BLACK in $N(v)$.

4: When all nodes are processed, all selected links $\{uv | v \in N(u), \forall v \in GG\}$ form the final network topology, denoted by $S\Theta GG$. Each node can shrink its transmission range as long as it sufficiently reaches its farthest neighbor in the final topology.

---

neighbors, where $k \geq 9$ is an adjustable parameter.

4. **Bounded average physical node degree**: the expected average physical node degree is at most a small constant. Here the physical degree of a node u in a structure H is defined as the number of nodes inside the disk centered at u with radius $\max_{uv \in H} \|uv\|$.

5. **Planar**: there are no edges crossing each other. This enables several localized routing algorithms, such as [2, 5, 7, 8], to be performed on top of this structure and guarantee the packet delivery without using the routing table.

6. **Neighbors Θ-separated**: the directions between any two logical neighbors of any node are separated by at least an angle θ, which reduces the communication interferences.

It is the *first* known *localized* topology control strategy for all nodes together to maintain such a *single* structure with these desired properties. Previously, only a centralized algorithm was reported in [1]. The first step is Algorithm 1 that can construct a power-efficient topology for unicast, then it extends to the final algorithm (Algorithm 2) that can support power-efficient broadcast at the same time.

Definition 1 (θ-Dominating Region) For each neighbor node v of a node u, the θ-*dominating region* of v is the 2θ-cone emanated from u, with the edge uv as its axis.

Let $N_{UDG}(u)$ be the set of neighbors of node u in UDG, and let $N(u)$ be the set of neighbors of node u in the final topology, which is initialized as the set of neighbor nodes in GG.

Algorithm 1 constructs a degree-$(k-1)$ planar power spanner.

Lemma 1 *Graph $S\Theta GG$ is connected if the underlying graph GG is connected. Furthermore, given any two nodes u and v, there exists a path $\{u, t_1, \ldots, t_r, v\}$ connecting them such that all edges have length less than $\sqrt{2}\|uv\|$.*

Theorem 1 *The structure $S\Theta GG$ has node degree at most $k-1$ and is planar power spanner with neighbors Θ-separated. Its power stretch factor is at most $\rho = \sqrt{2}^\beta / (1 - (2\sqrt{2}\sin\frac{\pi}{k})^\beta)$, where $k \geq 9$ is an adjustable parameter.*

Obviously, the construction is consistent for two endpoints of each edge: if an edge uv is kept by node u, then it is also kept by node v. It is worth mentioning that, the number 3 in criterion $\|xy\| > \max(\|uv\|, 3\|ux\|, 3\|vy\|)$ is carefully selected.

Theorem 2 *The structure $LS\Theta GG$ is a degree-bounded planar spanner. It has a constant power spanning ratio $2\rho + 1$, where ρ is the power spanning ratio of $S\Theta GG$. The node degree is bounded by $k-1$ where $k \geq 9$ is a customizable parameter in $S\Theta GG$.*

Algorithm 2 Construct $LS\Theta GG$: Planar Spanner with Bounded Degree and Low Weight

---

1: All nodes together construct the graph $S\Theta GG$ in a localized manner, as described in Algorithm 1. Then, each node marks its incident edges in $S\Theta GG$ *unprocessed*.

2: Each node u locally broadcasts its incident edges in $S\Theta GG$ to its one-hop neighbors and listens to its neighbors. Then, each node x can learn the existence of the set of 2-hop links $E_2(x)$, which is defined as follows: $E_2(x) = \{uv \in S\Theta GG \mid u \text{ or } v \in N_{\mathrm{UDG}}(x)\}$. In other words, $E_2(x)$ represents the set of edges in $S\Theta GG$ with at least one endpoint in the transmission range of node x.

3: Once a node x learns that its *unprocessed* incident edge xy has the smallest ID among all *unprocessed* links in $E_2(x)$, it will delete edge xy if there exists an edge $uv \in E_2(x)$ (here both u and v are different from x and y), such that $\|xy\| > \max(\|uv\|, 3\|ux\|, 3\|vy\|)$; otherwise it simply marks edge xy *processed*. Here assume that $uvyx$ is the convex hull of u, v, x and y. Then the link status is broadcasted to all neighbors through a message UPDATESTATUS(XY).

4: Once a node u receives a message UPDATESTATUS(XY), it records the status of link xy at $E_2(u)$.

5: Each node repeats the above two steps until all edges have been *processed*. Let $LS\Theta GG$ be the final structure formed by all remaining edges in $S\Theta GG$.

Theorem 3 *The structure $LS\Theta GG$ is low-weighted.*

Theorem 4 *Assuming that both the ID and the geometry position can be represented by $\log n$ bits each, the total number of messages during constructing the structure $LS\Theta GG$ is in the range of $[5n, 13n]$, where each message has at most $O(\log n)$ bits.*

Compared with previous known low-weighted structures [10, 12], $LS\Theta GG$ not only achieves more desirable properties, but also costs much less messages during construction. To construct $LS\Theta GG$, each node only needs to collect the information $E_2(x)$ which costs at most $6n$ messages for n nodes. The Algorithm 2 can be generally applied to any known degree-bounded planar spanner to make it low-weighted while keeping all its previous properties, except increasing the spanning ratio from ρ to $2\rho + 1$ theoretically.

In addition, the expected average node interference in the structure is bounded by a small constant. This is significant on its own due to the following reasons: it has been taken for granted that "*a network topology with small logical node degree will guarantee a small interference*" and recently Burkhart et al. [3] showed that this is not true generally. This work also shows that, although generally a small logical node degree cannot guarantee a small interference, the expected average interference is indeed small if the logical communication neighbors are chosen carefully.

Theorem 5 *For a set of nodes produced by a Poisson point process with density n, the expected maximum node interferences of EMST, GG, RNG, and Yao are at least $\Theta(\log n)$.*

Theorem 6 *For a set of nodes produced by a Poisson point process with density n, the expected average node interferences of EMST are bounded from above by a constant.*

This result also holds for nodes deployed with uniform random distribution.

Applications

Localized topology control in wireless ad hoc networks are critical mechanisms to maintain network connectivity and provide feedback to communication protocols. The major traffic in networks are unicast communications. There is a compelling need to conserve energy and improve network performance by maintaining an energy-efficient topology in localized ways. This algorithm achieves this by choosing relatively smaller power levels and size of communication neighbors for each node (e.g., reducing interference). Also, broadcasting is often necessary in MANET routing protocols. For example, many unicast routing protocols such as Dynamic Source Routing (DSR), Ad Hoc On Demand Distance Vector (AODV), Zone Routing Protocol (ZRP), and Location Aided Routing (LAR) use broadcasting or a derivation of it to establish routes. It is highly important to use power-efficient broadcast algorithms for such networks since wireless devices are often powered by batteries only.

Cross-References

▶ Applications of Geometric Spanner Networks
▶ Geometric Spanners
▶ Planar Geometric Spanners
▶ Sparse Graph Spanners

Recommended Reading

1. Bose P, Gudmundsson J, Smid M (2002) Constructing plane spanners of bounded degree and low weight. In: Proceedings of European symposium of algorithms, University of Rome, 17–21 Sept 2002
2. Bose P, Morin P, Stojmenovic I, Urrutia J (2001) Routing with guaranteed delivery in ad hoc wireless networks. ACM/Kluwer Wirel Netw 7(6):609–616, 3rd international workshop on discrete algorithms and methods for mobile computing and communications, pp 48–55 (1999)
3. Burkhart M, von Rickenbach P, Wattenhofer R, Zollinger A (2004) Does topology control reduce interference. In: ACM international symposium on mobile Ad-Hoc networking and computing (Mobi-Hoc), Tokyo, 24–26 May 2004
4. Gabriel KR, Sokal RR (1969) A new statistical approach to geographic variation analysis. Syst Zool 18:259–278
5. Karp B, Kung HT (2000) Gpsr: greedy perimeter stateless routing for wireless networks. In: Proceedings of the ACM/IEEE international conference on mobile computing and networking (MobiCom), Boston, 6–11 Aug 2000
6. Kleinrock L, Silvester J (1978) Optimum transmission radii for packet radio networks or why six is a magic number. In: Proceedings of the IEEE national telecommunications conference, Birmingham, pp 431–435, 4–6 Dec 1978
7. Kuhn F, Wattenhofer R, Zollinger A (2002) Asymptotically optimal geometric mobile ad-hoc routing. In: International workshop on discrete algorithms and methods for mobile computing and communications (DIALM), Atlanta, 28 Sept 2002
8. Kuhn F, Wattenhofer R, Zollinger A (2003) Worst-case optimal and average-case efficient geometric ad-hoc routing. In: ACM international symposium on mobile Ad-Hoc networking and computing (Mobi-Hoc), Anapolis, 1–3 June 2003
9. Li L, Halpern JY, Bahl P, Wang Y-M, Wattenhofer R (2001) Analysis of a cone-based distributed topology control algorithms for wireless multi-hop networks. In: PODC: ACM symposium on principle of distributed computing, Newport, 26–29 Aug 2001
10. Li X-Y (2003) Approximate MST for UDG locally. In: COCOON, Big Sky, 25–28 July 2003
11. Li X-Y, Wan P-J, Wang Y, Frieder O (2002) Sparse power efficient topology for wireless networks. In: IEEE Hawaii international conference on system sciences (HICSS), Big Island, 7–10 Jan 2002
12. Li X-Y, Wang Y, Song W-Z, Wan P-J, Frieder O (2004) Localized minimum spanning tree and its applications in wireless ad hoc networks. In: IEEE INFOCOM, Hong Kong, 7–11 Mar 2004
13. Song W-Z, Wang Y, Li X-Y, Frieder O (2004) Localized algorithms for energy efficient topology in wireless ad hoc networks. In: ACM international symposium on mobile Ad-Hoc networking and computing (MobiHoc), Tokyo, 24–26 May 2004
14. Toussaint GT (1980) The relative neighborhood graph of a finite planar set. Pattern Recognit 12(4):261–268
15. Wan P-J, Calinescu G, Li X-Y, Frieder O (2001) Minimum-energy broadcast routing in static ad hoc wireless networks. ACM wireless networks (2002), To appear, Preliminary version appeared in IEEE INFOCOM, Anchorage, 22–26 Apr 2001
16. Wang Y, Li X-Y (2003) Efficient construction of bounded degree and planar spanner for wireless networks. In: ACM DIALMPOMC joint workshop on foundations of mobile computing, San Diego, 19 Sept 2003
17. Yao AC-C (1982) On constructing minimum spanning trees in k-dimensional spaces and related problems. SIAM J Comput 11:721–736

Degree-Bounded Trees

Martin Fürer
Department of Computer Science and Engineering, The Pennsylvania State University, University Park, PA, USA

Keywords

Bounded degree spanning trees; Bounded degree Steiner trees

Years and Authors of Summarized Original Work

1994; Fürer, Raghavachari

Problem Definition

The problem is to construct a spanning tree of small degree for a connected undirected graph $G = (V, E)$. In the Steiner version of the problem, a set of distinguished vertices $D \subseteq V$ is given along with the input graph G. A Steiner tree is a tree in G which spans at least the set D.

As finding a spanning or Steiner tree of the smallest possible degree Δ^* is NP-hard, one is interested in approximating this minimization problem. For many such combinatorial optimization problems, the goal is to find an approximation in polynomial time (a constant or larger factor). For the spanning and Steiner tree problems, the iterative polynomial time approximation algorithms of Fürer and Raghavachari [8] (see also [14]) find much better solutions. The degree Δ of the solution tree is at most $\Delta^* + 1$.

There are very few natural NP-hard optimization problems for which the optimum can be achieved up to an additive term of 1. One such problem is coloring a planar graph, where coloring with four colors can be done in polynomial time. On the other hand, 3-coloring is NP-complete even for planar graphs. An other such problem is edge coloring a graph of degree Δ. While coloring with $\Delta + 1$ colors is always possible in polynomial time, Δ edge coloring is NP-complete.

Chvátal [3] has defined the *toughness* $\tau(G)$ of a graph as the minimum ratio $|X|/c(X)$ such that the subgraph of G induced by $V \setminus X$ has $c(X) \geq 2$ connected components. The inequality $1/\tau(G) \leq \Delta^*$ immediately follows. Win [17] has shown that $\Delta^* < \frac{1}{\tau(G)} + 3$; i.e., the inverse of the toughness is actually a good approximation of Δ^*.

A set X, such that the ratio $|X|/c(X)$ is the toughness $\tau(G)$, can be viewed as witnessing the upper bound $|X|/c(X)$ on $\tau(G)$ and therefore the lower bound $c(X)/|X|$ on Δ^*. Strengthening this notion, Fürer and Raghavachari [8] define X to be a witness set for $\Delta^* \geq d$ if d is the smallest integer greater or equal to $(|X| + c(X) - 1)/|X|$. Their algorithm not only outputs a spanning tree, but also a witness set X, proving that its degree is at most $\Delta^* + 1$.

Key Results

The minimum degree spanning tree and Steiner tree problems are easily seen to be NP-hard, as they contain the Hamiltonian path problem. Hence, we cannot expect a polynomial time algorithm to find a solution of minimal possible degree Δ^*. The same argument also shows that an approximation by a factor less than $3/2$ is impossible in polynomial time unless $P = NP$.

Initial approximation algorithms obtained solutions of degree $O(\Delta^* \log n)$ [6], where $n = |V|$ is the number of vertices. The optimal result for the spanning tree case has been obtained by Fürer and Raghavachari [7, 8].

Theorem 1 *Let Δ^* be the degree of an unknown minimum degree spanning tree of an input graph $G = (V, E)$. There is a polynomial time approximation algorithm for the minimum degree spanning tree problem that finds a spanning tree of degree at most $\Delta^* + 1$.*

Later this result has been extended to the Steiner tree case [8].

Theorem 2 *Assume a Steiner tree problem is defined by a graph $G = (V, E)$ and an arbitrary subset D of vertices V. Let Δ^* be the degree of an unknown minimum degree Steiner tree of G spanning at least the set D. There is a polynomial time approximation algorithm for the minimum degree Steiner tree problem that finds a Steiner tree of degree at most $\Delta^* + 1$.*

Both approximation algorithms run in time $O(mn \log n\, \alpha(m, n))$, where m is the number of edges and α is the inverse Ackermann function.

Applications

Some possible direct applications are in networks for noncritical broadcasting, where it might be desirable to bound the load per node, and in designing power grids, where the cost of splitting increases with the degree. Another major benefit of a small degree network is limiting the effect of node failure.

Furthermore, the main results on approximating the minimum degree spanning and Steiner tree problems have been the basis for approximating various network design problems, sometimes involving additional parameters.

Klein, Krishnan, Raghavachari and Ravi [11] find 2-connected subgraphs of approximately minimal degree in 2-connected graphs, as well as approximately minimal degree spanning trees (branchings) in directed graphs. Their algorithms run in quasi-polynomial time, and approximate the degree Δ^* by $(1 + \epsilon)\Delta^* + O(\log_{1+\epsilon} n)$.

Often the goal is to find a spanning tree that simultaneously has a small degree and a small weight. For a graph having an minimum weight spanning tree (MST) of degree Δ^* and weight w, Fischer [5] finds a spanning tree with degree $O(\Delta^* + \log n)$ and weight w, (i.e., an MST of small weight) in polynomial time.

Könemann and Ravi [12, 13] provide a bi-criteria approximation. For a given $B^* \geq \Delta^*$, let w be the minimum weight of any spanning tree of degree at most B^*. The polynomial time algorithm finds a spanning tree of degree $O(B^* + \log n)$ and weight $O(w)$. In the second paper, the algorithm adapts to the case of a different degree bound on each vertex. Chaudhuri et al. [2] further improved this result to approximate both the degree B^* and the weight w by a constant factor.

In another extension of the minimum degree spanning tree problem, Ravi and Singh [15] have obtained a strict generalization of the $\Delta^* + 1$ spanning tree approximation [8]. Their polynomial time algorithm finds an MST of degree $\Delta^* + k$ for the case of a graph with k distinct weights on the edges.

Recently, there have been some drastic improvements. Again, let w be the minimum cost of

a spanning tree of given degree B^*. Goemans [9] obtains a spanning tree of cost w and degree $B^* + 2$. Finally, Singh and Lau [16] decrease the degree to $B^* + 1$ and also handle individual degree bounds Δ_v^* for each vertex v in the same way.

Interesting approximation algorithms are also known for the 2-dimensional Euclidian minimum weight bounded degree spanning tree problem, where the vertices are points in the plane and edge weights are the Euclidian distances. Khuller, Raghavachari, and Young [10] show factor 1.5 and 1.25 approximations for degree bounds 3 and 4 respectively. These bounds have later been improved slightly by Chan [1]. Slightly weaker results are obtained by Fekete et al. [4], using flow-based methods, for the more general case where the weight function just satisfies the triangle inequality.

Open Problems

The time complexity of the minimum degree spanning and Steiner tree algorithms [8] is $O(mn\,\alpha(m, n) \log n)$. Can it be improved to $O(mn)$? In particular, what can be gained by initially selecting a reasonable Steiner tree with some greedy technique instead of starting the iteration with an arbitrary Steiner tree?

Is there an efficient parallel algorithm that can obtain a $\Delta^* + 1$ approximation in poly-logarithmic time? Fürer and Raghavachari [6] have obtained such an NC-algorithm, but only with a factor $O(\log n)$ approximation of the degree.

Cross-References

▶ Fully Dynamic Connectivity
▶ Graph Connectivity
▶ Minimum Energy Cost Broadcasting in Wireless Networks
▶ Minimum Spanning Trees
▶ Steiner Forest
▶ Steiner Trees

Recommended Reading

1. Chan TM (2004) Euclidean bounded-degree spanning tree ratios. Discret Comput Geom 32(2):177–194
2. Chaudhuri K, Rao S, Riesenfeld S, Talwar K (2006) A push-relabel algorithm for approximating degree bounded MSTs. In: Proceedings of the 33rd international colloquium on automata, languages and programming (ICALP 2006), Part I. LNCS, vol 4051. Springer, Berlin, pp 191–201
3. Chvátal V (1973) Tough graphs and Hamiltonian circuits. Discret Math 5:215–228
4. Fekete SP, Khuller S, Klemmstein M, Raghavachari B, Young N (1997) A network-flow technique for finding low-weightbounded-degree spanning trees. In: Proceedings of the 5th integer programming and combinatorial optimization conference (IPCO 1996), and J Algorithms 24(2):310–324
5. Fischer T (1993) Optimizing the degree of minimum weight spanning trees. Technical report TR93–1338. Computer Science Department, Cornell University
6. Fürer M, Raghavachari B (1990) An NC approximation algorithm for the minimum-degree spanning tree problem. In: Proceedings of the 28th annual Allerton conference on communication, control and computing, 1990, pp 174–281
7. Fürer M, Raghavachari B (1992) Approximating the minimum degree spanning tree to within one from the optimal degree. In: Proceedings of the third annual ACM-SIAM symposium on discrete algorithms (SODA 1992), pp 317–324
8. Fürer M, Raghavachari B (1994) Approximating the minimum-degree Steiner tree to within one of optimal. J Algorithms 17(3):409–423
9. Goemans MX (2006) Minimum bounded degree spanning trees. In: Proceedings of the 47th annual IEEE symposium on foundations of computer science (FOCS 2006), pp 273–282
10. Khuller S, Raghavachari B, Young N (1996) Low-degree spanning trees of small weight. SIAM J Comput 25(2):355–368
11. Klein PN, Krishnan R, Raghavachari B, Ravi R (2004) Approximation algorithms for finding low-degree subgraphs. Networks 44(3):203–215
12. Könemann J, Ravi R (2002) A matter of degree: improved approximation algorithms for degree-bounded minimum spanning trees. SIAM J Comput 31(6):1783–1793
13. Könemann J, Ravi R (2005) Primal-dual meets local search: approximating MSTs with nonuniform degree bounds. SIAM J Comput 34(3):763–773
14. Raghavachari B (1995) Algorithms for finding low degree structures. In: Hochbaum DS (ed) Approximation algorithms for NP-hard problems. PWS Publishing Company, Boston, pp 266–295
15. Ravi R, Singh M (2006) Delegate and conquer: an LP-based approximation algorithm for minimum degree MSTs. In: Proceedings of the 33rd international colloquiumon automata, languages and programming (ICALP 2006) Part I. LNCS, vol 4051. Springer, Berlin, pp 169–180
16. Singh M, Lau LC (2007) Approximating minimum bounded degree spanning trees to within one of optimal. In: Proceedings of the thirty-ninth annual ACM symposiumon theory of computing (STOC 2007), New York, pp 661–670
17. Win S (1989) On a connection between the existence of k-trees and the toughness of a graph. Graphs Comb 5(1):201–205

Delaunay Triangulation and Randomized Constructions

Olivier Devillers

Inria Nancy – Grand-Est, Villers-lès-Nancy, France

Keywords

Convex hull; Delaunay triangulation; Randomization; Voronoi diagram

Years and Authors of Summarized Original Work

1989; Clarkson, Shor
1993; Seidel
2002; Devillers
2003;Amenta, Choi, Rote

Problem Definition

The Delaunay triangulation and the Voronoi diagram are two classic geometric structures in the field of computational geometry. Their success can perhaps be attributed to two main reasons: Firstly, there exist practical, efficient algorithms to construct them; and secondly, they have an enormous number of useful applications ranging from meshing and 3D-reconstruction to interpolation.

Given a set S of n sites in some space $\mathbb{E}$, we define the Voronoi region $V_S(p)$ of $p \in S$ to be the set of points in $\mathbb{E}$ whose nearest neighbor in S is p (for some distance δ):

$$V(p) = \left\{ x \in \mathbb{E}, \forall q \in S \setminus \{p\} \delta(x, p) < \delta(x, q) \right\}.$$

It is easily seen that these regions form a partition of $\mathbb{E}$ into convex regions which we refer to as *cells*. These concepts may be extended into more exotic spaces such as periodic and hyperbolic spaces or metric spaces using convex distances, though we restrict ourselves to the case where $\mathbb{E}$ is the Euclidean space $\mathbb{E} = \mathbb{R}^d$ and the distance δ is the L_2 norm.

The Voronoi diagram $\mathbb{V}(S)$ may now be defined as the limit between the different Voronoi cells

$$\mathbb{V}(S) = \mathbb{E} \setminus \bigcup_{p \in S} V_S(p).$$

The Delaunay triangulation $\mathbb{D}(S)$ is the geometric dual of $\mathbb{V}(S)$. More formally, $\mathbb{D}(S)$ is a simplicial complex defined by

$$\sigma \in \mathbb{D}(S) \Longleftrightarrow \bigcap_{p \in \sigma} \overline{V_S}(p) \neq \emptyset,$$

where $\overline{V_S}(p)$ is the closure of the Voronoi cell $V_S(p)$ (see Fig. 1).

Voronoi diagrams and Delaunay triangulations have received a lot of attention in the literature

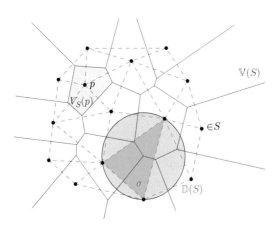

Delaunay Triangulation and Randomized Constructions, Fig. 1 The Voronoi diagram of a set S of 15 points and its dual Delaunay triangulation

with several surveys, books, or book chapters (e.g., [4, 14]) and hundreds of papers. In this article, we will focus on randomized construction algorithms for the Delaunay triangulation. Such algorithms use randomness to speed up their running time but do not assume any randomness in the data distribution.

Key Results

Delaunay Properties

Empty Ball Property
One crucial property of the Delaunay triangulation, which is the basis of many algorithms, is the *empty ball property*, which guarantees that a triangle is a Delaunay triangle of S if and only if the interior of its circumball does not contain any point of S.

Size of the Triangulation
In the plane, the combinatorial properties of a triangulation (not necessarily Delaunay) are completely fixed by the Euler relation. In particular, given n vertices, h of which on the convex hull, every triangulation must have $2n - 2 - h$ triangles and $3n - 3 - h$ edges. In dimension d, the Dehn-Sommerville relations yield a linear dependence for the number of simplices of all dimensions on the number of simplices of dimensions k for $k \leq \left\lceil \frac{d}{2} \right\rceil$; this gives an $O\left(n^{\lceil \frac{d}{2} \rceil}\right)$ upper bound for the number of simplices of all dimension. For both Delaunay and more general triangulations, these bounds are tight in the worst case.

These bounds can be tightened given some assumptions on the distribution of the input sites. If the points are uniformly distributed in a compact convex of fixed volume, then the triangulation size (its total number of simplices) is $\Theta(n)$, with a constant exponential in d [9]. In 3D, and for reconstruction purposes, it is convenient to assume that the points lie on a surface. It is known that the Delaunay triangulation of points uniformly distributed on a convex polyhedron has size $\Theta(n)$ (for a constant depending on the polyhedron complexity). For points uniformly distributed on a (non convex) polyhedron, the triangulation's size is between $\Omega(n)$ and $O(n \log n)$ [12]. If, instead of making a probabilistic assumption, we assume that the points are a "good sampling" of the surface such that every small ball centered on the surface contains between 1 and κ points (where κ is a constant), then the size of the Delaunay triangulation is $\Theta(n)$ for a polyhedron, $O(n \log n)$ for a *generic* smooth surface [3], and $\Omega(n \sqrt{n})$ for a nongeneric surface (e.g., a cylinder). In the case of the cylinder, a uniformly distributed point set has a triangulation of size $\Theta(n \log n)$. In dimension d, a p-dimensional polyhedron whose faces have a "good sampling" has size $O(n^k)$ where $k = \frac{d+1-\left\lceil \frac{d+1}{p+1} \right\rceil}{p}$ [2].

First Algorithms

Many classical techniques in algorithmic and computational geometry have been used to attack the problem of constructing the Delaunay triangulation and the Voronoi Diagram. The gift wrapping and the incremental approaches were introduced in the 1970s [11], followed by some worst case optimal algorithms in 2D, based on divide-and-conquer [13] and sweep line techniques [10]. In higher dimensions, the optimal worst case construction of Delaunay triangulation and convex hulls was solved in the 1990s.

In the remainder of the entry, we will describe some further algorithmic techniques that may be used to construct the Delaunay triangulation.

Randomized Construction

One popular and efficient method, applied to the Delaunay triangulation at the end of the 1980s [5], is Randomized Incremental Construction (RIC). The idea is to exploit the simplicity of an incremental algorithm while avoiding its worst case behavior by simply adjusting the order of insertion of the points.

Conflict Graph

Recalling that $\mathbb{D}(S)$ is the set of triangles with vertices in S whose circumballs are empty, the idea is to maintain for a sequence $\varnothing = S_0 \subset S_1 \subset S_2 \subset \ldots \subset S_n = S$, where $|S_i| = i$, a sequence of triangulations $\mathbb{D}(S_i)$ with associated *conflict graphs*. We define the conflict graph to be a bipartite graph that links a point p of $S \setminus S_i$ to a simplex σ in $\mathbb{D}(S_i)$ if the circumball of σ contains p (p *and σ are called in conflict*). The information contained in the conflict graph simplifies the construction of $\mathbb{D}(S_{i+1})$ from $\mathbb{D}(S_i)$ since it gives directly the simplices in $\mathbb{D}(S_i) \setminus \mathbb{D}(S_{i+1})$.

The key point comes from an analysis based on random sampling [6]; let's assume that S_i is a random sample of size i of S. We say that a simplex has *width* j if it has j points in conflict in S. In which case a Delaunay simplex is a simplex of width 0. Denote by Δ_j the number of simplices of width j and let $\Delta_{\leq k} = \sum_{j \leq k} \Delta_j$. We first bound $\Delta_{\leq k}$ using the following remark: a simplex of width j is a Delaunay simplex of a random sample R of size $\frac{n}{k}$ of S with probability $p_k = \frac{1}{k^{d+1}} \left(1 - \frac{1}{k}\right)^j$ (vertices of σ must be chosen in R and points in conflict must not). Notice that for $j \in [2 \ldots k]$, we have $\left(1 - \frac{1}{k}\right)^j \geq \left(1 - \frac{1}{k}\right)^k \geq \frac{1}{4}$ since $\left(1 - \frac{1}{x}\right)^x$ is an increasing function of value $\frac{1}{4}$ for $x = 2$. For $k \geq 2$, we have (using $\mathbb{P}$ to denote the probability measure)

$$|\mathbb{D}(R)| = O\left(\frac{n}{k}^{\left\lceil \frac{d}{2} \right\rceil}\right) = \sum_{\sigma \in S^{d+1}} \mathbb{P}(\sigma \in \mathbb{D}(R)) = \sum_{j \leq n} \Delta_j p_j \geq \frac{\Delta_0}{k^{d+1}} + \frac{\Delta_1 \frac{1}{2}}{k^{d+1}}$$

$$+ \sum_{j=2}^{k} \frac{\Delta_j \frac{1}{4}}{k^{d+1}} \geq \frac{\Delta_{\leq k}}{4k^{d+1}}, \Delta_{\leq k} \leq O\left(\frac{n}{k}^{\left\lceil \frac{d}{2} \right\rceil} k^{d+1}\right) = O\left(n^{\left\lceil \frac{d}{2} \right\rceil} k^{\left\lceil \frac{d+1}{2} \right\rceil}\right).$$

We can now analyze the incremental construction of $\mathbb{D}(S)$. The probability that a triangle of width j appears during the construction is

$$p'_j = \frac{\binom{d+1}{j+d+1}}{(j+d+1)!}$$

(the number of permutations that look at the vertices of σ before the points in conflict divided by the total number of permutations). Then the cost of the algorithm is given by the total number of conflicts occurring during the construction:

$$\sum_{\sigma \in S^{d+1}} \text{width}\,(\sigma)\mathbb{P}(\sigma \text{ appears}) = \sum_j \Delta_j \cdot j \cdot p'_j = \sum_j (\Delta_{\leq j} - \Delta_{\leq j-1}) j \cdot p'_j$$

$$= \sum_j \Delta_{\leq j}(j \cdot p'_j - (j+1)p'_{j+1}) \leq \sum_j n^{\lceil \frac{d}{2} \rceil} j^{\lceil \frac{d+1}{2} \rceil} O\left(\frac{1}{j^{d+2}}\right) \leq O\left(n^{\lceil \frac{d}{2} \rceil} \sum_j j^{-\lceil \frac{d}{2} \rceil}\right)$$

which gives $O(n \log n)$ for $d = 2$ and $O\left(n^{\lceil \frac{d}{2} \rceil}\right)$ for higher d.

Backward Analysis

A simpler way of analyzing RIC is backward analysis [15], and we will sketch it in 2D. The idea is quite simple and consists in asking: *what*

is the cost of the last step? The answer is that the cost of modifying the triangulation during last insertion is clearly proportional to the *degree* (number of simplices incident) of the last point inserted into the triangulation. Since the last point is a random point, its expected degree is

$$\mathbb{E}\text{xp}(\text{degree}(\text{last point})) = \mathbb{E}\text{xp}(\text{degree}(\text{random point})) = \frac{1}{n} \sum_{p \in S} \text{degree}(p) \leq \frac{6n}{n} = 6,$$

and summing over all insertion steps gives a linear cost for updating the triangulation. It remains to count the cost of updating the conflict graph. We remark that there is a conflict between the last point p_n and a triangle created by the insertion of the jth point p_j if and only if the edge $p_j p_n$ exists in $\mathbb{D}(S_j \cup \{p_n\})$. Since p_j and p_n are both random points of $S_j \cup \{p_n\}$, it happens with probability $O\left(\frac{1}{j}\right)$, the expected number of conflicts for p_n is thus $O\left(\sum_j \frac{1}{j}\right) = O(\log n)$, and the total number of conflicts is $O\left(\sum_k \log k\right) = O(n \log n)$.

Delaunay Hierarchy

The conflict graph approach assumes complete knowledge of S to initialize the conflict graph. Using a lazy approach and postponing the conflict determination, it is possible to obtain online algorithms [5].

Among the online schemes to construct the Delaunay triangulation, the Delaunay

hierarchy [7] gives good results both in theory and in practice. The Delaunay hierarchy constructs a sequence of random samples $S = S_0 \supset S_1 \supset \cdots \supset S_h$ such that $\mathbb{P}(p \in S_i \mid p \in S_{i-1}) = \alpha$. Then the Delaunay triangulations of $\mathbb{D}(S_i), 1 \leq i \leq h$ are maintained under point insertions. Pointers from a vertex of $\mathbb{D}(S_i)$ to the vertices at the same position in $\mathbb{D}(S_{i-1})$ (if $i < h$) and in $\mathbb{D}(S_{i+1})$ (if it actually belongs to S_{i+1}) are also computed.

When a new point p needs to be inserted, it is located by walking in $\mathbb{D}(S_h)$ (using neighborhood relations) to reach the closest vertex w_h of p in $\mathbb{D}(S_h)$. Then the hierarchy is descended, walking in $\mathbb{D}(S_i)$ from w_{i+1} to find w_i the closest neighbor in sample S_i. Using these neighbors, it is easy to insert p in $\mathbb{D}(S_h)$ and in the triangulation of other samples that the random process assigns to p.

In 2D, the expected cost of the walk at any level is $O(\alpha^{-1})$ and the expected value for h is

$O\left(\frac{\log n}{-\log\alpha}\right)$. Thus the theoretical complexity of the algorithm is $O(n\log n)$. The value of α can be optimized depending on the input distribution: for random points $\alpha = \frac{1}{30}$ gives good timings and a very low memory requirement in addition to the one needed for $\mathbb{D}(S)$.

A Less Randomized Construction

Constructing the Delaunay triangulation by inserting the points in a random order presents a drawback with respect to memory management. Since the inserted point is random in S, there is very little chance that the triangles needed are present in the cache memory. So, an idea is to sort the points using a space-filling curve (see Fig. 2-left) to ensure locality of the insertions. Unfortunately, when inserting the points in such an order, the randomized complexity results no longer apply and the number of created and destroyed triangles during the construction may explode on certain data sets.

A smart solution has been proposed: it is possible to use an insertion order random enough to apply randomized complexity results and allow some locality to benefit from cache memory. BRIO (Biased Randomized Insertion Order) [1] proposes to partition S in a set of random samples $S = \bigcup_{0\leq i\leq h} S_i$ such that $|S_i| = \alpha|S_{i+1}|$ for $\alpha \leq 1$, a small constant (e.g., $\alpha = \frac{1}{4}$), and to insert the samples by increasing size, each

sample being sorted using a spatial filling curve (see Fig. 2-right).

In the random setting, we have seen that the probability for a triangle of width j to appear in the conflict graph algorithm was $\frac{1\cdot2\cdot3}{(j+1)(j+2)(j+3)} = \Theta(j^{-3})$. Using BRIO, this probability is a bit less intricate to compute, but it can be bounded in terms of α and it can be shown that it is still $\Theta(j^{-3})$ and thus randomized complexity results still apply.

Experimental Results

On a 16 GB, 2.3 GHz desktop CGAL currently computes the Delaunay triangulation of up to 200M points in 2D and 50M points in 3D [8].

Static timings are almost constant with respect to the total number of points and are about 1 μs per point in 2D and 8 μs per point in 3D. In the dynamic setting, one million points are processed in 6s in 2D and 25s in 3D.

URLs to Code and Data Sets

CGAL, among a big collection of computational geometry algorithms, provides implementations for Delaunay triangulations in 2D, 3D, and general dimension. It computes the Delaunay triangulation in 2D and 3D using the Delaunay hierarchy in a dynamic setting and using BRIO for static computation (http://cgal.org).

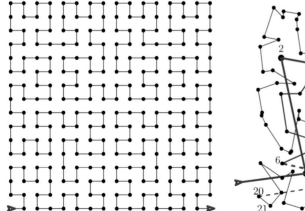 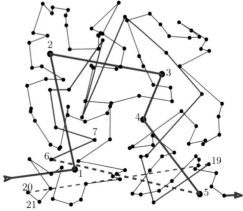

Delaunay Triangulation and Randomized Constructions, Fig. 2 *Right*: the Hilbert space-filling curve. *Left*: random points sorted with BRIO

Cross-References

▶ Triangulation Data Structures
▶ Voronoi Diagrams and Delaunay Triangulations

Recommended Reading

1. Amenta N, Choi S, Rote G (2003) Incremental constructions con BRIO. In: Proceedings of the 19th annual symposium on computational geometry, San Diego, pp 211–219. doi:10.1145/777792.777824, http://page.inf.fu-berlin.de/~rote/Papers/pdf/Incremental+constructions+con+BRIO.pdf
2. Amenta N, Attali D, Devillers O (2012) A tight bound for the Delaunay triangulation of points on a polyhedron. Discret Comput Geom 48:19–38. doi:10.1007/s00454-012-9415-7, http://hal.inria.fr/hal-00784900
3. Attali D, Boissonnat JD, Lieutier A (2003) Complexity of the Delaunay triangulation of points on surfaces: the smooth case. In: Proceedings of the 19th annual symposium on computational geometry, San Diego, pp 201–210. doi:10.1145/777792.777823, http://dl.acm.org/citation.cfm?id=777823
4. Aurenhammer F, Klein R (2000) Voronoi diagrams. In: Sack JR, Urrutia J (eds) Handbook of computational geometry. Elsevier/North-Holland, Amsterdam, pp 201–290. ftp://ftp.cis.upenn.edu/pub/cis610/public_html/ak-vd-00.ps
5. Boissonnat JD, Teillaud M (1986) A hierarchical representation of objects: the Delaunay tree. In: Proceedings of the 2nd annual symposium computational geometry, Yorktown Heights, pp 260–268. http://dl.acm.org/citation.cfm?id=10543
6. Clarkson KL, Shor PW (1989) Applications of random sampling in computational geometry, II. Discret Comput Geom 4:387–421. doi:10.1007/BF02187740, http://www.springerlink.com/content/b9n24vr730825p71/
7. Devillers O (2002) The Delaunay hierarchy. Int J Found Comput Sci 13:163–180. doi:10.1142/S0129054102001035, http://hal.inria.fr/inria-00166711
8. Devillers O (2012) Delaunay triangulations, theory vs practice. In: Abstracts 28th European workshop on computational geometry, Assisi, pp 1–4. http://hal.inria.fr/hal-00850561, invited talk
9. Dwyer R (1993) The expected number of k-faces of a Voronoi diagram. Int J Comput Math 26(5):13–21. doi:10.1016/0898-1221(93)90068-7, http://www.sciencedirect.com/science/article/pii/0898122193900687
10. Fortune SJ (1987) A sweepline algorithm for Voronoi diagrams. Algorithmica 2:153–174. doi:10.1007/BF01840357, http://www.springerlink.com/content/n88186tl165168rw/
11. Frederick CO, Wong YC, Edge FW (1970) Two-dimensional automatic mesh generation for structural analysis. Internat J Numer Methods Eng 2:133–144. doi:10.1002/nme.1620020112/abstract
12. Golin MJ, Na HS (2002) The probabilistic complexity of the Voronoi diagram of points on a polyhedron. In: Proceedings of the 18th annual symposium on computational geometry. Barcelona http://www.cse.ust.hk/~golin/pubs/SCG_02.pdf
13. Lee DT (1978) Proximity and reachability in the plane. PhD thesis, Coordinated Science Lab., University of Illinois, Urbana. http://oai.dtic.mil/oai/oai?verb=getRecord&metadataPrefix=html&identifier=ADA069764
14. Okabe A, Boots B, Sugihara K (1992) Spatial tessellations: concepts and applications of Voronoi diagrams. Wiley, Chichester
15. Seidel R (1993) Backwards analysis of randomized geometric algorithms. In: Pach J (ed) New trends in discrete and computational geometry, algorithms and combinatorics, vol 10. Springer, pp 37–68. http://ftp.icsi.berkeley.edu/ftp/pub/techreports/1992/tr-92-014.ps.gz

Derandomization of k-SAT Algorithm

Dominik Scheder
Institute for Interdisciplinary Information Sciences, Tsinghua University, Beijing, China
Institute for Computer Science, Shanghai Jiaotong University, Shanghai, China

Keywords

Algorithms for satisfiability; Derandomization; Local search

Years and Authors of Summarized Original Work

2011; Moser, Scheder

Problem Definition

Satisfiability is *the* central NP-complete problem. Given a Boolean formula in conjunctive normal form, for example, $(x \vee \bar{y} \vee z) \wedge (\bar{x} \vee \bar{z}) \wedge \ldots,$

decide whether there is a satisfying assignment. An important subclass is k-SAT, where the input is restricted to k-CNF formulas: CNF formulas in which every clause has at most k literals. In 1999, Uwe Schöning [6] gave an extremely simple randomized algorithm for k-SAT of running time

$$\left(\frac{2(k-1)}{k}\right)^n \text{poly}(n).$$

In particular this solves 3-SAT in time $O^*(1.334^n)$, 4-SAT in $O^*(1.5^n)$ for 4-SAT, and so on (we use O^* to suppress polynomial factors in n). Several authors have attempted to derandomize Schöning's algorithm, albeit at the cost of a greater running time: an algorithm of Dantsin, Goerdt, Hirsch, Kannan, Kleinberg, Papadimitriou, Raghavan, and Schöning [2] runs in time $O^*((2k/(k+1))^n)$, which for $k = 3$ is $O^*(1.5^n)$. For $k = 3$ Brueggemann and Kern [1] achieve $O(1.473^n)$; Scheder [5] achieves $O(1.465^n)$; Kutzkov and Scheder [3] reduced this to $O(1.439^n)$. All improvements suffer from two drawbacks: they fall short of achieving the running time of Schöning's randomized algorithm; most of them are tailored to $k = 3$; finally, they are all fairly complicated.

Key Results

We describe a simple deterministic algorithm due to Moser and Scheder [4] with a running time that matches that of Schöning's up to subexponential overhead. That is, we prove the following theorem:

Theorem 1 *There is a deterministic algorithm deciding satisfiability of k-CNF formulas over n variables in time $\left(\frac{2(k-1)}{k}\right)^{n+o(n)}$.*

Notation

For a CNF formula F we denote by $\text{vbl}(F)$ the set of variables appearing in F. Usually $n = |\text{vbl}(F)|$ denotes the number of variables in a formula. By $\{0, 1\}^{\text{vbl}(F)}$ (or $\{0, 1\}^n$), we denote the set of all truth assignments to these variables. We make frequent use of the notation $F^{[u \mapsto b]}$, which is the CNF formula created from F by replacing every occurrence of the literal u by b and of $\bar{u}$ by $1 - b$. For a literal u and a truth assignment α, we denote by $\alpha[u = b]$ the truth assignment that sets u to b and agrees with α otherwise.

A Promise Version of k-SAT

The heart of the proof will be an algorithm solving a promise version of k-SAT:

Theorem 2 *Let F be a k-CNF formula over n variables, $\alpha \in \{0, 1\}^n$ a (not necessarily satisfying) assignment, and $r \in \mathbb{N}_0$. There is a deterministic algorithm* sb-fast *with the following properties:*

1. *If F is unsatisfiable,* sb-fast (F, r) *returns* unsatisfiable.
2. *If F has a satisfying assignment α^* with $d_H(\alpha, \alpha^*) \leq r$, then* sb-fast$(F, \alpha, r)$ *returns* satisfiable.
3. *Otherwise (i.e., if F is satisfiable but all satisfying assignments of F are too far from α), then* sb-fast(F, α, r) *might return* unsatisfiable *or* satisfiable.

Furthermore, sb-fast *runs in time $(k-1)^{r+o(r)}\text{poly}(n)$.*

The "inner random walk" in Schöning's algorithm has all properties stated in Theorem 2, except that it is randomized (with a small error probability). Combining Theorem 2 with the covering code machinery of Dantsin, Goerdt, Hirsch, Kannan, Kleinberg, Papadimitriou, Raghavan, and Schöning [2] yields our main result, Theorem 1.

Theorem 3 *Suppose there is an algorithm A which satisfies properties 1–4 of Theorem 2 and runs in time $c^r\text{poly}(n)$. Then there is an algorithm B solving k-SAT in time $\left(\frac{2c}{c+1}\right)^{n+o(n)}$. Furthermore, B is deterministic if A is.*

Plugging in $c = k - 1$ yields Theorem 2.

Preliminaries: A Slower Algorithm

Dantsin et al. [2] give a deterministic algorithm, henceforth called sb-slow, satisfying Point 1–4 of Theorem 2 with a running time of $k^r \text{poly}(n)$. We will start by explaining and analyzing this algorithm because our algorithm sb-fast uses it as a subroutine in case the input formula F is "well behaved."

Algorithm sb-slow(F, α, r). F is a k-CNF formula over n variables, $\alpha \in \{0, 1\}^{\text{vbl}(F)}$ a truth assignment, and $r \in \mathbb{N}_0$.

1. If α satisfies F, return satisfiable.
2. Else if $r = 0$, return unsatisfiable.
3. Else:
   (a) Pick some clause $C = (u_1 \vee \cdots \vee u_\ell)$ unsatisfied by α. Note that $\ell \leq k$ holds in any case, but $\ell < k$ is possible.
   (b) Set $F_i := F^{[u_i=1]}$.
   (c) Call sb-slow$(F_i, \alpha, r - 1)$ for all $1 \leq i \leq \ell$.
   (d) If some of these ℓ recursive calls returns satisfiable, return satisfiable, otherwise return unsatisfiable.

It is obvious that sb-slow runs in time $k^r \text{poly}(n)$. For correctness, note that sb-slow returns unsatisfiable if F is unsatisfiable. If F has a satisfying assignment $\alpha^* \neq \alpha$, let $(u_1 \vee \ldots \vee u_k)$ be the clause picked in step (3a). Now α^* satisfies some literal u_i in that clause, and thus the formula $F_i := F^{[u_i=1]}$ is satisfiable, as well. Since neither u_i nor $\bar{u}_i$ appears in F_i, we see that $\alpha^*[u_i = 0]$ satisfies F_i. Since $d_H(\alpha^*[u_i = 0], \alpha) = d_H(\alpha^*, \alpha) - 1 \leq r - 1$, the call sb-slow$(F_i, \alpha, r - 1)$ will return satisfiable.

Speeding Up the Algorithm

Let $k \in \mathbb{N}_0$ be fixed from now on.

Definition 1 Let F be a k-CNF formula and $\alpha \in \{0, 1\}^{\text{vbl}(F)}$. We say that F is good (with respect to α) if α satisfies all k-clauses of F (it might still violate smaller clauses).

Observe that if F is good, then $F^{[u=1]}$ is good for every literal u. If F is good with respect to α, then sb-slow(F, α, r) picks a clause of size at most $k - 1$ in step (3a) and causes at most $k - 1$ recursive calls, each again with a good formula:

Lemma 1 *Suppose F is a k-CNF that is good with respect to α. Then* sb-slow (F, α, r) *runs in time* $(k - 1)^r \text{poly}(n)$.

Great! The only thing that is left is to do something smart for formulas that are not good, i.e., have some unsatisfied clauses of size k. We will now describe how sb-fast proceeds and give a precise pseudocode description later. The algorithm sb-fast greedily finds a maximal set of pairwise disjoint k-clauses that are unsatisfied by α: $C_1, \ldots, C_m$, all over disjoint sets of variables. This can be done in polynomial time. Let β be an assignment to these variables.

Proposition 1 $F^{[\beta]}$ *is good with respect to α.*

This is easy to see: consider a k-clause C in F that is not satisfied by α. Then C might or might not be among $C_1, \ldots, C_m$, but by maximality it shares at least one variable with some i. This variable disappears in $F^{[\beta]}$, and C is either satisfied or shrinks to something smaller than k.

Fix $t = \lfloor \log_k \log_2 n \rfloor$. If $m \leq t$ sb-fast can simply iterate over all $2^{km} \leq 2^{kt} \leq (\log_2 n)^k$ different assignments β to the variables in $C_1, \ldots, C_m$: each $F^{[\beta]}$ is good, and thus sb-fast $(F^{[\beta], \alpha, r})$ runs in time $(k-1)^r \text{poly}(n)$. Also, if F is unsatisfiable then all $F^{[\beta]}$ are. If F has a satisfying assignment α^* with $d_H(\alpha^*, \alpha) \leq r$, then at least one $F^{[\beta]}$ does, too. So this step is correct. We are left with the case that $m \geq t$. In this case we define $G := C_1, \ldots, C_t$.

k-ary Covering Codes

The set $[k]^t$ is endowed with a Hamming distance, just as the Hamming cube is: for

$w, w' \in [k]^t$, we define $d_H(w, w') := |\{i \in [t] \mid w_i \neq w'_i\}|$. The k-ary Hamming ball around w of radius s is the set $B_s^{(k)}(w) := \{w' \in [k]^t \mid d_H(w, w') \leq s\}$. The number of elements in such a ball is independent of w. We define and observe

$$\text{vol}^{(k)}(t, s) := \left| B_s^{(k)}(w) \right| = \sum_{i=0}^{s} \binom{t}{i} (k-1)^i.$$

If $C \subseteq [k]^t$ and $\cup_{w \in C} B_s^{(k)}(w) = [k]^t$, we call C a k-ary code of length t and covering radius s. Using the probabilistic method it is easy to show the following result:

Lemma 2 *Let* $t, k \in \mathbb{N}$, *and* $s \in \mathbb{N}_0$. *There exists a k-ary code of length t and covering radius s with at most*

$$\left\lceil \frac{t \ln(k) k^t}{\text{vol}^{(k)}(t, s)} \right\rceil$$

elements.

Observe that $k^t \leq \log_2 n$ and thus there are at most $2^{\log_2 n} = n$ subsets $C \subseteq [k]^t$. We iterate through all of them and find a smallest code of covering radius s.

Consider $G = (C_1, \ldots, C_t)$, our maximal set of pairwise disjoint k-clauses not satisfied by α. Any satisfying assignment α^* of F must satisfy at least one literal in each C_i. Since they are pairwise disjoint, this implies $d_H(\alpha, \alpha^*) \geq t$. There are exactly k^t assignments that satisfy G and have distance exactly t from α. Each such assignment can be represented by a $w \in [k]^t$ in the obvious way. To be more precise, for $w \in [k]^t$ we define $\alpha[G, w]$ to be the assignment which we obtain from α by flipping the w_i^{th} literal in C_i, for $1 \leq i \leq t$. If G is understood from the context, we write $\alpha[w]$ instead of $\alpha[G, w]$.

Example 1 Consider $G = ((x_1, y_1, z_1), (x_2, y_2, z_2), (x_3, y_3, z_3))$, $\alpha = (0, \ldots, 0)$ and $t = 3$. Let $w = (2, 3, 3)$. Then $\alpha[w]$ is the assignment that sets $y_1, z_2,$ and z_3 to 1 and all other variables to 0.

Proposition 2 *We observe the following facts about $\alpha[w]$:*

1. $d_H(\alpha, \alpha[w]) = t$ *for every* $w \in [k]^t$.
2. *If α^* satisfies F, then for some $w^* \in [k]^t$ we have $d_H(\alpha[w^*], \alpha^*) = d_H(\alpha, \alpha^*) - t$.*
3. *Let $w, w' \in [k]^t$. Then $d_H(\alpha[w], \alpha[w']) = 2d_H(w, w')$.*

Lemma 3 *Let t and G be defined as above, and let $C \subseteq [k]^t$ be a k-ary code of covering radius s. If α^* is a satisfying assignment of F, then there is some $w \in C$ such that $d_H(\alpha[w^*], \alpha^*) \leq d_H(\alpha, \alpha^*) - t + 2s$.*

In particular, if $B_r(\alpha)$ contains a satisfying assignment, then there is some $w \in C$ such that $B_{r-t+2s}(\alpha[w])$ contains it, too.

Proof (of Lemma 3) By Proposition 2, there is some $w^* \in [k]^t$ such that $d_H(\alpha[w^*], \alpha^*) = d_H(\alpha, \alpha^*) - t \leq r - t$. Since C has covering radius s, there is some $w \in C$ such that $d_H(ww^*) \leq s$, and by Observation 2, $d_H(\alpha[w], \alpha[w^*]) \leq 2s$. The lemma now follows from the triangle inequality. The proof is illustrated in Fig. 1.

We now state sb-fast formally. We compute an optimal k-ary code C of length t and covering radius $s = t/k$ that is fixed throughout the run of the algorithm.

Algorithm sb-fast(F, α, r).

1. If α satisfies F, return satisfiable.
2. Else if $r = 0$, return unsatisfiable.
3. Else let G be maximal set of pairwise disjoint k-clauses of F unsatisfied by α.
4. If $|G| \geq t := \log_k \log_2 n$: Call sb-slow($F^{[\beta]}, \alpha, r$) for every $\beta \in \{0, 1\}^{\text{vbl}(G)}$ and return satisfiable iff at least one call returns satisfiable.
5. Else, if $|G| \geq t$ set $G = (C_1, \ldots, C_t)$ and call sb-fast($F, \alpha[G, w], r - t + 2t/k$) for every $w \in C$ and return satisfiable iff at least one call returns satisfiable.

Derandomization of
k-SAT Algorithm, Fig. 1
The distance from α^* to
$\alpha[w]$ is at most the distance
from α^* to $\alpha[w^*]$ plus $2s$

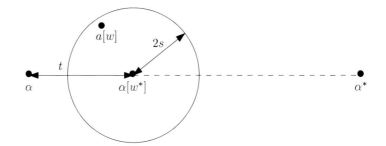

Correctness of sb-fast follows from the above discussion: If there is some $\alpha^* \in B_r(\alpha)$ that satisfies F, then for at least one $w \in \mathcal{C}$ it holds that $d_H(\alpha[w], \alpha^*) \leq r - t + 2t/k$, thus the corresponding recursive call to sb-fast will be successful.

What about the running time? If $|G| \leq t$, then every call to sb-slow($F^{[\beta]}, \alpha, r$) runs in time $O((k-l)^r \text{poly}(n))$. Otherwise, the procedure sb-fast calls itself recursively for each $w \in \mathcal{C}$. Every level further into the recursion, the parameter r decreases by $t - 2t/k$. The running time is therefore $|\mathcal{C}|^{r/(t-2t/k)} \text{poly}(n)$. To evaluate this we have to estimate the size of $\mathcal{C}$. Recall that $s = t/k$.

$$
|\mathcal{C}| \leq \frac{k^t \text{poly}(t)}{\text{vol}^{(k)}(t, s)} \leq \frac{k^t \text{poly}(t)}{\binom{t}{s}(k-1)^s}
$$

$$
\leq \frac{k^t \text{poly}(t)}{\left(\frac{t}{s}\right)^s \left(\frac{t}{t-s}\right)^{t-s}(k-1)^s}
$$

$$
= \frac{k^t \text{poly}(t)}{k^s \left(\frac{k}{k-1}\right)^{t-s}(k-1)^s}
$$

$$
= (k-1)^{t-2s} \text{poly}(t).
$$

Therefore,

$$
|\mathcal{C}|^{r/(t-2t/k)} \leq \left((k-1)^{t-2s} \text{poly}(t)\right)^{r/(t-2s)}
$$

$$
= (k-1)^r \text{poly}(t)^{r/(t-2s)}.
$$

Since t is a growing function in n, the term $\text{poly}(t)^{1/(t-2s)}$ converges to 1 as n grows, and the running time is at most $(k-1)^{r+o(r)\text{poly}(n)}$. This completes the proof of Theorem 2.

Cross-References

▶ Backtracking Based k-SAT Algorithms
▶ Exact Algorithms and Strong Exponential Time Hypothesis
▶ Exact Algorithms and Time/Space Tradeoffs
▶ Exact Algorithms for General CNF SAT
▶ Exact Algorithms for k SAT Based on Local Search
▶ Exact Algorithms for Maximum Two Satisfiability
▶ Exact Graph Coloring Using Inclusion-Exclusion
▶ Random Planted 3-SAT
▶ Thresholds of Random k-SAT
▶ Unique k-SAT and General k-SAT

Recommended Reading

1. Brueggemann T, Kern W (2004) An improved deterministic local search algorithm for 3-SAT. Theor Comput Sci 329(1–3):303–313
2. Dantsin E, Goerdt A, Hirsch EA, Kannan R, Kleinberg J, Papadimitriou C, Raghavan O, Schöning U (2002) A deterministic $(2 - 2/(k + 1))^n$ algorithm for k-SAT based on local search. Theor Comput Sci 289:69–83
3. Kutzkov K, Scheder D (2010) Using CSP to improve deterministic 3-SAT. CoRR abs/1007.1166
4. Moser RA, Scheder D (2011) A full derandomization of Schöning's k-SAT algorithm. In: Fortnow L, Vadhan SP (eds) Proceedings of the 43rd ACM symposium on theory of computing, STOC 2011, San Jose, 6–8 June 2011. ACM, pp 245–252
5. Scheder D (2008) Guided search and a faster deterministic algorithm for 3-SAT. In: Proceedings of the 8th Latin American symposium on theoretical informatics (LATIN'08), Búzios. Lecture notes in computer science, vol 4957, pp 60–71
6. Schöning U (1999) A probabilistic algorithm for k-SAT and constraint satisfaction problems. In: FOCS '99: proceedings of the 40th annual symposium on foundations of computer science, Washington, DC. IEEE Computer Society, p 410

Deterministic Broadcasting in Radio Networks

Leszek Gąsieniec
University of Liverpool, Liverpool, UK

Keywords

Dissemination of information; One-to-all communication; Wireless networks

Years and Authors of Summarized Original Work

2000; Chrobak, Gąsieniec, Rytter

Problem Definition

One of the most fundamental communication problems in wired as well as wireless networks is **broadcasting**, where one distinguished source node has a message that needs to be sent to all other nodes in the network.

The radio network abstraction captures the features of distributed communication networks with multi-access channels, with minimal assumptions on the channel model and processors' knowledge. Directed edges model unidirectional links, including situations in which one of two adjacent transmitters is more powerful than the other. In particular, there is no feedback mechanism (see, for example, [13]). In some applications, collisions may be difficult to distinguish from the noise that is normally present on the channel, justifying the need for protocols that do not depend on the reliability of the collision detection mechanism (see [9,10]). Some network configurations are subject to frequent changes. In other networks, topologies could be unstable or dynamic; for example, when mobile users are present. In such situations, algorithms that do not assume any specific topology are more desirable.

More formally a radio network is a directed graph where by n we denote the number of nodes in this graph. If there is an edge from u to v, then we say that v is an *out-neighbor* of u and u is an *in-neighbor* of v. Each node is assigned a unique identifier from the set $\{1, 2, \ldots, n\}$. In the broadcast problem, one node, for example, node 1, is distinguished as the *source node*. Initially, the nodes do not possess any other information. In particular, they do not know the network topology.

The time is divided into discrete time steps. All nodes start simultaneously, have access to a common clock, and work synchronously. A broadcasting algorithm is a protocol that for each identifier *id*, given all past messages received by *id*, specifies, for each time step t, whether *id* will transmit a message at time t, and if so, it also specifies the message. A message M transmitted at time t from a node u is sent instantly to all its out-neighbors. An out-neighbor v of u receives M at time step t only if no collision occurred, that is, if the other in-neighbors of v do not transmit at time t at all. Further, collisions cannot be distinguished from background noise. If v does not receive any message at time t, it knows that either none of its in-neighbors transmitted at time t or that at least two did, but it does not know which of these two events occurred. The *running time* of a broadcasting algorithm is the smallest t such that for any network topology, and any assignment of identifiers to the nodes, all nodes receive the source message no later than at step t.

All efficient radio broadcasting algorithms are based on the following purely combinatorial concept of selectors.

Selectors Consider subsets of $\{1, \ldots, n\}$. We say that a set S *hits* a set X iff $|S \cap X| = 1$, and that S *avoids* Y iff $S \cap Y = \emptyset$. A family $\mathcal{S}$ of sets is a *w-selector* if it satisfies the following property:

(*) For any two disjoint sets X, Y with $w / 2 \leq |X| \leq w$, $|Y| \leq w$, there is a set in $\mathcal{S}$ which hits X and avoids Y.

A complete layered network is a graph consisting of layers $L_0, \ldots, L_{m-1}$, in which each node in layer L_i is directly connected to every node in layer L_{i+1}, for all $i = 0, \ldots, m - 1$. The layer L_0 contains only the source node s.

Key Results

Theorem 1 ([5]) *For all positive integers w and n, s.t., $w \leq n$, there exists a w-selector $\bar{S}$ with $O(w \log n)$ sets.*

Theorem 2 ([5]) *There exists a deterministic $O(n \log^2 n)$-time algorithm for broadcasting in radio networks with arbitrary topology.*

Theorem 3 ([5]) *There exists a deterministic $O(n \log n)$-time algorithm for broadcasting in complete layered radio networks.*

Applications

Prior to this work, Bruschi and Del Pinto showed in [1] that radio broadcasting requires time $\Omega(n \log D)$ in the worst case. In [4], Chlebus et al. presented a broadcasting algorithm with time complexity $O(n^{11/6})$ – the first subquadratic upper bound. This upper bound was later improved to $O(n^{5/3} \log^3 n)$ by De Marco and Pelc [8] and by Chlebus et al. [3] to $O(n^{3/2})$ by application of finite geometries.

Recently, Kowalski and Pelc in [12] proposed a faster $O(n \log n \log D)$ – time radio broadcasting algorithm, where D is the eccentricity of the network. Later, Czumaj and Rytter showed in [6] how to reduce this bound to $O(n \log^2 D)$. The results presented in [5] (see Theorems 1–3, as well as further improvements in [6, 12]) are existential (non-constructive). The proofs are based on the probabilistic method. A discussion on efficient explicit construction of selectors was initiated by Indyk in [11] and then continued by Chlebus and Kowalski in [2].

More careful analysis and further discussion on selectors in the context of *combinatorial group testing* can be found in [7], where DeBonis et al. proved that the size of selectors is $\Theta(w \log \frac{n}{w})$.

Open Problems

The exact complexity of radio broadcasting remains an open problem, although the gap between the lower and upper bounds $\Omega(n \log D)$

and $O(n \log^2 D)$ is now only a factor of $\log D$. Another promising direction for further studies is improvement of efficient explicit construction of selectors.

Recommended Reading

1. Bruschi D, Del Pinto M (1997) Lower bounds for the broadcast problem in mobile radio networks. Distrib Comput 10(3):129–135
2. Chlebus BS, Kowalski DR (2005) Almost optimal explicit selectors. In: Proceedings of 15th international symposium on fundamentals of computation theory, Lübeck, pp 270–280
3. Chlebus M, Gąsieniec L, Östlin A, Robson JM (2000) Deterministic broadcasting in radio networks. In: Proceedings of 27th international colloquium on automata, languages and programming, Geneva. LNCS, vol 1853, pp 717–728
4. Chlebus BS, Gąsieniec L, Gibbons AM, Pelc A, Rytter W (2002) Deterministic broadcasting in unknown radio networks. Distrib Comput 15(1): 27–38
5. Chrobak M, Gąsieniec L, Rytter W (2002) Fast broadcasting and gossiping in radio networks,. In: Proceedings of 41st annual symposium on foundations of computer science, Redondo Beach, 2000, pp 575–581; Full version in J Algorithms 43(2):177–189 (2002)
6. Czumaj A, Rytter W (2006) Broadcasting algorithms in radio networks with unknown topology. J Algorithms 60(2):115 143
7. De Bonis A, Gąsieniec L, Vaccaro U (2005) Optimal two-stage algorithms for group testing problems. SIAM J Comput 34(5):1253–1270
8. De Marco G, Pelc A (2001) Faster broadcasting in unknown radio networks. Inf Process Lett 79(2): 53–56
9. Ephremides A, Hajek B (1998) Information theory and communication networks: an unconsummated union. IEEE Trans Inf Theory 44: 2416–2434
10. Gallager R (1985) A perspective on multiaccess communications. IEEE Trans Inf Theory 31: 124–142
11. Indyk P (2002) Explicit constructions of selectors and related combinatorial structures, with applications. In: Proceedings of 13th annual ACM-SIAM symposium on discrete algorithms, San Francisco, pp 697–704
12. Kowalski DR, Pelc A (2005) Broadcasting in undirected ad hoc radio networks. Distrib Comput 18(1):43–57
13. Massey JL, Mathys P (1985) The collision channel without feedback. IEEE Trans Inf Theory 31:192–204

Deterministic Searching on the Line

Spyros Angelopoulos
Sorbonne Universités, L'Université Pierre et
Marie Curie (UPMC), Université Paris 06, Paris,
France

Keywords

Online Searching; Pure strategies; Searching for a point on the infinite line; Searching in one dimension

Years and Authors of Summarized Original Work

1993; Baeza-Yates, Culberson, Rawlins
2001; Jaillet, Stafford

Problem Definition

In the *Linear Search Problem* (LSP), we seek efficient strategies for locating an immobile target on the infinite line. More formally, the search environment consists of the infinite (i.e., unbounded) line, with a point O designated as a specific start point. A mobile searcher is initially located at O, whereas the target may be hidden at any point on the line. The searcher's strategy S defines the movement of the searcher on the line; on the other hand, the hider's strategy H is defined as the precise placement of the target on the line, and we denote by $|H|$ the distance of the target from the start point. Given strategies S, H, the *cost* of locating the target, denoted by $c(S, H)$ is the total distance traversed by the searcher at the first time the target is located. The *normalized cost* of the strategies is defined as the quantity $\bar{c}(S, H) = \frac{c(S,H)}{|H|}$.

The objective of the linear search problem is to determine a strategy S for the searcher that minimizes the worst-case normalized cost, namely, the quantity $\sup_H \bar{c}(S, H)$; the latter is often referred to as the *competitive ratio* of the strategy S, due to similarities of this setup with the competitive analysis of online algorithms. In game-theoretic terms, the problem can be described as a zero-sum game between the searcher and the hider in which we seek the minimax strategy of the game.

Extensions

A natural extension of the linear search problem is the *m-ray search* problem, also known as the *star search* problem. Here, the search environment consists of m infinite rays, with the start point O being their common intersection point. Clearly, the linear search problem is precisely the 2-ray search problem.

Constraints

It must be noted that if $|H|$ is arbitrarily small, no strategy of constant competitive ratio exists. Hence, a frequent and natural assumption in the field is to assume that $|H| \geq 1$, i.e., that the target is hidden at least at some minimum allowed distance from the start point. A different assumption that can be made in order to circumvent this complication is that the search strategy must incorporate an infinite sequence of infinitesimal steps (i.e., depths of exploration). In this entry we assume the former, namely, that $|H| \geq 1$.

In addition, we assume only deterministic (i.e., pure) strategies for both the searcher and the hider. We note that a substantial amount of previous work has addressed mixed strategies under given probability distributions on the placement of the target. We refer the reader to the textbook of Alpern and Gal [1].

Key Results

We consider two variants of the problem. In the first variant, the searcher lacks any information concerning the hidden target. In the second variant, the searcher knows that the target is within distance $h = |H|$ from the start point O.

Note that for the linear search problem, the searcher's strategy is completely determined by the sequence of search depths $\{x_i\}_{i \geq 1}$, where x_i denotes the total distance from the start point in which the line is searched during the i-th exploration.

Searching with No Information

It has long been known that a *doubling* strategy, namely, the strategy $\{2^i\}_{i \geq 1}$ attains an optimal competitive ratio equal to 9 for this variant. The result is due to Beck and Newman [4] and redis-covered by Baeza-Yates et al. [2].

Calculating an upper bound on the competitive ratio is easy: if the target is at distance h from O, the doubling strategy will discover it at traversal $2k + 1$, where $2^{2k-1} < h$ and $2^{2k+1} \geq h$. The total distance traversed by the searcher is equal to $2 \sum_{i=1}^{2k} 2i + d = 4(2^{2k} - 1) + h$. The competitive ratio is maximized when $h \to \infty$ and converges (from below) to 9.

An elegant approach for proving the tightness of this bound is based on lower bounds on certain functionals over positive sequences [8]. Let $\{x_i\}_{i \geq 1}$ be an optimal search strategy, then it is easy to see that its competitive ratio is at least equal to $\sup_k 1 + 2\frac{\sum_{i=1}^{k+1} x_i}{x_k}$. In addition, it can be readily seen that an optimal search strategy $\{x_i\}_{i \geq 1}$ must be *monotone*, i.e., $x_i \geq x_j$ for $i > j$. Given the above, Gal shows that there exists $a > 1$ such that

$$\sup_k 1 + 2\frac{\sum_{i=1}^{k+1} x_i}{x_k} \geq \sup_a \lim_{k \to \infty} 1 + 2\frac{\sum_{i=1}^{k+1} a^i}{a^k},$$

which in turn is at least equal to 9, for all $a > 1$.

Informally, the above argument shows that *geometric* strategies of the form $\{a^i\}_{i \geq 1}$ comprise the space of optimal strategies, and by choosing $a = 2$ one obtains the best strategy.

Searching with an Upper Bound on the Target Distance

In the setting in which the searcher has an upper bound h on the distance of the target from the start point, it is possible to obtain improved competitive ratios. Jaillet and Stafford [9] approach this problem by solving the following "dual" problem: given a target competitive ratio r, and the upper bound h, what is the largest "extent" (i.e., the furthest one can go in both directions)

that can be searched while guaranteeing a competitive ratio at most r? A solution to this problem implies a solution to the (primal) search problem: it suffices to find the smallest r such that $e(r) \geq h$, where $e(h)$ is the best-possible extent.

The dual problem of finding $e(r)$ is addressed by means of a series of linear-program formulations. The solution to this series of linear programs defines a search strategy in which the search depths $\{x_i\}_{i \geq 1}$ are determined by an appropriate linear recurrence relation. As a last step, $e(r)$ is obtained as a particular element of the sequence that is generated by the linear recurrence in question. It should be noted that although the strategy is optimal, this technique does not yield a closed-form expression of the optimal competitive ratio (given h).

A similar approach leads to a solution for m-ray searching with an upper bound on the target distance. The crucial difficulty here, as opposed to the case $m = 2$, is in showing that an optimal strategy can be found in the class of *cyclic* strategies: these are strategies in which the searcher always visits the rays in some fixed round-robin order. This seemingly intuitive property is surprisingly hard to be established formally. To bypass this obstacle, one must first show that the property holds when the search depths form a nondecreasing sequence. Then one can argue that, once a searcher is at the start point, it will always choose to explore the ray that has been explored the least up to the current point. As noted in [9], this "least-extended-so-far discipline is the link between non-decreasing depths and the cyclic property that is sought." Once the optimality of cyclic strategies is established, a similar approach can be applied as in the case $m = 2$; namely, the search depths are determined by a (more complicated) linear recurrence.

Applications

The problem has obvious applications in the context of robotic navigation in an unknown environment. Strategies based on doubling are used in searching more complicated environments, e.g., a graph [11]. The linear search problem and its generalization have connections with the design of black-box strategies for obtaining *interruptible* algorithms. The latter class consists of algorithms with the property that they return efficient solutions even if interrupted during their execution. Such algorithms are very desirable in the context of real-time and anytime applications in artificial intelligence [5].

Cross-References

▶ Randomized Searching on Rays or the Line

Recommended Reading

1. Alpern S, Gal S (2003) The theory of search games and rendezvous. Kluwer Academic, Boston
2. Baeza-Yates R, Culberson J, Rawlins G (1993) Searching in the plane. Inf Comput 106(2): 234–252
3. Beck A (1964) On the linear search problem. Naval Res Logist 2:221–228
4. Beck A, Newman DJ (1970) Yet more on the linear search problem. Isr J Math 8:419–429
5. Bernstein DS, Finkelstein L, Zilberstein S (2003) Contract algorithms and robots on rays: unifying two scheduling problems. In: Proceedings of the 18th international joint conference on artificial intelligence (IJCAI), Acapulco, pp 1211–1217
6. Bose P, De Carufel J, Durocher S (2013) Revisiting the problem of searching on a line. In: Proceedings of the 21st European symposium on algorithms (ESA), Sophia Antipolis, pp 205–216
7. Gal S (1972) A general search game. Isr J Math 12:34–45
8. Gal S (1974) Minimax solutions for linear search problems. SIAM J Appl Math 27:17–30
9. Jaillet P, Stafford M (2001) Online searching. Oper Res 49:501–515
10. Kirkpatrick DG (2009) Hyperbolic dovetailing. In: Proceedings of the 17th annual European symposium on algorithms (ESA), Copenhagen, pp 616–627
11. Koutsoupias E, Papadimitriou C, Yannakakis M (1996) Searching a fixed graph. In: Proceedings of the 23rd international colloquium on automata, languages, and programming (ICALP), Paderborn, pp 280–289
12. López-Ortiz A, Schuierer S (2001) The ultimate strategy to search on m rays. Theor Comput Sci 261(2):267–295
13. Schuierer S (2001) Lower bounds in online geometric searching. Comput Geom Theory Appl 18(1): 37–53

Dictionary Matching

Moshe Lewenstein
Department of Computer Science, Bar-Ilan University, Ramat-Gan, Israel

Keywords

Approximate dictionary matching; Approximate text indexing

Years and Authors of Summarized Original Work

2004; Cole, Gottlieb, Lewenstein

Problem Definition

Indexing and *dictionary matching* are generalized models of pattern matching. These models have attained importance with the explosive growth of multimedia, digital libraries, and the Internet.

1. *Text Indexing:* In text indexing one desires to preprocess a text t, of length n, and to answer where subsequent queries p, of length m, appear in the text t.
2. *Dictionary Matching:* In dictionary matching one is given a dictionary D of strings $p_1, \ldots, p_d$ to be preprocessed. Subsequent queries provide a query string t, of length n, and ask for each location in t at which patterns of the dictionary appear.

Key Results

Text Indexing

The *indexing* problem assumes a large text that is to be preprocessed in a way that will allow the following efficient future queries. Given a query pattern, one wants to find all text locations that match the pattern in time proportional to the *pattern length and to the number of occurrences*.

To solve the indexing problem, Weiner [14] invented the *suffix tree* data structure (originally called a *position tree*), which can be constructed in linear time, and subsequent queries of length m are answered in time $O(m \log |\Sigma| + tocc)$, where $tocc$ is the number of pattern occurrences in the text.

Weiner's suffix tree in effect solved the indexing problem for exact matching of fixed texts. The construction was simplified by the algorithms of McCreight and, later, Chen and Seiferas. Ukkonen presented an online construction of the suffix tree. Farach presented a linear time construction for large alphabets (specifically, when the alphabet is $\{1, \ldots, n^c\}$, where n is the text size and c is some fixed constant). All results, besides the latter, work by handling one suffix at a time. The latter algorithm uses a divide-and-conquer approach, dividing the suffixes to be sorted to even-position suffixes and odd-position suffixes. See the entry on ▶ Suffix Tree Construction for full details. The standard query time for finding a pattern p in a suffix tree is $O(m \log |\Sigma|)$. By slightly adjusting the suffix tree, one can obtain a query time of $O(m + \log n)$; see [12].

Another popular data structure for indexing is suffix arrays. Suffix arrays were introduced by Manber and Myers. Others proposed linear time constructions for linearly bounded alphabets. All three extend the divide and conquer approach presented by Farach. The construction in [11] is especially elegant and significantly simplifies the divide-and-conquer approach, by dividing the suffix set into three groups instead of two. See the entry on ▶ Suffix Array Construction for full details. The query time for suffix arrays is $O(m + \log n)$ achievable by embedding additional lcp (longest common prefix) information into the data structure. See [11] for reference to other solutions. *Suffix Trays* were introduced in [6] as a merge between suffix trees and suffix arrays. The construction time of suffix trays is the same as for suffix trees and suffix arrays. The query time is $O(m + \log |\Sigma|)$.

Solutions for the indexing problem in *dynamic* texts, where insertions and deletions (of single characters or entire substrings) are allowed,

appear in several Papers; see [2] and references therein.

Dictionary Matching

Dictionary matching is, in some sense, the "inverse" of text indexing. The large body to be preprocessed is a set of patterns, called the *dictionary*. The queries are texts whose length is typically significantly smaller than the dictionary size. It is desired to find all (exact) occurrences of dictionary patterns in the text in time proportional to the *text length and to the number of occurrences*.

Aho and Corasick [1] suggested an automaton-based algorithm that preprocesses the dictionary in time $O(d)$ and answers a query in time $O(n + docc)$, where $docc$ is the number of occurrences of patterns within the text. Another approach to solving this problem is to use a generalized suffix tree. A *generalized suffix tree* is a suffix tree for a collection of strings. Dictionary matching is done for the dictionary of patterns. Specifically, a suffix tree is created for the generalized string $p_1 \$_1 p_2 \$_2 \ldots \$ p_d \$_d$, where the $\$_i$'s are not in the alphabet. A randomized solution using a fingerprint scheme was proposed in [3]. In [7] a parallel work-optimal algorithm for dictionary matching was presented. Ferragina and Luccio [8] considered the problem in the external memory model and suggested a solution based upon the String Btree data structure along with the notion of a certificate for dictionary matching. Two-dimensional dictionary matching is another fascinating topic which appears as a separate entry. See also the entry on ▶ Multidimensional String Matching.

Dynamic Dictionary Matching

Here one allows insertion and deletion of patterns from the dictionary D. The first solution to the problem was a suffix tree-based method for solving the dynamic dictionary matching problem. Idury and Schäffer [10] showed that the failure function (function mapping from one longest matching prefix to the next longest matching prefix; see [1]) approach and basic scanning loop of the Aho-Corasick algorithm can be adapted to dynamic dictionary matching for improved initial dictionary preprocessing time. They also showed that faster search time can be achieved at the expense of slower dictionary update time.

A further improvement was later achieved by reducing the problem to maintaining a sequence of well-balanced parentheses under certain operations. In [13] an optimal method was achieved based on a labeling paradigm, where labels are given to, sometimes overlapping, substrings of different lengths. The running times are $O(|D|)$ preprocessing time, $O(m)$ update time, and $O(n + docc)$ time for search. See [13] for other references.

Text Indexing and Dictionary Matching with Errors

In most real-life systems, there is a need to allow errors. With the maturity of the solutions for *exact* indexing and *exact* dictionary matching, the quest for *approximate* solutions began. Two of the classical measures for approximating closeness of strings, Hamming distance and Edit distance, were the first natural measures to be considered.

Approximate Text Indexing

For approximate text indexing, given a distance k, one preprocesses a specified text t. The goal is to find all locations ℓ of t within distance k of the query p, i.e., for the Hamming distance all locations ℓ such that the length m substring of t beginning at that location can be made equal to p with at most k character substitutions. (An analogous statement applies for the edit distance.) For $k = 1$ [4] one can preprocess in time $O(n \log^2 n)$ and answer subsequent queries p in time $O(m \sqrt{\log n} \log \log n + occ)$. For small $k \geq 2$, the following naive solutions can be achieved. The first possible solution is to traverse a suffix tree checking all possible configurations of k, or less, mismatches in the pattern. However, while the preprocessing needed to build a suffix tree is cheap, the search is expensive, namely, $O(m^{k+1}|\Sigma|^k + occ)$. Another possible solution, for the Hamming distance measure only, leads to data structures of size approximately $O(n^{k+1})$ embedding all mismatch possibilities into the tree. This can be slightly improved by using the

data structures for $k = 1$, which reduce the size to approximately $O(n^k)$.

Approximate Dictionary Matching

The goal is to preprocess the dictionary along with a threshold parameter k in order to support the following subsequent queries: Given a query text, seek all pairs of patterns (from the dictionary) and text locations which match within distance k. Here once again there are several algorithms for the case where $k = 1$ [4, 9]. The best solution for this problem has query time $O(m \log \log n + occ)$; the data structure uses space $O(n \log n)$ and can be built in time $O(n \log n)$.

The solutions for $k = 1$ in both problems (Approximate Text Indexing and Approximate Dictionary Matching) are based on the following, elegant idea, presented in Indexing terminology. Say a pattern p matches a text t at location i with one error at location j of p (and at location $i + j - 1$ of t). Obviously, the $j - 1$-length prefix of p matches the aligned substring of t and so does the $m - j - 1$ length suffix. If t and p are reversed, then the $j - 1$-th length prefix of p becomes a $j - 1$-th length suffix of p^R (that is p reverse). Notice that there is a match with, at most one error, if (1) the suffix of p starting at location $j + 1$ matches the (prefix of the) suffix of t starting at location $i + j$ and (2) the suffix of p^R starting at location $m - j + 1$ (the reverse of the $j - 1$-th length prefix of p) matches the (prefix of the) suffix of t^R starting at location $m - i - j + 3$. So, the problem now becomes a search for locations j which satisfy the above. To do so, the abovementioned solutions, naturally, use two suffix trees, one for the text and one for its reverse (with additional data structure tricks to answer the query fast). In dictionary matching the suffix trees are defined on the dictionary. The problem is that this solution does not carry over for $k \geq 2$. See the introduction of [5] for a full list of references.

Text Indexing and Dictionary Matching Within (Small) Distance k

Cole et al. [5] proposed a new method that yields a unified solution for approximate text indexing, approximate dictionary matching, and other related problems. However, since the solution is somewhat involved, it will be simpler to explain the ideas on the following problem. The desire is to index a text t to allow fast searching for all occurrences of a pattern containing, at most, k don't cares (don't cares are special characters which match all characters).

Once again, there are two possible, relatively straightforward, solutions to be elaborated. The first is to use a suffix tree, which is cheap to preprocess, but causes the search to be expensive, namely, $O(m|\Sigma|^k + occ)$ (if considering k mismatches this would increase to $O(m^{k+1}|\Sigma|^k + occ)$. To be more specific, imagine traversing a path in a suffix tree. Consider the point where a don't care is reached. If in the middle of an edge the only text suffixes (representing substrings) that can match the pattern with this don't care must also go through this edge, so simply continue traversing. However, if at a node, then all the paths leaving this node must be explored. This explains the mentioned time bound.

The second solution is to create a tree that contains all strings that are at Hamming distance k from a suffix. This allows fast search but leads to trees of size exponential in k, namely, $O(n^{k+1})$ size trees. To elaborate, the tree, called a k-error trie, is constructed as follows. First, consider the case for one don't care, i.e., a 1-error trie, and then extend it. At any node v a don't care may need to be evaluated. Therefore, create a special subtree branching off this node that represents a don't care at this node. To understand this subtree, note that the subtree (of the suffix tree) rooted at v is actually a compressed trie of (some of the) suffixes of the text. Denote the collection of suffixes S_v. The first character of all these suffixes has to be removed (or, perhaps better imagined as a replacement with a don't care character). Each will be a new suffix of the text. Denote the new collection as S'_v. Now, create a new compressed trie of suffixes for S'_v, calling this new subtree an *error tree*. Do so for every v. The suffix tree along with its error trees is a 1-error trie. Turning to queries in the 1-error trie, when traversing the 1-error trie, do so with the suffix tree up till the don't care at node v. Move into the error tree at node v and continue the traversal of the pattern.

To create a 2-error trie, simply take each error tree and construct an error tree for each node within. A $(k+1)$-error trie is created recursively from a k-error trie. Clearly the 1-error trie is of size $O(n^2)$, since any node u in the original suffix tree will appear in all the new subtrees of the 1-error trie created for each of the nodes v which are ancestors of u. Likewise, the k-error trie is of size $O(n^{k+1})$.

The method introduced in Cole et al. [5] uses the idea of the error trees to form a new data structure, which is called a k-errata trie. The k-errata trie will be much smaller than $O(n^{k+1})$. However, it comes at the cost of a somewhat slower search time. To understand the k-errata tries, it is useful to first consider the 1-errata tries and to extend. The 1-errata trie is constructed as follows. The suffix tree is first decomposed with a centroid path decomposition (which is a decomposition of the nodes into paths, where all nodes along a path have their subtree sizes within a range 2^r and 2^{r+1}, for some integer r). Then, as before, error trees are created for each node v of the suffix tree with the following difference. Namely, consider the subtree, T_v, at node v and consider the edge (v, x) going from v to child x on the centroid path. T_v can be partitioned into two subtrees, $T_x \cup (v, x)$, and T'_v all the rest of T_v. An error tree is created for the suffixes in T'_v. The 1-errata trie is the suffix tree with all of its error trees. Likewise, a $(k+1)$-errata trie is created recursively from a k-errata trie. The contents of a k-errata trie should be viewed as a collection of error trees, k levels deep, where error trees at each level are constructed on the error trees of the previous level (at level 0 there is the original suffix tree). The following lemma helps in obtaining a bound on the size of the k-errata trie.

Lemma 1 *Let C be a centroid decomposition of a tree T. Let u be an arbitrary node of T and π be the path from the root to u. There are at most $\log n$ nodes v on π for which v and v's parent on π are on different centroid paths.*

The implication is that every node u in the original suffix tree will only appear in $\log n$ error trees of the 1-*errata trie* because each ancestor v of u is on the path π from the root to u and only $\log n$ such nodes are on different centroid paths than their children (on π). Hence, u appears in only $\log^k n$ error trees in the k-*errata trie*. Therefore, the size of the k-*errata trie* is $O(n \log^k n)$. Creating the k-*errata tries* in $O(n \log^{k+1} n)$ can be done. To answer queries on a k-*errata trie*, given the pattern with (at most) k don't cares, the 0th level of the k-*errata trie*, i.e., the suffix tree, needs to be traversed. This is to be done until the first don't care, at location j, in the pattern is reached. If at node v in the 0th level of the k-*errata trie*, enter the (1st level) error tree hanging off of v and traverse this error tree from location $j + 2$ of the pattern (until the next don't care is met). However, the error tree hanging off of node v does not contain the subtree hanging off of v that is along the centroid path. Hence, continue traversing the pattern in the 0th level of the k-*errata trie*, starting along the edge on the centroid path leaving v (until the next don't care is met). The search is done recursively for k don't cares and, hence, yields an $O(2^k m)$ time search.

Recall that a solution for indexing text that supports queries of a pattern with k don't cares has been described. Unfortunately, when indexing to support k mismatch queries, not to mention k edit operation queries, the traversal down a k-*errata trie* can be very time consuming as frequent branching is required since an error may occur at any location of the pattern. To circumvent this problem, search many error trees in parallel. In order to do so, the error trees have to be grouped together. This needs to be done carefully; see [5] for the full details. Moreover, edit distance needs even more careful handling. The time and space of the algorithms achieved in [5] are as follows:

Approximate Text Indexing: The data structure for mismatches uses space $O(n \log^k n)$, takes time $O(n \log^{k+1} n)$ to build, and answers queries in time $O((\log^k n) \log \log n + m + occ)$. For edit distance, the query time becomes $O((\log^k n) \log \log n + m + 3^k \cdot occ)$. It must

be pointed out that this result is mostly effective for constant k.

Approximate Dictionary Matching: For k mismatches the data structure uses space $O(n + d \log^k d)$, is built in time $O(n + d \log^{k+1} d)$, and has a query time of $O((m + \log^k d) \cdot \log \log n + occ)$. The bounds for edit distance are modified as in the indexing problem.

Applications

Approximate Indexing has a wide array of applications in signal processing, computational biology, and text retrieval, among others. Approximate Dictionary Matching is important in digital libraries and text retrieval systems.

Cross-References

▶ Indexed Approximate String Matching
▶ Indexed Two-Dimensional String Matching
▶ Multidimensional String Matching
▶ Multiple String Matching
▶ Suffix Tree Construction
▶ Suffix Trees and Arrays

Recommended Reading

1. Aho AV, Corasick MJ (1975) Efficient string matching. Commun ACM 18(6):333–340
2. Alstrup S, Brodal GS, Rauhe T (2000) Pattern matching in dynamic texts. In: Proceedings of the symposium on discrete algorithms (SODA), San Francisco, pp 819–828
3. Amir A, Farach M, Matias Y (1992) Efficient randomized dictionary matching algorithms. In: Proceedings of the symposium on combinatorial pattern matching (CPM), Tucson, pp 259–272
4. Amir A, Keselman D, Landau GM, Lewenstein N, Lewenstein M, Rodeh M (1999) Indexing and dictionary matching with one error. In: Proceedings of the workshop on algorithms and data structures (WADS), Vancouver, pp 181–192
5. Cole R, Gottlieb L, Lewenstein M (2004) Dictionary matching and indexing with errors and don't cares. In: Proceedings of the symposium on theory of computing (STOC), Chicago, pp 91–100

6. Cole R, Kopelowitz T, Lewenstein M (2006) Suffix trays and suffix trists: structures for faster text indexing. In: Proceedings of the international colloquium on automata, languages and programming (ICALP), Venice, pp 358–369

7. Farach M, Muthukrishnan S (1995) Optimal parallel dictionary matching and compression. In: Symposium on parallel algorithms and architecture (SPAA), Santa Barbara, pp 244–253

8. Ferragina P, Luccio F (1998) Dynamic dictionary matching in external memory. Inf Comput 146(2): 85–99

9. Ferragina P, Muthukrishnan S, de Berg M (1999) Multi-method dispatching: a geometric approach with applications to string matching. In: Proceedings of the symposium on the theory of computing (STOC), Atlanta, pp 483–491

10. Idury RM, Schäffer AA (1992) Dynamic dictionary matching with failure functions. In: Proceedings of the 3rd annual symposium on combinatorial pattern matching, Tucson, pp 273–284

11. Karkkainen J, Sanders P, Burkhardt S (2006) Linear work suffix array construction. J ACM 53(6):918–936

12. Mehlhorn K (1979) Dynamic binary search. SIAM J Comput 8(2):175–198

13. Sahinalp SC, Vishkin U (1996) Efficient approximate and dynamic matching of patterns using a labeling paradigm. In: Proceedings of the foundations of computer science (FOCS), Burlington, pp 320–328

14. Weiner P (1973) Linear pattern matching algorithm. In: Proceedings of the symposium on switching and automata theory, Iowa City, pp 1–11

Dictionary-Based Data Compression

Travis Gagie[1,2] and Giovanni Manzini[1,3]
[1]Department of Computer Science, University of Eastern Piedmont, Alessandria, Italy
[2]Department of Computer Science, University of Helsinki, Helsinki, Finland
[3]Department of Science and Technological Innovation, University of Piemonte Orientale, Alessandria, Italy

Keywords

LZ compression; Lempel compression; Parsing-based compression; Ziv

Years and Authors of Summarized Original Work

1977; Ziv, Lempel

Problem Definition

The problem of lossless data compression is the problem of compactly representing data in a format that admits the faithful recovery of the original information. Lossless data compression is achieved by taking advantage of the redundancy which is often present in the data generated by either humans or machines.

Dictionary-based data compression has been "the solution" to the problem of lossless data compression for nearly 15 years. This technique originated in two theoretical papers of Ziv and Lempel [15, 16] and gained popularity in the "1980s" with the introduction of the Unix tool compress (1986) and of the gif image format (1987). Although today there are alternative solutions to the problem of lossless data compression (e.g., Burrows-Wheeler compression and Prediction by Partial Matching), dictionary-based compression is still widely used in everyday applications: consider for example the zip utility and its variants, the modem compression standards V.42bis and V.44, and the transparent compression of pdf documents. The main reason for the success of dictionary-based compression is its unique combination of compression power and compression/decompression speed. The reader should refer to [13] for a review of several dictionary-based compression algorithms and of their main features.

Key Results

Let T be a string drawn from an alphabet Σ. Dictionary-based compression algorithms work by parsing the input into a sequence of substrings (also called words) $T_1, T_2, \ldots, T_d$ and by encoding a compact representation of these substrings. The parsing is usually done incrementally and on-line with the following iterative procedure. Assume the encoder has already parsed the sub-strings $T_1, T_2, \ldots, T_{i-1}$. To proceed, the encoder maintains a dictionary of potential candidates for the next word T_i and associates a unique codeword with each of them. Then, it looks at the incoming data, selects one of the candidates, and emits the corresponding codeword. Different algorithms use different strategies for establishing which words are in the dictionary and for choosing the next word T_i. A larger dictionary implies a greater flexibility for the choice of the next word, but also longer codewords. Note that for efficiency reasons the dictionary is usually not built explicitly: the whole process is carried out implicitly using appropriate data structures.

Dictionary-based algorithms are usually classified into two families whose respective ancestors are two parsing strategies, both proposed by Ziv and Lempel and today universally known as LZ78 [16] and LZ77 [15].

The LZ78 Algorithm

Assume the encoder has already parsed the words $T_1, T_2, \ldots, T_{i-1}$, that is, $T = T_1 T_2 \cdots T_{i-1} \hat{T_i}$ for some text suffix $\hat{T_i}$. The LZ78 dictionary is defined as the set of strings obtained by adding a single character to one of the words $T_1, \ldots, T_{i-1}$ or to the empty word. The next word T_i is defined as the longest prefix of $\hat{T_i}$ which is a dictionary word. For example, for $T = aabbaaabaabaabba$ the LZ78 parsing is: $a, ab, b, aa, aba, abaa, bb, a$. It is easy to see that all words in the parsing are distinct, with the possible exception of the last one (in the example the word a). Let T_0 denote the empty word. If $T_i = T_j \alpha$, with $0 \leq j < i$ and $\alpha \in \Sigma$, the codeword emitted by LZ78 for T_i will be the pair (j, α). Thus, if LZ78 parses the string T into t words, its output will be bounded by $t \log t + t \log |\Sigma| + \Theta(t)$ bits.

The LZ77 Algorithm

Assume the encoder has already parsed the words $T_1, T_2, \ldots, T_{i-1}$, that is, $T = T_1 T_2 \cdots T_{i-1} \hat{T_i}$ for some text suffix $\hat{T_i}$. The LZ77 dictionary is defined as the set of strings of the form $w\alpha$ where $\alpha \in \Sigma$ and w is a substring of T starting in the already parsed portion of T. The next word T_i is defined as the longest prefix of $\hat{T_i}$ which is a dictionary word. For example, for $T = aabbaaabaabaabba$ the LZ77 parsing is: $a, ab, ba, aaba, abaabb, a$. Note that, in some sense, $T_5 = abaabb$ is defined in terms of itself: it is a copy of the dictionary word $w\alpha$ with w starting at the second a of T_4 and extending into T_5! It is easy to see that all words in the parsing are distinct, with the possible exception of the last one (in the example the word a), and that the number of words in the LZ77 parsing is smaller than in the LZ78 parsing. If $T_i = w\alpha$ with $\alpha \in \Sigma$, the codeword for T_i is the triplet (s_i, ℓ_i, α) where s_i is the distance from the start of T_i to the last occurrence of w in $T_1 T_2 \cdots T_{i-1}$, and $\ell_i = |w|$.

Entropy Bounds

The performance of dictionary-based compressors has been extensively investigated since their introduction. In [15] it is shown that LZ77 is optimal for a certain family of sources, and in [16] it is shown that LZ78 achieves asymptotically the best compression ratio attainable by a finite-state compressor. This implies that, when the input string is generated by an ergodic source, the compression ratio achieved by LZ78 approaches the entropy of the source. More recent work has established similar results for other Ziv–Lempel compressors and has investigated the rate of convergence of the compression ratio to the entropy of the source (see [14] and references therein).

It is possible to prove compression bounds without probabilistic assumptions on the input, using the notion of *empirical entropy*. For any string T, the order k empirical entropy $H_k(T)$ is the maximum compression one can achieve using a uniquely decodable code in which the codeword for each character may depend on the k characters immediately preceding it [6]. The following lemma is a useful tool for establishing upper bounds on the compression ratio of dictionary-based algorithms which hold pointwise on every string T.

Lemma 1 ([6, Lemma 2.3]) *Let* $T = T_1 T_2 \cdots T_d$ *be a parsing of T such that each word* T_i *appears at most M times. Then, for any* $k \geq 0$

$$d \log d \leq |T| H_k(T) + d \log(|T|/d)$$
$$+ d \log M + \Theta(kd + d),$$

where $H_k(T)$ *whereis the k-th order empirical entropy of T.* □

Consider, for example, the algorithm LZ78. It parses the input T into t distinct words (ignoring the last word in the parsing) and produces an output bounded by $t \log t + t \log |\Sigma| + \Theta(t)$ bits. Using Lemma 1 and the fact that $t = O(|T|/\log T)$, one can prove that LZ78's output is at most $|T| H_k(T) + o(|T|)$ bits. Note that the bound holds for any $k \geq 0$: this means that LZ78 is essentially "as powerful" as any compressor that encodes the next character on the basis of a finite context.

Algorithmic Issues
One of the reasons for the popularity of dictionary-based compressors is that they admit linear-time, space-efficient implementations. These implementations sometimes require non-trivial data structures: the reader is referred to [12] and references therein for further reading on this topic.

Greedy vs. Non-Greedy Parsing
Both LZ78 and LZ77 use a greedy parsing strategy in the sense that, at each step, they select the longest prefix of the unparsed portion which is in the dictionary. It is easy to see that for LZ77 the greedy strategy yields an optimal parsing; that is, a parsing with the minimum number of words. Conversely, greedy parsing is not optimal for LZ78: for any sufficiently large integer m there exists a string that can be parsed to $O(m)$ words and that the greedy strategy parses in $\Omega(m^{3/2})$ words. In [9] the authors describe an efficient algorithm for computing an optimal parsing for the LZ78 dictionary and, indeed, for any dictionary with the prefix-completeness property (a dictionary is prefix-complete if any prefix of a dictionary word is also in the dictionary).

Interestingly, the algorithm in [9] is a one-step lookahead greedy algorithm: rather than choosing the longest possible prefix of the unparsed portion of the text, it chooses the prefix that results in the longest advancement in the *next* iteration.

Applications

The natural application field of dictionary-based compressors is lossless data compression (see, for example [13]). However, because of their deep mathematical properties, the Ziv–Lempel parsing rules have also found applications in other algorithmic domains.

Prefetching
Krishnan and Vitter [7] considered the problem of prefetching pages from disk into memory to anticipate users' requests. They combined LZ78 with a pre-existing prefetcher P_1 that is asymptotically at least as good as the best memoryless prefetcher, to obtain a new algorithm P that is asymptotically at least as good as the best finite-state prefetcher. LZ78's dictionary can be viewed as a trie: parsing a string means starting at the root, descending one level for each character in the parsed string and, finally, adding a new leaf. Algorithm P runs LZ78 on the string of page requests as it receives them, and keeps a copy of the simple prefetcher P_1 for each node in the trie; at each step, P prefetches the page requested by the copy of P_1 associated with the node LZ78 is currently visiting.

String Alignment
Crochemore, Landau and Ziv-Ukelson [4] applied LZ78 to the problem of sequence alignment, i.e., finding the cheapest sequence of character insertions, deletions and substitutions that transforms one string T into another T' (the cost of an operation may depend on the character or characters involved). Assume, for simplicity, that $|T| = |T'| = n$. In 1980 Masek and Paterson proposed an $O(n^2/\log n)$-time algorithm with the restriction that the costs be rational; Crochemore et al.'s algorithm allows real-valued costs, has the same asymptotic cost

in the worst case, and is asymptotically faster for compressible texts.

The idea behind both algorithms is to break into blocks the matrix $A[1 \dots n, 1 \dots n]$ used by the obvious $O(n^2)$-time dynamic programming algorithm. Masek and Paterson break it into uniform-sized blocks, whereas Crochemore et al. break it according to the LZ78 parsing of T and T'. The rationale is that, by the nature of LZ78 parsing, whenever they come to solve a block $A[i \dots i', j \dots j']$, they can solve it in $O(i' - i + j' - j)$ time because they have already solved blocks identical to $A[i \dots i' - 1, j \dots j']$ and $A[i \dots i', j \dots j' - 1]$ [8]. Lifshits, Mozes, Weimann and Ziv-Ukelson [8] recently used a similar approach to speed up the decoding and training of hidden Markov models.

Compressed Full-Text Indexing

Given a text T, the problem of compressed full-text indexing is defined as the task of building an index for T that takes space proportional to the entropy of T and that supports the efficient retrieval of the occurrences of any pattern P in T. In [10] Navarro proposed a compressed full-text index based on the LZ78 dictionary. The basic idea is to keep two copies of the dictionary as tries: one storing the dictionary words, the other storing their reversal. The rationale behind this scheme is the following. Since any non-empty prefix of a dictionary word is also in the dictionary, if the sought pattern P occurs within a dictionary word, then P is a suffix of some word and easy to find in the second dictionary. If P overlaps two words, then some prefix of P is a suffix of the first word–and easy to find in the second dictionary–and the remainder of P is a prefix of the second word–and easy to find in the first dictionary. The case when P overlaps three or more words is a generalization of the case with two words. Recently, Arroyuelo et al. [1] improved the original data structure in [10]. For any text T, the improved index uses $(2 + \epsilon)|T|H_k(T) + o(|T| \log |\Sigma|)$ bits of space, where $H_k(T)$ is the k-th order empirical entropy of T, and reports all occ occurrences of P in T in $O(|P|^2 \log |P| + (|P| + occ) \log |T|)$ time.

Independently of [10], in [5] the LZ78 parsing was used together with the Burrows-Wheeler compression algorithm to design the first full-text index that uses $o(|T| \log |T|)$ bits of space and reports the occ occurrences of P in T in $O(|P| + occ)$ time. If $T = T_1 T_2 \cdots T_d$ is the LZ78 parsing of T, in [5] the authors consider the string $T_\$ = T_1 \$ T_2 \$ \cdots \$ T_d \$$ where $\$$ is a new character not belonging to Σ. The string $T_\$$ is then compressed using the Burrows-Wheeler transform. The $\$$'s play the role of anchor points: their positions in $T_\$$ are stored explicitly so that, to determine the position in T of any occurrence of P, it suffices to determine the position with respect to any of the $\$$'s. The properties of the LZ78 parsing ensure that the overhead of introducing the $\$$'s is small, but at the same time the way they are distributed within $T_\$$ guarantees the efficient location of the pattern occurrences.

Related to the problem of compressed full-text indexing is the compressed matching problem in which text and pattern are given together (so the former cannot be preprocessed). Here the task consists in performing string matching in a compressed text without decompressing it. For dictionary-based compressors this problem was first raised in 1994 by A. Amir, G. Benson, and M. Farach, and has received considerable attention since then. The reader is referred to [11] for a recent review of the many theoretical and practical results obtained on this topic.

Substring Compression Problems

Substring compression problems involve preprocessing T to be able to efficiently answer queries about compressing substrings: e.g., how compressible is a given substring s in T? what is s's compressed representation? or, what is the least compressible substring of a given length ℓ? These are important problems in bioinformatics because the compressibility of a DNA sequence may give hints as to its function, and because some clustering algorithms use compressibility to measure similarity. The solutions to these problems are often trivial for simple compressors, such as Huffman coding or run-length encoding, but they are open for more powerful algorithms, such as dictionary-based compressors, BWT compressors, and PPM compressors. Recently, Cormode and Muthukrishnan [3] gave some preliminary solutions for LZ77. For any string s, let $C(s)$ denote the number of words in the LZ77-parsing of s, and let LZ77(s) denote the LZ77-compressed representation of s. In [3] the authors show that, with $O(|T| \text{ polylog}(|T|))$ time preprocessing, for any substring s of T they can: $a)$ compute LZ77(s) in $O(C(s) \log |T| \log \log |T|)$ time, $b)$ compute an approximation of $C(s)$ within a factor $O(\log |T| \log^* |T|)$ in $O(1)$ time, $c)$ find a substring of length ℓ that is close to being the least compressible in $O(|T|\ell / \log \ell)$ time. These bounds also apply to general versions of these problems, in which queries specify another substring t in T as context and ask about compressing substrings when LZ77 starts with a dictionary already containing the words in the LZ77 parsing of t.

Grammar Generation

Charikar et al. [2] considered LZ78 as an approximation algorithm for the NP-hard problem of finding the smallest context-free grammar that generates only the string T. The LZ78 parsing of T can be viewed as a context-free grammar in which for each dictionary word $T_i = T_j \alpha$ there is a production $X_i \to X_j \alpha$. For example, for $T = aabbaaabaabaabba$ the LZ78 parsing is: a, ab, b, aa, aba, $abaa$, bb, a, and the corresponding grammar is: $S \to X_1 \ldots X_7 X_1$, $X_1 \to a$, $X_2 \to X_1 b$, $X_3 \to b$, $X_4 \to X_1 a$, $X_5 \to X_2 a$, $X_6 \to X_5 a$, $X_7 \to X_3 b$. Charikar et al. showed LZ78's approximation ratio is in $O((|T|/\log |T|)^{2/3}) \cap \Omega(|T|^{2/3}\log |T|)$; i.e., the grammar it produces has size at most $f(|T|) \cdot m^*$, where $f(|T|)$ is a function in this intersection and m^* is the size of the smallest grammar. They also showed m^* is at least the number of words output by LZ77 on T, and used LZ77 as the basis of a new algorithm with approximation ratio $O(\log(|T|/m^*))$.

URL to Code

The source code of the gzip tool (based on LZ77) is available at the page http://www.gzip.org/. An LZ77-based compression library zlib is available from http://www.zlib.net/. A more recent, and more efficient, dictionary-based compressor is LZMA (Lempel–Ziv Markov chain Algorithm), whose source code is available from http://www.7-zip.org/sdk.html.

Cross-References

► Arithmetic Coding for Data Compression
► Boosting Textual Compression
► Burrows-Wheeler Transform
► Compressed Text Indexing

Recommended Reading

1. Arroyuelo D, Navarro G, Sadakane K (2006) Reducing the space requirement of LZ-index. In: Proceedings of 17th combinatorial pattern matching conference (CPM). LNCS, vol 4009. Springer, pp 318–329
2. Charikar M, Lehman E, Liu D, Panigraphy R, Prabhakaran M, Sahai A, Shelat A (2005) The smallest grammar problem. IEEE Trans Inf Theory 51:2554–2576
3. Cormode G, Muthukrishnan S (2005) Substring compression problems. In: Proceedings of. 16th ACM-SIAM symposium on discrete algorithms (SODA '05), pp 321–330
4. Crochemore M, Landau G, Ziv-Ukelson M (2003) A subquadratic sequence alignment algorithm for unrestricted scoring matrices. SIAM J Comput 32:1654–1673
5. Ferragina P, Manzini G (2005) Indexing compressed text. J ACM 52:552–581
6. Kosaraju R, Manzini G (1999) Compression of low entropy strings with Lempel–Ziv algorithms. SIAM J Comput 29:893–911
7. Krishnan P, Vitter J (1998) Optimal prediction for prefetching in the worst case. SIAM J Comput 27:1617–1636
8. Lifshits Y, Mozes S, Weimann O, Ziv-Ukelson M (2007) Speeding up HMMdecoding and training by exploiting sequence repetitions. Springer, 2007
9. Matias Y, Sahinalp C (1999) On the optimality of parsing in dynamic dictionary based data compression. In: Proceedings 10th annual ACM-SIAM symposium on discrete algorithms (SODA'99), pp 943–944
10. Navarro G (2004) Indexing text using the Ziv–Lempel trie. J Discret Algorithm 2:87–114
11. Navarro G, Tarhio J (2005) LZgrep: a Boyer-Moore string matching tool for Ziv–Lempel compressed text. Softw Pract Exp 35:1107–1130
12. Sahinalp C, Rajpoot N (2003) Dictionary-based data compression: an algorithmic perspective. In: Sayood

K (ed) Lossless compression handbook. Academic Press, pp 153–167
13. Salomon D (2007) Data compression: the complete reference, 4th edn. Springer, London
14. Savari S (1997) Redundancy of the Lempel–Ziv incremental parsing rule. IEEE Trans Inf Theory 43:9–21
15. Ziv J, Lempel A (1977) A universal algorithm for sequential data compression. IEEE Trans Inf Theory 23:337–343
16. Ziv J, Lempel A (1978) Compression of individual sequences via variable-length coding. IEEE Trans Inf Theory 24:530–536

Differentially Private Analysis of Graphs

Sofya Raskhodnikova and Adam Smith
Computer Science and Engineering Department, Pennsylvania State University, University Park, State College, PA, USA

Keywords

Degree distribution; Graphs; Privacy; Subgraph counts

Years and Authors of Summarized Original Work

2013; Blum, Blocki, Datta, Sheffet
2013; Chen, Zhou
2013; Kasiviswanatan, Nissim, Raskhodnikova, Smith
2015; Borgs, Chayes, Smith
2015; Raskhodnikova, Smith

Problem Definition

Many datasets can be represented by graphs, where nodes correspond to individuals and edges capture relationships between them. On one hand, such datasets contain potentially sensitive information about individuals; on the other hand, there are significant public benefits from allowing access to *aggregate* information about the data. Thus, analysts working with such graphs are faced with two conflicting goals: protecting privacy of individuals and publishing accurate aggregate statistics. This article describes algorithms for releasing accurate graph statistics while preserving a rigorous notion of privacy, called *differential privacy*.

Differential privacy was introduced by Dwork et al. [6]. It puts a restriction on the algorithm that processes sensitive data and publishes the output. Intuitively, differential privacy requires that, for every individual, the output distribution of the algorithm is roughly the same whether or not this individual's data is present in the dataset. Next, we give a formal definition of differential privacy, specialized to datasets represented by graphs.

Two graphs are called *neighbors* if one can be obtained from the other by removing a node and its adjacent edges. Given a parameter $\epsilon > 0$, an algorithm A is ϵ-*node differentially private* if for all neighbor graphs G and G' and for all sets S of possible outputs produced by A:

$$\Pr[A(G) \subseteq S] \le e^{\epsilon} \cdot \Pr[A(G) \subseteq S].$$

This variant of differential privacy is called *node-differential privacy* because neighbor graphs are defined with respect to node removals. Analogously, we can define *edge differential privacy* by letting graphs be neighbors if they differ in exactly one edge. Intuitively, edge differential privacy protects edges (which represent connections between people), while node-differential privacy protects nodes together with their adjacent edges (i.e., all information pertaining to individuals). Node-differential privacy is a stronger privacy definition, but it is much harder to attain because it requires the output distribution of the algorithm to hide much larger differences in the input graph.

We would like to design differentially private algorithms (preferably, node-differentially private) that compute accurate graph statistics on a large family of realistic graphs. Typically, graphs that contain sensitive information, such as friendships, sexual relationships, and communication patterns, are sparse. Some examples of graph statistics we would like to compute on these graphs are the number of edges, small subgraph counts, and the degree distribution.

Most work on the topic considers an analyst who wants to evaluate a real-valued function f on the private input graph G (e.g., the number of triangles or the number of connected components in G). The goal is to release as good an approximation as possible to the true value $f(G)$. Differentially private algorithms must be randomized, so we try to minimize the expectation of the random variable $\text{error}_A(G) = |A(G) - f(G)|$. We will also discuss work on algorithms that release higher-dimensional summaries (i.e., output a real vector).

Bibliographical Notes

Edge privacy was first studied by Nissim et al. [16], and the distinction between node and edge privacy was laid out by Hay et al. [9]. Edge differentially private algorithms for a variety of tasks have been widely investigated. Examples include subgraph counts, degree distributions, and parameters of generative statistical models. Gehrke et al. [7] investigated a notion whose strength lies between edge and node privacy: node privacy for bounded-degree graphs. (The focus of their work is a generalization of differential privacy, called *zero-knowledge privacy*.)

Until recently, no node-differentially private algorithms (where privacy guarantees hold with respect to all graphs) were known that compute accurate graph statistics on realistic (namely, sparse) graphs. The first such algorithms were designed independently by Blocki et al. [3], Kasiviswanathan et al. [11], and Chen and Zhou [5]. Those algorithms look at releasing one real-valued statistic at a time. Two more recent works focus on higher-dimensional node-private releases: Raskhodnikova and Smith [17] and Borgs et al. [4].

This encyclopedia entry focuses on node-differentially private algorithms, since these offer the strongest privacy guarantees. Progress, however, continues on edge-private algorithms; see Lin and Kifer [13], Karwa and Slavkovic [10], Lu and Miklau [14], and Zhang et al. [18] for recent results.

Key Results

The main difficulty in the design of node-private algorithms is that techniques based on *local sensitivity* of a function (which are the basis of the best edge-private algorithms) yield node-private algorithms whose error on "typical" inputs swamps the statistic that one wants to release. The local sensitivity of a function f is a discrete analogue of the derivative of f – it measures how much the value of f can change when the input graph is replaced with its neighbor. On sparse graphs, the local sensitivity can be larger than the value of the function. Any method whose error is proportional to the local sensitivity will have large relative error.

Focus on a "Preferred Subset"

To get around the challenge of high local sensitivity, two works [3, 11] independently designed algorithms that are given a set S of "nice" graphs that hopefully contains G (e.g., graphs with an upper bound on the maximum degree). These algorithms are private on *all* graphs and return an accurate answer *on graphs in S*. What makes this approach work is that S is selected so that the sensitivity of f is small when restricted to inputs in S.

Let $\mathbb{G}$ denote the set of all labeled, undirected graphs. We will call $S \subseteq \mathbb{G}$ the "preferred" subset. Define the *Lipschitz constant* (also called the *restricted sensitivity*) of f on S to be

$$\Delta_f(S) = \sup_{G, G' \in S} \frac{\|f(G') - f(G)\|_1}{d_{\text{node}}(G, G')} \, ,$$

where d_{node} is the node distance between two graphs – the number of vertex insertions and deletions needed to go from G to G'. Blocki et al. [3] and Kasiviswanathan et al. [11] give methods for adding noise proportional to the Lipschitz constant of f on S.

Theorem 1 ([3, 11]) *For every $S \subseteq \mathbb{G}$, function $f : S \to \mathbb{R}$, and $\epsilon > 0$, there exists an algorithm A_S that is ϵ-differentially private (for all inputs) and such that, for all $G \in S$,*

$$\mathbb{E}|A_S(G) - f(G)| = O(\Delta_S(f)/\epsilon^2).$$

Moreover, for $S = \mathbb{G}_D$ (the set of D-bounded graphs), the running time of A is the running time for one evaluation of f plus a fixed polynomial in the size of G.

The same works [3, 11] also give generic reductions showing that given any algorithm that is ϵ-differentially private when restricted to graphs in S, one can design an algorithm A that has similar behavior on graphs in S but is ϵ'-differentially private for *all* inputs, for ϵ' not too much larger than ϵ.

"Down" Sensitivity

Rather than focusing on a single "nice" subset, some works [5, 17] sought to add noise proportional to a quantity related to, but usually much smaller than, the local sensitivity.

Define the *down sensitivity* (called *empirical global sensitivity* when first defined by Chen and Zhou [5]) of f at a graph G to be the Lipschitz constant of f *when restricted to the set of induced subgraphs of G*. Specifically, we write $G \preceq H$ to denote that G is an induced subgraph of H (i.e., G can be obtained by deleting a set of vertices from H) and define the down sensitivity to be

$$DS_f(G) = \max_{H, H' \text{neighbors}, H \preceq H' \preceq G} |f(G') - f(G)|.$$

By carefully (and privately) selecting the "preferred" subset based on the input, one can add noise essentially proportional to the down sensitivity.

Theorem 2 ([17]) *For every monotone function* $f : \mathbb{G} \to \mathbb{R}$ *and* $\epsilon > 0$, *there is an algorithm* A_f *that is* ϵ-*differentially private and such that, for all* $G \in \mathbb{G}$,

$$\mathbb{E}|A_f(G) - f(G)|$$

$$= \frac{DS_f(G) + 1}{\epsilon} \cdot O(\log \log \max_{G'} DS_f(G')).$$

Moreover, A_f can be made efficient when f is a generalized linear query (a class that includes counting occurrences of a fixed subgraph).

The down sensitivity is low for many commonly studied statistics in graphs that satisfy α-decay, a condition on the degree distribution that is satisfied by known generative models (including those that generate "scale-free"). (See [11] for a definition of α-decay.)

Lipschitz Extensions and Higher-Dimensional Releases

The main technical tool in the down-sensitivity-based results [5, 17] is the construction of *efficient* (i.e., polynomial time computable) *Lipschitz extensions* of the function f from subsets S of graphs to the space of all graphs. Kasiviswanathan et al. [11] and Chen and Zhou [5] give efficient Lipschitz extensions of several useful functions (including graph counts) that return a single real value. Raskhodnikova and Smith [17] give efficient Lipschitz extensions of higher-dimensional functions, namely, the degree distribution and adjacency matrix of a graph.

Borgs et al. [4] use the Lipschitz extension technique together with the *exponential mechanism* to provide the first node-differentially private algorithms for fitting high-dimensional statistical models to a given graph (specifically, they consider *stochastic block models* and generalizations thereof).

Applications

The algorithms discussed above address a real problem: datasets containing sensitive information about relationships among a collection of individuals are often valuable sources of information, but publishing useful summaries about such data without leaking individual information is difficult. Even when the graphs are "anonymized" by removing all obviously identifying information, such as names, addresses, birthdays, and zip codes, they present a privacy risk. For example, [1, 15] give de-anonymization attacks based only on unlabeled links. Node-differentially private algorithms offer a principled method for releasing information about a network while providing rigorous privacy guarantees (though some authors argue that even stronger notions may be needed [7, 12]).

Open Problems

Gupta et al. [8] and Blocki et al. [2] give edge differentially private algorithms for releasing a data structure that approximates the sizes of all cuts in the input graph in the following sense: for any cut, with high probability, the estimated cut size is accurate (the first reference gives weaker approximation guarantees with a stronger quantifier order: with high probability, all cut sizes are accurate). It is open whether a node-differentially private algorithm can obtain similar results.

For datasets that do not contain information about relationships, but only contain personal attributes that come from a relatively small set, differentially private algorithms can output a large number of statistics at once (see ▶ Query Release via Online Learning and ▶ Geometric Approaches to Answering Queries cross-referenced below). It is open how to do achieve similar results for graph statistics, even with edge differential privacy.

Finally, all algorithms we discussed release numerical graph statistics. The subject of differentially private synthetic graphs is largely unexplored. See [10, 13] for initial results.

Cross-References

- ▶ Beyond Worst Case Sensitivity in Private Data Analysis
- ▶ Geometric Approaches to Answering Queries
- ▶ Private Spectral Analysis
- ▶ Query Release via Online Learning

Acknowledgments The authors were supported in part by NSF award IIS-1447700, Boston University's Hariri Institute for Computing and Center for Reliable Information Systems and Cyber Security, and, while visiting the Harvard Center for Research on Computation & Society, by a Simons Investigator grant to Salil Vadhan.

Recommended Reading

1. Backstrom L, Dwork C, Kleinberg J (2007) Wherefore art thou r3579x? Anonymized social networks, hidden patterns, and structural steganography. In: Proceedings of the 16th international World Wide Web conference, Banff, pp 181–190
2. Blocki J, Blum A, Datta A, Sheffet O (2012) The Johnson-Lindenstrauss transform itself preserves differential privacy. In: 53rd annual IEEE symposium on foundations of computer science, FOCS 2012, New Brunswick, 20–23 Oct 2012. IEEE Computer Society, pp 410–419. doi:10.1109/FOCS.2012.67, http://dx.doi.org/10.1109/FOCS.2012.67
3. Blocki J, Blum A, Datta A, Sheffet O (2013) Differentially private data analysis of social networks via restricted sensitivity. In: Innovations in theoretical computer science (ITCS), Berkeley, pp 87–96
4. Borgs C, Chayes JT, Smith A (2015) Private graphon estimation for sparse graphs. arXiv:150606162 [mathST]
5. Chen S, Zhou S (2013) Recursive mechanism: towards node differential privacy and unrestricted joins. In: ACM SIGMOD international conference on management of data, New York, pp 653–664
6. Dwork C, McSherry F, Nissim K, Smith A (2006) Calibrating noise to sensitivity in private data analysis. In: Halevi S, Rabin T (eds) TCC, New York, vol 3876, pp 265–284
7. Gehrke J, Lui E, Pass R (2011) Towards privacy for social networks: a zero-knowledge based definition of privacy. In: Ishai Y (ed) TCC, Providence. Lecture notes in computer science, vol 6597. Springer, pp 432–449
8. Gupta A, Roth A, Ullman J (2012) Iterative constructions and private data release. In: TCC, Taormina
9. Hay M, Li C, Miklau G, Jensen D (2009) Accurate estimation of the degree distribution of private networks. In: International conference on data mining (ICDM), Miami, pp 169–178
10. Karwa V, Slavkovic A (2014) Inference using noisy degrees: differentially private ß-model and synthetic graphs. statME arXiv:1205.4697v3 [stat.ME]
11. Kasiviswanathan SP, Nissim K, Raskhodnikova S, Smith A (2013) Analyzing graphs with node-differential privacy. In: Theory of cryptography conference (TCC), Tokyo, pp 457–476
12. Kifer D, Machanavajjhala A (2011) No free lunch in data privacy. In: Sellis TK, Miller RJ, Kementsietsidis A, Velegrakis Y (eds) SIGMOD conference. ACM, Athens, Greece, pp 193–204
13. Lin BR, Kifer D (2013) Information preservation in statistical privacy and Bayesian estimation of unattributed histograms. In: ACM SIGMOD international conference on management of data, New York City, pp 677–688
14. Lu W, Miklau G (2014) Exponential random graph estimation under differential privacy. In: 20th ACM

SIGKDD international conference on knowledge discovery and data mining, New York City, pp 921–930

15. Narayanan A, Shmatikov V (2009) De-anonymizing social networks. In: IEEE symposium on security and privacy, Oakland, pp 173–187
16. Nissim K, Raskhodnikova S, Smith A (2007) Smooth sensitivity and sampling in private data analysis. In: Symposium on theory of computing (STOC), San Diego, pp 75–84, full paper on authors' web sites
17. Raskhodnikova S, Smith A (2015) High-dimensional Lipschitz extensions and node-private analysis of network data. arXiv:150407912
18. Zhang J, Cormode G, Procopiuc CM, Srivastava D, Xiao X (2015) Private release of graph statistics using ladder functions. In: ACM SIGMOD international conference on management of data, Melbourne, pp 731–745

Dilation of Geometric Networks

Rolf Klein
Institute for Computer Science, University of Bonn, Bonn, Germany

Keywords

Detour; Spanning ratio; Stretch factor

Years and Authors of Summarized Original Work

2005; Ebbers-Baumann, Grüne, Karpinski, Klein, Knauer, Lingas

Problem Definition

Notations

Let $G = (V, E)$ be a plane geometric network, whose vertex set V is a finite set of point sites in $\mathbb{R}^2$, connected by an edge set E of non-crossing straight line segments with endpoints in V. For two points $p \neq q \in V$, let $\xi_G(p, q)$ denote a shortest path from p to q in G. Then

$$\sigma(p, q) := \frac{|\xi_G(p, q)|}{|pq|} \quad (1)$$

is the detour one encounters when using network G, in order to get from p to q, instead of walking straight. Here, $|\,.\,|$ denotes the Euclidean length.

The *dilation* of G is defined by

$$\sigma(G) := \max_{p \neq q \in V} \sigma(p, q). \quad (2)$$

This value is also known as the spanning ratio or the stretch factor of G. It should, however, not be confused with the geometric dilation of a network, where the points on the edges are also being considered, in addition to the vertices.

Given a finite set S of points in the plane, one would like to find a plane geometric network $G = (V, E)$ whose dilation $\sigma(G)$ is as small as possible, such that S is contained in V. The value of

$$\Sigma(S) := \inf\{\sigma(G); G = (V, E) \text{ finite plane geometric network where } S \subset V\}$$

is called the *dilation of point set S*. The problem is in computing, or bounding, $\Sigma(S)$ for a given set S.

Related Work

If edge crossings were allowed, one could use spanners whose stretch can be made arbitrarily close to 1; see the monographs by Eppstein [6] or Narasimhan and Smid [12]. Different types of triangulations of S are known to have their stretch factors bounded from above by small constants, among them the Delaunay triangulation of stretch ≤ 2.42; see Dobkin et al. [3], Keil and Gutwin [10], and Das and Joseph [2]. Eppstein [5] has characterized all triangulations T of dilation $\sigma(T) = 1$; these triangulations are shown in Fig. 1. Trivially, $\Sigma(S) = 1$ holds for each point set S contained in the vertex set of such a triangulation T.

Key Results

The previous remark's converse also turns out to be true.

Dilation of Geometric Networks, Fig. 1 The triangulations of dilation 1

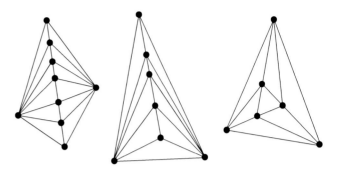

Theorem 1 ([11]) *If S is not contained in one of the vertex sets depicted in Fig. 1, then $\Sigma(S) > 1$.*

That is, if a point set S is not one of these special sets, then each plane network including S in its vertex set has a dilation larger than some lower bound $1 + \eta(S)$. The proof of Theorem 1 uses the following density result. Suppose one connects each pair of points of S with a straight line segment. Let S' be the union of S and the resulting crossing points. Now the same construction is applied to S' and repeated. For the limit point set S^∞, the following theorem holds. It generalizes work by Hillar and Rhea [8] and by Ismailescu and Radoičić [9] on the intersections of lines.

Theorem 2 ([11]) *If S is not contained in one of the vertex sets depicted in Fig. 1, then S^∞ lies dense in some polygonal part of the plane.*

For certain infinite structures can concrete lower bounds be proven.

Theorem 3 ([4]) *Let N be an infinite plane network all of whose faces have a diameter bounded from above by some constant. Then $\sigma(N) > 1.00156$ holds.*

Theorem 4 ([4]) *Let C denote the (infinite) set of all points on a closed convex curve. Then $\Sigma(C) > 1.00157$ holds.*

Theorem 5 ([4]) *Given n families $F_i, 2 \leq i \leq n$, each consisting of infinitely many equidistant parallel lines. Suppose that these families are in general position.*

Then their intersection graph G is of dilation at least $2/\sqrt{3}$.

The proof of Theorem 5 makes use of Kronecker's theorem on simultaneous approxima-

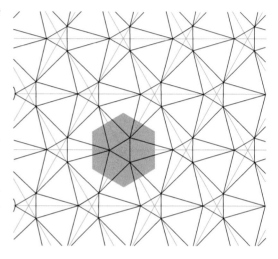

Dilation of Geometric Networks, Fig. 2 A network of dilation ~ 1.1247

tion. The bound is attained by the packing of equiangular triangles.

Finally, there is a general upper bound to the dilation of finite point sets.

Theorem 6 ([4]) *Each finite point set S is of dilation $\Sigma(S) < 1.1247$.*

To prove this upper bound, one can embed any given finite point set S in the vertex set of a scaled, and slightly deformed, finite part of the network depicted in Fig. 2. It results from a packing of equilateral triangles by replacing each vertex with a small triangle and by connecting neighboring triangles as indicated.

Applications

A typical university campus contains facilities like lecture halls, dorms, library, mensa, and supermarkets, which are connected by some path system. Students in a hurry are tempted to walk straight across the lawn, if the shortcut seems worth it. After a while, this causes new paths to appear. Since their intersections are frequented by many people, they attract coffee shops or other new facilities. Now, people will walk across the lawn to get quickly to a coffee shop, and so on.

D. Eppstein [5] has asked what happens to the lawn if this process continues. The above results show that (1) part of the lawn will be completely destroyed, and (2) the temptation to walk across the lawn cannot, in general, be made arbitrarily small by a clever path design.

Open Problems

For practical applications, upper bounds to the weight (= total edge length) of a geometric network would be valuable, in addition to upper dilation bounds. Some theoretical questions require further investigation, too. Is $\Sigma(S)$ always attained by a finite network? How to compute, or approximate, $\Sigma(S)$ for a given finite set S? Even for a set as simple as S_5, the corners of a regular 5-gon, is the dilation unknown. The smallest dilation value known, for a triangulation containing S_5 among its vertices, equals 1.0204; see Fig. 3. Finally, what is the precise value of $\sup\{\Sigma(S); S \text{ finite}\}$?

Cross-References

▶ Geometric Dilation of Geometric Networks

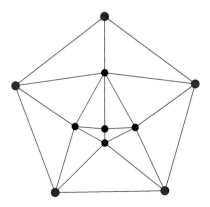

Dilation of Geometric Networks, Fig. 3 The best known embedding for S_5

Recommended Reading

1. Aronov B, de Berg M, Cheong O, Gudmundsson J, Haverkort H, Vigneron A (2005) Sparse geometric graphs with small dilation. In: Deng X, Du D (eds) Algorithms and computation: proceedings of the 16th international symposium (ISAAC 2005), Sanya. LNCS, vol 3827, pp 50–59. Springer, Berlin
2. Das G, Joseph D (1989) Which triangulations approximate the complete graph? In: Proceedings of the international symposium on optimal algorithms, Varna. LNCS, vol 401, pp 168–192. Springer, Berlin
3. Dobkin DP, Friedman SJ, Supowit KJ (1990) Delaunay graphs are almost as good as complete graphs. Discret Comput Geom 5:399–407
4. Ebbers-Baumann A, Gruene A, Karpinski M, Klein R, Knauer C, Lingas A (2007) Embedding point sets into plane graphs of small dilation. Int J Comput Geom Appl 17(3):201–230
5. Eppstein D, The geometry junkyard. http://www.ics.uci.edu/~eppstein/junkyard/dilation-free/
6. Eppstein D (1999) Spanning trees and spanners. In: Sack J-R, Urrutia J (eds) Handbook of computational geometry. Elsevier, Amsterdam, pp 425–461
7. Eppstein D, Wortman KA (2005) Minimum dilation stars. In: Proceedings of the 21st ACM symposium on computer geometry (SoCG), Pisa, pp 321–326
8. Hillar CJ, Rhea DL (2006) A result about the density of iterated line intersections. Comput Geom Theory Appl 33(3):106–114
9. Ismailescu D, Radoičić R (2004) A dense planar point set from iterated line intersections. Comput Geom Theory Appl 27(3):257–267
10. Keil JM, Gutwin CA (1992) The Delaunay triangulation closely approximates the complete Euclidean graph. Discret Comput Geom 7:13–28
11. Klein R, Kutz M, Penninger R (2015) Most finite point sets have dilation >1. Discret Comput Geom 53:80–106
12. Narasimhan G, Smid M (2007) Geometric spanner networks. Cambridge University Press, Cambridge/New York

Direct Routing Algorithms

Costas Busch
Department of Computer Science, Lousiana State University, Baton Rouge, LA, USA

Keywords

Bufferless packet switching; Collision-free packet scheduling; Hot-potato routing

Years and Authors of Summarized Original Work

2006; Busch, Magdon-Ismail, Mavronicolas, Spirakis

Problem Definition

The performance of a communication network is affected by the *packet collisions* which occur when two or more packets appear simultaneously in the same network node (router) and all these packets wish to follow the same outgoing link from the node. Since network links have limited available bandwidth, the collided packets wait on buffers until the collisions are resolved. Collisions cause delays in the packet delivery time and also contribute to the network performance degradation.

Direct routing is a packet delivery method which avoids packet collisions in the network. In direct routing, after a packet is injected into the network it follows a path to its destination without colliding with other packets, and thus without delays due to buffering, until the packet is absorbed at its destination node. The only delay that a packet experiences is at the source node while it waits to be injected into the network.

In order to formulate the direct routing problem, the network is modeled as a graph where all the network nodes are synchronized with a common time clock. Network links are bidirectional, and at each time step any link can be crossed by at most two packets, one packet in each direction. Given a set of packets, the *routing time* is defined to be the time duration between the first packet injection and the last packet absorbtion.

Consider a set of N packets, where each packet has its own source and destination node. In the *direct routing problem*, the goal is first to find a set of paths for the packets in the network, and second, to find appropriate injection times for the packets, so that if the packets are injected at the prescribed times and follow their paths they will be delivered to their destinations without collisions. The *direct scheduling problem* is a variation of the above problem, where the

paths for the packets are given a priori, and the only task is to compute the injection times for the packets.

A *direct routing algorithm* solves the direct routing problem (similarly, a *direct scheduling algorithm* solves the direct scheduling problem). The objective of any direct algorithm is to minimize the routing time for the packets. Typically, direct algorithms are *offline*, that is, the paths and the injection schedule are computed ahead of time, before the packets are injected into the network, since the involved computation requires knowledge about all packets in order to guarantee the absence of collisions between them.

Key Results

Busch, Magdon-Ismail, Mavronicolas, and Spirakis, present in [6] a comprehensive study of direct algorithms. They study several aspects of direct routing such as the computational complexity of direct problems and also the design of efficient direct algorithms. The main results of their work are described below.

Hardness of Direct Routing

It is shown in [Sect. 4 in 6] that the optimal direct scheduling problem, where the paths are given and the objective is to compute an optimal injection schedule (that minimizes the routing time) is an NP-complete problem. This result is obtained with a reduction from vertex coloring, where vertex coloring problems are transformed to appropriate direct scheduling problems in a 2-dimensional grid. In addition, it is shown in [6] that approximations to the direct scheduling problem are as hard to obtain as approximations to vertex coloring. A natural question is what kinds of approximations can be obtained in polynomial time. This question is explored in [6] for general and specific kinds of graphs, as described below.

Direct Routing in General Graphs

A direct algorithm is given in [Section 3 in 6] that solves approximately the optimal direct scheduling problem in general network topologies. Suppose that a set of packets and respective paths are given. The injection schedule is computed in polynomial time with respect to the size of the graph and the number of packets. The routing time is measured with respect to the *congestion* C of the packet paths (the maximum number of paths that use an edge), and the *dilation* D (the maximum length of any path).

The result in [6] establishes the existence of a simple greedy direct scheduling algorithm with routing time $rt = O(C \cdot D)$. In this algorithm, the packets are processed in an arbitrary order and each packet is assigned the smallest available injection time. The resulting routing time is worst-case optimal, since there exist instances of direct scheduling problems for which no direct scheduling algorithm can achieve a better routing time. A trivial lower bound on the routing time of any direct scheduling problem is $\Omega(C + D)$, since no algorithm can deliver the packets faster than the congestion or dilation of the paths. Thus, in the general case, the algorithm in [6] has routing time $rt = O((rt^*)^2)$, where rt^* is the optimal routing time.

Direct Routing in Specific Graphs

Several direct algorithms are presented in [6] for specialized network topologies. The algorithms solve the direct routing problem where first good paths are constructed and then an efficient injection schedule is computed. Given a set of packets, let C^* and D^* denote the optimal congestion and dilation, respectively, for all possible sets of paths for the packets. Clearly, the optimal routing time is $rt^* = \Omega(C^* + D^*)$. The upper bounds in the direct algorithm in [6] are expressed in terms of this lower bound. All the algorithms run in time polynomial to the size of the input.

Tree

The graph G is an arbitrary tree. A direct routing algorithm is given in [Section 3.1 in 6], where each packet follows the shortest path from its source to the destination. The injection schedule is obtained using the greedy algorithm with a particular ordering of the packets. The routing time of the algorithm is asymptotically optimal: $rt \leq 2C^* + D^* - 2 < 3 \cdot rt^*$.

Mesh

The graph G is a d-dimensional mesh (grid) with n nodes [10]. A direct routing algorithm is proposed in [Section 3.2 in 6], which first constructs efficient paths for the packets with congestion $C = O(d \log n \cdot C^*)$ and dilation $D = O(d^2 \cdot D^*)$ (the congestion is guaranteed with high probability). Then, using these paths the injection schedule is computed giving a direct algorithm with the routing time:

$$rt = O(d^2 \log^2 n \cdot C^* + d^2 \cdot D^*)$$
$$= O(d^2 \log^2 n \cdot rt^*).$$

This result follows from a more general result which is shown in [6], that says that if the paths contain at most b "bends", i.e., at most b dimension changes, then there is a direct scheduling algorithm with routing time $O(b \cdot C + D)$. The result follows because the constructed paths have $b = O(d \log n)$ bends.

Butterfly

The graph G is a butterfly network with n input and n output nodes [10]. In [Section 3.3 in 6] the authors examine permutation routing problems in the butterfly, where each input (output) node is the source (destination) of exactly one packet. An efficient direct routing algorithm is presented in [6] which first computes good paths for the packets using Valiant's method [14, 15]: two butterflies are connected back to back, and each path is formed by choosing a random intermediate node in the output of the first butterfly. The chosen paths have congestion $C = O(\lg n)$ (with high probability) and dilation $D = 2 \lg n = O(D^*)$. Given the paths, there is a direct schedule with routing time very close to optimal: $rt \leq 5 \lg n = O(rt^*)$.

Hypercube

The graph G is a hypercube with n nodes [10]. A direct routing algorithm is given in [Section 3.4 in 6] for permutation routing problems. The algorithm first computes good paths for the packets by selecting a single random intermediate node for each packet. Then an appropriate injection schedule gives routing time $rt < 14 \lg n$, which is worst-case optimal since there exist permutations for which $D^* = \Omega(\lg n)$.

Lower Bound for Buffering

In [Section 5 in 6] an additional problem has been studied about the amount of buffering required to provide small routing times. It is shown in [6] that there is a direct scheduling problem for which every direct algorithm requires routing time $\Omega(C \cdot D)$; at the same time, $C + D = \Theta(\sqrt{C \cdot D}) = o(C \cdot D)$. If buffering of packets is allowed, then it is well known that there exist packet scheduling algorithms [11, 12] with routing time very close to the optimal $O(C + D)$. In [6] it is shown that for the particular packet problem, in order to convert a direct injection schedule of routing time $O(C \cdot D)$ to a packet schedule with routing time $O(C + D)$, it is necessary to buffer packets in the network nodes in total $\Omega(N^{4/3})$ times, where a packet buffering corresponds to keeping a packet in an intermediate node buffer for a time step, and N is the number of packets.

Related Work

The only previous work which specifically addresses direct routing is for permutation problems on trees [3, 13]. In these papers, the resulting routing time is $O(n)$ for any tree with n nodes. This is worst-case optimal, while the result in [6] is asymptotically optimal for all routing problems in trees.

Cypher et al. [7] study an online version of direct routing in which a worm (packet of length L) can be re-transmitted if it is dropped (they also allow the links to have bandwidth $B \geq 1$). Adler et al. [1] study time constrained direct routing, where the task is to schedule as many packets as possible within a given time frame. They show that the time constrained version of the problem is NP-complete, and also study approximation algorithms on trees and meshes. Further, they discuss how much buffering could help in this setting.

Other models of bufferless routing are *matching routing* [2] where packets move to their desti-

nations by swapping packets in adjacent nodes, and *hot-potato routing* [4, 5, 8, 9] in which packets follow links that bring them closer to the destination, and if they cannot move closer (due to collisions) they are deflected toward alternative directions.

Applications

Direct routing represent collision-free communication protocols, in which packets spend the smallest amount of time possible time in the network once they are injected. This type of routing is appealing in power or resource constrained environments, such as optical networks, where packet buffering is expensive, or sensor networks where energy resources are limited. Direct routing is also important for providing quality of service in networks. There exist applications where it is desirable to provide guarantees on the delivery time of the packets after they are injected into the network, for example in streaming audio and video. Direct routing is suitable for such applications.

Cross-References

▶ Oblivious Routing
▶ Packet Routing

Recommended Reading

1. Adler M, Khanna S, Rajaraman R, Rosén A (2003) Timeconstrained scheduling of weighted packets on trees and meshes. Algorithmica 36: 123–152
2. Alon N, Chung F, Graham R (1994) Routing permutations on graphs via matching. SIAM J Discret Math 7(3):513–530
3. Alstrup S, Holm J, de Lichtenberg K, Thorup M (1998) Direct routing on trees. In: Proceedings of the ninth annual ACM-SIAM, symposium on discrete algorithms (SODA 98), San Francisco, pp 342–349
4. Ben-Dor A, Halevi S, Schuster A (1998) Potential function analysis of greedy hot-potato routing. Theory Comput Syst 31(1):41–61

5. Busch C, Herlihy M, Wattenhofer R (2000) Hard-potato routing. In: Proceedings of the 32nd annual ACM symposium on theory of computing, Portland, pp 278–285
6. Busch C, Magdon-Ismail M, Mavronicolas M, Spirakis P (2006) Direct routing: algorithms and complexity. Algorithmica 45(1):45–68
7. Cypher R, Meyer auf der Heide F, Scheideler C, Vöcking, B (1996) Universal algorithms for store-and-forward and wormhole routing. In: Proceedings of the 28th ACM symposium on theory of computing, Philadelphia, pp 356–365
8. Feige U, Raghavan P (1992) Exact analysis of hot-potato routing. In: IEEE (ed) Proceedings of the 33rd annual, symposium on foundations of computer science, Pittsburgh, pp 553–562
9. Kaklamanis C, Krizanc D, Rao S (1993) Hot-potato routing on processor arrays. In: Proceedings of the 5th annual ACM, symposium on parallel algorithms and architectures, Velen, pp 273–282
10. Leighton FT (1992) Introduction to parallel algorithms and architectures: arrays – trees – hypercubes. Morgan Kaufmann, San Mateo
11. Leighton FT, Maggs BM, Rao SB (1994) Packet routing and jobscheduling in O(congestion+dilation) steps. Combinatorica 14:167–186
12. Leighton T, Maggs B, Richa AW (1999) Fast algorithms for finding O(congestion + dilation) packet routing schedules. Combinatorica 19:375–401
13. Symvonis A (1996) Routing on trees. Inf Process Lett 57(4):215–223
14. Valiant LG (1982) A scheme for fast parallel communication. SIAM J Comput 11:350–361
15. Valiant LG, Brebner GJ (1981) Universal schemes for parallel communication. In: Proceedings of the 13th annual ACM, symposium on theory of computing, Milwaukee, pp 263–277

Directed Perfect Phylogeny (Binary Characters)

Jesper Jansson
Laboratory of Mathematical Bioinformatics,
Institute for Chemical Research, Kyoto
University, Gokasho, Uji, Kyoto, Japan

Keywords

Binary character; Character state matrix; Perfect phylogeny; Phylogenetic reconstruction; Phylogenetic tree

Years and Authors of Summarized Original Work

1991; Gusfield
1995; Agarwala, Fernández-Baca, Slutzki
2004; Pe'er, Pupko, Shamir, Sharan

Problem Definition

Let $S = \{s_1, s_2, \ldots, s_n\}$ be a set of elements called *objects* and let $C = \{c_1, c_2, \ldots, c_m\}$ be a set of functions from S to $\{0, 1\}$ called *characters*. For each object $s_i \in S$ and character $c_j \in C$, we say that s_i *has* c_j if $c_j(s_i) = 1$ or that s_i *does not have* c_j if $c_j(s_i) = 0$, respectively (in this sense, characters are *binary*). Then the set S and its relation to C can be naturally represented by a matrix M of size $(n \times m)$ satisfying $M[i, j] = c_j(s_i)$ for every $i \in \{1, 2, \ldots, n\}$ and $j \in \{1, 2, \ldots, m\}$. Such a matrix M is called a *binary character state matrix*.

Next, for each $s_i \in S$, define the set $C_{s_i} = \{c_j \in C : s_i \text{ has } c_j\}$. A *phylogeny for* S is a tree whose leaves are bijectively labeled by S, and a *directed perfect phylogeny for* (S, C) (if one exists) is a rooted phylogeny T for S in which each $c_j \in C$ is associated with exactly one edge of T in such a way that for any $s_i \in S$, the set of all characters associated with the edges on the path in T from the root to leaf s_i is equal to C_{s_i}. See Figs. 1 and 2 for two examples.

Now, define the following problem.

Problem 1 (The Directed Perfect Phylogeny Problem for Binary Characters)

INPUT: An $(n \times m)$-binary character state matrix M for some S and C.

OUTPUT: A directed perfect phylogeny for (S, C), if one exists; otherwise, *null*.

Key Results

In the presentation below, define a set S_{c_j} for each $c_j \in C$ by $S_{c_j} = \{s_i \in S : s_i \text{ has } c_j\}$. The next lemma is the key to solving the Directed

a

M	c_1	c_2	c_3	c_4	c_5	c_6	c_7	c_8
s_1	0	0	1	1	1	0	1	0
s_2	0	1	1	1	0	0	0	0
s_3	1	0	0	0	0	1	0	1
s_4	0	0	1	1	0	0	1	0
s_5	1	0	0	0	0	0	0	0

b

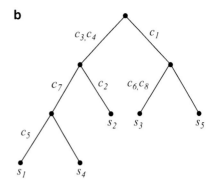

Directed Perfect Phylogeny (Binary Characters), Fig. 1 (a) A (5×8)-binary character state matrix M. (b) A directed perfect phylogeny for (S, C)

M	c_1	c_2
s_1	1	0
s_2	1	1
s_3	0	1

Directed Perfect Phylogeny (Binary Characters), Fig. 2 This binary character state matrix admits no directed perfect phylogeny

Perfect Phylogeny Problem for Binary Characters efficiently. It is also known in the literature as *the pairwise compatibility theorem* [5].

Lemma 1 *There exists a directed perfect phylogeny for* (S, C) *if and only if for each pair* $c_j, c_k \in C$, *it holds that* $S_{c_j} \cap S_{c_k} = \emptyset$, $S_{c_j} \subseteq S_{c_k}$, *or* $S_{c_k} \subseteq S_{c_j}$.

Short constructive proofs of the lemma can be found in, e.g., [8] and [14]. An algebraic proof of a slightly more general version of the lemma was given earlier by Estabrook, Johnson, and McMorris [3,4].

Using Lemma 1, it is trivial to construct a top-down algorithm for the problem that runs in $O(nm^2)$ time. As one might expect, a faster algorithm is possible. Gusfield [7] observed that after sorting the columns of M in nonincreasing lexicographic order, all duplicate copies of a column appear in a consecutive block of columns and column j is to the right of column k if S_{c_j} is a proper subset of S_{c_k}, and then exploited these two facts together with Lemma 1 to obtain the following result:

Theorem 1 ([7]) *The Directed Perfect Phylogeny Problem for Binary Characters can be solved in* $O(nm)$ *time.*

For a description of the original algorithm and a proof of its correctness, see [7] or [14]. A conceptually simplified version of the algorithm based on keyword trees can be found in Chapter 17.3.4 in [8]. Gusfield [7] also gave an adversary argument to prove a corresponding lower bound of $\Omega(nm)$ on the running time, showing that his algorithm is time optimal:

Theorem 2 ([7]) *Any algorithm that decides if a given binary character state matrix M admits a directed perfect phylogeny must, in the worst case, examine all entries of M.*

Agarwala, Fernández-Baca, and Slutzki [1] noted that the input binary character state matrix is often sparse, i.e., in general, most of the objects will not have most of the characters. In addition, they noted that for the sparse case, it is more efficient to represent the input (S, C) by all the sets S_{c_j} for $j \in \{1, 2, \ldots, m\}$, where each set S_{c_j} is defined as above and each S_{c_j} is specified as a linked list, than by using a binary character state matrix. Agarwala et al. [1] proved that with this alternative representation of S and C, the algorithm of Gusfield can be modified to run in time proportional to the total number of 1s in the corresponding binary character state matrix:

Theorem 3 ([1]) *The variant of the Directed Perfect Phylogeny Problem for Binary*

Characters in which the input is given as linked lists representing all the sets S_{c_j} for $j \in \{1, 2, \ldots, m\}$ can be solved in $O(h)$ time, where $h = \sum_{j=1}^{m} |S_{c_j}|$.

For a description of the algorithm, refer to [1] or [6]. Observe that Theorem 3 does not contradict Theorem 2; in fact, Gusfield's lower bound argument for proving Theorem 2 considers an input matrix consisting mostly of 1s.

When only a portion of an $(n \times m)$-binary character state matrix is available, an $\tilde{O}(nm)$-time algorithm by Pe'er et al. [13] can fill in the missing entries with 0s and 1s so that the resulting matrix admits a directed perfect phylogeny, if possible. A ZDD-based algorithm for enumerating all such solutions was recently developed by Kiyomi et al. [11].

Theorem 4 ([13]) *The variant of the Directed Perfect Phylogeny Problem for Binary Characters in which the input consists of an incomplete binary character state matrix can be solved in $\tilde{O}(nm)$ time.*

Applications

Directed perfect phylogenies for binary characters are used to describe the evolutionary history for a set of objects (e.g., biological species) that share some observable traits and that have evolved from a "blank" ancestral object which has none of the traits. Intuitively, the root of a directed perfect phylogeny corresponds to the blank ancestral object, and each directed edge $e = (u, v)$ corresponds to an evolutionary event in which the hypothesized ancestor represented by u gains the characters associated with e, transforming it into the hypothesized ancestor or object represented by v. For simplicity, it may be assumed that each character can emerge once only during the evolutionary history and is never lost after it has been gained, so that a leaf s_i is a descendant of the edge associated with a character c_j if and only if s_i has c_j. When this requirement is too strict, one can relax it to permit errors, for example, by letting each character be associated with more than one edge in the phylogeny

(i.e., allow each character to emerge many times) while minimizing the total number of such associations (*Camin-Sokal optimization*) or by keeping the requirement that each character emerges only once but allowing it to be lost multiple times (*Dollo parsimony*) [5, 6]. Such relaxations generally increase the computational complexity of the underlying computational problems; see, e.g., [2] and [15].

Binary characters are commonly used by biologists and linguists. Traditionally, morphological traits or directly observable features of species were employed by biologists as binary characters, and recently, binary characters based on genomic information such as substrings in DNA or protein sequences, SNP markers, protein regulation data, and shared gaps in a given multiple alignment have become more and more prevalent. Chapter 17.3.2 in [8] mentions several examples where phylogenetic trees have been successfully constructed based on such types of binary character data. In the context of reconstructing the evolutionary history of natural languages, linguists often use phonological and morphological characters with just two states [10].

The Directed Perfect Phylogeny Problem for Binary Characters is closely related to *the Perfect Phylogeny Problem*, a fundamental problem in computational evolutionary biology and phylogenetic reconstruction [5, 6, 14]. This problem (also described in more detail in Encyclopedia entry ▶ Perfect Phylogeny (Bounded Number of States)) introduces nonbinary characters so that each character $c_j \in C$ has a set of allowed states $\{0, 1, \ldots, r_j - 1\}$ for some integer r_j, and for each $s_i \in S$, character c_j is in one of its allowed states. Generalizing the notation used above, define the set $S_{c_j,\alpha}$ for every $\alpha \in \{0, 1, \ldots, r_j - 1\}$ by $S_{c_j,\alpha} = \{s_i \in S : \text{the state of } s_i \text{ on } c_j \text{ is } \alpha\}$. Then, the objective of *the Perfect Phylogeny Problem* is to construct (if possible) an *unrooted* phylogeny T for S such that the following holds: for each $c_j \in C$ and distinct states α, β of c_j, the minimal subtree of T that connects $S_{c_j,\alpha}$ and the minimal subtree of T that connects $S_{c_j,\beta}$ are vertex-disjoint. McMorris [12] showed that the special case with $r_j = 2$ for all $c_j \in C$ can be reduced to the

Directed Perfect Phylogeny Problem for Binary Characters in $O(nm)$ time: for each $c_j \in C$, if the number of 1s in column j of M is greater than the number of 0s, then set entry $M[i, j]$ to $1 - M[i, j]$ for all $i \in \{1, 2, \ldots, n\}$. Therefore, another application of Gusfield's algorithm [7] is as a subroutine for solving the Perfect Phylogeny Problem in $O(nm)$ time when $r_j = 2$ for all $c_j \in C$. Even more generally, the Perfect Phylogeny Problem for directed as well as undirected *cladistic* characters can be solved in polynomial time by a similar reduction to the Directed Perfect Phylogeny Problem for Binary Characters (see [6]).

In addition to the above, it is possible to apply Gusfield's algorithm to determine whether two given trees describe compatible evolutionary history, and if so, merge them into a single tree so that no branching information is lost (see [7] for details). Finally, Gusfield's algorithm has also been used by Hanisch, Zimmer, and Lengauer [9] to implement a particular operation on documents defined in their Protein Markup Language (ProML) specification.

Cross-References

▶ Directed Perfect Phylogeny (Binary Characters)
▶ Perfect Phylogeny Haplotyping

Acknowledgments JJ was funded by the Hakubi Project at Kyoto University and KAKENHI grant number 26330014.

Recommended Reading

1. Agarwala R, Fernández-Baca D, Slutzki G (1995) Fast algorithms for inferring evolutionary trees. J Comput Biol 2(3):397–407
2. Bonizzoni P, Braghin C, Dondi R, Trucco G (2012) The binary perfect phylogeny with persistent characters. Theor Comput Sci 454:51–63
3. Estabrook GF, Johnson CS Jr, McMorris FR (1976) An algebraic analysis of cladistic characters. Discret Math 16(2):141–147
4. Estabrook GF, Johnson CS Jr, McMorris FR (1976) A mathematical foundation for the analysis of cladistic character compatibility. Math Biosci 29(1–2): 181–187
5. Felsenstein J (2004) Inferring phylogenies. Sinauer Associates, Sunderland
6. Fernández-Baca D (2001) The perfect phylogeny problem. In: Cheng X, Du DZ (eds) Steiner trees in industry. Kluwer Academic, Dordrecht, pp 203–234
7. Gusfield DM (1991) Efficient algorithms for inferring evolutionary trees. Networks 21:19–28
8. Gusfield DM (1997) Algorithms on strings, trees, and sequences. Cambridge University Press, New York
9. Hanisch D, Zimmer R, Lengauer T (2002) ProML – the protein markup language for specification of protein sequences, structures and families. In Silico Biol 2:0029. http://www.bioinfo.de/isb/2002/02/0029/
10. Kanj IA, Nakhleh L, Xia G (2006) Reconstructing evolution of natural languages: complexity and parameterized algorithms. In: Proceedings of the 12th annual international computing and combinatorics conference (COCOON 2006). Lecture notes in computer science, vol 4112. Springer, Berlin/Heidelberg, pp 299–308
11. Kiyomi M, Okamoto Y, Saitoh T (2012) Efficient enumeration of the directed binary perfect phylogenies from incomplete data. In: Proceedings of the 11th international symposium on experimental algorithms (SEA 2012). Lecture notes in computer science, vol 7276. Springer, Berlin/Heidelberg, pp 248–259
12. McMorris FR (1977) On the compatibility of binary qualitative taxonomic characters. Bull Math Biol 39(2):133–138
13. Pe'er I, Pupko T, Shamir R, Sharan R (2004) Incomplete directed perfect phylogeny. SIAM J Comput 33(3):590–607
14. Setubal JC, Meidanis J (1997) Introduction to computational molecular biology. PWS Publishing Company, Boston
15. Sridhar S, Dhamdhere K, Blelloch GE, Halperin E, Ravi R, Schwartz R (2007) Algorithms for efficient near-perfect phylogenetic tree reconstruction in theory and practice. IEEE/ACM Trans Comput Biol Bioinform 4(4):561–571

Discrete Ricci Flow for Geometric Routing

Jie Gao[1], Xianfeng David Gu[1], and Feng Luo[2]
[1]Department of Computer Science, Stony Brook University, Stony Brook, NY, USA
[2]Department of Mathematics, Rutgers University, Piscataway, NJ, USA

Keywords

Geometric routing; Greedy routing; Greedy embedding; Virtual coordinates; Wireless networks

Years and Authors of Summarized Original Work

2009; Sarkar, Yin, Gao, Luo, Gu
2010; Sarkar, Zeng, Gao, Gu
2010; Zeng, Sarkar, Luo, Gu, Gao
2011; Jiang, Ban, Goswami, Zeng, Gao, Gu
2011; Yu, Ban, Sarkar, Zeng, Gu, Gao
2012; Yu, Yin, Han, Gao, Gu
2013; Ban, Goswami, Zeng, Gu, Gao
2013; Li, Zeng, Zhou, Gu, Gao

Problem Definition

The problem is concerned about computing virtual coordinates for greedy routing in a wireless ad hoc network. Consider a set of wireless nodes S densely deployed inside a geometric domain $\mathcal{R} \subseteq \mathbb{R}^2$. Nodes within communication range can directly communicate with each other. We ask whether one can compute a set of virtual coordinates for S such that greedy routing has guaranteed delivery. In particular, each node forwards the message to the neighbor whose distance to the destination, computed under the virtual coordinates and some metric function d, is the smallest. If such a neighbor can always be found, greedy routing successfully delivers the message to the destination. The problem can be phrased as finding a *greedy embedding* of S in some geometric space, such that greedy routing always succeeds.

In the setting of this entry, we assume that the nodes are a dense sample of the domain $\mathcal{R}$ such that the communication graph on S contains a triangulated mesh Σ as a discrete approximation of $\mathcal{R}$.

Key Results

The key result is a family of distributed algorithms for computing the greedy embedding using discrete Ricci flow. Given a triangular mesh Σ with vertex set V, edge set E, and face set F, we can define a piecewise linear metric by the edge lengths on Σ: $l : E \rightarrow \mathbb{R}^+$ that satisfies the triangle inequality for each triangle face. The piecewise linear metric determines the corner angles of the triangles on Σ, by the cosine law. The *discrete curvature* K_i at a vertex v_i is defined as the angle deficit on the mesh. If v_i is an interior vertex, $K_i = 2\pi - \sum_j \theta_j$, where θ_j's are the corner angles at v_i. If v_i is a vertex on the boundary, $K_i = \pi - \sum_j \theta_j$, where θ_j's are the corner angles at v_i. Thus, the curvature at an interior vertex v_i is 0 if the surface is flat at v_i. The curvature at a boundary vertex v_i is 0 if the boundary is locally a straight line at v_i (see Fig. 1). The famous Gauss-Bonnet theorem states that the total curvature is a topological invariant: $\sum_{v_i \in V} K_i = 2\pi \chi(\Sigma)$, where $\chi(\Sigma)$ is the Euler characteristic number (The Euler characteristics number of a surface is $2 - 2g - h$, where g is the genus or the number of handles and h is the number of holes.) of Σ. Ricci flow is a process that deforms the surface metric to meet any target curvature that is admissible by the Gauss-Bonnet theorem.

A conformal map in the continuous surface preserves the intersection angle of any two curves. In the discrete case, the "intersection angle" is defined using the circle packing metric [10, 11]. We place a circle at each vertex v_i with radius γ_i such that for each edge e_{ij}, the circles at v_i, v_j intersect or are tangent to each other. The intersection angle is denoted by $\phi(e_{ij})$. The pair of vertex radii and the intersection angles on a mesh Σ, (Γ, Φ), are called a *circle packing metric* of Σ (see Fig. 1). Two circle packing metrics (Γ_1, Φ_1) and (Γ_2, Φ_2) on the same mesh are *conformal equivalent*, if $\Phi_1 \equiv \Phi_2$. Therefore, a conformal deformation of a circle packing metric only modifies the vertex radii γ_i's and preserves the intersection angles. Note that the circle packing metric and the edge lengths (the piecewise linear metric) on one mesh can be converted to each other by using the cosine law.

Now we are ready to introduce the discrete Ricci flow algorithm. Let u_i be $\log \gamma_i$ for each vertex. Then the discrete Ricci flow, introduced in the work of [2], is defined as follows: $\frac{du_i(t)}{dt} = \bar{K}_i - K_i$, where $K_i, \bar{K}_i$ are the current and target curvature at vertex v_i, respectively. Discrete Ricci

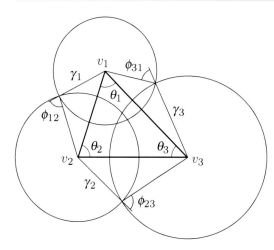

Discrete Ricci Flow for Geometric Routing, Fig. 1
The circle packing metric

flow can be formulated in the variational setting, namely, it is a negative gradient flow of some special energy form: $f(\mathbf{u}) = \int_{\mathbf{u_0}}^{\mathbf{u}} \sum_{i=1}^{n} (\bar{K}_i - K_i) du_i$, where $\mathbf{u_0}$ is an arbitrary initial metric and $\bar{K}$ is the prescribed target curvature. The integration above is well defined and called the *Ricci energy*. The discrete Ricci flow is the negative gradient flow of the discrete Ricci energy. The discrete metric which induces $\bar{K}$ is the minimizer of the energy. Computing the desired circle packing metric with prescribed curvature $\bar{K}$ is equivalent to minimizing the discrete Ricci energy. The discrete Ricci energy is strictly convex (namely, its Hessian is positive definite after a normalization). The global minimum uniquely exists, corresponding to the metric $\bar{\mathbf{u}}$, which induces $\bar{K}$. The discrete Ricci flow converges to this global minimum and the convergence is exponentially fast [2], i.e., $|\bar{K}_i - K_i(t)| < c_1 e^{-c_2 t}$, where c_1, c_2 are two positive constants. This represents a centralized algorithm for computing the discrete Ricci flow on Σ. In the following, we describe the distributed algorithm for different types of greedy routing scenarios.

Discrete Ricci Flow Algorithm

To apply discrete Ricci flow for greedy routing, we take a triangular mesh Σ as a subgraph from the communication graph. All non-triangular

faces are considered as network holes that will be mapped to circular holes in the embedding. All nodes not on hole boundaries have zero curvature under the mapping. Thus, the embedding is denoted as a circular domain. With the virtual coordinates and Euclidean distance metric, greedy routing guarantees delivery. (For a node in the interior of the triangulation, if the corner angle is greater than $2\pi/3$, we will adopt greedy routing on an edge that has provably guaranteed delivery.)

In particular, we set all edge lengths to be initially 1, which determines the initial curvature at each node. In particular, we choose the circle packing metric by placing a circle of initial radius $1/2$ on each node. The circles at adjacent nodes are tangent to each other. Thus, the intersection angle is kept at 0. We now set the target curvature at interior nodes to be zero and at hole boundary nodes to be $2\pi/k$ with k as the number of nodes on the hole boundary. The algorithms run in a gossip style. In each round, each node exchanges its radius with neighbors and computes its own Gaussian curvature. The algorithm stops when the current curvature is within error ϵ from the specified target curvature.

At each gossip round, node v_i is associated with a disk with radius e^{u_i}, where u_i is a scalar value. The length of the edge connecting v_i and v_j equals to $e^{u_i} + e^{u_j}$. The corner angles of each triangle can be estimated using cosine law by each node locally. That is, the angle θ_i^{jk} in triangle $[v_i, v_j, v_k]$ is

$$\theta_i^{jk} = \cos^{-1} \frac{l_{ij}^2 + l_{ki}^2 - l_{jk}^2}{2 l_{ij} l_{ki}}$$

The curvature k_i at v_i is

$$k_i = \begin{cases} 2\pi - \sum_{jk} \theta_i^{jk}, & v_i \notin \partial M \\ \pi - \sum_{jk} \theta_i^{jk}, & v_i \in \partial M \end{cases}$$

When the target curvature is not met, u_i is modified proportionally to the difference between target curvature and the current curvature.

$$u_i \Leftarrow u_i + \delta(\bar{k}_i - k_i)$$

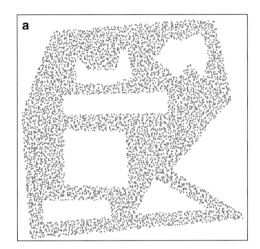
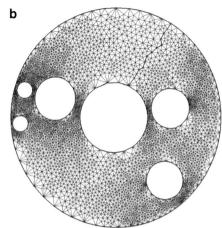

Discrete Ricci Flow for Geometric Routing, Fig. 2 (a) A network of 7,000 nodes with many holes; (b) virtual coordinates

Once the curvatures are computed, the triangulation is then flattened out by a simple flooding from a triangle root. Given three edge lengths of the root triangle $[v_0, v_1, v_2]$, the node coordinates can be constructed directly. Then the neighboring triangle of the root, e.g., $[v_1, v_0, v_i]$, can be flattened; the virtual coordinates of v_i are the intersection of two circles, one is centered at v_0 with radius l_{0i} and the other is centered at v_1 with radius l_{1i}. In a similar way, the neighbors of the newly flattened triangles can be further embedded. The virtual coordinates of the whole network are thus computed (Fig. 2).

Discrete Hyperbolic Ricci Flow

The key result in conformal geometry says that any surface with a Riemannian metric admits a Riemannian metric of constant Gaussian curvature, which is conformal to the original metric. Such metric is called the uniformization metric. Thus, depending on the surface topology, the uniformization metric has either positive constant, zero, or negative constant curvature everywhere. Simply connected surfaces with constant curvature are only of three canonical types: the sphere (constant positive curvature everywhere), the Euclidean plane (zero curvature everywhere), and the hyperbolic plane (negative curvature everywhere). Discrete Ricci flow is a powerful tool to compute the uniformization metric.

In our setting, when the triangulation Σ has two or more holes, it has negative total curvature. Thus, its uniformization metric is hyperbolic. To actually embed the surface and realize the uniformization metric, the holes in the network are cut open to get a simply connected triangulation T. Using discrete hyperbolic Ricci flow, we embed T in a convex region S in hyperbolic space. Each node is given a hyperbolic coordinate. Each edge uv has a length $d(u, v)$ as the geodesic between u, v in the hyperbolic space. In this way, greedy routing with the hyperbolic metric (i.e., send the message to the neighbor closer to the destination measured by hyperbolic distance) has *guaranteed delivery*.

The hyperbolic Ricci flow is very similar to the Euclidean version with a few modifications. First all metrics are hyperbolic. The edge length l_{ij} of e_{ij} is determined by the hyperbolic cosine law:

$$\cosh l_{ij} = \cosh \gamma_i \cosh \gamma_j + \sinh \gamma_i \sinh \gamma_j \cos \phi_{ij}. \tag{1}$$

Let $u_i = \log \tanh \frac{\gamma_i}{2}$; *the discrete Ricci flow is defined as*

$$\frac{d u_i(t)}{dt} = -K_i, \tag{2}$$

where K_i is the discrete Gaussian curvature at v_i. Once the hyperbolic metric is computed, we can embed the triangulation isometrically onto the Poincare disk.

Generalized Discrete Surface Ricci Flow

There are many schemes for discrete surface Ricci flow [14], including tangential circle packing, Thurston's circle packing, inversive distance circle packing, Yamabe flow, virtual radius circle packing, and mixed typed schemes. All of them can be unified as follows. The combinatorial structure of the triangulation is Σ; it is with one of three background geometries: Euclidean $\mathbb{E}^2$, hyperbolic $\mathbb{H}^2$, and spherical $\mathbb{S}^2$. Each vertex is associated with a circle; the vertex radii function is $\gamma : V \to \mathbb{R}^+$. Each vertex is also associated with a constant ϵ, which indicates the scheme. Each edge has a conformal structure coefficient $\eta : E \to \mathbb{R}$. So a circle packing metric is given by $(\Sigma, \gamma, \eta, \epsilon)$. The discrete conformal factor is given by

$$u_i = \begin{cases} \log \gamma_i & , \mathbb{E}^2 \\ \log \tanh \frac{\gamma_i}{2} & , \mathbb{H}^2 \\ \log \tan \frac{\gamma_i}{2} & , \mathbb{S}^2 \end{cases}$$

The length of $[v_i, v_j]$ is given by

$$\begin{cases} l_{ij}^2 = 2\eta_{ij}e^{u_i+u_j} + \epsilon_i e^{2u_i} + \epsilon_j e^{2u_j} & , \mathbb{E}^2 \\ \cosh l_{ij} = \frac{4\eta_{ij}e^{u_i+u_j}+(1+\epsilon_i e^{2u_i})(1+\epsilon_j e^{2u_j})}{(1-\epsilon_i e^{2u_i})(1-\epsilon_j e^{2u_j})} & , \mathbb{H}^2 \\ \cos l_{ij} = \frac{4\eta_{ij}e^{u_i+u_j}+(1-\epsilon_i e^{2u_i})(1-\epsilon_j e^{2u_j})}{(1+\epsilon_i e^{2u_i})(1+\epsilon_j e^{2u_j})} & , \mathbb{S}^2 \end{cases}$$

The discrete Ricci flow is given by

$$\frac{du_i(t)}{dt} = \bar{K}_i - K_i(t),$$

where $\bar{K} : V \to \mathbb{R}$ is the prescribed target curvature, which is the negative gradient flow of the discrete Ricci energy

$$E(u) = \int^{\mathbf{u}} \sum_i (\bar{K}_i - K_i) du_i.$$

For the discrete surfaces with Euclidean background geometry, the Ricci energy is convex on the space $\sum_i u_i = 0$. For those with hyperbolic background geometry, the energy is convex. For spherical case, the energy is indefinite.

For Yamabe scheme (where $\epsilon \equiv 0$), the combinatorial structure Σ is Delaunay, if for each edge $[v_i, v_j]$ share by two faces $[v_i, v_j, v_k]$ and $[v_j, v_i, v_l]$, $\theta_{ij}^k + \theta_{ji}^l \leq \pi$. If during the Yamabe flow, the combinatorial structure can be updated to ensure the Delaunay condition, then for any $\bar{K} : V \to (-\infty, 2\pi)$ satisfying the Gauss-Bonnet constraint $\sum_{v \in V} \bar{K}(v) = 2\pi\chi(\Sigma)$, the Yamabe flow with surgery can lead to the discrete metric that realizes the target curvature; the convergence is exponentially fast. This theorem implies the discrete uniformization theorem: any closed polyhedral surface admits a polyhedral metric discretely conformal to the original one, which induces constant Gaussian curvature everywhere [4,5] (Fig. 3).

Applications

The presented Ricci flow algorithms can be applied for a variety of routing primitives for large-scale wireless sensor networks with nonuniform node distribution. Besides guaranteed delivery [8], we can also achieve multiple additional desirable routing objectives, all derived from the unique property of a conformal mapping. For example, greedy routing on a circular domain may accumulate high traffic load on the interior hole boundaries. To alleviate that, we can reflect the network along a hole boundary using a Mobius transformation and map a copy of the network to cover the interior of the hole, recursively [9] (see Fig. 4). Routing on this covering space makes traffic load more balanced as hole boundaries essentially "disappear." In another case, when there are sudden link or node failures, we can apply a Mobius transformation to generate a different circular domain, with the sizes and positions of the holes rearranged, on which greedy routing generates a different path [6]. Thus, quick recovery from a spontaneous failure is possible. The hyperbolic Ricci flow can be used to map the domain with the holes cut open to a convex polygon that can tile up the entire hyperbolic

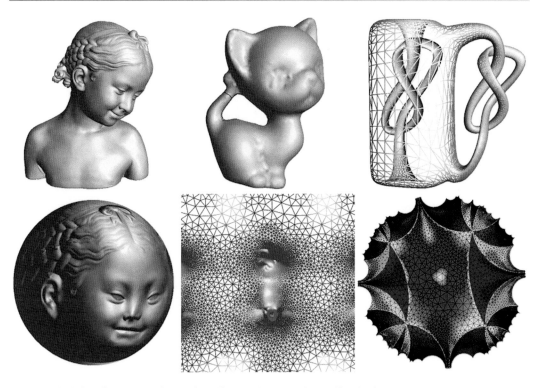

Discrete Ricci Flow for Geometric Routing, Fig. 3 Discrete surface uniformization

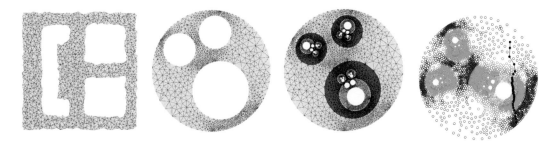

Discrete Ricci Flow for Geometric Routing, Fig. 4 Three-level circular reflections and a routing path

plane. This mapping supports greedy routing with specified "homotopy types," i.e., routes that go around holes in different ways [13] (see Fig. 5). Hyperbolic embedding can be generalized to 3D sensor networks with complex topology as in the case of monitoring underground tunnels [12]. Additional applications include generation of "space filling curves" for arbitrary domains [1], supporting greedy routing in mobile networks [7] and load balanced routing [3].

Open Problems

Given a smooth surface S with a Riemannian metric $\mathbf{g}$, the smooth Ricci flow leads to the uniformization metric $e^{2\lambda}\mathbf{g}$, where λ is the smooth conformal factor. If the surface is tessellated to get a discrete surface M_0 and discrete Ricci flow is performed on M_0, one obtains discrete conformal factor function u_0. When M is subdivided by n times, the discrete conformal factor is u_n, whether $\lim_{n \to \infty} u_n = \lambda$.

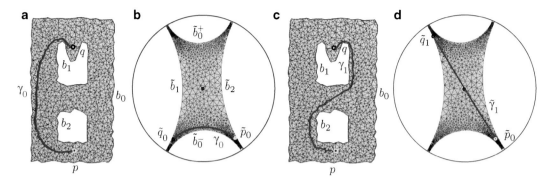

Discrete Ricci Flow for Geometric Routing, Fig. 5
Computing the shortest paths using the hyperbolic embedding of a 3-connected domain with 1,286 nodes. Two different paths are generated using greedy routing toward images of the destination in different patches. (**a**) Shortest path 1. (**b**) Geodesic of (**a**). (**c**) Shortest path 2. (**d**) Geodesic of (**c**)

Experimental Results

The convergence rate, i.e., the number of iterations is proportional to $O(\frac{log\,(1/\epsilon)}{\delta})$, where δ is the step size in the Ricci flow algorithm and ϵ is the error bound on the curvature. In our experiments we take ϵ to be $1e-6$. Routing with the virtual coordinates has 100 % delivery rate and the average path stretch (compared to the shortest path in the network) is no greater than 2.

URLs to Code and Data Sets

http://www.cs.sunysb.edu/~gu/tutorial/RicciFlow.html

Recommended Reading

1. Ban X, Goswami M, Zeng W, Gu XD, Gao J (2013) Topology dependent space filling curves for sensor networks and applications. In: Proceedings of 32nd annual IEEE conference on computer communications (INFOCOM'13), Turin
2. Chow B, Luo F (2003) Combinatorial Ricci flows on surfaces. J Differ Geom 63(1):97–129
3. Goswami M, Ni C-C, Ban X, Gao J, Xianfeng Gu D, Pingali V (2014) Load balanced short path routing in large-scale wireless networks using area-preserving maps. In: Proceedings of the 15th ACM international symposium on mobile ad hoc networking and computing (Mobihoc'14), Philadelphia, Aug 2014. pp 63–72
4. Gu X, Luo F, Sun J, Wu T (2013) A discrete uniformization theorem for polyhedral surfaces. arXiv:13094175
5. Gu X, Guo R, Luo F, Sun J, Wu T (2014) A discrete uniformization theorem for polyhedral surfaces ii. arXiv:14014594
6. Jiang R, Ban X, Goswami M, Zeng W, Gao J, Gu XD (2011) Exploration of path space using sensor network geometry. In: Proceedings of the 10th international symposium on information processing in sensor networks (IPSN'11), Chicago, pp 49–60
7. Li S, Zeng W, Zhou D, Gu XD, Gao J (2013) Compact conformal map for greedy routing in wireless mobile sensor networks. In: Proceedings of 32nd annual IEEE conference on computer communications (INFOCOM'13), Turin
8. Sarkar R, Yin X, Gao J, Luo F, Gu XD (2009) Greedy routing with guaranteed delivery using Ricci flows. In: Proceedings of the 8th international symposium on information processing in sensor networks (IPSN'09), San Francisco, pp 97–108
9. Sarkar R, Zeng W, Gao J, Gu XD (2010) Covering space for in-network sensor data storage. In: Proceedings of the 9th international symposium on information processing in sensor networks (IPSN'10), Stockholm, pp 232–243
10. Stephenson K (2005) Introduction to circle packing: the theory of discrete analytic functions. Cambridge University Press, New York
11. Thurston WP (1997) Three-dimensional geometry and topology, vol 1. Princeton University Press, Princeton
12. Yu X, Yin X, Han W, Gao J, Gu XD (2012) Scalable routing in 3d high genus sensor networks using graph embedding. In: Proceedings of the 31st annual IEEE conference on computer communications (INFOCOM'12), Orlando, pp 2681–2685
13. Zeng W, Sarkar R, Luo F, Gu XD, Gao J (2010) Resilient routing for sensor networks using hyperbolic embedding of universal covering space. In:

Proceedings of the 29th annual IEEE conference on computer communications (INFOCOM'10), San Diego, pp 1694–1702

14. Zhang M, Guo R, Zeng W, Luo F, Yau ST, Gu X (2014) The unified discrete surface Ricci flow. Graph Models. doi:. http://www.sciencedirect.com/science/article/pii/S1524070314000344

Distance Oracles for Sparse Graphs

Liam Roditty
Department of Computer Science, Bar-Ilan University, Ramat-Gan, Israel

Keywords

Data structures; Distance oracles; Graph algorithms; Shortest paths

Years and Authors of Summarized Original Work

2005; Thorup, Zwick
2012; Pătraşcu, Roditty
2012; Pătraşcu, Roditty, Thorup

Problem Definition

Let $G = (V, E)$ be a weighted undirected graph with n vertices and m edges. A distance oracle is a data structure capable of representing almost shortest paths efficiently, both in terms of space requirement and query time. Thorup and Zwick [7] showed that for any integer $k \geq 1$ it is possible to preprocess the graph in $\tilde{O}(mn^{1/k})$ time and generate a compact data structure of size $O(kn^{1+1/k})$ that answers approximate distance queries with $2k - 1$ multiplicative stretch in $O(k)$ time. This means that for every $u, v \in V$, it is possible to retrieve an estimate $\hat{d}(u, v)$ to the distance $d(u, v)$ in $O(k)$ time, such that $d(u, v) \leq \hat{d}(u, v) \leq (2k - 1)d(u, v)$. Recently, [8] showed, using a clever query algorithm, that the query time of Thorup and Zwick

can be reduced from $O(k)$ to $O(\log k)$. Even more recently, [1] showed that the query time of Thorup and Zwick can be reduced to $O(1)$. Thorup and Zwick [7] showed, based on the girth conjecture of [2], that there are dense enough graphs which cannot be represented by a data structure of size less than $n^{1+1/k}$ without increasing the stretch above $2k - 1$ for any integer k. Therefore, for dense graphs their distance oracle is optimal assuming the girth conjecture holds.

This suggests that the distance oracle of Thorup and Zwick can be improved only for sparse graphs and in particular, graphs with less than $n^{1+1/k}$ edges. Alternatively, it might be possible to get below the $2k - 1$ multiplicative stretch by allowing an additive stretch as well.

Notice, however, that we cannot gain from introducing also additive stretch without getting an improved multiplicative stretch distance oracles for sparse graphs (i.e., $m = O(n)$). A data structure with size $S(m, n)$ and stretch (α, β), where α is multiplicative stretch and β is additive stretch, implies a data structure with size $S((\beta + 1)m, n + \beta m)$ and multiplicative stretch of α, as if we divide every edge into $\beta + 1$ edges then all distances become a multiply of $\beta + 1$ and additive stretch of β is useless. For graphs with $m = O(n)$, the size of the data structure is asymptotically the same.

Key Results

Pătraşcu and Roditty [4] obtained a distance oracle for sparse unweighted graphs with $m = \tilde{O}(n)$ of size $\tilde{O}(m^{5/3})$ that can supply in $O(1)$ time an estimate of the distance with multiplicative stretch 2. For dense graphs, the distance oracle has size of $\tilde{O}(n^{5/3})$ and stretch $(2, 1)$.

Pătraşcu et al. [5] extended this result for weighted graphs and generalized it. In particular, they show that for any fixed positive integers k and ℓ, there is a distance oracle with stretch $\alpha = 2k + 1 \pm \frac{2}{\ell} = 2k + 1 - \frac{2}{\ell}, 2k + 1 + \frac{2}{\ell}$, that uses $\tilde{O}(m^{1+2/(\alpha+1)})$ space. The query time is $O(k + \ell)$.

D

Sommer et al. [6] proved a three-way trade-off between space, stretch, and query time of approximate distance oracles. They show that any distance oracle that can give stretch α answers to distance queries in time $O(t)$ must use $n^{1+\Omega(1/(t\alpha))}/\log n$ space. Their result is obtained by a reduction from lopsided set disjointness to distance oracles, using the framework introduced by [3]. Any improvement to this lower bound requires a major breakthrough in lower bounds techniques. In particular, it does not imply anything even for slightly non-constant query time as $\Omega(\log n)$ and slightly non-linear space as $n^{1.01}$.

Pătrașcu and Roditty [4] showed also a conditional lower bound for distance oracle that is based on a conjecture on the hardness of the set intersection problem. They showed that a distance oracle for unweighted graphs with $m = \tilde{O}(n)$ edges, which can distinguish between distances of 2 and 4 in constant time (as multiplicative stretch strictly less than 2 implies) requires $\tilde{\Omega}(n^2)$ space, assuming the conjecture holds. Thus, non-constant query time is essential to get stretch smaller than 2.

Pătrașcu et al. [5] showed, based also on a conjecture on the hardness of the set intersection problem, for any fixed positive integer ℓ, that there are graphs with m edges such that a distance oracle with constant query time and stretch below $3 - 2/(\ell + 1)$ must use space $\tilde{\Omega}(m^{1+1/(2-1/\ell)})$.

Open Problems

The conditional lower bounds of [5] for sparse graphs do not say anything on stretch 3. The best space upper bound for stretch 3 in sparse graphs and dense graphs is $\tilde{O}(n^{1.5})$. While in dense graphs this is tight due to the existence of graphs with $\Omega(n^{1.5})$ edges and girth 6, for sparse graphs nothing is known. Therefore, we have the following two open problems:

Can we get a $o(n^{1.5})$ space for stretch 3 in sparse graphs? Can we get stretch less than 3 for space $\tilde{O}(n^{1.5})$ in sparse graphs?

Recommended Reading

1. Chechik S (2014) Approximate distance oracles with constant query time. In: STOC, New York
2. Erdős P (1964) Extremal problems in graph theory. In: Simonovits M (ed) Theory of graphs and its applications, pp 29–36. https://www.renyi.hu/~p_erdos/1970-22.pdf
3. Pătrașcu M (2008) (Data) structures. In: Proceedings of 49th FOCS, Philadelphia, pp 434–443
4. Pătrașcu M, Roditty L (2014) Distance oracles beyond the Thorup-Zwick bound. SIAM J Comput 43(1):300–311
5. Pătrașcu M, Roditty L, Thorup M (2012) A new infinity of distance oracles for sparse graphs. In: FOCS, New Brunswick, pp 738–747
6. Sommer C, Verbin E, Yu W (2009) Distance oracles for sparse graphs. In: Proceedings of 50th FOCS, Atlanta, pp 703–712
7. Thorup M, Zwick U (2005) Approximate distance oracles. JACM 52(1):1–24
8. Wulff-Nilsen C (2013) Approximate distance oracles with improved query time. In: SODA, New Orleans, pp 539–549

Distance-Based Phylogeny Reconstruction (Fast-Converging)

Miklós Csűrös
Department of Computer Science, University of Montréal, Montréal, QC, Canada

Keywords

Learning an evolutionary tree

Years and Authors of Summarized Original Work

2003; King, Zhang, Zhou

Problem Definition

Introduction

From a mathematical point of view, a phylogeny defines a probability space for random sequences observed at the leaves of a binary tree T. The tree T represents the unknown hierarchy of common ancestors to the sequences. It is assumed that (unobserved) ancestral sequences are associated with the inner nodes. The tree along with the associated sequences models the evolution of a molecular sequence, such as the protein sequence of a gene. In the conceptually simplest case, each tree node corresponds to a species, and the gene evolves within the organismal lineages by vertical descent.

Phylogeny reconstruction consists of finding T from observed sequences. The possibility of such reconstruction is implied by fundamental principles of molecular evolution, namely, that random mutations within individuals at the genetic level spreading to an entire mating population are not uncommon, since often they hardly influence evolutionary fitness [15]. Such mutations slowly accumulate, and, thus, differences between sequences indicate their evolutionary relatedness.

The reconstruction is theoretically feasible in several known situations. In some cases, distances can be computed between the sequences, and used in a distance-based algorithm. Such an algorithm is fast-converging if it almost surely recovers T, using sequences that are polynomially long in the size of T. Fast-converging algorithms exploit statistical concentration properties of distance estimation.

Formal Definitions

An evolutionary *topology* $U(X)$ is an unrooted binary tree in which leaves are bijectively mapped to a set of species X. A *rooted topology* T is obtained by rooting a topology U on one of the edges uv: a new node ρ is added (the *root*), the edge uv is replaced by two edges ρv and ρu, and the edges are directed outwards on paths from ρ to the leaves. The edges, vertices, and leaves of a rooted or unrooted topology T are denoted by $\mathrm{E}(T)$, $\mathrm{V}(T)$ and $\mathrm{L}(T)$, respectively.

The edges of an unrooted topology U may be equipped with a a positive *edge length* function $d: \mathrm{E}(U) \mapsto (0, \infty)$. Edge lengths induce a *tree metric* $d: \mathrm{V}(U) \times \mathrm{V}(U) \mapsto [0, \infty)$ by the extension $d(u, v) = \sum_{e \in u \rightsquigarrow v} d(e)$, where $u \rightsquigarrow v$ denotes the unique path from u to v. The value $d(u,$ $v)$ is called the *distance* between u and v. The pairwise distances between leaves form a *distance matrix*.

An *additive tree metric* is a function $\delta: X \times X \mapsto [0, \infty)$ that is equivalent to the distance matrix induced by some topology $U(X)$ and edge lengths. In certain random models, it is possible to define an additive tree metric that can be estimated from dissimilarities between sequences observed at the leaves.

In a *Markov model of character evolution* over a rooted topology T, each node u has an associated *state*, which is a random variable $\xi(u)$ taking values over a fixed alphabet $\mathrm{A} = \{1, 2, \ldots r\}$. The vector of leaf states constitutes the *character* $\xi = (\xi(u): u \in \mathrm{L}(T))$. The states form a first-order Markov chain along every path. The joint distribution of the node states is specified by the marginal distribution of the root state, and the conditional probabilities $\mathbb{P}\{\xi(v) = b | \xi(u) = a\} = p_e(a \rightarrow b)$ on each edge e, called *edge transition probabilities*.

A *sample* of length ℓ consists of independent and identically distributed characters $\varXi = (\xi_i: i = 1, \ldots \ell)$. The *random sequence* associated with the leaf u is the vector $\varXi(u) = (\xi_i(u): i = 1, \ldots \ell)$.

A *phylogeny reconstruction algorithm* is a function F mapping samples to unrooted topologies. The *success probability* is the probability that $\mathrm{F}(\varXi)$ equals the true topology.

Popular Random Models

Neyman Model [14]
The edge transition probabilities are

$$p_e(a \rightarrow b) = \begin{cases} 1 - \mu_e & \text{if } a = b; \\ \frac{\mu_e}{r-1} & \text{if } a \neq b \end{cases}$$

with some edge-specific mutation probability $0 < \mu_e < 1 - 1/r$. The root state is uniformly distributed. A distance is usually defined by

$$d(u, v) = -\frac{r-1}{r} \ln\left(1 - \frac{r}{r-1}\mathbb{P}\{\xi(u) \neq \xi(v)\}\right).$$

General Markov Model

There are no restrictions on the edge transition probabilities in the general Markov model. For identifiability [1, 16], however, it is usually assumed that $0 < \det \mathbf{P}_e < 1$, where $\mathbf{P}_e$ is the stochastic matrix of edge transition probabilities. Possible distances in this model include the *paralinear* distance [1, 12] and the *LogDet* distance [13, 16]. This latter is defined by $d(u, v) = -\ln \det \mathbf{J}_{uv}$, where $\mathbf{J}_{uv}$ is the matrix of joint probabilities for $\xi(u)$ and $\xi(v)$.

It is often assumed in practice that sequence evolution is effected by a continuous-time Markov process operating on the edges. Accordingly, the edge length directly measures time. In particular, $\mathbf{P}_e = e^{\mathbf{Q} \cdot d(e)}$ on every edge e, where $\mathbf{Q}$ is the instantaneous rate matrix of the underlying process.

Key Results

It turns out that the hardness of reconstructing an unrooted topology U from distances is determined by its *edge depth* $\rho(U)$. Edge depth is defined as the smallest integer k for which the following holds. From each endpoint of every edge $e \in \mathrm{E}(U)$, there is a path leading to a leaf, which does not include e and has at most k edges.

Theorem 1 (Erdős, Steel, Székely, Warnow [6])
If U has n leaves, then $\rho(U) \leq 1 + \log_2(n - 1)$. Moreover, for almost all random n-leaf topologies under the uniform or Yule-Harding distributions, $\rho(U) \in O(\log \log n)$

Theorem 2 (Erdős, Steel, Székely, Warnow [6])
For the Neyman model, there exists a polynomial-time algorithm that has a success probability $(1 - \delta)$ for random samples of length

$$\ell = O\left(\frac{\log n + \log \frac{1}{\delta}}{f^2 (1 - 2g)^{4\rho + 6}} \right), \qquad (1)$$

where $0 < f = \min_e \mu_e$ and $g = \max_e \mu_e < 1/2$ are extremal edge mutation probabilities, and ρ is the edge depth of the true topology.

Theorem 2 can be extended to the general Markov model with analogous success rates for LogDet distances [7], as well as to a number of other Markov models [2].

Equation (1) shows that phylogenies can be reconstructed with high probability from polynomially long sequences. Algorithms with such sample size requirements were dubbed *fast-converging* [9]. Fast convergence was proven for the short quartet methods of Erdős et al. [6, 7], and for certain variants [11] of the so-called disk-covering methods introduced by Huson et al. [9]. All these algorithms run in $\Omega(n^5)$ time. Csűrös and Kao [3] initiated the study of computationally efficient fast-converging algorithms, with a cubic-time solution. Csűrös [2] gave a quadratic-time algorithm. King et al. [10] designed an algorithm with an optimal running time of $O(n \log n)$ for producing a phylogeny from a matrix of estimated distances.

The short quartet methods were revisited recently: [4] described an $O(n^4)$-time method that aims at succeeding even if only a short sample is available. In such a case, the algorithm constructs a forest of "trustworthy" edges that match the true topology with high probability.

All known fast-converging distance-based algorithms have essentially the same sample bound as in (1), but Daskalakis et al. [5] recently gave a twist to the notion of fast convergence. They described a polynomial-time algorithm, which outputs the true topology almost surely from a sample of size $O(\log n)$, given that edge lengths are not too large. Such a bound is asymptotically optimal [6]. Interestingly, the sample size bound does not involve exponential dependence on the edge depth: the algorithm does not rely on a distance matrix.

Applications

Phylogenies are often constructed in molecular evolution studies, from aligned DNA or protein sequences. Fast-converging algorithms have mostly a theoretical appeal at this point. Fast convergence promises a way to handle the increasingly important issue of constructing large-scale phylogenies: see, for example, the CIPRES project (http://www.phylo.org/).

Cross-References

Similar algorithmic problems are discussed under the heading

▶ Distance-Based Phylogeny Reconstruction: Safety and Edge Radius

Recommended Reading

Joseph Felsenstein wrote a definitive guide [8] to the methodology of phylogenetic reconstruction.

1. Chang JT (1996) Full reconstruction of Markov models on evolutionary trees: identifiability and consistency. Math Biosci 137:51–73
2. Csűrös M (2002) Fast recovery of evolutionary trees with thousands of nodes. J Comput Biol 9(2):277–297, Conference version at RECOMB 2001
3. Csűrös M, Kao M-Y (2001) Provably fast and accurate recovery of evolutionary trees through harmonic greedy triplets. SIAM J Comput 31(1):306–322, Conference version at SODA (1999)
4. Daskalakis C, Hill C, Jaffe A, Mihaescu R, Mossel E, Rao S (2006) Maximal accurate forests from distance matrices. In: Proceedings of research in computational biology (RECOMB), pp 281–295
5. Daskalakis C, Mossel E, Roch S (2006) Optimal phylogenetic reconstruction. In: Proceedings of ACM symposium on theory of computing (STOC), pp 159–168
6. Erdős PL, Steel MA, Székely LA, Warnow TJ (1999) A few logs suffice to build (almost) all trees (I). Random Struct Algorithm 14:153–184, Preliminary version as DIMACS TR97-71
7. Erdős PL, Steel MA, Székely LA, Warnow TJ (1999) A few logs suffice to build (almost) all trees (II). Theor Comput Sci 221:77–118, Preliminary version as DIMACS TR97-72
8. Felsenstein J (2004) Inferring pylogenies. Sinauer Associates, Sunderland
9. Huson D, Nettles S, Warnow T (1999) Disk-covering, a fast converging method of phylogenetic reconstruction. J Comput Biol 6(3–4):369–386, Conference version at RECOMB (1999)
10. King V, Zhang L, Zhou Y (2003) On the complexity of distance based evolutionary tree reconstruction. In: Proceedings of ACM-SIAM symposium on discrete algorithms (SODA), pp 444–453
11. Lagergren J (2002) Combining polynomial running time and fast convergence for the disk-covering method. J Comput Syst Sci 65(3):481–493
12. Lake JA (1994) Reconstructing evolutionary trees from DNA and protein sequences: paralinear distances. Proc Natl Acad Sci USA 91:1455–1459
13. Lockhart PJ, Steel MA, Hendy MD, Penny D (1994) Recovering evolutionary trees under a more realistic model of sequence evolution. Mol Biol Evol 11:605–612
14. Neyman J (1971) Molecular studies of evolution: a source of novel statistical problems. In: Gupta SS, Yackel J (eds) Statistical decision theory and related topics. Academic, New York, pp 1–27
15. Ohta T (2002) Near-neutrality in evolution of genes and gene regulation. Proc Natl Acad Sci USA 99:16134–16137
16. Steel MA (1994) Recovering a tree from the leaf colourations it generates under a Markov model. Appl Math Lett 7:19–24

Distance-Based Phylogeny Reconstruction: Safety and Edge Radius

Olivier Gascuel[1], Fabio Pardi[1], and Jakub Truszkowski[2,3]
[1] Institut de Biologie Computationnelle, Laboratoire d'Informatique, de Robotique et de Microélectronique de Montpellier (LIRMM), CNRS and Université de Montpellier, Montpellier cedex 5, France
[2] Cancer Research UK Cambridge Institute, University of Cambridge, Cambridge, UK
[3] European Molecular Biology Laboratory, European Bioinformatics Institute (EMBL-EBI), Wellcome Trust Genome Campus, Hinxton, Cambridge, UK

Keywords

Distance methods; Optimal radius; Performance analysis; Phylogeny reconstruction; Robustness; Safety radius approach

Years and Authors of Summarized Original Work

1999; Atteson
2005; Elias, Lagergren
2006; Dai, Xu, Zhu

2010; Pardi, Guillemot, Gascuel 2013; Bordewich, Mihaescu

Problem Definition

A phylogeny is an evolutionary tree tracing the shared history, including common ancestors, of a set of extant species or "taxa." Phylogenies are increasingly reconstructed on the basis of molecular data (DNA and protein sequences) using statistical techniques such as likelihood and Bayesian methods. Algorithmically, these techniques suffer from the discrete nature of tree topology space. Since the number of tree topologies increases exponentially as a function of the number of taxa, and each topology requires a separate likelihood calculation, it is important to restrict the search space and to design efficient heuristics. Distance methods for phylogeny reconstruction serve this purpose by inferring trees in a fraction of the time required for the more statistically rigorous methods. Distance methods also provide fairly accurate starting trees to be further refined by more sophisticated methods. Moreover, the input to a distance method is the matrix of pairwise evolutionary distances among taxa, which are estimated by maximum likelihood, so that distance methods also have sound statistical justifications.

Mathematically, a phylogenetic tree is a triple $T = (V, E, l)$ where V is the set of nodes (extant taxa correspond to leaves, ancestors to internal nodes), E is the set of edges (branches) representing relations of descent, and l is a function that assigns positive lengths to each edge in E, representing a measure of evolutionary divergence, for example, in terms of time, or amount of change between DNA and protein sequences. Any phylogenetic tree T defines a metric D_T on its leaf set L: let $P_T(u, v)$ define the unique path through T from u to v; then the distance from u to v is set to $D_T(u, v) = \sum_{e \in P_T(u,v)} l(e)$.

Distance methods for phylogeny reconstruction rely on the fundamental result [22] that the map $T \to D_T$ is reversible; i.e., a tree T can be reconstructed from its tree metric, a problem that

can be solved in $O(n \log n)$ time [14]. However, in practice D_T is not known, and one must use molecular sequence data to estimate a distance matrix D that approximates D_T [9]. As the amount of sequence data increases, D can be assumed to converge to D_T. A minimal requirement for any distance method is *consistency*: for any tree T, and for distance matrices D "close enough" to D_T, the algorithm should output a tree with the same topology as T (i.e., with the same underlying graph (V, E)). The present chapter deals with the question of when any distance algorithm for phylogeny reconstruction can be guaranteed to output the correct phylogeny as a function of the divergence between D and D_T. Atteson [1] demonstrated that this question can be precisely answered for neighbor joining (NJ) [18], one of the most cited algorithms in computational biology (with more than 35,000 citations up to 2014), and a number of NJ's variants.

The Neighbor Joining (NJ) Algorithm of Saitou and Nei [18]

NJ is *agglomerative*: it works by using the input matrix D to identify a pair of taxa $x, y \in L$ that are neighbors in T, i.e., there exists a node $u \in V$ such that $\{(u,x), (u,y)\} \subset E$. Then, the algorithm creates a node c that is connected to x and y, extends the distance matrix to c, and solves the reduced problem on $L \cup \{c\} \setminus \{x,y\}$. The pair (x, y) is chosen to minimize the following sum:

$$S_D(x, y) = (|L| - 2) \cdot D(x, y)$$
$$- \sum_{z \in L} (D(z, x) + D(z, y)).$$

The soundness of NJ is based on the observation that, if $D = D_T$ for a tree T, the value $S_D(x, y)$ will be minimized for a pair (x, y) that are neighbors in T.

Balanced Minimum Evolution and Algorithms Inspired by It

A number of papers (reviewed in [11]) have been dedicated to the various interpretations and properties of the S_D criterion. One of these interpretations consists of observing that agglomerating the pair of nodes that minimizes S_D is equivalent to choosing, among all the trees that can be obtained in this way, the one that minimizes a simple linear formula [16] to calculate the length of a tree from the distances between its leaves [11], thus connecting distance and parsimony methods [9]. As the optimization principle seeking the tree that minimizes this formula has been named balanced minimum evolution (BME) [6], NJ can then be seen as a greedy algorithm for BME.

This remarkable connection between NJ and BME naturally spurred the proposal of alternative algorithms for BME. One of these, GreedyBME, consists of iteratively adding taxa to a tree so that, at each step, the resulting tree is the one that minimizes BME among all the binary trees that can be obtained in this way [6]. More involved algorithms can be obtained by combining a simple tree construction algorithm such as NJ or GreedyBME, with a local search based on the traditional tree rearrangements used in phylogenetics [9], such as *nearest-neighbor interchange* (NNI) or *subtree pruning and regrafting* (SPR).

The Fast Neighbor Joining (FNJ) Algorithm of Elias and Lagergren [7]

Standard implementations of NJ require $O(n^3)$ computations, where n is the number of taxa in the data set. Since a distance matrix only has n^2 entries, many attempts have been made to construct a distance algorithm that would only require $O(n^2)$ computations while retaining the accuracy of NJ. To this end, one of the most interesting results is the fast neighbor joining (FNJ) algorithm of Elias and Lagergren [7].

Most of the computation of NJ is used in the recalculations of the sums $S_D(x, y)$ after each agglomeration step. Although each recalculation can be performed in constant time, and although it is not necessary to consider all pairs of taxa (x, y) in order to find the one that minimizes this sum [20], the number of pairs to consider remains, in the worst case, $O(k^2)$ when k nodes are left to agglomerate. Thus, summing over k, $O(n^3)$ computations are required in all.

Elias and Lagergren take a related approach to agglomeration, which does not exhaustively seek the minimum value of $S_D(x, y)$ at each step, but instead uses a heuristic to maintain a list of candidates of "visible pairs" (x, y) for agglomeration. At the $(n - k)$th step, when two neighbors are agglomerated from a k-taxa tree to form a $(k - 1)$-taxa tree, FNJ has a list of $O(k)$ visible pairs for which $S_D(x, y)$ is calculated. The pair joined is selected from this list. By trimming the number of pairs considered, Elias and Lagergren achieved an algorithm which requires only $O(n^2)$ computations. Other similar improvements to neighbor joining have also been proposed in recent years [8, 13, 20].

Safety Radius Performance Analysis (Atteson [1])

In order to provide accuracy guarantees for distance-based algorithms, Atteson [1] tackled the following question: if D is a distance matrix that approximates a tree metric D_T, can one have some confidence in the algorithm's ability to reconstruct T, or parts of it, given D, based on some measure of the distance between D and D_T? For two matrices, D_1 and D_2, the L_∞ distance between them is defined by $||D_1 - D_2||_\infty = \max_{i,j} |D_1(i, j) - D_2(i, j)|$. Moreover, let $\mu(T)$ denote the length of the shortest internal edge of a tree T. This is an important quantity, as short branches in a phylogeny are difficult to resolve, because of the relatively few (if any) molecular changes occurring on a short branch.

The *safety radius* of an algorithm A is then the greatest value of r with the property that given any phylogeny T, and any distance matrix D satisfying $||D - D_T||_\infty < r \cdot \mu(T)$, A will return a tree $\hat{T}$ with the same topology as T. Similarly, the *edge radius* of A is the greatest value of r, for which the presence in $\hat{T}$ of an edge $e \in E$ is guaranteed whenever $||D - D_T||_\infty < r \cdot l(e)$. As an easy consequence of these definitions, the safety radius is always at least as large as the edge radius. Moreover, both the safety radius and the edge radius can also be attributed to an optimization principle, assuming an exact optimization algorithm.

Key Results

Atteson [1] proved the following theorems:

Theorem 1 *The safety radius of NJ is* $1/2$.

Theorem 2 *The largest possible safety radius for any algorithm is* $1/2$.

Indeed, given any μ, one can find two different trees T_1, T_2 and a distance matrix D such that $\mu = \mu(T_1) = \mu(T_2)$, and $\|D - D_{T_1}\|_\infty = \mu/2 = \|D - D_{T_2}\|_\infty$. Since D is equidistant from two distinct tree metrics, no algorithm could assign it to the "closest" tree.

In their presentation of FNJ, Elias and Lagergren updated Atteson's results for their algorithm. They showed:

Theorem 3 *The safety radius of FNJ is* $1/2$.

An insight on the above results on neighbor-joining-type algorithms is provided by the fact that the optimization principle they are linked to, BME, has itself safety radius $1/2$ [15]. A simple consequence of this [15] is the fact that also GreedyBME has safety radius $1/2$, a result first proven by Shigezumi [19]. Finally, performing a local search guided by BME and based on SPR leads to an algorithm with safety radius greater or equal to $1/3$, regardless of the method used to construct the initial tree [2].

The edge radius of a number of algorithms has also been studied. As conjectured by Atteson [1] and formally proven by Dai et al. [5], the edge radius of NJ is $1/4$. Interestingly, other heuristics, related to NJ via the principle they seek to optimize (BME), perform better than NJ in terms of edge radius: GreedyBME has edge radius $1/3$ [3]; moreover, building an initial tree with GreedyBME and then performing a local search guided by BME and based on NNI or SPR operations constitute an algorithm with edge radius $1/3$ [3].

Finally, we note that the safety radius framework has also been applied to the ultrametric setting where the correct tree T is rooted and all tree leaves are at the same distance from the root [10]. These trees are called "molecular clock" trees in phylogenetics and "indexed hierarchies" in data analysis. In this setting, the optimal safety radius is equal to 1 (instead of $1/2$), and a number of standard algorithms (e.g., UPGMA, with time complexity in $O(n^2)$) have a safety radius of 1.

Open Problems

With increasing amounts of sequence data becoming available for an increasing number of species, distance algorithms such as NJ should be useful for quite some time. Currently, the bottleneck in the process of building phylogenies is estimating distances, rather than exploring tree topologies. Two algorithms were recently developed to reconstruct trees from incomplete distance matrices. These algorithms use character information as well as distances and hence cannot be categorized as pure distance methods.

FastTree [17] is an NJ-like heuristic that avoids computing the full distance matrix. For each taxon, FastTree computes the distances to $O(\sqrt{n})$ close neighbors. FastTree also uses sequence profiles to approximate $S_D(x, y)$ values in constant time. The overall algorithm takes $O(san\sqrt{n}\log n)$ time and $O(san + n\sqrt{n})$ memory, where s is the length of the input sequences and a is their alphabet size. FastTree has been shown to be highly accurate with simulated data [17], but no formal guarantee has yet been shown for this algorithm.

The only known $o(n^2)$ algorithm with theoretical guarantees is LSHTree [4]. It uses locality-sensitive hashing to rapidly find candidate pairs of close sequences for merging. After each merge, LSHTree reconstructs ancestral sequences at new internal nodes to ensure that a close pair of sequences can be found at each iteration. LSHTree is guaranteed to reconstruct the correct tree from sequences of logarithmic length under a Markov model of sequence evolution. The exact running time of LSHTree depends on the branch lengths.

As we have shown, a number of distance-based tree building algorithms have been analyzed in the safety radius framework. However, computer simulations (e.g., [6, 7]) have shown that not all algorithms with optimal safety radius achieve the same accuracy: for example, NJ is slightly more accurate than FNJ (both having safety radius $= 1/2$), but is beaten by heuristics based on NNI or SPR moves (with demonstrated safety radius $>= 1/3$, but possibly $= 1/2$). Moreover, some well-established methods (e.g., based on least squares [10, 21]) have safety radius converging to 0 when the number of taxa increases, which contradicts the common practice. These experimental observations indicate that the safety radius approach should be sharpened to provide better theoretical analysis of method performance (see [12] for a work in this direction). In particular, the choice of the L_∞ norm to measure the error in a distance matrix seems to have little statistical or biological justification.

An alternative analysis framework, strictly linked to the one presented here, is the one seeking to estimate the minimum sequence length required for accurate reconstruction of the correct tree. It is discussed in a separate entry of this encyclopedia [A].

Cross-References

▶ Distance-Based Phylogeny Reconstruction (Fast-Converging)

Recommended Reading

1. Atteson K (1999) The performance of neighbor-joining methods of phylogenetic reconstruction. Algorithmica 25:251–278
2. Bordewich M, Gascuel O, Huber KT, Moulton V (2009) Consistency of topological moves based on the balanced minimum evolution principle of phylogenetic inference. IEEE/ACM Trans Comput Biol Bioinformatics 6:110–117
3. Bordewich M, Mihaescu R (2013) Accuracy guarantees for phylogeny reconstruction algorithms based on balanced minimum evolution. IEEE/ACM Trans Comput Biol Bioinformatics 10:576–583
4. Brown DG, Truszkowski J (2012) Fast phylogenetic tree reconstruction using locality-sensitive hashing. Algorithms in bioinformatics. Springer, Berlin/Heidelberg, pp 14–29
5. Dai W, Xu Y, Zhu B (2006) On the edge l∞ radius of Saitou and Nei's method for phylogenetic reconstruction. Theor Comput Sci 369:448–455
6. Desper R, Gascuel O (2002) Fast and accurate phylogeny reconstruction algorithms based on the minimum- evolution principle. J Comput Biol 9:687–706
7. Elias I, Lagergren J (2005) Fast neighbor joining. In: Proceedings of the 32nd international colloquium on automata, languages, and programming (ICALP), Lisbon, pp 1263–1274
8. Evans J, Sheneman L, Foster J (2006) Relaxed neighbor joining: a fast distance-based phylogenetic tree construction method. J Mol Evol 62:785–792
9. Felsenstein J (2004) Inferring phylogenies. Sinauer Associates, Sunderland
10. Gascuel O, McKenzie A (2004) Performance analysis of hierarchical clustering algorithms. J Classif 21:3–18
11. Gascuel O, Steel M (2006) Neighbor-joining revealed. Mol Biol Evol 23:1997–2000
12. Gascuel O, Steel M (2014) A 'stochastic safety radius' for distance-based tree reconstruction. Algorithmica 1–18. http://dx.doi.org/10.1007/s00453-015-0005-y
13. Gronau I, Moran S (2007) Neighbor joining algorithms for inferring phylogenies via LCA distances. J Comput Biol 14:1–15
14. Hein J (1989) An optimal algorithm to reconstruct trees from additive distance data. Bull Math Biol 51:597–603
15. Pardi F, Guillemot S, Gascuel O (2010) Robustness of phylogenetic inference based on minimum evolution. Bull Math Biol 72:1820–1839
16. Pauplin Y (2000) Direct calculation of a tree length using a distance matrix. J Mol Evol 51:41–47
17. Price MN, Dehal PS, Arkin AP (2009) FastTree: computing large minimum evolution trees with profiles instead of a distance matrix. Mol Biol Evol 26:1641–1650
18. Saitou N, Nei M (1987) The neighbor-joining method: a new method for reconstructing phylogenetic trees. Mol Biol Evol 4:406–425
19. Shigezumi T (2006) Robustness of greedy type minimum evolution algorithms. In: Computational science–ICCS, Reading, pp 815–821
20. Simonsen M, Mailund T, Pedersen CNS (2011) Inference of large phylogenies using neighbour-joining. In: Fred A, Felipe J, Gamboa H (eds) Biomedical engineering systems and technologies. Communications in computer and information science, vol 127. Springer, Berlin/Heidelberg, pp 334–344
21. Willson S (2005) Minimum evolution using ordinary least-squares is less robust than neighbor-joining. Bull Math Biol 67:261–279
22. Zarestkii K (1965) Reconstructing a tree from the distances between its leaves. (In Russian) Usp Math Nauk 20:90–92

Distributed Algorithms for Minimum Spanning Trees

Sergio Rajsbaum
Instituto de Matemáticas, Universidad Nacional
Autónoma de México (UNAM), México City,
México

Keywords

Minimum weight spanning tree

Years and Authors of Summarized Original Work

1983; Gallager RG, Humblet PA, Spira PM

Problem Definition

Consider a communication network, modeled by an undirected weighted graph $G = (V, E)$, where $|V| = n$, $|E| = m$. Each vertex of V represents a processor of unlimited computational power; the processors have unique identity numbers (ids), and they communicate via the edges of E by sending messages to each other. Also, each edge $e \in E$ has associated a weight $w(e)$, known to the processors at the endpoints of e. Thus, a processor knows which edges are incident to it and their weights, but it does not know any other information about G. The network is *asynchronous*: each processor runs at an arbitrary speed, which is independent of the speed of other processors. A processor may wake up spontaneously or when it receives a message from another processor. There are no failures in the network. Each message sent arrives at its destination within a finite but arbitrary delay. A *distributed algorithm* A for G is a set of local algorithms, one for each processor of G, that include instructions for sending and receiving messages along the edges of the network. Assuming that A terminates (i.e., all the local algorithms eventually terminate), its *message complexity* is the total number of messages sent over any execution of the algo-

rithm, in the worst case. Its *time complexity* is the worst-case execution time, assuming processor steps take negligible time, and message delays are normalized to be at most 1 unit.

A *minimum spanning tree* (MST) of G is a subset E' of E such that the graph $T = (V, E')$ is a tree (connected and acyclic) and its total weight, $w(E') = \sum_{e \in E'} w(e)$, is as small as possible. The computation of an MST is a central problem in combinatorial optimization, with a rich history dating back to 1926 [2], and up to now, the book [12] collects properties, classical results, applications, and recent research developments.

In the *distributed MST problem*, the goal is to design a distributed algorithm A that terminates always and computes an MST T of G. At the end of an execution, each processor knows which of its incident edges belong to the tree T and which do not (i.e., the processor writes in a local output register the corresponding incident edges). It is remarkable that in the distributed version of the MST problem, a communication network is solving a problem where the input is the network itself. This is one of the fundamental starting points of network algorithms.

It is not hard to see that if all edge weights are different, the MST is unique. Due to the assumption that processors have unique ids, it is possible to assume that all edge weights are different: whenever two edge weights are equal, ties are broken using the processor ids of the edge endpoints. Having a unique MST facilitates the design of distributed algorithms, as processors can locally select edges that belong to the unique MST. Notice that if processors do not have unique ids and edge weights are not different, there is no deterministic MST (nor any spanning tree) distributed algorithm, because it may be impossible to break the symmetry of the graph, for example, in the case it is a cycle with all edge weights equal.

Key Results

The distributed MST problem has been studied since 1977, and dozens of papers have been written on the subject. In 1983, the fundamental distributed *GHS algorithm* in [5] was published, the first to solve the MST problem with $O(m + n \log n)$ message complexity. The paper has had a very significant impact on research in distributed computing and won the 2004 Edsger W. Dijkstra Prize in Distributed Computing.

It is not hard to see that any distributed MST algorithm must have $\Omega(m)$ message complexity (intuitively, at least one message must traverse each edge). Also, results in [3, 4] imply an $\Omega(n \log n)$ message complexity lower bound for the problem. Thus, the GHS algorithm is optimal in terms of message complexity.

The $\Omega(m + n \log n)$ message complexity lower bound for the construction of an MST applies also to the problem of finding an arbitrary spanning tree of the graph. However, for specific graph topologies, it may be easier to find an arbitrary spanning tree than to find an MST. In the case of a complete graph, $\Omega(n^2)$ messages are necessary to construct an MST [8], while an arbitrary spanning tree can be constructed in $O(n \log n)$ messages [7].

The time complexity of the GHS algorithm is $O(n \log n)$. In [1] it is described how to improve its time complexity to $O(n)$ while keeping the optimal $O(m + n \log n)$ message complexity. It is clear that $\Omega(D)$ time is necessary for the construction of a spanning tree, where D is the diameter of the graph. And in the case of an MST, the time complexity may depend on other parameters of the graph. For example, due to the need for information flow among processors residing on a common cycle, as in an MST construction, at least one edge of the cycle must be excluded from the MST. If messages of unbounded size are allowed, an MST can be easily constructed in $O(D)$ time, by collecting the graph topology and edge weights in a root processor. The problem becomes interesting in the more realistic model where messages are of size $O(\log n)$ and an edge weight can be sent in a single message. When the number of messages is not important, one can assume without loss of generality that the model is synchronous. For near-time optimal algorithms and lower bounds, see [10] and references herein.

Applications

The distributed MST problem is important to solve, both theoretically and practically, as an MST can be used to save on communication, in various tasks such as broadcast and leader election, by sending the messages of such applications over the edges of the MST.

Also, research on the MST problem, and in particular the MST algorithm of [5], has motivated a lot of work. Most notably, the algorithm of [5] introduced various techniques that have been in widespread use for multicasting, query and reply, cluster coordination and routing, protocols for handshake, synchronization, and distributed phases. Although the algorithm is intuitive and is easy to comprehend, it is sufficiently complicated and interesting that it has become a challenge problem for formal verification methods, e.g., [11].

Open Problems

There are many open problems in this area, and only a few significant ones are mentioned. As far as message complexity, although the asymptotically tight bound of $O(m + n \log n)$ for the MST problem in general graphs is known, finding the actual constants remains an open problem. There are smaller constants known for general spanning trees than for MST though [6].

As mentioned above, near-time optimal algorithms and lower bounds appear in [10] and references herein. The optimal time complexity remains an open problem. Also, in a synchronous model for overlay networks, where all processors are directly connected to each other, an MST can be constructed in sublogarithmic time, namely, $O(\log \log n)$ communication rounds [9], and no corresponding lower bound is known.

Cross-References

▶ Synchronizers, Spanners

Recommended Reading

1. Awerbuch B (1987) Optimal distributed algorithms for minimum weight spanning tree, counting, leader election and related problems (detailed summary). In: Proceedings of the 19th annual ACM symposium on theory of computing, New York City. ACM, New York, pp 230–240
2. Borůvka O (2001) Otakar Borůvka on minimum spanning tree problem (translation of both the 1926 papers, comments, history). Discret Math 233: 3–36
3. Burns JE (1980) A formal model for message-passing systems, TR-91. Indiana University, Bloomington
4. Frederickson G, Lynch N (1984) The impact of synchronous communication on the problem of electing a leader in a ring. In: Proceedings of the 16th annual ACM symposium on theory of computing, Washington, DC. ACM, New York, pp 493–503
5. Gallager RG, Humblet PA, Spira PM (1983) A distributed algorithm for minimum-weight spanning trees. ACM Trans Prog Lang Syst 5(1): 66–77
6. Johansen KE, Jorgensen UL, Nielsen SH (1987) A distributed spanning tree algorithm. In: Proceedings of the 2nd international workshop on distributed algorithms (DISC), Amsterdam. Lecture notes in computer science, vol 312. Springer, Berlin/Heidelberg, pp 1–12
7. Korach E, Moran S, Zaks S (1984) Tight upper and lower bounds for some distributed algorithms for a complete network of processors. In: Proceedings of the 3rd symposium on principles of distributed computing (PODC), Vancouver. ACM, New York, pp 199–207
8. Korach E, Moran S, Zaks S (1985) The optimality of distributive constructions of minimum weight and degree restricted spanning trees in a complete network of processors. In: Proceedings of the 4th symposium on principles of distributed computing (PODC), Minaki. ACM, New York, pp 277–286
9. Lotker Z, Patt-Shamir B, Pavlov E, Peleg D (2005) Minimum-weight spanning tree construction in $O(\log \log n)$ communication rounds. SIAM J Comput 35(1):120–131
10. Lotker Z, Patt-Shamir B, Peleg D (2006) Distributed MST for constant diameter graphs. Distrib Comput 18(6):453–460
11. Moses Y, Shimony B (2006) A new proof of the GHS minimum spanning tree algorithm. In: 20th international symposium on distributed computing (DISC), Stockholm, 18–20 Sept 2006. Lecture notes in computer science, vol 4167. Springer, Berlin/Heidelberg, pp 120–135
12. Wu BY, Chao KM (2004) Spanning trees and optimization problems. Discrete mathematics and its applications. Chapman Hall, Boca Raton

Distributed Computing for Enumeration

Alexandre Termier
IRISA, University of Rennes, 1, Rennes, France

Keywords

Cluster; Enumeration; Multicore; Parallelism; Work sharing; Work stealing

Years and Authors of Summarized Original Work

2006; Buehrer, Parthasarathy, Chen
2012; Martins, Manquiho, Lynce
2014; Negrevergne, Termier, Rousset, Mehaut

Problem Definition

This entry considers enumeration of combinatorial problems, which can be formulated as follows. Given a large search space C and a predicate of interest $P : C \mapsto$ {true, false} the goal is to enumerate the solutions $S \subseteq C$ such that $\forall ?s \in S \quad P(s) = $ true. In most settings, C is the complete set of combinations of an initial set G; hence, $|C| = 2^{|G|}$ and the problem is NP-hard. There are also cases where the elements to enumerate are not sets but other combinatorial structures such as sequences or graphs.

We restrict ourselves to the case where C can be organized as an *enumeration tree*:

- There exists a distinguished element of C called the root
- There exists (at least) a *parent* function *parent*: $C \setminus \{\text{root}\} \mapsto C$

Finding S implies, in the worst case, to enumerate all elements of C. A large body of research has exploited properties of P together with *parent* to avoid enumerating some parts of C, as described in the ▶ Reverse Search; Enumeration

Algorithms entry. The focus of the current entry is to exploit parallel computing devices in order to speedup the enumeration process. Due to the prevalence of multicore processors nowadays, they will be the main focus of this entry. However, many solutions presented also apply to a cluster setting.

Key Results

The main challenges of distributed enumeration, as well as existing solutions, are presented below.

Synchronization

For most distributed algorithms, one important challenge is to ensure the sharing of information between parallel processes, either by message passing in a cluster setting or by access to shared memory locations in a multicore setting.

This entry considers a tree-shaped enumeration, where all branches of the enumeration tree are independent. In such setting, complex synchronization mechanisms are not needed. The main difficulty is thus to guarantee that such tree-shaped enumeration can take place, i.e., find (at least) one *parent* function (cf. ▶ Reverse Search; Enumeration Algorithms entry).

Note that in some parallel SAT solvers [6] using *portfolio*-based approaches, each computing resource exploits a different strategy (and thus a different *parent* function) to explore the search space. Ideally these strategies are orthogonal: they explore different parts of the space, and they exchange information to help mutual pruning.

Load Balancing

Most recent algorithms explore the enumeration tree with Depth-First Search (DFS). The simplest way to perform distributed enumeration is to partition the enumeration tree and assign each subtree to a parallel process. A simple example is shown in Fig. 1. The figure shows the enumeration tree. Each subtree of the root is assigned to a different computing resource (either to a node in a cluster or to a core of a multicore processor).

However, depending on the choice of *parent* function and of pruning strategies, each subtree is likely to be of a different size, resulting on large running time differences for the parallel processes. Such phenomenon is called *load unbalance*. On the right of Fig. 1, the execution time is represented for both computing resources, assuming that computing for any node of the enumeration tree takes exactly one time unit. Computing resource T_1 computes for four time units, while computing resource T_2 computes for eight time units. This means that T_1 has been idle for four time units: for more than half of the execution time, the execution has only exploited

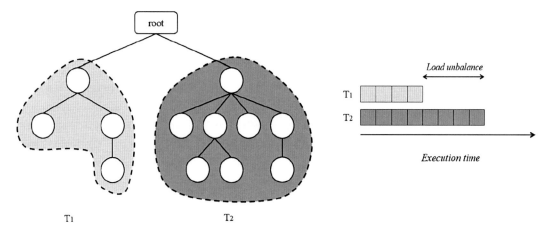

Distributed Computing for Enumeration, Fig. 1 Partitioning of the enumeration tree

one of the two available computing resources, resulting in a longer execution time (in an ideal case, as there are 12 nodes to explore, the execution time should be 6 time units). The solution to this problem is to use dynamic load balancing strategies such as *work sharing* or *work stealing*.

Work sharing is based on a simple producer/consumer principle: tasks (here nodes of the enumeration tree) are not assigned statically but made available in a single pool. Each idle computing resource can request one or more tasks from this pool. By reducing the granularity of the work each computing resource has to perform, this technique effectively reduces load unbalance. A disadvantage of this technique is that if computation for a single node of the enumeration tree is short, the computing resource will make frequent requests to the central work pool, with an increased risk of synchronization overheads over accesses to the pool. This can be limited by passing more than one enumeration tree node to each computing resource [1]; however, if the computation time of each of these nodes is unpredictable, a new load unbalance situation may arise.

Work stealing [2] is an "optimistic" improvement over work sharing in cases where the computation time of each task is unpredictable (which is often the case in distributed enumeration). The idea is that each computing resource enqueues a list of tasks to perform, which can either come from a central pool (as in work sharing) or from a partition of the search space (as in a static partitioning). When a computing resource finishes its tasks and becomes idle, it will query the work queues of the other resources and "steal" part of a nonempty queue (usually from the biggest queue).

Recent works from Hanusse et al. [3] have proposed an enumeration approach that mixes DFS and BFS. The parallelism that takes place in the BFS steps naturally lead to a good load balancing without needing work-stealing mechanisms. Their approach show promising results.

a large dataset $\mathcal{D}$. The enumeration tree is organized such that each branch explores a subset of the dataset, which allows to benefit from data locality effects and efficiently exploit the cache in case of multicore processors.

Techniques to limit load unbalance presented above tend to dispatch nodes from the same branch to different computing resources, which destroys data locality. This can lead to a vastly increased bandwidth usage (memory bus bandwidth for multicores and network bandwidth in clusters), resulting in a severe loss of performance.

Several solutions have been proposed by the pattern mining community in order to combine good load balance and good data locality. These solutions are designed for multicore processors. In the case of work stealing, Buehrer et al. [4] propose a method which dynamically decides, for a given node of the enumeration tree, if it has to be mined by the same thread as its parent (preserving locality) or if it can be enqueued and possibly stolen by another thread. This method takes into account the system load, which is a function of the size of the queues for the threads.

In the case of work sharing, Négrevergne et al. [5] divide the task pool in one queue per thread. Task assignement to threads prioritizes data locality.

Applications

The main applications of distributed enumeration are pattern mining (a field of data mining) and SAT problem solving [6]. Pattern mining consists in finding regularities in data, whereas solving SAT consists in finding if there exists a truth assignment for variables of a propositional formula. Both problems explore a huge search space and have numerous applications, they thus require to exploit as much as possible available parallel computing power.

Data Locality

In some classes of applications, such as pattern mining, testing the predicate P requires to access

Open Problems

Having an optimal parallel scaling for applications using distributed enumeration is still an open problem. The techniques presented in this entry allow to design algorithms having satisfying results on existing multicore computers with tens of cores. However as the number of cores will grow towards *many-core processors* (hundreds of cores), the current algorithms are unlikely to exhibit a good parallel scalability. Novel approaches with fewer overheads for handling the parallelism will be required.

Cross-References

▶ Reverse Search; Enumeration Algorithms

Recommended Reading

1. Zaki MJ (1999) Parallel and distributed association mining: a survey. IEEE Concurr 7(4):14–25
2. Blumofe RD, Leiserson CE (1999) Scheduling multithreaded computations by work stealing. J ACM 46(5):720–748
3. Hanusse N, Maabout S (2011) A parallel algorithm for computing borders. In: Proceedings of the 20th conference on information and knowledge management (CIKM'11), Glasgow, pp 1639–1648
4. Buehrer G, Parthasarathy S, Chen Y-K (2006) Adaptive parallel graph mining for CMP architectures. In: Proceedings of the 6th IEEE international conference on data mining (ICDM'06), Hong Kong, pp 97–106
5. Négrevergne B, Termier A, Rousset M-C, Méhaut J-F (2014) ParaMiner: a generic pattern mining algorithm for mutli-core architectures. Data Min Knowl Discov 28(3):593–633
6. Martins R, Manquiho VM, Lynce I (2012) An overview of parallel SAT solving. Constraints 17(3):304–347

Distributed Randomized Broadcasting in Wireless Networks under the SINR Model

Tomasz Jurdziński[1] and Dariusz R. Kowalski[2]
[1]Institute of Computer Science, University of Wrocław, Wrocław, Poland
[2]Department of Computer Science, University of Liverpool, Liverpool, UK

Keywords

Ad hoc network; Broadcasting; Distributed algorithms; SINR model; Wireless network

Years and Authors of Summarized Original Work

2012; Yu, Hua, Wang, Tan, Lau
2013; Daum, Gilbert, Kuhn, Newport
2013; Jurdziński, Kowalski, Różański, Stachowiak
2013; Jurdziński, Kowalski, Stachowiak

D

Problem Definition

Broadcasting is a fundamental problem in communication networks, where one distinguished node, called the *source*, holds a piece of information, and the goal is to disseminate this message to all other nodes in the network.

The *signal-to-interference-and-noise-ratio model*, *SINR* for short, generalizes the abstract radio networks model (RN) in the following way: nodes located in a metric space communicate by transmitting a signal to the wireless medium, and the quantitative accumulation of interference and signal attenuation are taken into account when deciding which nodes successfully receive the signal.

In more detail, a wireless network consists of *n nodes*, deployed into the Euclidean plane; each node v has its *transmission power P_v*, which is a positive real number. A network is *uniform* when transmission powers P_v are equal or *nonuniform* otherwise. In the following, the uniform networks are considered.

Nodes work synchronously in rounds; each node can either act as a transmitter or as a receiver during a round.

Interferences and collisions are determined by three fixed model parameters: path loss $\alpha > 2$, threshold $\beta \geq 1$, and ambient noise $\mathcal{N} > 0$. The $SINR(v, u, \mathcal{T})$ ratio, for given nodes u, v and a set of transmitting nodes $\mathcal{T}$, is defined as

$$SINR(v, u, \mathcal{T}) = \frac{P_v \text{dist}(v, u)^{-\alpha}}{\mathcal{N} + \sum_{w \in \mathcal{T} \setminus \{v\}} P_w \text{dist}(w, u)^{-\alpha}},$$

where $\text{dist}(\cdot, \cdot)$ is the distance function on the plane. A node u successfully receives a message from a node v in a round if $v \in \mathcal{T}$, $u \notin \mathcal{T}$, and

$$\text{SINR}(v, u, \mathcal{T}) \geq \beta, \qquad (1)$$

where $\mathcal{T}$ is the set of nodes transmitting at that round.

A single message sent in an execution of any algorithm can carry the source message and at most logarithmic, in the size of the network, number of control bits. A node other than the source starts executing the broadcast protocol after the first successful receipt of the source message; it is often called a *non-spontaneous wake-up model*.

In an *ad hoc network*, there is no central knowledge of network topology. A node v participating in an execution of a protocol knows the size of a network n or only a linear upper bound on the size. In the case that a network is deployed in the Euclidean space, one can distinguish between the case that each node knows its coordinates and the case that it does not know them.

Assuming that the transmission power of nodes is equal to P, the largest distance from the transmitter in which a message can be received is equal to $r = (P/(\mathcal{N}\beta))^{1/\alpha}$, provided only one node is transmitting in the whole network.

Sensitivity. Due to physical constraints, it is often assumed that the actual distance on which message can be received is smaller than $r = (P/(\mathcal{N}\beta))^{1/\alpha}$. This assumption is expressed by the *sensitivity parameter* $0 < \varepsilon_s < 1$ such that a message transmitted by v is received at a node u in a round with the set of transmitters $\mathcal{T}$ if $\text{SINR}(v, u, \mathcal{T}) \geq \beta$ and $\text{dist}(v, u) \leq (1 - \varepsilon_s)r$.

The setting with $\varepsilon_s = 0$ is called the model with *strong sensitivity*, and $\varepsilon_s > 0$ defines the model with *weak sensitivity*.

Connectivity. In order to determine which nodes are connected in a network, the notion of a *communication graph* is introduced. To this aim, the *connectivity parameter* ε_c is introduced such that $1 > \varepsilon_c \geq \varepsilon_s \geq 0$. An edge (u, v) appears in the communication graph iff $\text{dist}(u, v) \leq (1 - \varepsilon_c)r$. The setting with $\varepsilon_s = \varepsilon_c$ is called the model with *weak connectivity*, and the inequality $\varepsilon_s < \varepsilon_c$ defines the model with *strong connectivity*.

Time complexity of a randomized broadcasting algorithm is the number of rounds after which, for all communication networks defined by given SINR parameters α, β, and $\mathcal{N}$, and the parameters ε_s, ε_c, the source message is delivered to all nodes accessible from the source node in the communication graph with high probability (whp), i.e., with the probability at least $1 - 1/n$, where n is the number of nodes in the network.

Key Results

Complexity of the broadcasting problem significantly differs in various models obtained by constraints imposed on the sensitivity and connectivity parameters.

The Model with Strong Sensitivity and Strong Connectivity

In this setting reception of messages is determined merely by Eq. (1), and the communication graph does not contain links of distance close to r due to $\varepsilon_c > 0$.

Theorem 1 ([3]) *The broadcasting problem can be solved in time $O\left(D \log n + \log^2 n\right)$ with high probability in the model with strong connectivity and strong sensitivity for networks in Euclidean two-dimensional space with known coordinates.*

For the setting that nodes do not know their coordinates, Daum et al. [2] provided a broadcasting algorithm which relies on a parameter R_s equal to the maximum ratio among distances between pairs of nodes connected by an edge in the communication graph. That is, $R_s = \max\{\text{dist}(u, v)/\text{dist}(x, y) \mid (u, v), (x, y) \in E\}$, where $G(V, E)$ is the communication graph of a network.

Theorem 2 ([2]) *The broadcasting problem can be solved in time $O\left(D \log n \log^{\alpha+1} R_s\right)$ with high probability in the model with strong connectivity and strong sensitivity.*

As R_s might be even exponential with respect to n, the solution from Theorem 2 is inefficient for some networks. The following theorem gives a solution independent of geometric properties of a network.

Theorem 3 ([6]) *The broadcasting problem can be solved in time* $O\left(D \log^2 n\right)$ *with high probability in the model with strong sensitivity and strong connectivity.*

The Model with Strong Sensitivity and Weak Connectivity

In this setting reception of messages is determined merely by Eq. (1), and an edge (u, v) belongs to the communication graph iff $\text{SINR}(u, v, \emptyset) \geq \beta$.

Theorem 4 ([2]) *There exist families of networks with diameter 2 in the model with strong sensitivity and weak connectivity in which each broadcasting algorithm requires* $\Omega(n)$ *rounds to accomplish broadcast.*

Theorem 5 ([2]) *The broadcasting problem can be solved in time* $O(n \log^2 n)$ *with high probability in the model with strong sensitivity and weak connectivity.*

The Model with Weak Sensitivity and Strong Connectivity

In this setting, transmissions on unreliable links (i.e., on distance very close to r) are "filtered out." Moreover, the communication graph connects only nodes in distance at most $(1 - \varepsilon_c)r$, which is strictly smaller than r.

Theorem 6 ([7]) *The broadcasting problem can be solved in time* $O\left(D + \log^2 n\right)$ *with high probability in the model with weak sensitivity and strong connectivity for networks deployed on the Euclidean plane, provided* $0 < \varepsilon_s < \varepsilon_c = 2/3$. *This solution works in the model with* **spontaneous wake-up** *and requires* **power control mechanism**.

When applied directly to the case of non-spontaneous wake-up, the algorithm from [7] gives time bound $O\left(D \log^2 n\right)$.

Corollary 1 *The broadcasting problem can be solved in time* $O\left(D \log^2 n\right)$ *with high probability in the model with weak sensitivity and strong connectivity for networks deployed on the Euclidean*

plane, provided $0 < \varepsilon_s < \varepsilon_c = 2/3$. *This solution works in the model with* **non-spontaneous wake-up** *and requires* **power control mechanism**.

The Model with Weak Sensitivity and Weak Connectivity

In this setting, the maximal distances for a successful transmission and for connecting nodes by an edge are equal and both smaller than the largest theoretically possible range r following from Eq. (1).

Theorem 7 ([5]) *There exist (i) an infinite family of n-node networks requiring* $\Omega(n \log n)$ *rounds to accomplish broadcast whp and, (ii) for every* $D, \Delta = O(n)$, *an infinite family of n-node networks of diameter D and maximum degree of the communication graph* Δ *requiring* $\Omega(D\Delta)$ *rounds to accomplish broadcast whp in the model with weak sensitivity and weak connectivity (i.e.,* $0 < \varepsilon_s = \varepsilon_c < 1$) *for networks on the plane.*

Using appropriate combinatorial structures, deterministic broadcasting algorithms were obtained with complexities close to the above lower bounds, provided nodes know their coordinates.

Theorem 8 ([5]) *The broadcasting problem can be solved deterministically in time* $O\left(\min \left\{D\Delta \log^2 N, n \log N\right\}\right)$ *in the model with weak sensitivity and weak connectivity (i.e.,* $0 < \varepsilon_s = \varepsilon_c < 1$) *for networks in Euclidean two-dimensional space with known coordinates, with IDs in the range* $[1, N]$.

The above result translates to a randomized algorithm with complexity $O\left(\min \left\{D\Delta \log^2 n, n \log n\right\}\right)$, since nodes can choose unique IDs in the polynomial range with high probability.

Recently, Chlebus et al. [1] provided a randomized algorithm for the setting without knowledge of coordinates.

Theorem 9 ([1]) *The broadcasting problem can be solved in time* $O\left(n \log^2 n\right)$ *with high probability in the model with weak sensitivity and weak connectivity.*

Distributed Randomized Broadcasting in Wireless Networks under the SINR Model, Table 1 Complexity of randomized broadcasting with non-spontaneous wake-up for various sensitivity and connectivity settings. The result from [7] marked with $\star$ requires power control mechanism and $\varepsilon_c = 2/3$. The positive results requiring knowledge of coordinates apply only to the Euclidean space

	Coordinates	Strong connectivity: $\varepsilon_c > \varepsilon_s$	Weak connectivity: $\varepsilon_c = \varepsilon_s$
Strong sensitivity:	Known	$O\left(D \log n + \log^2 n\right)$	$\Omega(n)$
$\varepsilon_s = 0$	Unknown	$O\left(D \log^2 n\right), O\left(D \log n \log^{\alpha+1} R_s\right)$	$O\left(n \log^2 n\right)$
Weak sensitivity:	Known		$\Omega\left(\min\{D\Delta, n\}\right)$ $O\left(\min\left\{D\Delta \log^2 n, n \log n\right\}\right)$
$\varepsilon_s > 0$	Unknown	$O\left(D \log^2 n\right)\ \star$	$O\left(n \log^2 n\right)$

Applications

Using similar techniques to those in [5], an efficient deterministic broadcasting algorithm was obtained in the model with strong connectivity and strong sensitivity.

Theorem 10 ([4]) *The broadcasting problem can be solved deterministically in time $O\left(D \log^2 N\right)$ in the model with strong connectivity and strong sensitivity for networks in Euclidean two-dimensional space with known coordinates, with IDs in the range $[1, N]$.*

The solution in [7] applies to a more general problem of multi-broadcast and was further generalized in [8] to the setting in which nodes wake up in various time steps.

Positive results from [1, 2, 6] work also in a more general setting when nodes are deployed in a bounded-growth metric space.

Open Problems

In all considered models, there is at least $\log n$ gap between the established lower and upper bounds for the complexity of broadcasting. A natural open problem is to tighten these bounds.

As seen in Table 1, it is not known whether the complexity of broadcasting depends on sensitivity for each connectivity setting.

Another interesting research direction is to explore the impact of additional features such as power control or carrier sensing on the complexity of the problem.

Cross-References

▶ Broadcasting in Geometric Radio Networks

Recommended Reading

1. Chlebus BS, Kowalski DR, Vaya S (2014) Distributed communication in bare-bones wireless networks, manuscript
2. Daum S, Gilbert S, Kuhn F, Newport CC (2013) Broadcast in the ad hoc SINR model. In: DISC, Jerusalem, Lecture notes in computer science, vol 8070. Springer, pp 358–372
3. Jurdzinski T, Kowalski DR, Rozanski M, Stachowiak G (2013) Distributed randomized broadcasting in wireless networks under the SINR model. In: DISC, Jerusalem, Lecture notes in computer science, vol 8070. Springer, pp 373–387
4. Jurdzinski T, Kowalski DR, Stachowiak G (2013) Distributed deterministic broadcasting in uniform-power ad hoc wireless networks. In: Gasieniec L, Wolter F (eds) FCT, Liverpool. Lecture notes in computer science, vol 8070. Springer, pp 195–209
5. Jurdzinski T, Kowalski DR, Stachowiak G (2013) Distributed deterministic broadcasting in wireless networks of weak devices. In: ICALP (2), Riga. Lecture notes in computer science, vol 7966. Springer, pp 632–644
6. Jurdzinski T, Kowalski DR, Rozanski M, Stachowiak G (2014) On the impact of geometry on ad hoc communication in wireless networks. In: Halldórsson MM, Dolev S (eds) PODC, Paris. ACM, pp 357–366
7. Yu D, Hua QS, Wang Y, Tan H, Lau FCM (2012) Distributed multiple-message broadcast in wireless ad-hoc networks under the SINR model. In: Even G, Halldórsson MM (eds) SIROCCO, Reykjavik. Lecture notes in computer science, vol 7355. Springer, pp 111–122
8. Yu D, Hua QS, Wang Y, Yu J, Lau FCM (2013) Efficient distributed multiple-message broadcasting in unstructured wireless networks. In: INFOCOM, Turin, pp 2427–2435

Distributed Snapshots

Michel Raynal
Institut Universitaire de France and IRISA,
Université de Rennes, Rennes, France

Keywords

Asynchronous message-passing system; Consistent checkpointing; Consistent global state; Lattice of global states; Stable property

Years and Authors of Summarized Original Work

1985; Chandy, Lamport

Preliminary Remark

The presentation of this entry of the Encyclopedia follows Chapter 6 of [15], to which it borrows the presentation style and all figures. The reader interested on this topic will find developments in [15].

The Notion of a Global State

Modeling the Execution of a Process: The Event Point of View

A distributed computation involving n asynchronous sequential processes $p_1, \ldots, p_n$, communicating by directed channels (hence, a directional channel can be represented by two directed channels). Channels can be FIFO (first in first out) or non-FIFO.

A distributed computation can be modeled by a (reflexive) partial order on the events produced by the processes. An event corresponds to the sending of a message, the reception of a message, or a nonempty sequence of operations which does not involve the sending or the reception of a message. This partial order, due to Lamport and called *happened before* relation [12], is defined as follows. Let e_1 and e_2 be two events; $e_1 \xrightarrow{ev} e_2$ is the smallest order relation such that:

- Process order: e_1 and e_2 are the same event or have been produced by the same process, and e_1 was produced before e_2.
- Message order: e_1 is the sending of a message m and e_2 is its reception.
- Transitive closure: There is an event e such that $e_1 \xrightarrow{ev} e \wedge e \xrightarrow{ev} e_2$.

Modeling the Execution of a Process: The Local State Point of View

Let us consider a process p_i, which starts in the initial state σ_i^0. Let e_i^x be its xth event. The transition function $\delta_i()$ associated with p_i is consequently such that $\sigma_i^x = \delta_i(\sigma_i^x, e_i^x)$, where $x \geq 1$.

While a distributed computation can be modeled the partial order $\xrightarrow{ev}$ ("action" point of view), it follows from the previous definition that it can also be modeled by a partial order on its local states ("state" point of view). This partial order, denoted $\xrightarrow{\sigma}$, is defined as follows: $\sigma_i^x \xrightarrow{\sigma} \sigma_j^y \stackrel{\text{def}}{=} e_i^{x+1} \xrightarrow{ev} e_j^y$.

A two-process distributed execution is described in Fig. 1. The relation $\xrightarrow{ev}$ on event and the relation $\xrightarrow{\sigma}$ on local states can be easily extracted from it. As an example, we have $e_1^1 \xrightarrow{ev} e_2^3$ and $\sigma_2^0 \xrightarrow{\sigma} \sigma_1^2$. Two local states σ and σ' which are not related by $\xrightarrow{\sigma}$ are independent. This is denoted $\sigma \| \sigma'$.

Orphan and In-Transit Messages

Let us consider an ordered pair of local states $\langle \sigma_i, \sigma_j \rangle$ from different processes p_i and p_j and a message m sent by p_i to p_j.

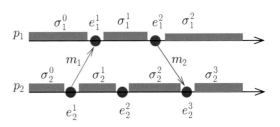

Distributed Snapshots, Fig. 1 A two-process distributed execution

- If m is sent by p_i after σ_i and received by p_j before σ_j, this message is *orphan* with respect to $\langle \sigma_i, \sigma_j \rangle$. This means that m is received and not sent with respect to the ordered pair $\langle \sigma_i, \sigma_j \rangle$.
- If m is sent by p_i before σ_i and received by p_j after σ_j, this message is *in-transit* with respect to $\langle \sigma_i, \sigma_j \rangle$. This means that m is sent and not yet received with respect to the ordered pair $\langle \sigma_i, \sigma_j \rangle$.

As an example, the message m_1 is orphan with respect to the ordered pair $\langle \sigma_2^0, \sigma_1^1 \rangle$, while the message m_2 is in-transit with respect to the ordered pair $\langle \sigma_1^2, \sigma_2^1 \rangle$.

Consistent Global State

A *global state* is a vector of n local states (one per process), plus a set of channel states (one per directed channel). The state of a channel is a sequence of messages if the channel is FIFO or a set of messages if the channel is non-FIFO.

A *consistent global state* (also called *snapshot*) is a global state in which the computation has passed or could have passed. More formally, a global state is a pair (Σ, M) where the vector of local states $\Sigma = [\sigma_1, \dots, \sigma_n]$ and the set of channel states $M = \cup_{\{(i,j)\}} cs(i,j)$ are such that for any directed pair (i, j) we have:

- $\sigma_i \| \sigma_j$. (This means that there is no orphan message with respect to the ordered pair $\langle \sigma_i, \sigma_j \rangle$.)
- $cs(i, j)$ contains all the messages which are in transit with respect to the ordered pair $\langle \sigma_i, \sigma_j \rangle$ and only them. (This means that $cs(i, j)$ contains all the messages (and only them) sent by p_i before σ_i and not yet received by p_j when it enters σ_j.)

As an example, when looking at Fig. 1, $([\sigma_1^2, \sigma_2^1], \{cs(1, 2) = m_2, \ cs(2, 1) = \emptyset\})$ is a consistent global state, while both $([\sigma_1^1, \sigma_2^0], \{cs(1, 2) = \dots, \ cs(2, 1) = \dots\})$ and $([\sigma_1^0, \sigma_2^2], \{cs(1, 2) = \emptyset, \ cs(2, 1) = \emptyset\})$ are not consistent (the first because, as the message m_1 is orphan with respect to the ordered pair $\langle \sigma_1^1, \sigma_2^0 \rangle$,

we do not have $\sigma_1^1 \| \sigma_2^0$; the second because the message m_1 does not belong to the channel state $cs(2, 1)$.).

The Lattice of Global States

Let us consider a vector of local states $\Sigma = [\sigma_1, \dots, \sigma_n]$ belonging to a consistent global state. The consistent global state, where $\Sigma' = [\sigma_1', \dots, \sigma_n']$, is directly reachable from Σ if there is a process p_i whose next event e_i, and we have $\forall j \neq j : \sigma_j' = \sigma_j$ and $\sigma_i = \delta_i(\sigma_i, e_i)$. This is denoted $\Sigma \xrightarrow{GS} \Sigma'$. By definition $\Sigma \xrightarrow{GS} \Sigma$. More generally, $\Sigma \xrightarrow{GS} \Sigma_a \xrightarrow{GS} \Sigma_b \cdots \Sigma_y \xrightarrow{GS} \Sigma_z$ is denoted $\Sigma \xrightarrow{GS^*} \Sigma_z$.

It can be shown that the set of all the vectors Σ associated with the consistent global states produced by a distributed computation is a lattice [3, 16]. The lattice obtained from the computation of Fig. 1 is described in Fig. 2. In this lattice, the notation $\Sigma = [a, b]$ means $\Sigma = [\sigma_1^a, \sigma_2^b]$.

Problem Definition

Specification of the Computation of a Consistent Global State

The problem to determine on-the-fly a consistent global state (in short CGS) was introduced, precisely defined, and solved by Chandy and Lamport in 1985 [2]. It can be defined by the following properties. The first is liveness property, while the last two are safety properties.

- Termination. If one or more processes launch the computation of a consistent global state, then this global state computation terminates.
- Consistency. If a global state is computed, then it is consistent.
- Validity. Let Σ_{start} be the global state of the computation when CGS starts, Σ_{end} be its global state when CGS terminates, and Σ be the global state returned by CGS. We have $\Sigma_{\text{start}} \xrightarrow{GS^*} \Sigma$ and $\Sigma \xrightarrow{GS^*} \Sigma_{\text{end}}$.

The validity property states that the global state which is returned depends on the time at

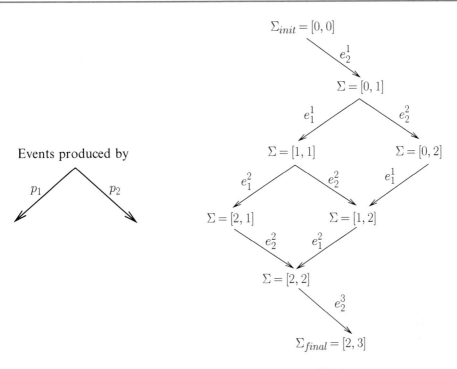

$\Sigma_{init} = [0, 0]$

e_2^1

$\Sigma = [0, 1]$

e_1^1 e_2^2

Events produced by

p_1 p_2

$\Sigma = [1, 1]$ $\Sigma = [0, 2]$

e_1^2 e_2^2 e_1^1

$\Sigma = [2, 1]$ $\Sigma = [1, 2]$

e_2^2 e_1^2

$\Sigma = [2, 2]$

e_2^3

$\Sigma_{final} = [2, 3]$

Distributed Snapshots, Fig. 2 The lattice associated with the computation of Fig. 2

which its computation is launched. Without this property, returning always the initial global state would be a correct solution.

Principles of CGS Algorithms

To compute a consistent global state, each process p_i is in charge of (a) recording a copy of its local state σ_i (sometimes called its local snapshot) and (b) the states of its input (or output) channels. In order that the computed global state satisfies the safety properties, in one way or another, all CGS algorithms have two things to do.

- Synchronization. In order to ensure that there is no orphan messages with respect to each ordered pair of local states $\langle \sigma_i, \sigma_j \rangle$, such that there is a directed channel from p_i to p_j, the processes must synchronize the recording of their local states which will define the consistent global state that is computed.
- Message recording. Each process has to record all the messages it receives (or

messages it sends) which are in transit with respect to the computed global state.

Key Result 1: Chandy-Lamport's Algorithm

Chandy and Lamport 's algorithm is denoted CL85 in the following.

Assumption

CL85 considers a failure-free asynchronous system. Asynchronous means that each process proceeds to its speed which can vary arbitrarily with time and remains always unknown to the other processes. Message transfer delays also are arbitrary (but finite).

CL85 assumes that the processes are connected by a directed communication graph, which is strongly connected (there is a directed path from any process to any other process). Each process has consequently a nonempty set of input channels and a nonempty set of output

D

channels. Moreover, each directed channel is a FIFO channel.

The Algorithm in Two Rules

CL85 requires that each process computes the state of its input channels. At a high abstraction level, it can formulated with two rules.

- "Local state recording" rule. When a process p_i records its local state σ_i, it sends a special control message (called *marker*) on each of its outgoing channels.

  It is important to notice that as channels are FIFO, a marker partitions the messages sent on a channel in two sets: the messages sent before the marker and the messages sent after the marker.

- "Input channel state recording" rule. When a process p_i receives a marker on one of its input channels $c(j, i)$, there are two cases.
  - If not yet done, it records its local state (i.e., it applies the first rule) and defines $cs(j, i)$ (the state of the input channel $c(j, i)$) as the empty sequence.
  - If it has already recorded its local state (i.e., executed the first rule), p_i defines $cs(j, i)$ as the sequence of application messages received on $c(j, i)$ between the recording of its local state and the reception of the marker on this input channel.

Properties of the Computed Global State

If one or more processes execute the first rule, it follows from this rule, and the fact that the communication graph is strongly connected, all the processes will execute this rule. Hence, a marker is sent on each directed channel, and each process records the state of its input channels. This proves the liveness property.

The consistency property follows from the following simple observation. Let us color processes and messages as follows. A process is initially green and becomes red when it sends a marker on each of its output channels. Moreover, a message has the color of its sender at the time the message is sent.

It is easy to see that the previous two rules guarantee that a green process turns red before

receiving a red message (hence, there is no orphan messages). Moreover, the green messages received on a channel $c(j, i)$ by a red process p_i are the messages that are in transit with respect to the ordered pair $\langle \sigma_j, \sigma_i \rangle$. Hence, all the *in-transit* messages are recorded and only them.

The Inherent Uncertainty on the Computed Global State

The proof of the validity property is a little bit more involved. The interested reader will consult [2, 15]. When looking at the lattice of Fig. 2, let us consider that the CL85 algorithm is launched when the observed computation is in the global state $\Sigma_{\text{start}} = [0, 1]$ and terminates when it is in the global state $\Sigma_{\text{end}} = [2, 2]$. The consistency property states that the global state Σ which is returned is one of the following global states: $[0, 1]$, $[1, 1]$, $[0, 2]$, $[2, 1]$, $[1, 2]$, or $[2, 2]$.

This uncertainty on the computed global state is intrinsic to the nature of distributed computing. (Eliminate would require to freeze the execution of the application we are observing, which in some sense forces it to execute sequentially.)

The main property of the consistent global state Σ that is computed is that the application has passed through it or could have passed through it. While an external omniscient observer can know if the application passed or not through Σ, no process can know it. This noteworthy feature characterizes the relativistic nature of the observation of distributed computations.

Message-Passing Snapshot Versus Shared Memory Snapshot

The notion of a shared memory snapshot has been introduced in [1]. A snapshot object is an object that consists of an array of atomic multi-reader/single-write atomic registers, one per process (a process can read any register but can write only the register it is associated with). A snapshot object provides the processes with two operations denoted update() and snapshot(). The update() operation allows the invoking process to store a new value in its register, while the snapshot() operation allows it to obtain the values of all the atomic read/write registers as if that operation was executed instantaneously.

More precisely, the invocations of the operations update() and snapshot() are linearizable [8].

Differently from the snapshot values returned in a message-passing system (whose global structure is a lattice), the arrays of values returned by the snapshot() operations of a shared memory system can be totally ordered. This is a fundamental difference, which is related to the communication medium. In one case, the underlying shared memory is a centralized component, while in the second case, the underlying message-passing system is inherently distributed, making impossible to totally order all the message-passing snapshots.

Other Assumptions and Algorithms

Algorithms that compute consistent global states in systems equipped with non-FIFO channels have been designed. Such algorithms are described in [11, 13, 15].

A communication-induced (CI) algorithm is a distributed algorithm that does not use additional control messages (such as markers). In these algorithms, control information (if needed) has to be carried by application messages. CI algorithms that compute consistent global states have been investigated in [6].

Global states computation in large-scale distributed systems is addressed in [9].

Key Result 2: A Necessary and Sufficient Condition

The Issue

An important question is the following one: Given a set of x, $1 \leq x \leq n$, local states from different processes do these local states belong to a consistent global state?

If $x = n$, the answer is easy: any ordered pair of local states $\langle \sigma_i, \sigma_j \rangle$ has to be such that $\sigma_i \| \sigma_j$ (none of them causally depends on the other). Hence, the question is interesting when $1 \leq x < n$. This problem was addressed and solved by Netzer and Xu [14] and generalized in [7].

As a simple example, let us consider the execution in Fig. 3, where there are three processes p_i, p_j, and p_k that have recorded the local states

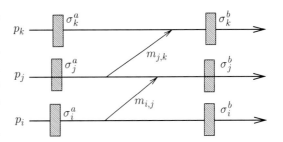

Distributed Snapshots, Fig. 3 A simple zigzag path

σ_x^a and σ_x^b, where $x \in \{i, j, k\}$. These local states produced by the computation are the only local states which have been recorded. Said another way, these recorded local states can be seen as local checkpoints.

An instance of the previous question is the following one: can the set $\{\sigma_i^a, \sigma_k^b\}$ be extended by the addition of a recorded local state σ_j of p_j (i.e., $\sigma_j = \sigma_j^a$ or $\sigma_j = \sigma_j^b$) such that the resulting global state $\Sigma = [\sigma_i^a, \sigma_j, \sigma_k^b]$ is a consistent?

It is easy to see that, despite the fact that the local states σ_i^a and σ_k^b are independent ($\sigma_i^a \| \sigma_k^b$), neither $[\sigma_i^a, \sigma_j^a, \sigma_k^b]$ nor $[\sigma_i^a, \sigma_j^b, \sigma_k^b]$ is consistent. More precisely, $[\sigma_i^a, \sigma_j^a, \sigma_k^b]$ is not consistent because, due the message $m_{j,k}$, the local states σ_j^a and σ_k^b are not independent, while $[\sigma_i^a, \sigma_j^b, \sigma_k^b]$ is not consistent because, due the message $m_{i,j}$, σ_i^a and σ_j^b are not independent.

The Result

The notion of a *zigzag* path has been introduced by Netzer and Xu in [14]. An example of a simple zigzag path is the sequence of messages $\langle m_{i,j}, m_{j,k} \rangle$ of Fig. 3, where we see that the local states σ_i^a and σ_k^b are related by this zigzag path. A zigzag path captures hidden dependencies linking recorded local states. These dependencies are hidden in the sense that not all of them can be captured by the relation $\xrightarrow{\sigma}$ defined on local states (as shown in the figure).

The main result due to Netzer and Xu [14] is the following: a set x local states, with $1 \leq x < n$ and at most one local state per process, can be extended to a consistent global state if and only if no two of them are related by a zigzag path. This result has been extended in [7].

D

Applications: Global Snapshots in Action

Distributed snapshots are a key element to understand and master the uncertainty created by asynchrony. From a practical point of view, they are important in distributed checkpointing, in the detection of stable properties defined on the set of global states, and in the debugging of distributed programs.

Detection of Stable Properties

A stable property is a property that, one true, remains true forever. In the distributed context, examples of distributed stable properties are deadlock (once deadlocked, an application remains forever deadlocked), termination (once terminated, an application remains forever terminated) [4,5], object inaccessibility, etc.

Algorithms that compute consistent global states satisfying the liveness, consistency, and validity properties previously stated can be used to detect stable properties. This follows from the observation that if the computed global state Σ satisfies a stable property P, then the global state Σ_{end} also satisfies P.

Checkpointing

A checkpoint is a global state from which a computation can be resumed. Trivially, checkpointing and consistent global states computation are problems which are very close [7]. The interested reader can consult [10, 15] for more details.

Cross-References

▶ Causal Order, Logical Clocks, State Machine Replication
▶ Distributed Computing for Enumeration
▶ Snapshots in Shared Memory

Recommended Reading

1. Afek Y, Attiya H, Dolev D, Gafni E, Merritt M, Shavit S (1993) Atomic snapshots of shared memory. J ACM 40(4):873–890
2. Chandy KM, Lamport L (1985) Distributed snapshots: determining global states of distributed systems. ACM Trans Comput Syst 3(1):63–75
3. Cooper R, Marzullo K (1991) Consistent detection of global predicates. In: Proceedings of the ACM/ONR workshop on parallel and distributed debugging, Santa Cruz. ACM, pp 163–173
4. Dijkstra EWD, Scholten CS (1980) Termination detection for diffusing computations. Inf Process Lett 11(1):1–4
5. Francez N (1980) Distributed termination. ACM Trans Program Lang Syst 2(1):42–55
6. Hélary J-M, Mostéfaoui A, Raynal M (1999) Communication-induced determination of consistent snapshots. IEEE Trans Parallel Distrib Syst 10(9):865–877
7. Hélary J-M, Netzer RHB, Raynal M (1999) Consistency criteria for distributed checkpoints. IEEE Trans Softw Eng 2(2):274–281
8. Herlihy MP, Wing JM (1990) Linearizability: a correctness condition for concurrent objects. ACM Trans Program Lang Syst 12(3):463–492
9. Kshemkalyani AD (2010) Fast and message-efficient global snapshot algorithms for large-scale distributed systems. IEEE Trans Parallel Distrib Syst 21(9):1281–1289
10. Kshemkalyani AD, Singhal M (2008) Distributed computing: principles, algorithms and systems. Cambridge University Press, Cambridge, 736p
11. Lai TH, Yang TH (1987) On distributed snapshots. Inf Process Lett 25:153–158
12. Lamport L (1978) Time, clocks, and the ordering of events in a distributed system. Commun ACM 21(7):558–565
13. Mattern F (1993) Efficient algorithms for distributed snapshots and global virtual time approximation. J Parallel Distrib Comput 18:423–434
14. Netzer RHB, Xu J (1995) Necessary and sufficient conditions for consistent global snapshots. IEEE Trans Parallel Distrib Syst 6(2):165–169
15. Raynal M (2013) Distributed algorithms for message-passing systems. Springer, 515p. ISBN:978-3-642-38122-5
16. Schwarz R, Mattern F (1994) Detecting causal relationships in distributed computations: in search of the Holy Grail. Distrib Comput 7(3):149–174

Distributed Vertex Coloring

Devdatt Dubhashi
Department of Computer Science, Chalmers University of Technology, Gothenburg, Sweden
Gothenburg University, Gothenburg, Sweden

Keywords

Distributed computation; Vertex coloring

Years and Authors of Summarized Original Work

2004; Finocchi, Panconesi, Silvestri

Problem Definition

The *vertex coloring problem* takes as input an undirected graph $G := (V, E)$ and computes a *vertex coloring*, i.e., a function, $c : V \to [k]$ for some positive integer k such that adjacent vertices are assigned different colors (that is, $c(u) \neq c(v)$ for all $(u, v) \in E$). In the $(\Delta + 1)$ vertex coloring problem, k is set equal to $\Delta + 1$ where Δ is the maximum degree of the input graph G. In general, $(\Delta + 1)$ colors could be necessary as the example of a clique shows. However, if the graph satisfies certain properties, it may be possible to find colorings with far fewer colors. Finding the minimum number of colors possible is a computationally hard problem: the corresponding decision problems are NP-complete [5]. In Brooks–Vizing colorings, the goal is to try to find colorings that are near optimal.

In this paper, the model of computation used is the synchronous, message passing framework as used in standard distributed computing [11]. The goal is then to describe very simple algorithms that can be implemented easily in this distributed model that simultaneously are *efficient* as measured by the number of rounds required and have good performance *quality* as measured by the number of colors used. For efficiency, the number of rounds is require to be poly-logarithmic in n, the number of vertices in the graph and for performance quality, the number of colors used is should be near-optimal.

Key Results

Key theoretical results related to distributed $(\Delta + 1)$-vertex coloring are due to Luby [9] and Johansson [7]. Both show how to compute a $(\Delta + 1)$-coloring in $O(\log n)$ rounds with high probability. For Brooks–Vizing colorings,

Kim [8] showed that if the graph is square or triangle free, then it is possible to color it with $O(\Delta / \log \Delta)$ colors. If, moreover, the graph is regular of sufficiently high degree ($\Delta \gg \log n$), then Grable and Panconesi [6] show how to color it with $O(\Delta / \log \Delta)$ colors in $O(\log n)$ rounds. See [10] for a comprehensive discussion of probabilistic techniques to achieve such colorings.

The present paper makes a comprehensive experimental analysis of distributed vertex coloring algorithms of the kind analyzed in these papers on various classes of graphs. The results are reported in section "Experimental Results" below and the data sets used are described in section "Data Sets."

Applications

Vertex coloring is a basic primitive in many applications: classical applications are *scheduling problems* involving a number of pairwise restrictions on which jobs can be done simultaneously. For instance, in attempting to schedule classes at a university, two courses taught by the same faculty member cannot be scheduled for the same time slot. Similarly, two course that are required by the same group of students also should not conflict. The problem of determining the minimum number of time slots needed subject to these restrictions can be cast as a vertex coloring problem. One very active application for vertex coloring is *register allocation*. The register allocation problem is to assign variables to a limited number of hardware registers during program execution. Variables in registers can be accessed much quicker than those not in registers. Typically, however, there are far more variables than registers so it is necessary to assign multiple variables to registers. Variables conflict with each other if one is used both before and after the other within a short period of time (for instance, within a subroutine). The goal is to assign variables that do not conflict so as to minimize the use of non-register memory. A simple approach to this is to create a graph where the nodes represent variables and an edge represents conflict between its nodes. A coloring is then a conflict-free assignment. If the number of colors used is less than the number of registers then a conflict-free register assignment is possible. Modern applications include *assigning frequencies* to mobile radios and other users of the electro-magnetic spectrum. In the simplest case, two customers that are sufficiently close must be assigned different frequencies, while those that are distant can share frequencies. The problem of minimizing the number of frequencies is then a vertex coloring problem. For more applications and references, see Michael Trick's coloring page [12].

Open Problems

The experimental analysis shows convincingly and rather surprisingly that the simplest, trivial, version of the algorithm actually performs best uniformly! In particular, it significantly outperforms the algorithms which have been analyzed rigorously. The authors give some heuristic recurrences that describe the performance of the trivial algorithm. It is a challenging and interesting open problem to give a rigorous justification of these recurrences. Alternatively, and less appealing, a rigorous argument that shows that the trivial algorithm dominates the ones analyzed by Luby and Johansson is called for. Other issues about how local structure of the graph impacts on the performance of such algorithms (which is hinted at in the paper) is worth subjecting to further experimental and theoretical analysis.

Experimental Results

All the algorithms analyzed start by assigning an initial *palette* of colors to each vertex, and then repeating the following simple iteration round:

1. *Wake up!*: Each vertex independently of the others wakes up with a certain probability to participate in the coloring in this round.
2. *Try!*: Each vertex independently of the others, selects a tentative color from its palette of colors at this round.

3. *Resolve conflicts!*: If no neighbor of a vertex selects the same tentative color, then this color becomes final. Such a vertex exits the algorithm, and the remaining vertices update their palettes accordingly. If there is a conflict, then it is resolved in one of two ways: Either all conflicting vertices are deemed unsuccessful and proceed to the next round, or an independent set is computed, using the so-called Hungarian heuristic, amongst all the vertices that chose the same color. The vertices in the independent set receive their final colors and exit. The Hungarian heuristic for independent set is to consider the vertices in random order, deleting all neighbors of an encountered vertex which itself is added to the independent set, see [1, p. 91] for a cute analysis of this heuristic to prove Turan's Theorem.
4. *Feed the Hungry!*: If a vertex runs out of colors in its palette, then fresh new colors are given to it.

Several parameters can be varied in this basic scheme: the wake up probability, the conflict resolution and the size of the initial palette are the most important ones.

In $(\Delta + 1)$-coloring, the initial palette for a vertex v is set to $[\Delta] := \{1, \cdots, \Delta + 1\}$ (global setting) or $[d(v) + 1]$ (where $d(v)$ is the degree of vertex v) (local setting). The experimental results indicate that (a) the best wake-up probability is 1, (b) the local palette version is as good as the global one in running time, but can achieve significant color savings and (c) the Hungarian heuristic can be used with vertex identities rather than random numbers giving good results.

In the Brooks–Vizing colorings, the initial palette is set to $[d(v)/s]$ where s is a *shrinking factor*. The experimental results indicate that uniformly, the best algorithm is the one where the wake-up probability is 1, and conflicts are resolved by the Hungarian heuristic. This is both with respect to the running time, as well as the number of colors used. Realistically useful values of s are between 4 and 6 resulting in Δ/s-colorings. The running time performance is excellent, with even graphs with a thousand vertices colored within 20–30 rounds. When compared to the best sequential algorithms, these algorithms use between twice or thrice as many colors, but are much faster.

Data Sets

Test data was both generated synthetically using various random graph models, and benchmark real life test sets from the second DIMACS implementation challenge [3] and Joe Culberson's web-site [2] were also used.

Cross-References

▶ Graph Coloring
▶ Randomization in Distributed Computing
▶ Randomized Gossiping in Radio Networks

Recommended Reading

1. Alon N, Spencer J (2000) The probabilistic method. Wiley, New York
2. Culberson JC. http://web.cs.ualberta.ca/~joe/Coloring/index.html
3. Ftp site of DIMACS implementation challenges. ftp://dimacs.rutgers.edu/pub/challenge/
4. Finocchi I, Panconesi A, Silvestri R (2004) An experimental analysis of simple distributed vertex soloring algorithms. Algorithmica 41:1–23
5. Garey M, Johnson DS (1979) Computers and intractability: a guide to the theory of NP-completeness. W.H. Freeman, San Francisco
6. Grable DA, Panconesi A (2000) Fast distributed algorithms for Brooks–Vizing colorings. J Algorithms 37:85–120
7. Johansson Ö (1999) Simple distributed $(\Delta + 1)$-coloring of graphs. Inf Process Lett 70:229–232
8. Kim J-H (1995) On Brook's theorem for sparse graphs. Comb Probab Comput 4:97–132
9. Luby M (1993) Removing randomness in parallel without processor penalty. J Comput Syst Sci 47(2):250–286
10. Molly M, Reed B (2002) Graph coloring and the probabilistic method. Springer, Berlin
11. Peleg D (2000) Distributed computing: a locality-sensitive approach. SIAM monographs on discrete mathematics and applications, vol 5. Society for Industrial and Applied Mathematics, Philadelphia
12. Trick M Michael Trick's coloring page. http://mat.gsia.cmu.edu/COLOR/color.html

Document Retrieval on String Collections

Rahul Shah
Department of Computer Science, Louisiana State University, Baton Rouge, LA, USA

Keywords

Geometric range searching; String matching; Suffix tree; Top-k retrieval

Years and Authors of Summarized Original Work

2002; Muthukrishan
2009; Hon, Shah, Vitter
2012; Navarro, Nekrich
2013; Shah, Sheng, Thankachan, Vitter

Problem Definition

Indexing data so that it can be easily searched is one of the most fundamental problems in computer science. Especially in the fields of databases and information retrieval, indexing is at the heart of query processing. One of the most popular indexes, used by all search engines, is the inverted index. However, in many cases like bioinformatics, eastern language texts, and phrase queries for Web, one may not be able to assume word demarcations. In such cases, these documents are to be seen as a string of characters. Thus, more sophisticated solutions are required for these string documents.

Formally, we are given a collection of D documents $\mathcal{D} = \{d_1, d_2, d_3, \ldots, d_D\}$. Each document d_i is a string drawn from the character set Σ of size σ and the total number of characters across all the documents is n. Our task is to preprocess this collection and build a data structure so that queries can be answered as quickly as possible. The query consists of a pattern string P, of length p, drawn from Σ. As the answer to the query, we are supposed output all the documents d_i in which this pattern P occurs as a substring. This is called the *document listing* problem. In a more advanced top-k version, the query consists of a tuple (P, k) where k is an integer. Now, we are supposed to output only the k most *relevant* documents. This is called the *top-k document retrieval* problem.

The notion of relevance is captured by a *score function*. The function score(P, d) denotes the score of the document d with respect to the pattern P. It can be the number of times P occurs in d, known as *term frequency*, or the distance between two closest occurrences of P in d, or any other function. Here, we will assume that score(P, d) is solely dependent on the set of occurrences of P in d and is known at the time of construction of the data structure.

Key Results

The first formal study of this problem was initiated by Muthukrishnan [4]. He took the approach of augmenting the generalized suffix tree with additional information. All subsequent works have used generalized suffix trees as their starting point. A generalized suffix tree GST is a compact trie of all the lexicographically sorted suffixes of all the documents. Thus, n total suffixes are stored and there are n leaves in this trie. Each edge in GST is labeled with a string and each root to leaf path (labels concatenated) represents some suffix of some document. The overall number of nodes in GST is $O(n)$. With each leaf, we associate a document id, which indicates the document to which that particular suffix belongs.

When the pattern P comes as a query, we first traverse from the root downward and find a vertex, which is known as locus(P). This can be done in $O(p)$ time. This is the first vertex below the edge where P finished in the root to leaf traversal. If v is locus(P) then all the leaves in the subtree of v represent the suffixes whose prefix is P. For any vertex v, let path(v) be the string obtained by concatenating all the labels from the root until v.

Document Listing Problem
Let us first see how the document listing problem is solved. One easy solution is to reach the locus

v of P and then visit all the leaves in the subtree of v. But this is costly as the number of leaves occ may be much more than number of unique document labels ndoc among these leaves. Optimally, we want to achieve $O(p + \text{ndoc})$ time.

To overcome this issue, Muthukrishnan first proposed to use a document array D_A. To construct this, he traverses all the leaves in GST from left-to-right and takes the corresponding document id. Thus, $D_A[i]$ = document id of ith lexicographically smallest suffix in GST. It is easy to find boundary points sp and ep such that the subtree of locus v corresponds to entries form $D_A[sp, \ldots, ep]$. To uniquely find documents in $D_A[sp, \ldots ep]$, we must not traverse the entire subarray as this would cost us O(occ). To avoid this, we construct another array C called a chain array. $C[i] = j$, where $j < i$ is the largest index for which $D_A[i] = D_A[j]$. If no such j exists then $C[i] = -1$. Thus, every document entry $D_A[i]$ links to the previous entry with the same document id. Now, to solve the document listing problem, one needs to get all the i's such that $sp \leq i \leq ep$ and $C[i] < sp$. The second constraint guarantees that every document is output only once. Muthukrishnan shows how to repeatedly apply range minimum queries (RMQ) to achieve constant time per output id.

Theorem 1 ([4]) *Given a collection of documents of total length n, we can construct a data structure taking $O(n)$ space, such that document listing queries for pattern P can be answered in optimal $O(p + \text{ndoc})$ time.*

Top-*k* Document Retrieval
Hon et al. [3] brought in an additional constraint of *score function* which can capture various notions of relevance like frequency and proximity. Instead of reporting all the documents, we only care to output the k highest scoring documents, as these would be the most relevant. Thus, $O(p + \text{ndoc})$ time is not optimal. We briefly describe their solution.

Let ST_d denote a suffix tree only for suffixes of document d. They augment the GST with *Links*. Link L is a 4-tuple: origin_node o, target_node t, document_id d, score_value s.

Essentially, $(L.o, L.t, L.d, L.s)$ is a link if and only if (t', o') is an edge in the suffix tree ST_d of the document d. Here, t' is the node in ST_d for which path(t') = path$(L.t)$. Similarly, path(o') = path$(L.o)$. The score value of the link $L.s$ = score(path$(o'), d)$. For $L.o$ and $L.t$, we use the preorder-id of those nodes in the GST. The total number of link entries is the same as the sum of number of edges in each individual suffix tree of all the documents, which is $O(n)$. They store these links in an array $\mathcal{L}$, sorted by target. In case of a tie among targets, we sort them further by their origin values.

When they execute the query (P, k), they first find the locus node v in GST. Now, the task is to find the top-k highest scoring documents within the subtree of v. Because the links are essentially edges in individual suffix trees, for any document d_i, there is at most one link whose origin o is within the subtree of v and whose target t is outside the subtree. Moreover, note that if the target t is outside the subtree then it must be one of the ancestors of v. The score of this link is exactly the score(P, d_i). Then the query is: Among all the links, whose origin starts within the subtree of locus v and whose target is outside the subtree of v, find the top-k highest scoring links. The documents are to be output in sorted order of score values. Let f_v be the preorder value of v and l_v be the preorder value of the last node in the subtree of v, then any qualifying link L would satisfy $f_v \leq L.o \leq l_v$ and $L.t \leq f_v$. And then, among all such links, only get k with the highest scores.

Their main idea is that one needs to look for at most p different target values – one for each ancestor of v. In the sorted array $\mathcal{L}$, these links come as at most p different subarray chunks. Moreover, within every target value the links originating from the subtree of v also come contiguously. Let $(l_1, r_1), (l_2, r_2), \ldots, (l_g, r_g)$ with $g < p$ be the intervals of the array $\mathcal{L}$ in which any link L satisfies $f_v \leq L.o \leq l_v$ and $L.t \leq f_v$. We skip here the description of how these intervals are found in $O(p)$ time. Now, the task is to get top-k highest ranking links from these intervals. For this, they construct a Range Maximum Query (RMQ) structure on score values of $\mathcal{L}$. They ap-

ply RMQs over each interval and put these values in a heap. Then they do extract-min from the heap, which at most maintains $O(k)$ elements. If they output an element from (l_a, r_a), the next greatest element from the same interval is put in the heap. They stop when the heap outputs k links. This takes $O(p + k \log k)$ time.

Theorem 2 ([3]) *Given a collection of documents of total length n, we can construct a data structure taking $O(n)$ space, such that top-k document retrieval queries (P, k) can be answered in $O(p + k \log k)$ time.*

Navarro and Nekrich [5] further improved the time to optimal $O(p + k)$. To achieve this, they first change the target attribute of the link to target_depth td. They model the links as two dimensional points (x_i, y_i) with weights w_i as the score. They maintain a global array of these points sorted by their x-coordinates, which are the preorders of origins of the links, while y stands for target_depth. If h is the depth of locus v and v spans the preorders $[a, b]$, then their query is to obtain points in $[a, b] \times [0, h]$ with top-k highest weights. First, they make a basic unit structure for $m \leq n$ points and answer these queries in $O(m^f + k)$ time. Here, $0 < f < 1$ is a constant. This is done by partitioning points by weights. Within each partition, the weights are disregarded. Then they start executing query $Q = [a, b] \times [0, h]$ from the highest partition. If less than $k' < k$ points qualify, then they output these points, change k to $k - k'$ and go to the next partition and so on. At some stage, in some partition more than k points will qualify. One cannot output all of them, and must get only the highest weighted points from that partition. For this, the partitions are further recursively divided into next level partitions. The depth of this recursion is constant and there are at most $O(m^f)$ partitions to be queried and each point is output in constant time. This gives $O(m^f + k)$ time. For sorted reporting, they show how Radix sort can be applied in a constant number of rounds.

Next, with this as a basic unit, they show how to create a layerwise data structure, so that we choose the appropriate layer according to h when the query comes. The parameters of that layer ensure that we get $O(h + k)$ time for the query.

Theorem 3 ([5]) *Given a collection of documents of total length n, we can construct a data structure taking $O(n)$ space, such that top-k document retrieval queries can be answered in $O(p + k)$ time.*

External Memory Document Retrieval

Shah et al. [6] obtained the first external memory results for this problem. In the external memory model, we take B as the block size and we count I/O (input/output) operations as the performance measure of the algorithm. They achieved optimal $O(p/B + \log_B n + k/B)$ I/Os for unsorted retrieval. They take $O(n \log^* n)$ space, which is slightly super linear. We briefly describe the structure here. They first make ranked components of GST. The rank of a node v with subtree size s_v is $\lfloor \log\lceil s_v/B \rceil \rfloor$. Ranked components are the contiguous set of vertices of the same rank. Apart from rank 0 vertices, all other components form downward paths (this is very similar to heavy path decomposition). Now, for links, instead of global array $\mathcal{L}$, they keep the set of links associated with each component. Basically, every link belongs to the component where its target is. They maintain two structures, one is a 3-sided structure in 2D [2] and the other is a 3D dominance structure [1].

The query processing first finds the locus v. Also, the query parameter k is converted into a score threshold τ using sketching structures. We are interested in links such that $f_v \leq L.o \leq l_v$, $L.t \leq f_v$, and $L.s \geq \tau$. These are four constraints. In external memory, three constraints are manageable, but not four. So they decompose the query into those with three constraints. They categorize the answer set into two kinds of links (i) the links whose targets are in the same component as v, and (ii) the links whose targets are in components ranked higher than v. By the property of rank components there are at most $\log n/B$ such higher components. For the second kind of links, we query all the higher-ranked 3-sided (2D) structures [2] with f_v, l_v, τ as the parameters. As long as the links are coming from

the subtree of v, the target values of links need not be checked. For the first kind of links, one cannot drop the target condition, i.e., $L.t \leq f_v$ must be satisfied. However, they show that a slight renumbering of pre-orders based on visiting the child in its own rank component allows condition $L.o \leq l_v$ to be dropped. Such queries are answered by 3D dominance structures. Using this, they obtain $O(p/B + \log^2(n/B) + k/B)$ query I/Os with linear space structure. They further bootstrap this to remove the middle term $\log \log^2(n/B)$ by doubling space requirements. This recursively leads to the following result.

Theorem 4 ([6]) *Given a document collection of size n, we can construct an $O(n \log^* n)$ space structure in external memory, which can answer the top-k document retrieval queries (P, k) in $O(p/B + \log_B n + k/B)$ I/Os. The output is unsorted.*

As a side effect of this result, they also obtain internal memory sorted top-k retrieval in $O(p + k)$ time like [5], and better, just $O(k)$ time if $locus(P)$ is given. This is because the answers come from at most $\log n$ different components. For dominance and 3-sided queries, one can get sorted outputs. And then, atomic heaps are used to merge at most $\log n$ sorted streams. Since atomic heaps only have $O(\log n)$ elements at a time, they can generate each output in $O(1)$ time.

Cross-References

▶ Compressed Document Retrieval on String Collections
▶ Suffix Trees and Arrays

Recommended Reading

1. Afshani P (2008) On dominance reporting in 3d. In: ESA, Karlsruhe, pp 41–51
2. Arge L, Samoladas V, Vitter JS (1999) On two-dimensional indexability and optimal range search indexing. In: PODS, Philadephia, pp 346–357
3. Hon WK, Shah R, Vitter JS (2009) Space-efficient framework for top-k string retrieval problems. In: FOCS, Atlanta, pp 713–722

4. Muthukrishnan S (2002) Efficient algorithms for document retrieval problems. In: SODA, San Francisco, pp 657–666
5. Navarro G, Nekrich Y (2012) Top-k document retrieval in optimal time and linear space. In: SODA, Kyoto, pp 1066–1077
6. Shah R, Sheng C, Thankachan SV, Vitter JS (2013) Top-k document retrieval in external memory. In: ESA, Sophia Antipolis, pp 803–814

Double Partition

Xiaofeng Gao
Department of Computer Science, Shanghai Jiao Tong University, Shanghai, China

Keywords

Approximation; Connected dominating set; Unit disk graph; Wireless network

Years and Authors of Summarized Original Work

2008; Gao, Huang, Zhang, Wu
2009; Huang, Gao, Zhang, Wu

Problem Definition

This problem deals with the design and analysis of a novel approximation algorithm for the minimum weight connected dominating set problem (MWCDS) under unit disk graph (UDG) model. The WCDS is proved to be NP-hard in 1970s, while for a long period researchers could not find a constant-factor approximation until 2006, when Ambühl et al. first introduced an approximation with ratio of 89 under UDG model. Inspired by their subroutines, we proposed a $(10 + \epsilon)$-approximation with double partition technique, which greatly improved the efficiency and effectiveness of the problem.

Given a homogeneous wireless ad hoc network, represented as an undirected graph $G = (V, E)$, where each vertex $v_i \in V$ in the network has the same communication range, we denote it as the *unit* 1 distance. Two nodes v_1, v_2 can communicate with each other when their Euclidean distance is smaller than 1, and correspondingly, the edge set $E = \{(v_i, v_j) \mid \text{dist}(v_i, v_j) \leq 1\}$. If v_i and v_j are connected, then we say that v_j is a neighbor of v_i (and vice versa). Such communication model is named as *unit disk graph* (UDG). Additionally, each vertex v_i has a weight w_i.

Objective

We hope to find a connected dominating subset $U \subseteq V$ in the given graph with the minimum weight, such that each vertex $v_i \in V$ is either in U or has a neighbor in U and the induced graph $G[U]$ is connected. In addition, the weight of U (defined as the sum of weights for elements in U, say, $W(U) = \sum_{v_i \in U} w_i$) is the minimum among all connected dominating subsets satisfying the above requirements.

Constraints

1. **Unit disk graph:** We restrict our discussion on two-dimensional space where each vertex has the same communication range, and edges between vertices are constructed according to the distance constraint.
2. **Weight minimization:** We focus on the weight version of minimum connected dominating set problem, which is much harder than the cardinality version, and thus providing a constant-factor approximation seems more difficult.

Problem 1 (Minimum Weight Connected Dominating Set in Unit Disk Graph)
INPUT: *A unit disk graph $G = (V, E)$ and a weight assigned on each vertex*
OUTPUT: *A minimum weight connected dominating vertex subset $U \subset V$ such that (1) wirelength is minimized and (2) the area-density constraints $D_{ij} \leq K$ are satisfied for all $B_{ij} \in B$*

Key Results

The minimum weight connected dominating set problem (MWCDS) can be divided into two parts: selecting a minimum weight dominating set (MWDS) and connecting the dominating set into a connected dominating set. In this chapter, we will focus on the former part, while the latter part is equivalent to solving a node-weighted Steiner tree problem in unit disk graphs.

The first constant-factor approximation algorithm for MWCDS under UDG was proposed by Ambühl C. et al. in 2006 [1], which is a polynomial-time 89-approximation. Later, Gao X. et al. [2] introduced a better approximation scheme with approximation ratio of $(6 + \epsilon)$ for MWDS, and Huang Y. et al. [3] further extended this idea to MWCDS with approximation ratio of $(10 + \epsilon)$. The main idea of their methods involves a double partition technique and a shifting strategy to reduce the redundant vertices selected through the algorithms.

In recent year, the approximation for MWDS in UDG received further improvements from $(6 + \epsilon)$ to 5 by Dai and Yu [4], to 4 by Erlebach T. et al. [5] and Zou F. et al. [6] independently, and to 3.63 by Willson J. et al. [7]. Meanwhile, to connect the dominating set in UDG, Ambühl C. et al. [1] gave a 12-approximation, Huang Y. et al. [3] provided a 4-approximation, and Zou F. et al. [8] constructed a 2.5ρ-approximation with a known ρ-approximation algorithm for the minimum network Steiner tree problem. Recently, the minimum approximation for network Steiner tree problem has an approximation ratio of 1.39 [9], so the best approximation ratio for MWCDS problem in UDG is 7.105 up to now.

Double Partition Technique

Given a UDG G containing n disks in the plane. Let $\mu < \frac{\sqrt{2}}{2}$ be a real number which is sufficiently close to $\frac{\sqrt{2}}{2}$, say, $\mu = 0.7$. Partition the area into squares with side length μ. If the whole area has boundary $P(n) \times Q(n)$, where $P(n)$ and $Q(n)$ are two polynomial functions on

n, then given the integer even constant K and letting $K \times K$ squares form a block, our partition will have at most $\left(\lceil \frac{P(n)}{K} \rceil + 1\right) \times \left(\lceil \frac{Q(n)}{K} \rceil + 1\right)$ blocks. We will discuss algorithm to compute MWDS for each block firstly and then combine them together.

MWDS in $K \times K$ Squares

Assume each block B has K^2 squares S_{ij}, for $i, j \in \{0, 1, \ldots, K-1\}$. Let V_{ij} be the set of disks in S_{ij}. If we have a dominating set D for this block, then for each square S_{ij}, its corresponding dominating set is (1) either a disk from inside S_{ij} (since $\text{dist}(d, d') \leq 1$ for any two disks within this square) or (2) a group of disks from neighbor region around S_{ij}, the union of which can cover all disk centers inside the square. Then if we want to select a minimum weight dominating set, for each square we will have two choices. However, instead of selecting dominating sets square by square, we hope to select them strip by strip to avoid repeated computation for some disks. For this purpose, we have the following lemmas.

Lemma 1 ([1]) *Let P be a set of points located in a strip between lines $y = y_1$ and $y = y_2$ for some $y_1 < y_2$. Let D be a set of weighted disks with uniform radius whose centers are above the line $y = y_2$ or below the line $y = y_1$. Furthermore, assume that the union of the disks in D covers all points in P. Then a minimum weight subset of D that covers all points in P can be computed in polynomial time.*

The proof of result for Lemma 1 is in fact constructive. It gives a polynomial-time algorithm by a dynamic programming. It says that as long as the set of centers P in a horizontal strip can be dominated by a set of centers D above and/or below the strip, then an optimal subset of D dominating P can be found in polynomial time.

Our next work is to select some disks for each square within a strip so that those disks can be covered by disks from the upper and lower strips. To better illuminate the strategy, we divide the neighbor parts of S_{ij} into eight regions $UL, UM, \ UR, CL, CR, LL, LM, and LR$ as shown in Fig. 1. The four lines forming S_{ij} are $x = x_1$, $x = x_2$, $y = y_1$, and $y = y_2$.

Double Partition, Fig. 1 S_{ij} and its neighbor regions

Denote by $Left = UL \cup CL \cup LL$, $Right = UR \cup CR \cup LR$, $Up = UL \cup UM \cup UR$, and $Down = LL \cup LM \cup LR$. After that, we will have Lemma 2.

Lemma 2 *Suppose* $p \in V_{ij}$ *is a disk in* S_{ij} *which can be dominated by a disk* $d \in LM$. *We draw two lines* p_l *and* p_r, *which intersect* $y = y_1$ *by angle* $\frac{\pi}{4}$ *and* $\frac{3\pi}{4}$. *Then the shadow* P_{LM} *surrounded by* $x = x_1$, $x = x_2$, $y = y_1$, p_l, *and* p_r *(shown in Fig. 2) can also be dominated by* d. *Similar results can be held for shadow* P_{UM}, P_{CL}, *and* P_{CR}, *which can be defined with a rotation.*

Proof We split shadow P_{LM} into two halves with vertical line $x = x_p$, where x_p is x-coordinate of disk p. Then we prove that the right half of P_{LM} can be covered by d. The left half can be proved symmetrically. Let o be intersection point of $x = x_p$ and $y = y_1$, a that of p_r and $x = x_2$ (or p_r and $y = y_1$), and b that of $x = x_2$ and $y = y_1$. Intuitively, the right half can be either a quadrangle $pabo$ or a triangle pao. We will prove both cases as follows:

Quadrangle case: Draw the perpendicular line of the line segment pa, namely, p_m. When d is under p_m as in Fig. 3a, we will have $\text{dist}(d, a) \leq \text{dist}(p, d) \leq 1$. Moreover, it is trivial that $\text{dist}(d, o)$ and $\text{dist}(d, b)$ are all < 1. Thus, d can cover the whole quadrangle. When d is above the line p_m as in Fig. 3b, we draw an auxiliary line $y = y_a$ parallel with $y = y_1$ and

$x = x_d$ intersecting $y = y_a$ at point c. Since d lies above p_m, $\angle cad \leq \pi/4$, and thus

$$\text{dist}(d, a) = \frac{\text{dist}(c, a)}{\cos \angle cad} < \frac{\sqrt{2}/2}{\cos \frac{\pi}{4}} = 1.$$

Note that both $\text{dist}(d, o)$ and $\text{dist}(d, b)$ are less than 1, so d can cover the whole quadrangle.

Triangle case: Similarly, draw p_m as described above. The proof remains when d is under p_m (see Fig. 3c). When d is above p_m as in Fig. 3d, we draw auxiliary line $x = x_d$ intersecting $y = y_1$ at c. Then we will get the same conclusion.

With the help of Lemma 2, we can select a region from S_{ij}, where the disks inside this region can be covered by disks from Up and $Down$ neighbor area. We name this region as "sandglass," with formal definition as follows:

Definition 1 (Sandglass) If D is a dominating set for square S_{ij} and $D \cap V_{ij} = \emptyset$, then there exists a subset $V_M \subseteq V_{ij}$ which can only be covered by disks from UM and LM (we can set $V_M = \emptyset$ if there are no such disks). Choose $V_{LM} \subseteq V_M$ the disks that can be covered by disks from LM, and draw p_l and p_r line for each $p \in V_{LM}$. Choose the leftmost p_l and rightmost p_r and form a shadow similar as that in Lemma 2. Symmetrically, choose V_{UM} and form a shadow with leftmost and rightmost lines. The union of the two shadows form a "sandglass" region Sand_{ij} of S_{ij} (see Fig. 4a, where solid circle represents V_{LM}, while hollow circle represents V_{UM}). Fig. 4b–4d gives other possible shapes of Sand_{ij}.

Lemma 3 *Suppose D is a dominating set for S_{ij} and Sand_{ij}s are chosen in the above way. Then any disks in Sand_{ij} can be dominated by disks only from neighbor region $Up \cup Down$, and disks from $S_{ij} \setminus \text{Sand}_{ij}$ can be dominated by disks only from neighbor region $Left \cup Right$.*

Proof Suppose to the contrary, there exists a disk $d \in \text{Sand}_{ij}$ which cannot be dominated by disks from $Up \cup Down$. Since D is a dominating set, there must be a $d' \in CL \cup CR$ which dominates

Double Partition, Fig. 2
Different shapes for
shadow P_{LM}

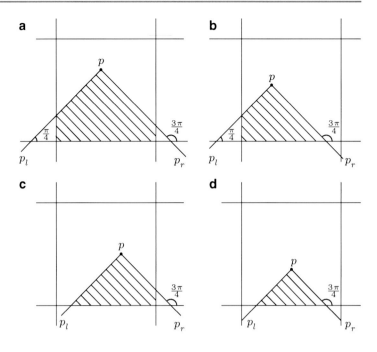

Double Partition, Fig. 3
Shape of shadow and
location of d. (**a**) d to left
of p_m. (**b**) d to left of p_m.
(**c**) d to left of p_m. (**d**) d
to left of p_m

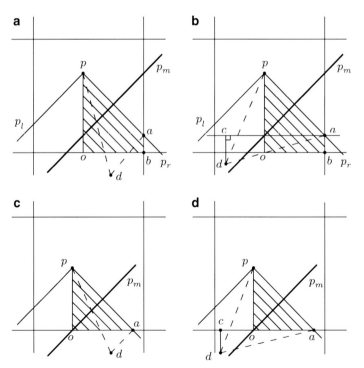

d. Without loss of generality, assume d belongs
to lower half of the sandglass which is formed by
p_1 and p_2, and let $d' \in CL$ (see Fig. 5). Based on
our assumption, d cannot locate in p_1's triangle
shadow to *Down* region (otherwise, since p_1 can

be dominated by a disk from LM, d can also
be dominated by this disk). We then draw d_l
and d_r to CL region and form a shadow to CL.
Then by Lemma 2 every disk from this shadow
can be dominated by d'. Obviously, p_1 belongs

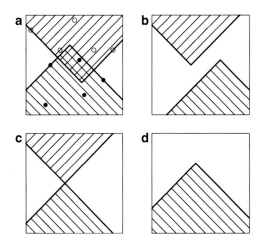

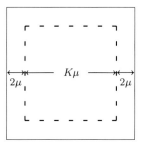

Double Partition, Fig. 6 Block selection

Double Partition, Fig. 4 Sandglass Sand$_{ij}$ for S_{ij}. (**a**) Form of sandglass. (**b**) Sandglass without intersection. (**c**) Sandglass with single disk. (**d**) Sandglass with half side

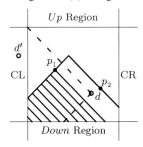

Double Partition, Fig. 5 Proof for sandglass

Algorithm 1 Calculate MWDS in $K \times K$ squares

Input: $K \times K$ squares with inner disks
Output: A local MWDS
1: For each S_{ij}, choose its sandglass or randomly select $d \in S_{ij}$.
2: If $d \in S_{ij}$ is selected, then remove d and all disks dominated by d.
3: For each strip $\cup_{j=1}^{K} S_{ij}$ from $i = 1$ to K, calculate a dominating set for the union of disks in the sandglasses.
4: For each strip $\cup_{i=1}^{K} S_{ij}$ from $j = 1$ to K, calculate a dominating set for the remaining disks not covered by Step 3.
Return the union of disks chosen in the above steps for $K \times K$ squares.

---

to this region, but p_1 is a disk which cannot be dominated by disks from CL, a contradiction.

Till now we already find "sandglass" region in which disks can be dominated by disks only from Up and $Down$ regions. In our algorithm, for each square S_{ij}, we can firstly decide whether to choose a disk inside this square as dominating set or to choose a dominating set from its neighbor region. If we choose the latter case, the algorithm will randomly select 4 disks d_1, d_2, d_3, and d_4 from S_{ij} and make corresponding sandglass (we can also choose less than 4 disks to form the sandglass). By enumeration of all possible sandglasses including the case of choosing one disk inside the square, for all squares within $K \times K$ area, there are at most $[\sum_{i=0}^{4} C_n^i \cdot 2^i]^{K^2}$ choices (n is the number of disks), which can be calculated within polynomial time. Moreover, when considering choosing a dominating set from neighbor

regions, we should also include regions around this $K \times K$ areas such that we will not miss disks outside the region. Therefore, we should consider $(K + 4) \times (K + 4)$ area, where the inner region is our selected block and the surrounding four strips are the assistance (shown as Fig. 6).

In all, we will have Algorithm 1 with four steps to calculate an MWDS for $K \times K$ squares. We enumerate all possible cases for each S_{ij} and choose the solution with minimum weight, which forms an MWDS for S_{ij}.

MWDC for the Whole Region

As discussed above, if our plane has size $P(n) \times Q(n)$, then there are at most $(\lceil \frac{P(n)}{K} \rceil + 1) \times (\lceil \frac{Q(n)}{K} \rceil + 1)$ blocks in the plane. We name each block B^{xy}, where $0 \leq x \leq \lceil \frac{P(n)}{K} \rceil + 1$ and $0 \leq y \leq \lceil \frac{Q(n)}{K} \rceil + 1$. Then, using Algorithm 1 to calculate dominating set for each block and by combining them together, we obtain a dominating set for our original partition.

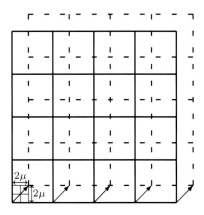

Double Partition, Fig. 7 Move blocks

Algorithm 2 Calculate MWDS for the whole plane

Input: G in region $P(n) \times Q(n)$
Output: A global MWDS
1: For a certain partition, calculate MWDS for each block B^{xy}, sum the weight of MWSD for each block, and form a solution.
2: Move each block to two squares to the right and two squares to the top of the original block.
3: Repeat Step 1 for new partition, and get a new solution.
4: Repeat Step 2 for $\lceil \frac{K}{2} \rceil$ times, and choose the minimum solution among those steps.
   Return the solution from Step 4 as our final result.

Next, we move our blocks to different positions by shifting policy. Move every block two squares right and two squares up to its original position, which can be seen from Fig. 7. Then calculate dominating set for each block again, and combine the solution together. We do this process $\frac{K}{2}$ times, choose the minimum solution as our final result. The whole process can be shown as Algorithm 2.

Performance Ratio

In the following, we extend our terminology "dominate" to points (a point is a location which is not necessarily a disk). A point p is *dominated* by a set of disks if the distance between p and at least one center of the disks is not more than 1. We say an area is *dominated* by a set of disks if every point in this area is dominated by the set of disks. Let OPT be optimal solution for

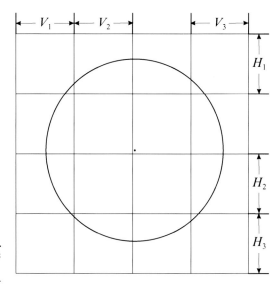

Double Partition, Fig. 8 An example for disk cover region

our problem and $w(OPT)$ the weight of optimal solution.

Theorem 1 *Algorithm 2 always outputs a dominating set with weight within $6 + \varepsilon$ times of the optimum one.*

Proof Our proof mainly has two phases. The first phase analyzes that our Algorithm 1 gives a 6-approximation for disks in $K \times K$ squares. The second phase proves that result from Algorithm 2 is less than $(6 + \varepsilon) \cdot w(OPT)$.

Phase 1: If a disk has radius 2 and our partition has side length $\mu < \frac{\sqrt{2}}{2}$, then a disk may dominate disks from at most 16 squares, which can be shown in Fig. 8. Simply, if a disk in OPT is used to dominate the square it belongs to, then we will remove this disk before calculating MWDS for strips. Therefore, it will be used only once. If a disk is not used to dominate the square containing it, then it may be used 3 times in calculating its 3 horizontal neighbor strips (H_1, H_2, and H_3 as shown in Fig. 8) and another 3 times in calculating its 3 vertical neighbor strips (V_1, V_2, and V_3 in Fig. 8). Therefore, Algorithm 1 is a 6-approximation for each block.

Phase 2: Now we consider the disks in side strips for a block. As discussed above, when calculating MWDS for a strip, we may use disks within $(K + 2) \times (K + 2)$ squares. Therefore, we can divide a block $B^{(xy)}$ into three kinds of squares, just as shown in Fig. 9 ($0 \leq x \leq \lceil \frac{P(n)}{K} \rceil$, and $0 \leq y \leq \lceil \frac{Q(n)}{K} \rceil$). If a disk belongs to inner part $\mathcal{A}$ of $B^{(xy)}$, it will be used at most 6 times during calculating process. We name those disks as d_{in}. If a disk belongs to side part $\mathcal{B}$ of $B^{(xy)}$, it may be used at most 5 times for calculating $B^{(xy)}$, but it may used at most 4 times when calculating $B^{(xy)}$'s neighbor block. We name those disks as d_{side}. If a disk belongs to corner squares $\mathcal{C}$ of $B^{(xy)}$, it may be used at most 4 times for calculating $B^{(xy)}$ and at most 8 times for neighbor blocks. We name those disks as d_{corner}. In addition, we know that during shifting process a node can stay at most 4 times in side or corner square. If we name l as the lth shifting, then our final solution will have the following inequality:

$$W(\text{Solution})$$

$$= \min_l \left\{ \sum_{\text{Sol}_l} [6w(d_{in}^l) + 9w(d_{\text{side}}^l) + 12w(d_{\text{corner}}^l)] \right\}$$

$$\leq \frac{1}{\frac{K}{2}} \sum_{l=0}^{\frac{K}{2}} \left\{ 6w(d_{in}^l) + 4 \cdot 12w(d_{\text{side}}^l + d_{\text{corner}}^l) \right\}$$

$$= 6w(OPT) + \frac{42}{\frac{K}{2}} w(OPT)$$

$$\leq (6 + \varepsilon)w(OPT)$$

where $\varepsilon = 42 / \frac{K}{2}$ can be arbitrarily small when K is sufficiently large.

Applications

Dominating set problem is widely used in network-related applications. For instance, in mobile and wireless ad hoc networks, it is implemented for communication virtual backbone selection to improve routing efficiency, for sensor coverage problem to extend network lifetime, and for clustering and data gathering problem to avoid flooding and energy waste. In optical network and data center networks, it is used for network management and switch-centric routing protocols. In social network applications, it is used for many cluster-related problems like positive influence, effective leader

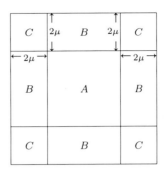

Double Partition, Fig. 9 Divide block $B^{(xy)}$ into 3 parts

group, etc. A weighted dominating set is a generalized heterogeneous network model to describe real-world applications, which is more realistic and practical.

Open Problems

There are two open problems for the minimum weighted connected dominating set (MWCDS) problem under unit disk graph (UDG). Firstly, there is another grid partition design to construct a constant approximation for domatic partition problem in unit disk graph [10]. Could this new technique result in an improvement in running time or performance ratio? Secondly, does MWCDS problem in UDG have a polynomial-time approximation scheme (PTAS)? Currently, no one has answered the above questions.

Cross-References

▶ Minimum Connected Sensor Cover

Recommended Reading

1. Ambühl C, Erlebach T, Mihalák M, Nunkesser M (2006) Constant-factor approximation for minimum weight (connected) dominating sets in unit disk graphs. In: Proceedings of the 9th international workshop on approximation algorithms for combinatorial optimization problems (APPROX 2006), Barcelona, 28–30 Aug 2006, pp 3–14
2. Gao X, Huang Y, Zhang Z, Wu W (2008) (6 + ϵ)-approximation for minimum weight dominating set in unit disk graphs. In: Proceedings of the 14th annual international conference on computing and combinatorics (COCOON), Dalian, 27–29 June 2008, pp 551–557

3. Huang Y, Gao X, Zhang Z, Wu W (2009) A better constant-factor approximation for weighted dominating set in unit disk graph. J Comb Optim 18(2):179–194

4. Dai D, Yu C (2009) A $(5 + \epsilon)$-approximation algorithm for minimum weight dominating set in unit disk graph. Theor Comput Sci 410(8–10):756–765

5. Erlebach T, Mihalák M (2009) A $(4 + \epsilon)$-approximation for the minimum weight dominating set problem in unit disk graph. In: Proceedings of the 7th workshop on approximation and online algorithms (WAOA), Copenhagen, 10–11 Sept 2009, pp 135–146

6. Zou F, Wang Y, Xu X, Du H, Li X, Wan P, Wu W (2011) New approximation for weighted dominating sets and connected dominating sets in unit disk graphs. Theor Comput Sci 412(3):198–208

7. Willson J, Wu W, Wu L, Ding L, Du D-Z (2014) New approximations for maximum lifetime coverage. Optim J Math Program oper Res. doi:10.1080/02331934.2014.883507

8. Zou F, Li X, Gao S, Wu W (2009) Node-weighted Steiner tree approximation in unit disk graphs. J Comb Optim 18(4):342–349

9. Byrka J, Grandoni F, Rothvoss T, Sanita L (2010) An improved LP-based approximation for Steiner tree. In: Proceedings of the 42th ACM symposium on theory of computing (STOC 2010), Cambridge, 6–8 June 2010, pp 583–592

10. Pandit S, Pemmaraju S, Varadarajan K (2009) Approximation algorithms for domatic partitions of unit disk graphs. In: Proceedings of the 12th international workshop on approximation algorithms for combinatorial optimization problems (APPROX 2009), Berkeley, 21–23 Aug 2009, pp 312–325

Dynamic Approximate All-Pairs Shortest Paths: Breaking the *O(mn)* Barrier and Derandomization

Monika Henzinger[1], Sebastian Krinninger[2], and Danupon Nanongkai[3]
[1] University of Vienna, Vienna, Austria
[2] Faculty of Computer Science, University of Vienna, Vienna, Austria
[3] School of Computer Science and Communication, KTH Royal Institute of Technology, Stockholm, Sweden

Keywords

Approximation algorithms; Data structures; Derandomization; Dynamic graph algorithms

Years and Authors of Summarized Original Work

2013; Henzinger, Krinninger, Nanongkai

Problem Definition

Given an undirected, unweighted graph with n nodes and m edges that is modified by a sequence of edge insertions and deletions, the problem is to maintain a data structure that quickly answers queries that ask for the length $d(u, v)$ of the shortest path between two arbitrary nodes u and v in the graph, called the *distance* of u and v. The fastest *exact* algorithm for this problem is randomized and takes amortized $O\left(n^2 \left(\log n + \log^2 ((m + n)/n)\right)\right)$ time per update and constant query time [6, 11]. In the *decremental* case, i.e., if only edge deletions are allowed, there exists a *deterministic* algorithm with amortized time $O(n^2)$ per deletion [7]. More precisely, its *total update time* for a sequence of up to m deletions is $O(mn^2)$. Additionally, there is a randomized algorithm with $O(n^3 \log^2 n)$ total update time and constant query time [1]. However, in the decremental case, when only α-*approximate* answers are required, i.e., when it suffices to output an estimate $\delta(u, v)$ such that $d(u, v) \leq \delta(u, v) \leq \alpha d(u, v)$ for all nodes u and v, the total update time can be significantly improved: Let $\epsilon > 0$ be a small constant. The fastest prior work was a class of randomized algorithms with total update time $\tilde{O}(mn)$ for $\alpha = 1 + \epsilon$ [10], $\tilde{O}\left(n^{5/2+O(1/\sqrt{\log n})}\right)$ for $\alpha = 3 + \epsilon$, and $\tilde{O}\left(n^{2+1/k+O(1/\sqrt{\log n})}\right)$ for $\alpha = 2k - 1 + \epsilon$ [4].

This leads to the question whether for $\alpha = 1 + \epsilon$ (a), a total update time of $o(nm)$ is possible and (b) a deterministic algorithm with total update time $\tilde{O}(nm)$ exists.

As pointed out in [3] and several other places, a *deterministic* algorithm is interesting due to the fact that deterministic algorithms can deal with an *adaptive offline adversary* (the strongest adversary model in online computation [2, 5]), while the randomized algorithms developed so far assume an *oblivious adversary* (the weakest adversary model) where the order of edge deletions must be fixed before an algorithm makes random choices.

Key Results

The paper of Henzinger, Krinninger, and Nanongkai [8] presents two algorithms for $\alpha = 1 + \epsilon$. The first one is a deterministic algorithm with total update time $\tilde{O}(mn)$. The second one studies a slightly relaxed version of the problem: Given a constant β, let $\delta(u, v)$ be an (α, β)-*approximation* if $d(u, v) \leq \delta(u, v) \leq \alpha d(u, v) + \beta$ for all nodes u and v. The second algorithm is a randomized algorithm with total update time $\tilde{O}(n^{5/2})$ that can guarantee both a $(1 + \epsilon, 2)$ and a $(2 + \epsilon, 0)$ approximation.

The results build on two prior techniques, namely, an exact decremental single-source shortest path data structure [7], called *ES-tree*, and the $(1 + \epsilon, 0)$-approximation algorithm of [10], called *RZ-algorithm*. The RZ-algorithm chooses for all integer i with $1 \leq i \leq \log n$, $\tilde{O}(n/(\epsilon 2^i))$ random nodes as *centers*, and maintains an ES-tree up to distance 2^{i+2} for each center. For correctness, it exploits the fact that the random choice of centers guarantees the following *invariant (I)*: For every pair of nodes u and v with distance $d(u, v)$, there exists with high probability a center c such that $d(u, c) \leq \epsilon d(u, v)$ and $d(c, v) \leq d(u, v)$. The total update time per center is $O(m2^i)$ resulting in a total update time of $\tilde{O}(mn)$. The deterministic algorithm of [8] derandomizes this algorithm by initially choosing centers fulfilling invariant (I) and after each update (a) greedily generating new centers to guarantee that (I) continues to hold and (b) moving the root of the existing ES-trees. To achieve a running time of $\tilde{O}(mn)$, the algorithm is not allowed to create more than $\tilde{O}(n/(\epsilon 2^i))$ many centers for each i. This condition is fulfilled by dynamically assigning each center a set of $\Omega(2^i)$ vertices such that no vertex is assigned to two centers.

The improved randomized algorithm uses the idea of an *emulator*, a sparser *weighted* graph that approximates the distances of the original graph. Emulators were used for dynamic shortest-path algorithms before [4]. The challenge when using an emulator is that edge deletions in the original graph might lead to edge deletions, edge insertions, or weight increases in the emulator, requiring in principle the use of a fully dynamic shortest-path algorithm on the emulator. Bernstein and Roditty [4] deal with this challenge by using an emulator where the number of *distance changes* between any two nodes can be bounded. However, the RZ-algorithm requires that the number of times that the distance between any two nodes changes is at most R before that distance exceeds R for any integer R with $1 \leq R \leq n$. As the emulator used by Bernstein and Roditty does not fulfill this property, they cannot run the RZ-algorithm on it. The new algorithm does not construct such an emulator either. Instead, it builds an emulator where the error introduced by edge insertions is limited and runs the RZ-algorithm with modified ES-trees, called *monotone ES-trees*, on this emulator. The analysis exploits the fact that the distance between any two nodes in the original graph can only *increase* after an edge deletion. Thus, even if an edge deletion leads to changes in the emulator that *decrease* their distance in the emulator, the corresponding ES-trees do not have to be updated, i.e., the distance of a vertex to its root in the ES-tree *never* decreases. The analysis shows that the error introduced through the use of monotone ES-trees in the RZ-algorithm is small so that the claimed approximation ratio is achieved. However, since the ES-trees are run on the sparse emulator the overall running time is $o(mn)$.

Open Problems

The main open problem is to find a similarly efficient algorithm in the *fully dynamic setting*, where both edge insertions and deletions are allowed. A further open problem is to extend the derandomization technique to the *exact* algorithm of [1].

Another challenge is to obtain similar results for *weighted, directed* graphs. We recently extended some of the above techniques to weighted, directed graphs and presented a randomized algorithm with $\tilde{O}\left(mn^{0.986}\right)$ total update time for $(1 + \epsilon)$-approximate *single-source* shortest paths [9].

Cross-References

▶ Decremental All-Pairs Shortest Paths
▶ Fully Dynamic All Pairs Shortest Paths

Recommended Reading

1. Baswana S, Hariharan R, Sen S (2007) Improved decremental algorithms for maintaining transitive closure and all-pairs shortest paths. J Algorithms 62(2):74–92. Announced at STOC, 2002
2. Ben-David S, Borodin A, Karp RM, Tardos G, Wigderson A (1994) On the power of randomization in on-line algorithms. Algorithmica 11(1):2–14. Announced at STOC, 1990
3. Bernstein A (2013) Maintaining shortest paths under deletions in weighted directed graphs. In: STOC, Palo Alto, pp 725–734
4. Bernstein A, Roditty L (2011) Improved dynamic algorithms for maintaining approximate shortest paths under deletions. In: SODA, San Francisco, pp 1355–1365
5. Borodin A, El-Yaniv R (1998) Online computation and competitive analysis. Cambridge University Press, Cambridge
6. Demetrescu C, Italiano GF (2004) A new approach to dynamic all pairs shortest paths. J ACM 51(6):968–992. Announced at STOC, 2003
7. Even S, Shiloach Y (1981) An on-line edge-deletion problem. J ACM 28(1):1–4
8. Henzinger M, Krinninger S, Nanongkai D (2013) Dynamic approximate all-pairs shortest paths: breaking the $O(mn)$ barrier and derandomization. In: FOCS, Berkeley
9. Henzinger M, Krinninger S, Nanongkai D (2014) Sublinear-time decremental algorithms for single-source reachability and shortest paths on directed graphs. In: STOC, New York
10. Roditty L, Zwick U (2012) Dynamic approximate all-pairs shortest paths in undirected graphs. SIAM J Comput 41(3):670–683. Announced at FOCS, 2004
11. Thorup M (2004) Fully-dynamic all-pairs shortest paths: faster and allowing negative cycles. In: SWAT, Humlebæk, pp 384–396

Dynamic Approximate-APSP

Aaron Bernstein
Department of Computer Science, Columbia University, New York, NY, USA

Keywords

Dynamic graph algorithms; Hop distances; Shortest paths

Years and Authors of Summarized Original Work

2009; Bernstein

Problem Definition

A dynamic graph algorithm maintains information about a graph that is changing over time. Given a property $\mathcal{P}$ of the graph (e.g., maximum matching), the algorithm must support an online sequence of query and update operations, where an update operation changes the underlying graph, while a query operation asks for the state of $\mathcal{P}$ in the current graph. In the typical model studied, each update affects a single edge, in which case the most general setting is the *fully dynamic* one, where an update can either insert an edge, delete an edge, or change the weight of an edge. Common restrictions of this include the *decremental* setting, where an update can only delete an edge or increase a weight, and the *incremental* setting where an update can insert an edge or decrease a weight.

This entry addresses the problem of maintaining α-approximate all-pairs shortest paths (APSP) in the fully dynamic setting in a weighted, undirected graph (the approximation factor α depends on the algorithm); the goal is to maintain an undirected graph G with real-valued nonnegative edge weights under an online intermixed sequence of the following operations:

- **delete**(u, v) (update): remove edge (u, v) from G.

- **insert**(u, v) (update): insert an edge (u, v) into G.
- **change weight**(u, v, w) (update): change the weight of edge (u, v) to w.
- **distance**(u, v) (query): return an α-approximation to the shortest $u - v$ distance in G.
- **path**(u, v) (query): return an α-approximate shortest path from u to v.

Approaches

The naive approach to the fully dynamic APSP problem is to recompute shortest paths from scratch after every update, allowing queries to be answered in optimal time. Letting n be the number of vertices and m the number of edges, computing APSP requires $O(mn + n^2 \log\log(n))$ time in sparse graphs [8] or slightly less than n^3 in dense graphs [9, 13]. If we allow approximation, a slightly better approach would be to construct an *approximate distance oracle* after each update, i.e., a static data structure for answering approximate distance queries quickly; an oracle for returning k-approximate distances ($k \geq 3$) can be constructed in time $O(\min\{n^2 \log(n), kmn^{1/k}\})$ [1, 11]. Another simple-minded approach would be to not perform any work during the updates and to simply compute the shortest $u-v$ path from scratch when a query arrived; using Dijkstra's algorithm with Fibonacci heaps [7], this would lead to a constant update time and a query time of $O(m + n \log(n))$.

The goal of a dynamic algorithm is to improve upon the above approaches by taking advantage of the fact that each update only affects a single edge, so one can reuse information between updates and thus avoid recomputing from scratch. In a breakthrough result, Demetrescu and Italiano showed that in the most general case of a directed graph with arbitrary real weights, one can answer updates in amortized time $O(n^2 \log^3(n))$ while maintaining optimal $O(1)$ time for distance queries [6]; Thorup improved the update time slightly to $O(n^2 (\log(n) + \log^2((m+n)/n)))$ [10]. This entry addresses a recent result of Bernstein [3] which shows that in undirected graphs, one can significantly improve upon this n^2 update time by settling for approximate distances.

Key Results

Bernstein's paper starts by showing that assuming integral weights, there is an algorithm for maintaining $(2 + \epsilon)$-approximate APSP *up to a bounded distance d* with amortized update time $O(md)$. This is efficient for small distances, but in the general case d can be very large, especially in weighted graphs. Bernstein overcomes this by introducing a high-level framework for extending shortest path-related algorithms that are efficient for small distances to ones that are efficient for general graphs; he later applied this approach to two other results [4, 5].

The first step of the approach is to show that with simple scaling techniques, an algorithm that is efficient for small (weighted) distances can be extended to an algorithm that is efficient for shortest paths with few edges (regardless of weight). Applying this technique to the above algorithm yields the following result:

Definition 1 Let the *hop distance* of a path be the number of edges it contains. A graph G is said to have *approximate hop diameter h* if for every pair of vertices (x, y), there is a $(1 + \epsilon)$-approximate shortest $x - y$ path with hop distance $\leq h$.

Theorem 1 ([3]) *Let G be an undirected graph with nonnegative real edge weights, and let R be the ratio of the heaviest to the lightest nonzero weight. One can maintain $(2 + \epsilon)$-APSP in the fully dynamic setting with amortized update time $O(mh \log(nR))$, where h is the approximate hop diameter of the graph.*

Shortcut Edges

Theorem 1 provides an efficient algorithm for small h, but on its own a result that is efficient for small hop diameter is not particularly powerful as even in unweighted graphs h can be $\Omega(n)$. The second step of Bernstein's approach is to show that regardless of whether the original graph is weighted, one can add weighted edges to reduce the hop diameter. A *shortcut edge* (x, y) is a new edge constructed by the algorithm that has weight $w(x, y)$ with $\delta(x, y) \leq w(x, y) \leq (1+\epsilon)\delta(x, y)$, where $\delta(x, y)$ is the shortest $x - y$ distance. It is clear that because shortcut weights are tethered to shortest distances, they do not change (weighted) distances in the graph. But a shortcut edge can greatly reduce hop distances; for example, in an unweighted graph where $\delta(x, y) = 1000$, adding a single shortcut edge (x, y) of weight 1000 decreases the $x - y$ hop distance to 1 while also decreasing the hop distance of paths that go through x and y. Bernstein adapts techniques from spanner and emulator theory (see in particular Thorup and Zwick's result on graph sparsification [12]) to show that in fact a small number of shortcut edges suffice to greatly reduce the hop diameter of a graph.

Theorem 2 ([3]) *Let G be an undirected graph with nonnegative real edge weights, and let R be the ratio of the heaviest to the lightest edge weight in the graph. There exists an algorithm that in time $O(m \cdot n^{O(1/\sqrt{\log(n)})} \cdot \log(nR))$ constructs a set S of $O(n^{1+O(1/\sqrt{\log(n)})} \cdot \log(nR))$ shortcut edges such that adding S to the edges of the graph reduces the approximate hop diameter to $n^{O(1/\sqrt{\log(n)})}$.*

Theorems 1 and 2 combined encapsulate the approach of Bernstein's algorithm: take an algorithm that works well for small distances, use scaling to transform it into an algorithm that is efficient for graphs of small hop diameter, and then add shortcut edges to the original graph to ensure a small hop diameter. For dynamic APSP, there are additional complications that arise from edges being inserted and deleted over time, but the basic approach remains the same.

Theorem 3 ([3]) *Let G be an undirected graph with real nonnegative edge weights, and let R be the ratio of the maximum edge weight appearing in the graph during any point in the update sequence to the minimum nonzero edge weight. There is an algorithm that maintains fully dynamic $(2 + \epsilon)$-approximate APSP in amortized update time $O(m \cdot n^{O(1/\sqrt{\log(n)})} \cdot \log(nR))$ and can answer distance queries in worst-case time $O(\log \log \log(n))$.*

Open Problems

The main open problem for fully dynamic approximate APSP is to develop an efficient algorithm for maintaining $(1 + \epsilon)$ approximate distances, possibly with a small additive error in the unweighted case. Another interesting problem would be to achieve $o(n^2)$ update times for dense graphs – this can already be done to some extent by combining the result of Bernstein discussed here with the fully dynamic spanner of Baswana et al. [2], but only for unweighted graphs and at the cost of a much worse approximation ratio. Other open problems include removing the dependence on $\log(R)$ and developing an efficient *deterministic* algorithm for the problem.

Cross-References

▶ Decremental All-Pairs Shortest Paths
▶ Decremental Approximate-APSP in Directed Graphs
▶ Dynamic Approximate All-Pairs Shortest Paths: Breaking the $O(mn)$ Barrier and Derandomization
▶ Trade-Offs for Dynamic Graph Problems

Recommended Reading

1. Baswana S, Sen S (2006) Approximate distance oracles for unweighted graphs in expected $o(n^2)$ time. ACM Trans Algorithms 2(4):557–577
2. Baswana S, Khurana S, Sarkar S (2012) Fully dynamic randomized algorithms for graph spanners. ACM Trans Algorithms 8(4):35
3. Bernstein A (2009) Fully dynamic approximate all-pairs shortest paths with query and close to linear update time. In: Proceedings of the 50th FOCS, Atlanta, pp 50–60
4. Bernstein A (2012) Near linear time $(1 + \epsilon)$-approximation for restricted shortest paths in undirected graphs. In: Proceedings of the 23rd SODA, San Francisco, pp 189–201
5. Bernstein A (2013) Maintaining shortest paths under deletions in weighted directed graphs: [extended abstract]. In: STOC, Palo Alto, pp 725–734
6. Demetrescu C, Italiano GF (2004) A new approach to dynamic all pairs shortest paths. J ACM 51(6):968–992. doi:http://doi.acm.org/10.1145/1039488.1039492

7. Fredman ML, Tarjan RE (1987) Fibonacci heaps and their uses in improved network optimization algorithms. J ACM 34(3):596–615
8. Pettie S (2004) A new approach to all-pairs shortest paths on real-weighted graphs. Theor Comput Sci 312(1):47–74. doi:10.1016/S0304-3975(03)00402-X
9. Takaoka T (1992) A new upper bound on the complexity of the all pairs shortest path problem. Inf Process Lett 43(4):195–199. doi:10.1016/0020-0190(92)90200-F
10. Thorup M (2004) Fully-dynamic all-pairs shortest paths: faster and allowing negative cycles. In: Proceedings of the 9th SWAT, Humlebaek, pp 384–396
11. Thorup M, Zwick U (2005) Approximate distance oracles. J ACM 52(1):1–24
12. Thorup M, Zwick U (2006) Spanners and emulators with sublinear distance errors. In: Proceedings of the 17th SODA, Miami, pp 802–809
13. Williams R (2014) Faster all-pairs shortest paths via circuit complexity. In: STOC, New York, pp 664–673

Dynamic Trees

Renato F. Werneck
Microsoft Research Silicon Valley, La Avenida, CA, USA

Keywords

Link-cut trees

Years and Authors of Summarized Original Work

2005; Tarjan, Werneck

Problem Definition

The *dynamic tree problem* is that of maintaining an arbitrary n-vertex forest that changes over time through edge insertions (*links*) and deletions (*cuts*). Depending on the application, one associates information with vertices, edges, or both. Queries and updates can deal with individual vertices or edges, but more commonly they refer to entire paths or trees. Typical operations include finding the minimum-cost edge along a path,

determining the minimum-cost vertex in a tree, or adding a constant value to the cost of each edge on a path (or of each vertex of a tree). Each of these operations, as well as *links* and *cuts*, can be performed in $O(\log n)$ time with appropriate data structures.

Key Results

The obvious solution to the dynamic tree problem is to represent the forest explicitly. This, however, is inefficient for queries dealing with entire paths or trees, since it would require actually traversing them. Achieving $O(\log n)$ time per operation requires mapping each (possibly unbalanced) input tree into a balanced tree, which is better suited to maintaining information about paths or trees implicitly. There are three main approaches to perform the mapping: path decomposition, tree contraction, and linearization.

Path Decomposition

The first efficient dynamic tree data structure was Sleator and Tarjan's *ST-trees* [13, 14], also known as *link-cut trees* or simply *dynamic trees*. They are meant to represent rooted trees, but the user can change the root with the *evert* operation. The data structure partitions each input tree into vertex-disjoint paths, and each path is represented as a binary search tree in which vertices appear in symmetric order. The binary trees are then connected according to how the paths are related in the forest. More precisely, the root of a binary tree becomes a *middle child* (in the data structure) of the parent (in the forest) of the topmost vertex of the corresponding path. Although a node has no more than two children (left and right) within its own binary tree, it may have arbitrarily many middle children. See Fig. 1. The path containing the root (*qlifcba* in the example) is said to be *exposed*, and is represented as the topmost binary tree. All path-related queries will refer to this path. The *expose* operation can be used to make any vertex part of the exposed path.

With standard balanced binary search trees (such as red-black trees), ST-trees support each dynamic tree operation in $O(\log^2 n)$ amortized

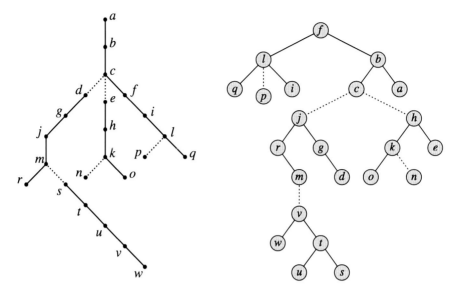

Dynamic Trees, Fig. 1 An ST-tree (Adapted from [14]). On the *left*, the original tree, rooted at *a* and already partitioned into paths; on the *right*, the actual data structure. *Solid edges* connect nodes on the same path; *dashed edges* connect different paths

time. This bound can be improved to $O(\log n)$ amortized with locally biased search trees, and to $O(\log n)$ in the worst case with globally biased search trees. Biased search trees (described in [5]), however, are notoriously complicated. A more practical implementation of ST-trees uses *splay trees*, a self-adjusting type of binary search trees, to support all dynamic tree operations in $O(\log n)$ amortized time [14].

Tree Contraction

Unlike ST-trees, which represent the input trees directly, Frederickson's *topology trees* [6, 7, 8] represent a *contraction* of each tree. The original vertices constitute level 0 of the contraction. Level 1 represents a partition of these vertices into *clusters*: a degree-one vertex can be combined with its only neighbor; vertices of degree two that are adjacent to each other can be clustered together; other vertices are kept as singletons. The end result will be a smaller tree, whose own partition into clusters yields level 2. The process is repeated until a single cluster remains. The topology tree is a representation of the contraction, with each cluster having as children its constituent clusters on the level below. See Fig. 2.

With appropriate pieces of information stored in each cluster, the data structure can be used to answer queries about the entire tree or individual paths. After a *link* or *cut*, the affected topology trees can be rebuilt in $O(\log n)$ time.

The notion of tree contraction was developed independently by Miller and Reif [11] in the context of parallel algorithms. They propose two basic operations, *rake* (which eliminates vertices of degree one) and *compress* (which eliminates vertices of degree two). They show that $O(\log n)$ rounds of these operations are sufficient to contract any tree to a single cluster. Acar et al. translated a variant of their algorithm into a dynamic tree data structure, *RC-trees* [1], which can also be seen as a randomized (and simpler) version of topology trees.

A drawback of topology trees and RC-trees is that they require the underlying forest to have vertices with bounded (constant) degree in order to ensure $O(\log n)$ time per operation. Similarly, although ST-trees do not have this limitation when aggregating information over paths, they require bounded degrees to aggregate over trees. Degree restrictions can be addressed by "ternarizing" the input forest (replacing high-degree vertices

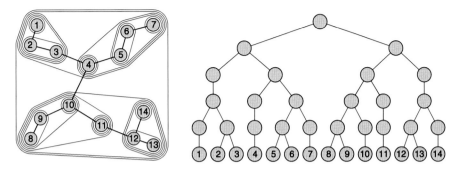

Dynamic Trees, Fig. 2 A topology tree (Adapted from [7]). On the *left*, the original tree and its multilevel partition; on the *right*, a corresponding topology tree

Dynamic Trees, Fig. 3
The *rake* and *compress*
operations, as used by top
trees (From [16]))

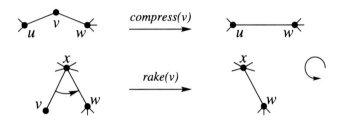

with a series of low-degree ones [9]), but this introduces a host of special cases.

Alstrup et al.'s *top trees* [3, 4] have no such limitation, which makes them more generic than all data structures previously discussed. Although also based on tree contraction, their clusters behave not like vertices, but like *edges*. A *compress* cluster combines two edges that share a degree-two vertex, while a *rake* cluster combines an edge with a degree-one endpoint with a second edge adjacent to its other endpoint. See Fig. 3.

Top trees are designed so as to completely hide from the user the inner workings of the data structure. The user only specifies what pieces of information to store in each cluster, and (through call-back functions) how to update them after a cluster is created or destroyed when the tree changes. As long as the operations are properly defined, applications that use top trees are completely independent of how the data structure is actually implemented, i.e., of the order in which *rakes* and *compresses* are performed.

In fact, top trees were not even proposed as stand-alone data structures, but rather as an interface on top of topology trees. For efficiency reasons, however, one would rather have a more direct implementation. Holm, Tarjan, Thorup

and Werneck have presented a conceptually simple stand-alone algorithm to update a top tree after a *link* or *cut* in $O(\log n)$ time in the worst case [17]. Tarjan and Werneck [16] have also introduced *self-adjusting top trees*, a more efficient implementation of top trees based on path decomposition: it partitions the input forest into edge-disjoint paths, represents these paths as splay trees, and connects these trees appropriately. Internally, the data structure is very similar to ST-trees, but the paths are edge-disjoint (instead of vertex-disjoint) and the ternarization step is incorporated into the data structure itself. All the user sees, however, are the *rakes* and *compresses* that characterize tree contraction.

Linearization

ET-trees, originally proposed by Henzinger and King [10] and later slightly simplified by Tarjan [15], use yet another approach to represent dynamic trees: *linearization*. It maintains an *Euler tour* of the each input tree, i.e., a closed path that traverses each edge twice–once in each direction. The tour induces a linear order among the vertices and arcs, and therefore can be represented as a balanced binary search tree. Linking and cutting edges from the forest corresponds to joining and splitting the affected binary trees, which can be done in $O(\log n)$ time. While linearization is arguably the simplest of the three approaches, it has a crucial drawback: because each edge appears twice, the data structure can only aggregate information over trees, not paths.

Lower Bounds

Dynamic tree data structures are capable of solving the *dynamic connectivity* problem on acyclic graphs: given two vertices v and w, decide whether they belong to the same tree or not. Pătraşcu and Demaine [12] have proven a lower bound of $\Omega(\log n)$ for this problem, which is matched by the data structures presented here.

Applications

Sleator and Tarjan's original application for dynamic trees was Dinic's blocking flow algorithm [13]. Dynamic trees are used to maintain a forest of arcs with positive residual capacity. As soon as the source s and the sink t become part of the same tree, the algorithm sends as much flow as possible along the s–t path; this reduces to zero the residual capacity of at least one arc, which is then *cut* from the tree. Several maximum flow and minimum-cost flow algorithms incorporating dynamic trees have been proposed ever since (some examples are [9, 15]). Dynamic tree data structures, especially those based on tree contraction, are also commonly used within dynamic graph algorithms, such as the dynamic versions of minimum spanning trees [6, 10], connectivity [10], biconnectivity [6], and bipartiteness [10]. Other applications include the evaluation of dynamic expression trees [8] and standard graph algorithms [13].

Experimental Results

Several studies have compared the performance of different dynamic-tree data structures; in most cases, ST-trees implemented with splay trees are the fastest alternative. Frederickson, for example, found that topology trees take almost 50 % more time than splay-based ST-trees when executing dynamic tree operations within a maximum flow algorithm [8]. Acar et al. [2] have shown that RC-trees are significantly slower than splay-based ST-trees when most operations are *links* and *cuts* (such as in network flow algorithms), but faster when queries and value updates are dominant. The reason is that splaying changes the structure of ST-trees even during queries, while RC-trees remain unchanged.

Tarjan and Werneck [17] have presented an experimental comparison of several dynamic tree data structures. For random sequences of *links* and *cuts*, splay-based ST-trees are the fastest alternative, followed by splay-based ET-trees, self-adjusting top trees, worst-case top trees, and RC-trees. Similar relative performance was observed in more realistic sequences of operations, except when queries far outnumber structural operations; in this case, the self-adjusting data structures are slower than RC-trees and worst-case top trees. The same experimental study also considered the "obvious" implementation of ST-trees, which represents the forest explicitly and require linear time per operation in the worst case. Its simplicity makes it significantly faster than the $O(\log n)$-time data structures for path-related queries and updates, unless paths are hundred nodes long. The sophisticated solutions are more useful when the underlying forest has high diameter or there is a need to aggregate information over trees (and not only paths).

Cross-References

Recommended Reading

1. Acar UA, Blelloch GE, Harper R, Vittes JL, Woo SLM (2004) Dynamizing static algorithms, with applications to dynamic trees and history independence. In: Proceedings of the 15th annual ACM-SIAM symposium on discrete algorithms (SODA). SIAM, pp 524–533

2. Acar UA, Blelloch GE, Vittes JL (2005) An experimental analysis of change propagation in dynamic trees. In: Proceedings of the 7th workshop on algorithm engineering and experiments (ALENEX). pp 41–54

3. Alstrup S, Holm J, de Lichtenberg K, Thorup M (1997) Minimizing diameters of dynamic trees. In: Proceedings of the 24th international colloquium on automata, languages and programming (ICALP), Bologna, 7–11 July 1997. Lecture notes in computer science, vol 1256. Springer, pp 270–280

4. Alstrup S, Holm J, Thorup M, de Lichtenberg K (2005) Maintaining information in fully dynamic trees with top trees. ACM Trans Algorithm 1(2):243–264

5. Bent SW, Sleator DD, Tarjan RE (1985) Biased search trees. SIAM J Comput 14(3):545–568

6. Frederickson GN (1985) Data structures for on-line update of minimum spanning trees, with applications. SIAM J Comput 14(4):781–798

7. Frederickson GN (1997) Ambivalent data structures for dynamic 2-edge-connectivity and k smallest spanning trees. SIAM J Comput 26(2):484–538

8. Frederickson GN (1997) A data structure for dynamically maintaining rooted trees. J Algorithms 24(1):37–65

9. Goldberg AV, Grigoriadis MD, Tarjan RE (1991) Use of dynamic trees in a network simplex algorithm for the maximum flow problem. Math Program 50:277–290

10. Henzinger MR, King V (1997) Randomized fully dynamic graph algorithms with polylogarihmic time per operation. In: Proceedings of the 27th annual ACM symposium on theory of computing (STOC), pp 519–527

11. Miller GL, Reif JH (1985) Parallel tree contraction and its applications. In: Proceedings of the 26th annual IEEE symposium on foundations of computer science (FOCS), pp 478–489

12. Pǎtraşcu M, Demaine ED (2004) Lower bounds for dynamic connectivity. In: Proceedings of the 36th annual ACM symposium on theory of computing (STOC), pp 546–553

13. Sleator DD, Tarjan RE (1983) A data structure for dynamic trees. J Comput Syst Sci 26(3):362–391

14. Sleator DD, Tarjan RE (1985) Self-adjusting binary search trees. J ACM 32(3):652–686

15. Tarjan RE (1997) Dynamic trees as search trees via Euler tours, applied to the network simplex algorithm. Math Program 78:169–177

16. Tarjan RE, Werneck RF (2005) Self-adjusting top trees. In: Proceedings of the 16th annual ACM-SIAM symposium on discrete algorithms (SODA), pp 813–822

17. Tarjan RE, Werneck RF (2007) Dynamic trees in practice. In: Proceedings of the 6th workshop on experimental algorithms (WEA). Lecture notes in computer science, vol 4525, pp 80–93

18. Werneck RF (2006) Design and analysis of data structures for dynamic trees. PhD thesis, Princeton University

E

Edit Distance Under Block Operations

S. Cenk Sahinalp
Laboratory for Computational Biology, Simon
Fraser University, Burnaby, BC, USA

Keywords

Block edit distance

Years and Authors of Summarized Original Work

2000; Cormode, Paterson, Sahinalp, Vishkin
2000; Muthukrishnan, Sahinalp

Problem Definition

Given two strings $S = s_1 s_2 \ldots s_n$ and $R = r_1 r_2 \ldots r_m$ (wlog let $n \geq m$) over an alphabet $\sigma = \{\sigma_1, \sigma_2, \ldots \sigma_\ell\}$, the *standard edit distance* between S and R, denoted $ED(S, R)$ is the minimum number of *single character edits*, specifically *insertions*, *deletions* and *replacements*, to transform S into R (equivalently R into S).

If the input strings S and R are permutations of the alphabet σ (so that $|S| = |R| = |\sigma|$) then an analogous *permutation edit distance* between S and R, denoted $PED(S, R)$ can be defined as the minimum number of single character *moves*, to transform S into R (or vice versa).

A generalization of the standard edit distance is *edit distance with moves*, which, for input strings S and R is denoted $EDM(S, R)$, and is defined as the minimum number of character edits and *substring (block) moves* to transform one of the strings into the other. A move of block $s[j, k]$ to position h transforms $S = s_1 s_2 \ldots s_n$ into $S' = s_1 \ldots s_{j-1} s_{k+1} s_{k+2} \ldots s_{h-1} s_j \ldots s_k s_h \ldots s_n$ [4].

If the input strings S and R are permutations of the alphabet σ (so that $|S| = |R| = |\sigma|$) then $EDM(S, R)$ is also called as the transposition distance and is denoted $TED(S, R)$ [1].

Perhaps the most general form of the standard edit distance that involves edit operations on blocks/substrings is the *block edit distance*, denoted $BED(S, R)$. It is defined as the minimum number of single character edits, block moves, as well as *block copies* and *block uncopies* to transform one of the strings into the other. Copying of a block $s[j, k]$ to position h transforms $S = s_1 s_2 \ldots s_n$ into $S' = s_1 \ldots s_j s_{j+1} \ldots s_k \ldots s_{h-1} s_j \ldots s_k s_h \ldots s_n$. A block uncopy is the inverse of a block copy: it deletes a block $s[j, k]$ provided there exists $s[j', k'] = s[j, k]$ which does not overlap with $s[j, k]$ and transforms S into $S' = s_1 \ldots s_{j-1} s_{k+1} \ldots s_n$.

Throughout this discussion all edit operations have unit cost and they may overlap; i.e., a character can be edited on multiple times.

Key Results

There are exact and approximate solutions to computing the edit distances described above with varying performance guarantees. As can be expected, the best available running times as well as the approximation factors for computing these edit distances vary considerably with the edit operations allowed.

Exact Computation of the Standard and Permutation Edit Distance

The fastest algorithms for exactly computing the standard edit distance have been available for more than 25 years.

Theorem 1 (Levenshtein [9]) *The standard edit distance ED(S, R) can be computed exactly in time $O(n \cdot m)$ via dynamic programming.*

Theorem 2 (Masek-Paterson [11]) *The standard edit distance ED(S, R) can be computed exactly in time $O(n + n \cdot m/\log_{|\sigma|}^2 n)$ via the "four-Russians trick".*

Theorem 3 (Landau-Vishkin [8]) *It is possible to compute ED(S, R) in time $O(n \cdot ED(S, R))$.*

Finally, note that if S and R are permutations of the alphabet σ, $PED(S, R)$ can be computed much faster than the standard edit distance for general strings: Observe that $PED(S, R) = n - LCS(S, R)$ where $LCS(S, R)$ represents the longest common subsequence of S and R. For permutations $S, R, LCS(S, R)$ can be computed in time $O(n \cdot \log \log n)$ [3].

Approximate Computation of the Standard Edit Distance

If some approximation can be tolerated, it is possible to considerably improve the $\tilde{O}(n \cdot m)$ time ($\tilde{O}$ notation hides polylogarithmic factors) available by the techniques above. The fastest algorithm that *approximately* computes the standard edit distance works by *embedding* strings S and R from alphabet σ into shorter strings S' and R' from a larger alphabet σ' [2]. The embedding is achieved by applying a general version of the *Locally Consistent Parsing* [13, 14] to partition the strings R and S into *consistent blocks* of size c to $2c - 1$; the partitioning is consistent in the sense that identical (long) substrings are partitioned identically. Each block is then replaced with a label such that identical blocks are identically labeled. The resulting strings S' and R' preserve the edit distance between S and R approximately as stated below.

Theorem 4 (Batu-Ergun-Sahinalp [2]) *ED(S, R) can be computed in time $\tilde{O}(n^{1+\epsilon})$ within an approximation factor of $\min\{n^{\frac{1-\epsilon}{3}+o(1)}, (ED(S, R)/n^\epsilon)^{\frac{1}{2}+o(1)}\}$.*

For the case of $\epsilon = 0$, the above result provides an $\tilde{O}(n)$ time algorithm for approximating $ED(S, R)$ within a factor of $\min\{n^{\frac{1}{3}+o(1)}, ED(S, R)^{\frac{1}{2}+o(1)}\}$.

Approximate Computation of Edit Distances Involving Block Edits

For all edit distance variants described above which involve blocks, there are no known polynomial time algorithms; in fact it is NP-hard to compute $TED(S, R)$ [1], $EDM(S, R)$ and $BED(S, R)$ [10]. However, in case S and R are permutations of σ, there are polynomial time algorithms that approximate transposition distance within a constant factor:

Theorem 5 (Bafna-Pevzner [1]) *TED(S, R) can be approximated within a factor of 1.5 in $O(n^2)$ time.*

Furthermore, even if S and R are arbitrary strings from σ, it is possible to approximately compute both $BED(S, R)$ and $EDM(S, R)$ in near linear time. More specifically obtain an embedding of S and R to binary vectors $f(S)$ and $f(R)$ such that:

Theorem 6 (Muthukrishnan-Sahinalp [12]) $\frac{\|f(S)-f(R)\|_1}{\log^* n} \leq BED(S, R) \leq \|f(S) - f(R)\|_1 \cdot \log n.$

In other words, the Hamming distance between $f(S)$ and $f(R)$ approximates $BED(S, R)$ within a factor of $\log n \cdot \log^* n$. Similarly for $EDM(S, R)$, it is possible to embed S and R to integer valued vectors $F(S)$ and $F(R)$ such that:

Theorem 7 (Cormode-Muthukrishnan [4])
$\frac{\|F(S)-F(R)\|_1}{\log^* n} \leq EDM(S, R) \leq \|F(S) - F(R)\|_1 \cdot \log n.$

In other words, the L_1 distance between $F(S)$ and $F(R)$ approximates $EDM(S, R)$ within a factor of $\log n \cdot \log^* n$.

The embedding of strings S and R into binary vectors $f(S)$ and $f(R)$ is introduced in [5] and is based on the Locally Consistent Parsing described above. To obtain the embedding, one needs to hierarchically partition S and R into growing size *core* blocks. Given an alphabet σ, Locally Consistent Parsing can identify only a limited number of substrings as core blocks. Consider the lexicographic ordering of these core blocks. Each dimension i of the embedding $f(S)$ simply indicates (by setting $f(S)[i] = 1$) whether S includes the ith core block corresponding to the alphabet σ as a substring. Note that if a core block exists in S as a substring, Locally Consistent Parsing will identify it.

Although the embedding above is exponential in size, the resulting binary vector $f(S)$ is very sparse. A simple representation of $f(S)$ and $f(R)$, exploiting their sparseness can be computed in time $O(n \log^* n)$ and the Hamming distance between $f(S)$ and $f(R)$ can be computed in linear time by the use of this representation [12].

The embedding of S and R into integer valued vectors $F(S)$ and $F(R)$ are based on similar techniques. Again, the total time needed to approximate $EDM(S, R)$ within a factor of $\log n \cdot \log^* n$ is $O(n \log^* n)$.

versions S_1 and S_2, and the parties communicate to reconcile the differences between the two versions. An information theoretic lower bound on the number of bits to communicate between the two parties is then $\Omega(BED(S, R)) \cdot \log n$. The embedding of S and R to binary strings $f(S)$ and $f(R)$ provides a simple protocol [5] which gives a near-optimal tradeoff between the number of rounds of communication and the total number of bits exchanged and works with high probability.

Another important application is to the Sequence Nearest Neighbors (SNN) problem, which asks to preprocess a set of strings $S_1, \ldots, S_k$ so that given an on-line query string R, the string S_i which has the lowest distance of choice to R can be computed in time polynomial with $|R|$ and polylogarithmic with $\sum_{j=1}^{k} |S_j|$. There are no known exact solutions for the SNN problem under any edit distance considered here. However, in [12], the embedding of strings S_i into binary vectors $f(S_i)$, combined with the Approximate Nearest Neighbors results given in [6] for Hamming Distance, provides an approximate solution to the SNN problem under block edit distance as follows.

Theorem 8 (Muthukrishnan-Sahinalp [12])
It is possible to preprocess a set of strings $S_1, \ldots, S_k$ from a given alphabet σ in $O(poly(\sum_{j=1}^{k} |S_j|))$ time such that for any on-line query string R from σ one can compute a string S_i in time $O(polylog(\sum_{j=1}^{k} |S_j|) \cdot poly(|R|))$ which guarantees that for all $h \in [1, k]$, $BED(S_i, R) \leq BED(S_h, R) \cdot \log(\max_j |S_j|) \cdot \log^*(\max_j |S_j|).$

Applications

Edit distances have important uses in computational evolutionary biology, in estimating the evolutionary distance between pairs of genome sequences under various edit operations. There are also several applications to the *document exchange problem* or *document reconciliation problem* where two copies of a text string S have been subject to edit operations (both single character and block edits) by two parties resulting in two

Open Problems

It is interesting to note that when dealing with permutations of the alphabet σ the problem of computing both character edit distances and block edit distances become much easier; one can compute $PED(S, R)$ exactly and $TED(S, R)$ within an approximation factor of 1.5 in $\tilde{O}(n)$ time. For arbitrary strings, it is an open question whether one can approximate $TED(S, R)$ or $BED(S, R)$ within a factor of $o(\log n)$ in polynomial time.

One recent result in this direction shows that it is not possible to obtain a polylogarithmic approximation to $TED(S, R)$ via a greedy strategy [7]. Furthermore, although there is a lower bound of $\Omega(n^{\frac{1}{3}})$ on the approximation factor that can be achieved for computing the standard edit distance in $\tilde{O}(n)$ time by the use of string embeddings, there is no general lower bound on how closely one can approximate $ED(S, R)$ in near linear time.

Cross-References

▶ Approximate String Matching

Recommended Reading

1. Bafna V, Pevzner PA (1998) Sorting by transpositions. SIAM J Discret Math 11(2):224–240
2. Batu T, Ergün F, Sahinalp SC (2006) Oblivious string embeddings and edit distance approximations. In: Proceedings of the ACM-SIAM SODA, pp 792–801
3. Besmaphyatnikh S, Segal M (2000) Enumerating longest increasing subsequences and patience sorting. Inf Process Lett 76(1–2):7–11
4. Cormode G, Muthukrishnan S (2002) The string edit distance matching problem with moves. In: Proceedings of the ACM-SIAM SODA, pp 667–676
5. Cormode G, Paterson M, Sahinalp SC, Vishkin U (2000) Communication complexity of document exchange. In: Proceedings of the ACM-SIAM SODA, pp 197–206
6. Indyk P, Motwani R (1998) Approximate nearest neighbors: towards removing the curse of fimensionality. In: Proceedings of the ACM STOC, pp 604–613
7. Kaplan H, Shafrir N (2005) The greedy algorithm for shortest superstrings. Info Process Lett 93(1):13–17
8. Landau G, Vishkin U (1989) Fast parallel and serial approximate string matching. J Algorithms 10:157–169
9. Levenshtein VI (1965) Binary codes capable of correcting deletions, insertions, and reversals. Dokl Akad Nauk SSSR 163(4):845–848 (Russian). (1966) Sov Phys Dokl 10(8):707–710 (English translation)
10. Lopresti DP, Tomkins A (1997) Block edit models for approximate string matching. Theor Comput Sci 181(1):159–179
11. Masek W, Paterson M (1980) A faster algorithm for computing string edit distances. J Comput Syst Sci 20:18–31
12. Muthukrishnan S, Sahinalp SC (2000) Approximate nearest neighbors and sequence comparison with block operations. In: Proceedings of the ACM STOC, pp 416–424
13. Sahinalp SC, Vishkin U (1994) Symmetry breaking for suffix tree construction. In: ACM STOC, pp 300–309
14. Sahinalp SC, Vishkin U (1996) Efficient approximate and dynamic matching of patterns using a labeling paradigm. In: Proceeding of the IEEE FOCS, pp 320–328

Efficient Decodable Group Testing

Hung Q. Ngo[1] and Atri Rudra[2]
[1]Computer Science and Engineering, The State University of New York, Buffalo, NY, USA
[2]Department of Computer Science and Engineering, State University of New York, Buffalo, NY, USA

Keywords

Coding theory; Nonadaptive group testing; Sublinear-time decoding

Years and Authors of Summarized Original Work

2011; Ngo, Porat, Rudra

Problem Definition

The basic group testing problem is to identify the unknown set of *positive items* from a large population of *items* using as few *tests* as possible. A test is a subset of items. A test returns positive if there is a positive item in the subset. The semantics of "positives," "items," and "tests" depend on the application.

In the original context [3], group testing was invented to solve the problem of identifying syphilis-infected blood samples from a large collection of WWII draftees' blood samples. In this case, items are blood samples, which are positive if they are infected. A test is a *pool* (group) of blood samples. Testing a group of samples at a time will save resources if the test outcome is negative. On the other hand, if the test outcome is positive, then all we know is that at least one sample in the pool is positive, but we do not know which one(s).

In *nonadaptive combinatorial group testing* (NACGT), we assume that the number of positives is at most d for some fixed integer d and that all tests have to be specified in advance before any test outcome is known. The NACGT paradigm has found numerous applications in many areas of mathematics, computer science, and computational biology [4, 9, 10].

A NACGT strategy with t tests on a universe of N items is represented by a $t \times N$ binary matrix $\mathbf{M} = (m_{ij})$, where $m_{ij} = 1$ iff item j belongs to test i. Let $\mathbf{M}_i$ and $\mathbf{M}^j$ denote row i and column j of $\mathbf{M}$, respectively. Abusing notation, we will also use $\mathbf{M}_i$ (respectively, $\mathbf{M}^j$) to denote the set of rows (respectively, columns) corresponding to the 1-entries of row i (respectively, column j). In other words, $\mathbf{M}_i$ is the ith pool, and $\mathbf{M}^j$ is the set of pools that item j belongs to.

Let $D \subset [N]$ be the unknown subset of positive items, where $|D| \leq d$. Let $\mathbf{y} = (y_i)_{i=1}^t \in \{0, 1\}^t$ denote the test outcome vector, i.e., $y_i = 1$ iff the ith test is positive. Then, the test outcome vector is precisely the (Boolean) union of the positive columns: $\mathbf{y} = \bigcup_{j \in D} \mathbf{M}^j$. The task of identifying the unknown subset D from the test outcome vector $\mathbf{y}$ is called *decoding*.

The main problem In many modern applications of NACGT, there are two key requirements for an NACGT scheme:

1. *Small number of tests.* "Tests" are computationally expensive in many applications.
2. *Efficient decoding.* As the item universe size N can be extremely large, it would be ideal for the decoding algorithm to run in time sublinear in N and more precisely in $\text{poly}(d, \log N)$ time.

Key Results

To be able to uniquely identify an arbitrary subset D of at most d positives, it is necessary and suffi-

cient for the test outcome vectors $\mathbf{y}$ to be different for distinct subsets D of at most d positives. An NACGT matrix with the above property is called *d-separable*. However, in general such matrices only admit the brute force $\Omega(N^d)$-time decoding algorithm. A very natural decoding algorithm called the *naïve decoding algorithm* runs much faster, in time $O(tN)$.

Definition 1 (Naïve decoding algorithm) Eliminate all items that participate in negative tests; return the remaining items.

This algorithm does not work for arbitrary d-separable matrices. However, if the test matrix $\mathbf{M}$ satisfies a slightly stronger property called d-*disjunct*, then the naïve decoding algorithm is guaranteed to work correctly.

Definition 2 (Disjunct matrix) A $t \times N$ binary matrix $\mathbf{M}$ is said to be d-disjunct iff $\mathbf{M}^j \setminus \bigcup_{k \in S} \mathbf{M}^k \neq \emptyset$ for any set S of d columns and any $j \notin S$. (See Fig. 1.)

Minimize Number of Tests

It is remarkable that d-disjunct matrices not only allow for linear time decoding, which is a vast improvement over the brute-force algorithm for separable matrices, but also have asymptotically the same number of tests as d-separable matrices [4]. Let $t(d, N)$ denote the minimum number of rows of an N-column d-disjunct matrix. It has been known for about 40 years [5] that $t(\Omega(\sqrt{N}), N) = \Theta(N)$, and for $d = O(\sqrt{N})$ we have

$$\Omega \left(\frac{d^2}{\log d} \log N \right) \leq t(d, N) \leq O(d^2 \log N).$$
(1)

A $t \times N$ d-disjunct matrix with $t = O(d^2 \log N)$, rows can be constructed randomly or even deterministically (see [11]). However, the decoding time $O(tN)$ of the naïve decoding algorithm is still too slow for modern applications, where in most cases $d \ll N$ and thus $t \ll N$.

Efficient Decodable Group Testing, Fig. 1 A d-disjunct matrix has the following property: for any subset S of d (not necessarily contiguous) columns, and any column j that is not present in S, there exists a row i that has a 1 in column j and all zeros in S

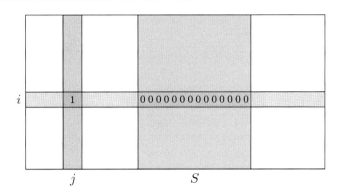

Efficient Decoding

An ideal decoding time would be in the order of poly$(d, \log N)$, which is sublinear in N for practical ranges of d. Ngo, Porat, and Rudra [10] showed how to achieve this goal using a couple of ideas: (a) two-layer test matrix construction and (b) code concatenation using a list recoverable code.

(a) Two-layer test matrix construction The idea is to construct **M** by stacking on top of one another two matrices: a "filtering" matrix **F** and an "identification" matrix **D**. (See Fig. 2.) The filtering matrix is used to quickly identify a "small" set of L candidate items including *all* the positives. Then, the identification matrix is used to pinpoint precisely the positives. For example, let **D** be any d-disjunct matrix, and that from the tests corresponding to the rows of **F**, we can produce a set S of $L = \text{poly}(d, \log N)$ candidate items in time poly$(d, \log N)$. Then, by running the naïve decoding algorithm on S using test results corresponding to the rows of **D**, we can identify all the positives in time poly$(d, \log N)$. To formalize the notion of "filtering matrix," we borrow a concept from coding theory, where producing a small list of candidate codewords is the *list decoding problem* [6].

Definition 3 (List-disjunct matrix) Let $d + \ell \leq N$ be positive integers. A matrix **F** is (d, ℓ)-list disjunct if and only if $\bigcup_{j \in T} \mathbf{M}^j \setminus \bigcup_{k \in S} \mathbf{M}^k \neq \emptyset$ for any two disjoint sets S and T of columns of **F** with $|S| = d$ and $|T| = \ell$. (See Fig. 3.)

Note that a matrix is d-disjunct matrix iff it is $(d, 1)$-list disjunct. However, the relaxation to $\ell = \Theta(d)$ allows the existence (and construction) of $(d, O(d))$-list-disjunct matrices with $\Theta(d \log(N/d))$ rows. The existence of such small list-disjunct matrices is crucially used in the second idea below.

(b) Code Concatenation with list recoverable codes A $t \times N$ (d, ℓ)-list-disjunct matrix admits $O(tN)$-decoding time using the naïve decoding algorithm. However, to achieve poly$(d, \log N)$ decoding time overall, we will need to construct list-disjunct matrices that allow for a poly$(d, \log N)$ decoding time. In particular, to use such a matrix as a filtering matrix, it is necessary that $\ell = \text{poly}(d)$. To construct efficiently decodable list-disjunct matrices, we need other ideas. Ngo, Porat, and Rudra [10] used a connection to list recoverable codes [6] to construct such matrices. This connection was used to construct $(d, O(d^{3/2}))$-list-disjunct matrices with $t = o(d^2 \log_d N)$ rows that can be decoded in poly(t) time. This along with the construction in Fig. 2 implies the following result:

Theorem 1 ([10]) *Given any d-disjunct matrix, it can be converted into another matrix with $1 + o(1)$ times as many rows that is also efficiently decodable (even if the original matrix was not).*

Other constructions of list-disjunct matrices with worse parameters were obtained earlier by Indyk, Ngo and Rudra [7], and Cheraghchi [1] using connections to expanders and randomness extractors.

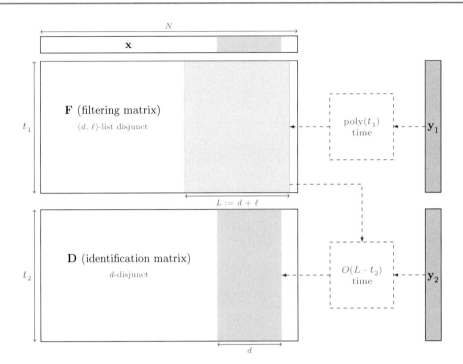

Efficient Decodable Group Testing, Fig. 2 The vector *x* denotes the characteristic vector of the *d* positives (illustrated by the *orange box*). The final matrix is the stacking of **F**, which is a (d, ℓ)-list-disjunct matrix, and **D**, which is a *d*-disjunct matrix. The result vector is naturally divided into y_1 (the part corresponding to **F** and denoted by the *red vector*) and y_2 (the part corresponding

to **D** and denoted by the *blue vector*). The decoder first uses y_1 to compute a superset of the set of positives (denoted by *green box*), which is then used with y_2 to compute the final set of positives. The first step of the decoding is represented by the *red-dotted box*, while the second step (naïve decoder) is denoted by the *blue-dotted box*

Efficient Decodable Group Testing, Fig. 3 A (d, ℓ)-list-disjunct matrix satisfies the following property: for any subset S of size d and any disjoint subset T of size ℓ, there exists a row i that has a 1 in at least one column in T and all zeros in S

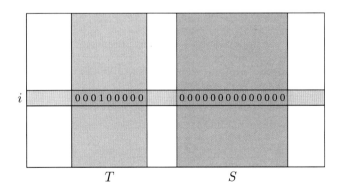

Applications

Heavy hitter is one of the most fundamental problems in data streaming [8]. Cormode and Muthukrishnan [2] showed that an NACGT scheme that is efficiently decodable and is also *explicit* solves a natural version of the heavy hitter problem. An explicit construction means

one needs an algorithm that outputs a column or a specific entry of **M** instead of storing the entire matrix **M** which can be extremely space consuming. This is possible with Theorem 1 by picking the filtering and decoding matrices to be explicit.

Another important generalization of NACGT matrices are those that can handle errors in the test outcomes. Again this is possible with the construction of Fig. 2 if the filtering and decoding matrices are also error tolerant. The list-disjunct matrices constructed by Cheraghchi are also error tolerant [1].

Open Problems

The outstanding open problem in group testing theory is to close the gap (1). An explicit construction of (d, d)-list-disjunct matrices is not known; solving this problem will lead to a scheme that is (near-)optimal in all desired objectives.

Recommended Reading

1. Cheraghchi M (2013) Noise-resilient group testing: Limitations and constructions. Discret Appl Math 161(1–2):81–95
2. Cormode G, Muthukrishnan S (2005) What's hot and what's not: tracking most frequent items dynamically. ACM Trans Database Syst 30(1):249–278
3. Dorfman R (1943) The detection of defective members of large populations. Ann Math Stat 14(4):436–440
4. Du DZ, Hwang FK (2000) Combinatorial group testing and its applications. Series on applied mathematics, vol 12, 2nd edn. World Scientific, River Edge
5. D'yachkov AG, Rykov VV (1982) Bounds on the length of disjunctive codes. Problemy Peredachi Informatsii 18(3):7–13
6. Guruswami V (2004) List decoding of error-correcting codes (Winning thesis of the 2002 ACM doctoral dissertation competition). Lecture notes in computer science, vol 3282. Springer
7. Indyk P, Ngo HQ, Rudra A (2010) Efficiently decodable non-adaptive group testing. In: Proceedings of the twenty first annual ACM-SIAM symposium on discrete algorithms (SODA'2010). ACM, New York, pp 1126–1142
8. Muthukrishnan S (2005) Data streams: algorithms and applications. Found Trends Theor Comput Sci 1(2)
9. Ngo HQ, Du DZ (2000) A survey on combinatorial group testing algorithms with applications to DNA library screening. In: Discrete mathematical problems with medical applications (New Brunswick, 1999). DIMACS series in discrete mathematics and theoretical computer science, vol 55. American Mathematical Society, Providence, pp 171–182
10. Ngo HQ, Porat E, Rudra A (2011) Efficiently decodable error-correcting list disjunct matrices and applications – (extended abstract). In: ICALP (1), pp 557–568
11. Porat E, Rothschild A (2011) Explicit nonadaptive combinatorial group testing schemes. IEEE Trans Inf Theory 57(12):7982–7989

Efficient Dominating and Edge Dominating Sets for Graphs and Hypergraphs

Andreas Brandstädt[1,2] and Ragnar Nevries[1]
[1]Computer Science Department, University of Rostock, Rostock, Germany
[2]Department of Informatics, University of Rostock, Rostock, Germany

Keywords

Efficient domination; Efficient edge domination; Exact cover

Years and Authors of Summarized Original Work

2010; Brandstädt, Hundt, Nevries
2011; Brandstädt, Mosca
2012; Brandstädt, Leitert, Rautenbach
2013; Brandstädt, Milanič, Nevries
2014; Brandstädt, Giakoumakis
2014; Nevries

Problem Definition

For a hypergraph $H = (V, \mathcal{E})$, a subset of edges $\mathcal{E}' \subseteq \mathcal{E}$ is an *exact cover* of H, if every vertex of V is contained in exactly one hyperedge of $\mathcal{E}'$, that is, for all $e, f \in \mathcal{E}'$ with $e \neq f$, $e \cap f = \emptyset$ and $\bigcup \mathcal{E}' = V$. The EXACT COVER (XC) problem asks for the existence of an exact cover in a given hypergraph H. Exact Cover is in Karp's famous list of 21 NP-complete problems; it is NP-complete even for 3-element hyperedges (problem X3C [SP2] in [14]).

Let G be a finite simple undirected graph with vertex set V and edge set E. A vertex *dominates* itself and all its neighbors, i.e., every vertex $v \in V$ dominates its closed neighborhood $N[v] = \{u \mid u = v$ or $uv \in E\}$. A vertex subset D of G is an *efficient dominating (e.d.)* set, if, for every vertex $v \in V$, there is exactly one $d \in D$ dominating v [1, 2]. An edge subset M of G is an *efficient edge dominating (e.e.d.)* set, if it is an efficient dominating set in the line graph $L(G)$ of G [15]. Efficient dominating sets are sometimes also called *independent perfect dominating sets*, and efficient edge dominating sets are also known as *dominating induced matchings*.

The EFFICIENT DOMINATION (ED) problem for a graph G asks for the existence of an e.d. set in G. The EFFICIENT EDGE DOMINATION (EED) problem asks for the existence of an e.d. set in the line graph $L(G)$.

For a graph G, let $\mathcal{N}(G)$ denote its closed neighborhood hypergraph, that is, for every vertex $v \in V$, the closed neighborhood $N[v]$ is a hyperedge in $\mathcal{N}(G)$; note that this is a multiset since distinct vertices may have the same closed neighborhood. For a graph G, the *square* G^2 has the same vertex set as G and two vertices, x and y, are adjacent in G^2, if and only if their distance in G is at most 2. Note that G^2 is isomorphic to $L(\mathcal{N}(G))$.

By definition, the ED problem on a graph G is the same as the Exact Cover problem on its closed neighborhood hypergraph $\mathcal{N}(G)$, and the EED problem is the same as the Exact Cover problem on $L(\mathcal{N}(G))$.

Key Results

ED and EED are NP-complete; their complexity on special graph classes was studied in various papers – see, e.g., [2, 3, 12, 16–18, 20, 22, 24, 25] for ED and [5, 7, 11, 15, 19, 21] for EED. In particular, ED remains NP-complete for chordal graphs as well as for (very restricted) bipartite graphs such as chordal bipartite graphs, and EED is NP-complete for bipartite graphs but solvable in linear time for chordal graphs.

ED for Graphs

A key tool in [8] is a reduction of ED for G to the maximum-weight independent set problem for G^2, which is based on the following observation:

For a hypergraph $H = (V, \mathcal{E})$ and $e \in \mathcal{E}$, let $\omega(e) := |e|$ be an edge weight function. For the line graph $L(H)$, let $\alpha_\omega(L(H))$ denote the maximum weight of an independent vertex set in $L(H)$. The weight of any independent vertex set in $L(H)$ is at most $|V|$, and H has an exact cover, if and only if $\alpha_\omega(L(H)) = |V|$. Using the fact that G^2 is isomorphic to $L(\mathcal{N}(G))$ and ED on G corresponds to Exact Cover on $\mathcal{N}(G)$, this means that ED on G can be reduced to the maximum weight of an independent vertex set in G^2, similarly for EED. This unified approach helps to answer some open questions on ED and EED for graph classes; one example is ED for strongly chordal graphs: Since for a dually chordal graph G, its square G^2 is chordal, ED is solvable in polynomial time for dually chordal graphs and thus for strongly chordal graphs [8] (recall that ED is NP-complete for chordal graphs). Similar properties of powers lead to polynomial time for ED on AT-free graphs using known results [8]. For P_5-free graphs having an e.d., G^2 is P_4-free [9].

ED is NP-complete for planar bipartite graphs of maximum degree 3 [9]. In [23], this is sharpened by adding a girth condition: ED is NP-complete for planar bipartite graphs of maximum degree 3 and girth at least g, for every fixed g.

From the known results, it follows that ED is NP-complete for F-free graphs whenever F contains a cycle or a claw. Thus, F can be assumed to be cycle- and claw-free (see, e.g., [9]); such graphs F are called *linear forests*. For $(P_3 + P_3)$-free graphs and thus for P_7-free graphs, ED is NP-complete. ED is robustly solvable in time $O(nm)$ for P_5-free graphs and for $(P_4 + P_2)$-free graphs [9, 23]. For every fixed $k \geq 1$, ED is solvable in polynomial time for $(P_5 + kP_2)$-free graphs [4]. For P_6-free graphs, the complexity of ED is an open problem, and correspondingly for $(P_6 + kP_2)$-free graphs; these are the only open cases for F-free graphs.

EED for Graphs

The fact that graphs having an e.e.d. are K_4-free leads to a simple linear time algorithm for EED on chordal graphs. More generally, EED is solvable in polynomial time for hole-free graphs and thus for weakly chordal graphs and for chordal bipartite graphs [7]. This also follows from the fact that, for a weakly chordal graph G, $L(G)^2$ is weakly chordal [10] and from the reduction of EED for G to the maximum-weight independent set problem for $L(G)^2$. In [23], this is improved to a robust $O(nm)$ time algorithm for EED on hole-free graphs. In [8], we show that EED is solvable in linear time for dually chordal graphs.

One of the open problems for EED was its complexity on P_k-free graphs. In [5], we show that EED is solvable in linear time for P_7-free graphs. The complexity of EED remains open for P_k-free graphs, $k \geq 8$. In [11], EED is solved in polynomial time on claw-free graphs. EED is NP-complete for planar bipartite graphs of maximum degree 3 [7]. In [23], it is shown that EED is NP-complete for planar bipartite graphs of maximum degree 3 and girth at least g, for every fixed g.

XC, ED, and EED for Hypergraphs

The notion of α-acyclicity [13] is one of the most important and most frequently studied hypergraph notions. Among the many equivalent conditions describing α-acyclic hypergraphs, we take the following: For a hypergraph $H = (V, \mathcal{E})$, a tree T with node set $\mathcal{E}$ and edge set E_T is a *join tree* of H, if, for all vertices $v \in V$, the set of hyperedges $\mathcal{E}_v := \{e \in \mathcal{E} \mid v \in e\}$ containing v induces a subtree of T. H is α-*acyclic*, if it has a join tree. Let $H^* := (\mathcal{E}, \{\mathcal{E}_v \mid v \in V\})$ be the *dual hypergraph* of H. The hypergraph $H = (V, \mathcal{E})$ is a *hypertree*, if there is a tree T with vertex set V such that, for all $e \in \mathcal{E}$, $T[e]$ is connected. Obviously, H is α-acyclic, if and only if its dual H^* is a hypertree.

By a result of Duchet, Flament, and Slater (see, e.g., [6]), it is known that H is a hypertree, if and only if H has the Helly property and its line graph $L(H)$ is chordal. In its dual version,

it says that H is α-acyclic, if and only if H is conformal and its 2-section graph is chordal. In [8], we show:

(i) ED and XC are NP-complete for α-acyclic hypergraphs but solvable in polynomial time for hypertrees.
(ii) EED is NP-complete for hypertrees but solvable in polynomial time for α-acyclic hypergraphs.

Recommended Reading

1. Bange DW, Barkauskas AE, Slater PJ (1988) Efficient dominating sets in graphs. In: Ringeisen RD, Roberts FS (eds) Applications of discrete mathematics. SIAM, Philadelphia, pp 189–199
2. Bange DW, Barkauskas AE, Host LH, Slater PJ (1996) Generalized domination and efficient domination in graphs. Discret Math 159:1–11
3. Biggs N (1973) Perfect codes in graphs. J Comb Theory (B) 15:289–296
4. Brandstädt A, Giakoumakis V (2014) Efficient domination for $(P_5 + kP_2)$-free graphs. Manuscript, arXiv:1407.4593v1
5. Brandstädt A, Mosca R (2011) Dominating induced matchings for P_7-free graphs in linear time. Technical report CoRR, arXiv:1106.2772v1; Extended abstract in: Asano T, Nakano S-I, Okamoto Y, Watanabe O (eds) Algorithms and computation. LNCS, vol 7074. Springer, pp 100–109
6. Brandstädt A, Le VB, Spinrad JP (1999) Graph classes: a survey. SIAM monographs on discrete mathematics and applications, vol 3. SIAM, Philadelphia
7. Brandstädt A, Hundt C, Nevries R (2010) Efficient edge domination on hole-free graphs in polynomial time. In: Lópes-Ortiz A (ed) LATIN 2010: theoretical informatics, Oaxaca. LNCS, vol 6034. Springer, pp 650–661
8. Brandstädt A, Leitert A, Rautenbach D (2012) Efficient dominating and edge dominating sets for graphs and hypergraphs. Technical report CoRR, arXiv:1207.0953v2 and in: Choa K-M, Hsu T-S, Lee D-T (eds) Algorithms and computation. LNCS, vol 7676. Springer, pp 267–277
9. Brandstädt A, Milanič M, Nevries R (2013) New polynomial cases of the weighted efficient domination problem. Technical report CoRR, arXiv:1304.6255 and in: Chatterjee K, Sgall J (eds) Mathematical foundations of computer science 2013. LNCS, vol 8087. Springer, pp 195–206

10. Cameron K, Sritharan R, Tang Y (2003) Finding a maximum induced matching in weakly chordal graphs. Discret Math 266:133–142

11. Cardoso DM, Korpelainen N, Lozin VV (2011) On the complexity of the dominating induced matching problem in hereditary classes of graphs. Discret Appl Math 159:521–531

12. Chang GJ, Pandu Rangan C, Coorg SR (1995) Weighted independent perfect domination on co-comparability graphs. Discret Appl Math 63:215–222

13. Fagin R (1983) Degrees of acyclicity for hypergraphs and relational database schemes. J ACM 30:514–550

14. Garey MR, Johnson DS (1979) Computers and intractability – a guide to the theory of NP-completeness. Freeman, San Francisco

15. Grinstead DL, Slater PL, Sherwani NA, Holmes ND (1993) Efficient edge domination problems in graphs. Inf Process Lett 48:221–228

16. Kratochvíl J (1991) Perfect codes in general graphs, Rozpravy Československé Akad. Věd Řada Mat. Přírod Vď 7. Akademia, Praha

17. Liang YD, Lu CL, Tang CY (1997) Efficient domination on permutation graphs and trapezoid graphs. In: Jiang T, Lee DT (eds) Computing and combinatorics, Shanghai. LNCS, vol 1276. Springer, pp 232–241

18. Lin Y-L (1998) Fast algorithms for independent domination and efficient domination in trapezoid graphs. In: Chwa K-Y, Ibarra OH (eds) Algorithms and computation, Taejon. LNCS, vol 1533. Springer, pp 267–275

19. Lu CL, Tang CY (1998) Solving the weighted efficient edge domination problem on bipartite permutation graphs. Discret Appl Math 87:203–211

20. Lu CL, Tang CY (2002) Weighted efficient domination problem on some perfect graphs. Discret Appl Math 117:163–182

21. Lu CL, Ko M-T, Tang CY (2002) Perfect edge domination and efficient edge domination in graphs. Discret Appl Math 119:227–250

22. Milanič M (2011) A hereditary view on efficient domination. Extended abstract In: Adacher L, Flamini M, Leo G, Nicosia G, Pacifici A, Piccialli V (eds) Proceedings of the 10th cologne-twente workshop on graphs and combinatorial optimization, Frascati, pp 203–206. Full version: Hereditary efficiently dominatable graphs. J Graph Theory 73:400–424

23. Nevries R (2014) Efficient domination and polarity. Ph.D. thesis, University of Rostock, Germany

24. Yen C-C (1992) Algorithmic aspects of perfect domination. Ph.D. thesis, National Tsing Hua University, Taiwan

25. Yen C-C, Lee RCT (1996) The weighted perfect domination problem and its variants. Discret Appl Math 66:147–160

Efficient Methods for Multiple Sequence Alignment with Guaranteed Error Bounds

S.M. Yiu and Francis Y.L. Chin
Department of Computer Science, University of Hong Kong, Hong Kong, China

Keywords

Multiple global alignment; Multiple string alignment

Years and Authors of Summarized Original Work

1993; Gusfield

Problem Definition

Multiple sequence alignment is an important problem in computational biology. Applications include finding highly conserved subregions in a given set of biological sequences and inferring the evolutionary history of a set of taxa from their associated biological sequences (e.g., see [9]). There are a number of measures proposed for evaluating the goodness of a multiple alignment, but prior to this work, no efficient methods are known for computing the optimal alignment for any of these measures. The work of Gusfield [7] gives two computationally efficient multiple alignment approximation algorithms for two of the measures with approximation ratio of less than 2. For one of the measures, they also derived a randomized algorithm, which is much faster and with high probability and reports a multiple alignment with small error bounds. To the best knowledge of the entry authors, this work is the first to provide approximation algorithms (with guarantee error bounds) for this problem.

Notations and Definitions

Let X and Y be two strings of alphabet Σ. The pairwise alignment of X and Y maps X and Y into strings X' and Y' that may contain spaces, denoted by '$\_$', where (1) $|X'| = |Y'| = \ell$ and (2) removing spaces from X' and Y' returns X and Y, respectively. The score of the alignment is defined as $d(X', Y') = \sum_{i=1}^{\ell} s(X'(i), Y'(i))$ where $X'(i)$ (and $Y'(i)$) denotes the ith character in X' (and Y') and $s(a, b)$ with $a, b \in \Sigma \cup$ '$\_$' is the distance-based scoring scheme that satisfies the following assumptions:

1. $s(\text{'}\_\text{'}, \text{'}\_\text{'}) = 0$;
2. Triangular inequality: for any three characters, $x, y, z, s(x, z) \leq s(x, y) + s(y, z))$.

Let $\chi = X_1, X_2, \ldots, X_k$ be a set of $k > 2$ strings of alphabet Σ. A multiple alignment A of these k strings maps $X_1, X_2, \ldots, X_k$ to $X'_1, X'_2, \ldots, X'_k$ that may contain spaces such that (1) $|X'_1| = |X'_2| = \cdots = |X'_k| = \ell$ and (2) removing spaces from X'_i returns X_i for all $1 \leq i \leq k$. The multiple alignment A can be represented as a $k \times \ell$ matrix.

The Sum of Pairs (SP) Measure

The score of a multiple alignment A, denoted by $SP(A)$, is defined as the sum of the scores of pairwise alignments induced by A, that is, $\sum_{i<j} d(X'_i, X'_j) = \sum_{i<j} \sum_{p=1}^{\ell} s(X'_i[p], X'_j[p])$ where $1 \leq i < j \leq k$.

Problem 1 (Multiple Sequence Alignment with Minimum SP Score)

INPUT: A set of k strings, a scoring scheme s.
OUTPUT: A multiple alignment A of these k strings with minimum $SP(A)$.

The Tree Alignment (TA) Measure

In this measure, the multiple alignment is derived from an evolutionary tree. For a given set χ of k strings, let $\chi' \supseteq \chi$. An evolutionary tree T'_χ for

χ is a tree with at least k nodes, where there is a one-to-one correspondence between the nodes and the strings in χ'. Let $X'_u \in \chi'$ be the string for node u. The score of T'_χ, denoted by TA (T'_χ), is defined as $\sum_{e=(u,v)} D(X'_u, X'_v)$ where e is an edge in T'_χ and $D(X'_u, X'_v)$ denotes the score of the optimal pairwise alignment for X'_u and X'_v. Analogously, the multiple alignment of χ under the TA measure can also be represented by a $|\chi'| \times \ell$ matrix, where $|\chi'| \geq k$, with a score defined as $\sum_{e=(u,v)} d(X'_u, X'_v)$ (e is an edge in T'_χ), similar to the multiple alignment under the SP measure in which the score is the summation of the alignment scores of all pairs of strings. Under the TA measure, since it is always possible to construct the $|\chi'| \times \ell$ matrix such that $d(X'_u, X'_v) = D(X'_u, X'_v)$ for all $e = (u, v)$ in T'_χ and we are usually interested in finding the multiple alignment with the minimum TA value, so $D(X'_u, X'_v)$ is used instead of $d(X'_u, X'_v)$ in the definition of TA (T'_χ).

Problem 2 (Multiple Sequence Alignment with Minimum TA Score)

INPUT: A set of k strings, a scoring scheme s.
OUTPUT: An evolutionary tree T for these k strings with minimum TA(T).

Key Results

Theorem 1 *Let A^* be the optimal multiple alignment of the given k strings with minimum SP score. They provide an approximation algorithm (the center star method) that gives a multiple alignment A such that $\frac{SP(A)}{SP(A^*)} \leq \frac{2(k-1)}{k} = 2 - \frac{2}{k}$.*

The center star method is to derive a multiple alignment which is consistent with the optimal pairwise alignments of a center string with all the other strings. The bound is derived based on the triangular inequality of the score function. The time complexity of this method is $O(k^2 \ell^2)$, where ℓ^2 is the time to solve the pairwise alignment by dynamic programming and k^2 is needed to find the center string, X_c, which gives the minimum value of $\sum_{i \neq c} D(X_c, X_i)$.

Theorem 2 *Let A^* be the optimal multiple alignment of the given k strings with minimum SP score. They provide a randomized algorithm that gives a multiple alignment A such that $\frac{SP(A)}{SP(A^*)} \leq 2 + \frac{1}{r-1}$ with probability at least $1 - \left(\frac{r-1}{r}\right)^p$ for any $r > 1$ and $p \geq 1$. Instead of computing $\binom{k}{2}$ optimal pairwise alignments to find the best center string, the randomized algorithm only considers p randomly selected strings to be candidates for the best center string; thus, this method needs to x compute only $(k-1)p$ optimal pairwise alignments in $O(kp\ell^2)$ time where $1 \leq p \leq k$.*

Theorem 3 *Let T^* be the optimal evolutionary tree of the given k strings with minimum TA score. They provide an approximation algorithm that gives an evolutionary tree T such that $\frac{TA(T)}{TA(T^*)} \leq \frac{2(k-1)}{k} = 2 - \frac{2}{k}$.*

In the algorithm, they first compute all the $\binom{k}{2}$ optimal pairwise alignments to construct a graph with every node representing a distinct string X_i and the weight of each edge (X_i, X_j) as $D(X_i X_j)$. This step determines the overall time complexity $O(k^2\ell^2)$. Then, they find a minimum spanning tree from the graph. The multiple alignment has to be consistent with the optimal pairwise alignments represented by the edges of this minimum spanning tree.

Applications

Multiple sequence alignment is a fundamental problem in computational biology. In particular, multiple sequence alignment is useful in identifying those common structures, which may only be weakly reflected in the sequence and not easily revealed by pairwise alignment. These common structures may carry important information for their evolutionary history, critical conserved motifs, and common 3D molecular structure, as well as biological functions.

More recently, multiple sequence alignment is also used in revealing noncoding RNAs (ncR-NAs) [2]. In this type of multiple alignment, we are not only align the underlying sequences but also the secondary structures of the RNAs. Researchers believe that ncRNAs that belong to the same family should have common components giving a similar secondary structure. The multiple alignment can help to locate and identify these common components.

Open Problems

A number of open problems related to the work of Gusfield remain open. For the SP measure, the center star method can be extended to the q-star method ($q > 2$) with approximation ratio of $2 - q/k$ [1, 10], sect. 7.5 of [11]). Whether there exists an approximation algorithm with better approximation ratio or with better time complexity is still unknown. For the TA measure, to be the best knowledge of the entry authors, the approximation ratio in Theorem 3 is currently the best result.

Another interesting direction related to this problem is the constrained multiple sequence alignment problem [12] which requires the multiple alignment to contain certain aligned characters with respect to a given constrained sequence. The best known result [6] is an approximation algorithm (also follows the idea of center star method) which gives an alignment with approximation ratio of $2 - 2/k$ for k strings.

For the complexity of the problem, Wang and Jiang [13] were the first to prove the NP-hardness of the problem with SP score under a *nonmetric* distance measure over a 4-symbol alphabet. More recently, in [5], the multiple alignment problem with SP score, star alignment, and TA score have been proved to be NP-hard for all binary or larger alphabets under *any metric*. Developing efficient approximation algorithms with good bounds for any of these measures is desirable.

Experimental Results

Two experiments have been reported in the paper showing that the worst-case error bounds in

Theorems 1 and 2 (for the SP measure) are pessimistic compared to the typical situation arising in practice.

The scoring scheme used in the experiments is $s(a, b) = 0$ if $a = b$; $s(a, b) = 1$ it either a or b is a space; otherwise $s(a, b) = 2$. Since computing the optimal multiple alignment with minimum SP score has been shown to be NP-hard, they evaluate the performance of their algorithms using the lower bound of $\sum_{i<j} D(X_i, X_j)$ (recall that $D(X_i, X_j)$ is the score of the optimal pairwise alignment of X_i and X_j). They have aligned 19 similar amino acid sequences with average length of 60 of homeoboxs from different species. The ratio of the scores of reported alignment by the center star method to the lower bound is only 1.018 which is far from the worst-case error bound given in Theorem 1. They also aligned 10 not-so-similar sequences near the homeoboxes, and the ratio of the reported alignment to the lower bound is 1.162. Results also show that the alignment obtained by the randomized algorithm is usually not far away from the lower bound.

Data Sets

The exact sequences used in the experiments are not provided.

Cross-References

▶ Statistical Multiple Alignment

Recommended Reading

1. Bafna V, Lawler EL, Pevzner PA (1997) Approximation algorithms for multiple sequence alignment. Theor Comput Sci 182:233–244
2. Dalli D, Wilm A, Mainz I, Stegar G (2006) STRAL: progressive alignment of non-coding RNA using base pairing probability vectors in quadratic time. Bioinformatics 22(13):1593–1599
3. Do C, Brudno M, Batzoglou S (2004) ProbCons: probabilistic consistency-based multiple alignment of amino acid sequences. In: Proceedings of the thirteenth national conference on artificial intelligence, pp 703–708, San Jose. AAAI Press
4. Edgar R (2004) MUSCLE: multiple sequence alignment with high accuracy and high throughput. Nucleic Acids Res 32:1792–1797
5. Elias I (2003) Setting the intractability of multiple alignment. In: Proceedings of the 14th annual international symposium on algorithms and computation (ISAAC 2003), Kyoto, pp 352–363
6. Francis YL, Chin NLH, Lam TW, Prudence WHW (2005) Efficient constrained multiple sequence alignment with performance guarantee. J Bioinform Comput Biol 3(1):1–18
7. Gusfield D (1993) Efficient methods for multiple sequence alignment with guaranteed error bounds. Bull Math Biol 55(1):141–154
8. Notredame C, Higgins D, Heringa J (2000) T-coffee: a novel method for multiple sequence alignments. J Mol Biol 302:205–217
9. Pevsner J (2003) Bioinformatics and functional genomics. Wiley, New York
10. Pevzner PA (1992) Multiple alignment, communication cost, and graph matching. SIAM J Appl Math 52:1763–1779
11. Pevzner PA (2000) Computational molecular biology: an algorithmic approach. MIT, Cambridge
12. Tang CY, Lu CL, Chang MDT, Tsai YT, Sun YJ, Chao KM, Chang JM, Chiou YH, Wu CM, Chang HT, Chou WI (2002) Constrained multiple sequence alignment tool development and its application to RNase family alignment. In: Proceedings of the first IEEE computer society bioinformatics conference (CSB 2002), Stanford, pp 127–137
13. Wang L, Jiang T (1994) On the complexity of multiple sequence alignment. J Comput Biol 1:337–48
14. Ye Y, Cheung D, Wang Y, Yiu SM, Qing Z, Lam TW, Ting HF GLProbs: aligning multiple sequences adaptively. IEEE/ACM Trans Comput Biol Bioinformatics. doi:http://doi.ieeecomputersociety.org/10.1109/TCBB.2014.2316820

Efficient Polynomial Time Approximation Scheme for Scheduling Jobs on Uniform Processors

Klaus Jansen
Department of Computer Science, University of Kiel, Kiel, Germany

Keywords

Approximation schemes; Integer linear programming; Scheduling; Sensitivity analysis

Years and Authors of Summarized Original Work

2009; Jansen
2011; Jansen, Robenek
2014; Chen, Jansen, Zhang

Problem Definition

We consider the following fundamental problem in scheduling theory. Suppose that there is a set $\mathcal{J}$ of n independent jobs J_j with processing time p_j and a set $\mathcal{P}$ of m nonidentical processors P_i that run at different speeds s_i. If job J_j is executed on processor P_i, then processor P_i needs p_j/s_i time units to complete the job. The goal is to find an assignment $a : \mathcal{J} \to \mathcal{P}$ for the jobs to the processors that minimizes the total length of the schedule $\max_{i=1,\dots,m} \sum_{J_j : a(J_j) = P_i} p_j/s_i$. This is the minimum time needed to complete all jobs on the processors. The problem is denoted $Q||C_{\max}$ and it is also called the minimum makespan problem on uniform parallel processors. By simplicity we may assume that the number m of processors is bounded by the number of jobs; otherwise select only the fastest n machines in $O(m)$ time.

Key Results

The scheduling problem on uniform and also identical processors is NP-hard [7] and the existence of a polynomial time algorithm for it would imply $P = NP$. Hochbaum and Shmoys [9, 10] presented a family of polynomial time approximation algorithms $\{A_\epsilon | \epsilon > 0\}$ for both scheduling problems, where each algorithm A_ϵ generates a schedule of length $(1 + \epsilon) OPT(I)$ for each instance I and has running time polynomial in the input size $|I|$. Such a family of algorithms is called a polynomial time approximation scheme (PTAS). It is allowed that the running time of each algorithm A_ϵ is exponential in $1/\epsilon$. The running time of the PTAS for uniform processors by Hochbaum and Shmoys [10] is $(n/\epsilon)^{O(1/\epsilon^2)}$.

Two restricted classes of approximation schemes were defined to classify different faster approximation scheme. An efficient polynomial time approximation scheme (EPTAS) is a PTAS with running time $f(1/\epsilon) \, poly(|I|)$ for some function f, while a fully polynomial time approximation scheme (FPTAS) runs in time $poly(1/\epsilon, |I|)$; polynomial in $1/\epsilon$ and the size $|I|$ of the instance. Since the scheduling problem on identical and also uniform processors is NP-hard in the strong sense (it contains bin packing as special case), we cannot hope for an FPTAS. For identical processors, Hochbaum and Shmoys (see [8]) and Alon et al. [1] gave an EPTAS with running time $f(1/\epsilon) + O(n)$, where f is doubly exponential in $1/\epsilon$.

Known Techniques

Hochbaum and Shmoys [9] introduced the dual approximation approach for identical and uniform processors and used the relationship between these scheduling problems and the bin packing problem. This relationship between scheduling on identical processors and bin packing problem had been exploited already by Coffman et al. [3]. Using the dual approximation approach, Hochbaum and Shmoys [9] proposed a PTAS for scheduling on identical processors with running time $(n/\epsilon)^{O(1/\epsilon^2)}$.

The main idea in the approach is to guess the length of the schedule by using binary search and to consider the corresponding bin packing instance with scaled identical bin size equal to 1. Then they distinguish between large items with size $> \epsilon$ and small items with size $\leq \epsilon$. For the large items they use a dynamic programming approach to calculate the minimum number of bins needed to pack them all. Afterward, they pack the remaining small items in a greedy way in enlarged bins of size $1 + \epsilon$ (i.e., they pack into any bin that currently contains items of total size at most 1; and if no such bin exists, then they open a new bin).

Furthermore, Hochbaum and Shmoys (see [8]) and Alon et al. [1] achieved an improvement to linear time by using an integer linear program for the cutting stock formulation of bin packing for the large items and a result on integer linear programming with a fixed number of variables by Lenstra [15]. This gives an EPTAS for identical

E

processors with running time $f(1/\epsilon) + O(n)$ where f is doubly exponential in $1/\epsilon$.

For uniform processors, the decision problem for the scheduling problem with makespan at most T can be viewed as a bin packing problem with different bin sizes. Using an ϵ-relaxed version of this bin packing problem, Hochbaum and Shmoys [10] were also able to obtain a PTAS for scheduling on uniform processors with running time $(n/\epsilon)^{O(1/\epsilon^2)}$. The main underlying idea in their algorithm is a clever rounding technique and a nontrivial dynamic programming approach over the different bins ordered by their sizes.

New Results

Recently, Jansen [11] proposed an EPTAS for scheduling jobs on uniform machines:

Theorem 1 ([11]) *There is an EPTAS (a family of algorithms* $\{A_\epsilon | \epsilon > 0\}$) *which, given an instance* I *of* $Q||C_{\max}$ *with* n *jobs and* m *processors with different speeds and a positive number* $\epsilon > 0$, *produces a schedule for the jobs of length* $A_\epsilon(I) \le (1 + \epsilon)OPT(I)$. *The running time of* A_ϵ *is*

$$2^{O(1/\epsilon^2 \log^3(1/\epsilon))} + poly(n).$$

Interestingly, the running time of the EPTAS is only single exponential in $1/\epsilon$.

Integer Linear Programming and Grouping Techniques

The new algorithm uses the dual approximation method by Hochbaum and Shmoys [10] to transform the scheduling problem into a bin packing problem with different bin sizes. Next, the input is structured by rounding bin sizes and processing times to values of the form $(1 + \delta)^i$ and $\delta(1 + \delta)^i$ with $i \in \mathbb{Z}$ where δ depends on ϵ. After sorting the bins according to their sizes, $c_1 \ge \ldots \ge c_m$, three groups of bins are built: $\mathcal{B}_1$ with the largest K bins (where K is constant). Let G be the smallest index such that capacity $c_{K+G+1} \le \gamma c_K$ where $\gamma < 1$ depends on ϵ; such an index G exists for $c_m \le \gamma c_K$. In this case $\mathcal{B}_2$ is the set of the next G largest bins where the maximum size $c_{\max}(\mathcal{B}_2) = c_{K+1}$ divided by the minimum size $c_{\min}(\mathcal{B}_2) = c_{K+G}$ is bounded by a constant $1/\gamma$ and $\mathcal{B}_3$ is the set with the remaining smaller bins

of size smaller than γc_K. This generates a gap of constant size between the capacities of bins in $\mathcal{B}_1$ and $\mathcal{B}_3$. If the rate c_m/c_K, where c_m is the smallest bin size, is larger the constant γ, then a simpler instance is obtained with only two groups $\mathcal{B}_1$ and $\mathcal{B}_2$ of bins.

For $\mathcal{B}_1$ all packings for the very large items are computed (those which only fit there). If there is a feasible packing, then a mixed integer linear program (MILP) or an integer linear program (ILP) in the simpler case is used to place the other items into the bins. The placement of the large items into the second group $\mathcal{B}_2$ is done via integral configuration variables; similar to the ILP formulation for bin packing by Fernandez de la Vega and Lueker [6]. Fractional configuration variables are used for the placement of large items into $\mathcal{B}_3$. Furthermore, additional fractional variables are taken to place small items into $\mathcal{B}_1$, $\mathcal{B}_2$, and $\mathcal{B}_3$. The MILP has only a constant number of integral variables and, therefore, can be solved via the algorithm by Lenstra or Kannan [14, 15].

In order to avoid that the running time is doubly exponential in $1/\epsilon$, a recent result by Eisenbrand and Shmonin [5] about integer cones is used. To apply their result a system of equalities for the integral configuration variables is considered and the corresponding coefficients are rounded. Then each feasible solution of the modified MILP contains at most $O(1/\delta \log^2(1/\delta))$ integral variables with values larger than zero. By choosing the strictly positive integral variables in the MILP, the number of integral configuration variables is reduced from $2^{O(1/\delta \log(1/\delta))}$ to $O(1/\delta \log^2(1/\delta))$. The number of choices is bounded by $2^{O(1/\delta^2 \log^3(1/\delta))}$.

Afterward, the fractional variables in the MILP solution are rounded to integral values using ideas from scheduling job shops [13] and scheduling on unrelated machines [16]. The effect of the rounding is that most of the items can be placed directly into the bins. Only a few of them cannot be placed this way, and here is where the K largest bins and the gap between $\mathcal{B}_1$ and $\mathcal{B}_3$ come into play. It can be proved that these items can be moved to the K largest bins by increasing their sizes only slightly.

Algorithm Avoiding the MILP

Recently an EPTAS for scheduling on uniform machines is presented by Jansen and Robenek [12] that avoids the use of an MILP or ILP solver. In the new approach instead of solving (M)ILPs, an LP-relaxation and structural information about the "closest" ILP solution is used.

In the following the main techniques are described for identical processors. For a given LP-solution x, the distance to the closest ILP solution y in the infinity norm is studied, i.e., $\|x - y\|_\infty$. For the constraint matrix A_δ of the considered LP, this distance is defined by

$$\text{max-gap}(A_\delta) := \max\{\min\{\|y^\star - x^\star\|_\infty : y^\star$$

$$\text{solution of ILP}\} : x^\star \text{solution of LP}\}.$$

Let $C(A_\delta)$ denote an upper bound for max-gap (A_δ). The running time of the algorithm is $2^{O(1/\epsilon \log(1/\epsilon) \log(C(A_\delta)))} + poly(n)$. The algorithm for uniform processors is more complex, but we obtain a similar running time $2^{O(1/\epsilon \log(1/\epsilon) \log(C(\tilde{A}_\delta)))} + poly(n)$, where the constraint matrix $\tilde{A}_\delta$ is slightly different. For the details we refer to [12].

It can be proved using a result by Cook et al. [4] that $C(A_\delta), C(\tilde{A}_\delta) \le 2^{O(1/\epsilon \log^2(1/\epsilon))}$. Consequently, the algorithm has a running time at most $2^{O(1/\epsilon^2 \log^3(1/\epsilon))} + poly(n)$, the same as in [11]. But, to our best knowledge, no instance is known to take on the value $2^{O(1/\epsilon \log^2(1/\epsilon))}$ for max-gap(A_δ). We conjecture $C(A_\delta) \le poly(1/\epsilon)$. If that holds, the running time of the algorithm would be $2^{O(1/\epsilon \log^2(1/\epsilon))} + poly(n)$ and thus improve the result in [11].

Lower Bounds

Recently, Chen, Jansen, and Zhang [2] proved the following lower bound on the running time: For scheduling on an arbitrary number of identical machines, denoted by $P||C_{\max}$, a polynomial time approximation scheme (PTAS) of running time $2^{O((1/\epsilon)^{1-\delta})} * poly(n)$ for any $\delta > 0$ would imply that the exponential time hypothesis (ETH) for 3-SAT fails.

Open Problems

The main open question is whether there is an EP-TAS for scheduling jobs on identical and uniform machines with a running time $2^{O(1/\epsilon \log^c(1/\epsilon))} * poly(n)$.

Experimental Results

None is reported.

Recommended Reading

1. Alon N, Azar Y, Woeginger GJ, Yadid T (1998) Approximation schemes for scheduling on parallel machines. J Sched 1:55–66
2. Chen L, Jansen K, Zhang G (2014) On the optimality of approximation schemes for the classical scheduling problem. In: Symposium on discrete algorithms (SODA 2014), Portland
3. Coffman EG, Garey MR, Johnson DS (1978) An application of bin packing to multiprocessor scheduling. SIAM J Comput 7:1–17
4. Cook W, Gerards AMH, Schrijver A, Tardos E (1986) Sensitivity theorems in integer linear programming. Math Program 34:251–264
5. Eisenbrand F, Shmonin G (2006) Caratheodory bounds for integer cones. Oper Res Lett 34:564–568
6. Fernandez de la Vega W, Lueker GS (1981) Bin packing can be solved within $1 + \epsilon$ in linear time. Combinatorica 1:349–355
7. Garey MR, Johnson DS (1979) Computers and intractability: a guide to the theory of NP-completeness. W.H. Freeman, San Francisco
8. Hochbaum DS (1997) Various notions of approximations: good, better, best, and more. In: Hochbaum DS (ed) Approximation algorithms for NP-hard problems, chap 9. Prentice Hall, Boston, pp 346–398
9. Hochbaum DS, Shmoys DB (1987) Using dual approximation algorithms for scheduling problems: practical and theoretical results. J ACM 34:144–162
10. Hochbaum DS, Shmoys DB (1988) A polynomial approximation scheme for scheduling on uniform processors: using the dual approximation approach. SIAM J Comput 17:539–551
11. Jansen K (2010) An EPTAS for scheduling jobs on uniform processors: using an MILP relaxation with a constant number of ntegral vasriables. SIAM J Discret Math 24(2):457–485
12. Jansen K, Robenek C (2011) Scheduling jobs on identical and uniform processors revisited. In: Workshop on approximation and online algorithms, WAOA 2011. LNCS, vol 7164, pp 109–122 and Technical report, University of Kiel, Saarbrücken, TR-1109

13. Jansen K, Solis-Oba R, Sviridenko M (2003) Makespan minimization in job shops: a linear time approximation acheme. SIAM J Discret Math 16:288–300
14. Kannan R (1987) Minkowski's convex body theorem and integer programming. Math Oper Res 12:415–440
15. Lenstra HW (1983) Integer programming with a fixed number of variables. Math Oper Res 8:538–548
16. Lenstra JK, Shmoys DB, Tardos E (1990) Approximation algorithms for scheduling unrelated parallel machines. Math Program 24:259–272

Engineering Algorithms for Computational Biology

David A. Bader
College of Computing, Georgia Institute of Technology, Atlanta, GA, USA

Keywords

High-performance computational biology

Years and Authors of Summarized Original Work

2002; Bader, Moret, Warnow

Problem Definition

In the 50 years since the discovery of the structure of DNA, and with new techniques for sequencing the entire genome of organisms, biology is rapidly moving towards a data-intensive, computational science. Many of the newly faced challenges require high-performance computing, either due to the massive-parallelism required by the problem, or the difficult optimization problems that are often combinatoric and NP-hard. Unlike the traditional uses of supercomputers for regular, numerical computing, many problems in biology are irregular in structure, significantly more challenging to parallelize, and integer-based using abstract data structures.

Biologists are in search of biomolecular sequence data, for its comparison with other genomes, and because its structure determines function and leads to the understanding of biochemical pathways, disease prevention and cure, and the mechanisms of life itself. Computational biology has been aided by recent advances in both technology and algorithms; for instance, the ability to sequence short contiguous strings of DNA and from these reconstruct the whole genome and the proliferation of high-speed microarray, gene, and protein chips for the study of gene expression and function determination. These high-throughput techniques have led to an exponential growth of available genomic data.

Algorithms for solving problems from computational biology often require parallel processing techniques due to the data- and compute-intensive nature of the computations. Many problems use polynomial time algorithms (e.g., all-to-all comparisons) but have long running times due to the large number of items in the input; for example, the assembly of an entire genome or the all-to-all comparison of gene sequence data. Other problems are compute-intensive due to their inherent algorithmic complexity, such as protein folding and reconstructing evolutionary histories from molecular data, that are known to be NP-hard (or harder) and often require approximations that are also complex.

Key Results

None

Applications

Phylogeny Reconstruction

A phylogeny is a representation of the evolutionary history of a collection of organisms or genes (known as taxa). The basic assumption of process necessary to phylogenetic reconstruction is repeated divergence within species or genes. A phylogenetic reconstruction is usually depicted as a tree, in which modern taxa are depicted at the leaves and ancestral taxa occupy internal nodes, with the edges of the tree denoting evolutionary relationships among the taxa. Reconstructing phylogenies is a major component of modern research programs in biology and medicine (as well as linguistics). Naturally, scientists are interested in phylogenies for the sake of knowledge, but such analyses also have many uses in applied research and in the commercial arena. Existing phylogenetic reconstruction techniques suffer from serious problems of running time (or, when fast, of accuracy). The problem is particularly serious for large data sets: even though data sets comprised of sequence from a single gene continue to pose challenges (e.g., some analyses are still running after 2 years of computation on medium-sized clusters), using whole-genome data (such as gene content and gene order) gives rise to even more formidable computational problems, particularly in data sets with large numbers of genes and highly-rearranged genomes.

To date, almost every model of speciation and genomic evolution used in phylogenetic reconstruction has given rise to NP-hard optimization problems. Three major classes of methods are in common use. Heuristics (a natural consequence of the NP-hardness of the problems) run quickly, but may offer no quality guarantees and may not even have a well-defined optimization criterion, such as the popular *neighbor-joining* heuristic [9]. Optimization based on the criterion of *maximum parsimony* (MP) [4] seeks the phylogeny with the least total amount of change needed to explain modern data. Finally, optimization based on the criterion of *maximum likelihood* (ML) [5] seeks the phylogeny that is the most likely to have given rise to the modern data.

Heuristics are fast and often rival the optimization methods in terms of accuracy, at least on datasets of moderate size. Parsimony-based methods may take exponential time, but, at least for DNA and amino acid data, can often be run to completion on datasets of moderate size. Methods based on maximum likelihood are very slow (the point estimation problem alone appears intractable) and thus restricted to very small instances, and also require many more assumptions than parsimony-based methods, but appear capable of outperforming the others in terms of the quality of solutions when these assumptions are met. Both MP- and ML-based analyses are often run with various heuristics to ensure timely

termination of the computation, with mostly unquantified effects on the quality of the answers returned.

Thus there is ample scope for the application of high-performance algorithm engineering in the area. As in all scientific computing areas, biologists want to study a particular dataset and are willing to spend months and even years in the process: accurate branch prediction is the main goal. However, since all exact algorithms scale exponentially (or worse, in the case of ML approaches) with the number of taxa, speed remains a crucial parameter – otherwise few datasets of more than a few dozen taxa could ever be analyzed.

Experimental Results

As an illustration, this entry briefly describes a high-performance software suite, GRAPPA (Genome Rearrangement Analysis through Parsimony and other Phylogenetic Algorithms) developed by Bader et al. *GRAPPA* extends Sankoff and Blanchette's breakpoint phylogeny algorithm [10] into the more biologically-meaningful inversion phylogeny and provides a highly-optimized code that can make use of distributed- and shared-memory parallel systems (see [1, 2, 6, 7, 8, 11] for details). In [3], Bader et al. gives the first linear-time algorithm and fast implementation for computing inversion distance between two signed permutations. *GRAPPA* was run on a 512-processor IBM Linux cluster with Myrinet and obtained a 512-fold speedup (linear speedup with respect to the number of processors): a complete breakpoint analysis (with the more demanding inversion distance used in lieu of breakpoint distance) for the 13 genomes in the Campanulaceae data set ran in less than 1.5 h in an October 2000 run, for a *million-fold* speedup over the original implementation. The latest version features significantly improved bounds and new distance correction methods and, on the same dataset, exhibits a speedup factor of *over one billion*. GRAPPA achieves this speedup through a combination of parallelism and high-performance algorithm engineering.

Although such spectacular speedups will not always be realized, many algorithmic approaches now in use in the biological, pharmaceutical, and medical communities may benefit tremendously from such an application of high-performance techniques and platforms.

This example indicates the potential of applying high-performance algorithm engineering techniques to applications in computational biology, especially in areas that involve complex optimizations: Bader's reimplementation did not require new algorithms or entirely new techniques, yet achieved gains that turned an impractical approach into a usable one.

Cross-References

► Distance-Based Phylogeny Reconstruction (Fast-Converging)
► Distance-Based Phylogeny Reconstruction: Safety and Edge Radius
► Efficient Methods for Multiple Sequence Alignment with Guaranteed Error Bounds
► High Performance Algorithm Engineering for Large-Scale Problems
► Local Alignment (with Affine Gap Weights)
► Local Alignment (with Concave Gap Weights)
► Multiplex PCR for Gap Closing (Whole-Genome Assembly)
► Peptide De Novo Sequencing with MS/MS
► Perfect Phylogeny Haplotyping
► Phylogenetic Tree Construction from a Distance Matrix
► Sorting Signed Permutations by Reversal (Reversal Distance)
► Sorting Signed Permutations by Reversal (Reversal Sequence)
► Sorting by Transpositions and Reversals (Approximate Ratio 1.5)
► Substring Parsimony

Recommended Reading

1. Bader DA, Moret BME, Warnow T, Wyman SK, Yan M (2001) High-performance algorithm engineering for gene-order phylogenies. In: DIMACS workshop on whole genome comparison. Rutgers University, Piscataway
2. Bader DA, Moret BME, Vawter L (2001) Industrial applications of high-performance computing for phylogeny reconstruction. In: Siegel HJ (ed) Proceedings of the SPIE commercial applications for high-performance computing, vol 4528. Denver, pp 159–168
3. Bader DA, Moret BME, Yan M (2001) A linear-time algorithm for computing inversion distance between signed permutations with an experimental study. J Comput Biol 8(5):483–491
4. Farris JS (1983) The logical basis of phylogenetic analysis. In: Platnick NI, Funk VA (eds) Advances in cladistics. Columbia University Press, New York, pp 1–36
5. Felsenstein J (1981) Evolutionary trees from DNA sequences: a maximum likelihood approach. J Mol Evol 17:368–376
6. Moret BME, Bader DA, Warnow T, Wyman SK, Yan M (2001) GRAPPA: a high performance computational tool for phylogeny reconstruction from gene-order data. In: Proceedings of the botany, Albuquerque
7. Moret BME, Bader DA, Warnow T (2002) High-performance algorithm engineering for computational phylogenetics. J Supercomput 22:99–111, Special issue on the best papers from ICCS'01
8. Moret BME, Wyman S, Bader DA, Warnow T, Yan M (2001) A new implementation and detailed study of breakpoint analysis. In: Proceedings of the 6th Pacific symposium biocomputing (PSB 2001), Hawaii, Jan 2001, pp 583–594
9. Saitou N, Nei M (1987) The neighbor-joining method: a new method for reconstruction of phylogenetic trees. Mol Biol Evol 4:406–425
10. Sankoff D, Blanchette M (1998) Multiple genome rearrangement and breakpoint phylogeny. J Comput Biol 5:555–570
11. Yan M (2004) High performance algorithms for phylogeny reconstruction with maximum parsimony. PhD thesis, Electrical and Computer Engineering Department, University of New Mexico, Albuquerque, Jan 2004

Engineering Algorithms for Large Network Applications

Christos Zaroliagis
Department of Computer Engineering and Informatics, University of Patras, Patras, Greece

Years and Authors of Summarized Original Work

2002; Schulz, Wagner, Zaroliagis

Problem Definition

Dealing effectively with applications in large networks, it typically requires the efficient solution of one ore more underlying algorithmic problems. Due to the size of the network, a considerable effort is inevitable in order to achieve the desired efficiency in the algorithm.

One of the primary tasks in large network applications is to answer queries for finding best routes or paths as efficiently as possible. Quite often, the challenge is to process a vast number of such queries on-line: a typical situation encountered in several real-time applications (e.g., traffic information systems, public transportation systems) concerns a query-intensive scenario, where a central server has to answer a huge number of on-line customer queries asking for their best routes (or optimal itineraries). The main goal in such an application is to reduce the (average) response time for a query.

Answering a best route (or optimal itinerary) query translates in computing a minimum cost (shortest) path on a suitably defined directed graph (digraph) with nonnegative edge costs. This in turn implies that the core algorithmic problem underlying the efficient answering of queries is the single-source single-target shortest path problem.

Although the straightforward approach of precomputing and storing shortest paths for all pairs of vertices would enabling the optimal answering of shortest path queries, the quadratic space requirements for digraphs with more than 10^5 vertices makes such an approach prohibitive for large and very large networks. For this reason, the main goal of almost all known approaches is to keep the space requirements as small as possible. This in turn implies that one can afford a heavy (in time) preprocessing, which does not blow up space, in order to speed-up the query time.

The most commonly used approach for answering shortest path queries employs Dijkstra's algorithm and/or variants of it. Consequently, the main challenge is how to reduce the algorithm's search-space (number of vertices visited), as this would immediately yield a better query time.

Key Results

All results discussed concern answering of *optimal* (or *exact* or *distance-preserving*) shortest paths under the aforementioned query-intensive scenario, and are all based on the following generic approach. A preprocessing of the input network $G = (V, E)$ takes place that results in a data structure of size $O(|V| + |E|)$ (i.e., linear to the size of G). The data structure contains additional information regarding certain shortest paths that can be used later during querying.

Depending on the pre-computed additional information as well as on the way a shortest path query is answered, two approaches can be distinguished. In the first approach, *graph annotation*, the additional information is attached to vertices or edges of the graph. Then, speed-up techniques to Dijkstra's algorithm are employed that, based on this information, decide quickly which part of the graph does not need to be searched. In the second approach, an *auxiliary graph G'* is constructed hierarchically. A shortest path query is then answered by searching only a small part of G', using Dijkstra's algorithm enhanced with heuristics to further speed-up the query time.

In the following, the key results of the first [3, 4, 9, 11] and the second approach [1, 2, 5, 7, 8, 10] are discussed, as well as results concerning modeling issues.

First Approach: Graph Annotation

The first work under this approach concerns the study in [9] on large railway networks. In that paper, two new heuristics are introduced: the *angle-restriction* (that tries to reduce the search space by taking advantage of the geometric layout of the vertices) and the *selection of stations* (a subset of vertices is selected among which all pairs shortest paths are pre-computed). These two heuristics along with a combination of the classical *goal-directed* or A^* *search* turned out to be rather efficient. Moreover, they motivated two important generalizations [10, 11] that gave further improvements to shortest path query times.

The full exploitation of geometry-based heuristics was investigated in [11], where both street and railway networks are considered. In that paper, it is shown that the search space of Dijkstra's algorithm can be significantly reduced (to 5–10 % of the initial graph size) by extracting geometric information from a given layout of the graph and by encapsulating pre-computed shortest path information in resulted geometric objects, called *containers*. Moreover, the dynamic case of the problem was investigated, where edge costs are subject to change and the geometric containers have to be updated.

A powerful modification to the classical Dijkstra's algorithm, called *reach-based routing*, was presented in [4]. Every vertex is assigned a so-called *reach value* that determines whether a particular vertex will be considered during Dijkstra's algorithm. A vertex is excluded from consideration if its reach value is small; that is, if it does not contribute to any path long enough to be of use for the current query.

A considerable enhancement of the classical A^* *search* algorithm using landmarks (selected vertices like in [9, 10]) and the triangle inequality with respect to the shortest path distances was shown in [3]. Landmarks and triangle inequality help to provide better lower bounds and hence boost A^* search.

Second Approach: Auxiliary Graph

The first work under this approach concerns the study in [10], where a new hierarchical decomposition technique is introduced called *multi-level graph*. A multi-level graph $\mathcal{M}$ is a digraph which is determined by a sequence of subsets of V and which extends E by adding multiple levels of edges. This allows to efficiently construct, during querying, a subgraph of $\mathcal{M}$ which is substantially smaller than G and in which the shortest path distance between any of its vertices is equal to the shortest path distance between the same vertices in G. Further improvements of this approach have been presented recently in [1]. A refinement of the above idea was introduced in [5], where the multi-level overlay graphs are introduced. In such

a graph, the decomposition hierarchy is not determined by application-specific information as it happens in [9, 10].

An alternative hierarchical decomposition technique, called *highway hierarchies*, was presented in [7]. The approach takes advantage of the inherent hierarchy possessed by real-world road networks and computes a hierarchy of coarser views of the input graph. Then, the shortest path query algorithm considers mainly the (much smaller in size) coarser views, thus achieving dramatic speed-ups in query time. A revision and improvement of this method was given in [8]. A powerful combination of the highway hierarchies with the ideas in [3] was reported in [2].

Modeling Issues

The modeling of the original best route (or optimal itinerary) problem on a large network to a shortest path problem in a suitably defined directed graph with appropriate edge costs also plays a significant role in reducing the query time. Modeling issues are thoroughly investigated in [6]. In that paper, the first experimental comparison of two important approaches (time-expanded versus time-dependent) is carried out, along with new extensions of them towards realistic modeling. In addition, several new heuristics are introduced to speed-up query time.

Applications

Answering shortest path queries in large graphs has a multitude of applications, especially in traffic information systems under the aforementioned scenario; that is, a central server has to answer, as fast as possible, a huge number of on-line customer queries asking for their best routes or itineraries. Other applications of the above scenario involve route planning systems for cars, bikes and hikers, public transport systems for itinerary information of scheduled vehicles (like trains or buses), answering queries in spatial databases, and web searching. All the above applications concern real-time systems in which users continuously enter their requests for finding their best connections or routes. Hence, the main goal is to reduce the (average) response time for answering a query.

Open Problems

Real-world networks increase constantly in size either as a result of accumulation of more and more information on them, or as a result of the digital convergence of media services, communication networks, and devices. This scaling-up of networks makes the scalability of the underlying algorithms questionable. As the networks continue to grow, there will be a constant need for designing faster algorithms to support core algorithmic problems.

Experimental Results

All papers discussed in section "Key Results" contain important experimental studies on the various techniques they investigate.

Data Sets

The data sets used in [6, 11] are available from http://lso-compendium.cti.gr/ under problems 26 and 20, respectively.

The data sets used in [1, 2] are available from http://www.dis.uniroma1.it/~challenge9/.

URL to Code

The code used in [9] is available from http://doi.acm.org/10.1145/351827.384254.

The code used in [6, 11] is available from http://lso-compendium.cti.gr/ under problems 26 and 20, respectively.

The code used in [3] is available from http://www.avglab.com/andrew/soft.html.

Cross-References

▶ Implementation Challenge for Shortest Paths
▶ Shortest Paths Approaches for Timetable Information

Recommended Reading

1. Delling D, Holzer M, Müller K, Schulz F, Wagner D (2006) High-performance multi-level graphs. In: 9th DIMACS challenge on shortest paths, Rutgers University, Nov 2006
2. Delling D, Sanders P, Schultes D, Wagner D (2006) Highway hierarchies star. In: 9th DIMACS challenge on shortest paths, Rutgers University, Nov 2006
3. Goldberg AV, Harrelson C (2005) Computing the shortest path: A^* search meets graph theory. In: Proceedings of the 16th ACM-SIAM symposium on discrete algorithms – SODA. ACM, New York/SIAM, Philadelphia, pp 156–165
4. Gutman R (2004) Reach-based routing: a new approach to shortest path algorithms optimized for road networks. In: Algorithm engineering and experiments – ALENEX (SIAM, 2004). SIAM, Philadelphia, pp 100–111
5. Holzer M, Schulz F, Wagner D (2006) Engineering multi-level overlay graphs for shortest-path queries. In: Algorithm engineering and experiments – ALENEX (SIAM, 2006). SIAM, Philadelphia, pp 156–170
6. Pyrga E, Schulz F, Wagner D, Zaroliagis C (2007) Efficient models for timetable information in public transportation systems. ACM J Exp Algorithmic 12(2.4):1–39
7. Sanders P, Schultes D (2005) Highway hierarchies hasten exact shortest path queries. In: Algorithms – ESA 2005. Lecture notes in computer science, vol 3669. pp 568–579
8. Sanders P, Schultes D (2006) Engineering highway hierarchies. In: Algorithms – ESA 2006. Lecture notes in computer science, vol 4168. pp 804–816
9. Schulz F, Wagner D, Weihe K (2000) Dijkstra's algorithm on-line: an empirical case study from public railroad transport. ACM J Exp Algorithmics 5(12): 1–23
10. Schulz F, Wagner D, Zaroliagis C (2002) Using multi-level graphs for timetable information in railway systems. In: Algorithm engineering and experiments – ALENEX 2002. Lecture notes in computer science, vol 2409. pp 43–59
11. Wagner D, Willhalm T, Zaroliagis C (2005) Geometric containers for efficient shortest path computation. ACM J Exp Algorithm 10(1.3):1–30

Engineering Geometric Algorithms

Dan Halperin
School of Computer Science, Tel-Aviv
University, Tel Aviv, Israel

Keywords

Certified and efficient implementation of geometric algorithms; Geometric computing with certified numerics and topology

Years and Authors of Summarized Original Work

2004; Halperin

Problem Definition

Transforming a theoretical geometric algorithm into an effective computer program abounds with hurdles. Overcoming these difficulties is the concern of *engineering geometric algorithms*, which deals, more generally, with the design and implementation of certified and efficient solutions to algorithmic problems of geometric nature. Typical problems in this family include the construction of Voronoi diagrams, triangulations, arrangements of curves and surfaces (namely, space subdivisions), two- or higher-dimensional search structures, convex hulls and more.

Geometric algorithms strongly couple topological/combinatorial structures (e.g., a graph describing the triangulation of a set of points) on the one hand, with numerical information (e.g., the coordinates of the vertices of the triangulation) on the other. Slight errors in the numerical calculations, which in many areas of science and engineering can be tolerated, may lead to detrimental mistakes in the topological structure, causing the computer program to crash, to loop infinitely, or plainly to give wrong results.

Straightforward implementation of geometric algorithms as they appear in a textbook, using standard machine arithmetic, is most likely to fail. This entry is concerned only with *certified* solutions, namely, solutions that are guaranteed to construct the exact desired structure or a good approximation of it; such solutions are often referred to as *robust*.

The goal of engineering geometric algorithms can be restated as follows: *Design and implement geometric algorithms that are at once robust and efficient in practice.*

Much of the difficulty in adapting in practice the existing vast algorithmic literature in computational geometry comes from the assumptions that are typically made in the theoretical study of geometric algorithms that (1) the input is in general position, namely, degenerate input is precluded, (2) computation is performed on an ideal computer that can carry out real arithmetic to infinite precision (so-called real RAM), and (3) the cost of operating on a small number of simple geometric objects is "unit" time (e.g., equal cost is assigned to intersecting three spheres and to comparing two integer numbers).

Now, in real life, geometric input is quite often degenerate, machine precision is limited, and operations on a small number of simple geometric objects within the same algorithm may differ 100-fold and more in the time they take to execute (when aiming for certified results). Just implementing an algorithm carefully may not suffice and often redesign is called for.

Key Results

Tremendous efforts have been invested in the design and implementation of robust computational-geometry software in recent years. Two notable large-scale efforts are the CGAL library [1] and the geometric part of the LEDA library [14]. These are jointly reviewed in the survey by Kettner and Näher [13]. Numerous other relevant projects, which for space constraints are not reviewed here, are surveyed by Joswig [12] with extensive references to papers and Web sites.

A fundamental engineering decision to take when coming to implement a geometric algorithm is what will the underlying arithmetic be, that is, whether to opt for exact computation or use the machine floating-point arithmetic. (Other less commonly used options exist as well.) To date, the CGAL and LEDA libraries are almost exclusively based on exact computation. One of the reasons for this exclusivity is that exact computation emulates the ideal computer (for restricted problems) and makes the adaptation of algorithms from theory to software easier. This is facilitated by major headway in developing tools for efficient computation with rational or algebraic numbers (GMP [3], LEDA [14], CORE [2] and more). On top of these tools, clever techniques for reducing the amount of exact computation were developed, such as floating-point filters and the higher-level geometric filtering.

The alternative is to use the machine floating-point arithmetic, having the advantage of being very fast. However, it is nowhere near the ideal infinite precision arithmetic assumed in the theoretical study of geometric algorithms and algorithms have to be carefully redesigned. See, for example, the discussion about imprecision in the manual of QHULL, the convex hull program by Barber et al. [5]. Over the years a variety of specially tailored floating-point variants of algorithms have been proposed, for example, the carefully crafted VRONI package by Held [11], which computes the Voronoi diagram of points and line segments using standard floating-point arithmetic, based on the topology-oriented approach of Sugihara and Iri. While VRONI works very well in practice, it is not theoretically certified. *Controlled perturbation* [9] emerges as a systematic method to produce certified approximations of complex geometric constructs while using floating-point arithmetic: the input is perturbed such that all predicates are computed accurately even with the limited-precision machine arithmetic, and a method is given to bound the necessary magnitude of perturbation that will guarantee the successful completion of the computation.

Another decision to take is how to represent the output of the algorithm, where the major issue is typically how to represent the coordinates of vertices of the output structure(s). Interestingly, this question is crucial when using exact computation since there the output coordinates can be prohibitively large or simply impossible to finitely enumerate. (One should note though that many geometric algorithms are *selective* only, namely, they do not produce new geometric entities but just select and order subsets of the input coordinates. For example, the output of an algorithm for computing the convex hull of a set of points in the plane is an ordering of a subset of the input points. No new point is computed. The discussion in this paragraph mostly applies to algorithms that output new geometric constructs, such as the intersection point of two lines.) But even when using floating-point arithmetic, one may prefer to have a more compact bit-size representation than, say, machine doubles. In this direction there is an effective, well-studied solution for the case of polygonal objects in the plane, called *snap rounding*, where vertices and intersection points are snapped to grid vertices while retaining certain topological properties of the exact desired structure. Rounding with guarantees is in general a very difficult problem, and already for polyhedral objects in 3-space the current attempts at generalizing snap rounding are very costly (increasing the complexity of the rounded objects to the third, or even higher, power).

Then there are a variety of engineering issues depending on the problem at hand. Following are two examples of engineering studies where the experience in practice is different from what the asymptotic resource measures imply. The examples relate to fundamental steps in many geometric algorithms: decomposition and point location.

Decomposition

A basic step in many geometric algorithms is to decompose a (possibly complex) geometric object into simpler subobjects, where each subobject typically has constant descriptive complexity. A well-known example is the triangulation of a polygon. The choice of decomposition may have a significant effect on the efficiency in practice of various algorithms that rely on decomposition. Such is the case when constructing Minkowski sums of polygons in the plane. The Minkowski sum of two sets A and B in $\mathbb{R}^d$ is the vector sum of the two sets $A \oplus B = \{a + b | a \in A, b \in B\}$. The simplest approach to computing Minkowski sums of two polygons in the plane proceeds in three steps: triangulate each polygon, then compute the sum of each triangle of one polygon with each triangle of the other, and finally take the union of all the subsums. In asymptotic measures, the choice of triangulation (over alternative decompositions) has no effect. In practice though, triangulation is probably the worst choice compared with other convex decompositions, even fairly simple heuristic ones (not necessarily optimal), as shown by experiments on a dozen different decomposition methods [4]. The explanation is that triangulation increases the overall complexity of the subsums and in turn makes the union stage more complex – indeed by a constant factor, but a noticeable factor in practice. Similar phenomena were observed in other situations as well. For example, when using the prevalent vertical decomposition of arrangements – often it is too costly compared with sparser decompositions (i.e., decompositions that add fewer extra features).

Point Location

A recurring problem in geometric computing is to process given planar subdivision (planar map), so as to efficiently answer *point-location* queries: Given a point q in the plane, which face of the map contains q? Over the years a variety of point-location algorithms for planar maps were implemented in CGAL, in particular, a hierarchical search structure that guarantees logarithmic query time after expected $O(n \log n)$ preprocessing time of a map with n edges. This algorithm is referred to in CGAL as the *RIC* point-location algorithm after the preprocessing method which uses randomized incremental construction. Several simpler, easier-to-program algorithms for point location were also implemented. None of the latter beats the RIC algorithm in query time. However, the RIC is by far the slowest of all the implemented

algorithms in terms of preprocessing, which in many scenarios renders it less effective. One of the simpler methods devised is a variant of the well-known *jump-and-walk* approach to point location. The algorithm scatters points (so-called *landmarks*) in the map and maintains the landmarks (together with their containing faces) in a nearest-neighbor search structure. Once a query q is issued it finds the nearest landmark ℓ to q, and "walks" in the map from ℓ toward q along the straight line segment connecting them. This landmark approach offers query time that is only slightly more expensive than the RIC method while being very efficient in preprocessing. The full details can be found in [10]. This is yet another consideration when designing (geometric) algorithms: the cost of preprocessing (and storage) versus the cost of a query. Quite often the effective (practical) tradeoff between these costs needs to be deduced experimentally.

Applications

Geometric algorithms are useful in many areas. Triangulations and arrangements are examples of basic constructs that have been intensively studied in computational geometry, carefully implemented and experimented with, as well as used in diverse applications.

Triangulations

Triangulations in two and three dimensions are implemented in CGAL [7]. In fact, CGAL offers many variants of triangulations useful for different applications. Among the applications where CGAL triangulations are employed are meshing, molecular modeling, meteorology, photogrammetry, and geographic information systems (GIS). For other available triangulation packages, see the survey by Joswig [12].

Arrangements

Arrangements of curves in the plane are supported by CGAL [15], as well as envelopes of surfaces in three-dimensional space. Forthcoming is support also for arrangements of curves on surfaces. CGAL arrangements have been used in motion planning algorithms, computer-aided design and manufacturing, GIS, computer graphics, and more (see Chap. 1 in [6]).

Open Problems

In spite of the significant progress in certified implementation of effective geometric algorithms, the existing theoretical algorithmic solutions for many problems still need adaptation or redesign to be useful in practice. One example where progress can have wide repercussions is devising effective decompositions for curved geometric objects (e.g., arrangements) in the plane and for higher-dimensional objects. As mentioned earlier, suitable decompositions can have a significant effect on the performance of geometric algorithms in practice.

Certified fixed-precision geometric computing lags behind the exact computing paradigm in terms of available robust software, and moving forward in this direction is a major challenge. For example, creating a certified floating-point counterpart to CGAL is a desirable (and highly intricate) task.

Another important tool that is largely missing is consistent and efficient rounding of geometric objects. As mentioned earlier, a fairly satisfactory solution exists for polygonal objects in the plane. Good techniques are missing for curved objects in the plane and for higher-dimensional objects (both linear and curved).

URL to Code

http://www.cgal.org

Cross-References

▶ LEDA: a Library of Efficient Algorithms
▶ Robust Geometric Computation

Recommended Reading

Conferences publishing papers on the topic include the ACM Symposium on Computational Geometry (SoCG), the Workshop on Algorithm Engineering and Experiments (ALENEX), the Engineering and Applications Track of the European Symposium on Algorithms (ESA), its predecessor and the Workshop on Experimental Algorithms (WEA). Relevant journals include the *ACM Journal on Experimental Algorithmics, Computational Geometry: Theory and Applications* and the *International Journal of Computational Geometry and Applications*. A wide range of relevant aspects are discussed in the recent book edited by Boissonnat and Teillaud [6], titled *Effective Computational Geometry for Curves and Surfaces*.

1. The CGAL project homepage. http://www.cgal.org/. Accessed 6 Apr 2008
2. The CORE library homepage. http://www.cs.nyu.edu/exact/core/. Accessed 6 Apr 2008
3. The GMP webpage. http://gmplib.org/. Accessed 6 Apr 2008
4. Agarwal PK, Flato E, Halperin D (2002) Polygon decomposition for efficient construction of Minkowski sums. Comput Geom Theory Appl 21(1–2):39–61
5. Barber CB, Dobkin DP, Huhdanpaa HT (2008) Imprecision in QHULL. http://www.qhull.org/html/qh-impre.htm. Accessed 6 Apr 2008
6. Boissonnat J-D, Teillaud M (eds) (2006) Effective computational geometry for curves and surfaces. Springer, Berlin
7. Boissonnat J-D, Devillers O, Pion S, Teillaud M, Yvinec M (2002) Triangulations in CGAL. Comput Geom Theory Appl 22(1–3):5–19
8. Fabri A, Giezeman G-J, Kettner L, Schirra S, Schönherr S (2000) On the design of CGAL a computational geometry algorithms library. Softw Pract Exp 30(11):1167–1202
9. Halperin D, Leiserowitz E (2004) Controlled perturbation for arrangements of circles. Int J Comput Geom Appl 14(4–5):277–310
10. Haran I, Halperin D (2006) An experimental study of point location in general planar arrangements. In: Proceedings of 8th workshop on algorithm engineering and experiments, Miami, pp 16–25
11. Held M (2001) VRONI: an engineering approach to the reliable and efficient computation of Voronoi diagrams of points and line segments. Comput Geom Theory Appl 18(2):95–123
12. Joswig M (2004) Software. In: Goodman JE, O'Rourke J (eds) Handbook of discrete and computational geometry, chapter 64, 2nd edn. Chapman & Hall/CRC, Boca Raton, pp 1415–1433

13. Kettner L, Näher S (2004) Two computational geometry libraries: LEDA and CGAL. In: Goodman JE, O'Rourke J (eds) Handbook of discrete and computational geometry, chapter 65, 2nd edn. Chapman & Hall/CRC, Boca Raton, pp 1435–1463
14. Mehlhorn K, Näher S (2000) LEDA: a platform for combinatorial and geometric computing. Cambridge University Press, Cambridge
15. Wein R, Fogel E, Zukerman B, Halperin D (2007) Advanced programming techniques applied to CGAL's arrangement package. Comput Geom Theory Appl 36(1–2):37–63

Enumeration of Non-crossing Geometric Graphs

Shin-ichi Tanigawa
Research Institute for Mathematical Sciences (RIMS), Kyoto University, Kyoto, Japan

Keywords

Enumeration; Non-crossing (crossing-free) geometric graphs; Triangulations

Years and Authors of Summarized Original Work

2009; Katoh, Tanigawa
2014; Wettstein

Problem Definition

Let P be a set of n points in the plane in general position, i.e., no three points are collinear. A *geometric graph on P* is a graph on the vertex set P whose edges are straight-line segments connecting points in P. A geometric graph is called *non-crossing* (or *crossing-free*) if any pair of its edges does not have a point in common except possibly their endpoints. We denote by $\mathcal{P}(P)$ the set of all non-crossing geometric graphs on P (which are also called *plane straight-line graphs* on P). A graph class $\mathcal{C}(P) \subseteq \mathcal{P}(P)$ can be defined by imposing additional properties such as connectivity, degree bound, or cycle-freeness. Examples of $\mathcal{C}(P)$ are the set of *triangulations* (i.e., inclusion-wise maximal graphs in $\mathcal{P}(P)$),

the set of *non-crossing perfect matchings*, the set of *non-crossing spanning k-connected graphs*, the set of *non-crossing spanning trees*, and the set of *non-crossing spanning cycles* (i.e., simple polygons). The problem is to enumerate all graphs in $\mathcal{C}(P)$ for a given set P of n points in the plane.

The following notations will be used to denote the cardinality of $\mathcal{C}(P)$: $\mathsf{tri}(P)$ for triangulations, $\mathsf{pg}(P)$ for plane straight-line graphs, $\mathsf{st}(P)$ for non-crossing spanning trees, and $\mathsf{cg}(P)$ for non-crossing spanning connected graphs.

Key Results

Enumeration of Triangulations

The first efficient enumeration algorithm for triangulations was given by Avis and Fukuda [3] as an application of their reverse search technique. The algorithm relies on well-known properties of Delaunay triangulations.

A triangulation T on P is called *Delaunay* if no point in P is contained in the interior of the circumcircle of a triangle in T. If it is assumed for simplicity that no four points in P lie on a circle, then the Delaunay triangulation on P exists and is unique. The Delaunay triangulation has the lexicographically largest angle vector among all triangulations on P, where the angle vector of a triangulation is the list of all the angles sorted in nondecreasing order.

For a triangulation T, a *Lawson edge* is an edge ab which is incident to two triangles, say abc and abd in T, and the circumcircle of abc contains d in its interior. *Flipping* a Lawson edge ab (i.e., replacing ab with another diagonal edge cd) always creates a triangulation having a lexicographically larger angle vector. Moreover a triangulation has a Lawson edge if and only if it is not Delaunay. In other words, any triangulation can be converted to the Delaunay triangulation by flipping Lawson edges.

In the algorithm by Avis and Fukuda, a rooted search tree on the set of triangulations is defined such that the root is the Delaunay triangulation and the parent of a non-Delaunay triangulation T

is a triangulation obtained by flipping the smallest Lawson edge in T (assuming a fixed total ordering on edges). Since the Delaunay triangulation can be computed in $O(n \log n)$ time, all the triangulations can be enumerated by tracing the rooted search tree based on the reverse search technique. A careful implementation achieves $O(n \cdot \text{tr}(P))$ time with $O(n)$ space.

An improved algorithm was given by Bespamyatnikh [5], which runs in $O(\log \log n \cdot \text{tr}(P))$ time with $O(n)$ space. His algorithm is also based on the reverse search technique, but the rooted search tree is defined by using the lexicographical ordering of edge vectors rather than angle vectors. This approach was also applied to the enumeration of pointed pseudo-triangulations [4]. See [6] for another approach.

Enumeration of Non-crossing Geometric Graphs

In [3], Avis and Fukuda also developed an enumeration algorithm for non-crossing spanning trees, whose running time is $O(n^3 \cdot \text{sp}(P))$. This was improved to $O(n \log n \cdot \text{sp}(P))$ by Aichholzer et al. [1]. They also gave enumeration algorithms for plane straight-line graphs and non-crossing spanning connected graphs with running time $O(n \log n \cdot \text{pg}(P))$ and $O(n \log n \cdot \text{sc}(P))$, respectively.

Katoh and Tangiawa [8] proposed a simple enumeration technique for wider classes of non-crossing geometric graphs. The same approach was independently given by Razen and Welzl [9] for counting the number of plane straight-line graphs, and the following description in terms of Delaunay triangulations is from [9].

Since each graph in $\mathcal{C}(P)$ is a subgraph of a triangulation, one can enumerate all graphs in $\mathcal{C}(P)$ by first enumerating all triangulations and then enumerating all graphs in $\mathcal{C}(P)$ in each triangulation. The output may contain duplicates, but one can avoid duplicates by enumerating only graphs in $\{G \in \mathcal{C}(P) \mid L(T) \subseteq E(G) \subseteq E(T)\}$ for each triangulation T, where $L(T)$ denotes the set of the Lawson edges in T. This enumeration framework leads to an algorithm with time complexity $O((t^{\text{pre}} + \log \log n)\text{tri}(P) + t \cdot \text{c}(P))$

and space complexity $O(n + s)$ provided that graphs in $\{G \in \mathcal{C}(G) \mid L(T) \subseteq E(G) \subseteq E(T)\}$ can be enumerated in $O(t)$ time per graph with $O(t^{\text{pre}})$ time preprocessing and $O(s)$ space for each triangulation T. For example, in the case of non-crossing spanning trees, one can use a fast enumeration algorithm for spanning trees in a given undirected graph to solve each subproblem, and the current best implementation gives an enumeration algorithm for non-crossing spanning trees with time complexity $O(n \cdot \text{tri}(P) + \text{st}(P))$.

For plane straight-line graphs and spanning connected graphs, $\text{pg}(P) \geq (\sqrt{8})^n \text{tri}(P)$ [9] and $\text{cs}(P) \geq 1.51^n \text{tri}(P)$ [8] hold for any P in general position. Hence $\text{tri}(P)$ is dominated by $\text{pg}(P)$ and $\text{cs}(P)$, respectively, and plane straight-line graphs or non-crossing spanning connected graphs can be enumerated in constant time on average with $O(n)$ space [8]. The same technique can be applied to the set of non-crossing spanning 2-connected graphs. It is not known whether there is a constant $c > 1$ such that $\text{st}(P) \geq c^n \text{tri}(P)$ for every P in general position.

In [8] an approach that avoids enumerating all triangulations was also discussed. Suppose that a nonempty subset $\mathcal{I}$ of $\mathcal{P}(P)$ satisfies a monotone property, i.e., for every $G, G' \in \mathcal{P}(P)$ with $G \subseteq G'$, $G' \in \mathcal{I}$ implies $G \in \mathcal{I}$, and suppose that $\mathcal{C}(P)$ is the set of all maximal elements in $\mathcal{I}$. Then all graphs in $\mathcal{C}(P)$ can be enumerated just by enumerating all triangulations T on P with $L(T) \in \mathcal{I}$, and this can be done efficiently based on the reverse search technique. This approach leads to an algorithm for enumerating non-crossing minimally rigid graphs in $O(n^2)$ time per output with $O(n)$ space, where a graph $G = (V, E)$ is called *minimally rigid* if $|E| = 2|V| - 3$ and $|E'| \leq 2|V'| - 3$ for any subgraph $G' = (V', E')$ with $|V'| \geq 2$.

Enumeration of Non-crossing Perfect Matchings

Wettstein [10] proposed a new enumeration (and counting) technique for non-crossing geometric graphs. This is motivated from a counting algorithm of triangulations by Alvarez and

Seidel [2] and can be used for enumerating, e.g., non-crossing perfect matchings, plane straight-line graphs, convex subdivisions, and triangulations. The following is a sketch of the algorithm for non-crossing perfect matchings.

A matching can be reduced to an empty graph by removing edges one by one. By fixing a rule for the removing edge in each matching, one can define a rooted search tree $\mathcal{T}$ on the set of non-crossing matchings, and the set of non-crossing matchings can be enumerated by tracing $\mathcal{T}$. To reduce time complexity, the first idea is to trace only a subgraph $\mathcal{T}'$ of $\mathcal{T}$ induced by a subclass of non-crossing matchings by a clever choice of removing edges. Another idea is a compression of the search tree $\mathcal{T}'$ by using an equivalence relation on the subclass of non-crossing matchings. The resulting graph $\mathcal{G}$ is a digraph on the set of equivalence classes, where there is a one-to-one correspondence between non-crossing perfect matchings and directed paths of length $n/2$ from the root. A crucial observation is that $\mathcal{G}$ has at most $2^n n^3$ edges while the number of non-crossing perfect matchings is known to be at least poly$(n) \cdot 2^n$ for any P in general position [7]. Hence non-crossing perfect matchings can be enumerated in polynomial time on average by first constructing $\mathcal{G}$ and then enumerating all the dipaths of length $n/2$ in $\mathcal{G}$. It was also noted in [10] that the algorithm can be polynomial-time delay, but still the space complexity is exponential in n.

Open Problems

A challenging open problem is to design an efficient enumeration algorithm for the set of non-crossing spanning cycles, the set of highly connected triangulations, or the set of degree-bounded triangulations or non-crossing spanning trees. It is also not known whether triangulations can be enumerated in constant time per output.

Cross-References

▶ Enumeration of Paths, Cycles, and Spanning Trees

▶ Reverse Search; Enumeration Algorithms

▶ Voronoi Diagrams and Delaunay Triangulations

Recommended Reading

1. Aichholzer O, Aurenhammer F, Huemer C, Krasser H (2007) Gray code enumeration of plane straight-line graphs. Graphs Comb 23(5):467–479
2. Alvarez V, Seidel R (2013) A simple aggregate algorithm for counting triangulations of planar point sets and related problems. In: Proceedings of the 29th annual symposium on computational geometry (SoCG2013), Rio de Janeiro. ACM
3. Avis D, Fukuda K (1996) Reverse search for enumeration. Discret Appl Math 65(1–3):21–46
4. Bereg S (2005) Enumerating pseudo-triangulations in the plane. Comput Geom Theory Appl 30(3):207–222
5. Bespamyatnikh S (2002) An efficient algorithm for enumeration of triangulations. Comput Geom Theory Appl 23(3):271–279
6. Brönnimann H, Kettner L, Pocchiola M, Snoeyink J (2006) Counting and enumerating pointed pseudotriangulations with the greedy flip algorithm. SIAM J Comput 36(3):721–739
7. García A, Noy M, Tejel J (2000) Lower bounds on the number of crossing-free subgraphs of K_n. Comput Geom Theory Appl 16(4):211–221
8. Katoh N, Tanigawa S (2009) Fast enumeration algorithms for non-crossing geometric graphs. Discret Comput Geom 42(3):443–468
9. Razen A, Welzl E (2011) Counting plane graphs with exponential speed-up. In: Calude C, Rozenberg G, Salomaa A, Maurer HA (eds) Rainbow of computer science. Springer, Berlin/Heidelberg, pp 36–46
10. Wettstein M (2014) Counting and enumerating crossing-free geometric graphs. In: Proceedings of the 30th annual symposium on computational geometry (SoCG2014), Kyoto. ACM

Enumeration of Paths, Cycles, and Spanning Trees

Roberto Grossi
Dipartimento di Informatica, Università di Pisa, Pisa, Italy

Keywords

Amortized analysis; Arborescences; Cycles; Elementary circuits; Enumeration algorithms; Graphs; Paths; Spanning trees

Years and Authors of Summarized Original Work

1975; Johnson
1975; Read and Tarjan
1994; Shioura, Tamura, Uno
1995; Kapoor, Ramesh
1999; Uno
2013; Birmelé, Ferreira, Grossi, Marino, Pisanti, Rizzi, Sacomoto, Sagot

Problem Definition

Let $G = (V, E)$ be a (directed or undirected) graph with $n = |V|$ vertices and $m = |E|$ edges. A *walk* of length k is a sequence of vertices $v_0, \ldots, v_k \in V$ such that v_i and v_{i+1} are connected by an edge of E, for any $0 \le i < k$. A *path* π of length k is a walk $v_0, \ldots, v_k$ such that any two vertices v_i and v_j are distinct, for $0 \le i < j \le k$: this is also called *st-path* where $s = v_0$ and $t = v_k$. A *cycle* (or, equivalently, *elementary circuit*) C of length $k + 1$ is a path $v_0, \ldots, v_k$ such that v_k and v_0 are connected by an edge of E.

We denote by $\mathcal{P}_{st}(G)$ the set of *st*-paths in G for any two given vertices $s, t \in V$ and by $\mathcal{C}(G)$ the set of cycles in G. Given a graph G, the problem of *st-path enumeration* asks for generating all the paths in $\mathcal{P}_{st}(G)$. The problem of *cycle enumeration* asks for generating all the cycles in $\mathcal{C}(G)$.

We denote by $\mathcal{S}(G)$ the set of spanning trees in a connected graph G, where a spanning tree $T \subseteq E$ is a set of $|T| = n - 1$ edges such that no cycles are contained in T and each vertex in V is incident to at least an edge of T. Given a connected graph G, the problem of *spanning tree enumeration* asks for generating all the spanning trees in $\mathcal{S}(G)$.

Typical costs of enumeration algorithms are proportional to the output size times a polynomial function of the graph size. Sometimes enumeration is meant with the stronger property of *listing*, where each solution is explicitly output. In the latter case, we define an algorithm for a listing problem to be *optimally output sensitive* if its

running time is $O(n + m + K)$ where K is the following output cost for the enumeration problem at hand, namely, $\mathcal{P}_{st}(G)$, $\mathcal{C}(G)$, or $\mathcal{S}(G)$.

- $K = \sum_{\pi \in \mathcal{P}_{st}(G)} |\pi|$ where $|\pi|$ is the number of nodes in the *st*-path π.
- $K = \sum_{C \in \mathcal{C}(G)} |C|$ where $|C|$ is the number nodes in the cycle C.
- $K = \sum_{T \in \mathcal{S}(G)} |T| = |\mathcal{S}(G)| \cdot (n - 1)$ for spanning trees.

Although the above is a notion of optimality for listing solutions explicitly, it is possible in some cases that the enumeration algorithm can efficiently encode the differences between consecutive solutions in the sequence produced by the enumeration. This is the case of spanning trees, where a cost of $K = |\mathcal{S}(G)|$ is possible when they are implicitly represented during enumeration. This is called CAT (constant amortized time) enumeration in [28].

Key Results

Some possible approaches to attack the enumeration problems are listed below, where the term "search" is meant as an exploration of the space of solutions.

Backtrack search. A backtracking algorithm finds the solutions for a listing problem by exploring the search space and abandoning a partial solution (thus the name "backtracking") that cannot be completed to a valid one.

Binary partition search. An algorithm divides the search space into two parts. In the case of graphs, this is generally done by taking an edge (or a vertex) and (i) searching for all solutions that include that edge (resp. vertex) and (ii) searching for all solutions that do not include that edge (resp. vertex). Point (i) can sometimes be implemented by contracting the edge, i.e., merging the endpoints of the edge and their adjacency list.

Differential encoding search. The space of solutions is encoded in such a way that consecutive solutions differ by a constant number of modifications. Although not every enumeration

E

problem has properties that allow such encoding, this technique leads to very efficient algorithms.

Reverse search. This is a general technique to explore the space of solutions by reversing a local search algorithm. This approach implicitly generates a tree of the search space that is traversed by the reverse search algorithm. One of the properties of this tree is that it has bounded height, a useful fact for proving the time complexity of the algorithm.

Although there is some literature on techniques for enumeration problems [38, 39, 41], many more techniques and "tricks" have been introduced when attacking particular problems. For a deep understanding of the topic, the reader is recommended to review the work of researchers such as David Avis, Komei Fukuda, Shin-ichi Nakano, and Takeaki Uno.

Path and Cycles

Listing all the cycles in a graph is a classical problem whose efficient solutions date back to the early 1970s. In particular, at the turn of the 1970s, several algorithms for enumerating all cycles of an undirected graph were proposed. There is a vast body of work, and the majority of the algorithms listing all the cycles can be divided into the following three classes (see [1, 23] for excellent surveys).

Search space algorithms. Cycles are looked for in an appropriate search space. In the case of undirected graphs, the *cycle vector space* [6] turned out to be the most promising choice: from a basis for this space, all vectors are computed, and it is tested whether they are a cycle. Since the algorithm introduced in [43], many algorithms have been proposed: however, the complexity of these algorithms turns out to be exponential in the dimension of the vector space and thus in n. For the special case of planar graphs, the paper in [34] describes an algorithm listing all the cycles in $O((|\mathcal{C}(G)| + 1)n)$ time.

Backtrack algorithms. All paths are generated by backtrack, and, for each path, it is tested whether it is a cycle. One of the first algorithms based on this approach is the one proposed in [37], which is however exponential in $|\mathcal{C}(G)|$. By adding a simple pruning strategy,

this algorithm has been successively modified in [36]: it lists all the cycles in $O(nm(|\mathcal{C}(G)| + 1))$ time. Further improvements were proposed in [16], [35], and [27], leading to $O((|\mathcal{C}(G)| + 1)(m + n))$ time algorithms that work for both directed and undirected graphs. Apart from the algorithm in [37], all the algorithms based on this approach are *polynomial-time delay*, that is, the time elapsed between the outputting of two cycles is polynomial in the size of the graph (more precisely, $O(nm)$ in the case of the algorithm of [36] and $O(m)$ in the case of the other three algorithms).

Algorithms using the powers of the adjacency matrix. This approach uses the so-called variable adjacency matrix, that is, the formal sum of edges joining two vertices. A nonzero element of the pth power of this matrix is the sum of all walks of length p: hence, to compute all cycles, we compute the nth power of the variable adjacency matrix. This approach is not very efficient because of the non-simple walks. All algorithms based on this approach (e.g., [26] and [45]) basically differ only on the way they avoid to consider walks that are neither paths nor cycles.

For directed graphs, the best known algorithm for listing cycles is Johnson's [16]. It builds upon Tarjan's backtracking search [36], where the search starts from the least vertex of each strongly connected component. After that, a new strongly connected component is discovered, and the search starts again from the least vertex in it. When exploring a strongly connected component with a recursive backtracking procedure, it uses an enhanced marking system to avoid visiting same cycle multiple times. A vertex is marked each time it enters the backtracking stack. Upon leaving the stack, if a cycle is found, then the vertex is unmarked. Otherwise, it remains marked until another vertex involved in a cycle is popped from the stack, and there exists a path of marked vertices (not in the stack) between these two vertices. This strategy is implemented using a collection of lists B, one list per vertex containing its marked neighbors not in the stack. Unmarking is done by a recursive procedure. The complexity of the algorithm is $O(n + m + |\mathcal{C}(G)|m)$ time and $O(n + m)$ space.

For undirected graphs, Johnson's bound can be improved with an optimal output-sensitive algorithm [2]. First of all, the cycle enumeration problem is reduced to the st-path enumeration by considering any spanning tree of the given graph G and its non-tree edges $b_1, b_2, \ldots, b_r$. Then, for $i = 1, 2, \ldots, r$, the cycles in $C(G)$ can be listed as st-paths in $G \setminus \{b_1, \ldots, b_i\}$, where s and t are the endpoint of non-tree edge b_i. Hence, the subproblem to be solved with an optimal output-sensitive algorithm is the st-path enumeration problem. Binary partition search is adopted to avoid duplicated output, but the additional ingredient is the notion of *certificate*, which is a suitable data structure that maintains the biconnected components of the residual graph and guarantees that each recursive call thus produces at least one solution. Its amortized analysis is based on a lower bound on the number of st-paths that can be listed in the residual graph, so as to absorb the cost of maintaining the certificate. The final cost is $O(m+n+\sum_{\pi \in \mathcal{P}_{st}(G)} |\pi|)$ time and $O(n+m)$ space, which is optimal for listing.

Spanning Trees

Listing combinatorial structures in graphs has been a long-time problem of interest. In his 1970 book [25], Moon remarks that "many papers have been written giving algorithms, of varying degrees of usefulness, for listing the spanning trees of a graph" (citation taken from [28]). Among others, he cites [7, 9, 10, 13, 42] – some of these early papers date back to the beginning of the twentieth century. More recently, in the 1960s, Minty proposed an algorithm to list all spanning trees [24].

The first algorithmic solutions appeared in the 1960s [24] and the combinatorial papers even much earlier [25]. Other results from Welch, Tiernan, and Tarjan for this and other problems soon followed [36, 37, 43] and used backtracking search. Read and Tarjan presented an algorithm taking $O(m + n + |S(G)| \cdot m)$ time and $O(m + n)$ space [27]. Gabow and Myers proposed the first algorithm [11] which is optimal when the spanning trees are explicitly listed, taking $O(m + n + |S(G)| \cdot n)$ time and $O(m + n)$ space.

When the spanning trees are implicitly enumerated, Kapoor and Ramesh [17] showed that an elegant incremental representation is possible by storing just the $O(1)$ information needed to reconstruct a spanning tree from the previously enumerated one, requiring a total of $O(m + n + |S(G)|)$ time and $O(mn)$ space [17], later reduced to $O(m)$ space by Shioura et al. [32]. These methods use the reverse search where the elements are the spanning trees. The rule for moving along these elements and for their differential encoding is based upon the observation that adding a non-tree edge and removing a tree edge of the cycle thus formed produces another spanning tree from the current one. Some machinery is needed to avoid duplicated spanning trees and to spend $O(1)$ amortized cost per generated spanning tree.

A simplification of the incremental enumeration of spanning trees is based on matroids and presented by Uno [39]. It is a binary partition search giving rise to a binary enumeration tree, where the two children calls generated by the current call correspond to the fact that the current edge is either contracted in $O(n)$ time or deleted in $O(m - n)$ time. There is a trimming and balancing phase in $O(n(m - n))$ time: trimming removes the edges that do not appear in any of the spanning trees that will be generated by the current recursive call and contracts the edges that appear in all of these spanning trees. Balancing splits the recursive calls as in the divide-and-conquer paradigm. A crucial property proved in [39] is that the residual graph will generate at least $\Omega(n(m - n))$ spanning trees, and thus the total cost per call, which is dominated by trimming and balancing, can be amortized as $O(1)$ per spanning tree. The method in [39] works also for directed spanning trees (arborescences) with an amortized $O(\log n)$ time cost per directed spanning tree.

Applications

The classical problem of listing all the cycles of a graph has been extensively studied for its many applications in several fields, ranging from the mechanical analysis of chemical structures [33] to the design and analysis of reliable communication networks and the graph isomorphism problem [43]. Almost 40 years after, the problem of efficiently listing all cycles of a graph is still an active area of research (e.g., [14, 15, 22, 29, 30, 44]). New application areas have emerged in the last decade, such as bioinformatics: for example, two algorithms for this problem have been proposed in [20] and [21] while studying biological interaction graphs, with important network properties derived for feedback loops, signaling paths, and dependency matrix, to name a few.

When considering weighted cycles, the paper in [19] proves that there is no polynomial total time algorithm (unless $P = NP$) to enumerate negative-weight (simple) cycles in directed weighted graphs. Uno [40] and Ferreira et al. [8] considered the enumeration of chordless cycles and paths. A chordless or induced cycle (resp., path) in an undirected graph is a cycle (resp., path) such that the subgraph induced by its vertices contains exactly the edges of the cycle (resp., path). Both chordless cycles and paths are very natural structures in undirected graphs with an important history, appearing in many papers in graph theory related to chordal graphs, perfect graphs, and co-graphs (e.g., [4, 5, 31]), as well as many NP-complete problems involving them (e.g., [3, 12, 18]).

As for spanning trees, we refer to the section "K-best enumeration" of this book.

Recommended Reading

1. Bezem G, Leeuwen Jv (1987) Enumeration in graphs. Technical Report RUU-CS-87-07, Utrecht University
2. Birmelé E, Ferreira R, Grossi R, Marino A, Pisanti N, Rizzi R, Sacomoto G, Sagot MF (2013) Optimal listing of cycles and st-paths in undirected graphs. In: Proceedings of the twenty-fourth annual ACM-SIAM symposium on discrete algorithms, New Orleans. SIAM, pp 1884–1896
3. Chen Y, Flum J (2007) On parameterized path and chordless path problems. In: IEEE conference on computational complexity, San Diego, pp 250–263
4. Chudnovsky M, Robertson N, Seymour P, Thomas R (2006) The strong perfect graph theorem. Ann Math 164:51–229
5. Conforti M, Rao MR (1992) Structural properties and decomposition of linear balanced matrices. Math Program 55:129–168
6. Diestel R (2005) Graph theory. Graduate texts in mathematics. Springer, Berlin/New York
7. Duffin R (1959) An analysis of the wang algebra of networks. Trans Am Math Soc 93:114–131
8. Ferreira RA, Grossi R, Rizzi R, Sacomoto G, Sagot M (2014) Amortized $\tilde{O}(|V|)$-delay algorithm for listing chordless cycles in undirected graphs. In: Proceedings of European symposium on algorithms. LNCS, vol 8737. Springer, Berlin/Heidelberg, pp 418–429
9. Feussner W (1902) Uber stromverzweigung in netzformigen leitern. Ann Physik 9:1304–1329
10. Feussner W (1904) Zur berechnung der stromstarke in netzformigen leitern. Ann Physik 15:385–394
11. Gabow HN, Myers EW (1978) Finding all spanning trees of directed and undirected graphs. SIAM J Comput 7(3):280–287
12. Haas R, Hoffmann M (2006) Chordless paths through three vertices. Theor Comput Sci 351(3):360–371
13. Hakimi S (1961) On trees of a graph and their generation. J Frankl Inst 272(5):347–359
14. Halford TR, Chugg KM (2004) Enumerating and counting cycles in bipartite graphs. In: IEEE Communication Theory Workshop, Cancun
15. Horváth T, Gärtner T, Wrobel S (2004) Cyclic pattern kernels for predictive graph mining. In: Proceedings of 10th ACM SIGKDD, Seattle, pp 158–167
16. Johnson DB (1975) Finding all the elementary circuits of a directed graph. SIAM J Comput 4(1):77–84
17. Kapoor S, Ramesh H (1995) Algorithms for enumerating all spanning trees of undirected and weighted graphs. SIAM J Comput 24:247–265
18. Kawarabayashi K, Kobayashi Y (2008) The induced disjoint paths problem. In: Lodi A, Panconesi A, Rinaldi G (eds) IPCO. Lecture notes in computer science, vol 5035. Springer, Berlin/Heidelberg, pp 47–61
19. Khachiyan L, Boros E, Borys K, Elbassioni K, Gurvich V (2006) Generating all vertices of a polyhedron is hard. In: Proceedings of the seventeenth annual ACM-SIAM symposium on discrete algorithm, society for industrial and applied mathematics, Philadelphia, SODA '06, Miami, pp 758–765
20. Klamt S et al (2006) A methodology for the structural and functional analysis of signaling and regulatory networks. BMC Bioinform 7:56
21. Klamt S, von Kamp A (2009) Computing paths and cycles in biological interaction graphs. BMC Bioinform 10:181
22. Liu H, Wang J (2006) A new way to enumerate cycles in graph. In: AICT and ICIW, Washington, DC, USA pp 57–59
23. Mateti P, Deo N (1976) On algorithms for enumerating all circuits of a graph. SIAM J Comput 5(1):90–99
24. Minty G (1965) A simple algorithm for listing all the trees of a graph. IEEE Trans Circuit Theory 12(1):120–120

25. Moon J (1970) Counting labelled trees. Canadian mathematical monographs, vol 1. Canadian Mathematical Congress, Montreal
26. Ponstein J (1966) Self-avoiding paths and the adjacency matrix of a graph. SIAM J Appl Math 14:600–609
27. Read RC, Tarjan RE (1975) Bounds on backtrack algorithms for listing cycles, paths, and spanning trees. Networks 5(3):237–252
28. Ruskey F (2003) Combinatorial generation. Preliminary working draft University of Victoria, Victoria
29. Sankar K, Sarad A (2007) A time and memory efficient way to enumerate cycles in a graph. In: Intelligent and advanced systems, Kuala Lumpur pp 498–500
30. Schott R, Staples GS (2011) Complexity of counting cycles using Zeons. Comput Math Appl 62:1828–1837
31. Seinsche D (1974) On a property of the class of n-colorable graphs. J Comb Theory, Ser B 16(2):191–193
32. Shioura A, Tamura A, Uno T (1994) An optimal algorithm for scanning all spanning trees of undirected graphs. SIAM J Comput 26:678–692
33. Sussenguth E (1965) A graph-theoretical algorithm for matching chemical structures. J Chem Doc 5:36–43
34. Syslo MM (1981) An efficient cycle vector space algorithm for listing all cycles of a planar graph. SIAM J Comput 10(4):797–808
35. Szwarcfiter JL, Lauer PE (1976) A search strategy for the elementary cycles of a directed graph. BIT Numer Math 16:192–204
36. Tarjan RE (1973) Enumeration of the elementary circuits of a directed graph. SIAM J Comput 2(3):211–216
37. Tiernan JC (1970) An efficient search algorithm to find the elementary circuits of a graph. Commun ACM 13:722–726
38. Uno T (1998) New approach for speeding up enumeration algorithms. Algorithms and computation. Springer, Berlin/Heidelberg, pp 287–296
39. Uno T (1999) A new approach for speeding up enumeration algorithms and its application for matroid bases. In: COCOON, Tokyo, pp 349–359
40. Uno T (2003) An output linear time algorithm for enumerating chordless cycles. In: 92nd SIGAL of information processing society Japan, Tokyo pp 47–53, (in Japanese)
41. Uno T (2003) Two general methods to reduce delay and change of enumeration algorithms. National Institute of Informatics, Technical Report NII-2003-004E, Tokyo, Apr. 2003
42. Wang K (1934) On a new method for the analysis of electrical networks. Nat Res Inst for Eng Academia Sinica Memoir (2):19
43. Welch JT Jr (1966) A mechanical analysis of the cyclic structure of undirected linear graphs. J ACM 13:205–210
44. Wild M (2008) Generating all cycles, chordless cycles, and hamiltonian cycles with the principle of exclusion. J Discret Algorithms 6:93–102
45. Yau S (1967) Generation of all hamiltonian circuits, paths, and centers of a graph, and related problems. IEEE Trans Circuit Theory 14:79–81

Equivalence Between Priority Queues and Sorting

Rezaul A. Chowdhury
Department of Computer Sciences, University of Texas, Austin, TX, USA
Stony Brook University (SUNY), Stony Brook, NY, USA

Keywords

AC^0 operation; Pointer machine; Priority queue; Sorting; Word RAM

Synonyms

Heap

Years and Authors of Summarized Original Work

2007 (2002); Thorup

Problem Definition

A *priority queue* is an abstract data structure that maintains a set Q of elements, each with an associated value called a *key*, under the following set of operations [5, 6]:

insert(Q, x, k): Inserts element x with key k into Q.

find-min(Q): Returns an element of Q with the minimum key but does not change Q.

delete(Q, x, k): Deletes element x with key k from Q.

Additionally, the following operations are often supported:

delete-min(Q): Deletes an element with the minimum key value from Q and returns it.

decrease-key(Q, x, k): Decreases the current key k' of x to k assuming $k < k'$.

meld(Q_1, Q_2): Given priority queues Q_1 and Q_2, returns the priority queue $Q_1 \cup Q_2$.

Observe that a *delete-min* can be implemented as a *find-min* followed by a *delete*, a *decrease-key* as a *delete* followed by an *insert*, and a *meld* as a series of *find-min*, *delete* and *insert*. However, more efficient implementations of *decrease-key* and *meld* often exist [5,6].

Priority queues have many practical applications including event-driven simulation, job scheduling on a shared computer, and computation of shortest paths, minimum spanning forests, minimum cost matching, optimum branching, etc. [5,6].

A priority queue can trivially be used for sorting by first inserting all keys to be sorted into the priority queue and then by repeatedly extracting the current minimum. The major contribution of Mikkel Thorup's 2002 article (Full version published in 2007) titled "Equivalence between Priority Queues and Sorting" [17] is a reduction showing that the converse is also true. Taken together, these two results imply that priority queues are computationally equivalent to sorting, that is, asymptotically, the per key cost of sorting is the update time of a priority queue.

A result similar to those in the current work [17] was presented earlier by the same author [14] which resulted in monotone priority queues (i.e., meaning that the extracted minimums are nondecreasing) with amortized time bounds only. In contrast, the current work [17] constructs general priority queues with worst-case bounds.

In addition to establishing the equivalence between priority queues and sorting, Thorup's reductions [17] are also used to translate several known sorting results into new results on priority queues.

Background

Some relevant background information is summarized below which will be useful in understanding the key results in section "Key Results."

- A standard *word RAM* models what one programs in a standard imperative programming language such as C. In addition to direct and indirect addressing and conditional jumps, there are functions, such as addition and multiplication, operating on a constant number of words. The memory is divided into words, addressed linearly starting from 0. The running time of a program is the number of instructions executed and the space is the maximal address used. The word length is a machine-dependent parameter which is big enough to hold a key and at least logarithmic in the number of input keys so that they can be addressed.

- A pointer machine is like the word RAM except that addresses cannot be manipulated.

- The AC^0 complexity class consists of constant-depth circuits with unlimited fan-in [18]. Standard AC^0 operations refer to the operations available via C but where the functions on words are in AC^0. For example, this includes addition but not multiplication.

- Integer keys will refer to nonnegative integers. However, if the input keys are signed integers, the correct ordering of the keys is obtained by flipping their sign bits and interpreting them as unsigned integers. Similar tricks work for floating point numbers and integer fractions [14].

- The atomic heaps of Fredman and Willard [7] are used in one of Thorup's reductions [17]. These heaps can support updates and searches in sets of $\mathcal{O}\left(\log^2 n\right)$ keys in $\mathcal{O}(1)$ worst-case time [20]. However, atomic heaps use multiplication operations which are not in AC^0.

Key Results

The main results in this paper are two reductions from priority queues to sorting. The stronger of the two, stated in Theorem 1, is for integer priority queues running on a standard word RAM.

Theorem 1 *If for some nondecreasing function S, up to n integer keys can be sorted in $S(n)$ time per key, an integer priority queue can be im-*

plemented supporting find-min *in constant time, and updates, i.e.,* insert *and* delete, *in* $\mathcal{O}(S(n))$ *time. Here n is the current number of keys in the queue. The reduction uses linear space. The reduction runs on a standard word RAM assuming that each integer key is contained in a single word.*

The reduction above provides the following new bounds for linear space integer priority queues improving previous bounds given by Han [8] and Thorup [14], respectively:

1. (**Deterministic**) $\mathcal{O}(\log\log n)$ update time using a sorting algorithm by Han [9].
2. (**Randomized**) $\mathcal{O}\left(\sqrt{\log\log n}\right)$ expected update time using a sorting algorithm given by Han and Thorup [10].

The reduction in Theorem 1 employs atomic heaps [7] which, in addition to being very complicated, use AC^0 operations. The following slightly weaker recursive reduction which does not restrict the domain of the keys is completely combinatorial.

Theorem 2 *If for some nondecreasing function S, up to n keys can be sorted in $S(n)$ time per key, a priority queue can be implemented supporting* find-min *in constant time, and updates in $T(n)$ time where n is the current number of keys in the queue and $T(n)$ satisfies the recurrence:*

$$T(n) = \mathcal{O}(S(n)) + T(\mathcal{O}(\log n))$$

The reduction runs on a pointer machine in linear space using only standard AC^0 operations.

This reduction implies the following new integer priority queue bounds not implied by Theorem 1, which improve previous bounds given by Thorup in 1998 [13] and 1997 [15], respectively:

1. (**Deterministic in AC^0**) $\mathcal{O}\left((\log\log n)^{1+\epsilon}\right)$ update time for any constant $\epsilon > 0$ using a standard AC^0 sorting algorithm given by Han and Thorup [10].

2. (**Randomized in AC^0**) $\mathcal{O}(\log\log n)$ expected update time using a randomized AC^0 sorting algorithm given by Thorup [15].

The Reduction in Theorem 1

Given a sorting routine that can sort up to n keys in $S(n)$ time per key, the priority queue is constructed as follows. All keys are assumed to be distinct.

The data structure has two major components: a partially sorted list of keys called a *base list* and a set of *level buffers* (also called *update buffers*). Most keys of the priority queue reside in the base list partitioned into logarithmic-sized disjoint sets called *base sets*. While the keys inside any given base set are not required to be sorted, each of those keys must be larger than every key in the base set (if any) appearing before it in the list. Keys inside each base set are stored in a doubly linked list allowing constant time updates. The first base set in the list containing the smallest key among all base sets is also maintained in an atomic heap so that the current minimum can be found in constant time. Each level buffer has a different capacity and accumulates updates (*insert/delete*) with key values in a different range. Smaller level buffers accept updates with smaller keys. An atomic heap is used to determine in constant time which level buffer collects a new update. When a level buffer accumulates enough updates, they first enter a sorting phase and then a merging phase. In the merging phase each update is applied on the proper base set in the key list, and invariants on base set size and ranges of level buffers are fixed. These phases are not executed immediately, instead they are executed in fixed time increments over a period of time. A level buffer continues to accept new updates, while some updates accepted by it earlier are still in the sorting phase, and some even older updates are in the merging phase. Every time it accepts a new update, $\mathcal{O}(S(n))$ time is spent on the sorting phase associated with it and $\mathcal{O}(1)$ time on its merging phase including rebalancing of base sets and scanning. This strategy allows the sorting and merging phases to complete execution by the time the level buffer becomes full again and thus keeping the movement of updates through different phases smooth while maintaining an $\mathcal{O}(S(n))$ worst-case time bound per update. Moreover, the size and capacity constraints ensure that the smallest key in the data structure is available in $\mathcal{O}(1)$ time. More details are given below.

The Base List: The base list consists of base sets $A_1, A_2, \ldots, A_k$, where $\frac{\Phi}{4} \leq |A_i| \leq \Phi$ for $i < k$, and $|A_k| \leq \Phi$ for some $\Phi = \Theta(\log n)$. The exact value of Φ is chosen carefully to make sure that it conforms with the requirements of the delicate worst-case base set rebalancing protocol used by the reduction. The base sets are partitioned by *base splitters* $s_0, s_1, \ldots, s_{k+1}$, where $s_0 = -\infty$, $s_{k+1} = \infty$, and for $i = 1, \ldots, k-1$, $\max A_{i-1} < s_i \leq \min A_i$. If a base set becomes too large or too small, it is split or joined with an adjacent set, respectively.

Level Buffers: Among the base splitters $l + 2 = \Theta(\log n)$ are chosen to become *level splitters* $t_0, t_1, \ldots, t_l, t_{l+1}$ with $t_0 = s_0 = -\infty$ and $t_{l+1} = s_{k+1} = \infty$, so that for $j > 0$, the number of keys in the base list below t_j is around $4^{j+1}\Phi$. These splitters are placed in an atomic heap. As the base list changes the level splitters are moved, as needed, in order to maintain their exponential distribution. Associated with each level splitter t_j, $1 \leq j \leq l$, is a *level buffer* B_j containing keys in $[t_{j-1}, t_{j+2})$, where $t_{l+2} = \infty$. Buffer B_j consists of an *entrance* buffer, a *sorter*, and a *merger*, each with capacity for 4^j keys. Level j works in a cycle of 4^j steps. The cycle starts with an empty entrance, at most 4^j updates in the sorter, and a sorted list of at most 4^j updates in the merger. In each step one may accept an update for the entrance, spend $S\left(4^j\right) = \mathcal{O}(S(n))$ time in the sorter and $\mathcal{O}(1)$ time in merging the sorted list in the merger with the $\mathcal{O}\left(4^j\right)$ base splitters in $[t_{j-1}, t_{j+2})$ and scanning for a new t_j among them. Therefore, after 4^j such steps, the sorted list is correctly merged with the base list, a new t_j is found, and a new sorted list is produced. The sorter then takes the role of the merger, the entrance becomes the

sorter, and the empty merger becomes the new entrance.

Handling Updates: When a new update key k (*insert*/*delete*) is received, the atomic heap of level splitters is used to find in $\mathcal{O}(1)$ time the t_j such that $k \in [t_{j-1}, t_j)$. If $k \in [t_0, t_1)$, its position is identified among the $\mathcal{O}(1)$ base splitters below t_1, and the corresponding base set is updated in $\mathcal{O}(1)$ time using the doubly linked list and the atomic heap (if exists) over the keys of that set. If $k \in [t_{j-1}, t_j)$ for some $j > 1$, the update is placed in the entrance of B_j, performing one step of the cycle of B_j in $\mathcal{O}(S(n))$ time. Additionally, during each update another splitter t_r is chosen in a round-robin fashion, and a step of a cycle of level r is executed in $\mathcal{O}(S(n))$ time. This additional work ensures that after every l updates some progress is made on moving each level splitter.

A *find-min* returns the minimum element of the base list which is available in $\mathcal{O}(1)$ time.

The Reduction in Theorem 2

This reduction follows from the previous reduction by replacing the atomic heap containing the level splitters with a data structure similar to a level buffer and the atomic heap over the keys of the first base set with a recursively defined priority queue satisfying the following recurrence for update time: $T(n) = \mathcal{O}(S(n)) + T(\mathcal{O}(\Phi))$.

Further Improvement

Alstrup et al. [1] presented a general reduction that transforms a priority queue to support *insert* in $\mathcal{O}(1)$ time while keeping the other bounds unchanged. This reduction can be used to reduce the cost of insertion to a constant in Theorems 1 and 2.

Applications

Thorup's equivalence results [17] can be used to translate known sorting results into new results on priority queues for integers and strings in different computational models (see section "Key Results"). These results can also be viewed as a new means of proving lower bounds for sorting via priority queues.

A new RAM priority queue that matches the bounds in Theorem 1 and also supports *decrease-key* in $\mathcal{O}(1)$ time is presented by Thorup [16]. This construction combines Andersson's exponential search trees [2] with the priority queues implied by Theorem 1. The reduction in Theorem 1 is also used by Pagh et al. [12] in order to develop an adaptive integer sorting algorithm for the word RAM and by Arge and Thorup [3] to develop a sorting algorithm that is simultaneously I/O efficient and internal memory efficient in the RAM model of computation. Cohen et al. [4] use a priority queue generated through this reduction to obtain a simple and fast amortized implementation of a reservoir sampling scheme that provides variance optimal unbiased estimation of subset sums. Reductions from meldable priority queues to sorting presented by Mendelson et al. [11] use the reductions from non-meldable priority queues to sorting given in [17].

An external-memory version of Theorem 1 has been proved by Wei and Yi [19].

Open Problems

One major open problem is to find a general reduction (if one exists) that allows us to decrease the value of a key in constant time. Another open question is whether the gap between the bounds implied by Theorems 1 and 2 can be reduced or removed. For example, for a hypothetical linear time-sorting algorithm, Theorem 1 implies a priority queue with an update time of $\mathcal{O}(1)$, while Theorem 2 implies only $\mathcal{O}(\log^* n)$-time updates.

Cross-References

► Cache-Oblivious Sorting
► External Sorting and Permuting
► Minimum Spanning Trees
► Single-Source Shortest Paths
► String Sorting
► Suffix Tree Construction

Recommended Reading

1. Alstrup S, Husfeldt T, Rauhe T, Thorup M (2005) Black box for constant-time insertion in priority queues (note). ACM TALG 1(1):102–106
2. Andersson A (1996) Faster deterministic sorting and searching in linear space. In: Proceedings of the 37th FOCS, Burlington, pp 135–141
3. Arge L, Thorup M (2013) RAM-efficient external memory sorting. In: Proceedings of the 24th ISAAC, Hong Kong, pp 491–501
4. Cohen E, Duffield N, Kaplan H, Lund C, Thorup M (2009) Stream sampling for variance-optimal estimation of subset sums. In: Proceedings of the 20th SODA, New York, pp 1255–1264
5. Cormen T, Leiserson C, Rivest R, Stein C (2009) Introduction to algorithms. MIT, Cambridge
6. Fredman M, Tarjan R (1987) Fibonacci heaps and their uses in improved network optimization algorithms. J ACM 34(3):596–615
7. Fredman M, Willard D (1994) Trans-dichotomous algorithms for minimum spanning trees and shortest paths. J Comput Syst Sci 48:533–551
8. Han Y (2001) Improved fast integer sorting in linear space. Inf Comput 170(8):81–94. Announced at STACS'00 and SODA'01
9. Han Y (2004) Deterministic sorting in $O(n \log \log n)$ time and linear space. J Algorithms 50(1):96–105. Announced at STOC'02
10. Han Y, Thorup M (2002) Integer sorting in $O(n \sqrt{\log \log n})$ expected time and linear space. In: Proceedings of the 43rd FOCS, Vancouver, pp 135–144
11. Mendelson R, Tarjan R, Thorup M, Zwick U (2006) Melding priority queues. ACM TALG 2(4):535–556. Announced at SODA'04
12. Pagh A, Pagh R, Thorup M (2004) On adaptive integer sorting. In: Proceedings of the 12th ESA, Bergen, pp 556–579
13. Thorup M (1998) Faster deterministic sorting and priority queues in linear space. In: Proceedings of the 9th SODA, San Francisco, pp 550–555
14. Thorup M (2000) On RAM priority queues. SIAM J Comput 30(1):86–109. Announced at SODA'96
15. Thorup M (2002) Randomized sorting in $O(n \log \log n)$ time and linear space using addition, shift, and bit-wise boolean operations. J Algorithms 42(2):205–230. Announced at SODA'97
16. Thorup M (2004) Integer priority queues with decrease key in constant time and the single source shortest paths problem. J Comput Syst Sci (special issue on STOC'03) 69(3):330–353
17. Thorup M (2007) Equivalence between priority queues and sorting. J ACM 54(6):28. Announced at FOCS'02
18. Vollmer H (1999) Introduction to circuit complexity: a uniform approach. Springer, Berlin/New York

E

19. Wei Z, Yi K (2014) Equivalence between priority queues and sorting in external memory. In: Proceedings of 22nd ESA, Wroclaw, pp 830–841
20. Willard D (2000) Examining computational geometry, van Emde Boas trees, and hashing from the perspective of the fusion tree. SIAM J Comput 29(3):1030–1049. Announced at SODA'92

Estimating Simple Graph Parameters in Sublinear Time

Oded Goldreich[1] and Dana Ron[2]
[1] Department of Computer Science, Weizmann Institute of Science, Rehovot, Israel
[2] School of Electrical Engineering, Tel-Aviv University, Ramat-Aviv, Israel

Keywords

Graph parameters; Sublinear-time algorithms

Years and Authors of Summarized Original Work

2008; Goldreich, Ron

Problem Definition

A *graph parameter* σ is a real-valued function over graphs that is invariant under graph isomorphism. For example, the average degree of the graph, the average distance between pairs of vertices, and the minimum size of a vertex cover are graph parameters. For a fixed graph parameter σ and a graph $G = (V, E)$, we would like to compute an estimate of $\sigma(G)$. To this end we are given query access to G and would like to perform this task in time that is *sublinear* in the size of the graph and with high success probability. In particular, this means that we do not read the entire graph but rather only access (random) parts of it (via the query mechanism). Our main focus here is on a very basic graph parameter: its average degree, denoted $\overline{d}(G)$.

The estimation algorithm is given an approximation parameter $\epsilon > 0$. It should output a value $\hat{d}$ such that with probability at least $2/3$

(over the random choices of the algorithm) it holds that $\overline{d}(G) \leq \hat{d} \leq (1 + \epsilon) \cdot \overline{d}(G)$. (The error probability can be decreased to 2^{-k} by invoking the algorithm $\Theta(k)$ times and outputting the median value.) For any vertex $v \in V = [n]$ of its choice, where $[n] \stackrel{\text{def}}{=} \{1, \ldots, n\}$, the estimation algorithm may query the degree of v, denoted $d(v)$. We refer to such queries as *degree queries*. In addition, the algorithm may ask for the ith neighbor of v, for any $1 \leq i \leq d(v)$. These queries are referred to as *neighbor queries*. We assume for simplicity that G does not contain any isolated vertices (so that, in particular, $\overline{d}(G) \geq 1$). This assumption can be removed.

Key Results

The problem of estimating the average degree of a graph in sublinear time was first studied by Feige [7]. He considered this problem when the algorithm is allowed only degree queries, so that the problem is a special case of estimating the average value of a function given query access to the function. For a general function $d : [n] \rightarrow [n - 1]$, obtaining a constant-factor estimate of the average value of the function (with constant success probability) requires $\Omega(n)$ queries to the function. Feige showed that when d is the degree function of a graph, for any $\gamma \in (0, 1]$, it is possible to obtain an estimate of the average degree that is within a factor of $(2 + \gamma)$ by performing only $O(\sqrt{n}/\gamma)$ (uniformly selected) queries. He also showed that in order to go below a factor of 2 in the quality of the estimate, $\Omega(n)$ queries are necessary.

However, given that the object in question is a graph, it is natural to allow the algorithm to query the neighborhood of vertices of its choice and not only their degrees; indeed, the aforementioned problem definition follows this natural convention. Goldreich and Ron [10] showed that by giving the algorithm this extra power, it is possible to break the factor-2 barrier. They provide an algorithm that, given $\epsilon > 0$, outputs a $(1 + \epsilon)$-factor estimate of the average degree (with probability at least $2/3$) after performing $O(\sqrt{n} \cdot$

poly($\log n$, $1/\epsilon$)) degree and neighbor queries. In fact, since a degree query to vertex v can be replaced by $O(\log d(v)) = O(\log n)$ neighbor queries, which implement a binary search, degree queries are not necessary. Furthermore, when the average degree increases, the performance of the algorithm improves, as stated next.

Theorem 1 *There exists an algorithm that makes only neighbor queries to the input graph and satisfies the following condition. On input $G = (V, E)$ and $\epsilon \in (0, 1)$, with probability at least $2/3$, the algorithm halts within $O\left(\sqrt{n/\overline{d}(G)} \cdot \text{poly}(\log n, 1/\epsilon)\right)$ steps and outputs a value in $[\overline{d}(G), (1 + \epsilon) \cdot \overline{d}(G)]$.*

The running time stated in Theorem 1 is essentially optimal in the sense that (as shown in [10]) a $(1 + \epsilon)$-factor estimate requires $\Omega(\sqrt{n/(\epsilon\overline{d}(G))})$ queries, for every value of n, for $\overline{d}(G) \in [2, o(n)]$, and for $\epsilon \in [\omega(n^{-1/4}), o(n/\overline{d}(G))]$.

The following is a high-level description of the algorithm and the ideas behind its analysis. For the sake of simplicity, we only show how to obtain a $(1 + \epsilon)$-factor estimate by performing $O\left(\sqrt{n} \cdot \text{poly}(\log n, 1/\epsilon)\right)$ queries (under the assumption that $\overline{d}(G) \geq 1$). For the sake of the presentation, we also allow the algorithm to perform degree queries. We assume that $\epsilon \leq 1/2$, or else we run the algorithm with $\epsilon = 1/2$. We first show how to obtain a $(2 + \epsilon)$-approximation by performing only degree queries and then explain how to improve the approximation by using neighbor queries as well.

Consider a partition of the graph vertices into buckets $B_1, \ldots, B_r$, where

$$B_i \stackrel{\text{def}}{=} \{v : (1 + \epsilon/8)^{i-1} \leq d(v) < (1 + \epsilon/8)^i\} \tag{1}$$

and $r = O(\log n/\epsilon)$. By this definition,

$$\frac{1}{n}\sum_{i=1}^{r} |B_i| \cdot (1 + \epsilon/8)^i \in \left[\overline{d}(G), (1 + \epsilon/8) \cdot \overline{d}(G)\right]. \tag{2}$$

Suppose we could obtain an estimate $\hat{b}_i$ of the size of each bucket B_i such that $\hat{b}_i \in [(1 - \epsilon/8)|B_i|, (1 + \epsilon/8)|B_i|]$. Then

$$\frac{1}{n}\sum_{i=1}^{r} \hat{b}_i \cdot (1 + \epsilon/8)^i$$

$$\in \left[(1 - \epsilon/8) \cdot \overline{d}(G), (1 + 3\epsilon/8) \cdot \overline{d}(G)\right]. \tag{3}$$

Now, for each B_i, if we uniformly at random select $\Omega\left(\frac{n}{|B_i|} \cdot \frac{\log r}{\epsilon^2}\right)$ vertices, then, by a multiplicative Chernoff bound, with probability $1 - O(1/r)$, the fraction of sampled vertices that belong to B_i is in the interval $\left[(1 - \epsilon/8)\frac{|B_i|}{n}, (1 + \epsilon/8)\frac{|B_i|}{n}\right]$. By querying the degree of each sampled vertex, we can determine to which bucket it belongs and obtain an estimate of $|B_i|$. Unfortunately, if B_i is much smaller than $\sqrt{n}$, then the sample size required to estimate $|B_i|$ is much larger than the desired $O\left(\sqrt{n} \cdot \text{poly}(\log n, 1/\epsilon)\right)$. Let $L \stackrel{\text{def}}{=} \{i : |B_i| \geq \sqrt{\epsilon n/8r}\}$ denote the set of indices of *large* buckets. The basic observation is that if, for each $i \in L$, we have an estimate $\hat{b}_i \in [(1 - \epsilon/8)|B_i|, (1 + \epsilon/8)|B_i|]$, then

$$\frac{1}{n}\sum_{i \in L} \hat{b}_i \cdot (1 + \epsilon/8)^i$$

$$\in \left[(1/2 - \epsilon/4) \cdot \overline{d}(G), (1 + 3\epsilon/8) \cdot \overline{d}(G)\right]. \tag{4}$$

The reasoning is essentially as follows. Recall that $\sum_v d(v) = 2|E|$. Consider an edge (u, v) where $u \in B_j$ and $v \in B_k$. If $j, k \in L$, then this edge contributes twice to the sum in Eq. (4): once when $i = j$ and once when $i = k$. If $j \in L$ and $k \notin L$ (or vice versa), then this edge contributes only once. Finally, if $j, k \notin L$, then the edge does not contribute at all, but there are at most $\epsilon n/8$ edges of this latter type. Since it is possible to obtain such estimates $\hat{b}_i$ for all $i \in L$ simultaneously, with constant success probability, by sampling $O\left((\sqrt{n} \cdot \text{poly}(\log n, 1/\epsilon)\right)$ vertices, we can get a

$(2+\epsilon)$-factor estimate by performing this number of degree queries. Recall that we cannot obtain an approximation factor below 2 by performing $o(n)$ queries if we use only degree queries.

In order to obtain the desired factor of $(1 + \epsilon)$, we estimate the number of edges (u, v) such that $u \in B_j$ and $v \in B_k$ with $j \in L$ and $k \notin L$, which are counted only once in Eq. (4). Here is where neighbor queries come into play. For each $i \in L$ (more precisely, for each i such that $\hat{b}_i$ is sufficiently large), we estimate

$$e_i \stackrel{\text{def}}{=} |\{(u, v) : u \in B_i, \ v \in B_k \text{ for } k \notin L\}|.$$

This is done by uniformly sampling *neighbors* of vertices in B_i, querying their degree, and therefore estimating the fraction of edges incident to vertices in B_i whose other endpoint belongs to B_k for $k \notin L$. If we denote the estimate of e_i by $\hat{e}_i$, then we can get that by performing $O\left((\sqrt{n} \cdot \text{poly}(\log n, 1/\epsilon)\right)$ neighbor queries, with high constant probability, the $\hat{e}_i$'s are such that

$$\frac{1}{n} \sum_{i \in L} \left(\hat{b}_i \cdot (1 + \epsilon/8)^i + \hat{e}_i \right)$$
$$\in \left[(1 - \epsilon/2) \cdot \overline{d}(G), (1 + \epsilon/2) \cdot \overline{d}(G) \right].$$
$$(5)$$

By dividing the left-hand side in Eq. (5) by $(1 - \epsilon/2)$, we obtain the $(1 + \epsilon)$-factor we sought.

Estimating the Average Distance

Another graph parameter considered in [10] is the average distance between vertices. For this parameter, the algorithm is given access to a *distance-query oracle*. Namely, it can query the distance between any two vertices of its choice. As opposed to the average degree parameter where neighbor queries could be used to improve the quality of the estimate (and degree queries were not actually necessary), distance queries are crucial for estimating the average distance, and neighbor queries are not of much use. The main (positive) result concerning the average distance parameter is stated next.

Theorem 2 *There exists an algorithm that makes only distance queries to the input graph*

and satisfies the following condition. On input $G = (V, E)$ and $\epsilon \in (0, 1)$, with probability at least $2/3$, the algorithm halts within $O\left(\sqrt{n/\overline{D}(G)} \cdot \text{poly}(1/\epsilon)\right)$ steps and outputs a value in $[\overline{D}(G), (1 + \epsilon) \cdot \overline{D}(G)]$, where $\overline{D}(G)$ is the average of the all-pairs distances in G. A corresponding algorithm exists for the average distance to a given vertex $s \in V$.

Comments for the Recommended Reading

The current entry falls within the scope of sublinear-time algorithms (see, e.g., [4]).

Other graph parameters that have been studied in the context of sublinear-time algorithms include the minimum weight of a spanning tree [2, 3, 5], the number of stars [11] and the number of triangles [6], the minimum size of a vertex cover [13–15, 17], the size of a maximum matching [14, 17], and the distance to having various properties [8, 13, 16]. Related problems over weighted graphs that represent distance metrics were studied in [12] and [1].

Recommended Reading

1. Bădoiu M, Czumaj A, Indyk P, Sohler C (2005) Facility location in sublinear time. In: Automata, languages and programming: thirty-second international colloquium (ICALP), Lisbon, pp 866–877
2. Chazelle B, Rubinfeld R, Trevisan L (2005) Approximating the minimum spanning tree weight in sublinear time. SIAM J Comput 34(6):1370–1379
3. Czumaj A, Sohler C (2009) Estimating the weight of metric minimum spanning trees in sublinear time. SIAM J Comput 39(3):904–922
4. Czumaj A, Sohler C (2010) Sublinear-time algorithms, in [9]
5. Czumaj A, Ergun F, Fortnow L, Magen A, Newman I, Rubinfeld R, Sohler C (2005) Approximating the weight of the euclidean minimum spanning tree in sublinear time. SIAM J Comput 35(1):91–109
6. Eden T, Levi A, Ron D, Seshadhri C (2015) Approximately counting triangles in sublinear time. In: Proceedings of FOCS 2015, Berkeley. (To appear) (see also arXiv:1504.00954 and arXiv:1505.01927)
7. Feige U (2006) On sums of independent random variables with unbounded variance, and estimating

the average degree in a graph. SIAM J Comput 35(4):964–984

8. Fischer E, Newman I (2007) Testing versus estimation of graph properties. SIAM J Comput 37(2):482–501

9. Goldreich O (ed) (2010) Property testing: current research and surveys. LNCS, vol 6390. Springer, Heidelberg

10. Goldreich O, Ron D (2008) Approximating average parameters of graphs. Random Struct Algorithms 32(4):473–493

11. Gonen M, Ron D, Shavitt Y (2011) Counting stars and other small subgraphs in sublinear time. SIAM J Discret Math 25(3):1365–1411

12. Indyk P (1999) Sublinear-time algorithms for metric space problems. In: Proceedings of the thirty-first annual ACM symposium on the theory of computing (STOC), Atlanta, pp 428–434

13. Marko S, Ron D (2009) Distance approximation in bounded-degree and general sparse graphs. Trans Algorithms 5(2):article number 22

14. Nguyen HN, Onak K (2008) Constant-time approximation algorithms via local improvements. In: Proceedings of the forty-ninth annual symposium on foundations of computer science (FOCS), Philadelphia, pp 327–336

15. Parnas M, Ron D (2007) Approximating the minimum vertex cover in sublinear time and a connection to distributed algorithms. Theor Comput Sci 381(1–3):183–196

16. Parnas M, Ron D, Rubinfeld R (2006) Tolerant property testing and distance approximation. J Comput Syst Sci 72(6):1012–1042

17. Yoshida Y, Yamamoto M, Ito H (2009) An improved constant-time approximation algorithm for maximum matchings. In: Proceedings of the fourty-first annual ACM symposium on the theory of computing (STOC), Bethesda, pp 225–234

Euclidean Traveling Salesman Problem

Artur Czumaj
Department of Computer Science, Centre for Discrete Mathematics and Its Applications, University of Warwick, Coventry, UK

Keywords

Approximation algorithms; Euclidean graphs; PTAS; TSP

Years and Authors of Summarized Original Work

1998; Arora
1999; Mitchell

Problem Definition

This entry considers geometric optimization $\mathcal{NP}$-hard problems like the Euclidean traveling salesman problem and the Euclidean Steiner tree problem. These problems are geometric variants of standard graph optimization problems, and the restriction of the input instances to geometric or Euclidean case arises in numerous applications (see [1,2]). The main focus of this entry is on the Euclidean traveling salesman problem.

The Euclidean Traveling Salesman Problem (TSP)

For a given set S of n points in the Euclidean space $\mathbb{R}^d$, find the minimum length path that visits each point exactly once. The cost $\delta(x, y)$ of an edge connecting a pair of points $x, y \in \mathbb{R}^d$ is equal to the Euclidean distance between points x and y, that is, $\delta(x, y) = \sqrt{\sum_{i=1}^{d}(x_i - y_i)^2}$, where $x = (x_1, \ldots, x_d)$ and $y = (y_1, \ldots, y_d)$. More generally, the distance could be defined using other norms, such as ℓ_p norms for any $p > 1$,

$$\delta(x, y) = \left(\sum_{i=1}^{p}(x_i - y_i)^p \right)^{1/p}.$$

For a given set S of points in Euclidean space $\mathbb{R}^d$, for a certain integer d, $d \geq 2$, a *Euclidean graph* (network) is a graph $G = (S, E)$, where E is a set of straight-line segments connecting pairs of points in S. If all pairs of points in S are connected by edges in E, then G is called a *complete Euclidean graph on S*. The cost of the graph is equal to the sum of the costs of the edges of the graph, $\mathrm{cost}(G) = \sum_{(x,y) \in E} \delta(x, y)$.

A *polynomial-time approximation scheme (PTAS)* is a family of algorithms $\{\mathcal{A}_\varepsilon\}$ such that, for each fixed $\varepsilon > 0$, $\mathcal{A}_\varepsilon$ runs in polynomial time in the size of the input and produces a $(1 + \varepsilon)$-approximation.

Related Work

The classical book by Lawler et al. [16] provides extensive information about the TSP. Also, the survey exposition of Bern and Eppstein [8] presents the state of the art for geometric TSP until 1995, and the survey of Arora [2] discusses the research after 1995.

Key Results

We begin with the hardness results. The TSP in general graphs is well known to be $\mathcal{NP}$-hard, and the same claim holds for the Euclidean TSP [14, 18].

Theorem 1 *The Euclidean TSP is $\mathcal{NP}$-hard.*

Perhaps rather surprisingly, it is still not known if the decision version of the problem is $\mathcal{NP}$-complete [14]. (The decision version of the Euclidean TSP: given a point set in the Euclidean space $\mathbb{R}^d$ and a number t, verify if there is a simple path of length smaller than t that visits each point exactly once.)

The approximability of TSP has been studied extensively over the last few decades. It is not hard to see that TSP is not approximable in polynomial time (unless $\mathcal{P} = \mathcal{NP}$) for arbitrary graphs with arbitrary edge costs. When the weights satisfy the triangle inequality (the so-called metric TSP), there is a polynomial-time 3/2-approximation algorithm due to Christofides [9], and it is known that no PTAS exists (unless $\mathcal{P} = \mathcal{NP}$). This result has been strengthened by Trevisan [21] to include Euclidean graphs in high dimensions (the same result holds also for any ℓ_p metric).

Theorem 2 (Trevisan [21]) *If $d \geq \log n$, then there exists a constant $\varepsilon > 0$ such that the Euclidean TSP in $\mathbb{R}^d$ is $\mathcal{NP}$-hard to approximate within a factor of $1 + \varepsilon$.*

In particular, this result implies that if $d \geq \log n$, then the Euclidean TSP in $\mathbb{R}^d$ has no PTAS unless $\mathcal{P} = \mathcal{NP}$.

The same result holds also for any ℓ_p metric. Furthermore, Theorem 2 implies that Euclidean TSP in $\mathbb{R}^{\log n}$ is *APX PB*-hard under E-reductions and *APX*-complete under AP-reductions.

It has been believed for some time that Theorem 2 might hold for smaller values of d, in particular even for $d = 2$, but this has been disproved independently by Arora [1] and Mitchell [17].

Theorem 3 (Arora [1] and Mitchell [17]) *The Euclidean TSP on the plane has a PTAS.*

The main idea of the algorithms of Arora and Mitchell is rather simple, but the details of the analysis are quite complicated. Both algorithms follow the same approach. One first proves a so-called structure theorem, which demonstrates that there is a $(1 + \varepsilon)$-approximation that has some local properties (in the case of the Euclidean TSP, there is a quadtree partition of the space containing all the points such that there is a $(1 + \varepsilon)$-approximation in which each cell of the quadtree is crossed by the tour at most a constant number of times and only in some prespecified locations). Then, one uses dynamic programming to find an optimal (or almost optimal) solution that obeys the local properties specified in the structure theorem.

The original algorithms presented in the first conference version of [1] and in the early version of [17] have the running times of the form $\mathcal{O}(n^{1/\epsilon})$ to obtain a $(1 + \varepsilon)$-approximation, but this has been subsequently improved. In particular, Arora's randomized algorithm in [1] runs in time $\mathcal{O}(n(\log n)^{1/\epsilon})$, and it can be derandomized with a slowdown of $\mathcal{O}(n)$. The result from Theorem 3 can be also extended to higher dimensions. Arora shows the following result.

Theorem 4 (Arora [1]) *For every constant d, the Euclidean TSP in $\mathbb{R}^d$ has a PTAS.*

For every fixed $c > 1$ and given any n points in $\mathbb{R}^d$, there is a randomized algorithm that finds a $\left(1 + \frac{1}{c}\right)$-approximation of the optimum traveling salesman tour in $\mathcal{O}\left(n(\log n)^{(\mathcal{O}(\sqrt{dc}))^{d-1}}\right)$ time. In particular, for any constant d and c, the running time is $\mathcal{O}\left(n(\log n)^{\mathcal{O}(1)}\right)$. The algorithm can be derandomized by increasing the running time by a factor of $\mathcal{O}(n^d)$.

This has been later extended by Rao and Smith [19], who proved the following.

Theorem 5 (Rao and Smith [19]) *There is a deterministic algorithm that computes a $\left(1 + \frac{1}{c}\right)$-approximation of the optimum traveling salesman tour in $\mathcal{O}\left(2^{(cd)^{\mathcal{O}(d)}} n + (cd)^{\mathcal{O}(d)} n \log n\right)$ time.*

There is a randomized algorithm that succeeds with probability at least $\frac{1}{2}$ and that computes a $\left(1 + \frac{1}{c}\right)$-approximation of the optimum traveling salesman tour in expected $\left(c\sqrt{d}\right)^{\mathcal{O}\left(d(c\sqrt{d})^{d-1}\right)} n + \mathcal{O}(d\, n \log n)$ time.

These results are essentially asymptotically optimal in the decision tree model thanks to a lower bound of $\Omega(n \log n)$ for any sublinear approximation for 1-dimensional Euclidean TSP due to Das et al. [12]. In the *real RAM* model, one can further improve the randomized results.

Theorem 6 (Bartal and Gottlieb [6]) *Given a set S of n points in d-dimensional grid $\{0, \ldots, \Delta\}^d$ with $\Delta = 2^{(cd)^{\mathcal{O}(d)}} n$, there is a randomized algorithm that with probability $1 - e^{-\mathcal{O}_d(n^{1/3d})}$ computes a $\left(1 + \frac{1}{c}\right)$-approximation of the optimum traveling salesman tour for S in time $2^{(cd)^{\mathcal{O}(d)}} n$ in the integer RAM model.*

If the data is not given in the integral form, then one may round the data into this form using the floor or mod functions, and assuming these functions are atomic operations, the rounding can be done in $\mathcal{O}(dn)$ total time, leading to the following theorem.

Theorem 7 (Bartal and Gottlieb [6]) *Given a set of n points in $\mathbb{R}^d$, there is a randomized algorithm that with probability $1 - e^{-\mathcal{O}_d(n^{1/3d})}$ computes a $\left(1 + \frac{1}{c}\right)$-approximation of the optimum traveling salesman tour in time $2^{(cd)^{\mathcal{O}(d)}} n$ in the real RAM model with atomic floor or mod operations.*

Applications

The techniques developed by Arora [1] and Mitchell [17] found numerous applications in the design of polynomial-time approximation schemes for geometric optimization problems.

Euclidean Minimum Steiner Tree

For a given set S of n points in the Euclidean space $\mathbb{R}^d$, find the minimum-cost network connecting all the points in S (where the cost of a network is equal to the sum of the lengths of the edges defining it).

Euclidean k-median

For a given set S of n points in the Euclidean space $\mathbb{R}^d$ and an integer k, find k-medians among the points in S so that the sum of the distances from each point in S to its closest median is minimized.

Euclidean k-TSP

For a given set S of n points in the Euclidean space $\mathbb{R}^d$ and an integer k, find the shortest tour that visits at least k points in S.

Euclidean k-MST

For a given set S of n points in the Euclidean space $\mathbb{R}^d$ and an integer k, find the shortest tree that visits at least k points in S.

Euclidean Minimum-Cost k-Connected Subgraph

For a given set S of n points in the Euclidean space $\mathbb{R}^d$ and an integer k, find the minimum-cost subgraph (of the complete graph on S) that is k-connected.

Theorem 8 *For every constant d, the following problems have a PTAS:*

- *Euclidean minimum Steiner tree problem in $\mathbb{R}^d$ [1, 19]*
- *Euclidean k-median problem in $\mathbb{R}^d$ [5]*
- *Euclidean k-TSP and the Euclidean k-MST problems in $\mathbb{R}^d$ [1]*
- *Euclidean minimum-cost k-connected subgraph problem in $\mathbb{R}^d$ (constant k) [10]*

The technique developed by Arora [1] and Mitchell [17] led also to some quasi-polynomial-time approximation schemes, that is, the algorithms with the running time of $n^{\mathcal{O}(\log n)}$. For example, Arora and Karokostas [4] gave a quasi-polynomial-time approximation scheme for the Euclidean minimum latency problem, Das and Mathieu [13] gave a quasi-polynomial-time approximation scheme for the Euclidean capacitated vehicle routing problem, and Remy and Steger [20] gave a quasi-polynomial-time approximation scheme for the minimum-weight triangulation problem.

For more discussion, see the survey by Arora [2] and Czumaj and Lingas [11].

Extensions to Planar Graphs and Metric Spaces with Bounded Doubling Dimension

The dynamic programming approach used by Arora [1] and Mitchell [17] is also related to the recent advances for a number of optimization problems for planar graphs and in graphs in metric spaces with bounded doubling dimension. For example, Arora et al. [3] designed a PTAS for the TSP in weighted planar graphs (cf. [15] for a linear-time PTAS), and there is a PTAS for metric spaces with bounded doubling dimension [7].

Open Problems

An interesting open problem is if the quasi-polynomial-time approximation schemes mentioned above (for the minimum latency, the capacitated vehicle routing, and the minimum-weight triangulation problems) can be extended to obtain PTAS. For more open problems, see Arora [2].

Experimental Results

The Web page of the 8th DIMACS Implementation Challenge, http://dimacs.rutgers.edu/Challenges/TSP/, contains a lot of instances.

URLs to Code and Data Sets

The Web page of the 8th DIMACS Implementation Challenge, http://dimacs.rutgers.edu/Challenges/TSP/, contains a lot of instances.

Cross-References

▶ Approximation Schemes for Geometric Network Optimization Problems
▶ Metric TSP
▶ Minimum k-Connected Geometric Networks
▶ Minimum Weight Triangulation

Recommended Reading

1. Arora S (1998) Polynomial time approximation schemes for Euclidean traveling salesman and other geometric problems. J Assoc Comput Mach 45(5):753–782
2. Arora S (2003) Approximation schemes for NP-hard geometric optimization problems: a survey. Math Program Ser B 97:43–69
3. Arora S, Grigni M, Karger D, Klein P, Woloszyn A (1998) A polynomial time approximation scheme for weighted planar graph TSP. In: Proceedings of the 9th annual ACM-SIAM symposium on discrete algorithms (SODA), San Francisco, pp 33–41
4. Arora S, Karakostas G (1999) Approximation schemes for minimum latency problems. In: Proceedings of the 31st annual ACM symposium on theory of computing (STOC), Atlanta, pp 688–693
5. Arora S, Raghavan P, Rao S (1998) Approximation schemes for Euclidean k-medians and related problems. In: Proceedings of the 30th annual ACM symposium on theory of computing (STOC), Dallas, pp 106–113
6. Bartal Y, Gottlieb LA (2013) A linear time approximation scheme for Euclidean TSP. In: Proceedings of the 54th IEEE symposium on foundations of computer science (FOCS), Berkeley, pp 698–706
7. Bartal Y, Gottlieb LA, Krauthgamer R (2012) The traveling salesman problem: low-dimensionality implies a polynomial time approximation scheme. In: Proceedings of the 44th annual ACM symposium on theory of computing (STOC), New York, pp 663–672
8. Bern M, Eppstein D (1996) Approximation algorithms for geometric problems. In: Hochbaum D (ed) Approximation algorithms for NP-hard problems. PWS Publishing, Boston
9. Christofides N (1976) Worst-case analysis of a new heuristic for the traveling salesman problem. Technical report, Graduate School of Industrial Administration, Carnegie-Mellon University, Pittsburgh
10. Czumaj A, Lingas A (1999) On approximability of the minimum-cost k-connected spanning subgraph problem. In: Proceedings of the 10th annual ACM-SIAM symposium on discrete algorithms (SODA), Baltimore, pp 281–290
11. Czumaj A, Lingas A (2007) Approximation schemes for minimum-cost k-connectivity problems in geometric graphs. In: Gonzalez TF (ed) Handbook of

approximation algorithms and metaheuristics. CRC, Boca Raton

12. Das G, Kapoor S, Smid M (1997) On the complexity of approximating Euclidean traveling salesman tours and minimum spanning trees. Algorithmica 19(4):447–462
13. Das A, Mathieu C (2010) A quasi-polynomial time approximation scheme for Euclidean capacitated vehicle routing. In: Proceedings of the 21st annual ACM-SIAM symposium on discrete algorithms (SODA), Austin, pp 390–403
14. Garey MR, Graham RL, Johnson DS (1976) Some NP-complete geometric problems. In: Proceedings of the 8th annual ACM symposium on theory of computing (STOC), Hershey, pp 10–22
15. Klein P (2008) A linear-time approximation scheme for TSP in undirected planar graphs with edge-weights. SIAM J Comput 37(6):1926–1952
16. Lawler EL, Lenstra JK, Rinnooy Kan AHG, Shmoys DB (1985) The traveling salesman problem: a guided tour of combinatorial optimization. Wiley, Chichester/New York
17. Mitchell JSB (1999) Guillotine subdivisions approximate polygonal subdivisions: a simple polynomial-time approximation scheme for geometric TSP, k-MST, and related problems. SIAM J Comput 28(4):1298–1309
18. Papadimitriou CH (1977) Euclidean TSP is NP-complete. Theor Comput Sci 4:237–244
19. Rao SB, Smith WD (1998) Approximating geometrical graphs via "spanners" and "banyans." In: Proceedings of the 30th annual ACM symposium on theory of computing (STOC), Dallas, pp 540–550
20. Remy J, Steger A (2006) A quasi-polynomial time approximation scheme for minimum weight triangulation. In: Proceedings of the 38th annual ACM symposium on theory of computing (STOC), Seattle, pp 316–325
21. Trevisan L (2000) When Hamming meets Euclid: the approximability of geometric TSP and Steiner tree. SIAM J Comput 30(2):475–485

Exact Algorithms and Strong Exponential Time Hypothesis

Joshua R. Wang and Ryan Williams
Department of Computer Science, Stanford University, Stanford, CA, USA

Keywords

Exact algorithms; Exponential-time hypothesis; Satisfiability; Treewidth

Years and Authors of Summarized Original Work

2010; Pǎtraşscu, Williams
2011; Lokshtanov, Marx, Saurabh
2012; Cygan, Dell, Lokshtanov, Marx, Nederlof, Okamoto, Paturi, Saurabh, Wahlström

Problem Definition

All problems in NP can be exactly solved in $2^{\text{poly}(n)}$ time via exhaustive search, but research has yielded faster exponential-time algorithms for many NP-hard problems. However, some key problems have not seen improved algorithms, and problems with improvements seem to converge toward $O(C^n)$ for some unknown constant $C > 1$.

The satisfiability problem for Boolean formulas in conjunctive normal form, CNF-SAT, is a central problem that has resisted significant improvements. The complexity of CNF-SAT and its special case k-SAT, where each clause has k literals, is the canonical starting point for the development of NP-completeness theory.

Similarly, in the last 20 years, two hypotheses have emerged as powerful starting points for understanding exponential-time complexity. In 1999, Impagliazzo and Paturi [5] defined the *exponential-time hypothesis* (ETH), which asserts that 3-*SAT cannot be solved in subexponential time.* Namely, it asserts there is an $\epsilon > 0$ such that 3-SAT cannot be solved in $O((1 + \epsilon)^n)$ time. ETH has been a surprisingly useful assumption for ruling out subexponential-time algorithms for other problems [2, 6]. A stronger hypothesis has led to more fine-grained lower bounds, which is the focus of this article. Many NP-hard problems are solvable in C^n time via exhaustive search (for some $C > 1$) but are not known to be solvable in $(C - \epsilon)^n$ time, for any $\epsilon > 0$. The *strong exponential-time hypothesis* (SETH) [1,5] asserts that for every $\epsilon > 0$, there exists a k such that k-SAT cannot be solved in time $O((2 - \epsilon)^n)$. SETH has been very useful in establishing tight (and exact) lower bounds for many problems. Here we survey some of these tight results.

Key Results

The following results are reductions from k-SAT to other problems. They can be seen either as new attacks on the complexity of SAT or as lower bounds for exact algorithms that are conditional on SETH.

Lower Bounds on General Problems
The following problems have lower bounds conditional on SETH. The reduction for the first problem is given to illustrate the technique.

k-Dominating Set
A *dominating set* of a graph $G = (V, E)$ is a subset $S \subseteq V$ such that every vertex is either in S or is a neighbor of a vertex in S. The k-DOMINATING SET problem asks to find a dominating set of size k. Assuming SETH, for any $k \geq 3$ and $\epsilon > 0$, k-DOMINATING SET cannot be solved in $O(n^{k-\epsilon})$ time [8].

The reduction from SAT to k-DOMINATING SET proceeds as follows. Fix some $k \geq 3$ and let F be a CNF formula on n variables; we build a corresponding graph G_F. Partition its variables into k equally sized parts of n/k variables. For each part, take all $2^{n/k}$ partial assignments and make a node for each partial assignment. Make each of the k parts into a clique (disjoint from the others). Add a dummy node for each partial assignment clique that is connected to every node in that clique but has no other edges. Add m more nodes, one for each clause. Finally, make an edge from a partial assignment node to a clause node iff the partial assignment satisfies the clause. We observe that there is a k-dominating set in G_F if F is satisfiable.

2Sat+2Clauses
The 2SAT+2CLAUSES problem asks whether a Boolean formula is satisfiable, given that it is a 2-CNF with two additional clauses of arbitrary length. Assuming SETH, for any $m = n^{1+o(1)}$ and $\epsilon > 0$, 2SAT+2CLAUSES cannot

be solved in $O(n^{2-\epsilon})$ time [8]. It is known that 2SAT+2CLAUSES can be solved in $O(mn + n^2)$ time [8].

HornSat+kClauses
The HORNSAT+kCLAUSES problem asks whether a Boolean formula is satisfiable, given that it is a CNF of clauses that contain at most one nonnegative literal per clause (a Horn CNF), conjoined with k additional clauses of arbitrary length but only positive literals. Assuming SETH, for any $k \geq 2$ and $\epsilon > 0$, HORNSAT+kCLAUSES cannot be solved in $O((n + m)^{k-\epsilon})$ time [8]. It can be trivially solved in $O(n^k \cdot (m + n))$ time by guessing a variable to set to true for each of the k additional clauses and checking if the remaining Horn CNF is satisfiable in linear time.

3-Party Set Disjointness
The 3-PARTY SET DISJOINTNESS problem is a communication problem with three parties and three subsets $S_1, S_2, S_3 \subseteq [m]$, where the ith party has access to all sets except for S_i. The parties wish to determine if $S_1 \cap \cdots \cap S_3 = \varnothing$. Clearly this can be done with $O(m)$ bits of communication. Assuming SETH, 3-PARTY SET DISJOINTNESS cannot be solved using protocols running in $2^{o(n)}$ time and communicating only $o(m)$ bits [8].

k-SUM
The k-SUM problem asks whether a set of n numbers contains a k-tuple that sums to zero. Assuming SETH, k-SUM on n numbers cannot be solved in $n^{o(k)}$ time for any $k < n^{0.99}$. (It is well known that k-SUM is in $O(n^{\lceil k/2 \rceil})$ time) [8].

For all the problems below, we can solve in $2^n n^{O(1)}$ time via exhaustive search.

k-Hitting Set
Given a set system $\mathcal{F} \subseteq 2^U$ in some universe U, a *hitting set* is a subset $H \subseteq U$ such that $H \cap S \neq \varnothing$ for every $S \in \mathcal{F}$. The k-HITTINGSET problem asks whether there is a hitting set of size at most t, given that each set $S \in \mathcal{F}$ has at most k elements. SETH is equivalent to the claim that for all $\epsilon > 0$,

there is a k for which k-HITTINGSET cannot be solved in time $O((2 - \epsilon)^n)$ [3].

k-Set Splitting

Given a set system $\mathcal{F} \subseteq 2^U$ in some universe U, a set splitting is a subset $X \subseteq U$ such that the first element of the universe is in X and for every $S \in \mathcal{F}$, neither $S \subseteq X$ nor $S \subseteq (U \setminus X)$. The k-SETSPLITTING problem asks whether there is a set splitting, given that each set $S \in \mathcal{F}$ has at most k elements. SETH is equivalent to the claim that for all $\epsilon > 0$, there is a k for which k-SETSPLITTING cannot be solved in time $O((2 - \epsilon)^n)$ [3].

k-NAE-Sat

The k-NAE-SAT problem asks whether a k-CNF has an assignment where the first variable is set to true and each clause has both a true literal and a false literal. SETH is equivalent to the claim that for all $\epsilon > 0$, there is a k for which k-NAE-SAT cannot be solved in time $O((2 - \epsilon)^n)$ [3].

c-VSP-Circuit-SAT

The c-VSP-CIRCUIT-SAT problem asks whether a cn-size Valiant series-parallel circuit over n variables has a satisfying assignment. SETH is equivalent to the claim that for all $\epsilon > 0$, there is a k for which c-VSP-CIRCUIT-SAT cannot be solved in time $O((2 - \epsilon)^n)$ [3].

Problems Parameterized by Treewidth

A variety of NP-complete problems have been shown to be much easier on graphs of bounded treewidth. Reductions starting from SETH given by Lokshtanov, Marx, and Saurabh [7] can also prove lower bounds that depend on the treewidth of an input graph, tw(G). The following are proven via analyzing the pathwidth of a graph, pw(G), and the fact that tw(G) $\leq$ pw(G).

Independent Set

An *independent set* of a graph $G = (V, E)$ is a subset $S \subseteq V$ such that the subgraph induced by S contains no edges. The INDEPENDENT SET problem asks to find an independent set of maximum size. Assuming SETH, for any $\epsilon > 0$,

INDEPENDENT SET cannot be solved in $(2 - \epsilon)^{\text{tw}(G)} n^{O(1)}$ time.

Dominating Set

A *dominating set* of a graph $G = (V, E)$ is a subset $S \subseteq V$ such that every vertex is either in S or is a neighbor of a vertex in S. The DOMINATING SET problem asks to find a dominating set of minimum size. Assuming SETH, for any $\epsilon > 0$, DOMINATING SET cannot be solved in $(3 - \epsilon)^{\text{tw}(G)} n^{O(1)}$ time.

Max Cut

A *cut* of a graph $G = (V, E)$ is a partition of V into S and $V \setminus S$. The size of a cut is the number of edges that have one endpoint in S and the other in $V \setminus S$. The MAX CUT problem asks to find a cut of maximum size. Assuming SETH, for any $\epsilon > 0$, MAX CUT cannot be solved in $(2 - \epsilon)^{\text{tw}(G)} n^{O(1)}$ time.

Odd Cycle Transversal

An *odd cycle transversal* of a graph $G = (V, E)$ is a subset $S \subseteq V$ such that the subgraph induced by $V \setminus S$ is bipartite. The ODD CYCLE TRANSVERSAL problem asks to, given an integer k, determine whether there is an odd cycle transversal of size k. Assuming SETH, for any $\epsilon > 0$, ODD CYCLE TRANSVERSAL cannot be solved in $(3 - \epsilon)^{\text{tw}(G)} n^{O(1)}$ time.

Graph Coloring

A *q-coloring* of a graph $G = (V, E)$ is a function $\mu : V \to [q]$. A q-coloring is *proper* if for all edges $(u, v) \in E$, $\mu(u) \neq \mu(v)$. The q-COLORING problem asks to decide whether the graph has a proper q-coloring. Assuming SETH, for any $q \geq 3$ and $\epsilon > 0$, q-COLORING cannot be solved in $(q - \epsilon)^{\text{tw}(G)} n^{O(1)}$ time.

Partition Into Triangles

A graph $G = (V, E)$ can be partitioned into triangles if there is a partition of the vertices into $S_1, S_2, \ldots, S_{n/3}$ such that each S_i induces a triangle in G. The PARTITION INTO TRIANGLES problem asks to decide whether the graph can be partitioned into triangles. Assuming SETH, for any $\epsilon > 0$, PARTITION INTO TRIANGLES cannot be solved in $(2 - \epsilon)^{\text{tw}(G)} n^{O(1)}$ time.

All of the above results are tight, in the sense that when $\epsilon = 0$, there is an algorithm for each of them.

Showing Difficulty Via Set Cover

Given a set system $\mathcal{F} \subseteq 2^U$ in some universe U, a set cover is a subset $\mathcal{C} \subseteq \mathcal{F}$ such that $\bigcup_{S \in \mathcal{C}} S = U$. The SET COVER problem asks whether there is a set cover of size at most t.

Cygan et al. [3] also gave reductions from SET COVER to several other problems, showing lower bounds conditional on the assumption that for all $\epsilon > 0$, there is a k such that SET COVER where sets in $\mathcal{F}$ have size at most k cannot be computed in time $O^*((2 - \epsilon)^n)$.

It is currently unknown how SET COVER is related to SETH; if there is a reduction from CNF-SAT to SET COVER, then all of these problems would have conditional lower bounds as well.

Steiner Tree

Given a graph $G = (V, E)$ and a set of terminals $T \subseteq V$, a *Steiner Tree* is a subset $X \subseteq V$ such that the graph induced by X is connected and $T \subseteq X$. The STEINER TREE problem asks whether G has a Steiner tree of size at most t. With the above SET COVER assumption, for all $\epsilon > 0$, STEINER TREE cannot be solved in $O^*((2 - \epsilon)^t)$ time.

Connected Vertex Cover

A *connected vertex cover* of a graph $G = (V, E)$ is a subset $X \subseteq V$ such that the subgraph induced by X is connected and every edge contains at least one endpoint in X. The CONNECTED VERTEX COVER problem asks whether G has a connected vertex cover of size at most t. With the above SET COVER assumption, for all $\epsilon > 0$, CONNECTED VERTEX COVER cannot be solved in $O^*((2 - \epsilon)^t)$ time.

Set Partitioning

Given a set system $\mathcal{F} \subseteq 2^U$ in some universe U, a set partitioning is a set cover $\mathcal{C}$ where pairwise disjoint elements have an empty intersection. The SET PARTITIONING problem asks whether there

is a set partitioning of size at most t. With the above SET COVER assumption, for all $\epsilon > 0$, SET PARTITIONING cannot be solved in $O^*((2 - \epsilon)^n)$ time.

Subset Sum

The SUBSET SUM problem asks whether a set of n positive numbers contains a subset that sums to a target t. With the above SET COVER assumption, for all $\delta < 1$, SUBSET SUM cannot be solved in $O^*(t^\delta)$ time. Note that there is a dynamic programming solution that runs in $O(nt)$ time.

Open Problems

- Does ETH imply SETH?
- Does SETH imply SET COVER requires $O^*((2 - \epsilon)^n)$ time for all $\epsilon > 0$?
- Does SETH imply that the Traveling Salesman Problem in its most general, weighted form requires $O^*((2 - \epsilon)^n)$ time for all $\epsilon > 0$?
- Given two graphs F and G, on k and n nodes, respectively, the SUBGRAPH ISOMORPHISM problem asks whether a (noninduced) subgraph of G is isomorphic to F. Does SETH imply that SUBGRAPH ISOMORPHISM cannot be solved in $2^{O(n)}$?

Cross-References

▶ Backtracking Based k-SAT Algorithms
▶ Exact Algorithms for General CNF SAT
▶ Exact Algorithms for Treewidth

Recommended Reading

1. Calabro C, Impagliazzo R, Paturi R (2009) The complexity of satisfiability of small depth circuits. In: Chen J, Fomin F (eds) Parameterized and exact computation. Lecture notes in computer science, vol 5917. Springer, Berlin/Heidelberg, pp 75–85. doi:10.1007/978-3-642-11269-0_6. http://dx.doi.org/10.1007/978-3-642-11269-0_6
2. Chen J, Chor B, Fellows M, Huang X, Juedes D, Kanj IA, Xia G (2005) Tight lower bounds for certain parameterized np-hard problems. Inf Comput 201(2):216–231. doi:http://dx.doi.org/10.1016/j.ic.

2005.05.001. http://www.sciencedirect.com/science/article/pii/S0890540105000763

3. Cygan M, Dell H, Lokshtanov D, Marx D, Nederlof J, Okamoto Y, Paturi R, Saurabh S, Wahlstrom M (2012) On problems as hard as cnf-sat. In: Proceedings of the 2012 IEEE conference on computational complexity (CCC '12), Washington, DC. IEEE Computer Society, pp 74–84. doi:10.1109/CCC.2012.36. http://dx.doi.org/10.1109/CCC.2012.36

4. Fomin F, Kratsch D (2010) Exact exponential algorithms. Texts in theoretical computer science, an EATCS series. Springer, Berlin/Heidelberg

5. Impagliazzo R, Paturi R (2001) On the complexity of k-sat. J Comput Syst Sci 62(2):367–375. doi:http://dx.doi.org/10.1006/jcss.2000.1727. http://www.sciencedirect.com/science/article/pii/S0022000000917276

6. Impagliazzo R, Paturi R, Zane F (2001) Which problems have strongly exponential complexity? J Comput Syst Sci 63(4):512–530. doi:http://dx.doi.org/10.1006/jcss.2001.1774. http://www.sciencedirect.com/science/article/pii/S002200000191774X

7. Lokshtanov D, Marx D, Saurabh S (2011) Known algorithms on graphs of bounded treewidth are probably optimal. In: Proceedings of the twenty-second Annual ACM-SIAM symposium on discrete algorithms (SODA '11), San Francisco. SIAM, pp 777–789. http://dl.acm.org/citation.cfm?id=2133036.2133097

8. Pătraşcu M, Williams R (2010) On the possibility of faster sat algorithms. In: Proceedings of the twenty-first annual ACM-SIAM symposium on discrete algorithms (SODA '10), Philadelphia. Society for Industrial and Applied Mathematics, pp 1065–1075. http://dl.acm.org/citation.cfm?id=1873601.1873687

9. Woeginger G (2003) Exact algorithms for np-hard problems: a survey. In: Jünger M, Reinelt G, Rinaldi G (eds) Combinatorial optimization Eureka, you shrink! Lecture notes in computer science, vol 2570. Springer, Berlin/Heidelberg, pp 185–207. doi:10.1007/3-540-36478-1_17. http://dx.doi.org/10.1007/3-540-36478-1_17

Exact Algorithms and Time/Space Tradeoffs

Jesper Nederlof
Technical University of Eindhoven, Eindhoven, The Netherlands

Keywords

Dynamic programming; Knapsack; Space efficiency; Steiner tree; Subset discrete Fourier transform

Years and Authors of Summarized Original Work

2010; Lokshtanov, Nederlof

Problem Definition

In the *subset sum problem*, we are given integers $a_1, \ldots, a_n, t$ and are asked to find a subset $X \subseteq \{1, \ldots, n\}$ such that $\sum_{i \in X} a_i = t$. In the *Knapsack problem*, we are given $a_1, \ldots, a_n, b_1, \ldots, b_n, t, u$ and are asked to find a subset $X \subseteq \{1, \ldots, n\}$ such that $\sum_{i \in X} a_i \leq t$ and $\sum_{i \in X} b_i \geq u$. It is well known that both problems can be solved in $O(nt)$ time using dynamic programming. However, as is typical for dynamic programming, these algorithms require a lot of working memory and are relatively hard to execute in parallel on several processors: the above algorithms use $O(t)$ space which may be *exponential* in the input size.

This raises the question: when can we avoid these disadvantages and still be (approximately) as fast as dynamic programming algorithms? It appears that by (slightly) loosening the time budget, space usage and parallelization can be significantly improved in many dynamic programs.

Key Results

A Space Efficient Algorithm for Subset Sum

In this article, we will use $\tilde{O}(\cdot)$ to suppress factors that are poly-logarithmic in the input size. In what follows, we will discuss how to prove the following theorem:

Theorem 1 (Lokshtanov and Nederlof, [7]) *There is an algorithm counting the number of solutions of a subset sum instance in $\tilde{O}(n^2 t (n + \log t))$ time and $(n + lg(t))(\lg nt)$ space.*

The Discrete Fourier Transform

We use Iverson's bracket notation: given a Boolean predicate b, $[b]$ denotes 1 if b is true and 0 otherwise. Let $P(x)$ be a polynomial of degree $N - 1$, and let $p_0, \ldots, p_{N-1}$ be its coefficients. Thus, $P(x) = \sum_{i=0}^{N-1} p_i x^i$. Let ω_N denote the N'th root of unity, that is, $\omega_N = e^{\frac{2\pi i}{N}}$. Let k, t be integers such that $k \neq t$. By the summation formula for geometric progressions $\left(\sum_{\ell=0}^{N-1} r^\ell = \frac{1-r^N}{1-r} \text{ for } r \neq 1 \right)$, we have:

$$\sum_{\ell=0}^{N-1} \omega_N^{\ell(k-t)} = \frac{1 - \omega_N^{(k-t)N}}{1 - \omega_N^{k-t}} = \frac{1 - \left(\omega_N^N\right)^{k-t}}{1 - \omega_N^{k-t}}$$

$$= \frac{1 - (1)^{k-t}}{1 - \omega_N^{k-t}} = 0.$$

On the other hand, if $k = t$, then $\sum_{\ell=0}^{N-1} \omega_N^{\ell(k-t)} = \sum_{\ell=0}^{N-1} 1 = N$. Thus, both cases can be compactly summarized as $\sum_{\ell=0}^{N-1} \omega_N^{\ell(k-t)} = [k = t]N$. As a consequence, we can express a coefficient p_t of $P(x)$ directly in terms of its evaluations:

$$p_t = \sum_{k=0}^{N-1} [k = t] p_k$$

$$= \sum_{k=0}^{N-1} \frac{1}{N} \sum_{\ell=0}^{N-1} \omega_N^{\ell(k-t)} \sum p_k$$

$$= \frac{1}{N} \sum_{\ell=0}^{N-1} \omega_N^{-\ell t} \sum_{k=0}^{N-1} p_k \left(\omega_N^\ell\right)^k \qquad (1)$$

$$= \frac{1}{N} \sum_{\ell=0}^{N-1} \omega_N^{-\ell t} P(\omega_N^\ell)$$

Using the Discrete Fourier Transform for Subset Sum

Given an instance $a_1, \ldots, a_n, t$ of subset sum, define the polynomial $P(x)$ to be $P(x) = \prod_{i=1}^{n}(1 + x^{a_i})$. Clearly, we can discard integers a_i larger than t, and assume that $P(x)$ has degree at most $N = nt$. If we expand the products in this polynomial to get rid of the parentheses, we get a sum of 2^n products and each of these products is of the type x^k and corresponds to a subset

$X \subseteq \{1, \ldots, n\}$ such that $\sum_{i \in X} a_i = k$. Thus, if we aggregate these products, we obtained the normal form $P(x) = \sum_{k=0}^{N-1} p_k x^k$, where p_k equals the number of subsets $X \subseteq [n]$ such that $\sum_{i \in X} a_i = k$. Plugging this into Eq. 1, we have that the number of subset sum solutions equals

$$p_t = \frac{1}{N} \sum_{\ell=0}^{N-1} \omega_N^{-\ell t} \prod_{i=1}^{n} \left(1 + \omega_N^{\ell a_i}\right). \qquad (2)$$

Given Eq. 2, the algorithm suggests itself: evaluation of the right-hand side gives the number of solutions of the subset sum instance. Given ω_N, this would be a straightforward on the unit-cost RAM model (recall that in this model arithmetic instructions as $+, -, *$ and $/$ are assumed to take constant time): the required powering operations are performed in $\log(N)$ arithmetic operations so an overall upper bound would be $O(n^2 t \log(nt))$ time.

However, still the value of this algorithm is not clear yet: for example, ω_N may be irrational, so it is not clear how to perform the arithmetic efficiently. This is an issue that also arises for the folklore fast Fourier transform (see, e.g., [3] for a nice exposition), and this issue is usually not addressed (a nice exception is Knuth [6]). Moreover in our case we should also be careful on the space budget: for example, we cannot even store 2^t in the usual way within our space budget. But, as we will now see, it turns out that we can simply evaluate Eq. 2 with finite precision and round to the nearest integer within the resource bounds claimed in Theorem 1.

Evaluating Equation 2 with Finite Precision

The algorithm establishing Theorem 1 is presented in Algorithm 1. Here, ρ represents the amount of precision the algorithm works with. The procedure tr_ρ truncates ρ bits after the decimal point. The procedure $\text{apxr}_\rho(z)$ returns an estimate of ω_N^z. In order to do this, estimates of ω_N^z with z being powers of 2 are precomputed in Lines 3–4. We omit an explicit implementation of the right-hand side of Line 4 since this is very standard; for example, one can use an approximation of π together with a binary splitting approach

Algorithm 1 Approximate evaluation of Eq. 2

Algorithm: SSS($a_1, \ldots, a_n, t$)
Require: for every $1 \le i \le n, a_i < t$.
1: $\rho \leftarrow 3n + 6 \log nt$
2: $s \leftarrow 0$
3: **for** $0 \le q \le \log N - 1$ **do**
4:     //store roots of unity for powers of two
5:     $r_q \leftarrow \mathrm{tr}_\rho(e^{\frac{2\pi 2^q}{N_i}})$
6: **end for**
7: **for** $0 \le \ell \le N - 1$ **do**
8:     $p \leftarrow \mathrm{apxr}_\rho(-\ell t \% N)$
9:     **for** $1 \le i \le n$ **do**
10:        $p \leftarrow \mathrm{tr}_\rho(p * (1 + \mathrm{apxr}_\rho(\ell a_i \% N)))$
11:    **end for**
12:    $s \leftarrow s + p$
13: **end for**
14: **return** $\mathrm{rnd}(\mathrm{tr}_\rho(\frac{s}{N}))$ //round to nearest int

Algorithm: $\mathrm{apxr}_\rho(z)$
Require: $z < N$.
15: $p' \leftarrow 1$
16: **for** $1 \le q \le \log N - 1$ **do**
17:    **if** 2^q divides z **then**
18:        $p \leftarrow \mathrm{tr}_\rho(p' * r_q)$
19:    **end if**
20: **end for**
21: **return** p'

(see [1, Section 4.9.1]) or a Taylor expansion-based approach (see [2]). Crude upper bounds on the time and space usage of both approaches are $O(\rho^2 \log N)$ time and $O(n \log nt + \log nt)$ space.

Let us proceed with verifying whether Algorithm 1 satisfies the resource bounds of Theorem 1. It is easy to see that all intermediate numbers have modulus at most $2^n N$, so their estimates can be represented with $O(\rho)$ bits. For all multiplications we will use an asymptotically fast algorithm running in $\tilde{O}(n)$ time (e.g., [5]). Then, Line 3–4 take $\tilde{O}(\rho^2 \log N)$ time. Line 6 takes $\tilde{O}(\rho \lg N)$ time; Lines 7–8 take $n\rho \lg N$ time, which is the bottleneck. So overall, the algorithm uses $\tilde{O}(Nn\rho) = \tilde{O}(n^2 t(n + \log t))$ time. The space usage is dominated by the precomputed values which use $O(\log N\rho) = O(n + \log t(\log nt))$ space.

For the correctness of Algorithm 1, let us first study what happens if we work with infinite precision (i.e., $\rho = \infty$). Note that $\mathrm{apxr}_\infty(z) = \omega_N^z$ since it computes

$$\prod_{q=1}^{\log N - 1} [2^q \text{ divides } z] r_q$$

$$= \prod_{q=1}^{\log N - 1} [2^q \text{ divides } z] \omega_N^{2^q}$$

$$= \omega_N^{\sum_{q=1}^{\log N - 1} [2^q \text{ divides } z] 2^q} = \omega^z.$$

Moreover, note that on iteration ℓ of the for-loop of Line 5, we will have on Line 9 that $p = P(\omega_N^\ell)$ by the definition of $P(x)$. Then, it is easy to see that Algorithm 1 indeed evaluates the right-hand side of Eq. 2.

Now, let us focus on the finite precision. The algorithm computes an N-sized sum of $(n + 2 \log N)$-sized products of precomputed values, (increased by one). Note that it is sufficient to guarantee that on Line 9 in every iteration ℓ, $|p - \omega_N^{-\ell t} \prod_{i=1}^n (1 + \omega_N^{\ell a_i})| \le 0.4$, since then the total error of s on Line 10 is at most $0.4N$ and the total error of s/N is 0.4, which guarantees rounding to the correct integer. Recall that p is the result of an $(n + 2 \log N)$-sized product, so let us analyze how the approximation error propagates in this situation. If $\hat{a}, \hat{b}$ are approximations of a, b and we approximate c by $\mathrm{tr}_\rho \hat{a} * \hat{b}$, we have

$$|c - \hat{c}| \le |a - \hat{a}||b| + |b - \hat{b}||a| + |a - \hat{a}||b - \hat{b}| + 2^{-\rho}.$$

Thus, if a is the result of a (i-1)-sized product, and using an upper bound of 2 for the modulus of any of the product terms in the algorithm, we can upper bound the error of E_i estimating an i-length product as follows: $E_1 \le 2^{-\rho}$ and for $i > 1$:

$$E_i \le 2E_{i-1} + 2^{-\rho}2^{i-1} + 2^{-\rho}E_{i-1} + 2^{-\rho}$$

$$\le 3E_{i-1} + 2^{-\rho}2^i.$$

Using straightforward induction we have that $E_i \le 6^i 2^{-\rho}$ So indeed, setting $\rho = 3n + 6 \log nt$ suffices.

A Generic Framework

For Theorem 1, we only used that the to be determined value is a coefficient of a (relatively) small degree polynomial that we can evaluate efficiently. Whether this is the case for other problems solved by dynamic programming can be seen from the structure of the used recurrence: when the recurrence can be formulated over a polynomial ring where the polynomials have small degree, we can evaluate it fast and interpolate with the same technique as above to find a required coefficient. For example, for Knapsack, one can use the polynomial $P(x, y) = \prod_{i=1}^{n}(x^{a_i} y^{b_i})$ and look for a nonzero coefficient of $x^{t'} y^{u'}$ where $t' \leq t$ and $u' \geq u$ to obtain a pseudo-polynomial time and polynomial space algorithm as well.

Naturally, this technique does not only apply to the polynomial ring. In general, if the ring would be $R \subseteq C^{N \times N}$ equipped with matrix addition and multiplication, we just need a matrix that simultaneously diagonalizes all matrices of R (in the above case, R are all circulant matrices which are simultaneously diagonalized by the Fourier matrix).

Applications

The framework applies to many dynamic programming algorithms. A nice additional example is the algorithm of Dreyfus and Wagner for Steiner tree [4, 7, 8].

Cross-References

▶ Knapsack

Recommended Reading

1. Brent R, Zimmermann P (2010) Modern computer arithmetic. Cambridge University Press, New York
2. Chudnovsky DV, Chudnovsky GV (1997) Approximations and complex multiplication according to ramanujan. In: Pi: a source book. Springer, New York, pp 596–622
3. Dasgupta S, Papadimitriou CH, Vazirani U (2006) Algorithms. McGraw-Hill, Boston
4. Dreyfus SE, Wagner RA (1972) The Steiner problem in graphs. Networks 1:195–207
5. Fürer M (2009) Faster integer multiplication. SIAM J Comput 39(3):979–1005
6. Knuth DE (1997) The art of computer programming, vol 2 (3rd edn.): seminumerical algorithms. Addison-Wesley Longman, Boston
7. Lokshtanov D, Nederlof J (2010) Saving space by algebraization. In: Proceedings of the forty-second ACM symposium on theory of computing, STOC '10, Cambridge. ACM, New York, pp 321–330
8. Nederlof J (2013) Fast polynomial-space algorithms using inclusion-exclusion. Algorithmica 65(4):868–884

Exact Algorithms for Bandwidth

Marek Cygan
Institute of Informatics, University of Warsaw, Warsaw, Poland

Keywords

Bandwidth; Exponential time algorithms; Graph ordering

Years and Authors of Summarized Original Work

2000; Feige
2008; Cygan, Pilipczuk
2010; Cygan, Pilipczuk
2012; Cygan, Pilipczuk

Problem Definition

Given a graph G with n vertices, an *ordering* is a bijective function $\pi : V(G) \to \{1, 2, \ldots, n\}$. The bandwidth of π is a maximal length of an edge, i.e.,

$$\text{bw}(\pi) = \max_{uv \in E(G)} |\pi(u) - \pi(v)|.$$

The bandwidth problem, given a graph G and a positive integer b, asks if there exists an ordering of bandwidth at most b.

Key Results

An exhaustive search for the bandwidth problem enumerates all the $n!$ orderings, trying to find one of bandwidth at most b. The first single exponential time algorithm is due to Feige and Kilian [6], which we are going to describe now.

Bucketing

Definition 1 For a positive integer k, let $\mathcal{I}_k$ be the collection of $\lceil n/k \rceil$ sets obtained by splitting the set $\{1, \ldots, n\}$ into equal parts (except the last one), i.e., $\mathcal{I}_k = \{\{1, \ldots, k\}, \{k+1, \ldots, 2k\}, \ldots\}$. A function $f : V(G) \to \mathcal{I}_k$ is called a k-bucket assignment, if for every edge $uv \in E(G)$ at least one of the following conditions is satisfied:

- $f(u) = f(v)$,
- $|\max f(u) - \min f(v)| \le b$,
- $|\min f(u) - \max f(v)| \le b$.

Clearly, if a function $f : V(G) \to \mathcal{I}_k$ is not a k-bucket assignment, then there is no ordering π of bandwidth at most b consistent with f, where π is *consistent* with f iff $\pi(v) \in f(v)$ for each $v \in V(G)$. A bucket function can be seen as a rough assignment – instead of assigning vertices to their final positions in the ordering, we assign them to intervals.

The $\mathcal{O}(10^n \mathrm{poly}(n))$ time algorithm of [6] is based on two ideas, both related to the notion of bucket assignments. For the sake of presentation, let us assume that n is divisible by b, whereas b is a power of two. Moreover, we assume that G is connected, as otherwise it is enough to consider each connected component of G separately.

First, one needs to show that there is a family of at most $n3^{n-1}$ b-bucket assignments $\mathcal{F}$, such that any ordering of bandwidth at most b is consistent with some b-bucket assignment from $\mathcal{F}$. We create $\mathcal{F}$ recursively by branching. First, fix an arbitrary vertex v_0, and assign it to some interval from $\mathcal{I}_k$ (there are at most n choices here). Next, consider any vertex v without assigned interval, which has a neighbor u with already assigned interval. By the assumption that G is connected, v always exists. Note that in order to create a valid bucket assignment, v has to be assigned either to the same interval as u or to one of its two neighboring intervals. This gives at most three branches to be explored.

In the second phase, consider some b-bucket assignment $f \in \mathcal{F}$. We want to check whether there exists some ordering of bandwidth at most b consistent with f. To do this, for each vertex v, we branch into two choices, deciding whether v should be assigned to the left half of $f(v)$ or to the right half of $f(v)$. This leads to at most 2^n $b/2$-bucket assignments to be processed. The key observation is that each of those assignments can be naturally split into two independent subproblems. This is because each edge within an interval of length $b/2$ and each edge between two neighboring intervals of length $b/2$ will be of length at most $b - 1$. Additionally, each edge connecting two vertices with at least two intervals of length $b/2$ in between would lead to violating the constraint of being a valid $b/2$-bucket assignment. Therefore, it is enough to consider vertices in even and odd intervals separately (see Fig. 1). Such routine of creating more and more refined bucket assignments can be continued, where the running time used for n vertices satisfies

$$T(n) = 2^n \cdot 2 \cdot T\left(\frac{n}{2}\right)$$

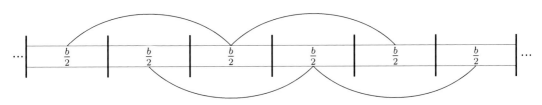

Exact Algorithms for Bandwidth, Fig. 1 *Thick vertical lines* separate subsequent intervals from $\mathcal{I}_{b/2}$. Meaningful edges connect vertices with exactly one interval in between

which in turn gives $T(n) = 4^n \text{poly}(n)$. Since we have $|\mathcal{F}| \leq n3^{n-1}$, we end up with $\mathcal{O}(12^n \text{poly}(n))$ time algorithm. If instead of generating b-bucket assignments one uses $b/2$-bucket assignments (there are at most $n5^{n-1}$ of them), then the running time can be improved to $10^n \text{poly}(n)$.

Dynamic Programming

In [2, 5], Cygan and Pilipczuk have shown that for a single $(b + 1)$-bucket assignment, one can check in time and space $\mathcal{O}(2^n \text{poly}(n))$ whether there exists an ordering of bandwidth at most b consistent with it. Since there are at most $n3^{n-1}$ $(b + 1)$-bucket assignments, this leads to $\mathcal{O}(6^n \text{poly}(n))$ time algorithm.

The key idea is to assign vertices to their final positions consistent with some $f \in \mathcal{F}$ in a very specific order. Let us color the set of positions $\{1, \ldots, n\}$ with $\text{color}(i) = (i - 1) \bmod (b + 1)$. Define a *color order* of positions, where positions from $\{1, \ldots, n\}$ are sorted by their color values, breaking ties with position values (see Fig. 2).

A lemma that proves usefulness of the color order shows that if we assign vertices to positions in the color order, then we can use the standard Held-Karp dynamic programming over subsets approach. In particular, in a state of

$b+1$			$b+1$			$b+1$							
1	5	9	12	2	6	10	13	3	7	11	14	4	8

Exact Algorithms for Bandwidth, Fig. 2 An index of each position in a color order for $n = 14$ and $b = 3$

dynamic programming, it is enough to store the subset $S \subseteq V(G)$ of vertices already assigned to the first $|S|$ positions in the color order (see Fig. 3).

Further Improvements

Instead of upper bounding running time of the algorithm for each $(b + 1)$-bucket assignment separately, one can count the number of states of the dynamic programming routine used by the algorithm throughout the processing of all the bucket assignments. As shown in [2], this leads to $\mathcal{O}(5^n \text{poly}(n))$ running time, which with more insights and more technical analysis can be further improved to $O(4.83^n)$ [5] and $O(4.383^n)$ [3]. If only polynomial space is allowed, then the best known algorithm needs $O(9.363^n)$ running time [4].

Related Work

Concerning small values of b, Saxe [8] presented a nontrivial $O(n^{b+1})$ time and space dynamic programming, consequently proving the problem to be in XP. However, Bodlaender et al. [1] have shown that bandwidth is hard for any fixed level of the W hierarchy.

For a related problem of minimum distortion embedding, Fomin et al. [7] obtained a $\mathcal{O}(5^n \text{poly}(n))$ time algorithm, improved by Cygan and Pilipczuk [4] to running times same as for the best known bandwidth algorithms.

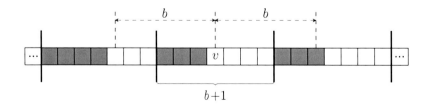

Exact Algorithms for Bandwidth, Fig. 3 When a vertex v is to be assigned to the next position in the color order, then all its neighbors from the left interval cannot be yet assigned a position, whereas all its neighbors from the right interval have to be already assigned in order to obtain an ordering of bandwidth at most b

Open Problems

Many vertex ordering problems admit $\mathcal{O}(2^n$ poly$(n))$ time and space algorithms, like Hamiltonicity, cutwidth, pathwidth, optimal linear arrangement, etc. In [2], Cygan and Pilipczuk have shown that a dynamic programming routine with such a running time is possible, provided a $(b + 1)$-bucket assignment is given. A natural question to ask is whether it is possible to obtain $\mathcal{O}(2^n$ poly$(n))$ without the assumption of having a fixed assignment to be extended.

Cross-References

▶ Graph Bandwidth

Recommended Reading

1. Bodlaender HL, Fellows MR, Hallett MT (1994) Beyond np-completeness for problems of bounded width: hardness for the W hierarchy. In: Leighton FT, Goodrich MT (eds) Proceedings of the twenty-sixth annual ACM symposium on theory of computing, Montréal, 23–25 May 1994. ACM, pp 449–458, doi:10.1145/195058.195229. http://doi.acm.org/10.1145/195058.195229
2. Cygan M, Pilipczuk M (2008) Faster exact bandwidth. In: Broersma H, Erlebach T, Friedetzky T, Paulusma D (eds) Graph-theoretic concepts in computer science, 34th international workshop, WG 2008, Durham, June 30–July 2, 2008. Revised papers, Lecture notes in computer science, vol 5344, pp 101–109. doi:10.1007/978-3-540-92248-3_10. http://dx.doi.org/10.1007/978-3-540-92248-3_10
3. Cygan M, Pilipczuk M (2010) Exact and approximate bandwidth. Theor Comput Sci 411(40–42):3701–3713. doi:10.1016/j.tcs.2010.06.018. http://dx.doi.org/10.1016/j.tcs.2010.06.018
4. Cygan M, Pilipczuk M (2012) Bandwidth and distortion revisited. Discret Appl Math 160(4–5):494–504. doi:10.1016/j.dam.2011.10.032. http://dx.doi.org/10.1016/j.dam.2011.10.032
5. Cygan M, Pilipczuk M (2012) Even faster exact bandwidth. ACM Trans Algorithms 8(1):8. doi:10.1145/2071379.2071387. http://doi.acm.org/10.1145/2071379.2071387
6. Feige U (2000) Coping with the np-hardness of the graph bandwidth problem. In: Halldórsson MM (ed) Proceedings of the 7th Scandinavian workshop on algorithm theory (SWAT), Bergen. Lecture notes in computer science, vol 1851. Springer, pp 10–19
7. Fomin FV, Lokshtanov D, Saurabh S (2011) An exact algorithm for minimum distortion embedding. Theor Comput Sci 412(29):3530–3536. doi:10.1016/j.tcs.2011.02.043. http://dx.doi.org/10.1016/j.tcs.2011.02.043
8. Saxe J (1980) Dynamic programming algorithms for recognizing small-bandwidth graphs in polynomial time. SIAM J Algebr Methods 1:363–369

Exact Algorithms for Dominating Set

E

Dieter Kratsch
UFM MIM – LITA, Université de Lorraine, Metz, France

Keywords

Branch and reduce; Dominating set; Exact exponential algorithm; Measure and conquer

Years and Authors of Summarized Original Work

2005; Fomin, Grandoni, Kratsch
2008; van Rooij, Bodlaender
2011; Iwata

Problem Definition

The dominating set problem is a classical NP-hard optimization problem which fits into the broader class of covering problems. Hundreds of papers have been written on this problem that has a natural motivation in facility location.

Definition 1 For a given undirected, simple graph $G = (V, E)$, a subset of vertices $D \subseteq V$ is called a *dominating set* if every vertex $u \in V - D$ has a neighbor in D. The minimum dominating set problem (abbr. MDS) is to find a *minimum dominating set* of G, i.e., a dominating set of G of minimum cardinality.

Problem 1 (MDS)

INPUT: Undirected simple graph $G = (V, E)$.
OUTPUT: A minimum dominating set D of G.

Various modifications of the dominating set problem are of interest, some of them obtained by putting additional constraints on the dominating set as, e.g., requesting it to be an independent set or to be connected. In graph theory, there is a huge literature on domination dealing with the problem and its many modifications. In graph algorithms, the MDS problem and some of its modifications like independent dominating set and connected dominating set have been studied as benchmark problems for attacking NP-hard problems under various algorithmic approaches.

Known Results

The algorithmic complexity of MDS and its modifications when restricted to inputs from a particular graph class has been studied extensively. Among others, it is known that MDS remains NP-hard on bipartite graphs, split graphs, planar graphs, and graphs of maximum degree 3. Polynomial time algorithms to compute a minimum dominating set are known, e.g., for permutation, interval, and k-polygon graphs. There is also a $O(3^k n^{O(1)})$ time algorithm to solve MDS on graphs of treewidth at most k.

The dominating set problem is one of the basic problems in parameterized complexity; it is W[2]-complete and thus it is unlikely that the problem is fixed parameter tractable. On the other hand, the problem is fixed parameter tractable on planar graphs. Concerning approximation, MDS is equivalent to MINIMUM SET COVER under L-reductions. There is an approximation algorithm solving MDS within a factor of $1 + \log|V|$, and it cannot be approximated within a factor of $(1 - \epsilon) \ln|V|$ for any $\epsilon > 0$, unless NP $\subset$ DTIME($n^{\log \log n}$).

Exact Exponential Algorithms

If P $\neq$ NP, then no polynomial time algorithm can solve MDS. Even worse, it has been observed in [5] that unless SNP$\subseteq$ SUBEXP (which is considered to be highly unlikely), there is not even a subexponential time algorithm solving the dominating set problem.

The trivial $O(2^n (n + m))$ algorithm, which simply checks all the 2^n vertex subsets whether they are dominating, clearly solves MDS. Three

faster algorithms have been established in 2004. The algorithm of Fomin et al. [5] uses a deep graph-theoretic result due to B. Reed, stating that every graph on n vertices with minimum degree at least three has a dominating set of size at most $3n/8$, to establish an $O(2^{0.955n})$ time algorithm solving MDS. The $O(2^{0.919n})$ time algorithm of Randerath and Schiermeyer [9] uses very nice ideas including matching techniques to restrict the search space. Finally, Grandoni [6] established an $O(2^{0.850n})$ time algorithm to solve MDS.

Key Results

Branch and Reduce and Measure and Conquer

The work of Fomin, Grandoni, and Kratsch presents a simple and easy way to implement a recursive branch and reduce algorithm to solve MDS. It was first presented at ICALP 2005 [2] and later published in 2009 in [3]. The running time of the algorithm is significantly faster than the ones stated for the previous algorithms. This is heavily based on the analysis of the running time by measure and conquer, which is a method to analyze the worst case running time of (simple) branch and reduce algorithms based on a sophisticated choice of the measure of a problem instance.

Theorem 1 *There is a branch and reduce algorithm solving* MDS *in time* $O(2^{0.610n})$ *using polynomial space.*

Theorem 2 *There is an algorithm solving* MDS *in time* $O(2^{0.598n})$ *using exponential space.*

The algorithms of Theorems 1 and 2 are simple consequences of a transformation from MDS to MINIMUM SET COVER (abbr. MSC) combined with new exact exponential time algorithms for MSC.

Problem 2 (MSC)

INPUT: Finite set $\mathcal{U}$ and a collection $\mathcal{S}$ of subsets $S_1, S_2, \ldots S_t$ of $\mathcal{U}$.

OUTPUT: A minimum set cover $\mathcal{S}'$, where $\mathcal{S}' \subseteq \mathcal{S}$ is a set cover of $(\mathcal{U}, \mathcal{S})$ if $\bigcup_{S_i \in \mathcal{S}'} S_i = \mathcal{U}$.

Theorem 3 *There is a branch and reduce algorithm solving* MSC *in time* $O(2^{0.305(|\mathcal{U}|+|\mathcal{S}|)})$ *using polynomial space.*

Applying memorization to the polynomial space algorithm of Theorem 3, the running time can be improved as follows.

Theorem 4 *There is an algorithm solving* MSC *in time* $O(2^{0.299(|\mathcal{S}|+|\mathcal{U}|)})$ *needing exponential space.*

The analysis of the worst case running time of the simple branch and reduce algorithm solving MSC (of Theorem 3) is done by a careful choice of the measure of a problem instance which allows to obtain an upper bound that is significantly smaller than the one that could be obtained using the standard measure. The refined analysis leads to a collection of recurrences. Then, random local search was used to compute the weights, used in the definition of the measure, aiming at the best achievable upper bound of the worst case running time. By now various other methods to do these time-consuming computations are available; see, e.g., [1].

Getting Faster MDS Algorithms

There is a lot of interest in exact exponential algorithms for solving MDS and in improving their best known running times. Two important improvements on the running times of the original algorithm stated in Theorems 1 and 2 have been achieved. To simplify the comparison, let us mention that in [4] those running times are stated as $O(1.5259^n)$ using polynomial space and $O(1.5132^n)$ needing exponential space.

Van Rooij and Bodlaender presented faster exact exponential algorithms solving MDS that are strongly based on the algorithms of Fomin et al. and the methods of their analysis. By introducing new reduction rules in the algorithm and a refined analysis, they achieved running time $O(1.5134^n)$ using polynomial space and time $O(1.5063^n)$, presented at STACS 2008. This analysis has been further improved in [11] to achieve a running time of $O(1.4969^n)$ using polynomial space, which

was published in 2011. It should be emphasized that memorization cannot be applied to the latter algorithm.

The currently best known algorithms solving MDS have been obtained by Ywata [7] and presented at IPEC 2011.

Theorem 5 *There is a branch and reduce algorithm solving* MDS *in time* $O(1.4864^n)$ *using polynomial space.*

Theorem 6 *There is an algorithm solving* MDS *in time* $O(1.4689^n)$ *needing exponential space.*

Ywata's polynomial space branch and reduce algorithm is also strongly related to the algorithm of Fomin et al. and its analysis. The improvement in the running time is achieved by some crucial change in the order of branchings in the algorithm solving MSC, i.e., the algorithm branches on the same element consecutively. These consecutive branchings can then be exploited by a refined analysis using global weights called potentials. Thus, such an analysis is dubbed "potential method." By a variant of memorization where dynamic programming memorizes only solutions of subproblems with small number of elements, an algorithm of running time $O(1.4689^n)$ needing exponential space has been obtained.

Counting Dominating Sets

A strongly related problem is #DS that asks to determine for a given graph G the number of dominating sets of size k, for any k. In [8], Nederlof, van Rooij, and van Dijk show how to combine inclusion/exclusion and a branch and reduce algorithm while using measure and conquer, as to obtain an algorithm (needing exponential space) of running time $O(1.5002^n)$. Clearly, this also solves MDS.

Applications

There are various other NP-hard domination-type problems that can be solved by exact exponential algorithms based on an algorithm solving

MINIMUM SET COVER: any instance of the initial problem is transformed to an instance of MSC, and then an algorithm solving MSC is applied and thus the initial problem is solved. Examples of such problems are TOTAL DOMINATING SET, k-DOMINATING SET, k-CENTER, and MDS on split graphs. Measure and conquer and the strongly related quasiconvex analysis of Eppstein [1] have been used to design and analyze a variety of exact exponential branch and reduce algorithms for NP-hard problems, optimization, counting, and enumeration problems; see [4].

Open Problems

While for many algorithms it is easy to show that the worst case analysis is tight, this is not the case for the nowadays time analysis of branch and reduce algorithms. For example, the worst case running times of the branch and reduce algorithms of Fomin et al. [3] solving MDS and MSC remain unknown; a lower bound of $\Omega(3^{n/4})$ for the MDS algorithm is known. The situation is similar for many other branch and reduce algorithms. Consequently, there is a strong need for new and better tools to analyze the worst case running time of branch and reduce algorithms.

Cross-References

▶ Connected Dominating Set
▶ Exact Algorithms for Induced Subgraph Problems
▶ Exact Algorithms for Maximum Independent Set
▶ Minimal Dominating Set Enumeration

Recommended Reading

1. Eppstein D (2006) Quasiconvex analysis of backtracking algorithms. ACM Trans Algorithms 2(4):492–509
2. Fomin FV, Grandoni F, Kratsch D (2005) Measure and conquer: domination – a case study. In: Proceedings of ICALP 2005, Lisbon
3. Fomin FV, Grandoni F, Kratsch D (2009) A measure & conquer approach for the analysis of exact algorithms. J ACM 56(5)
4. Fomin FV, Kratsch D (2010) Exact exponential algorithms. Springer, Heidelberg
5. Fomin FV, Kratsch D, Woeginger GJ (2004) Exact (exponential) algorithms for the dominating set problem. In: Proceedings of WG 2004, Bonn. LNCS, vol 3353. Springer, pp 245–256
6. Grandoni F (2004) Exact algorithms for hard graph problems. PhD thesis, Università di Roma "Tor Vergata", Roma, Mar 2004
7. Iwata Y (2011) A faster algorithm for Dominating Set analyzed by the potential method. In: Proceedings of IPEC 2011, Saarbrücken. LNCS, vol 7112. Springer, pp 41–54
8. Nederlof J, van Rooij JMM, van Dijk TC (2014) Inclusion/exclusion meets measure and conquer. Algorithmica 69(3):685–740
9. Randerath B, Schiermeyer I (2004) Exact algorithms for MINIMUM DOMINATING SET. Technical Report, zaik-469, Zentrum für Angewandte Informatik Köln, Apr 2004
10. van Rooij JMM (2011) Exact exponential-time algorithms for domination problems in graphs. PhD thesis, University Utrecht
11. van Rooij JMM, Bodlaender HL (2011) Exact algorithms for dominating set. Discret Appl Math 159(17):2147–2164
12. Woeginger GJ (2003) Exact algorithms for NP-hard problems: a survey. Combinatorial optimization – Eureka, you shrink. LNCS, vol 2570. Springer, Berlin/Heidelberg, pp 185–207

Exact Algorithms for General CNF SAT

Edward A. Hirsch
Laboratory of Mathematical Logic, Steklov Institute of Mathematics, St. Petersburg, Russia

Keywords

Boolean satisfiability; Exponential-time algorithms; SAT

Years and Authors of Summarized Original Work

1998; Hirsch
2003; Schuler

Problem Definition

The satisfiability problem (*SAT*) for Boolean formulas in conjunctive normal form (*CNF*) is one of the first **NP**-complete problems [2, 13]. Since its **NP**-completeness currently leaves no hope for polynomial-time algorithms, the progress goes by decreasing the exponent. There are several versions of this parametrized problem that differ in the parameter used for the estimation of the running time.

Problem 1 (SAT)

INPUT: Formula F in CNF containing n variables, m clauses, and l literals in total.
OUTPUT: "Yes" if F has a *satisfying assignment*, i.e., a substitution of Boolean values for the variables that makes F true. "No" otherwise.

The bounds on the running time of SAT algorithms can be thus given in the form $|F|^{O(1)} \cdot \alpha^n$, $|F|^{O(1)} \cdot \beta^m$, or $|F|^{O(1)} \cdot \gamma^l$, where $|F|$ is the length of a reasonable bit representation of F (i.e., the formal input to the algorithm). In fact, for the present algorithms, the bases β and γ are constants, while α is a function $\alpha(n, m)$ of the formula parameters (because no better constant than $\alpha = 2$ is known).

Notation

A formula in conjunctive normal form is a set of clauses (understood as the conjunction of these clauses), a clause is a set of literals (understood as the disjunction of these literals), and a literal is either a Boolean variable or the negation of a Boolean variable. A truth assignment assigns Boolean values (*false* or *true*) to one or more variables. An assignment is abbreviated as the list of literals that are made true under this assignment (e.g., assigning *false* to x and *true* to y is denoted by $\neg x, y$). The result of the application of an assignment A to a formula F (denoted $F[A]$) is the formula obtained by removing the clauses containing the true literals from F and removing the falsified literals from the remaining clauses. For example, if $F = (x \vee \neg y \vee z) \wedge (y \vee \neg z)$, then $F[\neg x, y] = (z)$. A *satisfying* assignment for F is an assignment A such that $F[A] =$ true. If such an assignment exists, F is called *satisfiable*.

Key Results

Bounds for β and γ

General Approach and a Bound for β
The trivial brute-force algorithm enumerating all possible assignments to the n variables runs in 2^n polynomial-time steps. Thus $\alpha \leq 2$, and by trivial reasons also $\beta, \gamma \leq 2$. In the early 1980s, Monien and Speckenmeyer noticed that β could be made smaller. (They and other researchers also noticed that α could be made smaller for a special case of the problem where the length of each clause is bounded by a constant; the reader is referred to another entry (*Local search algorithms for k-SAT*) of the *Encyclopedia* for relevant references and algorithms.) Then Kullmann and Luckhardt [12] set up a framework for divide-and-conquer (Also called *DPLL* due to the papers of Davis and Putnam [6] and Davis, Logemann, and Loveland [7].) algorithms for SAT that split the original problem into several (yet usually a constant number of) subproblems by substituting the values of some variables and simplifying the obtained formulas. This line of research resulted in the following upper bounds for β and γ:

Theorem 1 (Hirsch [8]) *SAT can be solved in time*

1. $|F|^{O(1)} \cdot 2^{0.30897m}$;
2. $|F|^{O(1)} \cdot 2^{0.10299l}$.

A typical divide-and-conquer algorithm for SAT consists of two phases: splitting of the original problem into several subproblems (e.g., reducing SAT(F) to SAT(F[x]) and SAT(F[\neg x])) and simplification of the obtained subproblems using polynomial-time transformation rules that do not affect the satisfiability of the subproblems (i.e., they replace a formula by an equisatisfiable one). The subproblems $F_1, \ldots, F_k$ for splitting are chosen so that the corresponding recurrent inequality using the simplified problems

$F'_1, \ldots, F'_k,$

$$T(F) \leq \sum_{i=1}^{k} T\left(F'_i\right) + const,$$

gives a desired upper bound on the number of leaves in the recurrence tree and, hence, on the running time of the algorithm. In particular, in order to obtain the bound $|F|^{O(1)} \cdot 2^{0.30897m}$ one takes either two subproblems $F[x], F[\neg x]$ with recurrent inequality

$$t_m \leq t_{m-3} + t_{m-4}$$

or four subproblems $F[x, y], F[x, \neg y], F[\neg x, y], F[\neg x, \neg y]$ with recurrent inequality

$$t_m \leq 2t_{m-6} + 2t_{m-7}$$

where $t_i = max_{m(G) \leq i} T(G)$. The simplification rules used in the $|F|^{O(1)} \cdot 2^{0.30897m}$-time and the $|F|^{O(1)} \cdot 2^{0.10299l}$-time algorithms are as follows:

Simplification Rules

Elimination of 1-Clauses If F contains a 1-clause (a), replace F by $F[a]$.

Subsumption If F contains two clauses C and D such that $C \subseteq D$, replace F by $F \backslash \{D\}$.

Resolution with Subsumption Suppose a literal a and clauses C and D are such that a is the only literal satisfying both conditions $a \in C$ and $\neg a \in D$. In this case, the clause $(C \cup D) \backslash \{a, \neg a\}$ is called the *resolvent by the literal a* of the clauses C and D and denoted by $R(C, D)$.
    The rule is: if $R(C, D) \subseteq D$, replace F by $(F \backslash \{D\}) \cup \{R(C, D)\}$.

Elimination of a Variable by Resolution [6] Given a literal a, construct the formula $DP_a(F)$ by

1. Adding to F all resolvents by a
2. Removing from F all clauses containing a or $\neg a$

The rule is: if $DP_a(F)$ is not larger in m (resp., in l) than F, then replace F by $DP_a(F)$.

Elimination of Blocked Clauses A clause C is *blocked* for a literal a w.r.t. F if C contains the literal a, and the literal $\neg a$ occurs only in the clauses of F that contain the negation of at least one of the literals occurring in $C \backslash \{a\}$. For a CNF-formula F and a literal a occurring in it, the assignment $I(a, F)$ is defined as

$$\{a\} \cup \{\text{literals } x \notin \{a, \neg a\} \mid \text{the clause}$$
$$\{\neg a, x\} \text{ is blocked for } \neg a \text{ w.r.t. } F\}.$$

Lemma 2 (Kullmann [11])

(1) If a clause C is blocked for a literal a w.r.t. F, then F and $F \backslash \{C\}$ are equi-satisfiable.
(2) Given a literal a, the formula F is satisfiable iff at least one of the formulas $F[\neg a]$ and $F[I(a, F)]$ is satisfiable.

The first claim of the lemma is employed as a simplification rule.

Application of the Black and White Literals Principle Let P be a binary relation between literals and formulas in CNF such that for a variable v and a formula F, at most one of $P(v, F)$ and $P(\neg v, F)$ holds.

Lemma 3 *Suppose that each clause of F that contains a literal w satisfying $P(w, F)$ contains also at least one literal b satisfying $P(\neg b, F)$. Then F and $F[\{l \mid P(\neg l, F)\}]$ are equi-satisfiable.*

A Bound for γ
To obtain the bound $|F|^{O(1)} \cdot 2^{0.10299l}$, it is enough to use a pair $F[\neg a], F[I(a, F)]$ of subproblems (see Lemma 2(2)) achieving the desired recurrent inequality $t_l \leq t_{l-5} + t_{l-17}$ and to switch to the $|F|^{O(1)} \cdot 2^{0.30897m}$-time algorithm if there are none. A recent (much more technically involved) improvement to this algorithm [16] achieves the bound $|F|^{O(1)} \cdot 2^{0.0926l}$.

A Bound for α

Currently, no non-trivial constant upper bound for α is known. However, starting with [14] there was an interest to non-constant bounds. A series of randomized and deterministic algorithms showing successive improvements was developed, and at the moment the best possible bound is achieved by a deterministic divide-and-conquer algorithm employing the following recursive procedure. The idea behind it is a dichotomy: either each clause of the input formula can be shortened to its first k literals (then a k-CNF algorithm can be applied), or all these literals in one of the clauses can be assumed false. (This clause-shortening approach can be attributed to Schuler [15], who used it in a randomized fashion. The following version of the deterministic algorithm achieving the best known bound both for deterministic and randomized algorithms appears in [5].)

Procedure S

Input: a CNF formula F and a positive integer k.

1. Assume F consists of clauses $C_1, \ldots, C_m$. Change each clause C_i to a clause D_i as follows: If $|C_i| > k$ then choose any k literals in C_i and drop the other literals; otherwise leave C_i as is, i.e., $D_i = C_i$. Let F' denote the resulting formula.
2. Test satisfiability of F' using the $m \cdot \text{poly}\,(n) \cdot (2 - 2/(k + 1))^n$-time k-CNF algorithm defined in [4].
3. If F' is satisfiable, output "satisfiable" and halt. Otherwise, for each i, do the following:
   1. Convert F to F_i as follows:
      1. Replace C_j by D_j for all $j < i$.
      2. Assign *false* to all literals in D_i.
   2. Recursively invoke *Procedure S* on (F_i, k).
4. Return "unsatisfiable".

The algorithm just invokes *Procedure S* on the original formula and the integer parameter $k = k * (m, n)$. The most accurate analysis of this family of algorithms by Calabro, Impagli-

azzo, and Paturi [1] implies that, assuming that $m > n$, one can obtain the following bound by taking $k(m, n) = 2\log(m/n) + \text{const}$. (This explicit bound is not stated in [1] and is inferred in [3].)

Theorem 4 (Dantsin, Hirsch [3]) *Assuming $m > n$, SAT can be solved in time*

$$|F|^{O(1)} \cdot 2^n \left(1 - \frac{1}{O\left(\log\left(m/n\right)\right)}\right).$$

Applications

While SAT has numerous applications, the presented algorithms have no direct effect on them.

Open Problems

Proving a constant upper bound on $\alpha < 2$ remains a major open problem in the field, as well as the hypothetic existence of $(1 + \varepsilon)^l$-time algorithms for arbitrary small $\varepsilon > 0$.

It is possible to perform the analysis of a divide-and-conquer algorithm and even to generate simplification rules automatically [10]. However, this approach so far led to new bounds only for the (NP-complete) optimization version of 2-SAT [9].

Experimental Results

Jun Wang has implemented the algorithm yielding the bound on β and collected some statistics regarding the number of applications of the simplification rules [17].

Cross-References

▶ Exact Algorithms for k SAT Based on Local Search

Recommended Reading

1. Calabro C, Impagliazzo R, Paturi R (2006) A duality between clause width and clause density for SAT. In: Proceedings of the 21st annual IEEE conference on computational complexity (CCC 2006), Prague. IEEE Computer Society, pp 252–260
2. Cook SA (2006) The complexity of theorem proving procedures. In: Proceedings of the third annual ACM symposium on theory of computing, Shaker Heights, May 1971. ACM, pp 151–158
3. Dantsin E, Hirsch EA (2008) Worst-case upper bounds. In: Biere A, van Maaren H, Walsh T (eds) Handbook of satisfiability. IOS. In: Biere A, Heule MJH, van Maaren H, Walsh T (eds) Handbook of satisfiability. Frontiers in artificial intelligence and applications, vol 185. IOS Press, Amsterdam/Washington, DC, p 980. ISBN:978-1-58603-929-5, ISSN:0922-6389
4. Dantsin E, Goerdt A, Hirsch EA, Kannan R, Kleinberg J, Papadimitriou C, Raghavan P, Schöning U (2002) A deterministic $(2 - 2/(k + 1))^n$ algorithm for k-SAT based on local search. Theor Comput Sci 289(1):69–83
5. Dantsin E, Hirsch EA, Wolpert A (2006) Clause shortening combined with pruning yields a new upper bound for deterministic SAT algorithms. In: Proceedings of CIAC-2006, Rome. Lecture notes in computer science, vol 3998. Springer, Berlin, pp 60–68
6. Davis M, Putnam H (1960) A computing procedure for quantification theory. J ACM 7: 201–215
7. Davis M, Logemann G, Loveland D (1962) A machine program for theorem-proving. Commun ACM 5:394–397
8. Hirsch EA (2000) New worst-case upper bounds for SAT. J Autom Reason 24(4):397–420
9. Kojevnikov A, Kulikov A (2006) A new approach to proving upper bounds for MAX-2-SAT. In: Proceedings of the seventeenth annual ACM-SIAM symposium on discrete algorithms (SODA 2006), Miami. ACM/SIAM, pp 11–17
10. Kulikov A (2005) Automated generation of simplification rules for SAT and MAXSAT. In: Proceedings of the eighth international conference on theory and applications of satisfiability testing (SAT 2005), St. Andrews. Lecture notes in computer science, vol 3569. Springer, Berlin, pp 430–436
11. Kullmann O (1999) New methods for 3-{SAT} decision and worst-case analysis. Theor Comput Sci 223(1–2):1–72
12. Kullmann O, Luckhardt H (1998) Algorithms for SAT/TAUT decision based on various measures, 71pp. Preprint, http://cs-svr1.swan.ac.uk/csoliver/papers.html
13. Levin LA (1973) Universal search problems. Probl Inf Trans 9(3):265–266. In Russian. English translation Trakhtenbrot BA (1984) A survey of russian approaches to perebor (Brute-force Search) algorithms. Ann Hist Comput 6(4):384–400
14. Pudlák P (1998) Satisfiability – algorithms and logic. In: Proceedings of the 23rd international symposium on mathematical foundations of computer science, MFCS'98. Lecture notes in computer science, Brno, vol 1450. Springer, Berlin, pp 129–141
15. Schuler R (2005) An algorithm for the satisfiability problem of formulas in conjunctive normal form. J Algorithms **54**(1), 40–44
16. Wahlström M (2005) An algorithm for the SAT problem for formulae of linear length. In: Proceedings of the 13th annual European symposium on algorithms, ESA 2005. Lecture notes in computer science, Mallorca, vol 3669. Springer, Berlin, pp 107–118
17. Wang J (2002) Generating and solving 3-SAT. MSc thesis, Rochester Institute of Technology, Rochester

Exact Algorithms for Induced Subgraph Problems

Michał Pilipczuk
Institute of Informatics, University of Warsaw, Warsaw, Poland
Institute of Informatics, University of Bergen, Bergen, Norway

Keywords

Exact algorithms; Hereditary property; Induced subgraph; 2^n barrier

Years and Authors of Summarized Original Work

2010; Fomin, Villanger
2013; Bliznets, Fomin, Pilipczuk, Villanger

Problem Definition

A graph class Π is a set of simple graphs. One can also think of Π as a *property*: Π comprises all the graphs that satisfy a certain condition. We say that class (property) Π is *hereditary* if it is closed under taking induced subgraphs. More precisely, whenever $G \in \Pi$ and H is an induced subgraph of G, then also $H \in \Pi$.

We shall consider the MAXIMUM INDUCED Π-SUBGRAPH problem: given a graph G, find the largest (in terms of the number of vertices) induced subgraph of G that belongs to Π. Suppose now that class Π is polynomial-time recognizable: there exists an algorithm that decides whether a given graph H belongs to Π in polynomial time. Then MAXIMUM INDUCED Π-SUBGRAPH on an n-vertex graph G can be solved by brute force in time (The $\mathcal{O}^\star(\cdot)$ notation hides factors polynomial in the input size.) $\mathcal{O}^\star(2^n)$: we iterate through all the induced subgraphs of G, and on each of them, we run a polynomial-time test deciding whether it belongs to Π.

Can we do anything smarter? Of course, this very much depends on the class Π we are working with. MAXIMUM INDUCED Π-SUBGRAPH is a generic problem that encompasses many other problems as special cases; examples include CLIQUE (Π = complete graphs), INDEPENDENT SET (Π = edgeless graphs), or FEEDBACK VERTEX SET (Π = forests). It is convenient to assume that Π is also hereditary; this assumption is satisfied in many important examples, including the aforementioned special cases.

So far, the MAXIMUM INDUCED Π-SUBGRAPH problem has been studied for many graph classes Π, and basically in all the cases it turned out that it is possible to find an algorithm with running time $\mathcal{O}(c^n)$ for some $c < 2$. Obtaining a result of this type is often informally called *breaking the 2^n barrier*. While the algorithms share a common general methodology, vital details differ depending on the structural properties of the class Π. This makes each and every algorithm of this type contrived to a particular scenario. However, it is tempting to formulate the following general conjecture.

Conjecture 1 ([1]) *For every hereditary, polynomial-time recognizable class of graphs* Π, *there exists a constant* $c_\Pi < 2$ *for which there is an algorithm solving* MAXIMUM INDUCED Π-SUBGRAPH *in time* $\mathcal{O}(c_\Pi^n)$.

On one hand, current partial progress on this conjecture consists of scattered results exploiting different properties of particular classes Π,

without much hope for proving more general statements. On the other hand, finding a counterexample refuting Conjecture 1 based, e.g., on the Strong Exponential Time Hypothesis seems problematic: the input to MAXIMUM INDUCED Π-SUBGRAPH consists only of $\binom{n}{2}$ bits of information about adjacencies between the vertices, and it seems difficult to model the search space of a general k-SAT using such input under the constraint that Π has to be hereditary and polynomial-time recognizable.

It can be that Conjecture 1 is either false or very difficult to prove, and therefore, one can postulate investigating its certain subcases connected to well-studied classes of graphs. For instance, one could assume that graphs from Π have constant treewidth or that Π is a subclass of chordal or interval graphs. Another direction is to strengthen the assumption about the description of the class Π by requiring that belonging to Π can be expressed in some formalism (e.g., some variant of logic). Finally, one can investigate the algorithms for MAXIMUM INDUCED Π-SUBGRAPH where Π is not required to be hereditary; here, natural nonhereditary properties are connectivity and regularity.

Key Results

Table 1 presents a selection of results on the MAXIMUM INDUCED Π-SUBGRAPH problem. Since the algorithms are usually quite technical when it comes to details, we now present an overview of the general methodology and most important techniques. In the following, we assume that Π is hereditary and polynomial-time recognizable.

Most often, the general approach is to examine the structure of the input instance and of a fixed, unknown optimum solution. The goal is to identify as broad spectrum of situations as possible where the solution can be found by examining $\mathcal{O}((2 - \varepsilon)^n)$ candidates, for some $\varepsilon > 0$. By checking the occurrence of each of these situations, we eventually narrow down our investigations to the case where we have a well-defined structure of the input instance and

Exact Algorithms for Induced Subgraph Problems, Table 1 Known results for MAXIMUM INDUCED Π-SUBGRAPH. The first part of the table presents results for problems for which breaking the 2^n barrier follows directly from branching on forbidden subgraphs. The second part contains results for which breaking the barrier requires a nontrivial insight into the structure of Π. Finally, the last part contains results for nonhereditary classes Π. Here, ε denotes a small, positive constant, and its index specifies a parameter on which the value of this constant depends

Property	Time complexity	Reference
Edgeless	$\mathcal{O}(1.2109^n)$	Robson [10]
Biclique	$\mathcal{O}(1.3642^n)$	Gaspers et al. [6]
Cluster graph	$\mathcal{O}(1.6181^n)$	Fomin et al. [3]
Bipartite	$\mathcal{O}(1.62^n)$	Raman et al. [9]
Acyclic	$\mathcal{O}(1.7347^n)$	Fomin et al. [2]
Constant treewidth	$\mathcal{O}(1.7347^n)$	Fomin et al. [2]
Planar	$\mathcal{O}(1.7347^n)$	Fomin et al. [4]
d-degenerate	$\mathcal{O}((2 - \varepsilon_d)^n)$	Pilipczuk×2 [8]
Chordal	$\mathcal{O}((2 - \varepsilon)^n)$	Bliznets et al. [1]
Interval	$\mathcal{O}((2 - \varepsilon)^n)$	Bliznets et al. [1]
r-regular	$\mathcal{O}((2 - \varepsilon_r)^n)$	Gupta et al. [7]
Matching	$\mathcal{O}(1.6957^n)$	Gupta et al. [7]

a number of assumptions about how the solution looks like. Then, hopefully, a direct algorithm can be devised.

Let us consider a very simple example of this principle, which is also a technique used in many algorithms for breaking the 2^n barrier. Suppose the input graph has n vertices and assume the optimum solution is of size larger that $(1/2 + \delta)n$, for some $\delta > 0$. Then, as candidates for the optimum solution, we can consider all the vertex subsets of at least this size: there is only $(2 - \varepsilon)^n$ of them, where $\varepsilon > 0$ depends on δ. Similarly, if the optimum solution has size smaller than $(1/2 - \delta)n$, then we can identify this situation by iterating through all the vertex subsets of size $(1/2 - \delta)n$ (whose number is again $(2 - \varepsilon)^n$ for some $\varepsilon > 0$) and verifying that none of them induces a graph belonging to Π; note that here we use the assumption that Π is hereditary. In this case we can solve the problem by looking at all vertex subsets of size

at most $(1/2 - \delta)n$. All in all, we can solve the problem faster than $\mathcal{O}^\star(2^n)$ provided that the number of vertices in the optimum solution differs by at least δn from $n/2$, for some $\varepsilon > 0$. More precisely, for every $\delta > 0$ we will obtain a running time of the form $\mathcal{O}((2 - \varepsilon)^n)$, where ε tends to 0 when δ tends to 0. Hence, we can focus only on the situation when the number of vertices in the optimum solution is very close to $n/2$.

We now give an overview of some other important techniques.

Branching on Forbidden Induced Subgraphs

Every hereditary graph class Π can be characterized by giving a minimal set of forbidden induced subgraphs $\mathcal{F}$: a graph belongs to Π if and only if it does not contain any graph from $\mathcal{F}$ as an induced subgraph, and $\mathcal{F}$ is inclusion-wise minimal with this property. For instance, the class of forests is characterized by $\mathcal{F}$ being the family of all the cycles, whereas taking $\mathcal{F}$ to be the family of all the cycles of length at least 4 gives the class of chordal graphs. For many important classes the family $\mathcal{F}$ is infinite, but there are notable examples where it is finite, like cluster, trivially perfect, or split graphs.

If Π is characterized by a finite set of forbidden subgraphs $\mathcal{F}$, then already a simple branching strategy yields an algorithm working in time $\mathcal{O}((2 - \varepsilon)^n)$, for some $\varepsilon > 0$ depending on $\mathcal{F}$. Without going into details, we iteratively find a forbidden induced subgraph that is not yet removed by the previous choices and branch on the fate of all the undecided vertices in this subgraph, omitting the branch where all of them are included in the solution. Since this forbidden induced subgraph is of constant size, a standard analysis shows that the running time of this algorithm is $\mathcal{O}((2 - \varepsilon)^n)$ for some $\varepsilon > 0$ depending on $\max_{H \in \mathcal{F}} |V(H)|$. This simple observation can be combined with more sophisticated techniques in case when $\mathcal{F}$ is infinite. We can namely start the algorithm by branching on forbidden induced subgraphs that are of constant size and, when their supply is exhausted, turn to some other algorithms. The following lemma provides a formalization of this concept; a graph is called $\mathcal{F}$-*free* if it does not contain any graph from $\mathcal{F}$ as an induced subgraph.

Lemma 1 ([1]) *Let $\mathcal{F}$ be a finite set of graphs and let ℓ be the maximum number of vertices in a graph from $\mathcal{F}$. Let Π be a hereditary graph class that is polynomial-time recognizable. Assume that there exists an algorithm $\mathcal{A}$ that for a given $\mathcal{F}$-free graph G on n vertices, in time $\mathcal{O}((2-\varepsilon)^n)$ finds a maximum induced subgraph of G that belongs to Π, for some $\varepsilon > 0$. Then there exists an algorithm $\mathcal{A}'$ that for a given graph G on n vertices, in time $\mathcal{O}((2-\varepsilon')^n)$ finds a maximum induced subgraph of G that is $\mathcal{F}$-free and belongs to Π, where $\varepsilon' > 0$ is a constant depending on ε and ℓ.*

Thus, for the purpose of breaking the 2^n barrier, it is sufficient to focus on the case when no constant-size forbidden induced subgraph is present in the input graph.

Exploiting a Large Substructure

Here, the general idea is to look for a large substructure in the graph that can be leveraged to design an algorithm breaking the barrier. Let us take as an example the MAXIMUM INDUCED CHORDAL SUBGRAPH problem, considered by Bliznets et al. [1]. Suppose that in the input graph G one can find a clique Q of size δn, for some $\delta > 0$; recall that the largest clique in a graph can be found as fast as in time $\mathcal{O}(1.2109^n)$ [10]. Then consider the following algorithm: guess, by considering $2^{n-|Q|}$ possibilities, the intersection of the optimum solution with $V(G) \setminus Q$. Then observe that, since Q is a clique, every induced cycle in G can have only at most two vertices in common with Q. Hence, the problem of optimally extending the choice on $V(G) \setminus Q$ to Q essentially boils down to solving a VERTEX COVER instance on $|Q|$ vertices, which can be

done in time $\mathcal{O}(1.2109^{|Q|})$. As Q constitutes a linear fraction of all the vertices, the overall running time is $\mathcal{O}(1.2109^{|Q|} \cdot 2^{n-|Q|})$, which is $\mathcal{O}((2-\varepsilon)^n)$ for some $\varepsilon > 0$ depending on δ. Thus, one can focus on the case where the largest clique in the input graph, and hence also in any maximum-sized induced chordal subgraph, has less than δn vertices.

Potential Maximal Cliques

A *potential maximal clique (PMC)* in a graph G is a subset of vertices that becomes a clique in some inclusion-wise minimal triangulation (By a triangulation of a graph we mean any its chordal supergraph.) of G. Fomin and Villanger in [2] observed two facts. Firstly, whenever H is an induced subgraph of G of treewidth t, then there exists a minimal triangulation TG of G that captures H in the following sense: every clique of TG intersects $V(H)$ only at a subset of some bag of a fixed width-t tree decomposition of H. Secondly, a graph G on n vertices can have only $\mathcal{O}(1.734601^n)$ PMCs, which can be enumerated in time $\mathcal{O}(1.734601^n)$. Intuitively, this means that we can effectively search the space of treewidth-t induced subgraphs of G in time $\mathcal{O}(1.734601^n \cdot n^{\mathcal{O}(t)})$ using dynamic programming. Slightly more precisely, treewidth-t induced subgraphs of G can be assembled in a dynamic programming manner using states of the form (Ω, X), where Ω is a PMC in G and X is a subset of Ω of size at most $t + 1$, corresponding to $\Omega \cap V(H)$. In this manner one can obtain an algorithm with running time $\mathcal{O}(1.734601^n \cdot n^{\mathcal{O}(t)})$ for finding the maximum induced treewidth-t subgraph, which in particular implies a $\mathcal{O}(1.734601^n)$-time algorithm for MAXIMUM INDUCED FOREST, equivalent to FEEDBACK VERTEX SET. Recently, Fomin et al. [5] extended this framework to encapsulate also problems where the induced subgraph H is in addition required to satisfy a property expressible in *Monadic Second-Order Logic*.

Recommended Reading

1. Bliznets I, Fomin FV, Pilipczuk M, Villanger Y (2013) Largest chordal and interval subgraphs faster than 2^n. In: Bodlaender HL, Italiano GF (eds) ESA, Sophia Antipolis. Lecture Notes in Computer Science, vol 8125. Springer, pp 193–204
2. Fomin FV, Villanger Y (2010) Finding induced subgraphs via minimal triangulations. In: Marion JY, Schwentick T (eds) STACS, Nancy. LIPIcs, vol 5. Schloss Dagstuhl – Leibniz-Zentrum fuer Informatik, pp 383–394
3. Fomin FV, Gaspers S, Kratsch D, Liedloff M, Saurabh S (2010) Iterative compression and exact algorithms. Theor Comput Sci 411(7–9):1045–1053
4. Fomin FV, Todinca I, Villanger Y (2011) Exact algorithm for the maximum induced planar subgraph problem. In: Demetrescu C, Halldórsson MM (eds) ESA, Saarbrücken. Lecture notes in computer science, vol 6942. Springer, pp 287–298
5. Fomin FV, Todinca I, Villanger Y (2014) Large induced subgraphs via triangulations and CMSO. In: Chekuri C (ed) SODA, Portland. SIAM, pp 582–583
6. Gaspers S, Kratsch D, Liedloff M (2012) On independent sets and bicliques in graphs. Algorithmica 62(3–4):637–658
7. Gupta S, Raman V, Saurabh S (2012) Maximum *r*-regular induced subgraph problem: fast exponential algorithms and combinatorial bounds. SIAM J Discr Math 26(4):1758–1780
8. Pilipczuk M, Pilipczuk M (2012) Finding a maximum induced degenerate subgraph faster than 2^n. In: Thilikos DM, Woeginger GJ (eds) IPEC, Ljubljana. Lecture notes in computer science, vol 7535. Springer, pp 3–12
9. Raman V, Saurabh S, Sikdar S (2007) Efficient exact algorithms through enumerating maximal independent sets and other techniques. Theory Comput Syst 41(3):563–587
10. Robson JM (1986) Algorithms for maximum independent sets. J Algorithms 7(3):425–440

Exact Algorithms for *k* SAT Based on Local Search

Kazuo Iwama
Computer Engineering, Kyoto University, Sakyo, Kyoto, Japan
School of Informatics, Kyoto University, Sakyo, Kyoto, Japan

Keywords

CNF satisfiability; Exponential-time algorithm; Local search

Years and Authors of Summarized Original Work

1999; Schöning

Problem Definition

The CNF satisfiability problem is to determine, given a CNF formula F with n variables, whether or not there exists a satisfying assignment for F. If each clause of F contains at most k literals, then F is called a k-CNF formula and the problem is called k-SAT, which is one of the most fundamental NP-complete problems. The trivial algorithm is to search 2^n 0/1-assignments for the n variables. But since [6], several algorithms which run significantly faster than this $O(2^n)$ bound have been developed. As a simple exercise, consider the following straightforward algorithm for 3-SAT, which gives us an upper bound of 1.913^n: choose an arbitrary clause in F, say, $(x_1 \vee \overline{x_2} \vee x_3)$. Then generate seven new formulas by substituting to these x_1, x_2, and x_3 all the possible values except $(x_1, x_2, x_3) = (0, 1, 0)$ which obviously unsatisfies F. Now one can check the satisfiability of these seven formulas and conclude that F is satisfiable iff at least one of them is satisfiable. (Let $T(n)$ denote the time complexity of this algorithm. Then one can get the recurrence $T(n) \leq 7 \times T(n-3)$ and the above bound follows.)

Key Results

In the long history of k-SAT algorithms, the one by Schöning [11] is an important breakthrough. It is a standard local search and the algorithm itself is not new (see, e.g., [7]). Suppose that y is the current assignment (its initial value is selected uniformly at random). If y is a satisfying assignment, then the algorithm answers yes and terminates. Otherwise, there is at least one clause whose three literals are all false under y. Pick an arbitrary such clause and select one of the three literals in it at random. Then flip (true to false and vice versa) the value of that variable, replace y with that new assignment, and then repeat the same procedure. More formally:

SCH(CNF formula F, integer I)

repeat I times

$y = $ uniformly random vector $\in \{0, 1\}^n$

$z = \textbf{RandomWalk}(F, y)$;

if z satisfies F

then output(z); exit;

end

output('Unsatisfiable');

RandomWalk(CNF formula $G(x_1, x_2, \ldots, x_n)$, assignment y);

$y' = y$;

for $3n$ times

if y' satisfies G

then return y'; exit;

$C \leftarrow$ an arbitrary clause of G that is not satisfied by y';

Modify y' as follows:

select one literal of C uniformly at random and flip the assignment to this literal;

end

return y'

Schöning's analysis of this algorithm is very elegant. Let $d(a, b)$ denote the Hamming distance between two binary vectors (assignments) a and b. For simplicity, suppose that the formula F has only one satisfying assignment y^* and the current assignment y is far from y^* by Hamming distance d. Suppose also that the currently false clause C includes three variables, x_i, x_j, and x_k. Then y and y^* must differ in at least one of these three variables. This means that if the value of x_i, x_j, or x_k is flipped, then the new assignment gets closer to y^* by Hamming distance one with probability at least 1/3. Also, the new assignment gets farther by Hamming distance one with probability at most 2/3. The argument can be generalized to the case that F has multiple satisfying assignments. Now here comes the key lemma:

Lemma 1 *Let F be a satisfiable formula and y^* be a satisfying assignment for F. For each assignment y, the probability that a satisfying assignment (that may be different from y^*) is found by **RandomWalk** (F, y) is at least $(1/(k - 1))^{d(y, y^*)}/p(n)$, where $p(n)$ is a polynomial in n.*

By taking the average over random initial assignments, the following theorem follows:

Theorem 1 *For any satisfiable formula F on n variables, the success probability of **RandomWalk** (F, y) is at least $(k/2(k - 1))^n/p(n)$ for some polynomial p. Thus, by setting $I = (2(k - 1)/k)^n \cdot p(n)$, **SCH** finds a satisfying assignment with high probability. When $k = 3$, this value of I is $O(1.334^n)$.*

Applications

The Schöning's result has been improved by a series of papers [1, 3, 9] based on the idea of [3]. Namely, RandomWalk is combined with the (polynomial time) 2SAT algorithm, which makes it possible to choose better initial assignments. For derandomization of **SCH**, see [2]. Iwama and Tamaki [4] developed a nontrivial combination of **SCH** with another famous, backtrack-type algorithm by [8], resulting in the then fastest algorithm with $O(1.324^n)$ running time. The current fastest algorithm is due to [10], which is based on the same approach as [4] and runs in time $O(1.32216^n)$.

Open Problems

k-SAT is probably the most popular NP-complete problem for which numerous researchers are competing for its fastest algorithm. Thus, improving its time bound is always a good research target.

Experimental Results

AI researchers have also been very active in SAT algorithms including local search; see, e.g., [5].

Cross-References

▶ Exact Algorithms for General CNF SAT
▶ Random Planted 3-SAT

Recommended Reading

1. Baumer S, Schuler R (2003) Improving a probabilistic 3-SAT algorithm by dynamic search and independent clause pairs. ECCC TR03-010. Also presented at SAT
2. Dantsin E, Goerdt A, Hirsch EA, Kannan R, Kleinberg J, Papadimitriou C, Raghavan P, Schöning U (2002) A deterministic $(2 - 2/(k + 1))^n$ algorithm for k-SAT based on local search. Theor Comput Sci 289(1):69–83
3. Hofmeister T, Schöning U, Schuler R, Watanabe O (2002) Probabilistic 3-SAT algorithm further improved. In: Proceedings 19th symposium on theoretical aspects of computer science, Juan-les-Pins. LNCS, vol 2285, pp 193–202
4. Iwama K, Tamaki S (2004) Improved upper bounds for 3-SA T. In: Proceedings 15th annual ACM-SIAM symposium on discrete algorithms, New Orleans, pp 321–322
5. Kautz H, Selman B (2003) Ten challenges redux: recent progress in propositional reasoning and search. In: Proceedings 9th international conference on principles and practice of constraint programming, Kinsale, pp 1–18
6. Monien B, Speckenmeyer E (1985) Solving satisfiability in less than 2^n steps. Discret Appl Math 10:287–295
7. Papadimitriou CH (1991) On selecting a satisfying truth assignment. In: Proceedings 32nd annual symposium on foundations of computer science, San Juan, pp 163–169
8. Paturi R, Pudlák P, Saks ME, Zane F (1998) An improved exponential-time algorithm for k-SAT. In: Proceedings 39th annual symposium on foundations of computer science, Palo Alto, pp 628–637; J ACM 52(3):337–364 (2006)
9. Rolf D (2003) 3-SAT $\in$ $RTIME(O(1.32793^n))$. ECCC TR03-054
10. Rolf D (2006) Improved bound for the PPSZ/Schöning-algorithm for 3-SAT. J Satisf Boolean Model Comput 1:111–122
11. Schöning U (1999) A probabilistic algorithm for k-SAT and constraint satisfaction problems. In: Proceedings 40th annual symposium on foundations of computer science, New York, pp 410–414

Exact Algorithms for Maximum Independent Set

Fabrizio Grandoni
IDSIA, USI-SUPSI, University of Lugano, Lugano, Switzerland

Keywords

Exact exponential algorithms; Maximum independent set

Years and Authors of Summarized Original Work

1977; Tarjan, Trojanowski
1985; Robson
1999; Beigel
2009; Fomin, Grandoni, Kratsch

Problem Definition

Let $G = (V, E)$ be an n-node undirected, simple graph without loops. A set $I \subseteq V$ is called an *independent set* of G if the nodes of I are pairwise not adjacent. The *maximum independent set* (MIS) problem asks to determine the maximum cardinality $\alpha(G)$ of an independent set of G. MIS is one of the best studied NP-hard problems.

We will need the following notation. The (open) *neighborhood* of a vertex v is $N(v) = \{u \in V : uv \in E\}$, and its *closed neighborhood* is $N[v] = N(v) \cup \{v\}$. The degree $\deg(v)$ of v is $|N(v)|$. For $W \subseteq V$, $G[W] = (W, E \cap \binom{W}{2})$ is the *graph induced* by W. We let $G - W = G[V - W]$.

Key Results

A very simple algorithm solves MIS (exactly) in $O^*(2^n)$ time: it is sufficient to enumerate all the subsets of nodes, check in polynomial time whether each subset is an independent set or not, and return the maximum cardinality independent set. We recall that the O^* notation suppresses polynomial factors in the input size. However, much faster (though still exponential-time) algorithms are known. In more detail, there exist algorithms that solve MIS in worst-case time $O^*(c^n)$ for some constant $c \in (1, 2)$. In this section, we will illustrate some of the most relevant techniques that have been used in the design and analysis of exact MIS algorithms. Due to space constraints, our description will be slightly informal (please see the references for formal details).

Bounding the Size of the Search Tree

All the nontrivial exact MIS algorithms, starting with [7], are recursive branching algorithms. As an illustration, consider the following simple MIS algorithm Alg1. If the graph is empty, output $\alpha(G) = 0$ (base instance). Otherwise, choose any node v of maximum degree, and output

$$\alpha(G) = \max\{\alpha(G - \{v\}), 1 + \alpha(G - N[v])\}.$$

Intuitively, the subgraph $G - \{v\}$ corresponds to the choice of not including v in the independent set (v is *discarded*), while the subgraph $G - N[v]$ to the choice of including v in the independent set (v is *selected*). Observe that, when v is selected, the neighbors of v have to be discarded. We will later refer to this branching as a *standard branching*.

The running time of the above algorithm, and of branching algorithms more in general, can be bounded as follows. The recursive calls induce a *search tree*, where the root is the input instance and the leaves are base instances (that can be solved in polynomial time). Observe that each branching step can be performed in polynomial time (excluding the time needed to solve subproblems). Furthermore, the height of the search tree is bounded by a polynomial. Therefore, the running time of the algorithm is bounded by $O^*(L(n))$, where $L(n)$ is the maximum number of leaves of any search tree that can be generated by the considered algorithm on an input instance with n nodes. Let us assume that $L(n) \leq c^n$ for some constant $c \geq 1$. When we branch at node v,

we generate two subproblems containing $n - 1$ and $n - |N[v]|$ nodes, respectively. Therefore, c has to satisfy $c^n \geq c^{n-1} + c^{n-|N[v]|}$. Assuming pessimistically $|N[v]| = 1$, one obtains $c^n \geq 2c^{n-1}$ and therefore $c \geq 2$. We can conclude that the running time of the algorithm is $O^*(2^n)$. Though the running time of Alg1 does not improve on exhaustive search, much faster algorithms can be obtained by branching in a more careful way and using a similar type of analysis. This will be discussed in the next subsections.

Refined Branching Rules

Several refined branching rules have been developed for MIS. Let us start with some *reduction rules*, which reduce the problem without branching (alternatively, by branching on a single subproblem). An isolated node v can be selected w.l.o.g.:

$$\alpha(G) = 1 + \alpha(G - N[v]).$$

Observe that if $N[u] \subseteq N[v]$, then node v can be discarded w.l.o.g. (*dominance*):

$$\alpha(G) = \alpha(G - \{v\}).$$

This rule implies that nodes of degree 1 can always be selected.

Suppose that we branch at a node v, and in the branch where we discard v we select exactly one of its neighbors, say w. Then by replacing w with v, we obtain a solution of the same cardinality including v: this means that the branch where we select v has to provide the optimal solution. Therefore, we can assume w.l.o.g. that the optimal solution either contains v or at least 2 of its neighbors. This idea is exploited in the *folding* operation [1], which we next illustrate only in the case of degree-2 nodes. Let $N[v] = \{w_1, w_2\}$. Remove $N[v]$. If $w_1 w_2 \notin E$, create a node v' and add edges between v' and nodes in $N(w_1) \cup N(w_2) - \{v\}$. Let $G_{\text{fold}}(v)$ be the resulting graph. Then, one has

$$\alpha(G) = 1 + \alpha(G_{\text{fold}}(v)).$$

Intuitively, including node v' in the optimal solution to $G_{\text{fold}}(v)$ corresponds to selecting both w_1 and w_2, while discarding v' corresponds to selecting v.

Let Alg2 be the algorithm that exhaustively applies the mentioned reduction rules and then performs a standard branching on a node of maximum degree. Reduction rules reduce the number of nodes at least by 1; hence, we have the constraint $c^n \geq c^{n-1}$. If we branch at node v, $\deg(v) \geq 3$. This gives $c^n \geq c^{n-1} + c^{n-4}$, which is satisfied by $c \geq 1.380\ldots$. Hence, the running time is in $O^*(1.381^n)$.

Let us next briefly sketch some other useful ideas that lead to refined branchings. A *mirror* [3] of a node v is a node u at distance 2 from v such that $N(v) - N(u)$ induces a clique. By the above discussion, if we branch by discarding v, we can assume that we select at least two neighbors of v and therefore we have also to discard the mirrors $M(v)$ of v. In other terms, we can use the refined branching

$$\alpha(G) = \max\{\alpha(G-\{v\}-M(v)),\, 1+\alpha(G-N[v])\}.$$

A *satellite* [5] of a node v is a node u at distance 2 from v such that there exists a node $u' \in N(v) \cap N(u)$ that satisfies $N[u'] - N[v] = \{u\}$. Observe that if an optimal solution discards u, then we can discard v as well by dominance since $N[u'] \subseteq N[v]$ in $G - \{u\}$. Therefore, we can assume that in the branch where we select v, we also select its satellites $S(v)$. In other terms,

$$\alpha(G) = \max\{\alpha(G-\{v\}),\, 1 + |S(v)|$$
$$+ \alpha(G - N[v] - \cup_{u \in S(v)} N[u])\}.$$

Another useful trick [4] is to branch on nodes that form a *small separator* (of size 1 or 2 in the graph), hence isolating two or more connected components that can be solved independently (see also [2,5]).

Measure and Conquer

Above we always used the number n of nodes as a *measure* of the size of subproblems. As observed in [3], much tighter running time bounds can be achieved by using smarted measures. As an illustration, we will present a refined bound on the running time of Alg2.

Let us measure the size of subproblems with the number n_3 of nodes of degree at least 3 (*large nodes*). Observe that, when $n_3 = 0$, G is a collection of isolated nodes, paths, and cycles. Therefore, in that case, Alg2 only applies reduction rules, hence solving the problem in polynomial time. In other terms, $L(n_3) = L(0) = 1$ in this case. If the algorithm applies any reduction rule, the number of large nodes cannot increase and we obtain the trivial inequality $c^{n_3} \geq c^{n_3}$. Suppose next that Alg2 performs a standard branching at a node v. Note that at this point all nodes in the graph are large. If $\deg(v) \geq 4$, then we obtain the inequality $c^{n_3} \geq c^{n_3-1} + c^{n_3-5}$ which is satisfied by $c \geq 1.324\ldots$. Otherwise ($\deg(v) = 3$), observe that the neighbors of v have degree 3 in G and at most 2 in $G - \{v\}$. Therefore, the number of large nodes is at most $n_3 - 4$ in both subproblems $G - \{v\}$ and $G - N[v]$. This gives the inequality $c^{n_3} \geq 2c^{n_3-4}$ which is satisfied by $c \geq 2^{1/4} < 1.1893$. We can conclude that the running time of the algorithm is in $O^*(1.325^n)$. In [3], each node is assigned a weight which is a growing function of its degree, and the measure is the sum of node weights (a similar measure is used also in [2,5]).

In [2], it is shown how to use a fast MIS algorithm for graphs of maximum degree Δ to derive faster MIS algorithms for graphs of maximum degree $\Delta + 1$. Here the measure used in the analysis is a combination of the number of nodes and edges.

Memorization

So far we described algorithms with polynomial space complexity. *Memorization* [6] is a technique to speed up exponential-time branching algorithms at the cost of an exponential space complexity. The basic idea is to store the optimal solution to subproblems in a proper (exponential size) data structure. Each time a new subproblem is generated, one first checks (in polynomial time) whether that subproblem was already solved before. This way one avoids to solve the same subproblem several times.

In order to illustrate this technique, it is convenient to consider the variant `Alg3` of `Alg2` where we do not apply folding. This way, each subproblem corresponds to some induced subgraph $G[W]$ of the input graph. We will also use the standard measure though memorization is compatible with measure and conquer. By adapting the analysis of `Alg2`, one obtains the constraint $c^n \geq c^{n-1} + c^{n-3}$ and hence a running time of $O^*(1.466^n)$. Next, consider the variant `Alg3mem` of `Alg3` where we apply memorization. Let $L_k(n)$ be the maximum number of subproblems on k nodes generated by `Alg3mem` starting from an instance with n nodes. A slight adaptation of the standard analysis shows that $L_k(n) \leq 1.466^{n-k}$. However, since there are at most $\binom{n}{k}$ induced subgraphs on k nodes and we never solve the same subproblem twice, one also has $L_k(n) \leq \binom{n}{k}$. Using Stirling's formula, one obtains that the two upper bounds are roughly equal for $k = \alpha n$ and $\alpha = 0.107\ldots$. We can conclude that the running time of `Alg3mem` is in $O^*(\sum_{k=0}^n L_k(n)) = O^*(\sum_{k=0}^n \min\{1.466^{n-k}, \binom{n}{k}\}) = O^*(\max_{k=0}^n \min\{1.466^{n-k}, \binom{n}{k}\}) = O^*(1.466^{(1-0.107)n}) = O^*(1.408^n)$. The analysis can be refined [6] by bounding the number of *connected* induced subgraphs with k nodes in graphs of small maximum degree.

Cross-References

▶ Exact Algorithms for Dominating Set

Recommended Reading

1. Beigel R (1999) Finding maximum independent sets in sparse and general graphs. In: ACM-SIAM symposium on discrete algorithms (SODA), Baltimore, pp 856–857
2. Bourgeois N, Escoffier B, Paschos VT, van Rooij JMM (2012) Fast algorithms for max independent set. Algorithmica 62(1–2):382–415
3. Fomin FV, Grandoni F, Kratsch D (2009) A measure & conquer approach for the analysis of exact algorithms. J ACM 56(5):Article no. 25
4. Fürer M (2006) A faster algorithm for finding maximum independent sets in sparse graphs. In:
5. Kneis J, Langer A, Rossmanith P (2009) A fine-grained analysis of a simple independent set algorithm. In: Foundations of software technology and theoretical computer science (FSTTCS), Kanpur, pp 287–298
6. Robson JM (1986) Algorithms for maximum independent sets. J Algorithms 7(3):425–440
7. Tarjan R, Trojanowski A (1977) Finding a maximum independent set. SIAM J Comput 6(3):537–546

Latin American theoretical informatics symposium (LATIN), Valdivia, pp 491–501

E

Exact Algorithms for Maximum Two-Satisfiability

Ryan Williams
Department of Computer Science, Stanford University, Stanford, CA, USA

Keywords

Max 2-SAT

Years and Authors of Summarized Original Work

2004; Williams

Problem Definition

In the maximum 2-satisfiability problem (abbreviated as MAX 2-SAT), one is given a Boolean formula in conjunctive normal form, such that each clause contains at most two literals. The task is to find an assignment to the variables of the formula such that a maximum number of clauses are satisfied.

MAX 2-SAT is a classic optimization problem. Its decision version was proved *NP*-complete by Garey, Johnson, and Stockmeyer [7], in stark contrast with 2-SAT which is solvable in linear time [2]. To get a feeling for the difficulty of the problem, the *NP*-completeness reduction is sketched here. One can transform any 3-SAT instance F into a MAX 2-SAT instance F', by replacing each clause of F such as

$$c_i = (\ell_1 \vee \ell_2 \vee \ell_3),$$

where ℓ_1, ℓ_2, and ℓ_3 are arbitrary literals, with the collection of 2-CNF clauses

$$(\ell_1), (\ell_2), (\ell_3), (c_i), (\neg\ell_1 \vee \neg\ell_2), (\neg\ell_2 \vee \neg\ell_3),$$
$$(\neg\ell_1 \vee \neg\ell_3), (\ell_1 \vee c_i), (\ell_2 \vee c_i), (\ell_3 \vee c_i),$$

where c_i is a new variable. The following are true:

- If an assignment satisfies c_i, then exactly seven of the ten clauses in the 2-CNF collection can be satisfied.
- If an assignment does not satisfy c_i, then exactly six of the ten clauses can be satisfied.

If F is satisfiable then there is an assignment satisfying 7/10 of the clauses in F', and if F is not satisfiable, then no assignment satisfies more than 7/10 of the clauses in F'. Since 3-SAT reduces to MAX 2-SAT, it follows that MAX 2-SAT (as a decision problem) is NP-complete.

Notation

A CNF formula is represented as a set of clauses.

The letter ω denotes the smallest real number such that for all $\epsilon > 0$, n by n matrix multiplication over a field can be performed in $O(n^{\omega+\epsilon})$ field operations. Currently, it is known that $\omega < 2.373$ [4, 16]. The field matrix product of two matrices A and B is denoted by $A \times B$.

Let A and B be matrices with entries from $\mathbb{R} \cup \{\infty\}$. The *distance product* of A and B (written in shorthand as $A \circledast B$) is the matrix C defined by the formula

$$C[i, j] = \min_{k=1,\dots,n} \{A[i, k] + B[k, j]\}.$$

A word on m's and n's: in reference to graphs, m and n denote the number of edges and the number of nodes in the graph, respectively. In reference to CNF formulas, m and n denote the number of clauses and the number of variables, respectively.

Key Result

The primary result of this entry is a procedure solving Max 2-Sat in $O(m \cdot 2^{\omega n/3})$ time. The method can be generalized to *count* the number of solutions to *any* constraint optimization problem with at most two variables per constraint. Indeed, in the same running time, one can find a Boolean assignment that maximizes any given degree-two polynomial in n variables [18, 19]. In this entry, we shall restrict attention to be Max 2-Sat, for simplicity. There are several other known exact algorithms for Max 2-Sat that are more effective in special cases, such as sparse instances [3, 8, 9, 11–13, 15, 17]. The procedure described below is the only one known (to date) that runs in c^n steps for a constant $c < 2$.

Key Idea

The algorithm gives a reduction from MAX 2-SAT to the problem MAX TRIANGLE, in which one is given a graph with integer weights on its nodes and edges, and the goal is to output a 3-cycle of maximum weight. At first, the existence of such a reduction sounds strange, as MAX TRIANGLE can be trivially solved in $O(n^3)$ time by trying all possible 3-cycles. The key is that the reduction exponentially increases the problem size, from a MAX 2-SAT instance with m clauses and n variables to a MAX TRIANGLE instance having $O(2^{2n/3})$ edges, $O(2^{n/3})$ nodes, and weights in the range $\{-m, \dots, m\}$.

Note that if MAX TRIANGLE required $\Theta(n^3)$ time to solve, then the resulting MAX 2-SAT algorithm would take $\Theta(2^n)$ time, rendering the above reduction pointless. However, it turns out that the brute-force search of $O(n^3)$ for MAX TRIANGLE is not the best one can do: using fast matrix multiplication, there is an algorithm for MAX TRIANGLE that runs in $O(Wn^\omega)$ time on graphs with weights in the range $\{-W, \dots, W\}$.

Main Algorithm

First, a reduction from MAX 2-SAT to MAX TRIANGLE is described, arguing that each triangle of weight K in the resulting graph is in one-to-one correspondence with an assignment that satisfies K clauses of the MAX 2-SAT instance. Let a, b be reals, and let $\mathbb{Z}[a, b] := [a, b] \cap \mathbb{Z}$.

Lemma 1 *If* MAX TRIANGLE *on graphs with n nodes and weights in $\mathbb{Z}[-W, W]$ is solvable in $O(f(W) \cdot g(n))$ time, for polynomials f and g, then* MAX 2-SAT *is solvable in $O(f(m) \cdot g(2^{n/3}))$ time, where m is the number of clauses and n is the number of variables.*

Proof Let C be a given 2-CNF formula. Assume without loss of generality that n is divisible by 3. Let F be an instance of MAX 2-SAT. Arbitrarily partition the n variables of F into three sets P_1, P_2, P_3, each having $n/3$ variables. For each P_i, make a list L_i of all $2^{n/3}$ assignments to the variables of P_i.

Define a graph $G = (V, E)$ with $V = L_1 \cup L_2 \cup L_3$ and $E = \{(u, v) | u \in P_i, v \in P_j, i \neq j\}$. That is, G is a complete tripartite graph with $2^{n/3}$ nodes in each part, and each node in G corresponds to an assignment to $n/3$ variables in C. Weights are placed on the nodes and edges of G as follows. For a node v, define $w(v)$ to be the number of clauses that are satisfied by the partial assignment denoted by v. For each edge $\{u, v\}$, define $w(\{u, v\}) = -W_{uv}$, where W_{uv} is the number of clauses that are satisfied by *both* u and v.

Define the weight of a triangle in G to be the total sum of all weights and nodes in the triangle.

Claim 1 There is a one-to-one correspondence between the triangles of weight K in G and the variable assignments satisfying exactly K clauses in F.

Proof Let a be a variable assignment. Then there exist unique nodes $v_1 \in L_1, v_2 \in L_2$, and $v_3 \in L_3$ such that a is precisely the concatenation of v_1, v_2, v_3 as assignments. Moreover, any triple of nodes $v_1 \in L_1, v_2 \in L_2$, and $v_3 \in L_3$ corresponds to an assignment. Thus, there is a one-to-one correspondence between triangles in G and assignments to F.

The number of clauses satisfied by an assignment is exactly the weight of its corresponding

triangle. To see this, let $T_a = \{v_1, v_2, v_3\}$ be the triangle in G corresponding to assignment a. Then

$$
\begin{aligned}
w(T_a) &= w(v_1) + w(v_2) + w(v_3) + w(\{v_1, v_2\}) \\
&\quad + w(\{v_2, v_3\}) + w(\{v_1, v_3\}) \\
&= \sum_{i=1}^{3} |\{c \in F | v_i \text{ satisfies } F\}| \\
&\quad - \sum_{i, j : i \neq j} |\{c \in F | v_i \text{ and } v_j \text{ satisfy } F\}| \\
&= |\{c \in F | a \text{ satisfies } F\}|,
\end{aligned}
$$

where the last equality follows from the inclusion-exclusion principle.

Notice that the number of nodes in G is $3 \cdot 2^{n/3}$, and the absolute value of any node and edge weight is m. Therefore, running a MAX TRIANGLE algorithm on G, a solution to MAX 2-SAT, is obtained in $O(f(m) \cdot g(3 \cdot 2^{n/3}))$, which is $O(f(m) \cdot g(2^{n/3}))$ since g is a polynomial. This completes the proof of Lemma 1.

Next, a procedure is described for finding a maximum triangle faster than brute-force search, using fast matrix multiplication. Alon, Galil, and Margalit [1] (following Yuval [22]) showed that the distance product for matrices with entries drawn from $\mathbb{Z}[-W, W]$ can be computed using fast matrix multiplication as a subroutine.

Theorem 1 (Alon, Galil, Margalit [1]) *Let A and B be $n \times n$ matrices with entries from $\mathbb{Z}[-W, W] \cup \{\infty\}$. Then $A \circledast B$ can be computed in $O(W n^\omega \log n)$ time.*

Proof (Sketch) One can replace ∞ entries in A and B with $2W + 1$ in the following. Define matrices A' and B', where

$$
A'[i, j] = x^{3W - A[i,j]}, \quad B'[i, j] = x^{3W - B[i,j]},
$$

and x is a variable. Let $C = A' \times B'$. Then

$$
C[i, j] = \sum_{k=1}^{n} x^{6W - A[i,k] - B[k,j]}.
$$

The next step is to pick a number x that makes it easy to determine, from the sum of arbitrary powers of x, the largest power of x appearing in the sum; this largest power immediately gives the minimum $A[i, k] + B[k, j]$. Each $C[i, j]$ is a polynomial in x with coefficients from $\mathbb{Z}[0, n]$. Suppose each $C[i, j]$ is evaluated at $x = (n + 1)$. Then each entry of $C[i, j]$ can be seen as an $(n + 1)$-ary number, and the position of this number's most significant digit gives the minimum $A[i, k] + B[k, j]$.

In summary, $A \otimes_d B$ can be computed by constructing

$$A'[i, j] = (n + 1)^{3W - A[i,j]},$$

$$B'[i, j] = (n + 1)^{3W - B[i,j]}$$

in $O(W \log n)$ time per entry, computing $C = A' \times B'$ in $O(n^\omega \cdot (W \log n))$ time (as the sizes of the entries are $O(W \log n)$), then extracting the minimum from each entry of C, in $O(n^2 \cdot W \log n)$ time. Note if the minimum for an entry $C[i, j]$ is at least $2W + 1$, then $C[i, j] = \infty$.

Using the fast distance product algorithm, one can solve MAX TRIANGLE faster than brute force. The following is based on an algorithm by Itai and Rodeh [10] for detecting if an unweighted graph has a triangle in less than n^3 steps. The result can be generalized to *counting* the number of k-*cliques*, for arbitrary $k \geq 3$. (To keep the presentation simple, the counting result is omitted. Concerning the k-clique result, there is unfortunately no asymptotic runtime benefit from using a k-clique algorithm instead of a triangle algorithm, given the current best algorithms for these problems.)

Theorem 2 MAX TRIANGLE *can be solved in* $O(W n^\omega \log n)$, *for graphs with weights drawn from* $\mathbb{Z}[-W, W]$.

Proof First, it is shown that a weight function on nodes and edges can be converted into an equivalent weight function with weights on only edges. Let w be the weight function of G, and redefine the weights to be:

$$w'(\{u, v\}) = \frac{w(u) + w(v)}{2} + w(\{u, v\}),$$

$$w'(u) = 0.$$

Note the weight of a triangle is unchanged by this reduction.

The next step is to use a fast distance product to find a maximum weight triangle in an edge-weighted graph of n nodes. Construe the vertex set of G as the set $\{1, \ldots, n\}$. Define A to be the $n \times n$ matrix such that $A[i, j] = -w(\{i, j\})$ if there is an edge $\{i, j\}$, and $A[i, j] = \infty$ otherwise. The claim is that there is a triangle through node i of weight at least K if and only if $(A \circledast A \circledast A)[i, i] \leq -K$. This is because $(A \circledast A \circledast A)[i, i] \leq -K$ if and only if there are distinct j and k such that $\{i, j\}, \{j, k\}, \{k, i\}$ are edges and $A[i, j] + A[j, k] + A[k, i] \leq -K$, i.e., $w(\{i, j\}) + w(\{j, k\}) + w(\{k, i\}) \geq K$.

Therefore, by finding an i such that $(A \circledast A \circledast A)[i, i]$ is minimized, one obtains a node i contained in a maximum triangle. To obtain the actual triangle, check all m edges $\{j, k\}$ to see if $\{i, j, k\}$ is a triangle.

Theorem 3 MAX 2-SAT *can be solved in* $O(m \cdot 1.732^n)$ *time.*

Proof Given a set of clauses C, apply the reduction from Lemma 1 to get a graph G with $O(2^{n/3})$ nodes and weights from $\mathbb{Z}[-m, m]$. Apply the algorithm of Theorem 2 to output a max triangle in G in $O(m \cdot 2^{\omega n/3} \log(2^{n/3})) = O(m \cdot 1.732^n)$ time, using the $O(n^{2.376})$ matrix multiplication of Coppersmith and Winograd [4].

Applications

By modifying the graph construction, one can solve other problems in $O(1.732^n)$ time, such as Max Cut, Minimum Bisection, and Sparsest Cut. In general, any constraint optimization problem for which each constraint has at most two variables can be solved faster using the above approach. For more details, see [18] and the survey by Woeginger [21]. Techniques similar to the above algorithm have also been used by Dorn

[6] to speed up dynamic programming for some problems on planar graphs (and in general, graphs of bounded branchwidth).

Open Problems

- Improve the space usage of the above algorithm. Currently, $\Theta(2^{2n/3})$ space is needed. A very interesting open question is if there is a $O(1.99^n)$ time algorithm for MAX 2-SAT that uses only *polynomial* space. This question would have a positive answer if one could find an algorithm for solving the k-CLIQUE problem that uses polylogarithmic space and $n^{k-\delta}$ time for some $\delta > 0$ and $k \geq 3$.
- Find a faster-than-2^n algorithm for MAX 2-SAT that does not require fast matrix multiplication. The fast matrix multiplication algorithms have the unfortunate reputation of being impractical.
- Generalize the above algorithm to work for MAX k-SAT, where k is any positive integer. The current formulation would require one to give an efficient algorithm for finding a small hyperclique in a hypergraph. However, no general results are known for this problem. It is conjectured that for all $k \geq 2$, MAX k-SAT is in $\bar{O}(2^{n(1-\frac{1}{k+1})})$ time, based on the conjecture that matrix multiplication is in $n^{2+o(1)}$ time [17].

Cross-References

▶ All Pairs Shortest Paths via Matrix Multiplication

Recommended Reading

1. Alon N, Galil Z, Margalit O (1997) On the exponent of the all-pairs shortest path problem. J Comput Syst Sci 54:255–262
2. Aspvall B, Plass MF, Tarjan RE (1979) A linear-time algorithm for testing the truth of certain quantified boolean formulas. Inf Proc Lett 8(3):121–123
3. Bansal N, Raman V (1999) Upper bounds for Max Sat: further improved. In: Proceedings of ISAAC, Chennai. LNCS, vol 1741. Springer, Berlin, pp 247–258
4. Coppersmith D, Winograd S (1990) Matrix multiplication via arithmetic progressions. JSC 9(3):251–280
5. Dantsin E, Wolpert A (2006) Max SAT for formulas with constant clause density can be solved faster than in $O(2^n)$ time. In: Proceedings of the 9th international conference on theory and applications of satisfiability testing, Seattle. LNCS, vol 4121. Springer, Berlin, pp 266–276
6. Dorn F (2006) Dynamic programming and fast matrix multiplication. In: Proceedings of 14th annual European symposium on algorithms, Zurich. LNCS, vol 4168. Springer, Berlin, pp 280–291
7. Garey M, Johnson D, Stockmeyer L (1976) Some simplified NP-complete graph problems. Theor Comput Sci 1:237–267
8. Gramm J, Niedermeier R (2000) Faster exact solutions for Max2Sat. In: Proceedings of CIAC. LNCS, vol 1767, Rome. Springer, Berlin, pp 174–186
9. Hirsch EA (2000) A $2^{m/4}$-time algorithm for Max 2-SAT: corrected version. Electronic colloquium on computational complexity report TR99-036
10. Itai A, Rodeh M (1978) Finding a minimum circuit in a graph. SIAM J Comput 7(4):413–423
11. Kneis J, Mölle D, Richter S, Rossmanith P (2005) Algorithms based on the treewidth of sparse graphs. In: Proceedings of workshop on graph theoretic concepts in computer science, Metz. LNCS, vol 3787. Springer, Berlin, pp 385–396
12. Kojevnikov A, Kulikov AS (2006) A new approach to proving upper bounds for Max 2-SAT. In: Proceedings of the seventeenth annual ACM-SIAM symposium on discrete algorithms, Miami, pp 11–17
13. Mahajan M, Raman V (1999) Parameterizing above guaranteed values: MAXSAT and MAXCUT. J Algorithms 31(2):335–354
14. Niedermeier R, Rossmanith P (2000) New upper bounds for maximum satisfiability. J Algorithms 26:63–88
15. Scott A, Sorkin G (2003) Faster algorithms for MAX CUT and MAX CSP, with polynomial expected time for sparse instances. In: Proceedings of RANDOM-APPROX 2003, Princeton. LNCS, vol 2764. Springer, Berlin, pp 382–395
16. Vassilevska Williams V (2012) Multiplying matrices faster than Coppersmith-Winograd. In: Proceedings of the 44th annual ACM symposium on theory of computing, New York, pp 887–898
17. Williams R (2004) On computing k-CNF formula properties. In: Theory and applications of satisfiability testing. LNCS, vol 2919. Springer, Berlin, pp 330–340
18. Williams R (2005) A new algorithm for optimal 2-constraint satisfaction and its implications. Theor Comput Sci 348(2–3):357–365
19. Williams R (2007) Algorithms and resource requirements for fundamental problems. PhD thesis, Carnegie Mellon University

E

20. Woeginger GJ (2003) Exact algorithms for NP-hard problems: a survey. In: Combinatorial optimization – Eureka! You shrink! LNCS, vol 2570. Springer, Berlin, pp 185–207
21. Woeginger GJ (2004) Space and time complexity of exact algorithms: some open problems. In: Proceedings of 1st international workshop on parameterized and exact computation (IWPEC 2004), Bergen. LNCS, vol 3162. Springer, Berlin, pp 281–290
22. Yuval G (1976) An algorithm for finding all shortest paths using $N^{2.81}$ infinite-precision multiplications. Inf Process Lett 4(6):155–156

Exact Algorithms for Treewidth

Ioan Todinca
INSA Centre Val de Loire, Universite d'Orleans, Orléans, France

Keywords

Extremal combinatorics; Potential maximal cliques; Treewidth

Years and Authors of Summarized Original Work

2008; Fomin, Kratsch, Todinca, Villanger
2012; Bodlaender, Fomin, Koster, Kratsch, Thilikos
2012; Fomin, Villanger

Problem Definition

The *treewidth* parameter intuitively measures whether the graph has a "treelike" structure. Given an undirected graph $G = (V, E)$, a *tree decomposition* of G is a pair $(\mathcal{X}, T)$, where $T = (I, F)$ is a tree and $\mathcal{X} = \{X_i \mid i \in I\}$ is a collection of subsets of V called *bags* satisfying:

1. $\bigcup_{i \in I} X_i = V$,
2. For each edge uv of G, there is a bag X_i containing both endpoints,

3. For all $v \in V$, the set $\{i \in I \mid v \in X_i\}$ induces a connected subtree of T.

The *width* of a tree decomposition $(\mathcal{X}, T)$ is the size of its largest bag, minus one. The *treewidth* of G, denoted by tw(G), is the minimum width over all possible tree decompositions. One can easily observe that n-vertex graphs have treewidth at most $n - 1$ and that the graphs of treewidth at most one are exactly the forests.

Given a graph G and a number k, the TREEWIDTH problem consists in deciding if tw(G) $\leq k$. Arnborg, Corneil, and Proskurowski show that the problem is NP-hard [1]. On the positive side, Bodlaender [2] gives an algorithm solving the problem in time $2^{\mathcal{O}(k^3)} n$. Bouchitté and Todinca [4,5] prove that the problem is polynomial on classes of graphs with polynomially many minimal separators, with an algorithm based on the notion of *potential maximal clique*. This latter technique is also employed by several exact, moderately exponential algorithms for TREEWIDTH.

Key Results

TREEWIDTH can be solved in $\mathcal{O}^*(2^n)$ time by adapting the $\mathcal{O}(n^k)$ algorithm of Arnborg et al. [1] or the Held-Karp technique initially designed for the TRAVELING SALESMAN problem [12]. (We use here the $\mathcal{O}^*$ notation that suppresses polynomial factors.) Fomin et al. [9] break this "natural" 2^n barrier with an algorithm running in time $\mathcal{O}^*(1.8135^n)$, using the same space complexity. Bodlaender et al. [3] present a polynomial-space algorithm running in $\mathcal{O}^*(2.9512^n)$ time. A major improvement for both results is due to Fomin and Villanger [8].

Theorem 1 ([8]) *The* TREEWIDTH *problem can be solved in* $\mathcal{O}^*(1.7549^n)$ *time using exponential space and in* $\mathcal{O}^*(2.6151^n)$ *time using polynomial space.*

These algorithms use an alternative definition for treewidth. A graph $H = (V, E)$ is *chordal* or *triangulated* if it has no induced cycle with four or more vertices. It is well-known that a chordal graph has tree decompositions whose bags are exactly its maximal cliques. Given an arbitrary graph $G = (V, E)$, a chordal graph $H = (V, F)$ on the same vertex set is called a *minimal triangulation* of G if H contains G as a subgraph and no chordal subgraph of H contains G. The treewidth of G can be defined as the minimum clique size of H minus one, over all minimal triangulations H of G.

A vertex subset S of graph G is a *minimal separator* if there are two distinct components $G[C]$ and $G[D]$ of the graph $G[V \setminus S]$ such that $N_G(C) = N_G(D) = S$ ($N_G(C)$ denotes the neighborhood of C in graph G).

A vertex subset Ω of G is a *potential maximal clique* if there exists some minimal triangulation H of G such that Ω induces a maximal clique in H. Potential maximal cliques are characterized as follows [4]: Ω is a potential maximal clique of G if and only if (i) for each pair of vertices $u, v \in \Omega$, u and v are adjacent or see a same component of $G[V \setminus \Omega]$, and (ii) no component of $G[V \setminus \Omega]$ sees the whole set Ω. As an example, when G is a cycle, its minimal separators are exactly the pairs of nonadjacent vertices, and the potential maximal cliques are exactly the triples of vertices.

A *block* is pair (S, C) such that S is a minimal separator of G and $G[C]$ is a component of $G[V \setminus S]$. Denote by $R_G(S, C)$ the graph obtained from $G[S \cup C]$ by turning S into a clique, i.e., by adding all missing edges with both endpoints in S. The treewidth of G can be obtained as follows:

$$\mathrm{tw}(G) = \min_S \left(\max_C \mathrm{tw}(R(S, C)) \right) \quad (1)$$

where the minimum is taken over all minimal separators S and the maximum is taken over all connected components $G[C]$ of $G[V \setminus S]$.

All quantities $\mathrm{tw}(R_G(S, C))$ can be computed by dynamic programming over blocks (S, C), by increasing the size of $S \cup C$. We only consider here blocks (S, C) such that $S = N_G(C)$ (see [4] for more details).

$$\mathrm{tw}(R_G(S, C))$$

$$= \min_{S \subset \Omega \subseteq S \cup C} \left(\max_{1 \le i \le p} (|\Omega| - 1, \mathrm{tw}(R_G(S_i, C_i))) \right)$$
$$(2)$$

where the minimum is taken over all potential maximal cliques Ω with $S \subset \Omega \subseteq S \cup C$ and the maximum is taken over all pairs (S_i, C_i), where $G[C_i]$ is a component of $G[C \setminus \Omega]$ and $S_i = N_G(C_i)$. Let Π_G denote the set of all potential maximal cliques of graph G. It was pointed in [9] that the number of triples (S, Ω, C) like in Eq. 2 is at most $n|\Pi_G|$, which proves that TREEWIDTH can be computed in $\mathcal{O}^*(|\Pi_G|)$ time and space, if Π_G is given in the input.

Therefore, it remains to give a good upper bound for the number $|\Pi_G|$ of potential maximal cliques of G, together with efficient algorithms for listing these objects. Based on the previously mentioned characterization of potential maximal cliques, Kratsch et al. provide an algorithm listing them in time $\mathcal{O}^*(1.8135^n)$. Fomin and Villanger [8] improve this result, thanks to the following combinatorial theorem:

Theorem 2 ([8]) *Let* $G = (V, E)$ *be an n-vertex graph, let* v *be a vertex of* G, *and* b, f *be two integers. The number of vertex subsets* B *containing* v *such that* $G[B]$ *is connected,* $|B| = b + 1$, *and* $|N_G(B)| = f$ *is at most* $\binom{b+f}{f}$.

The elegant inductive proof also leads to an $\mathcal{O}^*(\binom{b+f}{f})$ time algorithm listing all such sets B. Eventually, the potential maximal cliques of an input graph G can be listed in $\mathcal{O}^*(1.7549^n)$ time [8]. This bound was further improved to $\mathcal{O}^*(1.7347^n)$ in [7].

In order to obtain polynomial-space algorithms for TREEWIDTH, Bodlaender et al. [3] provide a relatively simple divide-and-conquer algorithm, based on the Held-Karp approach, running in $\mathcal{O}^*(4^n)$ time. They also observe that Eq. 1 can be used for recursive, polynomial-space algorithms, by replacing the minimal separators S by *balanced* separators, in the sense that each component of $G[V \setminus S]$ contains at most $n/2$ vertices. This leads to polynomial-space algorithm with $\mathcal{O}^*(2.9512^n)$ running time.

Fomin and Villanger [8] restrict the balanced separators to a subset of the potential maximal cliques, and based on Theorem 2 they obtain, still using polynomial space, a running time of $\mathcal{O}^*(2.6151^n)$.

We refer to the book of Fomin and Kratsch [6] for more details on the TREEWIDTH problem and more generally on exact algorithms.

Applications

Exact algorithms based on potential maximal cliques have been extended to many other problems like FEEDBACK VERTEX SET, LONGEST INDUCED PATH, or MAXIMUM INDUCED SUBGRAPH WITH A FORBIDDEN PLANAR MINOR. More generally, for any constant t and any property $\mathcal{P}$ definable in counting monadic second-order logic, consider the problem of finding, in an arbitrary graph G, a maximum-size induced subgraph $G[F]$ of treewidth at most t and with property $\mathcal{P}$. This generic problem can be solved in $\mathcal{O}^*(|\Pi_G|)$ time, if Π_G is part of the input [7, 10]. Therefore, there is an algorithm in $\mathcal{O}^*(1.7347^n)$ time for the problem, significantly improving the $\mathcal{O}^*(2^n)$ time for exhaustive search.

Open Problems

Currently, the best known upper bound on the number of potential maximal cliques in n-vertex graphs is of $\mathcal{O}^*(1.7347^n)$ and does not seem to be tight [7]. Simple examples show that this bound is of at least $3^{n/3} \sim 1.4425^n$. A challenging question is to find a tight upper bound and efficient algorithms enumerating all potential maximal cliques of arbitrary graphs.

Experimental Results

Several experimental results are reported in [3], especially on an "engineered" version of the $\mathcal{O}^*(2^n)$ time and space algorithm based on the Held-Karp approach. This dynamic programming algorithm is compared with the branch and bound approach of Gogate and Dechter [11] on instances of up to 50 vertices. The results are relatively similar. Bodlaender et al. [3] also observe that the polynomial-space algorithms become too slow even for small instances.

Cross-References

▶ Kernelization, Preprocessing for Treewidth

Recommended Reading

1. Arnborg S, Corneil DG, Proskurowski A (1987) Complexity of finding embeddings in a k-tree. SIAM J Algebr Discret Methods 8(2):277–284
2. Bodlaender HL (1996) A linear-time algorithm for finding tree-decompositions of small treewidth. SIAM J Comput 25(6):1305–1317
3. Bodlaender HL, Fomin FV, Koster AMCA, Kratsch D, Thilikos DM (2012) On exact algorithms for treewidth. ACM Trans Algorithms 9(1):12
4. Bouchitté V, Todinca I (2001) Treewidth and minimum fill-in: grouping the minimal separators. SIAM J Comput 31(1):212–232
5. Bouchitté V, Todinca I (2002) Listing all potential maximal cliques of a graph. Theor Comput Sci 276(1–2):17–32
6. Fomin FV, Kratsch D (2010) Exact exponential algorithms, 1st edn. Springer, New York
7. Fomin FV, Villanger Y (2010) Finding induced subgraphs via minimal triangulations. In: Marion JY, Schwentick T (eds) STACS, Nancy. LIPIcs, vol 5. Schloss Dagstuhl – Leibniz-Zentrum fuer Informatik, pp 383–394
8. Fomin FV, Villanger Y (2012) Treewidth computation and extremal combinatorics. Combinatorica 32(3):289–308
9. Fomin FV, Kratsch D, Todinca I, Villanger Y (2008) Exact algorithms for treewidth and minimum fill-in. SIAM J Comput 38(3):1058–1079
10. Fomin FV, Todinca I, Villanger Y (2014) Large induced subgraphs via triangulations and cmso. In: Chekuri C (ed) SODA, Portland. SIAM, pp 582–583
11. Gogate V, Dechter R (2004) A complete anytime algorithm for treewidth. In: Chickering DM, Halpern JY (eds) UAI, Banff. AUAI Press, pp 201–208
12. Held M, Karp RM (1962) A dynamic programming approach to the sequencing problem. J Soc Ind Appl Math 10(1):196–210

Exact Algorithms on Graphs of Bounded Average Degree

Marcin Pilipczuk
Institute of Informatics, University of Bergen,
Bergen, Norway
Institute of Informatics, University of Warsaw,
Warsaw, Poland

Keywords

Bounded average degree graphs; Bounded degree graphs; Chromatic number; Counting perfect matchings; Traveling salesman problem; TSP

Years and Authors of Summarized Original Work

2010; Björklund, Husfeldt, Kaski, Koivisto
2012; Björklund, Husfeldt, Kaski, Koivisto
2013; Cygan, Pilipczuk
2014; Golovnev, Kulikov, Mihajlin

Problem Definition

We focus on the following question: how an assumption on the sparsity of an input graph, such as bounded (average) degree, can help in designing exact (exponential-time) algorithms for NP-hard problems. The following classic problems are studied:

Traveling Salesman Problem Find a minimum-length Hamiltonian cycle in an input graph with edge weights.

Chromatic Number Find a minimum number k for which the vertices of an input graph can be colored with k colors such that no two adjacent vertices receive the same color.

Counting Perfect Matchings Find the number of perfect matchings in an input graph.

Key Results

The classic algorithms of Bellman [1] and Held and Karp [10] for traveling salesman problem run in $2^n n^{\mathcal{O}(1)}$ time for n-vertex graphs. Using the inclusion-exclusion principle, the chromatic number of an input graph can be determined within the same running time bound [4]. Finally, as long as counting perfect matchings is concerned, a half-century-old $2^{n/2} n^{\mathcal{O}(1)}$-time algorithm of Ryser for bipartite graphs [12] has only recently been transferred to arbitrary graphs by Björklund [2].

In all three aforementioned cases, it is widely open whether the 2^n or $2^{n/2}$ factor in the running time bound can be improved. In 2008, Björklund, Husfeldt, Kaski, and Koivisto [5,6] observed that such an improvement can be made if we restrict ourselves to bounded degree graphs. Further work of Cygan and Pilipczuk [8] and Golovnev, Kulikov, and Mihajlin [9] extended these results to graphs of bounded average degree.

Bounded Degree Graphs

Traveling Salesman Problem

Let us present the approach of Björklund, Husfeldt, Kaski, and Koivisto on the example of traveling salesman problem. Assume we are given an n-vertex edge-weighted graph G. The classic dynamic programming algorithm picks a root vertex r and then, for every vertex $v \in V(G)$ and every set $X \subseteq V(G)$ containing v and r, computes $T[X, v]$: the minimum possible length of a path in G with vertex set X that starts in r and ends in v. The running time bound $2^n n^{\mathcal{O}(1)}$ is dominated by the number of choices of the set X.

The simple, but crucial, observation is as follows: if a set X satisfies $X \cap N_G[u] = \{u\}$ for some $u \in V(G) \setminus \{r\}$, then the values $T[X, v]$ are essentially useless, as no path starting in r can visit the vertex u without visiting any neighbor of u (here $N_G[u] = N_G(u) \cup \{u\}$ stands for the closed neighborhood of u). Let us call a set $X \subseteq V(G)$ *useful* if $X \cap N_G[u] \neq \{u\}$ for every $u \in V(G) \setminus \{r\}$. The argumentation so far proved that we may skip the computation of $T[X, v]$ for all sets X that are not useful. The natural question is how many different useful sets may exist in an n-vertex graph?

Consider the following greedy procedure: initiate $A = \emptyset$ and, as long as there exists a vertex $u \in V(G)$ such that $N_G[u] \cap N_G[A] = \emptyset$, add an arbitrarily chosen vertex u to the set A. By construction, the set A satisfies the following property: for every $u_1, u_2 \in A$, we have $N_G[u_1] \cap N_G[u_2] = \emptyset$. An interesting fact is that $|A| = \Omega(n)$ for graphs of bounded degree: whenever we insert a vertex u into the set A, we cannot later insert into A any neighbor of u nor any neighbor of a neighbor of u. However, if the maximum degree of G is bounded by d, then there are at most d neighbors of u, and every such neighbor has at most $d-1$ further neighbors. Consequently, when we insert a vertex u into A, we prohibit at most $d + d(d-1) = d^2$ other

vertices from being inserted into A, and $|A| \geq n/(1 + d^2)$.

It is easy to adjust the above procedure such that the root vertex r does not belong to A. Observe that for every useful set X and every $u \in A$, we have $X \cap N_G[u] \neq \{u\}$ and, furthermore, the sets $N_G[u]$ for $u \in A$ are pairwise disjoint. We can think of choosing a useful set X as follows: first, for every $u \in A$, we choose the intersection $X \cap N_G[u]$ (there are $2^{|N_G[u]|} - 1$ choices, as the choice $\{u\}$ is forbidden), and, second, we choose the set $X \setminus N_G[A]$. Hence, the number of useful sets is bounded by

$$\left(\prod_{u \in A} 2^{|N_G[u]|} - 1 \right) \cdot 2^{n-|N_G[A]|} = 2^n \cdot \prod_{u \in A} \left(1 - 2^{-|N_G[u]|} \right) \leq 2^n \cdot \prod_{u \in A} (1 - 2^{-d-1})$$

$$= 2^n \cdot (1 - 2^{-d-1})^{|A|} \leq 2^n \cdot (1 - 2^{-d-1})^{\frac{n}{1+d^2}}$$

$$= \left(2 \cdot \sqrt[1+d^2]{1 - 2^{-d-1}} \right)^n .$$

Thus, for every degree bound d, there exists a constant $\varepsilon_d > 0$ such that the number of useful sets in an n-vertex graph of maximum degree d is bounded by $(2 - \varepsilon_d)^n$, yielding a $(2 - \varepsilon_d)^n n^{\mathcal{O}(1)}$-time algorithm for traveling salesman problem. A better dependency on d in the formula for ε_d can be obtained using a projection theorem of Chung, Frankl, Graham, and Shearer [7] (see [5]).

Chromatic Number

A similar reasoning can be performed for the problem of determining the chromatic number of an input graph. Here, it is useful to rephrase the problem as follows: find a minimum number k such that the vertex set of an input graph can be covered by k maximal independent sets; note that we do not insist that the independent sets are disjoint. Observe that if X is a set of vertices covered by one or more such maximal independent sets, we have $X \cap N_G[u] \neq \emptyset$ for every $u \in V(G)$, as otherwise the vertex u should have been included into one of the covering sets. Hence, we can call a set $X \subseteq V(G)$ *useful* if it intersects every closed neighborhood in G,

and we obtain again a $(2 - \varepsilon_d)^n$ bound on the number of useful sets. An important contribution of Björklund, Husfeldt, Kaski, and Koivisto [5] can be summarized as follows: using the fact that the useful sets are upward-closed (any superset of a useful set is useful as well), we can trim the fast subset convolution algorithm of [3] to consider useful sets only. Consequently, we obtain a $(2 - \varepsilon_d)^n n^{\mathcal{O}(1)}$-time algorithm for computing the chromatic number of an input graph of maximum degree bounded by d.

Bounded Average Degree

Generalizing Algorithms for Bounded Degree Graphs

The above approach for traveling salesman problem has been generalized to graphs of bounded average degree by Cygan and Pilipczuk [8] using the following observation. Assume a graph G has n vertices and average degree bounded by d. Then, a simple Markov-type inequality implies that for every $\zeta > 1$ there are at most n/ζ vertices of degree larger than ζd. However, this bound cannot be tight for all values of ζ at once, and one can prove the following: if we want at most $n/(\alpha\zeta)$ vertices of degree larger than ζd for some $\alpha > 1$, then we can always find such a constant ζ of order roughly exponential in α.

An appropriate choice of α and the corresponding value of ζ allow us to partition the vertex set of an input graph into a large part of bounded degree and a very small part of unbounded degree. The extra multiplicative gap of α in the size bound allows us to hide the cost of extensive branching on the part with unbounded degree in the gains obtained by considering only (appropriately defined) useful sets in the bounded degree part.

With this line of reasoning, Cygan and Pilipczuk [8] showed that for every degree bound d, there exists a constant $\varepsilon_d > 0$ such that traveling salesman problem in graphs of bounded average degree by d can be solved in $(2 - \varepsilon_d)^n n^{\mathcal{O}(1)}$ time. It should be noted that the constant ε_d depends here doubly exponentially on d, as opposed to single-exponential dependency in the works for bounded degree graphs.

Furthermore, Cygan and Pilipczuk showed how to express the problem of counting perfect matchings in an n-vertex graph as a specific variant of a problem of counting Hamiltonian cycles in an $n/2$-vertex graph. This reduction not only gives a simpler $2^{n/2} n^{\mathcal{O}(1)}$-time algorithm for counting perfect matchings, as compared to the original algorithm of Björklund [2], but since the reduction does not increase the number of edges in a graph, it also provides a $(2 - \varepsilon_d)^{n/2} n^{\mathcal{O}(1)}$-time algorithm in the case of bounded average degree.

In a subsequent work, Golovnev, Kulikov, and Mihajlin [9] showed how to use the aforementioned multiplicative gap of α to obtain a $(2 - \varepsilon_d)^n n^{\mathcal{O}(1)}$-time algorithm for computing the chromatic number of a graph with average degree bounded by d. Furthermore, they expressed all previous algorithms as the task of determining one coefficient in a carefully chosen polynomial, obtaining polynomial space complexity without any significant loss in time complexity.

Counting Perfect Matchings in Bipartite Graphs

A somewhat different line of research concerns counting perfect matchings in bipartite graphs. Here, a $2^{n/2} n^{\mathcal{O}(1)}$-time algorithm is known for several decades [12]. Cygan and Pilipczuk presented a very simple $2^{(1-1/(3.55d))n/2} n^{\mathcal{O}(1)}$-time algorithm for this problem in graphs of average degree at most d, improving upon the previous works of Servedio and Wan [13] and Izumi and Wadayama [11]. Furthermore, this result generalizes to the problem of computing the permanent of a matrix over an arbitrary commutative ring with the number of nonzero entries linear in the dimension of the matrix.

Cross-References

▸ Exact Graph Coloring Using Inclusion-Exclusion
▸ Fast Subset Convolution

Recommended Reading

1. Bellman R (1962) Dynamic programming treatment of the travelling salesman problem. J ACM 9:61–63
2. Björklund A (2012) Counting perfect matchings as fast as ryser. In: Rabani Y (ed) SODA, Kyoto. SIAM, pp 914–921
3. Björklund A, Husfeldt T, Kaski P, Koivisto M (2007) Fourier meets möbius: fast subset convolution. In: Johnson DS, Feige U (eds) STOC, San Diego. ACM, pp 67–74
4. Björklund A, Husfeldt T, Koivisto M (2009) Set partitioning via inclusion-exclusion. SIAM J Comput 39(2):546–563
5. Björklund A, Husfeldt T, Kaski P, Koivisto M (2010) Trimmed moebius inversion and graphs of bounded degree. Theory Comput Syst 47(3):637–654
6. Björklund A, Husfeldt T, Kaski P, Koivisto M (2012) The traveling salesman problem in bounded degree graphs. ACM Trans Algorithms 8(2):18
7. Chung FRK, Frankl P, Graham RL, Shearer JB (1986) Some intersection theorems for ordered sets and graphs. J Comb Theory Ser A 43(1):23–37
8. Cygan M, Pilipczuk M (2013) Faster exponential-time algorithms in graphs of bounded average degree. In: Fomin FV, Freivalds R, Kwiatkowska MZ, Peleg D (eds) ICALP (1). Lecture notes in computer science, vol 7965. Springer, Berlin/Heidelberg, pp 364–375

9. Golovnev A, Kulikov AS, Mihajlin I (2014) Families with infants: a general approach to solve hard partition problems. In: ICALP (1). Lecture notes in computer science. Springer, Berlin/Heidelberg, pp 551–562. Available at http://arxiv.org/abs/1311.2456
10. Held M, Karp RM (1962) A dynamic programming approach to sequencing problems. J Soc Ind Appl Math 10:196–210
11. Izumi T, Wadayama T (2012) A new direction for counting perfect matchings. In: FOCS, New Brunswick. IEEE Computer Society, pp 591–598
12. Ryser H (1963) Combinatorial mathematics. The Carus mathematical monographs. Mathematical Association of America, Buffalo
13. Servedio RA, Wan A (2005) Computing sparse permanents faster. Inf Process Lett 96(3):89–92

Exact Graph Coloring Using Inclusion-Exclusion

Andreas Björklund and Thore Husfeldt
Department of Computer Science, Lund
University, Lund, Sweden

Keywords

Vertex coloring

Years and Authors of Summarized Original Work

2006; Björklund, Husfeldt

Problem Definition

A *k-coloring* of a graph $G = (V, E)$ assigns one of k colors to each vertex such that neighboring vertices have different colors. This is sometimes called *vertex coloring*.

The smallest integer k for which the graph G admits a k-coloring is denoted $\chi(G)$ and called the *chromatic number*. The number of k-colorings of G is denoted $P(G; k)$ and called the *chromatic polynomial*.

Key Results

The central observation is that $\chi(G)$ and $P(G; k)$ can be expressed by an inclusion-exclusion formula whose terms are determined by the number of independent sets of induced subgraphs of G. For $X \subseteq V$, let $s(X)$ denote the number of nonempty independent vertex subsets disjoint from X, and let $s_r(X)$ denote the number of ways to choose r nonempty independent vertex subsets $S_1, \ldots, S_r$ (possibly overlapping and with repetitions), all disjoint from X, such that $|S_1| + \cdots + |S_r| = |V|$.

Theorem 1 ([1]) *Let G be a graph on n vertices.*

1.

$$\chi(G) = \min_{k \in \{1, \ldots, n\}} \left\{ k : \sum_{X \subseteq V} (-1)^{|X|} s(X)^k > 0 \right\}.$$

2. For $k = 1, \ldots, n$,

$$P(G; k) = \sum_{r=1}^{k} \binom{k}{r} \left(\sum_{X \subseteq V} (-1)^{|X|} s_r(X) \right).$$

The time needed to evaluate these expressions is dominated by the 2^n evaluations of $s(X)$ and $s_r(X)$, respectively. These values can be precomputed in time and space within a polynomial factor of 2^n because they satisfy

$$s(X) = \begin{cases} 0, & \text{if } X = V, \\ s(X \cup \{v\}) + s(X \cup \{v\} \cup N(v)) + 1, & \text{for } v \notin X, \end{cases}$$

where $N(v)$ are the neighbors of v in G. Alternatively, the values can be computed using exponential-time, polynomial-space algorithms from the literature.

This leads to the following bounds:

Theorem 2 ([3]) *For a graph G on n vertices, $\chi(G)$ and $P(G;k)$ can be computed in*

1. *Time and space $2^n n^{O(1)}$.*
2. *Time $O(2.2461^n)$ and polynomial space*

The space requirement can be reduced to $O(1.292^n)$ [4].

The techniques generalize to arbitrary families of subsets over a universe of size n, provided membership in the family can be decided in polynomial time [3, 4], and to the Tutte polynomial and the Potts model [2].

Applications

In addition to being a fundamental problem in combinatorial optimization, graph coloring also arises in many applications, including register allocation and scheduling.

Recommended Reading

1. Björklund A, Husfeldt T (2008) Exact algorithms for exact satisfiability and number of perfect matchings. Algorithmica 52(2):226–249
2. Björklund A, Husfeldt T, Kaski P, Koivisto M (2007) Fourier meets Möbius: fast subset convolution. In: Proceedings of the 39th annual ACM symposium on theory of computing (STOC), San Diego, 11–13 June 2007. Association for Computing Machinery, New York, pp 67–74
3. Björklund A, Husfeldt T, Koivisto M (2009) Set partitioning via inclusion-exclusion. SIAM J Comput 39(2):546–563
4. Björklund A, Husfeldt T, Kaski P, Koivisto M (2011) Covering and packing in linear space. Inf Process Lett 111(21–22):1033–1036

Exact Quantum Algorithms

Ashley Montanaro
Department of Computer Science, University of Bristol, Bristol, UK

Keywords

Exact algorithms; Quantum algorithms; Quantum query complexity

Years and Authors of Summarized Original Work

2013; Ambainis

Problem Definition

Many of the most important known quantum algorithms operate in the query complexity model. In the simplest variant of this model, the goal is to compute some Boolean function of n input bits by making the minimal number of queries to the bits. All other resources (such as time and space) are considered to be free. In the model of *exact* quantum query complexity, one insists that the algorithm succeeds with certainty on every allowed input. The aim is then to find quantum algorithms which satisfy this constraint and still outperform any possible classical algorithm. This can be a challenging task, as achieving a probability of error equal to zero requires delicate cancellations between the amplitudes in the quantum algorithm. Nevertheless, efficient exact quantum algorithms are now known for certain functions.

Some basic Boolean functions which we will consider below are:

- Parity^n: $f(x_1, \ldots, x_n) = x_1 \oplus x_2 \oplus \cdots \oplus x_n$.
- Threshold$_k^n$: $f(x_1, \ldots, x_n) = 1$ if $|x| \geq k$, and $f(x) = 0$ otherwise, where $|x| := \sum_i x_i$

is the Hamming weight of x. The special case $k = n/2$ is called the majority function.

- Exact$_k^n$: $f(x_1, \ldots, x_n) = 1$ if $|x| = k$, and $f(x) = 0$ otherwise.
- NE ("not-all-equal") on 3 bits: $f(x_1, x_2, x_3) = 0$ if $x_1 = x_2 = x_3$, and $f(x_1, x_2, x_3) = 1$ otherwise.

Key Results

Early Results

One of the earliest results in quantum computation was that the parity of 2 bits can be computed with certainty using only 1 quantum query [6], implying that Parity^n can be computed using $\lceil n/2 \rceil$ quantum queries. By contrast, any classical algorithm which computes this function must make n queries. The quantum algorithm for Parity^n can be used as a subroutine to obtain speedups over classical computation for other problems. For example, based on this algorithm the majority function on n bits can be computed exactly using $n + 1 - w(n)$ quantum queries, where $w(n)$ is the number of 1s in the binary expansion of n [8]; this result has recently been improved (see below).

If the function to be computed is partial, i.e., some possible inputs are disallowed, the separation between exact quantum and classical query complexity can be exponential. For example, in the Deutsch-Jozsa problem we are given query access to an n-bit string x (with n even) such that either all the bits of x are equal or exactly half of them are equal to 1. Our task is to determine which is the case. Any exact classical algorithm must make at least $n/2 + 1$ queries to bits of x to solve this problem, but it can be solved with only one quantum query [7]. An exponential separation is even known between exact quantum and *bounded-error* classical query complexity for a different partial function [5].

Recent Developments

For some years, the best known separation between exact quantum and classical query complexity of a total Boolean function (i.e., a function $f : \{0, 1\}^n \rightarrow \{0, 1\}$ with all possible n-bit strings allowed as input) was the factor of 2 discussed above. However, recently the first example has been presented of an exact quantum algorithm for a family of total Boolean functions which achieves a lower asymptotic query complexity than the best possible classical algorithm [1].

The family of functions used can be summarized as a "not-all-equal tree of depth d." It is based around the recursive use of the NE function. Define the function $NE^0(x_1) = x_1$ and then for $d > 0$

$$NE^d(x_1, \ldots, x_{3^d})$$
$$= NE(NE^{d-1}(x_1, \ldots, x_{3^{d-1}}), NE^{d-1}(x_{3^{d-1}+1}, \ldots, x_{2 \cdot 3^{d-1}}), NE^{d-1}(x_{2 \cdot 3^{d-1}+1}, \ldots, x_{3^d})).$$

Then the following separation is known:

Theorem 1 (Ambainis [1]) *There is an exact quantum algorithm which computes* NE^d *using* $O(2.593\ldots^d)$ *queries. Any classical algorithm which computes* NE^d *must make* $\Omega(3^d)$ *queries, even if it is allowed probability of failure 1/3.*

In addition, Theorem 1 implies the first known asymptotic separation between exact quantum and classical communication complexity for a total function. Improvements over the best possible

classical algorithms are also known for the other basic Boolean functions previously mentioned.

Theorem 2 (Ambainis, Iraids, and Smotrovs [2]) *There is an exact quantum algorithm which computes* Exact$_k^n$ *using* $\max\{k, n - k\}$ *queries and an exact quantum algorithm which computes* Threshold$_k^n$ *using* $\max\{k, n - k + 1\}$ *queries. Both of these complexities are optimal.*

By contrast, it is easy to see that any exact classical algorithm for these functions must make

n queries. An optimal exact quantum algorithm for the special case Exact_2^4 had already been found prior to this, in work which also gave optimal exact quantum query algorithms for all Boolean functions on up to 3 bits [9].

Methods

We briefly describe the main ingredients of the efficient quantum algorithm for NE^d [1]. The basic idea is to fix some small d_0, start with an exact quantum algorithm which computes NE^{d_0} using fewer queries than the best possible classical algorithm, and then amplify the separation by using the algorithm recursively. A difficulty with this approach is that the standard approach for using a quantum algorithm recursively incurs a factor of 2 penalty in the number of queries with each recursive call. This factor of 2 is required to "uncompute" information left over after the algorithm has completed. Therefore, a query complexity separation by a factor of 2 or less does not immediately give an asymptotic separation.

This problem can be addressed by introducing the notion of p-computation. Let $p \in [-1, 1]$. A quantum algorithm $\mathcal{A}$ is said to p-compute a function $f(x_1, \ldots, x_n)$ if, for some state $|\psi_{\text{start}}\rangle$:

- Whenever $f(x_1, \ldots, x_n) = 0$, $\mathcal{A}|\psi_{\text{start}}\rangle = |\psi_{\text{start}}\rangle$.
- Whenever $f(x_1, \ldots, x_n) = 1$, $\mathcal{A}|\psi_{\text{start}}\rangle = p|\psi_{\text{start}}\rangle + \sqrt{1-p^2}|\psi\rangle$ for some $|\psi\rangle$, which may depend on x, such that $\langle \psi | \psi_{\text{start}}\rangle = 0$.

It can be shown that if there exists an algorithm which p-computes some function f for some $p \leq 0$, there exists an exact quantum algorithm which computes f using the same number of queries. Further, if an algorithm (-1)-computes some function f, the same algorithm can immediately be used recursively, without needing any additional queries at each level of recursion. Thus, to obtain an asymptotic quantum-classical separation for NE^d, it suffices to obtain an algorithm which (-1)-computes NE^{d_0} using strictly fewer than 3^{d_0} queries, for some d_0.

The NE^d problem also behaves particularly well with respect to p-computation for general values of p:

Lemma 1 *If there is an algorithm $\mathcal{A}$ which p-computes NE^{d-1} using k queries, there is an algorithm $\mathcal{A}'$ which p'-computes NE^d with $2k$ queries, for $p' = 1 - 4(1-p)^2/9$.*

This lemma allows algorithms for NE^{d-1} to be lifted to algorithms for NE^d, at the expense of making the value of p worse. Nevertheless, given that it is easy to write down an algorithm which (-1)-computes NE^0 using one query, the lemma is sufficient to obtain an exact quantum algorithm for NE^2 using 4 queries. This is already enough to prove an asymptotic quantum-classical separation, but this separation can be improved using the following lemma (a corollary of a variant of amplitude amplification):

Lemma 2 *If there is an algorithm $\mathcal{A}$ which p-computes NE^d using k queries, there is an algorithm $\mathcal{A}'$ which p'-computes NE^d with $2k$ queries, for $p' = 2p^2 - 1$.*

Interleaving Lemmas 1 and 2 allows one to derive an algorithm which (-1)-computes NE^8 using 2,048 queries, which implies an exact quantum algorithm for NE^d using $O(2{,}048^{d/8}) = O(2.593\ldots^d)$ queries.

Experimental Results

It is a difficult task to design exact quantum query algorithms, even for small functions, as these algorithms require precise cancellations between amplitudes. One way to gain numerical evidence for what the exact quantum query complexity of a function should be is to use the formulation of quantum query complexity as a semidefinite programming (SDP) problem [4]. This allows one to estimate the optimal success probability of any quantum algorithm using a given number of queries to compute a given function. If this success probability is very close to 1, this gives numerical evidence that there exists an exact quantum algorithm using that number of queries.

This approach has been applied for all Boolean functions on up to 4 bits, giving strong evidence that the only function on 4 bits which requires 4 quantum queries is the AND function and functions equivalent to it [9]. This has led to the conjecture that, for any n, the only function on n bits which requires n quantum queries to be computed exactly is the AND function and functions equivalent to it. This would be an interesting contrast with the classical case where most functions on n bits require n queries. This conjecture has recently been proven for various special cases: symmetric functions, monotone functions, and functions with formula size n [3].

Cross-References

▶ Quantum Algorithm for the Parity Problem
▶ Quantum Search

Recommended Reading

1. Ambainis A (2013) Superlinear advantage for exact quantum algorithms. In: Proceedings of the 45th annual ACM symposium on theory of computing, pp 891–900. arXiv:1211.0721
2. Ambainis A, Iraids J, Smotrovs J (2013) Exact quantum query complexity of EXACT and THRESHOLD. In: Proceedings of the 8th conference on the theory of quantum computation, communication, and cryptography (TQC'13), pp 263–269. arXiv:1302.1235
3. Ambainis A, Gruska J, Zheng S (2014) Exact query complexity of some special classes of Boolean functions. arXiv:1404.1684
4. Barnum H, Saks M, Szegedy M (2003) Quantum query complexity and semi-definite programming. In: Proceedings of the 18th annual IEEE conference on computational complexity, Aarhus, pp 179–193
5. Brassard G, Høyer P (1997) An exact quantum polynomial-time algorithm for Simon's problem. In: Proceedings of the fifth Israeli symposium on theory of computing and systems, Aarhus, Denmark pp 12–23. quant-ph/9704027
6. Cleve R, Ekert A, Macchiavello C, Mosca M (1998) Quantum algorithms revisited. Proc R Soc Lond A 454(1969):339–354. quant-ph/9708016
7. Deutsch D, Jozsa R (1992) Rapid solution of problems by quantum computation. Proc R Soc Lond Ser A 439(1907):553–558
8. Hayes T, Kutin S, van Melkebeek D (2002) The quantum black-box complexity of majority. Algorithmica 34(4):480–501. quant-ph/0109101
9. Montanaro A, Jozsa R, Mitchison G (2013) On exact quantum query complexity. Algorithmica 71(4):775–796

Experimental Implementation of Tile Assembly

Constantine G. Evans
Division of Biology and Bioengineering,
California Institute of Technology, Pasadena, CA, USA

Keywords

DNA tiles; Experimental tile self-assembly

Years and Authors of Summarized Original Work

2007; Schulman, Winfree
2008; Fujibayashi, Hariadi, Park, Winfree, Murata
2009; Barish, Schulman, Rothemund, Winfree
2012; Schulman, Yurke, Winfree

Problem Definition

From the earliest works on tile self-assembly, abstract theoretical models and experimental implementations have been linked. In 1998, in addition to developing the abstract and kinetic Tile Assembly Models (aTAM and kTAM) [14], Winfree et al. demonstrated the use of DNA tiles to construct a simple, periodic lattice [16]. Periodic lattices and "uniquely addressed" assemblies, where each tile type appears once in each assembly, have been widely studied, with systems employing up to a thousand unique tiles in three dimensions [8, 13]. While these systems provide insight into the behavior of DNA tile systems, *algorithmic* tile systems of more theoretical interest pose specific challenges for experimental implementation.

In the aTAM, abstract tiles attach individually to empty lattice sites if bonds of a sufficient total strength b (at least abstract "temperature" τ) can be made, and once attached, never detach. Experimentally, free tiles and assemblies of bound tiles are in solution. Tiles have short single-stranded "sticky ends" regions that form bonds with complementary regions on other tiles. Tiles attach to assemblies at rates dependent only upon their concentrations, regardless of the strength of bonds that can be made. Once attached, tiles can detach and do so at a rate that is exponentially dependent upon the total strength of the bonds [6]. Thus, for a tile t_i with concentration $[t_i]$ binding by a total abstract bond strength b, we have attachment and detachment rates of

$$r_f = k_f[t_i] \qquad r_b = k_f e^{-b \Delta G_{se}^\circ / RT + \alpha} \qquad (1)$$

where k_f is an experimentally determined rate constant, α is a constant binding free energy change (e.g., from entropic considerations), ΔG_{se}° is the free energy change of a single-strength bond, and T is the (physical) temperature. Using the substitutions $[t_i] = e^{-G_{mc} + \alpha}$, $G_{se} = -\Delta G_{se}^\circ / RT$, and $\hat{k}_f = k_f e^{\alpha}$, these can be simplified to

$$r_f = \hat{k}_f e^{-G_{mc}} \qquad r_b = \hat{k}_f e^{-b G_{se}} \qquad (2)$$

where G_{se} is a (positive) unitless free energy for a single-strength bond (larger values correspond to stronger bonds), G_{mc} is a free energy analogue of concentration (larger values correspond to lower concentrations), and $\hat{k}_f$ is an adjusted rate constant.

These rates are the basis of the kinetic Tile Assembly Model (kTAM), which is widely used as a physical model of tile assembly [14]. Tiles that attach faster than they detach will tend to remain attached and allow further growth: for example, if $G_{mc} < 2G_{se}$, tile attachments by $b \geq 2$ will be favorable. Tiles that detach faster than they attach will tend to remain detached and not allow further growth. Since G_{mc} is dependent upon tile concentration, and G_{se} is dependent

upon physical temperature (lower temperatures result in larger G_{se} values), the attachment and detachment rates can be tuned such that attachment is slightly more favorable than detachment for tiles attaching by a certain total bond strength and less favorable for less strongly bound tiles. In this way, in the limit of low concentrations and slow growth, the kTAM approximates the aTAM at a given abstract temperature τ. When moving away from this limit and toward experimentally feasible conditions, however, the kTAM provides insight into many of the challenges faced in experimental implementation of algorithmic tile assembly:

Growth errors: While tile assembly in the aTAM is error-free, tiles can attach in erroneous locations in experiments. Even ignoring the possibility of lattice defects, malformed tiles, and other experimental peculiarities, errors can arise in the kTAM via tiles that attach by less than the required bond strength (e.g., one single-strength bond for a $\tau = 2$ system) and are then "frozen" in place by further attachments [4]. As the further growth of algorithmic systems depends on the tiles already present in an assembly, a single erroneously incorporated tile can propagate undesired growth via further, valid attachments. These errors can arise both in growth sites where another tile could attach correctly ("growth errors") and lattice sites where no correct tile could attach ("facet nucleation errors") [3, 14].

Seeding: Tile assembly in the aTAM is usually initiated from a designated "seed" tile. In solution, however, tiles are free to attach to all other tiles and can form assemblies without starting from a seed, even if this requires several unfavorable attachments to form a stable structure that can allow further growth. Depending upon the tile system, these "spuriously nucleated" structures can potentially form easily. For example, a $T = 2$ system with boundaries of identical tiles that attach by double bonds on both sides can readily form long strings of boundary tiles [10, 11].

Tile depletion: As free tiles in solution are incorporated into assemblies, their concentrations are correspondingly reduced. This depletion lowers the attachment rates for those tiles and in turn changes the favorability of growth. If different tile types are incorporated in different quantities, their attachment rates will become unequal, and at some point in assembly, attachment by two single-strength bonds may be favorable for one tile type and unfavorable for another.

Tile design: While theoretical constructions may employ an arbitrary number of sticky ends types, this number is limited by tile designs in practice. Most tiles use short single-stranded DNA regions of 5–10 nucleotides (nt), limiting the number of possible sticky ends to 4^5–4^{10} at best. However, since partial bonds can form between subsequences of the sticky ends, sequences with sufficient orthogonality are required, and since DNA binding strength is sequence dependent, sequences with similar binding energies are required [5]. Both of these effects place considerably more stringent limits on the number of sticky ends and change the behavior of experimental systems.

Key Results

Winfree and Bekbolatov developed a tileset transformation, "uniform proofreading," that reduced per-site *growth* error rates from $r_{err} \approx me^{-G_{se}}$ (where m is the number of possible errors) to $\approx me^{-KG_{se}}$ by scaling each tile into a $K \times K$ block of individually attaching tiles with unique internal bonds [15]. However, this transformation did not reduce facet nucleation errors. Chen and Goel later created a modified transformation, "snaked proofreading," that reduced both growth and facet nucleation errors by changing the strengths of the internal bonds used [3]. These and other proofreading methods have the potential to drastically reduce error rates in experimental systems.

Schulman et al. analyzed tile system nucleation through the consideration of "critical nuclei," tile assemblies where melting and further growth are equally favorable, and showed that by ensuring a sufficient number of unfavorable attachments would be required for a critical nucleus to form, the rate of spurious nucleation can be kept arbitrarily low [11]. Using this analysis, Schulman et al. constructed the "zigzag" ribbon system, which forms a ribbon where each row must assemble completely before the next can begin growth, as an example of a system where spurious nucleation can be made arbitrarily low by increasing ribbon width. To nucleate desired structures, this system makes use of a large, preformed seed structure to allow the growth of the first ribbon row.

Schulman et al. also devised a "constant-temperature" growth technique where the concentrations of assemblies, controlled by the concentration of initial seeds in a nucleation-controlled system, are kept small enough in comparison to the concentrations of free tiles that growth does not significantly deplete tile concentrations, which thus remain approximately constant [12]. After growth is completed, the remaining free tiles are "deactivated" by adding an excess of DNA strands complementary to specific sticky ends sequences.

In analyzing the effects of DNA sequences on tile assembly, Evans and Winfree showed an exponential increase of error rates in the kTAM for partial binding between different sticky ends sequences and for differing sequence-dependent binding energies and developed algorithms for sequence design and assignment to reduce these effects [5]. With reasonable design constraints, their algorithms suggested limits of around 80 sticky ends types for tiles using 5 nt sticky ends and around 360 for tiles using 10 nt sticky ends before significant sequence effects begin to become unavoidable and must be incorporated into tile system design.

Experimental Results

While numerous designs exist for tile structures, experimental implementations have usually used either double-crossover (DX) tiles with 5 or 6 nt sticky ends [16] or single-stranded tiles (SST) with 10 and 11 nt sticky ends [17]. SSTs potentially offer a significantly larger sequence

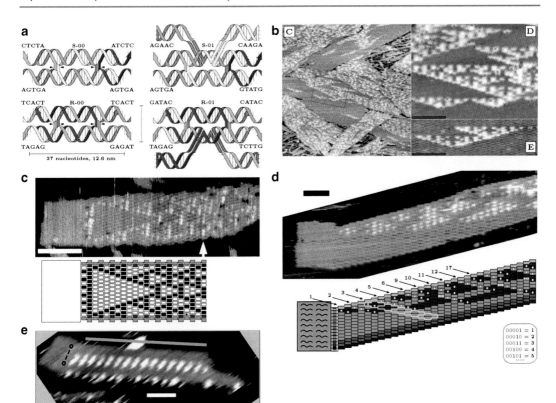

Experimental Implementation of Tile Assembly, Fig. 1 Experimental results for algorithmic tile assembly. (**a**) and (**b**) show the Rothemund et al. XOR system's DX tiles and resulting structures, with (**b**) illustrating the high error rates and seeding problems of the system [9]. (**c**) shows the Fujibayashi et al. fixed-width XOR ribbon [7], while (**d**) shows the Barish et al. binary counter ribbon with partial 2 × 2 proofreading [2]; the rectangular structures on the left of both systems are preformed DNA origami seeds. (**e**) shows an example bit-copying ribbon from Schulman et al. [12]

space and have been employed in large, non-algorithmic systems [8, 13] but have not yet been used for complex algorithmic systems.

Early experiments in algorithmic tile assembly using DX tiles did not employ any of the key results discussed above. Rothemund et al. implemented a simple XOR system of four logical tiles (eight tiles were needed owing to structural considerations), using DNA hairpins on "one-valued" tiles as labels [9] and flexible, one-dimensional seeds (Fig. 1a,b). While assemblies grew, and Sierpinski triangle patterns were visible, error rates were between 1 and 10 % per tile. Barish et al. implemented more complex bit-copying and binary counting systems in a similar way, finding per-tile error rates of around 10 % [1].

More recently, Fujibayashi et al. used rigid DNA origami structures to serve as seeds for the growth of a fixed-width XOR ribbon system and, in doing so, reduced error rates to 1.4 % per tile without incorporating proofreading [7] (Fig. 1c). This seeding mechanism was also used by Barish et al. to seed zigzag bit-copying and binary counting ribbon systems that implemented 2 × 2 uniform proofreading [2]. With nucleation control and proofreading, these systems resulted in dramatically reduced error rates of 0.26 % per proofreading block for copying and 4.1 % for the more algorithmically complex binary counting, which only partially implemented uniform proofreading (Fig. 1d).

A similar bit-copying ribbon was later implemented by Schulman et al., with the addition of the constant-temperature, constant-concentration growth method and the use of biotin-streptavidin labels rather than DNA hairpins. The result was a decrease in error rates by almost a factor of ten to 0.034 % per block [12] (Fig. 1e). At this error rate, structures of around 2,500 error-free blocks, or 10,000 individual tiles, could be grown with reasonable yields, suggesting that with the incorporation of proofreading, nucleation control and constant-concentration growth methods, low-error experimental implementations of increasingly complex algorithmic tile systems may be feasible up to sequence space limitations.

Cross-References

▶ Robustness in Self-Assembly

Recommended Reading

1. Barish RD, Rothemund PWK, Winfree E (2005) Two computational primitives for algorithmic self-assembly: copying and counting. Nano Lett 5(12):2586–2592. doi:10.1021/nl052038l
2. Barish RD, Schulman R, Rothemund PWK, Winfree E (2009) An information-bearing seed for nucleating algorithmic self-assembly. PNAS 106:6054–6059. doi:10.1073/pnas.0808736106
3. Chen HL, Goel A (2005) Error free self-assembly using error prone tiles. In: DNA 10, Milan. LNCS, vol 3384. Springer, pp 702–707
4. Doty D (2012) Theory of algorithmic self-assembly. Commun ACM 55(12):78–88. doi:10.1145/2380656.2380675
5. Evans CG, Winfree E (2013) DNA sticky end design and assignment for robust algorithmic self-assembly. In: DNA 19, Tempe. LNCS, vol 8141. Springer, pp 61–75. doi:10.1007/978-3-319-01928-4_5
6. Evans CG, Hariadi RF, Winfree E (2012) Direct atomic force microscopy observation of DNA tile crystal growth at the single-molecule level. J Am Chem Soc 134:10,485–10,492. doi:10.1021/ja301026z
7. Fujibayashi K, Hariadi R, Park SH, Winfree E, Murata S (2008) Toward reliable algorithmic self-assembly of DNA tiles: a fixed-width cellular automaton pattern. Nano Lett 8(7):1791–1797. doi:10.1021/nl0722830
8. Ke Y, Ong LL, Shih WM, Yin P (2012) Three-dimensional structures self-assembled from DNA bricks. Science 338(6111):1177–1183. doi:10.1126/science.1227268
9. Rothemund PWK, Papadakis N, Winfree E (2004) Algorithmic self-assembly of DNA Sierpinski triangles. PLoS Biol 2(12):e424. doi:10.1371/journal.pbio.0020424
10. Schulman R, Winfree E (2007) Synthesis of crystals with a programmable kinetic barrier to nucleation. PNAS 104(39):15,236–15,241. doi:10.1073/pnas.0701467104
11. Schulman R, Winfree E (2010) Programmable control of nucleation for algorithmic self-assembly. SIAM J Comput 39(4):1581–1616. doi:10.1137/070680266
12. Schulman R, Yurke B, Winfree E (2012) Robust self-replication of combinatorial information via crystal growth and scission. PNAS 109(17):6405–6410. doi:10.1073/pnas.1117813109
13. Wei B, Dai M, Yin P (2012) Complex shapes self-assembled from single-stranded DNA tiles. Nature 485(7400):623–626
14. Winfree E (1998) Simulations of computing by self-assembly. Technical report CaltechCSTR:1998.22, Pasadena
15. Winfree E, Bekbolatov R (2004) Proofreading tile sets: error correction for algorithmic self-assembly. In: DNA 9, Madison. Wisconson, LNCS, vol 2943. Springer, pp 126–144
16. Winfree E, Liu F, Wenzler LA, Seeman NC (1998) Design and self-assembly of two-dimensional DNA crystals. Nature 394(6693):539–544
17. Yin P, Hariadi RF, Sahu S, Choi HMT, Park SH, LaBean TH, Reif JH (2008) Programming DNA tube circumferences. Science 321(5890):824–826. doi:10.1126/science.1157312

Experimental Methods for Algorithm Analysis

Catherine C. McGeoch
Department of Mathematics and Computer Science, Amherst College, Amherst, MA, USA

Keywords

Algorithm engineering; Empirical algorithmics; Empirical analysis of algorithms; Experimental algorithmics

Years and Authors of Summarized Original Work

2001; McGeoch

Problem Definition

Experimental analysis of algorithms describes not a specific algorithmic problem, but rather an approach to algorithm design and analysis. It complements, and forms a bridge between, traditional *theoretical analysis*, and the application-driven methodology used in *empirical analysis*.

The traditional theoretical approach to algorithm analysis defines algorithm efficiency in terms of counts of dominant operations, under some abstract model of computation such as a RAM; the input model is typically either worst-case or average-case. Theoretical results are usually expressed in terms of asymptotic bounds on the function relating input size to number of dominant operations performed.

This contrasts with the tradition of empirical analysis that has developed primarily in fields such as operations research, scientific computing, and artificial intelligence. In this tradition, the efficiency of implemented programs is typically evaluated according to CPU or wall-clock times; inputs are drawn from real-world applications or collections of benchmark test sets, and experimental results are usually expressed in comparative terms using tables and charts.

Experimental analysis of algorithms spans these two approaches by combining the sensibilities of the theoretician with the tools of the empiricist. Algorithm and program performance can be measured experimentally according to a wide variety of *performance indicators*, including the dominant cost traditional to theory, bottleneck operations that tend to dominate running time, data structure updates, instruction counts, and memory access costs. A researcher in experimental analysis selects performance indicators most appropriate to the scale and scope of the specific research question at hand. (Of course time is not the only metric of interest in algorithm studies; this approach can be used to analyze other properties such as solution quality or space use.)

Input instances for experimental algorithm analysis may be randomly generated or derived from application instances. In either case, they typically are described in terms of a small-to medium-sized collection of *controlled parameters*. A primary goal of experimentation is to investigate the cause-and-effect relationship between input parameters and algorithm/program performance indicators.

Research goals of experimental algorithmics may include discovering functions (not necessarily asymptotic) that describe the relationship between input and performance, assessing the strengths and weaknesses of different algorithm/data structures/programming strategies, and finding best algorithmic strategies for different input categories. Results are typically presented and illustrated with graphs showing comparisons and trends discovered in the data.

The two terms "empirical" and "experimental", are often used interchangeably in the literature. Sometimes the terms "old style" and "new style" are used to describe, respectively, the empirical and experimental approaches to this type of research. The related term "algorithm engineering" refers to a systematic design process that takes an abstract algorithm all the way to an implemented program, with an emphasis on program efficiency. Experimental and empirical analysis is often used to guide the algorithm engineering process. The general term *algorithmics* can refer to both design and analysis in algorithm research.

Key Results

None

Applications

Experimental analysis of algorithms has been used to investigate research problems originating in theoretical computer science. One example arises in the average-case analysis of algorithms for the One-Dimensional Bin Packing problem. Experimental analyses have led to new theorems about the performance of the optimal algorithm; new asymptotic bounds on average-case performance of approximation algorithms; extensions of theoretical results to new models of inputs; and to new algorithms with tighter approximation guarantees. Another example is the experimental discovery of a type of phase-transition behavior for random instances of the 3CNF-Satisfiabilty problem, which has led to new ways to characterize the difficulty of problem instances.

A second application of experimental algorithmics is to find more realistic models of computation, and to design new algorithms that perform better on these models. One example is found in the development of new memory-based models of computation that give more accurate time predictions than traditional unit-cost models. Using these models, researchers have found new cache-efficient and I/O-efficient algorithms that exploit properties of the memory hierarchy to achieve significant reductions in running time.

Experimental analysis is also used to design and select algorithms that work best in practice, algorithms that work best on specific categories of inputs, and algorithms that are most robust with respect to bad inputs.

Data Sets

Many repositories for data sets and instance generators to support experimental research are available on the Internet. They are usually organized according to specific combinatorial problems or classes of problems.

URL to Code

Many code repositories to support experimental research are available on the Internet. They are usually organized according to specific combinatorial problems or classes of problems. Skiena's *Stony Brook Algorithm Repository* (www.cs.sunysb.edu/~algorith/) provides a comprehensive collection of problem definitions and algorithm descriptions, with numerous links to implemented algorithms.

Recommended Reading

The algorithmic literature containing examples of experimental research is much too large to list here. Some articles containing advice and commentary on experimental methodology in the context of algorithm research appear in the list below.

The workshops and journals listed below are specifically intended to support research in experimental analysis of algorithms. Experimental work also appears in more general algorithm research venues such as SODA (ACM/IEEE Symposium on Data Structures and Algorithms), *Algorithmica*, and *ACM Transactions on Algorithms*.

1. ACM Journal of Experimental Algorithmics. Launched in 1996, this journal publishes contributed articles as well as special sections containing selected papers from ALENEX and WEA. Visit www.jea. acm.org, or visit portal.acm.org and click on ACM Digital Library/Journals/Journal of Experimental Algorithmics
2. ALENEX. Beginning in 1999, the annual workshop on Algorithm Engineering and Experimentation is sponsored by SIAM and ACM. It is co-located with SODA, the SIAM Symposium on Data Structures and Algorithms. Workshop proceedings are published in the Springer LNCS series. Visit www.siam.org/meetings/ for more information
3. Barr RS, Golden BL, Kelly JP, Resende MGC, Stewart WR (1995) Designing and reporting on computational experiments with heuristic methods. J Heuristics 1(1):9–32
4. Cohen PR (1995) Empirical methods for artificial intelligence. MIT, Cambridge
5. DIMACS Implementation Challenges. Each DIMACS Implementation Challenge is a year-long cooperative research event in which researchers cooperate to find the most efficient algorithms and strategies for selected algorithmic problems. The DIMACS Challenges since 1991 have targeted a variety of optimization problems on graphs; advanced data structures; and scientific application areas involving computational biology and parallel computation. The DIMACS Challenge proceedings are published by AMS as part of the DIMACS Series in Discrete Mathematics and Theoretical Computer Science. Visit dimacs.rutgers.edu/Challenges for more information
6. Johnson DS (2002) A theoretician's guide to the experimental analysis of algorithms. In: Goodrich MH, Johnson DS, McGeoch CC (eds) Data structures, near neighbors searches, and methodology: fifth and sixth DIMACS implementation challenges, vol 59, DIMACS series in discrete mathematics and theoretical

computer science. American Mathematical Society, Providence
7. McGeoch CC (1996) Toward an experimental method for algorithm simulation. INFORMS J Comput 1(1):1–15
8. WEA. Beginning in 2001, the annual Workshop on Experimental and Efficient Algorithms is sponsored by EATCS. Workshop proceedings are published in the Springer LNCS series

Exponential Lower Bounds for *k*-SAT Algorithms

Dominik Scheder
Institute for Interdisciplinary Information Sciences, Tsinghua University, Beijing, China
Institute for Computer Science, Shanghai Jiaotong University, Shanghai, China

Keywords

Exponential algorithms; *k*-SAT; Lower bounds; PPSZ algorithm; Proof complexity

Years and Authors of Summarized Original Work

2013; Shiteng Chen, Dominik Scheder, Navid Talebanfard, Bangsheng Tang

Problem Definition

Given a propositional formula in conjunctive normal form, such as $(x \vee y) \wedge (\bar{x} \vee \bar{y} \vee z) \wedge (\bar{z})$, one wants to find an assignment of truth values to the variables that makes the formula evaluate to true. Here, $[x \mapsto 1, y \mapsto 0, z \mapsto 0]$ does the job. We call such formulas *CNF formulas* and such assignments *satisfying assignments*. SAT is the problem of deciding whether a given CNF formula is satisfiable. If every clause (such as $(\bar{x} \vee \bar{y} \vee z)$ above) has at most *k* literals, we call this a *k*-CNF formula. The above example is a 3-CNF formula. The problem of deciding whether a given *k*-CNF formula is satisfiable is called *k*-SAT. This is one of the most fundamental NP-complete problems.

Several clever algorithms have been developed for *k*-SAT. In this note we are mostly concerned with the PPSZ algorithm [3]. This is itself an improved version of the older PPZ algorithm [4]. Another prominent SAT algorithm is Schöning's random walk algorithm [6], which is slower than PPSZ, but has the benefit that it can be turned into a deterministic algorithm [5].

Given that we currently cannot prove P $\neq$ NP, all super-polynomial lower bounds on the running time of *k*-SAT algorithms must be either conditional, that is, rest on widely believed but yet unproven assumptions, or must be for a particular family of algorithms. In this note we sketch exponential lower bounds for the PPSZ algorithm, which is the currently fastest algorithm for *k*-SAT. We measure the running time of a SAT algorithm in terms of *n*, the number of variables. Often probabilistic algorithms for *k*-SAT (like PPSZ) have polynomial running time and success probability p^n for some $p < 1$. One can turn this into a Monte Carlo algorithm with success probability at least $1/2$ by repeating it $(1/p)^n$ times. We prefer the formulation of PPSZ as having polynomial running time, and we are interested in the worst-case success probability p^n.

Key Results

The worst-case behavior of PPSZ is exponential. That is, there are satisfiable *k*-CNF formulas on *n* variables, for which PPSZ finds a satisfying assignment with probability at most $2^{-\Omega(n)}$. More precisely, there is a constant C and a sequence $\epsilon_k \leq \frac{C \log^2 k}{k}$ such that the worst-case success probability of PPSZ for *k*-SAT is at most $2^{-(1-\epsilon_k)n}$. See Theorem 3 below for a formal statement.

The PPSZ Algorithm

The PPSZ algorithm, named after its inventors Paturi, Pudlák, Saks, and Zane [3], is the fastest known algorithm for k-SAT. We now give a brief description of it: Choose a random ordering σ of the n variables $x_1, \ldots, x_n$ of F. Choose random truth values $b = (b_1, \ldots, b_n) \in \{0, 1\}$. Iterate through the variables in the ordering given by σ. When processing x_i check whether it is "obvious" what the correct value of x_i should be. If so, fix x_i to that value. Otherwise, fix x_i to b_i. By fixing we mean replacing each occurrence of x_i in F by that value (and each occurrence of $\bar{x}_i$ by the negation of that value). After all variables have been processed, the algorithm returns the satisfying assignment it has found or returns failure if it has run into a contradiction.

It remains to specify what "obvious" means: Given a CNF formula F and a variable x_i, we say that the correct value of x_i is *obviously* b if the statement $x_i = b$ can be derived from F by width-w resolution, where w is some large constant (think of $w = 1,000$). This can be checked in time $O(n^w)$, which is polynomial.

Let $\text{ppsz}(F, \sigma, b)$ be the return value of ppsz. That is, $\text{ppsz}(F, \sigma, b) \in \text{sat}(F) \cup \{\text{failure}\}$, where $\text{sat}(F)$ is the set of satisfying assignments of F.

A Very Brief Sketch of the Analysis of PPSZ

Let σ be a permutation of $x_1, \ldots, x_n$ and let $b = (b_1, \ldots, b_n) \subset \{0, 1\}^n$. Suppose we run PPSZ on F using this permutation σ and the truth values b. For $1 \leq i \leq n$, define Z_i to be 1 if PPSZ did not find it obvious what the correct value of x_i should be. Let $Z = Z_1 + \cdots + Z_n$. To underline the dependence on F, σ, and b, we sometimes write $Z(F, \sigma, b)$. It is not difficult to show the following lemma.

Lemma 1 ([3]) *Let F be a satisfiable CNF formula over n variables. Let σ be a random permutation of its variables and let $a \in \{0, 1\}^n$ be satisfying assignment of F. Then*

$$\Pr_{\sigma, b}[\text{ppsz}(F, \sigma, b) = a] = \mathbb{E}_\sigma[2^{-Z(F, \sigma, a)}] \quad (1)$$

Since $x \mapsto 2^x$ is a convex function, Jensen's inequality implies that $\mathbb{E}_\sigma[2^{-Z}] \geq 2^{-\mathbb{E}[Z]}$, and by linearity of expectation, it holds that $\mathbb{E}[Z] = \sum_{i=1}^n \mathbb{E}[Z_i]$.

Lemma 2 ([3]) *There are numbers $c_k \in [0, 1]$ such that the following holds: If F is a k-CNF formula over n variables with a unique satisfying assignment a, then $\mathbb{E}_\sigma[Z_i(F, \sigma, a)] \leq c_k$ for all $1 \leq i \leq n$. Furthermore, for large k we have $c_k \approx 1 - \frac{\pi^2}{6k}$, and in particular $c_3 = 2\ln(2) - 1 \approx 0.38$.*

Combining everything, Paturi, Pudlák, Saks, and Zane obtain their main result:

Theorem 1 ([3]) *Let F be a k-CNF formula with a unique satisfying assignment a. Then PPSZ finds this satisfying assignment with probability at least $2^{-c_k n}$.*

It takes a considerable additional effort to show that the same bound holds also if F has multiple satisfying assignments:

Theorem 2 ([2]) *Let F be a satisfiable k-CNF formula. Then PPSZ finds a satisfying assignment with probability at least $2^{-c_k n}$.*

We sketch the intuition behind the proof of Lemma 2. It turns out that in the worst case the event $Z_i = 1$ can be described by the following random experiment: Let $T = (V, E)$ be the infinite rooted $(k - 1)$-ary tree. For each node $v \in V$ choose $\tau(v) \in [0, 1]$ randomly and independently. Call a node v *alive* if $\tau(v) \geq \tau(\text{root})$. Then $\Pr[Z_i = 1]$ is (roughly) equal to the probability that T contains an infinite path of alive vertices, starting with the root. Call this probability c_k. A simple calculation shows that $c_3 = 2\ln(2) - 1$. For larger values of c_k, there is not necessarily a closed form, but Paturi, Pudlák, Saks, and Zane show that $c_k \approx 1 - \frac{\pi^2}{6k}$ for large k.

Hard Instances for the PPSZ Algorithm

One can construct instances on which the success probability of PPSZ is exponentially small. The construction is probabilistic and rather simple. Its analysis is quite technical, so we can only sketch it here. We start with some easy estimates. By Lemma 1 we can write the success probability of PPSZ as

$$\Pr_{\sigma, b}[\text{ppsz}(F, \sigma, b) \in \text{sat}(F)]$$

$$= \sum_{a \in \text{sat}(F)} \mathbb{E}_\sigma[2^{-Z(F, \sigma, a)}]. \quad (2)$$

Above we used Jensen's inequality to prove $\mathbb{E}[2^{-Z}] \geq 2^{-\mathbb{E}[Z]}$. In this section we want to construct *hard instances*, that is, instances on which the success probability (2) is exponentially small. Thus, we cannot use Jensen's inequality, as it gives a lower bound, not an upper. Instead, we use the following trivial estimate:

$$\sum_{a \in \text{sat}(F)} \mathbb{E}_\sigma[2^{-Z(F, \sigma, a)}] \leq \sum_{a \in \text{sat}(F)} \max_\sigma 2^{-Z(F, \sigma, a)}$$

$$\leq |\text{sat}(F)| \cdot \max_{a \in \text{sat}(F), \sigma} 2^{-Z(F, \sigma, a)}. \quad (3)$$

We would like to construct a satisfiable k-CNF formula F for which (i) $|\text{sat}(F)|$ is small, i.e., F has few satisfying assignments, and (ii) $Z(F, \sigma, a)$ is large for every permutation and every satisfying assignment a. It turns out there are formulas satisfying both requirements:

Theorem 3 *There are numbers ϵ_k converging to 0 such that the following holds: For every k, there is a family $(F_n)_{n \geq 1}$, where each F_n is a satisfiable k-CNF formula over n variables such that*

1. $|\text{sat}(F_n)| \leq 2^{\epsilon_k n}$.
2. $Z(F, \sigma, a) \geq (1 - \epsilon_k)n$ *for all σ and all $a \in$* $\text{sat}(F_n)$.

Thus, the probability of PPSZ finding a satisfying assignment of F_n is at most $2^{-(1-2\epsilon_k)n}$. Furthermore, $\epsilon_k \leq \frac{C \log^2(k)}{k}$ for some universal constant C.

This theorem shows that PPSZ has exponentially small success probability. Also, it shows that the *strong exponential time hypothesis* (SETH) holds for PPSZ: As k grows, the advantage over the trivial success probability 2^{-n} becomes negligible.

The Probabilistic Construction

Let $A \in \mathbb{F}_2^{n \times n}$. The system $Ax = 0$ defines a Boolean function $f_A : \{0, 1\}^n \rightarrow \{0, 1\}$ as follows: $f_A(x) = 1$ if and only if $A \cdot x = 0$. Say A is k-*sparse* if every row of A has at most k nonzero entries. If A is k-sparse, then f_A can be written as a k-CNF formula with n variables and $2^{k-1} n$ clauses. Our construction will be probabilistic. For this, we define a distribution over k-sparse matrices in $\mathbb{F}_2^{n \times n}$. Our distribution will have the form $\mathcal{D}^n$, where $\mathcal{D}$ is a distribution over row vectors from $\mathbb{F}_2^n$. That is, we sample each row of A independently from $\mathcal{D}$. Let us describe $\mathcal{D}$. Define $\mathbf{e}_i \in \mathbb{F}_2^n$ to be the vector with a 1 at the ith position and 0 elsewhere. Sample $i_1, \ldots, i_k \in \{1, \ldots, n\}$ uniformly and independently and let $X = \mathbf{e}_{i_1} + \cdots + \mathbf{e}_{i_k}$. Clearly, $X \in \mathbb{F}_2^n$ has at most k nonzero entries. This is our distribution $\mathcal{D}$.

Let A be a random matrix sampled as described, and write f_A as a k-CNF formula F. Note that $\text{sat}(F) = \ker A$. The challenge is to show that F satisfies the two conditions of Theorem 3.

Lemma 3 (A has high rank) *With probability $1 - o(1)$, $|\ker(A)| \leq 2^{\epsilon_k n}$.*

This shows that F satisfies the first condition of the theorem, i.e., it has few satisfying assignments. Lemma 3 is quite straightforward to prove, though not trivial. The next lemma shows that $Z(F, \sigma, a)$ is large.

Lemma 4 *With probability $1 - o(1)$, it holds that $Z(F, \sigma, a) \geq (1 - \epsilon_k)n$ for all permutations σ and all $a \in \text{sat}(F)$.*

Proving this lemma is the main technical challenge. The proof uses ideas from proof complexity (indeed, the above construction is inspired by constructions in proof complexity).

Open Problems

Suppose the true worst-case success probability of PPSZ on k-CNF formulas is $2^{-r_k n}$. Paturi, Pudlák, Saks, and Zane have proved that $r_k \leq 1 - \Omega(1k)$. Chen, Scheder, Talebanfard, and

Tang showed that $r_k \geq 1 - O\left(\frac{\log^2 k}{k}\right)$. Can one close this gap by construction harder instances or maybe even improve the analysis of PPSZ?

What is the average-case success probability of PPSZ on F when we sample A from $\mathcal{D}^n$? Note that F is exponentially hard with probability $1 - o(1)$, but this might leave a $1/n$ probability that F is very easy for PPSZ.

The construction of [1] is probabilistic. Can one make it explicit? The proof of Lemma 4 uses (implicit in [1]) a nonstandard notion of expansion. We do not know of explicit construction of those expanders.

Cross-References

▶ Backtracking Based k-SAT Algorithms
▶ Derandomization of k-SAT Algorithm
▶ Exact Algorithms and Strong Exponential Time Hypothesis
▶ Exact Algorithms and Time/Space Tradeoffs
▶ Exact Algorithms for General CNF SAT
▶ Exact Algorithms for k-SAT Based on Local Search
▶ Exact Algorithms for Maximum Two-Satisfiability
▶ Exact Graph Coloring Using Inclusion-Exclusion
▶ Random Planted 3-SAT
▶ Thresholds of Random k-Sat
▶ Unique k-SAT and General k-SAT

Recommended Reading

1. Chen S, Scheder D, Tang B, Talebanfard N (2013) Exponential lower bounds for the PPSZ k-SAT algorithm. In Khanna S (ed) SODA, New Orleans. SIAM, pp 1253–1263
2. Hertli T (2011) 3-SAT faster and simpler – unique-SAT bounds for PPSZ hold in general. In Ostrovsky R (ed) FOCS, Palm Springs. IEEE, pp 277–284
3. Paturi R , Pudlák P, Saks ME, Zane F (2005) An improved exponential-time algorithm for k-SAT. J ACM 52(3):337–364
4. Paturi R, Pudlák P, Zane F (1999) Satisfiability coding lemma. Chicago J Theor Comput Sci 11(19). (electronic)
5. Robin RA, Scheder D (2011) A full derandomization of Schöning's k-SAT algorithm. In Fortnow L Salil P, Vadhan (eds) STOC, San Jose. ACM, pp 245–252
6. Schöning U (1999) A probabilistic algorithm for k-SAT and constraint satisfaction problems. In FOCS '99: Proceedings of the 40th annual symposium on foundations of computer science, New York. IEEE Computer Society, Washington, DC, p 410

External Sorting and Permuting

Jeffrey Scott Vitter
University of Kansas, Lawrence, KS, USA

Keywords

Disk; External memory; I/O; Out-of-core; Permuting; Secondary storage; Sorting

Synonyms

Out-of-core sorting

Years and Authors of Summarized Original Work

1988; Aggarwal, Vitter

Problem Definition

Notations The main properties of magnetic disks and multiple disk systems can be captured by the commonly used *parallel disk model* (PDM), which is summarized below in its current form as developed by Vitter and Shriver [22]:

N = problem size (in units of data items);

M = internal memory size(inunitsofdata items);

B = block transfer size (in units of data items);

D = number of independent disk drives;

P = number of CPUs,

where $M < N$, and $1 \leq DB \leq M/2$. The data items are assumed to be of fixed length. In a single I/O, each of the D disks can simultaneously transfer a block of B contiguous data items. (In the original 1988 article [2], the D blocks per I/O were allowed to come from the same disk, which is not realistic.) If $P \leq D$, each of the P processors can drive about D/P disks; if $D < P$, each disk is shared by about P/D processors. The internal memory size is M/P per processor, and the P processors are connected by an interconnection network.

It is convenient to refer to some of the above PDM parameters in units of disk blocks rather than in units of data items; the resulting formulas are often simplified. We define the lowercase notation

$$n = \frac{N}{B}, \qquad m = \frac{M}{B}, \qquad q = \frac{Q}{B}, \qquad z = \frac{Z}{B} \tag{1}$$

to be the problem input size, internal memory size, query specification size, and query output size, respectively, in units of disk blocks.

The primary measures of performance in PDM are:

1. The number of I/O operations performed
2. The amount of disk space used
3. The internal (sequential or parallel) computation time

For reasons of brevity in this survey, focus is restricted onto only the first two measures. Most of the algorithms run in $O(N \log N)$ CPU time with one processor, which is optimal in the comparison model, and in many cases are optimal for parallel CPUs. In the word-based RAM model, sorting can be done more quickly in $O(N \log \log N)$ CPU time. Arge and Thorup [5] provide sorting algorithms that are theoretically optimal in terms of both I/Os and time in the word-based RAM model. In terms of auxiliary storage in external memory, algorithms and data structures should ideally use linear space, which means

$O(N/B) = O(n)$ disk blocks of storage. Vitter [20] gives further details about the PDM model and provides optimal algorithms and data structures for a variety of problems. The content of this chapter comes largely from an abbreviated form of [19].

Problem 1 External sorting
INPUT: The input data records R_0, R_1, R_2, ... are initially "striped" across the D disks, in units of blocks, so that record R_i is in block $\lfloor i/B \rfloor$ and block j is stored on disk $j \bmod D$.
OUTPUT: A striped representation of a permuted ordering $R_{\sigma(0)}$, $R_{\sigma(1)}$, $R_{\sigma(2)}$, ... of the input records with the property that $key(R_{\sigma(i)}) \leq key(R_{\sigma(i+1)})$ for all $i \geq 0$.

Permuting is the special case of sorting in which the permutation that describes the final position of the records is given explicitly and does not have to be discovered, for example, by comparing keys.

Problem 2 Permuting
INPUT: Same input assumptions as in external sorting. In addition, a permutation σ of the integers $\{0, 1, 2, \ldots, N-1\}$ is specified.
OUTPUT: A striped representation of a permuted ordering $R_{\sigma(0)}$, $R_{\sigma(1)}$, $R_{\sigma(2)}$, ... of the input records.

Key Results

Theorem 1 ([2, 15]) *The average-case and worst-case number of I/Os required for sorting $N = nB$ data items using D disks is*

$$Sort(N) = \Theta\left(\frac{n}{D} \log_m n\right). \tag{2}$$

Theorem 2 ([2]) *The average-case and worst-case number of I/Os required for permuting N data items using D disks is*

$$\Theta\left(\min\left\{\frac{N}{D}, Sort(N)\right\}\right). \tag{3}$$

A more detailed lower bound is provided in (9) in section "Lower Bounds on I/O."

Matrix transposition is the special case of permuting in which the permutation can be represented as a transposition of a matrix from row-major order into column-major order.

Theorem 3 ([2]) *With D disks, the number of I/Os required to transpose a $p \times q$ matrix from row-major order to column-major order is*

$$\Theta\left(\frac{n}{D}\log_m \min\{M, p, q, n\}\right), \qquad (4)$$

where $N = pq$ and $n = N/B$.

Matrix transposition is a special case of a more general class of permutations called *bit-permute/complement* (BPC) permutations, which in turn is a subset of the class of *bit-matrix-multiply/complement* (BMMC) permutations. BMMC permutations are defined by a $\log N \times \log N$ nonsingular 0-1 matrix A and a $(\log N)$-length 0-1 vector c. An item with binary address x is mapped by the permutation to the binary address given by $Ax \oplus c$, where $\oplus$ denotes bitwise exclusive-or. BPC permutations are the special case of BMMC permutations in which A is a permutation matrix, that is, each row and each column of A contain a single 1. BPC permutations include matrix transposition, bit-reversal permutations (which arise in the FFT), vector-reversal permutations, hypercube permutations, and matrix re-blocking. Cormen et al. [8] characterize the optimal number of I/Os needed to perform any given BMMC permutation solely as a function of the associated matrix A, and they give an optimal algorithm for implementing it.

Theorem 4 ([8]) *With D disks, the number of I/Os required to perform the BMMC permutation defined by matrix A and vector c is*

$$\Theta\left(\frac{n}{D}\left(1 + \frac{\text{rank}(\gamma)}{\log m}\right)\right), \qquad (5)$$

where γ is the lower-left $\log n \times \log B$ submatrix of A.

The two main paradigms for external sorting are *distribution* and *merging*, which are discussed in the following sections for the PDM model.

Sorting by Distribution

Distribution sort [12] is a recursive process that uses a set of $S - 1$ partitioning elements to partition the items into S disjoint buckets. All the items in one bucket precede all the items in the next bucket. The sort is completed by recursively sorting the individual buckets and concatenating them together to form a single fully sorted list.

One requirement is to choose the $S - 1$ partitioning elements so that the buckets are of roughly equal size. When that is the case, the bucket sizes decrease from one level of recursion to the next by a relative factor of $\Theta(S)$, and thus there are $O(\log_S n)$ levels of recursion. During each level of recursion, the data are scanned. As the items stream through internal memory, they are partitioned into S buckets in an online manner. When a buffer of size B fills for one of the buckets, its block is written to the disks in the next I/O, and another buffer is used to store the next set of incoming items for the bucket. Therefore, the maximum number of buckets (and partitioning elements) is $S = \Theta(M/B) = \Theta(m)$, and the resulting number of levels of recursion is $\Theta(\log_m n)$. How to perform each level of recursion in a linear number I/Os is discussed in [2, 14, 22].

An even better way to do distribution sort, and deterministically at that, is the BalanceSort method developed by Nodine and Vitter [14]. During the partitioning process, the algorithm keeps track of how evenly each bucket has been distributed so far among the disks. It maintains an invariant that guarantees good distribution across the disks for each bucket.

The distribution sort methods mentioned above for parallel disks perform write operations in complete stripes, which make it easy to write parity information for use in error correction and recovery. But since the blocks written in each stripe typically belong to multiple buckets, the buckets themselves will not be striped on the disks, and thus the disks must be used independently during read operations. In the write phase, each bucket must therefore keep track of the last block written to each disk so that the blocks for the bucket can be linked together.

An orthogonal approach is to stripe the contents of each bucket across the disks so that read operations can be done in a striped manner. As a result, the write operations must use disks independently, since during each write, multiple buckets will be writing to multiple stripes. Error correction and recovery can still be handled efficiently by devoting to each bucket one block-sized buffer in internal memory. The buffer is continuously updated to contain the exclusive-or (parity) of the blocks written to the current stripe, and after $D - 1$ blocks have been written, the parity information in the buffer can be written to the final (Dth) block in the stripe.

Under this new scenario, the basic loop of the distribution sort algorithm is, as before, to read one memory load at a time and partition the items into S buckets. However, unlike before, the blocks for each individual bucket will reside on the disks in contiguous stripes. Each block therefore has a predefined place where it must be written. With the normal round-robin ordering for the stripes (namely, $\ldots, 1, 2, 3, \ldots, D, 1, 2, 3, \ldots, D, \ldots$), the blocks of different buckets may "collide," meaning that they need to be written to the same disk, and subsequent blocks in those same buckets will also tend to collide. Vitter and Hutchinson [21] solve this problem by the technique of *randomized cycling*. For each of the S buckets, they determine the ordering of the disks in the stripe for that bucket via a random permutation of $\{1, 2, \ldots, D\}$. The S random permutations are chosen independently. If two blocks (from different buckets) happen to collide during a write to the same disk, one block is written to the disk and the other is kept on a write queue. With high probability, subsequent blocks in those two buckets will be written to different disks and thus will not collide. As long as there is a small pool of available buffer space to temporarily cache the blocks in the write queues, Vitter and Hutchinson [21] show that with high probability the writing proceeds optimally.

The randomized cycling method or the related merge sort methods discussed at the end of section "Sorting by Merging" are the methods of choice for sorting with parallel disks. Distribution sort algorithms may have an advantage over the merge approaches presented in section "Sorting by Merging" in that they typically make better use of lower levels of cache in the memory hierarchy of real systems, based upon analysis of distribution sort and merge sort algorithms on models of hierarchical memory.

Sorting by Merging

The *merge* paradigm is somewhat orthogonal to the distribution paradigm of the previous section. A typical merge sort algorithm works as follows [12]: In the "run formation" phase, the n blocks of data are scanned, one memory load at a time; each memory load is sorted into a single "run," which is then output onto a series of stripes on the disks. At the end of the run formation phase, there are $N/M = n/m$ (sorted) runs, each striped across the disks. (In actual implementations, "replacement selection" can be used to get runs of $2M$ data items, on the average, when $M \gg B$ [12].) After the initial runs are formed, the merging phase begins. In each pass of the merging phase, R runs are merged at a time. For each merge, the R runs are scanned and its items merged in an online manner as they stream through internal memory. Double buffering is used to overlap I/O and computation. At most $R = \Theta(m)$ runs can be merged at a time, and the resulting number of passes is $O(\log_m n)$.

To achieve the optimal sorting bound (2), each merging pass must be done in $O(n/D)$ I/Os, which is easy to do for the single-disk case. In the more general multiple-disk case, each parallel read operation during the merging must on the average bring in the next $\Theta(D)$ blocks needed for the merging. The challenge is to ensure that those blocks reside on different disks so that they can be read in a single I/O (or a small constant number of I/Os). The difficulty lies in the fact that the runs being merged were themselves formed during the previous merge pass. Their blocks were written to the disks in the previous pass without knowledge of how they would interact with other runs in later merges.

The Greed Sort method of Nodine and Vitter [15] was the first optimal deterministic EM algorithm for sorting with multiple disks. It works by relaxing the merging process with a final pass to fix the merging. Aggarwal and Plaxton [1] developed an optimal deterministic merge sort based upon the Sharesort hypercube parallel sorting algorithm. To guarantee even distribution during the merging, it employs two high-level merging schemes in which the scheduling is almost oblivious. Like Greed Sort, the Sharesort algorithm is theoretically optimal (i.e., within a constant factor of optimal), but the constant factor is larger than the distribution sort methods.

One of the most practical methods for sorting is based upon the *simple randomized merge sort* (SRM) algorithm of Barve et al. [7], referred to as "randomized striping" by Knuth [12]. Each run is striped across the disks, but with a random starting point (the only place in the algorithm where randomness is utilized). During the merging process, the next block needed from each disk is read into memory, and if there is not enough room, the least needed blocks are "flushed" (without any I/Os required) to free up space.

Further improvements in merge sort are possible by a more careful prefetching schedule for the runs. Barve et al. [6], Kallahalla and Varman [11], Shah et al. [17], and others have developed competitive and optimal methods for prefetching blocks in parallel I/O systems.

Hutchinson et al. [10] have demonstrated a powerful duality between parallel writing and parallel prefetching, which gives an easy way to compute optimal prefetching and caching schedules for multiple disks. More significantly, they show that the same duality exists between distribution and merging, which they exploit to get a provably optimal and very practical parallel disk merge sort. Rather than use random starting points and round-robin stripes as in SRM, Hutchinson et al. [10] order the stripes for each run independently, based upon the randomized cycling strategy discussed in section "Sorting by Distribution" for distribution sort. These approaches have led to successfully faster external memory sorting algorithms [9]. Clever algorithm engineering optimizations on multicore architectures have won recent big data sorting competitions [16].

Handling Duplicates: Bundle Sorting

For the problem of *duplicate removal*, in which there are a total of K distinct items among the N items, Arge et al. [4] use a modification of merge sort to solve the problem in $O(n \max\{1, \log_m(K/B)\})$ I/Os, which is optimal in the comparison model. When duplicates get grouped together during a merge, they are replaced by a single copy of the item and a count of the occurrences. The algorithm can be used to sort the file, assuming that a group of equal items can be represented by a single item and a count.

A harder instance of sorting called *bundle sorting* arises when there are K distinct key values among the N items, but all the items have different secondary information that must be maintained, and therefore items cannot be aggregated with a count. Matias et al. [13] develop optimal distribution sort algorithms for bundle sorting using

$$O(n \max\{1, \log_m \min\{K, n\}\}) \qquad (6)$$

I/Os and prove the matching lower bound. They also show how to do bundle sorting (and sorting in general) *in place* (i.e., without extra disk space).

Permuting and Transposition

Permuting is the special case of sorting in which the key values of the N data items form a permutation of $\{1, 2, \ldots, N\}$. The I/O bound (3) for permuting can be realized by one of the optimal sorting algorithms except in the extreme case $B \log m = o(\log n)$, where it is faster to move the data items one by one in a nonblocked way. The one-by-one method is trivial if $D = 1$, but with multiple disks, there may be bottlenecks on individual disks; one solution for doing the permuting in $O(N/D)$ I/Os is to apply the randomized balancing strategies of [22].

Matrix transposition can be as hard as general permuting when B is relatively large (say, $\frac{1}{2}M$) and N is $O(M^2)$, but for smaller B, the special structure of the transposition permutation makes transposition easier. In particular, the matrix can be broken up into square submatrices of B^2 elements such that each submatrix contains B

blocks of the matrix in row-major order and also B blocks of the matrix in column-major order. Thus, if $B^2 < M$, the transpositions can be done in a simple one-pass operation by transposing the submatrices one at a time in internal memory. Thonangi and Yang [18] discuss other types of permutations realizable with fewer I/Os than sorting.

Fast Fourier Transform and Permutation Networks

Computing the fast Fourier transform (FFT) in external memory consists of a series of I/Os that permit each computation implied by the FFT directed graph (or butterfly) to be done while its arguments are in internal memory. A permutation network computation consists of an oblivious (fixed) pattern of I/Os such that any of the $N!$ possible permutations can be realized; data items can only be reordered when they are in internal memory. A permutation network can be realized by a series of three FFTs.

The algorithms for FFT are faster and simpler than for sorting because the computation is nonadaptive in nature, and thus the communication pattern is fixed in advance [22].

Lower Bounds on I/O

The following proof of the permutation lower bound (3) of Theorem 2 is due to Aggarwal and Vitter [2]. The idea of the proof is to calculate, for each $t \geq 0$, the number of distinct orderings that are realizable by sequences of t I/Os. The value of t for which the number of distinct orderings first exceeds $N!/2$ is a lower bound on the average number of I/Os (and hence the worst-case number of I/Os) needed for permuting.

Assuming for the moment that there is only one disk, $D = 1$, consider how the number of realizable orderings can change as a result of an I/O. In terms of increasing the number of realizable orderings, the effect of reading a disk

block is considerably more than that of writing a disk block, so it suffices to consider only the effect of read operations. During a read operation, there are at most B data items in the read block, and they can be interspersed among the M items in internal memory in at most $\binom{M}{B}$ ways, so the number of realizable orderings increases by a factor of $\binom{M}{B}$. If the block has never before resided in internal memory, the number of realizable orderings increases by an extra $B!$ factor, since the items in the block can be permuted among themselves. (This extra contribution of $B!$ can only happen once for each of the N/B original blocks.) There are at most $n + t \leq N \log N$ ways to choose which disk block is involved in the tth I/O (allowing an arbitrary amount of disk space). Hence, the number of distinct orderings that can be realized by all possible sequences of t I/Os is at most

$$(B!)^{N/B} \left(N(\log N) \binom{M}{B} \right)^t . \qquad (7)$$

Setting the expression in (7) to be at least $N!/2$, and simplifying by taking the logarithm, the result is

$$N \log B + t \left(\log N + B \log \frac{M}{B} \right) = \Omega(N \log N). \qquad (8)$$

Solving for t gives the matching lower bound $\Omega(n \log_m n)$ for permuting for the case $D = 1$. The general lower bound (3) of Theorem 2 follows by dividing by D.

Hutchinson et al. [10] derive an asymptotic lower bound (i.e., one that accounts for constant factors) from a more refined argument that analyzes both input operations and output operations. Assuming that $m = M/B$ is an increasing function, the number of I/Os required to sort or permute n indivisible items, up to lower-order terms, is at least

$$\frac{2N}{D} \frac{\log n}{B \log m + 2 \log N} \sim \begin{cases} \dfrac{2n}{D} \log_m n & \text{if } B \log m = \omega(\log N); \\[2mm] \dfrac{N}{D} & \text{if } B \log m = o(\log N). \end{cases} \qquad (9)$$

For the typical case in which $B \log m = \omega(\log N)$, the lower bound, up to lower order terms, is $2n \log_m n$ I/Os. For the pathological in which $B \log m = o(\log N)$, the I/O lower bound is asymptotically N/D.

Permuting is a special case of sorting, and hence the permuting lower bound applies also to sorting. In the unlikely case that $B \log m = o(\log n)$, the permuting bound is only $\Omega(N/D)$, and in that case the comparison model must be used to get the full lower bound (2) of Theorem 1 [2]. In the typical case in which $B \log m = \Omega(\log n)$, the comparison model is not needed to prove the sorting lower bound; the difficulty of sorting in that case arises not from determining the order of the data but from permuting (or routing) the data.

The proof used above for permuting also works for permutation networks, in which the communication pattern is oblivious (fixed). Since the choice of disk block is fixed for each t, there is no $N \log N$ term as there is in (7), and correspondingly there is no additive $\log N$ term in the inner expression as there is in (8). Hence, solving for t gives the lower bound (2) rather than (3). The lower bound follows directly from the counting argument; unlike the sorting derivation, it does not require the comparison model for the case $B \log m = o(\log n)$. The lower bound also applies directly to FFT, since permutation networks can be formed from three FFTs in sequence. The transposition lower bound involves a potential argument based upon a togetherness relation [2].

For the problem of bundle sorting, in which the N items have a total of K distinct key values (but the secondary information of each item is different), Matias et al. [13] derive the matching lower bound.

The lower bounds mentioned above assume that the data items are in some sense "indivisible," in that they are not split up and reassembled in some magic way to get the desired output. It is conjectured that the sorting lower bound (2) remains valid even if the indivisibility assumption is lifted. However, for an artificial problem related to transposition, removing the indivisibility assumption can lead to faster algorithms.

Whether the conjecture is true is a challenging theoretical open problem.

Applications

Sorting and sorting-like operations account for a significant percentage of computer use [12], with numerous database applications. In addition, sorting is an important paradigm in the design of efficient EM algorithms, as shown in [20], where several applications can be found. With some technical qualifications, many problems that can be solved easily in linear time in internal memory, such as permuting, list ranking, expression tree evaluation, and finding connected components in a sparse graph, require the same number of I/Os in PDM as does sorting.

Open Problems

Several interesting challenges remain. One difficult theoretical problem is to prove lower bounds for permuting and sorting without the indivisibility assumption. Another question is to determine the I/O cost for each individual permutation, as a function of some simple characterization of the permutation, such as number of inversions. A continuing goal is to develop optimal EM algorithms and to translate theoretical gains into observable improvements in practice.

Many interesting challenges and opportunities in algorithm design and analysis arise from new architectures being developed. For example, Arge et al. [3] propose the *parallel external memory (PEM)* model for the design of efficient algorithms for chip multiprocessors, in which each processor has a private cache and shares a larger main memory with the other processors. The paradigms described earlier form the basis for efficient algorithms for sorting, selection, and prefix sums. Further architectures to explore include other forms of multicore architectures, networks of workstations, hierarchical storage devices, disk drives with processing capabilities, and storage devices based upon microelectromechanical systems (MEMS). Active (or intelligent) disks, in which disk drives have some processing capability and can filter information sent to the host, have been proposed to further reduce the I/O bottleneck, especially in large database applications. MEMS-based nonvolatile storage has the potential to serve as an intermediate level in the memory hierarchy between DRAM and disks. It could ultimately provide better latency and bandwidth than disks, at less cost per bit than DRAM.

URL to Code

Two systems for developing external memory algorithms are TPIE and STXXL, which can be downloaded from http://www.madalgo. au.dk/tpie/ and http://stxxl.sourceforge.net/, respectively. Both systems include subroutines for sorting and permuting and facilitate development of more advanced algorithms.

Cross-References

▶ I/O-Model

Recommended Reading

1. Aggarwal A, Plaxton CG (1994) Optimal parallel sorting in multi-level storage. In: Proceedings of the 5th ACM-SIAM symposium on discrete algorithms, Arlington, vol 5, pp 659–668
2. Aggarwal A, Vitter JS (1988) The input/output complexity of sorting and related problems. Commun ACM 31:1116–1127
3. Arge L, Goodrich MT, Nelson M, Sitchinava N (2008) Fundamental parallel algorithms for private-cache chip multiprocessors. In: Proceedings of the 20th symposium on parallelism in algorithms and architectures, Munich, pp 197–206
4. Arge L, Knudsen M, Larsen K (1993) A general lower bound on the I/O-complexity of comparison-based algorithms. In: Proceedings of the workshop on algorithms and data structures, Montréal. Lecture notes in computer science, vol 709, pp 83–94
5. Arge L, Thorup M (2013) RAM-efficient external memory sorting. In: Proceedings of the 24th international symposium on algorithms and computa-

tion, Hong Kong. Lecture notes in computer science, vol 8283, pp 491–501
6. Barve RD, Kallahalla M, Varman PJ, Vitter JS (2000) Competitive analysis of buffer management algorithms. J Algorithms 36:152–181
7. Barve RD, Vitter JS (2002) A simple and efficient parallel disk mergesort. ACM Trans Comput Syst 35:189–215
8. Cormen TH, Sundquist T, Wisniewski LF (1999) Asymptotically tight bounds for performing BMMC permutations on parallel disk systems. SIAM J Comput 28:105–136
9. Dementiev R, Sanders P (2003) Asynchronous parallel disk sorting. In: Proceedings of the 15th ACM symposium on parallelism in algorithms and architectures, San Diego, pp 138–148
10. Hutchinson DA, Sanders P, Vitter JS (2005) Duality between prefetching and queued writing with parallel disks. SIAM J Comput 34:1443–1463
11. Kallahalla M, Varman PJ (2005) Optimal read-once parallel disk scheduling. Algorithmica 43:309–343
12. Knuth DE (1998) Sorting and searching. The art of computer programming, vol 3, 2nd edn. Addison-Wesley, Reading
13. Matias Y, Segal E, Vitter JS (2006) Efficient bundle sorting. SIAM J Comput 36(2):394–410
14. Nodine MH, Vitter JS (1993) Deterministic distribution sort in shared and distributed memory multiprocessors. In: Proceedings of the 5th ACM symposium on parallel algorithms and architectures, Velen, vol 5. ACM, pp 120–129
15. Nodine MH, Vitter JS (1995) Greed sort: an optimal sorting algorithm for multiple disks. J ACM 42:919–933
16. Rahn M, Sanders P, Singler J (2010) Scalable distributed-memory external sorting. In: Proceedings of the 26th IEEE international conference on data engineering, Long Beach, pp 685–688
17. Shah R, Varman PJ, Vitter JS (2004) Online algorithms for prefetching and caching on parallel disks. In: Proceedings of the 16th ACM symposium on parallel algorithms and architectures, Barcelona, pp 255–264
18. Thonangi R, Yang J (2013) Permuting data on random-access block storage. Proc VLDB Endow 6(9):721–732
19. Vitter JS (2001) External memory algorithms and data structures: dealing with massive data. ACM Comput Surv 33(2):209–271
20. Vitter JS (2008) Algorithms and data structures for external memory. Series on foundations and trends in theoretical computer science. Now Publishers, Hanover. (Also referenced as Volume 2, Issue 4 of Foundations and trends in theoretical computer science, Now Publishers)
21. Vitter JS, Hutchinson DA (2006) Distribution sort with randomized cycling. J ACM 53:656–680
22. Vitter JS, Shriver EAM (1994) Algorithms for parallel memory I: two-level memories. Algorithmica 12:110–147

F

Facility Location

Karen Aardal[1,2], Jaroslaw Byrka[1,2], and
Mohammad Mahdian[3]
[1]Centrum Wiskunde & Informatica (CWI),
Amsterdam, The Netherlands
[2]Department of Mathematics and Computer
Science, Eindhoven University of Technology,
Eindhoven, The Netherlands
[3]Yahoo! Research, Santa Clara, CA, USA

Keywords

Plant location; Warehouse location

Years and Authors of Summarized Original Work

1997; Shmoys, Tardos, Aardal

Problem Definition

Facility location problems concern situations
where a planner needs to determine the location
of facilities intended to serve a given set of
clients. The objective is usually to minimize
the sum of the cost of opening the facilities and
the cost of serving the clients by the facilities,
subject to various constraints, such as the number
and the type of clients a facility can serve. There
are many variants of the facility location problem,
depending on the structure of the cost function
and the constraints imposed on the solution.
Early references on facility location problems
include Kuehn and Hamburger [35], Balinski and
Wolfe [8], Manne [40], and Balinski [7]. Review
works include Krarup and Pruzan [34] and
Mirchandani and Francis [42]. It is interesting to
notice that the algorithm that is probably one of
the most effective ones to solve the uncapacitated
facility location problem to optimality is the
primal-dual algorithm combined with branch-
and-bound due to Erlenkotter [16] dating back
to 1978. His primal-dual scheme is similar to
techniques used in the modern literature on
approximation algorithms.

More recently, extensive research into approximation algorithms for facility location problems
has been carried out. Review articles on this
topic include Shmoys [49, 50] and Vygen [55].
Besides its theoretical and practical importance,
facility location problems provide a showcase of
common techniques in the field of approximation
algorithms, as many of these techniques such as
linear programming rounding, primal-dual methods, and local search have been applied successfully to this family of problems. This entry
defines several facility location problems, gives
a few historical pointers, and lists approximation algorithms with an emphasis on the results
derived in the paper by Shmoys, Tardos, and
Aardal [51]. The techniques applied to the *uncapacitated facility location* (UFL) problem are
discussed in some more detail.

In the UFL problem, a set $\mathcal{F}$ of n_f facilities and a set C of n_c clients (also known as cities, or demand points) are given. For every facility $i \in \mathcal{F}$, the facility opening cost is equal to f_i. Furthermore, for every facility $i \in \mathcal{F}$ and client $j \in C$, there is a connection cost c_{ij}. The objective is to open a subset of the facilities and connect each client to an open facility so that the total cost is minimized. Notice that once the set of open facilities is specified, it is optimal to connect each client to the open facility that yields smallest connection cost. Therefore, the objective is to find a set $S \subseteq \mathcal{F}$ that minimizes $\sum_{i \in S} f_i + \sum_{j \in C} \min_{i \in S}\{c_{ij}\}$. This definition and the definitions of other variants of the facility location problem in this entry assume unit demand at each client. It is straightforward to generalize these definitions to the case where each client has a given demand. The UFL problem can be formulated as the following integer program due to Balinski [7]. Let y_i, $i \in \mathcal{F}$ be equal to 1 if facility i is open, and equal to 0 otherwise. Let x_{ij}, $i \in \mathcal{F}$, $j \in C$ be the fraction of client j assigned to facility i.

$$\min \sum_{i \in \mathcal{F}} f_i y_i + \sum_{i \in \mathcal{F}} \sum_{j \in C} c_{ij} x_{ij} \qquad (1)$$

$$\text{subject to} \sum_{i \in \mathcal{F}} x_{ij} = 1, \quad \text{for all } j \in C, \qquad (2)$$

$$x_{ij} - y_i \leq 0, \quad \text{for all } i \in \mathcal{F}, \; j \in C \qquad (3)$$

$$x \geq 0, \; y \in \{0, 1\}^{n_f} \qquad (4)$$

In the linear programming (LP) relaxation of UFL the constraint $y \in \{0, 1\}^{n_f}$ is substituted by the constraint $y \in [0, 1]^{n_f}$. Notice that in the uncapacitated case, it is not necessary to require $x_{ij} \in \{0, 1\}$, $i \in \mathcal{F}$, $j \in C$ if each client has to be serviced by precisely one facility, as $0 \leq x_{ij} \leq 1$ by constraints (2) and (4). Moreover, if x_{ij} is not integer, then it is always possible to create an integer solution with the same cost by assigning client j completely to one of the facilities currently servicing j.

A γ-approximation algorithm is a polynomial algorithm that, in case of minimization, is guar-

anteed to produce a feasible solution having value at most γz^*, where z^* is the value of an optimal solution, and $\gamma \geq 1$. If $\gamma = 1$ the algorithm produces an optimal solution. In case of maximization, the algorithm produces a solution having value at least γz^*, where $0 \leq \gamma \leq 1$.

Hochbaum [25] developed an $O(\log n)$-approximation algorithm for UFL. By a straightforward reduction from the Set Cover problem, it can be shown that this cannot be improved unless $NP \subseteq DTIME[n^{O(\log \log n)}]$ due to a result by Feige [17]. However, if the connection costs are restricted to come from distances in a metric space, namely $c_{ij} = c_{ji} \geq 0$ for all $i \in \mathcal{F}, j \in C$ (nonnegativity and symmetry) and $c_{ij} + c_{ji'} + c_{i'j'} \geq c_{ij'}$ for all $i, i' \in \mathcal{F}, j, j' \in C$ (triangle inequality), then constant approximation guarantees can be obtained. In all results mentioned below, except for the maximization objectives, it is assumed that the costs satisfy these restrictions. If the distances between facilities and clients are Euclidean, then for some location problems approximation schemes have been obtained [5].

Variants and Related Problems

A variant of the uncapacitated facility location problem is obtained by considering the objective coefficients c_{ij} as the per unit profit of servicing client j from facility i. The maximization version of UFL, max-UFL is obtained by maximizing the profit minus the facility opening cost, i.e., $\max \sum_{i \in \mathcal{F}} \sum_{j \in C} c_{ij} x_{ij} - \sum_{i \in \mathcal{F}} f_i y_i$. This variant was introduced by Cornuéjols, Fisher, and Nemhauser [15].

In the k-median problem the facility opening cost is removed from the objective function (1) to obtain $\min \sum_{i \in M} \sum_{j \in N} c_{ij} x_{ij}$, and the constraint that no more than k facilities may be opened, $\sum_{i \in M} y_i \leq k$, is added. In the k-center problem the constraint $\sum_{i \in M} y_i \leq k$ is again included, and the objective function here is to minimize the maximum distance used on a link between an open facility and a client.

In the capacitated facility location problem a capacity constraint $\sum_{j \in C} x_{ij} \leq u_i y_i$ is added for all $i \in \mathcal{F}$. Here it is important to distinguish between the splittable and the unsplittable case, and also between *hard capacities* and *soft capacities*. In the splittable case one has $x \geq 0$, allowing for a client to be serviced by multiple depots, and in the unsplittable case one requires $x \in \{0, 1\}^{n_f \times n_c}$. If each facility can be opened at most once (i.e., $y_i \in \{0, 1\}$), the capacities are called hard; otherwise, if the problem allows a facility i to be opened any number r of times to serve ru_i clients, the capacities are called soft.

In the k-level facility location problem, the following are given: a set C of clients, k disjoint sets $\mathcal{F}_1, \ldots, \mathcal{F}_k$ of facilities, an opening cost for each facility, and connection costs between clients and facilities. The goal is to connect each client j through a path $i_1, \ldots, i_k$ of open facilities, with $i_\ell \in \mathcal{F}_\ell$. The connection cost for this client is $c_{ji_1} + c_{i_1 i_2} + \cdots + c_{i_{k-1} i_k}$. The goal is to minimize the sum of connection costs and facility opening costs.

The problems mentioned above have all been considered by Shmoys, Tardos, and Aardal [51], with the exceptions of max-UFL, and the k-center and k-median problems. The max-UFL variant is included for historical reasons, and k-center and k-median are included since they have a rich history and since they are closely related to UFL. Results on the capacitated facility location problem with hard capacities are mentioned as this, at least from the application point of view, is a more realistic model than the soft capacity version, which was treated in [51]. For k-level facility location, Shmoys et al. considered the case $k = 2$. Here, the problem for general k is considered.

There are many other variants of the facility location problem that are not discussed here. Examples include K-facility location [33], universal facility location [24, 38], online facility location [3, 18, 41], fault tolerant facility location [28, 30, 54], facility location with outliers [12, 28], multicommodity facility location [48], priority facility location [37, 48], facility location with hierarchical facility costs [52], stochastic facility location [23, 37, 46], connected facility location [53], load-balanced facility location [22, 32, 37], concave-cost facility location [24], and capacitated-cable facility location [37, 47].

Key Results

Many algorithms have been proposed for location problems. To begin with, a brief description of the algorithms of Shmoys, Tardos, and Aardal [51] is given. Then, a quick overview of some key results is presented. Some of the algorithms giving the best values of the approximation guarantee γ are based on solving the LP-relaxation by a polynomial algorithm, which can actually be quite time consuming, whereas some authors have suggested fast combinatorial algorithms for facility location problems with less competitive γ-values. Due to space restrictions the focus of this entry is on the algorithms that yield the best approximation guarantees. For more references the survey papers by Shmoys [49, 50] and by Vygen [55] are recommended.

The Algorithms of Shmoys, Tardos, and Aardal

First the algorithm for UFL is described, and then the results that can be obtained by adaptations of the algorithm to other problems are mentioned.

The algorithm solves the LP relaxation and then, in two stages, modifies the obtained fractional solution. The first stage is called *filtering* and it is designed to bound the connection cost of each client to the most distant facility fractionally serving him. To do so, the facility opening variables y_i are scaled up by a constant and then the connection variables x_{ij} are adjusted to use the closest possible facilities.

To describe the second stage, the notion of *clustering*, formalized later by Chudak and Shmoys [13] is used. Based on the fractional solution, the instance is cut into pieces called *clusters*. Each cluster has a distinct client called the *cluster center*. This is done by iteratively choosing a client, not covered by the previous clusters, as the next cluster center, and adding to this cluster the facilities that serve the cluster center in the fractional solution, along with other clients served by these facilities. This construction of clusters guarantees that the facilities in each cluster are open to a total extent of one, and therefore after opening the facility with the smallest opening cost in each cluster, the total facility opening cost that is paid does not exceed the facility opening cost of the fractional solution. Moreover, by choosing clients for the cluster centers in a greedy fashion, the algorithm makes each cluster center the minimizer of a certain cost function among the clients in the cluster. The remaining clients in the cluster are also connected to the opened facility. The triangle inequality for connection costs is now used to bound the cost of this connection. For UFL, this filtering and rounding algorithm is a 4-approximation algorithm. Shmoys et al. also show that if the filtering step is substituted by *randomized filtering*, an approximation guarantee of 3.16 is obtained.

In the same paper, adaptations of the algorithm, with and without randomized filtering, was made to yield approximation algorithms for the soft-capacitated facility location problem, and for the 2-level uncapacitated problem. Here, the results obtained using randomized filtering are discussed.

For the problem with soft capacities two versions of the problem were considered. Both have equal capacities, i.e., $u_i = u$ for all $i \in \mathcal{F}$. In the first version, a solution is "feasible" if the y-variables either take value 0, or a value between 1 and $\gamma' \geq 1$. Note that γ' is not required to be integer, so the constructed solution is not necessarily integer. This can be interpreted as allowing for each facility i to expand to have capacity $\gamma' u$ at a cost of $\gamma' f_i$. A (γ, γ')-approximation algorithm is a polynomial algorithm that produces such a feasible solution having a total cost within a factor of γ of the true optimal cost, i.e., with $y \in \{0, 1\}^{n_f}$. Shmoys et al. developed a $(5.69, 4.24)$-approximation algorithm for the splittable case of this problem, and a $(7.62, 4.29)$-approximation algorithm for the unsplittable case.

In the second soft-capacitated model, the original problem is changed to allow for the y-variables to take nonnegative integer values, which can be interpreted as allowing multiple facilities of capacity u to be opened at each location. The approximation algorithms in this case produces a solution that is feasible with respect to this modified model. It is easy to show that the approximation guarantees

obtained for the previous model also hold in this case, i.e., Shmoys et al. obtained a 5.69-approximation algorithm for splittable demands and a 7.62-approximation algorithm for unsplittable demands. This latter model is the one considered in most later papers, so this is the model that is referred to in the paragraph on soft capacity results below.

UFL

The first algorithm with constant performance guarantee was the 3.16-approximation algorithm by Shmoys, Tardos, and Aardal, see above. Since then numerous improvements have been made. Guha and Khuller [19, 20] proved a lower bound on approximability of 1.463, and introduced a *greedy augmentation procedure*. A series of approximation algorithms based on LP-rounding was then developed (see e.g., [10, 13]). There are also greedy algorithms that only use the LP-relaxation implicitly to obtain a lower bound for a primal-dual analysis. An example is the JMS 1.61-approximation algorithm developed by Jain, Mahdian, and Saberi [29]. Some algorithms combine several techniques, like the 1.52-approximation algorithm of Mahdian, Ye, and Zhang [39], which uses the JMS algorithm and the greedy augmentation procedure. Currently, the best known approximation guarantee is 1.5 reported by Byrka [10]. It is obtained by combining a randomized LP-rounding algorithm with the greedy JMS algorithm.

max-UFL

The first constant factor approximation algorithm was derived in 1977 by Cornuéjols et al. [15] for max-UFL. They showed that opening one facility at a time in a greedy fashion, choosing the facility to open as the one with highest marginal profit, until no facility with positive marginal profit can be found, yields a $(1 - 1/e) \approx 0.632$-approximation algorithm. The current best approximation factor is 0.828 by Ageev and Sviridenko [2].

k-Median, *k*-Center

The first constant factor approximation algorithm for the k-median problem is due to Charikar,

Guha, Tardos, and Shmoys [11]. This LP-rounding algorithm has the approximation ratio of $6\frac{2}{3}$. The currently best known approximation ratio is $3 + \epsilon$ achieved by a local search heuristic of Arya, et al. [6] (see also a separate entry *k-median and Facility Location*).

The first constant factor approximation algorithm for the k-center problem was given by Hochbaum and Shmoys [26], who developed a 2-approximation algorithm. This performance guarantee is the best possible unless $P = NP$.

Capacitated Facility Location

For the soft-capacitated problem with equal capacities, the first constant factor approximation algorithms are due to Shmoys et al. [51] for both the splittable and unsplittable demand cases, see above. Recently, a 2-approximation algorithm for the soft capacitated facility location problem with unsplittable unit demands was proposed by Mahdian et al. [39]. The integrality gap of the LP relaxation for the problem is also 2. Hence, to improve the approximation guarantee one would have to develop a better lower bound on the optimal solution.

In the hard capacities version it is important to allow for splitting the demands, as otherwise even the feasibility problem becomes difficult. Suppose demands are splittable, then we may to distinguish between the equal capacity case, where $u_i = u$ for all $i \in \mathcal{F}$, and the general case. For the problem with equal capacities, a 5.83-approximation algorithm was given by Chudak and Wiliamson [14]. The first constant factor approximation algorithm, with $\gamma = 8.53 + \epsilon$, for general capacities was given by Pál, Tardos, and Wexler [44]. This was later improved by Zhang, Chen, and Ye [57] who obtained a 5.83-approximation algorithm also for general capacities.

k-Level Problem

The first constant factor approximation algorithm for $k = 2$ is due to Shmoys et al. [51], with $\gamma = 3.16$. For general k, the first algorithm, having $\gamma = 3$, was proposed by Aardal, Chudak, and Shmoys [1]. For $k = 2$, Zhang [56] developed a 1.77-approximation algorithm. He also showed

that the problem for $k = 3$ and $k = 4$ can be approximated by $\gamma = 2.523$ (This value of γ deviates slightly from the value 2.51 given in the paper. The original argument contained a minor calculation error.) and $\gamma = 2.81$ respectively.

Applications

Facility location has numerous applications in the field of operations research. See the book edited by Mirchandani and Francis [42] or the book by Nemhauser and Wolsey [43] for a survey and a description of applications of facility location in problems such as plant location and locating bank accounts. Recently, the problem has found new applications in network design problems such as placement of routers and caches [22, 36], agglomeration of traffic or data [4, 21], and web server replications in a content distribution network [31, 45].

Open Problems

A major open question is to determine the exact approximability threshold of UFL and close the gap between the upper bound of 1.5 [10] and the lower bound of 1.463 [20]. Another important question is to find better approximation algorithms for k-median. In particular, it would be interesting to find an LP-based 2-approximation algorithm for k-median. Such an algorithm would determine the integrality gap of the natural LP relaxation of this problem, as there are simple examples that show that this gap is at least 2.

Experimental Results

Jain et al. [28] published experimental results comparing various primal-dual algorithms. A more comprehensive experimental study of several primal-dual, local search, and heuristic algorithms is performed by Hoefer [27]. A collection of data sets for UFL and several other location problems can be found in the OR-library maintained by Beasley [9].

Cross-References

▶ Assignment Problem
▶ Bin Packing (hardness of Capacitated Facility Location with unsplittable demands)
▶ Circuit Placement
▶ Greedy Set-Cover Algorithms (hardness of a variant of UFL, where facilities may be built at all locations with the same cost)
▶ Local Approximation of Covering and Packing Problems
▶ Local Search for *K*-medians and Facility Location

Recommended Reading

1. Aardal K, Chudak FA, Shmoys DB (1999) A 3-approximation algorithm for the *k*-level uncapacitated facility location problem. Inf Process Lett 72:161–167
2. Ageev AA, Sviridenko MI (1999) An 0.828-approximation algorithm for the uncapacitated facility location problem. Discret Appl Math 93:149–156
3. Anagnostopoulos A, Bent R, Upfal E, van Hentenryck P (2004) A simple and deterministic competitive algorithm for online facility location. Inf Comput 194(2):175–202
4. Andrews M, Zhang L (1998) The access network design problem. In: Proceedings of the 39th annual IEEE symposium on foundations of computer science (FOCS). IEEE Computer Society, Los Alamitos, pp 40–49
5. Arora S, Raghavan P, Rao S (1998) Approximation schemes for Euclidean *k*-medians and related problems. In: Proceedings of the 30th annual ACM symposium on theory of computing (STOC). ACM, New York, pp 106–113
6. Arya V, Garg N, Khandekar R, Meyerson A, Munagala K, Pandit V (2001) Local search heuristics for k-median and facility location problems. In: Proceedings of the 33rd annual ACM symposium on theory of computing (STOC). ACM, New York, pp 21–29
7. Balinski ML (1966) On finding integer solutions to linear programs. In: Proceedings of the IBM scientific computing symposium on combinatorial problems. IBM, White Plains, pp 225–248
8. Balinski ML, Wolfe P (1963) On Benders decomposition and a plant location problem. In: ARO-27. Mathematica Inc., Princeton
9. Beasley JE (2008) Operations research library. http://people.brunel.ac.uk/~mastjjb/jeb/info.html. Accessed 2008
10. Byrka J (2007) An optimal bifactor approximation algorithm for the metric uncapacitated facility location problem. In: Proceedings of the 10th international workshop on approximation algorithms for combinatorial optimization problems (APPROX). Lecture notes in computer science, vol 4627. Springer, Berlin, pp 29–43
11. Charikar M, Guha S, Tardos E, Shmoys DB (1999) A constant factor approximation algorithm for the k-median problem. In: Proceedings of the 31st annual ACM symposium on theory of computing (STOC). ACM, New York, pp 1–10
12. Charikar M, Khuller S, Mount D, Narasimhan G (2001) Facility location with outliers. In: Proceedings of the 12th annual ACM-SIAM symposium on discrete algorithms (SODA). SIAM, Philadelphia, pp 642–651
13. Chudak FA, Shmoys DB (2003) Improved approximation algorithms for the uncapacitated facility location problem. SIAM J Comput 33(1):1–25
14. Chudak FA, Wiliamson DP (1999) Improved approximation algorithms for capacitated facility location problems. In: Proceedings of the 7th conference on integer programing and combinatorial optimization (IPCO). Lecture notes in computer science, vol 1610. Springer, Berlin, pp 99–113
15. Cornuéjols G, Fisher ML, Nemhauser GL (1977) Location of bank accounts to optimize float: an analytic study of exact and approximate algorithms. Manag Sci 8:789–810
16. Erlenkotter D (1978) A dual-based procedure for uncapacitated facility location problems. Oper Res 26:992–1009
17. Feige U (1998) A threshold of ln n for approximating set cover. J ACM 45:634–652
18. Fotakis D (2003) On the competitive ratio for online facility location. In: Proceedings of the 30th international colloquium on automata, languages and programming (ICALP). Lecture notes in computer science, vol 2719. Springer, Berlin, pp 637–652
19. Guha S, Khuller S (1998) Greedy strikes back: improved facility location algorithms. In: Proceedings of the 9th ACM-SIAM symposium on discrete algorithms (SODA). SIAM, Philadelphia, pp 228–248
20. Guha S, Khuller S (1999) Greedy strikes back: improved facility location algorithms. J Algorithm 31:228–248
21. Guha S, Meyerson A, Munagala K (2001) A constant factor approximation for the single sink edge installation problem. In: Proceedings of the 33rd annual ACM symposium on theory of computing (STOC). ACM, New York, pp 383–388
22. Guha S, Meyerson A, Munagala K (2000) Hierarchical placement and network design problems. In: Proceedings of the 41st annual IEEE symposium on foundations of computer science (FOCS). IEEE Computer Society, Los Alamitos, pp 603–612
23. Gupta A, Pál M, Ravi R, Sinha A (2004) Boosted sampling: approximation algorithms for stochastic optimization. In: Proceedings of the 36st annual ACM symposium on theory of computing (STOC). ACM, New York, pp 417–426

24. Hajiaghayi M, Mahdian M, Mirrokni VS (2003) The facility location problem with general cost functions. Networks 42(1):42–47

25. Hochbaum DS (1982) Heuristics for the fixed cost median problem. Math Program 22(2):148–162

26. Hochbaum DS, Shmoys DB (1985) A best possible approximation algorithm for the k-center problem. Math Oper Res 10:180–184

27. Hoefer M (2003) Experimental comparison of heuristic and approximation algorithms for uncapacitated facility location. In: Proceedings of the 2nd international workshop on experimental and efficient algorithms (WEA). Lecture notes in computer science, vol 2647. Springer, Berlin, pp 165–178

28. Jain K, Mahdian M, Markakis E, Saberi A, Vazirani VV (2003) Approximation algorithms for facility location via dual fitting with factor-revealing LP. J ACM 50(6):795–824

29. Jain K, Mahdian M, Saberi A (2002) A new greedy approach for facility location problems. In: Proceedings of the 34st annual ACM symposium on theory of computing (STOC). ACM, New York, pp 731–740

30. Jain K, Vazirani VV (2000) An approximation algorithm for the fault tolerant metric facility location problem. In: Approximation algorithms for combinatorial optimization, proceedings of APPROX. Lecture notes in computer science, vol 1913. Springer, Berlin, pp 177–183

31. Jamin S, Jin C, Jin Y, Raz D, Shavitt Y, Zhang L (2000) On the placement of internet instrumentations. In: Proceedings of the 19th annual joint conference of the IEEE computer and communications Societies (INFOCOM), vol 1. IEEE Computer Society, Los Alamitos, pp 295–304

32. Karger D, Minkoff M (2000) Building Steiner trees with incomplete global knowledge. In: Proceedings of the 41st annual IEEE symposium on foundations of computer science (FOCS). IEEE Computer Society, Los Alamitos, pp 613–623

33. Krarup J, Pruzan PM (1990) Ingredients of locational analysis. In: Mirchandani P, Francis R (eds) Discrete location theory. Wiley, New York, pp 1–54

34. Krarup J, Pruzan PM (1983) The simple plant location problem: survey and synthesis. Eur J Oper Res 12:38–81

35. Kuehn AA, Hamburger MJ (1963) A heuristic program for locating warehouses. Manag Sci 9:643–666

36. Li B, Golin M, Italiano G, Deng X, Sohraby K (1999) On the optimal placement of web proxies in the internet. In: Proceedings of the 18th annual joint conference of the IEEE computer and communications societies (INFOCOM). IEEE Computer Society, Los Alamitos, pp 1282–1290

37. Mahdian M (2004) Facility location and the analysis of algorithms through factor-revealing programs. PhD thesis, MIT, Cambridge

38. Mahdian M, Pál M (2003) Universal facility location. In: Proceedings of the 11th annual european symposium on algorithms (ESA). Lecture notes in computer science, vol 2832. Springer, Berlin, pp 409–421

39. Mahdian M, Ye Y, Zhang J (2006) Approximation algorithms for metric facility location problems. SIAM J Comput 36(2):411–432

40. Manne AS (1964) Plant location under economies-of-scale – decentralization and computation. Manag Sci 11:213–235

41. Meyerson A (2001) Online facility location. In: Proceedings of the 42nd annual IEEE symposium on foundations of computer science (FOCS). IEEE Computer Society, Los Alamitos, pp 426–431

42. Mirchandani PB, Francis RL (1990) Discrete location theory. Wiley, New York

43. Nemhauser GL, Wolsey LA (1990) Integer and combinatorial optimization. Wiley, New York

44. Pál M, Tardos E, Wexler T (2001) Facility location with nonuniform hard capacities. In: Proceedings of the 42nd annual IEEE symposium on foundations of computer science (FOCS). IEEE Computer Society, Los Alamitos, pp 329–338

45. Qiu L, Padmanabhan VN, Voelker G (2001) On the placement of web server replicas. In: Proceedings of the 20th annual joint conference of the IEEE computer and communications societies (INFOCOM). IEEE Computer Society, Los Alamitos, pp 1587–1596

46. Ravi R, Sinha A (2006) Hedging uncertainty: approximation algorithms for stochastic optimization problems. Math Program 108(1):97–114

47. Ravi R, Sinha A (2002) Integrated logistics: approximation algorithms combining facility location and network design. In: Proceedings of the 9th conference on integer programming and combinatorial optimization (IPCO). Lecture notes in computer science, vol 2337. Springer, Berlin, pp 212–229

48. Ravi R, Sinha A (2004) Multicommodity facility location. In: Proceedings of the 15th annual ACM-SIAM symposium on discrete algorithms (SODA). SIAM, Philadelphia, pp 342–349

49. Shmoys DB (2000) Approximation algorithms for facility location problems. In: Jansen K, Khuller S (eds) Approximation algorithms for combinatorial optimization, vol 1913, Lecture notes in computer science. Springer, Berlin, pp 27–33

50. Shmoys DB (2004) The design and analysis of approximation algorithms: facility location as a case study. In: Thomas RR, Hosten S, Lee J (eds) Proceedings of symposia in applied mathematics, vol 61. AMS, Providence, pp 85–97

51. Shmoys DB, Tardos E, Aardal K (1997) Approximation algorithms for facility location problems. In: Proceedings of the 29th annual ACM symposium on theory of computing (STOC). ACM, New York, pp 265–274

52. Svitkina Z, Tardos E (2006) Facility location with hierarchical facility costs. In: Proceedings of the 17th annual ACM-SIAM symposium on discrete algorithm (SODA). SIAM, Philadelphia, pp 153–161

53. Swamy C, Kumar A (2004) Primal-dual algorithms for connected facility location problems. Algorithmica 40(4):245–269

54. Swamy C, Shmoys DB (2003) Fault-tolerant facility location. In: Proceedings of the 14th annual ACM-SIAM symposium on discrete algorithms (SODA). SIAM, Philadelphia, pp 735–736

55. Vygen J (2005) Approximation algorithms for facility location problems (lecture notes). Technical report No. 05950-OR, Research Institute for Discrete Mathematics, University of Bonn. http://www.or.uni-bonn.de/~vygen/fl.pdf

56. Zhang J (2004) Approximating the two-level facility location problem via a quasi-greedy approach. In: Proceedings of the 15th annual ACM-SIAM symposium on discrete algorithms (SODA). SIAM, Philadelphia, pp 808–817. Also, Math Program 108:159–176 (2006)

57. Zhang J, Chen B, Ye Y (2005) A multiexchange local search algorithm for the capacitated facility location problem. Math Oper Res 30(2):389–403

Failure Detectors

Rachid Guerraoui
School of Computer and Communication
Sciences, EPFL, Lausanne, Switzerland

Keywords

Distributed oracles; Failure information; Partial synchrony; Time-outs

Years and Authors of Summarized Original Work

1996; Chandra, Toueg

Problem Definition

A distributed system is comprised of a collection of processes. The processes typically seek to achieve some common task by communicating through message passing or shared memory. Most interesting tasks require, at least at certain points of the computation, some form of *agreement* between the processes. An abstract form of such agreement is *consensus* where processes need to agree on a single value among a set of proposed values. Solving this seemingly elementary problem is at the heart of reliable distributed computing and, in particular, of distributed database commitment, total ordering of messages, and emulations of many shared object types.

Fischer, Lynch, and Paterson's seminal result in the theory of distributed computing [13] says that consensus cannot be deterministically solved in an *asynchronous* distributed system that is prone to process failures. This impossibility holds consequently for all distributed computing problems which themselves rely on consensus.

Failures and *asynchrony* are fundamental ingredients in the consensus impossibility. The impossibility holds even if only *one* process *fails*, and it does so only by *crashing*, i.e., stopping its activities. Tolerating crashes is the least one would expect from a distributed system for the goal of distribution is in general to avoid single points of failures in centralized architectures. Usually, actual distributed applications exhibit more severe failures where processes could deviate arbitrarily from the protocol assigned to them.

Asynchrony refers to the absence of assumptions on process speeds and communication delays. This absence prevents any process from distinguishing a crashed process from a correct one and this inability is precisely what leads to the consensus impossibility. In practice, however, distributed systems are not completely asynchronous: some timing assumptions can typically be made. In the best case, if precise lower and upper bounds on communication delays and process speeds are assumed, then it is easy to show that consensus and related impossibilities can be circumvented despite the crash of any number of processes [20].

Intuitively, the way that such timing assumptions circumvent asynchronous impossibilities is by providing processes with *information about failures*, typically through *time-out* (or *heart-beat*) mechanisms, usually underlying actual distributed applications. Whereas certain information about failures can indeed be obtained in distributed systems, the accuracy of such information might vary from a system to another, depending on the underlying network, the load of the application, and the mechanisms used to detect failures. A crucial problem in this context is to characterize such information, in an abstract and precise way.

Key Results

The Failure Detector Abstraction

Chandra and Toueg [5] defined the *failure detector* abstraction as a simple way to capture failure information that is needed to circumvent asynchronous impossibilities, in particular the consensus impossibility. The model considered in [5] is a message passing one where processes can fail by *crashing*. Processes that crash stop their activities and do not recover. Processes that do not crash are said to be *correct*. At least one process is supposed to be correct in every execution of the system.

Roughly speaking, a failure detector is an oracle that provides processes with information about failures. The oracle is accessed in each computation step of a process and it provides the process with a value conveying some failure information. The value is picked from some set of values, called the *range* of the failure detector. For instance, the range could be the set of subsets of processes in the system, and each subset could depict the set of processes detected to have crashed, or considered to be correct. This would correspond to the situation where the failure detector is implemented using a time-out: every process q that does not communicate within some time period with some process p, would be included in subset of processes suspected of having crashed by p.

More specifically, a failure detector is a function, D, that associates to each *failure pattern*, F, a set of *failure detector histories* $\{H_i\} = D(F)$. Both the failure pattern and the failure detector history are themselves functions.

- A failure pattern F is a function that associates to each time t, the set of processes $F(t)$ that have indeed crashed by time t. This notion assumes the existence of a global clock, outside the control of the processes, as well as

a specific concept of *crash* event associated with time. A set of failure pattern is called an *environment*.

- A failure detector history H is also a function, which associates to each process p and time t, some value v from the range of failure detector values. (The range of a failure detector D is denoted R_D.) This value v is said to be output by the failure detector D at process p and time t.

Two observations are in order.

- By construction, the output of a failure detector does not depend on the computation, i.e., on the actual steps performed by the processes, on their algorithm or the input of such algorithm. The output of the failure detector depends solely on the failure pattern, namely on whether and when processes crashed.
- A failure detector might associate several histories to each failure pattern. Each history represents a suite of possible combinations of outputs for the same given failure pattern. This captures the inherent non-determinism of a failure detection mechanism. Such a mechanism is typically itself implemented as a distributed algorithm and the variations in communication delays for instance could lead the same mechanism to output (even slightly) different information for the same failure pattern.

To illustrate these concepts, consider two classical examples of failure detectors.

1. The *perfect* failure detector outputs a subset of processes, i.e., the range of the failure detector is the set of subsets of processes in the system. When a process q is output at some time t at a process p, then q is said to be *detected* (of having crashed) by p. The *perfect* failure detector guarantees the two following properties:

   - Every process that crashes is eventually permanently detected;
   - No correct process is ever detected.

2. The *eventually strong* failure detector outputs a subset of processes: when a process q is output at some time t at a process p, then q is said to be *suspected* (of having crashed) by p. An *eventually strong* failure detector ensures the two following properties:
   - Every process that crashes is eventually suspected;
   - Eventually, some correct process is never suspected.

The *perfect* failure detector is *reliable*: if a process q is detected, then q has crashed. An *eventually strong* failure detector is *unreliable*: there never is any guarantee that the information that is output is accurate. The use of the term *suspected* conveys that idea. The distinction between *unreliability* and *reliability* was precisely captured in [14] for the general context where the range of the failure detector can be arbitrary.

Consensus Algorithms

Two important results were established in [5].

Theorem 1 (Chandra-Toueg [5]) *There is an algorithm that solves consensus with a perfect failure detector.*

The theorem above implicitly says that if the distributed system provides means to implement perfect failure detection, then the consensus impossibility can be circumvented, even if all but one process crashes. In fact, the result holds for any failure pattern, i.e., in any environment.

The second theorem below relates the existence of a consensus algorithm to a resilience assumption. More specifically, the theorem holds in the *majority* environment, which is the set of failure patterns where more than half of the processes are correct.

Theorem 2 (Chandra-Toueg [5]) *There is an algorithm that implements consensus with an eventually strong failure detector in the majority environment.*

The algorithm underlying the result above is similar to *eventually synchronous* consensus algorithms [10] and share also some similarities with the *Paxos* algorithm [18]. It is shown in [5] that no algorithm using solely the *eventually strong* failure detector can solve consensus without the majority assumption. (This result is generalized to any unreliable failure detector in [14].) This resilience lower bound is intuitively due to the possibility of partitions in a message passing system where at least half of the processes can crash and failure detection is unreliable. In shared memory for example, no such possibility exists and consensus can be solved with the *eventually strong* failure [19].

Failure Detector Reductions

Failure detectors can be compared. A failure detector D_2 is said to be *weaker* than a failure detector D_1 if there is an asynchronous algorithm, called a *reduction* algorithm, which, using D_1, can emulate D_2. Three remarks are important here.

- The fact that the reduction algorithm is asynchronous means that it does not use any other source of failure information, besides D_1.
- Emulating failure detector D_2 means implementing a distributed variable that mimics the output that could be provided by D_2.
- The existence of a reduction algorithm depends on environment. Hence, strictly speaking, the fact that a failure detector is weaker than another one depends on the environment under consideration.

If failure detector D_1 is weaker than D_2, and vice et versa, then D_1 and D_2 are said to be *equivalent*. Else, if D_1 is weaker than D_2 and D_2 is not weaker than D_1, then D_1 is said to be *strictly weaker* than D_2. Again, strictly speaking, these notions depend on the considered environment.

The ability to compare failure detectors help define a notion of *weakest* failure detector to solve a problem. Basically, a failure detector D is the weakest to solve a problem P if the two following properties are satisfied:

- There is an algorithm that solves P using D.
- If there is an algorithm that solves P using some failure detector D', then D is weaker than D'.

Theorem 3 (Chandra-Hadzilacos-Toueg [4])
The eventually strong failure detector is the weakest to solve consensus in the majority environment.

The weakest failure detector to implement consensus in any environment was later established in [8].

Applications

A Practical Perspective
The identification of the failure detector concept had an impact on the design of reliable distributed architectures. Basically, a failure detector can be viewed as a first class service of a distributed system, at the same level as a name service or a file service. Time-out and heartbeat mechanisms can thus be hidden under the failure detector abstraction, which can then export a unified interface to higher level applications, including consensus and state machine replication algorithms [2, 11, 21].

Maybe more importantly, a failure detector service can encapsulate synchrony assumptions: these can be changed without impact on the rest of the applications. Minimal synchrony assumptions to devise specific failure detectors could be explored leading to interesting theoretical results [1, 7, 12].

A Theoretical Perspective
A second application of the failure detector concept is a theory of distributed computability. Failure detectors enable to classify problems. A problem A is *harder* (resp. *strictly harder*) than problem B if the weakest failure detector to solve B is weaker (resp. strictly weaker) than the weakest failure detector to solve A. (This notion is of course parametrized by a specific environment.)

Maybe surprisingly, the induced failure detection reduction between problems does not exactly match the classical *black-box* reduction notion. For instance, it is well known that there is no asynchronous distributed algorithm that can use a *Queue* abstraction to implement a *Compare-*

Swap abstraction in a system of $n > 2$ processes where $n − 1$ can fail by crashing [15]. In this sense, a *Compare-Swap* abstraction is strictly more powerful (in a *black-box* sense) than a *Queue* abstraction. It turns out that:

Theorem 4 (Delporte-Fauconnier-Guerraoui [9]) *The weakest failure detector to solve the Queue problem is also the weakest to solve the Compare-Swap problem in a system of $n > 2$ processes where $n − 1$ can fail by crashing.*

In a sense, this theorem indicates that reducibility as induced by the failure detector notion is different from the traditional *black-box* reduction.

Open Problems

Several issues underlying the failure detector notion are still open. One such issue consists in identifying the weakest failure detector to solve the seminal *set-agreement* problem [6]: a decision task where processes need to agree on up to k values, instead of a single value as in consensus. Three independent groups of researchers [3, 16, 22] proved the impossibility of solving this problem in an asynchronous system with k failures, generalizing the consensus impossibility [13]. Determining the weakest failure detector to circumvent this impossibility would clearly help understand the fundamentals of failure detection reducibility.

Another interesting research direction is to relate the complexity of distributed algorithm with the underlying failure detector [17]. Clearly, failure detectors circumvents asynchronous impossibilities, but to what extend do they boost the complexity of distributed algorithms? One would of course expect the complexity of a solution to a problem to be higher if the failure detector is weaker. But to what extend?

Cross-References

▸ Asynchronous Consensus Impossibility
▸ Atomic Broadcast

▶ Causal Order, Logical Clocks, State Machine Replication
▶ Linearizability

Recommended Reading

1. Aguilera MK, Delporte-Gallet C, Fauconnier H, Toueg S (2003) On implementing omega with weak reliability and synchrony assumptions. In: 22th ACM symposium on principles of distributed computing, pp 306–314
2. Bertier M, Marin O, Sens P (2003) Performance analysis of a hierarchical failure detector. In: Proceedings 2003 international conference on dependable systems and networks (DSN 2003), San Francisco, 22–25 June 2003, pp 635–644
3. Boroswsky E, Gafni E (n.d.) Generalized FLP impossibility result for t-resilient asynchronous computations. In: Proceedings of the 25th ACM symposium on theory of computing. ACM, pp 91–100
4. Chandra TD, Hadzilacos V, Toueg S (1996) The weakest failure detector for solving consensus. J ACM 43(4):685–722
5. Chandra TD, Toueg S (1996) Unreliable failure detectors for reliable distributed systems. J ACM 43(2):225–267
6. Chauduri S (1993) More choices allow more faults: set consensus problems in totally asynchronous systems. Inf Comput 105(1):132–158
7. Chen W, Toueg S, Aguilera MK (2002) On the quality of service of failure detectors. IEEE Trans Comput 51(1):13–32
8. Delporte-Gallet C, Fauconnier H, Guerraoui R (2002) Failure detection lower bounds on registers and consensus. In: Proceedings of the 16th international symposium on distributed computing, LNCS, vol 2508
9. Delporte-Gallet C, Fauconnier H, Guerraoui R (2005) Implementing atomic objects in a message passing system. Technical report, EPFL Lausanne
10. Dwork C, Lynch NA, Stockmeyer L (1988) Consensus in the presence of partial synchrony. J ACM 35(2):288–323
11. Felber P, Guerraoui R, Fayad M (1999) Putting oo distributed programming to work. Commun ACM 42(11):97–101
12. Fernández A, Jiménez E, Raynal M (2006) Eventual leader election with weak assumptions on initial knowledge, communication reliability and synchrony. In: Proceedings of the international symposium on dependable systems and networks (DSN), pp 166–178
13. Fischer MJ, Lynch NA, Paterson MS (1985) Impossibility of distributed consensus with one faulty process. J ACM 32(2):374–382
14. Guerraoui R (2000) Indulgent algorithms. In: Proceedings of the 19th annual ACM symposium on principles of distributed computing, ACM, Portland, pp 289–297
15. Herlihy M (1991) Wait-free synchronization. ACM Trans Program Lang Syst 13(1):123–149
16. Herlihy M, Shavit N (1993) The asynchronous computability theorem for t-resilient tasks. In: Proceedings of the 25th ACM symposium on theory of computing, pp 111–120
17. Keidar I, Rajsbaum S (2002) On the cost of fault-tolerant consensus when there are no faults-a tutorial. In: Tutorial 21st ACM symposium on principles of distributed computing
18. Lamport L (1998) The part-time parliament. ACM Trans Comput Syst 16(2):133–169
19. Lo W-K, Hadzilacos V (1994) Using failure detectors to solve consensus in asynchronous shared memory systems. In: Proceedings of the 8th international workshop on distributed algorithms. LNCS, vol 857, pp 280–295
20. Lynch N (1996) Distributed algorithms. Morgan Kauffman
21. Michel R, Corentin T (2006) In search of the holy grail: looking for the weakest failure detector for wait-free set agreement. Technical Report TR 06-1811, INRIA
22. Saks M, Zaharoglou F (1993) Wait-free k-set agreement is impossible: the topology of public knowledge. In: Proceedings of the 25th ACM symposium on theory of computing, ACM, pp 101–110

False-Name-Proof Auction

Makoto Yokoo
Department of Information Science and Electrical Engineering, Kyushu University, Nishi-ku, Fukuoka, Japan

Keywords

False-name-proof auctions; Pseudonymous bidding; Robustness against false-name bids

Years and Authors of Summarized Original Work

2004; Yokoo, Sakurai, Matsubara

Problem Definition

In Internet auctions, it is easy for a bidder to submit multiple bids under multiple identifiers (e.g., multiple e-mail addresses). If only one item/good is sold, a bidder cannot make any additional profit by using multiple bids. However, in combinatorial auctions, where multiple items/goods are sold simultaneously, submitting multiple bids under fictitious names can be profitable. A bid made under a fictitious name is called a *false-name bid*.

Here, use the same model as the GVA section. In addition, false-name bids are modeled as follows.

- Each bidder can use multiple identifiers.
- Each identifier is unique and cannot be impersonated.
- Nobody (except the owner) knows whether two identifiers belongs to the same bidder or not.

The goal is to design a *false-name-proof protocol*, i.e., a protocol in which using false-names is useless, thus bidders voluntarily refrain from using false-names.

The problems resulting from collusion have been discussed by many researchers. Compared with collusion, a false-name bid is easier to execute on the Internet since obtaining additional identifiers, such as another e-mail address, is cheap. False-name bids can be considered as a very restricted subclass of collusion.

Key Results

The Generalized Vickrey Auction (GVA) protocol is (dominant strategy) incentive compatible, i.e., for each bidder, truth-telling is a dominant strategy (a best strategy regardless of the action of other bidders) if there exists no false-name bids. However, when false-name bids are possible, truth-telling is no longer a dominant strategy, i.e., the GVA is not false-name-proof.

Here is an example, which is identical to Example 1 in the GVA section.

Example 1 Assume there are two goods a and b, and three bidders, bidder 1, 2, and 3, whose types are θ_1, θ_2, and θ_3, respectively. The evaluation value for a bundle $v(B, \theta_i)$ is determined as follows.

	$\{a\}$	$\{b\}$	$\{a, b\}$
θ_1	$6	$0	$6
θ_2	$0	$0	$8
θ_3	$0	$5	$5

As shown in the GVA section, good a is allocated to bidder 1, and b is allocated to bidder 3. Bidder 1 pays $3 and bidder 3 pays $2.

Now consider another example.

Example 2 Assume there are only two bidders, bidder 1 and 2, whose types are θ_1 and θ_2, respectively. The evaluation value for a bundle $v(B, \theta_i)$ is determined as follows.

	$\{a\}$	$\{b\}$	$\{a, b\}$
θ_1	$6	$5	$11
θ_2	$0	$0	$8

In this case, the bidder 1 can obtains both goods, but he/she requires to pay $8, since if bidder 1 does not participate, the social surplus would have been $8. When bidder 1 does participate, bidder 1 takes everything and the social surplus except bidder 1 becomes 0. Thus, bidder 1 needs to pay the decreased amount of the social surplus, i.e., $8.

However, bidder 1 can use another identifier, namely, bidder 3 and creates a situation identical to Example 1. Then, good a is allocated to bidder 1, and b is allocated to bidder 3. Bidder 1 pays $3 and bidder 3 pays $2. Since bidder 3 is a false-name of bidder 1, bidder 1 can obtain both goods by paying $3 + $2 = $5. Thus, using a false-name is profitable for bidder 1.

The effects of false-name bids on combinatorial auctions are analyzed in [4]. The obtained results can be summarized as follows.

- As shown in the above example, the GVA protocol is not false-name-proof.
- There exists no false-name-proof combinatorial auction protocol that satisfies Pareto efficiency.
- If a surplus function of bidders satisfies a condition called *concavity*, then the GVA is guaranteed to be false-name-proof.

F

Also, a series of protocols that are false-name-proof in various settings have been developed: combinatorial auction protocols [2, 3], multi-unit auction protocols [1], and double auction protocols [5].

Furthermore, in [2], a distinctive class of combinatorial auction protocols called a Price-oriented, Rationing-free (PORF) protocol is identified. The description of a PORF protocol can be used as a guideline for developing strategy/false-name proof protocols.

The outline of a PORF protocol is as follows:

1. For each bidder, the price of each bundle of goods is determined independently of his/her own declaration, while it depends on the declarations of other bidders. More specifically, the price of bundle (a set of goods) B for bidder i is determined by a function $p(B, \Theta_X)$, where Θ_X is a set of declared types by other bidders X.
2. Each bidder is allocated a bundle that maximizes his/her utility independently of the allocations of other bidders (i.e., rationing-free). The prices of bundles must be determined so that *allocation feasibility* is satisfied, i.e., no two bidders want the same item.

Although a PORF protocol appears to be quite different from traditional protocol descriptions, surprisingly, it is a sufficient and necessary condition for a protocol to be strategy-proof. Furthermore, if a PORF protocol satisfies the following additional condition, it is guaranteed to be false-name-proof.

Definition 1 (No Super-Additive price increase (NSA)) For any subset of bidders $S \subseteq N$ and $N' = N \setminus S$, and for $i \in S$, denote B_i as a bundle that maximizes i's utility, then $\sum_{i \in S} p(B_i, \bigcup_{j \in S \setminus \{i\}} \{\theta_j\} \cup \Theta_{N'}) \geq p(\bigcup_{i \in S} B_i, \Theta_{N'})$.

An intuitive description of this condition is that the price of buying a combination of bundles (the right side of the inequality) must be smaller than or equal to the sum of the prices for buying these bundles separately (the left side). This condition

is also a necessary condition for a protocol to be false-name-proof, i.e., any false-name-proof protocol can be described as a PORF protocol that satisfies the NSA condition.

Here is a simple example of a PORF protocol that is false-name-proof. This protocol is called the Max Minimal-Bundle (M-MB) protocol [2]. To simplify the protocol description, a concept called a *minimal* bundle is introduced.

Definition 2 (minimal bundle) Bundle B is called minimal for bidder i, if for all $B' \subset B$ and $B' \neq B$, $v(B', \theta_i) < v(B, \theta_i)$ holds.

In this new protocol, the price of bundle B for bidder i is defined as follows:

- $p(B, \Theta_X) = \max_{B_j \subseteq M, j \in X} v(B_j, \theta_j)$, where $B \cap B_j \neq \emptyset$ and B_j is minimal for bidder j.

How this protocol works using Example 1 is described here. The prices for each bidder is determined as follows.

	$\{a\}$	$\{b\}$	$\{a, b\}$
bidder 1	$8	$8	$8
bidder 2	$6	$5	$6
bidder 3	$8	$8	$8

The minimal bundle for bidder 1 is $\{a\}$, the minimal bundle for bidder 2 is $\{a, b\}$, and the minimal bundle for bidder 3 is $\{b\}$. The price of bundle $\{a\}$ for bidder 1 is equal to the largest evaluation value of conflicting bundles. In this case, the price is $8, i.e., the evaluation value of bidder 2 for bundle $\{a, b\}$. Similarly, the price of bidder 2 for bundle $\{a, b\}$ is 6, i.e., the evaluation value of bidder 1 for bundle $\{a\}$. As a result, bundle $\{a, b\}$ is allocated to bidder 2.

It is clear that this protocol satisfies the allocation feasibility. For each good l, choose bidder j^* and bundle B_j^* that maximize $v(B_j, \theta_j)$ where $l \in B_j$ and B_j is minimal for bidder j. Then, only bidder j^* is willing to obtain a bundle that contains good l. For all other bidders, the price of a bundle that contains l is higher than (or equal to) his/her evaluation value.

Furthermore, it is clear that this protocol satisfies the NSA condition. In this pricing scheme,
$p(B \cup B', \Theta_X) = \max(p(B, \Theta_X), p(B', \Theta_X))$
holds for all B, B', and Θ_X. Therefore, the following formula holds

$$p\left(\bigcup_{i \in S} B_i, \Theta_X\right) = \max_{i \in S} p(B_i, \Theta_X) \le \sum_{i \in S} p(B_i, \Theta_X).$$

Furthermore, in this pricing scheme, prices increase monotonically by adding opponents, i.e., for all $X' \supseteq X$, $p(B, \Theta_{X'}) \ge p(B, \Theta_X)$ holds. Therefore, for each i, $p(B_i, \bigcup_{j \in S \setminus \{i\}} \{\theta_j\} \cup \Theta_{N'}) \ge p(B_i, \Theta_{N'})$ holds. Therefore, the NSA condition, i.e., $\sum_{i \in S} p(B_i, \bigcup_{j \in S \setminus \{i\}} \{\theta_j\} \cup \Theta_{N'}) \ge p(\bigcup_{i \in S} B_i, \Theta_{N'})$ holds.

Applications

In Internet auctions, using multiple identifiers (e.g., multiple e-mail addresses) is quite easy and identifying each participant on the Internet is virtually impossible. Combinatorial auctions have lately attracted considerable attention. When combinatorial auctions become widely used in Internet auctions, false-name-bids could be a serious problem.

Open Problems

It is shown that there exists no false-name-proof protocol that is Pareto efficient. Thus, it is inevitable to give up the efficiency to some extent. However, the theoretical lower-bound of the efficieny loss, i.e., the amount of the efficiency loss that is inevitabe for any false-name-proof protocol, is not identified yet. Also, the efficiency loss of existing false-name-proof protocols can be quite large. More efficient false-name-proof protocols in various settings are needed.

Cross-References

▶ Generalized Vickrey Auction

Recommended Reading

1. Iwasaki A, Yokoo M, Terada K (2005) A robust open ascending-price multi-unit auction protocol against false-name bids. Decis Support Syst 39:23–39
2. Yokoo M (2003) The characterization of strategy/false-name proof combinatorial auction protocols: price-oriented, rationing-free protocol. In: Proceedings of the 18th international joint conference on artificial intelligence, pp 733–739
3. Yokoo M, Sakurai Y, Matsubara S (2001) Robust combinatorial auction protocol against false-name bids. Artif Intell 130:167–181
4. Yokoo M, Sakurai Y, Matsubara S (2004) The effect of false-name bids in combinatorial auctions: new fraud in internet auctions. Game Econ Behav 46:174–188
5. Yokoo M, Sakurai Y, Matsubara S (2005) Robust double auction protocol against false-name bids. Decis Support Syst 39:23–39

Fast Minimal Triangulation

Yngve Villanger
Department of Informatics, University of Bergen, Bergen, Norway

Keywords

Minimal fill problem

Years and Authors of Summarized Original Work

2005; Heggernes, Telle, Villanger

Problem Definition

Minimal triangulation is the addition of an inclusion minimal set of edges to an arbitrary undirected graph, such that a chordal graph is obtained. A graph is *chordal* if every cycle of length at least 4 contains an edge between two nonconsecutive vertices of the cycle.

More formally, Let $G = (V, E)$ be a simple and undirected graph, where $n = |V|$ and $m = |E|$. A graph $H = (V, E \cup F)$, where

$E \cap F = \emptyset$ is a *triangulation* of G if H is chordal, and H is a *minimal* triangulation if there exists no $F' \subset F$, such that $H' = (V, E \cup F')$ is chordal. Edges in F are called *fill edges*, and a triangulation is minimal if and only if the removal of any single fill edge results in a chordless four cycle [10].

Since minimal triangulations were first described in the mid-1970s, a variety of algorithms have been published. A complete overview of these along with different characterizations of chordal graphs and minimal triangulations can be found in the survey of Heggernes et al. [5] on minimal triangulations. Minimal triangulation algorithms can roughly be partitioned into algorithms that obtain the triangulation through elimination orderings, and those that obtain it through vertex separators. Most of these algorithms have an $O(nm)$ running time, which becomes $O(n^3)$ for dense graphs. Among those that use elimination orderings, Kratsch and Spinrad's $O(n^{2.69})$-time algorithm [8] is currently the fastest one. The fastest algorithm is an $o(n^{2.376})$-time algorithm by Heggernes et al. [5]. This algorithm is based on vertex separators, and will be discussed further in the next section. Both the algorithm of Kratsch and Spinrad [8] and the algorithm of Heggernes et al. [5] use the matrix multiplication algorithm of Coppersmith and Winograd [3] to obtain an $o(n^3)$-time algorithm.

Key Results

For a vertex set $A \subset V$, the subgraph of G induced by A is $G[A] = (A, W)$, where $uv \in W$ if $u, v \in A$ and $uv \in E\}$). The closed neighborhood of A is $N[A] = U$, where $u, v \in U$ for every $uv \in E$, where $u \in A\}$ and $N(A) = N[A] \setminus A$. A is called a *clique* if $G[A]$ is a complete graph. A vertex set $S \subset V$ is called a *separator* if $G[V \setminus S]$ is disconnected, and S is called a *minimal* separator if there exists a pair of vertices $a, b \in V \setminus S$ such that a, b are contained in different connected components of $G[V \setminus S]$, and in the same connected component of $G[V \setminus S']$ for any $S' \subset S$. A vertex set $\Omega \subseteq V$ is a *potential maximal clique* if there

exists no connected component of $G[V \setminus \Omega]$ that contains Ω in its neighborhood, and for every vertex pair $u, v \in \Omega$, uv is an edge or there exists a connected component of $G[V \setminus \Omega]$ that contains both u and v in its neighborhood.

From the results in [1, 7], the following recursive minimal triangulation algorithm is obtained. Find a vertex set A which is either a minimal separator or a potential maximal clique. Complete $G[A]$ into a clique. Recursively for each connected component C of $G[V \setminus A]$ where $G[N[C]]$ is not a clique, find a minimal triangulation of $G[N[C]]$. An important property here is that the set of connected components of $G[V \setminus A]$ defines independent minimal triangulation problems.

The recursive algorithm just described defines a tree, where the given input graph G is the root node, and where each connected component of $G[V \setminus A]$ becomes a child of the root node defined by G. Now continue recursively for each of the subproblems defined by these connected components. A node H which is actually a subproblem of the algorithm is defined to be at *level* i, if the distance from H to the root in the tree is i. Notice that all subproblems at the same level can be triangulated independently. Let k be the number of levels. If this recursive algorithm can be completed for every subgraph at each level in $O(f(n))$ time, then this trivially provides an $O(f(n) \cdot k)$-time algorithm.

The algorithm in Fig. 1 uses queues to obtain this level-by-level approach, and matrix multiplication to complete all the vertex separators at a given level in $O(n^\alpha)$ time, where $\alpha < 2.376$ [3]. In contrast to the previously described recursive algorithm, the algorithm in Fig. 1 uses a partitioning subroutine that either returns a *set* of minimal separators or a potential maximal clique.

Even though all subproblems at the same level can be solved independently they may share vertices and edges, but no nonedges (i.e., pair of vertices that are not adjacent). Since triangulation involves edge addition, the number of nonedges will decrease for each level, and the sum of nonedges for all subproblems at the same level will never exceed n^2. The partitioning algorithm

---

Algorithm FMT - Fast Minimal Triangulation
Input: An arbitrary graph $G = (V,E)$.
Output: A minimal triangulation G' of G.

Let Q_1, Q_2 and Q_3 be empty queues; Insert G into Q_1; $G' = G$;
repeat
  Construct a zero matrix M with a row for each vertex in V (columns are added later);
  while Q_1 is nonempty **do**
    Pop a graph $H = (U, D)$ from Q_1;
    Call **Algorithm Partition**(H) which returns a vertex subset $A \subset U$;
    Push vertex set A onto Q_3;
    for each connected component C of $H[U \setminus A]$ **do**
      Add a column in M such that $M(v, C) = 1$ for all vertices $v \in N_H(C)$;
      if there exists a non-edge uv in $H[N_H[C]]$ with $u \in C$ **then**
        Push $H_C = (N_H[C], D_C)$ onto Q_2, where $uv \notin D_C$ if $u \in C$ and $uv \notin D$;
  Compute MM^T;
  Add to G' the edges indicated by the nonzero elements of MM^T;
  while Q_3 is nonempty **do**
    Pop a vertex set A from Q_3;
    if $G'[A]$ is not complete **then** Push $G'[A]$ onto Q_2;
  Swap names of Q_1 and Q_2;
until Q_1 is empty

Fast Minimal Triangulation, Fig. 1 Fast minimal triangulation algorithm

in Fig. 2 exploits this fact and has an $O(n^2 - m)$ running time, which sums up to $O(n^2)$ for each level. Thus, each level in the fast minimal triangulation algorithm given in Fig. 1 can be completed in $O(n^2 + n^\alpha)$ time, where $O(n^\alpha)$ is the time needed to compute MM^T. The partitioning algorithm in Fig. 2 actually finds a set A that defines a set of minimal separators, such that no subproblem contains more than four fifths of the nonedges in the input graph. As a result, the number of levels in the fast minimal triangulation algorithm is at most $\log_{4/5}(n^2) = 2\log_{4/5}(n)$, and the running time $O(n^\alpha \log n)$ is obtained.

Applications

The first minimal triangulation algorithms were motivated by the need to find good pivotal orderings for Gaussian elimination. Finding an optimal ordering is equivalent to solving the minimum triangulation problem, which is a nondeterministic polynomial-time hard problem. Since any minimum triangulation is also a minimal triangulation, and minimal triangulations can be found in polynomial time, then the set of minimal triangulations can be a good place to search for a pivotal ordering.

Probably because of the desired goal, the first minimal triangulation algorithms were based on orderings, and produced an ordering called a minimal elimination ordering. The problem of computing a minimal triangulation has received increasing attention since then, and several new applications and characterizations related to the vertex separator properties have been published. Two of the new applications are computing the tree-width of a graph, and reconstructing evolutionary history through phylogenetic trees [6]. The new separator-based characterizations of minimal triangulations have increased the knowledge of minimal triangulations [1, 7, 9]. One result based on these characterizations is an algorithm that computes

Algorithm Partition
Input: A graph $H = (U, D)$ (a subproblem popped from Q_1).
Output: A subset A of U such that either $A = N[K]$ for some connected $H[K]$
                     or A is a potential maximal clique of H (and G').

Part I: defining P
Unmark all vertices of H; $k = 1$;
while there exists an unmarked vertex u **do**
      if $\mathcal{E}_{\bar{H}}(U \setminus N_H[u]) < \frac{2}{5}|\bar{E}(H)|$ **then** Mark u as an **s**-vertex (stop vertex);
      else
              $C_k = \{u\}$; Mark u as a **c**-vertex (component vertex);
              while there exists a vertex $v \in N_H[C_k]$ which is unmarked or marked as an **s**-vertex **do**
                    if $\mathcal{E}_{\bar{H}}(U \setminus N_H[C_k \cup \{v\}]) \geq \frac{2}{5}|\bar{E}(H)|$ **then**
                          $C_k = C_k \cup \{v\}$; Mark v as a **c**-vertex (component vertex);
                    else
                          Mark v as a **p**-vertex (potential maximal clique vertex); Associate v with C_k;
              $k = k + 1$;
P = the set of all **p**-vertices and **s**-vertices;

Part II: defining A
if $H[U \setminus P]$ has a full component C **then** $A = N_H[C]$;
else if there exist two non-adjacent vertices u,v such that u is an **s**-vertex
              and v is an **s**-vertex or a **p**-vertex **then** $A = N_H[u]$;
else if there exist two non-adjacent **p**-vertices u and v, where u is associated with C_i
              and v is associated with C_j and $u \notin N_H(C_j)$ and $v \notin N_H(C_i)$ **then** $A = N_H[C_i \cup \{u\}]$;
else $A = P$;

Fast Minimal Triangulation, Fig. 2 Partitioning algorithm. Let $\bar{E}(H) = W$, where $uv \in W$ if $uv \notin D$ be the set of nonedges of H. Define $\mathcal{E}_{\bar{H}}(S)$ to be the sum of degrees in $\bar{H} = (U, \bar{E})$ of vertices in $S \subseteq U = V(H)$

the tree-width of a graph in polynomial time if the number of minimal separators is polynomially bounded [2]. A second application is faster exact (exponential-time) algorithms for computing the tree-width of a graph [4].

Open Problems

The algorithm described shows that a minimal triangulation can be found in $O((n^2 + n^\alpha) \log n)$ time, where $O(n^\alpha)$ is the time required to preform an $n \times n$ binary matrix multiplication. As a result, any improved binary matrix multiplication algorithm will result in a faster algorithm for computing a minimal triangulation. An interesting question is whether or not this relation goes the other way as well. Does there exist an $O((n^2 + n^\beta) f(n))$ algorithm for binary matrix multiplication, where $O(n^\beta)$ is the time required to find a minimal triangulation and $f(n) = o(n^{\alpha-2})$ or at least $f(n) = O(n)$. A possibly simpler and related question previously asked in [8] is: Is it at least as hard to compute a minimal triangulation as to determine whether a graph contains at least one triangle? A more algorithmic question is if there exists an $O(n^2 + n^\alpha)$-time algorithm for computing a minimal triangulation.

Cross-References

▶ Treewidth of Graphs

Recommended Reading

1. Bouchitté V, Todinca I (2001) Treewidth and minimum fill-in: grouping the minimal separators. SIAM J Comput 31:212–232
2. Bouchitté V, Todinca I (2002) Listing all potential maximal cliques of a graph. Theor Comput Sci 276(1–2):17–32
3. Coppersmith D, Winograd S (1990) Matrix multiplication via arithmetic progressions. J Symb Comput 9(3):251–280
4. Fomin FV, Kratsch D, Todinca I (2004) Exact (exponential) algorithms for treewidth and minimum fill-in. In: ICALP of LNCS, vol 3142. Springer, Berlin, pp 568–580
5. Heggernes P, Telle JA, Villanger Y (2005) Computing minimal triangulations in time $O(n^\alpha log \, n) = o(n^{2.376})$. SIAM J Discret Math 19(4):900–913
6. Huson DH, Nettles S, Warnow T (1999) Obtaining highly accurate topology estimates of evolutionary trees from very short sequences. In: RECOMB, pp 198–207
7. Kloks T, Kratsch D, Spinrad J (1997) On treewidth and minimum fill-in of asteroidal triple-free graphs. Theor Comput Sci 175:309–335
8. Kratsch D, Spinrad J (2006) Minimal fill in $O(n^{2.69})$ time. Discret Math 306(3):366–371
9. Parra A, Scheffler P (1997) Characterizations and algorithmic applications of chordal graph embeddings. Discret Appl Math 79:171–188
10. Rose D, Tarjan RE, Lueker G (1976) Algorithmic aspects of vertex elimination on graphs. SIAM J Comput 5:146–160

Fast Subset Convolution

Petteri Kaski
Department of Computer Science, School of Science, Aalto University, Helsinki, Finland
Helsinki Institute for Information Technology (HIIT), Helsinki, Finland

Keywords

Convolution; Min-sum semiring; Möbius inversion; Set covering; Set packing; Set partitioning; Sum-product ring; Zeta transform

Years and Authors of Summarized Original Work

2007; Björklund, Husfeldt, Kaski, Koivisto

Problem Definition

A basic strategy to solve hard problems by dynamic programming is to express the partial solutions using a recurrence over the 2^n subsets of an n-element set U. Our interest here is in recurrences that have the following structure:

> For each subset $S \subseteq U$, in order to obtain the partial solution at S, we consider all possible ways to *partition* S into two disjoint parts, T and $S \setminus T$, with $T \subseteq S$.

Fast subset convolution [1] is a technique to speed up the evaluation of such recurrences, assuming the recurrence can be reduced to a suitable algebraic form. In more precise terms, let R be an algebraic ring, such as the integers equipped with the usual arithmetic operations (addition, negation, multiplication). We seek a fast solution to:

Problem (Subset Convolution)

INPUT: *Two functions* $f : 2^U \to R$ *and* $g : 2^U \to R$.

OUTPUT: *The function* $f * g : 2^U \to R$, *defined for all* $S \subseteq U$ *by*

$$(f * g)(S) = \sum_{T \subseteq S} f(T)g(S \setminus T). \quad (1)$$

Here, we may view the output $f * g$ and the inputs f and g each as a table with 2^n entries, where each entry is an element of R. If we evaluate the sum (1) directly for each $S \subseteq U$ in turn, in total we will execute $\Theta\left(\sum_{s=0}^{n} \binom{n}{s} 2^s\right) = \Theta(3^n)$ arithmetic operations in R to obtain $f * g$ from f and g.

Key Results

We can considerably improve on the $\Theta(3^n)$ direct evaluation by taking advantage of the ring structure of R (i.e., the possibility to form and, after multiplication, *cancel* linear combinations of the entries in f and g):

Theorem 1 (Fast Subset Convolution [1])
There exists an algorithm that solves SUBSET CONVOLUTION *in* $O(2^n n^2)$ *arithmetic operations in* R.

In what follows, we present an algorithm that proceeds via reduction to the union product and fast Möbius inversion; an alternative proof is possible via reduction to the symmetric difference product and fast Walsh–Hadamard (Fourier) transforms.

Fast Evaluation via the Union Product

Let us start with a relaxed version of subset convolution. Namely, instead of *partitioning* S, we split S in all possible ways into a *cover* (A, B) with $A \cup B = S$; this cover need not be disjoint (i.e., we need not have $A \cap B = \emptyset$) as would be required by subset convolution. For f and g as earlier, define the *union product* (*covering product*) $f \cup g : 2^U \rightarrow R$ for all $S \subseteq U$ by

$$(f \cup g)(S) = \sum_{\substack{A, B \subseteq U \\ A \cup B = S}} f(A)g(B).$$

The union product diagonalizes into a pointwise product via a pair of mutually inverse linear transforms. For a given $f : 2^U \rightarrow R$, the *zeta transform* $f\zeta : 2^U \rightarrow R$ is defined for all $S \subseteq U$ by $f\zeta(S) = \sum_{T \subseteq S} f(T)$, and its inverse the *Möbius transform* $f\mu : 2^U \rightarrow R$ is defined for all $S \subseteq U$ by $f\mu(S) = (-1)^{|S|} \sum_{T \subseteq S}(-1)^{|T|} f(T)$. Using the zeta and Möbius transforms to diagonalize into a pointwise product, the union product can be evaluated as

$$f \cup g = ((f\zeta) \cdot (g\zeta))\mu. \tag{2}$$

We can now reduce subset convolution to a union product over a polynomial ring with coefficients in the original ring R. Denote by $R[w]$ the univariate polynomial ring with indeterminate w and coefficients in the ring R. Let $f, g : 2^U \rightarrow R$

be the given input to subset convolution. Extend the input $f : 2^U \rightarrow R$ to the input $f_w : 2^U \rightarrow R[w]$ defined for all $S \subseteq U$ by $f_w(S) = f(S)w^{|S|}$. Extend g similarly to g_w. Compute the union product $f_w \cup g_w$ using (2) over $R[w]$. For all $S \subseteq U$, it now holds that the coefficient of the monomial $w^{|S|}$ in the polynomial $(f_w \cup g_w)(S)$ is equal to $(f * g)(S)$.

To compute (2) fast, we require algorithms that evaluate zeta and Möbius transforms over an arbitrary ring in $O(2^n n)$ arithmetic operations. We proceed via the following recurrence for $j = 1, 2, \ldots, n$. Let us assume that $U = \{1, 2, \ldots, n\}$. Let $z_0 = f$. Suppose $z_{j-1} : 2^U \rightarrow R$ is available. Then, we compute $z_j : 2^U \rightarrow R$ for all $S \subseteq U$ by

$$z_j(S) = \begin{cases} z_{j-1}(S) & \text{if } j \notin S; \\ z_{j-1}(S) + z_{j-1}(S \setminus \{j\}) & \text{if } j \in S. \end{cases}$$

We have $f\zeta = z_n$. The recurrence carries out exactly $2^{n-1}n$ additions in R. To compute the Möbius transform $f\mu$ of a given input f, first transform the input by negating the values of all sets that have odd size, then run the previous recurrence with the transformed input as z_0, and transform the output z_n by negating the values of all sets that have odd size. The result is $f\mu$.

Remarks The fast algorithm (2) for the union product (in a dual form that considers intersections instead of unions) is due to Kennes [9], who used the algorithm to speed up an implementation of the Dempster–Shafer theory of evidence. The fast recurrences for the zeta and Möbius transforms are special cases of an algorithm of Yates [12] for multiplying a vector with an iterated Kronecker product; see Knuth [10, §4.6.4].

Extensions and Variations

A number of extensions and variations of the basic framework are possible [1]. Iterated subset convolution (union product) enables one to solve set partitioning and packing (covering) problems. Assuming the input is sparse, more careful control over the space usage of the framework can be obtained by splitting the fast zeta transform into two parts [5]. Similarly, the running time can be controlled by *trimming* [4] the transforms, for example, to the down-closure (subset-closure) of the desired outputs and/or the up-closure (superset-closure) of the supports of the inputs in 2^U. A trimmed complementary dual to the union product is investigated in [3].

Beyond the subset lattice $(2^U, \subseteq, \cup, \cap)$, fast algorithms are known for the zeta and Möbius transforms of lattices $(L, \leq, \vee, \wedge)$ with few join-irreducible elements [6]. This implies fast analogs of the union product (the *join product*) for such lattices.

Applications

Fast subset convolution and its variants are applied to speed up dynamic programming algorithms that build up a solution from partitions of smaller solutions such that *there is little or no interaction between the parts*. Connectivity, partitioning, and subgraph counting problems on graphs are natural examples [1–3, 8, 11].

To apply fast subset convolution, it is necessary to reduce the recurrence at hand into the algebraic form (1). Let us briefly discuss two types of recurrences as examples.

Boolean Subset Convolution

Suppose that f and g are $\{0, 1\}$-valued, and we are seeking to decide whether there exists a valid partition of S into two parts so that one part is valid by f and the other part valid by g. This can be modeled as a Boolean (OR–AND) subset convolution:

$$(f *_{\vee,\wedge} g)(S) = \bigvee_{T \subseteq S} f(T) \wedge g(S \setminus T).$$

Boolean subset convolutions can be efficiently reduced into a subset convolution (1) over the integers simply by replacing the OR with a sum and the AND with multiplication.

Min-Sum Subset Convolution

Another common situation occurs when we are seeking the *minimum cost* to partition S so that

the cost of one part is measured by f and the other by g, where both f and g take nonnegative integer values. This can be modeled as a min-sum subset convolution:

$$(f *_{\min,+} g)(S) = \min_{T \subseteq S} f(T) + g(S \setminus T).$$

A min-sum subset convolution over nonnegative integers can be reduced to a subset convolution (1) over a univariate polynomial ring $\mathbb{Z}[x]$ with integer coefficients. Extend $f : 2^U \to \mathbb{Z}_{\geq 0}$ to $f_x : 2^U \to \mathbb{Z}[x]$ by setting $f_x(S) = x^{f(S)}$ for all $S \subseteq U$. Extend g similarly to g_x. Now observe that the degree of the least-degree monomial with a nonzero coefficient in the polynomial $(f_x * g_x)(S)$ equals $(f *_{\min,+} g)(S)$. This reduction requires computation with polynomials of degree $O(D)$ with $D = \max\{\max_{S \subseteq U} f(S), \max_{S \subseteq U} g(S)\}$, which may not be practical compared with the $O(3^n)$ baseline if D is large.

We refer to [1] for a more detailed discussion and examples.

Cross-References

▶ Enumeration of Paths, Cycles, and Spanning Trees
▶ Exact Algorithms and Time/Space Tradeoffs
▶ Exact Graph Coloring Using Inclusion-Exclusion
▶ Steiner Trees

Recommended Reading

1. Björklund A, Husfeldt T, Kaski P, Koivisto M (2007) Fourier meets Möbius: fast subset convolution. In: Johnson DS, Feige U (eds) STOC. ACM, pp 67–74. doi:10.1145/1250790.1250801, http://doi.acm.org/10.1145/1250790.1250801
2. Björklund A, Husfeldt T, Kaski P, Koivisto M (2008) Computing the Tutte polynomial in vertex-exponential time. In: FOCS. IEEE Computer Society, pp 677–686. doi:10.1109/FOCS.2008.40, http://doi.ieeecomputersociety.org/10.1109/FOCS.2008.40
3. Björklund A, Husfeldt T, Kaski P, Koivisto M (2009) Counting paths and packings in halves. In: [7], pp 578–586. doi:10.1007/978-3-642-04128-0_52, http://dx.doi.org/10.1007/978-3-642-04128-0_52

4. Björklund A, Husfeldt T, Kaski P, Koivisto M (2010) Trimmed Moebius inversion and graphs of bounded degree. Theory Comput Syst 47(3):637–654. doi:10.1007/s00224-009-9185-7, http://dx.doi.org/10.1007/s00224-009-9185-7

5. Björklund A, Husfeldt T, Kaski P, Koivisto M (2011) Covering and packing in linear space. Inf Process Lett 111(21–22):1033–1036. doi:10.1016/j.ipl.2011.08.002, http://dx.doi.org/10.1016/j.ipl.2011.08.002

6. Björklund A, Koivisto M, Husfeldt T, Nederlof J, Kaski P, Parviainen P (2012) Fast zeta transforms for lattices with few irreducibles. In: Rabani Y (ed) SODA. SIAM, pp 1436–1444. doi:10.1137/1.9781611973099.113, http://dx.doi.org/10.1137/1.9781611973099.113

7. Fiat A, Sanders P (eds) (2009) Algorithms – ESA 2009, 17th annual european symposium, Copenhagen, 7–9 Sept 2009. Proceedings, lecture notes in computer science, vol 5757. Springer

8. Fomin FV, Lokshtanov D, Raman V, Saurabh S, Rao BVR (2012) Faster algorithms for finding and counting subgraphs. J Comput Syst Sci 78(3):698–706. doi:10.1016/j.jcss.2011.10.001, http://dx.doi.org/10.1016/j.jcss.2011.10.001

9. Kennes R (1992) Computational aspects of the mobius transformation of graphs. IEEE Trans Syst Man Cybern 22(2):201–223. doi:10.1109/21.148425, http://dx.doi.org/10.1109/21.148425

10. Knuth DE (1998) The art of computer programming, vol 2. Seminumerical algorithms, 3rd edn. Addison-Wesley, Upper Saddle River

11. van Rooij JMM, Bodlaender HL, Rossmanith P (2009) Dynamic programming on tree decompositions using generalised fast subset convolution. In: [7], pp 566–577. doi:10.1007/978-3-642-04128-0_51, http://dx.doi.org/10.1007/978-3-642-04128-0_51

12. Yates F (1937) The design and analysis of factorial experiments. Imperial Bureau of Soil Science, Harpenden

Faster Deterministic Fully-Dynamic Graph Connectivity

Christian Wulff-Nilsen
Department of Computer Science, University of Copenhagen, Copenhagen, Denmark

Keywords

Connectivity; Data structures; Dynamic graphs

Years and Authors of Summarized Original Work

2013; Wulff-Nilsen

Problem Definition

We consider the fully dynamic graph connectivity problem. Here we wish to maintain a data structure for a simple graph that changes over time. We assume a fixed set of n vertices and updates consist of adding and deleting single edges. The data structure should support queries for vertex pairs (u, v) of whether u and v are connected in the current graph.

Key Results

The first nontrivial data structure is due to Frederickson [2] who showed how to achieve deterministic worst-case update time $O(\sqrt{m})$ and query time $O(1)$, where m is the current number of edges of the graph. Using a sparsification technique, Eppstein et al. [1] obtained $O(\sqrt{n})$ update time. Much faster amortized bounds can be achieved. Henzinger and King [3] gave a data structure with $O(\log^3 n)$ randomized expected amortized update time and $O(\log n / \log \log n)$ query time. Update time was improved to $O(\log^2 n)$ by Henzinger and Thorup [4]. A deterministic data structure with $O(\log^2 n)$ amortized update time and $O(\log n / \log \log n)$ query time was given by Holm et al. [5]. Thorup [8] achieved a randomized expected amortized update time of $O(\log n (\log \log n)^3)$ and a query time of $O(\log n / \log \log \log n)$. The fastest known deterministic amortized data structure was given in [9]. Its update time is $O(\log^2 n / \log \log n)$ and query time is $O(\log n / \log \log n)$. Kapron et al. [6] gave a Monte Carlo algorithm with polylogarithmic worst-case operation time. A general cell-prove lower bound of $\Omega(\log n)$ was shown by Pătraşcu and Demaine [7].

In the following, we sketch the main ideas in the data structure presented in [9].

A Simple Data Structure

We start with a simple data structure similar to that of Thorup [8] (which is based on the data structure of Holm et al. [5]) that achieves $O(\log^2 n)$ update time and $O(\log n)$ query time. In the subsection below, we give the main ideas for improving these bounds by a factor of $\log \log n$.

In the following, denote by $G = (V, E)$ the current graph. The data structure maintains for each edge $e \in E$ a *level* $\ell(e)$ between 0 and $\ell_{max} = \lfloor \log n \rfloor$. Initially, an edge has level 0 and its level can only increase over time. For the amortization, we can think of $\ell_{max} - \ell(e)$ as the amount of credits associated with edge e, and every time $\ell(e)$ increases, e pays one credit (which may correspond to more than one unit of time). Let G_i denote the subgraph of G with vertex set V and containing the edges of level at least i and refer to each connected component of G_i as a *level i cluster*. The following invariant is maintained:

Invariant: For each i, any level i cluster contains at most $n/2^i$ vertices.

The clusters nest and thus have a forest representation. More specifically, the *cluster forest* of G is a forest $\mathcal{C}$ of rooted trees where a node u at depth i corresponds to a level i cluster $C(u)$ and the children of u correspond to level $i + 1$ clusters contained in $C(u)$. Note that roots of $\mathcal{C}$ correspond to components of $G_0 = G$ and leaves correspond to vertices of G. Hence, if we can maintain $\mathcal{C}$, we can answer a connectivity query (u, v) in $O(\log n)$ time by traversing the leaf-to-root paths from u and v, respectively, and checking whether the roots are distinct.

In the following, for each node $w \in \mathcal{C}$, denote by $n(w)$ the number of vertices of G contained in $C(w)$; equivalently, $n(w)$ is the number of leaves in the subtree of $\mathcal{C}$ rooted at w.

Edge insertions: When a new edge e is inserted into G, its level $\ell(e)$ is initialized to 0. Updating $\mathcal{C}$ amounts to merging the roots corresponding to the endpoints of e if these roots are distinct.

Edge deletions: Handling the deletion of an edge $e = (u, v)$ is more involved. Let $i = \ell(e)$, let $C(w_i)$ be the level i cluster containing e, and let $C(u_{i+1})$ and $C(v_{i+1})$ be the level $i + 1$ clusters containing u and v, respectively. If $C(u_{i+1}) = C(v_{i+1})$, no changes occur in $\mathcal{C}$. Otherwise, consider the multigraph M obtained from $C(w_i)$ by contracting its level $i + 1$ child clusters to single vertices. We search in parallel in M from $C(u_{i+1})$ and $C(v_{i+1})$, respectively, using a standard search procedure like BFS or DFS. Note that all edges visited have level i. If a search procedure visits a vertex a already visited by the other procedure, the removal of e does not disconnect the level i cluster containing e. In this case, we terminate both search procedures. Consider the two sets V_u and V_v of vertices of M visited by the procedures from $C(u_{i+1})$ resp. $C(v_{i+1})$ where we only include a in one of the sets. Then $V_u \cap V_v = \emptyset$, and since $n(w_i) \leq n/2^i$ by our invariant, either $\sum_{w \in V_u} n(w) \leq n/2^{i+1}$ or $\sum_{w \in V_v} n(w) \leq n/2^{i+1}$. Assume w.l.o.g. the former. Then we increase the level of all visited edges between vertices in V_u to $i + 1$ without violating the invariant. These level increases pay for the search procedure from $C(u_{i+1})$, and since we ran the two procedures in parallel, they also pay for the search procedure from $C(v_{i+1})$.

If the two search procedures do not meet, we increase edge levels on one side as above but now $C(w_i)$ is split into two subclusters since we did not manage to reconnect it with level i edges. In this case, we recursively try to connect these two subclusters in the level $i - 1$ cluster containing them. If we are in this case at level 0, it means that a connected component of $G_0 = G$ is split in two.

Performance: To show how to implement the above with $O(\log^2 n)$ update time, let us assume for now that $\mathcal{C}$ is a forest of *binary* trees. In order for a search procedure to visit a level i edge from a level $i + 1$ cluster $C(a_{i+1})$ to a level $i + 1$ cluster $C(b_{i+1})$, it identifies the start point a of this edge (a, b) in G by traversing the path in $\mathcal{C}$ from a_{i+1} down to leaf a. It then visits (a, b) and traverses the path in $\mathcal{C}$ from leaf b up to

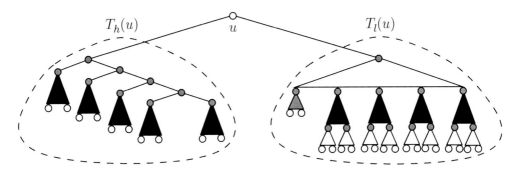

Faster Deterministic Fully-Dynamic Graph Connectivity, Fig. 1 Hybrid local tree for u. Left subtree $T_h(u)$ is a simple local tree for the heavy children and the black subtrees are rank trees attached to a rank path ending in

the root of $T_h(u)$. Right subtree $T_l(u)$ is a lazy local tree for the light children; see [8,9] for details on the structure of this tree

b_{i+1}. To guide the downward searches for level i edges, we maintain for every node w of $\mathcal{C}$ a bitmap whose ith bit is 1 iff there is a level i-edge incident to a leaf in the subtree of $\mathcal{C}$ rooted at w. Since trees in $\mathcal{C}$ are binary, the start point a of (a, b) can be identified from a_{i+1} in $O(\log n)$ time using these bitmaps. Hence, each edge level increase costs $O(\log n)$. Since edge levels can only increase, an edge pays a total of $O(\log^2 n)$. Hence, we achieve an amortized update time of $O(\log^2 n)$.

Above, we assumed that trees in $\mathcal{C}$ are binary. To handle the general case, we modify $\mathcal{C}$ to a different forest $\mathcal{C}_L$ by adding a *simple local tree* $L(u)$ between each non-leaf node u and its children. Associate with each node $v \in \mathcal{C}$ a *rank* $\text{rank}(v) = \lfloor \log n(v) \rfloor$. To form $L(u)$, let C be the set of its children. As long as there are nodes in C with the same rank r, we give them a common parent with rank $r + 1$ and replace them by this parent in C. When this procedure terminates, we have at most $\log n$ *rank trees* whose roots have pairwise distinct ranks and we attach these roots to a *rank path* whose root is u; rank tree roots with bigger rank are attached closer to u than rank tree roots of smaller rank. The resulting tree $L(u)$ is binary; see the left subtree in Fig. 1 for an illustration. Hence, the trees in $\mathcal{C}_L$ are binary as well, and it is easy to see that they have height $O(\log n)$. The performance analysis above for $\mathcal{C}$ then carries through to $\mathcal{C}_L$, and we still have an amortized update time of $O(\log^2 n)$.

A Faster Data Structure

The above data structure has update time $O(\log^2 n)$ and query time $O(\log n)$. We now sketch how to speed up both bounds by a factor of $\log \log n$. We can get the speedup for query time by adding an upward shortcutting system in $\mathcal{C}_L$ where for each leaf-to-root path, we have shortcuts each skipping $\Theta(\log \log n)$ vertices. Maintaining this shortcutting system can be done efficiently. This system also gives a factor $\log \log n$ speedup for each of the upward searches performed by the search procedures described earlier. Speeding up downward searches can be done using a variant of a downward shortcutting system of Thorup [8].

These two shortcutting systems alone do not suffice to improve update time to $O(\log^2 n / \log \log n)$. The data structure needs to support merges and splits of clusters, and with the simple local trees defined above, this costs $O(\log n)$ time per merge/split, and each edge needs to pay a total of $O(\log^2 n)$ for this over all its level increases. Thorup [8] considered *lazy local trees* which can be maintained much more efficiently under cluster merges/splits. However, using these trees to form $\mathcal{C}_L$ may increase the height of trees in this forest to order $\log n \log \log n$ which will slow down our upward shortcutting system by a factor of $\log \log n$. To handle this, consider a hybrid of the simple local tree and Thorup's lazy local tree. For a non-leaf node u in $\mathcal{C}$, a child v is called *heavy* if $n(v) \geq n(u) / \log^{\epsilon} n$ where $\epsilon > 0$ is a constant that we can pick arbitrarily small. A child that is not heavy is called *light*. Now, the *hybrid local tree* of u consists of a simple local tree $T_h(u)$ for the heavy children and a lazy local tree $T_l(u)$ for the light children; see Fig. 1 for an illustration. It can be shown that trees in $\mathcal{C}_L$ have height $O(\frac{1}{\epsilon}\log n)$ if we use hybrid local trees. Furthermore, $\mathcal{C}_L$ can be maintained efficiently under cluster merges and splits. The reason is that, although the hybrid local trees contain simple local trees which are expensive to maintain, these simple local trees are very small as each of them has at most $\log^\epsilon n$ leaves. Hence, maintaining them is not a bottleneck in the data structure.

Combining hybrid local trees with the two shortcutting systems suffice to obtain a factor $\log\log n$ speedup for updates and queries. This gives a deterministic data structure with $O(\log^2 n/\log\log n)$ update time and $O(\log n/\log\log n)$ query time.

Open Problems

Two main open problems for dynamic connectivity are:

- Is there a data structure with $O(\log n)$ operation time (which would be optimal by the lower bound in [7])?
- Is there a data structure with worst-case polylogarithmic operation time which is not Monte Carlo?

Recommended Reading

1. Eppstein D, Galil Z, Italiano GF, Nissenzweig A (1997) Sparsification – a technique for speeding up dynamic graph algorithms. J ACM 44(5):669–696. See also FOCS'92
2. Frederickson GN (1985) Data structures for on-line updating of minimum spanning trees, with applications. SIAM J Comput 14(4):781–798. See also STOC'83
3. Henzinger MR, King V (1995) Randomized dynamic graph algorithms with polylogarithmic time per operation. In: Proceedings of twenty-seventh annual ACM symposium on theory of computing (STOC), Las Vegas, pp 519–527
4. Henzinger MR, Thorup M (1997) Sampling to provide or to bound: with applications to fully dynamic graph algorithms. Random Struct Algorithms 11(4):369–379. See also ICALP'96
5. Holm J, de Lichtenberg K, Thorup M (2001) Polylogarithmic deterministic fully-dynamic algorithms for connectivity, minimum spanning tree, 2-edge, and biconnectivity. J ACM 48(2):723–760. See also STOC'98
6. Kapron BM, King V, Mountjoy B (2013) Dynamic graph connectivity in polylogarithmic worst case time. In: Proceedings of twenty-fourth annual ACM-SIAM symposium on discrete algorithms (SODA), New Orleans, pp 1131–1142
7. Pătraşcu M, Demaine E (2006) Logarithmic lower bounds in the cell-probe model. SIAM J Comput 35(4). Special issue 36th ACM symposium on theory of computing (STOC 2004)
8. Thorup M (2000) Near-optimal fully-dynamic graph connectivity. In: Proceedings of thirty-second annual ACM symposium on theory of computing (STOC), Portland, pp 343–350
9. Wulff-Nilsen C (2013) Faster deterministic fully-dynamic graph connectivity. In: Proceedings of twenty-fourth annual ACM-SIAM symposium on discrete algorithms (SODA), New Orleans, pp 1757–1769

Fault-Tolerant Connected Dominating Set

Donghyun Kim[1], Wei Wang[2], Weili Wu[3,4,5], and Alade O. Tokuta[1]
[1]Department of Mathematics and Physics, North Carolina Central University, Durham, NC, USA
[2]School of Mathematics and Statistics, Xi'an Jiaotong University, Xi'an, Shaanxi, China
[3]College of Computer Science and Technology, Taiyuan University of Technology, Taiyuan, Shanxi Province, China
[4]Department of Computer Science, California State University, Los Angeles, CA, USA
[5]Department of Computer Science, The University of Texas at Dallas, Richardson, TX, USA

Keywords

Approximation algorithm; Connected dominating set; Fault tolerance; Graph algorithm; Vertex connectivity; Virtual backbone

Years and Authors of Summarized Original Work

2010; Kim, Wang, Li, Zhang, Wu
2013; Wang, Kim, An, Gao, Li, Zhang, Wu

Problem Definition

The problem of interest is to find a virtual backbone with a certain level of fault tolerance. Virtual backbone is a subset of nodes to be in charge of routing messages among the other nodes and is a very effective tool to improve the communication efficiency of various wireless networks such as mobile ad hoc networks and wireless sensor networks [3]. It is known that a virtual backbone with smaller cardinality works more efficiently. Without the fault-tolerance consideration, the problem of computing minimum cardinality virtual backbone can be formulated as a minimum connected dominating set problem [1], which is a well-known NP-hard problem [2]. To improve the fault tolerance of a connected dominating set C in homogenous wireless networks, C needs to exhibit two additional properties [4]:

- k-connectivity: C has to be k-vertex-connected so that the virtual backbone can survive even after $k - 1$ backbone nodes fail.
- m-domination: each node v has to be adjacent to at least m nodes in C so that v can be still connected even after $m - 1$ neighboring backbone nodes fail.

The actual value of the two integers, k and m, can be determined by a network operator based on the degree of fault tolerance desired. The majority of the results on this topic consider homogenous wireless networks which is a wireless network of uniform hardware functionality. In this case, the network can be abstracted using the unit disk graph model [6].

Mathematical Formulation

Given a unit disk graph $G = (V, E)$, a subset $C \subseteq V$ is a dominating set in G if for each node $v \in V \setminus C$, v has a neighboring node in C. C is an m-dominating set in G if for each node $v \in V \setminus C$, v has at least m neighboring nodes in C. C is a connected dominating set in G if C is a dominating set in G and if $G[C]$, the subgraph of G induced by C, is connected. C is a k-connected dominating set in G if $G[C]$ is a dominating set in G and $G[C]$ is k-vertex-connected. Finally, C is a k-connected m-dominating set if (a) $G[C]$ is k-vertex-connected and (b) C is an m-dominating set in G. Given $G = (V, E)$, the minimum k-connected m-dominating set problem is to find a minimum cardinality subset C of V satisfying those two requirements.

Key Results

The initial discussion about the need of fault tolerance in virtual backbones has been made by Dai and Wu [4]. Since the minimum k-connected m-dominating set problem in NP-hard, many efforts are made to design a constant factor approximation algorithm for the problem. In [7], Wang et al. proposed a constant factor approximation algorithm for the problem with $k = 2$ and $m = 1$. In [8], Shang et al. introduced a constant factor approximation algorithm for arbitrary integer m and $k = 1, 2$. Later, lots of efforts are made to introduce a constant factor approximation algorithm for arbitrary k and m pairs [9–13]. However, all of them do not work or lose the claimed constant approximation bound in some instances when $k \geq 3$ [14, 15].

In [16], the authors introduce an $O(1)$ approximation algorithm, Fault-Tolerant Connected Dominating Sets Computation Algorithm (FT-CDS-CA), which computes $(3, m)$-CDSs in UDGs. The core part of the algorithm is for computing a $(3, 3)$-CDS, and then it can be easily adapted to compute $(3, m)$-CDS for any $m \geq 1$. The following sections will introduce some key ideas and results of this work.

Constant Approximation for 3-Connected m-Dominating Set

Core Idea
The algorithm starts from a 2-connected 3-dominating set $Y_0 := C_{2,3}$, which can be done

by the algorithm in [8]. Then, it augments the connectivity of the subset by adding a set of nodes $C_0 \subset V \setminus Y_0$ into Y_0 while guaranteeing the number of the newly added nodes C_0 is within a constant factor of $|Y_0|$. In order to do so, the entry introduces the concept of a good node and a bad node. A node u in a 2-connected graph G_2 is called *a good node* if $G_2 \setminus \{u\}$ is still 2-connected, that is, it cannot constitute a separator with any other node in G_2; otherwise it is *a bad node*. An important observation is that a 2-connected graph without bad nodes is 3-connected. Then the entry shows that one can always convert a bad node into a good node by adding a constant number of nodes into Y_0 while not introducing new bad nodes, and they gave an efficient way to achieve this goal. By repeatedly changing bad nodes in Y_0 into good nodes until no bad node is left, Y_0 eventually becomes 3-connected whose size is guaranteed to be within a constant factor of the optimal solution.

Brief Description

A. *Removing Separators* If a 2-connected graph G_2 is not 3-connected, then there exists a pair of nodes u and v, called separator of G_2, such that $G_2 \setminus \{u, v\}$ splits into several parts. It can be shown, due to the properties of UDG, that by adding the internal nodes of at most a constant number of H_3-paths (by an H_3-path we mean a path with length at most three connecting two nodes of a subgraph H of G_2, the internal nodes of which do not belong to H) into Y_0, $\{u, v\}$ is no longer a separator of Y_0, and the nodes newly added are good nodes because $Y_0 = C_{2,3}$ is a 3-dominating set.

B. *Decomposition of a Connected Graph into a Leaf-Block Tree* In graph theory, a block of a graph is a maximal 2-connected subgraph [5]. Given a 2-connected subgraph Y_0 (initially, this is a $C_{2,3}$) and the set X of bad points in Y_0, we select $v \in X$ as a root and compute a leaf-block tree T_0 of $Y_0 \setminus \{v\}$ [5]. Then, T_0 constitutes of a set of blocks $\{B_1, B_2, \ldots, B_s\}$ and a set of cut vertices $\{c_1, \ldots, c_t\}$. An important fact is that v can constitute a separator only with another node in $\{c_1, \ldots, c_t\}$.

C. *Good Blocks vs Bad Blocks* In the process of decomposing a block B with root v into a leaf-block tree, it is important to identify those blocks B_i containing *internal bad nodes*, that is, those bad nodes in B_i that cannot be connected with nodes outside B_i directly without going through v (otherwise, it is called *external bad nodes*). We call such a block B_i with (resp. without) an internal bad node *a bad block* (resp. *a good block*). A key fact is that an internal bad node in B_i can only constitute a separator of Y_0 with another node inside B_i, while this may not be true for external bad nodes.

D. *Multilevel Decomposition* The purpose of the multilevel decomposition is to find a block B with root v such that $B \setminus \{v\}$ contains only good blocks. We assume that $X \neq \emptyset$, since otherwise Y_0 is already 3-connected. After setting $B \leftarrow Y_0$, FT-CDS-CA first picks one $v \in X$ and starts the initial decomposition process (say level-0 decomposition). Then, $B \setminus \{v\}$ is decomposed into a (level-0) leaf-block graph T_0, which is a tree whose vertices consist of a set of blocks $\{B_1, \ldots, B_s\}$ and a set of cut vertices $\{c_1, c_2, \ldots, c_t\}$ ($s \geq 2$ and $t \geq 1$). Now, FT-CDS-CA examines each block in T_0 to see if there is a block B_i having an internal bad node in it. If all blocks are good blocks, then we are done in this step. Otherwise, there must exist some B_i having an internal node $w \in B_i$ which constitutes a separator $\{w, u\}$ of Y_0 with another node $u \in B_i \subset Y_0$. Now, set $v \leftarrow w$ and $B \leftarrow B_i$, start next level (level-1) decomposition. By repeating such process, we can keep making our problem smaller and eventually can find a block B with root v such that $B \setminus \{v\}$ contains only good blocks.

E. *Merging Blocks (Reconstructing the Leaf-Block Tree with a New Root)* After the multilevel decomposition process, we obtain a series of blocks: $Y_l \subset Y_{l-1} \subset \cdots Y_1 \subset Y_0$, where $Y_l = B$ is the final block with a root v such that there is no bad block in the leaf-block tree of $Y_l \setminus \{v\}$. In the induced subgraph $G[Y_l]$, v constitutes a separator with any of $c_1, c_2, \ldots, c_t$, but in Y_0 this is not necessarily true, since there exist some blocks B_i having external nodes that are adjacent to some nodes in $Y_0 \setminus Y_l$ (otherwise, Y_l and $Y_0 \setminus Y_l$ cannot be connected with each other). So these blocks that can be connected directly with $Y_0 \setminus Y_l$ without going through v should be merged together with $Y_0 \setminus Y_l$ into a larger block. After merging all possible blocks into one bigger block, we obtain a modified leaf-block tree T'_l in which one bigger block VB (we call it a *virtual block*) is added representing all the merged blocks and $Y_0 \setminus Y_l$, and all the cut vertices c_i which do not constitute a separator with v have to be removed. Moreover, we mark every remaining cut vertex of VB as a *virtual cut vertex*. In essence, the above merging process can be considered as a process to generate a leaf-block tree directly from $Y_0 \setminus \{v\}$ with all blocks being good except possibly for the virtual block.

F. One Bad-Node Elimination At this point, we have a leaf-block tree T'_l with $V(T'_l) = \{B_1, B_2, \ldots, B_s, VB\} \cup \{c_1, \ldots, c_t\}$, which is obtained through the decomposition of $Y_0 \setminus \{v\}$ (or, equivalently, through the merging process), where v is the internal bad node chosen as root in $B = Y_l$. Note in T'_l, every B_i is a good block except possibly for the virtual block VB. The key point here is that we must have $s \geq 1$; otherwise v would be a good node. In this step, a simple process is employed to make either v or one of the cut vertices in $\{v_1, v_2, \ldots, v_t\} \setminus C$ (C is the set of cut vertices in the virtual block VB) to be a good node. Consider two cases: (i) if the leaf-block tree T'_l has only virtual cut vertices (i.e., T_l is a star centered at VB), then the bad node v becomes a good node by removing the separators consisting of v and the virtual cut vertices, and (ii) if the leaf-block tree T'_l has a cut vertex which is not a virtual cut vertex, then we can find a path $P = (\tilde{B}_0, \tilde{c}_1, \ldots, R)$ in T'_l with one endpoint $\tilde{B}_0$ being a leaf in the tree T'_l and the other endpoint R being a block (cut vertex) with degree larger than two or the virtual block VB (a virtual cut vertex c), if the former does not exist. In this case, two consecutive blocks $\tilde{B}_i$ and $\tilde{B}_{i+1}$ can be found which share a common cut vertex $\tilde{c}_i$. Then it can be shown that at most five H_3-paths are needed such that $\tilde{c}_i$ cannot constitute a pair

of separator of Y_0 with any of the remaining cut vertex or v. Meanwhile, it is still possible that $\tilde{c}_i$ may constitute separators of Y_0 with the external nodes $\tilde{B}_i$ and $\tilde{B}_{i+1}$ (clearly $\tilde{c}_i$ cannot constitute separators of Y_0 with the internal nodes $\tilde{B}_i$ and $\tilde{B}_{i+1}$). It can be proved that the total number of external nodes in $\tilde{B}_i$ and $\tilde{B}_{i+1}$ that may constitute separators of Y_0 with $\tilde{c}_i$ is at most five. In both cases, the number of H_3-paths added to change one bad node (v or $\tilde{c}_i$) into a good node is at most a constant.

Open Problems

While Wang et al. [16] manage to introduce a constant factor approximation algorithm for the minimum k-connected m-dominating set problem in unit disk graph with $k = 3$ and arbitrary integer $m \geq 1$, it is still open to design an approximation algorithm for the case with $k \geq 4$.

Experimental Results

Wang et al.'s work [16] presents some simulation results. The results show that when a 2-connected 3-dominating set computed by Shang et al.'s approach [8] is augmented to a 3-connected 3-dominating set using their algorithm, the size of the connected dominating set will modestly increase roughly less than 25 %. Their algorithm is also compared with an optimal solution using an exhaustive computation within small-scale random unit disk graphs. The result shows the performance gap between exact algorithm and their algorithm is no greater than 39.27 %.

Cross-References

► Connected Dominating Set
► Efficient Dominating and Edge Dominating Sets for Graphs and Hypergraphs
► Exact Algorithms for Dominating Set
► Strongly Connected Dominating Set

Recommended Reading

1. Guha S, Khuller S (1998) Approximation algorithms for connected dominating sets. Algorithmica 20:374–387
2. Garey MR, Johnson DS (1979) Computers and intractability: a guide to the theory of NP-completeness. Freeman, San Francisco
3. Sinha P, Sivakumar R, Bharghavan V (2001) Enhancing ad hoc routing with dynamic virtual infrastructures. In: Proceedings of the 20th annual joint conference of the IEEE computer and communications societies, Anchorage vol 3, pp 1763–1772
4. Dai F, Wu J (2005), On constructing k-connected k-dominating set in wireless network. In: Proceedings of the 19th IEEE international parallel and distributed processing symposium, Denver
5. Diestel R (2005) Graph theory, 3rd edn. Springer, Heidelberg
6. Clark BN, Colbourn CJ, Johnson DS (1990) Unit disk graphs. Discret Math 86:165–177
7. Wang F, Thai MT, Du DZ (2009) 2-connected virtual backbone in wireless network. IEEE Trans Wirel Commun 8(3):1230–1237
8. Shang W, Yao F, Wan P, Hu X (2007) On minimum m-connected k-dominating set problem in unit disc graphs. J Comb Optim 16(2):99–106
9. Li Y, Wu Y, Ai C, Beyah R (2012) On the construction of k-connected m-dominating sets in wireless networks. J Comb Optim 23(1):118–139
10. Thai MT, Zhang N, Tiwari R, Xu X (2007) On approximation algorithms of k-connected m-dominating sets in disk graphs. Theor Comput Sci 358:49–59
11. Wu Y, Wang F, Thai MT, Li Y (2007) Constructing k-connected m-dominating sets in wireless sensor networks. In: Proceedings of the 2007 military communications conference, Orlando
12. Wu Y, Li Y (2008) Construction algorithms for k-connected m-dominating sets in wireless sensor networks. In: Proceedings of 9th ACM international symposium on mobile ad hoc networking and computing, Hong Kong
13. Zhang N, Shin I, Zou F, Wu W, Thai MT (2008) Trade-off scheme for fault tolerant connected dominating sets on size and diameter. In: Proceedings of the 1st ACM international workshop on foundations of wireless ad hoc and sensor networking and computing (FOWANC '08), Hong Kong
14. Kim D, Gao X, Zou F, Du DZ (2011) Construction of fault-tolerant virtual backbones in wireless networks. Handbook on security and networks. World Scientific, Hackensack, pp 488–509
15. Kim D, Wang W, Li X, Zhang Z, Wu W (2010) A new constant factor approximation for computing 3-connected m-dominating sets in homogeneous wireless networks. In: Proceedings of the 29th IEEE conference on computer communications, San Diego
16. Wang W, Kim D, An MK, Gao W, Li X, Zhang Z, Wu W (2013) On construction of quality fault-tolerant virtual backbone in wireless networks. IEEE/ACM Trans Netw 21(5):1499–1510

Fault-Tolerant Quantum Computation

Ben W. Reichardt
Electrical Engineering Department, University of Southern California (USC), Los Angeles, CA, USA

Keywords

Quantum noise threshold

Years and Authors of Summarized Original Work

1996; Shor, Aharonov, Ben-Or, Kitaev

Problem Definition

Fault tolerance is the study of reliable computation using unreliable components. With a given noise model, can one still reliably compute? For example, one can run many copies of a classical calculation in parallel, periodically using majority gates to catch and correct faults. Von Neumann showed in 1956 that if each gate fails independently with probability p, flipping its output bit $0 \leftrightarrow 1$, then such a fault tolerance scheme still allows for arbitrarily reliable computation provided that p is below some constant threshold (whose value depends on the model details) [10].

In a quantum computer, the basic gates are much more vulnerable to noise than classical transistors – after all, depending on the implementation, they are manipulating single electron spins, photon polarizations, and similarly fragile subatomic particles. It might not be possible to engineer systems with noise rates less than

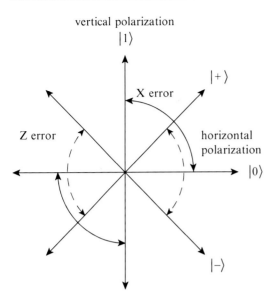

Fault-Tolerant Quantum Computation, Fig. 1 Bit-flip X errors flip 0 and 1. In a qubit, $|0\rangle$ and $|1\rangle$ might be represented by *horizontal* and *vertical polarization* of a photon, respectively. Phase-flip Z errors flip the $\pm45°$ polarized states $|+\rangle$ and $|-\rangle$

10^{-2}, or perhaps 10^{-3}, per gate. Additionally, the phenomenon of entanglement makes quantum systems *inherently* fragile. For example, in Schrödinger's cat state – an equal superposition between a living cat and a dead cat, often idealized as $1/\sqrt{2}(|0^n\rangle + |1^n\rangle)$ – an interaction with just one quantum bit ("qubit") can collapse, or decohere, the entire system. Fault tolerance techniques will therefore be essential for achieving the considerable potential of quantum computers. Practical fault tolerance techniques will need to control high noise rates and do so with low overhead, since qubits are expensive.

Quantum systems are continuous, not discrete, so there are many possible noise models. However, the essential features of quantum noise for fault tolerance results can be captured by a simple discrete model similar to the one Von Neumann used. The main difference is that, in addition to bit-flip X errors which swap 0 and 1, there can also be phase-flip Z errors which swap $|+\rangle \equiv 1/\sqrt{2}(|0\rangle + |1\rangle)$ and $|-\rangle \equiv 1/\sqrt{2}(|0\rangle - |1\rangle)$ (Fig. 1). A noisy gate is modeled as a perfect gate followed by independent introduction of X, Z, or Y (which is both X and Z) errors with respective probabilities p_X, p_Z, p_Y. One popular model is independent depolarizing noise ($p_X = p_Z = p_Y \equiv p/3$); a depolarized qubit is completely randomized.

Faulty measurements and preparations of single-qubit states must additionally be modeled, and there can be memory noise on resting qubits. It is often assumed that measurement results can be fed into a classical computer that works perfectly and dynamically adjusts the quantum gates, although such control is not necessary. Another common, though unnecessary, assumption is that any pair of qubits in the computer can interact; this is called a *nonlocal* gate. In many proposed quantum computer implementations, however, qubit mobility is limited so gates can be applied only locally, between physically nearby qubits.

Key Results

The key result in fault tolerance is the existence of a noise *threshold*, for certain noise and computational models. The noise threshold is a positive, constant noise rate (or set of model parameters) such that with noise below this rate, reliable computation is possible. That is, given an inputless quantum circuit $\mathcal{C}$ of perfect gates, there exists a "simulating" circuit FTC of faulty gates such that with probability at least 2/3, say, the measured output of $\mathcal{C}$ agrees with that of FTC. Moreover, FTC should be only polynomially larger than $\mathcal{C}$.

A quantum circuit with N gates can a priori tolerate only $O(1/N)$ error per gate, since a single failure might randomize the entire output. In 1996, Shor showed how to tolerate $O(1/\text{poly}(\log N))$ error per gate by encoding each qubit into a poly($\log N$)-sized quantum error-correcting code and then implementing each gate of the desired circuit directly on the encoded qubits, alternating computation and error correction steps (similar to Von Neumann's scheme) [8]. Shor's result has two main technical pieces:

1. The discovery of quantum error-correcting codes (QECCs) was a major result. Remarkably, even though quantum errors can be continuous, codes that correct discrete errors suffice. (Measuring the syndrome of a code block projects into a discrete error event.) The first quantum code, discovered by Shor, was a nine-qubit code consisting of the concatenation of the three-qubit repetition code $|0\rangle \rightarrow |000\rangle$, $|1\rangle \rightarrow |111\rangle$ to protect against bit-flip errors, with its dual $|+\rangle \mapsto |+++\rangle$, $|-\rangle \mapsto |---\rangle$ to protect against phase-flip errors. Since then, many other QECCs have been discovered. Codes like the nine-qubit code that can correct bit- and phase-flip errors separately are known as Calderbank-Shor-Steane (CSS) codes and have quantum code words which are simultaneously superpositions over code words of classical codes in both the $|0/1\rangle$ and $|+/-\rangle$ bases.

2. QECCs allow for quantum memory or for communicating over a noisy channel. For computation, however, it must be possible to compute on encoded states without first decoding. An operation is said to be *fault tolerant* if it cannot cause correlated errors within a code block. With the n-bit majority code, all classical gates can be applied *transversely* – an encoded gate can be implemented by applying the unencoded gate to bit i of each code block, $1 \le i \le n$. This is fault tolerant because a single failure affects at most 1 bit in each block, and thus, failures can't spread too quickly. For CSS quantum codes, the controlled-NOT gate CNOT, $|a,b\rangle \rightarrow |a, a \oplus b\rangle$, can similarly be applied transversely. However, the CNOT gate by itself is not universal, so Shor also gave a fault-tolerant implementation of the Toffoli gate $|a,b,c\rangle \rightarrow |a,b,c \oplus (a \wedge b)\rangle$. Procedures are additionally needed for error correction using faulty gates and for the initial preparation step. The encoding of $|0\rangle$ will be a highly entangled state and difficult to prepare (unlike 0^n for the classical majority code).

However, Shor did not prove the existence of a *constant* tolerable noise rate, a noise threshold. Several groups – Aharonov/Ben-Or, Kitaev, and Knill/Laflamme/Zurek – each had the idea of using smaller codes and *concatenating* the procedure repeatedly on top of itself. Intuitively, with a distance-three code (i.e., code that corrects any one error), one expects the "effective" logical error rate of an encoded gate to be at most cp^2 for some constant c, because one error can be corrected but two errors cannot. The effective error rate for a twice-encoded gate should then be at most $c(cp^2)^2$; and since the effective error rate is dropping doubly exponentially fast in the number of levels of concatenation, the overhead in achieving a $1/N$ error rate is only poly(log N). The threshold for improvement, $cp^2 < p$, is $p < 1/c$. However, this rough argument is not rigorous, because the effective error rate is ill defined, and logical errors need not fit the same model as physical errors (e.g., they will not be independent).

Aharonov and Ben-Or and Kitaev gave independent rigorous proofs of the existence of a positive constant noise threshold, in 1997 [1,5].

Broadly, there has since been progress on two fronts of the fault tolerance problem:

1. First, work has proceeded on extending the set of noise and computation models in which a fault tolerance threshold is known to exist. For example, correlated or even adversarial noise, leakage errors (where a qubit leaves the $|0\rangle$, $|1\rangle$ subspace), and non-Markovian noise (in which the environment has a memory) have all been shown to be tolerable in theory, even with only local gates.

2. Threshold existence proofs establish that building a working quantum computer is possible *in principle*. Physicists need only engineer quantum systems with a low enough constant noise rate. But realizing the potential of a quantum computer will require *practical* fault tolerance schemes. Schemes will have to tolerate a high noise rate (not just some constant) and do so with low overhead (not just polylogarithmic).

However, rough estimates of the noise rate tolerated by the original existence proofs are not promising – below 10^{-6} noise per gate. If the true threshold is only 10^{-6}, then building a quantum computer will be next to impossible. Therefore, second, there has been substantial work on optimizing fault tolerance schemes primarily in order to improve the tolerable noise rate. These optimizations are typically evaluated with simulations and heuristic analytical models. Recently, though, Aliferis, Gottesman, and Preskill have developed a method to prove reasonably good threshold lower bounds, up to 2×10^{-4}, based on counting "malignant" sets of error locations [3].

In a breakthrough, Knill has constructed a novel fault tolerance scheme based on very efficient distance-*two* codes [6]. His codes cannot correct any errors, and the scheme uses extensive postselection on no detected errors – i.e., on detecting an error, the enclosing subroutine is restarted. He has estimated a threshold above 3% per gate, an order of magnitude higher than previous estimates. Reichardt has proved a threshold lower bound of 10^{-3} for a similar scheme [7], somewhat supporting Knill's high estimate. However, reliance on postselection leads to an enormous overhead at high error rates, greatly limiting practicality. (A classical fault tolerance scheme based on error detection could not be efficient, but quantum teleportation allows Knill's scheme to be at least theoretically efficient.) There seems to be tradeoff between the tolerable noise rate and the overhead required to achieve it.

There are several complementary approaches to quantum fault tolerance. For maximum efficiency, it is wise to exploit any known noise structure before switching to general fault tolerance procedures. Specialized techniques include careful quantum engineering, techniques from nuclear magnetic resonance (NMR) such as dynamical decoupling and composite pulse sequences, and decoherence-free subspaces. For very small quantum computers, such techniques may give sufficient noise protection.

It is possible that an inherently reliable quantum-computing device will be engineered or discovered, like the transistor for classical computing, and this is the goal of *topological quantum computing* [4].

Applications

As quantum systems are noisy and entanglement fragile, fault tolerance techniques will probably be essential in implementing any quantum algorithms – including efficient factoring and quantum simulation.

The quantum error-correcting codes originally developed for fault tolerance have many other applications, including quantum key distribution.

Open Problems

Dealing with noise may turn out to be the most daunting task in building a quantum computer. Currently, physicists' low-end estimates of achievable noise rates are only slightly below theorists' high-end (mostly simulation based) estimates of tolerable noise rates, at reasonable levels of overhead. However, these estimates are made with different noise models – most simulations are based on the simple independent depolarizing noise model, and threshold lower bounds for more general noise are much lower. Also, both communities may be being too optimistic. Unanticipated noise sources may well appear as experiments progress. The probabilistic noise models used by theorists in simulations may not match reality closely enough, or the overhead/threshold tradeoff may be impractical. It is not clear if fault-tolerant quantum computing will work in practice, unless inefficiencies are wrung out of the system. Developing more efficient fault tolerance techniques is a major open problem. Quantum system engineering, with more realistic simulations, will be required to understand better various tradeoffs and strategies for working with gate locality restrictions.

The gaps between threshold upper bounds, threshold estimates, and rigorously proven threshold lower bounds are closing, at least for simple noise models. Our understanding of what to expect with more realistic noise models is less developed, though. One current line of research is in extending threshold proofs to more realistic noise models – e.g., [2]. A major open question here is whether a noise threshold can be shown to even *exist* where the bath Hamiltonian is unbounded – e.g., where system qubits are coupled to a non-Markovian, harmonic oscillator bath. Even when a threshold is known to exist, rigorous threshold lower *bounds* in more general noise models may still be far too conservative (according to arguments, mostly intuitive, known as "twirling") and, since simulations of general noise models are impractical, new ideas are needed for more efficient analyses.

Theoretically, it is of interest what is the best asymptotic overhead in the simulating circuit FTC versus C? Overhead can be measured in terms of size N and depth/time T. With concatenated coding, the size and depth of FTC are $O(N \text{polylog } N)$ and $O(T \text{polylog } N)$, respectively. For classical circuit C, however, the depth can be only $O(T)$. It is not known if the quantum depth overhead can be improved.

Experimental Results

Fault tolerance schemes have been simulated for large quantum systems, in order to obtain threshold estimates. For example, extensive simulations including geometric locality constraints have been run by Thaker et al. [9].

Error correction using very small codes has been experimentally verified in the lab.

URL to Code

Andrew Cross has written and distributes code for giving Monte Carlo estimates of and rigorous lower bounds on fault tolerance thresholds: http://web.mit.edu/awcross/www/qasm-tools/. Emanuel Knill has released *Mathematica* code for estimating fault tolerance thresholds for certain postselection-based schemes: http://arxiv.org/e-print/quant-ph/0404104.

Cross-References

▶ Quantum Error Correction

Recommended Readings

1. Aharonov D, Ben-Or M (1997) Fault-tolerant quantum computation with constant error rate. In: Proceedings 29th ACM symposium on theory of computing (STOC), pp 176–188. quant-ph/9906129
2. Aharonov D, Kitaev AY, Preskill J (2006) Fault-tolerant quantum computation with long-range correlated noise. Phys Rev Lett 96:050504. quant-ph/0510231
3. Aliferis P, Gottesman D, Preskill J (2006) Quantum accuracy threshold for concatenated distance-3 codes. Quantum Inf Comput 6:97–165. quant-ph/0504218
4. Freedman MH, Kitaev AY, Larsen MJ, Wang Z (2002) Topological quantum computation. Bull AMS 40(1):31–38
5. Kitaev AY (1997) Quantum computations: algorithms and error correction. Russ Math Surv 52:1191–1249
6. Knill E (2005) Quantum computing with realistically noisy devices. Nature 434:39–44
7. Reichardt BW (2006) Error-detection-based quantum fault tolerance against discrete Pauli noise. Ph.D. thesis, University of California, Berkeley. quant-ph/0612004
8. Shor PW (1996) Fault-tolerant quantum computation. In: Proceedings of the 37th symposium on foundations of computer science (FOCS). quant-ph/9605011
9. Thaker DD, Metodi TS, Cross AW, Chuang IL, Chong FT (2006) Quantum memory hierarchies: efficient designs to match available parallelism in quantum computing. In: Proceedings of the 33rd international symposium on computer architecture (ISCA), pp 378–390. quant-ph/0604070
10. von Neumann J (1956) Probabilistic logic and the synthesis of reliable organisms from unreliable components. In: Shannon CE, McCarthy J (eds) Automata studies. Princeton University Press, Princeton, pp 43–98

Finding Topological Subgraphs

Paul Wollan
Department of Computer Science, University of Rome La Sapienza, Rome, Italy

Keywords

Disjoint paths; Fixed-parameter tractability; Topological minor; Topological subgraph

Years and Authors of Summarized Original Work

2011; Grohe, Kawarabayashi, Marx, Wollan

Problem Definition

To *subdivide* an edge e in a graph G with endpoints u and v, delete the edge from the graph and add a path of length two connecting the vertices u and v. A graph G is a *subdivision* of graph H if G can be obtained from H by repeatedly subdividing edges. A graph H is a *topological subgraph* (or *topological minor*) of graph G if a subdivision of H is a subgraph of G. Equivalently, H is a topological subgraph of G if H can be obtained from G by deleting edges, deleting vertices, and suppressing vertices of degree 2 (to *suppress* a vertex of degree 2, delete the vertex and add an edge connecting its two neighbors). The notion of topological subgraphs appears in the classical result of Kuratowski in 1935 stating that a graph is planar if and only if it does not have a topological subgraph isomorphic to K_5 or $K_{3,3}$. This entry considers the problem of determining, given a graph G and H, whether G contains H as a topological minor.

Topological Subgraph Testing

Input: Graphs G and H
Output: Determine if H is a topological subgraph of G

Observe that a graph G on n vertices contains the cycle of length n as a topological subgraph if and only if G contains a Hamiltonian cycle. Thus, it is NP-complete to decide if H is a topological subgraph of a graph G with no further restrictions on G or H.

Previous Work

The algorithmic problem of testing for topological subgraphs was already studied in the 1970s by Lapaugh and Rivest [12] (also see [7]). Fortune, Hopcroft, and Wyllie [6] showed that the analogous problem in directed graphs is NP-complete even when H is a fixed small graph. Robertson and Seymour, as a consequence of their seminal work on graphs minors, showed that for a fixed graph H, there exists a polynomial time algorithm to check whether H is a topological subgraph of a graph G given in input. However, the running time of the Robertson-Seymour algorithm is $|V(G)|^{O(|V(H)|)}$. Following this, Downey and Fellows [4] (see also [5]) conjectured that the problem of topological subgraph testing is fixed parameter tractable: they conjectured that there exists a function f and a constant c such that there exists an algorithm for testing whether a graph H is a topological subgraph of G which runs in time $f(|V(H)|) \cdot |V(G)|^c$.

The problem of topological subgraph testing is closely related to that of minor testing and the k-disjoint paths problem. A graph H is a *minor* of G if H can be obtained from a subgraph of G by contracting edges. The k-disjoint paths problem instead takes as input k pairs of vertices $(s_1, t_1), \ldots, (s_k, t_k)$ of vertices in a graph G and asks if there exist pairwise internally vertex disjoint paths $P_1, \ldots, P_k$ such that the endpoints of P_i are s_i and t_i for all $1 \leq i \leq k$. Robertson and Seymour [13] considered a model of *labeled minor containment* that unites these two problems and showed that there is an $O(|V(G)|^3)$ time algorithm for both H-minor testing for a fixed graph H and the k-disjoint paths problem for a fixed value k.

For every H, there exists a finite list $H_1, \ldots, H_t$ of graphs such that a graph G contains H as a minor if and only if G contains H_i as a topological minor for some index i; this follows from the definition of minor and topological minor. Thus, the problem of minor testing reduces to the harder problem of topological minor testing. It is not difficult to reduce the problem of topological subgraph containment for a fixed graph H to the k-disjoint paths problem. For each vertex v of H, guess a vertex v' of G, and then for each edge uv of H, and seek to find a path connecting u' and v' in G such that these $|E(H)|$ paths are pairwise internally vertex disjoint. This approach yields the $|V(G)|^{O(|V(H)|)}$ time algorithm for topological subgraph testing mentioned above.

Key Results

The following theorem of Grohe, Kawarabayashi, Marx, and Wollan [8] shows that topological subgraph testing is fixed parameter tractable, confirming the conjecture of Downey and Fellows.

Theorem 1 *For every fixed, undirected graph H, there is an $O(|V(G)|^3)$ time algorithm that decides if H is a topological subgraph of G.*

Outline of the Proof

The algorithm given by Theorem 1 builds on the techniques first developed by Robertson and Seymour in their algorithm for minor testing and the k-disjoint paths problem. Fix a graph H and let G be a graph given in input. The algorithm separately considers each of the following three cases:

1. The tree-width of G is bounded (by an appropriate function on $|V(H)|$);
2. G has large tree-width, but the size of the largest clique minor is bounded (again by an appropriate function on $|V(H)|$);
3. G has a large clique minor.

Note that in the third case, the existence of a large clique minor necessarily forces the graph G to have large tree-width. We do not use any technical aspects of the parameter tree-width here and direct interested readers to [1, 2] for further discussion of this topic.

The Robertson-Seymour algorithm for minor testing offers a roadmap for the proof of Theorem 1; the discussion of the proof of Theorem 1 highlights where the proof builds on the tools of Robertson and Seymour and where new techniques are required. As in Robertson-Seymour's algorithm for minor testing, the algorithm considers a rooted version of the problem.

G has Bounded Tree-Width

Numerous problems can be efficiently solved when the input graph is restricted to have bounded tree-width (see [1, 3] for examples). For example, the k-disjoint paths problem can be solved in linear time in graphs of bounded tree-width [15]. Standard dynamic programming techniques can be used to solve the more general rooted version of the topological subgraph problem which the algorithm considers.

G has Large Tree-Width, but no Large Clique Minor

Robertson and Seymour showed that graphs of large tree-width which do not contain a fixed clique minor must contain a large, almost planar subgraph [11, 13]; this result is sometimes known as the flat-wall theorem. The proof of correctness for their disjoint paths algorithm hinges upon this theorem by showing that a vertex in the planar subgraph can be deleted without affecting the feasibility of a given disjoint paths problem. The proof of Theorem 1 builds on this approach. Given graphs G and H, say a vertex $v \in V(G)$ is *irrelevant* for the problem of topological subgraph testing if G contains H as a topological minor if and only if $G - v$ contains H as a topological minor. If the algorithm can efficiently find an irrelevant vertex v, then it can proceed by recursing on the graph $G - v$. In order to apply a similar irrelevant vertex argument to that developed by Robertson and Seymour for the disjoint paths problem, the proof of Theorem 1 shows that a large flat wall contains an irrelevant vertex for a given topological subgraph testing problem by first generalizing several technical results on rerouting systems of paths in graphs [10, 13] as well as deriving a stronger version of the flat-wall theorem.

G has a Large Clique Minor

In the Robertson and Seymour algorithm for minor testing, once the graph can be assumed to have a large clique minor, the algorithm trivially terminates. When considering the k-disjoint paths problem, again it is a relatively straightforward matter to find an irrelevant vertex for a given disjoint paths problem assuming the existence of a large clique minor. Instead, if we are considering the problem of testing topological subgraph containment, the presence of a large clique minor does not yield an easy recursion. Consider the case where we are testing for the existence of a topological subgraph of a 4-regular graph H in a graph G which contains a subcubic subgraph G' such that G' has a large clique minor. Whether or not G' will prove useful in finding a topological subgraph of H in G will depend entirely on whether or not it is possible to link many vertices of degree four (in G) to the clique minor in G' and not on the size itself of the clique minor in G'. Similar issues arise in [9] when developing structure theorems for excluded topological subgraphs.

The proof of Theorem 1 proceeds by considering separately the case when the large-degree vertices can be separated from the clique minor by a bounded sized separator or not. If they cannot, one can find the necessary rooted topological minors. Alternatively, if they can, the algorithm recursively calculates the rooted topological minors in subgraphs of G and replaces a portion of the graph with a bounded size gadget. This portion of the argument is substantially different from the approach of Robertson and Seymour to minor testing and comprises the major new development in the proof.

Applications

An *immersion* of a graph H into a graph G is defined like a topological embedding, except that the paths in G corresponding to the edges of H are only required to be pairwise edge disjoint instead of pairwise internally vertex disjoint. Formally, an immersion of H into G is a mapping v that associates with each vertex $v \in V(H)$ a distinct vertex $v(v) \in V(G)$ and with each edge $e = vw \in E(H)$ a path $v(e)$ in G with endpoints $v(v)$ and $v(w)$ in such a way that the paths $v(e)$ for $e \in E(H)$ are mutually edge disjoint. Robertson and Seymour [14] showed that graphs are well quasi-ordered under the immersion relation, proving a conjecture of Nash-Williams. In [8], the authors give a construction which implies the following corollary of Theorem 1.

Corollary 1 *For every fixed undirected graph H, there is an $O(|V(G)|^3)$ time algorithm that decides if there is an immersion of H into G.*

Again, the algorithm is uniform in H, which implies that the immersion problem is fixed parameter tractable. This answers another open question by Downey and Fellows [4, 5]. Corollary 1 also holds for the more restrictive "strong immersion"

version, where $v(v)$ cannot be the internal vertex of the path $v(e)$ for any $v \in V(G)$ and $e \in E(G)$.

Recommended Readings

1. Bodlaender H (2006) Treewidth: characterizations, applications, and computations. In: Graph-theoretic concepts in computer science. Lecture notes in computer science, vol 4271. Springer, Berlin/Heidelberg, pp 1–14
2. Bodlaender H, Koster A (2008) Combinatorial optimization on graphs of bounded treewidth. Comput J 51(3): 255–269
3. Courcelle B (1990) The monadic second-order logic of graphs I: Recognizable sets of finite graphs. Info Comput 85:12–75
4. Downey RG, Fellows MR (1992) Fixed-parameter intractability. In: Proceedings of the seventh annual structure in complexity theory conference, Boston, npp 36–49
5. Downey RG, Fellows MR (1999) Parameterized complexity. Monographs in computer science. Springer, New York
6. Fortune S, Hopcroft JE, Wyllie J (1980) The directed subgraph homeomorphism problem. Theor Comput Sci 10:111–121
7. Garey MR, Johnson DS (1979) Computers and intractability. W.H. Freeman, San Francisco
8. Grohe M, Kawarabayashi K, Marx D, Wollan P (2011) Finding topological subgraphs is fixed parameter tractable. In: STOC'11 proceedings of the 43rd ACM symposium on theory of computing, San Jose, pp 479–488
9. Grohe M, Marx D (2012) Structure theorem and isomorphism test for graphs with excluded topological subgraphs. In: STOC '12 proceedings of the forty-fourth annual ACM symposium on theory of computing, New York, p 173–192
10. Kawarabayashi K, Wollan P (2011) A shorter proof of the graph minor algorithm: the unique linkage theorem. In: STOC'10 proceedings of the 42rd ACM symposium on theory of computing, Cambridge, pp 687–694
11. Kawarabayashi K, Thomas R, Wollan P(2013) A new proof of the flat wall theorem. http://arxiv.org/abs/1207.6927
12. LaPaugh AS, Rivest RL (1980) The subgraph homeomorphism problem. J Comput Syst Sci 20(2):133–149
13. Robertson N, Seymour PD (1995) Graph minors XIII: the disjoint paths problem. J Combin Theory Ser B 63:65–110
14. Robertson N, Seymour PD (2010) Graph minors XXIII: Nash Williams' immersion conjecture. J Combin Theory Ser B 100(2):181–205
15. Scheffler P (1989) Linear time algorithms for graphs of bounded tree-width: the disjoint paths problem. Dresdner Reihe Forschung 5(9):49–52

First Fit Algorithm for Bin Packing

Gyorgy Dosa
University of Pannonia, Veszprém, Hungary

Keywords

Approximation ratio; Bin packing; First fit;
Tight result

Years and Authors of Summarized Original Work

1974; Johnson, Demers, Ullman, Garey, Graham
2013; Dósa, Sgall

Problem Definition

In the classical bin packing (BP) problem, we are given a set of items with rational sizes between 0 and 1, and we try to pack them into a minimum number of bins of unit size so that no bin contains items with total size more than 1. The problem definition originates in the early 1970s: Johnson's thesis [10] on bin packing together with Graham's work on scheduling [8, 9] (among other pioneering works) started and formed the whole area of approximation algorithms. The First Fit (FF) algorithm is one among the first algorithms which were proposed to solve the BP problem and analyzed in the early works. FF performs as follows: The items are first given in some list L and then are handled by the algorithm in this given order. Then, algorithm FF packs each item into the first bin where it fits; in case the item does not fit into any already opened bin, the algorithm opens a new bin and puts the actual item there. A closely related algorithm is Best Fit (BF); it packs the items also according to a given list, but each item is packed into the most full bin where it fits or the item is packed into a new bin only if it does not fit into any open bin. If the items are ordered in the list by decreasing sizes, the algorithms are called as FFD (first fit decreasing) and BFD (best fit decreasing).

Applications

There are many applications for bin packing (in industry, computer science, etc), and BP has many different versions. It is worth noting that BP has a strong relationship to the area of scheduling. So the scientific communities of packing and scheduling are almost the same.is a major

Key Results

It was immediately shown in the early works [6,12,15] that the asymptotic approximation ratio of FF and BF bin packing is 1.7. It means that if the optimum packing needs OPT bins, algorithm FF never uses more than $1.7 \cdot OPT + C$ bins, where C is a fixed constant (The same holds for the BF algorithm). It is easy to see that the multiplicative factor, i.e., 1.7, cannot be smaller. But the minimum value of the C constant, for which the statement remains valid, is not a simple issue.

First, Ullman in 1971 [15] showed that C can be chosen to be 3. But this is not the best choice.

Soon, the additive term was decreased in [6] to 2 and then in [7] to $FF \leq \lceil 1.7 \cdot OPT \rceil$; since both FF and OPT denote integer numbers, this is the same as $FF \leq 1.7 \cdot OPT + 0.9$.

Then, for many years, no new results were published regarding the possible decreasing of the additive term.

Another direction is considered in the many-times-cited work of Simchi-Levy [14]. He showed that the absolute approximation ratio of FF (and BF) is at most 1.75. It means that if we do not use an additive term in the inequality, then $FF \leq 1.75 \cdot OPT$ is valid.

Now, if we are interested in the tight result, we have two options. One is that we can try to decrease the multiplicative factor in the inequality of the absolute approximation ratio, i.e., the question is the following: What is the smallest number, say α, that can be substituted in the place of 1.75 such that the inequality $FF \leq \alpha \cdot OPT$ is valid for any input? The other direction is the following: What is the smallest possible value of the additive constant C such that the $FF \leq 1.7 \cdot OPT + C$ inequality is true for every input?

The next step was made independently from each other in two works. Xia and Tan [17] and Boyar et al. [1] proved that the absolute approximation ratio of FF is not larger than $12/7 \approx 1.7143$.

Moreover, [17] also dealt with the other direction and decreased the value of C to $FF \le 1.7 \cdot OPT + 0.7$.

If we are interested in how much the additive term (or the α factor) can be decreased, we must also deal with the lower bound of the algorithm. Regarding this, the early works give examples for both the asymptotic and absolute ratios. For the asymptotic bound, there exists such input for which $FF = 17k$ holds whenever $OPT = 10k + 1$; thus, the asymptotic upper bound 1.7 is tight, see [6, 12, 15]. For the absolute ratio, an example is given with $FF = 17$ and $OPT = 10$, i.e., an instance with approximation ratio exactly 1.7 [6, 12]. But no example was shown for large values of OPT.

It means that soon, it turned out that the value of the multiplicative factor of the absolute approximation ratio (i.e., α) cannot be smaller than 1.7 or, regarding the another measure, the additive constant cannot be chosen to be smaller than zero. But this remained an open question for 40 years whether the smallest possible choice of α is really 1.7 or, in other words, the smallest possible choice of the additive term is really zero.

Finally, the papers [3, 4] answered the question. Lower-bound instances are given with $FF = BF = \lfloor 1.7 \cdot OPT \rfloor$ for any value of OPT, and it is also shown that $FF = BF \le \lfloor 1.7 \cdot OPT \rfloor$ holds for any value of OPT. So this is the tight bound which was looked for 40 years.

Methods

To prove the upper bound, the main technique is the usage of a weighting function. Any item gets some weight according to its size. Then, to get the asymptotic ratio, it is only needed that any optimal bin has a weight at most 1.7 and any bin in the FF packing (with bounded exception) has a weight at least 1.

In the recent paper [13], a nice and surprising idea is presented: The same weight function that

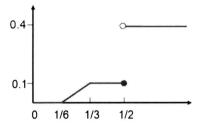

First Fit Algorithm for Bin Packing, Fig. 1 The bonus function

was used traditionally in the analysis is divided into two parts: scaled size and bonus. Thus, the weight of any item a is $w(a) = s(a) + b(a)$, where $s(a) = 6/5 \cdot a$ is the scaled size of the item and the remaining part $b(a)$ is the bonus of the item, which is defined as follows:

$b(a)$ is zero if the size of the item is below 1/6. The bonus is just 0.1 if a is between 1/3 and 1/2, and it is 0.4 if the size is above 1/2. Between 1/6 and 1/3, the bonus function is continuous and linear. We emphasize that this is the *same* old weighting function, only in a new costume. The bonus function can be seen in Fig. 1.

By this separation, it is easy to show that the weight of any optimal bin is at most 1.7, and this implies that the weight of the whole instance is at most $1.7 \cdot OPT$.

The key part is to show that on average, the weight of each FF bin is at least 1. This property trivially holds if the total size of the items in the bin is at least 5/6. It is not hard to handle the bins with single items; here, almost all of them must be bigger than 1/2, and such items have huge bonus (i.e., 0.4), together with the scaled size, that is, at least 0.6, we are again done. In the remaining bins, the next tricky calculation is used: The scaled size of the bin plus the bonus of the *following* bin is at least 1. By this trick, the proof will be almost done, but several further examinations are also needed for completing the tightness result.

New Lower Bound Construction

The lower bound construction works in the following way. Suppose, for the sake of simplicity, that OPT = 10k for some integer k, and let ε > 0, a small value.

The input consists of OPT small items of size approximately 1/6, followed by OPT medium-sized items of size approximately 1/3, followed by OPT large items of size exactly $1/2 + ε$. The optimum packs in each bin one item from each group. FF packs the small items into 2k bins with 5 items with the exception of the first and last of these bins, which will have 6 and 4 items, respectively. The sizes of items differ from 1/3, or differ from 1/6, in both directions by a small amount δ_i. Finally, every large item will occupy its own bin.

In the original construction, the choice of the small and medium-sized items is a bit difficult, so one could think that the construction *must* be so difficult, and thus, the construction cannot be tightened. It turns out, however, that this is not the case. The construction can be modified in the way that δ_i is exponentially decreasing but remains greater than ε for all i. This guarantees that only the item with the largest δ_i in a bin is relevant for its final size, and this in turn enables us to order the items so that no additional later item fits into these bins. Thus, by the modification, not only the construction is simpler but it also makes possible to prove the tightness.

Open Problems

There are many open problems regarding bin packing. For example, the tight absolute approximation ratio of BFD is an open question (For FFD, it was recently proved that $FFD \leq (11/9) \cdot OPT + 6/9$ and this is the tight result, see [5].).

Cross-References

▶ Current Champion for Online Bin Packing
▶ Harmonic Algorithm for Online Bin Packing
▶ Lower Bounds for Online Bin Packing
▶ Selfish Bin Packing Problems
▶ Subset Sum Algorithm for Bin Packing

Recommended Readings

1. Boyar J, Dósa G, Epstein L (2012) On the absolute approximation ratio for First Fit and related results. Discret Appl Math 160:1914–1923
2. Coffman EG, Garey MR, Johnson DS (1997) Approximation algorithms for bin packing: a survey. In: Hochbaum D (ed) Approximation algorithms. PWS Publishing Company, Boston
3. Dósa G, Sgall J (2013) First Fit bin packing: a tight analysis. In: Proceedings of the 30th symposium on theoretical aspects of computer science (STACS), LIPIcs vol 3. Schloss Dagstuhl, Kiel, Germany, pp 538–549
4. Dósa G, Sgall J (2014) Optimal analysis of Best Fit bin packing. In: Esparza J et al (eds) ICALP 2014. LNCS, part I, vol 8572. Springer, Heidelberg, Copenhagen, Denmark, pp 429–441
5. Dósa G, Li R, Han X, Tuza Zs (2013) Tight absolute bound for First Fit Decreasing bin-packing: FFD(L) <= 11/9 OPT(L) + 6/9. Theor Comput Sci 510:13–61
6. Garey MR, Graham RL, Ullman JD (1973) Worst-case analysis of memory allocation algorithms. In: Proceedings of the 4th symposium on theory of computing (STOC). ACM, Denver, Colorado, USA, pp 143–150
7. Garey MR, Graham RL, Johnson DS, Yao ACC (1976) Resource constrained scheduling as generalized bin packing. J Combin Theory Ser A 21:257–298
8. Graham RL (1966) Bounds for certain multiprocessing anomalies. Bell Syst Tech J 45:1563–1581
9. Graham RL (1969) Bounds on multiprocessing timing anomalies. SIAM J Appl Math 17:263–269
10. Johnson DS (1973) Near-optimal bin packing algorithms. PhD thesis, MIT, Cambridge, MA
11. Johnson DS (1974) Fast algorithms for bin packing. J Comput Syst Sci 8:272–314
12. Johnson DS, Demers A, Ullman JD, Garey MR, Graham RL (1974) Worst-case performance bounds for simple one-dimensional packing algorithms. SIAM J Comput 3:256–278
13. Sgall J (2012) A new analysis of Best Fit bin packing. In: Kranakis et al (eds) Proceedings of the 6th international conference FUN with algorithms. LNCS, vol 7288. Springer, Venice, Italy, pp 315–321
14. Simchi-Levi D (1994) New worst case results for the bin-packing problem. Naval Res Logist 41:579–585
15. Ullman JD (1971) The performance of a memory allocation algorithm. Technical report 100, Princeton University, Princeton
16. Williamson DP, Shmoys DB (2011) The design of approximation algorithms. Cambridge University Press, Cambridge
17. Xia B, Tan Z (2010) Tighter bounds of the First Fit algorithm for the bin-packing problem. Discret Appl Math 158:1668–1675

F

Fixed-Parameter Approximability and Hardness

Guy Kortsarz
Department of Computer Science, Rutgers University, Camden, NJ, USA

Keywords

Approximation; Exponential time conjecture; Fixed parameter; Inapproximability

Years and Authors of Summarized Original Work

1995; Alon, Yuster, Zwick
2005; Marx
2006; McCartin
2013; Hajiaghayi, Kortsarz

Problem Definition

NP-hard problems are believed to be intractable. This is the widely believed assumption that $P \neq NP$. For all our problems, the size of their input is denoted by n. In parameterized complexity, the input is refined to (I, k) with k a parameter related to the input, and the goal is to find an exact algorithm for the problem that runs in time $f(k) \cdot n^{O(1)}$, for some function f. In this survey, we parameterize by the optimum value of the instance unless stated otherwise. In addition, the optimum is always integral. In approximation algorithms, a ρ approximation for a minimization (maximization) problem P is a polynomial time algorithm A, such that for any instance I, A returns a solution of value $A(I)$ and $A(I)/\text{OPT}(I) \leq \rho$ ($\text{OPT}(I)/A(I) \leq \rho$) with $\text{OPT}(I)$ the optimum value for the instance. In both subjects, there are intractability results. The class FPT are the problems that admit an $f(k)n^{O(1)}$ time, exact solution for some function f. The classes $W[i]$ for every integer $i \geq 1$

Supported in part by NSF grant number 1218620.

satisfy FPT $\subseteq W[1] \subseteq W[2] \subseteq \ldots$. It is widely believed that all inclusions are strict. Consider the CLIQUE problem. Given a graph $G(V, E)$, a subset $U \subseteq V$, forms a *clique*, if for every $u, v \in U, (u, v) \in E$. The problem is

Input: A graph G and a parameter k.
Question: Is there in G a clique U of size $|U| \geq k$?

In [21], it is proved that CLIQUE admits no $n^{1-\epsilon}$ approximation unless $P = NP$. It is known that CLIQUE is $W[1]$-complete. Thus it is considered highly unlikely that CLIQUE $\in$ FPT. The SETCOVER problem is defined as follows:

Input: A universe U and a collection $\mathcal{S} = \{S_i\}$ of subsets of $\mathcal{U}$ and a parameter k.
Question: Is there a subcollection $\mathcal{S}' \subseteq \mathcal{S}$ containing at most k sets so that $\bigcup_{S_i \in \mathcal{S}'} S_i = \mathcal{U}$?

SETCOVER is $W[2]$-complete. In addition, Raz and Safra [27] show that unless $P = NP$, SETCOVER admits no $c \ln n$ algorithm for some constant c, almost matching the simple greedy $\ln n + 1$ ratio approximation algorithm.

Our Subject

Formally, we deal with the following subject: An algorithm for a minimization (resp., maximization) problem P is called an (r, t)-FPT-approximation algorithm for P with input parameter k, if the algorithm takes as input an instance I with value OPT and an integer parameter k and either computes a feasible solution to I with value at most $k \cdot r(k)$ (resp., at least $k / r(k)$ and $k / r(h) = o(k)$) or computes a certificate that $k < $ OPT (resp., $k > $ OPT) in time $t(k) \cdot |I|^{O(1)}$. The requirement that $k / r(k) = o(k)$ avoids returning a single vertex in the clique problem, claiming OPT approximation.

A problem is called (r, t)-FPT-inapproximable (or, (r, t)-FPT-hard) if it does not admit an (r, t)-FPT-approximation algorithm. An FPT approximation is mainly interesting if the problem is $W[1]$-hard and allowing running time $f(k) \cdot n^{O(1)}$ gives improved approximation. We restrict our attention to this scenario. Thus, we do not discuss many subjects such as approximations in OPT that run in *polynomial time* in n and upper and lower bounds, on algorithms with sub exponential time in n, for several conbinatorial problems.

Our Complexity Assumption

We assume the following conjecture throughout Impagliazzo et al. [4] conjectured the following:

> *Exponential Time Hypothesis* (ETH)
>
> 3-SAT cannot be solved in $2^{o(q)}(q + m)^{O(1)}$ time where q is the number of variables and m is the number of clauses.

The following is due to [4].

Lemma 1 *Assuming* ETH, *3*-SAT *cannot be solved in* $2^{o(m)}(q + m)^{O(1)}$ *time where* q *is the number of variables and* m *is the number of clauses.*

It is known that the ETH implies that $W[1] \neq$ FPT. This implies that $W[2] \neq$ FPT as well.

Key Results

We survey some FPT-approximability and FPT-inapproximability results. Our starting point is a survey by Marx [23], and we also discuss recent results. The simplest example we are aware of in which combining FPT running time and FPT-approximation algorithm gives an improved result is for the strongly connected directed subgraph (SCDS) problem.

Input: A directed graph $G(V, E)$, a set $T = \{t_1, t_2, \ldots, t_p\}$ of terminals, and an integer k.

Question: Is there a subgraph $G'(V, E')$ so that $|E'| \leq k$ and for every $t_i, t_j \in T$, there is a directed path in G' from t_i to t_j and vice versa?

The problem is in $W[1]$-hard. The best approximation algorithm known for this problem is n^ϵ for any constant ϵ. See Charikar et al. [5].

The following is due to [7].

Theorem 1 *The SCDS problem admits an FPT time 2 approximation ratio.*

Proof The directed Steiner tree problem is given a directed edge-weighted graph and a root r and a set $T = \{t_1, t_2, \ldots, t_p\}$ of terminals; find a minimum cost-directed tree rooted by r containing T. This problem belongs to FPT. See Dreyfus and Wagner [12]. Note that for every terminal t_i, any feasible solution contains a directed tree from t_i to T and a a reverse-directed Steiner tree from T to t_i. These two problems can be solved optimally in FPT time. In the second application, we reverse the direction of edges before we find the directed Steiner tree. Moreover, two such trees give a feasible solution for the SCDS problem as every two terminals t_j, t_k have a path via t_i. Clearly, the solution has value at most $2 \cdot$ OPT with OPT the optimum value for the SCDS instance. The claim follows.

Definition 1 A *polynomial time approximation scheme* (PTAS) for a problem P is a $1 + \epsilon$ approximation for any constant ϵ that runs in time $n^{f(1/\epsilon)}$. An *EPTAS* is such an algorithm that runs in time $f(1/\epsilon)n^{O(1)}$.

The *vertex cover* problem is to select the smallest possible subset U of V so that for every edge (u, v), either $u \in U$ or $v \in U$ (or both). In the *partial vertex cover problem*, a graph $G(V, E)$ and an integer k are given. The goal is to find a set U of k vertices that is touched by the largest number of edges. An edge (u, v) is touched by a set U if $u \in U$ or $v \in U$ or both. It is known that this problem admits no *PTAS* unless $P = NP$ (see Dinur and Safra [10]). The corresponding minimum partial vertex cover problem requires a set of k vertices touched by the *least number of edges*. This problem admits no better than 2-ratio, under the *small set expansion conjecture*. See [15]. Both problems belong to $W[1]$-hard. The following theorem of [23] relies on a technique called *color coding* [1].

Theorem 2 ([23]) *For every constant ϵ, the partial vertex cover problem (and in a similar proof the minimum partial vertex cover problem) admits an EPTAS that runs in time $f(k, 1/\epsilon) \cdot n^{O(1)}$ with n the number of vertices in the graph.*

Proof Let $D = \binom{k}{2}/\epsilon$. Sort the vertices $v_1, v_2, \ldots, v_n$ by nonincreasing degrees. If for

F

the largest degree, $d(v_1)$ satisfies $d(v_1) \geq D$, the algorithm outputs the set $\{v_1, v_2, \ldots, v_k\}$. These k vertices cover at least $\sum_{i=1}^{k} \deg(v_i) - \binom{k}{2}$ edges. Clearly, OPT $\leq \sum_{i=1}^{k} \deg(v_i)$. Hence, the value of the constructed solution is at least

$$\frac{\sum_{i=1}^{k} \deg(v_i) - \binom{k}{2}}{\sum_{i=1}^{k} \deg(v_i)} \geq 1 - \frac{\binom{k}{2}}{D} \geq 1 - \frac{\epsilon}{2} \geq \frac{1}{1 + \epsilon}$$

times the optimum for a $1 + \epsilon$ approximation. In the other case, the optimum OPT $\leq k \cdot D$. We guess the correct value of the optimum by trying all values between $1, \ldots, k \cdot D$. Fix the run with the correct OPT. Let E^* be the set of OPT edges that are touched by the optimum. An OPT labeling is an assignment of a label in $\{1, \ldots, \text{OPT}\}$ to the edges of E. We show that if the labels of E^* are pairwise distinct, we can solve the problem in time $h(k, 1/\epsilon)$. Let $\{u_1, u_2, \ldots, u_k\}$ be the optimum set. Let L_i be the labels of the edges of u_i. As all labels of E^* are pairwise distinct, $\{L(u_i)\}$ is a disjoint partition of all labels (as otherwise there is a labeling with less than OPT labels). The number of possible partitions of the labels into k sets is at most k^{OPT}. Given the *correct* partition $\{L_i\}$, we need to match every L_i with a vertex u_i so that the labels of u_i are L_i. This can be done in polynomial time by matching computation. To get a labeling with different pairwise labels on E^*, we draw for every edges a label between 1 and OPT, randomly and independently. The probability that the labels of E^* are disjoint is more than $1/\text{OPT}^{\text{OPT}}$. Repeating the random experiment for OPT^{OPT} times implies that with probability at least $1 - 1/e$, one of the labeling has different pairwise labels for E^*. This result can be derandomized [1].

We consider one example in which OPT is not the parameter [23]. Consider a graph that contains a set $X = \{x_1, \ldots, x_k\}$. so that $G \setminus X$ is a planar graph. Thus the parameter here is the number of vertices that need to be removed to make the graph. Consider the minimum coloring problem on G. We can determine the best coloring of X in time k^k. Then we can color $G \setminus X$ by four (different) colors. A simple calculation shows

that this algorithm has approximation ratio at most $7/3$.

The following is a simple relation exist between *EPTAS* and FPT theory.

Proposition 1 *If an optimization problem P admits an EPTAS, then $P \in$ FPT.*

Proof We prove the theorem for minimization problems. For maximization problems, the proof is similar. Assume that P has a $1 + \epsilon$ approximation that runs in time $f(1/\epsilon) \cdot n^{O(1)}$. Set $\epsilon = 1/(2k)$. Using the EPTAS algorithm gives an $f(2k)n^{O(1)}$ time $(1 + \epsilon)$ approximation. If the optimum is at most k, we get a solution of size at most $(1 + \epsilon)k = k + 1/2 < k + 1$. As the solution is integral, the cost is at most k. If the minimum is $k + 1$, the approximation will not return a better than $k + 1$ size solution. Thus the approximation returns cost at most k if and only if there is a solution of size at most k.

Thus we can rule out the possibility of an *EPTAS* if a problem is $W[1]$-hard. For example, this shows that the maximum independent set for unit disks graphs admits no *EPTAS* as it belongs to $W[1]$. See many more examples in [23]. Chen Grohe and Grüber [6] provide an early discussion of our topic. Lange wrote a PDF presentation for recent FPT approximation. The following theorem is due to Grohe and Grüber (see [19]).

Theorem 3 *If a maximization problem admits an FPT-approximation algorithm with performance ratio $\rho(k)$, then for some function ρ', there exists a $\rho'(k)$ polynomial time approximation algorithm for the problem.*

In the traveling salesperson with a deadline, the input is a metric on n points and a set $D \subseteq V$ with each $v \in D$ having a deadline t_v. A feasible solution is a simple path containing all vertices, so that *for every $v \in D$, the length of the tour until v is at most t_v*. The problem admits no constant approximation and is not in FPT when parameterized by $|D|$. See Bockenhauer, Hromkovic, Kneis, and Kupke [2]. In this entry, the authors give a 2.5 approximation that runs in time $n^{O(1)} + |D|! \cdot |D|$. The parameterized undirected multicut problem is given an undirected graph and a collection $\{s_i, t_i\}_{i=1}^m$ of pairs, and a parameter k is possible to remove at most k edges and disconnect all pairs. Garg, Vazirani, and Yannakakis give an $O(\log n)$ approximation for the problem [16]. In 2009, it was given a ratio 2 fixed-parameter approximation (Marx and Razgon) algorithm. However, Marx and Razgon [25] and Bousquet et al. [3] show that this problem is in fact in FPT. Fellows, Kulik, Rosamond, and Shachnai give the following tradeoff (see [14]). The best known exact time algorithm for the vertex cover problem has running time 1.273^k. The authors show that if we settle for an approximation result, then the running time can be improved. Specifically, they gave $\alpha \geq 1$ approximation for vertex cover that runs in time $1.237^{(2-\alpha)k}$. The minimum edge dominating set problem is given a graph and a parameter k, and there is a subset $E' \subset E$ of size at most k so that every edge in $E \setminus E'$ is adjacent to at least one edge in E'. Escoffier, Monnot, Paschos, and Mingyu Xiao (see [13]) prove that the problem admits a $1 + \epsilon$ ratio for any $0 \leq \epsilon \leq 1$ that runs in time $2^{(2-\epsilon)\cdot k}$. A kernel for a problem P is a reduction from an instance I to an instance I' whose size is $g(k)$, namely, a function of k, so that a yes answer for I implies a yes answer for I' and a no answer for I implies a no answer for I'. If a kernel exists, it is clear that $P \in$ FPT. However, the size of the kernel may determine what is the function of k in the $f(k) \cdot n^{O(1)}$ exact solution. The following result seems interesting because it may not be intuitive. In the *tree deletion* problem, we are given a graph $G(V, E)$ and a number k and the question is if we can delete up to k vertices and get a tree. Archontia Giannopoulou, Lokshtanov, Saket, and Suchy prove (see [17]) that the tree deletion problem admits a kernel of size $O(k^4)$. However, the problem does not admit an approximation ratio of OPT^c for any constant c.

Other Parameters

An *independent set* is a set vertices so that no two vertices in the set share an edge. In parameterized version given k, the question is if there is an independent set of size at least k. Clearly, the problem is $W[1]$-complete. Grohe [18] show that the maximum independent set admits a FPT-approximation scheme if the parameter is the genus of the graph. E. D. Demaine, M. Hajiaghayi, and K. Kawarabayashi [9] showed that vertex coloring has a ratio 2 approximation when parameterized by the *genus* of a graph. The tree augmentation problem is given an edge-weighted graph and a spanning tree whose edges have cost 0; find a minimum cost collection of edges to add to the tree, so that the resulting graph is 2-edge connected. The problem admits several polynomial time, ratio 2, and approximation algorithms. Breaking the 2 ratio for the problem is an important challenge in approximation algorithms. Cohen and Nutov parameterized the problem by the diameter D of the tree and gave an $f(D) \cdot n^{O(1)}$ time, $1 + \ln 2 < 1.7$ approximation algorithm for the problem [8].

Fixed-Parameter Inapproximability

The following inapproximability is from [11]. The additive maximum independent set problem is given a graph and a parameter k and a constant c, and the question is if the problem admits an independent set of size at least $k - c$ or no independent set of size k exists.

It turns out that the problem is equivalent to the independent set problem.

Theorem 4 *Unless $W[1] =$ FPT (hence, under the* ETH*), the independent set problem admits no additive c approximation.*

Proof Let I be the instance. Find the smallest d so that

$$\left\lceil \frac{dk - c}{d} \right\rceil \geq k.$$

Output d copies of G and let $k \cdot d$ be the parameter of the new instance I'. We show that the new graph has independent set of size $dk - c$ if and only if the original instance has an independent set of size k. If the original instance has an independent set of size k, taking union of d independent sets, we get an independent set of size $k \cdot d$.

Now say that I' has an independent set of size $dk - c$. The average size of an independent set in a graph in I' is then $(dk - c)/d$. Since the size of the independent set is integral, there is a copy that admits an independent set of size

$$\left\lceil \frac{dk - c}{d} \right\rceil \geq k.$$

An independent set I is *maximal* if for every $v \notin I$, $v + I$ is not an independent set. The problem of minimum size maximal independent set (MSDIS) is shown to be completely inapproximable in [11]. Namely, this problem is $(r(k), t(k))$-FPT-hard for any r, t, unless FPT $= W[2]$ (hence, under the ETH). The problem admits no $n^{1-\epsilon}$ approximation (see [20]). In the min-WSAT problem, a Boolean circuit is given and the task is to find a satisfying assignment of minimum weight. The weight of an assignment is the number of true variables. Min-WSAT was given a complete inapproximability by Chen, Grohe, Grüber (see [6]) 2006.

The above two problems are not monotone. This implies that the above results are non-surprising. The most meaningful complete inapproximability is given by Marx [24] who shows that the weighted circuit satisfiability for monotone or antimonotone circuits is completely FPT inapproximable.

Of course, if the problem has almost no gap, namely, the instance can have value k or $k - 1$, it is hard to get a strong hardness.

A natural question is if we can use gap reductions from approximation algorithms theory to get some strong lower bounds, in particular for clique and setcover. It turns out that this is very difficult even under the ETH conjecture. This subject is related to almost linear PCP (see [26]). In this entry, Moshkovitz poses a conjecture called *the projection game conjecture* (PGC). M. Hajiaghayi, R. Khandekar, and G. Kortsarz show the following theorem.

Theorem 5 *Under the* ETH *and* PGC *conjectures,* SETCOVER *is* (r, t)-FPT-*hard for* $r(k) = (\log k)^{\gamma}$ *and* $t(k) = \exp(\exp((\log k)^{\gamma})) \cdot$ poly$(n) = \exp\left(k^{(\log^f k)}\right) \cdot$ poly(n) *for some constant* $\gamma > 1$ *and* $f = \gamma - 1$.

Cross-References

▶ Color Coding

Recommended Readings

1. Alon N, Yuster R, Zwick U (1995) Color coding. J ACM 42(4):844–856
2. Böckenhauer HJ, Hromkovic J, Kneis J, Kupke J (2007) The parameterized approximability of TSP with deadlines. Theory Comput Syst 41(3):431–444
3. Bousquet N, Daligault J, Thomassé S (2011) Multicut is FPT. In: STOC'11, New York, pp 459–468
4. Calabro C, Impagliazzo R, Paturi R (1995) A duality between clause width and clause density for SAT. In: Computational Complexity, Prague, pp 252–260
5. Charikar M, Chekuri C, Cheung TY, Dai Z, Goel A, Guha S, Li M (1999) Approximation algorithms for directed steiner problems. J Algorithms 33(1):73–91
6. Chen Y, Grohe M, Grüber M (2006) On parameterized approximability. In: IWPEC 2006, Zurich, pp 109–120
7. Chitnis R, Hajiaghayi MT, Kortsarz G (2013) Fixed parameter and approximation algorithms: a new look. In: IPEC, Sophia Antipolis, pp 110–122
8. Cohen N, Nutov Z (2013) A (1+ln2)-approximation algorithm for minimum-cost 2-edge-connectivity augmentation of trees with constant radius. Theor Comput Sci 489–490: 67–74
9. Demaine E, Hajiaghayi MT, Kawarabayashi KI (2005) Algorithmic graph minor theory: decomposition, approximation, and coloring. In: FOCS 2005, Pittsburgh, pp 637–646
10. Dinur I, Safra S (2002) On the importance of being biased. In: STOC 2002, Québec, pp 33–42
11. Downey R, Fellows M (1995) Parameterized approximation of dominating set problems. Inf Process Lett 109(1):68–70
12. Dreyfus SE, Wagner RA (1971) The steiner problem in graphs. Networks 1(3):195–207
13. Escoffier B, Monnot J, Paschos VT, Xiao M (2012) New results on polynomial inapproximability and fixed parameter approximability of edge dominating set. In: IPEC, Ljubljana, pp 25–36
14. Fellows MR, Kulik A, Rosamond FA, Shachnai H (2012) Parameterized approximation via fidelity preserving transformations. In: ICALP, Warwick, pp 351–362
15. Gandhi R, Kortsarz G (2014) On set expansion problems and the small set expansion conjecture. In: WG, Nouan-le-Fuzelier, pp 189–200
16. Garg N, Vazirani VV, Yannakakis M (1996) Approximate max-flow min-(multi)cut theorems and their applications. SIAM J Comput 25(2):235–251

17. Giannopoulou A, Lokshtanov D, Saket S, Suchy O (2013) Tree deletion set has a polynomial kernel (but no OPT^O(1) approximation). arXiv:1309.7891
18. Grohe M (2003) Local tree-width, excluded minors, and approximation algorithms. Combinatorica 23(4):613–632
19. Grohe M, Grüber M (2007) Parameterized approximability of the disjoint cycle problem. In: ICALP 2007, Wroclaw, pp 363–374
20. Halldórsson MM (1993) Approximating the minimum maximal independence number. Inf Process Lett 46(4):169–172
21. Hastad J (1996) Clique is hard to approximate within $n^{1-\epsilon}$. In: FOCS, Burlington, pp 627–636
22. Impagliazzo R, Paturi R, Zane F (1998) Which problems have strongly exponential complexity? J Comput Syst Sci 63(4):512–530
23. Marx D (2005) Parameterized complexity and approximation algorithms. Comput J 51(1): 60–78
24. Marx D (2013) Completely inapproximable monotone and antimonotone parameterized problems. J Comput Syst Sci 79(1):144–151
25. Marx D, Razgon I (2014) Fixed-parameter tractability of multicut parameterized by the size of the cutset. SIAM J Comput 43(2):355–388
26. Moshkovitz D (2012) The projection games conjecture and the NP-hardness of ln n-approximating set-cover. In: APPROX-RANDOM 2012, Boston, pp 276–287
27. Raz R, Safra S (1997) A sub-constant error-probability low-degree test, and a sub-constant error-probability PCP characterization of NP. In: STOC 1997, San Diego, pp 475–484

Floorplan and Placement

Yoji Kajitani
Department of Information and Media Sciences, The University of Kitakyushu, Kitakyushu, Japan

Keywords

Alignment; Dissection; Layout; Packing

Years and Authors of Summarized Original Work

1994; Kajitani, Nakatake, Murata, Fujiyoshi

Problem Definition

The problem is concerned with efficient coding of the constraint that defines the placement of objects on a plane without mutual overlapping. This has numerous motivations, especially in the design automation of integrated semiconductor chips, where almost hundreds of millions of rectangular modules shall be placed within a small rectangular area (chip). Until 1994, the only known coding efficient in computer-aided design was *Polish-Expression* [1]. However, this can only handle a limited class of placements of the *slicing structure*. In 1994 Nakatake, Fujiyoshi, Murata, and Kajitani [2] and Murata, Fujiyoshi, Nakatake, and Kajitani [3] were finally successful to answer this long-standing problem in two contrasting ways. Their code names are *Bounded-Sliceline-Grid* (BSG) for floorplanning and *Sequence-Pair* (SP) for placement.

Notations

1. *Floorplanning, placement, compaction, packing, layout:* Often they are used as exchangeable terms. However, they have their own implications to be used in the following context. *Floorplanning* concerns the design of the plane by restricting and partitioning a given area on which objects are able to be properly *placed*. *Packing* tries a placement with an intention to reduce the area occupied by the objects. *Compaction* supports packing by pushing objects to the center of the placement. The result, including other environments, is the *layout*. BSG and SP are paired concepts, the former for "floorplanning," the latter for "placement."

2. *ABLR-relation:* The objects to be placed are assumed rectangles in this entry though they could be more general depending on the problem. For two objects p and q, p is said to be *above* q (denoted as pAq) if the bottom edge (boundary) of p is above the top edge of q. Other relations with respect to "*below*"

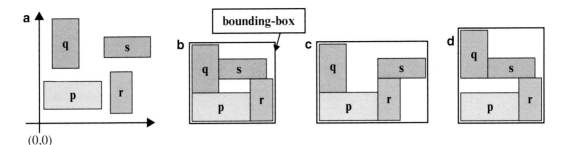

Floorplan and Placement, Fig. 1 (**a**) A feasible placement whose ABLR-relations could be observed differently. (**b**) Compacted placement if ABLR-relations are (qLr), (sAp), …. Its sequence-pair is SP = (qspr,pqrs) and single-sequence is SS = (2413). (**c**) Compacted placement for (qLr), (sRp), …. SP = (qpsr,pqrs). SS = (2143). (**d**) Compacted placement if (qAr), (sAp), …. SP = (qspr,prqs). SS = (3412)

(pBq), "*left-of*" (pLq), and "*right-of*" (pRq) are analogously defined. These four relations are generally called *ABLR-relations*.

A placement without mutual overlapping of objects is said to be *feasible*. Trivially, a placement is feasible if and only if every pair of objects is in one of ABLR-relations. The example in Fig. 1 will help these definitions.

It must be noted that a pair of objects may satisfy two ABLR-relations simultaneously, but not three. Furthermore, an arbitrary set of ABLR-relations is not necessarily *consistent* for any feasible placement. For example, any set of ABLR-relations including relations (pAq), (qAr), and (rAp) is not consistent.

3. *Compaction:* Given a placement, its *bounding-box* is the minimum rectangle that encloses all the objects. A placement of objects is evaluated by the smallness of the bounding-box's area, abbreviated as the *bb-area*. An ABLR-relation set is also evaluated by the minimum bb-area of all the placements that satisfy the set. However, given a consistent ABLR-relation set, the corresponding placement is not unique in general. Still, the minimum bb-area is easily obtained by a common technique called the "Longest-Path Algorithm." (See, e.g., [4].)

Consider the placement whose objects are all inside the 1st quadrant of the xy-coordinate system, without loss of generality with respect to minimizing the bb-area. It is evident that if a given ABLR-relation set is feasible, there is an object that has no object left or below it.

Place it such that its left-bottom corner is at the origin. From the remaining objects, take one that has no object left of or below it. Place it as leftward and downward as long as any ABLR-relation with already fixed objects is not violated. See Fig. 1 to catch the concept, where the ABLR-relation set is the one obtained the placement in (a) (so that it is trivially feasible). It is possible to obtain different ABLR-relation sets, according to which compaction would produce different placements.

4. *Slice-line:* If it is possible to draw a straight horizontal line or vertical line to separate the objects into two groups, the line is said a *slice-line*. If each group again has a slice-line, and so does recursively, the placement is said to be a *slicing structure*. Figure 2 shows placements of slicing and non-slicing structures.

5. *Spiral:* Two structures each consisting of four line segments connected by a *T-junction* as shown in Fig. 3a are *spirals*. Their regular alignment in the first quadrant as shown in (b) is the *Bounded-Sliceline-Grid* or BSG. A BSG is a *floorplan*, or a *T-junction dissection*, of the rectangular area into rectangular regions called *rooms*. It is denoted as an $n \times m$ BSG if the numbers of rows and columns of its rooms are n and m, respectively. According to the left-bottom room being p-type or q-type, the BSG is said to be p-type or q-type, respectively.

In a BSG, take two rooms x and y. The ABLR-relations between them are all that is defined by

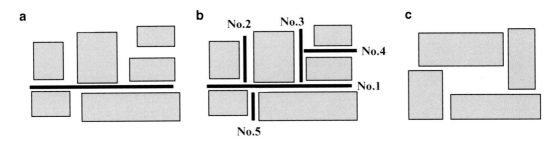

Floorplan and Placement, Fig. 2 (a) A placement with a slice-line. (b) A slicing structure since a slice-line can be found in each ith hierarchy No. $k(k = 1, 2, 3, 4)$. (c) A placement that has no slice-line

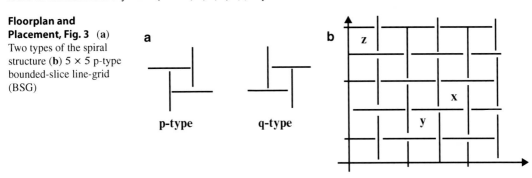

Floorplan and Placement, Fig. 3 (a) Two types of the spiral structure (b) 5 × 5 p-type bounded-slice line-grid (BSG)

the rule: If the bottom segment of x is the top segment of y (Fig. 3), room x is above room y. Furthermore, *Transitive Law* is assumed: If "x is above y" and "z is above x," then "z is above y."

Other relations are analogously defined.

Lemma 1 *A room is in a unique ABLR-relation with every other room.*

An $n \times n$ BSG has n^2 rooms. A BSG-assignment is a *one-to-one mapping* of n objects into the rooms of $n \times n$ BSG. ($n^2 - n$ rooms remain vacant.)

After a BSG-assignment, a pair of two objects inherits the same ABLR-relation as the ABLR-relation defined between corresponding rooms. In Fig. 3, if x, y, and z are the names of objects, the ABLR-relations among them are {(xAy), (xRz), (yBx), (yBz), (zLx), (zAy)}.

Key Results

The input is n objects that are rectangles of arbitrary sizes. The main concern is the *solution space*, the collection of distinct consistent ABLR-relation sets, to be generated by BSG or SP.

Theorem 1 ([4,5])

1. *For any feasible ABLR-relation set, there is a BSG-assignment into $n \times n$ BSG of any type that generates the same ABLR-relation set.*
2. *The size $n \times n$ is a minimum: if the number of rows or columns is less than n, there is a feasible ABLR-relation set that is not obtained by any BSG-assignment.*

The proof to (1) is not trivial [5] (Appendix). The number of solutions is $_{n^2}C_n$. A remarkable feature of an $n \times n$ BSG is that any ABLR-relation set of n objects is generated by a proper BSG-assignment. By this property, BSG is said to be *universal* [11].

In contrast to the BSG-based generation of consistent ABLR-relation sets, SP directly imposes the ABLR-relations on objects.

A pair of permutations of object names, represented as (Γ^+, Γ^-), is called the sequence-pair, or SP. See Fig. 1. An SP is decoded to a unique ABLR-relation set by the rule:

Consider a pair (x, y) of names such that x is before y in Γ^-. Then (xLy) or (xAy) if x is before or after y in Γ^+, respectively. ABLR-relations

"B" and "R" can be derived as the inverse of "A" and "L." Examples are given in Fig. 1.

A remarkable feature of sequence-pair is that its generation and decoding are both possible by simple operations. The question is what the solution space of all SPs is.

Theorem 2 *Any feasible placement has a corresponding SP that generates an ABLR-relation set satisfied by the placement. On the other hand, any SP has a corresponding placement that satisfies the ABLR-relation set derived from the SP.*

Using SP, a common compaction technique mentioned before is described in a very simple way:

Minimum Area Placement from $SP = (\Gamma^+, \Gamma^-)$

1. Relabel the objects such that $\Gamma^- = (1, 2, \ldots, n)$. Then $\Gamma^+ = (p_1, p_2, \ldots, p_n)$ will be a permutation of numbers $1, 2, \ldots, n$. It is simply a kind of normalization of SP [6]. But Kajitani [11] considers it a concept derived from Q-sequence [10] and studies its implication by the name of *single-sequence* or SS. In the example in Fig. 1b, p, q, r, and s are labeled as 1, 2, 3, and 4 so that SS = (2413).
2. Take object 1 and place it at the left-bottom corner in the 1st quadrant.
3. For $k = 2, 3, \ldots, n$, place k such that its left edge is at the rightmost edge of the objects with smaller numbers than k and lie before k in SS, and its bottom edge is at the topmost edge of the objects with smaller numbers than k and lie after k in SS.

Applications

Many ideas followed after BSG and SP [2–5] as seen in the reference. They all applied a common methodology of a stochastic heuristic search, called simulated annealing, to generate feasible placements one after another based on some evaluation (with respect to the smallness of the bb-area) and to keep the best-so-far as the output. This methodology has become practical

by the speed achieved due to their simple data structure. The first and naive implementation of BSG [2] could output the layout of sufficiently small area placement of 500 rectangles in several minutes. (Finding a placement with the minimum bb-area is NP-hard [3].) Since then many ideas followed, including currently widely used codes such as O-tree [7], B*-tree [8], corner block list [9], Q-sequence [10], single-sequence [11], and others. Their common feature is in coding the nonoverlapping constraint along horizontal and vertical directions, which is the inheritant property of rectangles.

As long as applications are concerned with the rectangle placement in the minimum area and do not mind mutual interconnection, the problem can be solved practically enough by BSG, SP, and those related ideas. However, in an integrated circuit layout problem, mutual connection is a major concern. Objects are not restricted to rectangles, even soft objects are used for performance. Many efforts have been devoted with a certain degree of success. For example, techniques concerned with rectilinear objects, rectilinear chip, insertion of small but numerous elements like buffers and decoupling capacitors, replacement for design change, symmetric placement for analog circuit design, three-dimensional placement, etc. have been developed. Here few of them is cited but it is recommended to look at proceedings of ICCAD (International Conference on Computer-Aided Design), DAC (Design Automation Conference), ASPDAC (Asia and South Pacific Design Automation Conference), DATE (Design Automation and Test in Europe), and journals TCAD (IEEE Trans. on Computer-Aided Design) and TCAS (IEEE Trans. on Circuit and Systems), particularly those that cover VLSI (Very Large Scale Integration) physical design.

Open Problems

BSG

The claim of Theorem 1 that a BSG needs n rows to provide any feasible ABLR-relation set is reasonable if considering a placement of all objects

Floorplan and Placement, Fig. 4
Octagonal BSG of size n, p-type: (**a**) If n is odd, it has $(n^2 + 1)/2$ rooms. (**b**) If n is even, it has $(n^2 + 2n)/2$ rooms

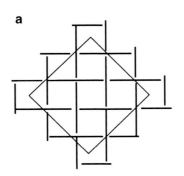

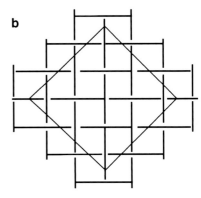

aligned vertically. This is due to the rectangular framework of a BSG. However, experiments have been suggesting a question if from the beginning [5] if we need such big BSGs. The octagonal BSG is defined in Fig. 4. It is believed to hold the following claim expecting a drastic reduction of the solution space.

Conjecture (BSG): For any feasible ABLR-relation set, there is an assignment of n objects into octagonal BSG of size n, any type, that generates the same ABLR-relation set.

If this is true, then the size of the solution space needed by a BSG reduces to $_{(n^2+1)/2}C_n$ or $_{(n^2+2n)/2}C_n$.

SP or SS

It is possible to define the universality of SP or SS in the same manner as defined for BSG. In general, two sequences of arbitrary k numbers $P = (p_1, p_2, \ldots, p_k)$ and $Q = (q_1, q_2, \ldots, q_k)$ are said *similar* with each other if $\text{ord}(p_i) = \text{ord}(q_i)$ for every i where $\text{ord}(p_i) = j$ implies that p_i is the jth smallest in the sequence. If they are single-sequences, two similar sequences generate the same set of ABLR-relations under the natural one-to-one correspondence between numbers.

An SS of length m (necessarily $\geq n$) is said *universal of order n* if SS has a subsequence (a sequence obtained from SS by deleting some of the numbers) that is similar to any sequence of length n. Since rooms of a BSG are considered n^2 objects, Theorem 1 implies that there is a

universal SS of order n whose length is n^2. The known facts about smaller universal SS are:

1. For $n = 2, 132, 231, 213$, and 312 are the shortest universal SS. Note that 123 and 321 are not universal.
2. For $n = 3$, SS $= 41352$ is the shortest universal SP.
3. For $n = 4$, the shortest length of universal SS 10 or less.
4. The size of universal SS is $\Omega(n^2)$ (Imahori S, Dec 2005, Private communication).

Open Problem (SP)

It is still an open problem to characterize the universal SP. For example, give a way to (1) certify a sequence as universal and (2) generate a minimum universal sequence for general n.

Cross-References

▶ Circuit Placement
▶ Slicing Floorplan Orientation

Recommended Readings

1. Wong DF, Liu CL (1985) A new algorithm for floorplan design. In: ACM/IEEE design automation conference (DAC), 23 Nov 1985, pp 101–107
2. Nakatake S, Murata H, Fujiyoshi K, Kajitani Y (1994) Bounded sliceline grid (BSG) for module packing. IEICE Technical report, Oct 1994, VLD94-66, vol 94, no 313, pp 19–24 (in Japanese)

3. Murata H, Fujiyoshi K, Nakatake S, Kajitani Y (1995) A solution space of size $(n!)^2$ for optimal rectangle packing. In: 8th Karuizawa workshop on circuits and systems, April 1995, pp 109–114
4. Murata H, Nakatake S, Fujiyoshi K, Kajitani Y (1996) VLSI module placement based on rectangle-packing by sequence-pair. IEEE Trans Comput Aided Des (TCAD) 15(12):1518–1524
5. Nakatake S, Fujiyoshi K, Murata H, Kajitani Y (1998) Module packing based on the BSG-structure and IC layout applications. IEEE Trans Comput Aided Des (TCAD) 17(6):519–530
6. Kodama C, Fujiyoshi K (2003) Selected sequence-pair: an efficient decodable packing representation in linear time using sequence-pair. In: Proceedings of Asia and South Pacific design automation conference (ASP-DAC), Bangalore, 2003, pp 331–337
7. Guo PN, Cheng CK, Yoshimura T (1998) An O-tree representation of non-slicing floorplan and its applications. In: 36th design automation conference (DAC), June 1998, pp 268–273
8. Chang Y-C, Chang Y-W, Wu G-M, Wu S-W (2000) B*-trees: a new representation for non-slicing floorplans. In: 37th design automation conference (DAC), June 2000, pp 458–463
9. Hong X, Dong S, Ma Y, Cai Y, Cheng CK, Gu J (2000) Corner block list: an efficient topological representation of non-slicing floorplan. In: International computer aided design (ICCAD) '00, San Jose, Nov 2000, pp 8–12
10. Sakanushi K, Kajitani Y, Mehta D (2003) The quarter-state-sequence floorplan representation. IEEE TCAS-I 50(3):376–386
11. Kajitani Y (2006) Theory of placement by single-sequence related with DAG, SP, BSG, and O-tree. In: International symposium on circuits and systems, May 2006

Flow Time Minimization

Luca Becchetti[1], Stefano Leonardi[1], Alberto Marchetti-Spaccamela[1], and Kirk Pruhs[2]
[1]Department of Information and Computer Systems, University of Rome, Rome, Italy
[2]Department of Computer Science, University of Pittsburgh, Pittsburgh, PA, USA

Keywords

Flow time: response time

Years and Authors of Summarized Original Work

2001; Becchetti, Leonardi, Marchetti-Spaccamela, Pruhs

Problem Definition

Shortest-job-first heuristics arise in sequencing problems, when the goal is minimizing the perceived latency of users of a multiuser or multitasking system. In this problem, the algorithm has to schedule a set of jobs on a pool of m identical machines. Each job has a release date and a processing time, and the goal is to minimize the average time spent by jobs in the system. This is normally considered a suitable measure of the quality of service provided by a system to interactive users. This optimization problem can be more formally described as follows:

Input

A set of m identical machines and a set of n jobs $1, 2, \ldots, n$. Every job j has a release date r_j and a processing time p_j. In the sequel, $\mathcal{I}$ denotes the set of feasible input instances.

Goal

The goal is minimizing the *average flow* (also known as *average response*) time of the jobs. Let C_j denote the time at which job j is completed by the system. The flow time or response time F_j of job j is defined by $F_j = C_j - r_j$. The goal is thus minimizing

$$\min \frac{1}{n} \sum_{j=1}^{n} F_j .$$

Since n is part of the input, this is equivalent to minimizing the *total* flow time, i.e., $\sum_{j=1}^{n} F_j$.

Off-line versus On-line

In the *off-line setting*, the algorithm has full knowledge of the input instance. In particular, for every $j = 1, \ldots, n$, the algorithm knows r_j and p_j.

Conversely, in the *on-line setting*, at any time t, the algorithm is only aware of the set of jobs released up to time t.

In the sequel, A and OPT denote, respectively, the algorithm under consideration and the optimal, off-line policy for the problem. $A(I)$ and $OPT(I)$ denote the respective costs on a specific input instance I.

Further Assumptions in the On-line Case

Further assumptions can be made as to the algorithm's knowledge of processing times of jobs. In particular, in this survey an important case is considered, realistic in many applications, i.e., that p_j is completely unknown to the on-line algorithms until the job eventually completes (*non-clairvoyance*) [1, 3].

Performance Metric

In all cases, as is common in combinatorial optimization, the performance of the algorithm is measured with respect to its optimal, off-line counterpart. In a minimization problem such as those considered in this survey, the competitive ratio ρ_A is defined as:

$$\rho_A = \max_{I \in \mathcal{I}} \frac{A(I)}{OPT(I)}.$$

In the off-line case, ρ_A is the *approximation ratio* of the algorithm. In the on-line setting, ρ_A is known as the *competitive ratio* of A.

Preemption

When *preemption* is allowed, a job that is being processed may be interrupted and resumed later after processing other jobs in the interim. As shown further, preemption is necessary to design efficient algorithms in the framework considered in this survey [5, 6].

Key Results

Algorithms

Consider any job j in the instance and a time t in A's schedule, and denote by $w_j(t)$ the amount of time spent by A on job j until t. Denote by $x_j(t) = p_j - w_j(t)$ its *remaining processing time* at t.

The best known heuristic for minimizing the average flow time when preemption is allowed is *shortest remaining processing time* (SRPT). At any time t, SRPT executes a pending job j such that $x_j(t)$ is minimum. When preemption is not allowed, this heuristic translates to *shortest job first* (SJF): at the beginning of the schedule, or when a job completes, the algorithm chooses a pending job with the shortest processing time and runs it to completion.

Complexity

The problem under consideration is polynomially solvable on a single machine when preemption is allowed [9, 10]. When preemption is allowed, SRPT is optimal for the single-machine case. On parallel machines, the best known upper bound for the preemptive case is achieved by SRPT, which was proven to be $O(\log \min n/m, P)$-approximate [6], P being the ratio between the largest and smallest processing times of the instance. Notice that SRPT is an on-line algorithm, so the previous result holds for the on-line case as well. The authors of [6] also prove that this lower bound is tight in the on-line case. In the off-line case, no non-constant lower bound is known when preemption is allowed.

In the non-preemptive case, no off-line algorithm can be better than $\Omega(n^{1/3-\epsilon})$-approximate, for every $\epsilon > 0$, the best upper bound being $O(\sqrt{n/m} \log(n/m))$ [6]. The upper and lower bound become $O(\sqrt{n})$ and $\Omega(n^{1/2-\epsilon})$ for the single machine case [5].

Extensions

Many extensions have been proposed to the scenarios described above, in particular for the preemptive, on-line case. Most proposals concern the power of the algorithm or the knowledge of the input instance. For the former aspect, one interesting case is the one in which the algorithm is equipped with faster machines than its optimal counterpart. This aspect has been considered in [4]. There the authors prove that even a moderate increase in speed makes some very simple heuristics have performances that can be very close to the optimum.

As to the algorithm's knowledge of the input instance, an interesting case in the on-line setting, consistent with many real applications, is the non-clairvoyant case described above. This aspect has been considered in [1, 3]. In particular, the authors of [1] proved that a randomized variant of the MLF heuristic described above achieves a competitive ratio that in the average is at most a polylogarithmic factor away from the optimum.

Applications

The first and traditional field of application for scheduling policies is resource assignment to processes in multitasking operating systems [11]. In particular, the use of shortest-job-like heuristics, notably the MLF heuristic, is documented in operating systems of wide use, such as UNIX and WINDOWS NT [8, 11]. Their application to other domains, such as access to Web resources, has been considered more recently [2].

Open Problems

Shortest-job-first-based heuristics such as those considered in this survey have been studied in depth in the recent past. Still, some questions remain open. One concerns the off-line, parallel-machine case, where no non-constant lower bound on the approximation is known yet. As to the on-line case, there still is no tight lower bound for the non-clairvoyant case on parallel machines. The current $\Omega(\log n)$ lower bound was achieved for the single-machine case [7], and there are reasons to believe that it is below the one for the parallel case by a logarithmic factor.

Cross-References

▶ Minimum Flow Time
▶ Minimum Weighted Completion Time

▶ Multilevel Feedback Queues
▶ Shortest Elapsed Time First Scheduling

Recommended Reading

1. Becchetti L, Leonardi S (2004) Nonclairvoyant scheduling to minimize the total flow time on single and parallel machines. J ACM 51(4):517–539
2. Crovella ME, Frangioso R, Harchal-Balter M (1999) Connection scheduling in web servers. In: Proceedings of the 2nd USENIX symposium on Internet technologies and systems (USITS-99), pp 243–254
3. Kalyanasundaram B, Pruhs K (2003) Minimizing flow time nonclairvoyantly. J ACM 50(4):551–567
4. Kalyanasundaram B, Pruhs K (2000) Speed is as powerful as clairvoyance. J ACM 47(4):617–643
5. Kellerer H, Tautenhahn T, Woeginger GJ (1996) Approximability and nonapproximability results for minimizing total flow time on a singlemachine. In: Proceedings of 28th annual ACM symposium on the theory of computing (STOC'96), pp 418–426
6. Leonardi S, Raz D (1997) Approximating total flow time on parallel machines. In: Proceedings of the annual ACM symposium on the theory of computing (STOC), pp 110–119
7. Motwani R, Phillips S, Torng E (1994) Nonclairvoyant scheduling. Theor Comput Sci 130(1):17–47
8. Nutt G (1999) Operating system projects using Windows NT. Addison-Wesley, Reading
9. Schrage L (1968) A proof of the optimality of the shortest remaining processing time discipline. Oper Res 16(1):687–690
10. Smith DR (1976) A new proof of the optimality of the shortest remaining processing time discipline. Oper Res 26(1):197–199
11. Tanenbaum AS (1992) Modern operating systems. Prentice-Hall, Englewood Cliffs

Force-Directed Graph Drawing

Ulrik Brandes
Department of Computer and Information Science, University of Konstanz, Konstanz, Germany

Keywords

Force-directed placement; Graph drawing; MDS; Spring embedder

Years and Authors of Summarized Original Work

1963; Tutte
1984; Eades

Problem Definition

Given a connected undirected graph, the problem is to determine a straight-line layout such that the structure of the graph is represented in a readable and unbiased way. Part of the problem is the definition of readable and unbiased.

Formally, we are given a simple, undirected graph $G = (V, E)$ with vertex set V and edge set $E \subseteq \binom{V}{2}$. Let $n = |V|$ be the number of vertices and $m = |E|$ the number of edges. The *neighbors* of a vertex v are defined as $N(v) = \{u : \{u, v\} \in E\}$, and $\deg(v) = |N(v)|$ is its *degree*. We assume that G is connected, for otherwise the connected components can be treated separately.

A (two-dimensional) layout for G is a vector $p = (p_v)_{v \in V}$ of vertex positions $p_v = \langle x_y, y_v \rangle \in \mathbb{R}^2$. Since edges are drawn as line segments, the drawing is completely determined by these vertex positions. All approaches in this chapter generalize to higher-dimensional layouts, and there are variants for various graph classes and desired layout features; we only discuss some of them briefly at the end.

The main idea is to make use of physical analogies. A graph is likened to a system of objects (the vertices) that are subject to varying forces (derived from structural features). Forces cause the objects to move around until those pushing and pulling into different directions cancel each other out and the graph layout reaches an equilibrium state. Equivalently, states might be described by an energy function so that forces are not specified directly, but derived from gradients of the energy function.

As a reference model, consider the layout energy function

$$A(p) = \sum_{\{u,v\} \in E} \|p_u - p_v\|^2 \qquad (1)$$

where $\|p_u - p_v\|^2 = (x_u - x_v)^2 + (y_u - y_v)^2$ is the squared Euclidean distance of the endpoints of edge $\{u, v\}$. It associates with a layout the sum of squared edge lengths, so that its minimization marks an attempt to position adjacent vertices close to each other. Because of its straightforward physical analogy, we refer to the minimization of Eq. (1) as the *attraction model*.

Note that minimum-energy layouts of this pure attraction model are degenerate in that all vertices are placed in the same position, since such layouts p are exactly those for which $A(p) = 0$ for a connected graph.

Even if it has not been the starting point of any of the approaches sketched in the next section, it is instructive to think of them as different solutions to the degeneracy problem inherent in the attraction model.

Key Results

We present force-directed layout methods as variations on the attraction model. The first two variants retain the objective but introduce constraints, whereas the other two modify the objective (Fig. 1).

For the constraint-based variants, it is more convenient to analyze the attraction model in matrix form. A necessary condition for a (local) minimum of any objective function is that all partial derivatives vanish. For the attraction model (1), this amounts to

$$\frac{\partial}{\partial x_v} A(p) = \frac{\partial}{\partial y_v} A(p) = 0 \quad \text{for all } v \in V.$$

For any $v \in V$,

$$\frac{\partial}{\partial x_v} A(p) = \sum_{u \in N(v)} 2(x_v - x_u) \stackrel{!}{=} 0,$$

and likewise for $\frac{\partial}{\partial y_v} A(p)$. The necessary conditions can therefore be translated into

$$x_v = \frac{\sum_{u \in N(v)} x_u}{\deg(v)} \quad \text{and} \quad y_v = \frac{\sum_{u \in N(v)} y_u}{\deg(v)}$$

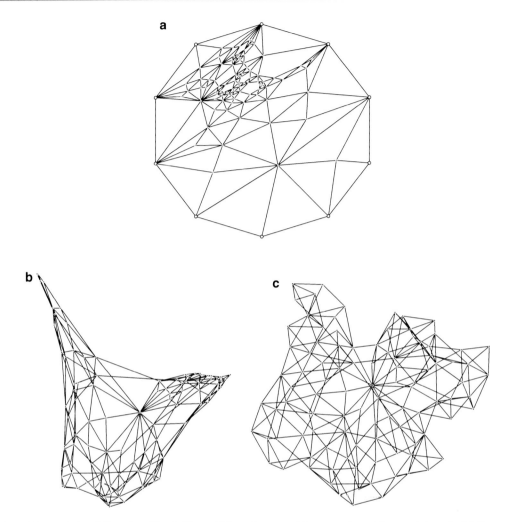

Force-Directed Graph Drawing, Fig. 1 Three different layouts of the same planar triconnected graph. (**a**) Barycentric. (**b**) Spectral. (**c**) Stress

for all $v \in V$, i.e., every vertex must lie in the barycenter of its neighbors. Bringing all variables to the left-hand side, we obtain a system of linear equations whose coefficients form an eminent graph-related matrix, the *Laplacian matrix* $L(G) = D(G) - A(G)$, where $D(G)$ is a diagonal matrix with diagonal entries $\deg(v)$, $v \in V$, and $A(G)$ is the adjacency matrix of G. The entries of $L = L(G) = (\ell_{uv})_{u,v \in V}$ are thus

$$\ell_{uv} = \begin{cases} \deg(v) & \text{if } u = v \\ -1 & \text{if } u \neq v \text{ and } \{u, v\} \in E \\ 0 & \text{otherwise.} \end{cases}$$

so that the optimality conditions can be written as

$$L \cdot p = \mathbf{0}, \qquad (2)$$

where $\mathbf{0}$ is an $n \times 2$-matrix of zeros. As discussed above, the solutions to this system of linear equations are given by those layouts p in which all x- and all y-coordinates are identical.

Fixed Boundary

An intuitive approach to prevent attraction from collapsing an entire graph onto a single point is to grab a few of its vertices and drag them apart. Technically, this corresponds to constraining the layout by fixing select vertices to distinct positions.

Let $B \subseteq V$ be a nonempty subset of *boundary vertices* for which positions $\hat{p}_v = \langle \hat{x}_v, \hat{y}_v \rangle$, $v \in B$ are pre-specified. A layout is called *barycentric* (with respect to these constraints) if it satisfies

$$
p_v = \begin{cases} \hat{p}_v & \text{if } v \in B \\ \frac{1}{\deg(v)} \sum_{u \in N(v)} p_u & \text{otherwise.} \end{cases}
$$

We next show that the solution of the attraction model with a proper boundary constraint is unique by showing that the reduced system of linear equations has a coefficient matrix with a nonzero determinant. Let L^B denote the matrix obtained by striking out the rows and columns of L indexed by B. Then, a barycentric layout is a solution of

$$
L^B \cdot p_{V \setminus B} = \left(\sum_{u \in N(v) \cap B} \hat{p}_u \right)_{V \setminus B}. \qquad (3)
$$

Different from the pure attraction model, the barycentric model (with a nondegenerate boundary) has a nondegenerate solution that is uniquely defined. Recall that a system of linear equations has a unique solution if and only if the determinant of its matrix of coefficients is positive. The Matrix Tree Theorem [9] asserts that the determinant of every principal minor of a Laplacian matrix equals the number of spanning trees of its associated multigraph, and L^B is a principal minor of the Laplacian of the graph obtained from G by contracting the vertices in B. Since this graph has at least one spanning tree, the determinant of L^B is positive and the solution of (3) is thus unique.

The barycentric approach was introduced in Tutte [15]. The main result shown in this paper is, in fact, that barycentric layouts of a triconnected planar graph with one face constrained to a convex polygon are planar. For the purpose of graph drawing, less desirable properties are exponentially small resolution of angles and edge lengths as evidenced by a family of triangular graphs obtained by starting from a triangle and adding

vertices adjacent to the same two initial vertices and the most recently added one. For non-triconnected graphs, degenerate subgraph layouts are possible because components lacking boundary vertices are mapped to a line if between a separation pair, or to a point if hinging on a cut vertex.

Orthogonality

Barycentric layouts are systematically biased by the choice of boundary. An alternative constraint avoiding the single-point collapse is to constrain the coordinate vector of each dimension to be orthogonal to the degenerate layout.

Observe that the one-dimensional version of Eq. (2) can also be read as a special case of the eigenequation $Lx = \lambda x$, since $\lambda = 0$ is, in fact, an eigenvalue of L associated with eigenvector $x = \mathbf{1}$. The Laplacian of a simple undirected graph is a real, symmetric, and positive semi-definite matrix so that the eigenvalues are real and nonnegative, and eigenvectors associated with different eigenvalues are orthogonal.

Rearranging the eigenequation yields $\lambda = \frac{x^T L x}{x^T x}$, where $x^T x$ only normalizes for scale. Since $x^T L x = A(x)$, eigenvectors $x \perp y$ associated with the smallest positive eigenvalues yield the best layout in the attraction model subject to orthogonality also with the degenerate layout $\mathbf{1}$. Note that $\mathbf{1} \perp x$ implies that the average of all coordinates x_v is zero, so that the layouts are centered on the origin.

Spectral drawings based on the Laplacian have been proposed by Hall [7], but can be also be defined via other matrices [11]. It is interesting to note that Laplacian spectral layout corresponds to classical multidimensional scaling using the square root of effective resistance as a measure of distance between vertices. While spectral layouts display symmetries, they are highly cluttered and imbalanced for graphs of low algebraic connectivity as measured by their smallest positive eigenvalue.

Distances

The terms in the objective function of the attraction model correspond to the potential energy of a spring with ideal length zero. To avoid

F

collapse, one can thus replace them by springs of some nonzero ideal length. While this takes care of the adjacent pairs of vertices, vertices that are more than one edge away from each other can be connected by springs of different length, say proportional to their shortest-path distance. Down-weighting the influence of distant pairs, we obtain the *stress-minimization model* with objective

$$S(p) = \sum_{u,v \in V} \frac{1}{d(u,v)^2} \left(\| p_u - p_v \| - d(u,v) \right)^2 ,$$

constituting another special case of MDS [12] with graph-theoretic distances $d(u,v)$ as input and inverse quadratic weights. This instantiation has been proposed as a graph drawing method by Kamada and Kawai [8] using gradient descent to determine locally optimal layouts. The use of majorization [13] was shown to be superior by Gansner, Koren, and North [6]. A comprehensive survey of variant layout objective functions is given in Chen and Buja [3].

Repulsion

Instead of springs with nonzero ideal length, a dual physical analogy motivates another approach to counter the collapse caused by attraction, namely, repulsion.

The classic *spring embedder* of Eades [4] specifies forces rather than an energy function. While there is a logarithmic force $\log \frac{\| p_u - p_v \|}{l}$. $(p_u - p_v)$ between adjacent vertices that is neutral if their distance equals a desired value l, nonadjacent vertices push each other apart with quadratically with $\frac{(p_u - p_v)}{\| p_u - p_v \|}$. Both forces are up to scaling constants. A layout is obtained by iteratively evaluating the forces exerted on a vertex by all others and then moving it in the direction of the resulting force, until an approximate equilibrium is obtained.

Many, many variants of the spring embedder have been proposed. The most widely used from Fruchterman and Reingold [5] replaces the forces by quadratically declining repulsion between all pairs of vertices and additional quadratic attraction between adjacent pairs and also introduces

several pragmatic improvements. Brandes and Pich [2] find that suitably initialized stress MDS yields superior results, though.

More force-directed methods are surveyed in Brandes [1] and Kobourov [10], and forces have been used very creatively to realize different layout objectives such as common direction of edges, edge curvatures, angles between incident edges, preferred locations, and many more. A relation with graph clustering is pointed out in Noack [14].

Recommended Reading

1. Brandes U (2001) Drawing on physical analogies. In: Kaufmann M, Wagner D (eds) Drawing graphs: methods and models. Lecture notes in computer science, vol 2025. Springer, Berlin/Heidelberg, pp 71–86
2. Brandes U, Pich C (2009) An experimental study on distance-based graph drawing. In: Proceedings of the 16th international symposium on graph drawing (GD'08), Heraklion. Lecture notes in computer science, vol 5417. Springer, pp 218–229
3. Chen L, Buja A (2013) Stress functions for nonlinear dimension reduction, proximity analysis, and graph drawing. J Mach Learn Res 14:1145–1173
4. Eades P (1984) A heuristic for graph drawing. Congr Numerantium 42:149–160
5. Fruchterman TMJ, Reingold EM (1991) Graph drawing by force-directed placement. Softw Pract Exp 21(11):1129–1164
6. Gansner ER, Koren Y, North SC (2005) Graph drawing by stress majorization. In: Proceedings of the 12th international symposium on graph drawing (GD'04), New York. Lecture notes in computer science, vol 3383. Springer, New York, pp 239–250. doi: 10.1007/978-3-540-31843-9_25
7. Hall KM (1970) An r-dimensional quadratic placement algorithm. Manag Sci 17(3):219–229
8. Kamada T, Kawai S (1989) An algorithm for drawing general undirected graphs. Inf Process Lett 31:7–15
9. Kirchhoff GR (1847) Über die Auflösung der Gleichungen, auf welche man bei der Untersuchung der linearen Verteilung galvanischer Ströme geführt wird. Ann Phys Chem 72:497–508
10. Kobourov SG (2013) Force-directed drawing algorithms. In: Tamassia R (ed) Handbook of graph drawing and visualization. CRC, Boca Raton, pp 383–408
11. Koren Y (2005) Drawing graphs by eigenvectors: theory and practice. Comput Math Appl 49(11–12):1867–1888. doi:10.1016/j.camwa.2004.08.015
12. Kruskal JB (1964) Multidimensional scaling for optimizing goodness of fit to a nonmetric hypothesis. Psychometrika 29(1):1–27

13. de Leeuw J (1977) Applications of convex analysis to multidimensional scaling. In: Barra JR, Brodeau F, Romier G, van Cutsem B (eds) Recent developments in statistics. North-Holland, Amsterdam, pp 133–145
14. Noack A (2009) Modularity clustering is force-directed layout. Phys Rev E 79:026,102
15. Tutte WT (1963) How to draw a graph. Proc Lond Math Soc 13(3):743–768

FPGA Technology Mapping

Jason Cong[1] and Yuzheng Ding[2]
[1]Department of Computer Science, UCLA, Los Angeles, CA, USA
[2]Xilinx Inc., Longmont, CO, USA

Keywords

FlowMap; Lookup-table mapping; LUT mapping

Years and Authors of Summarized Original Work

1992; Cong, Ding

Problem Definition

Introduction

Field-programmable gate array (FPGA) is a type of integrated circuit (IC) device that can be (re)programmed to implement custom logic functions. A majority of FPGA devices use lookup table (LUT) as the basic logic element, where a LUT of K logic inputs (K-LUT) can implement any Boolean function of up to K variables. An FPGA also contains other logic elements, such as registers, programmable interconnect resources, dedicated logic resources such as memory blocks and digital signal processing (DSP) blocks, and input/output resources [6].

The programming of an FPGA involves the transformation of a logic design into a form suitable for implementation on the target FPGA device. This generally takes multiple steps. For LUT-based FPGAs, *technology mapping* is to transform a general Boolean logic network (obtained from the design specification through earlier transformations) into a functionally equivalent K-LUT network that can be implemented by the target FPGA device. The objective of a technology mapping algorithm is to generate, among many possible solutions, an optimized one according to certain criteria, some of which are timing optimization, which is to make the resultant implementation operable at faster speed; area minimization, which is to make the resultant implementation compact in size; and power minimization, which is to make the resultant implementation low in power consumption. The algorithm presented here, named *FlowMap* [2], is for timing optimization; it was the first provably optimal polynomial time algorithm for technology mapping problems on general Boolean networks, and the concepts and approach it introduced have since generated numerous useful derivations and applications.

Data Representation and Preliminaries

The input data to a technology mapping algorithm for LUT-based FPGA is a *general Boolean network*, which can be modeled as a direct acyclic graph $N = (V, E)$. A node $v \in V$ can either represent a logic signal source from outside of the network, in which case it has no incoming edge and is called a *primary input* (PI) node, or it can represent a *logic gate*, in which case it has incoming edge(s) from PIs and/or other gates, which are its logic input(s). If the logic output of the gate is also used outside of the network, its node is a *primary output* (PO), which can have no outgoing edge if it is only used outside.

If edge $\langle u, v \rangle \in E$, u is said to be a *fanin* of v and v a *fanout* of u. For a node v, $input(v)$ denotes the set of its fanins; similarly, for a subgraph H, $input(H)$ denotes the set of distinct nodes outside of H that are fanins of nodes in H. If there is a direct path in N from a node u to a node v, u is said to be a *predecessor* of v and v a *successor* of u. The *input network* of a node v, denoted N_v, is the subgraph containing v and all of its predecessors. A *cone* of a non-PI node

v, denoted C_v, is a subgraph of N_v containing v and possibly some of its *non-PI* predecessors, such that for any node $u \in C_v$, there is a path from u to v in C_v. If $|input(C_v)| \leq K, C_v$ is called a *K-feasible* cone. The network N is *K-bounded* if every non-PI node has a K-feasible cone. A *cut* of a non-PI node v is a bipartition (X, X') of nodes in N_v such that X' is a cone of v; $input(X')$ is called the *cut-set* of (X, X') and $n(X, X') = |input(X')|$ the *size* of the cut. If $n(X, X') \leq K, (X, X')$ is a *K-feasible* cut. The *volume* of (X, X') is $vol(X, X') = |X'|$.

A *topological order* of the nodes in the network N is a linear ordering of the nodes in which each node appears after all of its predecessors and before any of its successors. Such an order is always possible for an acyclic graph.

Problem Formulation

A *K-cover* of a given Boolean network N is a network $N_M = (V_M, E_M)$, where V_M consists of the PI nodes of N and some K-feasible cones of nodes in N, such that for each PO node v of N, V_M contains a cone C_v of v; and if $C_u \in V_M$, then for each non-PI node $v \in input(C_u)$, V_M also contains a cone C_v of v. Edge $\langle u, C_v \rangle \in E_M$ if and only if PI node $u \in input(C_v)$; edge $\langle C_u, C_v \rangle \in E_M$ if and only if non-PI node $u \in input(C_v)$. Since each K-feasible cone can be implemented by a K-LUT, a K-cover can be implemented by a network of K-LUTs. Therefore, the *technology mapping* problem for

K-LUT based FPGA, which is to transform N into a network of K-LUTs, is to find a K-cover N_M of N.

The *depth* of a network is the number of edges in its longest path. A technology mapping solution N_M is *depth optimal* if among all possible mapping solutions of N it has the minimum depth. If each level of K-LUT logic is assumed to contribute a constant amount of logic delay (known as the *unit delay* model), the minimum depth corresponds to the smallest logic propagation delay through the mapping solution, or in other words, the fastest K-LUT implementation of the network N. The problem solved by the *FlowMap* algorithm is *depth-optimal technology mapping for K-LUT based FPGAs*.

A Boolean network that is not K-bounded may not have a mapping solution as defined above. To make a network K-bounded, *gate decomposition* may be used to break larger gates into smaller ones. The *FlowMap* algorithm applies, as preprocessing, an algorithm named *DMIG* [3] that converts all gates into 2-input ones in a depth-optimal fashion, thus making the network K-bounded for $K \geq 2$. Different decomposition schemes may result in different K-bounded networks and consequently different mapping solutions; the optimality of *FlowMap* is with respect to a given K-bounded network Fig. 1 illustrates a Boolean network, its DAG, a covering with 3-feasible cones, and the resultant 3-LUT network. As illustrated, the cones in

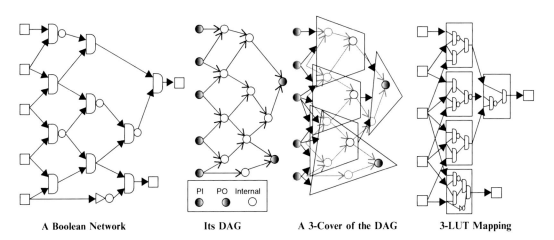

A Boolean Network   **Its DAG**   **A 3-Cover of the DAG**   **3-LUT Mapping**

FPGA Technology Mapping, Fig. 1 A Boolean network, its DAG. a 3-feasible cone covering, and a 3-LUT mapping

the covering may overlap; this is allowed and often beneficial. (When the mapped network is implemented, the overlapped portion of logic will be replicated into each of the K-LUTs that contain it)

Key Results

The *FlowMap* algorithm takes a two-phase approach. In the first phase, it determines for each non-PI node a preferred K-feasible cone as a candidate for the covering; the cones are computed such that if used, they will yield a depth-optimal mapping solution. This is the central piece of the algorithm. In the second phase, the cones necessary to form a cover are chosen to generate a mapping solution.

Structure of Depth-Optimal K-Covers

Let $M(v)$ denote a K-cover (or equivalently, K-LUT mapping solution) of the input network N_v of v. If v is a PI, $M(v)$ consists of v itself. (For simplicity, in the rest of the article, $M(v)$ shall be referred as *a K-cover of v.*) With that defined, first there is

Lemma 1 *If C_v is the K-feasible cone of v in a K-cover $M(v)$, then $M(v) = \{C_v\} + \cup\{M(u) : u \in input(C_v)\}$ where $M(u)$ is a certain K-cover of u. Conversely, if C_v is a K-feasible cone of v, and for each $u \in input(C_v)$, $M(u)$ a K-cover of u, then $M(v) = \{C_v\} + \cup\{M(u) : u \in input(C_v)\}$ is a K-cover of v.*

In other words, a K-cover of a node consists of a K-feasible cone of the node and a K-cover of each input of the cone. Note that for $u_1 \in input(C_v), u_2 \in input(C_v), M(u_1)$ and $M(u_2)$ may overlap, and an overlapped portion may or may not be covered the same way; the union above includes all *distinct* cones from all parts. Also note that for a given C_v, there can be different K-covers of v containing C_v, varying by the choice of $M(u)$ for each $u \in input(C_v)$.

Let $d(M(v))$ denote the depth of $M(v)$. Then

Lemma 2 *For K-cover $M(v) = \{C_v\} + \cup\{M(u) : u \in input(C_v)\}$,*

$$d(M(v)) = max\{d(M(u)) : u \in input(C_v)\} + 1.$$

In particular, let $M^*(u)$ denote a K-cover of u with minimum depth, then $d(M(v)) \geq max\{d(M^*(u)) : u \in input(C_v)\} + 1$; the equality holds when every $M(u)$ in $M(v)$ is of minimum depth.

Recall that C_v defines a K-feasible cut (X, X') where $X' = C_v, X = N_v - C_v$. Let $H(X, X')$ denote the *height* of the cut (X, X'), defined as $H(X, X') = max\{d(M^*(u)) : u \in input(X')\} + 1$. Clearly, $H(X, X')$ gives the minimum depth of any K-cover of v containing $C_v = X'$. Moreover, by properly choosing the cut, $H(X, X')$ height can be minimized, which leads to a K-cover with minimum depth:

Theorem 1 *If K feasible cut (X, X') of v has the minimum height among all K-feasible cuts of v, then the K-cover $M^*(v) = \{X'\} + \cup\{M^*(u) : u \in input(X')\}$ is of minimum depth among all K-covers of v.*

That is, a minimum height K-feasible cut defines a minimum depth K-cover. So the central task for depth-optimal technology mapping becomes the computation of a minimum height K-feasible cut for each PO node.

By definition, the height of a cut depends on the (depths of) minimum depth K-covers of nodes in $N_v - \{v\}$. This suggests a *dynamic programming* procedure that follows topological order, so that when the minimum depth K-cover of v is to be determined, a minimum depth K-cover of each node in $N_v - \{v\}$ is already known and the height of a cut can be readily determined. This is how the first phase of the *FlowMap* algorithm is carried out.

Minimum Height K-Feasible Cut Computation

The first phase of *FlowMap* was originally called the *labeling phase,* as it involves the computation of a *label* for each node in the K-bounded graph. The label of a non-PI node v, denoted $l(v)$, is defined as the minimum height of any cut of v. For convenience, the labels of PI nodes are defined to be 0.

The so-defined label has an important *monotonic* property.

Lemma 3 *Let* $p = max\{l(u) : u \in input(v)\}$, *then* $p \le l(v) \le p + 1$.

Note that this also implies that for any node $u \in N_v - \{v\}, l(u) \le p$. Based on this, in order to find a minimum height K-feasible cut, it is sufficient to check if there is one of height p; if not, then any K-feasible cut will be of minimum height $(p + 1)$, and one always exists for a K-bounded graph.

The search for a K-feasible cut of a height p $(p > 0; p = 0$ is trivial) in *FlowMap* is done by transforming N_v into a *flow network* F_v and computing a *network flow* [5] on it (hence the name). The transformation is as follows. For each node $u \in N_v - \{v\}, l(u) < p, F_v$ has two nodes u_1 and u_2, linked by a *bridge* edge $\langle u_1, u_2 \rangle$; F_v has a single *sink* node t for all other nodes in N_v, and a single *source* node s. For each PI node u of N_v, which corresponds to a bridge edge $\langle u_1, u_2 \rangle$ in F_v, F_v contains edge $\langle s, u_1 \rangle$; for each edge $\langle u, w \rangle$ in N_v, if both u and w have bridge edges in F_v, then F_v contains edge $\langle u_2, w_1 \rangle$; if u has a bridge edge but w does not, F_v contains edge $\langle u_2, t \rangle$; otherwise (neither has bridge) no corresponding edge is in F_v. The bridging edges have unit capacity; all others have infinite capacity. Noting that each edge in F_v with finite (unit) capacity corresponds to a node $u \in N_v$ with $l(u) < p$ and vice versa, and according to the max-flow min-cut theorem [5], it can be shown that.

Lemma 4 *Node* v *has a* K-*feasible cut of height* p *if and only if* F_v *has a maximum network flow of size no more than* K.

On the flow network F_v, a maximum flow can be computed by running the augmenting path algorithm [5]. Once a maximum flow is obtained, the *residual graph* of the flow network is disconnected, and the corresponding *min-cut* (X, X') can be identified as follows: $v \in X'$; for $u \in N_v - \{v\}$, if it is bridged in F_v, and u_1 can be reached in a depth-first search of the residual graph from s, then $u \in X$; otherwise $u \in X'$.

Note that as soon as the flow size exceeds K, the computation can stop, knowing there will not be a desired K-feasible cut. In this case, one can modify the flow network by bridging all nodes in $N_v - \{v\}$ allowing the inclusion of nodes u with $l(u) = p$ in the cut computation, and find a K-feasible cut with height $p + 1$ the same way.

An augmenting path is found in linear time to the number of edges, and there are at most K augmentations for each cut computation. Applying the algorithm to every node in topological order, one would have the following result.

Theorem 2 *In a* K-*bounded Boolean network of* n *nodes and* m *edges, the computation of a minimum height* K-*feasible cut for every node can be completed in* $O(Kmn)$ *time.*

The cut found by the algorithm has another property:

Lemma 5 *The cut* (X, X') *computed as above is the unique maximum volume min-cut; moreover, if* (Y, Y') *is another min-cut, then* $Y' \subseteq X'$.

Intuitively, a cut of larger volume defines a larger cone which covers more logic, therefore a cut of larger volume is preferred. Note however that Lemma 5 only claims maximum among min-cuts; if $n(X, X') < K$, there can be other cuts that are still K-feasible but with larger cut size and larger cut volume. A post-processing algorithm used by *FlowMap* tries to grow (X, X') by collapsing all nodes in X', plus one or more in the cut-set, into the sink, and repeat the flow computation; this will force a cut of larger volume, an improvement if it is still K-feasible.

K-Cover Construction

Once minimum height K-feasible cuts have been computed for all nodes, each node v has a K-feasible cone C_v defined by its cut, which has minimum depth. From here, constructing the K-cover $N_M = (V_M, E_M)$ is straight-forward. First, the cones of all PO nodes are included in V_M. Then, for any cone $C_v \in V_M$, cone C_u for each non-PI node $u \in input(C_v)$ is also included in V_M; so is every PI node $u \in input(C_v)$. Similarly, an edge $\langle C_u, C_v \rangle \in E_M$ for each non-

PI node $u \in input(C_v)$; an edge $\langle u, C_v \rangle \in E_M$ for each PI node $u \in input(C_v)$.

Lemma 6 *The K-cover constructed as above is depth optimal.*

This is a linear time procedure, therefore

Theorem 3 *The problem of depth-optimal technology mapping for K-LUT based FPGAs on a Boolean network of n nodes and m edges can be solved in $O(Kmn)$ time.*

Applications

The *FlowMap* algorithm has been used as a centerpiece or a framework for more complicated FPGA logic synthesis and technology mapping algorithms. There are many possible variations that can address various needs in its applications. Some are briefed below; details of such variations/applications can be found in [1,3].

Complicated Delay Models
With minimal change, the algorithm can be applied where non-unit delay model is used, allowing delay of the nodes and/or the edges to vary, as long as they are static. Dynamic delay models, where the delay of a net is determined by its post-mapping structure, cannot be applied to the algorithm. In fact, delay-optimal mapping under dynamic delay models is NP-hard [3].

Complicated Architectures
The algorithm can be adapted to FPGA architectures that are more sophisticated than homogeneous K-LUT arrays. For example, mapping for FPGA with two LUT sizes can be carried out by computing a cone for each size and dynamically choosing the best one.

Multiple Optimization Objectives
While the algorithm is for delay minimization, area minimization (in terms of the number of cones selected) as well as other objectives can also be incorporated, by adapting the criteria for cut selection. The original algorithm considers area minimization by maximizing the volume of the cuts; substantially, more minimization can be achieved by considering more K-feasible cuts

and making smart choices to, e.g., increase sharing among input networks, allow cuts of larger heights along no-critical paths, etc. [4] Achieving area optimality, however, is NP-hard [3].

Integration with Other Optimizations
The algorithm can be combined with other types of optimizations, including retiming, logic resynthesis, and physical synthesis.

Cross-References

▸ Circuit Partitioning: A Network-Flow-Based Balanced Min-Cut Approach
▸ Performance-Driven Clustering
▸ Sequential Circuit Technology Mapping

Recommended Reading

The *FlowMap* algorithm, with more details and experimental results, was published in [2]. General information about FPGA can be found in [6]. A good source of concepts and algorithms of network flow is [5]. Comprehensive surveys of FPGA design automation, including many variations and applications of the *FlowMap* algorithm, as well as other algorithms, are presented in [1,3]. A general approach based on the relation ship among K-feasible cuts, K-covers and technology mapping solutions, enabling optimization for complicated objectives, is given in [4].

1. Chen D, Cong J, Pan P (2006) FPGA design automation: a survey. In: Foundations and trends in electronic design automation, vol 1, no 3. Now Publishers, Hanover
2. Cong J, Ding Y (1992) An optimal technology mapping algorithm for delay optimization in lookup-table based FPGA designs. In: Proceedings of IEEE/ACM international conference on computer-aided design, San Jose, pp 48–53
3. Cong J, Ding Y (1996) Combinational logic synthesis for LUT based field programmable gate arrays. ACM Trans Design Autom Electron Syst 1(2):145–204
4. Cong J, Wu C, Ding Y (1999) Cut ranking and pruning: enabling a general and efficient FPGA mapping solution. In: Proceedings of ACM/SIGDA seventh international symposium on FPGAs, Monterey, pp 29–35
5. Tarjan R (1983) Data structures and network algorithms. SIAM, Philadelphia
6. Trimberger S (1994) Field-programmable gate array technology. Springer, Boston

Fractional Packing and Covering Problems

George Karakostas
Department of Computing and Software,
McMaster University, Hamilton, ON, Canada

Keywords

Approximation; Covering; Dual; FPTAS; Packing

Years and Authors of Summarized Original Work

1991; Plotkin, Shmoys, Tardos
1995; Plotkin, Shmoys, Tardos

Problem Definition

This entry presents results on fast algorithms that produce approximate solutions to problems which can be formulated as linear programs (LP) and therefore can be solved exactly, albeit with slower running times. The general format of the family of these problems is the following: Given a set of m inequalities on n variables, and an oracle that produces the solution of an appropriate optimization problem over a convex set $P \in \mathbb{R}^n$, find a solution $x \in P$ that satisfies the inequalities, or detect that no such x exists. The basic idea of the algorithm will always be to start from an infeasible solution x, and use the optimization oracle to find a direction in which the violation of the inequalities can be decreased; this is done by calculating a vector y that is a *dual solution corresponding to x*. Then, x is carefully updated towards that direction, and the process is repeated until x becomes "approximately" feasible. In what follows, the particular problems tackled, together with the corresponding optimization oracle, as well as the different notions of "approximation" used are defined.

- The **fractional packing problem** and its oracle are defined as follows:

PACKING: Given an $m \times n$ matrix A, $b > 0$, and a convex set P in $\mathbb{R}^n$ such that $Ax \geq 0$, $\forall x \in P$, is there $x \in P$ such that $Ax \leq b$?

PACK_ORACLE: Given m-dimensional vector $y \geq 0$ and P as above, return $\bar{x} :=$ arg $\min\{y^\top Ax : x \in P\}$.

- The **relaxed fractional packing problem** and its oracle are defined as follows:

RELAXED PACKING: Given $\varepsilon > 0$, an $m \times n$ matrix A, $b > 0$, and convex sets P and $\hat{P}$ in $\mathbb{R}^n$ such that $P \subseteq \hat{P}$ and $Ax \leq 0$, $\forall x \in \hat{P}$, find $x \in \hat{P}$ such that $Ax \leq (1 + \varepsilon)b$, or show that $\nexists x \in P$ such that $Ax \leq b$.

REL_PACK_ORACLE: Given m-dimensional vector $y \geq 0$ and P, $\hat{P}$ as above, return $\bar{x} \in \hat{P}$ such that $y^\top A\bar{x} \leq \min\{y^\top Ax : x \in P\}$.

- The **fractional covering problem** and its oracle are defined as follows:

COVERING: Given an $m \times n$ matrix A, $b > 0$, and a convex set P in $\mathbb{R}^n$ such that $Ax \geq 0$, $\forall x \in P$, is there $x \in P$ such that $Ax \geq b$?

COVER_ORACLE: Given m-dimensional vector $y \geq 0$ and P as above, return $\bar{x} :=$ arg $\max\{y^\top Ax : x \in P\}$.

- The **simultaneous packing and covering problem** and its oracle are defined as follows:

SIMULTANEOUS PACKING AND COVERING: Given $\hat{m} \times n$ and $(m - \hat{m}) \times n$ matrices $\hat{A}$, A respectively, $b > 0$ and $\hat{b} > 0$, and a convex set P in $\mathbb{R}^n$ such that $Ax \geq 0$ and $\hat{A}x \leq 0$, $\forall x \in P$, is there $x \in P$ such that $Ax \leq b$, and $\hat{A}x \geq \hat{b}$?

SIM_ORACLE: Given P as above, a constant v and a dual solution $(y, \hat{y})$, return $\bar{x} \in P$ such that $A\bar{x} \leq vb$, and $y^\top A\bar{x} - \sum_{i \in I(v,\bar{x})} \hat{y}_i \hat{a}_i \bar{x} = \min\{y^\top Ax - \sum_{i \in I(v,x)} \hat{y}_i \hat{a}_i x : x$ a vertex of P such that $Ax \leq vb\}$, where $I(v, x) := \{i : \hat{a}_i x \leq vb_i\}$.

- The **general problem** and its oracle are defined as follows:

GENERAL: Given an $m \times n$ matrix A, an arbitrary vector b, and a convex set P in $\mathbb{R}^n$, is there $x \in P$ such that $Ax \leq b$,?

GEN_ORACLE: Given m-dimensional vector $y \geq 0$ and P as above, return $\bar{x} := \arg\min\{y^{\mathsf{T}}Ax : x \in P\}$.

Definitions and Notation

For an error parameter $\varepsilon > x0$, a point $x \in P$ is an *ε-approximation solution* for the fractional packing (or covering) problem if $Ax \leq (1 + \varepsilon)b$ (or $Ax \geq (1 - \varepsilon)b$). On the other hand, if $x \in P$ satisfies $Ax \leq b$ (or $Ax \geq b$), then x is an *exact solution*. For the GENERAL problem, given an error parameter $\varepsilon > 0$ and a positive tolerance vector d, $x \in P$ is an *ε-approximation solution* if $Ax \leq b + \varepsilon d$ and an *exact solution* if $Ax \leq b$. An *ε-relaxed decision procedure* for these problems either finds an ε-approximation solution or correctly reports that no exact solution exists. In general, for a minimization (maximization) problem, an $(1 + \varepsilon)$-approximation $((1-\varepsilon)$-approximation$)$ algorithm returns a solution at most $(1 + \varepsilon)$ (at least $(1 - \varepsilon)$) times the optimal.

The algorithms developed work within time that depends polynomially on ε^{-1}, for any error parameter $\varepsilon > 0$. Their running time will also depend on the *width* ρ of the convex set P relative to the set of inequalities $Ax \leq b$ or $Ax \geq b$ defining the problem at hand. More specifically, the width ρ is defined as follows for each one of the problems considered here:

- PACKING: $\rho := \max_i \max_{x \in P} \frac{a_i x}{b_i}$.
- RELAXED PACKING: $\hat{\rho} := \max_i \max_{x \in \hat{P}} \frac{a_i x}{b_i}$.
- COVERING: $\rho := \max_i \max_{x \in P} \frac{a_i x}{b_i}$.
- SIMULTANEOUS PACKING AND COVERING: $\rho := \max_{x \in P} \max\{\max_i \frac{a_i x}{b_i}, \max_i \frac{\hat{a}_i x}{b_i}\}$.

- GENERAL: $\rho := \max_i \max_{x \in P} \frac{|a_i x - b_i|}{d_i} + 1$, where d is the tolerance vector defined above.

Key Results

Many of the results below were presented in [8] by assuming a model of computation with exact arithmetic on real numbers and exponentiation in a single step. But, as the authors mention [8], they can be converted to run on the RAM model by using approximate exponentiation, a version of the oracle that produces a *nearly* optimal solution, and a limit on the numbers used that is polynomial in the input length similar to the size of numbers used in exact linear programming algorithms. However, they leave as an open problem the construction of ε-approximate solutions using polylogarithmic precision for the general case of the problems they consider (as can be done, e.g., in the multicommodity flow case [4]).

Theorem 1 *For $0 < \varepsilon \leq 1$, there is a deterministic ε-relaxed decision procedure for the fractional packing problem that uses $O(\varepsilon^{-2}\rho \log(m\varepsilon^{-1}))$ calls to* PACK_ORACLE, *plus the time to compute Ax for the current iterate x between consecutive calls.*

For the case of P being written as a product of smaller-dimension polytopes, i.e., $P = P^1 \times \ldots \times P^k$, each P^l with width ρ^l $\left(\text{obviously} \rho \leq \sum_l \rho^l\right)$, and a separate PACK_ORACLE for each P^l, A^l, then randomization can be used to potentially speed up the algorithm. By using the notation $PACK\_ORACLE_l$ for the P^l, A^l oracle, the following holds:

Theorem 2 *For $0 < \varepsilon \leq 1$, there is a randomized ε-relaxed decision procedure for the fractional packing problem that is expected to use*

$$O\left(\varepsilon^{-2}\left(\sum_l \rho^l\right)\log(m\varepsilon^{-1}) + k\log(\rho\varepsilon^{-1})\right)$$

calls to $PACK\_ORACLE_l$ for some $l \in \{1, \ldots, k\}$ (possibly a different l in every call), plus the time to compute $\sum_l A^l x^l$ for the

F

current iterate $x = (x^1, x^2, \ldots, x^k)$ *between consecutive calls.*

Theorem 2 holds for RELAXED PACKING as well, if ρ is replaced by $\hat{\rho}$ and PACK_ORACLE by REL_PACK_ORACLE.

In fact, one needs only an *approximate* version of PACK_ORACLE. Let $C_\mathcal{P}(y)$ be the minimum cost $y^\mathsf{T} Ax$ achieved by PACK_ORACLE for a given y.

Theorem 3 *Let* PACK_ORACLE *be replaced by an oracle that, given vector* $y \geq 0$, *finds a point* $\bar{x} \in P$ *such that* $y^\mathsf{T} A\bar{x} \leq (1 + \varepsilon/2)C_\mathcal{P}(y) + (\varepsilon/2)\lambda y^\mathsf{T} b$, *where* λ *is minimum so that* $Ax \leq \lambda b$ *is satisfied by the current iterate* x. *Then, Theorems 1 and 2 still hold.*

Theorem 3 shows that even if no efficient implementation exists for an oracle, as in, e.g., the case when this oracle solves an NP-hard problem, a fully polynomial approximation scheme for it suffices.

Similar results can be proven for the fractional covering problem ($COVER\_OrACLE_l$ is defined similarly to $PACK\_ORACLE_l$ above):

Theorem 4 *For* $0 < \varepsilon < 1$, *there is a deterministic* ε*-relaxed decision procedure for the fractional covering problem that uses* $O(m + \rho \log^2 m + \varepsilon^{-2}\rho \log(m\varepsilon^{-1}))$ *calls to* COVER_ORACLE, *plus the time to compute* Ax *for the current iterate* x *between consecutive calls.*

Theorem 5 *For* $0 < \varepsilon < 1$, *there is a randomized* ε*-relaxed decision procedure for the fractional packing problem that is expected to use* $O\left(mk + \left(\sum_i \rho^l\right) \log^2 m + k \log \varepsilon^{-1} + \varepsilon^{-2}\left(\sum_l \rho^l\right) \log(m\varepsilon^{-1})\right)$ *calls to* $COVER\_ORACLE_l$ *for some* $l \in \{1,\ldots,k\}$ *(possibly a different* l *in every call), plus the time to compute* $\sum_l A^l x^l$ *for the current iterate* $x = (x^1, x^2, \ldots, x^k)$ *between consecutive calls.*

Let $C_\mathcal{C}(y)$ be the maximum cost $y^\mathsf{T} Ax$ achieved by COVER_ORACLE for a given y.

Theorem 6 *Let* COVER_ORACLE *be replaced by an oracle that, given vector* $y \geq 0$, *finds a point* $\bar{x} \in P$ *such that* $y^\mathsf{T} A\bar{x} \geq (1 - \varepsilon/2)C_\mathcal{C}(y) - (\varepsilon/2)\lambda y^\mathsf{T} b$, *where* λ *is maximum so that* $Ax \geq \lambda b$ *is satisfied by the current iterate* x. *Then, Theorems 4 and 5 still hold.*

For the simultaneous packing and covering problem, the following is proven:

Theorem 7 *For* $0 < \varepsilon \leq 1$, *there is a randomized* ε*-relaxed decision procedure for the simultaneous packing and covering problem that is expected to use* $O(m^2(\log^2 \rho)\varepsilon^{-2} \log(\varepsilon^{-1}m \log \rho))$ *calls to* SIM_ORACLE, *and a deterministic version that uses a factor of* $\log \rho$ *more calls, plus the time to compute* $\hat{A}x$ *for the current iterate* x *between consecutive calls.*

For the GENERAL problem, the following is shown:

Theorem 8 *For* $0 < \varepsilon < 1$, *there is a deterministic* ε*-relaxed decision procedure for the GENERAL problem that uses* $O(\varepsilon^{-2}\rho^2 \log(m\rho\varepsilon^{-1}))$ *calls to* GEN_ORACLE, *plus the time to compute* Ax *for the current iterate* x *between consecutive calls.*

The running times of these algorithms are proportional to the width ρ, and the authors devise techniques to reduce this width for many special cases of the problems considered. One example of the results obtained by these techniques is the following: If a packing problem is defined by a convex set that is a product of k smaller-dimension convex sets, i.e., $P = P^1 \times \ldots \times P^k$, and the inequalities $\sum_l A^l x^l \leq b$, then there is a randomized ε-relaxed decision procedure that is expected to use $O(\varepsilon^{-2}k \log(m\varepsilon^{-1}) + k \log k)$ calls to a subroutine that finds a minimum-cost point in $\hat{P}^l = \{x^l \in P^l : A^l x^l \leq b\}$, $l = 1, \ldots, k$ and a deterministic version that uses $O(\varepsilon^{-2}k^2 \log(m\varepsilon^{-1}))$ such calls, plus the time to compute Ax for the current iterate x between consecutive calls. This result can be applied to the multicommodity flow problem, but the required subroutine is a single-source minimum-cost flow computation, instead of a shortest-path calculation needed for the original algorithm.

Applications

The results presented above can be used in order to obtain fast approximate solutions to linear programs, even if these can be solved exactly by LP algorithms. Many approximation algorithms are based on the rounding of the solution of such programs, and hence one might want to solve them approximately (with the overall approximation factor absorbing the LP solution approximation factor), but more efficiently. Two such examples, which appear in [8], are mentioned here.

Theorems 1 and 2 can be applied for the improvement of the running time of the algorithm by Lenstra, Shmoys, and Tardos [5] for the scheduling of unrelated parallel machines without preemption ($R||C_{max}$): N jobs are to be scheduled on M machines, with each job i scheduled on exactly one machine j with processing time p_{ij}, so that the maximum total processing time over all machines is minimized. Then, for any fixed $r > 1$, there is a deterministic $(1 + r)$-approximation algorithm that runs in $O(M^2 N \log^2 N \log M)$ time and a randomized version that runs in $O(MN \log M \log N)$ expected time. For the version of the problem with preemption, there are polynomial-time approximation schemes that run in $O(MN^2 \log^2 N)$ time and $O(MN \log N \log M)$ expected time in the deterministic and randomized case, respectively.

A well-known lower bound for the metric Traveling Salesman Problem (metric TSP) on N nodes is the Held-Karp bound [2], which can be formulated as the optimum of a linear program over the *subtour elimination polytope*. By using a randomized minimum-cut algorithm by Karger and Stein [3], one can obtain a randomized approximation scheme that computes the Held-Karp bound in $O(N^4 \log^6 N)$ expected time.

Open Problems

The main open problem is the further reduction of the running time for the approximate solution of the various fractional problems. One direction would be to improve the bounds for specific problems, as has been done very successfully for the multicommodity flow problem in a series of papers starting with Shahrokhi and Matula [9]. Currently, the best running times for several versions of the multicommodity flow problems are achieved by Madry [6]. Shahrokhi and Matula [9] also led to a series of results by Grigoriadis and Khachiyan developed independently to [8], starting with [1] which presents an algorithm with a number of calls smaller than the one in Theorem 1 by a factor of $\log(m\varepsilon^{-1})/\log m$. Considerable effort has been dedicated to the reduction of the dependence of the running time on the width of the problem or the reduction of the width itself (e.g., see [10] for sequential and parallel algorithms for mixed packing and covering), so this can be another direction of improvement.

A problem left open by [8] is the development of approximation schemes for the RAM model that use only *polylogarithmic in the input length* precision and work for the general case of the problems considered.

Cross-References

▶ Approximation Schemes for Makespan Minimization

Recommended Reading

1. Grigoriadis MD, Khachiyan LG (1994) Fast approximation schemes for convex programs with many blocks and coupling constraints. SIAM J Optim 4:86–107
2. Held M, Karp RM (1970) The traveling-salesman problem and minimum cost spanning trees. Oper Res 18:1138–1162
3. Karger DR, Stein C (1993) An $\tilde{O}(n^2)$ algorithm for minimum cut. In: Proceeding of 25th annual ACM symposium on theory of computing (STOC), San Diego, pp 757–765
4. Leighton FT, Makedon F, Plotkin SA, Stein C, Tardos É, Tragoudas S (1995) Fast approximation algorithms for multicommodity flow problems. J Comput Syst Sci 50(2):228–243
5. Lenstra JK, Shmoys DB, Tardos É (1990) Approximation algorithms for scheduling unrelated parallel machines. Math Program A 24:259–272
6. Madry A (2010) Faster approximation schemes for fractional multicommodity flow problems via

dynamic graph algorithms. In: Proceedings of 42nd ACM symposium on theory of computing (STOC), Cambridge, pp 121–130

7. Plotkin SA, Shmoys DB, Tardos É (1991) Fast approximation algorithms for fractional packing and covering problems. In: Proceedings of 32nd annual IEEE symposium on foundations of computer science (FOCS), San Juan, pp 495–504

8. Plotkin SA, Shmoys DB, Tardos É (1995) Fast approximation algorithms for fractional packing and covering problems. Math Oper Res 20(2):257–301. Preliminary version appeared in [6]

9. Shahrokhi F, Matula DW (1990) The maximum concurrent flow problem. J ACM 37:318–334

10. Young NE (2001) Sequential and parallel algorithms for mixed packing and covering. In: Proceedings of 42nd annual IEEE symposium on foundations of computer science (FOCS), Las Vegas, pp 538–546

Frequent Graph Mining

Takashi Washio
The Institute of Scientific and Industrial Research, Osaka University, Ibaraki, Osaka, Japan

Keywords

Anti-monotonicity; Canonical depth sequence; Canonical graph representation; Data-driven enumeration; DFS code; Frequent subgraph; Pattern growth

Years and Authors of Summarized Original Work

2000; Inokuchi, Washio
2002; Yan, Han
2004; Nijssen, Kok

Problem Definition

This problem is to enumerate all subgraphs appearing with frequencies not less than a threshold value in a given graph data set. Let $G(V, E, L, \ell)$ be a labeled graph where V is a set of vertices, $E \subseteq V \times V$ a set of edges, L a set of labels and $\ell : V \cup E \rightarrow L$ a labeling function. A labeled graph $g(v, e, L, \ell)$ is a subgraph of $G(V, E, L, \ell)$, i.e., $g \sqsubseteq G$, if and only if a mapping $f : v \rightarrow V$ exists such that $\forall u_i \in v, f(u_i) \in V, \ell(u_i) = \ell(f(u_i))$, and $\forall (u_i, u_j) \in e, (f(u_i), f(u_j)) \in E, \ell(u_i, u_j) = \ell(f(u_i), f(u_j))$. Given a graph data set $D = \{G_i | i = 1, \ldots, n\}$, a support of g in D is a set of all G_i involving g in D, i.e., $D(g) = \{G_i | g \sqsubseteq G_i \in D\}$. Under a given threshold frequency called a minimum support $minsup > 0$, g is said to be frequent, if the size of $D(g)$ i.e., $|D(g)|$, is greater than or equal to $minsup$. Generic frequent graph mining is a problem to enumerate all frequent subgraphs g of D, while most algorithms focus on connected and undirected subgraphs. Some focus on induced subgraphs or limit the enumeration to closed frequent subgraphs where each of them is maximal in the frequent subgraphs having an identical support.

Key Results

Study of the frequent graph mining was initiated in the mid-1990s under motivation to analyze complex structured data acquired and accumulated in our society. Their major issue has been principles to efficiently extract frequent subgraphs embedded in a given graph data set. They invented many original canonical graph representations adapted to the data-driven extraction, which are different from these proposed in studies of efficient isomorphism checking [1] and graph enumeration without duplications [2].

Pioneering algorithms of frequent graph mining, SUBDUE [3] and GBI [4], did not solve the aforementioned standard problem but greedily extracted subgraphs concisely describing the original graph data under some measures such as minimum description length (MDL). The earliest algorithms for deriving a complete set of the frequent subgraphs are WARMR [5] and its extension FARMER [6]. They can flexibly focus on various types of frequent subgraphs for the enumeration by applying inductive logic programming (ILP) in artificial intelligence, while they are not very scalable in the size of the enumerated subgraphs.

AGM proposed in 2000 [7, 8] was an epoch-making study in the sense that it combined frequent item set mining [9] and the graph enumeration and enhanced the scalability for practical applications. It introduced technical strategies of (1) incremental candidate enumeration based on anti-monotonicity of the subgraph frequency, (2) canonical graph representation to avoid duplicated candidate subgraph extractions, and (3) data-driven pruning of the candidates by the minimum support. The anti-monotonicity is a fundamental nature of the subgraph frequency that $|D(g_1)| \leq |D(g_2)|$ for any subgraphs g_1 and g_2 in D if $g_2 \sqsubseteq g_1$. Dozens of frequent graph mining algorithms have been studied along this line after 2000. In the rest of this entry, gSpan [10] and Gaston [11], considered to be the most efficient up to date, are explained.

gSpan

gSpan derives all frequent connected subgraphs in a given data set of connected and undirected graphs [10]. For the aforementioned strategy (1), it applies a pattern growth technique which is data-driven enumeration of candidate frequent subgraphs. It is performed by tracing vertices and edges of each data graph G in a DFS manner. Figure 1b is an example search tree generated by starting from the vertex labeled as Y in the graph (a). In a search tree T, the vertex for the next visit is the one reachable from the current vertex by passing through an edge untraced yet in G. If the vertex for the next visit is the one visited earlier in G, the edge is called a backward edge otherwise a forward edge. They are depicted by dashed and solid lines, respectively, in Fig. 1b. When no more untraced edges are available from the current vertex, the search backtracks to the nearest vertex having the untraced edges. Any subtree of T represents a subgraph of G. We denote the sets of the forward and the backward edges in T as $E_{f,T} = \{e | \forall i, j, i < j, e = (v_i, v_j) \in E\}$ and $E_{b,T} = \{e | \forall i, j, i > j, e = (v_i, v_j) \in E\}$, respectively, where i and j are integer indices numbered at the vertices in their visiting order in T.

There exist many trees T representing an identical graph G as another tree of the graph (a) shown in Fig. 1c. This ambiguity causing duplication and miss in the candidate graph enumeration is avoided by introducing the strategy (2). gSpan applied the following three types of partial orders of the edges in T. Given $e_1 = (v_{i_1}, v_{j_1})$ and $e_2 = (v_{i_2}, v_{j_2})$,

$e_1 \prec_{f,T} e_2$ if and only if $j_1 < j_2$ for $e_1, e_2 \in E_{f,T}$,

$e_1 \prec_{b,T} e_2$ if and only if (i) $i_1 < i_2$ or (ii) $i_1 = i_2$ and $j_1 < j_2$, for $e_1, e_2 \in E_{b,T}$,

$e_1 \prec_{bf,T} e_2$ if and only if (i) $i_1 < j_2$ for $e_1 \in E_{b,T}, e_2 \in E_{f,T}$

or (ii) $j_1 \leq i_2$ for $e_1 \in E_{f,T}, e_2 \in E_{b,T}$.

The combination of these partial orders is known to give a linear order of the edges. We also assume a total order of the labels in L and define a representation of T, a DFS code, as a sequence of 5-tuples $((v_i, v_j), \ell(v_i), \ell((v_i, v_j)), \ell(v_j))$ following the trace order of the DFS in T. A DFS code is smaller if smaller edges and smaller labels appear in earlier 5-tuples in the sequence. Accordingly, we define the search tree T having the minimum DFS code as a canonical representation of G. The search tree in Fig. 1c is canonical, since its DFS code is the minimum. As any subtree of a canonical T has its minimum DFS code, it is also canonical for its corresponding subgraph.

Moreover, gSpan applies the DFS which chooses the untraced edge having the smallest 5-tuples at the current vertex for visiting the next vertex. This focuses on the minimum DFS code and ensures to enumerate the canonical

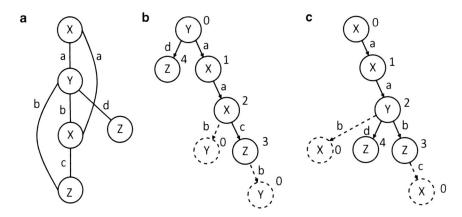

Frequent Graph Mining, Fig. 1 A data graph and its search trees

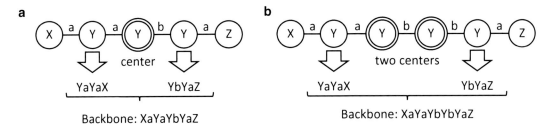

Frequent Graph Mining, Fig. 2 Examples of paths and their backbones

subtree of every subgraph before finding its other noncanonical subtrees. This efficiently prunes an infrequent subgraph without matching its multiple DSF codes in (3). In this manner, the canonical graph representation of gSpan is fully adapted to its search algorithm.

Gaston

Gaston also derives all frequent connected subgraphs in a given data set of connected and undirected graphs [11]. It uses the polynomial time complexity of the enumeration of paths and free trees. Gaston uses a canonical path representation, a backbone, in (2). Two sequences of the labels of the vertices and edges starting from a center, which is a middle vertex, in the path to the both terminals are derived as shown in Fig. 2a, and the reverse of the lexicographically smaller sequence with the appended larger sequence is defined to be the backbone. In case of a path having an even number of vertices, two centers, which are two middle vertices, are used as shown

in Fig. 2b. Starting from a single vertex, Gaston extends a path by adding a vertex to one of the terminals in the strategy (1). Finally, it efficiently counts the frequency of the extended path in the data set by using its backbone and prunes the infrequent paths in (3).

Gaston further enumerates free trees involving a frequent path as the longest path by iteratively adding vertices to the vertices in the free trees except for the terminal vertices of the path. Since the set of the free trees having a distinct backbone as its longest path is also distinct, the set does not intersect each other. This reduces the complexity of the enumeration in (1). Moreover, Gaston derives a canonical representation of a free tree, a canonical depth sequence, for (2) by transforming the tree to a rooted and ordered tree where the root is the center of its longest path, and its vertices and edges are arranged in a lexicographically descending order of the labels. If the two center exists in the path, the free tree is partitioned for each center, and each free tree is represented by its canonical depth sequence. Similarly to the DFS code of gSpan, any subtree involving the root in this rooted and ordered tree is a canonical depth sequence. This is beneficial for (1), since all canonical depth sequences are incrementally obtained in the depth first search. Gaston efficiently prunes the infrequent free trees in the data set by using the canonical depth sequence in (3).

Gaston further enumerates cyclic subgraphs from a frequent free tree by iteratively adding edges bridging vertex pairs in the tree in (1). For (2), Gaston avoids duplicated enumerations of the cyclic subgraphs by using Nauty algorithm for the graph isomorphism checking [1]. It prunes the infrequent cyclic subgraphs of the data set in (3) and finally derives the frequent subgraphs. Gaston works very efficiently for the sparse data graphs, since the candidate cyclic subgraphs is less in such graphs.

URLs to Code and Data Sets

gSpan suite (http://www.cs.ucsb.edu/~xyan/software/gSpan.htm), Gaston suite (http://www.liacs.nl/~snijssen/gaston/). Other common suites can be found for various frequent substructure mining (http://hms.liacs.nl/index.html).

Cross-References

▶ Enumeration of Non-crossing Geometric Graphs
▶ Enumeration of Paths, Cycles, and Spanning Trees
▶ Frequent Pattern Mining
▶ Graph Isomorphism
▶ Matching in Dynamic Graphs
▶ Tree Enumeration

Recommended Reading

1. McKay BD, Piperno A (2013) Practical graph isomorphism, II. J Symb Comput 60:94–112
2. Roberts F, Tesman B (2011) Applied combinatorics, 2nd edn. CRC, New York
3. Cook J, Holder L (1994) Substructure discovery using minimum description length and background knowledge. J Artif Intell Res 1:231–255
4. Yoshida K, Motoda H, Indurkhya N (1994) Graph-based induction as a unified learning framework. J Appl Intell 4:297–328
5. Dehaspe L, Toivonen H (1999) Discovery of frequent datalog patterns. Data Min Knowl Discov 3(1):7–36
6. Nijssen S, Kok J (2001) Faster association rules for multiple relations. In: Proceedings of the IJCAI2001: 17th international joint conference on artificial intelligence, Seattle, vol 2, pp 891–896
7. Inokuchi A, Washio T, Motoda H (2000) An Apriori-based algorithm for mining frequent substructures from graph data. In: Proceedings of the PKDD2000: principles of data mining and knowledge discovery, 4th Europian conference, Lyon. Lecture notes in artificial intelligence (LNAI), vol 1910, pp 13–23
8. Inokuchi A, Washio T, Motoda H (2003) Complete mining of frequent patterns from graphs: mining graph data. Mach Learn 50:321–354
9. Agrawal R, Srikant R (1994) Fast algorithms for mining association rules. In: Proceedings of the VLDB1994: 20th very large dada base conference, Santiago de Chile, pp 487–499
10. Yan X, Han J (2002) gspan: graph-based substructure pattern mining. In: Proceedings of the ICDM2002: 2nd IEEE conference on data mining, Maebashi City, pp 721–724
11. Nijssen S, Kok JN (2004) A quickstart in frequent structure mining can make a difference. In: Proceedings of the 10th ACM SIGKDD international conference on knowledge discovery and data mining, Seattle, pp 647–652

Frequent Pattern Mining

Takeaki Uno
National Institute of Informatics, Chiyoda, Tokyo, Japan

Keywords

Apriori; Backtrack; Database; Data mining; FIMI; FP-growth; FP-tree; Itemset mining; Frequency; Monotone property

Years and Authors of Summarized Original Work

2004; Uno, Kiyomi, Arimura

Problem Definition

Pattern mining is a fundamental problem in data mining. The problem is to find all the patterns appearing in the given database frequently. For a set $E = \{1, \ldots, n\}$ of items, an *itemset* (also called a *pattern*) is a subset of E. Let $\mathcal{D}$ be a given *database* composed of *transactions* $R_1, \ldots, R_m$, $R_i \subseteq E$. For an itemset P, an *occurrence* of P is a transaction of $\mathcal{D}$ such that $P \subseteq R$, and the *occurrence set* $\mathrm{Occ}(P)$ is the set of occurrences of P. The *frequency* of P, also called *support*, is $|\mathrm{Occ}(P)|$ and denoted by $\mathrm{frq}(P)$. For a given constant σ called *minimum support*, an itemset P is *frequent* if $\mathrm{frq}(P) \geq \sigma$. For given a database and a minimum support, *frequent itemset mining* is the problem of enumerating all frequent itemsets in $\mathcal{D}$.

Key Results

The enumeration is solved in output polynomial time [1], and the space complexity is input polynomial [5]. Many algorithms have been proposed for practical efficiency on real-world data [7, 8, 10, 11] that are drastically fast. We show algorithm LCM that is the winner in the competition [6], and several techniques used in LCM.

Algorithms

There have been a lot of algorithms for this problem. The itemsets satisfy the following *monotone property*, and this is used in almost all existing algorithms.

Lemma 1 *For any itemsets P and Q such that $P \supseteq Q$, there holds $\mathrm{frq}(P) \leq \mathrm{frq}(Q)$. In particular, $\mathrm{Occ}(P) \subseteq \mathrm{Occ}(Q)$.* □

Using the monotone property, we can enumerate all frequent itemsets from $\emptyset$ by recursively adding items.   □

Lemma 2 *Any frequent itemset P of size k is generated by adding an item to a frequent itemset of size $k - 1$.*

Apriori

Apriori algorithm was proposed in the first paper of frequent pattern mining, by Agrawal et al. in 1993 [1]. The computational resources are not enough in the era, and the database could not fit memory, thus stored in an HDD. Apriori is designed to be efficient in such environments so that it scans the database only few times. Apriori is a breadth-first search algorithm that iteratively generates all frequent itemsets of size 1, size 2, and so on. Apriori generates candidate itemsets by adding an item to each frequent itemset of size $k - 1$. From the monotone property, any frequent itemset of size k is in the candidate itemsets. Apriori then checks the inclusion relation between a transaction and all candidates. By doing this for all transactions, the frequencies of candidates are computed and infrequent candidates are removed. The algorithm is written as follows:

Algorithm $\mathrm{Apriori}(\mathcal{D}, \sigma)$:

1. $\mathcal{P}_0 = \{\emptyset\}$; $k := 0$;
2. **while** $\mathcal{P}_k \neq \emptyset$ **do**
3. $\mathcal{P}_{k+1} := \emptyset$;
4.   **for each** $P \in \mathcal{P}_k$, $\mathrm{frq}(P) := 0$; $\mathcal{P}_{k+1} := \mathcal{P}_{k+1} \cup \{P \cup \{i\} \mid i \in E\}$
5.   **for each** $R \in \mathcal{D}$, $\mathrm{frq}(P) := \mathrm{frq}(P) + 1$ for all $P \in \mathcal{P}_{k+1}, P \subseteq R$.
6.   remove all P from $\mathcal{P}_{k+1}$ satisfying $\mathrm{frq}(P) < \sigma$
7.   **output** all $P \in \mathcal{P}_{k+1}$; $k := k + 1$
8. **end while**

The space complexity of Apriori is $O(n\mu)$ where μ is the number of frequent itemsets of $\mathcal{D}$. The time complexity is $O(n||\mathcal{D}||\mu)$ where $||\mathcal{D}|| = \sum_{R \in \mathcal{D}} |R|$ is the size of $\mathcal{D}$. Hence Apriori is output polynomial time.

Backtrack Algorithm

Backtrack algorithm is a depth-first search-based frequent itemset mining algorithm that is first proposed Bayardo et al. [5] in 1998. The amount of memory in a computer was rapidly increasing in the era, and thus the databases began to fit the memory. We can then reduce the memory space for storing candidate itemsets and thus huge amount of itemsets can be enumerated. Moreover, a technique so-called down project accelerates the computation. According to the monotone property, we can see that $P \cup \{i\}$ is included in a transaction R only if $R \in \mathrm{Occ}(P)$ holds. By using this, down project reduces the checks only with transactions in $\mathrm{Occ}(P)$. Particularly, we can see that $\mathrm{Occ}(P \cup \{i\}) = \mathrm{Occ}(P) \cap \mathrm{Occ}(\{i\})$. Moreover, we can reduce the check for the duplication by using a technique so-called tail extension. We denote the maximum item in P by $tail(P)$. Tail extension generates itemsets $P \cup \{i\}$ only with $i, i > tail(P)$. In this way, any frequent itemset P is generated uniquely from another frequent itemset; thus duplications are efficiently avoided, by recursively generating with tail extensions.

Algorithm BackTrack $(P, \mathrm{Occ}(P), \sigma)$:

---

1. **output** P
2. **for each item** $i > tail(P)$ **do**
3. **if** $|\mathrm{Occ}(P) \cap \mathrm{Occ}(\{i\})| \geq \sigma$, then **call** BackTrack $(P \cup \{i\}, \mathrm{Occ}(P) \cap \mathrm{Occ}(\{i\}), \sigma)$
4. **end for**

The space complexity of BackTrack is $O(\|\mathcal{D}\|)$; thereby BackTrack is polynomial space. The time complexity is $O(\|\mathcal{D}\|\mu)$, since step 3 is done by marking transactions in $\mathrm{Occ}(P)$ and checking whether each transaction of $\mathrm{Occ}(\{i\})$ is marked in constant time. Moreover, since the depth of recursion is at most n, the delay of BackTrack is $O(n\|\mathcal{D}\|)$. BackTrack with down project reduces practical computation time in order of magnitude, in implementation competitions FIMI03 and FIMI04 [6].

Database Reduction

The technique of database reduction was first developed in FP-growth by Han et al. [8] and modified in LCM by Uno et al. [11]. Database reduction drastically reduces practical computation time, as shown in the experiments in [6].

We observe that down project removes the unnecessary transactions from the database given to a recursive call, where unnecessary transactions are those never used in the recursion, and this fastens the computation. The idea of database reduction is to further remove unnecessary items from the database. The unnecessary items are (1) items i satisfying $i < tail(P)$ and (2) items i such that $P \cup \{i\}$ is not frequent. Items of (1) are never used because of the rule of tail extension. (2) comes from that $P \cup \{i\} \cup \{j\}$ is not frequent for any item j, by the monotone property. Thus, the removal of these items never disturbs the enumeration. The database obtained by removing unnecessary items from each transaction of $\mathrm{Occ}(P)$ is called the *conditional database*.

In the deep of recursion, the conditional database tends to have few items since $tail(P)$ is large and $\mathrm{frq}(P)$ is small. In such cases, several transactions would be identical. The computation for the identical transactions is the same, and thus we unify these transactions and put a mark of their quantity to the unified transaction as the representation of the multiplicity. For example, three transactions $R_1, R_2, R_3 = \{100, 105, 110\}$ are replaced by $R_j = \{100, 105, 110\}$ and a mark "three" is put to R_j. By this, the computation time on the bottom levels of the recursion is drastically shortened when σ is large. This is because conditional databases usually have k items in the bottom levels where k is a small constant and thus can have at most 2^k different transactions. The obtained database is called the *reduced database* and is denoted by $\mathcal{D}^*(P, \sigma)$.

The computation for the unification of identical transactions can be done by, for example, radix sort in $O(\|\mathcal{D}^*(P, \sigma)\|)$ time [11]. FP-growth further reduces by representing the database by a trie [7, 8]. However, experiments in [6] show that the overheads of trie are often larger than the gain; thus in many cases FP-growth is slower than LCM. The computation of frequent itemset mining generates recursions widely spread as the depth, thus so-called bottom expanded. In such case, the computation time on

F

the bottom levels dominates the total computation time [9]; thus database reduction performs very well.

Delivery

Delivery [10, 11] is a technique to compute $Occ(P \cup \{i\})$ for all $i > tail(P)$ at once. Down project computes $Occ(P \cup \{i\})$ in $O(|Occ(\{i\})|)$ time and thus takes $O(\|\mathcal{D}\|)$ time for all i. Delivery computes $Occ(P \cup \{i\})$ for all i at once in $(\|Occ(P)\|)$ time. The idea is to find all $P \cup \{i\}$ that are included in R, for each transaction $R \in Occ(P)$. Actually, $P \cup \{i\} \subseteq R$ iff $i \in R$; thus this is done by just scanning items $i > tail(P)$. The algorithm is described as follows:

---

1. $Occ(P \cup \{i\}) := \emptyset$ for each $i > tail(P)$
2. **for** each $R \in Occ(P)$ **do**
3.   **for** each item $i \in R, i > tail(P)$, insert R to $Occ(P \cup \{i\})$
4. **end for**

By using the reduced database $\mathcal{D}^*(P, \sigma)$, delivery is done in $O(\|\mathcal{D}^*(P, \sigma)\|)$ time. Note that the frequency is the sum of multiplications of transactions in the reduced database.

Generalizations and Extensions

The frequent itemset mining problem is extended by varying patterns and databases, such as trees in XML databases, labeled graphs in chemical compound databases, and so on. Let $\mathcal{L}$ be a class of structures, and $\preceq$ be an binary relation on $\mathcal{L}$. A member of $\mathcal{L}$ is called a *pattern*. Suppose that we are given a *database* $\mathcal{D}$ composed of records $R_1, \ldots, R_m$, $R_i \in \mathcal{L}$. Itemset mining is the case that $\mathcal{L} = 2^E$ and $a \preceq b$ holds iff $a \subseteq b$. For a pattern $P \in \mathcal{L}$, an *occurrence* of P is a record $R \in \mathcal{D}$ such that $P \preceq R$, and the other notations are defined in the same way. For given a database and a minimum support, *frequent pattern mining* is the problem of enumerating all the frequent patterns in $\mathcal{D}$.

When $\mathcal{L}$ is arbitrary, the frequent pattern mining is hard. Thus, we often assume that $(\mathcal{L}, \preceq)$ is a lattice, and there is an element $\perp$ of $\mathcal{L}$ such that $\perp \preceq P$ holds for any $P \in \mathcal{L}$. We then have the monotone property.

Lemma 3 *For any $P, Q \in \mathcal{L}$ satisfying $P \preceq Q$, there holds* $frq(P) \geq frq(Q)$. $\square$

Let $suc(P)$ (resp., $prc(P)$) be the set of elements $Q \in \mathcal{L} \setminus \{P\}$ such that $P \preceq Q$ (resp., $Q \preceq P$) holds and no $X \in \mathcal{L} \setminus \{P, Q\}$ satisfies $P \preceq X \preceq Q$ (resp., $Q \preceq X \preceq P$). Using the monotone property, we can enumerate all frequent patterns from $\perp$ by recursively generating all elements of $suc(P)$.

In this general setting, Apriori needs an assumption that $(\mathcal{L}, \preceq)$ is modular; thus for any P, Q such that $P \preceq Q$, the length of any maximal chain $P \preceq X_1 \cdots X_k \preceq Q$ is identical. By this assumption, we can define the size of a pattern P by the length of the maximal chain from $\perp$ to P. Apriori then works by replacing $\{P \cup \{i\} \mid i \in E\}$ of step 4 by $suc(P)$.

Let T be the time to generate a pattern in $suc(P)$ and T' be the time to evaluate $a \preceq b$. Note that T' may be large, for example, in the case that $\mathcal{L}$ is the set of graphs and T' is the time for graph isomorphism. Apriori generates $|suc(P)|$ patterns for each pattern P, and we have to check whether each generated pattern is already in $\mathcal{P}_{i+1}$ by comparing P and each member of $\mathcal{P}_{i+1}$. Thus, the total computation time is $O(s(T + T'(\mu + |\mathcal{D}|)))$ where s is the maximum size of $suc(P)$.

The depth-first search algorithm needs an alternative for tail extension. The alternative is given by reverse search technique proposed by Avis and Fukuda [4]. A pattern Q is generated from many patterns in $prc(Q)$, and this makes duplications. We avoid this by defining the parent $P(Q)$ by one of $prc(Q)$ and allow to generate Q only from $P(Q)$, so that Q is uniquely generated. For example, the same as tail extension, we define an order in $prc(Q)$ and define $P(Q)$ by the minimum one in the order.

Algorithm Backtrack2 $(P, \text{Occ}(P), \sigma)$:

---

1. **output** P
2. **for each** $Q \in \text{suc}(P)$ **do**
3. compute $\text{Occ}(Q)$ from $\text{Occ}(P)$
4. **if** $\text{frq}(Q) \geq \sigma$ and $P = P(Q)$ **then call** Backtrack2 $(Q, \text{Occ}(Q), \sigma)$
5. **end for**

The time complexity of BackTrack2 is $O(s(T + T'' + T'|\mathcal{D}|)\mu)$ where T'' is the time to compute the parent of a pattern. The heaviest part of $T'|\mathcal{D}|$ is usually reduced by down project. The algorithm will be efficient if all patterns Q satisfying $P = P(Q)$ are efficiently enumerated. Such examples are sequences [12], trees [3], and motifs with wildcards [2].

Frequent Sequence Mining

$\mathcal{L}$ is composed of strings on alphabet Σ, and $a, b \in \mathcal{L}$ satisfy $a \preceq b$ iff a is a subsequence of b, i.e., a is obtained from b by deleting some letters. For a pattern P, $\text{suc}(P)$ is the set of strings obtained by inserting a letter to P at some position.

We define the parent of P by the string obtained by removing the last letter from P. Then, the children of P is generated by appending a letter to the tail of P. Since $\preceq$ can be tested in linear time, BackTrack2 runs in $O(|\Sigma| \times ||\mathcal{D}||)$ time for each frequent sequence.

Frequent Ordered Tree Mining

$\mathcal{L}$ is composed of rooted trees such that each vertex has a label and an ordering of children. Such a tree is called a labeled ordered tree. $a, b \in \mathcal{L}$ satisfy $a \preceq b$ iff a is a subtree of b with correspondence keeping the children orders and vertex labels; a vertex of label "A" has to be mapped to a vertex having label "A," and children orders do not change. For a pattern P, $\text{suc}(P)$ is the set of labeled ordered trees obtained by inserting a vertex as a leaf.

The rightmost path of an ordered tree is $\{v_1, \ldots, v_k\}$ where v_1 is the root, and v_i is the last child of v_{i-1}. We define the parent of P by that obtained by removing the rightmost leaf v_k. Then, the children of P is generated by appending a vertex with a label so that the vertex is the last child of a vertex in the rightmost path. Since $\preceq$ can be tested in linear time, BackTrack2 runs in $O(t|\Sigma| \times ||\mathcal{D}||)$ time for each frequent ordered tree, where t is the maximum height of the pattern tree.

Recommended Reading

1. Agrawal R, Mannila H, Srikant R, Toivonen H, Verkamo AI (1996) Fast discovery of association rules. In: Fayyad UM (ed) Advances in knowledge discovery and data mining. MIT, Menlo Park, pp 307–328
2. Arimura H, Uno T (2007) An efficient polynomial space and polynomial delay algorithm for enumeration of maximal motifs in a sequence. J Comb Optim 13:243–262
3. Asai T, Arimura T, Uno T, Nakano S (2003) Discovering frequent substructures in large unordered trees. LNAI 2843:47–61
4. Avis D, Fukuda K (1996) Reverse search for enumeration. Discret Appl Math 65:21–46
5. Bayardo RJ Jr (1998) Efficiently mining long patterns from databases. SIGMOD Rec 27:85–93
6. Goethals B (2003) The FIMI repository. http://fimi.cs.helsinki.fi/
7. Grahne G, Zhu J (2003) Efficiently using prefix-trees in mining frequent itemsets. In: IEEE ICDM'03 workshop FIMI'03, Melbourne
8. Han J, Pei J, Yin Y (2000) Mining frequent patterns without candidate generation. ACM SIGMOD Rec 29:1–12
9. Uno T (1998) New approach for speeding up enumeration algorithms. In: Chwa K-Y, Ibarra OH (eds) Algorithms and computation. LNCS, vol 1533. Springer, Berlin/Heidelberg, pp 287–296
10. Uno T, Asai T, Uchida Y, Arimura H (2004) An efficient algorithm for enumerating closed patterns in transaction databases. In: Suzuki E, Arikawa S (eds) Discovery science. LNCS, vol 3245. Springer, Berlin/Heidelberg, pp 16–31
11. Uno T, Kiyomi M, Arimura H (2004) LCM ver.2: efficient mining algorithms for frequent/closed/maximal itemsets. In: IEEE ICDM'04 workshop FIMI'04, Brighton
12. Wang J, Han J (2004) BIDE: efficient mining of frequent closed sequences. In: ICDE'04, Boston, pp 79–90

F

Fully Dynamic All Pairs Shortest Paths

Giuseppe F. Italiano
Department of Computer and Systems Science,
University of Rome, Rome, Italy
Department of Information and Computer
Systems, University of Rome, Rome, Italy

Years and Authors of Summarized Original Work

2004; Demetrescu, Italiano

Problem Definition

The problem is concerned with efficiently maintaining information about all-pairs shortest paths in a dynamically changing graph. This problem has been investigated since the 1960s [17, 18, 20], and plays a crucial role in many applications, including network optimization and routing, traffic information systems, databases, compilers, garbage collection, interactive verification systems, robotics, dataflow analysis, and document formatting.

A dynamic graph algorithm maintains a given property $\mathcal{P}$ on a graph subject to dynamic changes, such as edge insertions, edge deletions and edge weight updates. A dynamic graph algorithm should process queries on property $\mathcal{P}$ quickly, and perform update operations faster than recomputing from scratch, as carried out by the fastest static algorithm. An algorithm is said to be *fully dynamic* if it can handle both edge insertions and edge deletions. A *partially dynamic* algorithm can handle either edge insertions or edge deletions, but not both: it is *incremental* if it supports insertions only, and *decremental* if it supports deletions only. In this entry, fully dynamic algorithms for maintaining shortest paths on general directed graphs are presented.

In the *fully dynamic All Pairs Shortest Path (APSP) problem* one wishes to maintain a directed graph $G = (V, E)$ with real-valued edge weights under an intermixed sequence of the following operations:

Update(x, y, w): update the weight of edge (x, y) to the real value w; this includes as a special case both edge insertion (if the weight is set from $+\infty$ to $w < +\infty$) and edge deletion (if the weight is set to $w = +\infty$);

Distance(x, y): output the shortest distance from x to y.

Path(x, y): report a shortest path from x to y, if any.

More formally, the problem can be defined as follows.

Problem 1 (Fully Dynamic All-Pairs Shortest Paths)

INPUT: A weighted directed graph $G = (V, E)$, and a sequence σ of operations as defined above.
OUTPUT: A matrix D such entry $D[x, y]$ stores the distance from vertex x to vertex y throughout the sequence σ of operations.

Throughout this entry, m and n denotes respectively the number of edges and vertices in G.

Demetrescu and Italiano [3] proposed a new approach to dynamic path problems based on maintaining classes of paths characterized by local properties, i.e., properties that hold for all proper subpaths, even if they may not hold for the entire paths. They showed that this approach can play a crucial role in the dynamic maintenance of shortest paths.

Key Results

Theorem 1 *The fully dynamic shoretest path problem can be solved in* $O(n^2 \log^3 n)$ *amortized time per update during any intermixed sequence of operations. The space required is $O(mn)$.*

Using the same approach, Thorup [22] has shown how to slightly improve the running times:

Theorem 2 *The fully dynamic shoretest path problem can be solved in* $O(n^2(\log n +$

$\log^2(m/n)))$ *amortized time per update during any intermixed sequence of operations. The space required is O(mn).*

Applications

Dynamic shortest paths find applications in many areas, including network optimization and routing, transportation networks, traffic information systems, databases, compilers, garbage collection, interactive verification systems, robotics, dataflow analysis, and document formatting.

Open Problems

The recent work on dynamic shortest paths has raised some new and perhaps intriguing questions. First, can one reduce the space usage for dynamic shortest paths to $O(n^2)$? Second, and perhaps more importantly, can one solve efficiently fully dynamic *single-source* reachability and shortest paths on general graphs? Finally, are there any general techniques for making increase-only algorithms fully dynamic? Similar techniques have been widely exploited in the case of fully dynamic algorithms on undirected graphs [11–13].

Experimental Results

A thorough empirical study of the algorithms described in this entry is carried out in [4].

Data Sets

Data sets are described in [4].

Cross-References

▶ Dynamic Trees
▶ Fully Dynamic Connectivity
▶ Fully Dynamic Higher Connectivity
▶ Fully Dynamic Higher Connectivity for Planar Graphs
▶ Fully Dynamic Minimum Spanning Trees
▶ Fully Dynamic Planarity Testing
▶ Fully Dynamic Transitive Closure

Recommended Reading

1. Ausiello G, Italiano GF, Marchetti-Spaccamela A, Nanni U (1991) Incremental algorithms for minimal length paths. J Algorithm 12(4):615–638
2. Demetrescu C (2001) Fully dynamic algorithms for path problems on directed graphs. PhD thesis, Department of Computer and Systems Science, University of Rome "La Sapienza", Rome
3. Demetrescu C, Italiano GF (2004) A new approach to dynamic all pairs shortest paths. J Assoc Comput Mach 51(6):968–992
4. Demetrescu C, Italiano GF (2006) Experimental analysis of dynamic all pairs shortest path algorithms. ACM Trans Algorithms 2(4):578–601
5. Demetrescu C, Italiano GF (2005) Trade-offs for fully dynamic reachability on dags: breaking through the O(n2) barrier. J Assoc Comput Mach 52(2):147–156
6. Demetrescu C, Italiano GF (2006) Fully dynamic all pairs shortest paths with real edge weights. J Comput Syst Sci 72(5):813–837
7. Even S, Gazit H (1985) Updating distances in dynamic graphs. Method Oper Res 49:371–387
8. Frigioni D, Marchetti-Spaccamela A, Nanni U (1998) Semi-dynamic algorithms for maintaining single source shortest paths trees. Algorithmica 22(3):250–274
9. Frigioni D, Marchetti-Spaccamela A, Nanni U (2000) Fully dynamic algorithms for maintaining shortest paths trees. J Algorithm 34:351–381
10. Henzinger M, King V (1995) Fully dynamic biconnectivity and transitive closure. In: Proceedings of the 36th IEEE symposium on foundations of computer science (FOCS'95), Los Alamos. IEEE Computer Society, pp 664–672
11. Henzinger M, King V (2001) Maintaining minimum spanning forests in dynamic graphs. SIAM J Comput 31(2):364–374
12. Henzinger MR, King V (1999) Randomized fully dynamic graph algorithms with polylogarithmic time per operation. J ACM 46(4):502–516
13. Holm J, de Lichtenberg K, Thorup M (2001) Polylogarithmic deterministic fully-dynamic algorithms for connectivity, minimum spanning tree, 2-edge, and biconnectivity. J ACM 48:723–760
14. King V (1999) Fully dynamic algorithms for maintaining all-pairs shortest paths and transitive closure in digraphs. In: Proceedings of the 40th IEEE symposium on foundations of computer science (FOCS'99), Los Alamos. IEEE Computer Society, pp 81–99

F

15. King V, Sagert G (2002) A fully dynamic algorithm for maintaining the transitive closure. J Comput Syst Sci 65(1):150–167
16. King V, Thorup M (2001) A space saving trick for directed dynamic transitive closure and shortest path algorithms. In: Proceedings of the 7th annual international computing and combinatorics conference (CO-COON). LNCS, vol 2108. Springer, Berlin, pp 268–277
17. Loubal P (1967) A network evaluation procedure. Highw Res Rec 205:96–109
18. Murchland J (1967) The effect of increasing or decreasing the length of a single arc on all shortest distances in a graph. Technical report, LBS-TNT-26, London Business School, Transport Network Theory Unit, London
19. Ramalingam G, Reps T (1996) An incremental algorithm for a generalization of the shortest path problem. J Algorithm 21:267–305
20. Rodionov V (1968) The parametric problem of shortest distances. USSR Comput Math Math Phys 8(5):336–343
21. Rohnert H (1985) A dynamization of the all-pairs least cost problem. In: Proceedings of the 2nd annual symposium on theoretical aspects of computer science, (STACS'85). LNCS, vol 182. Springer, Berlin, pp 279–286
22. Thorup M (2004) Fully-dynamic all-pairs shortest paths: faster and allowing negative cycles. In: Proceedings of the 9th Scandinavian workshop on algorithm theory (SWAT'04), Humlebaek. Springer, Berlin, pp 384–396
23. Thorup M (2005) Worst-case update times for fully-dynamic all-pairs shortest paths. In: Proceedings of the 37th ACM symposium on theory of computing (STOC 2005). ACM, New York

Fully Dynamic Connectivity

Valerie King
Department of Computer Science, University of Victoria, Victoria, BC, Canada

Keywords

Fully dynamic graph algorithm for maintaining connectivity; Incremental algorithms for graphs

Years and Authors of Summarized Original Work

2001; Holm, de Lichtenberg, Thorup

Problem Definition

Design a data structure for an undirected graph with a fixed set of nodes which can process queries of the form "Are nodes i and j connected?" and updates of the form "Insert edge $\{i,j\}$"; "Delete edge $\{i,j\}$." The goal is to minimize update and query times, over the worst-case sequence of queries and updates. Algorithms to solve this problem are called "fully dynamic" as opposed to "partially dynamic" since both insertions and deletions are allowed.

Key Results

Holm et al. [4] gave the first deterministic fully dynamic graph algorithm for maintaining connectivity in an undirected graph with polylogarithmic amortized time per operation, specifically, $O(\log^2 n)$ amortized cost per update operation and $O(\log n / \log \log n)$ worst-case per query, where n is the number of nodes. The basic technique is extended to maintain minimum spanning trees in $O(\log^4 n)$ amortized cost per update operation and 2-edge connectivity and biconnectivity in $O(\log^5 n)$ amortized time per operation.

The algorithm relies on a simple novel technique for maintaining a spanning forest in a graph which enables efficient search for a replacement edge when a tree edge is deleted. This technique ensures that each nontree edge is examined no more than $\log_2 n$ times. The algorithm relies on previously known tree data structures, such as top trees or ET-trees to store and quickly retrieve information about the spanning trees and the nontree edges incident to them.

Algorithms to achieve a query time $O(\log n / \log \log \log n)$ and expected amortized update time $O(\log n (\log \log n)^3)$ for connectivity and $O(\log^3 n \log \log n)$ expected amortized update time for 2-edge and biconnectivity were given in [6]. Lower bounds showing a continuum of tradeoffs for connectivity between query and update times in the cell probe model which match the known upper bounds were proved in [5]. Specifically, if t_u and t_q are the amortized update and query time, respectively, then $t_q \cdot \log(t_u/t_q) = \Omega(\log n)$ and $t_u \cdot \log(t_q/t_u) = \Omega(\log n)$.

A previously known, somewhat different, randomized method for computing dynamic connectivity with $O(\log^3 n)$ amortized expected update time can be found in [2], improved to $O(\log^2 n)$ in [3]. A method which minimizes worst-case rather than amortized update time is given in [1] $O(\sqrt{n})$ time per update for connectivity as well as 2-edge connectivity and bipartiteness.

Open Problems

Can the worst-case update time be reduced to $O(n^{1/2})$, with polylogarithmic query time?

Can the lower bounds on the trade-offs in [6] be matched for all possible query costs?

Applications

Dynamic connectivity has been used as a subroutine for several static graph algorithms, such as the maximum flow problem in a static graph [7], and for speeding up numerical studies of the Potts spin model.

URL to Code

See http://www.mpi-sb.mpg.de/LEDA/friends/dyngraph.html for software which implements the algorithm in [2] and other older methods.

Cross-References

▶ Fully Dynamic All Pairs Shortest Paths
▶ Fully Dynamic Transitive Closure

Recommended Reading

1. Eppstein D, Galil Z, Italiano GF, Nissenzweig A (1997) Sparsification-a technique for speeding up dynamic graph algorithms. J ACM 44(5):669–696.1
2. Henzinger MR, King V (1999) Randomized fully dynamic graph algorithms with polylogarithmic time per operation. J ACM 46(4):502–536. (Presented at ACM STOC 1995)
3. Henzinger MR, Thorup M (1997) Sampling to provide or to bound: with applications to fully dynamic graph algorithms. Random Struct Algorithms 11(4):369–379. (Presented at ICALP 1996)
4. Holm J, De Lichtenberg K, Thorup M (2001) Polylogarithmic deterministic fully-dynamic algorithms for connectivity, minimum spanning tree, 2-Edge, and biconnectivity. J ACM 48(4):723–760. (Presented at ACM STOC 1998)
5. Iyer R, Karger D, Rahul H, Thorup M (2001) An experimental study of poly-logarithmic fully-dynamic connectivity algorithms. J Exp Algorithmics 6(4). (Presented at ALENEX 2000)
6. Pătraşcu M, Demaine E (2006) Logarithmic lower bounds in the cell-probe model. SIAM J Comput 35(4):932–963. (Presented at ACM STOC 2004)
7. Thorup M (2000) Near-optimal fully-dynamic graph connectivity. In: Proceedings of the 32nd ACM symposium on theory of computing, Portland. ACM STOC, pp 343–350
8. Thorup M (2000) Dynamic Graph Algorithms with Applications. In: Halldórsson MM (ed) 7th Scandinavian workshop on algorithm theory (SWAT), Norway, 5–7 July 2000, pp 1–9
9. Zaroliagis CD (2002) Implementations and experimental studies of dynamic graph algorithms. In: Experimental algorithmics, Dagstuhl seminar, Sept 2000. Lecture notes in computer science, vol 2547. Springer. Journal article: J Exp Algorithmics 229–278 (2000)

F

Fully Dynamic Connectivity: Upper and Lower Bounds

Giuseppe F. Italiano
Department of Computer and Systems Science, University of Rome, Rome, Italy
Department of Information and Computer Systems, University of Rome, Rome, Italy

Keywords

Dynamic connected components; Dynamic spanning forests

Years and Authors of Summarized Original Work

2000; Thorup

Problem Definition

The problem is concerned with efficiently maintaining information about connectivity in a dynamically changing graph. A dynamic graph algorithm maintains a given property $\mathcal{P}$ on a graph subject to dynamic changes, such as edge insertions, edge deletions and edge weight updates. A dynamic graph algorithm should process queries on property $\mathcal{P}$ quickly, and perform update operations faster than recomputing from scratch, as carried out by the fastest static algorithm. An algorithm is said to be *fully dynamic* if it can handle both edge insertions and edge deletions. A *partially dynamic* algorithm can handle either edge insertions or edge deletions, but not both: it is *incremental* if it supports insertions only, and *decremental* if it supports deletions only.

In the fully dynamic connectivity problem, one wishes to maintain an undirected graph $G = (V, E)$ under an intermixed sequence of the following operations:

Connected(u, v): Return *true* if vertices u and v are in the same connected component of the graph. Return *false* otherwise.

Insert(x, y): Insert a new edge between the two vertices x and y.

Delete(x, y): Delete the edge between the two vertices x and y.

Key Results

In this section, a high level description of the algorithm for the fully dynamic connectivity problem in undirected graphs described in [11] is presented: the algorithm, due to Holm, de Lichtenberg and Thorup, answers connectivity queries in $O(\log n / \log \log n)$ worst-case running time while supporting edge insertions and deletions in $O(\log^2 n)$ amortized time.

The algorithm maintains a spanning forest F of the dynamically changing graph G. Edges in F are referred to as *tree edges*. Let e be a tree edge of forest F, and let T be the tree of F

containing it. When e is deleted, the two trees T_1 and T_2 obtained from T after the deletion of e can be reconnected if and only if there is a non-tree edge in G with one endpoint in T_1 and the other endpoint in T_2. Such an edge is called a *replacement edge* for e. In other words, if there is a replacement edge for e, T is reconnected via this replacement edge; otherwise, the deletion of e creates a new connected component in G.

To accommodate systematic search for replacement edges, the algorithm associates to each edge e a level $\ell(e)$ and, based on edge levels, maintains a set of sub-forests of the spanning forest F: for each level i, forest F_i is the sub-forest induced by tree edges of level $\geq i$. Denoting by L denotes the maximum edge level, it follows that:

$$F = F_0 \supseteq F_1 \supseteq F_2 \supseteq \cdots \supseteq F_L .$$

Initially, all edges have level 0; levels are then progressively increased, but never decreased. The changes of edge levels are accomplished so as to maintain the following invariants, which obviously hold at the beginning.

Invariant (1): F is a maximum spanning forest of G if edge levels are interpreted as weights.

Invariant (2): The number of nodes in each tree of F_i is at most $n/2^i$.

Invariant (1) should be interpreted as follows. Let (u, v) be a non-tree edge of level $\ell(u, v)$ and let $u \cdots v$ be the unique path between u and v in F (such a path exists since F is a spanning forest of G). Let e be any edge in $u \cdots v$ and let $\ell(e)$ be its level. Due to (1), $\ell(e) \geq \ell(u, v)$. Since this holds for each edge in the path, and by construction $F_{\ell(u,v)}$ contains all the tree edges of level $\geq \ell(u, v)$, the entire path is contained in $F_{\ell(u,v)}$, i.e., u and v are connected in $F_{\ell(u,v)}$.

Invariant (2) implies that the maximum number of levels is $L \leq \lfloor \log_2 n \rfloor$.

Note that when a new edge is inserted, it is given level 0. Its level can be then increased at most $\lfloor \log_2 n \rfloor$ times as a consequence of edge deletions. When a tree edge $e = (v, w)$ of level $\ell(e)$ is deleted, the algorithm looks for a replacement edge at the highest possible level, if any.

Due to invariant (1), such a replacement edge has level $\ell \leq \ell(e)$. Hence, a replacement subroutine Replace$((u, w), \ell(e))$ is called with parameters e and $\ell(e)$. The operations performed by this subroutine are now sketched.

Replace$((u, w), \ell)$ finds a replacement edge of the highest level $\leq \ell$, if any. If such a replacement does not exist in level ℓ, there are two cases: if $\ell > 0$, the algorithm recurses on level $\ell - 1$; otherwise, $\ell = 0$, and the deletion of (v, w) disconnects v and w in G.

During the search at level ℓ, suitably chosen tree and non-tree edges may be promoted at higher levels as follows. Let T_v and T_w be the trees of forest F_ℓ obtained after deleting (v, w) and let, w.l.o.g., T_v be smaller than T_w. Then T_v contains at most $n/2^{\ell+1}$ vertices, since $T_v \cup T_w \cup \{(v, w)\}$ was a tree at level ℓ and due to invariant (2). Thus, edges in T_v of level ℓ can be promoted at level $\ell + 1$ by maintaining the invariants. Non-tree edges incident to T_v are finally visited one by one: if an edge does connect T_v and T_w, a replacement edge has been found and the search stops, otherwise its level is increased by 1.

Trees of each forest are maintained so that the basic operations needed to implement edge insertions and deletions can be supported in $O(\log n)$ time. There are few variants of basic data structures that can accomplish this task, and one could use the Euler Tour trees (in short ET-tree), first introduced in [17], for this purpose.

In addition to inserting and deleting edges from a forest, ET-trees must also support operations such as finding the tree of a forest that contains a given vertex, computing the size of a tree, and, more importantly, finding tree edges of level ℓ in T_v and non-tree edges of level ℓ incident to T_v. This can be done by augmenting the ET-trees with a constant amount of information per node: the interested reader is referred to [11] for details.

Using an amortization argument based on level changes, the claimed $O(\log^2 n)$ bound on the update time can be proved. Namely, inserting an edge costs $O(\log n)$, as well as increasing its level. Since this can happen $O(\log n)$ times, the total amortized insertion cost, inclusive of level

increases, is $O(\log^2 n)$. With respect to edge deletions, cutting and linking $O(\log n)$ forest has a total cost $O(\log^2 n)$; moreover, there are $O(\log n)$ recursive calls to Replace, each of cost $O(\log n)$ plus the cost amortized over level increases. The ET-trees over $F_0 = F$ allows it to answer connectivity queries in $O(\log n)$ worst-case time. As shown in [11], this can be reduced to $O(\log n/\log\log n)$ by using a $\Theta(\log n)$-ary version of ET-trees.

Theorem 1 *A dynamic graph G with n vertices can be maintained upon insertions and deletions of edges using $O(\log^2 n)$ amortized time per update and answering connectivity queries in $O(\log n/\log\log n)$ worst-case running time.*

Later on, Thorup [18] gave another data structure which achieves slightly different time bounds:

Theorem 2 *A dynamic graph G with n vertices can be maintained upon insertions and deletions of edges using $O(\log n \cdot (\log\log n)^3)$ amortized time per update and answering connectivity queries in $O(\log n/\log\log\log n)$ time.*

The bounds given in Theorems 1 and 2 are not directly comparable, because each sacrifices the running time of one operation (either query or update) in order to improve the other.

The best known lower bound for the dynamic connectivity problem holds in the bit-probe model of computation and is due to Pătraşcu and Tarniţă [16]. The bit-probe model is an instantiation of the cell-probe model with one-bit cells. In this model, memory is organized in cells, and the algorithms may read or write a cell in constant time. The number of cell probes is taken as the measure of complexity. For formal definitions of this model, the interested reader is referred to [13].

Theorem 3 *Consider a bit-probe implementation for dynamic connectivity, in which updates take expected amortized time t_u, and queries take expected time t_q. Then, in the average case of an input distribution, $t_u = \Omega\left(log^2 n/log^2(t_u + t_q)\right)$. In particular*

$$\max\{t_u, t_q\} = \Omega\left(\left(\frac{\log n}{\log\log n}\right)^2\right).$$

In the bit-probe model, the best upper bound per operation is given by the algorithm of Theorem 2, namely it is $O(\log^2 n / \log\log\log n)$. Consequently, the gap between upper and lower bound appears to be limited essentially to doubly logarithmic factors only.

Applications

Dynamic graph connectivity appears as a basic subproblem of many other important problems, such as the dynamic maintenance of minimum spanning trees and dynamic edge and vertex connectivity problems. Furthermore, there are several applications of dynamic graph connectivity in other disciplines, ranging from Computational Biology, where dynamic graph connectivity proved to be useful for the dynamic maintenance of protein molecular surfaces as the molecules undergo conformational changes [6], to Image Processing, when one is interested in maintaining the connected components of a bitmap image [3].

Open Problems

The work on dynamic connectivity raises some open and perhaps intruiguing questions. The first natural open problem is whether the gap between upper and lower bounds can be closed. Note that the lower bound of Theorem 3 seems to imply that different trade-offs between queries and updates could be possible: can we design a data structure with $o(\log n)$ time per update and $O(\text{poly}(\log n))$ per query? This would be particulary interesting in applications where the total number of queries is substantially larger than the number of updates.

Finally, is it possible to design an algorithm with matching $O(\log n)$ update and query bounds for general graphs? Note that this is possible in the special case of plane graphs [5].

Experimental Results

A thorough empirical study of dynamic connectivity algorithms has been carried out in [1, 12].

Data Sets

Data sets are described in [1, 12].

Cross-References

▶ Dynamic Trees
▶ Fully Dynamic All Pairs Shortest Paths
▶ Fully Dynamic Higher Connectivity
▶ Fully Dynamic Higher Connectivity for Planar Graphs
▶ Fully Dynamic Minimum Spanning Trees
▶ Fully Dynamic Planarity Testing
▶ Fully Dynamic Transitive Closure

Recommended Reading

1. Alberts D, Cattaneo G, Italiano GF (1997) Anempirical study of dynamic graph algorithms. ACM J Exp Algorithms 2
2. Beame P, Fich FE (2002) Optimal bounds for the predecessor problem and related problems. J Comput Syst Sci 65(1):38–72
3. Eppstein D (1997) Dynamic connectivity in digital images. Inf Process Lett 62(3):121–126
4. Eppstein D, Galil Z, Italiano GF, Nissenzweig A (1997) Sparsification – a technique for speeding up dynamic graph algorithms. J Assoc Comput Mach 44(5):669–696
5. Eppstein D, Italiano GF, Tamassia R, Tarjan RE, Westbrook J, Yung M (1992) Maintenance of a minimum spanning forest in a dynamic plane graph. J Algorithms 13:33–54
6. Eyal E, Halperin D (2005) Improved maintenance of molecular surfacesusing dynamic graph connectivity. In: Proceedings of the 5th international workshop on algorithms in bioinformatics (WABI 2005), Mallorca, pp 401–413
7. Frederickson GN (1985) Data structures for on-line updating of minimum spanning trees. SIAM J Comput 14:781–798
8. Frederickson GN (1991) Ambivalent data structures for dynamic 2-edge-connectivity and k smallest spanning trees. In: Proceedings of the 32nd symposium on foundations of computer science, pp 632–641

9. Henzinger MR, Fredman ML (1998) Lower bounds for fully dynamic connectivity problems in graphs. Algorithmica 22(3):351–362

10. Henzinger MR, King V (1999) Randomized fully dynamic graph algorithms with polylogarithmic time per operation. J ACM 46(4):502–516

11. Holm J, de Lichtenberg K, Thorup M (2001) Polylogarithmic deterministic fully-dynamic algorithms for connectivity, minimum spanning tree, 2-edge, and biconnectivity. J ACM 48:723–760

12. Iyer R, Karger D, Rahul H, Thorup M (2001) An experimental study of polylogarithmic, fully dynamic, connectivity algorithms. ACM J Exp Algorithmics 6

13. Miltersen PB (1999) Cell probe complexity – a survey. In: 19th conference on the foundations of software technology and theoretical computer science (FSTTCS). Advances in Data Structures Workshop

14. Miltersen PB, Subramanian S, Vitter JS, Tamassia R (1994) Complexity models for incremental computation. In: Ausiello G, Italiano GF (eds) Special issue on dynamic and on-line algorithms. Theor Comput Sci 130(1):203–236

15. Pătraşcu M, Demain ED (2004) Lower bounds for dynamic connectivity. In: Proceedings of the 36th ACM symposium on theory of computing (STOC), pp 546–553

16. Pätrascu M, Tarnita C (2007) On dynamic bit-probe complexity. Theor Comput Sci Spec Issue ICALP'05 380:127–142; In: Italiano GF, Palamidessi C (eds) A preliminary version in proceedings of the 32nd international colloquium on automata, languages and programming (ICALP'05), pp 969–981

17. Tarjan RE, Vishkin U (1985) An efficient parallel biconnectivity algorithm. SIAM J Comput 14:862–874

18. Thorup M (2000) Near-optimal fully-dynamic graph connectivity. In: Proceedings of the 32nd ACM symposium on theory of computing (STOC), pp 343–350

Fully Dynamic Higher Connectivity

Giuseppe F. Italiano
Department of Computer and Systems Science, University of Rome, Rome, Italy
Department of Information and Computer Systems, University of Rome, Rome, Italy

Keywords

Fully dynamic edge connectivity; Fully dynamic vertex connectivity

Years and Authors of Summarized Original Work

1997; Eppstein, Galil, Italiano, Nissenzweig

Problem Definition

The problem is concerned with efficiently maintaining information about edge and vertex connectivity in a dynamically changing graph. Before defining formally the problems, a few preliminary definitions follow.

Given an undirected graph $G = (V, E)$, and an integer $k \geq 2$, a pair of vertices $\langle u, v \rangle$ is said to be *k-edge-connected* if the removal of any $(k - 1)$ edges in G leaves u and v connected. It is not difficult to see that this is an equivalence relationship: the vertices of a graph G are partitioned by this relationship into equivalence classes called *k-edge-connected components*. G is said to be *k-edge-connected* if the removal of any $(k - 1)$ edges leaves G connected. As a result of these definitions, G is k-edge-connected if and only if any two vertices of G are k-edge-connected. An edge set $E' \subseteq E$ is an *edge-cut for vertices x and y* if the removal of all the edges in E' disconnects G into two graphs, one containing x and the other containing y. An edge set $E' \subseteq E$ is an *edge-cut for G* if the removal of all the edges in E' disconnects G into two graphs. An edge-cut E' for G (for x and y, respectively) is *minimal* if removing any edge from E' reconnects G (for x and y, respectively). The cardinality of an edge-cut E', denoted by $|E'|$, is given by the number of edges in E'. An edge-cut E' for G (for x and y, respectively) is said to be a *minimum cardinality edge-cut* or in short a *connectivity edge-cut* if there is no other edge-cut E'' for G (for x and y respectively) such that $|E''| < |E'|$. Connectivity edge-cuts are of course minimal edge-cuts. Note that G is k-edge-connected if and only if a connectivity edge-cut for G contains at least k edges, and vertices x and y are k-edge-connected if and only if a connectivity edge-cut for x and y contains at least k edges. A connectivity edge-cut of cardinality 1 is called a *bridge*.

The following theorem due to Ford and Fulkerson, and Elias, Feinstein and Shannon (see [7]) gives another characterization of k-edge connectivity.

Theorem 1 (Ford and Fulkerson, Elias, Feinstein and Shannon) *Given a graph G and two vertices x and y in G, x and y are k-edge-connected if and only if there are at least k edge-disjoint paths between x and y.*

In a similar fashion, a vertex set $V' \subseteq V - \{x, y\}$ is said to be a *vertex-cut* for vertices x and y if the removal of all the vertices in V' disconnects x and y. $V' \subset V$ is a *vertex-cut* for vertices G if the removal of all the vertices in V' disconnects G.

The cardinality of a vertex-cut V', denoted by $|V'|$, is given by the number of vertices in V'. A vertex-cut V' for x and y is said to be a *minimum cardinality vertex-cut* or in short a *connectivity vertex-cut* if there is no other vertex-cut V'' for x and y such that $|V''| < |V'|$. Then x and y are k-vertex-connected if and only if a connectivity vertex-cut for x and y contains at least k vertices. A graph G is said to be k-vertex-connected if all its pairs of vertices are k-vertex-connected. A connectivity vertex-cut of cardinality 1 is called an *articulation point*, while a connectivity vertex-cut of cardinality 2 is called a *separation pair*. Note that for vertex connectivity it is no longer true that the removal of a connectivity vertex-cut splits G into two sets of vertices.

The following theorem due to Menger (see [7]) gives another characterization of k-vertex connectivity.

Theorem 2 (Menger) *Given a graph G and two vertices x and y in G, x and y are k-vertex-connected if and only if there are at least k vertex-disjoint paths between x and y.*

A dynamic graph algorithm maintains a given property $\mathcal{P}$ on a graph subject to dynamic changes, such as edge insertions, edge deletions and edge weight updates. A dynamic graph algorithm should process queries on property $\mathcal{P}$ quickly, and perform update operations faster than recomputing from scratch, as carried out by the fastest static algorithm. An algorithm is *fully dynamic* if it can handle both edge insertions and edge deletions. A *partially dynamic* algorithm can handle either edge insertions or edge deletions, but not both: it is *incremental* if it supports insertions only, and *decremental* if it supports deletions only.

In the *fully dynamic k-edge connectivity problem* one wishes to maintain an undirected graph $G = (V, E)$ under an intermixed sequence of the following operations:

- *k-EdgeConnected(u, v):* Return *true* if vertices u and v are in the same k-edge-connected component. Return *false* otherwise.
- *Insert(x, y):* Insert a new edge between the two vertices x and y.
- *Delete(x, y):* Delete the edge between the two vertices x and y.

In the *fully dynamic k-vertex connectivity problem* one wishes to maintain an undirected graph $G = (V, E)$ under an intermixed sequence of the following operations:

- *k-VertexConnected(u, v):* Return *true* if vertices u and v are k-vertex-connected. Return *false* otherwise.
- *Insert(x, y):* Insert a new edge between the two vertices x and y.
- *Delete(x, y):* Delete the edge between the two vertices x and y.

Key Results

To the best knowledge of the author, the most efficient fully dynamic algorithms for k-edge and k-vertex connectivity were proposed in [3, 12]. Their running times are characterized by the following theorems.

Theorem 3 *The fully dynamic k-edge connectivity problem can be solved in:*

1. $O(\log^4 n)$ *time per update and* $O(\log^3 n)$ *time per query, for* $k = 2$

2. $O(n^{2/3})$ time per update and query, for $k = 3$
3. $O(n\alpha(n))$ time per update and query, for $k = 4$
4. $O(n \log n)$ time per update and query, for $k \geq 5$.

Theorem 4 *The fully dynamic k-vertex connectivity problem can be solved in:*

1. $O(\log^4 n)$ time per update and $O(\log^3 n)$ time per query, for $k = 2$
2. $O(n)$ time per update and query, for $k = 3$
3. $O(n\alpha(n))$ time per update and query, for $k = 4$.

Applications

Vertex and edge connectivity problems arise often in issues related to network reliability and survivability. In computer networks, the vertex connectivity of the underlying graph is related to the smallest number of nodes that might fail before disconnecting the whole network. Similarly, the edge connectivity is related to the smallest number of links that might fail before disconnecting the entire network. Analogously, if two nodes are k-vertex-connected then they can remain connected even after the failure of up to $(k - 1)$ other nodes, and if they are k-edge-connected then they can survive the failure of up to $(k - 1)$ links. It is important to investigate the dynamic versions of those problems in contexts where the networks are dynamically evolving, say, when links may go up and down because of failures and repairs.

Open Problems

The work of Eppstein et al. [3] and Holm et al. [12] raises some intriguing questions. First, while efficient dynamic algorithms for k-edge connectivity are known for general k, no efficient fully dynamic k-vertex connectivity is known for $k \geq 5$. To the best of the author's knowledge, in this case even no static algorithm is known. Second, fully dynamic 2-edge and 2-vertex connectivity can be solved in polylogarithmic time per update, while the best known update bounds for higher edge and vertex connectivity are polynomial: Can this gap be reduced, i.e., can one design polylogarithnmic algorithms for fully dynamic 3-edge and 3-vertex connectivity?

Cross-References

▶ Dynamic Trees
▶ Fully Dynamic All Pairs Shortest Paths
▶ Fully Dynamic Connectivity
▶ Fully Dynamic Higher Connectivity for Planar Graphs
▶ Fully Dynamic Minimum Spanning Trees
▶ Fully Dynamic Planarity Testing
▶ Fully Dynamic Transitive Closure

Recommended Reading

1. Dinitz EA (1993) Maintaining the 4-edge-connected components of a graph on-line. In: Proceedings of the 2nd Israel symposium on theory of computing and systems, Natanya, pp 88–99
2. Dinitz EA, Karzanov AV, Lomonosov MV (1990) On the structure of the system of minimal edge cuts in a graph. In: Fridman AA (ed) Studies in discrete optimization. Nauka, Moscow, pp 290–306 (in Russian)
3. Eppstein D, Galil Z, Italiano GF, Nissenzweig A (1997) Sparsification – a technique for speeding up dynamic graph algorithms. J Assoc Comput Mach 44(5):669–696
4. Frederickson GN (1997) Ambivalent data structures for dynamic 2-edge-connectivity and k smallest spanning trees. SIAM J Comput 26(2):484–538
5. Galil Z, Italiano GF (1992) Fully dynamic algorithms for 2-edge connectivity. SIAM J Comput 21:1047–1069
6. Galil Z, Italiano GF (1993) Maintaining the 3-edge-connected components of a graph on-line. SIAM J Comput 22:11–28
7. Harary F (1969) Graph theory. Addison-Wesley, Reading
8. Henzinger MR (1995) Fully dynamic biconnectivity in graphs. Algorithmica 13(6):503–538
9. Henzinger MR (2000) Improved data structures for fully dynamic biconnectivity. SIAM J Comput 29(6):1761–1815
10. Henzinger M, King V (1995) Fully dynamic biconnectivity and transitive closure. In: Proceedings of the 36th IEEE symposium on foundations of computer science (FOCS'95), Milwaukee, pp 664–672

11. Henzinger MR, King V (1999) Randomized fully dynamic graph algorithms with polylogarithmic time per operation. J ACM 46(4):502–516
12. Holm J, de Lichtenberg K, Thorup M (2001) Polylogarithmic deterministic fully-dynamic algorithms for connectivity, minimum spanning tree, 2-edge, and biconnectivity. J ACM 48:723–760
13. Karzanov AV, Timofeev EA (1986) Efficient algorithm for finding all minimal edge cuts of a nonoriented graph. Cybernetics 22:156–162
14. La Poutré JA (1992) Maintenance of triconnected components of graphs. In: La Poutré JA (ed) Proceedings of the 19th international colloquium on automata, languages and programming. Lecture notes in computer science, vol 623. Springer, Berlin, pp 354–365
15. La Poutré JA (2000) Maintenance of 2- and 3-edge-connected components of graphs II. SIAM J Comput 29(5):1521–1549
16. La Poutré JA, van Leeuwen J, Overmars MH (1993) Maintenance of 2- and 3-connected components of graphs, part I: 2- and 3-edge-connected components. Discr Math 114:329–359
17. La Poutré JA, Westbrook J (1994) Dynamic two-connectivity with backtracking. In: Proceedings of the 5th ACM-SIAM symposium on discrete algorithms, Arlington, pp 204–212
18. Westbrook J, Tarjan RE (1992) Maintaining bridge-connected and biconnected components on-line. Algorithmica 7:433–464

Fully Dynamic Higher Connectivity for Planar Graphs

Giuseppe F. Italiano
Department of Computer and Systems Science, University of Rome, Rome, Italy
Department of Information and Computer Systems, University of Rome, Rome, Italy

Keywords

Fully dynamic edge connectivity; Fully dynamic vertex connectivity

Years and Authors of Summarized Original Work

1998; Eppstein, Galil, Italiano, Spencer

Problem Definition

In this entry, the problem of maintaining a dynamic planar graph subject to edge insertions and edge deletions that preserve planarity but that can change the embedding is considered. In particular, in this problem one is concerned with the problem of efficiently maintaining information about edge and vertex connectivity in such a dynamically changing planar graph. The algorithms to solve this problem must handle insertions that keep the graph planar without regard to any particular embedding of the graph. The interested reader is referred to the chapter ▶ Fully Dynamic Planarity Testing of this encyclopedia for algorithms to learn how to check efficiently whether a graph subject to edge insertions and deletions remains planar (without regard to any particular embedding).

Before defining formally the problems considered here, a few preliminary definitions follow.

Given an undirected graph $G = (V, E)$, and an integer $k \geq 2$, a pair of vertices $\langle u, v \rangle$ is said to be *k-edge-connected* if the removal of any $(k - 1)$ edges in G leaves u and v connected. It is not difficult to see that this is an equivalence relationship: the vertices of a graph G are partitioned by this relationship into equivalence classes called *k-edge-connected components*. G is said to be *k-edge-connected* if the removal of any $(k - 1)$ edges leaves G connected. As a result of these definitions, G is k-edge-connected if and only if any two vertices of G are k-edge-connected. An edge set $E' \subseteq E$ is an *edge-cut for vertices x and y* if the removal of all the edges in E' disconnects G into two graphs, one containing x and the other containing y. An edge set $E' \subseteq E$ is an *edge-cut for G* if the removal of all the edges in E' disconnects G into two graphs. An edge-cut E' for G (for x and y, respectively) is *minimal* if removing any edge from E' reconnects G (for x and y, respectively). The cardinality of an edge-cut E', denoted by $|E'|$, is given by the number of edges in E'. An edge-cut E' for G (for x and y, respectively) is said to be a *minimum cardinality edge-cut* or in short a *connectivity edge-cut* if there is no other edge-cut E'' for G (for x and y, respectively) such that $|E''| < |E'|$. Connectivity edge-cuts are of course minimal edge-cuts. Note that G is k-edge-connected if and only if a connectivity edge-cut for G contains at least k edges, and vertices x and y are k-edge-connected if and only if a connectivity edge-cut for x and y contains at least k edges. A connectivity edge-cut of cardinality 1 is called a *bridge*.

In a similar fashion, a vertex set $V' \subseteq V - \{x, y\}$ is said to be a *vertex-cut* for vertices x and y if the removal of all the vertices in V' disconnects x and y. $V' \subset V$ is a *vertex-cut* for vertices G if the removal of all the vertices in V' disconnects G.

The cardinality of a vertex-cut V', denoted by $|V'|$, is given by the number of vertices in V'. A vertex-cut V' for x and y is said to be a *minimum cardinality vertex-cut* or in short a *connectivity vertex-cut* if there is no other vertex-cut V'' for x and y such that $|V''| < |V'|$. Then x and y are k-vertex-connected if and only if a connectivity vertex-cut for x and y contains at least k vertices. A graph G is said to be k-*vertex-connected* if all its pairs of vertices are k-vertex-connected. A connectivity vertex-cut of cardinality 1 is called an *articulation point*, while a connectivity vertex-cut of cardinality 2 is called a *separation pair*. Note that for vertex connectivity it is no longer true that the removal of a connectivity vertex-cut splits G into two sets of vertices.

A dynamic graph algorithm maintains a given property $\mathcal{P}$ on a graph subject to dynamic changes, such as edge insertions, edge deletions and edge weight updates. A dynamic graph algorithm should process queries on property $\mathcal{P}$ quickly, and perform update operations faster than recomputing from scratch, as carried out by the fastest static algorithm. An algorithm is *fully dynamic* if it can handle both edge insertions and edge deletions. A *partially dynamic* algorithm can handle either edge insertions or edge deletions, but not both: it is *incremental* if it supports insertions only, and *decremental* if it supports deletions only.

In the *fully dynamic k-edge connectivity problem for a planar graph* one wishes to maintain an undirected planar graph $G = (V, E)$ under an intermixed sequence of edge insertions, edge deletions and queries about the k-edge connectiv-

ity of the underlying planar graph. Similarly, in the *fully dynamic k-vertex connectivity problem for a planar graph* one wishes to maintain an undirected planar graph $G = (V, E)$ under an intermixed sequence of edge insertions, edge deletions and queries about the k-vertex connectivity of the underlying planar graph.

Key Results

The algorithms in [2, 3] solve efficiently the above problems for small values of k:

Theorem 1 *One can maintain a planar graph, subject to insertions and deletions that preserve planarity, and allow queries that test the 2-edge connectivity of the graph, or test whether two vertices belong to the same 2-edge-connected component, in $O(\log n)$ amortized time per insertion or query, and $O(\log^2 n)$ per deletion.*

Theorem 2 *One can maintain a planar graph, subject to insertions and deletions that preserve planarity, and allow testing of the 3-edge and 4-edge connectivity of the graph in $O(n^{1/2})$ time per update, or testing of whether two vertices are 3- or 4-edge-connected, in $O(n^{1/2})$ time per update or query.*

Theorem 3 *One can maintain a planar graph, subject to insertions and deletions that preserve planarity, and allow queries that test the 3-vertex connectivity of the graph, or test whether two vertices belong to the same 3-vertex-connected component, in $O(n^{1/2})$ amortized time per update or query.*

Note that these theorems improve on the bounds known for the same problems on general graphs, reported in the chapter ▶ Fully Dynamic Higher Connectivity

Applications

The interest reader is referred to the chapter ▶ Fully Dynamic Higher Connectivity for applications of dynamic edge and vertex connectivity.

The case of planar graphs is especially important, as these graphs arise frequently in applications.

Open Problems

A number of problems related to the work of Eppstein et al. [2, 3] remain open. First, can the running times per operation be improved? Second, as in the case of general graphs, also for planar graphs fully dynamic 2-edge connectivity can be solved in polylogarithmic time per update, while the best known update bounds for higher edge and vertex connectivity are polynomial: Can this gap be reduced, i.e., can one design polylogarithnmic algorithms at least for fully dynamic 3-edge and 3-vertex connectivity? Third, in the special case of planar graphs can one solve fully dynamic k-vertex connectivity for general k?

Cross-References

► Dynamic Trees
► Fully Dynamic All Pairs Shortest Paths
► Fully Dynamic Connectivity
► Fully Dynamic Higher Connectivity
► Fully Dynamic Minimum Spanning Trees
► Fully Dynamic Planarity Testing
► Fully Dynamic Transitive Closure

Recommended Reading

1. Galil Z, Italiano GF, Sarnak N (1999) Fully dynamic planarity testing with applications. J ACM 48:28–91
2. Eppstein D, Galil Z, Italiano GF, Spencer TH (1996) Separator based sparsification I: planarity testing and minimum spanning trees. J Comput Syst Sci Spec Issue STOC 93 52(1):3–27
3. Eppstein D, Galil Z, Italiano GF, Spencer TH (1999) Separator based sparsification II: edge and vertex connectivity. SIAM J Comput 28:341–381
4. Giammarresi D, Italiano GF (1996) Decremental 2- and 3-connectivity on planar graphs. Algorithmica 16(3):263–287
5. Hershberger J, Rauch M, Suri S (1994) Data structures for two-edge connectivity in planar graphs. Theor Comput Sci 130(1):139–161

Fully Dynamic Minimum Spanning Trees

Giuseppe F. Italiano
Department of Computer and Systems Science, University of Rome, Rome, Italy
Department of Information and Computer Systems, University of Rome, Rome, Italy

Keywords

Dynamic minimum spanning forests

Years and Authors of Summarized Original Work

2000; Holm, de Lichtenberg, Thorup

Problem Definition

Let $G = (V, E)$ be an undirected weighted graph. The problem considered here is concerned with maintaining efficiently information about a minimum spanning tree of G (or minimum spanning forest if G is not connected), when G is subject to dynamic changes, such as edge insertions, edge deletions and edge weight updates. One expects from the dynamic algorithm to perform update operations faster than recomputing the entire minimum spanning tree from scratch.

Throughout, an algorithm is said to be *fully dynamic* if it can handle both edge insertions and edge deletions. A *partially dynamic* algorithm can handle either edge insertions or edge deletions, but not both: it is *incremental* if it supports insertions only, and *decremental* if it supports deletions only.

Key Results

The dynamic minimum spanning forest algorithm presented in this section builds upon the dynamic connectivity algorithm described in the entry ► Fully Dynamic Connectivity. In particular, a few simple changes to that algorithm are sufficient to maintain a minimum spanning forest of a weighted undirected graph upon deletions of edges [13]. A general reduction from [11] can then be applied to make the deletions-only algorithm fully dynamic.

This section starts by describing a decremental algorithm for maintaining a minimum spanning forest under deletions only. Throughout the sequence of deletions, the algorithm maintains a minimum spanning forest F of the dynamically changing graph G. The edges in F are referred to as *tree edges* and the other edges (in $G - F$) are referred to as *non-tree edges*. Let e be an edge being deleted. If e is a non-tree edge, then the minimum spanning forest does not need to change, so the interesting case is when e is a tree edge of forest F. Let T be the tree of F containing e. In this case, the deletion of e disconnects the tree T into two trees T_1 and T_2: to update the minimum spanning forest, one has to look for the minimum weight edge having one endpoint in T_1 and the other endpoint in T_2. Such an edge is called a *replacement edge* for e.

As for the dynamic connectivity algorithm, to search for replacement edges, the algorithm associates to each edge e a level $\ell(e)$ and, based on edge levels, maintains a set of sub-forests of the minimum spanning forest F: for each level i, forest F_i is the sub-forest induced by tree edges of level $\geq i$. Denoting by L the maximum edge level, it follows that:

$$F = F_0 \supseteq F_1 \supseteq F_2 \supseteq \cdots \supseteq F_L.$$

Initially, all edges have level 0; levels are then progressively increased, but never decreased. The changes of edge levels are accomplished so as to maintain the following invariants, which obviously hold at the beginning.

Invariant (1): F is a maximum spanning forest of G if edge levels are interpreted as weights.

Invariant (2): The number of nodes in each tree of F_i is at most $n/2^i$.

Invariant (3): Every cycle C has a non-tree edge of maximum weight and minimum level among all the edges in C.

Invariant (1) should be interpreted as follows. Let (u,v) be a non-tree edge of level $\ell(u, v)$ and let $u \cdots v$ be the unique path between u and v in F (such a path exists since F is a spanning forest of G). Let e be any edge in $u \cdots v$ and let $\ell(e)$ be its level. Due to (1), $\ell(e) \geq \ell(u, v)$. Since this holds for each edge in the path, and by construction $F_{\ell(u,v)}$ contains all the tree edges of level $\geq \ell(u, v)$, the entire path is contained in $F_{\ell(u,v)}$, i.e., u and v are connected in $F_{\ell(u,v)}$.

Invariant (2) implies that the maximum number of levels is $L \leq \lfloor \log_2 n \rfloor$.

Invariant (3) can be used to prove that, among all the replacement edges, the lightest edge is on the maximum level. Let e_1 and e_2 be two replacement edges with $w(e_1) < w(e : 2)$, and let C_i be the cycle induced by e_i in F, $i = 1, 2$. Since F is a minimum spanning forest, e_i has maximum weight among all the edges in C_i. In particular, since by hypothesis $w(e_1) < w(e : 2)$, e_2 is also the heaviest edge in cycle $C = (C_1 \cup C_2) \setminus (C_1 \cap C_2)$. Thanks to Invariant (3), e_2 has minimum level in C, proving that $\ell(e_2) \leq \ell(e_1)$. Thus, considering non-tree edges from higher to lower levels is correct.

Note that initially, an edge is is given level 0. Its level can be then increased at most $\lfloor \log_2 n \rfloor$ times as a consequence of edge deletions. When a tree edge $e = (v, w)$ of level $\ell(e)$ is deleted, the algorithm looks for a replacement edge at the highest possible level, if any. Due to invariant (1), such a replacement edge has level $\ell \leq \ell(e)$. Hence, a replacement subroutine `Replace`$((u, w), \ell(e))$ is called with parameters e and $\ell(e)$. The operations performed by this subroutine are now sketched.

`Replace`$((u, w), \ell)$ finds a replacement edge of the highest level $\leq \ell$, if any, considering edges in order of increasing weight. If such a replacement does not exist in level ℓ, there are two cases: if $\ell > 0$, the algorithm recurses on level $\ell - 1$; otherwise, $\ell = 0$, and the deletion of (v,w) disconnects v and w in G.

It is possible to show that `Replace` returns a replacement edge of minimum weight on the highest possible level, yielding the following lemma:

Lemma 1 *There exists a deletions-only minimum spanning forest algorithm that can be initialized on a graph with n vertices and m edges and supports any sequence of edge deletions in $O(m \log^2 n)$ total time.*

The description of a fully dynamic algorithm which performs updates in $O(\log^4 n)$ time now follows. The reduction used to obtain a fully dynamic algorithm is a slight generalization of the construction proposed by Henzinger and King [11] and works as follows.

Lemma 2 *Suppose there is a deletions-only minimum spanning tree algorithm that, for any k and ℓ, can be initialized on a graph with k vertices and ℓ edges and supports any sequence of $\Omega(\ell)$ deletions in total time $O(\ell \cdot t(k, \ell))$, where t is a nondecreasing function. Then there exists a fully-dynamic minimum spanning tree algorithm for a graph with n nodes starting with no edges, that, for m edges, supports updates in time*

$$O\left(\log^3 n + \sum_{i=1}^{3+\log_2 m} \sum_{j=1}^{i} t\left(min\{n, 2^j\}, 2^j\right)\right).$$

The interested reader is referred to references [11] and [13] for the description of the construction that proves Lemma 2. From Lemma 1 one gets $t(k, \ell) = O(\log^2 k)$. Hence, combining Lemmas 1 and 2, the claimed result follows:

Theorem 3 *There exists a fully-dynamic minimum spanning forest algorithm that, for a graph with n vertices, starting with no edges, maintains a minimum spanning forest in $O(\log^4 n)$ amortized time per edge insertion or deletion.*

There is a lower bound of $\Omega(\log n)$ for dynamic minimum spanning tree, given by Eppstein et al. [6], which uses the following argument. Let A be an algorithm for maintaining a minimum spanning tree of an arbitrary (multi)graph G. Let A be such that change weight(e, Δ) returns the

edge f that replace e in the minimum spanning tree, if e is replaced. Clearly, any dynamic spanning tree algorithm can be modified to return f. One can use algorithm A to sort n positive numbers $x_1, x_2, \ldots, x_n$, as follows. Construct a multigraph G consisting of two nodes connected by $(n + 1)$ edges $e_0, e_1, \ldots, e_n$, such that edge e_0 has weight 0 and edge e_i has weight x_i. The initial spanning tree is e_0. Increase the weight of e_0 to $+\infty$. Whichever edge replaces e_0, say e_i, is the edge of minimum weight. Now increase the weight of e_i to $+\infty$: the replacement of e_i gives the second smallest weight. Continuing in this fashion gives the numbers sorted in increasing order. A similar argument applies when only edge decreases are allowed. Since Paul and Simon [14] have shown that any sorting algorithm needs $\Omega(n \log n)$ time to sort n numbers on a unit-cost random access machine whose repertoire of operations include additions, subtractions, multiplications and comparisons with 0, but not divisions or bit-wise Boolean operations, the following theorem follows.

Theorem 4 *Any unit-cost random access algorithm that performs additions, subtractions, multiplications and comparisons with 0, but not divisions or bit-wise Boolean operations, requires $\Omega(\log n)$ amortized time per operation to maintain a minimum spanning tree dynamically.*

Applications

Minimum spanning trees have applications in many areas, including network design, VLSI, and geometric optimization, and the problem of maintaining minimum spanning trees dynamically arises in such applications.

Algorithms for maintaining a minimum spanning forest of a graph can be used also for maintaining information about the connected components of a graph. There are also other applications of dynamic minimum spanning trees algorithms, which include finding the k smallest spanning trees [3–5, 8, 9], sampling spanning trees [7] and dynamic matroid intersection problems [10]. Note that the first two problems are not necessarily dynamic: however, efficient solutions for these problems need dynamic data structures.

Open Problems

The first natural open question is to ask whether the gap between upper and lower bounds for the dynamic minimum spanning tree problem can be closed. Note that this is possible in the special case of plane graphs [6].

Second, the techniques for dynamic minimum spanning trees can be extended to dynamic 2-edge and 2-vertex connectivity, which indeed can be solved in polylogarithmic time per update. Can one extend the same technique also to higher forms of connectivity? This is particularly important, since the best known update bounds for higher edge and vertex connectivity are polynomial, and it would be useful to design polylogarithnmic algorithms at least for fully dynamic 3-edge and 3-vertex connectivity.

Experimental Results

A thorough empirical study on the performance evaluation of dynamic minimum spanning trees algorithms has been carried out in [1, 2].

Data Sets

Data sets are described in [1, 2].

Cross-References

▶ Dynamic Trees
▶ Fully Dynamic All Pairs Shortest Paths
▶ Fully Dynamic Connectivity
▶ Fully Dynamic Higher Connectivity
▶ Fully Dynamic Higher Connectivity for Planar Graphs

▶ Fully Dynamic Planarity Testing
▶ Fully Dynamic Transitive Closure

Recommended Reading

1. Alberts D, Cattaneo G, Italiano GF (1997) An empirical study of dynamic graph algorithms. ACM J Exp Algorithm 2
2. Cattaneo G, Faruolo P, Ferraro Petrillo U, Italiano GF (2002) Maintaining dynamic minimum spanning trees: an experimental study. In: Proceeding 4th workshop on algorithm engineering and experiments (ALENEX 02), 6–8 Jan, pp 111–125
3. Eppstein D (1992) Finding the k smallest spanning trees. BIT 32:237–248
4. Eppstein D (1994) Tree-weighted neighbors and geometric k smallest spanning trees. Int J Comput Geom Appl 4:229–238
5. Eppstein D, Galil Z, Italiano GF, Nissenzweig A (1997) Sparsification – a technique for speeding up dynamic graph algorithms. J Assoc Comput Mach 44(5):669–696
6. Eppstein D, Italiano GF, Tamassia R, Tarjan RE, Westbrook J, Yung M (1992) Maintenance of a minimum spanning forest in a dynamic plane graph. J Algorithms 13:33–54
7. Feder T, Mihail M (1992) Balanced matroids. In: Proceeding 24th ACM symposium on theory of computing, Victoria, 04–06 May, pp 26–38
8. Frederickson GN (1985) Data structures for on-line updating of minimum spanning trees. SIAM J Comput 14:781–798
9. Frederickson GN (1991) Ambivalent data structures for dynamic 2-edge-connectivity and k smallest spanning trees. In: Proceeding 32nd symposium on foundations of computer science, San Juan, 01–04 Oct, pp 632–641
10. Frederickson GN, Srinivas MA (1989) Algorithms and data structures for an expanded family of matroid intersection problems. SIAM J Comput 18:112–138
11. Henzinger MR, King V (2001) Maintaining minimum spanning forests in dynamic graphs. SIAM J Comput 31(2):364–374
12. Henzinger MR, King V (1999) Randomized fully dynamic graph algorithms with polylogarithmic time per operation. J ACM 46(4):502–516
13. Holm J, de Lichtenberg K, Thorup M (2001) Polylogarithmic deterministic fully-dynamic algorithms for connectivity, minimum spanning tree, 2-edge, and biconnectivity. J ACM 48:723–760
14. Paul J, Simon W (1980) Decision trees and random access machines. In: Symposium über Logik und Algorithmik; See also Mehlhorn K (1984) Sorting and searching. Springer, Berlin, pp 85–97
15. Tarjan RE, Vishkin U (1985) An efficient parallel biconnectivity algorithm. SIAM J Comput 14:862–874

Fully Dynamic Planarity Testing

Giuseppe F. Italiano
Department of Computer and Systems Science,
University of Rome, Rome, Italy
Department of Information and Computer
Systems, University of Rome, Rome, Italy

Years and Authors of Summarized Original Work

1999; Galil, Italiano, Sarnak

Problem Definition

In this entry, the problem of maintaining a dynamic planar graph subject to edge insertions and edge deletions that preserve planarity but that can change the embedding is considered. Before formally defining the problem, few preliminary definitions follow.

A graph is *planar* if it can be embedded in the plane so that no two edges intersect. In a dynamic framework, a planar graph that is committed to an embedding is called *plane*, and the general term *planar* is used only when changes in the embedding are allowed. An edge insertion that preserves the embedding is called *embedding-preserving*, whereas it is called *planarity-preserving* if it keeps the graph planar, even though its embedding can change; finally, an edge insertion is called *arbitrary* if it is not known to preserve planarity. Extensive work on dynamic graph algorithms has used ad hoc techniques to solve a number of problems such as minimum spanning forests, 2-edge-connectivity and planarity testing for plane graphs (with embedding-preserving insertions) [5–7, 9–12]: this entry is concerned with more general planarity-preserving updates.

The work of Galil et al. [8] and of Eppstein et al. [3] provides a general technique for dynamic planar graph problems, including those mentioned above: in all these problems, one can deal with either arbitrary or planarity-preserving

insertions and therefore allow changes of the embedding.

The *fully dynamic planarity testing problem* can be defined as follows. One wishes to maintain a (not necessarily planar) graph subject to *arbitrary* edge insertions and deletions, and allow queries that test whether the graph is currently planar, or whether a potential new edge would violate planarity.

Key Results

Eppstein et al. [3] provided a way to apply the sparsification technique [2] to families of graphs that are already sparse, such as planar graphs.

The new ideas behind this technique are the following. The notion of a certificate can be expanded to a definition for graphs in which a subset of the vertices are denoted as *interesting*; these *compressed certificates* may reduce the size of the graph by removing uninteresting vertices. Using this notion, one can define a type of sparsification based on *separators*, small sets of vertices the removal of which splits the graph into roughly equal size components. Recursively finding separators in these components gives a *separator tree* which can also be used as a *sparsification tree*; the interesting vertices in each certificate will be those vertices used in separators at higher levels of the tree. The notion of a *balanced separator tree*, which also partitions the interesting vertices evenly in the tree, is introduced: such a tree can be computed in linear time, and can be maintained dynamically. Using this technique, the following results can be achieved.

Theorem 1 *One can maintain a planar graph, subject to insertions and deletions that preserve planarity, and allow queries that test whether a new edge would violate planarity, in amortized time $O(n^{1/2})$ per update or query.*

This result can be improved, in order to allow arbitrary insertions or deletions, even if they might let the graph become nonplanar, using the following approach. The data structure above can be used to maintain a planar subgraph of the given graph. Whenever one attempts to insert a new edge, and the resulting graph would be nonplanar, the algorithm does not actually perform the insertion, but instead adds the edge to a list of *nonplanar edges*. Whenever a query is performed, and the list of nonplanar edges is nonempty, the algorithm attempts once more to add those edges one at a time to the planar subgraph. The time for each successful addition can be charged to the insertion operation that put that edge in the list of nonplanar edges. As soon as the algorithm finds some edge in the list that can not be added, it stops trying to add the other edges in the list. The time for this failed insertion can be charged to the query the algorithm is currently performing. In this way the list of nonplanar edges will be empty if and only if the graph is planar, and the algorithm can test planarity even for updates in nonplanar graphs.

Theorem 2 *One can maintain a graph, subject to arbitrary insertions and deletions, and allow queries that test whether the graph is presently planar or whether a new edge would violate planarity, in amortized time $O(n^{1/2})$ per update or query.*

Applications

Planar graphs are perhaps one of the most important interesting subclasses of graphs which combine beautiful structural results with relevance in applications. In particular, planarity testing is a basic problem, which appears naturally in many applications, such as VLSI layout, graphics, and computer aided design. In all these applications, there seems to be a need for dealing with dynamic updates.

Open Problems

The $O(n^{1/2})$ bound for planarity testing is amortized. Can we improve this bound or make it worst-case?

Finally, the complexity of the algorithms presented here, and the large constant factors involved in some of the asymptotic time bounds, make some of the results unsuitable for practical applications. Can one simplify the methods while retaining similar theoretical bounds?

Cross-References

▸ Dynamic Trees
▸ Fully Dynamic All Pairs Shortest Paths
▸ Fully Dynamic Connectivity
▸ Fully Dynamic Higher Connectivity
▸ Fully Dynamic Higher Connectivity for Planar Graphs
▸ Fully Dynamic Minimum Spanning Trees
▸ Fully Dynamic Transitive Closure

Recommended Reading

1. Cimikowski R (1994) Branch-and-bound techniques for the maximum planar subgraph problem. Int J Comput Math 53:135–147
2. Eppstein D, Galil Z, Italiano GF, Nissenzweig A (1997) Sparsification – a technique for speeding up dynamic graph algorithms. J Assoc Comput Mach 44(5):669–696
3. Eppstein D, Galil Z, Italiano GF, Spencer TH (1996) Separator based sparsification I: planarity testing and minimum spanning trees. J Comput Syst Sci Spec Issue STOC 93 52(1):3–27
4. Eppstein D, Galil Z, Italiano GF, Spencer TH (1999) Separator based sparsification II: edge and vertex connectivity. SIAM J Comput 28: 341–381
5. Eppstein D, Italiano GF, Tamassia R, Tarjan RE, Westbrook J, Yung M (1992) Maintenance of a minimum spanning forest in a dynamic plane graph. J Algorithms 13:33–54
6. Frederickson GN (1985) Data structures for on-line updating of minimum spanning trees, with applications. SIAM J Comput 14:781–798
7. Frederickson GN (1997) Ambivalent data structures for dynamic 2-edge-connectivity and k smallest spanning trees. SIAM J Comput 26(2):484–538
8. Galil Z, Italiano GF, Sarnak N (1999) Fully dynamic planarity testing with applications. J ACM 48:28–91
9. Giammarresi D, Italiano GF (1996) Decremental 2- and 3-connectivity on planar graphs. Algorithmica 16(3):263–287
10. Hershberger J, Suri MR, Suri S (1994) Data structures for two-edge connectivity in planar graphs. Theor Comput Sci 130(1):139–161
11. Italiano GF, La Poutré JA, Rauch M (1993) Fully dynamic planarity testing in planar embedded graphs.

     In: 1st annual European symposium on algorithms, Bad Honnef, 30 Sept–2 Oct
12. Tamassia R (1988) A dynamic data structure for planar graph embedding. In: 15th international colloquium automata, languages, and programming. LNCS, vol 317. Springer, Berlin, pp 576–590

Fully Dynamic Transitive Closure

Valerie King
Department of Computer Science, University of Victoria, Victoria, BC, Canada

Keywords

All-pairs dynamic reachability; Fully dynamic graph algorithm for maintaining transitive closure; Incremental algorithms for digraphs;

Years and Authors of Summarized Original Work

1999; King

Problem Definition

Design a data structure for a directed graph with a fixed set of node which can process queries of the form "Is there a path from i to j ?" and updates of the form: "Insert edge (i, j)"; "Delete edge (i, j)". The goal is to minimize update and query times, over the worst case sequence of queries and updates. Algorithms to solve this problem are called "fully dynamic" as opposed to "partially dynamic" since both insertions and deletions are allowed.

Key Results

This work [4] gives the first deterministic fully dynamic graph algorithm for maintaining the transitive closure in a directed graph. It uses $O(n^2 \log n)$ amortized time per update and $O(1)$ worst case query time where n is number of nodes in the graph. The basic technique is extended to give fully dynamic algorithms for approximate and exact all-pairs shortest paths problems.

The basic building block of these algorithms is a method of maintaining all-pairs shortest paths with insertions and deletions for distances up to d. For each vertex v, a single-source shortest path tree of depth d which reach v ("In_v") and another tree of vertices which are reached by v ("Out_v") are maintained during any sequence of deletions. Each insert of a set of edges incident to v results in the rebuilding of In_v and Out_v I. For each pair of vertices x, y and each length, a count is kept of the number of v such that there is a path from x in In_v to y in Out_v of that length.

To maintain transitive closure, $\log n$ levels of these trees are maintained for trees of depth 2, where the edges used to construct a forest on one level depend on the paths in the forest of the previous level.

Space required was reduced from $O(n^3)$ to $O(n^2)$ in [6]. A $\log n$ factor was shaved off [7, 10]. Other tradeoffs between update and query time are given in [1, 7–10]. A deletions only randomized transitive closure algorithm running in $O(mn)$ time overall is given by [8] where m is the initial number of edges in the graph. A simple monte carlo transitive closure algorithm for acyclic graphs is presented in [5]. Dynamic single source reachability in a digraph is presented in [8, 9]. All-pairs shortest paths can be maintained with nearly the same update time [2].

Applications

None

Open Problems

Can reachability from a single source in a directed graph be maintained in $o(mn)$ time over a worst case sequence of m deletions?

Can strongly connected components be maintained in $o(mn)$ time over a worst case sequence of m deletions?

Experimental Results

Experimental results on older techniques can be found in [3].

Cross-References

▶ All Pairs Shortest Paths in Sparse Graphs
▶ All Pairs Shortest Paths via Matrix Multiplication
▶ Fully Dynamic All Pairs Shortest Paths
▶ Fully Dynamic Connectivity

Recommended Reading

1. Demestrescu C, Italiano GF (2005) Trade-offs for fully dynamic transitive closure on DAG's: breaking through the $O(n^2)$ barrier, (presented in FOCS 2000). J ACM 52(2):147–156
2. Demestrescu C, Italiano GF (2004) A new approach to dynamic all pairs shortest paths, (presented in STOC 2003). J ACM 51(6):968–992
3. Frigioni D, Miller T, Nanni U, Zaroliagis CD (2001) An experimental study of dynamic algorithms for transitive closure. ACM J Exp Algorithms 6(9)
4. King V (1999) Fully dynamic algorithms for maintaining all-pairs shortest paths andtransitive closure in digraphs. In: Proceedings of the 40th annual IEEE symposium on foundation of computer science (ComiIEEE FOCS). IEEE Computer Society, New York, pp 81–91
5. King V, Sagert G (2002) A fully dynamic algorithm for maintaining the transitive closure (presented in FOCS 1999). JCCS 65(1):150–167
6. King V, Thorup M (2001) A space saving trick for dynamic transitive closure and shortest path algorithms. In: Proceedings of the 7th annual international conference of computing and cominatorics (COCOON). Lecture notes computer science, vol 2108/2001. Springer, Heidelberg, pp 269–277
7. Roditty L (2003) A faster and simpler fully dynamic transitive closure. In: Proceedings of the 14th annual ACM-SIAM symposium on discrete algorithms (ACMIEEE SODA). ACM, Baltimore, pp 404–412
8. Roditty L, Zwick U (2002) Improved dynamic reachability algorithms for directed graphs. In: Proceedings of the 43rd annual symposium on foundation of computer science (IEEE FOCS). IEEE Computer Society, Vancouver, pp 679–688
9. Roditty L, Zwick U (2004) A fully dynamic reachability algorithm for directed graphs with an almost linear update time. In: Proceedings of the 36th ACM symposium on theory of computing (ACMSTOC). ACM, Chicago, pp 184–191
10. Sankowski S (2004) Dynamic transitive closure via dynamic matrix inverse. In: Proceedings of the 45th annual symposium on foundations of computer science (IEEE FOCS). IEEE Computer Society, Rome, pp 509–517